Control Systems
Engineering

Control Systems Engineering

FIFTH EDITION

I.J. Nagrath

M. Gopal

ANSHAN LTD
6 Newlands Road
Tunbridge Wells, Kent.
TN4 9AT. UK

Co-published in the U. K. by

ANSHAN LTD, 6 Newlands Road, Tunbridge Wells, Kent TN4 9AT
In 2008

Tel/Fax: +44(0)1892557767
e-mail: info@anshan.co.uk
Web Site: www.anshan.co.uk

ISBN: 978-1848-290-037

British Library Cataloguing in Publication Data
A Catalogue record for this book is available from the British Library

PREFACE TO THE FIFTH EDITION

The primary objective of the Fifth Edition is to provide adequate breadth and depth for two courses in control systems at the undergraduate level. With this objective in view, a chapter on optimal control was brought in the last edition. In this edition a full chapter on nonlinear control systems has been included which has detailed coverage of various types of nonlinearities, the phase-plane analysis and describing function analysis. The Liapunov stability analysis is covered in a preceding chapter.

The material for first undergraduate course spans chapters one to nine and good part of chapter ten on control system design, and the early part of Chapter 12, state variable formulations. This material is well selected and well designed to meet the needs of the students of all engineering disciplines.

Some specific improvements, based on student/teacher feedback, carried out in this edition are:

- Additional examples in Chapters 2, 3, 5, 8 and 10. These examples are designed to illustrate not only the chapter concepts but also what has preceded and a glimpse of what is beyond.
- Chapter 10 on control system design has been further enriched by the inclusion of PID controller design, comparison table of lag/lead compensators, industrial OPAMP compensating networks and a design example to illustrate step by step design procedure to meet stringent specifications using root-locus technique.

A second and elective course in control systems at undergraduate level is provided for by coverage of robust control, digital control, controllability and observability, observer and pole-placement design, Liapunov stability analysis, optimal control, nonlinear control and introduction to adaptive, fuzzy logic and neural network control. This material can also be used for a first graduate course for those students who do not take the elective at undergraduate level.

The authors welcome any constructive criticism and will be grateful for any appraisal by the readers.

I.J. NAGRATH
M. GOPAL

PREFACE TO THE THIRD EDITION

This book was first published in 1975 and a second edition came out in 1982. The present edition is a fully revised version in the context of the fast changing technology.

This text is designed for an undergraduate course in control systems for engineering students. The basic concepts of the course cut across disciplines, like electrical, mechanical, chemical, and aerospace. Therefore, some balance is maintained between engineering disciplines by drawing examples from all over. Here and there are some examples of nonengineering systems. Examples and problems, which in the previous edition were designed around now obsolete hardware, have been modified. With automation of heavy earth moving machinery, the course is acquiring relevance to civil engineering as well.

While the well written parts of the fundamentals of time and frequency domain are retained, newer aspects and modern examples have been brought in. A glimpse has been given to state variables in the modelling chapter (No. 2). Sensitivity issues have been highlighted in chapter 3 which aid in the robust system design later in chapter 10. Feedforward control and PID controllers have been introduced early in chapter 5. Robotic modelling and robotic control (decentralized) are some of the latest topics introduced.

In chapters of root locus, polar plots and Bode plots minor alterations are carried out. MATLAB examples on these are illustrated in Appendix III.

The Routh algebraic stability test and stability in frequency domain through Nyquist plot are now strengthened with well designed additional examples.

Chapter 10 on design and compensation has been pruned of some of the graphical procedures but has been considerably enlarged to include (*i*) PID controller design with tuning aided by a tuning tree and (*ii*) design of robust controllers. Robust control system design is based on simple procedure to minimize dominant root sensitivity; prefilter advantages and design is added. Also for some of the systems designed by the frequency-domain method, time-domain step response obtained by MATLAB is presented and comparison carried out.

Digital control systems (Ch. 11) is as in the previous edition restricted to sampled-data type inluding the basics of the z-transform.

Chapter 12 on state variable formulations and design has been considerably modified. Several concepts and procedures are now illustrated through simple examples rather than generalized formulations. Computational procedure for continuous-time systems in state variable form through discretization is added. Pole placement design is illustrated through simple examples before giving the generalized method (heavy derivation is relegated to Appendix V). More examples, are brought here from the chemical discipline. Of course controllability, observability and observer design stay with minor improvements in presentation.

Nonlinear control system chapter has now been brought in 13th location. It has been sufficiently pruned, thereby enhancing the stress on conceptual issues and procedures. Generalized approach to system stability is clarified and only Liapunov stability criterion and methods of formulating Liapunov function are retained, while the frequency-domain method of Popov, which was not finding any application in later chapter has been dropped.

Chapter on Optimal Control (which truly should form a full-fledged independent course) has been dropped in favour of chapter 14 on 'Advances in Control Systems' comprising three parts namely Adaptive control, Neural Network control and Fuzzy Logic control systems. The chapter is so written that in a short space all the issues of these control schemes, which are finding many practical applications, particularly the Adaptive and Fuzzy logic control are exposed to the reader; Robust control systems having been dealt with earlier in the design chapter (10). Those who proceed for higher studies can then choose any one area and study further through full courses these topics. This chapter has been mainly contributed by Dr. L.D. Bahera and K.K.N. Sastry. The results reported are obtained in the Center for Robotics and Intelligent Systems, BITS.

This book is already known for having given an exhaustive chapter (No. 4) on control system components. It has been updated by eliminating the obsolete components like amplidyne etc. and inclusion of dc servomotors, more on stepper motors, potentiometers and encoders. The topic of gyros is removed as optical gyros have become available and these are beyond the scope of this core level book. Hydraulic and pneumatic components have been strengthened as these still find applications; hydraulic components in particular, like in earth moving machinery.

Appendix III gives MATLAB: Tool For Design and Analysis of Control Systems. The basic commands and procedures alongwith several examples in design and analysis are included. The reader can supplement these with instructional material available with MATLAB software and solve any problem in any chapter of the book using this versatile tool. At several places in the book results obtained through MATLAB are presented.

Authors are grateful to Dr. L.D. Bahera and K.K.N. Sastry of BITS for contributing chapter 14 on Advances in Control Systems. Further, K.K.N Sastry has been very helpful in checking examples, problems and write-ups of some more chapters.

Authors are grateful to the Director and authorities of BITS, Pilani and IIT, Delhi for providing all the help and facilities for the second revision of this book.

The authors are highly thankful to K.N. Sharma for excellent preparation of illustrations.

The authors welcome any constructive criticism of the book and will be grateful for any appraisal by the readers.

I.J. NAGRATH
M. GOPAL

CONTENTS

1

INTRODUCTION

1

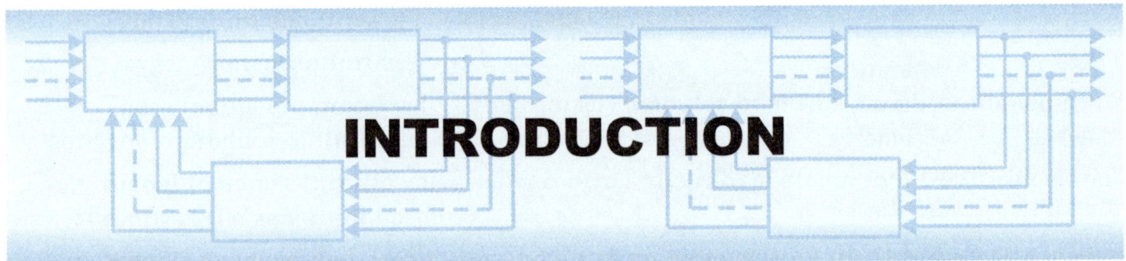

INTRODUCTION

1.1 THE CONTROL SYSTEM

The control system is that means by which any quantity of interest in a machine, mechanism or other equipment is maintained or altered in accordance with a desired manner. Consider, for example, the driving system of an automobile. Speed of the automobile is a function of the position of its accelerator. The desired speed can be maintained (or a desired change in speed can be achieved) by controlling pressure on the accelerator pedal. This automobile driving system (accelerator, carburettor and engine-vehicle) constitutes a control system. Figure 1.1 shows the general diagrammatic representation of a typical control system. For the automobile driving system the input (command) signal is the force on the accelerator pedal which through linkages causes the carburettor valve to open (close) so as to increase or decrease fuel (liquid form) flow to the engine bringing the engine-vehicle speed (controlled variable) to the desired value.

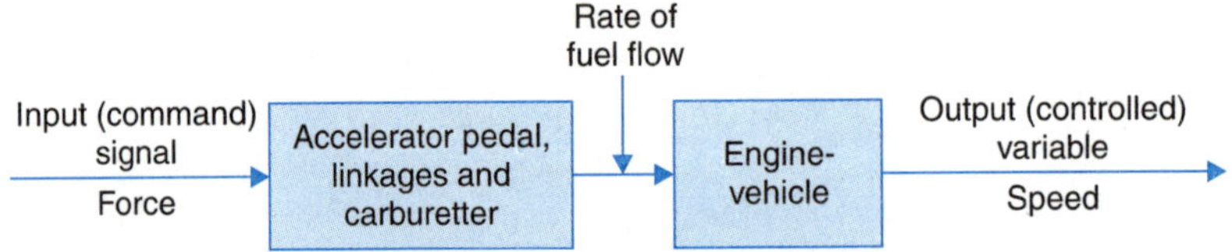

Fig. 1.1. The basic control system.

The diagrammatic representation of Fig. 1.1 is known as *block diagram* representation wherein each block represents an element, a plant, mechanism, device etc., whose inner details are not indicated. Each block has an input and output signal which are linked by a relationship characterizing the block. It may be noted that the signal flow through the block is unidirectional.

Closed-Loop Control

Let us reconsider the automobile driving system. The route, speed and acceleration of the automobile are determined and controlled by the driver by observing traffic and road conditions and by properly manipulating the accelerator, clutch, gear-lever, brakes and steering wheel, etc. Suppose the driver wants to maintain a speed of 50 km per hour (desired output). He accelerates the automobile to this speed with the help of the accelerator and then maintains it by holding the accelerator steady. No error in the speed of the automobile occurs so long as there are no gradients or other disturbances along the road. The actual speed of the automobile is measured by the speedometer and indicated on its dial. The driver reads the speed dial visually and compares the actual speed with the desired one mentally. If there is a deviation of speed from the desired speed, accordingly he takes the decision to increase or decrease the speed. The decision is executed by change in pressure of his foot (through muscular power) on the accelerator pedal.

These operations can be represented in a diagrammatic form as shown in Fig. 1.2. In contrast to the sequence of events in Fig. 1.1, the events in the control sequence of Fig. 1.2 follow a closed-loop, *i.e.*, the information about the instantaneous state of the output is feedback to the input and is used to modify it in such a manner as to achieve the desired output. It is on account of this basic difference that the system of Fig. 1.1 is called an *open-loop system,* while the system of Fig. 1.2 is called a *closed-loop system.*

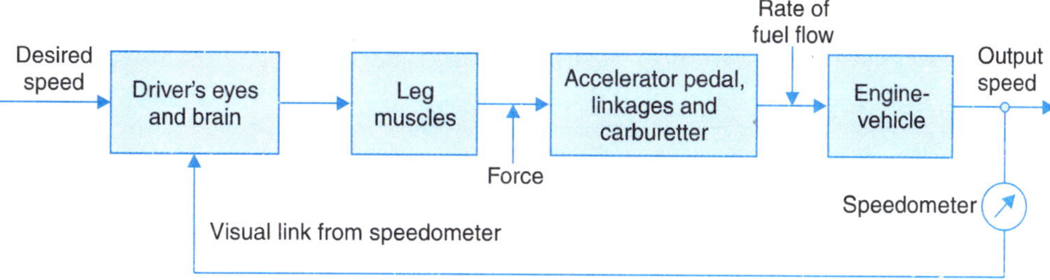

Fig. 1.2. Schematic diagram of a manually controlled closed-loop system.

Let us investigate another control aspect of the above example of an automobile (engine vehicle) say its steering mechanism. A simple block diagram of an automobile steering mechanism is shown in Fig. 1.3(a). The driver senses visually and by tactile means (body movement) the error between the actual and desired directions of the automobile as in Fig. 1.3(b). Additional information is available to the driver from the feel (sensing) of the steering wheel through his hand(s), these informations constitute the feedback signal(s) which are interpreted by driver's brain, who then signals his hand to adjust the steering wheel accordingly. This again is an example of a closed-loop system where human visual and tactile measurements constitute the feedback loop.

In fact unless human being(s) are not left out of in a control system study practically all control systems are a sort of closed-loop system (with intelligent measurement and sensing loop or there may indeed by several such loops).

Systems of the type represented in Figs. 1.2 and 1.3 involve continuous manual control by a human operator. These are classified as *manually controlled systems.* In many complex and fast-acting systems, the presence of human element in the control loop is undersirable because the system response may be too rapid for an operator to follow or the demand on operator's skill may be unreasonably high. Furthermore, some of the systems. *e.g.,* missiles, are self-destructive and in such systems human element must be excluded. Even in situations where manual control could be possible, an economic case can often be made out for reduction of human supervision. Thus in most situations the use of some equipment which performs the same intended function as a continuously employed human operator is preferred. A system incorporating such an equipment is known as *automatic control system.* In fact in most situations an automatic control system could be made to perform intended functions better than a human operator, and could further be made to perform such functions as would be impossible for a human operator.

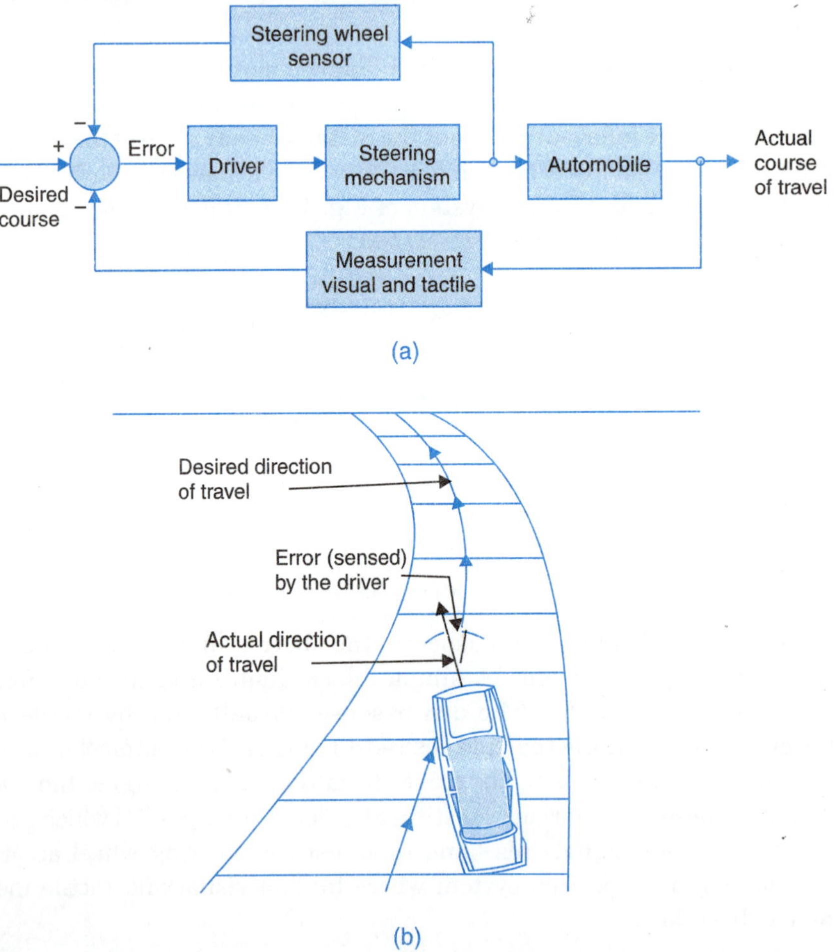

Fig. 1.3. (a) Automobile steering control system. (b) The driver uses the difference between the actual and desired direction of travel to adjust the steering wheel accordingly.

The general block diagram of an automatic control system which is characterised by a feedback loop, is shown in Fig. 1.4. An error detector compares a signal obtained through feedback elements, which is a function of the output response, with the reference input. Any difference between these two signals constitutes an error or actuating signal, which actuates the control elements. The control elements in turn alter the conditions in the plant (controlled member) in such a manner as to reduce the original error.

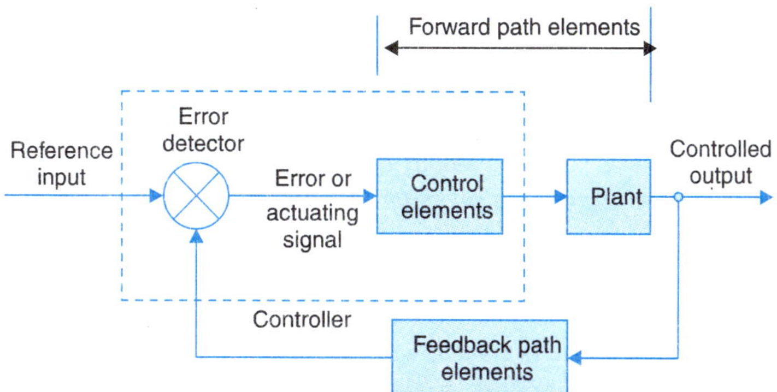

Fig. 1.4. General block diagram of an automatic control system.

In order to gain a better understanding of the interactions of the constituents of a control system, let us discuss a simple tank level control system shown in Fig. 1.5. This control system can maintain the liquid level h (controlled output) of the tank within accurate tolerance of the

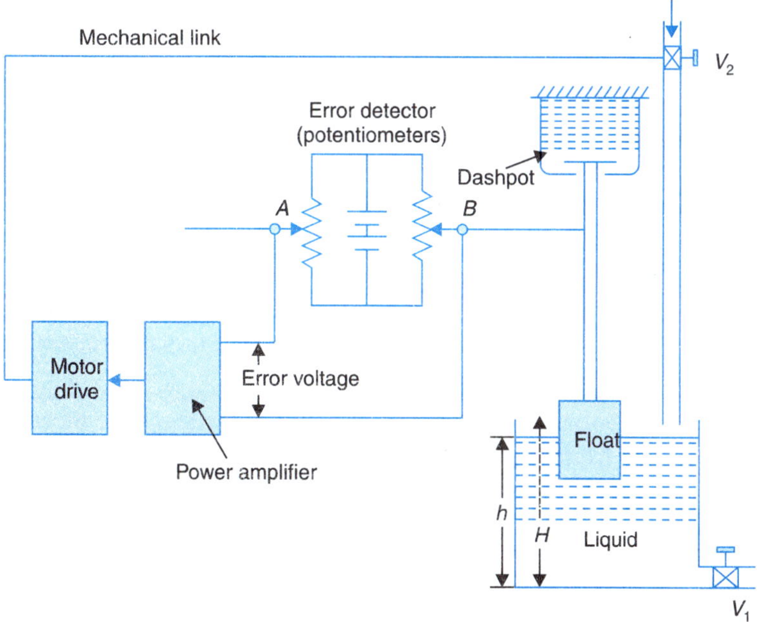

Fig. 1.5. Automatic tank-level control system.

desired liquid level even though the output flow rate through the valve V_1 is varied. The liquid level is sensed by a float (feedback path element), which positions the slider arm B on a potentiometer. The slider arm A of another potentiometer is positioned corresponding to the desired liquid level H (the reference input). When the liquid level rises or falls, the potentiometers (error detector) give an error voltage (error or actuating signal) proportional to the change in liquid level. The error voltage actuates the motor through a power amplifier (control elements) which in turn conditions the plant (*i.e.,* decreases or increases the opening of the valve V_2) in order to restore the desired liquid level. Thus the control system automatically attempts to correct any deviation between the actual and desired liquid levels in the tank.

Open-Loop Control

As stated already, any physical system which does not automatically correct for variation in its output, is called an open-loop system. Such a system may be represented by the block diagram of Fig. 1.6. In these systems the output remains constant for a constant input signal provided the external conditions remain unaltered. The output may be changed to any desired value by appropriately changing the input signal but variations in external conditions or internal parameters of the system may cause the output to vary from the desired value in an uncontrolled fashion. The open-loop control is, therefore, satisfactory only if such fluctuations can be tolerated or system components are designed and constructed so as to limit parameter variations and environmental conditions are well-controlled.

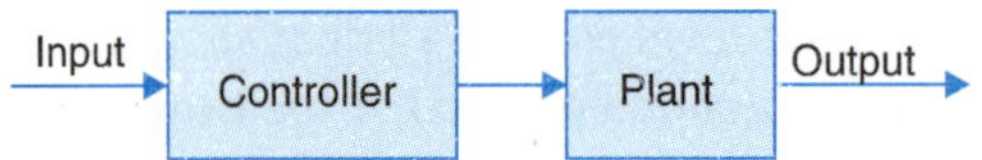

Fig. 1.6. General block diagram of open-loop system.

It is important to note that the fundamental difference between an open and closed-loop control system is that of feedback action. Consider, for example, a traffic control system for regulating the flow of traffic at the crossing of two roads. The system will be termed open-loop if red and green lights are put on by a timer mechanism set for predetermined fixed intervals of time. It is obvious that such an arrangement takes no account of varying rates of traffic flowing to the road crossing from the two directions. If on the other hand a scheme is introduced in which the rates of traffic flow along both directions are measured (some distance ahead of the crossing) and are compared and the difference is used to control the timings of red and green lights, a closed-loop system (feedback control) results. Thus the concept of feedback can be usefully employed to traffic control.

Unfortunately, the feedback which is the underlying principle of most control systems, introduces the possibility of undersirable system oscillations (hunting). Detailed discussion of feedback principle and the linked problem of stability are dealt with later in the book.

1.2 SERVOMECHANISMS

In modern usage the term *servomechanism* or servo is restricted to feedback control systems in which the controlled variable is mechanical position or time derivatives of position, *e.g.,* velocity and acceleration.

A servo system used to position a load shaft is shown in Fig. 1.7 in which the driving motor is geared to the load to be moved. The output (controlled) and desired (reference) positions θ_C and θ_R respectively are measured and compared by a potentiometer pair whose output voltage v_E is proportional to the error in angular position $\theta_E = \theta_R - \theta_C$. The voltage $v_E = K_P\theta_E$ is amplified and is used to control the field current (excitation) of a dc generator which supplies the armature voltage to the drive motor.

To understand the operation of the system assume K_P = 100 volts/rad and let the output shaft position be 0.5 rad. Corresponding to this condition, the slider arm B has a voltage of +50 volts. Let the slider arm A be also set at +50 volts. This gives zero actuating signal (v_E = 0). Thus the motor has zero output torque so that the load stays stationary at 0.5 rad.

Assume now that the new desired load position is 0.6 rad. To achieve this, the arm A is placed at +60 volts position, while the arm B remains instantaneously at +50 volts position. This creates an actuating signal of +10 volts, which is a measure of lack of correspondence between the actual load position and the commanded position. The actuating signal is amplified and fed to the servo motor which in turn generates an output torque which repositions the load. The system comes to a standstill only when the actuating signal becomes zero, *i.e.*, the arm B and the load reach the position corresponding to 0.6 rad (+60 volts position).

Consider now that a load torque T_L is applied at the output as indicated in Fig. 1.7. This will require a steady value of error voltage v_E which acting through the amplifier, generator, motor and gears will counterbalance the load torque. This would mean that a steady error will exist between the input and output angles. This is unlike the case when there is no load torque and consequently the angle error is zero. In control terminology, such loads are known as load disturbances and system has to be designed to keep the error to these disturbances within specified limits.

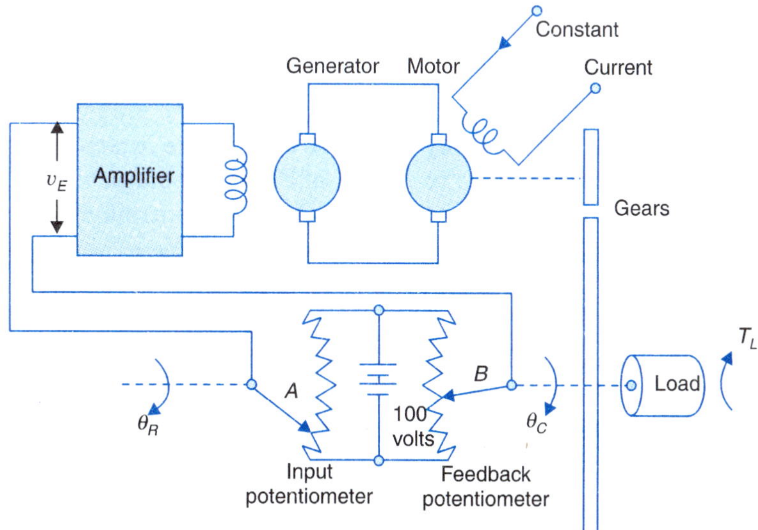

Fig. 1.7. A position control system.

By opening the feedback loop *i.e.,* disconnecting the potentiometer B, the reader can easily verify that any operator acting as part of feedback loop will find it very difficult to adjust θ_C to a desired value and to be able to maintain it. This further demonstrates the power of a negative feedback (hardware) loop.

The position control systems have innumerable applications, namely, machine tool position control, constant-tension control of sheet rolls in paper mills, control of sheet metal thickness in hot rolling mills, radar tracking systems, missile guidance systems, inertial guidance, roll stabilization of ships, etc. Some of these applications will be discussed in this book.

Robotics

Advances in servo mechanism has led to the development of the new field of control and automation, the robots and robotology. A robot is a mechanism devised to perform repetitive tasks which are tiresome for a human being or tasks to be performed in a hazardous environment say in a radioactive area. Robots are as varied as the tasks that can be imagined to be performed by them. Great strides are being made in this field with the explosion in the power of digital computer, interfacing and software tools which have brought to reality the application of vision and artificial intelligent for devising more versatile robots and increased applications of robotology in industrial automation. In fact in replacing a human being for a repetitive and/or hazardous task the robots can perform the task at a greater speed (so increased productivity) and higher precision (better quality and higher reliability of the product or service).

We shall describe here a robot manipulator arm as an example. The arm is devised to preform some of the tasks performed by a human arm (shoulder, elbow and wrist). Imitation of some of the elementary functions of hand is carried out by an end effector with three degrees of freedom in general (roll, yaw and pitch). The robot arm is a set of serial links with the beginning of each link jointed with the end of the preceeding link in form of a revolute joint (for relative rotary motion between the two links) or a prismatic joint (for relative translatory motion). The number of joints determine the degrees of freedom of the arm.

Figures 1.8 (*a*) and (*b*) show the schematic diagrams of two kinds of manipulator arms. To reduce joint inertia and gravity loading the drive motors are located in the base and the joints are belt driven. For a programmed trajectory of the manipulator tip, each joint requires not only a controlled angular (or translatory) motion but also controlled velocity, acceleration and torque. Further the mechanism complexity is such that the effective joint inertia may change by as much as 300% during a trajectory traversal. The answer to such control complexity is the computer control. The versatility of high-speed on-line computer further permits the sophistication of control through computer vision, learning of new tasks and other intelligent functions. Manipulators can perform delicate (light) as well as heavy tasks; for example, manipulator can pick up objects weighing hundreds of kilograms and position them with an accuracy of a centimeter or better.

Using robots (specially designed for broken-down tasks) an assembly line in a manufacturing process can be speeded up with added quality and reliability of the end product. Example can be cited of watch industry in Japan where as many as 150 tasks on the assembly are robot executed.

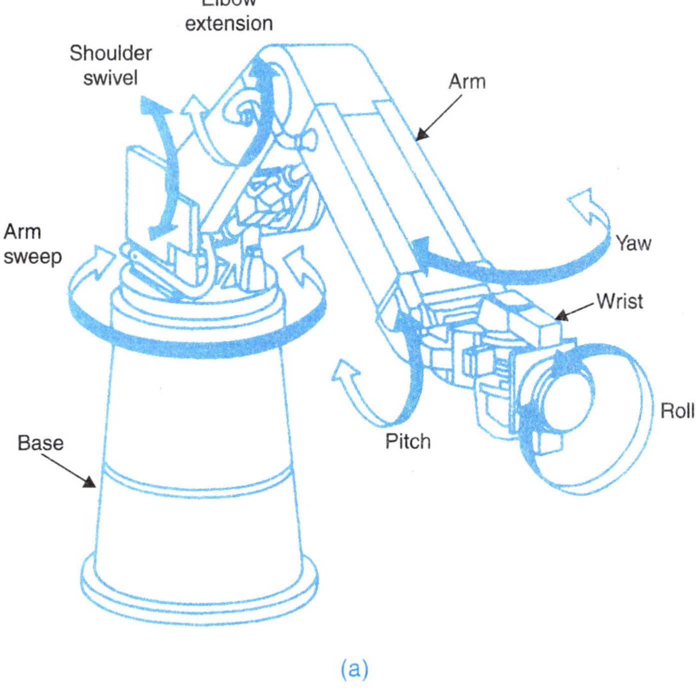

(a)

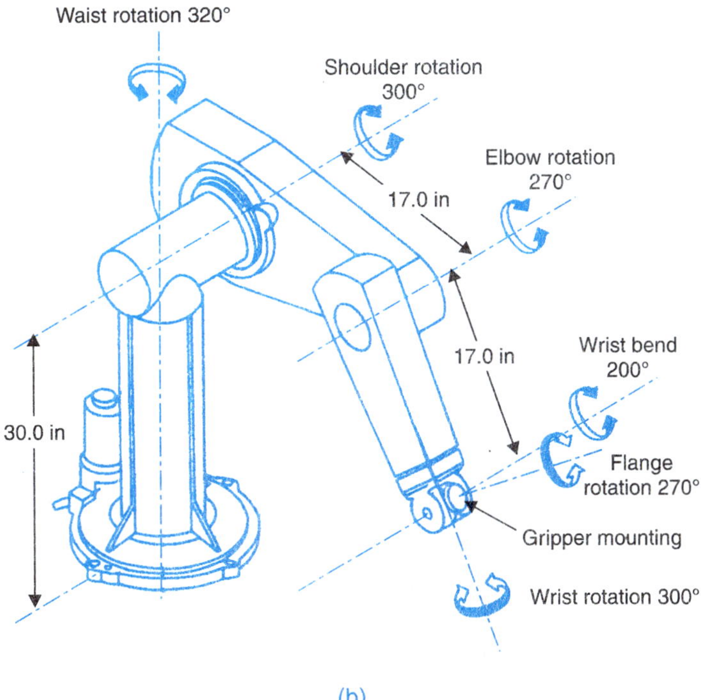

(b)

Fig. 1.8. (a) Cincinnati Milacron T^3 robot arm (b) PUMA 560 series robot arm.

For flexible manufacturing units mobile automations (also called AGV (automated guided vehicle)) have been devised and implemented which are capable of avoiding objects while travelling through a room or industrial plant.

1.3 HISTORY AND DEVELOPMENT OF AUTOMATIC CONTROL

It is instructive to trace brief historical development of automatic control. Automatic control systems did not appear until the middle of eighteenth century. The first automatic control system, the fly-ball governor, to control the speed of steam engines, was invented by James Watt in 1770. This device was usually prone to hunting. It was about hundred years later that Maxwell analyzed the dynamics of the fly-ball governor.

The schematic diagram of a speed control system using a fly-ball governor is shown in Fig. 1.9. The governor is directly geared to the output shaft so that the speed of the fly-balls is proportional to the output speed of the engine. The position of the throttle lever sets the desired speed. The lever pivoted as shown in Fig. 1.9 transmits the centrifugal force from the fly-balls to the bottom of the lower seat of the spring. Under steady conditions, the centrifugal force of the fly-balls balances the spring force* and the opening of flow control valve is just sufficient to maintain the engine speed at the desired value.

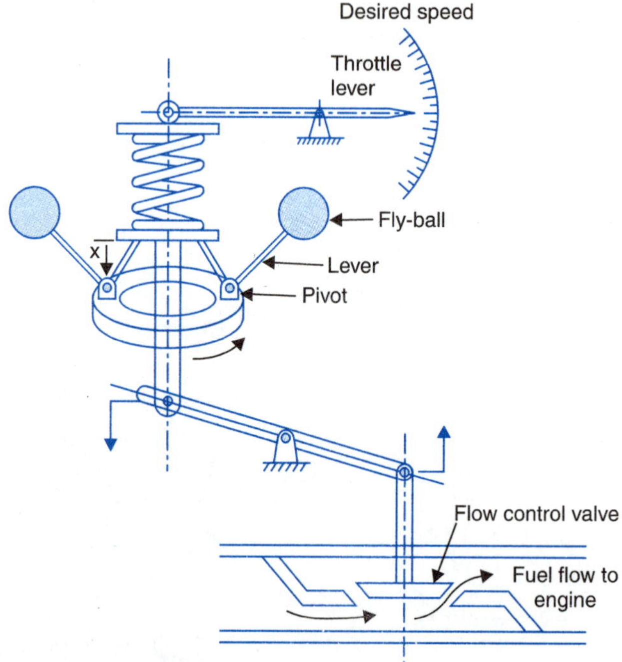

Fig. 1.9. Speed control system.

* The gravitational forces are normally negligible compared to the centrifugal force.

If the engine speed drops below the desired value, the centrifugal force of the fly-balls decreases, thus decreasing the force exerted on the bottom of the spring, causing x to move downward. By lever action, this results in wider opening of the control valve and hence more fuel supply which increases the speed of the engine until equilibrium is restored. If the speed increases, the reverse action takes place.

The change in desired engine speed can be achieved by adjusting the setting of throttle lever. For a higher speed setting, the throttle lever is moved up which in turn causes x to move downward resulting in wider opening of the fuel control valve with consequent increase of speed. The lower speed setting is achieved by reverse action.

The importance of positioning heavy masses like ships and guns quickly and precisely was realized during the World War I. In early 1920, Minorsky performed the classic work on the automatic steering of ships and positioning of guns on the shipboards.

A date of significance in automatic control systems in that of Hazen's work in 1934. His work may possibly be considered as a first struggling attempt to develop some general theory for servomechanisms. The word 'servo' has originated with him.

Prior to 1940 automatic control theory was not much developed and for most cases the design of control systems was indeed an art. During the decade of 1940's, mathematical and analytical methods were developed and practised and control engineering was established as an engineering discipline in its own rights. During the World War II it became necessary to design and construct automatic aeroplane pilots, gun positioning systems, radar tracking systems and other military equipments based on feedback control principle. This gave a great impetus to the automatic control theory.

The missile launching and guidance system of Fig. 1.10 is a sophisticated example of military applications of feedback control. The target plane is sited by a rotating radar antenna which then locks in and continuously tracks the target. Depending upon the position and velocity of the plane as given by the radar output data, the launch computer calculates the firing angle in terms of a launch command signal, which when amplified through a power amplifier drives the launcher (drive motor). The launcher angular position is feedback to the launch computer and the missile is triggered as soon as the error between the launch command signal and the missile firing angle becomes zero. After being fired the missile enters the radar beam which is tracking the target. The control system contained within the missile now receives a guidance signal from the beam which automatically adjusts the control surface of the missile such that the missile rides along the beam, finally homing on to the target.

It is important to note that the actual missile launching and guidance system is far more complex requiring control of gun's bearing as well as elevation. The simplified case discussed above illustrates the principle of feedback control.

The industrial use of automatic control has tremendously increased since the World War II. Modern industrial processes such as manufacture and treatment of chemicals and metals are now automatically controlled.

A simple example of an automatically controlled industrial process is shown in Fig. 1.11. This is a scheme employed in paper mills for reeling paper sheets. For best results the paper sheet must be pulled on to the wind-up roll at nearly constant tension. A reduction in tension will produce a loose roll, while an increase in tension may result in tearing of the paper sheet. If reel speed is constant, the linear velocity of paper and hence its tension increases, as the wind-up roll diameter increases. Tension control may be achieved by suitably varying the reel speed.

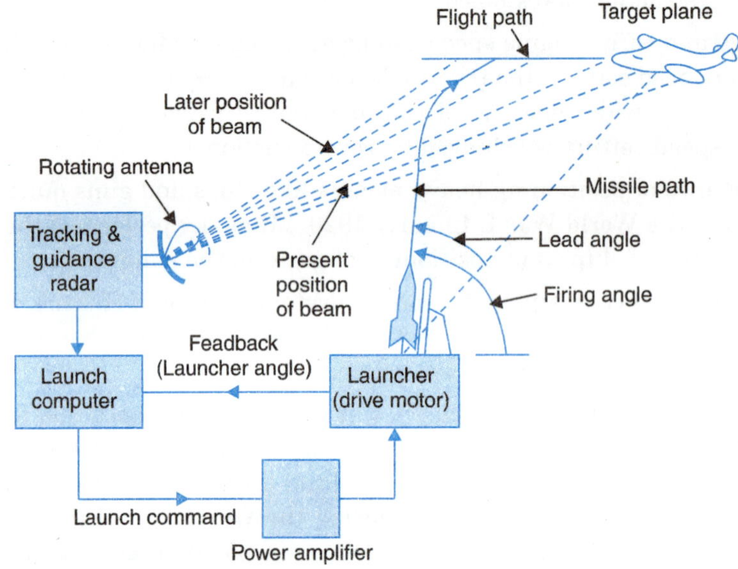

Fig. 1.10. Missile launching and guidance system.

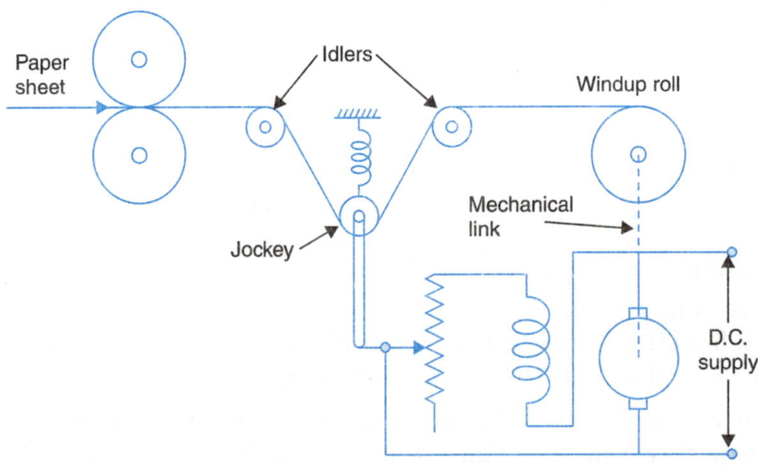

Fig. 1.11. A constant tension reeling system.

In the scheme shown in Fig. 1.11 the paper sheet passes over two idling and one jockey roll. The jockey roll is constrained to vertical motion only with its weight supported by paper tension and spring. Any change in tension moves the jockey in vertical direction, upward for increased tension and downward for decreased tension. The vertical motion of the jockey is used to change the field current of the drive motor and hence the speed of wind-up roll which adjusts the tension.

Another example of controlled industrial processes is a batch chemical reactor shown in Fig. 1.12. The reactants are initially charged into the reaction vessel of the batch reactor and are then agitated for a certain period of time to allow the reaction to take place. Upon completion of the reaction, the products are discharged.

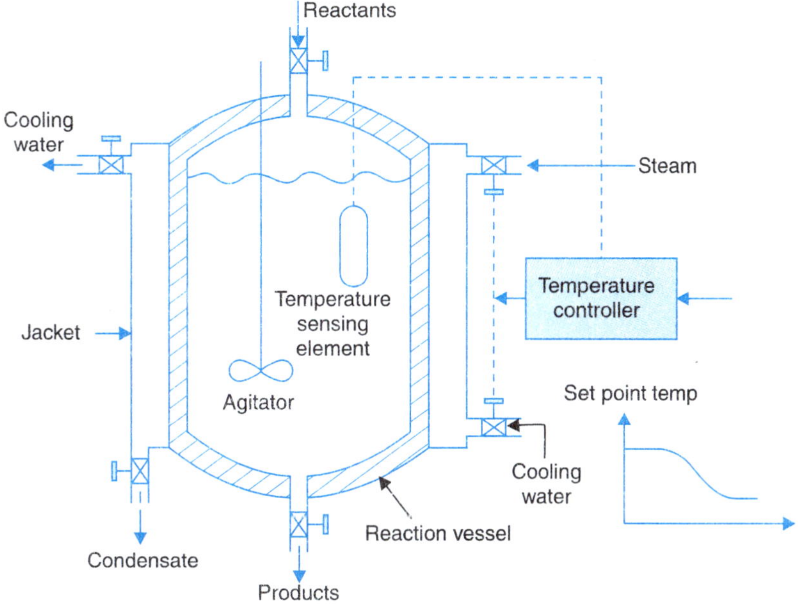

Fig. 1.12. A batch chemical process.

For a specific reaction there is an optimum temperature profile according to which the temperature of the reactor mass should be varied to obtain best results. Automatic temperature control is achieved by providing both steam and cooling water jackets for heating or cooling the reactor mass (cooling is required to remove exothermic heat of reaction during the period the reaction proceeds vigorously). During the heating phase, the controller closes the water inlet valve and opens and controls the steam inlet valve while the condensate valve is kept open. Reverse action takes place during the cooling phase.

Control engineering has enjoyed tremendous growth during the years since 1955. Particularly with the advent of analog and digital computers and with the perfection achieved in computer field, highly sophisticated control schemes have been devised and implemented. Furthermore, computers have opened up vast vistas for applying control concepts to non-engineering fields like business and management. On the technological front fully automated computer control schemes have been introduced for electric utilities and many complex industrial processes with several interacting variables particularly in the chemical and metallurgical processes.

A glorious future lies ahead for automation wherein computer control can run our industries and produce our consumer goods provided we can tackle with equal vigour and success the socio-economic and resource depletion problems associated with such sophisticated degree of automation.

1.4 DIGITAL COMPUTER CONTROL

In some of the examples of control systems of high level of complexity (robot manipulator of Fig. 1.9 and missile launching and guidance system of Fig. 1.11) it is seen that such control systems need a digital computer as a control element to digitally process a number of input signals to generate a number of control signals so as to manipulate several plant variables. In these control systems signals in certain parts of the plant are in analog form *i.e.,* continuous functions of the time variable, while the control computer handles data only in digital (or discrete) form. This requires signal discretization and analog-to-digital interfacing in form of *A/D* and *D/A* converters.

To begin with we will consider a simple form the digital control system knows as sampled-data control system. The block diagram of such a system with single feedback loop is illustrated in Fig. 1.13 wherein the sampler samples the error signal $e(t)$ every T seconds. The sampler is an electronic switch whose output is the discritized version of the analog error signal and is a train of pulses of the sampling frequency with the strength of each pulse being that the error signal at the beginning of the sampling period. The sampled signal is passed through a data hold and is then filtered by a digital filter in accordance with the control algorithm. The smoothed out control signal $u(t)$ is then used to manipulate the plant.

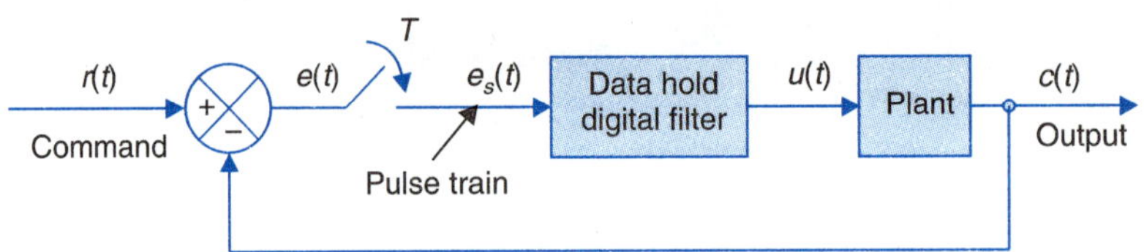

Fig. 1.13. Block diagram of a sampled-data control system.

It is seen above that computer control is needed in large and complex control schemes dealing with a number of input, output variables and feedback channels. This is borne out by the examples of Fig. 1.9 and 1.11. Further in chemical plants, a number of variables like temperatures, pressures and fluid flows have to be controlled after the information on throughput, its quality and its constitutional composition has been analyzed on-line. Such systems are referred to as multivariable control systems whose general block diagram is shown in Fig. 1.14.

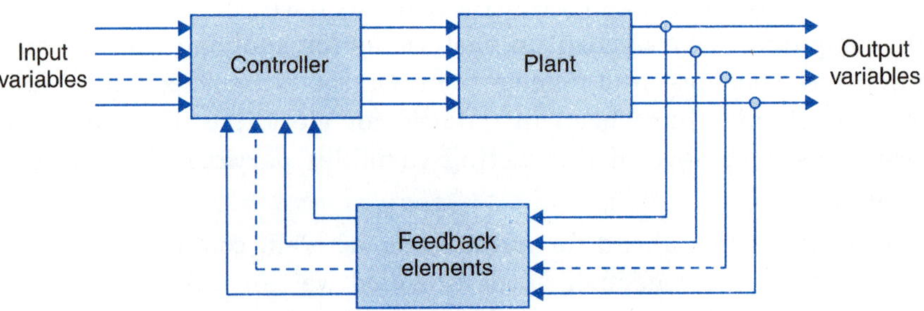

Fig. 1.14. General block diagram of a multivariable control system.

Where a few variable are to be controlled with a limited number of commands and the control algorithm is of moderate complexity and the plant process to be controlled is at a given physical location, a general purpose computer chip, the microprocessor (μP) is commonly employed. Such systems are known as μP-based control systems. Of course at the input/output interfacing A/D and D/A converter chips would be needed.

For large systems a central computer is employed for simultaneous control of several subsystems wherein certain hierarchies are maintained keeping in view the overall system objectives. Additional functions like supervisory control, fault recording, data logging etc, also become possible. We shall advance three examples of central computer control.

Automatic Aircraft Landing System

The automatic aircraft landing system in a simplified form is depicted in Fig. 1.15(a). The system consists of three basic parts: the aircraft, the radar unit and the controlling unit. The radar unit measures the approximate vertical and lateral positions of the aircraft, which are then transmitted to the controlling unit. From these measurements, the controlling unit calculates appropriate pitch and bank commands. These commands are then transmitted to the aircraft autopilots which in turn cause the aircraft to respond.

Assuming that the lateral control system and the vertical control system are independent (decoupled), we shall consider only the lateral control system whose block diagram is given in Fig. 1.15(b). The aircraft lateral position, $y(t)$, is the lateral distance of the aircraft from the extended centerline of the landing area on the deck of the aircraft carrier. The control system attempts to force $y(t)$ to zero. The radar unit measures $y(kT)$ is the sampled value of $y(t)$, with

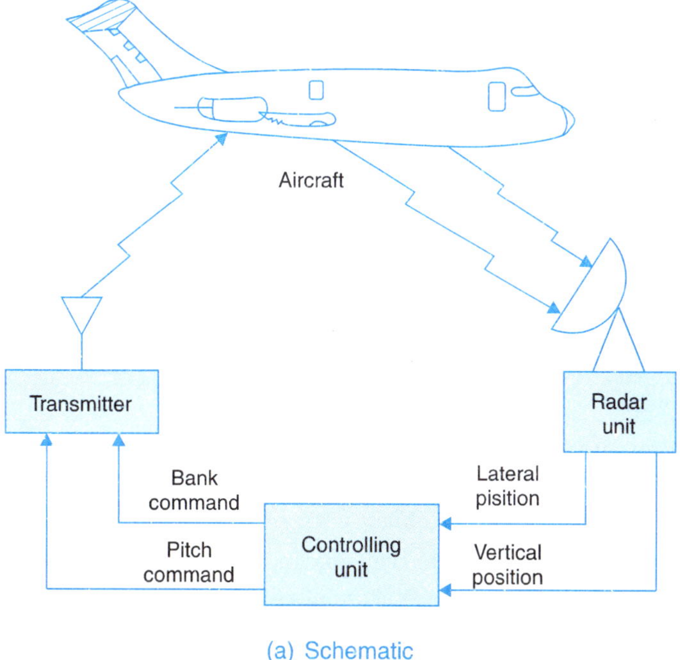

(a) Schematic

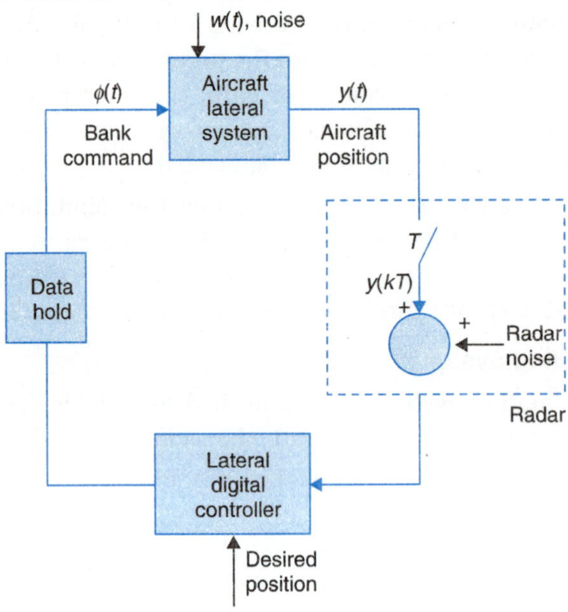

(b) Lateral landing system.

Fig. 1.15. Automatic aircraft landing system.

$T = 0.05$s and $k = 0, 1, 2, 3$ ... The digital controller processes these sampled values and generates the discrete bank command constant at the last value received until the next value is received. Thus the bank commands is updated every $t = 0.05$s, which is called the sampling period. The aircraft responds to the bank command, which changes the lateral position $y(t)$.

It may be noted here that the lateral digital controller must be able to compute the control signal within one sampling period. This is the computational stringency imposed on the central computer in all on-line computer control schemes.

Two unwanted inputs called **disturbances** appear into the system. These are (*i*) wind gust affecting the position of the aircraft and (*ii*) radar noise present in measurement of aircraft position. These are labelled as disturbance input in Figure 1.15(*b*). The system has to be designed to mitigate the effects of disturbance input so that the aircraft lands within acceptable limits of lateral accuracy.

Rocket Autopilot System

As another illustration of computer control, let us discuss an autopilot system which steers a rocket vehicle in response to radioed command. Figure 1.16 shows a simplified block diagram representation of the system.

The state of motion of the vehicle (velocity, acceleration) is fed to the control computer by means of motion sensors (gyros, accelerometers). A position pick-off feeds the computer with the information about rocket engine angle displacement and hence the direction in which the vehicle is heading. In response to heading-commands from the ground, the computer generates a signal which controls the hydraulic actuator and in turn moves the engine.

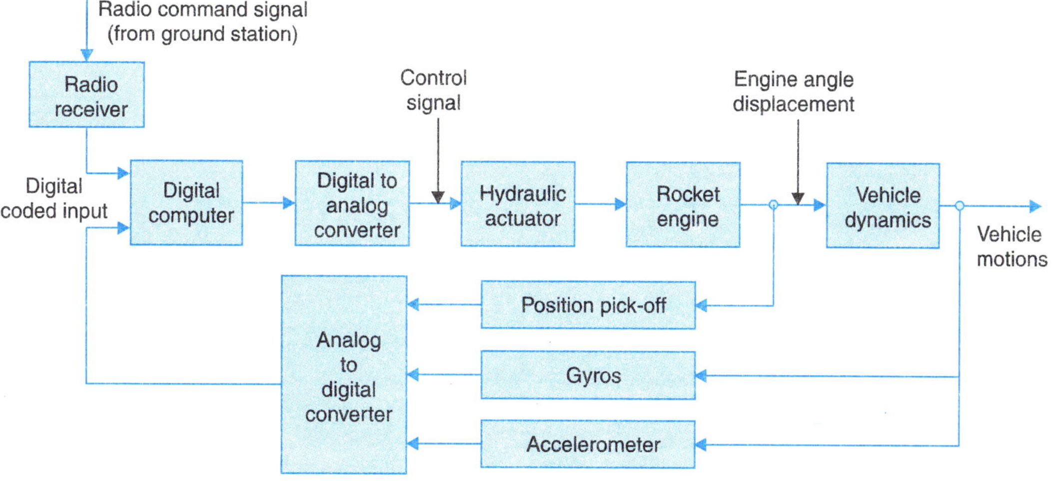

Fig. 1.16. A typical autopilot system.

Coordinated Boiler-Generator Control

Coordinated control system for a boiler-generator unit by a central computer is illustrated by the simplified schematic block diagram of Fig. 1.17. Various signal inputs to the control computer from suitable sensor blocks are:

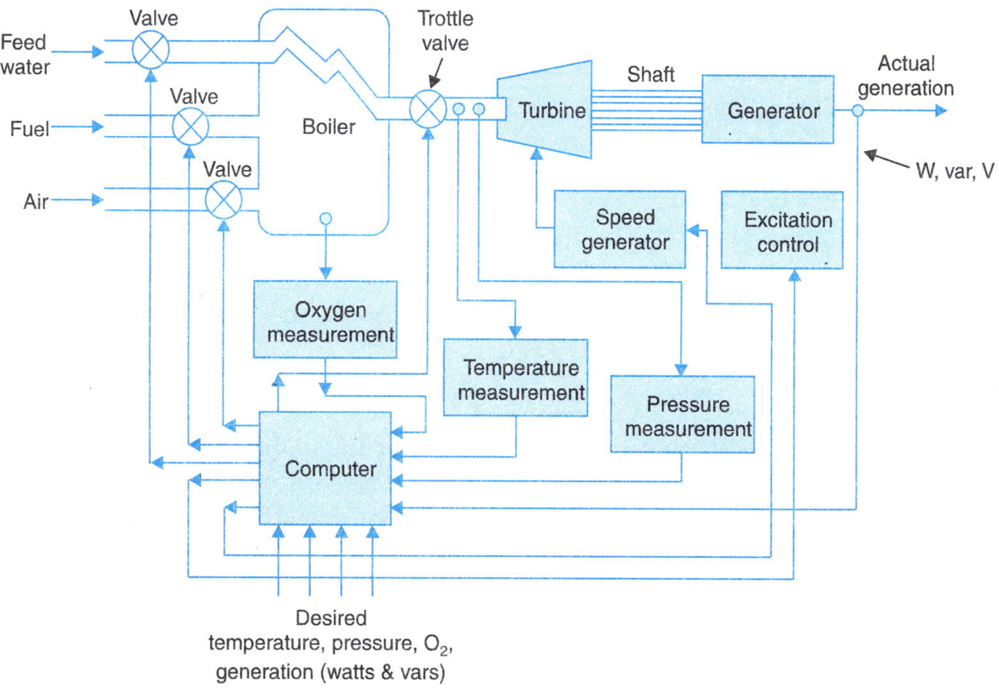

Fig. 1.17. Coordinated control for a boiler-generator.

- Watts, vars, line voltage.
- Temperature and pressure of steam inlet to turbine.
- Oxygen content in furnace air.

These inputs are processed by the control computer by means of a coordinated control algorithm to produce control signals as below:

- Signal to adjust throttle valve. This controls the rate of steam input to turbine and so controls the generator output.
- Signals to adjust fuel, feed water and air in accordance with the throttle valve opening.
- Signals which adjust generator excitation so as to control its var output (which indirectly controls the terminal voltage of the generator).

1.5 APPLICATION OF CONTROL THEORY IN NON-ENGINEERING FIELDS

We have considered in previous sections a number of applications which highlight the potentialities of automatic control to handle various engineering problems. Although control theory originally evolved as an engineering discipline, due to universality of the principles involved it is no longer restricted to engineering confines in the present state of art. In the following paragraphs we shall discuss some examples of control theory as applied to fields like economics, sociology and biology.

Consider an economic inflation problem which is evidenced by continually rising prices. A model of the vicious price-wage inflationary cycle, assuming simple relationship between wages, product costs and cost of living is shown in Fig 1.18. The economic system depicted in this figure is found to be a positive feedback system.

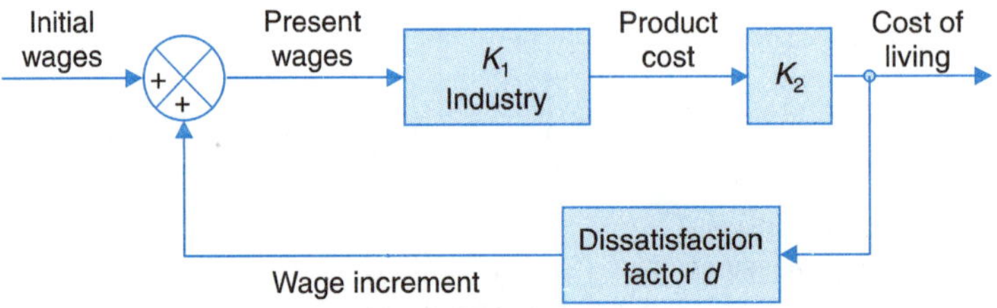

Fig. 1.18. Economic inflation dynamics.

To introduce yet another example of non-engineering application of control principles, let us discuss the dynamics of epidemics in human beings and animals. A normal healthy community has a certain rate of daily contracts C. When an epidemic disease affects this community the social pattern is altered as shown in Fig. 1.19. The factor K_1 contains the

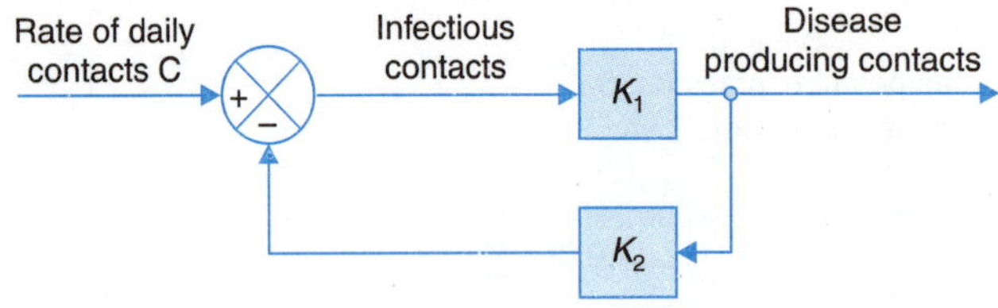

Fig. 1.19. Block diagram representation of epidemic dynamics.

statistical fraction of infectious contacts that actually produce the disease, while the factor K_2 accounts for the isolation of the sick people and medical immunization. Since the isolation and immunization reduce the infectious contacts, the system has a negative feedback loop.

In medical field, control theory has wide applications, such as temperature regulation, neurological, respiratory and cardiovascular controls. A simple example is the automatic anaesthetic control. The degree of anaesthesia of a patient undergoing operation can be measured from encephalograms. Using control principles anaesthetic control can be made completely automatic, thereby freeing the anaesthetist from observing constantly the general condition of the patient and making manual adjustments.

The examples cited above are somewhat over-simplified and are introduced merely to illustrate the universality of control principles. More complex and complete feedback models in various non-engineering fields are now available. This area of control is under rapid development and has a promising future.

1.6 THE CONTROL PROBLEM

In the above account the field of control systems has been surveyed with a wide variety of illustrative examples including those of some nonphysical systems. The basic block diagram of a control system given in Fig. 1.3 is reproduced in Fig. 1.20 wherein certain alternative block and signal nomenclature are introduced.

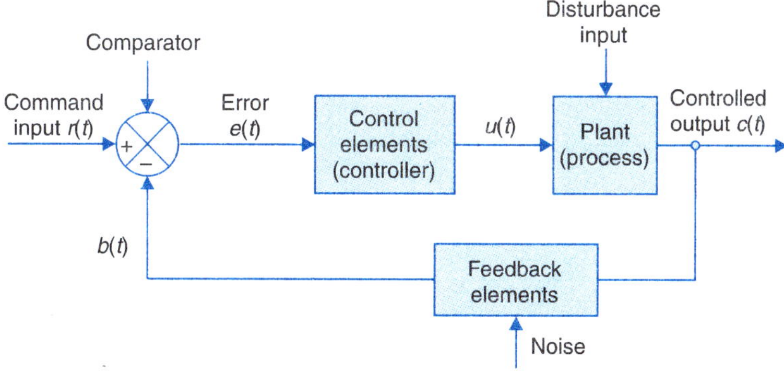

Fig. 1.20. The basic control loop.

Further the figure also indicates the presence of the disturbance input (load disturbance) in the plant and noise input in feedback element (noise enters in the measurement process; see example of automatic aircraft landing system in Fig. 1.15). This basic control loop with negative feedback responds to reduce the error between the command input (desired output) and the controlled output.

Further as we shall see in later chapters that negative feedback has several benefits like reduction in effects of disturbances input, plant nonlinearities and changes in plant parameters. A multivariable control system with several feedback loops essentially follows the same logic. In some mechanical systems and chemical processes a certain signal also is directly input to the controller elements particularly to counter the effect of load disturbance (not shown in the figure).

Generally, a controller (or a filter) is required to process the error signal such that the overall system statisfies certain criteria specifications. Some of these criteria are:

1. Reduction in effect of disturbance signal.
2. Reduction in steady-state errors.
3. Transient response and frequency response performance.
4. Sensitivity to parameter changes.

Solving the control problem in the light of the above criteria will generally involve following steps:

1. Choice of feedback sensor(s) to get a measure of the controlled output.
2. Choice of actuator to drive (manipulate) the plant like opening or closing a valve, adjusting the excitation or armature voltage of a motor.
3. Developing mathematical models of plant, sensor and actuator.
4. Controller design based on models developed in step 3 and the specified criteria.
5. Simulating system performance and fine tuning.
6. Iterate the above steps, if necessary.
7. Building the system or its prototype and testing.

The criteria and steps involved in system design and implementation and tools of analysis needed of this, form the subject matter of the later chapters.

2

MATHEMATICAL MODELS OF PHYSICAL SYSTEMS

2

MATHEMATICAL MODELS OF PHYSICAL SYSTEMS

2.1 INTRODUCTION

A *physical system* is a collection of physical objects connected together to serve an objective. Examples of a physical system may be cited from laboratory, industrial plant or utility services– an electronic amplifier composed of many components, the governing mechanism of a steam turbine or a communications statellite orbiting the earth are all examples of physical systems. A more general term *system* is used to describe a combination of components which may not all be physical, *e.g.*, biological, economic, socio-economic or management systems. Study in this book will be mainly restricted to physical systems though a few examples of general type systems will also be introduced.

No physical system can be represented in its full physical intricacies and therefore idealizing assumptions are always made for the purpose of analysis and synthesis of systems. An idealized physical system is called a *physical model.* A physical system can be modelled in a number of ways depending upon the specific problem to be dealt with and the desired accuracy. For example, an electronic amplifier may be modelled as an interconnection of linear lumped elements, or some of these may be pictured as nonlinear elements in case the stress is on distortion analysis. A communication satellite may be modelled as a point, a rigid body or a flexible body depending upon the type of study to be carried out. As idealizing assumptions are gradually removed for obtaining a more accurate model, a point of diminishing return is reached, *i.e.,* the gain in accuracy of representation is not commensurate with the increased complexity of the computation required. In fact, beyond a certain point there may indeed be an undetermined loss in accuracy of representation due to flow of errors in the complex computations.

Once a physical model of a physical system is obtained, the next step is to obtain a *mathematical model* which is the mathematical representation of the physical model through use of appropriate physical laws. Depending upon the choice of variables and the coordinate system, a given physical model may lead to different mathematical models. A network, for example, may be modelled as a set of nodal equations using Kirchhoff's current law or a set of mesh equations using Kirchhoff's voltage law. A control system may be modelled as a scalar differential equation describing the system or state variable vector-matrix differential equation. The particular mathematical model which gives a greater insight into the dynamic behaviour of physical system is selected.

When the mathematical model of a physical system is solved for various input conditions, the result represents the dynamic response of the system. The mathematical model of a system is *linear,* if it obeys the *principle of superposition and homogeneity.* This principle implies that if a system model has responses $y_1(t)$ and $y_2(t)$ to any two inputs $x_1(t)$ and $x_2(t)$ respectively, then the system response to the linear combination of these inputs

$$\alpha_1 x_1(t) + \alpha_2 x_2(t)$$

is given by the linear combination of the individual outputs, *i.e.,*

$$\alpha_1 y_1(t) + \alpha_2 y_2(t)$$

where α_1 are α_2 are constants.

Mathematical models of most physical systems are characterized by differential equations. A mathematical model is linear, if the differential equation describing it has coefficients, which are either functions only of the independent variable or are constants. If the coefficients of the describing differential equations are functions of time (the independent variable), then the mathematical model is *linear time-varying.* On the other hand, if the coefficients of the describing differential equations are constants, the model is *linear time-invariant.*

The differential equation describing a linear time-invariant system can be reshaped into different forms for the convenience of analysis. For example, for transient response or frequency response analysis of single-input-single-output linear systems, *the transfer function representation* (to be discussed later in this chapter) forms a useful model. On the other hand, when a system has multiple inputs and outputs, the *vector-matrix notation* (discussed in Chapter 12) may be more convenient. The mathematical model of a system having been obtained, the available mathematical tools can then be utilized for analysis or synthesis of the system.

Powerful mathematical tools like the Fourier and Laplace transforms are available for use in linear systems. Unfortunately no physical system in nature is perfectly linear. Therefore certain assumptions must always be made to get a linear model which, as pointed out earlier, is a compromise between the simplicity of the mathematical model and the accuracy of results obtained from it. However, it may not always be possible to obtain a valid linear model, for example, in the presence of a strong nonlinearity or in presence of distributive effects which can not be represented by lumped parameters.

A commonly adopted approach for handling a new problem is: first build a simplified model, linear as far as possible, by ignoring certain nonlinearities and other physical properties which may be present in the system and thereby get an approximate idea of the dynamic response of a system; a more complete model is then built for more complete analysis.

2.2 DIFFERENTIAL EQUATIONS OF PHYSICAL SYSTEMS

This section presents the method of obtaining differential equation models of physical systems by utilizing the physical laws of the process. Depending upon the system well-known physical laws like Newton's laws, Kirchhoff's laws, etc. will be used to build mathematical models.

We shall in first step build the physical model of the system as interconnection of idealized system elements and describe these in form of elemental laws. These idealized elements are sort of building blocks of the system. An ideal element results by making two basic assumptions.

1. Spatial distribution of the element is ignored and it is regarded as a point phenomen. Thus mass which has physical dimensions, is considered concentrated at a point and temperature in a room which is distributed out into the whole room space is replaced by a representative temperature as if of a single point in the room.

   The process of ignoring the spatial dependence by choosing a representative value is called lumping and the corresponding modelling is known as lumped-parameter modelling as distinguished from the distributed parameter modelling which accounts for space distribution.

2. We shall assume that the variables associated with the elements lie in the range that the element can be described by simple linear law of (*i*) a constant of proportionality or (*ii*) a first-order derivative or (*iii*) a first-order integration.

The last two forms are in fact alternatives and can be interconverted by a single differentiation or integration.

To begin with we shall consider ideal elements which have a single-port or two-terminal representation and so have two variables associated with it as shown in Fig. 2.1. These variables are indentified as

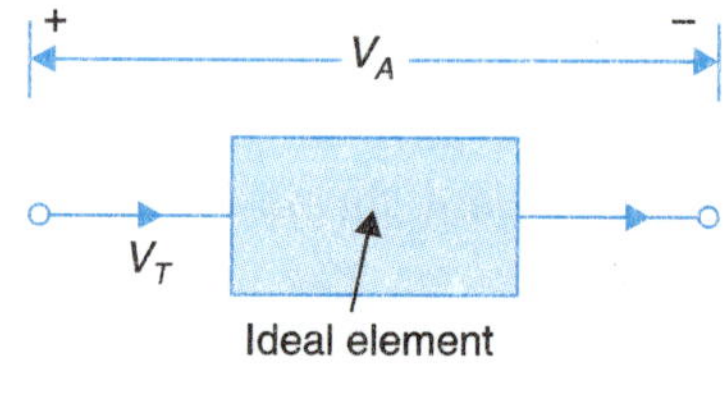

Fig. 2.1

1. **Through variable V_T** which sort of passes through the element and so has the same value in at one port and out at the other. For example, current through an electrical resistance.
2. **Across variable V_A** which appears across the two terminals of the element. For example, voltage across an electrical resistance.

Another classification of the element variables is

1. **Input variable** or independent variable (V_i)
2. **Output variable** or dependent (response) variable (V_o)

Thus V_i could be V_T or V_A and corresponding V_o would be V_A or V_T. The element is then represented in *block diagram* form (**cause-effect form**) as in Fig. 2.2 wherein the signal V_i flows into the block and flows out of it (V_0) suitably modified by the law represented by the block.

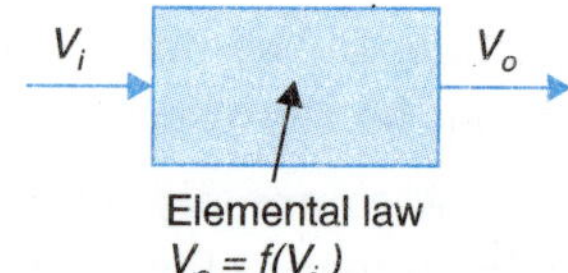

Fig. 2.2. Block diagram of element.

Mechanical Systems

Mechanical systems and devices can be modelled by means of three* ideal translatory and three ideal rotary elements. Their diagrammatic representation and elemental relationships are given in Fig. 2.3. In case of mass/inertia elements it may be noted that one terminal is always the inertial reference frame with respect to which the free terminal moves/rotates.

Through and Across variables for ideal mechanical elements* of Fig. 2.3 are indentified in Table 2.1 along with their units.

Table 2.1. Variables of Mechanical Elements

	Through variable	***Integrated through variable***	***Across variable***	***Integrated across variable***
Translational elements	Force, F $N = \text{kg-m/s}^2$	Translational momentum $p = \int_{-\infty}^{t} F dt$ $(N\text{-}s)$	Velocity difference $v = v_1 - v_2$ (m/s)	Displacement difference $x = x_1 - x_2$ (m)
Rotational elements	Torque, T $(N\text{-}m)$	Angular momentum $h = \int_{-\infty}^{t} T dt$ $(N\text{-}ms)$	Angular velocity difference $\omega = \omega_1 - \omega_2$ (rad/s)	Angular Displacement difference $\theta = \theta_1 - \theta_2$ (rad)

Mass/inertia and the two kinds of springs are the energy storage elements where in energy can be stored and retrieved without loss and so these are called conservative elements. Energy stored in these elements in expressed as:

Mass : $E = (1/2)\, Mv^2$ = kinetic energy (J) ; motional energy

Inertia : $E = (1/2)\, J\omega^2$ = kinetic energy (J) ; motional energy

Spring (translatory) : $E = 1/2\, Kx^2$ = potential energy (J) ; deformation energy

Spring (torsional) : $E = 1/2\, K\theta^2$ = potential energy (J) ; deformation energy

Damper is a **dissipative** element and power it consumes (lost in form of heat) is given as

$$P = fv^2 \, (W)$$
$$= f\omega^2 \, (W)$$

The elemental relationships in Fig. 2.3 are not expressed in momentum form (which is used mainly in **impulse excitation**). For illustration the relationship of mass can be integrated and expressed as

$$\int_{-\infty}^{t} F dt = M \int_{-\infty}^{t} (dv / dt)\, dt \qquad \text{or} \qquad p = Mv \; ; \text{if } v(-\infty) = 0$$

*Another element generally needed is the gear train which will be considered later in this Section.

(1) The mass element

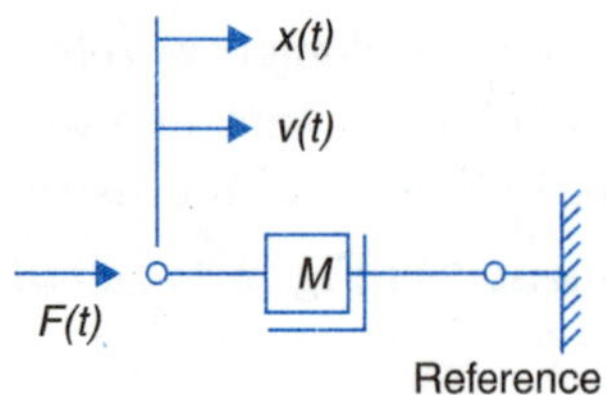

$$F = M\frac{dv}{dt} = M\frac{d^2x}{dt^2}$$

(2) The spring element

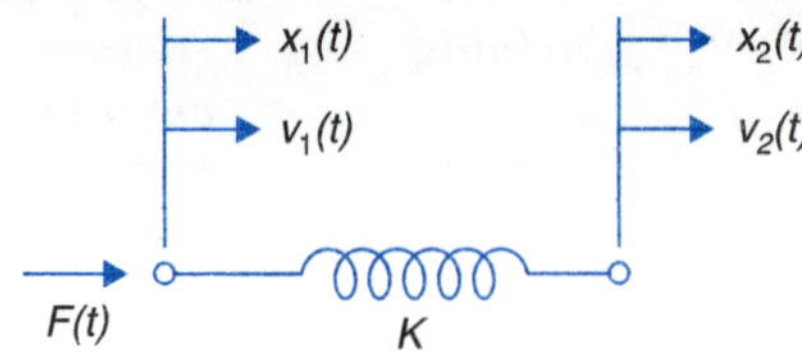

$$F = K(x_1 - x_2) = Kx = K\int_{-\infty}^{t}(v_1 - v_2)dt = K\int_{-\infty}^{t} v\,dt$$

(3) The damper element

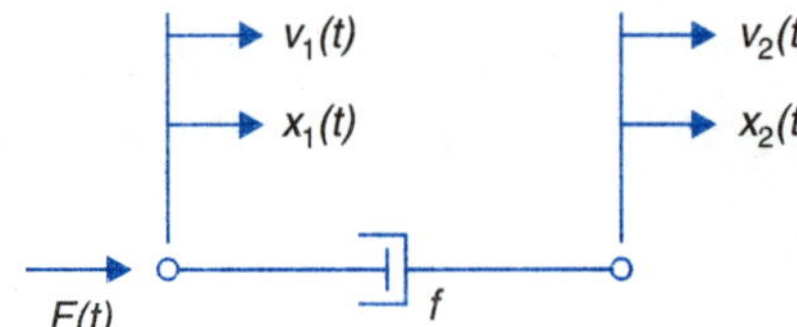

$$F = f(v_1 - v_2) = fv = f(\dot{x}_1 - \dot{x}_2) = f\dot{x}$$

x(m), v(m/sec), M(kg), F(newton), K(newton/m), f(newton per m/sec)

Rotational Elements

(4) The inertia element

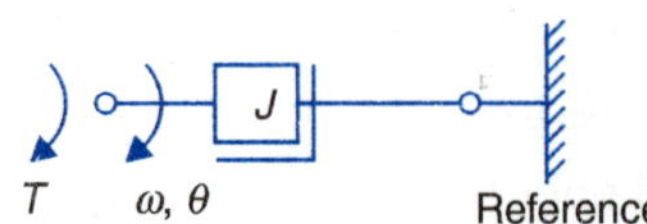

$$T = J\frac{d\omega}{dt} = J\frac{d^2\theta}{dt^2}$$

(5) The torsional spring element

K

T θ_1, ω_1 θ_2, ω_2

$$T = K(\theta_1 - \theta_2) = K\theta = K\int_{-\infty}^{t}(\omega_1 - \omega_2)dt = K\int_{-\infty}^{t}\omega\,dt$$

(6) The damper element

f

T ω_1, θ_1 ω_2, θ_2

$$T = f(\omega_1 - \omega_2) = f\omega = f(\dot{\theta}_1 - \dot{\theta}_2) = f\dot{\theta}$$

θ(rad), ω(rad/sec), J(kg-m^2), T(newton-m)

K(newton-m/rad), f(newton-m per rad/sec)

Fig. 2.3. Ideal elements for mechanical systems.

A mechanical system which is modelled using the three ideal elements presented above would yield a mathematical model which is an ordinary differential equation. Before we advance examples of this type of modelling, we will examine in some detail the friction which has been modelled as a linear element, the damper.

Friction

The friction exists in physical systems whenever mechanical surfaces are operated in sliding contact. The friction encountered in physical systems may be of many types:

(*i*) **Coulomb friction force:** The force of sliding friction between dry surfaces. This force is substantially constant.

(*ii*) **Viscous friction force:** The force of friction between moving surfaces separated by viscous fluid or the force between a solid body and a fluid medium. This force is approximately linearly proportional to velocity over a certain limited velocity range.

(*iii*) **Stiction:** The force required to initiate motion between two contacting surfaces (which is obviously more than the force required to maintain them in relative motion).

In most physical situations of interest, the viscous friction predominates. The ideal relation given in Fig. 2.3 is based on this assumption.

The friction force acts in a direction opposite to that of velocity. However, it should be realised that friction is not always undesirable in physical systems. Sometimes it may even be necessary to introduce friction intentionally to improve the dynamic response of the system (discussed in Chapter 5). Friction may be introduced intentionally in a system by use of a dashpot shown in Fig. 2.4. It consists of a piston and oil filled cylinder with a narrow annular passage between piston and cylinder. Any relative motion between piston and cylinder is resisted by oil with a friction force (fv).

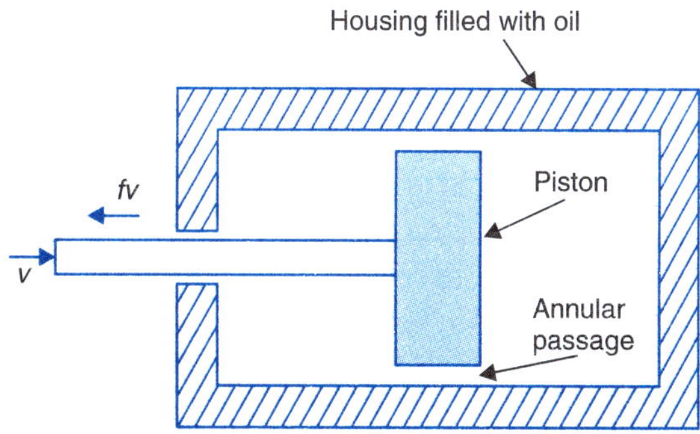

Fig. 2.4. Dashpot construction.

Translational Systems

Let us consider now the mechanical system shown in Fig. 2.5 (a). It is simply a mass M attached to a spring (stiffness K) and a dashpot (viscous friction coefficient f) on which the force F acts. Displacement x is positive in the direction shown. The zero position is taken to be at the point where the spring and mass are in static equilibrium.*

---

*Note that the gravitational effect is eliminated by this choice of zero position.

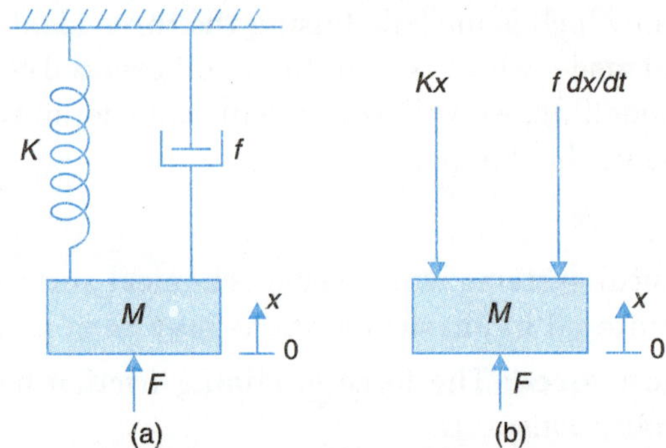

Fig. 2.5. (a) A mass-spring-dashpot system; (b) Free-body diagram.

The systematic way of analyzing such a system is to draw a free-body diagram* as shown in Fig. 2.5 (*b*). Then by applying Newton's law of motion to the free-body diagram, the force equation can be written as

$$F - f\frac{dx}{dt} - Kx = M\frac{d^2x}{dt^2} \qquad \text{or} \qquad F = M\frac{d^2x}{dt^2} + f\frac{dx}{dt} + Kx \qquad ...(2.1)$$

Equation (2.1) is a linear, constant coefficient differential equation of second-order. Also observe that the system has two storage elements (mass M and spring K).

Mechanical Accelerometer

In this simplest form, an accelerometer consists of a spring-mass-dashpot system shown in Fig. 2.6. The frame of the accelerometer is attached to the moving vehicle.

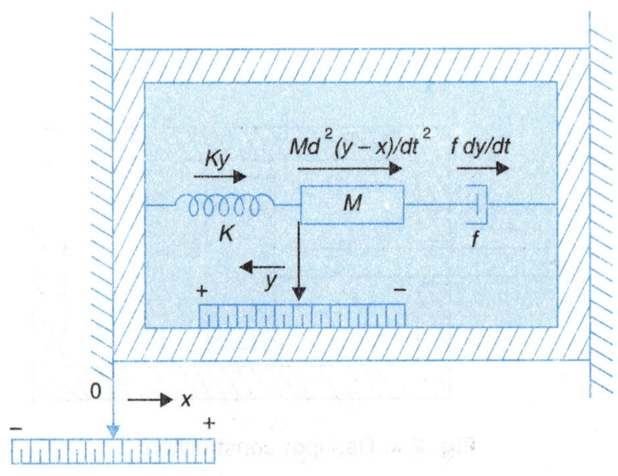

Fig. 2.6. Simplified diagram of an accelerometer.

Whenever the moving vehicle and hence the frame of the accelerometer is accelerated, the spring deflects until it produces enough force to accelerate the mass at the same rate as the

* In example (2.1), we shall see that there is one to one correspondence between free-body diagram approach and nodal method of analysis.

frame. The deflection of the spring which may be measured by a linear-motion potentiometer is a direct measure of acceleration.

Let

x = displacement of the moving vehicle (or frame) with respect to a fixed reference frame.

y = displacement of the mass M with respect to the accelerometer frame.

The positive directions for x and y are indicated on the diagram. Since y is measured with respect to the frame, the force on the mass due to spring is $-Ky$ and due to viscous friction is $-f\frac{dy}{dt}$. The motion of the mass with respect to the fixed reference frame in the positive direction of y is $(y - x)$.

The force equation for the system becomes

$$M\frac{d^2(y-x)}{dt^2} + f\frac{dy}{dt} + Ky = 0$$

or

$$M\frac{d^2y}{dt^2} + f\frac{dy}{dt} + Ky = M\frac{d^2x}{dt^2} = Ma \qquad \text{...(2.2)}$$

where a is the input acceleration.

If a constant acceleration is applied to the accelerometer, the output displacement y becomes constant under steady-state as the derivatives by y become zero, *i.e.*,

$$Ma = Ky \qquad \text{or} \qquad a = \left(\frac{K}{M}\right)y$$

The steady-state displacement y is thus a measure of the constant input acceleration. This instrument can also be used for displacement measurements as explained later in Section 2.3.

Nonlinear Spring

No spring is linear over an arbitrary range of extensions–in fact that is true of all physical as well as nonphysical systems. The linear spring elemental law

$$F = Ky$$

is applicable within a limited range of extension y measured beyond the unstretched end of the spring as in Fig. 2.7. Where large extensions are encountered the spring law changes to

$$F = Ky^2 \qquad \text{...(2.3)}$$

which is graphically represented in Fig. 2.8. This law does not obey the principle of superposition as shown below:

$$F_{S_1} = Ky_1 \; ; y_1 = \text{spring extension is linear range}$$

$$F_{S_2} = Ky_2^2 \; ; y_2 = \text{spring extension in nonlinear range}$$

$$F_{S_2} = F_{S_1} + F_{S_2}, y = y_1 + y_2 = \text{total spring extension}$$

$$F_S = Ky_1 + Ky_2^2 \neq Ky^2$$

So this behaviour (response) of the spring is not linear **(nonlinear)**

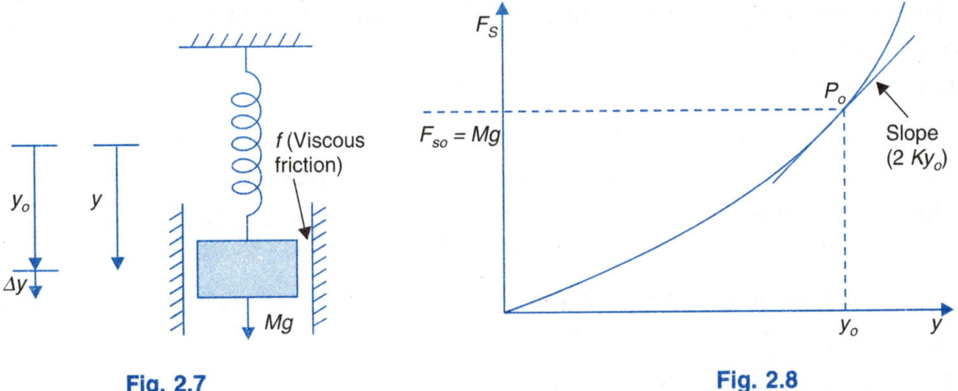

Fig. 2.7

Fig. 2.8

Linearization

Consider the mass-spring system of Fig. 2.7, under gravitational force Mg which when large pushes the spring to nonlinear region of operation. From the freebody diagram we can write the describing equation of the system as

$$Mg = M\ddot{y} + f\dot{y} + Ky^2 \qquad ...(2.4)$$

Under steady condition (rest position of the mass) all derivatives of y are zero. So eqn. (2) gives

$$Mg = Ky_o^{\,2} \text{ or } y_o = \sqrt{(Mg/K)} \qquad ...(2.5)$$

Consider now that the system moves through a small value Δy about y_0. Equation (2.4) can then be written as

$$Mg = M\frac{d^2}{dt^2}(y_o + \Delta y) + f\frac{d}{dt}(y_o + \Delta y) + K(y_o + \Delta y)^2 \qquad ...(2.6)$$

The nonlinear spring term can be approximated by retaining the first derivative term in Taylor series, *i.e.,*

$$y = K(y_o + \Delta y)^2 = Ky_o^2 + \frac{d}{dy}(Ky^2)\Big|_{y = y_o} \Delta y = Ky_o^2 + (2Ky_o^2)\Delta y$$

Substituting this approximated value we get

$$Mg = M\frac{d^2}{dt^2}(y_o + \Delta y) + f\frac{d}{dt}(y_o + \Delta y) + Ky_o + (2Ky_o^2)\Delta y$$

or

$$0 = M\frac{d^2}{dt^2}(\Delta y) + f\frac{d}{dt}(\Delta y) + (2Ky_o^2)\Delta y \qquad ...(2.7)$$

It is seen from eqn. (2.7) that the spring behaviour for small movement around $P_o(F_o = Mg, y_o)$ in Fig. 2.8 is linear with spring constant modified to $(2Ky_0)$, the slope of the spring characteristic at the point P_0, called the **operating point.**

If we relabel $x = \Delta y$ and also apply an external force F in positive direction of x (*i.e.,* downwards), eqn. (2.7) becomes

$$F = M\ddot{x} + f\dot{x} + K_o x \; ; \; K_o = 2Ky_o \qquad ...(2.8)$$

The technique of linearization presented above is also known as **small-signal** modelling. This is commonly used in **automatic regulating systems** which operate in a narrow range around the set point. It will be used in this section in modelling of hydraulic and penumatic systems which are otherwise nonlinear. The technique of small-signal linearization will be elaborated in Section 2.4 for multivariable components/devices.

Levered Systems

An ideal (mass and friction less) lever is shown in Fig. 2.9 (*a*) so long as the rotation θ about the axis of the lever is small

$$x = a\theta \text{ and } y = b\theta$$

so that $$\frac{x}{y} = \frac{a}{b}\text{; displacement ratio} \qquad \text{...(2.9}a\text{)}$$

also $$aF_1 = bF_2 \qquad \text{...(2.9}b\text{)}$$

or $$\frac{F_1}{F_2} = \frac{b}{a}\text{ ; force advantage}$$

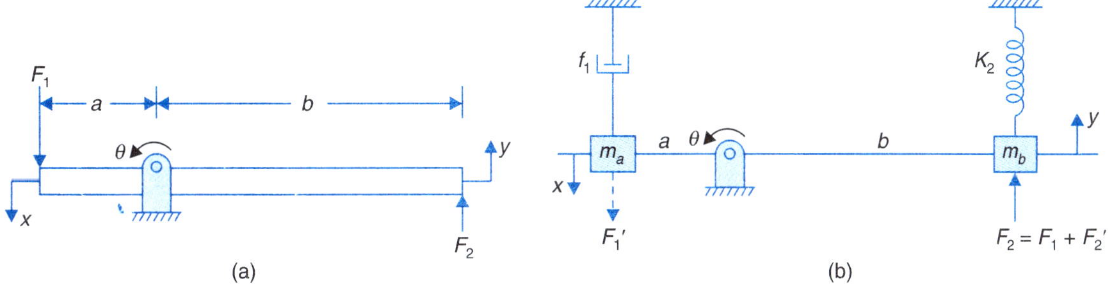

Fig. 2.9. Levered systems.

Consider now the levered system of Fig. 2.9 (*b*). External force F_2 acting on m_b comprises two components, *i.e.*,

$$F_2 = F_1 + F_2'$$

F_2' acts on (m_b, K_2) subsystem and F_1' the reflection of F_1 at *a*-end of the lever acts on (m_a, f_1) subsystem. The dynamical equations for the two systems are written as

$$F_2' = m_b\ddot{y} + K_2 y \qquad \text{...(2.10}a\text{)}$$

$$F_1' = m_a\ddot{x} + f_1\dot{x} \qquad \text{...(2.10}b\text{)}$$

But $F_1' = (b/a)\, F_1$ and $x = (a/b)y$. Substituting in eqn. (2.10*b*), we get

$$(b/a)\, F_1 = (a/b)\, m_a\ddot{y} + (a/b)\, f_1\dot{y}$$

or $$F_1 = (a/b)^2\, m_a\ddot{y} + (a/b)^2 f_1 y = m_a'\ddot{y} + f_1\ \dot{y}. \qquad \text{...(2.11)}$$

where $m_a' = (a/b)^2\, m_a$; mass at end '*a*' reflected at end '*b*' of the lever.

$f_1' = (a/b)^2 f_1$ friction at end '*a*' reflected at end '*b*' of the lever

Adding eqns. (2.10*a*) and (2.10*b*), we have

$$F_2 = F_1 + F_2' = (m_a' + m_b)\ \ddot{y} + f_1'\ \dot{y} + K_2 y \qquad \text{...(2.12)}$$

Equation (2.12) can be written down directly by reflecting the parameters from one end of the lever to the other in the inverse square displacement ratio of the lever.

Rotational Systems

Mechanical systems involving fixed-axis rotation occur in the study of machinery of many types and are very important. The modelling procedure is very close to that used in translation. In these systems, the variables of interest are the torque and angular velocity (or displacement). The three basic components for rotational systems are: moment of inertia, torsional spring and viscous friction.

The three ideal rotational elements with their relevant properties and conventions are shown in Fig. 2.3.

Let us consider now, the rotational mechanical system shown in Fig. 2.10 (*a*) which consists of a rotatable disc of moment of inertia J and a shaft of stiffness K. The disc rotates in a viscous medium with viscous friction coefficient f.

Let T be the applied torque which tends to rotate the disc. The free-body diagram is shown in Fig. 2.10 (*b*).

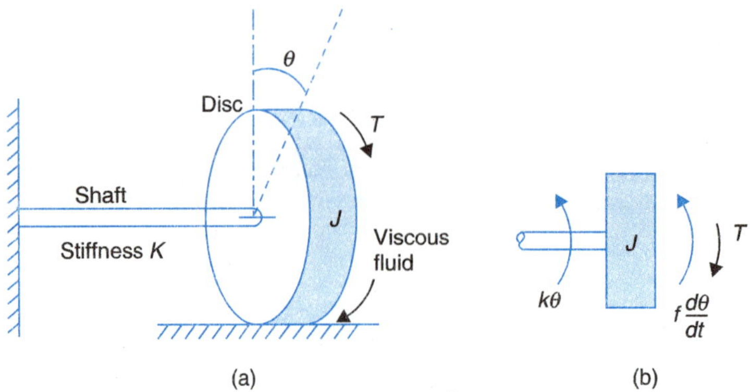

Fig. 2.10. (a) Rotational mechanical system; (b) Free-body diagram.

The torque equation obtained from the free-body diagram is

$$T - f\frac{d\theta}{dt} - K\theta = J\frac{d^2\theta}{dt^2} \quad \text{or} \quad T = J\frac{d^2\theta}{dt^2} + f\frac{d\theta}{dt} + K\theta \qquad ...(2.13)$$

Equation (2.13) is a linear constant coefficient differential equation describing the dynamics of the system shown in Fig. 2.10 (*a*). Again observe that the system has two storage elements, inertia J and shaft of stiffness K.

Gear Trains

Gear trains are used in control systems to attain the mechanical matching of motor to load. Usually a servomotor operates at high speed but low torque. To drive a load with high torque and low speed by such a motor, the torque magnification and speed reduction are achieved by gear trains. Thus in mechanical systems gear trains act as matching devices like transformers in electrical systems.

Figure 2.11 shows a motor driving a load through a gear train which consists of two gears coupled together. The gear with N_1 teeth is called the primary gear (analogous to primary winding of a transformer) and gear with N_2 teeth is called the secondary gear.

Angular displacements of shafts 1 and 2 are denoted by θ_1 and θ_2 respectively. The moment of inertia and viscous friction of motor and gear 1 are denoted by J_1 and f_1 and those of gear 2 and load are denoted by J_2 and f_2 respectively.

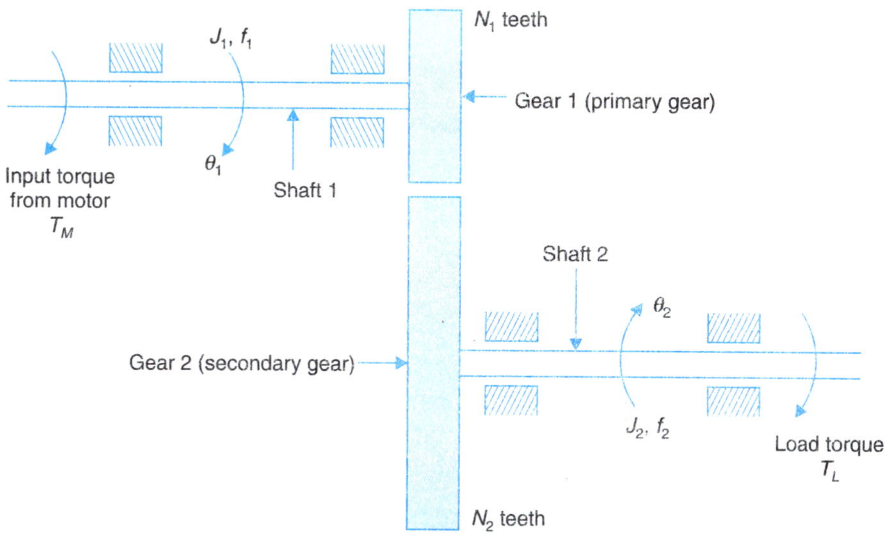

Fig. 2.11. Gear train system.

For the first shaft, the differential equation is

$$J_1\ddot{\theta}_1 + f_1\dot{\theta}_1 + T_1 = T_M \quad \text{...(2.14)}$$

where T_m is the torque developed by the motor and T_1 is the load torque on gear 1 due to the rest of the gear train.

For the second shaft $J_2\ddot{\theta}_2 + f_2\dot{\theta}_2 + T_L = T_2$ ...(2.15)

where T_2 is the torque transmitted to gear 2 and T_L is the load torque.

Let r_1 be the radius of gear 1 and r_2 be that of gear 2. Since the linear distance travelled along the surface of each gear is same, $\theta_1 r_1 = \theta_2 r_2$. The number of teeth on gear surface being proportional to gear radius, we obtain

$$\frac{\theta_2}{\theta_1} = \frac{N_1}{N_2} \quad \text{...(2.16)}$$

Here the stiffness of the shafts of the gear train is assumed to be infinite. In an ideal case of no loss in power transfer, the work done by gear 1 is equal to that of gear 2. Therefore,

$$T_1\theta_1 = T_2\theta_2 \quad \text{...(2.17)}$$

Combining eqns. (2.16) and (2.17) we have

$$\frac{T_1}{T_2} = \frac{\theta_2}{\theta_1} = \frac{N_1}{N_2} \quad \text{...(2.18)}$$

Differentiating θ_1 and θ_2 in eqn. (2.18) twice, we have the following relation for speed and acceleration.

$$\frac{\ddot{\theta}_2}{\ddot{\theta}_1} = \frac{\dot{\theta}_2}{\dot{\theta}_1} = \frac{N_1}{N_2} \quad \text{...(2.19)}$$

Thus if $N_1/N_2<1$, from eqns. (2.18) and (2.19) it is found that the gear train reduces the speed and magnifies the torque.

Eliminating T_1 and T_2 from eqns. (2.14) and (2.15) with the help of eqns. (2.18) and (2.19), we obtain

$$J_1\ddot{\theta}_1 + f_1\dot{\theta}_1 + \frac{N_1}{N_2}(J_2\ddot{\theta}_2 + f_2\dot{\theta}_2 + T_L) = T_M \qquad ...(2.20)$$

Elimination of θ_2 from eqn. (2.20) with the help of eqn. (2.19) yields,

$$\left[J_1 + \left(\frac{N_1}{N_2}\right)^2 J_2\right]\ddot{\theta}_1 + \left[f_1 + \left(\frac{N_1}{N_2}\right)^2 f_2\right]\dot{\theta}_1 + \left(\frac{N_1}{N_2}\right)T_L = T_M \qquad ...(2.21)$$

Thus the equivalent moment of inertia and viscous friction of gear train referred to shaft 1 are

$$J_{1eq} = J_1 + \left(\frac{N_1}{N_2}\right)^2 J_2 \,;\, f_{1eq} = f_1 + \left(\frac{N_1}{N_2}\right)^2 f_2$$

In terms of equivalent moment of inertia and friction, eqn. (2.21) may be written as

$$J_{1eq}\ddot{\theta}_1 + f_{1eq}\ddot{\theta}_1 + \left(\frac{N_1}{N_2}\right)T_L = T_M$$

Here $(N_1/N_2)\,T_L$ is the load torque referred to shaft 1.

Similarly, expressing θ_1 in terms of θ_2 in eqn. (2.20) with the help of eqn. (2.19), the equivalent moment of inertia and viscous friction of gear train referred to load shaft are

$$J_{2eq} = J_2 + \left(\frac{N_2}{N_1}\right)^2 J_1 \,;\, f_{2eq} = f_2 + \left(\frac{N_2}{N_1}\right)^2 f_1$$

Torque equation referred to the load shaft may then be expressed as

$$J_{2eq}\ddot{\theta}_2 + f_{2eq}\ddot{\theta}_2 + T_L = \left(\frac{N_2}{N_1}\right)T_M$$

It is observed that inertia and friction parameters are referred from one shaft of the gear train to the other in the direct square ratio of the gear teeth. The same will hold for shaft stiffness when present.

Electrical Systems

The resistor, inductor and capacitor are the three basic elements of electrical circuits. These circuits are analyzed by the application of Kirchhoff's voltage and current laws.

Let us analyze the *L-R-C* series circuit shown in Fig. 2.7 by using Kirchhoff's voltage law. The governing equations of the system are

$$L\frac{di}{dt} + Ri + \frac{1}{C}\int_{\infty}^{t} i\, dt = e \qquad ...(2.22)$$

$$\frac{1}{C}\int_{\infty}^{t} i\, dt = e_o \,;\, e_o = e_c \qquad ...(2.23)$$

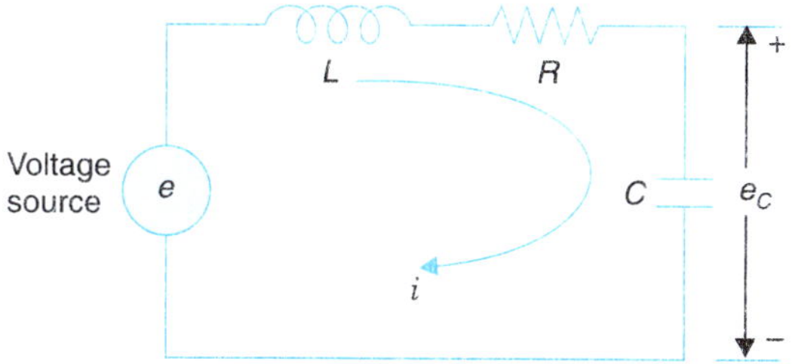

Fig. 2.12. L-R-C series circuit.

Elemental relationships are obvious from these equations. It is also to be noted that inductor and capacitor are the storage elements and resistor is the dissipative element. In terms of electric charge $q = \int i dt$, eqn. (2.22) becomes

$$L\frac{d^2q}{dt^2} + R\frac{dq}{dt} + \frac{1}{C}q = e \qquad ...(2.24)$$

Similarly, using Kirchhoff's current law, we obtain the following equations for *L-R-C* parallel circuit shown in Fig. 2.13.

$$C\frac{de}{dt} + \frac{1}{L}\int_{-\infty}^{t} e\,dt + \frac{e}{R} = i \qquad ...(2.25)$$

In terms of magnetic flux linkage $\phi = \int e dt$, eqn. (2.25) may be written as

$$C\frac{d^2\phi}{dt^2} + \frac{1}{R}\frac{d\phi}{dt} + \frac{1}{L}\phi = i \qquad ...(2.26)$$

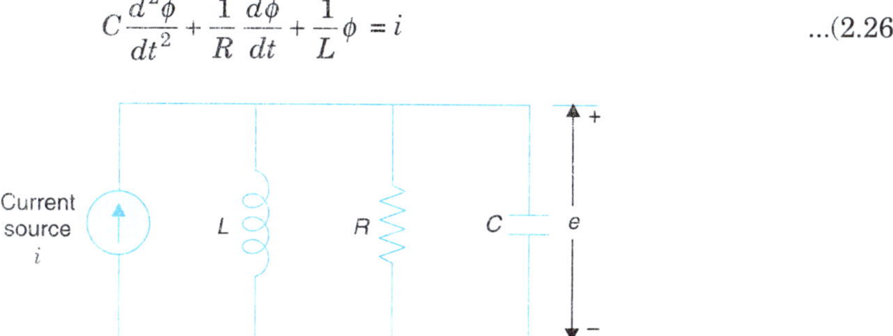

Fig. 2.13. L-R-C parallel circuit.

Analogous System

Comparing eqn. (2.1) for the mechanical translational system shown in Fig. 2.5 (*a*) or eqn. (2.13) for the mechanical rotational system shown in Fig. 2.10 (*a*) and eqn. (2.24) for the electrical system shown in Fig. 2.11, it is seen that they are of identical form. Such systems whose differential equations are of identical form are called *analogous systems*. The force F (torque T) and voltage e are the analogous variables here. This is called the Force (Torque)-Voltage analogy. A list of analogous variables in this analogy is given in Table 2.2.

Table 2.2. Analogous Quantities In Force (Torque)-Voltage Analogy

Mechanical translational systems	***Mechanical rotational systems***	***Electrical systems***
Force F	Torque T	Voltage e
Mass M	Moment of inertia J	Inductance L
Viscous friction coefficient f	Viscous friction coefficient f	Resistance R
Spring stiffness K	Torsional spring stiffness K	Reciprocal of capacitance $1/C$
Displacement x	Angular displacement θ	Charge q
Velocity $\dot{x}$	Angular velocity $\dot{\theta}$	Current i

Similarly eqns. (2.1) and (2.13) referred above and eqn. (2.26) for the electrical system shown in Fig. 2.13 are also identical. In this case force F (torque T) and current i are the analogous variables. This is called the Force (Torque)-Current analogy. A list of analogous quantities in this analogy is given in Table 2.3.

Table 2.3. Analogous Quantities In Force (Torque)-current Analogy

Mechanical translational systems	***Mechanical rotational systems***	***Electrical systems***
Force F	Torque T	Current i
Mass M	Moment of inertia J	Capacitance C
Viscous friction coefficient f	Viscous friction coefficient f	Reciprocal of resistance $1/R$
Spring stiffness K	Torsional spring stiffness K	Reciprocal of inductance $1/L$
Displacement x	Angular displacement θ	Magnetic flux linkage λ
Velocity $\dot{x}$	Angular velocity $\dot{\theta}$	Voltage e

The concept of analogous system is a useful technique for the study of various systems like electrical, mechanical, thermal, liquid-level, etc. If the solution of one system is obtained, it can be extended to all other systems analogous to it. Generally it is convenient to study a non-electrical system in terms of its electrical analog as electrical systems are more easily amenable to experimental study.

Thermal Systems

The basic requirement for the representation of thermal systems by linear models is that the temperature of the medium be uniform which is generally not the case. Thus for precise analysis a distributed parameter model must be used. Here, however, in order to simplify the analysis, uniformity of temperature is assumed and thereby the system is represented by a lumped parameter model.

Consider the simple thermal system shown in Fig. 2.14. Assume that the tank is insulated to eliminate heat loss to the surrounding air, there is no heat storage in the insulation and liquid in the tank is kept at uniform temperature by perfect mixing with the help of a stirrer. Thus a single temperature may be used to describe the thermal state of the entire liquid. (If complete mixing is not present, there is a complex temperature distribution throughout the liquid and the problem becomes one of the distributed parameters, requiring the use of partial differential equations). Assume that the steady-state temperature of the inflowing liquid is θ_i and that of the outflowing liquid is θ. The steady-state heat input rate from the heater is H. The liquid flow rate is of course assumed constant. To obtain a linear model we shall use small-signal analysis already illustrated for a nonlinear spring.

Let ΔH (*J*/min) be a small increase in the heat input rate from its steady-state value. This increase in heat input rate will result in increase of the heat outflow rate by an amount ΔH_1 and a heat storage rate of the liquid in the tank by an amount ΔH_2. Consequently the temperature of the liquid in the tank and therefore of the outflowing liquid rises by $\Delta\theta$(°C). Since the insulation has been regarded as perfect, the increase in heat outflow rate is only due to the rise in temperature of the outflowing liquid and is given by

$$\Delta H_1 = Qs\ \Delta\theta$$

where Q = steady liquid flow rate in kg/min; and s = specific heat of the liquid in *J*/kg °C.

The above relationship can be written in the form

$$\Delta H_1 = \Delta\theta/R \qquad \text{...(2.27)}$$

where $R = 1/Qs$, is defined as the *thermal resistance* and has the units of °C/J/min.

The rate of heat storage in the tank is given by

$$\Delta H_2 = Ms\frac{d(\Delta\theta)}{dt}$$

where M = mass of liquid in the tank in kg; and $\frac{d(\Delta\theta)}{dt}$ = rate of rise of temperature in the tank.

The above equation can be expressed in the form

$$\Delta H_2 = C\frac{d(\Delta\theta)}{dt} \qquad \text{...(2.28)}$$

where $C = Ms$, is defined as the *thermal capacitance* and has the units of *J*/°C. For the system of Fig. 2.14, the heat flow balance equation is

$$\Delta H = \Delta H_1 + \Delta H_2 = \frac{\Delta\theta}{R} + C\frac{d(\Delta\theta)}{dt}$$

or

$$RC\frac{d(\Delta\theta)}{dt} + \Delta\theta = R(\Delta H) \qquad \text{...(2.29)}$$

Equation (2.29) describes the dynamics of the thermal system with the assumption that the temperature of the inflowing liquid is constant.

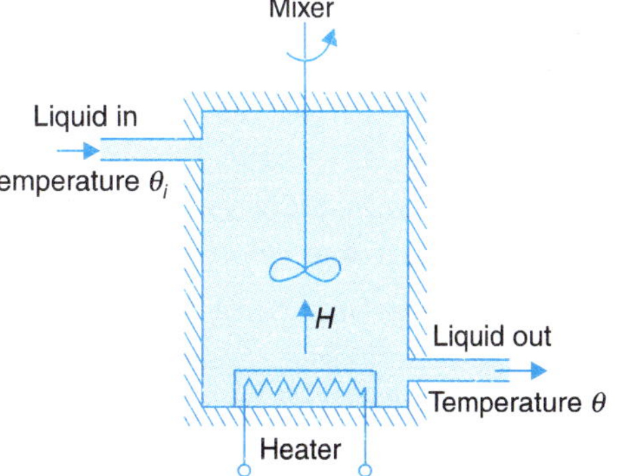

Fig. 2.14. Thermal system.

In practice, the temperature of the inflowing liquid fluctuates. Thus along with a heat input signal from the heater, there is an additional signal due to change in the temperature of the inflowing liquid which is known as the *disturbance signal.*

Let $\Delta\theta_i$ be change in the temperature of the inflowing liquid from its steady-state value. Now in addition to the change in heat input from the heater, there is a change in heat carried by the inflowing liquid. The heat flow equation, therefore, becomes

$$\Delta H + \frac{\Delta\theta_i}{R} = \frac{\Delta\theta}{R} + C\frac{d}{dt}(\Delta\theta)$$

or

$$RC\frac{d}{dt}(\Delta\theta) + \Delta\theta = \Delta\theta_i + R(\Delta H) \qquad ...(2.30)$$

Let us now relax the assumption that the tank insulation is perfect. As the liquid temperature increases by $\Delta\theta$, the rate of heat flow through the tank walls to the ambient medium increases by

$$\Delta H_3 = \frac{\Delta\theta}{R_t}$$

where R_t is the thermal resistance of the tank walls. Equation (2.30) is then modified to

$$\Delta H + \frac{\Delta\theta_i}{R} = \left(\frac{\Delta\theta}{R} + \frac{\Delta\theta}{R_t}\right) + C\frac{d}{dt}(\Delta\theta)$$

or

$$R'C\frac{d}{dt}(\Delta\theta) + \Delta\theta = \left(\frac{R'}{R}\right)\Delta\theta_i + R'(\Delta H)$$

where $R' = \frac{RR_t}{R + R_t}$ = effective thermal resistance due to liquid outflow and tank walls (it is a parallel combination of R and R_t).

It is still being assumed above that there is no heat storage in the tank walls. Relaxing this assumption will simply add to the thermal capacitance C.

Fluid Systems

The dynamics of the fluid systems can be represented by ordinary linear differential equations only if the fluid is incompressible and fluid flow is laminar. Industrial processes often involve fluid flow through connecting pipes and tanks where the flow is usually turbulent resulting in nonlinear equations describing the system.

Velocity of sound is a key parameter in fluid flow to determine the compressibility property. If the fluid velocity is much less than the velocity of sound, compressibility effects are usually small. As the velocity of sound in liquids is about 1500 m/s, compressibility* effects are rarely of importance in liquids and the treatment of compressibility is generally restricted to gases, where the velocity of sound is about 350 m/s.

Another important fluid property is the type of fluid flow-laminar or turbulent. Laminar flow is characterized by smooth motion of one laminar of fluid past another, while turbulent flow is characterized by an irregular and nearly random motion superimposed on the main

*The tendency of so-called incompressible fluids to compress slightly under pressure is called *fluid compliance*. This type of effect is accounted for in hydraulic pumps and motors discussed in Section 4.5.

motion of fluid. The transition from laminar to turbulent flow was first investigated by Osborne Reynolds, who after experimentation found that for pipe flow the transition conditions could be correlated by a dimensionless group which is now known as *Reynolds number,* Re.

From his experiments, Reynolds found that pipe flow will be laminar for Re less that 2,000 and turbulent for Re greater than 3,000. When Re is between 2,000 and 3,000, the type of flow is unpredictable and often changes back and forth between the laminar and turbulent states because of flow disturbances and pipe vibrations.

The pressure drop across a pipe section is given by

$$P = \frac{128 l\mu}{\pi D^4} Q \text{ ; for laminar flow} \qquad ...(2.31a)$$

$$= RQ$$

$$P = \frac{8K_t \rho l}{\pi^2 D^5} Q^2 \text{ ; for turbulent flow} \qquad ...(2.31b)$$

$$= K_T Q^2$$

where l = length of pipe section (m); D = diameter of pipe (m); μ = viscosity (Ns/m^2); Q = volumetric flow rate (m^3/s) ; K_t = a constant (to be determined experimentally); and ρ = mass density (kg/m^3).

Equation (2.31*a*) representing laminar flow is linear, *i.e.,*

$$P = RQ$$

where $R = \frac{128 l\mu}{\pi D^4}\left(\frac{N/m^2}{m^3/s}\right)$ is the *fluid resistance.*

Equation (2.31*b*) representing turbulent flow is nonlinear, *i.e.,*

$$P = K_T Q^2$$

where $K_T = \frac{8K_t \rho l}{\pi^2 D^5}$.

This equation can be linearized about the operating point (P_0, Q_0) by techniques discussed earlier in this section (*See* Nonlinear Spring). At the operating point,

$$P_o = K_T Q_o^{\,2}$$

Expanding the turbulent flow equation (2.31*b*) in Taylor series about the operating point and retaining first-order term only, we have

$$P = P_o + \left.\frac{dP}{dQ}\right|_{(P_o, Q_o)} (Q - Q_o)$$

It follows that $\quad P - P_o = 2K_T Q_o (Q - Q_o)$

or $\quad \Delta P = R\Delta Q \qquad ...(2.32)$

where $R = 2K_T Q_o$ is the *turbulent flow resistance.*

Equation (2.32) relates the incremental fluid flow to incremental pressure around the operating point in the case of turbulent flow.

Large pipes even when long offer small resistance while short devices that contain some contractions (orifices, nozzles, valves, etc.) offer large resistance to fluid flow. For these dissipation devices, head loss

$$P = \frac{8K\rho}{\pi^2 D^4} Q^2$$

where K is a constant. Experimentally determined values of K for various dissipation devices can be found in handbooks. This equation is analogous to eqn. (2.31*b*) and can be linearized about the operating point to obtain resistance offered by a dissipation device.

The other ideal element used in modelling fluid systems is the *fluid capacitance*. Consider a tank with cross-sectional area = $A(m^2)$.

$$\text{The rate of fluid storage in the tank} = A\frac{dH}{dt} = \frac{A}{\rho g}\frac{dP}{dt} = C\frac{dP}{dt} \quad \text{...(2.33)}$$

where H = fluid head in the tank (m); $P = \rho g H$ (N/m^2) = pressure at tank bottom; and $C = \frac{A}{\rho g}\left(\frac{m^3}{N/m^2}\right)$ = *capacitance* of the tank.

Inertial effect of fluid in a pipe line is modelled as *inertance* defined below.

$$\Delta P = L\frac{dQ}{dt}$$

where ΔP = pressure drop as on the pipe

Q = rate of fluid flow through pipe

$L = \frac{\rho l}{A}$ = inertance (Ns²/m⁵).

For small fluid accelerations, the inertance effect is usually neglected to obtain a simple mathematical model of the system. This is really true of hydraulic components used in control systems.

Liquid Level Systems

In terms of head $H(m)$, the fluid pressure is given by

$$P = \rho g H$$

The pressure-flow rate relations given by eqns. (2.32*a*) and (2.32*b*) may be expressed as the following head-flow rate relations:

$$H = RQ;\ \text{for laminar flow} \quad (2.34a)$$

where $R = \frac{128 l u}{\pi D^2 \rho g}$

$$\Delta H = R\Delta Q;\ \text{for turbulent flow} \quad (2.34b)$$

where $R = \frac{2K_T Q_o}{\rho g}$

The parameter R in eqns. (2.34) is referred to as *hydraulic resistance.*

$$\text{The rate of fluid storage in a tank} = A\frac{dH}{dt} = C\frac{dH}{dt}$$

where $C = A$ (m^2) = *hydraulic capacitance* of the tank.

Consider a simple liquid-level system shown in Fig. 2.15 where a tank is supplying liquid through an outlet. Under steady conditions, let Q_i be the liquid flow rate into the tank and Q_o be the outflow rate, while H_o is the steady liquid head in the tank. Obviously $Q_i = Q_o$.

Let ΔQ_i be a small increase in the liquid inflow rate from its steady-state value. This increase in liquid inflow rate causes increase of head of the liquid in the tank by ΔH, resulting in increase of liquid outflow rate by

$$\Delta Q_o = \Delta H/R$$

The system dynamics is described by the liquid flow rate balance equation:

Rate of liquid storage in the tank = rate of liquid inflow-rate of liquid outflow

Therefore

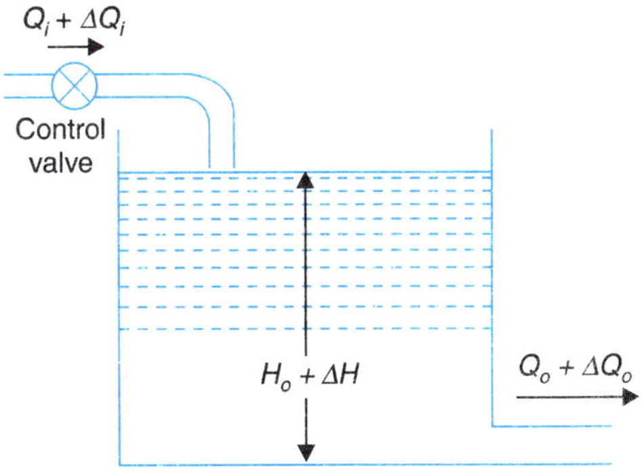

$$C\frac{d(\Delta H)}{dt} = \Delta Q_i - \frac{\Delta H}{R}$$

or
$$RC\frac{d(\Delta H)}{dt} + \Delta H = R(\Delta Q_i) \tag{2.35}$$

where C is the capacitance of the tank and R is the total resistance offered by the tank outlet and pipe.

Fig. 2.15. Liquid-level system.

Pneumatic Systems

We shall assume in our discussion that velocities of gases are a small fraction of the velocity of sound, which is true in a number of engineering applications. With this assumption, we treat pneumatic flow also as nearly incompressible. Therefore, the results presented earlier are directly applicable to this class of pneumatic systems.

Consider a simple pneumatic system shown in Fig. 2.16. A pneumatic source is supplying air to the pressure vessel through a pipe line.

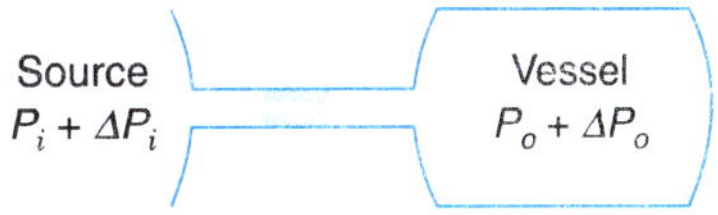

Fig. 2.16. Simple pneumatic system.

Let us define:

P_i = air pressure of the source at steady-state (N/m^2).

P_o = air pressure in the vessel at steady-state (N/m^2).

ΔP_i = small change in air pressure of the source from its steady-state value.

ΔP_o = small change in air pressure of the vessel from its steady-state value.

System dynamics is described by the equation:

Rate of gas storage in vessel = rate of gas inflow

or
$$C\frac{d(\Delta P_o)}{dt} = \frac{\Delta P}{R} = \frac{\Delta P_i - \Delta P_o}{R}$$

or
$$RC\frac{d(\Delta P_o)}{dt} + \Delta P_o = \Delta P_i \qquad \text{...(2.36)}$$

Table 2.4 summarizes the variables and parameters of the thermal, liquid level and pneumatic systems which are analogous to those of electrical systems.

Table 2.4. Analogous Quantities

Electrical systems	***Thermal systems***	***Liquid-level systems***	***Pneumatic systems***
Charge, q	Heat flow, J	Liquid flow, m^3	Air flow, m^3
Current, A	Heat flow rate, J/min	Liquid flow rate, m^3/min	Air flow rate, m^3/min
Voltage, V	Temperature, °C	Head, m	Pressure, N/m^2
Resistance, Ω	Resistance, $\frac{°C}{J/min}$	Resistance, $\frac{N/m^2}{m^3/min}$	Resistance, $\frac{N/m^2}{m^3/min}$
Capacitance, C	Capacitance, J/°C	Capacitance, m^2	Capacitance, $\frac{m^3}{N/m^2}$

2.3 DYNAMICS OF ROBOTIC MECHANISMS

Dynamical equations for robotic serial links will be illustrated here by means of two simple examples of two-link mechanisms.

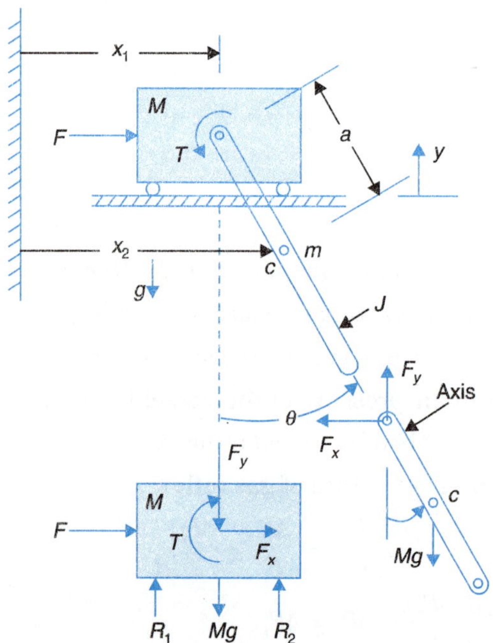

Fig. 2.17. Simplied model of a gantry robot.

Gantry Robot

A simple gantry robot mechanism is shown in Fig. 2.17(a) wherein the main body the crane is propelled by traction force, F. To the body is attached a rotating arm (mass m, and moment of inertia J about the axis of rotation) driven by an actuator located on the crane body. At the end of the arm a hand (**end effector**) would be attached for picking up objects (not shown in figure).

The free-body diagrams of the masses M and m are drawn in Figs. 2.17(*b*) and (*c*).

For mass *M*

$$M\ddot{x}_1 = Fx + F \quad ...(i)$$

$$M\ddot{y}_1 = R_1 + R_2 - F_y - Mg \quad ...(ii)$$

Equation (*ii*) is needed to be used only if reaction R_1 and R_2 are to be calculated.

For mass *m*

$$m\ddot{x}_2 = -F_x \quad ...(iii)$$

$$m\ddot{y}_2 = F_y - mg \quad ...(iv)$$

$$J\ddot{\theta} = aF_x \cos\theta - F_y \sin\theta + T \quad ...(v)$$

Eliminating F_x in eqns. (*i*) and (*iii*), we get

$$M\ddot{x}_1 + m\ddot{x}_2 = F \quad ...(vi)$$

Eliminating F_y in eqns. (*iv*) and (*v*), we have

$$J\ddot{\theta} = m\ddot{x}_2\, a \cos\theta + ma\,(\ddot{y}_2 + g) = T \quad ...(vii)$$

The displacements x_2 and y_2 are related to θ as follows:

$$x_2 = x_1 + a \sin\theta$$

$$y_2 = -a \cos\theta$$

Differentiating twice

$$\ddot{x}_2 = \ddot{x}_1 - a \sin\theta\,\dot{\theta}^2 + \cos\theta\,\ddot{\theta} \quad ...(viii)$$

$$\ddot{y}_2 = a \cos\theta\,\dot{\theta}^2 + a \sin\theta\,\ddot{\theta} \quad ...(ix)$$

Substituting $\ddot{x}_2$ and $\ddot{y}_2$ into eqns. (*vi*) and (*vii*), we get

$$(J + ma^2)\ddot{\theta} + ma \cos\theta\,\ddot{x}_1 + mga \sin\theta = T \quad ...(x)$$

$$(M + m)\ddot{x}_1 + ma \cos\theta\,\ddot{\theta} - ma \sin\theta\,\dot{\theta}^2 = F \quad ...(xi)$$

It is observed that the dynamic equations are nonlinear and coupled. Serial link manipulator is too complex a mechanism to be modelled by the free-body technique. It is much simpler to use energy method which employs generalized coordinates.

Lagrangian Mechanics

The Lagrangian L is defined as the difference between the kinetic energy K and the potential energy P of the system

$$L = K - P \quad ...(2.37)$$

The kinetic potential energy of the system may be expressed in any convenient coordinate system that will simplify the problem. It is not necessary to use Cartesian coordinates.

The dynamics equations, in terms of the coordinates used to express the kinetic and potential energy, are obtained as

$$F_i = \frac{d}{dt}\frac{\partial L}{\partial q_i} - \frac{\partial L}{\partial q_i} \quad ...(2.38)$$

where q_i are the coordinates in which the kinetic and potential energy are expressed q_i is the corresponding velocity, and F_i the corresponding force or torque; F_i is either a force or a torque i, depending upon whether q_i is a linear or an angular coordinate. These factors, torques, and coordinates are referred to as generalized forces, torques, and coordinates.

Illustrative Example

We shall derive the dynamic equations for a two-link serial manipulator as shown in Fig. 2.18. Here the link masses are represented by point masses at the end of the link. The manipulator hangs down in a field of granity g. As indicated in the figure θ_1 and θ_2 are chosen as the generalized coordinates.

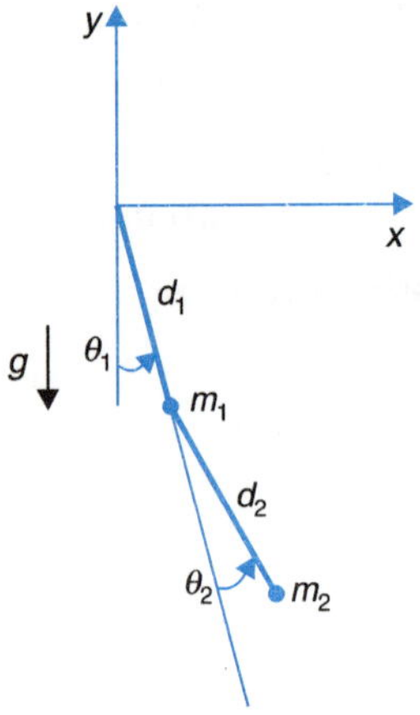

Fig. 2.18. A two link manipulator.

The Kinetic and Potential Energy

The kinetic energy of a mass is $K = 1/2\ mv^2$. So far the mass m_1 the kinetic energy is expressed as

$$K_1 = (1/2)m_1 d_1^2 \dot{\theta}_1^2 \qquad ...(i)$$

The potential energy with reference to the coordinate frame is expressed by the y-coordinate as

$$P_1 = -m_1 g d_1 \cos(\theta_1) \qquad ...(ii)$$

In the case of the second mass let us first write expressions for its Cartesian position coordinates. These are

$$x_2 = d_1 \sin(\theta_1) + d_2 \sin(\theta_1 + \theta_2) \qquad ...(iii)$$

$$y_2 = -d_1 \cos(\theta_1) - d_2 \cos(\theta_1 + \theta_2) \qquad ...(iv)$$

Differentiating these we get the velocity components of the mass m_2 as

$$\dot{x}_2 = d_1 \cos(\theta_1)\dot{\theta}_1 + d_2 \cos(\theta_1 + \theta_2)(\dot{\theta}_1 + \dot{\theta}_2)$$

$$\dot{y}_2 = d_1 \sin(\theta_2)\dot{\theta}_1 + d_2 \sin(\theta_1 + \theta_2)(\dot{\theta}_1 + \dot{\theta}_2)$$

The magnitude of the velocity squared is then

$$v_2^2 = d_1^2\dot{\theta}_1^2 + d_2^2(\dot{\theta}_1^2 + 2\dot{\theta}_1\dot{\theta}_2 + \dot{\theta}_1^2)$$

$$+ 2d_1d_2 \cos(\theta_1)\cos(\theta_1 + \theta_2)(\dot{\theta}_1^2 + \dot{\theta}_1\dot{\theta}_2)$$

$$+ 2d_1d_2 \sin(\theta_1)\sin(\theta_1 + \theta_2)(\dot{\theta}_1^2 + \dot{\theta}_1\dot{\theta}_2)$$

$$= d_1^2\dot{\theta}_1^2 + d_2^2(\dot{\theta}_1^2 + 2\dot{\theta}_1\dot{\theta}_2 + \dot{\theta}_2^2) + 2d_1d_2\cos(\theta_2)(\dot{\theta}_1^2 + \dot{\theta}_1\dot{\theta}_2)$$

and the kinetic energy of the mass m_2 is then

$$K_2 = 1/2m_2d_1^2\dot{\theta}_1^2 + 1/2m_2d_2^2(\dot{\theta}_1^2 + 2\dot{\theta}_1\dot{\theta}_2 + \theta_2^2) + m_2d_1d_2\cos(\theta_2)(\dot{\theta}_1^2 + \dot{\theta}_1\dot{\theta}_2)$$

From eqn. (iv) the potential energy of the mass m_2 is

$$P_2 = -m_2gd_1\cos(\theta_1) - m_2gd_2\cos(\theta_1 + \theta_2) \qquad ...(v)$$

The Lagrangian

The Lagrangian for the two-link system is

$$L = (K_1 + K_2) - (P_1 + P_2)$$

Substituting the values we get

$$L = 1/2(m_1 + m_2)d_1^2\dot{\theta}_1^2 + 1/2m_2d_2^2(\dot{\theta}_1^2 + 2\dot{\theta}_1\dot{\theta}_2 + \dot{\theta}_2^2)$$
$$+ m_2d_1d_2\cos(\theta_2)(\dot{\theta}_1^2 + \dot{\theta}_1\dot{\theta}_2)$$
$$+ (m_1 + m_2)gd_1\cos(\theta_1) + m_2gd_2\cos(\theta_1 + \theta_2) \qquad ...(vi)$$

The Dynamic Equations

The dynamic equations are derived below using the Lagrangian of eqn. (*vi*)

$$\frac{\partial L}{\partial \dot{\theta}_1} = (m_1 + m_2)d_1^2\dot{\theta}_1^2 + m_2d_2{}^2\dot{\theta}_1 + m_2d_2{}^2\dot{\theta}_2$$
$$+ 2m_2d_1d_2\cos(\theta_2)\dot{\theta}_1 + m_2d_1d_2\cos(\theta_2)\dot{\theta}_2$$

$$\frac{d}{dt}\frac{\partial L}{\partial \dot{\theta}_1} = [(m_1 + m_2)d_1{}^2 + m_2d_2{}^2 + 2m_2d_1d_2\cos(\dot{\theta}_2)]$$
$$+ [m_2d_2{}^2 + m_2d_1d_2\cos(\theta_2)]\ddot{\theta}_2$$
$$- 2m_2d_1d_2\sin(\theta_2)\dot{\theta}_1\dot{\theta}_2 - m_2d_1d_2\sin(\theta_2)\dot{\theta}_2^2$$

$$\frac{\partial L}{\partial \theta_1} = -(m_1 + m_2)gd_1\sin(\theta_1) - m_2gd_2\sin(\theta_1 + \theta_2)$$

These equations yield the torque at joint 1 as

$$T_1 = [(m_1 + m_2)d_1{}^2 + m_2d_2{}^2 + 2m_2d_1d_2\cos(\theta_2)]\ddot{\theta}_1$$
$$+ [m_2d_2{}^2 + m_2d_1d_2\cos(\theta_2)]\ddot{\theta}_2$$
$$- 2m_2d_1d_2\sin(\theta_2)\dot{\theta}_1\dot{\theta}_2 - m_2d_1d_2\sin(\theta_2)\dot{\theta}_2^2$$
$$+ (m_1 + m_2)gd_1\sin(\theta_1) + m_2gd_2\sin(\theta_1 + \theta_2) \qquad ...(vii)$$

Performing similar operations at joint 2,we have

$$\frac{\partial L}{\partial \dot{\theta}_2} = m_2d_1^2\dot{\theta}_1 + m_2d_2^2\dot{\theta}_2 + m_2d_1d_2\cos(\theta_2)\dot{\theta}_1$$

$$\frac{d}{dt}\frac{\partial L}{\partial \dot{\theta}_2} = m_2d_2^2\ddot{\theta}_1 + m_2d_2^2\ddot{\theta}_2 + m_2d_1d_2\cos(\theta_2)\ddot{\theta}_1 - m_2d_1d_2\sin(\theta_2)\dot{\theta}_1\dot{\theta}_2$$

$$\frac{\partial L}{\partial \theta_2} = -m_2d_1d_2\sin(\theta_2)\dot{\theta}_1\dot{\theta}_2 - m_2gd_2\sin(\theta_1 + \theta_2)$$

The torque at joint 2 is then given by

$$T_2 = [m_1d_2^2 + m_2d_1d_2\cos(\theta_2)]\ddot{\theta}_1 + m_2d_2^2\ddot{\theta}_2 - 2m_2d_1d_2\sin(\theta_2)\dot{\theta}_1\ddot{\theta}_2$$
$$- m_2d_1d_2\sin(\theta_2)\dot{\theta}_1^2 + m_2gd_2\sin(\theta_1 + \theta_2) \qquad ...(viii)$$

The torques at joint 1 and 2 (eqns. (*vii*) and (*viii*)), can be rewritten in the general form

$$T_1 = D_{11}\ddot{\theta}_1 + D_{12}\ddot{\theta}_2 + D_{111}\dot{\theta}_1^2 + D_{122}\dot{\theta}_2^2 + D_{112}\dot{\theta}_1\dot{\theta}_2 + D_{121}\dot{\theta}_2\dot{\theta}_1 + D_1 \qquad ...(ix)$$

$$T_2 = D_{12}\ddot{\theta}_1 + D_{22}\ddot{\theta}_2 + D_{211}\dot{\theta}_1^2 + D_{222}\dot{\theta}_2^2 + D_{212}\dot{\theta}_1\dot{\theta}_2 + D_{221}\dot{\theta}_2\dot{\theta}_1 + D_2 \qquad ...(x)$$

Various coefficients in the torque expressions of eqns. (*x*) and (*xi*) are defined below:

$$D_{11} = [(m_1 + m_2)d_1^2 + m_2d_2^2 + 2m_2d_1d_2\cos(\theta_2)]$$
$$D_{22} = m_2d_2^2$$

Coupling inertias

$$D_{12} = D_{21} = m_2 d_2^2 + m_2 d_1 d_2 \cos(\theta_2)$$

Centripetal acceleration coefficients

$$D_{111} = 0$$
$$D_{122} = D_{211} = -m_2 d_1 d_2 \sin(\theta_2)$$
$$D_{222} = 0$$

Coriolis acceleration coefficients

$$D_{112} = D_{121} = -m_2 d_1 d_2 \sin(\theta_2)$$
$$D_{212} = D_{221} = -m_2 d_1 d_2 \sin(\theta_2)$$

Gravity terms

$$D_1 = (m_1 + m_2) g d_1 \sin(\theta_1) + m_2 g d_2 \sin(\theta_1 + \theta_2)$$
$$D_2 = m_2 g d_2 \sin(\theta_1 + \theta_2).$$

2.4 TRANSFER FUNCTIONS

The transfer function of a linear time-invariant system is defined to be the ratio of the Laplace transform of the output variable to the Laplace transform of the input variable under the assumption that all initial condition are zero.

Consider the mass-spring-dashpot system shown in Fig. 2.5 (a), whose dynamics is described by the second-order differential equation (2.1).

Taking the Laplace transform of each term of this equation (assuming zero initial conditions), we obtain

$$F(s) = Ms^2X(s) + fsX(s) + KX(s)$$

Then the transfer function is

$$G(s) = \frac{X(s)}{F(s)} = \frac{1}{Ms^2 + fs + K} \quad ...(2.39)$$

The highest power of the complex variable s in the denominator of the transfer function determines the *order of the system*. The mass-spring-dashpot system under consideration is thus a second-order system, a fact which is already recognized from its differential equation.

The transfer function of the L-R-C circuits shown in Fig. 2.12 is similarly obtained by taking the Laplace transform of eqns. (2.22) and (2.23), with zero initial conditions. The resulting equations are

$$sLI(s) + RI(s) + \frac{1}{s}\frac{I(s)}{C} = E(s)$$
$$\frac{1}{s}\frac{I(s)}{C} = E_0(s)$$

If e is assumed to be the input variable and e_0 the output variable, the transfer function of the system is

$$\frac{E_0(s)}{E(s)} = \frac{1}{LCs^2 + RCs + 1} \quad ...(2.40)$$

Equation (2.39) and (2.40) reveal that the transfer function is an expression in s-domain, relating the output and input of the linear time-invariant system in terms of the system parameters and is independent of the input. It describes the input output behaviour of the system and does not give any information concerning the internal structure of the system. Thus, when the transfer function of a physical system is determined, the system can be represented by a *block*, which is a shorthand pictorial representation of the cause and effect relationship between input and output of the system. The signal flowing into the block (called input) flows out of it (called output) after being processed by the transfer function characterizing the block, see Fig. 2.19 (a). Functional operation of a system can be more readily visualized by examination of a block diagram rather than by the examination of the equations describing the physical system. Therefore, when working with a linear time-invariant system, we can think of a system or its sub-systems simply as interconnected blocks with each block described by a transfer function.

Laplace transforming eqn. (2.29), the transfer function of the thermal system shown in Fig. 2.14 is

$$\frac{\Delta\theta(s)}{\Delta H(s)} = \frac{R}{RCs + 1} \qquad \text{...(2.41)}$$

The block diagram representation of the system is shown in Fig. 2.19 (a). When this system is subjected to a disturbance, the dynamics is described by eqn. (2.30). Taking the Laplace transformation of this equation, we get

$$(RCs + 1)\,\Delta\theta(s) = \Delta\theta_i(s) + R\Delta H(s) \qquad \text{...(2.42)}$$

The corresponding block diagram representation is given in Fig. 2.19 (b).

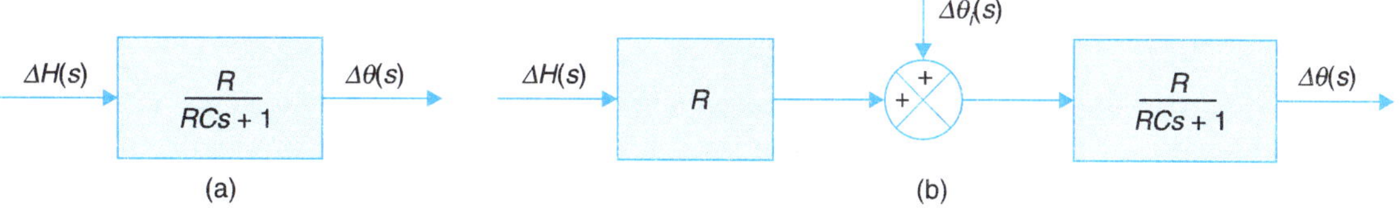

Fig. 2.19. Block diagram of the thermal system shown in Fig. 2.14.

Sinusoidal Transfer Functions

The steady-state response of a control system to a sinusoidal input is obtained by replacing s with $j\omega$ in the transfer function of the system.

Transfer function of the mechanical accelerometer shown in Fig. 2.6, obtained from eqn. (2.2), is

$$\frac{Y(s)}{X(s)} = \frac{Ms^2}{Ms^2 + fs + K} = \frac{s^2}{s^2 + \frac{f}{M}s + \frac{K}{M}}$$

Its sinusoidal transfer function becomes

$$\frac{Y(j\omega)}{X(j\omega)} = \frac{(j\omega)^2}{(j\omega)^2 + \frac{f}{M}(j\omega) + \frac{K}{M}} \qquad \text{...(2.43)}$$

Equation (2.43) represents the behaviour of the accelerometer when used as a device to measure sinusoidally varying displacement. If the frequency of the sinusoidal input signal $X(j\omega)$ is very low, *i.e.*, $\omega << \omega_n = \sqrt{(K/M)}$, then the transfer function given by eqn. (2.43) may be approximated by

$$\frac{Y(j\omega)}{X(j\omega)} \approx \frac{-\omega^2}{K/M}$$

The output signal is very weak for values of frequency $\omega << \omega_n$. Weak output signal coupled with the fact that some inherent noise may always be present in the system, makes the displacement measurement by the accelerometer in the low frequency range as quite unreliable.

For $\omega >> \omega_n$, the transfer function given by eqn. (2.43) may be approximated by

$$\frac{Y(j\omega)}{X(j\omega)} \approx 1$$

Thus, at very high frequencies the accelerometer output follows the sinusoidal displacement input. For this range of frequencies the basic accelerometer system can be used for displacement measurement particularly in seismographic studies.

For a sinusoidal input acceleration, the steady-state sinusoidal response of the accelerometer is given by

$$\frac{Y(j\omega)}{A(j\omega)} = \frac{1}{(j\omega)^2 + \frac{f}{M}(j\omega) + \frac{K}{M}}$$

As long as $\omega << \omega_n = \sqrt{(K/M)}$,

$$\frac{Y(j\omega)}{A(j\omega)} \approx \frac{M}{K}$$

The accelerometer is thus suitable for measurement of sinusoidally varying acceleration from zero frequency (constant acceleration) to a frequency which depends upon the choice of ω_n for the accelerometer. The sinusoidal behaviour of this type of transfer functions will be studied in greater details in Chapter 8.

Procedure for Deriving Transfer Functions

The following assumptions are made in deriving transfer functions of physical systems.

1. It is assumed that there is no loading, *i.e.,* no power is drawn at the output of the system. If the system has more than one nonloading elements in tandem, then the transfer function of each element can be determined independently and the overall transfer function of the physical system is determined by multiplying the individual transfer functions. In case of systems consisting of elements which load each other, the overall transfer function should be derived by basic analysis without regard to the individual transfer functions.
2. The system should be approximated by a linear lumped constant parameters model by making suitable assumptions.

To illustrate the point (1) above, let us consider two identical *RC* circuits connected in cascade so that the output from the first circuits is fed as input to the second as shown in Fig. 2.20.

The describing equations for this system are

$$\frac{1}{C}\int_{-\infty}^{t}(i_1 - i_2)dt + Ri_1 = e_i \quad ...(2.44a)$$

$$\frac{1}{C}\int_{-\infty}^{t}(i_2 - i_1)dt + Ri_2 = -\frac{1}{C}\int_{-\infty}^{t} i_2 dt = -e_o \quad ...(2.44b)$$

Taking the Laplace transforms of eqns. (2.44 (a)) and (2.44 (b)), assuming zero initial conditions, we obtain

$$\frac{1}{sC}[I_1(s) - I_2(s)] + RI_1(s) = E_i(s)$$

$$\frac{1}{sC}[I_2(s) - I_1(s)] + RI_2(s) = -\frac{1}{sC}[I_2(s)] = -E_o(s)$$

The transfer function obtained by eliminating $I_1(s)$ and $I_2(s)$ from the above equation is

$$\frac{E_o(s)}{E_i(s)} = \frac{1}{\tau^2 s^2 + 3\tau s + 1} \quad ...(2.45)$$

where $\tau = RC$.

The transfer function of each of the individual RC circuits is $1/(1 + s\tau)$. From eqn. (2.45) it is seen that overall transfer function of the two RC circuits connected in cascades is not equal to $[1/(\tau s + 1)]\,[1/(\tau s + 1)]$ but instead it is $1/(\tau^2 s^2 + 3\tau s + 1)$.

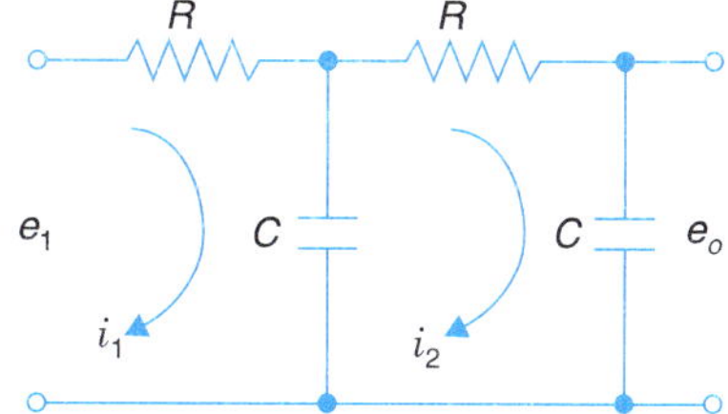

Fig. 2.20. RC circuits in cascade.

This difference is explained by the fact that while deriving the transfer function of a single RC circuits, it is assumed that the output is unloaded. However, when the input of second circuit is obtained from the output of first, a certain amount of energy is drawn from the first circuit and hence its original transfer function is no longer valid. The degree to which the overall transfer function is modified from the product of individual transfer functions depends upon the amount of loading.

As an example to illustrate the point (2) above, let us derive the transfer function of a d.c. servomotor. In servo applications, a d.c. motor is required to produce rapid accelerations from standstill. Therefore the physical requirements of such a motor are low inertia and high starting torque. Low inertia is attained with reduced armature diameter with a consequent increase in armature length such that the desired power output is achieved. Thus, except for minor differences in constructional features, a d.c. servometer is essentially an ordinary d.c. motor.

In control systems, the d.c. motors are used in two different control modes: armature-control mode with fixed field current, and field-control mode with fixed armature current.

Armature-control

Consider the armature-controlled d.c. motor shown in Fig. 2.21.

In this system,

R_a = resistance of armature (Ω).

L_a = inductance of armature winding (H).

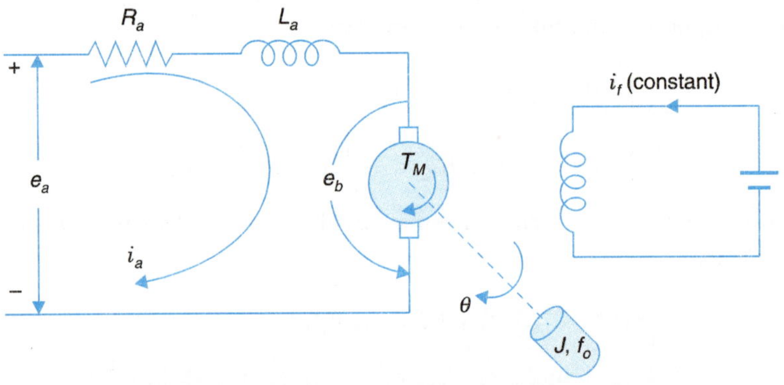

Fig. 2.21. Armature-controlled d.c. motor.

i_a = armature current (A).

i_f = field current (A).

e_a = applied armature voltage (V).

e_b = back emf (volts).

T_M = torque developed by motor (Nm).

θ = angular displacement of motor-shaft (rad).

J = equivalent moment of inertia of motor and load referred to motor shaft (kg-m^2).

f_0 = equivalent viscous friction coefficient of motor and load referred to motor shaft $\left(\frac{\text{Nm}}{\text{rad/s}}\right)$.

In servo applications, the d.c. motors are generally used in the linear range of the magnetization curve. Therefore, the air gap flux ϕ is proportional of the field current, *i.e.*,

$$\phi = K_f i_f \qquad ...(2.46)$$

where K_f is a constant.

The torque T_M developed by the motor is proportional to the product of the armature current and air gap flux, *i.e.*,

$$T_M = K_1 K_f i_f i_a \qquad ...(2.47)$$

where K_1 is a constant.

In the armature-controlled d.c. motor, the field current is kept constant, so that eqn. (2.46) can be written as

$$T_M = K_T i_a \qquad ...(2.48)$$

where K_T is known as the motor torque constant.

The motor back emf being proportional to speed is given as

$$e_b = K_b \frac{d\theta}{dt} \qquad ...(2.49)$$

where K_b is the back emf constant.

The differential equation of the armature circuit is

$$L_a \frac{di_a}{dt} + R_a i_a + e_b = e_a \qquad ...(2.50)$$

The torque equation is

$$J\frac{d^2\theta}{dt^2} + f_0\frac{d\theta}{dt} = T_M = K_T i_a \qquad ...(2.51)$$

Taking the Laplace transforms of eqns. (2.48) to (2.50), assuming zero initial conditions, we get

$$E_b(s) = K_b s\theta(s) \qquad ...(2.52)$$

$$(L_a s + R_a)I_a(s) = E_a(s) - E_b(s) \qquad ...(2.53)$$

$$(Js^2 + f_0 s)\,\theta(s) = T_M(s) = K_T I_a(s) \qquad ...(2.54)$$

From eqns. (2.51) to (2.53), the transfer function of the system is obtained as

$$G(s) = \frac{\theta(s)}{E_a(s)} = \frac{K_T}{s[(R_a + sL_a)(Js + f_0) + K_T K_b]} \qquad ...(2.55)$$

The block diagram representation of eqn. (2.53) is shown in Fig. 2.22 (*a*) where the circular block representing the differencing action is known as the *summing point*. Equation (2.54) is represented by a block shown in Fig. 2.22 (*b*).

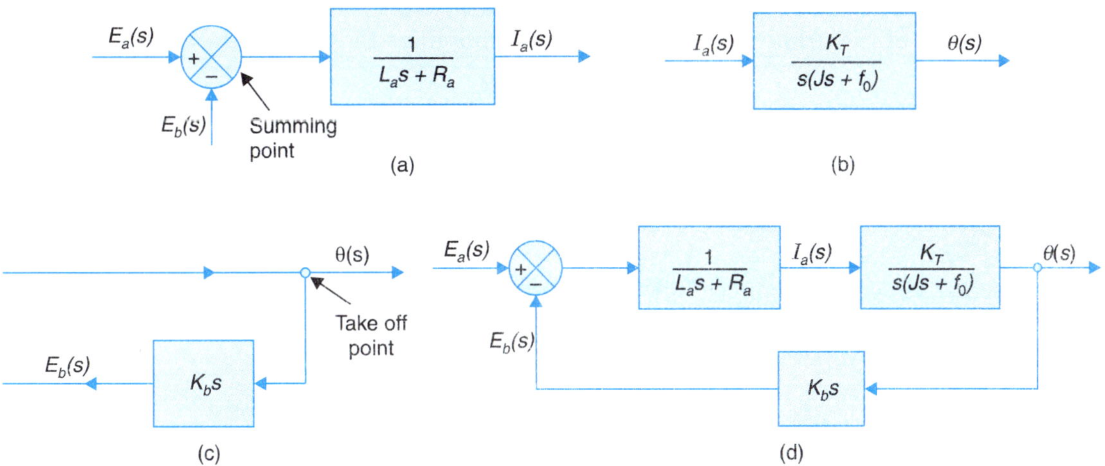

Fig. 2.22. Block diagram of armature-controlled d.c. motor.

Figure 2.22 (*c*) represents eqn. (2.52) where a signal is taken off from a *take-off* point and fed to the feedback block ($K_b s$). Fig. 2.22 (*d*) is the complete block diagram of the system under consideration, obtained by connecting the block diagram shown in Fig. 2.22 (*a*), (*b*) and (*c*). It may be pointed out here that when a signal is taken from the output of a block, this does not affect the output as per assumption 1 of the procedure for driving transfer functions advanced earlier.

However, it should be noted that the block diagram of the system under consideration can be directly obtained from the physical system of Fig. 2.21 by using the transfer functions of simple electrical and mechanical networks derived already. The voltage applied to the armature circuit is $E_a(s)$ which is opposed by the back emf ($E_b(s)$). The net voltage ($E_a - E_b$) acts on a linear circuit comprised of resistance and inductance in series, having the transfer function $1/(sL_a + R_a)$. The result is an armature current $I_a(s)$. For fixed field, the torque developed by the motor is $K_T I_a(s)$. This torque rotates the load at a speed $\theta(s)$ against the moment of inertia J and viscous friction with coefficient f_0 [the transfer function is $1/(Js + f_0)$]. The back emf signal $E_b = K_b\ \theta(s)$ is taken off from the shaft speed and fedback negatively to the summing point. The angle signal $\theta(s)$ is obtained by integrating (*i.e.,* $1/s$) the speed $\dot{\theta}(s)$. This results in the block diagram of Fig. 2.23, which is equivalent to that of Fig. 2.22 as can be seen by shifting the take off point from $\dot{\theta}(s)$ to $\theta(s)$ and replacing $K_b s$ by K_b.

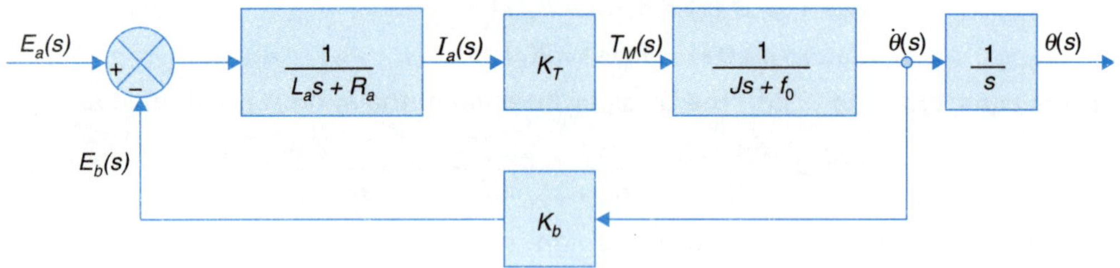

Fig. 2.23. Block diagram of armature-controlled d.c. motor.

The *armature circuit inductance* L_a *is usually negligible.* Therefore from eqn. (2.55), the transfer function of the armature controlled motor simplifies to

$$\frac{\theta(s)}{E_a(s)} = \frac{K_T/R_a}{Js^2 + s(f_0 + K_T K_b/R_a)} \qquad ...(2.56)$$

The term $(f_0 + K_T K_b/R_a)$ indicates that the back emf of the motor effectively increases the viscous friction of the system. Let

$$f = f_0 + K_T\, K_b/R_a$$

be the effective viscous friction coefficient. Then from eqn. (2.56)

$$\frac{\theta(s)}{E_a(s)} = \frac{K_T\ /\ R_a}{s(Js + f)} \qquad ...(2.57)$$

The transfer function given by eqn. (2.56) may be written in the form

$$\frac{\theta(s)}{E_a(s)} = \frac{K_m}{s(s\tau_m + 1)} \qquad ...(2.58)$$

where $K_m = K_T/R_a f$ = motor gain constant, and $\tau_m = J/f$ = motor time constant.

The motor torque and back emf constants K_T, K_b are interrelated. Their relationship is deduced below. In metric units, K_b is in V/rad/s and K_T is in *Nm/A*.

Electrical power converted to mechanical form = $e_b i_a = K_b \dot{\theta} i_a$ W

Power at shaft (in mechanical form) = $T\dot{\theta} = K_T i_a \dot{\theta}$ W

At steady speed these two powers balance. Hence

$$K_b \dot{\theta} i_a = K_T\, i_a \dot{\theta} \qquad \text{or} \qquad K_b = K_T \text{ (in MKS units)}$$

This result can be used to advantage in practice as K_b can be measured more easily and with greater accuracy than K_T.

Field-control

A field-controlled d.c. motor is shown in Fig. 2.24 (*a*).

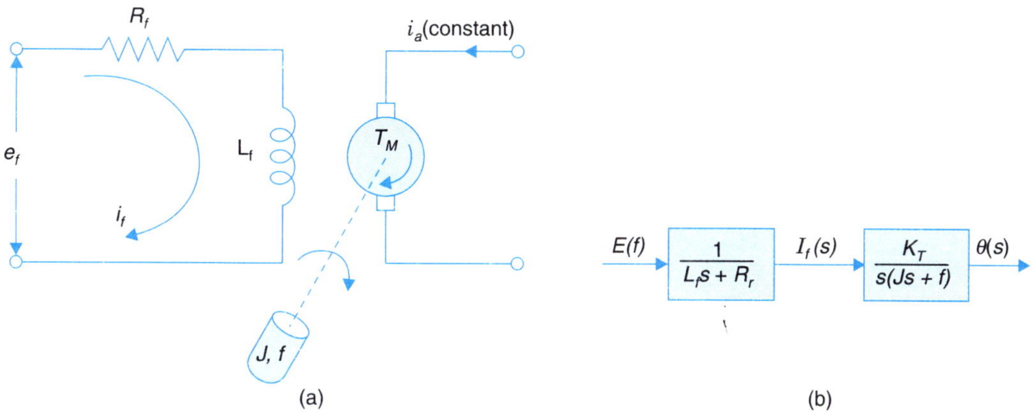

Fig. 2.24. (a) Field-controlled d.c. motor, (b) Block diagram of field-controlled motor.

In this system,

R_f = field winding resistance (Ω).

L_f = field winding inductance (H).

e = field control voltage (V).

i_f = field current (A).

T_M = torque developed by motor (Nm).

J = equivalent moment of inertia of motor and load referred to motor shaft (kg-m^2).

f = equivalent viscous friction coefficient of motor and load referred to motor shaft $\frac{Nm}{\text{rad / s}}$.

θ = angular displacement of motor shaft (rad).

In the field-controlled motor, the armature current is fed from a constant current source. Therefore, from eqn. (2.47)

$$T_M = K_1 K_f i_f i_a = K_T' i_f$$

where K_T' is a constant.

The equation for the field circuit is

$$L_f = \frac{di_f}{dt} + R_f i_f = e_f \quad \text{...(2.59)}$$

The torque equation is

$$J\frac{d^2\theta}{dt^2} = f\frac{d\theta}{dt} = T_M = K_T' i_f \quad \text{... (2.60)}$$

Taking the Laplace transform of eqns. (2.48) and (2.49), assuming zero initial conditions, we get

$$(L_f s + R_f) I_f(s) = E_f(s) \quad \text{...(2.61)}$$

$$(Js^2 + fs)\ \theta(s) = T_M(s) = K_T' I_f(s) \quad \text{...(2.62)}$$

From the above equations, the transfer function of the motor is obtained as

$$\frac{\theta(s)}{E_f(s)} = \frac{K_T'}{s(L_f s + R_f)(Js + f)} = \frac{K_m}{s(\tau_f s + 1)(\tau_{me} s + 1)} \qquad ...(2.63)$$

where $K_m = K_T'/R_f f$ = motor gain constant; $\tau_f = L_f/R_f$ = time constant of field circuit; and $\tau_{me} = J/f$; mechanical time constant.

The block diagram of the field-controlled d.c. motor obtained from eqns. (2.61) and (2.62) is given in Fig. 2.24 (*b*).

For small size motors field-control is advantageous because only a low power servo amplifier is required while the armature current which is not large can be supplied from an expensive constant current amplifier. For large size motors it is on the whole cheaper to use armature-control scheme. Further in armature-controlled motor, back emf contributes additional damping over and above that provided by load friction. With the advances made in permanent magnet materials, permanent magnet armature-controlled d.c. servomotor are now universally adopted (see Chapter 4).

2.5 BLOCK DIAGRAM ALGEBRA

As introduced earlier, the input-output behaviour of a linear system or element of a linear system is given by its transfer function

$$G(s) = C(s)/R(s)$$

where $R(s)$ = Laplace transformation of the input variable; and $C(s)$ = Laplace transform of the output variable.

A convenient graphical representation of this behaviour is the *block diagram* as shown in Fig. 2.25 (*a*) wherein the signal into the block represents the input $R(s)$ and the signal out of the block represents the output $C(s)$, while the block itself stands for the transfer function $G(s)$. The flow of information (signal) is unidirectional from the input to the output with the output being equal to the input multiplied by the transfer function of the block. A complex system comprising of several non-loading elements is represented by the interconnection of the blocks for individual elements. The blocks are connected by lines with arrows indicating the unidirectional flow of information from the output of one block to the input of the other. In addition to this, *summing* or *differencing* of signals is indicated by the symbols shown in Fig. 2.25 (*b*), while the *take-off* point of a signal is represented by Fig. 2.25 (*c*).

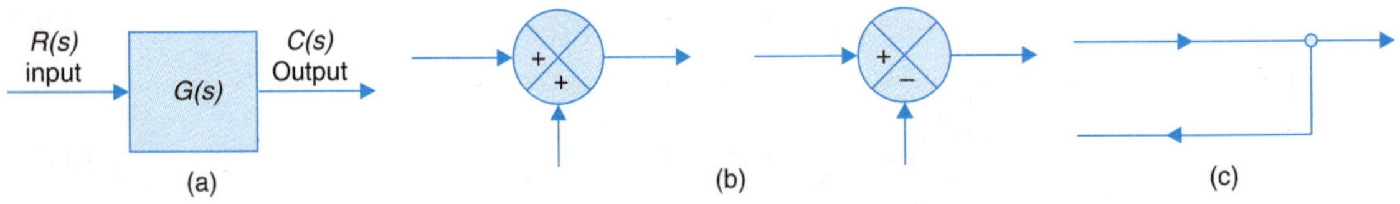

Fig. 2.25

Block diagrams of some of the control systems turn out to be very complex such that the evaluation of their performance requires simplification (or reduction) of block diagrams which is carried out by block diagram rearrangements. Some of the important block diagram rearrangements are discussed in this section.

Block Diagram of a Closed-loop System

Fig. 2.26 (*a*) shows the block diagram of a negative feedback system. With reference to this figure, the terminology used in block diagrams of control systems is given below.

$R(s)$ = reference input.

$C(s)$ = output signal or controlled variable.

$B(s)$ = feedback signal.

$E(s)$ = actuating signal.

$G(s) = C(s)/E(s)$ = forward path transfer function.

$H(s)$ = transfer function of the feedback elements.

$G(s)\, H(s) = B(s)/E(s)$ = loop transfer function.

$T(s) = C(s)/R(s)$ = closed-loop transfer function.

From Fig. 2.26 (*a*) we have

$$C(s) = G(s)/E(s) \quad \text{...(2.64)}$$

$$E(s) = R(s) - B(s) = R(s) - H(s)C(s) \quad \text{...(2.65)}$$

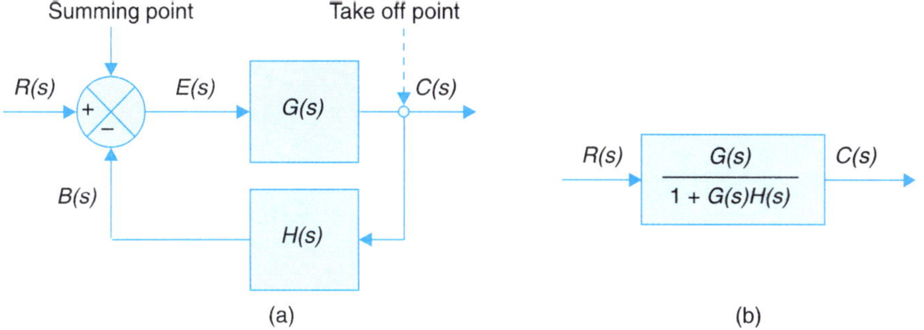

Fig. 2.26. (a) Block diagram of closed-loop system; (b) Reduction of block diagram shown in Fig. 2.26 (a).

Eliminating $E(s)$ from eqns. (2.64) and (2.65) we have

$$C(s) = G(s)R(s) - G(s)H(s)C(s)$$

or

$$\frac{C(s)}{R(s)} = T(s) = \frac{G(s)}{1+G(s)H(s)} \quad \text{...(2.66)}$$

Therefore the system shown in Fig. 2.26 (*a*) can be reduced to a single block shown in Fig. 2.26 (*b*).

Multiple-input-multiple-output Systems

When multiple inputs are present in a linear system, each input can be treated independently of the others. Complete output of the system can then be obtained by superposition, *i.e.,* outputs corresponding to each input along are added together.

Consider a two-input linear system shown in Fig. 2.27 (*a*). The response to the reference input can be obtained by assuming $U(s) = 0$. The corresponding block diagram shown in Fig. 2.27 (*b*) gives

$$C_R(s) = \text{output due to } R(s) \text{ acting alone}$$

$$= \frac{G_1(s)G_2(s)}{1+G_1(s)G_2(s)H(s)} R(s) \quad ...(2.67)$$

Similarly the response to the input $U(s)$ is obtained by assuming $R(s) = 0$. The block diagram corresponding to this case is shown in Fig. 2.27 (*c*), which gives

$$C_U(s) = \text{input due to } U(s) \text{ acting along}$$

$$= \frac{G_2(s)}{1+G_1(s)G_2(s)H(s)} U(s) \quad ...(2.68)$$

The response to the simultaneous application of $R(s)$ and $U(s)$ can be obtained by adding the two individual responses.

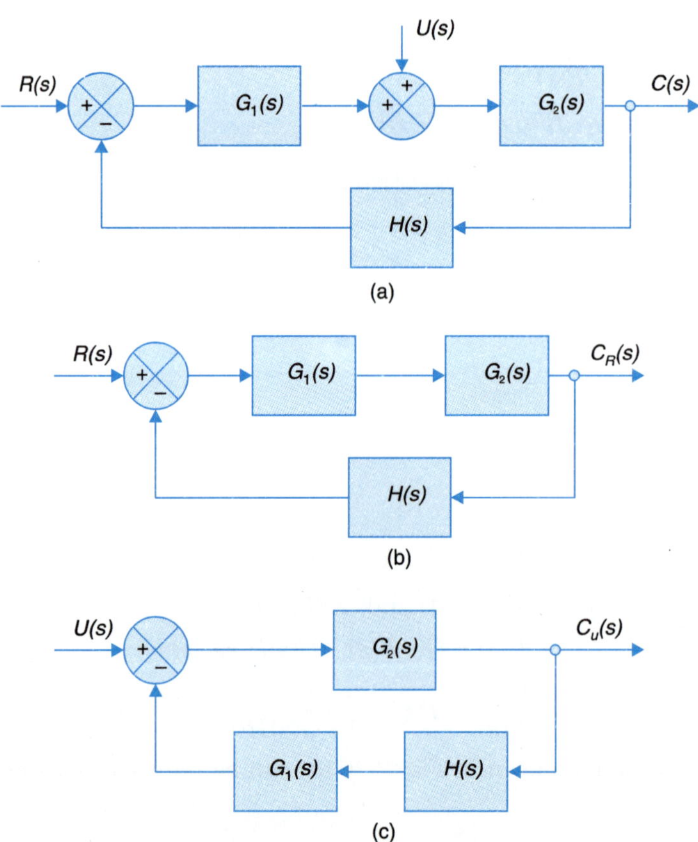

Fig. 2.27. Block diagram of two-input system.

Adding eqns. (2.67) and (2.68), we get

$$C(s) = C_R(s) + C_U(s) = \frac{G_2(s)}{1+G_1(s)G_2(s)H(s)}[G_1(s)R(s) + U(s)] \quad ...(2.69)$$

In case of multiple-input multiple-output system shown in Fig. 2.28 (r inputs and m outputs), the i-th output $C_i(s)$ is given by the principle of superposition as

$$C_i(s) = \sum_{j=1}^{r} G_{ij}(s) R_j(s);\ i = 1, 2, \ldots, m \qquad \ldots(2.70)$$

where $R_j(s)$ is the j-th input and $G_{ij}(s)$ is the transfer function between the i-th output and j-th input with all other inputs reduced to zero.

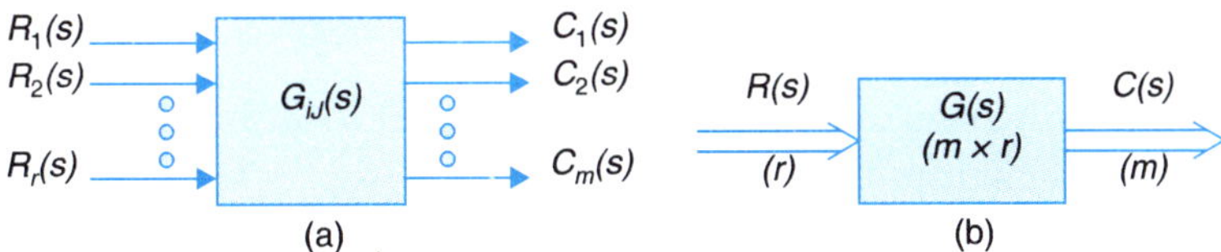

Fig. 2.28. Multiple-input-multiple-output systems.

Equation (2.69) can be expressed in matrix form as

$$\begin{bmatrix} C_1(s) \\ C_2(s) \\ \\ C_m(s) \end{bmatrix} = \begin{bmatrix} G_{11}(s) & G_{12}(s) & \ldots & G_{1r}(s) \\ \vdots & \vdots & & \vdots \\ G_{m1}(s) & G_{m2}(s) & \ldots & G_{mr}(s) \end{bmatrix} \begin{bmatrix} R_1(s) \\ R_2(s) \\ \vdots \\ R_r(s) \end{bmatrix} \qquad \ldots(2.71)$$

This can be expressed in compressed matrix notation as

$$\mathbf{C}(s) = \mathbf{G}(s)\mathbf{R}(s) \qquad \ldots(2.72)$$

where $\mathbf{R}(s)$ = vector of inputs (in Laplace transform), dimension r

$\mathbf{G}(s)$ = matrix transfer function ($m \times r$)

$\mathbf{C}(s)$ = vector of output (m)

The corresponding block diagram can be drawn as in Fig. 2.28(b) where thick arrows represent multi inputs and outputs.

When feedback loop is present each feedback signal is obtained by processing in general all the outputs. Thus for ith feedback signal we can write

$$B_i(s) = \sum_{j=1}^{m} H_{ij}(s) C_j(s) \qquad \ldots(2.73)$$

The ith error signal is then

$$E_i(s) = R_i(s) - B_i(s) \qquad \ldots(2.74)$$

Generalizing in matrix form we can write

$$\mathbf{C}(s) = \mathbf{G}(s)\ \mathbf{E}(s) \qquad \ldots(i)$$

$$\mathbf{E}(s) = \mathbf{R}(s) - \mathbf{B}(s) \qquad \ldots(ii)$$

$$\mathbf{B}(s) = \mathbf{H}(s)\ \mathbf{C}(s) \qquad \ldots(iii)$$

Substituting eqns. (ii) and (iii) in eqn. (i), we get

$$\mathbf{C}(s) = \mathbf{G}(s)\ [\mathbf{R}(s) - \mathbf{H}(s)\ \mathbf{C}(s)] = \mathbf{G}(s)\ \mathbf{R}(s) - \mathbf{G}(s)\ \mathbf{H}(s)\ \mathbf{C}(s) \qquad \ldots(iv)$$

This can be simplified as

$$\mathbf{C}(s) + \mathbf{G}(s)\,\mathbf{H}(s)\,\mathbf{C}(s) = \mathbf{G}(s)\,\mathbf{R}(s)$$

or $$\mathbf{C}(s)\,[\mathbf{I} + \mathbf{G}(s)\,\mathbf{H}(s)] = \mathbf{G}(s)\,\mathbf{R}(s)$$

or $$\mathbf{C}(s) = [\mathbf{I} + \mathbf{G}(s)\,\mathbf{H}(s)]^{-1}\,\mathbf{G}(s)\,\mathbf{R}(s) \quad ...(2.75)$$

wherein closed loop matrix transfer function as

$$\mathbf{T}(s) = \mathbf{G}(s)\,[\mathbf{I} + \mathbf{G}(s)\,\mathbf{H}(s)]^{-1} \quad ...(2.76)$$

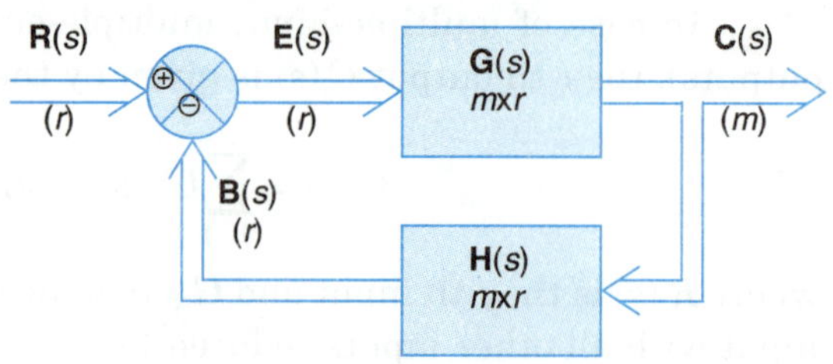

Fig. 2.29. Block diagram of multi-input multi-output closed-loop system.

The results are represented in thick arrow block diagram of Fig. 2.29. The reader may compare this block diagram and matrix equation (2.76) with the block diagram of Fig. 2.26 and eqn. (2.66) of the single-input-single-output case.

Block Diagram Reduction

As indicated earlier, a complex block diagram configuration can be simplified by certain rearrangements of block diagram using the rules of block diagram algebra. Some of the important rules are given in Table 2.5. All these rules are derived by simple algebraic manipulations of the equations representing the blocks.

As an example, let us consider the liquid-level system shown in Fig. 2.30 (note that because of interaction of the tanks, the complete transfer function cannot be obtained by multiplying individual transfer functions of the tanks).

In this system, a tank having liquid capacitance C_1 is supplying liquid through a pipe of resistance R_1 to another tank of liquid capacitance C_2, which delivers this liquid through a pipe of resistance R_2. The steady-state outflow rates of tank 1 and that of tank 2 are Q_1 and Q_2 and heads are H_1 and H_2 respectively.

Let ΔQ be a small deviation in the inflow rate Q. This results in

ΔH_1 = small deviation of the head of tank 1 from its steady-state value.

ΔH_2 = small deviation of the head of the tank 2 from its steady-state value.

ΔQ_1 = small deviation of the outflow rate of tank 1 from its steady-state value.

ΔQ_2 = small deviation of the outflow rate of tank 2 from its steady-state value.

Table 2.5. Rules of Block Diagram Algebra

Rule	*Original diagram*	*Equivalent diagram*
1. Combining blocks in cascade	X_1, G_1, X_1G_1, G_2, $X_1G_1G_2$	X_1, G_1G_2, $X_1G_1G_2$
2. Moving a summing point after a block	X_1, +, ±, X_2, $(X_1 \pm X_2)$, G, $G(X_1 \pm X_2)$	X_1, G, X_1G, +, ±, $G(X_1 \pm X_2)$, G, X_2

3. Moving a summing point ahead of a block	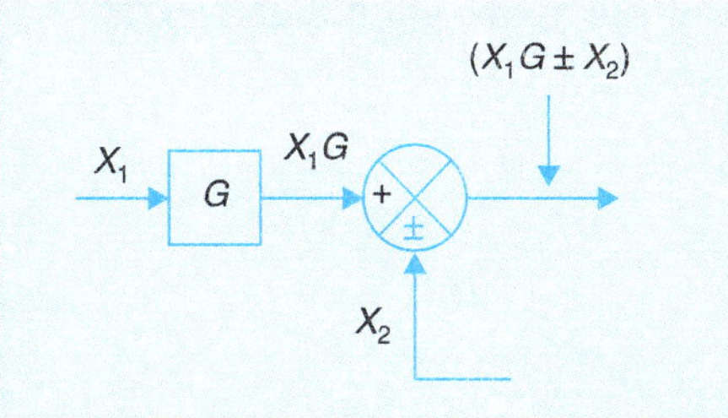 X_1, G, X_1G, $+$, $\pm$, X_2, $(X_1G \pm X_2)$	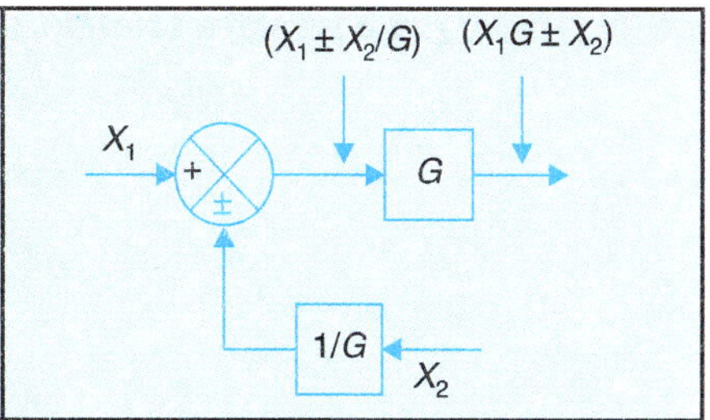 X_1, $+$, $\pm$, $(X_1 \pm X_2/G)$, G, $(X_1G \pm X_2)$, $1/G$, X_2
4. Moving a take off point after a block	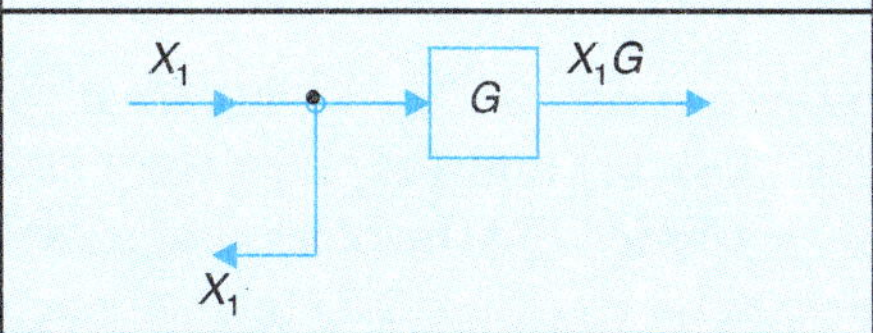 X_1, G, X_1G, X_1	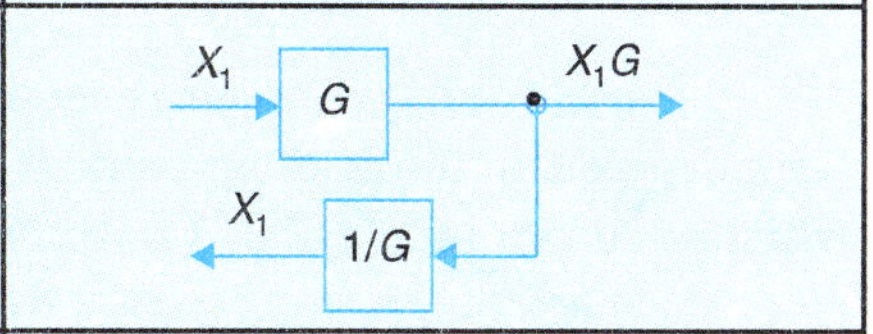 X_1, G, X_1G, X_1, $1/G$
5. Moving a take off point ahead of a block	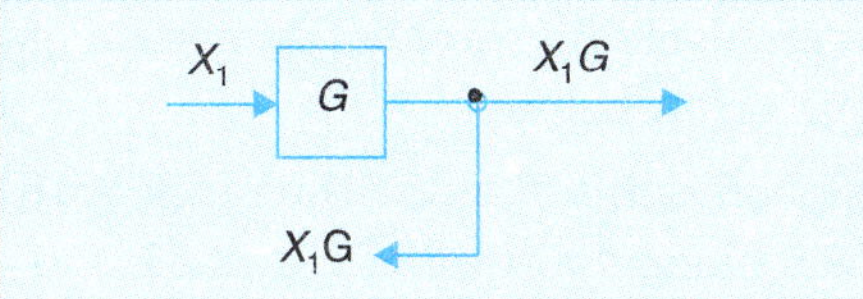 X_1, G, X_1G, X_1G	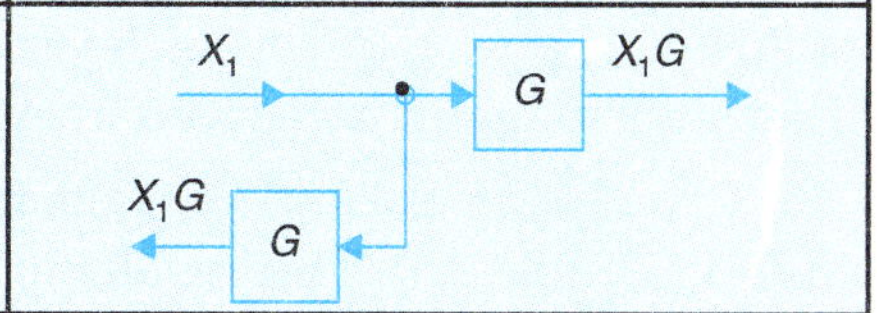 X_1, G, X_1G, X_1G, G
6. Eliminating a feedback loop	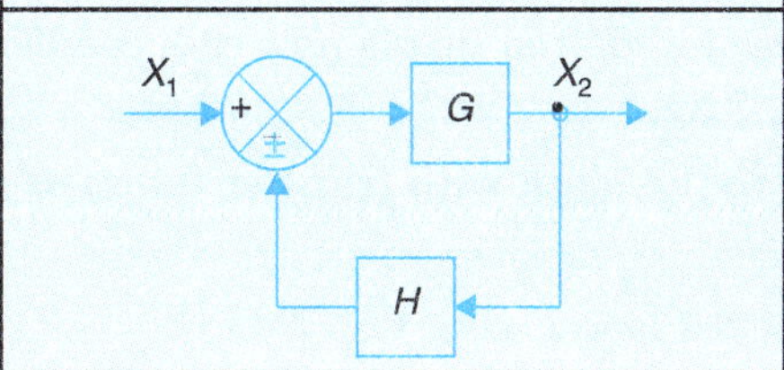 X_1, $+$, $\pm$, G, X_2, H	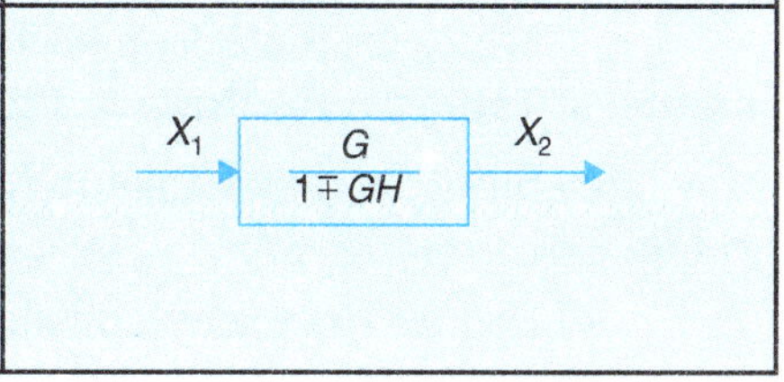 X_1, $\dfrac{G}{1 \mp GH}$, X_2

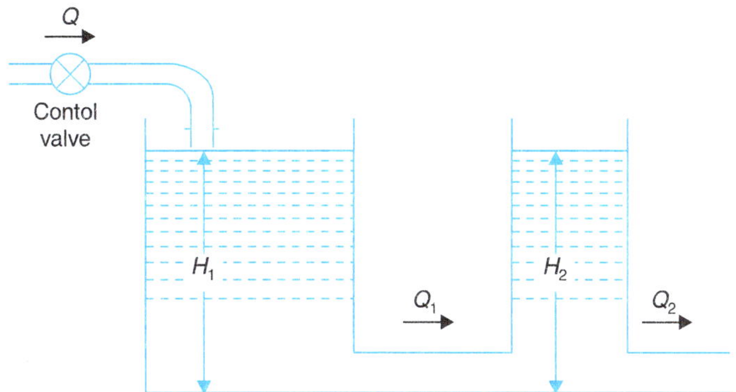

Fig. 2.30. Liquid-level system.

The flow balance equation for tank 1 is

$$\Delta Q = \Delta Q_1 + C_1 \frac{d}{dt}(\Delta H_1)$$

Similarly for tank 2

$$\Delta Q_1 = \Delta Q_2 + C_2 \frac{d}{dt}(\Delta H_2)$$

where $$\Delta Q_1 = \frac{\Delta H_1 - \Delta H_2}{R_1} \quad \text{and} \quad \Delta Q_2 = \frac{\Delta H_2}{R_2}$$

Taking the Laplace transform of the above equations we get

$$\Delta Q(s) - \Delta Q_1(s) = sC_1 \Delta H_1(s) \qquad ...(2.77)$$

$$\Delta Q_1(s) - \Delta Q_2(s) = sC_2 \Delta H_2(s) \qquad ...(2.78)$$

$$\Delta Q_1(s) = \frac{\Delta H_1(s) - \Delta H_2(s)}{R_1} \qquad ...(2.79)$$

$$\Delta Q_2(s) = \frac{\Delta H_2(s)}{R_2} \qquad ...(2.80)$$

The block diagram corresponding to eqns. (2.77) – (2.80) are given in Figs. 2.31 (*a*)-(*d*). Connecting the block diagrams of Fig. 2.31 (*a*) and (*b*) gives the block diagram for tank 1, which is shown in Fig. 2.31 (*e*). Similarly connecting the block diagrams of Fig. 2.31 (*c*) and (*d*) gives the block diagram for tank 2 which is shown in Fig. 2.31 (*f*). Connecting the block diagrams of Figs. 2.31 (*e*) and (*f*), gives the overall block diagram of the system as shown in Fig. 2.31 (*g*). This block diagram is reduced in steps given below.

(*i*) In Fig. 2.31 (*g*) shift the take off point T_1 after the block with transfer function $1/R_2$ (rule 4 of Table 2.5). This results in the block diagram of Fig. 2.31.

(*ii*) Minor feedback loop enclosed in dotted line is now reduced to a single block by rules 1 and 6 of Table 2.5 resulting in Fig. 2.31.

(*iii*) Shift the take off point T_2 to the block with transfer function $1/(R_2C_2s + 1)$ resulting in Fig. 2.31.

(*iv*) Reduce the encircled feedback loop giving Fig. 2.31 (*k*).

(*v*) Reduce Fig. 2.31 (*k*) to the single block of Fig. 2.31, which gives the overall transfer function of the system.

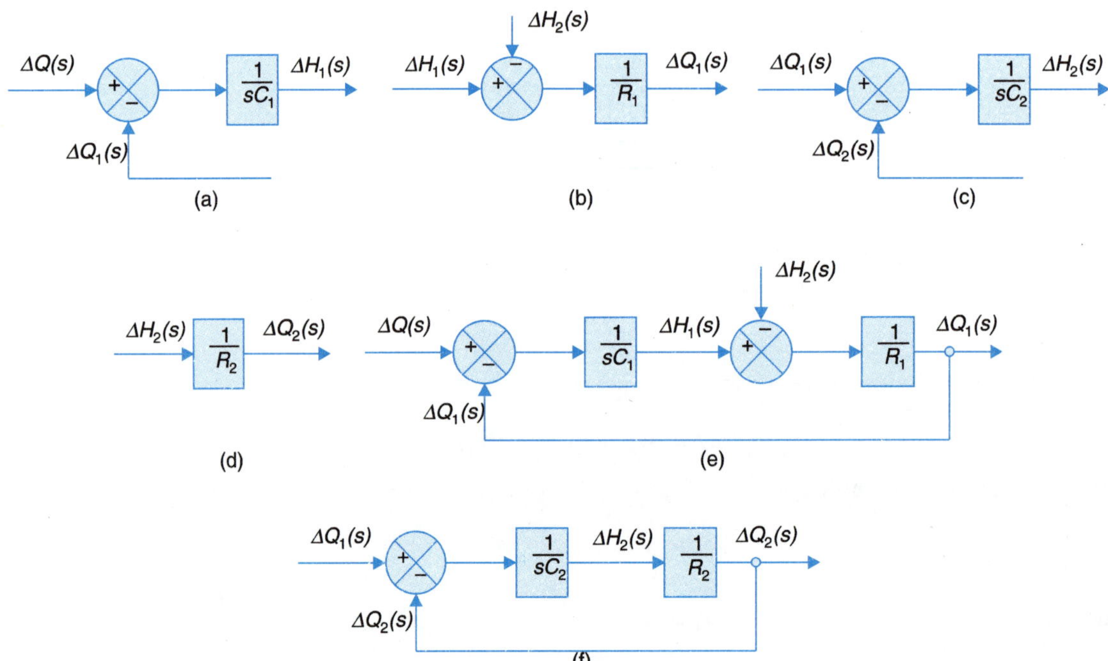

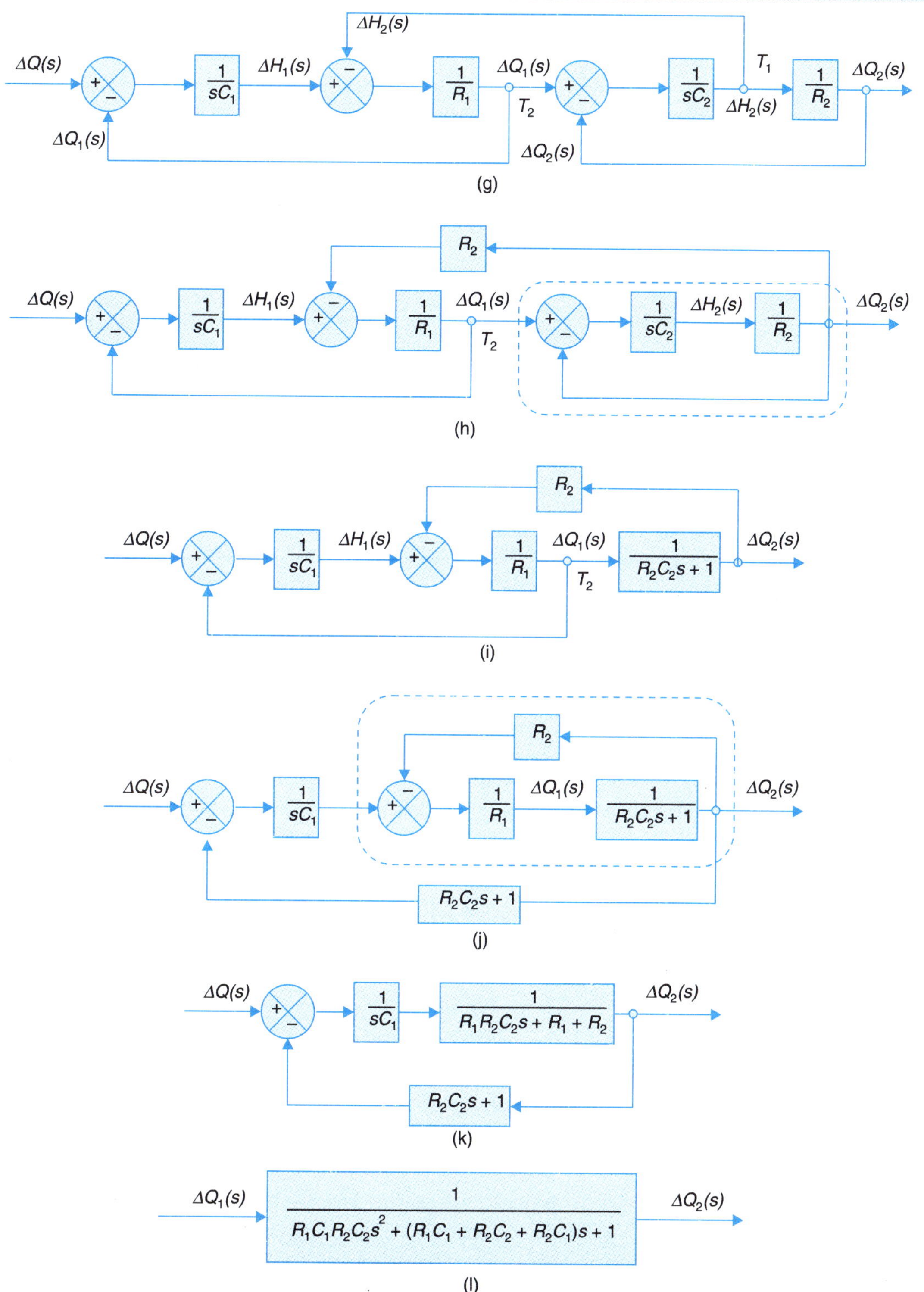

Fig. 2.31. Formation and reduction of block diagram of the system shown in Fig. 2.30.

Feedforward Compensation

Let us consider the example of Fig. 2.27 (a). The input $U(s)$ in this block diagram represents **Disturbance input** in control systems. Such an example will be considered later in this section 2.5 speed control system of Fig. 2.40. The effect of such an input is to introduce error into the system performance which needs to be kept low, within acceptable limits this is known as compensation. In several systems where the disturbance input can be predicted (or computed before hand), its effect can be eliminated by a feedforward compensation technique illustrated in the modified block diagram of Fig. 2.32.

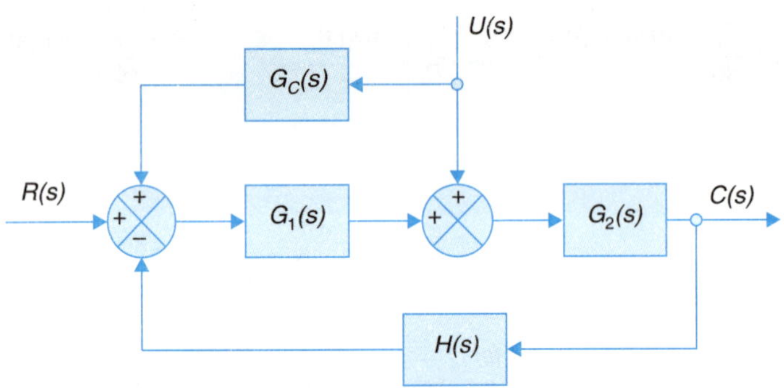

Fig. 2.32. Forward compensation.

The compensating block $G_c(s)$ causes additional input of $G_c(s)U(s)$ alongwith $U(s)$. It then follows from eqn. (2.67) that this would contribute to output a term

$$C_U(s) = \frac{G_2(s) + G_1(s)G_2(s)G_c(s)}{1 + G_1(s)G_2(s)H(s)} U(s) \quad \text{...(2.81)}$$

For this to cancel out the output component $C_U(s)$ due to disturbance input $U(s)$, the following condition has to be met.

$$G_2(s) + G_1(s)G_2(s)G_c(s) = 0 \quad \text{or} \quad G_c(s) = -\frac{1}{G_1(s)} \quad \text{...(2.82)}$$

The issues of this type of compensation will be considered further in Chapter 10.

2.6 SIGNAL FLOW GRAPHS

Block diagrams are very successful for representing control systems, but for complicated systems, the block diagram reduction process is tedious and time consuming. An alternate approach is that of signal flow graphs developed by S.J. Mason, which does not require any reduction process because of availability of a flow graph gain formula which relates the input and output system variables.

A signal flow graph is a graphical representation of the relationships between the variables of a set of linear algebraic equations. It consists of a network in which nodes representing each of the system variables are connected by directed branches. The closed-loop system whose block diagram is shown in Fig. 2.26 (a) has the signal flow representation given in Fig. 2.33 (*a*). The formulation of this signal flow graph is explained through the various signal flow terms defined below.

1. *Node.* It represents a system variable which is equal to the sum of all incoming signals at the node. Outgoing signals from the node do not affect the value of the node variable. For example, *R, E, B* and *C* are nodes in Fig. 2.33 (*a*). These symbols are also represent the corresponding node variables.

2. *Branch.* A signal travels along a branch from one node to another in the direction indicated by the branch arrow and in the process gets multiplied by the gain or transmittance of the branch. For example, the signal reaching the node *C* from the node *E* is given by *GE* where *G* is the branch transmittance and the branch is directed from the node *E* to the node *C* in Fig. 2.33 (*a*). Thus the value of the node variable *C* = *GE*.

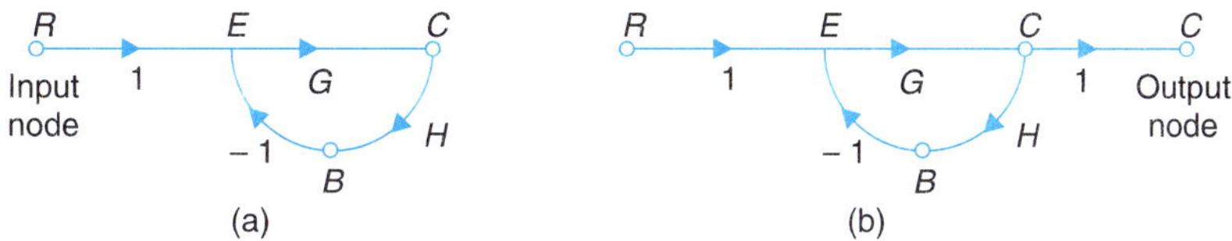

Fig. 2.33. Signal flow graph of a closed-loop system.

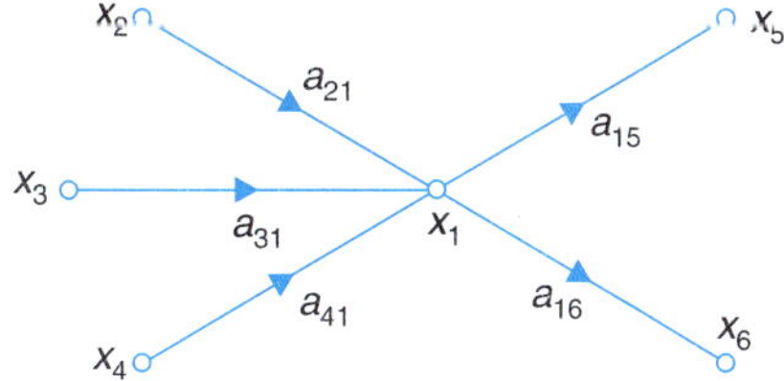

Fig. 2.34. Node as a summing point and as a transmitting point.

(*a*) *Node as a summing point*

With reference to the signal flow graph of Fig. 2.34, the node variable x_1 is expressed

$$x_1 = a_{21}\, x_2 + a_{31}\, x_3 + a_{41}\, x_4 = \text{sum of all incoming signals.}$$

(*b*) *Node as a transmitting point*

A node variable is transmitted through all branches outgoing from the node. Thus in the signal flow graph of Fig. 2.34.

$$x_5 = a_{15}\, x_1 \qquad \text{...(2.83)}$$

$$x_6 = a_{16}\, x_1$$

As already stated in (1) above the value of the node variable is not affected by the outgoing branches.

3. *Notation.* a_{ij} is the transmittance of the branch directed from node x_i to node x_j.

4. *Input node or source.* It is a node with only outgoing branches. For example, *R* is an input node in Fig. 2.33 (*a*).

5. *Output node or sink.* It is a node with only incoming branches. However, this condition is not always met. An additional branch with unit gain may be introduced in order to meet this specified condition. For example, the node C in Fig. 2.33 (a) has one outgoing branch but after introducing an additional branch with unit transmittance as shown in Fig. 2.33 (b) the node becomes an output node.

6. *Path.* It is the traversal of connected branches in the direction of the branch arrows such that no node is traversed more than once.

7. *Forward path.* It is a path from the input node to the output node. For example, R-E-C is a forward path in Fig. 2.33 (a).

8. *Loop.* It is a path which originates and terminates at the same node. For example, E-C-B-E is a loop in Fig. 2.33 (a).

9. *Non-touching loops.* Loops are said to be non-touching if they do not possess any common node.

10. *Forward path gain.* It is the product of the branch gains encountered in traversing a forward path. For example, forward path gain of the path R-E-C in Fig. 2.33 (a) is G.

11. *Loop gain.* It is the product of branch gains encountered in traversing the loop. For example, loop gain of the loop E-C-B-E in Fig. 2.33 (a) is $-GH$.

Construction of Signal Flow Graphs

The signal flow graph of a system is constructed from its describing equations. To outline the procedure, let us consider a system described by the following set of equations:

$$\begin{aligned} x_2 &= a_{12}x_1 + a_{32}x_3 + a_{42}x_4 + a_{52}x_5 \\ x_3 &= a_{23}x_2 \\ x_4 &= a_{34}x_3 + a_{44}x_4 \\ x_5 &= a_{35}x_3 + a_{45}x_4 \end{aligned} \qquad \text{...(2.84)}$$

where x_1 is the input variable and x_5 is the output variable.

The signal flow graph for this system is constructed as shown in Fig. 2.35. First the nodes are located as shown in Fig. 2.35 (a). The first equation in (2.84) states that x_2 is equal to sum of four signals and its signal flow graph is shown in Fig. 2.35 (b). Similarly, the signal flow graphs for the remaining three equations in (2.84) are constructed as shown in Figs. 2.35 (c), (d) and (e) respectively giving the complete signal flow graph of Figs. 2.35 (f).

The overall gain from input to output may be obtained by Mason's gain formula.

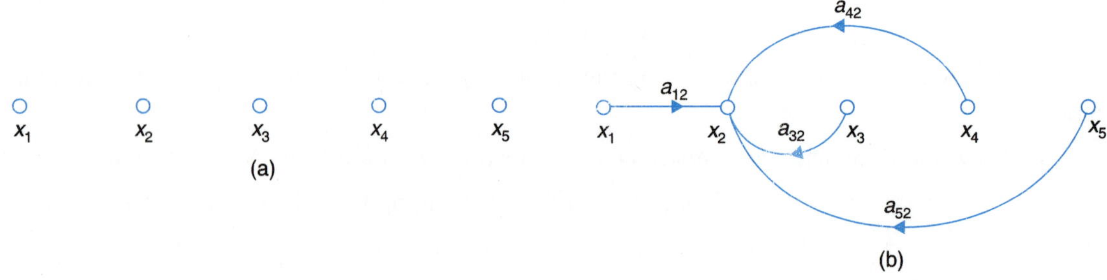

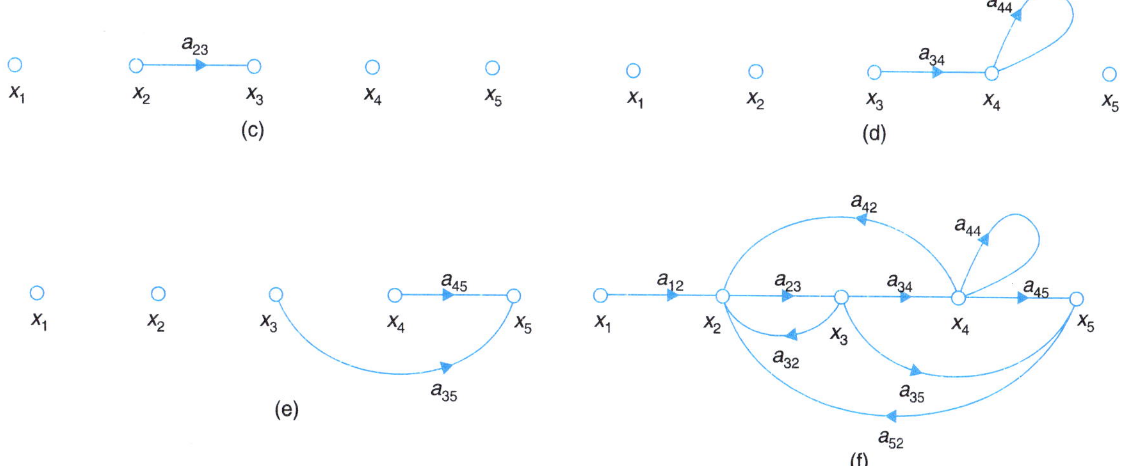

Fig. 2.35. Construction of signal flow graph for eqns. (2.84).

Mason's Gain Formula

The relationship between an input variable and an output variable of a signal flow graph is given by the net gain between the input and output nodes and is known as the overall gain of the system. Mason's gain formula for the determination of the overall system gain is given by:

$$T = \frac{1}{\Delta} \sum_K P_K \Delta_K \qquad \text{...(2.85)}$$

where P_K = path gan of K-th forward path; Δ = determinant of the graph = 1 – (sum of loop gains of all individual loops) + (sum of gain products of all possible combinations of two non-touching loops) – (sum of gain products of all possible combinations of three non-touching loops) + ... , *i.e.*,

$$\Delta = 1 - \sum_m P_{m_1} + \sum_m P_{m_2} - \sum_m P_{m_3} + \ldots \qquad \text{...(2.86)}$$

where P_{mr} = gain product m-th possible combination of r non-touching* loos; Δ_K = the value of Δ for the part of the graph not touching the K-th forward path; and T = overall gain of the system.

Let us illustrate the use of Mason's formula by finding the overall gain of the signal flow graph shown in Fig. 2.35. The following conclusions are drawn by inspection of this signal flow graph.

1. There are two forward paths with path gains

$$P_1 = a_{12}a_{23}a_{34}a_{45} \qquad \text{Fig. 2.36 }(a)$$

$$P_2 = a_{12}a_{23}a_{35} \qquad \text{Fig. 2.36 }(b)$$

2. There are five individual loops with loop gains

$$P_{11} = a_{23}a_{32} \qquad \text{Fig. 2.36 }(c)$$

*Non-touching implies that no node is common between the two.

$$P_{21} = a_{23}a_{34}a_{42} \qquad \text{Fig. 2.36 } (d)$$

$$P_{31} = a_{44} \qquad \text{Fig. 2.36 } (e)$$

$$P_{41} = a_{23}a_{34}a_{45}a_{52} \qquad \text{Fig. 2.36 } (f)$$

$$P_{51} = a_{23}a_{35}a_{52} \qquad \text{Fig. 2.36 } (g)$$

3. There are two possible combinations of two non-touching loops with loop gain products

$$P_{12} = a_{23}a_{32}a_{44} \qquad \text{Fig. 2.36 } (h)$$

$$P_{22} = a_{23}a_{35}a_{52}a_{44} \qquad \text{Fig. 2.36 } (i)$$

4. There are no combinations of three non-touching loops, four non-touching loops, etc. Therefore

$$P_{m3} = P_{m4} = \ldots = 0$$

Hence from eqn. (2.85)

$$\Delta = 1 - (a_{23}a_{32} + a_{23}a_{34}a_{42} + a_{44} + a_{23}a_{34}a_{45}a_{52} + a_{23}a_{35}a_{52}) + (a_{23}a_{32}a_{44} + a_{23}a_{35}a_{52}a_{44})$$

5. First forward path is in touch with all the loops. Therefore, $\Delta_1 = 1$. The second forward path is not in touch with one loop (Fig. 2.36 (*j*)). Therefore, $\Delta_2 = 1 - a_{44}$.

From eqn. (2.85), the

$$T = \frac{x_5}{x_1} = \frac{P_1\Delta_1 + P_2\Delta_2}{\Delta}$$

$$= \frac{a_{12}a_{23}a_{34}a_{45} + a_{12}a_{23}a_{35}(1 - a_{44})}{1 - a_{23}a_{32} - a_{23}a_{34}a_{42} - a_{44} - a_{23}a_{34}a_{45}a_{52} + a_{23}a_{32}a_{44} + a_{23}a_{35}a_{52}a_{44}} \qquad \ldots(2.87)$$

State Variable Formulation

So far we have considered the transfer function approach (single/multi input, output) using both block diagram algebra and signal flow graph. Before we consider further examples of signal flow graphs in control systems, we will consider an alternate organization of a system's differential equations as a set of first-order differential equations. This is known as the state variable formulation and will studies in detail in Chapter 12. Here we shall introduce the technique through a simple example by using the insight acquired into the physical systems considered so far.

Consider a simple system described by the first-order differential equation

$$\dot{x} = ax \; ; x(t = 0) = x(0) \qquad \ldots(2.88)$$

As x defines the system *state* for $t \geq 0$, it is called a *state variable* and eqn. (2.88) is the *state equation.*

Observe that the system has no input but has an initial condition (at $t = 0$). Integrating $\dot{x}$ we have

$$x = \int \dot{x}\, dt + x(0) \qquad \ldots(2.89)$$

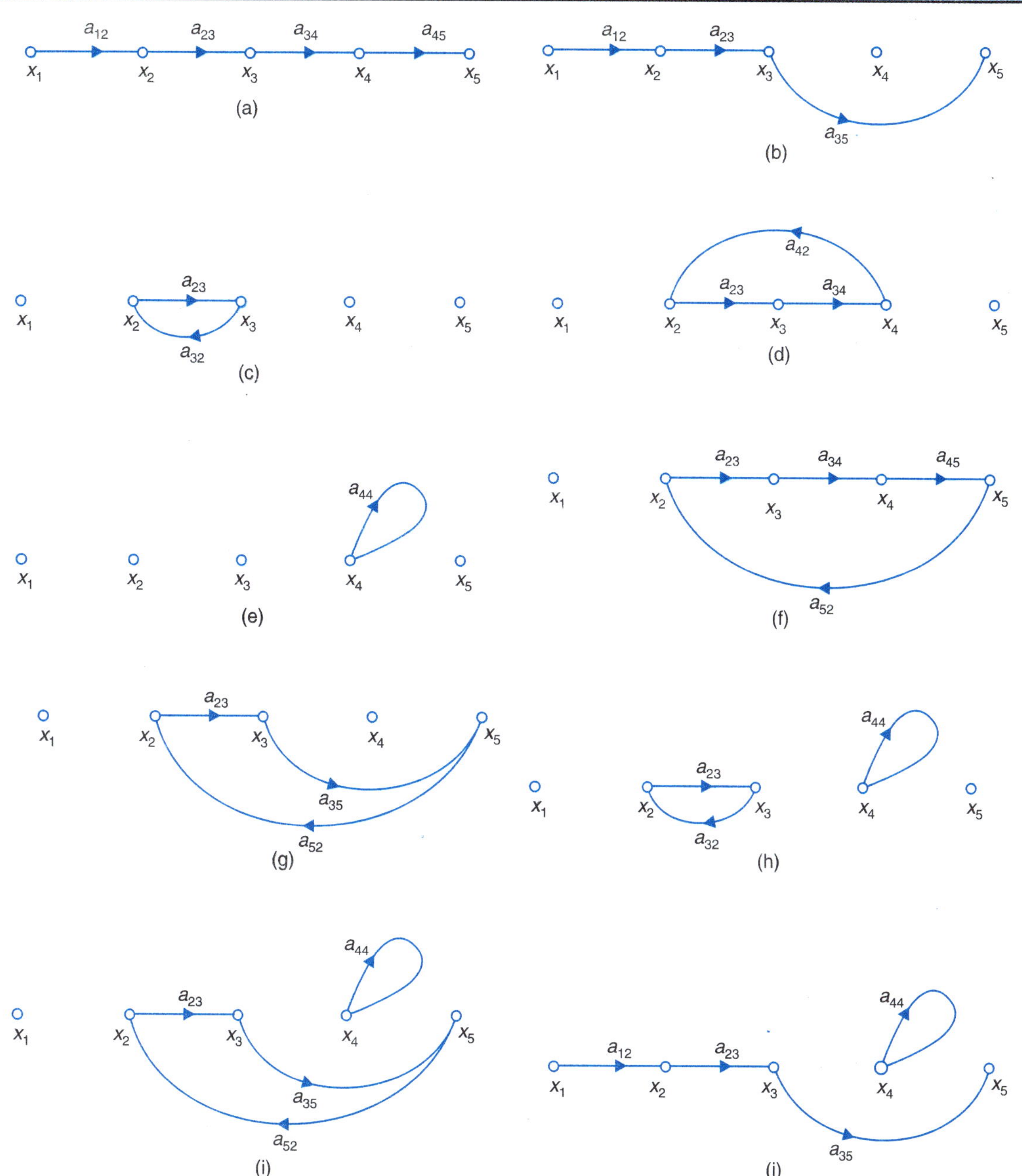

Fig. 2.36. Application of Mason's formula to the signal flow graph shown in Fig. 2.35.

The signal flow diagram for eqns. (2.88) and (2.89) can be drawn as in Fig. 2.37 (*a*) with integration symbol introduced as a transmittance of the branch from node $\dot{x}$ to x. This is the time-domain representation.

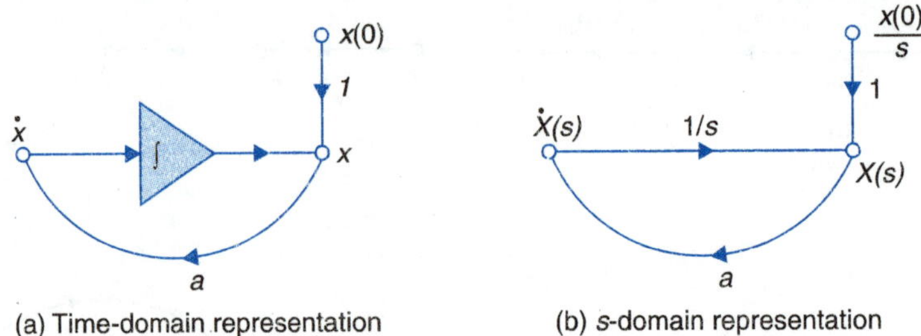

Fig. 2.37. Signal flow graph of first-order system.

Take now the Laplace transform of eqn. (2.88)

$$\dot{X}(s) = aX(s) \qquad \text{...(2.90}a\text{)}$$

But

$$\dot{X}(s) = s\,X(s) - x(0) \qquad \text{...(2.90}b\text{)}$$

or

$$X(s) = \frac{1}{s}\dot{X}(s) + \frac{1}{s}x(0) \qquad \text{...(2.91)}$$

Equations (2.90*a*) and (2.91) would give the signal flow graph in *s*-domain, as in Fig. 2.37 (*b*).

Consider now a first-order system with input *u*

$$\dot{x} = ax + bu;\ x(t = 0) = x(0) \qquad \text{...(2.92)}$$

whose *s*-domain signal flow diagram is drawn in Fig. 2.38 (*a*).

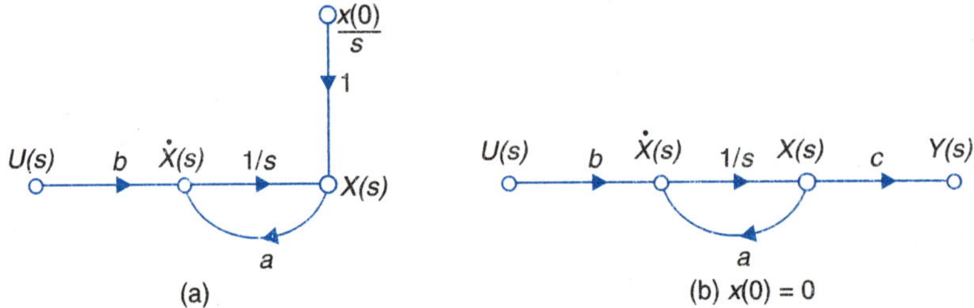

Fig. 2.38

Assume that the system output is given as

$$y = cx \qquad \text{...(2.93)}$$

The modified signal flow graph is drawn in Fig. 2.38 (*b*) wherein it is assumed that $x(0) = 0$. Observing that there is one forward path and one loop and applying the Mason's gain formula, we have

$$P_1 = bc/s$$

$$\Delta_1 = 1$$

$$\Delta = 1 - \Delta_{11} = 1 - (a/s) = (s - a)/s$$

$$T(s) = \frac{Y(s)}{U(s)} = \frac{P_1\Delta_1}{\Delta} = \frac{(bc/s)}{(s-a)/s} = \frac{bc}{s-a} \qquad \text{...(2.94)}$$

In case of second-order system two first-order state variable equations would be needed and in general *n* equations for *n*-th-order system. Let us consider the example of armature-controlled *dc* motor of Fig. 2.21 wherein we shall now regard the output of interest as motor speed $\omega = \dot{\theta}$. The block diagram of Fig. 2.22 is then redrawn as in Fig. 2.38 from which we can write

$$K_T I_a(s) = Js\omega(s) + f_0\,\omega(s)$$
$$E(s) - K_b\omega(s) = L_a s I_a(s) + R_0 I_a(s)$$

Taking the inverse Laplace transform (initial condition being zero)

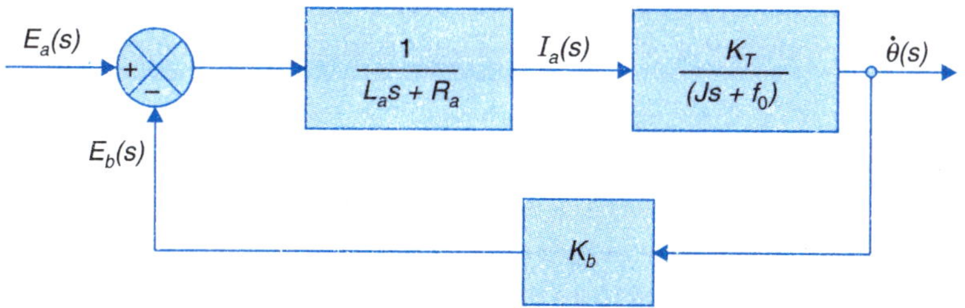

Fig. 2.38. Armature-controlled d.c. motor.

$$K_T i_a(t) = J\frac{d\omega(t)}{dt} + f_0\omega(t)$$
$$e_a(t) - K_b\omega(t) = L_a\frac{di_a(t)}{dt} + R_a i_a(t)$$

We can recognize these equations as first-order equations in ω and i_a which are then recognized as state variables, while $e(t)$ is the system input. We thus have

$$\frac{d\omega(t)}{dt} = (f_0/J)\omega(t) + (K_T/J)i_a(t) \qquad ...(2.95a)$$

$$\frac{di_a(t)}{dt} = -(K_b/L_a)\omega(t) - (R_a/L_a)\,i_a(t) + (1/L_a)e_a(t) \qquad ...(2.95b)$$

Droping the bracketed variable t for convenience of writing we can then rewrite these state equations in vector-matrix form as

$$\begin{bmatrix} \frac{d\omega}{dt} \\ \frac{di_a}{dt} \end{bmatrix} = \begin{bmatrix} -(f_0/J) & (K_T/J) \\ -(K_b/L_a) & -(R_a/L_a) \end{bmatrix}\begin{bmatrix} \omega \\ i_a \end{bmatrix} + \begin{bmatrix} 0 \\ 1/L_a \end{bmatrix} e_a \qquad ...(2.96a)$$

We will reidentify the state variables as

$$x_1 = \omega,\ x_2 = i_a$$

and input as $u = e$ and output as

$$y = \omega = [1 \quad 0]\begin{bmatrix} \omega \\ i_a \end{bmatrix} \qquad ...(2.96b)$$

We can now write eqns. (2.96*a*) and (*b*) in standard form as

$$\begin{bmatrix} x_1 \\ x_2 \end{bmatrix} = \begin{bmatrix} a_{11} & a_{12} \\ a_{21} & a_{22} \end{bmatrix}\begin{bmatrix} x_1 \\ x_2 \end{bmatrix} + \begin{bmatrix} b_1 \\ b_2 \end{bmatrix} u\,;\, u = e_a \qquad ...(2.97a)$$

$$y = [c_1 \quad c_2]\begin{bmatrix} x_1 \\ x_2 \end{bmatrix} \qquad ...(2.97b)$$

In compact vector-matrix rotation

$$\dot{\mathbf{x}} = \mathbf{Ax} + \mathbf{bu} \quad ...(2.98a)$$

$$\mathbf{y} = \mathbf{Cx} \quad ...(2.98b)$$

This is the state variable representation of a single-input-single-output (SISO) system. In multi-input multi-output (MIMO) system u and y will acquire the vector form.

The state variables identified in the above example are physical variables and are directly available for measurement. Indeed the state variables for a given system are not unique and these can be defined in a number of other ways. These and associated topics and will be discussed at length in Chapter 12.

Speed Control System

As an example let us consider a feedback speed control system whose objective is move the load at desired speed. This is easily achieved using the armature controlled *dc* motor of Fig. 2.21 by providing a feedback control loop as in Fig. 2.39 wherein the voltage signal e_t proportional to input speed (ω) generated by a *dc* tachometer coupled the motor armature is fed back negatively and is substracted from the reference voltage e_r creating the difference (error) signal e. This error signal e is then amplified to control the armature current i_a such that the motor acquires the desired speed, while driving the load (J, f and torque T_D).

A *dc* tachometer is just a conventional *dc* generator usually of permanent magnet kind, whose output voltage is a measure of speed. Thus

$$e_t = K_t\omega;\ \text{excitation being constant} \quad ...(2.99)$$

where K_t (V/rad/s) is the techometer constant.

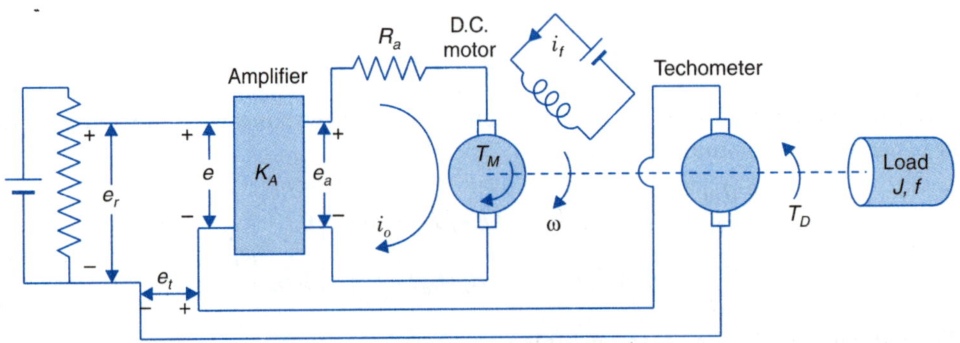

Fig. 2.39. A speed control system.

The block diagram of the speed control system of Fig. 2.39 can be easily drawn as in Fig. 2.40 (a) by modifying the block diagram of Fig. 2.23 of the armature-controlled *dc* motor. The outer feedback loop accounts for the speed feedback which is basic to the speed regulation action. The load torque T_D which opposes the motor torque enters negatively in the block diagram so that the torque applied to load (and motor) inertia and fraction is ($T_M - T_D$).

Let us convert this block diagram into a signal flow graph while making the simplifying assumption that $L_a \approx 0$. The signal flow graph is drawn in Fig 2.40 (b). Various system transfer functions are derived below using the Mason's gain formula.

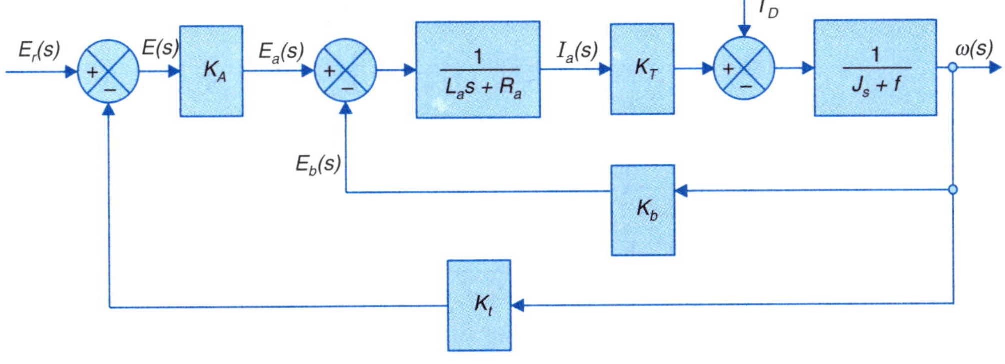

(a) Block diagram

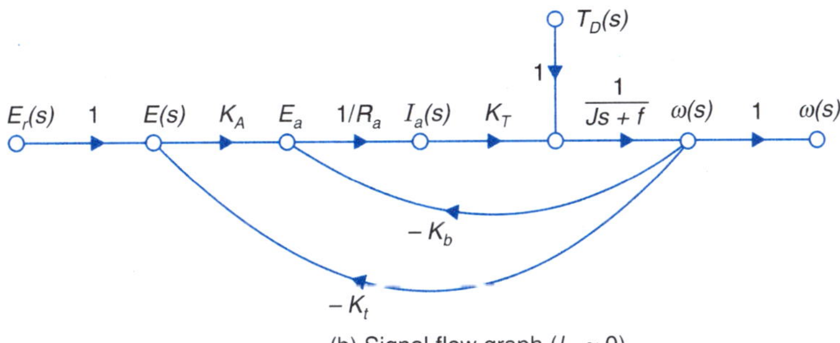

(b) Signal flow graph ($L_a \approx 0$)

Fig. 2.40. Speed control system.

Consider first the case with zero disturbance torque. By inspection of the signal flow graph, with $T_D(s) = 0$, it is found that:

1. There is only one forward path with path gain

$$P_1 = \frac{K_A K_T}{R_a(Js + f)}$$

2. There are two individual loops with loop gains

$$P_{11} = \frac{-K_T K_b}{R_a(Js + f)}$$

$$P_{21} = \frac{-K_A K_T K_t}{R_a(Js + f)}$$

3. There are no combinations of two non-touching loops, three non-touching loops, etc. Therefore

$$P_{m2} = P_{m3} = \dots = 0$$

Hence from eqn. (2.86)

$$\Delta = 1 + \left[\frac{K_T K_b}{R_a(Js + f)} + \frac{K_A K_T K_t}{R_a(Js + f)}\right] = 1 + \frac{K_T K_b + K_A K_T K_t}{R_a(Js + f)}$$

4. The forward path is in touch with both the loops. Therefore

$$\Delta_1 = 1$$

From *the Mason's gain formula of* eqn. (2.84), the overall gain is

$$T(s) = \frac{\omega(s)}{E_r(s)} = \frac{P_1 \Delta_1}{\Delta} = \frac{K_A K_T}{R_a(Js+f) + K_T K_b + K_A K_T K_t} \qquad ...(2.100)$$

With $K_t = 0$, the system is reduced to open-loop with the transfer function

$$G(s) = \frac{K_A K_T}{R_a(Js+f) + K_b K_T} = \frac{K}{(\tau s + 1)} \qquad ...(2.101)$$

where $$K = \frac{K_A K_T}{R_a f + K_T K_b}; \tau = \frac{R_a J}{R_a f + K_T K_b}$$

From eqn. (2.100), the closed-loop transfer function of the system is given by

$$T(s) = \frac{K/\tau}{s + \left(\frac{1+KK_t}{\tau}\right)} \qquad ...(2.102)$$

When the load (*disturbance*) torque $T_D(s)$ is present, the only change in the graph is the additional input $T_D(s)$.

Applying Mason's gain formula to the graph, the following transfer function is obtained between output speed and disturbance torque with zero reference voltage, *i.e.*, $E_r(s) = 0$

$$\left.\frac{\omega(s)}{T_D(s)}\right|_{E_r(s)=0} = \frac{\omega_D(s)}{T_D(s)} = \frac{-1}{Js + f + \frac{K_T}{R_a}(K_A K_t + K_b)} \qquad ...(2.103)$$

When there is no feedback ($K_t = 0$), eqn. (2.102) modifies to

$$\left.\frac{\omega(s)}{T_D(s)}\right|_{E_r(s)=0} = \frac{\omega_D(s)}{T_D(s)} = \frac{-1}{Js + f + \frac{K_T K_b}{R_a}} \qquad ...(2.104)$$

The additional term $\left(\frac{K_T K_A}{R_a}\right) K_t$ in the denominator of eqn. (2.102) (compared to eqn. (2.103)) arises on account of output (speed) feedback. As we shall see in Chapter 5 that this term reduces the effect of load (disturbance) torque on motor speed.

2.7 ILLUSTRATIVE EXAMPLES

Example 2.1 : Consider the mechanical system shown in Fig. 2.41 (*a*). A force $F(t)$ is applied to mass M_2. The free-body diagrams for the two masses are shown in Fig. 2.41 (*b*). From this figure, we have the following differential equations describing the dynamics of the system.

Solution.

$$F(t) - f_2(\dot{y}_2 - \dot{y}_1) - K_2(y_2 - y_1) = M_2 \ddot{y}_2$$

$$f_2(\dot{y}_2 - \dot{y}_1) + K_2(y_2 - y_1) - f_1\dot{y}_1 - K_1 y_1 = M_1 \ddot{y}_1$$

Rearranging we get

$$M_2 \ddot{y}_2 + f_2(\dot{y}_2 - \dot{y}_1) + K_2(y_2 - y_1) = F(t) \qquad ...(2.105)$$

$$M_1\ddot{y}_1 + f_1\dot{y}_1 - f_2(\dot{y}_2 - \dot{y}_1) + K_1y_1 - K_2(y_2 - y_1) = 0 \qquad ...(2.106)$$

These are two simultaneous second-order linear differential equations. Manipulation of these equations will result in a single differential equation (fourth-order) relating the response y_2 (or y_1) to input $F(t)$.

A spring-mass-damper system may be schematically represented as a network by showing the inertial reference frame as the second terminal of every mass (or inertia) element. As an example, the mechanical system of Fig. 2.41 (*a*) is redrawn in Fig. 2.42 which may be referred to as the *mechanical network*. Analogous electrical circuit based on force-current analogy (Table 2.2) is shown in Fig. 2.43. A look at Fig. 2.43 (electrical analog of Fig. 2.41 (*a*)) and Fig. 2.42 reveals that they are alike topologically.

The dynamical equations of the system [eqns. (2.105)-(2.106)] could also be obtained by writing nodal equations for the electrical network of Fig. 2.43 or for the mechanical network of Fig. 2.42 (with force and velocity analogous to current and voltage respectively) since the two are alike topologically. The result is:

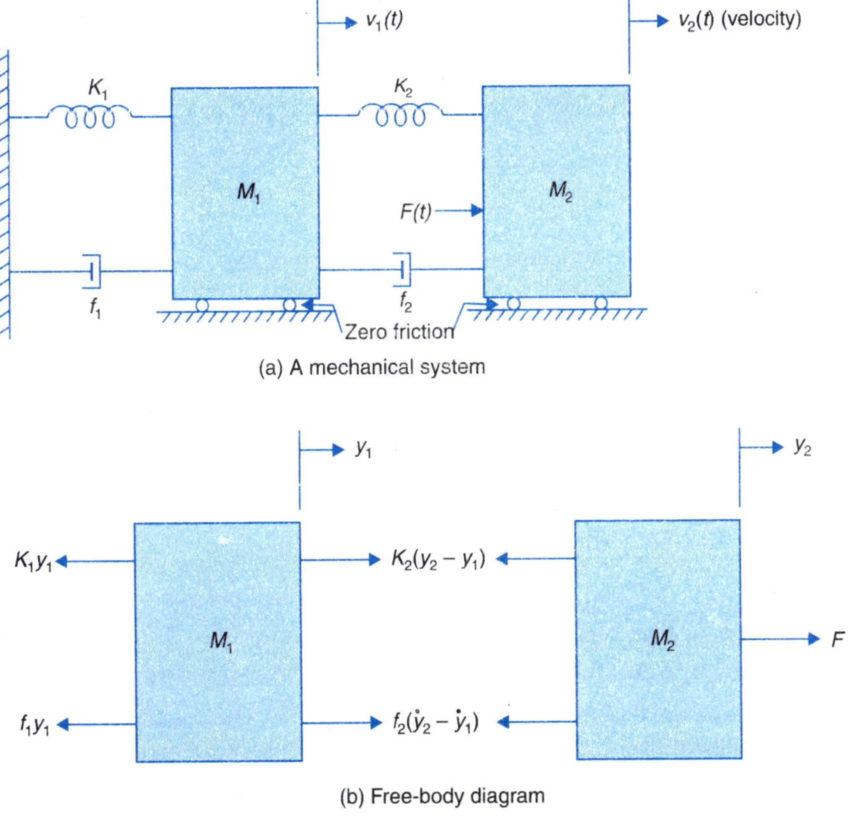

Fig. 2.41. (a) A mechanical system; (b) Free-body diagram.

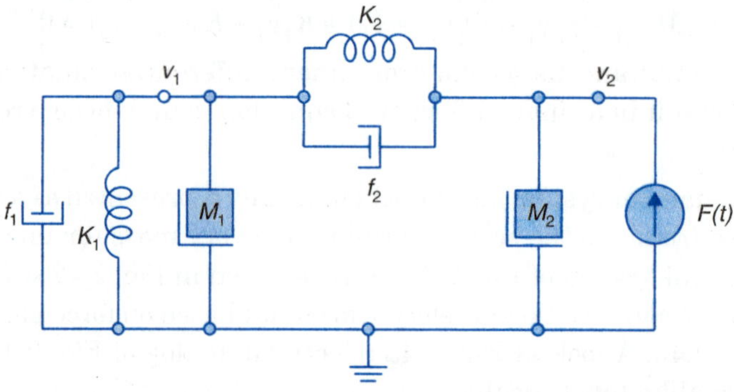

Fig. 2.42. Mechanical network for the system of Fig. 2.41.

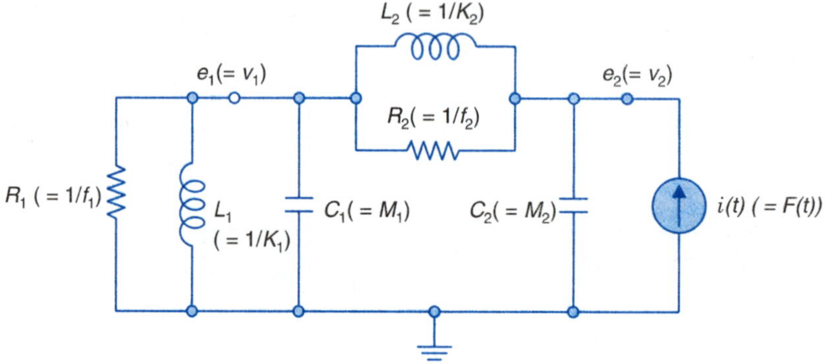

Fig. 2.43. Electrical analog for the system of Fig. 2.30 (a).

$$f_1 v_1 + K_1 \int_{-\infty}^{t} v_1 dt + M_1 \dot{v}_1 + K_2 \int_{-\infty}^{t} (v_1 - v_2) dt + f_2 (v_1 - v_2) = 0$$

$$M_2 \dot{v}_2 + K_2 \int_{-\infty}^{t} (v_2 - v_1) dt + f_2 (v_2 - v_1) = F(t)$$

The result is same as obtained earlier (with $y = \int_{-\infty}^{t} v dt$, $\dot{y} = v$ and $\ddot{y} = \dot{v}$) in eqns. (2.105) and (2.106) using the free-body diagram approach.

Signal Flow Graph

The mechanical system of Fig. 2.41(*a*) has four storage elements so it is a fourth order system and would be identified by four state variables. These can be defined as $x_1 = y_1$, $x_2 = \dot{y}_1$, $x_3 = y_2$, $x_4 = \dot{y}_2$. With reference to the free-body diagram of Fig. 2.41(*b*), the signal flow graph of Fig. 2.44 can be immediately drawn. The state variable equations can be written directly from the signal flow graph. The reader should write out these equations and organize them in matrix form. Here again the state variables are defined as physical variables, but this is not a unique choice.

Using Mason's gain formula the transfer function between any output and input $F(s)$ can be derived. It is identified here that there are seven loops but no combinations of two or more loops. Further there are two forward paths but there are no loops non-touching these paths. The reader should find the transfer function $y_1(s)/F(s)$.

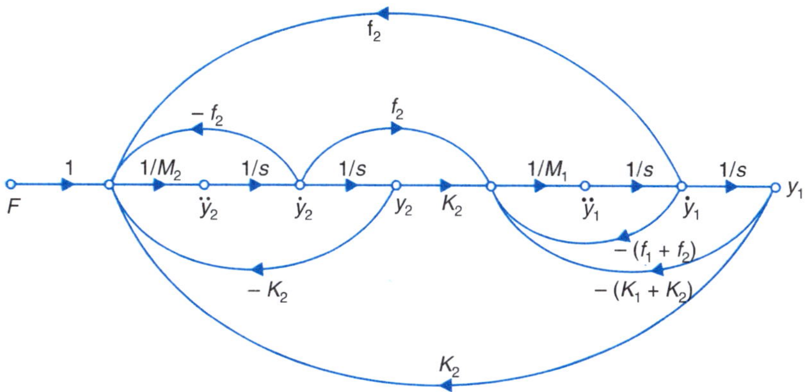

Fig. 2.44. Signal flow graph of the mechanical system of Fig. 2.41 (a).

Example 2.2 : Consider the circuit (electrical) of Fig. 2.45.

(*a*) Identify a set of state variables (physical variables).

(*b*) Draw the signal flow graph of the circuit in terms of the state variables identified in part (*a*).

(*c*) From the signal flow graph, write the state variable equations of the circuit.

(*d*) From the signal flow graph, determine the transfer function $E_C(s)/E(s)$.

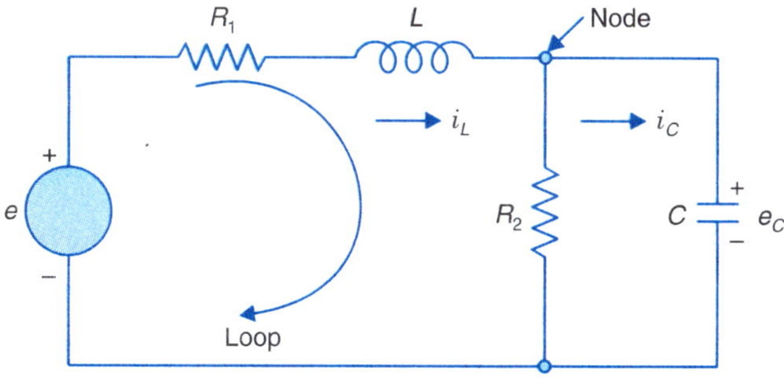

Fig. 2.45

Solution.

(*a*) This circuit has two storage elements, so these shall be two state variables. We shall identify these as the inductor current i_L and capacitor voltage e_C; both these are associated with energy storage. Remember that the state variables do not form a unique set.

From the elemental laws of inductor and capacitor we can draw the signal flow graph as in Fig. 2.46(*a*). The complete signal flow graph is then constructed by the KCL equation at the node and the KVL equation round the loop. These equations are:

$$i_L = \frac{e_C}{R_2} + i_C \quad \text{or} \quad i_C = i_L - \frac{e_C}{R_2} \qquad ...(i)$$

and

$$e = R_1 i_L + e_L + e_C \quad \text{or} \quad e_L = e - R_1 i_L - e_C \qquad ...(ii)$$

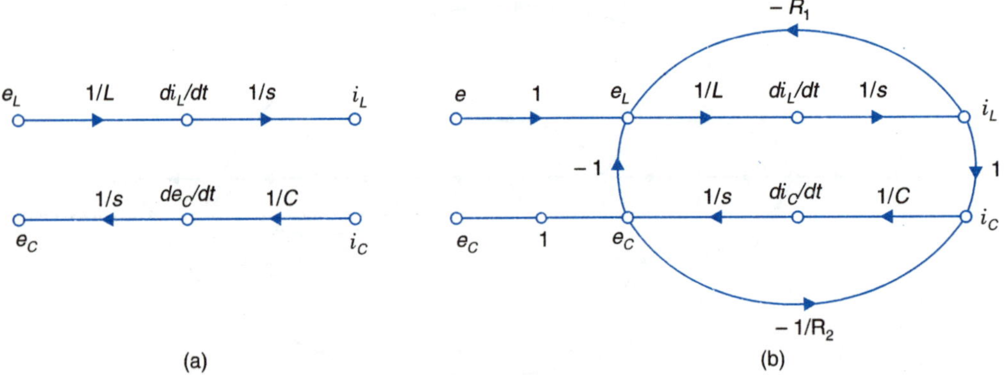

Fig. 2.46

The complete signal flow graph is drawn in Fig. 2.47(*b*).

(*c*) From the signal flow graph the two state variable equations can be written as below.

$$\frac{di_L}{dt} = \frac{1}{L}e_L = \frac{1}{L}(-e_C - R_1 i_L + e) = -\frac{R_1}{L}i_L - \frac{1}{L}e_C + \frac{1}{L}e \qquad ...(iii)$$

and
$$\frac{de_C}{dt} = \frac{1}{C}i_C = \frac{1}{C}\left(i_L + \frac{e_C}{R_2}\right) = \frac{1}{C}i_L - \frac{1}{R_2 C}e_C \qquad ...(iv)$$

Equations (*iii*) and (*iv*) can be written in matrix form

$$\begin{bmatrix} di_L/dt \\ di_C/dt \end{bmatrix} = \begin{bmatrix} -R_1/L & -1/L \\ 1/C & 1/R_2C \end{bmatrix}\begin{bmatrix} i_L \\ e_C \end{bmatrix} + \begin{bmatrix} 1/L \\ 0 \end{bmatrix} e$$

These are also known as **state space** equations.

(*d*) Input $E(s)$, output $E_C(s)$. From the singal flow graph we have,

Forward path $\qquad P_1 = \frac{1}{sL} \times \frac{1}{sC} = \frac{1}{s^2 LC}; \Delta_1 = 1$

Single loops $\qquad P_{11} = -\frac{R_1}{sL}, P_{21} = -\frac{1}{sR_2C}, P_{31} = -\frac{1}{s^2 LC}$

$$\Delta = 1 + \frac{R_1}{sL} + \frac{1}{sR_2C} + \frac{1}{s^2 LC}$$

Hence
$$\frac{E_C(s)}{E(s)} = \frac{P_1\Delta_1}{\Delta} = \frac{\frac{1}{s^2 LC}}{1 + \frac{R_1}{sR_2C} + \frac{1}{s^2 LC}} = \frac{1}{1 + s\left(R_1 C + \frac{L}{R_2}\right) + s^2 LC}$$

Example 2.3 : Consider a salt mixing tank shown in Fig. 2.48. A solution of salt in water at a concentration C_f(moles* of salt/m^3 of solution) is mixed with pure water to obtain an outflow

*A mole of a substance is defined as the amount of substance whose mass numerically equals its molecular weight. For example, a gram-mole of helium would have a mass of 4.003 g (molecular weight of helium = 4.003).

stream with salt concentration C_o. The water flow rate is assumed fixed at Q_w and the solution flow rate may be varied to achieve the desired concentration C_o (also see Problem 2.6). Volumetric hold-up of the tank is V, which is held constant. Let us assume that stirring causes perfect mixing so that composition of the liquid in the tank is uniform throughout.

Solution. For this system,

$$Q_f = K_v x_v \text{ ; } K_v \text{ is valve coefficient}$$

$$Q_o = Q_w + Q_f$$

The rate of salt inflow in the tank

$$m_i = Q_f C_f \text{ ; } \left(\frac{\text{m}^3}{\text{s}} \cdot \frac{\text{moles}}{\text{m}^3} = \text{moles / s}\right)$$

The rate of salt outflow from the tank

$$m_o = Q_o C_o \text{ ; (moles/s)}$$

The rate of salt accumulation in the tank

$$m_a = \frac{d}{dt}[VC_o(t)] = V\frac{dC_o}{dt}$$

where $VC_o(t)$ the salt hold-up of the tank at time t.

From the law of conversation of mass, we have

$$m_1 = m_a + m_o \qquad \text{or} \qquad Q_f C_f = V\frac{dC_o}{dt} + Q_o C_o$$

or

$$\tau\frac{dC_o}{dt} + C_o = Kx_v$$

where $K = (C_f K_v)/Q_o$ and $\tau = V/Q_o$ is the tank hold-up time.

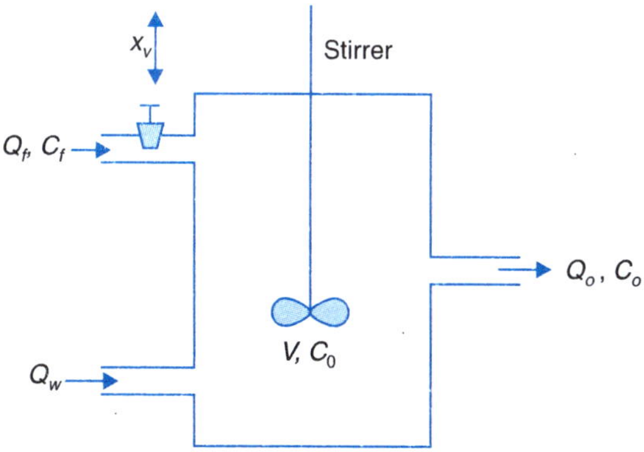

Fig. 2.48. A salt mixing tank.

The transfer function of the system is

$$\frac{C_o(s)}{X_v(s)} = \frac{K}{s\tau + 1}$$

Example 2.4 : The manipulator shown in Fig. 2.49 has a rotating joint followed by a linear (prismatic) joint. The whole mass of links is concentrated at the centre of mass. Derive the dynamic equations for the manipulator.

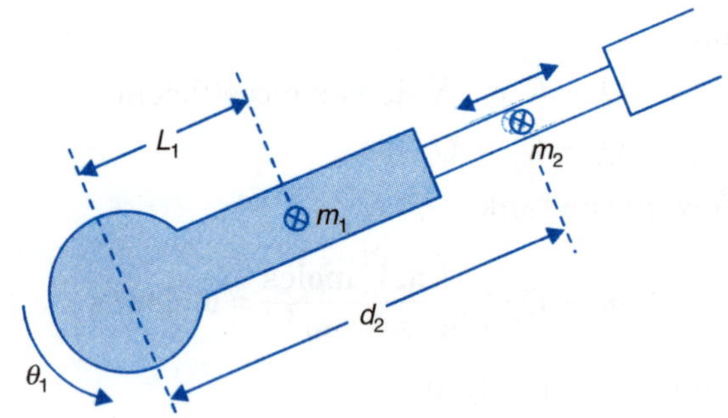

Fig. 2.49

Solution. Linear velocity of first centre of mass, $v_{c1} = l_1\dot{\theta}_1$ ...(i)

Kinetic energy of link 1 is

$$K_1 = (1/2)m_1 l_1^2 \dot{\theta}_1^2 \quad ...(ii)$$

Linear velocity of second centre of mass is

$$v_{c2} = \sqrt{(d_2^2\dot{\theta}_1^2 + \dot{d}_2^2)} \quad ...(iii)$$

Kinetic energy of link 2 is

$$K_2 = \frac{1}{2} m_2 (d_2^2\dot{\theta}_1^2 + \dot{d}_2^2) \quad ...(iv)$$

Hence the total kinetic energy of manipulator is

$$K = K_1 + K_2 \quad ...(v)$$

or

$$K(\theta_1, \dot{\theta}_1, d_2, \dot{d}_2) = (1/2)(m_1 l_1^2 + m_2 d_2^2)\dot{\theta}_1^2 + \frac{1}{2} m_2 \dot{d}_2^2 \quad ...(vi)$$

Potential energy of link 1 is

$$u_1 = m_1 l_1 g \sin \theta_1 \quad ...(vii)$$

Potential energy of link 2 is

$$u_2 = m_2 d_2 g \sin \theta_1 \quad ...(viii)$$

Hence total potential energy of manipulator is

$$U(\theta_1, d_2) = u_1 + u_2 = g(m_1 l_1 + m_2 d_2) \sin \theta_1 \quad ...(ix)$$

Taking partial derivatives, we have

$$\begin{bmatrix} \dfrac{\partial K}{\partial \dot{\theta}_1} \\ \dfrac{\partial K}{\partial \dot{d}_2} \end{bmatrix} = \begin{bmatrix} (m_1 l_1^2 + m_2 d_2^2)\dot{\theta}_1 \\ m_2 d_2 \end{bmatrix} \quad ...(x)$$

$$\begin{bmatrix} \dfrac{\partial K}{\partial \theta_1} \\ \dfrac{\partial K}{\partial d_2} \end{bmatrix} = \begin{bmatrix} 0 \\ m_2 d_2 \theta_1^2 \end{bmatrix} \quad ...(xi)$$

$$\begin{bmatrix} \dfrac{\partial U}{\partial \theta_1} \\ \dfrac{\partial U}{\partial d_2} \end{bmatrix} = \begin{bmatrix} g(m_1 l_1 + m_2 d_2)\cos\theta_1 \\ g m_2 \sin\theta_1 \end{bmatrix} \quad ...(xii)$$

From Lagrangian mechanics we know that

$$\begin{bmatrix} \tau_1 \\ \tau_2 \end{bmatrix} = \begin{bmatrix} \dfrac{d}{dt}\left(\dfrac{\partial K}{\partial \dot{\theta}_1}\right) \\ \dfrac{d}{dt}\left(\dfrac{\partial K}{\partial \dot{d}_2}\right) \end{bmatrix} - \begin{bmatrix} \dfrac{\partial K}{\partial \theta_1} \\ \dfrac{\partial K}{\partial d_2} \end{bmatrix} + \begin{bmatrix} \dfrac{\partial U}{\partial \theta_1} \\ \dfrac{\partial U}{\partial d_2} \end{bmatrix} \quad ...(xiii)$$

Thus from (x), (xi), (xii), and $(xiii)$ we have

$$\tau_1 = (m_1 l_1^2 + m_2 d_2^2)\ddot{\theta}_1 + 2m_2 d_2 \dot{\theta}_1 \dot{d}_2 + (m_1 l_1 + m_2 d_2)\, g \cos\theta_1 \quad ...(xiv)$$

$$\tau_2 = m_2 \dot{d}_2 - m_2 d_2 \dot{\theta}_1^2 + m_2 g \sin\theta_1 \quad ...(xv)$$

Equations (xiv) and (xv) are the dynamic equations of the manipulator.

Example 2.5 : The schematic diagram of a position control system is drawn in Fig. 2.50.

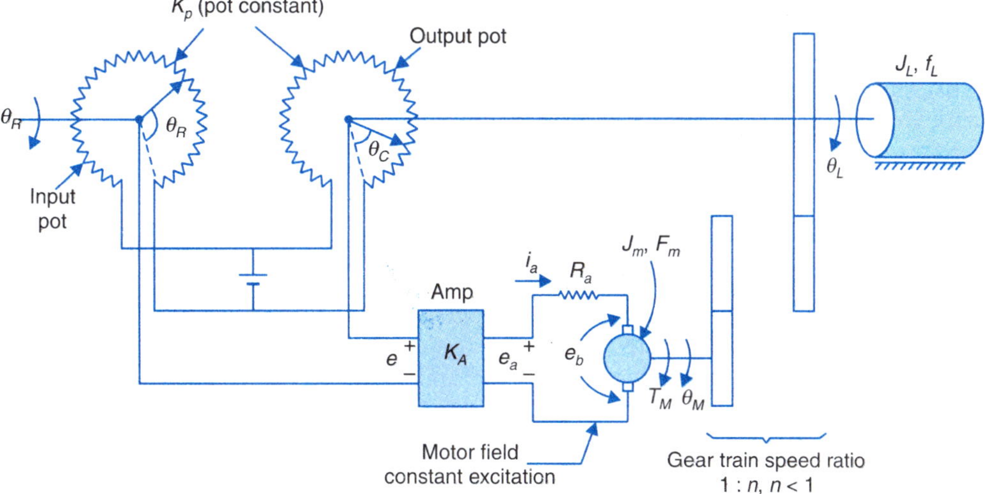

Fig. 2.50

— Various constituents of the system are:

- Drive motor-dc armature controlled.
- Load is driven through a reduction gearing, to amplify torque for moving the load.

- Load angular position is sensed by a circular potentiometer.
- Angle reference input is sensed by an identical potentiometer.
- Position error (in form of voltage e) is fed to amplifier with a power stage to feed the motor armature at voltage e_a.

From physical reasoning it is easily seen that at steady position of load, voltage e should be zero so the motor armature is stationary. It means $\theta_C = \theta_R$ *i.e.* steady-state error is zero. Any change in command θ_R introduces error e and the motor drives the load till $\theta_C = \theta_R$ once again. The nature of this moment is an important performance measure.

Various parameters and variables are indicated in Fig. 2.50. Motor parameters are:

Torque constant $= K_T$ kg-m/Amp. (armature)

Back emf $= K_b$ V/rad/sec

It is reminded here that numerically $K_T = K_b$.

Motor armature inductance L_a is negligible.

(*a*) Draw the block diagram of the motor and reduce it to the form $\omega_M(s)/E_a(s)$; ω_M = motor speed.

(*b*) Draw the complete system block diagram.

(*c*) Determine the overall transfer function $T(s) = \theta_C(s)/\theta_R(s)$.

(*d*) Briefly discuss the nature of $T(s)$.

(*e*) For numerical data given within the solution determine $T(s)$; K_A is variable gain.

Solution. From the control system knowledge acquired so far the block of the position control system (Fig. 2.50) can be drawn directly. For this purpose the load is reflected to the motor shaft giving the effective inertia and friction at motor shaft as

$$J = J_M + n^2 J_L \quad \text{...(i)}$$

$$f = f_M + n^2 f_L \quad \text{...(ii)}$$

(*a*) To begin with let us first draw the motor block diagram linking ω_M, motor speed, to armature voltage e_a. It is given in Fig. 2.51 (*a*) (this has been presented earlier and is repeated here for the reader to become fully conversant with it).

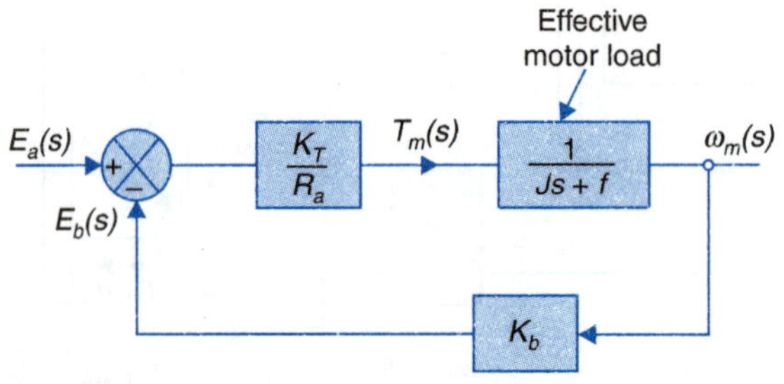

Fig. 2.51

Reducing the block diagram of Fig. 2.51, we get

$$\frac{\omega_M(s)}{E_a(s)} = \frac{K_m}{\tau' s + 1} \quad \text{...(iii)}$$

$$\text{where } K_m = \frac{K_T}{R_a f + K_T K_b} \quad \text{...(iv)}$$

$$\tau' = \frac{J}{f + (K_T K_b)/R_a} \quad ...(v)$$

This is represented in Fig. 2.52

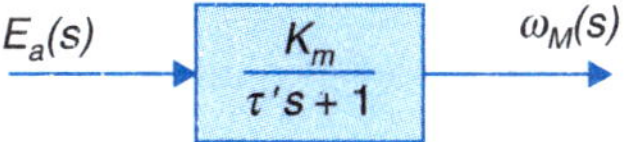

Fig. 2.52. Motor and load.

Note. τ' is the effective motor time constant including friction f and equivalent frictional effect caused by motor back emf $(K_b\ K_T)/R_a$. Symbol τ is reserved for motor's mechanical time constant *i.e.* $\tau = J/f$.

(*b*) The complete block diagram of the speed control system is now drawn in Fig. 2.53.

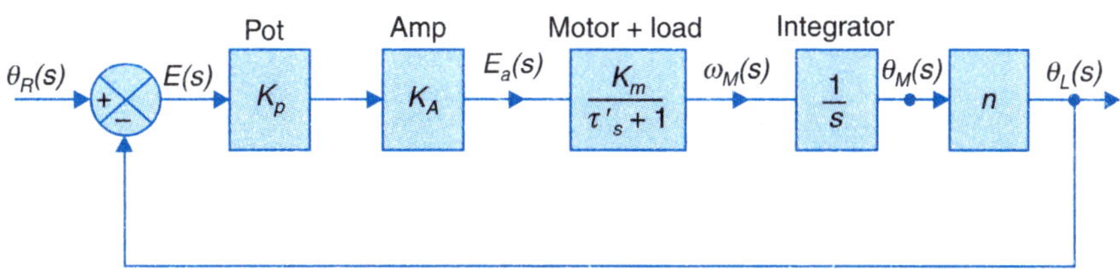

Fig. 2.53

(*c*) Forward path transfer function

$$G(s) = \frac{K_p K_A K_m n}{s(\tau' s + 1)} \quad ...(vi)$$

Overall transfer function

$$T(s) = \frac{\theta_L(s)}{\theta_R(s)} = \frac{G(s)}{1 + G(s)}$$

Substituting $G(s)$, we get

$$\frac{\theta_L(s)}{\theta_R(s)} = \frac{K_p K_A K_m n}{\tau' s^2 + s + K_p K_A K_m n} \quad ...(vii)$$

It is to be noticed that it is a second-order system compared to the first-order speed control system of Fig. 2.39. Increase in order has been caused by the integrator to get angle θ_M from speed $\dot{\theta}_M$.

(*d*) Data of system components

DC Motor

$$K_T = 60 \times 10^{-3} \text{ kg-m/A}$$
$$J_M = 1 \times 10^{-3} \text{ kg-m}^2$$
$$f_M = 10 \times 10^{-3} \text{ kg-m/rad/sec}$$
$$R_a = 1\ \Omega$$

Potentiometer

$$K_p = 1 \text{ V/rad}$$

Gear Train

$$n = 1/10 = 0.1$$

Load

$$J_L = 900 \times 10^{-3} = \text{kg-m}^2$$
$$f_L = 9{,}000 \times 10^{-3} \text{ kg-m/rad/sec.}$$

Effective load at motor shaft

$$J = 1 \times 10^{-3} + 900 \times 10^{-3} \times (0.1)^2$$
$$= 10 \times 10^{-3} \text{ kg-m}^2$$
$$f = 10 \times 10^{-3} + 9{,}000 \times 10^{-3} \times (0.1)^2$$
$$= 100 \times 10^{-3} \text{ kg-m/rad/sec.}$$

From Eqs. (*iv*) and (*v*)

$$K_m = \frac{60 \times 10^{-3}}{1 \times 100 \times 10^{-3} + (60 \times 10^{-3})^2} \; ; K_b = K_T$$
$$= \frac{60}{10 + 3.6} = 4.41$$
$$\tau' = \frac{10 \times 10^{-3}}{100 \times 10^{-3} + (60 \times 10^{-3})^2/1} = \frac{10}{100 + 3.6} = 0.097 \text{ sec.}$$

Substituting (values in Eq. (*vii*)

$$\frac{\theta_L(s)}{\theta_R(s)} = \frac{1 \times K_A \times 4.41 \times 0.1}{0.097 s^2 + s + 1 \times K_A \times 4.41 \times 0.1}$$

or
$$\frac{\theta_L(s)}{\theta_R(s)} = \frac{4.55\, K_A}{s^2 + 10.31 s + 4.55\, K_A} \qquad \text{...}(viii)$$

Choice of gain K_A in this transfer function depends on the system dynamics desired. The issues involved will be discussed in Chapter 5.

Example 2.6 : The block diagram of a speed control system is drawn in Fig. 2.54. Define the state variables and write the state and output equations of the system in vector-matrix form.

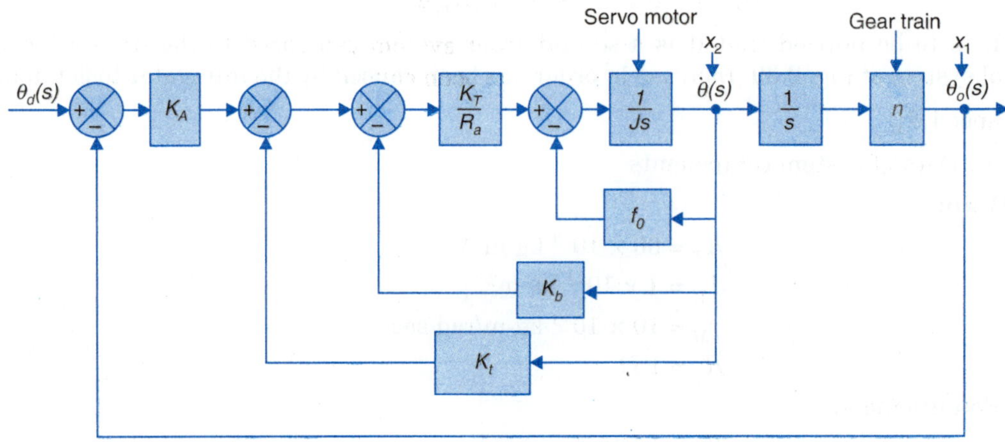

Fig. 2.54

Solution. There will be two state variables as this is a second-order system. We define the state variables as:

$$x_1 = \theta_o, \quad x_2 = \dot{\theta}$$

State equations written from the block diagram. For each block with one denominator s, input-output relations are written. In this block diagram there are two such blocks (2nd order system)

- Block $\left(\frac{n}{s}\right)$

$$x_1 = \left(\frac{n}{s}\right) x_2$$

In time domain $\quad \dot{x}_1 = n x_2$

- Block $\left(\frac{1}{Js}\right)$

$$\left\{(\theta_d - x_1)K_A - \left[(K_t + K_b)\frac{K_T}{R_a} + f_0\right]x_2\right\}\frac{1}{Js} = x_2$$

In time domain it can be expressed as

$$\dot{x}_2 = -\left(\frac{K_A}{J}\right)x_1 - \frac{f_0 + (K_t + K_b)\left(\frac{K_T}{R_a}\right)}{J}x_2 + \left(\frac{K_A}{J}\right)\theta_d$$

These two equation are written below as state equation

$$\begin{bmatrix}\dot{x}_1 \\ \dot{x}_2\end{bmatrix} = \begin{bmatrix} 0 & n \\ -\left(\frac{K_A}{J}\right) & -\frac{f_0 + (K_t + K_b)\left(\frac{K_T}{R_a}\right)}{J}\end{bmatrix}\begin{bmatrix}x_1 \\ x_2\end{bmatrix} + \begin{bmatrix}0 \\ \frac{K_A}{J}\end{bmatrix}\theta_d \quad \ldots(iii)$$

Output equation

$$y = \theta_o = x_1 \quad \text{or} \quad y = [1 \quad 0]\begin{bmatrix}x_1 \\ x_2\end{bmatrix}. \quad \ldots(iv)$$

PROBLEMS

2.1. Obtain the transfer functions of the mechanical systems shown in Figs. P-2.1(a) and (b).

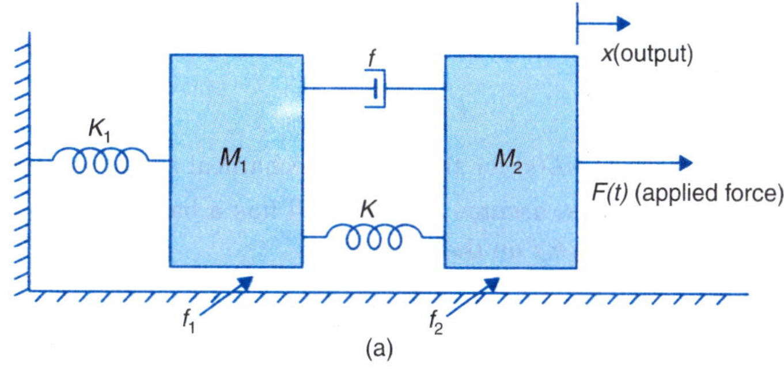

(a)

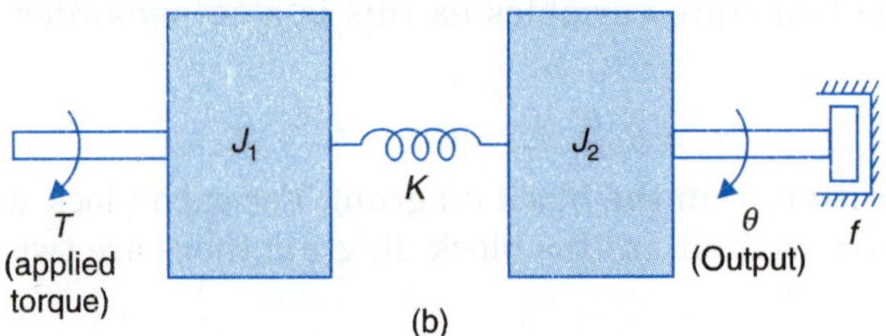

Fig. P-2.1.

2.2. Write the differential equations governing the behaviour of the mechanical system shown in Fig. P-2.2. Also obtain an analogous electrical circuit based on force-current analogy.

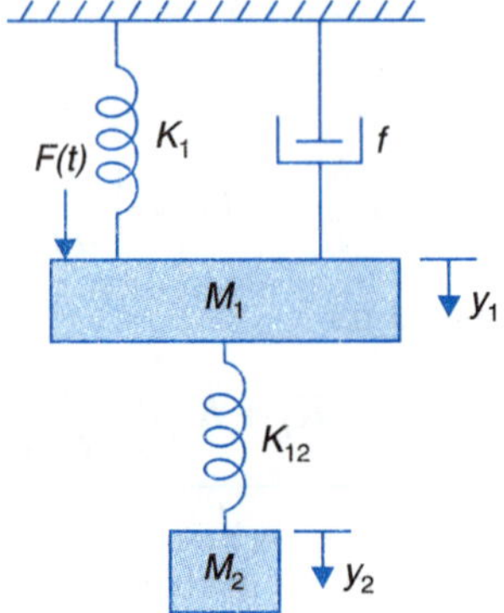

Fig. P-2.2.

2.3. Write the differential equations for the mechanical system shown in Fig. P-2.3. Also obtain an analogous electrical circuit based on force-current analogy.

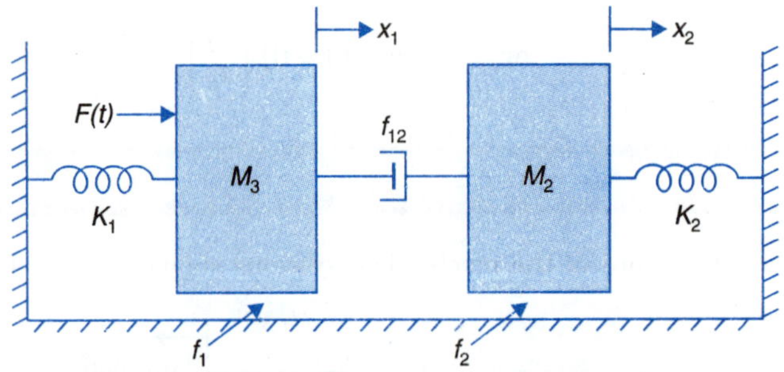

Fig. P-2.3.

2.4. Find the transfer function $X(s)/E(s)$ for the electromechanical system shown in Fig. P-2.4.
[**Hint:** for a simplified analysis, assume that the coil has a back emf $e_b = k_1\, dx/dt$ and the coil current i produces a force $F_C = k_2 i$ on the mass M]

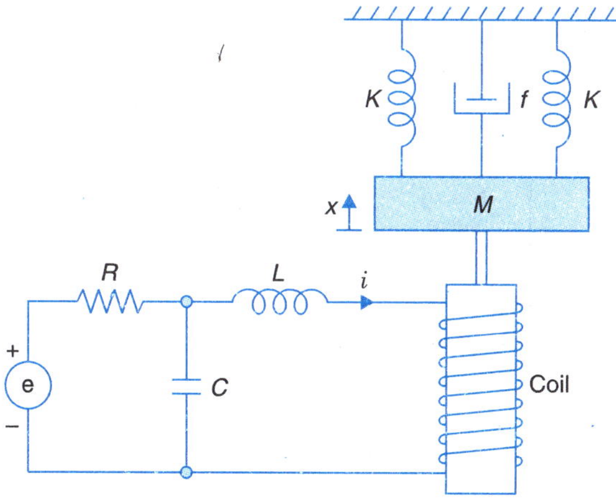

Fig. P-2.4.

2.5. Fig. P-2.5 shows a thermometer plugged into a bath of temperature θ_i. Obtain the transfer function $\theta(s)/\theta_i(s)$ of the thermometer and its electrical analogue. (The thermometer may be considered to have a thermal capacitance C which stores heat and a thermal resistance R which limits the heat flow). How the temperature indication of the thermometer will vary with time after the thermometer is suddenly plugged in ?

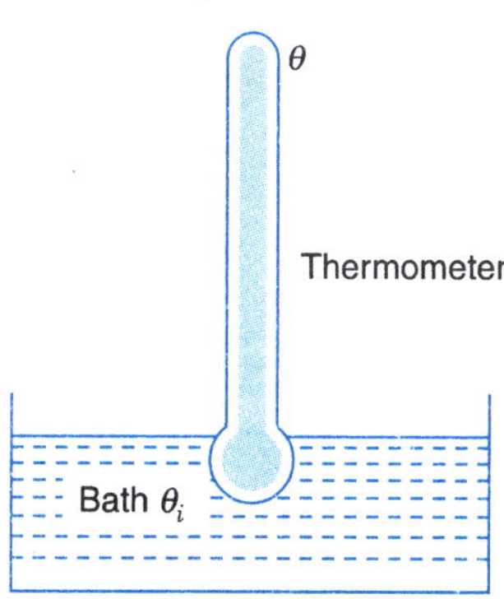

Fig. P-2.5.

2.6. The scheme of Fig. P-2.6 produces a steady stream flow of dilute salt solution with controlled concentration C_0. A concentrated solution of salt with concentration C_i is continuously mixed with pure water in a mixing valve. The valve characteristic is such that the total flow rate Q_0 through it is maintained constant but the inflow Q_i of concentrated salt solution may be linearly varied by controlling valve stem position x. The outflow rate from the salt mixing tank is the same as the flow rate into it from the mixing valve, such that the level of the dilute salt solution in the tank is maintained constant. Obtain the transfer function $C_0(s)/X(s)$. If from fully closed position, the valve stem is suddenly opened by x_0, determine the outstream salt concentration C_0 as a function of time.

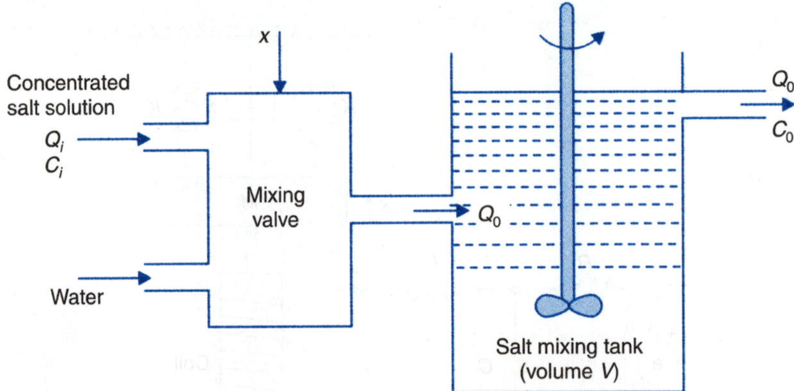

Fig. P-2.6.

2.7. In the speed control system shown in Fig. P-2.7, the generator field time constant is negligible. It is driven at constant speed giving a generated voltage of K_g volts/field amp. The motor is separately excited so as to have a counter emf of K_b volts per rad/sec. It produces a torque of K_T newton-m/amp. The motor and its load have a combined moment of inertia J kg-m^2 and negligible friction. The tachometer has K_t volts per rad/sec and the amplifier gain is K_A amps/volt. Draw the block diagram of this system and determine therefrom the transfer function $\omega(s)/E_i(s)$, where ω is the load speed.

With the system originally at rest, a control voltage e_i = 100 volts is suddenly applied. Determine how the load speed will change with time.

Given:

J = 6 kg-m^2 $\qquad K_A$ = 4 amp/volt $\qquad K_T$ = 1.5 newton-m/amp

K_g = 50 volts/amp $\qquad R_a$ = 1 ohm

K_t = 0.2 volts per rad/sec

[**Hint:** $K_b = K_T$ in MKS units]

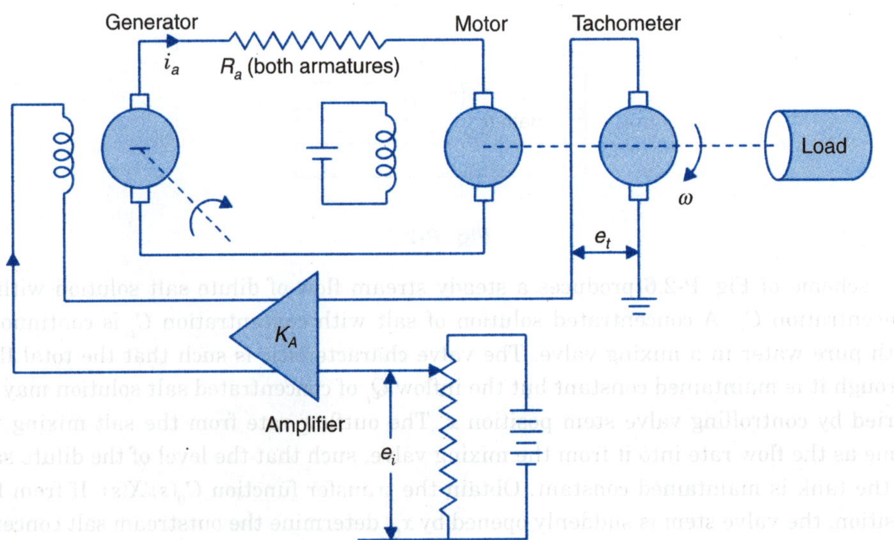

Fig. P-2.7.

2.8. Consider the positional servomechanism shown in Fig. P-2.8. Assume that the input to the system is the reference shaft position θ_R and the system output is the load shaft position θ_L. Draw the block diagram of the system indicating the transfer function of each block. Simplify the block diagram to obtain $\theta_L(s)/\theta_R(s)$ for the closed-loop system and also when the loop is open (in opening the loop, the lead from the output potentiometer driven by θ_C is disconnected and grounded). The parameters of the system are given below:

Sensitivity of error detector,	K_p = 10 volt/rad
Gain of d.c. amplifier,	K_A = 50 volts/volt
Motor field resistance,	R_f = 100 ohms
Motor field inductance,	L_f = 20 henrys
Motor torque constant,	K_T = 10 newton-m/amp
Moment of inertia of load,	J_L = 250 kg-m^2
Coefficient of viscous friction of load,	f_L = 2,500 newton-m per rad/sec
Motor to load gear ratio,	$(\dot{\theta}_L/\dot{\theta}_M) = 1/50$
Load to potentiometer gear ratio,	$\left(\dot{\theta}_C/\dot{\theta}_L\right) = 1$

Motor inertia and friction are negligible.

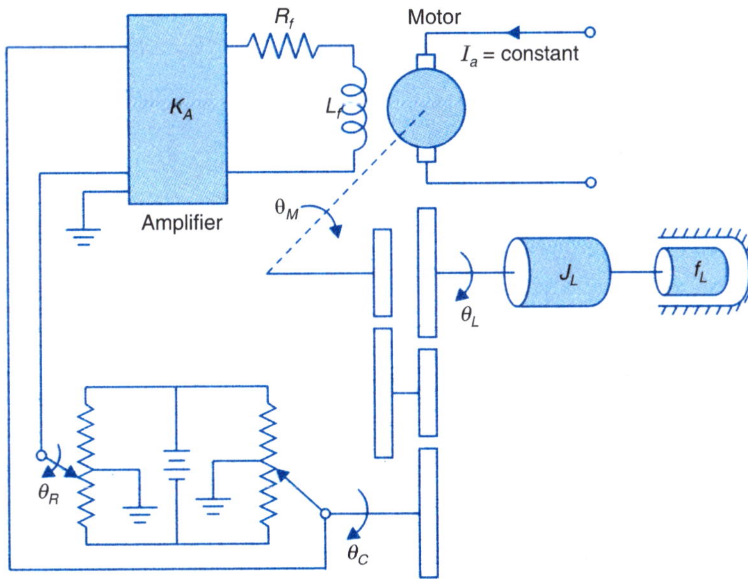

Fig. P-2.8.

2.9. Using block diagram reduction techniques, find the closed-loop transfer functions of the systems whose block diagrams are given in Figs. P-2.9(*a*) and (*b*).

2.10. For the system represented by the block diagram shown in Fig. P-2.10, evaluate the closed-loop transfer function, when the input R is (*i*) at station I, (*ii*) at station II.

2.11. From the block diagrams shown in Fig. P-2.11, determine C_1/R_1 and C_2/R_1 (assuming $R_2 = 0$).

2.12. Fig. P-2.12 shows a schematic diagram of liquid-level system. The flow of liquid Q_i into the tank is controlled by the pressure P of the incoming liquid and valve opening V_x (note that this is a more realistic model than the one shown in Fig. 2.15) through a nonlinear relationship.

$$Q_i = f(P, V_x)$$

Linearized liquid-level model and about the operating point $(P_0, Q_i = Q_0 H_0)$ is given as $\Delta Q_i = K_1 \Delta P + K_2 \Delta V_x$. Draw the signal flow graph and obtain therefrom the transfer function $\left.\frac{\Delta Q_0(s)}{\Delta V_x(s)}\right|_{\Delta Q_D = 0}$ with pressure remaining constant.

(The tank and output pipe may be considered to have liquid capacitance C and flow resistance R respectively).

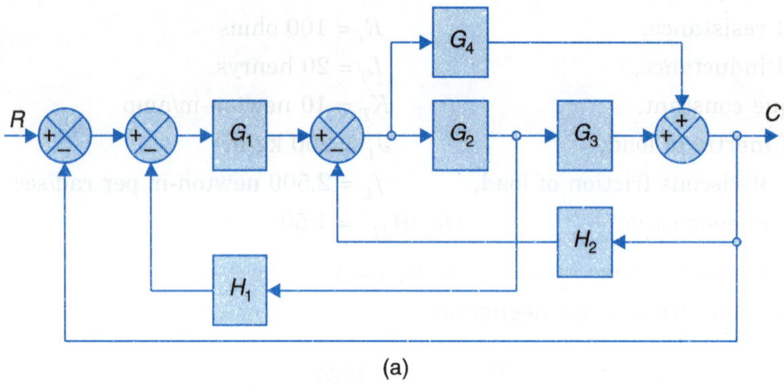

(a)

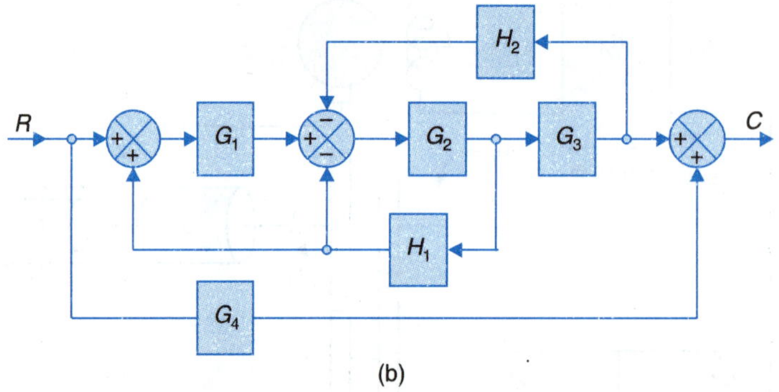

(b)

Fig. P-2.9.

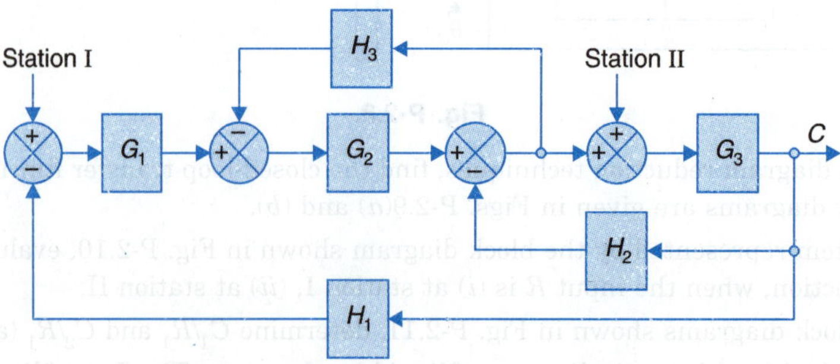

Fig. P-2.10.

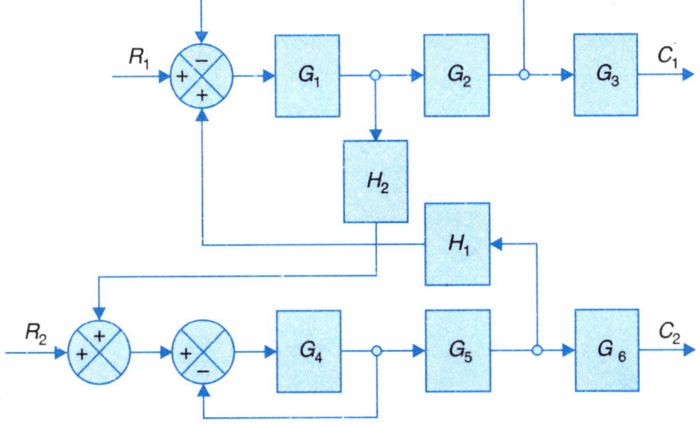

Fig. P-2.11.

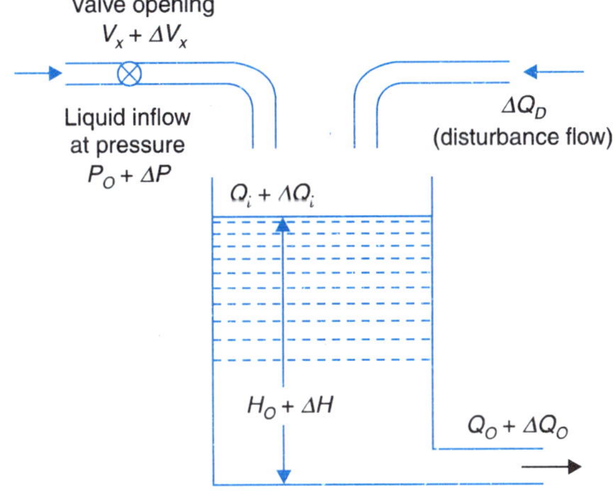

Fig. P-2.12.

2.13. Draw a signal flow graph and evaluate the closed-loop transfer function of a system whose block diagram is given in Fig. P-2.13.

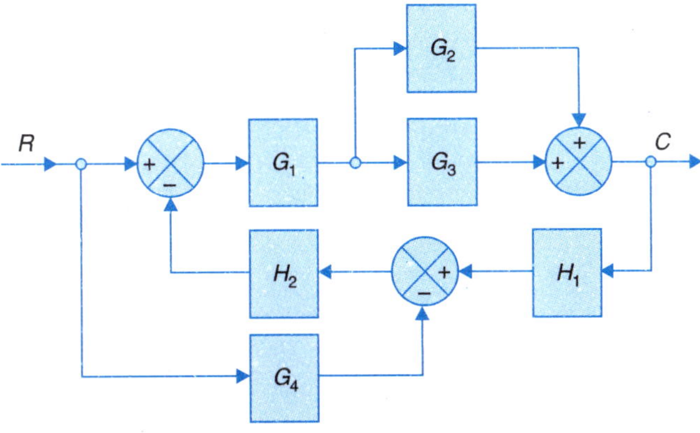

Fig. P-2.13.

2.14. Obtain the overall transfer function *C*/*R* from the signal flow graph shown in Fig. P-2.14.

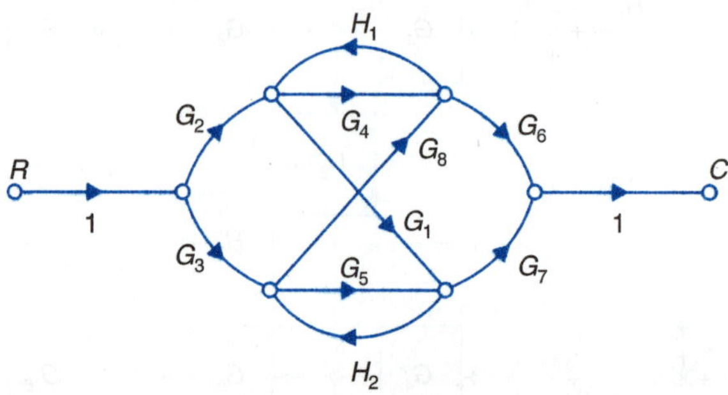

Fig. P-2.14.

2.15. Fig. P-2.15 gives the signal flow graph of a system with two inputs and two outputs. Find expressions for the outputs C_1 and C_2. Also determine the condition that makes C_1 independent of R_2 and C_2 independent of R_1.

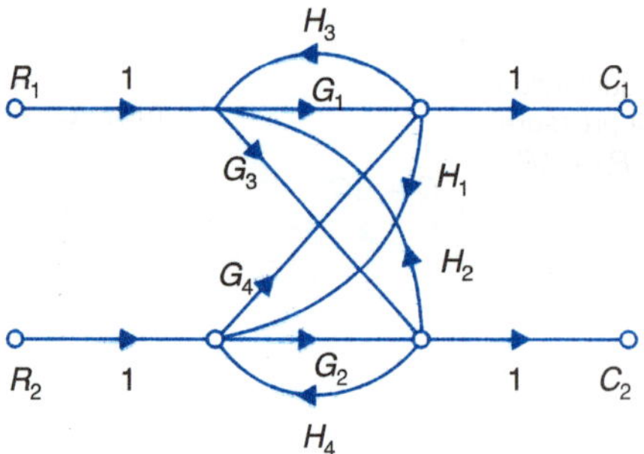

Fig. P-2.15.

2.16. For the system represented by the following equations, find the transfer function *X*(*s*)*U*(*s*) by signal flow graph technique.

$$x = x_1 + \beta_3 u$$

$$\dot{x}_1 = -a_1 x_1 + x_2 + \beta_2 u$$

$$\dot{x}_2 = -a_2 x_1 + \beta_1 u.$$

3

FEEDBACK CHARACTERISTICS OF CONTROL SYSTEMS

3

FEEDBACK CHARACTERISTICS OF CONTROL SYSTEMS

3.1 FEEDBACK AND NON-FEEDBACK SYSTEMS

Feedback systems play an important role in modern engineering practice because they have the possibility for being adopted to perform their assigned tasks automatically. A *non-feedback (open-loop) system* represented by the block diagram and signal flow graph in Fig. 3.1 (*a*), is activated by a single signal at the input (for single-input systems). There is no provision within this system for supervision of the output and no mechanism is provided correct (or compensate) the system behaviour for any lack of proper performance of system components, changing environment, loading or ignorance of the exact value of process parameters. On the other hand, a *feedback (closed-loop) system* represented by the block diagram and signal flow graph in Fig. 3.1 (*b*) is driven by two signals (more signals could be employed), one the input signal and the other, a signal called the feedback signal derived from the output of the system. The feedback signal gives this system the capability to act as self-correcting mechanism as explained below.

The output signal c is measured by a sensor $H(s)$, which produces a feedback signal b. The comparator compares the feedback signal b with the input (command) signal r generating the actuating signal e, which is as measure of discrepancy between r and b. The actuating signal is applied to the process $G(s)$ so as to influence the output c in a manner which tends to reduce the errors.

Feedback as a means of automatic regulation and control is, in fact, inherent in nature and can be noticed in many physical, biological and soft systems. For example, the body temperature of any living being is automatically regulated through a process which is essentially a feedback process, only it is far more complex than the diagram of Fig. 3.1 (*b*).

The beneficial effects of feedback in feedback systems with high loop gain, which will be elaborated in this Chapter, are enumerated below (discussion will not follow this order).

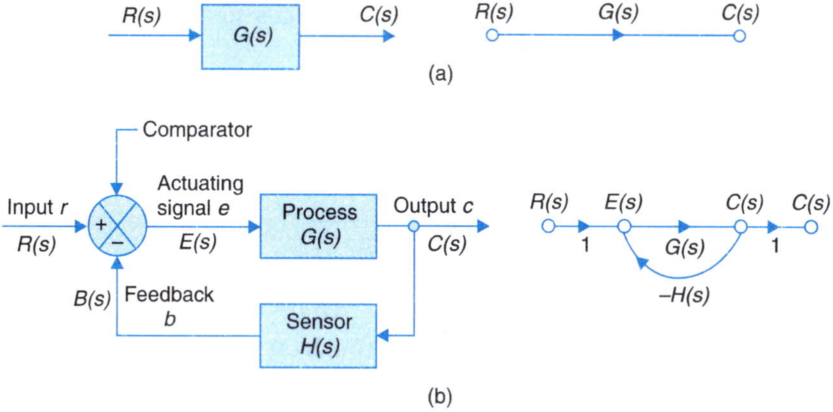

Fig. 3.1. (a) A non-feedback (open-loop) system
(b) A feedback (closed-loop) system

1. The controlled variable accurately follows the desired value.
2. Effect on the controlled variable of external disturbances other than those associated with the feedback sensor are greatly reduced.
3. Effect of variation in controller and process parameters (the forward path) on system performance is reduced to acceptable levels. These variations occur due to wear, aging, environmental changes etc.

   Feedback in the control loop allows accurate control of the output (by means of the input signal) even when process or controlled plant parameters are not known accurately.
4. Feedback in a control system greatly improves the speed of its response compared to the response speed capability of the plant/components composing the system (forward path).

The cost of achieving these improvements in system's performance through feedback will be discussed alongwith. These are: greater system complexity, need for much larger forward path gain and possibility of system instability (it means undesired/persistent oscillations of the output variable).

3.2 REDUCTION OF PARAMETER VARIATIONS BY USE OF FEEDBACK

One of the primary purpose of using feedback in control systems is to reduce the sensitivity of the system to parameter variations. The parameters of a system may vary with age, with changing environment (*e.g.,* ambient temperature), etc. Conceptually, *sensitivity* is a measure of the effectiveness of feedback in reducing the influence of these variations on system performance.

Let us define sensitivity on a quantitative basis. In the open-loop case

$$C(s) = G(s)\, R(s)$$

Suppose due to parameter variations $G(s)$ changes to $[G(s) + \Delta G(s)]$ where $|G(s)| >> |\Delta G(s)|$. The output of the open-loop system then changes to

$$C(s) + \Delta C(s) = [G(s) + \Delta G(s)]\, R(s)$$

or

$$\Delta C(s) = \Delta G(s)\, R(s) \qquad \text{...(3.1)}$$

Similarly, in the closed-loop case, the output

$$C(s) = \frac{G(s)}{1+G(s)H(s)} R(s)$$

changes to

$$C(s) + \Delta C(s) = \frac{G(s) + \Delta G(s)}{1+G(s)H(s) + \Delta G(s)H(s)} R(s)$$

due to the variation $\Delta G(s)$ in $G(s)$, the forward path transfer function. Since $|G(s)| >> |\Delta G(s)|$ we have from the above, the variation in the output as

$$\Delta C(s) \approx \frac{\Delta G(s)}{1+G(s)H(s)} R(s) \qquad \text{...(3.2)}$$

From eqns. (3.1) and (3.2) it is seen that in comparison to the open-loop system, the change in the output of the closed-loop system due to variation $G(s)$ is reduced by a factor of $[1 + G(s) H(s)]$ which is much greater than unity in most practical cases over the frequency $(s = j\omega)$ of interest.

The term *system sensitivity* is used to describe the relative variation in the overall transfer function $T(s) = C(s)/R(s)$ due to variation in $G(s)$ and is defined below:

$$\text{Sensitivity} = \frac{\text{percentage change in } T(s)}{\text{percentage change in } G(s)}$$

For small incremental variation in $G(s)$, the sensitivity is written in the quantitative form as

$$S_G^T = \frac{\partial T / T}{\partial G / G} = \frac{\partial LnT}{\partial LnG} \qquad \text{...(3.3)}$$

where S_G^T denotes the sensitivity of T with respect to G.

In accordance with the above definition, the sensitivity of the closed-loop system is

$$S_G^T = \frac{\partial T}{\partial G} \times \frac{G}{T} = \frac{(1+GH) - GH}{(1+GH)^2} \times \frac{G}{G/(1+GH)} = \frac{1}{1+GH} \qquad \text{...(3.4)}$$

Similarly, the sensitivity of the open-loop system is

$$S_G^T = \frac{\partial T}{\partial G} \times \frac{G}{T} = 1 \text{ (in this case } T = G) \qquad \text{...(3.5)}$$

Thus, the sensitivity of a closed-loop system with respect to variation in G is reduced by a factor $(1 + GH)$ as compared to that of an open-loop system.

The sensitivity of T with respect to H, the feedback sensor, is given as

$$S_H^T = \frac{\partial T}{\partial H} \times \frac{H}{T} = G\left[\frac{-G}{(1+GH)^2}\right] \frac{H}{G/(1+GH)} = \frac{-GH}{1+GH} \qquad \text{...(3.6)}$$

The above equation shows that for large values of GH, sensitivity of the feedback system with respect to H approaches unity. Thus, we see that the changes in H directly affect the system output. Therefore, it is important to use feedback elements which do not vary with environmental changes or can be maintained constant.

Very often the system's sensitivity is to be determined with a particular parameter (or parameters), with the transfer function expressed in ratio of polynomial form *i.e.,*

$$T(s) = \frac{N(s,\alpha)}{D(s,\alpha)}; \ \alpha = \text{parameter under consideration}$$

From eqn.(3.3)

$$S_\alpha^T = \left.\frac{\partial LnN}{\partial Ln\alpha}\right|_{\alpha_0} - \left.\frac{\partial LnD}{\partial Ln\alpha}\right|_{\alpha_0} \qquad \text{...(3.7}a\text{)}$$

$$= S_\alpha^N - S_\alpha^D \qquad \text{...(3.7}b\text{)}$$

where α_0 is the nominal value of the parameter around which the variation occurs.

The use of feedback in reducing sensitivity to parameter variations is an important advantage of feedback control systems. To have a highly accurate open-loop system, the components of $G(s)$ must be selected to meet the specifications rigidly in order to fulfil the overall goals of the system. On the other hand, in a closed-loop system $G(s)$ may be less rigidly specified, since the effects of parameter variations are mitigated by the use of feedback. However, a closed-loop system requires careful selection of the components of the feedback sensor $H(s)$. Since $G(s)$ is made up of power elements and $H(s)$ is made up of measuring elements which operate at low power levels, the selection of accurate $H(s)$ is far less costly than that of $G(s)$ to meet the exact specifications.

The price for improvement in sensitivity by use of feedback is paid in terms of *loss of system gain.* The open-loop system has a gain $G(s)$, while the gain of the closed-loop system is $G(s)/[1 + G(s)\,H(s)]$. Hence by use of feedback, the system gain is reduced by the same factor as by which the sensitivity of the system to parameter variations is reduced. Sufficient open-loop gain can, however, be easily built into a system so that we can afford to lose some gain to achieve improvement in sensitivity.

As a first example of feedback in reducing the system's sensitivity to parameter variations consider the feedback amplifier of Fig. 3.2(a) with negative feedback provided through a potential divider (feedback gain less than unity). Assuming the input impedance of the amplifier

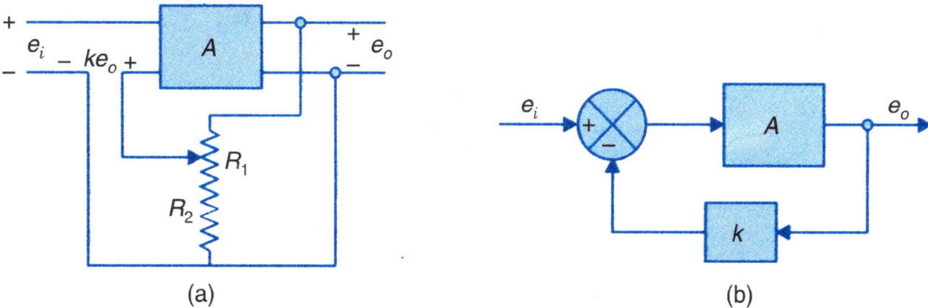

Fig. 3.2. Feedback amplifier.

to be infinite and its output impedance as zero, the equivalent block diagram of the system is drawn in Fig. 3.2 (*b*). Observe that both the forward gain A and feedback gain $k(< 1)$ are independent of frequency (in the range of frequencies of interest here).

It easily follows from the block diagram of Fig. 3.2 (*b*) that the overall gain of the amplifier circuit is

$$\frac{e_0}{e_i} = T = \frac{A}{1+kA};\ k = \frac{R_2}{R_1} \le 1 \qquad \text{...(3.8}a\text{)}$$

$$S_A^T = \frac{dT}{dA}\cdot\frac{A}{T} = \frac{1}{1+kA};\ \text{see eqn. (3.4)} \qquad \text{...(3.8}b\text{)}$$

For $A = 10^4$, $k = 0.1$

$$S_A^T = \frac{1}{1+10^3} = 0.001$$

While the feedback reduces the sensitivity to variation in forward gain to a very low figure (0.001), it also reduces the overall gain to

$$T = \frac{10^4}{1+10^3} = 10\ ;\ \text{compare with forward gain of } 10^4$$

Now sensitivity to feedback gain is given by

$$S_k^T = \frac{dT}{dk}\cdot\frac{k}{T} = \frac{-kA}{1+kA};\ \text{eqn. (3.6)} \qquad \text{...(3.8}c\text{)}$$

$$= \frac{-10^3}{1+10^3} = -1$$

S_k^T being equal to unity, the feedback constant $k = R_2/R_1$ must not vary *i.e.,* the resistor ratio R_2/R_1 must be accurate and stable.

In fact for such large A (= 10^4), $kA >> 1$ and so from eqn. (3.8*a*)

$$T = \frac{1}{k} = \frac{R_1}{R_2} = 10\ ;\ \text{independent of } A$$

As an example of control of system sensitivity, let us consider the speed control system of Fig. 2.39 which may be operated in open-or-closed-loop mode. The signal flow graph of this system is given in Fig. 2.40(*b*). The reduced signal flow graph of this system with $T_D = 0$, is drawn in Fig. 3.3

$$K = \frac{K_A K_T}{R_a f + K_T K_b};\ \tau = \frac{R_a f}{R_a f + K_T K_b}$$

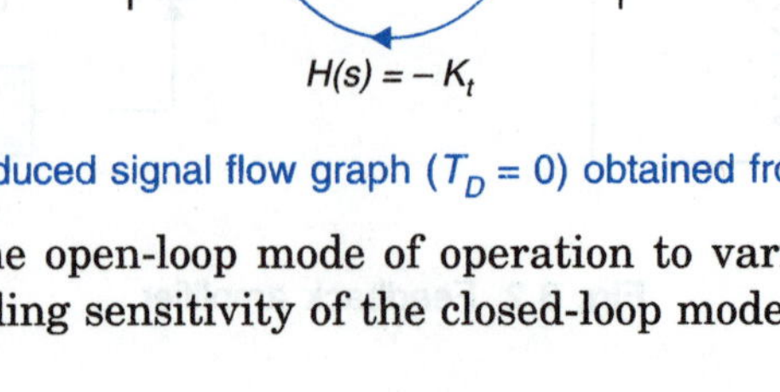

Fig. 3.3. Reduced signal flow graph ($T_D = 0$) obtained from Fig. 2.29.

The sensitivity of the open-loop mode of operation to variation-in the constant K is unity, while the corresponding sensitivity of the closed-loop mode is evaluated below.

From the signal flow graph of Fig. 3.3

$$T(s) = \frac{K}{\tau s + (1 + KK_t)} \tag{3.9a}$$

$$S_K^T = \frac{\partial T}{\partial K} \times \frac{K}{T} = \frac{s + \dfrac{1}{\tau}}{s + \left(\dfrac{1 + KK_t}{\tau}\right)} \tag{3.9b}$$

The expression (3.9*b*) can also be obtained by substituting $G(s) = K/(\tau s + 1)$ and $H(s) = K_t$ in eqn. (3.4).

For a typical application of this system, we might have $1/\tau = 0.1$ and $(1 + KK_t)\tau = 10$. Therefore from eqn, (3.4) we obtain

$$S_K^T = \frac{s + 0.1}{s + 10}$$

It follows from above that the sensitivity is a function of *s* and must be evaluated over the complete frequency band within which input has significant components. Our interest is to determine the upper limit for the sensitivity function $| S_K^T |$ over the frequency band and the frequency at which the maximum value occurs.

At a particular frequency, *e.g.*, $s = j\omega = j1$, the magnitude of the sensitivity is approximately:

$$\left| S_K^T \right| = 0.1$$

Thus the sensitivity of the closed-loop speed control system at this frequency is reduced by a factor of ten compared to that of the open-loop case.

Sensitivity studies in the frequency domain will be taken up in Chapter 9.

3.3 CONTROL OVER SYSTEM DYNAMICS BY USE OF FEEDBACK

Let us consider the elementary single-loop feedback system of Fig. 3.4. The open-loop transfer function of this system is

$$\frac{C(s)}{R(s)} = G(s) = \frac{K'}{s + \alpha} \qquad \text{...(3.10 (a))}$$

$$= \frac{K}{\tau s + 1};\ K = K'/\alpha,\ \tau = 1/\alpha \qquad \text{...(3.10 (b))}$$

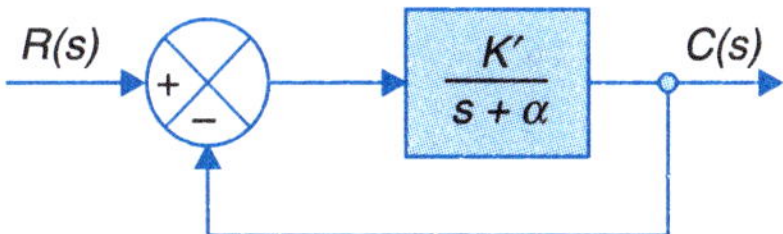

Fig. 3.4. A simple feedback system.

These are two alternative forms of expressing a transfer function. At $s = -\alpha$, $G(s)$ tends to infinity so is known as the **pole** of the system, while $\tau = 1/\alpha$ is known as its **time constant.** The *dc* gain of the system is

$$G(0) = K = K'/\alpha$$

With the feedback loop closed the closed-loop transfer function is

$$\frac{C(s)}{R(s)} = \frac{K}{\tau s + (1+K)} = \frac{K'}{s + (\alpha + K')} \qquad \text{...(3.11 (a))}$$

$$= \frac{K/(1+K)}{\tau_c s + 1}\ ;\ \tau_c = \tau/(1+K) \qquad \text{...(3.11 (b))}$$

We find from eqns (3.11 (*a*)) and (3.11 (*b*)) that the effect of closing the loop (that is introduction of negative feedback) is to shift the system's pole from $-\alpha$ to $-(\alpha + K')$ or $-\alpha(1 + K)$; alternatively to reduce the system time constant from τ to $\tau/(1 + K)$. Of course in the mean time the *dc* gain has reduced to $K/(1 + K)$.

We shall now examine the effect of these changes in system transfer function on its dynamic response.

For this purpose we shall assume that the system is excited (disturbed) by an impulse input $r(t) = \delta(t)$ (infinitely large input lasting for infinity short time). The Laplace transform of an impulse excitation is $R(s) = 1$ (this will be further discussed in Chapter 5). Taking the inverse Laplace transform of eqn. (3.10 (*a*)) and (3.11 (*a*)) with $R(s) = 1$, we have

$$c(t) = \mathcal{L}^{-1}\frac{K'}{s+\alpha} = K'e^{-\alpha t}$$

$$= K' e^{-t/\tau} \text{ (for non-feedback (open-loop) system)} \qquad \text{...(3.12)}$$

and

$$c(t) = \mathcal{L}^{-1}\frac{K'}{s+\alpha(1+K)} = K'e^{-\alpha(1+K)t}$$

$$= K' e^{-t/\tau_c} \text{ (for feedback (closed-loop) system)} \qquad \text{...(3.13)}$$

The location of pole and the dynamic response of non-feedback (open-loop) and feedback (closed-loop) system are shown in Fig. 3.5. These responses decay in accordance with respective

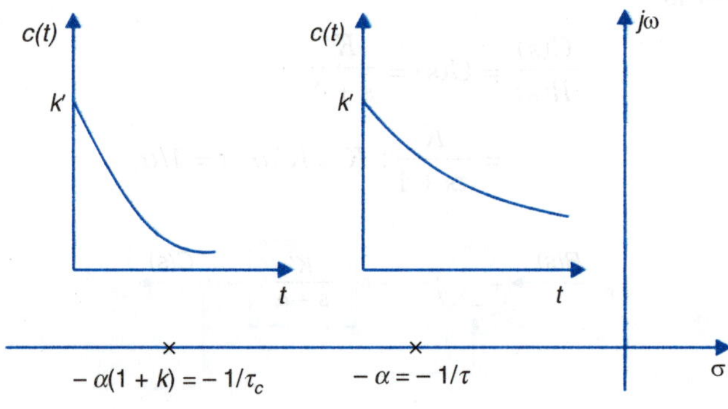

Fig. 3.5. Impulse response of open and closed-loop system (Fig. 3.4).

time constants. As the closed-loop time constant $\tau_c = \tau/(1 + K)$, its response decays much faster* which means that the speed of system's response with loop closed is faster by a factor of $(1 + K)$ compared to the open-loop system.

From this example, it is concluded that feedback controls the dynamics of the system by adjusting the location of its poles. It is, however, important to note here that feedback introduces the possibility of instability, that is, a closed-loop system may be unstable even though the open-loop is stable. The question of stability of control systems is treated in details in Chapter 6.

Consider once again the speed control system of Fig. 2.40 (*a*). Let the system be subjected to a suddenly applied constant input called step** input for which $E_r(s) = A/s$, where A is a constant. The output response of the system obtained by reference to the signal flow graph of Fig. 3.3 or directly from eqn. (2.101) is given by

$$\omega(s) = \frac{K/\tau}{s\left(s + \dfrac{1}{\tau}\right)} \quad \text{(for open-loop operation, } i.e., K_t = 0\text{)}$$

$$= \frac{K/\tau}{s\left(s + \dfrac{1 + KK_t}{\tau}\right)} \quad \text{(for closed-loop operation)}$$

Taking the inverse Laplace transform of the above equations, we get

$$\omega(t) = K(1 - e^{-t/\tau}) \text{ (for open-loop operation)} \qquad \text{...(3.14)}$$

$$= \frac{K}{1 + KK_t}(1 - e^{-t/\tau_c}) \text{ (for closed-loop operation)} \qquad \text{...(3.15)}$$

where τ_c (closed-loop time constant) $= \tau/(1 + KK_t)$.

It is seen from above that if the open-loop time constant τ is large, the transient response is poor and one choice is to replace the motor by another one with a lower time constant. Such a motor will obviously be more expensive and further due to physical limitations it is not possible to design and manufacture motor of a given size with time constant lower than a certain minimum value. Under such circumstances the closed-loop mode provides a lower time constant τ_c which can be conveniently adjusted by a suitable choice of KK_t. Unlimited reduction in τ_c is of course not practicable.

It is seen from eqn (2.102) and also eqn. (3.15) that the dc gain $T(0)$ of the closed-loop system is reduced by a factor of $(1 + KK_t)$ on account of the feedback loop. However, this only needs a scaling of $r(t)$ to obtain the desired $c(t)$.

From the above illustration we conclude that feedback is a powerful technique for control of system dynamics.

Effect of feedback on bandwidth

A control system is a low-pass filter—it responds to frequencies from *dc* up to a certain value ω_b at which the gain drops to $1/\sqrt{2}$ of its *dc* value. This frequency ω_b is the bandwidth of the

*In time 5τ the response decay to $e^{-5} = 0.0067$ or 0.67% of the value immediately after application of impulse.

**Discussed in detail in Chapter 5.

system. A large bandwidth implies that the system responds accurately to higher frequencies *i.e.,* fast changing signals, which is another way (frequency domain) of looking at the speed of response of a control system.

Consider once again the system of Fig. 3.4. The open- and closed-loop transfer functions (eqns (3.10*b*) and 3.11*b*) can be expressed in the frequency domain ($s = j\omega$) as

$$G(s) = \frac{K}{j\omega\tau + 1} \text{ ; open-loop} \qquad \text{...(3.16 (a))}$$

$$G(s) = \frac{K/(1+K)}{j\omega\tau_c + 1} \text{ ; closed-loop} \qquad \text{...(3.16 (b))}$$

Their bandwidth is determined as below

$$(\omega_b\tau)^2 + 1 = (\sqrt{2})^2 \text{ or } \omega_b(OL) = 1/\tau$$

$$(\omega_b\tau_c)^2 + 1 = (\sqrt{2})^2 \text{ or } \omega_b(CL) = 1/\tau_c$$

$$\frac{\omega_b(CL)}{\omega_b(OL)} = \frac{\tau}{\tau_c} = (1 + K)$$

Thus the closed-loop system has a bandwidth $(1 + K)$ times the bandwidth of the open-loop system this implies increased speed to response.

3.4 CONTROL OF THE EFFECTS OF DISTURBANCE SIGNALS BY USE OF FEEDBACK

Figure 3.6 shows the signal flow graph of a closed-loop system with the disturbance signal T_D in the forward path.

The ratio of the output $C(s)$ to the disturbance signal $T_D(s)$, when $R(s) = 0$, is obtained by applying the signal flow gain formula to the graph of Fig. 3.6, and is given by

$$\frac{C_D(s)}{T_D(s)} = \frac{-G_2(s)}{1 + G_1(s)G_2(s)H(s)} \qquad \text{...(3.18)}$$

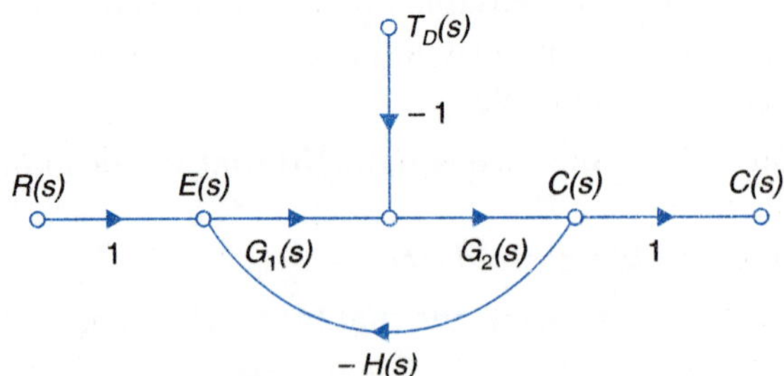

Fig. 3.6. A closed-loop system with a disturbance signal.

If $|G_1G_2H(s)| >> 1$ over the working range of s, then from eqn. (3.9)

$$\frac{C_D(s)}{T_D(s)} \approx \frac{-1}{G_1(s)H(s)}$$

Therefore, it is seen that if $G_1(s)$ is made sufficiently large, the effect of disturbance can be decreased by feedback.

Let us discuss the effect of load (disturbance) torque in the control system shown in Fig. 2.39. Assume that the load torque is a step signal for which $T_D(s) = A/s$ where A is a constant.

With reference to the signal flow graph of Fig. 2.40 (*b*) or directly from eqns. (2.103) and (2.104), the change in speed due to load torque along $[E_r(s) = 0]$ is given by

$$\omega_D(s) = \frac{-A}{s\left(Js + f + \dfrac{K_T K_b}{R_a}\right)} \qquad ...(3.19)$$

(for open-loop operation, $K_t = 0$)

$$= \frac{-A}{s\left[Js + f + \dfrac{K_T}{R_a}(K_A K_t + K_b)\right]} \qquad ...(3.20)$$

(for closed-loop operation)

The steady-state error in speed due to load is given by the *final value theorem* as

$$e_{ss} = \lim_{t\to\infty} \omega_D(t) = \lim_{s\to 0} s\omega_D(s) \qquad ...(3.21)$$

Using eqns. (3.19), (3.20) and (3.21), the steady-state speed errors in open and closed-loop cases are obtained as

$$e_{ss}(OL) = \frac{-AR_a}{R_a f + K_T K_b} \qquad ...(3.22)$$

and

$$e_{ss}(CL) = \frac{-AR_a}{R_a f + K_T(K_b + K_A K_t)} \qquad ...(3.23)$$

The ratio of steady-state error in output speed due to load torque; in open-loop and closed-loop cases obtained from eqns. (3.22) and (3.23), is

$$\frac{e_{ss}(CL)}{e_{ss}(OL)} = \frac{R_a f + K_T K_b}{R_a f + K_T(K_b + K_A K_t)}$$

Because of the additional term $K_T K_A K_t$ in the denominator, the effect of disturbance on the response of the system under consideration can be considerably reduced in closed-loop operation compared to that of the open-loop.

From the above analysis, it is seen that introduction of feedback decreases the effects of disturbances and noise signals in the forward path of the feedback loop. Feedback is introduced by a set of additional elements, called the measurement sensor H which may itself generate some noise. Let us evaluate the effect of this noise on the system performance.

Figure 3.7 shows the signal flow graph of a system with the noise signal $N(s)$ in the feedback path. Using the gain formula for the signal flow graph, the following result is obtained.

$$\left.\frac{C(s)}{N(s)}\right|_{R(s)=0} = \frac{C_n(s)}{N(s)} = \frac{-G_1(s)G_2(s)H_2(s)}{1 + G_1(s)G_2(s)H_1(s)H_2(s)}$$

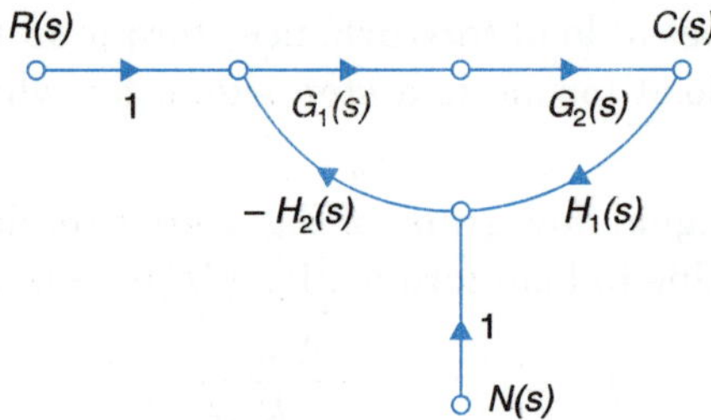

Fig. 3.7. Closed-loop system with measurement noise.

For large values of loop gain ($|G_1G_2H_1H_2(s)|>>1$), the above equation reduces to

$$\frac{C_n(s)}{N(s)} \approx -\frac{1}{H_1(s)}$$

Therefore, the effect of noise on output is

$$C_n(s) = -\frac{N(s)}{H_1(s)} \qquad ...(3.24)$$

Thus, for optimum performance of the system, the measurement sensor should be designed such that $H_1(s)$ is maximum, which is equivalent to maximizing the signal-to-noise ratio of the sensor.

The design specifications of the feedback sensor are far more stringent than those of the forward path transfer function. The feedback sensor must have low parameter variations as these are directly reflected in system response (the sensitivity $S_H^T \approx -1$). Further the signal-to-noise ratio for the sensor must be high as explained above. Usually it is possible to design and construct the sensor with such stringent specifications and at reasonable cost because the feedback elements operate at low power level.

To conclude, the use of feedback has the advantages of reducing sensitivity, improving transient response and minimizing the effects of disturbance signals in control systems. On the other hand, the use of feedback increases the number of components of the system, thereby increasing its complexity. Further it reduces the gain of the system and also introduces the possibility of instability. However, in most cases the advantages outweigh the disadvantages and therefore the feedback systems are commonly employed in practice.

3.5 LINEARIZING EFFECT OF FEEDBACK

Yet another property of feedback is its linearizing effect which is illustrated by means of the simple single-loop static system of Fig. 3.8 (a). In a static system various gains (transmittances) are independent of time. We shall assume that the forward block function is nonlinear expresses as

$$e = f(e) = e^2 \text{ ; square low function}$$

When the feedback loop is open

$$e = r \;\Rightarrow\; c = r^2$$

which is plotted in Fig. 3.8 (*b*) [graph (*i*)]. On the other hand when the loop is closed, we have

$$e = r - c$$

and so $$e = f(e) = (r - c)^2$$

which is plotted in graph (*ii*) of Fig. 3.8 (*b*).

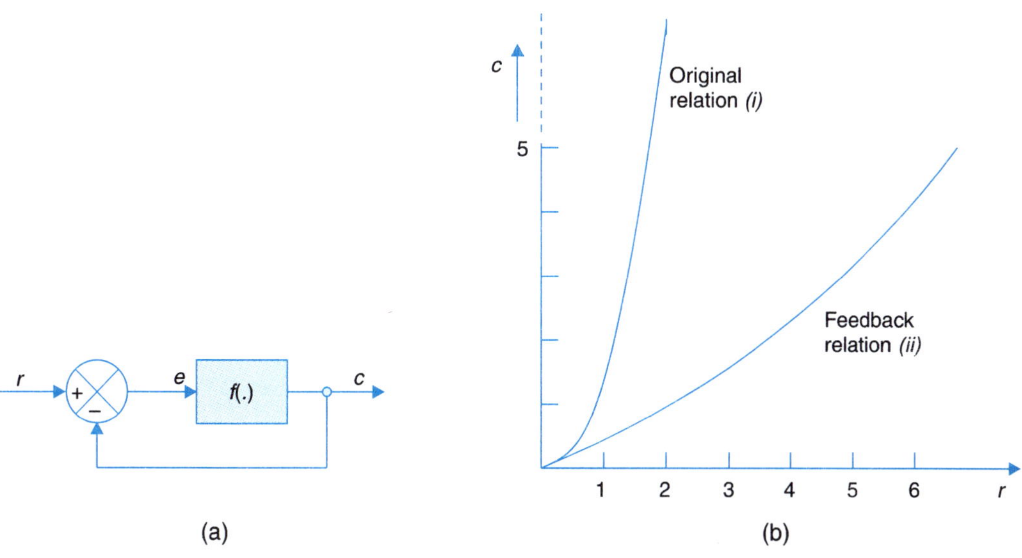

Fig. 3.8

It is easily seen by comparison of the graphs (*i*) and (*ii*) that the input-output relation [*c*(*r*)] is approximately linear over a much wider range for the closed-loop system compared to its open-loop behaviour.

3.6 REGENERATIVE FEEDBACK

The preceding material in this chapter has emphasized a negative or degenerative type of feedback. In regenerative feedback, the outputs is feedback with positive sign as shown in Fig. 3.9

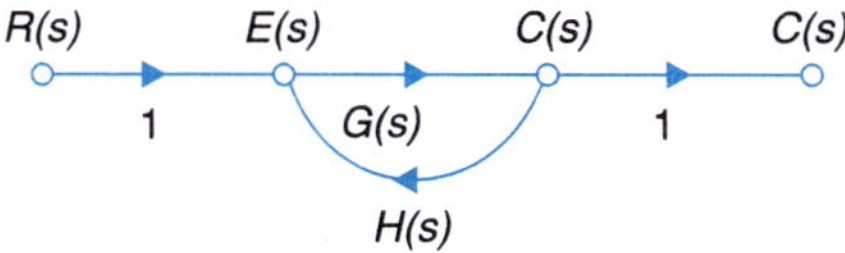

Fig. 3.9. A regenerative feedback control system.

In this case, the transfer function is given by

$$\frac{C(s)}{R(s)} = \frac{G(s)}{1 - G(s)H(s)} \quad \text{...(3.25)}$$

There is a negative sign in the denominator of eqn. (3.25) which indicates the possibility of denominator becoming equal to zero thereby giving an infinite output for a finite input which is the condition of instability.

The regenerative feedback is sometimes used for increasing the loop gain of feedback systems. Figure 3.10 shows a feedback system with an inner loop having regenerative feedback. This signal flow graph reduces to a single loop graph whose loop gain is

$$\frac{-G(s)H(s)}{1-G_f(s)}$$

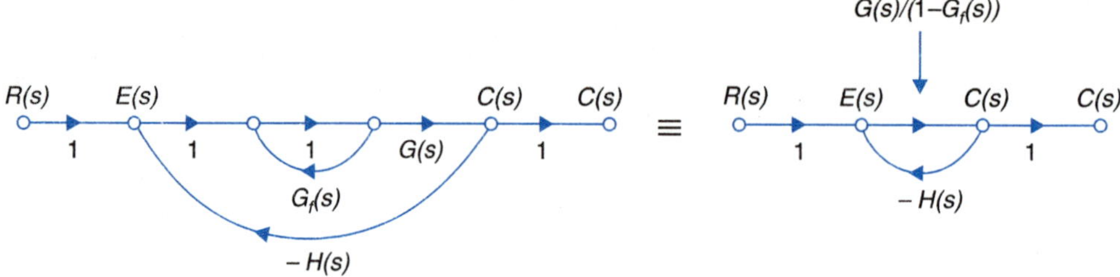

Fig. 3.10. Increasing loop gain by regenerative feedback.

If $G_f(s)$ is selected to be nearly unity, the loop gain becomes very high and the closed-loop transfer function approximates to

$$\frac{C(s)}{R(s)} = \frac{G(s)}{1-G_f(s)+G(s)H(s)} \approx \frac{1}{H(s)} \qquad ...(3.26)$$

Thus due to high loop gain provided by the inner regenerative feedback loop, the closed-loop transfer function becomes insensitive to $G(s)$.

3.7 ILLUSTRATIVE EXAMPLES

The effect of feedback on the performance of control systems is further illustrated with the help of following examples.

Example 3.1 : Consider the temperature control system of Fig. 3.11 which is set up to produce a steady stream flow of hot liquid at a controlled temperature. The temperature of the outflowing liquid is regulated automatically by means of a feedback sensor (say a *thermocouple*) which produces an output voltage e_t proportional to the temperature of the over flowing liquid. This voltage is subtracted from the reference voltage e_r to generate the error signal, $e = (e_r - e_t)$, *which* in turn regulates the current i_c through the heater element (and therefore the rate of heat input to the liquid) by means of silicon controlled rectifiers (SCRs) connected in full-wave operation with suitable logic circuitry.

To reduce the complexity of this problem, certain simplifying assumptions given below are made at this stage.

(*i*) The liquid inflow and outflow rates for the tank are equal so that the liquid level in the tank is maintained constant during the operation.

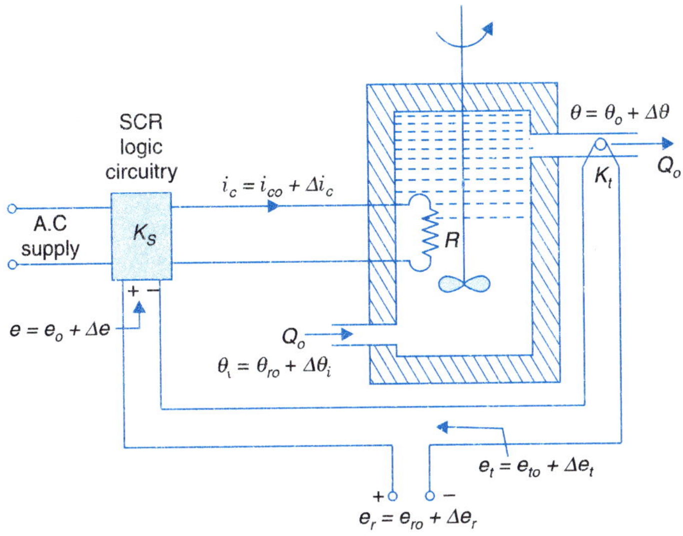

Fig. 3.11. A temperature control system.

(*ii*) The liquid in the tank is well-stirred so that its state can be described by the temperature θ of the outflowing liquid.

(*iii*) The tank is well-lagged so that the heat loss through its walls is negligible. Also the heat storage capacity of the tank walls is negligible.

(*iv*) The operation of the *SCR* circuit is linear, *i.e.*, $i_c = K_s e$, where K_s is the circuit gain in amps/volt.

Solution. The heat flow balance equation is

Rate of heat generated by the heater = rate of heat storage in the tank + rate of heat removed by the outflowing liquid

Mathematically this can be expressed as (in consistent units)

$$i_c^2 R = Mc\frac{d\theta}{dt} + Q_o \rho c(\theta - \theta_i) = C\frac{d\theta}{dt} + \frac{1}{R_t}(\theta - \theta_i)$$

where M = mass of the liquid in the tank; c = specific heat of the liquid; ρ = density of the liquid; Q_o = volume flow rate of the liquid θ_i = temperature of the inflowing liquid; R = resistance of the heater element; $C = Mc$; thermal capacitance of the liquid in the tank; and $R_t = \dfrac{1}{Q_o \rho c}$ = thermal resistance of the heat transfer process.

Substituting all the variables in terms of their steady plus incremental values, the above equation can be written as

$$K_s^2(e_o^2 + 2e_o\Delta e)R = Cd\Delta\theta/dt + (\theta_o - \theta_{io})/R_i + (\Delta\theta - \Delta\theta_i)/R_i \qquad ...(3.27)$$

wherein it has been assumed that $(\Delta e)^2 \approx 0$. Under steady operation, *i.e.*, with incremental values put as zero, we have

$$K_s^2 e_o^2 = (\theta_o - \theta_{io})/R_i \qquad ...(3.28)$$

Subtracting eqn. (3.28) from eqn. (3.27), we have the describing equation in terms of the incremental values about the operating point as

$$2K_s^2 e_o R\Delta e = C(d\Delta\theta/dt) + (\Delta\theta - \Delta\theta_i)/R_t \quad ...(3.29)$$

The incremental error is given by

$$\Delta e = \Delta e_r - \Delta e \quad ...(3.30)$$

Now $$\Delta e = K_t \Delta\theta \quad ...(3.31)$$

K_t being the constant of the temperature sensor.

Taking the Laplace transform of eqns. (3.29), (3.30) and (3.31) and reorganizing we get

$$\Delta\theta(s) = \frac{K\Delta E(s)}{\tau s + 1} + \frac{\Delta\theta_i(s)}{\tau s + 1} \quad ...(3.32)$$

$$\Delta E(s) = \Delta E_r(s) - \Delta E_i(s) \quad ...(3.33)$$

$$\Delta E_t(s) = K_t \Delta\theta(s) \quad ...(3.34)$$

where $$K = 2K_s^2 e_o RR_t \; ; \; \tau = R_t C$$

From eqns. (3.32), (3.33) and (3.34) we can draw the block diagram of the system as shown in Fig. 3.12 where the open-loop transfer function is

$$G(s) = \frac{K}{\tau s + 1}$$

and $\Delta\theta_i(s)$ is the change in the temperature of the inflowing liquid which can be regarded as a disturbance input entering the system through $1/(\tau s + 1)$.

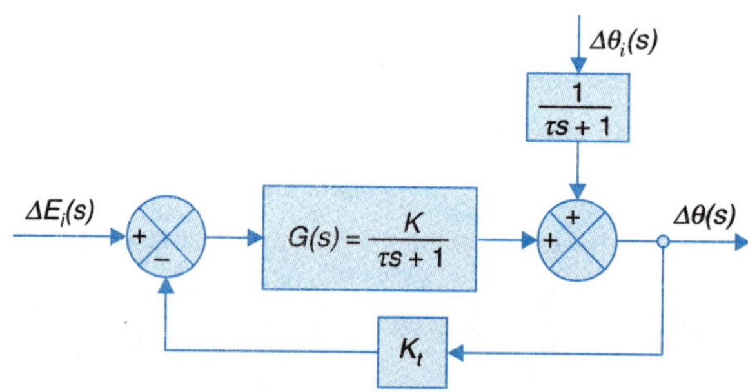

Fig. 3.12. Block diagram of the system shown in Fig. 3.9.

Assuming the disturbance signal $\Delta\theta_i$ to be zero, the steady change in the temperature of the outflowing liquid caused by an unwanted step change ΔE_r in the reference voltage is given by

$$\underset{t\to\infty}{\Delta\theta(t)} = \underset{s\to 0}{s}\left(\frac{\Delta E_r}{s}\right)\frac{K}{\tau s + 1 + KK_t} = \frac{\Delta E_r K}{1 + KK_t} \quad \text{(for closed-loop)} \quad ...(3.35)$$

$$= \Delta E_r K \quad \text{(for open-loop ; } K_t = 0) \quad ...(3.36)$$

It is easily observed from above that the steady change in the temperature of the outflowing liquid caused by an unwanted change in reference voltage is reduced by the factor $1/(1 + KK_t)$ in the closed-loop compared to the open-loop case.

Similarly, if the reference input is held fixed, *i.e.*, $\Delta e_r = 0$, a step change $\Delta\theta_i$ in the temperature of the inflowing liquid, causes a steady change in the output temperature of

$$\Delta\theta(t)_{t\to\infty} = \lim_{s\to 0} s\left(\frac{\Delta\theta_i}{s}\right)\frac{1}{\tau s + 1 + KK_t}$$

$$= \frac{\Delta\theta_i}{1 + KK_t} \quad \text{(closed-loop case)} \qquad ...(3.37)$$

$$= \Delta\theta_i \quad \text{(open-loop } K_t = 0) \qquad ...(3.38)$$

Thus, the change in temperature of the outflowing liquid caused by a change in temperature of the inflowing liquid can be reduced to any prescribed value by suitable choice of the loop gain KK_t.

If desired, the heat loss through the tank walls can be accounted for by introducing the term

$$Ah(\Delta\theta - \Delta\theta_e)$$

on the right hand side of eqn. (3.29), where h is the heat transfer constant and A is the surface area of the tank. This results in the additional disturbance signal $\Delta\theta_e$ caused by a change in the environmental temperature θe_0.

Example 3.2 : Consider the feedback control system shown in Fig. 3.13. The normal value of process parameter K is 1. Let us evaluate the sensitivity of transfer function $T(s) - C(s)/R(s)$ to variations in parameter K.

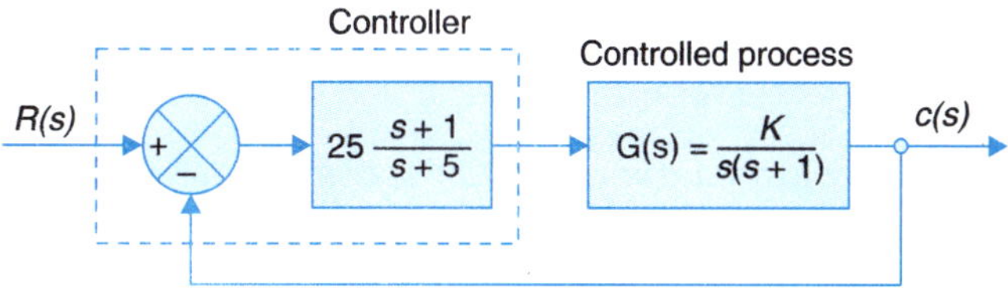

Fig. 3.13. A signal-loop configuration for $T(s) = \dfrac{25}{s^2 + 5s + 25}$.

Solution. From Fig. 3.13.

$$T(s) = \frac{C(s)}{R(s)} = \frac{25K}{s^2 + 5s + 25K} = \frac{1}{1 + \dfrac{s^2 + 5s}{25K}}$$

Therefore
$$S_K^T = \frac{\partial T}{\partial K} \times \frac{K}{T} = \frac{s(s+5)}{s^2 + 5s + 25K}$$

Since the normal value of K is 1, we have

$$S_K^T = \frac{s(s+5)}{s^2 + 5s + 25} \qquad ...(3.39)$$

S_K^T may be evaluated at various values of frequency (discussed in Chapter 9). At a particular frequency $\omega = 5$, the magnitude of sensitivity is approximately $|S_K^T| = 1.41$.

In a typical design problem, the process transfer function $G(s)$ is given. We select $T(s)$ which would meet the given specifications. $T(s)$ may be achieved by different feedback structures. Sensitivity function then allows quantitative comparison of these structures.

For example, the transfer function $T(s)$ of system of Fig. 3.13 may be realized by a two-loop structure of Fig. 3.14 (normal value of $K = 1$). For this structure,

$$T(s) = \frac{25K}{s^2 + (1+4K)s + 25K} = \frac{25K}{s(s+1) + K(4s+25)}$$

$$S_K^T = \frac{s(s+1)}{s(s+1) + K(4s+25)} = \frac{s(s+1)}{s^2 + 5s + 25} \quad \text{(for } K = 1\text{)} \qquad \text{...(3.40)}$$

The magnitude of sensitivity at $\omega = 5$ is $|S_K^T| \approx 1$.

This shows the superiority of the two-loop system of Fig. 3.14 over a single-loop system of Fig. 3.13. Thus the sensitivity function gives a firm basis for comparison of alternative designs.

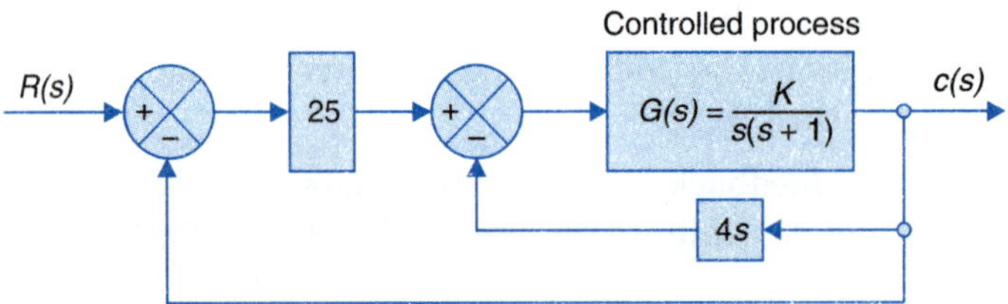

Fig. 3.14. A two-loop configuration for $T(s) = \dfrac{25}{s^2 + 5s + 25}$.

Example 3.3 : The model for automatic control of speed for an automobile on a straight run (assumed clear) is proposed below with its block diagram in Fig. 3.15.

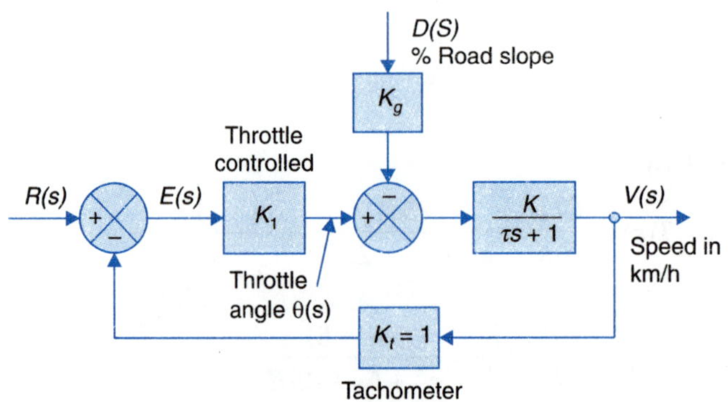

Fig. 3.15

(1) Engine-which is modelled as a single time constant $t = 20s$ and a gain of K(km/h/deg) with input throttle angle θ in degrees and output speed V in km/h.

K for various automobile models ranges from 1-2; a typical value being 1.5.

(2) Speed feedback is obtained by a tachometer coupled to the engine shaft which is then differenced from the reference input to obtain the error signal for manipulating the throttle angle.

(3) Throttle control has a time constant much less than τ and is ignored. So it is modelled as again K_1 in units of throttle/value angle in deg/unit vehicle speed error. Its typical/ value is $K_1 = 50$ deg/km/h.

(4) Disturbance (load) signal appears when the automobile runs on down a slope expressed in % (elevation per unit distance covered). This signal is converted into equivalent throttle angle by constant K_g deg (or throttle angle) per unit % road slope: $K_g = 100°$/unit % slope.

(*a*) Find the sensitivity of $T(s) = V(s)/R(s)$ to changes in K. Calculate its steady value ($s = 0$) for a nominal value of $K = 1.5$.

(*b*) Find the steady error e_{ss} for vehicle speed of 60 km/h.

(*c*) Calculate the up slope that will cause the vehicle to stall if it was running at 60 km/h on level road and reference input signal in assumed to be remain constant.

(*d*) What is the maximum permissible down slope for the vehicle speed not to exceed 100 km/h if it was running on level road at a speed (steady) of 10 km/h ?

(*e*) Calculate the ratio K_g/K_1 for the vehicle to stall on an up slope of 50%, if it is running on level road at a speed of 60 km/h.

(*f*) For a steady speed of 60 km/h, find R (input) with system loop (*i*) open and (*ii*) closed.

(*g*) For open-and closed-loop system with input as calculated in part (*f*), find the dynamic response. In what time the vehicle speed reaches 90% of its steady value for open- and closed-loop system ?

Solution. (*a*) From the block diagram of Fig. 3.15

$$T(s) = \frac{V(s)}{R(s)} = \frac{K_1 K}{\tau s + 1 + K_1 K} \qquad ...(i)$$

$$S_K^T = \frac{\partial T}{\partial K} \cdot \frac{K}{T} = \frac{\tau s + 1}{\tau s + 1 + K_1 K}$$

$$S_K^T \text{ (steady state)} = S_K^T(s = 0) = \frac{1}{1 + K_1 K} = \frac{1}{1 + 50 \times 1.5} = 0.0132 \text{ or } 1.32\%$$

(*b*) Steady state speed of vehicle = 60 km/h

Let $R(s) = \dfrac{A}{s}$

Then from eqn. (*i*)

$$V(s) = \frac{K_1 K}{\tau s + 1 + K_1 K} \cdot \frac{A}{s}$$

$$v(ss) = \lim_{s \to 0} sV(s) = \frac{AK_1 K}{1 + K_1 K}$$

$$\frac{A \times 50 \times 1.5}{1 + 50 \times 1.5} = 60 \qquad \text{or} \qquad A = 60.8 \text{ km/h}$$

$$\therefore \qquad e_{ss} = 60.8 - 60 = 0.8 \text{ km/h}$$

(*c*) Under vehicle stalled conditions:

Vehicle speed, tachometer output (feedback signal), input to engine-vehicle block are all = 0

Hence $AK_1 = K_g D$ or $D = \dfrac{AK_1}{K_g} = \dfrac{60.8 \times 50}{100} = 30.4\%$

(*d*) Steady vehicle speed on level road = 10 km/h

Then $A = 60.8 \times \dfrac{10}{60} = 10.133$ km/h

$$[(10.133 - 10)\, K_1 - (-D)\, K_g]\, K = 100$$

$$(6.65 + 100\, D) \times 1.5 = 100 \quad \text{or} \quad D = 0.6\% \text{ (down)}$$

(*e*) As in part (*c*) $AK_1 = K_g D$ or $K_g/K_1 = A/D = \dfrac{60.4}{50} = 1.21$

(*f*) Open-loop system $R \times K_1 \times K = 60$ or $R = \dfrac{60}{50 \times 1.5} = 0.8$ km/h

Closed-loop system $R \times \dfrac{K_1 K}{1 + K_1 K} = 60$ or $R = 60.8$ km/h

(*g*) Open-loop system $V(s) = \dfrac{K_1 K}{\tau s + 1} \cdot \dfrac{0.8}{s}$

Taking inverse Laplace transform

$$v(t) = 0.8\, K_1 K(1 - e^{-t/}\tau)\ ;\ \tau = 20s \quad \text{(given)}$$
$$= 60\,(11 - e^{-t/20})$$

If 90% speed, $t = t_1$

$$0.9 = 1 - e^{-t_1/20} \quad \text{or} \quad t_1 = 46\ s$$

Closed-loop system

$$V(s) = \frac{K_1 K}{\tau s + 1 + K_1 K} \cdot \frac{60.8}{s} = \frac{60.8 K'}{s(\tau' s + 1)}$$

Taking the inverse Laplace transform

$$v(t) = 0.8\, k'(1 - e^{-t/\tau'})$$

$$K' = \frac{K_1 K}{1 + K_1 K} = \frac{50 \times 1.5}{1 + 50 \times 1.5} = \frac{75}{76}$$

$$\tau' = \frac{\tau}{1 + K_1 K} = \frac{20}{76} = 0.263\ s$$

Substituting value $v(t) = 60\,(1 - e^{-t/0.263})$

From which we get time at 90% speed as

$$t_1 = 0.6\ s$$

Remarks : Observe that the closed-loop system reaches 90% of steady state speed in 0.6*s* compared to the corresponding time 46*s* for the open-loop system *i.e.*, speeding up of dynamic response by a faster of 46/0.6 ≈ 77 times.

Example 3.4 : The block diagram of a position control system is drawn in Fig. 3.16. It employs a field-controlled motor with negligible field line constant.

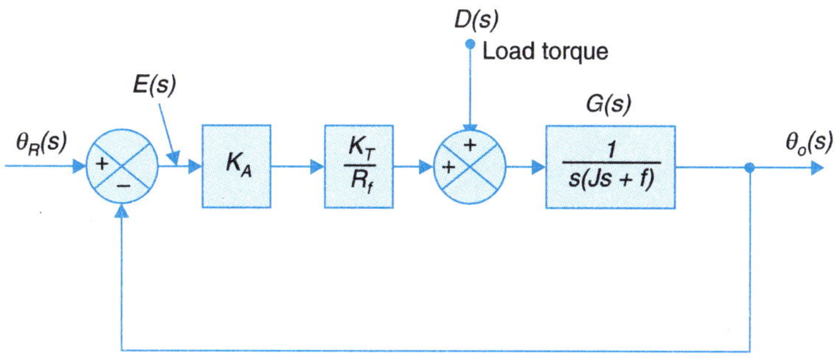

Fig. 3.16

Determine

(*a*) $T_D(s) = \theta_o(s)/D(s)$; $\theta_R(s) = 0$

(*b*) $T(s) = \theta_o(s)/\theta_R(s)$; $D(s) = 0$

(*c*) Sensitivity $S_{K_A}^{T_D}$

(*d*) Sensitivity $S_{K_A}^{T}$

(*e*) $E(s)/D(s)$; $\theta_R(s) = 0$; steady-rate error e_{ss} for unit step $D(s)$

(*f*) $E(s)/\theta_R(s)$; $D(s) = 0$; steady-rate error e_{ss} for unit step $\theta_R(s)$.

Find the numerical value of each part given

$$J = 0.1, f = 1, R_f = 1, K_T = 2.5, K_A = \text{adjustable gain.}$$

(*g*) Determine the value of K_A for e_{ss} in part (*e*) to be less than 1%. For this value of K_A find the numerical values of $S_{K_A}^{T_D}$ and $S_{K_A}^{T}$.

Give your comments on the results.

Solution. Before starting solution we identify

$$G(s) = \frac{1}{s(Js + f)} \quad \text{...(i)}$$

$$G'(s) = \frac{K}{s(Js + f)} \;;\; K = K_A K_T / R_f \quad \text{....(ii)}$$

(*a*) $\theta_R(s) = 0$

Forward path transfer function $= G(s)$

Feedback path gain $= k$(negative feedback)

We now find

$$T_D(s) = \frac{\theta_0(s)}{D(s)} = \frac{G(s)}{1 + KG(s)}$$

Substituting values

$$T_D(s) = \frac{1}{s(Js + f) + K_A K_T / R_f} \quad \text{...(iii)}$$

or $$T_D(s) = \frac{1}{s(0.1s+1)+2.5\,K_A} \qquad ...(iv)$$

(b) $D(s) = 0$

Forward path $= G'(s) = KG(s)$

Feedback path = unity (negative)

$$T(s) = \frac{\theta_0(s)}{\theta_R(s)} = \frac{KG(s)}{1+KG(s)}$$

or $$T(s) = \frac{K_A K_T/R_f}{s(sJ+f)+K_A K_T/R_f} \qquad ...(v)$$

substituting values

$$T(s) = \frac{2.5\,K_A}{s(0.1s+1)+2.5\,K_A}$$

(c) To find $S_{K_A}^{T_D}$, from Eq. (iii)

$$S_{K_A}^{T_D} = \frac{\partial \ln(1)}{\partial \ln(K_A)} - \frac{\partial \ln[s(Js+f)+K_A K_T/R_f]}{\partial \ln(K_A)}$$

or $$S_{K_A}^{T_D} = -\frac{K_A K_T/R_f}{s(Js+f)+K_A K_T/R_f} \qquad ...(vii)$$

Numerically

$$S_{K_A}^{T_D} = -\frac{2.5\,K_A}{s(0.1s+1)+2.5K_A} \qquad ...(viii)$$

(d) To find $S_{K_A}^{T}$; from Eq. (v)

$$S_{K_A}^{T} = \frac{\partial \ln(K_A K_T/R_f)}{\partial \ln(K_A)} - \frac{\partial \ln[s(Js+f)+K_A K_T/R_f]}{\partial \ln(K_A)}$$

$$= \frac{K_T/R_f}{K_A K_T/R_f}.K_A - \frac{K_A K_T/R_f}{s(Js+f)+K_A K_T/R_f}$$

or $$S_{K_A}^{T} = \frac{1}{[s(Js+f)+K_A K_T/R_f]} \qquad ...(ix)$$

Numerically $$S_{K_A}^{T} = \frac{1}{[s(0.1s+1)+2.5K_A]} \qquad ...(x)$$

Comment. If low frequencies $s = j\omega,\ \omega \to 0$

$$S_{K_A}^{T_D} = -1, \quad S_{k_A}{}^{T} = \frac{1}{2.5K_A}$$

$T_D(s)$ has K_A in feedback and so it is 100% sensitive to K_A.

However $T(s)$ has K_A in forward path and its sensitivity to K_A can be reduced to any desired value by choice of K_A.

(e) To find $E(s)/D(s)$; $\theta_R(s) = 0$

From Fig. 3.16

$$E(s) = \theta_R(s) - \theta_0(s) = -\theta_0(s), \text{ as } \theta_R(s) = 0$$

$$\frac{E(s)}{D(s)} = -\frac{1}{s(Js+f) + K_A K_T/R_f}$$

$$D(s) = \frac{1}{s}$$

Then $$E(s) = -\frac{1}{s[s(Js+f) + K_A K_T/R_f]} \quad \text{...}(xi)$$

$$e_{ss} = \lim_{s\to 0} sE(s)$$

or $$e_{ss} = -\frac{1}{K_A K_T/R_f} \quad \text{...}(xii)$$

or $$e_{ss} = -\frac{1}{2.5 K_A} \quad \text{...}(xiii)$$

(*f*) To find $E(s)/\theta_R(s)$; $D(s) = 0$

$$E(s) = \theta_R(s) - \theta_0(s)$$

$$\theta_o(s) = \frac{(K_A K_T/R_f)\,\theta_R(s)}{s(Js+f) + K_A K_T/R_f}$$

Then $$E(s) = \left[1 - \frac{K_A K_T/R_f}{s(Js+f) + K_A K_T/R_f}\right]\theta_R(s) \quad \text{...}(xiv)$$

or $$E(s) = \frac{s(sJ+f)}{s(Js+f) + K_A K_T/R_f} \cdot \frac{1}{s} \; ; \; \theta_R(s) = \frac{1}{s}$$

Therefore $$e_{ss} = \lim_{s\to 0} sE(s) = 0$$

(*g*) For unit step load torque (from Eq. (*xiii*))

$$e_{ss} = \frac{1}{2.5 K_A} = 0.01 \text{ ; 1\% or less}$$

which gives $$K_A = \frac{100}{2.5} = 40$$

$$S_{K_A}^{T_D}\,(K_A = 40) = -\frac{2.5 \times 40}{s(0.1s+f) + 2.5 \times 40} \quad \text{...}(xv)$$

Under dc conditions ($s = 0$)

$$S_{40}^{T_D}\,(\text{dc}) = -1 \quad (100\%)$$

Now $$S_{K_A}^{T}\,(K_A = 40) = -\frac{1}{[s(0.1s+1) + 2.5 \times 40]} \quad \text{...}(xvi)$$

Under dc condition $$S_{40}^{T} = \frac{1}{100} = 0.01 \quad (1\%)$$

Example 3.5 : Consider the speed control system of Fig. 3.17 wherein the inner loop corresponds to motor back emf. The controller is an integrater with gain K observe that the load is inertia only.

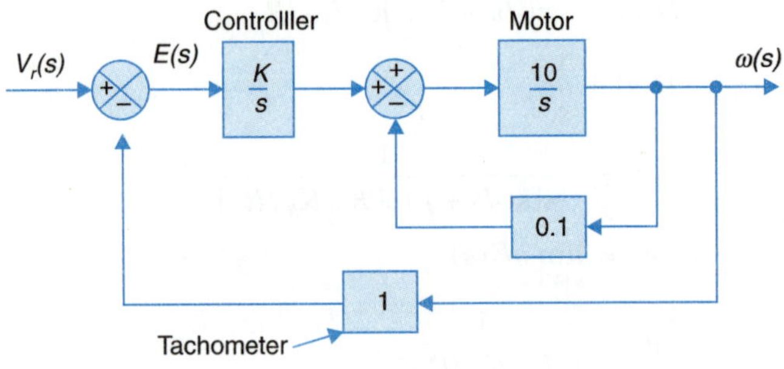

Fig. 3.17

(*a*) Determine the value of K for which steady-state error to unit ramp input ($V_r(s) = 1/s^2$) is less than 0.01 rad/sec.

(*b*) For the value of K found in part (*a*) determine, the sensitivity S_K^T, $T(s) = \omega(s)/V_r(s)$. What will be the limiting value of S_K^T at low frequencies ?

Solution. (*a*) Reducing the inner loop

$$G_{mot}(s) = \frac{10/s}{1 + 0.1 \times 10/s} = \frac{10}{s+1} \qquad ...(i)$$

From the forward path

$$\omega(s) = \left[\frac{10K}{s(s+1)}\right] E(s) \qquad ...(ii)$$

$$E(s) = V_r(s) - \omega(s) \qquad ...(iii)$$

$$E(s) = V_r(s) - \left[\frac{10K}{s(s+1)}\right] E(s)$$

or

$$E(s) = \left[\frac{s(s+1)}{s(s+1) + 10K}\right] V_r(s) \qquad ...(iv)$$

Input unit ramp, $V_r(s) = 1/s^2$. Then

$$E(s) = \frac{s(s+1)}{s(s+1) + 10K} \cdot \frac{1}{s^2} \qquad ...(v)$$

Steady-state error $e(ss) = \lim_{s \to 0} sE(s) = \lim_{s \to 0} \frac{(s+1)}{s(s+1) + 10K}$

or

$$e(ss) = \frac{1}{10K} = 0.01$$

which gives $K = 10$

(*b*)

$$T(s) = \frac{\omega(s)}{V_r(s)} = \frac{\dfrac{10K}{s(s+1)}}{1 + \dfrac{10K}{s(s+1)}}$$

or $$T(s) = \frac{10K}{s(s+1)+10K} \qquad ...(vi)$$

To find S_k^T

$$\frac{\partial \ln (10K)}{\partial \ln (K)} = \frac{10K}{10K} = 1$$

$$\frac{\partial \ln [s(s+1)+10K]}{\partial \ln (K)} = \frac{10K}{s(s+1)+10K}$$

Then $$S_K^T = 1 - \frac{10K}{s(s+1)+10K}$$

or $$S_K^T = \frac{s(s+1)}{s(s+1)+10K} \qquad ...(vii)$$

At $K = 10$

$$S_{10}^T = \frac{s(s+1)}{s(s+1)+100} \qquad ...(viii)$$

At low frequencies $s = j\omega$, ω tending to zero

$$S_{10}^T \approx \frac{1}{100} = 0.01 \text{ or } 1\%.$$

Example 3.6: The transfer function of an armature-controlled is employed to control the speed of J, f load in closed-loop with tachometer feedback as shown in Fig. 3.18. The motor-load transfer function is

$$G(s) = \frac{K_m}{(\tau' s + 1)}$$

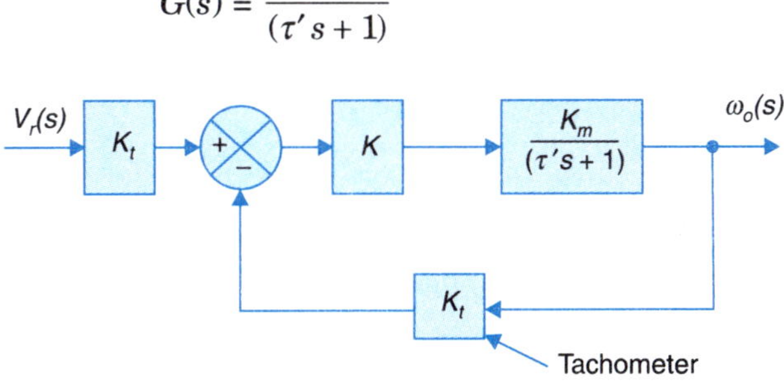

Fig. 3.18

Determine the time constant τ'' of the overall transfer function $T(l) = \omega_0(s)/V_r(s)$. What would be the value of gain K for $\tau'' = \tau'/10$? Calculate its numerical value for

$$\tau' = 0.1, K_m = 1.25, K_t = 0.2$$

Find $S_{K_t}^T$ and its limiting value for low frequencies.

Solution. For the block diagram given

$$T(s) = \frac{\omega_0(s)}{V_r(s)} = \frac{KG(s)}{1 + K_t K G(s)} \qquad ...(i)$$

Substituting for $G(s)$, we get

$$T(s) = \frac{K}{\tau' s + 1 + K_t K\, K_m} \qquad ...(ii)$$

or
$$T(s) = \frac{K/(1 + K_t K K_m)}{\tau'' s + 1} \qquad ...(iii)$$

Closed-loop time constant, $\tau'' = \tau'/(1 + K_t K K_m)$ ...(iv)

$$\tau'' = \tau'/10 = \tau'/(1 + K_t K K_m)$$

which gives $1 + K_t K K_m = 10$

Substituting values $K = \dfrac{10 - 1}{0.2 \times 1.25} = 36$

To find $S_{K_t}^T$

From Eq. (*ii*)

$$\frac{\partial \ln K}{\partial \ln K_t} = 0$$

$$\frac{\partial \ln [\tau' s + 1 + K_t K K_m]}{\partial \ln K_t} = \frac{K K_m}{\tau' s + 1 + K_t K K_m} . K_t$$

We then get
$$S_{K_t}^T = -\frac{K_t K K_m}{\tau' s + 1 + K_t K K_m} \qquad ...(v)$$

Substituting values
$$S_{0.2}^T = -\frac{0.2 \times 36 \times 1.25}{0.1s + 1 + 0.2 \times 36 \times 1.25} = -\frac{9}{0.1s + 9}$$

For low values of $s = j\omega$

$$S_{0.2}^T \approx -1 \quad (100\%)$$

The sensitivity is 100% because tachometer (K_t) is in the feedback path.

Example 3.7 : For the block diagram of Fig. 3.19 determine the sensitivity S_α^T, $T(s) = C(s)/R(s)$. Evaluate it at $\omega = 0.1$ and s rad (sec. for $\alpha = 2$).

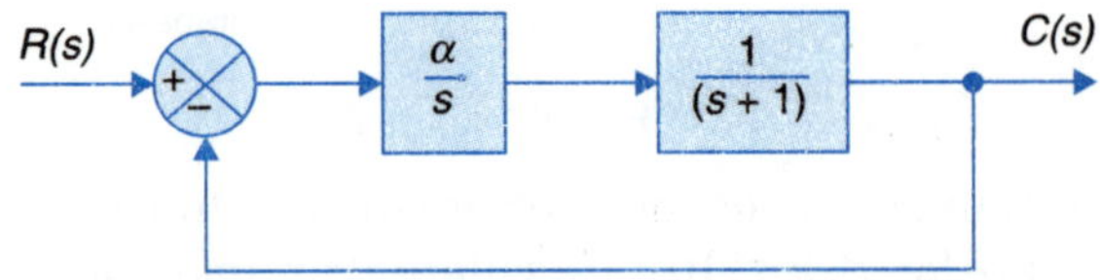

Fig. 3.19

Solution.
$$T(s) = \frac{\alpha}{s(s+1) + \alpha}$$

To find S_α^T
$$\frac{\partial \ln (\alpha)}{\partial \ln (\alpha)} = 1$$

$$\frac{\partial \ln [s(s+1) + \alpha]}{\partial \ln (\alpha)} = \frac{\alpha}{s(s+1) + \alpha}$$

Then $S_\alpha^T = 1 - \dfrac{\alpha}{s(s+1)+\alpha} = \dfrac{s(s+1)}{s(s+1)+\alpha}$

To evaluate S_α^T at $s = j\omega = j0.1$ and $j1$

$$s(s + 1) = j0.1(j0.1 + 1) = -0.01 + j0.1$$

$$S_2^T(j0.1) = \frac{-0.01 + j0.1}{(2-0.01)+j0.2} = 0.05 \angle 93°$$

$$S_2^T(j2) = \frac{-1 + j1}{(2-1)+j1} = 1\angle 90°$$

In terms of absolute values

$$| S_2^T(j0.1) | = 0.05, \; | S_2^T(j2) | = 1$$

It is observed that sensitivity increases with frequency.

Example 3.8 : The schematic of a liquid level control system is shown in Fig. 3.19. What of feedback it should ? Draw the block diagram of the system and write the transfer function of each block. Small letters represent incremental quantities.

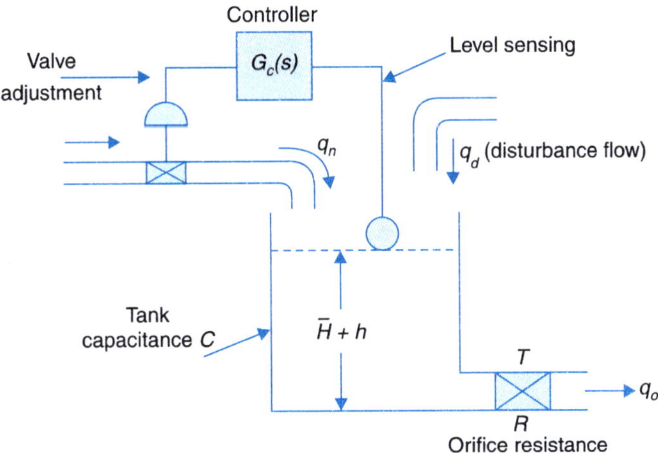

Fig. 3.20

Consider two types of controllers (*i*) $G_c(s) = K$ (proportional) and (*ii*) $G_e(s) = K/s$ *(integral)*. Determine the steady-state value of incremental height of liquid in tank, $h_{ss}(t)$ for unit step disturbance q_d.

Determine the sensitivity S_R^T in both cases.

Solution. Flow balance equation for the tank level system

$$q_i + q_d = C\frac{dh}{dt} + q_0 \qquad ...(i)$$

$$q_0 = \frac{h}{R} \qquad ...(ii)$$

Equation (*i*) can be expressed as

$$q_i + q_d = C\frac{dh}{dt} + \frac{h}{R} \qquad ...(iii)$$

Laplace transforming yields

$$Q_i(s) + Q_d(s) = \frac{(RCs+1)}{R} H(s) \quad ...(iv)$$

or

$$H(s) = \frac{R}{(RCs+1)} [Q_i(s) + Q_d(s)] \quad ...(v)$$

According to level sensing feedback controller

$$Q_i(s) = -G_c(s)\, H(s) \text{ ; negative feedback} \quad ...(vi)$$

From Eqs. (*v*) and (*vi*) the block diagram of Fig. 3.20 can be drawn.

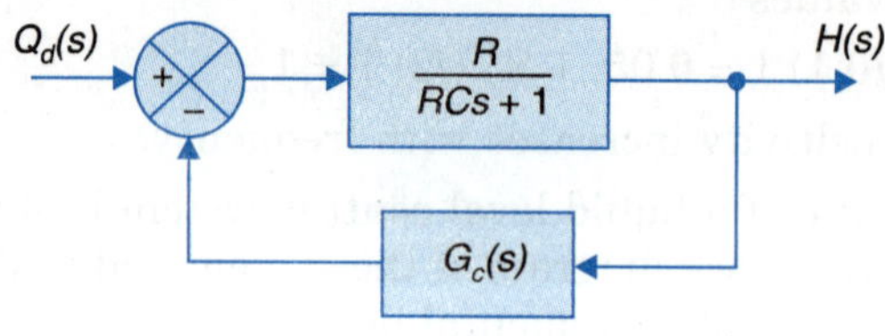

Fig. 3.21

(*i*) $G_c(s) = K$

From Fig. 3.20

$$\frac{H(s)}{Q_d(s)} = \frac{\dfrac{R}{RCs+1}}{1+\dfrac{KR}{RCs+1}}$$

or

$$\frac{H(s)}{Q_d(s)} = \frac{R}{RCs+(1+KR)} \quad ...(vii)$$

For unit step $Q_d(s) = \frac{1}{s}$. Then

$$h_{ss} = \lim_{s\to 0} sH(s) = \lim_{s\to 0} \frac{sR}{s[RCs+(1+KR)]}$$

or

$$h_{ss} = \frac{R}{1+KR} \quad ...(viii)$$

(*ii*) $G_c(s) = \frac{K}{s}$

From Fig. 3.20

$$\frac{H(s)}{Q_d(s)} = \frac{Rs}{RCs+(1+KR)}$$

For unit step disturbance $\theta_d(s) = \frac{1}{s}$ ......(*ix*)

$$h_{ss} = \lim_{s\to 0} sH(s) = \lim_{s\to 0} \frac{s\times sR}{s[RCs+(1+KR)]}$$

or

$$h_{ss} = 0 \quad ...(x)$$

Sensitivities

(*i*) $$T(s) = \frac{H(s)}{Q_d(s)} = \frac{R}{RCs + (1 + kR)}$$

To find S_R^T

$$\frac{\partial \ln (R)}{\partial \ln (R)} = 1$$

$$\frac{\partial \ln [RCs + (1 + kR)]}{\partial \ln (R)} = \frac{(Cs + K)\, R}{RCs + (1 + KKR)}$$

which gives

$$S_R^T = 1 - \frac{RCs + KR}{RCs + (1 + KR)}$$

or $$S_R^T = \frac{1}{RCs + (1 + KR)} \qquad ...(xi)$$

(*ii*) $$T(s) = \frac{Rs}{RCs + (1 + KR)}$$

To find S_R^T

$$\frac{\partial \ln (Rs)}{\partial \ln (R)} = 1$$

$$\frac{\partial \ln [RCs + (1 + KR)]}{\partial \ln (R)} = \frac{(Cs + K) R}{RCs + (1 + KR)}$$

which gives

$$S_R^T = 1 - \frac{RCs + KR}{RCs + (1 + KR)}$$

$$= \frac{1}{RCs + (1 + KR)} \qquad ...(xii)$$

Under low frequency conditions $s = j\omega$, $\omega \to 0$

(*i*) $S_R^T = \dfrac{1}{1 + KR}$ (*ii*) $S_R^T = \dfrac{1}{1 + KR}$ ...(*xiii*)

PROBLEMS

3.1. Consider the OPAM of Fig. P-3.1 connected as an inverting amplifier by means of the feedback resistance. Assuming that the OPAM offers infinite input resistance and has zero output resistance, find the expression for amplifier gain (T).

Draw the block diagram representing the amplifier as a feedback structure. Is the feedback negative or positive ?

Find expressions for S_A^T and $S_K^T (K = R_i/R_f)$. Calculate their values for $A = 10^4$ and $K = 0.1$.

Discuss the significance of these and calculate also the value of T.

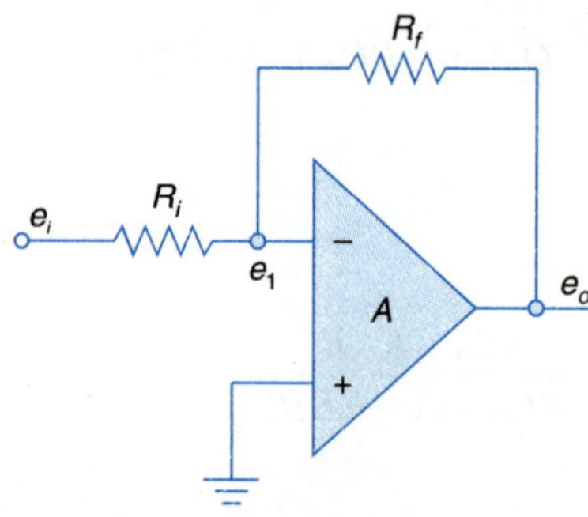

Fig. P-3.1

3.2. A simple voltage regulator is shown in Fig. P-3.2. A potentiometer is used at the output terminals of the generator to give a feedback voltage KV_o where K is constant ($K \leq 1$). The potentiometer resistance is high enough that it may be assumed to draw negligible current. The amplifier has a gain of 20 volts/volt. The generator gain is 50 volts/field amp. Reference voltage v_r = 50V.

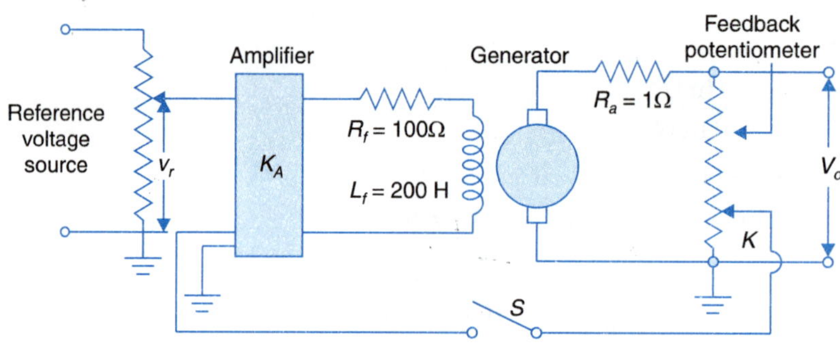

Fig. P-3.2

(*a*) Draw the block diagram of the system when the generator is supplying load current. Indicate the transfer function of each block.

(*b*) The system is operated closed-loop (S closed). Determine the value of K in order to give a steady no-load generator terminal voltage of 250 volts. What is the change in terminal voltage caused by a steady load current of 30 A ? What reference voltage would be required to restore the generator terminal voltage of 250 V ?

(*c*) The system is operated open-loop (S open), determine the reference voltage needed to obtain a steady no-load voltage of 250 V. What would be the change in terminal voltage for a load of 30 amps ?

(*d*) Compare the changes in terminal voltage in parts (*b*) and (*c*) and comment upon the effect of feedback in countering the changes in terminal voltage caused by load current.

3.3. The block diagram of a position control system is shown in Fig. P-3.2. Determine the sensitivity of closed-loop transfer function T which respect to G and H, the forward and feedback path transfer functions respectively, for ω = 1 rad/sec.

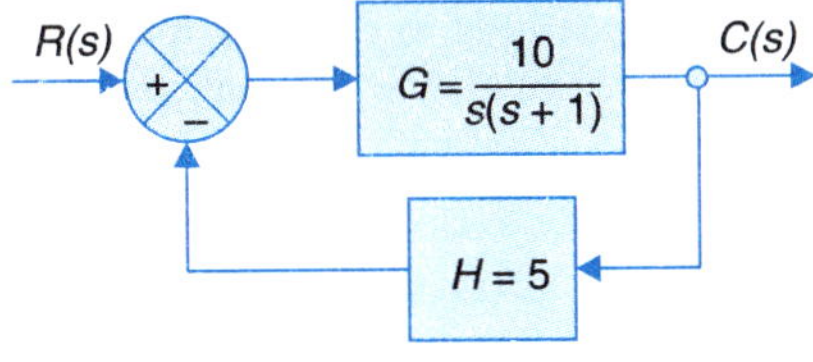

Fig. P-3.3

3.4. Consider the open-and closed-loop systems of Figs. P-3.4 (*a*) and (*b*).

(*a*) Fig. P-3.4(*a*) (open-loop): $K = 10$. For $R(s) = A/s$, find A for $c(ss)$, = 1.

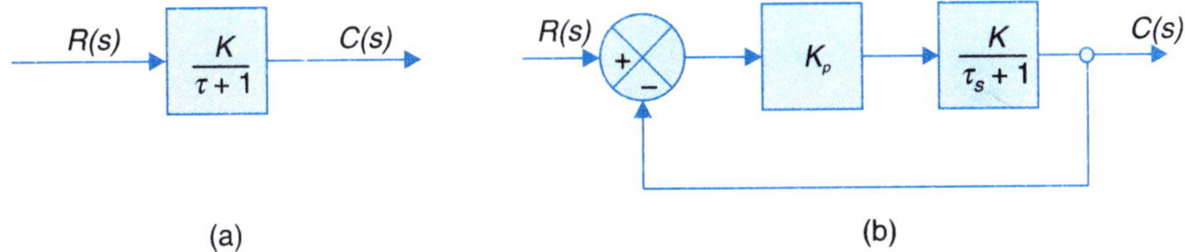

Fig. P-3.4

K now changes (increases) by 10% with A remaining constant find the % change in $c(ss)$.

(*b*) Fig. P-3.4 (*b*) (closed-loop): $K = 10$, $K_p = 100$. For $R(s) = A/s$, find A for $c(ss) = 1$. With this value of A find the % change in $c(ss)$ if K increase by 10%.

If a 2% change in $c(ss)$ is acceptable, which of the two systems does not require any monitoring and readjustment for environment causing 10% change in gain K of the plant.

3.5. In the system of Fig. P-3.5, $f(.)$ is an √function. Plot c vs r for $K = 10$ and observe the linearizing effect of feedback.

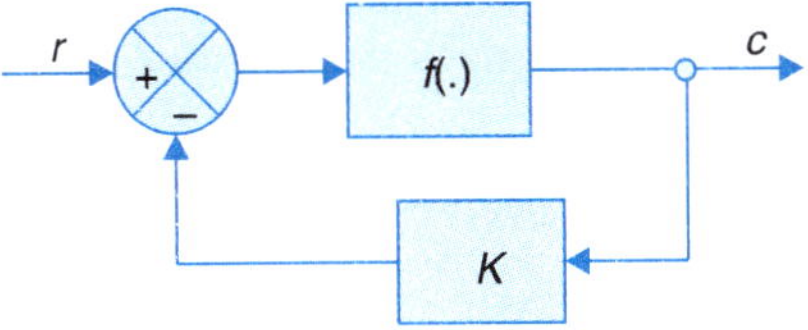

Fig. P-3.5

How should K be chosen to increase the range of r (positive) over which the input-output relationship is linear ? How does this affect the overall system gain (shape) ?

3.6. A servo system is represented by the signal flow graph shown in Fig. P-3.6. The variable T is the torque and E is the error. Determine:

(*a*) the overall transmission if, $K_1 = 1$, $K_2 = 5$ and $K_3 = 5$;

(*b*) the sensitivity of the system to changes in K_1 for $\omega = 0$.

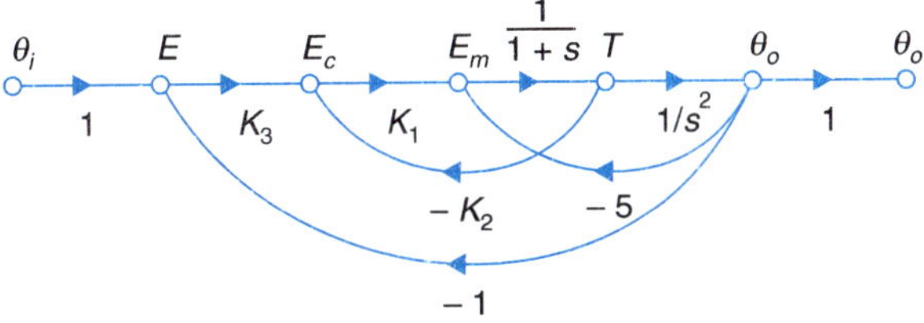

Fig. P-3.6

3.7. The field of a d.c. servomotor is separately excited by means of a d.c. amplifier of gain $K_A = 90$ (see Fig. P-3.7). The field has an inductance of 2 henrys and a resistance of 50 ohms. Calculate the effective field time constant.

A voltage proportional to the field current is now fed back negatively to the amplifier input. Determine the value of the feedback constant K to reduce the field time constant to 4 milliseconds.

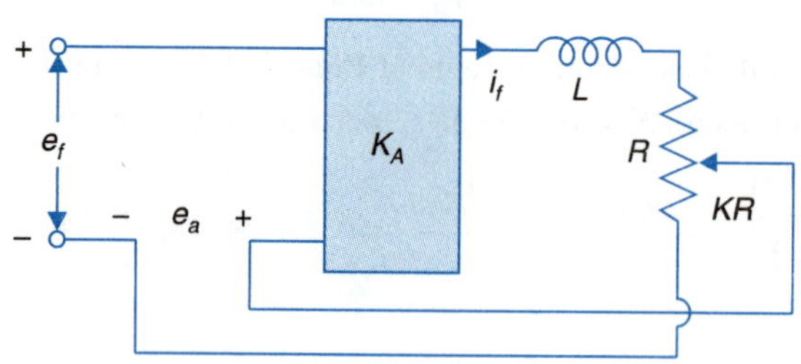

Fig. P-3.7

3.8. For the speed control system shown in Fig. P-3.8 assume that

(*i*) the reference and feedback tachometer are identical;

(*ii*) generator field time constant is negligible and its generated voltage is K^g volts/amp;

(*iii*) friction of motor and mechanical load is negligible.

(*a*) Find the time variation of output speed (ω_0) for a sudden reference input of 10 rad/sec. Find also the steady-state output speed.

(*b*) If the feedback loop is opened and gain K_A adjusted to give the same steady-state speed as in the case of the closed-loop, determine how the output speed varies with time and compare the speed of response in the two cases (closed- and open-loop).

(*c*) Compare the sensitivity of ω_o to changes in amplifier gain K_A and generator speed ω_g, with and without feedback.

[**Hint :** The generator gain constant K_g changes in direct proportion to generator speed *i.e.*, $K_g = K_g' \, \omega_g$ where K_g' is constant.]

The system constants are given below:

Moment of inertia of motor and load	$J = 5$ kg–m^2
Motor back emf constant	$K_t = 5$ volts per rad/sec
Total armature resistance of motor and generator	$R_a = 1$ ohm
Generator gain constant	$K_g = 50$ volts/amp
Amplifier gain	$K_A = 5$ amps/volt
Tachometer constant	$K_t = 0.5$ volts per rad/sec

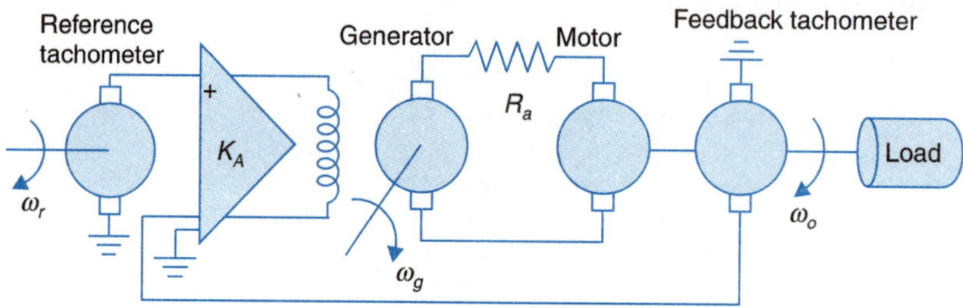

Fig. P-3.8

3.9. Figure P-3.9 shows a control system for supplying a steady flow of oil. The volume flow rate Q_0 of the oil from the tank is measured and the supply rate Q of oil to the tank is increased by an amount $K(\overline{Q} - Q_o)$, when the out flow differs from the desired flow Q_o. Under steady conditions $Q_o = \overline{Q}$ and the corresponding head in the tank is $\overline{H}$. The tank and output pipe may be considered to have a liquid capacitance C and flow resistance R.

For small deviations in flow rate from the steady-state condition, determine the closed-loop transfer function. Compare the sensitivity of this transfer function to changes in R with and without feedback. Also compare the system time constants in the two cases.

Suggest suitable hardware to implement the control scheme.

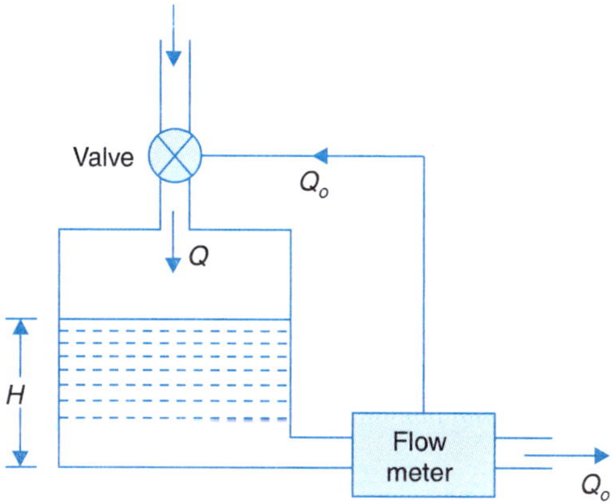

Fig. P-3.9

3.10. The scheme of Fig. P-2.6 is modified for automatic regulation of the concentration of outstream salt solution as shown in Fig. 3.10. The controller continuously monitors the output concentration C_o with the help of a conductivity cell located in the exit pipe and compares it with the desired concentration C_d to generate the error signal $(C_d - C_o)$. The controller manipulates the valve stem position according to the linear law, $x = K_c(C_d - C_o)$.

(*i*) Draw the signal flow graph of the system and obtain therefrom the transfer function $C_o(s)/C_d(s)$.

(*ii*) Compare the open-loop and closed-loop systems of Fig. P-2.6 and Fig. P-3.10 respectively for steady-state error in outstream salt concentration for unit-step disturbances in the inflow rate Q_i of the concentrated salt solution.

(*iii*) The system described above is operating around the point defined by (C_d^0, C_o^0, O_i^0). Compare the steady-state error caused in the outstream salt concentration to unit-step disturbance (ΔC_i) in the concentration of the inflowing concentrated salt solution for the open-loop case of Fig. P-2.6 and the closed-loop case of Fig. P-3.9.

[**Hint :** For part (*iii*) : Change in salt (mass) inflow rate is

$$(C_i^o + \Delta C_i)(Q_i^o + \Delta Q_i) - C_i^o Q_i^o \approx C_i^o[\Delta Q_i + (Q_i^o/C_i^o)\Delta C_i]$$

on the assumption that $\Delta C_i \Delta Q_i \approx 0$. The disturbance ΔC_i in C_i can thus be equivalently regarded as a disturbance $(Q_i^o/C_i^o)\,\Delta C_i$ in Q_i. Now

$$V\frac{d(\Delta C_o)}{dt} + Q_o(\Delta C_o) = C_i^o[\Delta Q_i + \underbrace{(Q_i^o / C_i^o)\Delta C_i}_{\text{Disturbance signal}}\ ]$$

$$\Delta Q_i = K\Delta x \ ; \ \Delta x = K_c(-\Delta C_o) \text{ since } (\Delta C_d = 0)$$

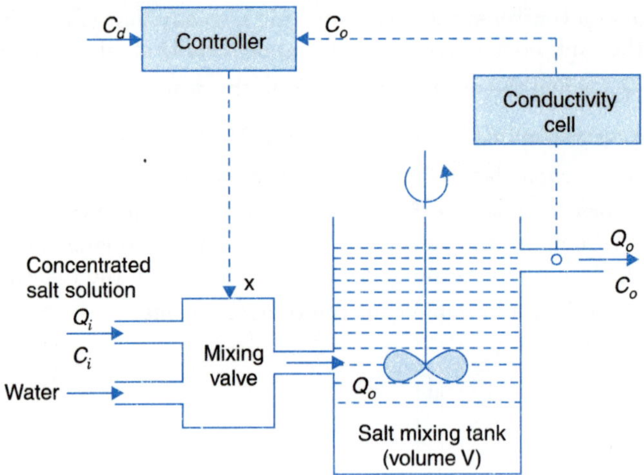

Fig. P-3.10

3.11. A temperature control industrial heating system is shown in schematic form in Fig. P-3.11. Modelling information on various constituents of the system is given below:

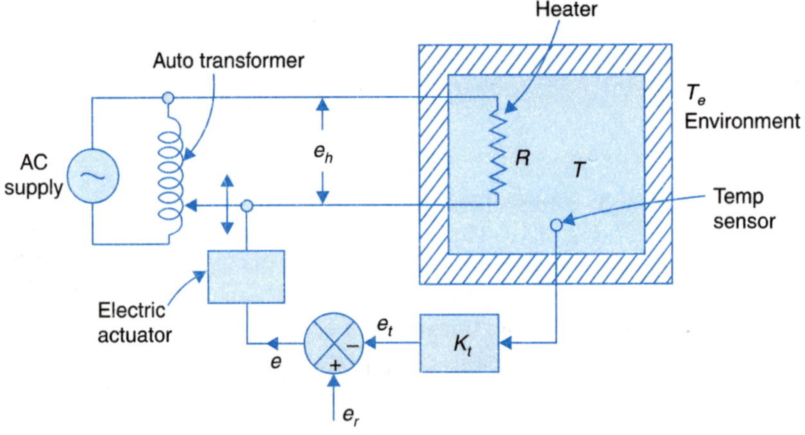

Fig. P-3.11

(*i*) Autotransformer: It's tapping point is adjusted by mechanically moving the tap changing gear by means of an electric actuator. Changes from a steady position in rms value of output voltage (e_h) is related to the input voltage (dc) of the actuator as

$$e_h = K_a e \text{ ; } e = \text{error signal (actuator input)}$$

(*ii*) For small changes in output voltage (rms) of the autotransformer the electric heater can be modelled as a linear relationship:

Rate of heat produced by the heating element $= K_h e$

(*iii*) Thermal capacitance of mass in the tank $= C$ J/°C

(*iv*) Thermal resistance of tank walls $= R_T$ °C/W

(*v*) Constant of temperature sensor $= K_t$ V/°C

Note : All variables are increments around their steady values.

Model the system in form of a block diagram giving the transfer function of each block.

(*a*) Derive the expressions for $S_{K_a}^{T}$ in open-and closed-loop models. Reduce this expression for dc condition *i.e.*, $s = 0$.

(*b*) Find $S_{K_t}^{T}$ and its expression for $s = 0$.

(*c*) Find the steady change in temperature of the mass for a step change in environment temperature in open-and closed-loop modes.

(*d*) If a dc servo motor is used as an actuator the governing relationship will be

$$e_h = K_a \int e\,dt$$

Show that in this case the steady state error (e) is zero for step input e_r and for step input disturbance T_e.

3.12. A ship has six degrees of freedom, three of them are translational motions and three are rotational. One of the rotational motions, the rolling motion, contributes much discomfort to the passengers. Different kinds of antiroll stabilizers have been employed in ships, probably the most modern being through power driven fins. The fins which look like small airplane wings protrude from each side of the ship below the water line. Either the area or the angle of attack of the fins is varied by power actuators in such a way as to produce controlling torque that opposes the torque of the sea.

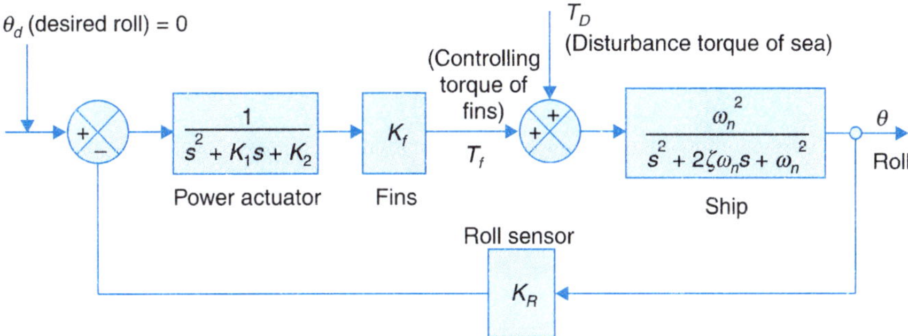

Fig. P-3.12

The rolling dynamics of the unstabilized ship is approximately represented by the transfer function given below:

$$G(s) = \frac{\omega_n^2}{s^2 + 2\zeta\omega_n s + \omega_n^2}$$

This is a highly underdamped system with ζ as low as 0.05. The roll sensor senses the roll and provides a signal to the power actuator which drives the fins to produce controlling torque $T_f(s)$. The block diagram of the system is shown in Fig. P-3.12.

(*a*) Determine the sensitivity of the system to changes in K_f (for $\omega = 0$).

(*b*) Discuss the effect of feedback in stabilizing the ship against rolling motion.

3.13. For the system whose signal flow graph is drawn in Fig. P-3.13

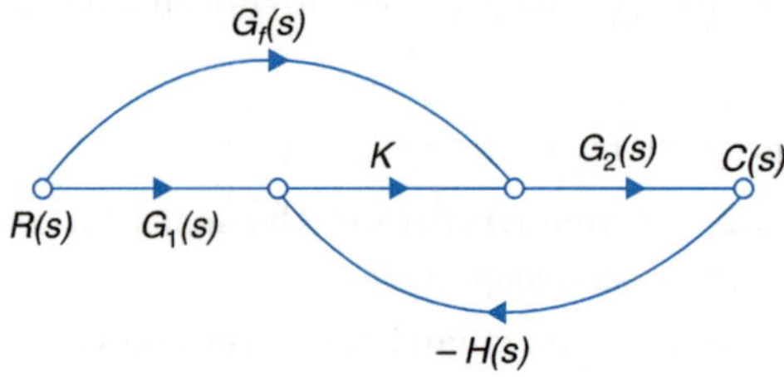

Fig. P-3.13

Find the transfer function $T(s) = C(s)/R(s)$

Find the system sensitivity S_K^T using eqn. (3.7*a*)

3.14 Fig. P-3.14 shows the block diagram of a speed control system with an integrator (*K/s*) in the forward path for a desired speed of 100 rad/s, show that steady output speed will be 100 rad/s (indicative of zero steady state error).

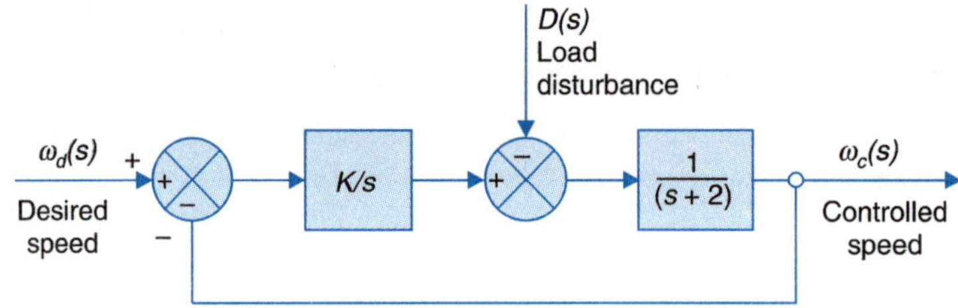

Fig. P-3.14

For a unit load disturbance $D(s) = 1/s$ find the *s*-domain expression for change in controlled speed, $\omega_c^D(s)$. Find the change in speed, caused by the disturbance at $t = 0$ and $t = \infty$ (steady state change).

Find the expression for $\omega_c^D(t)$ for $K = 0.5$, 1, 2 and 4. Sketch the nature of response. Which value of *K* should be preferred and why ? You may use Table 1.3 (Appendix 1) for finding Laplace inverse.

3.15 In a radar system, an electromagnetic pulse is radiated from an antenna into space. An echo pulse is received back if a conducting surface such as an airplane appears in the path of the signal. When the radar is in search of target, the antenna is continuously rotated. When target is located, the antenna is stopped and pointed towards the target by varying its angular direction until a maximum echo is heard. If energy is radiated in a narrow directional beam, accurate information about target location can be obtained. Narrow beam can be realised if the antenna size is large (*e.g.,* 20 m diameter). To drive this size of antenna, hydraulic or electric motors are used. One of the schemes utilizing electric motor is depicted in Fig. P-3.15. Determine:

(*a*) Sensitivity to changes in SCR gain, K_s for $\omega = 0.1$.

(*b*) The steady-state error of motor shaft, *i.e.,* $(\theta_R - \theta_M)$ when antenna is subjected to a constant wind gust torque of 100 newton-m.

The system constants are given below

Moment of inertia of load,	$J_L = 400$ kg–m^2
Potentiometer sensitivity,	$K_p = 1$ volts/rad
SCR gain,	$K_s = 500$ volt/volt
Back emf constant of motor,	$K_b = 1$ volt per rad/sec
	$R_a = 1$ ohm
	$N_1 = 40$
	$N_2 = 800 = N_3$

Load friction, motor inertia and friction are assumed to be negligible.

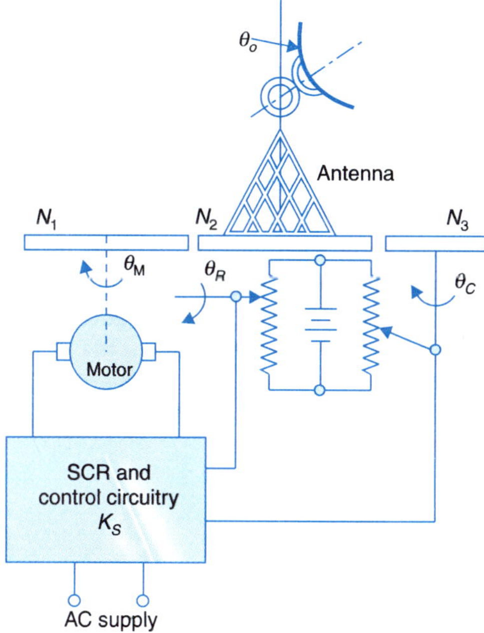

Fig. P-3.15

3.16. The liquid-level system shown in Fig. P-3.16 is modified for the automatic regulation of liquid head H (this means indirectly regulating the liquid outflow Q_0) by use of a float level sensor and a lever mechanism to adjust the opening of the valve as shown in Fig. P-3.16. For a change ΔH of head the lever-mechanism adjusts the valve opening by $\Delta v_x (\Delta v_x = -K_x \Delta H)$.

(*i*) Draw the signal flow graph of the system and obtain therefrom the transfer functions.

$$\left.\frac{\Delta H(s)}{\Delta P(s)}\right|_{\Delta Q_D = 0} \quad \text{and} \quad \left.\frac{\Delta Q_0(s)}{\Delta P(s)}\right|_{\Delta Q_D = 0}$$

(*ii*) Compare the open-loop and closed-loop systems of Fig. P-2.12 and Fig. P.3.16 respectively for the steady-state error in liquid outflow to the unit-step disturbance ΔQ_D.

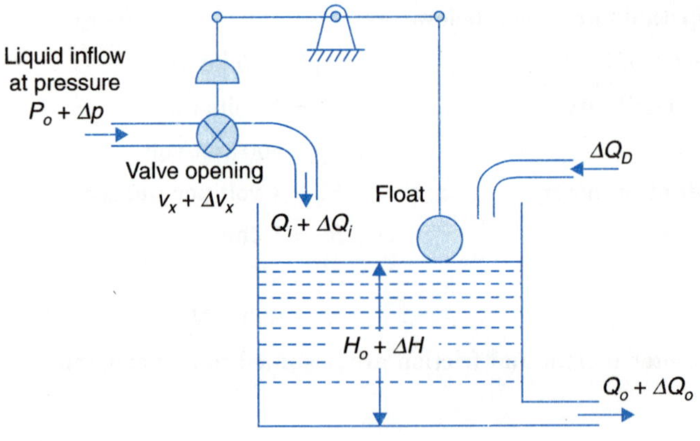

Fig. P-3.16

3.17. National economy control may be represented by the block diagram of Fig. 3.17. Private business investment fluctuates and represents disturbance to the system. Government senses the deviation of actual national income from the desired one and changes the business investments in order to hold the national income within tolerable limits.

(*a*) Determine the transfer function $C(s)/R(s)$ when the private business investment is zero.

(*b*) Determine the steady-state error when private business investment is zero and desired national income is represented by a unit step function.

(*c*) Determine the transfer function $C(s)/D(s)$ when the government does not act as a controller (loop involving government is open). Comment upon the stability of this system.

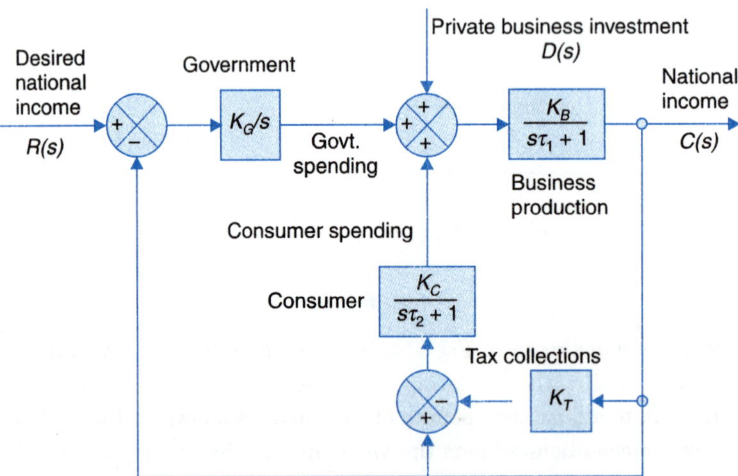

Fig. P-3.17. [From William A.Lynch and John G. Truxal, *Signals and Systems in Electrical Engineering,* 1962, McGraw Hill, New York, Reproduced with permission.]

3.18. (*a*) From the block diagram Fig. 3.12 for the temperature control system of Fig. 3.11 find the transfer function $\Delta\theta(s)/\Delta E_r(s)$ when $\Delta\theta_i = 0$ for open-and closed-loop conditions. Determine also the sensitivity of these transfer functions to change in K under steady d.c. conditions, *i.e.*, $s = 0$, (*b*) Modify the block diagram of Fig. 3.12 to include the effect of the disturbance signal $\Delta\theta_e$ caused by changes in the environmental temperature θ_{e0}. Find the steady change in the temperature of the outflowing liquid for open-and closed-loop cases due to a step change $\Delta\theta_e$ (assume $\Delta e_r = \Delta\theta_i = 0$).

3.19. A unity feedback system has forward path transfer function $G(s) = 20/(s + 1)$. Determine and compare the response of open-and closed-loop systems for unit-step input.

Suppose now that parameter variation occurring during operating conditions causes $G(s)$ to modify $G'(s) = 20/(s + 0.4)$. What will be the effect on unit-step response of open-and closed-loop systems ? Comment upon the sensitivity of the two systems to parameter variations.

4

CONTROL SYSTEMS AND COMPONENTS

4

CONTROL SYSTEMS AND COMPONENTS

4.1 INTRODUCTION

A closed-loop control system can be represented by the general block diagram shown in Fig. 4.1. Such a system is composed of three basic elements: the feedback element, controller and controlled system.

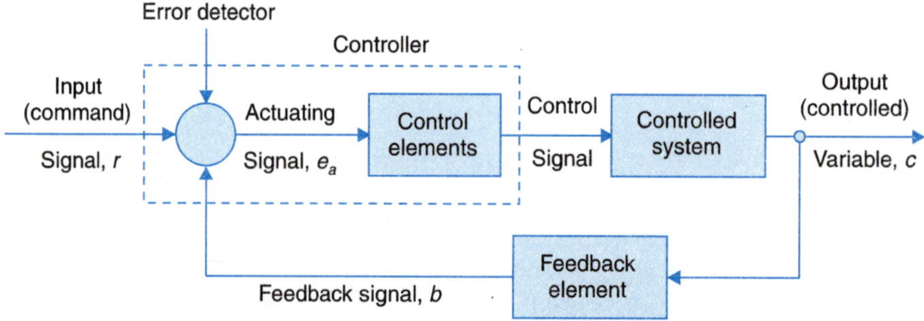

Fig. 4.1. General block diagram of a closed-loop control system.

The *feedback element* is a device which converts the *output* (*controlled*) *variable* c into another suitable variable, the feedback signal b, which then is compared with the input (command) signal.

The *controller* consists of an *error detector* and *control elements*. The error detector compares the feedback signal obtained from the plant output with the *input* (*command*) *signal* and determines therefrom the deviation known as the *actuating signal*. The actuating signal is usually at low power level. It is suitably manipulated by the control elements to produce a *control signal*. The manipulation may involve amplification, generation of a suitable function of the actuating signal and a power stage. The power stage in control elements is essential so that the control signal can drive the *controlled system* (a plant or process) to produce the desired output variable. Large power amplification may be involved in the plant or process being controlled.

This chapter is devoted to a general study of control system components, the emphasis being on those commonly used in servo-systems.

4.2 LINEAR APPROXIMATION OF NONLINEAR SYSTEMS

Practically none of the physical systems in nature are perfectly linear. Since most powerful mathematical tools are available for linear analysis, it is desirable to make linearizing assumptions whenever a compromise can be obtained between the simplicity of analysis and accuracy of results. As the components to be discussed in this chapter have somewhat nonlinear behaviour, it is worthwhile discussing a general linearizing technique before entering into a detailed discussion of components. In fact the technique has already been employed in the previous chapters where some of the components introduced were essentially nonlinear. The intuitive basis of linearization is that a smooth curve differs very little from its tangent line so long as the variable does not wander far from the point of tangency. Thus if the region of operation is restricted to a narrow range, a nonlinear system can be treated approximately as a linear system. This kind of approximation appears to be valid in most control systems since their purpose is to keep the controlled variable very close to the desired value. However, if the system is required to follow a varying desired value, one can analyze the system by linearizing it at several points along the curve.

Consider a general element with input $x(t)$ and output $y(t)$ whose input-output relationship $y = f(x)$ is shown graphically in Fig. 4.2. The relationship may be *nonlinear* but is assumed to be *continuous*.

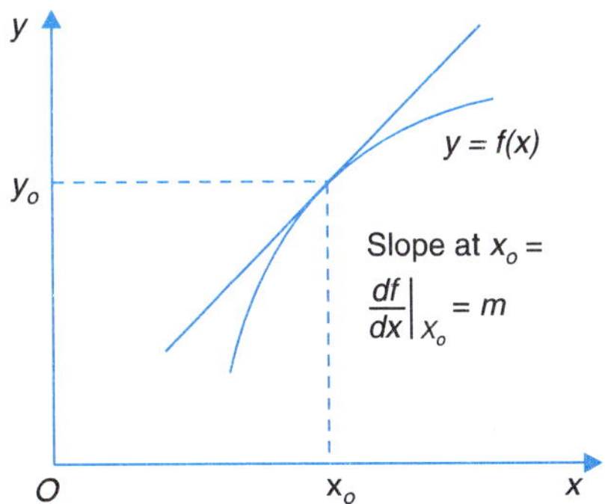

Fig. 4.2. Linearization of a nonlinear element.

Expansion of the equation $y = f(x)$ into a Taylor's series about the normal operating point (x_o, y_o) gives

$$y = f(x) = f(x_o) + \left.\frac{df}{dx}\right|_{x = x_o} \frac{x - x_o}{1!} + \left.\frac{d^2f}{dx^2}\right|_{x = x_o} \frac{(x - x_o)^2}{2!} + \ldots \qquad \ldots(4.1)$$

If the variation $(x - x_o)$ of the input about the normal operating point is small, higher than first-order terms in $(x - x_o)$ can be neglected yielding a linear approximation,

$$y = y_o + m(x - x_o) \quad \text{...(4.2)}$$

where $m = \left.\dfrac{df}{dx}\right|_{x = x_o}$ is the slope at the operating point.

Equation (4.2) may be rewritten as

$$y - y_o = m(x - x_o) \qquad \text{or} \qquad \Delta y = m\Delta x$$

If the output variable y depends on several input variables $x_1, x_2, ..., x_n$, *i.e.*, $y = f(x_1, x_2, ..., x_n)$, then the linear approximation for y can be obtained similarly by expanding this equation into Taylor's series about the operating point $(x_{1o}, x_{2o}, ..., x_{no}, y_o)$ and neglecting all terms of second and higher derivatives. This approximation gives

$$y = f(x_{1o}, x_{2o}, ..., x_{no}) + \left.\frac{\partial f}{\partial x_1}\right|_{\substack{x_1 = x_{1o}\\ \cdots\cdots \\ x_n = x_{no}}} (x_1 - x_{1o}) + \left.\frac{\partial f}{\partial x_2}\right|_{\substack{x_1 = x_{1o}\\ \cdots\cdots \\ x_n = x_{no}}} (x_2 - x_{2o}) + ...$$

$$+ \left.\frac{\partial f}{\partial x_n}\right|_{\substack{x_1 = x_{1o}\\ \cdots\cdots \\ x_n = x_{no}}} (x_n - x_{no})$$

This technique is employed in several examples in the following sections.

4.3 CONTROLLER COMPONENTS

Controller components are of three kinds:

1. *Sensors.* These are low-power transducers which produce output signal as a measure of the controlled variable; a linear (proportional) relationship is generally preferred though it could be a suitable functional relationship. Sensors are employed for a variety of measurements-position, velocity, acceleration, pressure, temperature or a quantity representing the chemical state of a reactor, neutron flux level in an atomic reactor etc. The output signal of the sensor is invariably in electrical form; analog or digital.

2. *Differencing and amplification.* Differencing to get the error signal and its amplification to suitable level in magnitude and power are most conveniently carried out electronically. Availability of OPAMs for differencing and stable amplification over a wide frequency range and large range of voltage (or current) and power levels by use of power transistors or SCR's at the final (power) stage are the chief advantages offered by the electronic systems. These will not be discussed here as they form part of standard courses in electronics.

3. *Actuators.* These are devices whose output is mechanical motion (translatory/rotary; though rotary motion is convenient and is preferred). The actuators are characterised by power output and speed-torque relationship to match the load. These could be electrical, hydraulic or pneumatic. An actuator in a control system performs a variety of tasks to manipulate the controlled process or plant. For example, it may open/close a valve in a hydraulic/steam/chemical

process (plant), turn a robot link w.r.t. to its neighbouring link, move a transformer tap (up or down), move up/down the control rods of a nuclear reactor etc.

Because of the flexibility inherent in transmitting electrical power (through cables) and desirable speed-torque characteristics which are linear, electric actuators (motors) are now widely adopted in control systems except in low speed but high torque applications where hydraulic actuators are still in use. Pneumatic actuators are not as messy as hydraulic ones but suffer from leakages and inherent inaccuracies.

4. *Electric system.* DC and ac motors are the two kinds of electric actuators; in low power ratings these are known as servomotors. DC motors are costlier than ac motors because of the additional cost of commutation gear. These have, however, the important advantages of linearity of characteristics and higher **stalled torque/inertia** ratio; this being an important figure of merit for a servomotor. Stalled torque is the torque developed by motor when stationary with full applied voltage (and full field in case of a DC motor). It may be pointed out here that high torque/inertia ratio means lower motor time constant and so faster dynamic response.

With advanced manufacturing techniques, low brush commutator friction and still higher torque/inertia ratios have been achieved in dc servomotors, such that these have practically taken over from ac servomotors in most control applications.

Electric actuators for stepped motion are known as stepper motors which will be dealt in details in section 4.

DC Servomotors

Modelling of dc motors, armature controlled and field controlled, has been considered at length in Chapter 2. Here we shall consider some of the constructional features of dc servomotors. With recent development in rare earth permanent magnets (PM) which have high residual flux density and a very high coercivity, dc servomotors are now constructed with PMs resulting in much higher torque/inertia ratio and also higher operating efficiency as these motor have no field losses. The speed of a permanent magnet dc (PMDC) motor is nearly directly proportional to armature voltage at a given load torque. Also the speed-torque characteristic at a given voltage is more flat than in a wound field motor as the effect of armature reaction is less pronounced in a PM motor.

Three types of constructions employed in PMDC servomotors are illustrated in Fig. 4.3 (*a*), (*b*) and (*c*). In Fig. 4.3 (*a*) the armature is slotted with dc winding placed inside these slots (as in a normal dc motor). Though quite reliable and rugged this type of construction has high inertia to reduce which the construction of Fig. 4.3 (*b*) is adopted where the winding is placed on the armature surface. Because of the larger air gap, stronger PMs are needed in this construction. A much lower inertia is achieved by placing the winding on a nonmagnetic cylinder which rotates in annular space between the PM stator and stationary rotor as illustrated in Fig. 4.3 (*c*). The air gap has to be still larger with consequent need of much stronger PMs. The constructional details of this low inertia motor are further brought out in Fig. 4.3 (*d*).

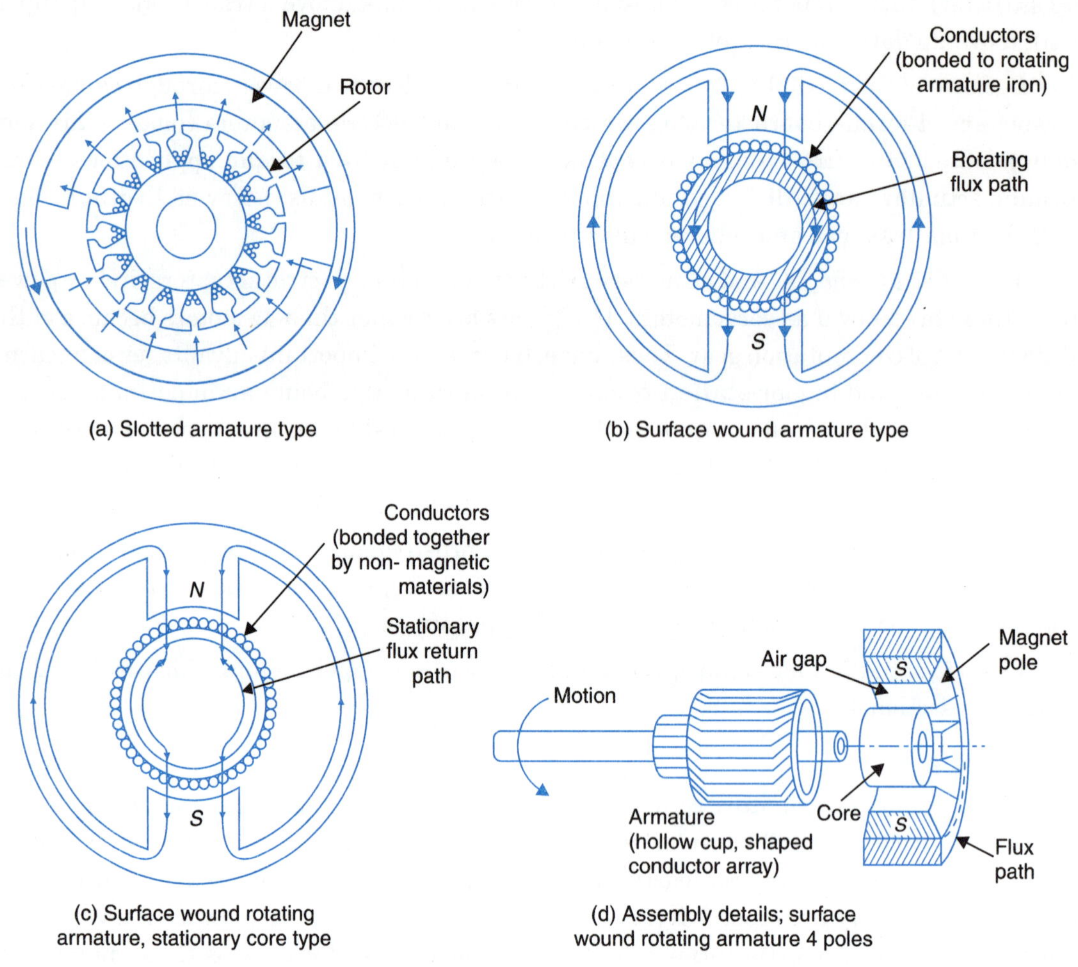

Fig. 4.3. Permanent magnet dc motor (PMDC).
Cross-sectional views of various types.

DC Tachogenerator

PMDC when coupled to a rotating shaft would generate a voltage proportional to speed and would thus serve as a tachometer. The current rating of such a tachometer is very small as it is to be connected to the input end of an OPAM. This device is commonly employed for speed feedback in speed control system or in the internal velocity feedback loop of a position control system.

Brushless DC motor

With advances in inverter technology commutation process of a dc motor can be accomplished by solid-state devices (transistor/SCR) thereby eliminating brush wear and friction problems of normal dc motor. Such an arrangement dictates reverse construction with armature winding placed on stator and PMs on the rotor. Rotor shaft position is sensed by means of suitable sensors (Hall effect sensors). The sensor pulses through suitable logic circuitry control the switching times of the switching matrix feeding inverted current to the stator windings from the dc source. The schematic diagram is shown in Fig. 4.3. In the absence of a regulator the motor speed is approximately directly proportional to the primary dc voltage.

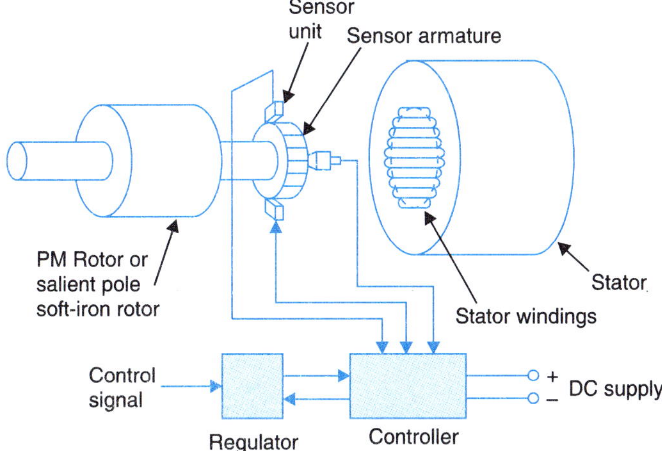

Fig. 4.4. Components of a brushless dc motor.

The switching circuit of a 3 phase motor connected in delta (the connection could be star as well) is shown in Fig. 4.4. The winding is energized with six pulses from 6 power transistors in cyclic sequence (16, 15, 35, 34, 24, 26). This sequence of turning on the transistors is controlled by the pulses obtained from three Hall effect sensors placed on the rotor at 120° elect spacing. Thus the stator and motor fields (PMs) remain locked in synchronism independent of rotor speed.

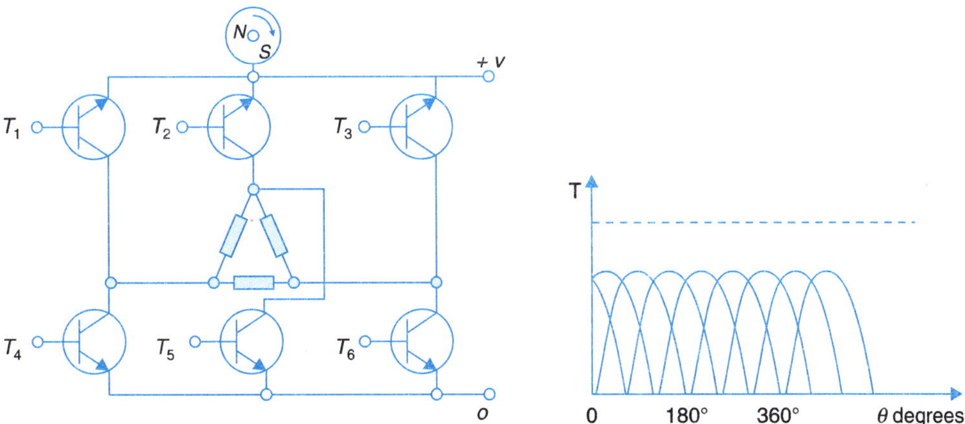

Fig. 4.5. Three phase, six pulse brushless dc motor.

A brushless dc motor with 3 phases delivers an even torque output and also the exploitation of the material is optimum. Though the electronic circuitry because of 3 phases is somewhat complex but reliable components and circuits are now available.

A.C. SERVOMOTORS

An a.c. servomotor is basically a two-phase induction motor except for certain special design features. A two-phase induction motor consists of two stator windings* oriented 90°. Figure 4.6 shows the schematic diagram for balanced operation of the motor, *i.e.*, voltages of equal rms magnitude and 90° phase difference are applied to the two stator phases, thus making their respective fields 90° apart in both time and space, resulting in a magnetic field of constant magnitude rotating at synchronous speed. The direction of rotation depends upon phase relationship of voltages V_1 and V_2. As the field sweeps over the rotor, voltages are induced in it producing current in the short-circuited rotor. The rotating magnetic field interacts with these currents producing a torque on the rotor in the direction of field rotation. The general shape of the torque-speed characteristics of a two-phase induction motor is shown in Fig. 4.7.

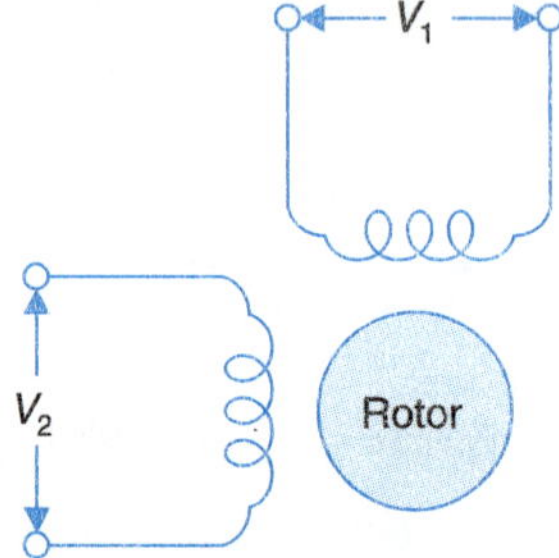

Fig. 4.6. Schematic diagram of a two-phase induction motor.

It is seen from this figure that the shape of the characteristic depends upon the ratio of the rotor reactance X to the rotor resistance R. In normal induction motors, X/R ratio is generally kept high so as to obtain the maximum torque close to the operating region which is usually around 5% slip.

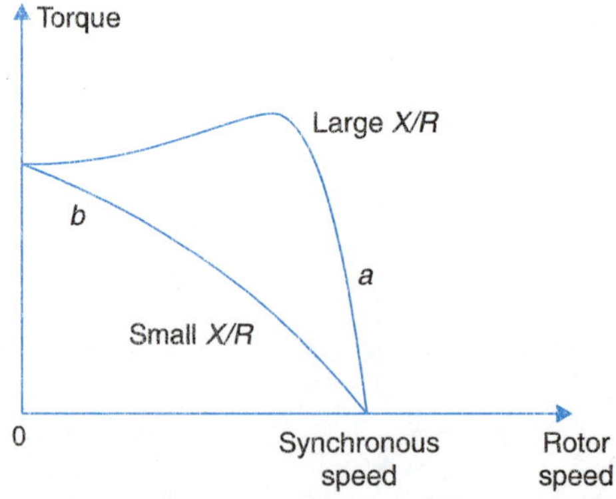

Fig. 4.7. Torque-speed characteristics of induction motor.

A two-phase servomotor differs in two ways from a normal induction motor.

* A two pole construction is commonly used.

1. The rotor of the servomotor is built with high resistance so that its X/R ratio is small and the torque-speed characteristic, as shown by the curve b of Fig. 4.7, is nearly linear in contrast to the highly nonlinear characteristic when large X/R ratio is used for servo applications, then because of the positive slope for part of the characteristic, the system using such a motor becomes unstable.

The rotor construction is usually squirrel cage or drag-cup type. The diameter of the rotor is kept small in order to reduce inertia and thus to obtain good accelerating characteristics. Drag-cup construction is used for very low inertia applications.

2. In servo applications, the voltages applied to the two stator windings are seldom balanced. As shown in Fig. 4.8, one of the phases known as the *reference phase* is excited by a constant voltage and the other phase, known as the *control phase* is energized by a voltage which is 90° out of phase with respect to the voltage of the reference phase. The control phase voltage is supplied from a servo amplifier and it has a variable magnitude and polarity (± 90° phase angle with respect to the reference phase). The direction of rotation of the motor reverses as the polarity of the control phase signal changes sign.

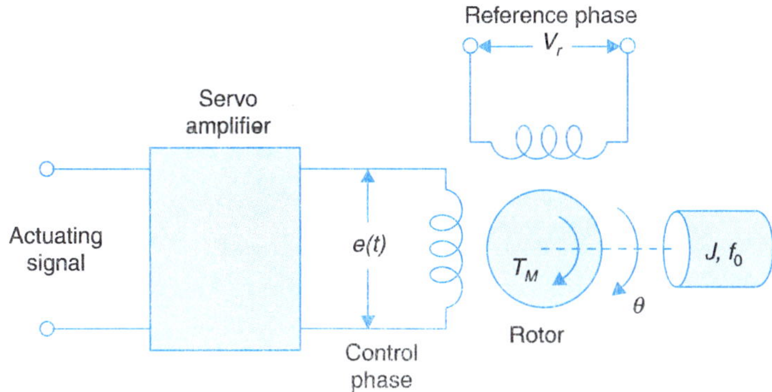

Fig. 4.8. Schematic diagram of a two-phase servomotor.

It can be proved using symmetrical components that starting torque of servomotor under unbalanced operation, is proportional to E, the rms value of the sinusoidal control voltage $e(t)$. A family of torque-speed curves with variable rms control voltage is shown in Fig. 4.9 (a). All these curves have negative slope. Note that the curve for zero control voltage goes through the origin and the motor develops a decelerating torque.

As seen from Fig. 4.9 (a), the torque-speed curves are still somewhat nonlinear. However, in the low-speed region, the curves are nearly linear and equidistant, *i.e.*, the torque varies linearly with speed as well as with control voltage. Since a servomotor seldom operates at high speeds, these curves can be linearized about the operating point.

The torque generated by the motor is a function of both the speed $\dot{\theta}$ and rms control, voltage E, *i.e.*, $T_M = f(\dot{\theta}, E)$. Expanding this equation into Taylor's series about the normal operating point (T_{Mo}, E_o, $\dot{\theta}_o$) and dropping off the terms of second- and higher-order derivatives, we get

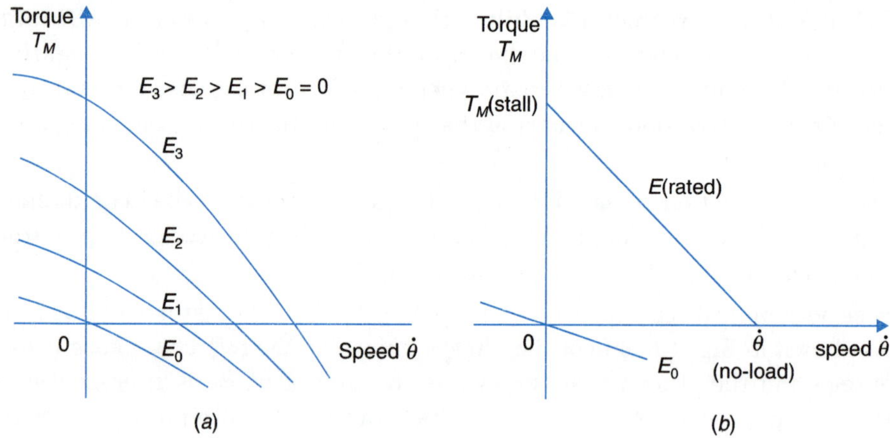

Fig. 4.9. Servomotor characteristics.

$$T_M = T_{M_o} + \left.\frac{\partial T_M}{\partial E}\right|_{\substack{E = E_o \\ \dot{\theta} = \dot{\theta}_o}} (E - E_o) + \left.\frac{\partial T_M}{\partial \dot{\theta}}\right|_{\substack{E = E_o \\ \dot{\theta} = \dot{\theta}_o}} (\dot{\theta} - \dot{\theta}_o) \quad \text{...(4.3)}$$

or
$$T_M - T_{M_o} = K(E - E_o) - f(\dot{\theta} - \dot{\theta}_o)$$

or
$$\Delta T_M = K\Delta E - f\Delta\dot{\theta} \quad \text{...(4.4)}$$

where
$$K = \left.\frac{\partial T_M}{\partial E}\right|_{\substack{E = E_o \\ \dot{\theta} = \dot{\theta}_o}} \text{ and } f = -\left.\frac{\partial T_M}{\partial \dot{\theta}}\right|_{\substack{E = E_o \\ \dot{\theta} = \dot{\theta}_o}} \text{ are constants.}$$

If the load consists of inertia J and viscous friction f_0, the torque equation, expressed in incremental notation, becomes

$$\Delta T_M = J\Delta\ddot{\theta} + f_0\Delta\dot{\theta} = K\Delta E - f\Delta\dot{\theta}$$

The above equation is valid even when ΔE varies with time so long as this variation is slow compared to sin $\omega_c t$; in frequency domain, it means that ω (cut-off) << ω_c where ω is the frequency of the signal ΔE, and ω_c is the carrier frequency. Control systems using such devices are called *carrier control systems.*

Taking the Laplace transform of above equation, we have

$$(Js^2 + f_0 s)\Delta\theta(s) = K\Delta E(s) - fs\Delta\theta(s)$$

The incremental motor transfer function about an operating point is then given by

$$G_M(s) = \frac{\Delta\theta(s)}{\Delta E(s)} = \frac{K}{Js^2 + (f_0 + f)s} = \frac{K_m}{s(\tau_m s + 1)} \quad \text{...(4.5)}$$

where $K_m = \dfrac{K}{f_0 + f}$ = motor gain constant; and $\tau_m = \dfrac{J}{f_0 + f}$ = motor time constant.

From eqn. (4.5) it is seen that the negative slope of the torque-speed characteristic of the servomotor contributes the viscous friction term f to the load friction f_0. If the slope

$$\left.\frac{\partial T_M}{\partial \dot{\theta}}\right|_{\substack{E = E_o \\ \dot{\theta} = \dot{\theta}_o}} = -f$$

is positive (as in part of torque-speed characteristic of a normal induction motor), then the effective viscous friction $(f_0 + f)$ may become negative. The control system using such a motor may have negative damping and may therefore be unstable. This is one of the reasons why the conventional induction motors are not used as servomotors.

When a.c. servomotor is used in a position control system, the operating point is $(E_0 = 0, \dot{\theta}_0 = 0)$, so that

$$\Delta\theta = \theta \text{ and } \Delta E = E$$

Therefore $$G(s) = \frac{\theta(s)}{E(s)} = \frac{K_m}{s(\tau_m s + 1)}$$

The motor constants K and f can be easily estimated from two well known tests: no-load and stall torque tests. The no-load speed $(T_M = 0)$ and stall torque $(\dot{\theta} = 0$, *i.e.*, motor rotor held stationary) at rated voltage applied to control phase are plotted in Fig. 4.9 (*b*). The straight line joining $\dot{\theta}$ (no-load speed) and T_M (stall torque) points approximately represents the torque-speed characteristic of a servomotor. Therefore,

$$K = \frac{\text{Stall torque (rated voltage)}}{\text{Rated control phase voltage}} \qquad \text{...(4.6(a))}$$

$$f = \frac{\text{Stall torque (rated voltage)}}{\text{No load speed (rated voltage)}} \qquad \text{...(4.6(b))}$$

It is further observed from Fig. 4.9 (*a*) that the slope of the torque-speed characteristic reduces as control phase voltage decreases. In motors used in practice, the torque-speed slope in low speed region is nearly one-half the slope at rated voltage. Therefore, for servomotors used in position control systems,

$$f = \frac{1}{2}\frac{\text{Stall torque (rated voltage)}}{\text{No load speed (rated voltage)}} \qquad \text{...(4.7)}$$

approximately.

AC Tachometer

The principle of operation of an ac tachometer or drag-up generator can be easily understood by referring to the schematic diagram of Fig. 4.10. As in the case of an ac servomotor, the two stator field coils are mounted at right angles to each other, *i.e.* in space quadrature. The tachometer is merely a thin aluminum cup that rotates in the air-gap between a fixed magnetic structure. This low inertia rotor of highly conducting material provides a uniformly short-circuited secondary conductors.

An approximate analysis of the ac tachometer is presented below:

Let the reference coil be excited with voltage $V_r \cos \omega_c t$, ω_c being the carrier frequency. It produces a reference flux of $\phi_r \sin \omega_c t$, as the resistance and leakage reactance voltage drops of the coil are of negligible order. This flux being alternating is equivalent to two rotating fluxes, ϕ_f and ϕ_b of equal magnitude rotating at synchronous speed ω_s in opposite directions. When the rotor (drag up) is stationary, ϕ_f and ϕ_b induce equal and opposite speed voltages in the open-circuited quadrature coil (placed at 90° elect in space w.r.t. the reference coil) so that the voltage at the quadrature coil terminals is zero. Both fields induce currents of frequency ω_c in the short circuited rotor producing reaction flux. In fact ϕ_f and ϕ_b (components of ϕ_r) are the resulting rotating air-gap fluxes.

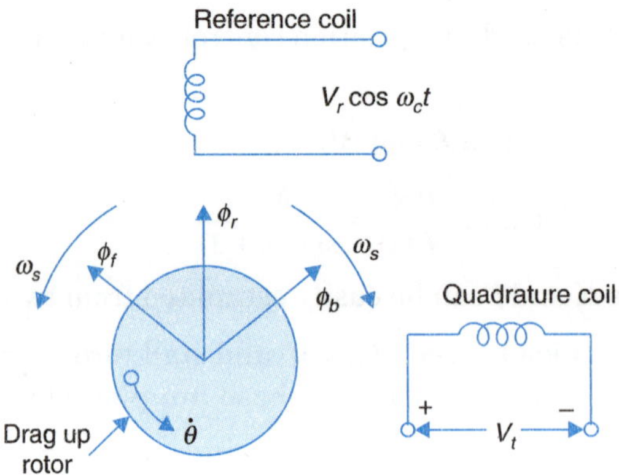

Fig. 4.10. AC tachometer.

Assume now that the rotor rotates at speed $\dot{\theta}$ (say in the direction of ϕ_f). As a result the relative speed of rotor w.r.t. ϕ_f decreases and that w.r.t. ϕ_b increases. Induced rotor currents due to ϕ_f tend to reduce so that the net ϕ_f strengthens, while, the induced rotor currents due to ϕ_b tend to increase so that net ϕ_b weakens. As a consequence of this unbalanced strengths of the fields ϕ_f and ϕ_b, a net voltage is induced in the quadrature coil. Because of 90° space displacement of this coil, its voltage is in phase quadrature to the voltage applied to the reference coil; the voltage magnitude being proportional (approximately) to the rotor speed. Reverse phenomenon happens if the rotor rotates in opposite direction. Thus the tachometer voltage (voltage of quadrature coil (open circuited)) is approximately given as

$$v_t(t) = K_t \dot{\theta} \sin \omega_c t \qquad ...(4.8(a))$$

This equation would also apply when $\dot{\theta}$ varies with time so long as the frequency of this variation ($a\ \omega_c$) << ω_c. Thus eqn. (4.8(a)) generalizes to

$$v_t(t) = K_t \dot{\theta}(t) \sin \omega_c t \qquad ...(4.8(b))$$

As $\dot{\theta}(t)$ is varying slowly w.r.t. sin $\omega_c t$, we can introduce the concept of instantaneous rms value as

$$V_t(t) = K_t \dot{\theta}(t) \qquad ...(4.8(c))$$

where K_t is the tachometer constant.

Potentiometers

These are simple reliable devices for measuring mechanical displacement (translatory linear or angular). A potentiometer is a simple voltage divider with three terminals-two fixed terminals and third a movable terminal attached to a jockey–symbolically represented in Fig. 4.11. The total resistance of the potentiometer (between fixed terminals) is spread out uniformly, linearly for translatory measurement or in helical from for angular measurement. If a fixed reference voltage is applied at the fixed terminals, it is easy to see that the voltage output at the movable terminal (w.r.t. one of the fixed terminals (the ground)) is proportional to displacement. This linear relationship is affected by the magnitude of the load resistance. The load resistance in fact is the input resistance of the device (usually amplifier) to which the potentiometer output is connected.

With reference to Fig. 4.11

R_p = total potentiometer resistance

x_t = total displacement of the potential over which R_p is spread out (uniformly)

x_i = displacement of the jockey from the ground terminal G (input)

R_L = load resistance

V_{REF} = reference voltage

V_o = output voltage (V_{MG})

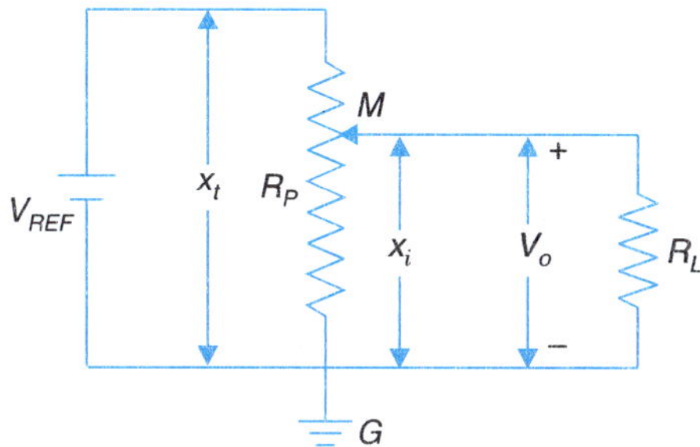

Fig. 4.11. A potentiometer with load.

As the resistance is spread out uniformly

$$R_i = (x_i/x_t)R_p$$

$$R_{eq} = R_i \parallel R_L = \frac{1}{\left[\dfrac{1}{R_L} + \left(\dfrac{x_t}{x_i}\right)\dfrac{1}{R_P}\right]}$$

Then by voltage dividing action

$$V_o = V_{REF}\frac{R_{eq}}{R_{eq} + (R_p - R_i)}$$

Substituting values and manipulating it is easily established that

$$V_o = \frac{V_{REF}}{(x_t/x_i) + (R_P/R_L)[1 - (x_i/x_t)]} \quad ...(4.10)$$

If R_L = (jockey open *i.e.*, no load)

$$V_o = V_{REF}\,(x_i/x_t); \text{ a linear relationship} \quad ...(4.11)$$

Finite value of R_L causes the output voltage to be nonlinear function of (x_i/x_t) (Eq. (4.10)). This relationship is plotted as V_o/V_{REF} vs x_i/x_t in Fig. 4.12. The figure also shows the actual relationship without loading which can vary above or below the ideal case because of manufacturing tolerances. Potentiometer linearity is defined later with reference to this.

It is seen from Fig. 4.12 that loading causes the potentiometer output to be a nonlinear function of (x_i), the mechanical displacement. This can be remedied by feeding the potentiometer output V_o to an analog buffer; a JFET input OPAM used as a noninverting unity gain amplifier. These have an input impedance of around 1000 M which eliminates loading.

Potentiometer specifications

1. *Total resistance* (R_p)

2. *Resolution*: It is the smallest incremental change that is possible in a potentiometer. For a wire wound potentiometer it is the per turn voltage. However for potentiometers where resistance material is spread out uniformally (single-turn, carbon filament), it is infinite.

3. *Linearity*: It is the variation of output vs input displacement from the linear (proportional) relationship.

4. *Precision*: It is an indication of linearity; say a linearity of 1%.

5. *Equivalent noise resistance*: It represents the electrical noise generated when moving the jockey of a wire wound potentiometer. It is not specified for non-wire wound potentiometer.

6. *Power handling capacity*: Power (P) in watts that the potentiometer can dissipate continuously. Thus the maximum reference voltage that can be applied across a potentiometer is given as

$$V(\text{max}) = \sqrt{PR_p}$$

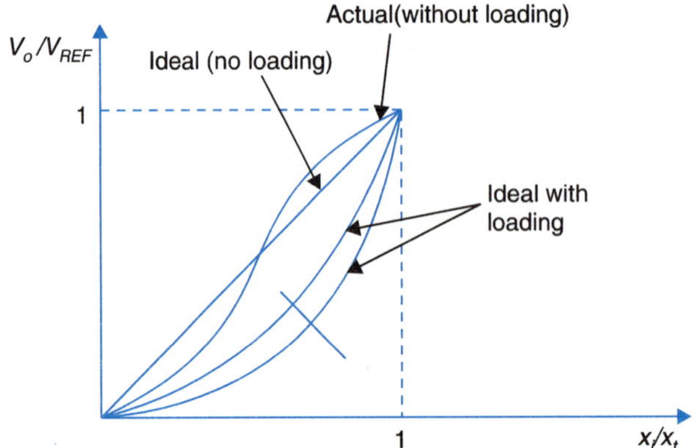

Fig. 4.12. Potentiometer characteristics.

Potentiometer Classification

1. *Wire wound potentiometers*: These are made of resistance wire wound on an insulating former. These could be:

- *Linear (translatory)*—available in shaft displacement of 7.5, 10, 15 cm.
- *Rotary*—single-turn (340°), three turn (1080°), five turn (1800°) and ten turn (3600°)

2. *Non-wire wound potentiometers*: These are divided into three groups:

- *Carbon composition*—coated film type and moulded type
- *Conductive plastic*—resistance element is in form of a cavity (coated with carbon resin mixture) in plastic base and the unit is hermatically sealed

- *Ceramic potentiometer*—resistance material comprises of a hybrid of ceramic and metal; immune to humidity.

Typical ranges and specifications of commercially available potentiometer are given in Table 4.1:

Specifications	*Potentiometer type*		
	Wire wound	*Conductive plastic*	*Ceramic*
Resistance	5 Ω to 50 k	1 k to 100 k	500 Ω to 2 M
Resolution	0.06% to 0.4%	Infinite	Infinite
Linearity	±0.25% to + 0.5%	±0.25 to + 0.5%	±0.25% to + 0.5%
Equivalent noise resistance	100 Ω	–	–
Maximum power rating	5 W	0.5 W	5 W

Potentiometer as a Feedback Element

Potentiometers are often used for position feedback in position control systems like robots. Wire wound potentiometers are not used as feedback elements because of noise and finite resolution. Instead moulded composition potentiometers are widely used for this kind of application as these have infinite resolution, low internal noise and are very robust. The block diagram of a position feedback circuit in a digital control system is given in Fig. 4.13. The signal from the buffer is fed to sample and hold circuit which holds constant value of the signal for the sampling period (this may not be needed for a slowly time varying signal). This is necessitated so that the analog to digital converter (ADC) can convert the hold signal to digital form. Both these blocks would need a finite amount of time for signal processing (unlike the analog blocks).

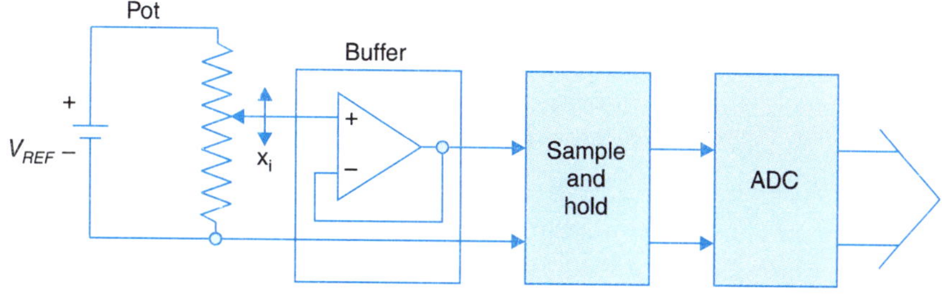

Fig. 4.13

For the system of Fig. 4.13

$$\text{Resolution (ADC)} = \frac{V_{REF}\ \text{(ADC)}}{2^N};\ N = \text{number of bits}$$

$$\text{Resolution (Pot)} = \frac{V_{REF}(\text{Pot})}{x_t} \text{ V/degree, } x_t = \text{total pot turns (in degrees)}$$

Hence,

Mechanical resolution of the system

$$= \frac{V_{REF}\ (\text{ADC})}{2^N} \frac{V_{REF}\ (\text{Pot})}{x_t} = \frac{V_{REF}\ (\text{ADC})\ x_t}{2^N\ V_{REF}\ (\text{Pot})}$$

For a single turn pot (340°), $N = 8$ bit, V_{REF} (ADC) = V_{REF} (Pot) = 5 V

$$\text{Resolution (ADC)} = \frac{5}{2^8} = 20 \text{ mV}$$

$$\text{Mechanical resolution (system)} = \frac{340°}{2^8} = 1.36°/\text{bit}$$

It may also be noted that V_{REF} (Pot) should not be more than V_{REF} (ADC), otherwise ADC would get damaged.

Optical Encoders

Optical encoders are frequently used in control systems (robots in particular) to convert linear or rotary displacement into digital code or pulse signals, Encoders are of two types:

Absolute encoders: Their output is a digitally coded signal with distinct digital code indicative of each particular least significant increment of resolution.

Incremental encoders: Their output is a pulse for each increment of resolution but these make no distinction between increments.

Because of their simple construction, low cost, ease of application and versatility incremental encoders are by far one of the most popular encoders.

Incremental Encoder

An incremental encoder typically has four parts: a light source (LED), a rotary (or translatory) disc, a stationary mask and a sensor (photodiode) as shown in Fig. 4.14. The disc has alternate opaque and transparent sectors (of equal width) which are etched by means of a photographic process on to a plastic disc (slots are cut out in case a metal disc is used). As the disc rotates during half of the increment cycle the transparent sectors of rotating and stationary discs come in alignment permitting the light from LED to reach the sensor thereby generating an electrical pulse (see Fig. 4.15). For fine resolution encoders (upto thousands of increments/revolution), multi-slit mask is often used to maximize the reception of shutter light.

The wave form of the sensor output of an encoder is generally triangular or sinusoidal depending upon the resolution required. Square wave signal compatible with digital logic are obtained from it by means of linear OPAM and comparator. Alternate transparent/opaque sectors of the disc (in developed form) and the square wave pulse train (obtained after signal processing) in synchronous with the disc are shown in Fig 4.15. The resolution of such an incremental encoder is give as:

$$\text{Basic resolution} = (360°/N)$$

where N = no. of sectors of disc; each sector is half transparent and half opaque.

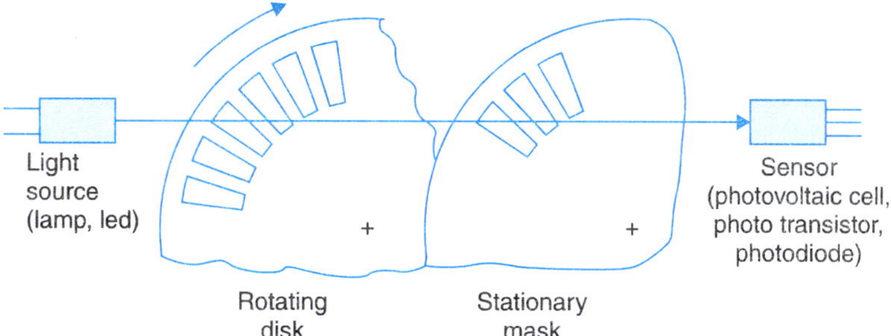

Fig. 4.14. Typical incremental optomechanics.

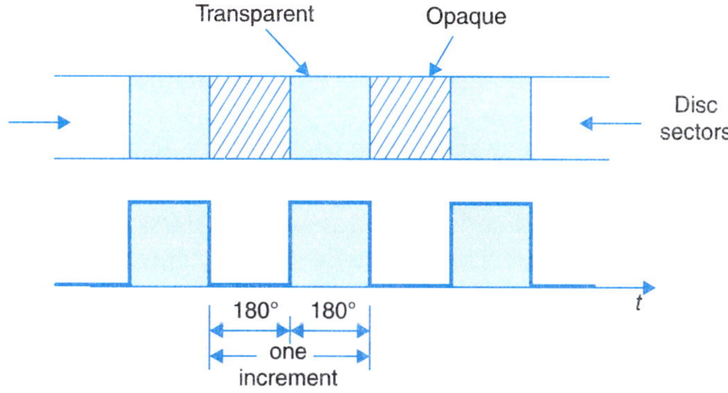

Fig. 4.15

In a dual channel encoder two optoelectronic channels are employed. These are installed on the same rotating disc and the mask but displaced at 90° to each other such that the two pulse output signals have a relative times phase displacement of 90° elect as shown in Fig. 4.16. The output of channel 2 is called the quadrature output. A circuit that senses the relative time phase of the outputs of the two channels determines the direction of rotation of the disc or the encoder shaft.

Considerable improvement in resolution of the encoder with two channels is achieved by processing these output by means of a digital circuit which generate a short duration pulse (impulse) at each change of state. Reader can easily imagine that this can be done by differentiating the signals and reversing the polarity of the negative impulses. The processed output is also shown in Fig. 4.16. It is easy to observe that there are now four impulses/disc increment so that the basic resolution becomes (360°/4N) *i.e.*, an improvement by a factor 4.

The output of the encoder is fed to a counter which counts the number of pulses; the count being a measure of angle (or translation) through which the encoder shaft has rotated. By sampling the counter at regular intervals by means of clock pulses it is possible to compute the speed of the encoder shaft.

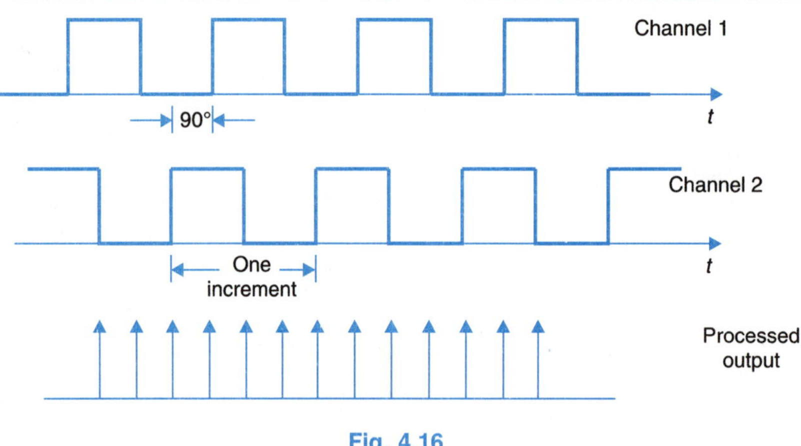

Fig. 4.16

Absolute Encoders

Absolute encoders have binary code etched on to the rotating disc which has as many track as the number bits in the code. Each particular least significant increment of resolution has a unique code; transparency regarded as 1 and opaqueness as 0. As many photodiode -LED pairs as the number of tracks are needed. These may be suitably spread round the tracks to avoid signal interference.

As absolute encoder is a digital transducer in the true sense as its output does not require any conditioning but can be directly read to get the position information.

For reasons mentioned already incremental transducer are most common in use. These are available with 1024 sectors and two channels giving a resolution is 5.27 minutes. Their chief disadvantage is that the output count is volatile when power is off. There are ways to overcome this.

Synchros

A *synchro* is an electromagnetic *transducer* commonly used a convert angular position of a shaft into an electric signal. It is commercially known as a *selsyn* or an *autosyn*.

The basic synchro unit is usually called a *synchro transmitter*. Its construction is similar to that of a three-phase alternator. The stator (stationary member) is of laminated silicon steel and is slotted to accommodate a balanced three-phase winding which is usually of concentric coil type (three identical coils are placed in the stator with their axis 120° apart) and is Y-connected. The rotor is of dumbbell construction and is wound with a concentric coil. An ac voltage is applied to the rotor winding through slip rings. The constructional features and schematic diagram of a synchro transmitter are shown in Figs. 4.17 and 4.18, respectively.

Let an ac voltage

$$v_r(t) = V_r \sin \omega_c t$$

be applied to the rotor of the *synchro* transmitter as shown in Fig. 4.18. This voltage causes a flow of magnetizing current in the rotor coil which produces a sinusoidally time varying flux directed along its axis and distributed nearly sinusoidally in the air gap along the stator periphery. Because of transformer action, voltages are induced in each of the stator coils. As

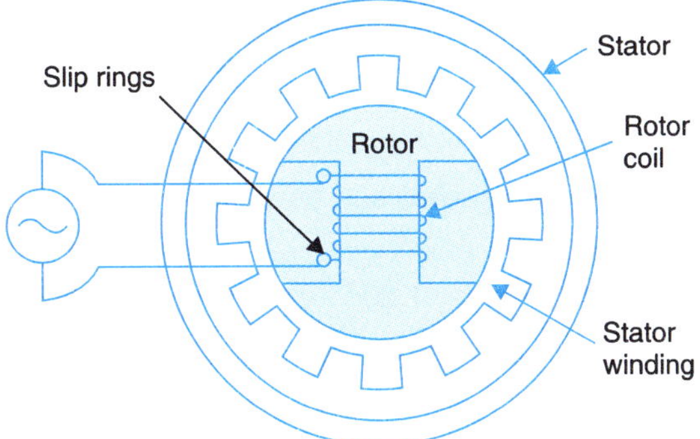

Fig. 4.17. Constructional features of synchro transmitter.

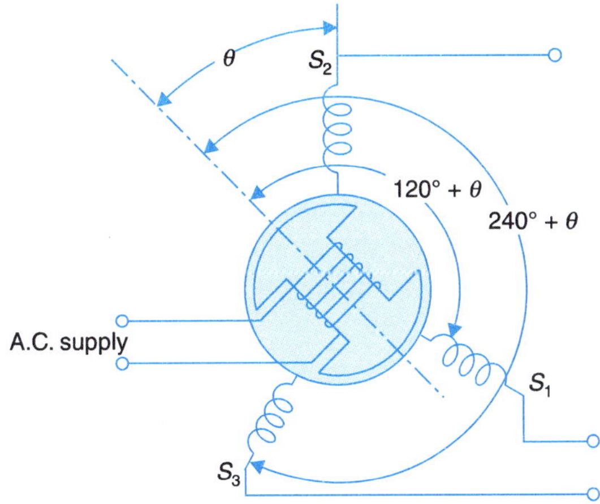

Fig. 4.18. Schematic diagram of synchro transmitter.

the air gap flux is sinusoidally distributed, the flux linking any stator coil is proportional to the cosine of the angle between the rotor and stator coil axes and so is the voltage induced in each stator coil. The stator coil voltages are of course in time phase with each other. Thus we see that the synchro transmitter acts like a single-phase transformer in which the rotor coil is the primary and the stator coils form the three secondaries.

Let v_{s_1n}, v_{s_2n} and v_{s_3n} respectively be the voltages induced in the stator coils S_1, S_2, S_3 with respect to the neutral. Then, for the rotor position of the synchro transmitter shown in Fig. 4.17, where the rotor axis makes an angle θ with the axis of the stator coils S_2

$$v_{s_1n} = KV_r \sin \omega_c t \cos (\theta + 120°) \quad ...(4.12)$$

$$v_{s_2n} = KV_r \sin \omega_c t \cos \theta \quad ...(4.13)$$

$$v_{s_3n} = KV_r \sin \omega_c t \cos (\theta + 240°) \quad ...(4.14)$$

The three terminal voltages of the stator are

$$v_{s_1s_2} = v_{s_1n} - v_{s_2n} = \sqrt{3}KV_r \sin (\theta + 240°) \sin \omega_c t \quad ...(4.15)$$

$$v_{s_2s_3} = v_{s_2n} - v_{s_3n} = \sqrt{3}\,KV_r \sin(\theta + 120°) \sin \omega_c t \qquad ...(4.16)$$

$$v_{s_3s_1} = v_{s_3n} - v_{s_1n} = \sqrt{3}\,KV_r \sin\theta \sin \omega_c t \qquad ...(4.17)$$

when $\theta = 0$, from eqns. (4.12)—(4.14) it is seen that maximum voltage is induced in the stator coil S_2, while it follows from eqn. (4.17) that the terminal voltage $v_{s_3s_1}$ is zero. This position of the rotor is defined as the *electrical zero* of the transmitter and is used as reference for specifying the angular position of the rotor (see Fig. 4.20).

Thus it is seen that the input to the synchro transmitter is the angular position of its rotor shaft and the output is a set of three single-phase voltages given by eqns. (4.15)—(4.17); the magnitudes of these voltages are functions of the shaft position.

The output of the synchro transmitter is applied to the stator windings of a *synchro control transformer*. The control transformer is similar in construction to a synchro transmitter except for the fact that the rotor of the control transformer is made cylindrical in shape so that the air gap is practically uniform. The system (transmitter-control transformer pair) acts as an error detector. Circulating currents of the same phase but of different magnitudes flow through the two sets of stator coils. The result is the establishment of an identical flux pattern in the air gap of the control transformer as the voltage drops in resistances and leakage reactances of the two sets of stator coils are usually small. The control transformer flux axis thus being in the same position as that of the synchro transmitter rotor, the voltage induced in the control transformer rotor is proportional to the cosine of the angle between the two rotors and is given by

$$e(t) = K'V_r \cos\phi \sin \omega_c t \qquad ...(4.18)$$

where ϕ is the angular displacement between the two rotors. When $\phi = 90°$, i.e., the two rotors are at right angles, then the voltage induced in the control transformer rotor is zero. This position is known as the *electrical zero* position of the control transformer. In Fig. 4.19, the transmitter and control transformer rotors are shown in their respective electrical zero positions.

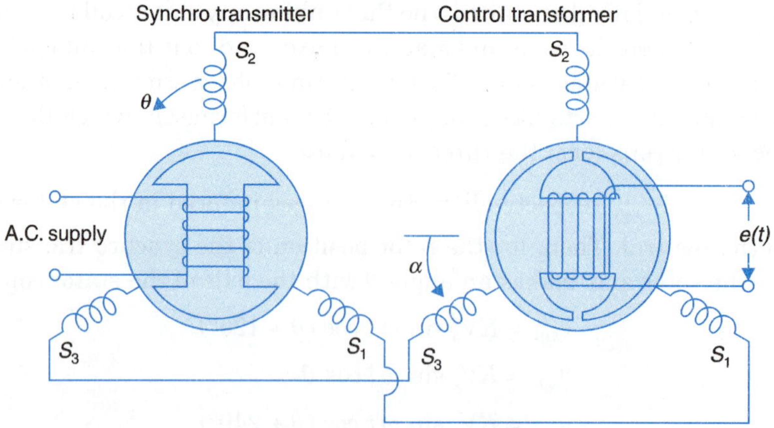

Fig. 4.19. Synchro error detector.

Let the rotor of the transmitter rotate through an angle θ in the direction indicated and let the control transformer rotor rotate in the same direction through an angle α resulting in a net angular separation of $\phi = (90° - \theta + \alpha)$ between the two rotors. From eqn. (4.18), the voltage at the rotor terminals of the control transformer is then

$$e(t) = KV_r \sin(\theta - \alpha) \sin \omega_c t \quad ...(4.19)$$

For small angular displacement between the two rotor positions,

$$e(t) = KV_r(\theta - \alpha) \sin \omega_c t \quad ...(4.20)$$

The synchro transmitter-control transformer pair thus acts as an error detector giving a voltage signal at the rotor terminals of the control transformer proportional to the angular difference between the transmitter and control transformer shaft positions. Equation (4.20), though derived for constant $(\theta - \alpha)$, is valid for varying conditions as well, so long as the rate of angle change is small enough for the speed voltages induced in the device to be negligible.

Equation (4.20) is represented graphically in Fig. (4.20) for an arbitrary time variation of $(\theta - \alpha)$.

We see from this diagram that the output of synchro error detector is a modulated signal, the modulating wave has the information regarding the lack of correspondence between the two rotor positions and the carrier wave is the ac input to the rotor of the synchro transmitter. This type of modulation is known as *suppressed-carrier modulation*. From eqn. (4.20) the modulating signal representing the discrepancy between the two shaft positions is

$$e_m(t) = K_s(\theta - \alpha) \quad ...(4.21)$$

where K_s is known as *sensitivity* of error detector and has the units of volts (rms)/rad angular difference of the shafts of the synchro pair.

As pointed out earlier the rotor of the control transformer is made cylindrical in shape so that the air gap is practically uniform. This is essential for a control transformer, since its rotor terminals are usually connected to an amplifier, therefore the change in the rotor output impedance with rotation of the shaft must be minimized. Another distinguishing feature is that the stator winding of the control transformer has a higher impedance per phase. This feature permits several control transformers to be fed from a single transmitter.

A.C. Position Control System

Consider the system shown in Fig. 4.21 in which the position of the mechanical load is controlled in accordance with the position of the reference shaft. This system employs ac components and all the signals other than the input and output shaft positions are suppressed-carrier modulated signals. Such systems are known as *carrier control systems* and are designed so that the signal cut-off frequency is much less than the carrier frequency. It is then sufficiently accurate to analyze these systems on the basis of modulating signals only.

The components used in this system are: synchro transmitter-control transformer pair as error detector; ac amplifier for signals amplification; ac servomotor to drive the load shaft through a gearing; ac tachometer for providing rate feedback. The servomotor for proper operation has to be provided with carrier voltages on the two phases 90° out of phase with respect to each other. This is achieved easily by exciting the reference motor phase and the synchro transmitter rotor coil directly from the carrier supply, while the carrier voltage driving the control phase of the motor is obtained by amplifying the error signal. In the process of this amplification, the carrier phase is shifted through 90° by use of two RC networks.*

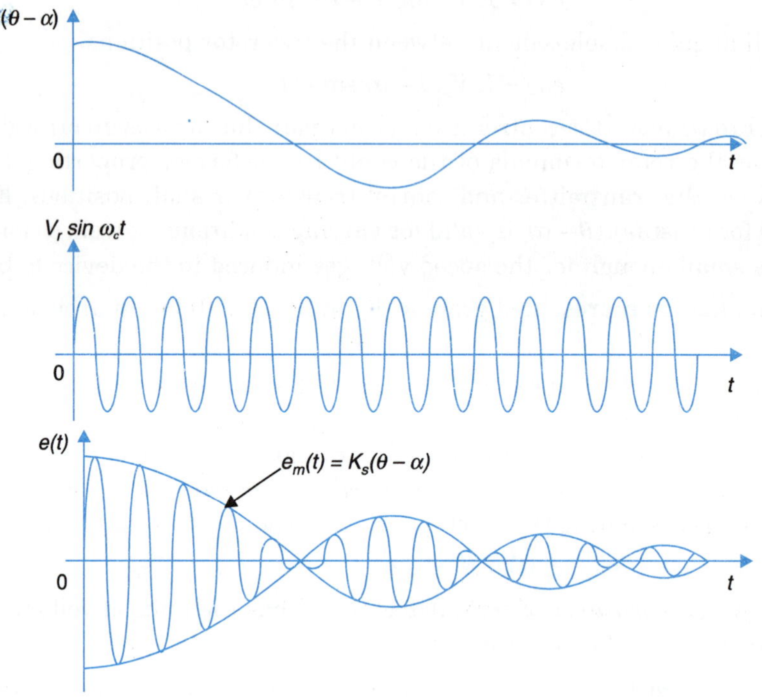

Fig. 4.20. Typical wave forms of synchro errors detector.

Linearized transfer functions derived earlier in this chapter are valid for the various components of this system as the signals are always small, it being a position tracking system. The operating point of the ac servo motor is ($v_C = 0$, $\dot{\theta}_M = 0$) as the motor is stationary under steady conditions. Referring to the component transfer functions, the signal flow graph of Fig. 4.22 can be at once drawn. The overall transfer function of the system obtained therefrom is

$$\frac{\theta_C(s)}{\theta_R(s)} = \frac{K_m K_A K_s n}{\tau_m s^2 + (1 + K_m K_A K_t)s + K_m K_A K_s n} \quad \text{...(4.22)}$$

* The parameters of the two *L*-type *RC* networks given below are selected to give a phase shift of 90° at the carrier frequency. Loading effects are reduced by proper choice of impedance level of the networks.

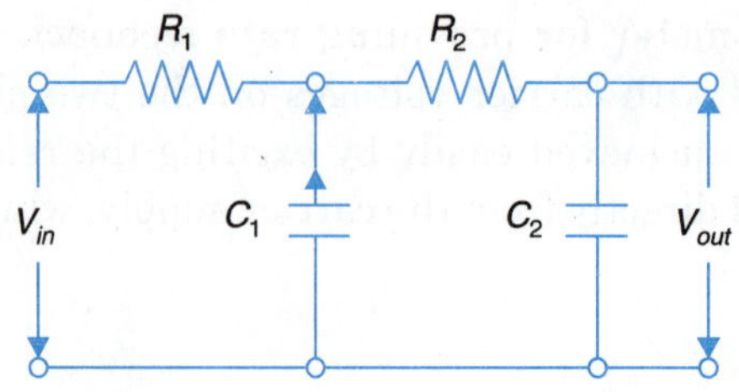

where K_m = motor gain constant in rad/volt; τ_m = motor time constant in sec; K_s = synchro sensitivity in volts/rad; K_t = tachometer constant in volts/rad/sec; K_A = amplifier gain in volts/volt; and n (gear ratio) = $\theta_C/\theta_M = \theta_L/\theta_M$.

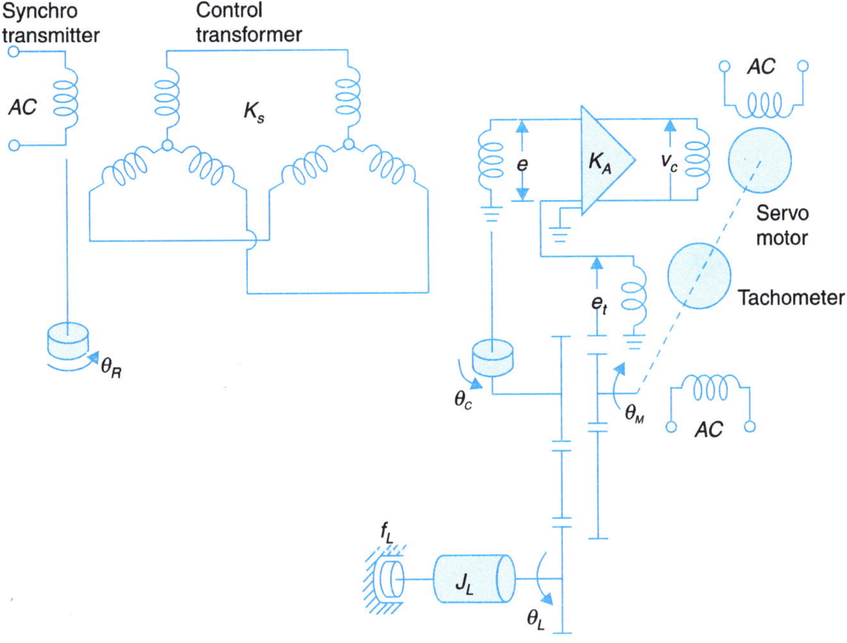

Fig. 4.21. A.C. position control system.

The tachometer feedback is of course employed negatively and as will be seen in Chapter 5, it improves the damping factor for the closed-loop system.

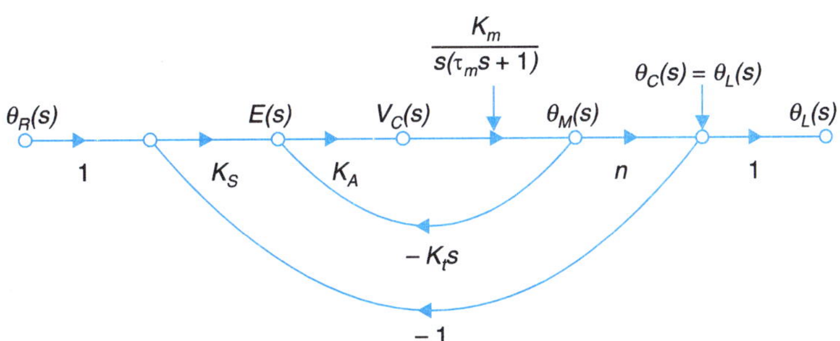

Fig. 4.22. Signal flow graph of the system shown in Fig. 4.21.

The AC/DC Control Systems

The control systems discussed in this text so far, can be broadly classified into dc control systems and ac or carrier control systems based on the kind of components employed. A dc component is characterized by a dc (constant) signal under steady state conditions with zero signal frequency (s = 0). On the other hand, steady conditions in ac components with zero modulating frequency mean the presence of sinusoidal signal of carrier frequency and constant amplitude. The speed control system of Fig. 2.39 is typically a dc control system employing all dc components wherein a varying signal carries the information instantaneously. The position control system of Fig. 4.21. may be cited as an example of an ac system where in ac components have been employed and the signals are in the form of suppressed carrier modulation and the signal information is not the instantaneous value of the signal, but rather the varying rms (or maximum) value of the signal. AC components are not all of the type which accept ac input signal and produce an ac output signal. Such a component is of course an ac amplifier. However, ac components like *synchros*, accept a dc input signal (shaft angular position), are provided with a carrier supply and produce a carrier modulated signal; while components like ac servomotor accept carrier modulated input signal at the control phase, are provided with carrier supply at the reference phase and produce a demodulated (dc) output signal in the form of an angular position or velocity.

The choice between dc and ac systems is chiefly based on economic considerations, ease of operations and maintenance and consideration of size and weight particularly for air-borne systems. One of the advantages of ac systems is the use of an ac amplifier which is inherently stable unlike the dc amplifier which suffers from the possibility of drift. Another major advantage of ac systems is the smaller frame size for ac components particularly for high carrier frequency. Even though network compensation is not very convenient, tachometer feedback compensation can be very effective in ac systems. On account of weight and size considerations, ac systems are preferred for aircraft and missile systems and also for instrumentation and certain low power ground applications.

With advances in integrated circuit OPAMPs (operational amplifiers), high-torque PMDCs employing rare-earth permanent magnets and SCR circuitry for feeding power to dc armature, dc servomotor have almost superseded ac servomotors in most applications. Further improvements in commutation technology has reduced commutator brush friction and wear. Higher sustained torque for a given frame size and elimination of commutator brush gear have become possible with brushless dc motors, elaborated earlier in this section. In dc servo systems OPAMPs offer the advantage of highly stable dc coupled amplification in feedback mode and ease in implementation of analog filtering in forward path of a control system.

For large power applications like heavy drives, etc., dc systems are an ideal choice on account of the availability of rugged high power amplification rotating and static amplifiers. Furthermore, dc network compensation is easily carried out and is very effective.

Control systems need not necessarily be restricted to use all dc or all ac components. Hybrid systems can be put together which combine the advantages of both type of components. The dc and ac components in a system have to be coupled together through modulators or demodulators so that they receive and put out the kind of signal desired.

4.4 STEPPER MOTORS

The stepper motor is special type of synchronous motor which is designed to rotate through a specific angle (called a step) for each electrical pulse received from its control unit. Typical step sizes are 7.5°, 15° or larger. The stepper motor is used in digitally controlled position control systems in open-loop mode. The input command is in form of a train of pulses to turn a shaft through specified angle.

There are two advantages in using stepper motors. One is their compatibility with digital systems and secondly no sensors are needed for position and speed sensing as these are directly obtained by counting input pulses and periodic counting if speed information is needed. Stepper motors have a wide range of applications: paper feed motors in typewriters and printer, positioning of print heads, pens in XY-plotters, recording heads in computer disk drives and in positioning of work tables and tools in numerically controlled machining equipment. The range of applications of these motors is increasing as these motors are becoming available in larger power ratings and with reducing cost.

Elementary operation of a four phase stepper motor with a two-pole PM rotor can be illustrated through the diagram of Fig. 4.23. Let us assume that the rotor is permanent magnet excited. Such a rotor aligns with the axis of the stator field with torque being proportion to the sin θ, θ being the angle of displacement between the rotor axis and stator field axis. The torque-angle characteristics is drawn in Fig. 4.24 (*a*) which phase '*a*' excited and also with phase *b* excited. The rotor thus assumes positions $\theta = 0°, 45°, 90°$ as phases $a, a + b, b, ...$ are excited. It is easily observed that the stable position of the rotor corresponds to that angle at which the torque is zero and is positive for smaller angles and negative for larger angles. Thus with phase '*a*' excited the stable (or locked) position is $\theta = 0°$ but not $\theta = 180°$ (unstable) and the torque has a maximum value at $\theta = 90°$. It is therefore easily concluded that each excitation pattern of phases corresponds to a unique position of the rotor. Patterns for phase windings excitations can be easily visualized for steps of 22.5°, 11.25°. Another feature of a PM stepper motor is that when excited it seeks a preferred position offers advantage in certain application.

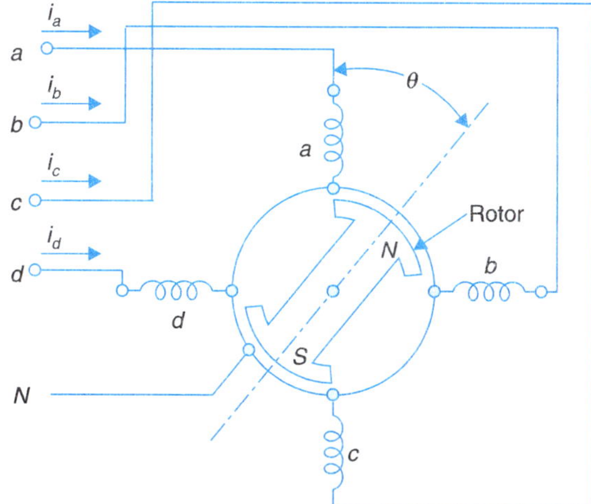

Fig. 4.23. A 4-phase 2-pole stepper motor-elementary diagram.

Consider now that the rotor (projecting pole) is made of just ferromagnetic material (no permanent magnets). The device now behaves as a variable-reluctance motor. The ferromagnetic rotor seeks the position in which it presents minimum reluctance to the stator field *i.e.*, the rotor axis aligns itself to the stator field axis. In Fig. 4.24 with phase '*a*' excited this happens at $\theta = 0°$ as well as $\theta = 180°$ in which positions the torque on the rotor is zero. At $\theta = 90°$ the rotor presents infinite reluctance to the phase '*a*' axis and so the torque has also a zero there. Thus the rotor torque is a function of the sin 2θ as drawn in Fig. 4.24 (*b*). Seen from stator there are thus two possible excitation for a given rotor position. For example, for rotor $\theta = 0°$ locking takes place when either '*a*' or '*b*' phase is excited. This fact is in contrast to the PM stepper motor and has to be kept in mind in using this type of motor.

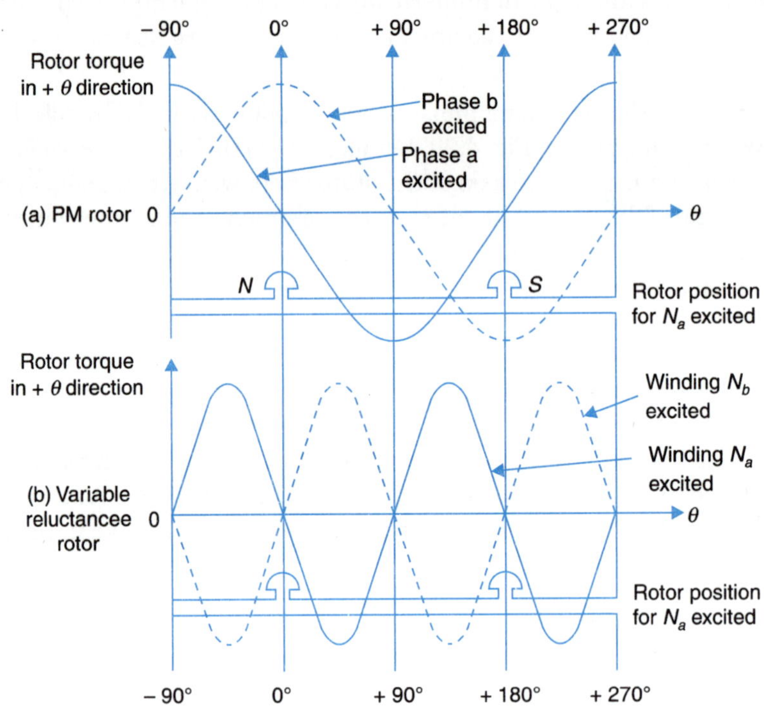

Fig. 4.24. Torque-angle characteristics of stepper motor (*a*) PM motor (*b*) variable-reluctance motor.

Having illustrated the operation of an elementary stepper motor and its two types, we shall now consider some further details.

Variable-reluctance Stepper Motor

A variable-reluctance stepper motor consists of a single or several stacks of stators and rotors—stators have a common frame and rotors have a common shaft as shown in the longitudinal cross-sectional view of Fig. 4.25 for a 3-stack motor. Both stators and rotors have toothed structure as shown in the end view of Fig. 4.26. The stator and rotor teeth are of same size and therefore can be aligned as shown in this figure. The stators are pulse excited, while the rotors are unexcited.

Consider a particular stator and rotor set shown in the developed diagram of Fig. 4.27. As the stator is excited, the rotor is pulled into the nearest minimum reluctance position–the position where stator and rotor teeth are aligned. The stator torque acting on the rotor is a function of the angular misalignment θ. There are two positions of zero torque: $\theta = 0$, rotor and stator teeth aligned and $\theta = 360°/(2 \times T) = 180°/T$ (T = number of rotor teeth), rotor teeth

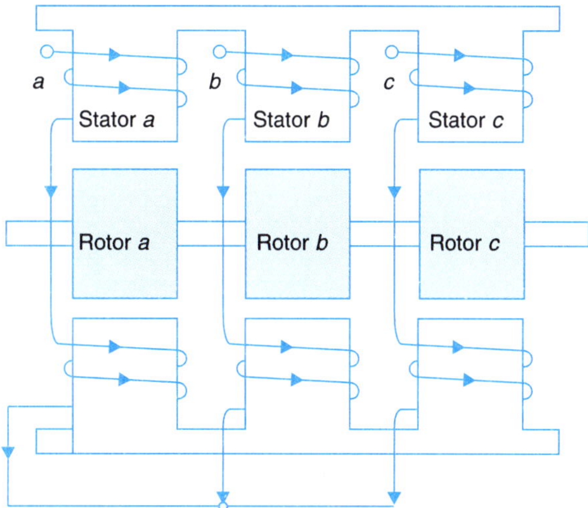

Fig. 4.25. Longitudinal cross-sectional view of 3-stack variable reluctance stepper motor.

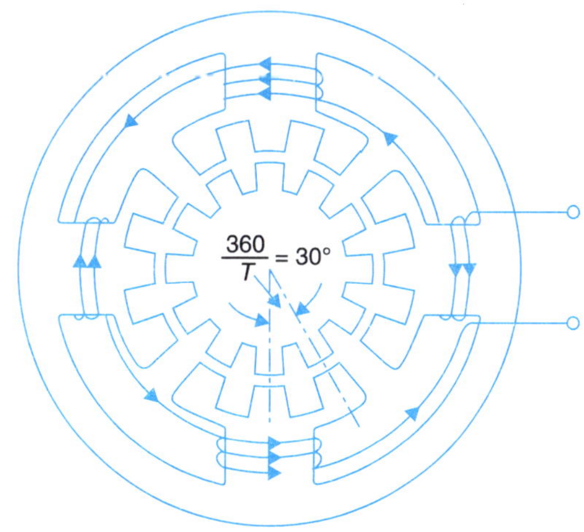

Fig. 4.26. End view of stator and rotor (12-teeth) of a multistack variable reluctance stepper motor.

aligned with stator slots. The shape of the static torque-angle curve of one stack of a stepper motor is shown in Fig. 4.28. It is nearly sinusoidal. Teeth aligned position ($\theta = 0$) is a stable position, *i.e.*, slight disturbance from this position in either direction brings the rotor back to it. Tooth-slot aligned position ($\theta = 180°/T$) is unstable, *i.e.*, slight disturbance from this position in either direction makes the rotor move away from it. The rotor thus locks into stator in position $\theta = 0$ (or multiple of $360°/T$). The dynamic torque-speed characteristic will differ from this due to speed emf induced in stator coils.

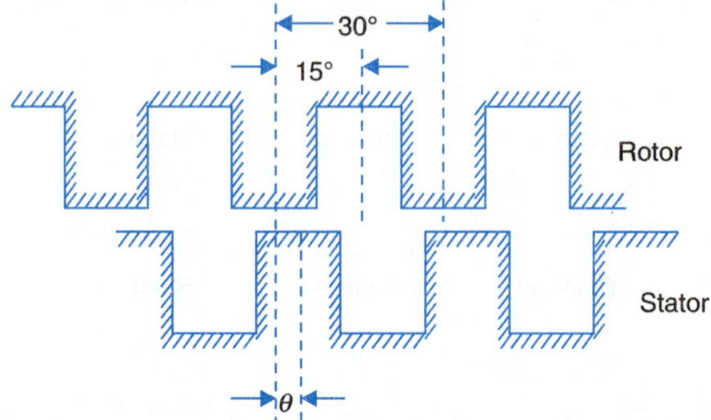

Fig. 4.27. Developed view of teeth of a pair of stator-rotor.

While the teeth on all the rotors are perfectly aligned, stator teeth of various stacks differ by an angular displacement of

$$\alpha = \frac{360°}{nT} \qquad ...(4.23)$$

where n = number of stacks.

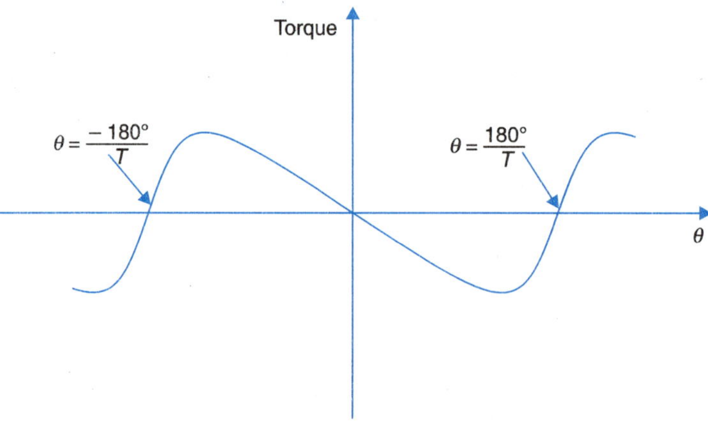

Fig. 4.28. Static torque-angle curve of a stepper motor stack.

Fig. 4.29 shows the developed diagram of a 3-stack stepper motor with rotor in such a position that stack c rotor teeth are aligned with its stator. Here

$$\alpha = \frac{360°}{3 \times 12} = 10° \text{ (number of rotor teeth = 12)}$$

In a multiple stack rotor, number of phases equals number of stacks. If phase a stator is pulse excited, the rotor will move by 10° in the direction indicated. If instead phase b is excited, the rotor will move by 10° opposite to the direction indicated. Pulse train with sequence *abcab* will make the rotor go through incremental motion in indicated direction, while sequence *bacba* will make it move in opposite direction. Directional control is only possible with three or more phases.

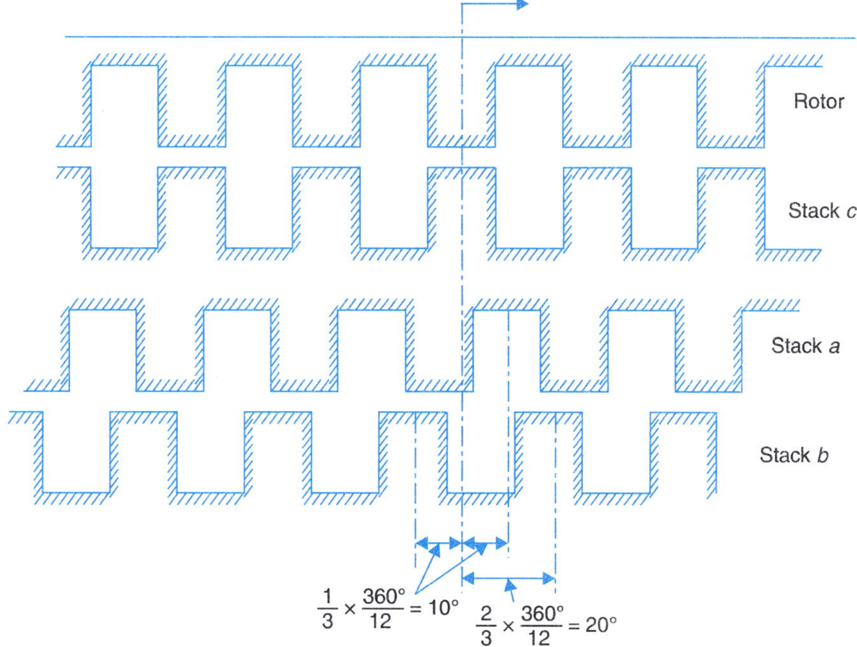

Fig. 4.29. Developed view of rotor and stator stacks of a 3-stack stepper motor.

Permanent-magnet Motor

Figure 4.30 shows the phases (stacks) of a 2 phase, 4-pole permanent-magnet stepper motor. The rotor is made of ferrite or rare earth material which is permanently magnetized. The stator

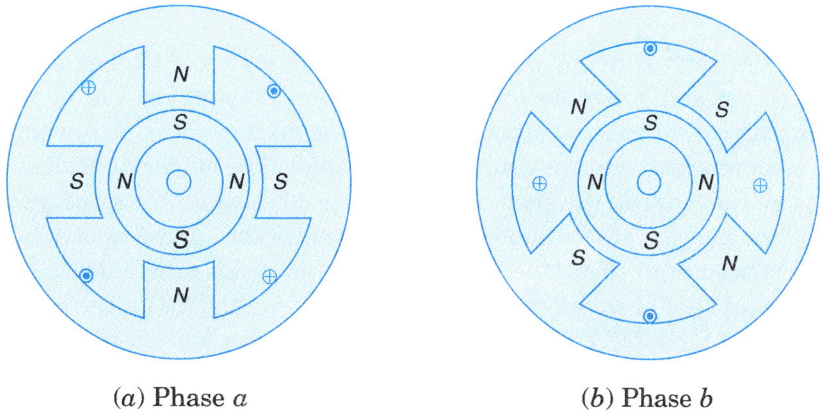

(*a*) Phase *a* (*b*) Phase *b*

Fig. 4.30. Permanent-magnet stepper motor.

stack of phase *b* is staggered from that of phase *a* by an angle of 90°. When the phase *a* is excited, the rotor is aligned as shown in Fig. 4.30 (*a*). If now the phase *b* is also excited, the effective stator poles shift counter-clockwise by $22\frac{1}{2}^\circ$ [Fig. 4.30 (*b*)] causing the rotor to move accordingly. If the phase *a* is now de-energized (with *b* still energized), the rotor will move another step of $22\frac{1}{2}^{\circ}$. The reversal of phase a winding current will produce a farther forward movement of $22\frac{1}{2}^{\circ}$, and so on. It is easy to visualize as to how the direction of movement can be reversed.

To simplify the switching arrangement, which is accomplished electronically double coils are provided for each phase. The schematic diagram of the switching circuit is shown in Fig. 4.31.

Compared to variable-reluctance motors, typical permanent-magnet stepper motors operate at larger steps up to 90°, and at maximum response rates of 300 pps.

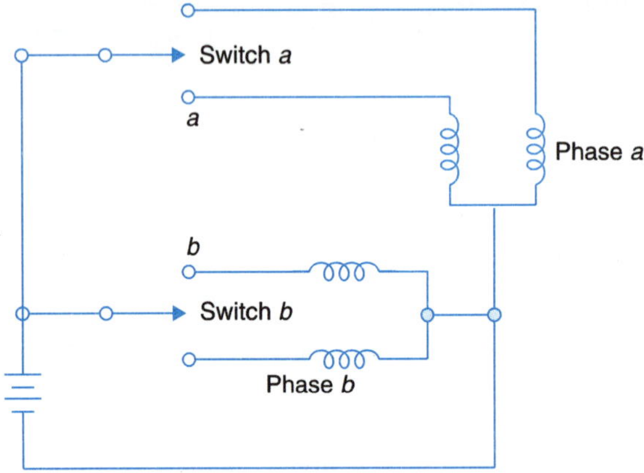

Fig. 4.31. Switching arrangement 2-phases permanent-magnet stepper motor.

Hybrid Stepper Motor

This is in fact a PM stepper motor with constructional features of toothed and stacked rotor adopted from the variable reluctance motor. The stator has only one set of winding-excited poles which interact with two rotor stacks. The permanent magnet is placed axially along the rotor in form of an annular cylinder over the motor shaft. The stacks at each end of the rotor are toothed. So all the teeth on the stack at one of the rotor acquire the same polarity while all the teeth of the stack at the others end of the rotor require the opposite polarity. The two sets of teeths are displaced from each other by one half of the tooth pitch (also called pole pitch). These constructional details are brought out by Figs. 4.32 (a) and (b) for the case of three teeth on each stack so that tooth pitch γ_t = 360°/3 = 120°. This motor has a 2-phase, 4-pole stator.

Consider now that the stator phase 'a' is excited such that the top stator pole acquires north polarity while the bottom stator pole acquires south polarity. As a result the nearest tooth of the front stack (assumed to be of north polarity) is pulled into locking position with the stator south pole (top) and the diametrically opposite tooth of the rear stack (south polarity) is simultaneously locked into the stator north pole (bottom). The repulsive forces on the remaining two front stack teeth balance out as these are symmetrically located w.r.t. the bottom stator pole and so do the repulsive forces due to the top stator pole on the remaining two rear stack teeth. This rotor position is thus a stable position with net torque on rotor being zero.

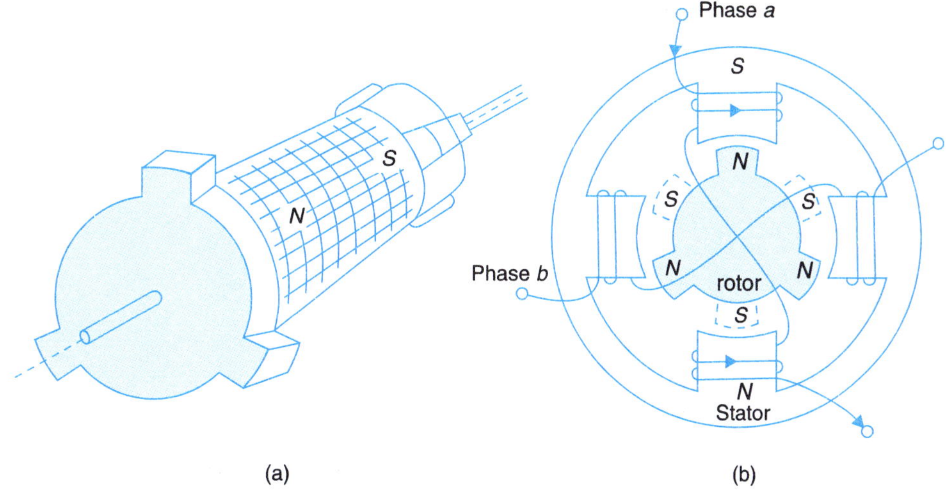

Fig. 4.32. Schematic views of hybrid stepper motor.

If the excitation is shifted from phase a to phase b such the right-hand pole becomes south and left hand pole becomes north, this would cause the rotor to turn anti-clockwise by $\gamma/4 = 30°$ into the new locking position. The rotor will turn clockwise by 30° if the phase b were oppositely excited.

If the stator excitation were to be removed, the rotor will continue to remain locked into the same position as it is prevented to move in either direction by torque caused by the permanent magnet excitation. This fact is in favour of hybrid (also PM) motors. Compared to PM motor finers steps for better resolution are easily obtained in hybrid motor by increasing the number of stack teeth and also be adding additional stack pairs on the rotor for example for a seven teeth stack the steps size is $(360°/7)/4 = 12.85°$. Also compared to variable-reluctance motor a hybrid motor requires less excitation because of the PM excited rotor.

Half stepping can be achieved in a hybrid motor by exciting phase a and then exciting phase b before switching off the excitation of phase a and so on. In fact any fractional step can be obtained by suitably proportioning the excitation of the two phases. Such stepping is known as microstepping.

As the stepping rate is increased, the rotor has less time to drive the load from one position to the next as the stator winding current pattern is shifted. Beyond a certain pulsing rate [(pulse/s) = pps)] the rotor cannot follow the command and would begin to miss pulses. The start range as shown in Fig. 4.33 is that in which the load position follows the pulses without losing steps. The slew-range is one in which the load velocity follows the pulse rate without losing a step, but cannot start, stop or reverse on command. Obviously the maximum slew rate would increase as the load is lightened. Typical variable-reluctance motors can operate at maximum position response rates upto 1200 pps.

Step skipping may also occur if motor oscillation amplitude about the locking position is too large.

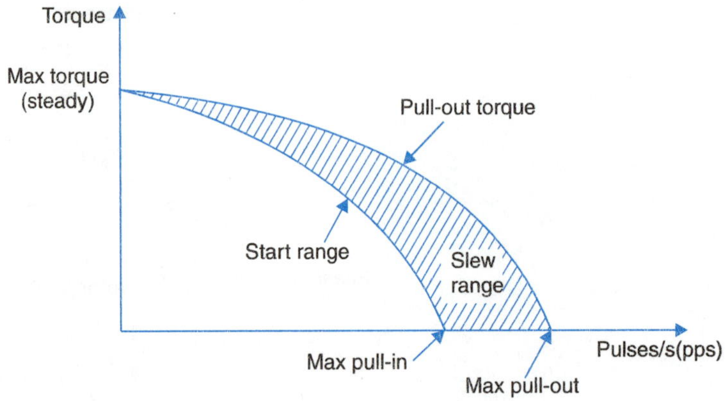

Fig. 4.33. Torque versus pulse/s for a stepper motor.

Use of Stepper Motors in Control Systems

There are two distinctly different ways of using stepper motors in control systems. One is the open-loop mode and other is the closed-loop mode.

As has been discussed earlier, the stepper motor is a digital device whose output in shaft angular displacement is completely determined by the number of input pulses. Consequently, there is no need for a feedback device to determine the position of motor shaft and, therefore, of the load connected to the motor shaft. This means that an open-loop step servo system can be designed to yield the same accuracy as a closed-loop analog system. This leads to simplified system design. Figure 4.34 illustrates the use of stepper motor in open-loop mode.

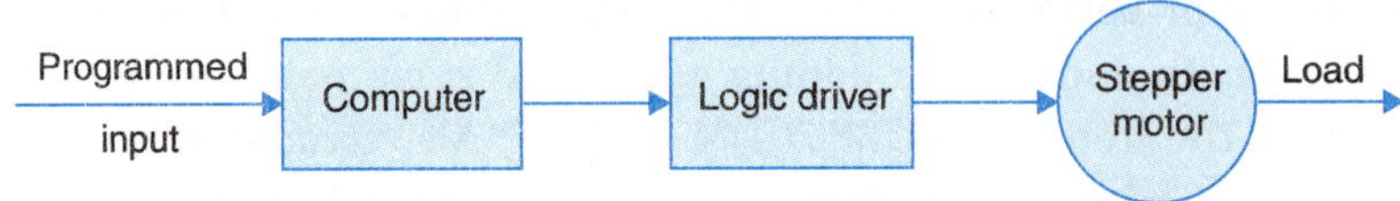

Fig. 4.34. Use of stepper motor in open-loop mode,

In the closed-loop (positional feedback) mode, the motor is used like conventional servo motor. A signal from the output is fed back and is used to operate a gate controlling the pulses from a pulse generator (Fig. 4.35).

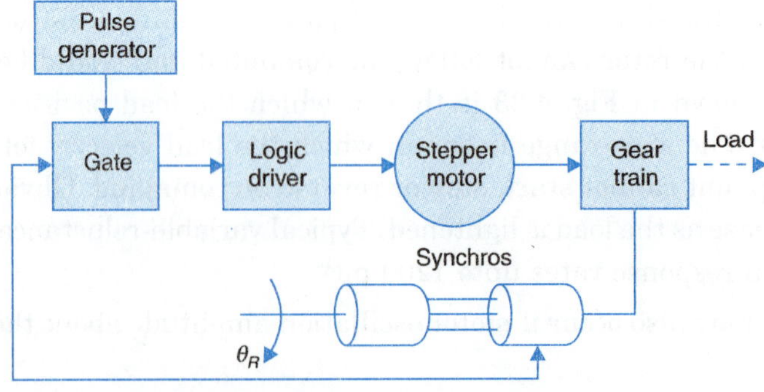

Fig. 4.35. Use of stepper motor in closed-loop mode.

Since stepper motors are essentially digital actuators, there is no need to use digital-to-analog and analog-to-digital converters in constructing digital control systems (refer Chapter 11).

4.5 HYDRAULIC SYSTEMS

Hydraulically operated components are frequently used in hydraulic feedback systems and in combined electromechanical-hydraulic systems. In these elements, power is transmitted through the action of fluid flow under pressure. The fluid used is relatively incompressible such as petroleum-base oils or certain non-inflammable synthetic fluids.

The main advantage of the hydraulic systems lies in the hydraulic motor which can be made much smaller in physical size than an electric motor for the same power output. In addition to this, hydraulic components are rapid acting and more rugged compared to the corresponding electrical components. On the other hand, hydraulic systems have the inherent problems of leak and of sealing them against foreign particles, operating noise and the tendency to become sluggish at low temperatures because of increase viscosity of the fluid. Furthermore, hydraulic lines are not as flexible as electric cables.

Common hydraulic control applications are power steering and brakes in automobiles, the steering mechanisms of large ships, the control of large machine tools, etc.

The hydraulic output devices used in control systems are generally of two types—those intended to produce rotatory motion and those whose output is translational. The first type of devices are known as hydraulic motors and the second ones are known as hydraulic linear actuators.

Some of the important hydraulic components are discussed in this section.

Hydraulic Pumps and Motors

A device frequently used in control systems giving large output torque and short response time is the *hydraulic transmission* shown in Fig. 4.36. It consists of a variable stroke hydraulic

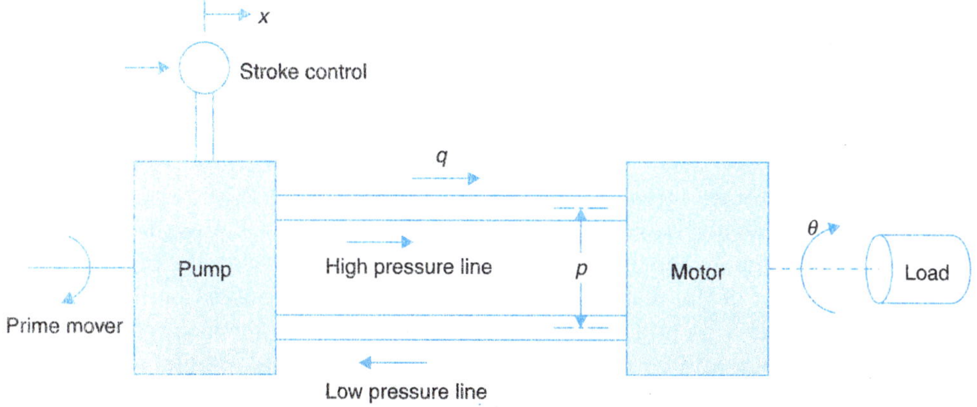

Fig. 4.36. Hydraulic transmission.

pump and a fixed stroke hydraulic motor. Control of the motor is exercised by varying the amount of oil delivered by the pump. This is carried out by mechanically changing the pump stroke. Like in a dc generator and motor, there is no essential difference between hydraulic pump and motor. In a pump, the input is mechanical power (torque at a certain speed) and output, hydraulic power (flow at a certain pressure) and in a motor, the input is hydraulic and output mechanical.

Figure 4.37 shows the constructional features of hydraulic pump and motor. The pistons in hydraulic pump bear against the stationary wobble plate and are carried round by the cylinder

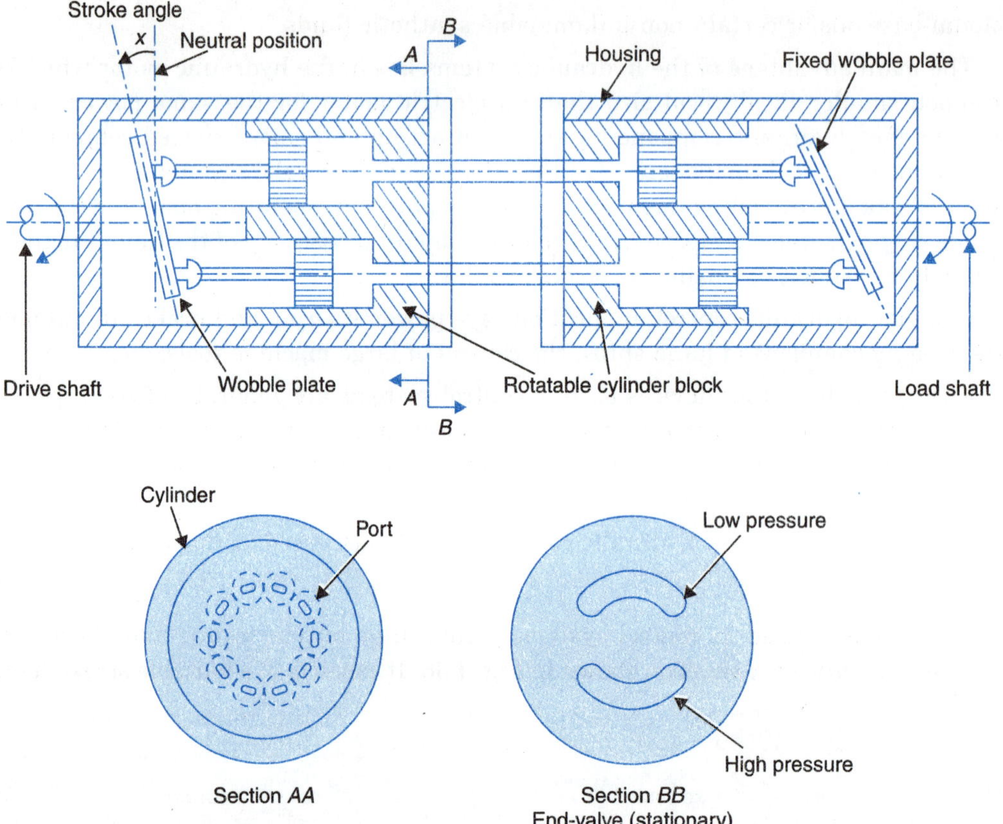

Fig. 4.37. Constructional details of hydraulic transmission system.

block which is made to rotate by the prime mover. (In some designs, the cylinder block is kept stationary and the wobble plate rotated). First consider that the wobble plate of the pump is in the neutral position. Now as the shaft is driven, the cylinder block with pistons rotates, but there is no displacement of the pistons into the cylinders and hence no pumping action takes place. Let the wobble plate be now tilted (Fig. 4.37). As the cylinder block is now rotated, the pistons rotate along with and are guided along the ridge of the wobble plate. Each piston therefore moves outwards in its cylinder as it comes in the top position and moves inwards in its cylinder as it comes in the bottom position. The result is a reciprocating pumping action with the high pressure oil being collected by the end-valve plate (fixed member) from cylinders in bottom position and delivered to the high pressure line. The oil is drawn in from the low pressure line through valve plate into the cylinders in top position. The oil flow rate can be varied by adjusting the stroke angle of the wobble plate.

From the pump, the oil is forced under pressure into the high pressure inlet of the end-valve plate of the motor. It exerts a force against the position that may be resolved into a component normal to the fixed wobble plate and other in the plane of the wobble plate. The normal force is opposed by bearings but the other force causes a torque around the output shaft of the hydraulic motor; thus the output shaft velocity is approximately proportional to flow of oil and hence stroke angle of the wobble plate of the pump. The pump constantly pushes oil into the high pressure line while it constantly receives oil from the low pressure line. Thus oil is merely circulated from pump to motor and back again.

The direction of rotation of the output shaft can be reversed by reversing the direction of the stroke angle of the wobble plate of the pump. The pump pistons having forward stroke in the pervious case, now have backward stroke and vice-versa. The high pressure line therefore becomes the low pressure line and vice-versa. Consequently the direction of the motor reverses.

Let us derive the transfer function of the hydraulic pump-motor system described above. The rate q_p at which the oil flows from the pump is accounted for as follows:

$$q_p = q_m + q_i + q_c \qquad \text{...(4.24)}$$

where q_m = oil flow rate through the motor; q_i = leakage flow rate; and q_c = compressibility flow rate.

The ideal rate q_p at which the oil flows from the pump is given by

$$q_p = K_p x \qquad \text{...(4.25)}$$

where K_p = a constant, the rate of oil flow per unit stroke angle; x = the stroke angle.

The rate of oil flow through the motor is proportional to motor speed, *i.e.*,

$$q_m = K_m \frac{d\theta}{dt} \qquad \text{...(4.26)}$$

where θ = angle through which motor turns; and K_m = motor displacement constant.

All the oil from the pump does not flow through the motor in the proper channels. Due to back pressure in the motor, which is built up by the hydraulic flow to overcome the resistance to movement offered by load on motor shaft, a portion of the ideal flow from the pump leaks back past the pistons of motor and pump. It is usually assumed that the leakage flow is proportional to motor pressure, *i.e.*,

$$q_l = K_l p \qquad \text{...(4.27)}$$

where K_l = constant; p = pressure drop across the motor = pressure difference between high and low pressure lines.

The back pressure built up by the motor not only causes leakage flow in the motor and pump as well, but also causes the oil in the lines to compress. Compressibility of the oil may usually be considered negligible except when the oil is at very high pressure and when there is a large quantity of oil in the system.

Volume compressibility flow is essentially proportional to pressure and therefore the rate of compressibility flow is proportional to the rate of change of pressure, *i.e.*,

$$q_c = K_c \frac{dp}{dt} \qquad \text{...(4.28)}$$

where K_c = coefficient of compressibility.

Substituting the values of various flow rates in eqn. (4.24), the flow rate equation can be written as

$$K_p x = K_m \frac{d\theta}{dt} + K_l p = K_e \frac{dp}{dt} \qquad \text{...(4.29)}$$

The torque T_M developed by the motor is proportional to the pressure drop and balances the load torque. If the load is assumed to consist of moment of inertia J and viscous friction with coefficient f, the torque equation is

$$T_M = K_T p = J \frac{d^2\theta}{dt^2} + f \frac{d\theta}{dt} \qquad \text{...(4.30)}$$

where K_T = motor torque constant.

If the motor losses are included as part of load, the hydraulic input power ($q_m p$) must equal the mechanical output power $[T_M(d\theta/dt)]$, *i.e.*,

$$q_m p = T_M \frac{d\theta}{dt} \qquad \text{...(4.31)}$$

or
$$K_m \frac{d\theta}{dt} p = K_T p \frac{d\theta}{dt}$$

or
$$K_m = K_T \text{ (in MKS units)} \qquad \text{...(4.32)}$$

Equation (4.30), now becomes

$$K_m p = J\ddot{\theta} + f\ddot{\theta} \qquad \text{...(4.33)}$$

From eqns. (4.29) and (4.33), we get

$$K_p x = \frac{K_c J}{K_m} \dddot{\theta} + \left(\frac{K_c f}{K_m} + \frac{K_l J}{K_m}\right)\ddot{\theta} + \left(\frac{K_l f}{K_m} + K_m\right)\dot{\theta} \qquad \text{...(4.34)}$$

Taking the Laplace transform of eqn. (4.34) we get

$$G(s) = \frac{\theta(s)}{X(s)} = \frac{K_p}{s\left[\frac{K_c J}{K_m} s^2 + \left(\frac{K_c J}{K_m} + \frac{K_l J}{K_m}\right)s + \left(K_m + \frac{K_l f}{K_m}\right)\right]} \qquad \text{...(4.35)}$$

Normally $K_c << K_m$ and therefore eqn. (4.35) may be simplified as

$$G(s) = \frac{\theta(s)}{X(s)} = \frac{K_p}{s\left[\frac{K_l J}{K_m} s + \left(K_m + \frac{K_l f}{K_m}\right)\right]} = \frac{K}{s(\tau s + 1)} \qquad \text{...(4.36)}$$

where
$$K = \frac{K_p}{K_m + \frac{K_l f}{K_m}}; \; \tau = \frac{JK_l}{K_m^{\,2} + K_l f}$$

It is observed that the transfer function of hydraulic transmission is of the same form as that of an electric motor given by eqn. (4.5). Further, the simplified transfer function of eqn. (4.36) can be obtained directly by neglecting the compressibility flow in eqn. (4.24).

Hydraulic Valves

In place of the variable displacement pump, one frequently used device is a constant pressure source and a valve to control the flow of oil to the hydraulic motor. This system has the advantage that the movable parts of the valve can be made much lighter than those of the stroke control mechanism in the pump, as a result of which, the time constants are reduced and the system becomes fast acting. The disadvantage of this system is that the valve has nonlinear characteristics and introduces problems of its own.

Figure 4.38 shows a three-way spool valve controlling the oil flow to the hydraulic motor. When the spool is in the neutral position ($x = 0$), the oil flow is blocked completely. As the spools is moved downwards (x positive), the upper pipe line to the motor gets connected to the high pressure source and the lower pipe line to the sump causing the motor to rotate in a particular direction. If the spools is moved upward (x negative) the lower pipe line to the motor gets connected to the high pressure source and the upper pipe line to the sump. The direction of oil flow in the motor and hence direction of its rotation is reversed.

Because turbulent flow exists at a sharp-edged orifice such as a valve, the flow equation is (refer section 2.2)

$$q_1 = Kx\sqrt{P_0 - P_1}\ ;\ x > 0$$

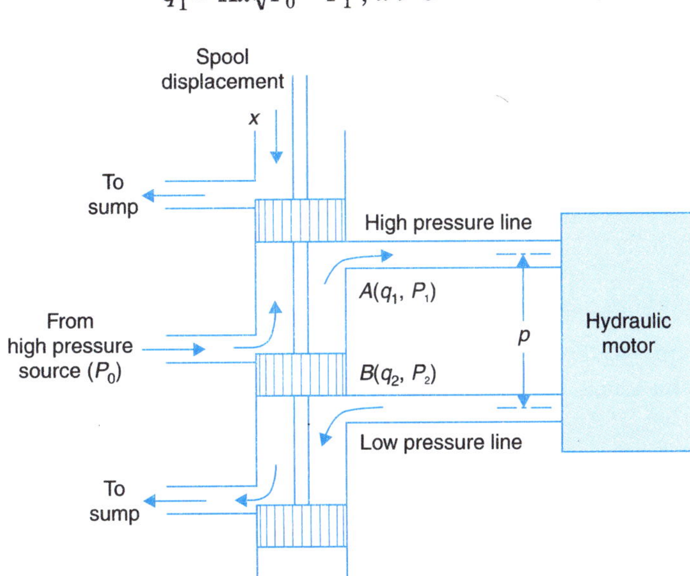

Fig. 4.38. Hydraulic valve.

where q_1 is the rate of flow through port A and K is flow coefficient which is a function of valve and fluid properties.

In addition, it follows that

$$q_2 = Kx\sqrt{P_2}\ ;\ x > 0$$

where q_2 is the flow rate at port B.

Since $q_1 = q_2 = q$, we have from above equations

$$q = \frac{Kx}{\sqrt{2}} \sqrt{P_0 - p}\ ; x > 0$$

where $p = P_1 - P_2$.

For $x < 0$, the relationship between p, q and x may be obtained in a similar manner.

The valve characteristics are shown in Fig. 4.39 which relate the volumetric oil flow rate q to the motor and the differential pressure p across the motor for different values of spool displacement x. Although the valve characteristics are nonlinear, for small values of x, these can be linearized. The relationship between q, x and p may be written as

$$q = f(x, p) \qquad ...(4.37)$$

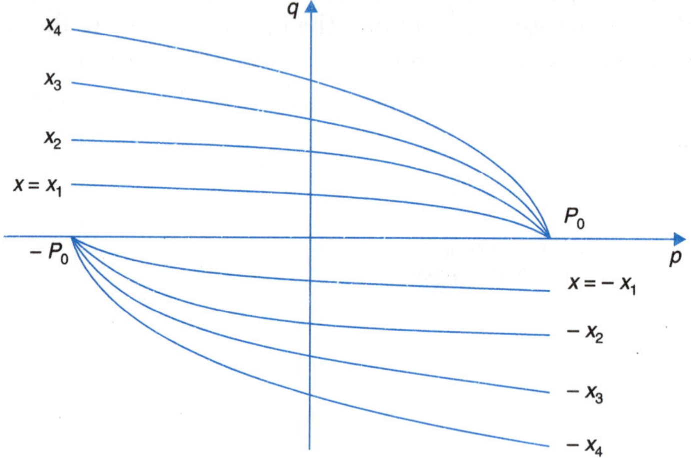

Fig. 4.39. Typical valve characteristics.

Expanding eqn. (4.37) into Taylor's series about the normal operating point (q_0, x_0, p_0) and neglecting all the terms of second and higher derivatives, we get

$$q = q_o + \left.\frac{\partial q}{\partial x}\right|_{\substack{x = x_o \\ p = p_o}} (x - x_o) + \left.\frac{\partial q}{\partial p}\right|_{\substack{x = x_o \\ p = p_o}} (p - p_o) \qquad ...(4.38)$$

For this system, the normal operation point corresponds to $q_o = 0, p_o = 0, x_o = 0$, therefore, from eqn. (4.38)

$$q = K_1 x - K_2 p \qquad ...(4.39)$$

where $$K_1 = \left.\frac{\partial q}{\partial x}\right|_{\substack{x = o \\ p = o}} ; K_2 = -\left.\frac{\partial q}{\partial p}\right|_{\substack{x = o \\ p = o}}$$

Equation (4.39) gives a linearized relationship among q, x and p.

The response equation of the valve-motor combination can now be written down from eqn. (4.29) of the pump-motor system by replacing its left-hand side by $q = K_1 x - K_2 p$, the inflow rate to the pipe-line-motor combination. Furthermore the compressibility is neglected right at this stage. Thus

$$q = K_1 x - K_2 p = K_m \dot{\theta} + K_l p \qquad \text{...(4.40)}$$

From eqn. (4.40) and the torque equation (4.30) the transfer function of the system is obtained (analogous to the pump-motor system) as

$$G(s) = \frac{\theta(s)}{X(s)} = \frac{K_1}{s\left[\frac{J(K_2 + K_l)}{K_m} s + K_m + \left(\frac{K_2 + K_l}{K_m}\right) f\right]} = \frac{K}{s(\tau s + 1)}$$

where $$K = \frac{K_1}{K_m + \left(\frac{K_2 + K_l}{K_m}\right) f}; \quad \tau = \frac{J(K_2 + K_l)}{K_m^2 + (K_2 + K_l) f} \qquad \text{...(4.41)}$$

Hydraulic Linear Actuator

The linear actuators are piston devices. A simple hydraulic actuator is shown in Fig. 4.40, in which the motion of the spool regulates the flow of oil to either side of the power cylinder.

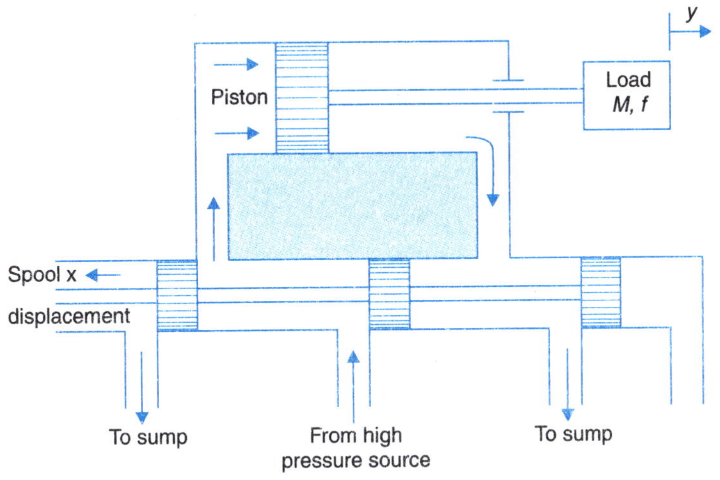

Fig. 4.40. Hydraulic linear actuator.

When the spool moves to the left, the oil from the high pressure source enters into the power cylinder, on the left of the power piston. This creates a differential pressure across the piston which causes the power piston to move to the right, pushing the oil is front of it to the sump. The oil is pressurized by a pump and is recirculated in the system. The load rigidly coupled to the piston moves a distance y from its reference position in response to the displacement x of the valve spool from its neutral position.

Let us first carry out a *simplified analysis* of the linear actuator. The relationship between the volumetric oil flow rate q into the power piston and the differential pressure p across the piston for small values of spool displacement x, has already been given in eqn. (4.40) and can be rewritten in the form given below:

$$K_2 p = K_1 x - q$$

Assuming leakage and compressibility flows to be negligible, the rate of oil flow into the piston is proportional to the rate at which the piston moves, *i.e.*,

$$q = A\frac{dy}{dt} \qquad ...(4.42)$$

where A is the area of the piston.

The force on the piston is (Ap) which moves the load consisting of mass M and viscous friction with coefficient f, according to the differential equation

$$Ap = M\ddot{y} + f\dot{y} \qquad ...(4.43)$$

From eqns. (4.39), (4.42) and (4.43) we can draw the signal flow graph given in Fig. 4.40.

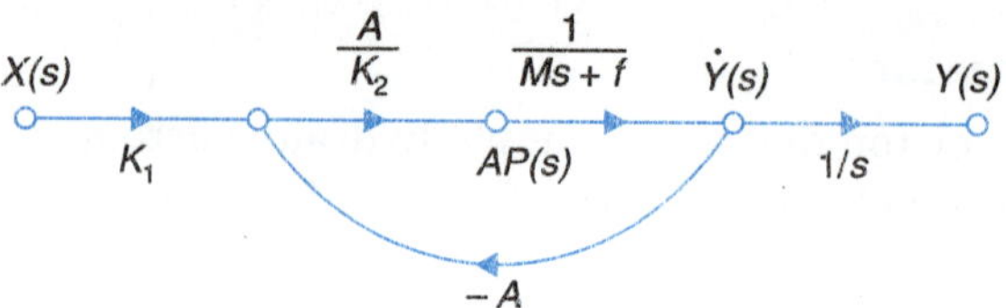

Fig. 4.41. Signal flow graph of hydraulic actuator.

The transfer function, as obtained from the signal flow graph, is

$$\frac{Y(s)}{X(s)} = \frac{A(K_1/K_2)}{s\left[Ms + \left(f + \frac{A^2}{K_2}\right)\right]} = \frac{K}{s(\tau s + 1)} \qquad ...(4.44)$$

where $$K = \frac{AK_1}{A^2 + fK_2}; \quad \tau = \frac{MK_2}{A^2 + fK_2}$$

It is found that the transfer function of the linear actuator is similar to that of hydraulic and electric motors and contains in it the term $1/s$ which accounts for the integrating action of the device.

A more exact analysis can now be carried out by taking the leakage and compressibility flows into account. The oil flow rate in the power cylinder is then given by

$$q = q_p + q_l + q_c \qquad ...(4.45)$$

where q_p = incompressible component of flow rate to the power cylinder; q_l = leakage flow rate around the power piston; and q_c = compressibility flow rate.

The rate of the incompressible component of flow into the power cylinder is proportional to the rate at which the power piston moves, *i.e.*,

$$q_p = K_p\frac{dy}{dt} \qquad ...(4.46)$$

where K_p = a constant.

From eqns. (4.27) and (4.28) the leakage flow rate and compressibility flow rate are given as

$$q_l = K_l p \qquad ...(4.47)$$

$$q_c = K_c\frac{dp}{dt} \qquad ...(4.48)$$

Substituting the various flow rates in eqn. (4.45) we have

$$q = K_1 x - K_2 p = K_p \frac{dy}{dt} + K_l p + K_c \frac{dp}{dt}$$

or

$$K_p \frac{dy}{dt} + K_c \frac{dp}{dt} + (K_l + K_2)p = K_1 x \qquad \text{...(4.49)}$$

The differential equation governing the load remains the same as in eqn. (4.43). Substituting for p from eqn. (4.43) in (4.49), we obtain

$$\frac{KM}{A}\dddot{y} + \left[\frac{K_c f}{A} + (K_l + K_2)\frac{M}{A}\right]\ddot{y} + \left[K_p + (K_l + K_2)\frac{f}{A}\right]\dot{y} = K_l x$$

From the above equation, the transfer function of the system is

$$\frac{Y(s)}{X(s)} = \frac{K_l}{s\left[\frac{K_c M}{A}s^2 + \left\{\frac{K_c f}{A} + (K_l + K_2)\frac{M}{A}\right\}s + \left\{K_p + (K_l + K_2)\frac{f}{A}\right\}\right]}$$

As previously stated, K_c is very small, therefore $K_c M/A$ and $K_c f/A$ can be neglected. The transfer function then reduces to

$$\frac{Y(s)}{X(s)} = \frac{K}{s(\tau s + 1)} \qquad \text{...(4.50)}$$

where $$K = \frac{K_l}{K_p + (K_l + K_2)f/A}; \quad \tau = \frac{(K_l + K_2)M/A}{K_p + (K_l + K_2)f/A}$$

With $K_l = 0$ and $K_p = A$, this transfer function reduces to the simplified case transfer function given in eqn. (4.44)

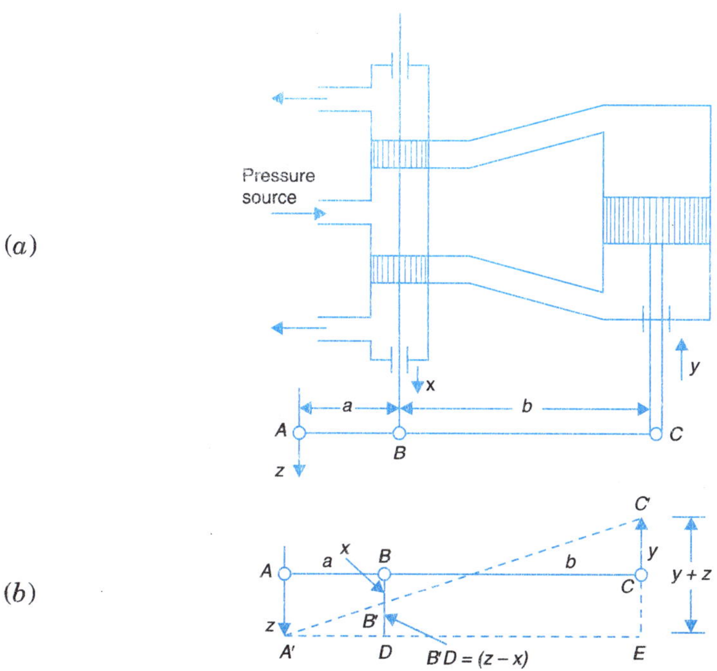

Fig. 4.42. Hydraulic actuator with negative feedback.

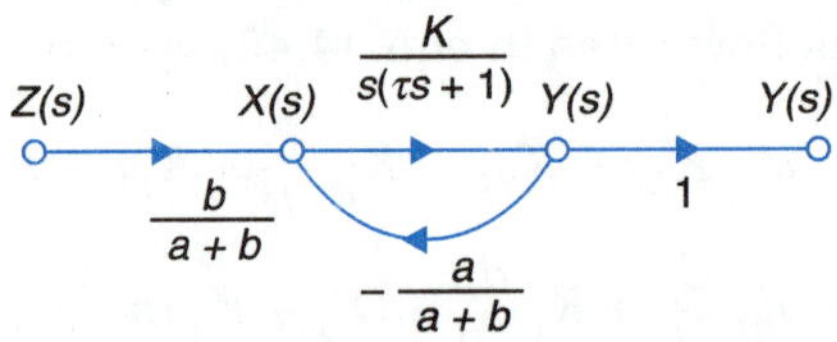

Fig. 4.43. Signal flow graph of system shown in Fig. 4.42.

Hydraulic linear actuator can also be modified to act as hydraulic amplifier (proportional controller) by providing a negative feedback through a link mechanism as shown in Fig. 4.42 (*a*). For small movements *x*, *y* and *z* can be regarded as linear. From the link geometry in Fig. 4.42 (*b*),

$$\frac{C'E}{A'E} = \frac{B'D}{A'D} \quad \text{or} \quad \frac{y+z}{a+b} = \frac{z-x}{a} \quad \text{or} \quad x = \frac{b}{a+b}z - \frac{a}{a+b}y \qquad \text{...(4.51)}$$

The transfer function $Y(s)/X(s)$ of the actuator already derived, is give in eqn, (4.44). Combining eqns. (4.44) and (4.51) we draw the signal flow graph of Fig. 4.42. The overall transfer function is obtained therefrom as

$$\frac{Y(s)}{Z(s)} = \frac{bK}{(a+b)\,s(\tau s+1)+Ka} \qquad \text{...(4.52)}$$

In the normal frequency range of hydraulic control systems

$$|(a + b)s(\tau s + 1)| << Ka$$

Therefore,

$$\frac{Y(s)}{Z(s)} \approx \frac{b}{a} \qquad \text{...(4.53)}$$

Thus the hydraulic actuator can be made to function as a linear amplifier over the frequency range of interest. The amplifier gain can be adjusted by a suitable choice of the link mechanism lever ratio *b*/*a*.

Hydraulic Feedback System

Let us discus a hydraulic power steering mechanisms whose simplified schematic diagram is shown in Fig. 4.44. The input to the system is the rotation of the steering wheel by the driver and the output is positioning of the car wheels in accordance with the input signal.

When the steering wheel is in the zero position, *i.e.*, the cross bar is horizontal, the wheels are directed parallel to the longitudinal axis of the car. For this condition, the spool is in the neutral position and the oil supply to the power cylinder is cut-off. When the steering wheel is turned anticlock-wise through an angle θ_i, the spool is made to move towards right by an amount *x* with the help of the gear mechanism. The high pressure oil enters on the left hand side of the power cylinder causing the power piston and hence the power ram to move towards right by an amount *y*. Through a proper drive linkage, a torque is applied to the wheels causing the desired displacement θ_o of the wheels.

A rigid linkage bar connects the power ram and the moveable valve housing. When the power ram moves towards right, the linkage moves as shown in Fig. 4.45. It is seen from this figure that movement of the power ram towards right causes a movement of the movable valve housing in such a direction as to seal off the high pressure side. The system then operates with a fixed θ_o for a given input *x* under steady conditions.

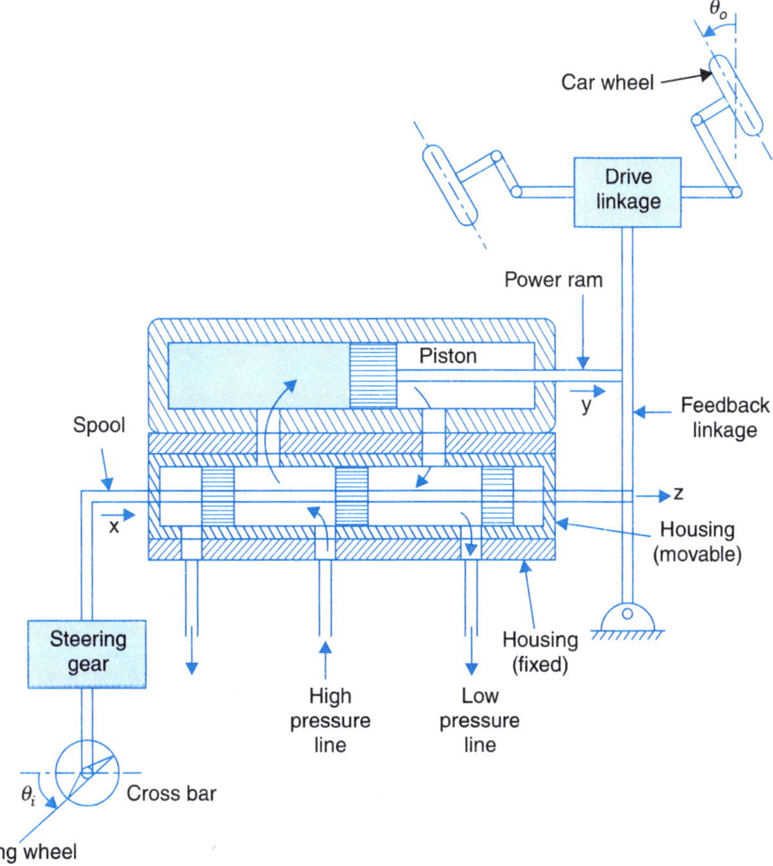

Fig. 4.44. Power steering mechanism. [From V. Del Toro and S. Parker, *Principles of Control systems Engineering,* 1960, McGraw-Hill, New York, Reproduced with permission].

The geometry of the linkage is shown in Fig. 4.45, from which we obtain for small movements.

$$\frac{AA'}{AC} = \frac{AA'}{AB + BC} = \frac{BB'}{BC}$$

or
$$\frac{y}{a+b} = \frac{z}{a} \qquad ...(4.54)$$

Therefore,
$$Z(s) = \frac{a}{a+b} Y(s) \qquad ...(4.55)$$

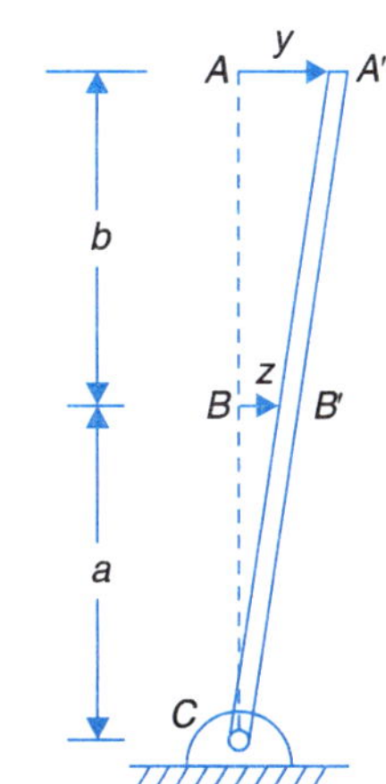

Fig. 4.45. Geometry of feedback linkage for small values of movement y.

Assuming the compressibility and leakage flows to be negligible the flow rate of oil into power cylinder due to spool movement is [refer eqn. 4.39] given by

$$q = K_1 e - K_2 p$$

where $e = (x - z)$ is the net displacement of spool relative to the valve-housing.

Therefore, $$Q(s) = K_1E(s) - K_2P(s) \quad ...(4.56)$$

and $$E(s) = X(s) - Z(s) \quad ...(4.57)$$

The flow rate of oil into the power cylinder is proportional to the rate at which the power piston displaces the oil, *i.e.*,

$$Q(s) = K_p s\, Y(s) \quad ...(4.58)$$

Assuming the load on power piston to consist of mass M and viscous friction with coefficient f, the force equation is given by

$$AP(s) = Ms^2Y(s) + fsY(s) \quad ...(4.59)$$

where A is the area of the piston. From eqns, (4.56) and (4.59) we get

$$Q(s) = K_1E(s) - K_2\frac{Ms^2 + fs}{A}Y(s) \quad ...(4.60)$$

The block diagram of the power steering mechanism obtained from eqns. (4.55)-(4.60) is shown in Fig. 4.46.

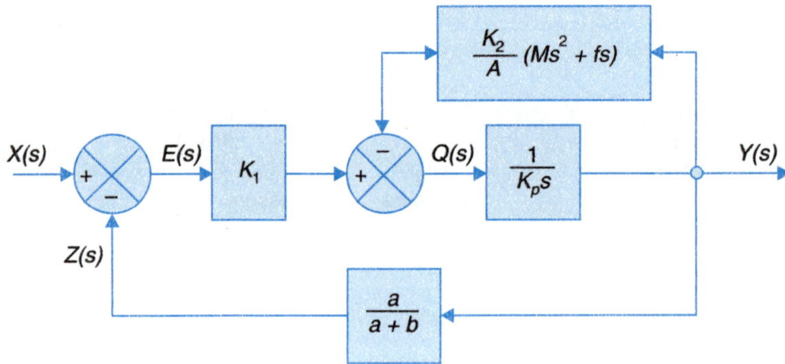

Fig. 4.46. Block diagram of power steering mechanism shown in Fig. 4.44 (a).

Example 4.1 : Consider the hydraulic actuator of Fig. 4.47 with lever feedback. Assuming the actuator to be an ideal integrator draw the system block diagram identifying various transfer functions.

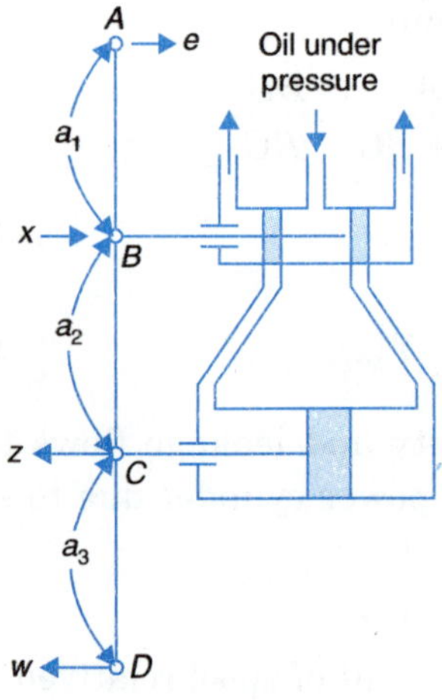

Fig. 4.47

Solution. It can be easily shown that

$$x = \left(\frac{a_2}{a_1 + a_2}\right)e - \left(\frac{a_1}{a_1 + a_2}\right)z \qquad ...(i)$$

and

$$w = \left(\frac{a_3}{a_2 + a_3}\right)e + \left(\frac{a_1 + a_2 + a_3}{a_2 + a_3}\right)z \qquad ...(ii)$$

For the actuator

$$z = K\int x\,dt \qquad ...(iii)$$

The block diagram with transfer functions indicated therein is drawn in Fig. 4.48.

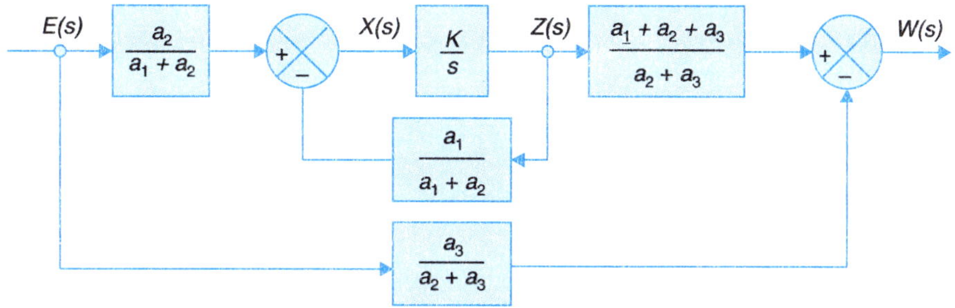

Fig. 4.48

Example 4.2 : Consider the fly-ball speed governing system of Fig. 4.48. Explain the operation of the system. Find the transfer function of each component and draw the overall block diagram and the transfer function $Y(s)/E(s)$. Show that under high gain condition the speed controller has proportional plus-integral action.

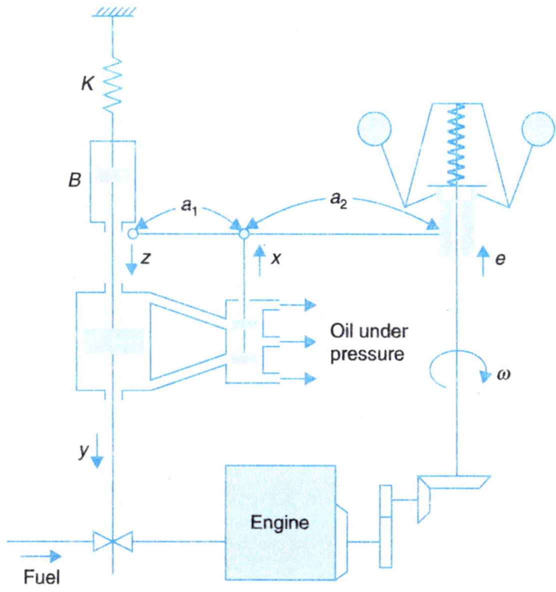

Fig. 4.49. Engine speed governing system.

Solution. The fly-ball governor is coupled to engine shaft via speed reduction gearing, thereby sensing engine speed. As engine speed increases the fly balls move outwards under centrifugal force. So that the governor sleeve moves upwards (*e*) corresponding to the speed error with reference to the speed setting of the governor. This movement via the lever and hydraulic actuator causes the power piston to move downward which in turn reduces the fuel valve opening therefore reducing engine speed. The reverse happens when engine speed decreases. The system thus has inherent negative feedback regulating action.

Using the lever eqn. (4.51) with appropriate symbols, we have

$$X(s) = \left(\frac{a_2}{a_1 + a_2}\right) E(s) - \left(\frac{a_1}{a_1 + a_2}\right) Z(s) \qquad ...(i)$$

If the hydraulic actuator's time constant is assumed negligible (refer eqn (4.41)), it acts as an integrator *i.e.*,

$$\frac{Y(s)}{X(s)} = \frac{K_A}{s} \;; K_A = \text{actuator gain} \qquad ...(ii)$$

Assuming the force input at z to be negligible

$$K\dot{z} + B(\dot{z} - \dot{y}) = 0 \qquad \text{or} \qquad \frac{Z(s)}{Y(s)} = \frac{sB}{sB + K} \qquad ...(iii)$$

Using component transfer functions of eqns. (*i*)-(*iii*), the system block diagram is drawn in Fig. 4.50. The overall transfer can be written down as

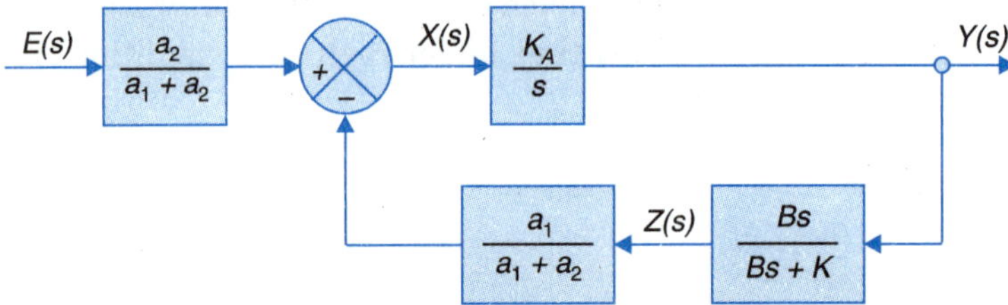

Fig. 4.50. Block diagram of the system of Fig. 4.48 transfer function can be written down as

$$\frac{Y(s)}{X(s)} = \left(\frac{a_2}{a_1 + a_2}\right) \frac{K_A/s}{1 + \left(\frac{a_1}{a_1 + a_2}\right)\left(\frac{BK_A}{Bs + K}\right)} \qquad ...(iv)$$

Let the gain be so adjusted that for the frequency range of interest (in hydraulic systems bandwidth is very low)

$$\left|\left(\frac{a_1}{a_1 + a_2}\right)\left(\frac{BK_A}{Bs + K}\right)\right| >> 1$$

The transfer function of eqn. (*iv*) then simplifies to

$$\frac{Y(s)}{X(s)} = \frac{a_2}{a_1}\left[1 + \frac{K}{Bs}\right] = \text{proportional plus integral terms} \qquad ...(v)$$

4.6 PNEUMATIC SYSTEMS

As different from hydraulic fluid, air medium (or other gases in special situations) is used in pneumatic control systems. Air medium has the advantage of being non-inflammable and having almost negligible viscosity compared to the high viscosity of hydraulic fluids, which also varies considerably with temperature causing a marked effect on the performance of control systems.

On the other hand, the high incompressibility of a hydraulic fluid causes the force wave to travel faster and therefore the hydraulic systems have a shorter response time; while in pneumatic systems there is a considerable amount of compressibility flow so that such systems are characterized by longer time delays.

Pneumatic systems find considerable application in the process control field. These are sometimes used in guided missiles and aircraft systems.

Pneumatic Bellows

It consists of a hollow chamber with thin metallic walls. The side walls of bellows are corrugated, while the input and output surfaces are flat as shown in Fig. 4.51. The action of the bellows is similar to that of a spring. An increase in the pressure within the bellows results in an increase in the separation between the input and output surfaces.

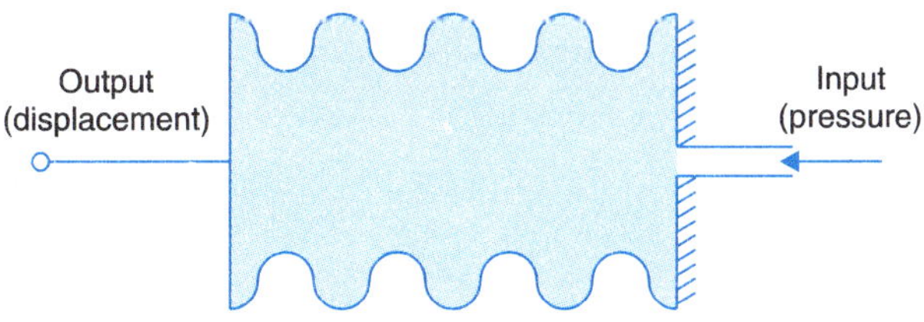

Fig. 4.51. Pneumatic bellows.

The force acting to separate two surfaces = $(\Delta P)A$, where A = area of each flat surface; and ΔP = differential pressure (internal pressure minus external pressure).

The force opposing the separation = $K(\Delta x)$, where K = stiffness of the belows; and Δx = displacement of movable surface from the reference.

Therefore, in equilibrium

$$K(\Delta x) = (\Delta P)A \qquad \text{or} \qquad \frac{\Delta X(s)}{\Delta P(s)} = \frac{A}{K}$$

This is the transfer function of the bellows.

Pneumatic Flapper Valve

This is an important component of many pneumatic systems. The power source for this device is the supply of air at constant pressure. A schematic diagram of the valve is shown in Fig. 4.52.

Pressurized air is fed through the orifice and is ejected from the nozzle towards the flapper. The flapper is positioned against the nozzle opening and the nozzle back pressure P_b is controlled by the nozzle-flapper distance e. As the flapper approaches the nozzle, the resistance to the flow of air through the nozzle increases with the result that the nozzle back pressure increases. If the nozzle is completely closed by the flapper, the nozzle back pressure P_b becomes equal to the supply pressure P_s. If the flapper is moved away from the nozzle so that nozzle flapper distance is large, then there is practically no restriction to flow and the nozzle back pressure takes on a minimum value close to the ambient pressure.

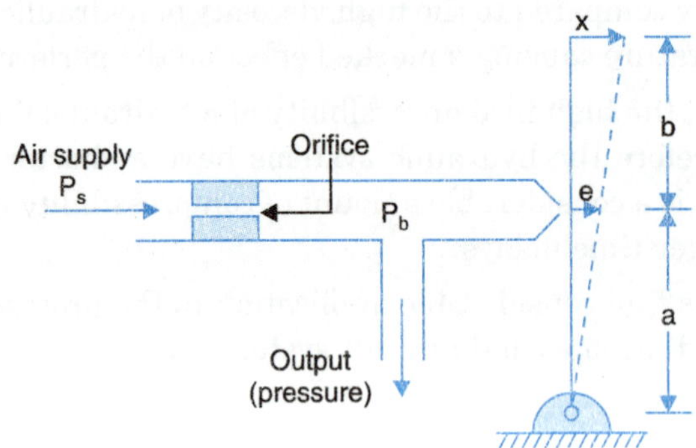

Fig. 4.52. A flapper valve.

Thus the flapper valve converts small changes in the position of the flapper into larger changes in the back pressure. A typical curve relating the nozzle back pressure P_b to the nozzle flapper distance e is shown in Fig. 4.53. The linear part of the curve is utilized in the valve operation. For this operating region, the transfer function of the valve is

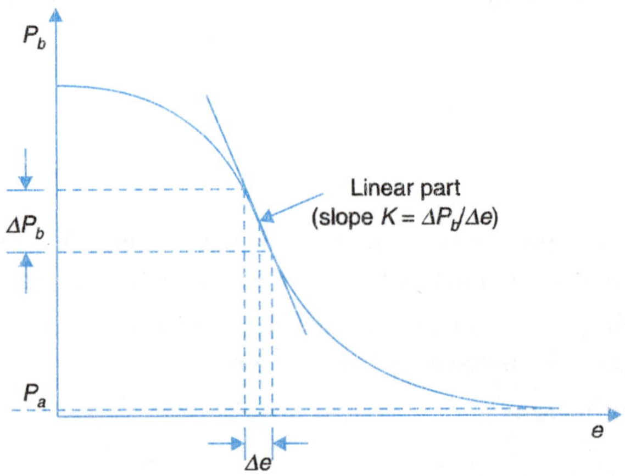

Fig. 4.53. Flapper valve characteristic.

$$\frac{\Delta P_b(s)}{\Delta X(s)} = \left(\frac{a}{a+b}\right)K \qquad \text{...(4.62)}$$

where $K < 0$ is the slope of the linear part of the curve and $e = [a/(a + b)]x$ is the nozzle-flapper distance as shown in Fig. 4.52.

Pneumatic Relay

For the flapper valve of the type discussed above, it is necessary to restrict the range of flapper displacement to very small value so that the linearity of operation is maintained. Because of this, the change in the output pressure is very small and therefore the need of a pneumatic amplifier in cascade with this device arises. This pneumatic amplifier is commonly known as pneumatic relay. A typical combination of flapper valve and relay is shown in Fig. 4.54. A ball is attached to the lower bellows surface. When the ball rests on its upper seat, the atmospheric opening is closed and the output pressure P becomes equal to the supply pressure P_s. When the ball rests on its lower seat, it shuts off the air supply and output pressure drops to ambient pressure. The output pressure can thus be made to vary from ambient to full supply pressure.

The movement of the flapper away from the nozzle causes the nozzle back pressure P_b to decrease, thus the bellows contracts, moving the ball upwards. The atmospheric opening is partially closed and the output pressure increases. When the flapper moves towards the nozzle, the opposite action takes place.

It is thus seen that an increase in separation between nozzle and flapper results in a decrease in output pressure in the case of the flapper valve and an increase in output pressure when the flapper valve is used with the pneumatic relay. The transfer function of the combination is therefore

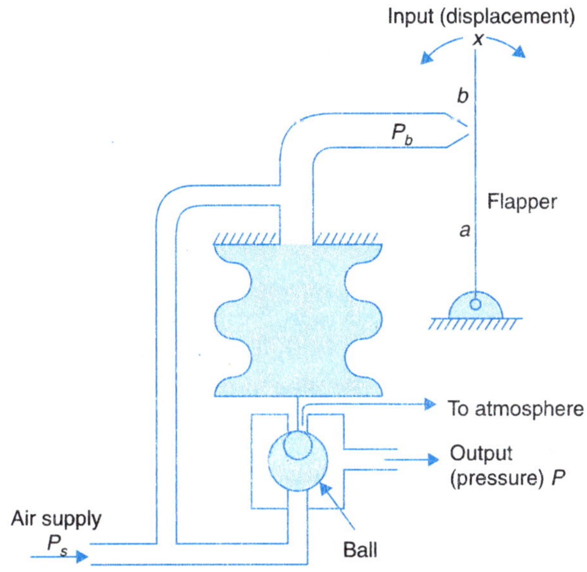

Fig. 4.54. A flapper valve with relay.

$$\frac{\Delta P(s)}{\Delta X(s)} = \frac{a}{a+b}K \qquad \text{...(4.63)}$$

where $K > 0$.

Pneumatic Actuator

The majority of the modern pneumatic control systems require translatory output motion. This motion is achieved by pneumatic actuators responding to changes in input air pressure. A schematic diagram of the pneumatic actuator is shown in Fig. 4.55. Assume that the actuator is used to position a load consisting of a spring with stiffness K, viscous friction with coefficient *f* and mass *M*. For a small variation ΔP in the input pressure, the force acting on the diaphragm is $A(\Delta P)$ where *A* is the area of the diaphragm. If Δy is the displacement of the actuator stem because of this force, then the force balance equation is

$$A(\Delta P) = M\Delta\ddot{y} + f\Delta\dot{y} + K\Delta y$$

Therefore the transfer function of the actuator is

$$\frac{\Delta Y(s)}{\Delta P(s)} = \frac{A}{Ms^2 + fs + K} \quad ...(4.64)$$

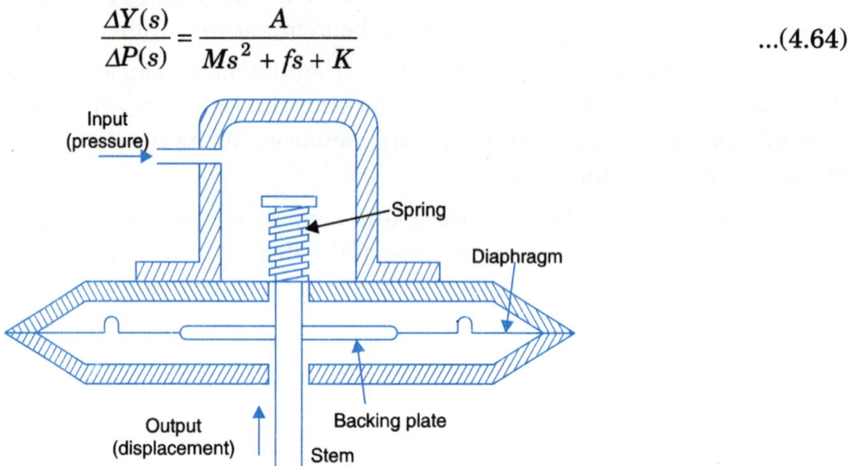

Fig. 4.55. Pneumatic actuator.

It may be noted that the stiffness and mass of the diaphragm and backing plate, are considered to be negligible.

Pneumatic Position Control System

The pneumatic position control system shown in Fig. 4.56 is used to position a fluid flow valve. The principle of operation of this system is that an input signal is applied to the flapper corresponding to the required opening of the plug valve. The nozzle back pressure operates the actuator and the actuator stem moves to give the required opening of the plug valve.

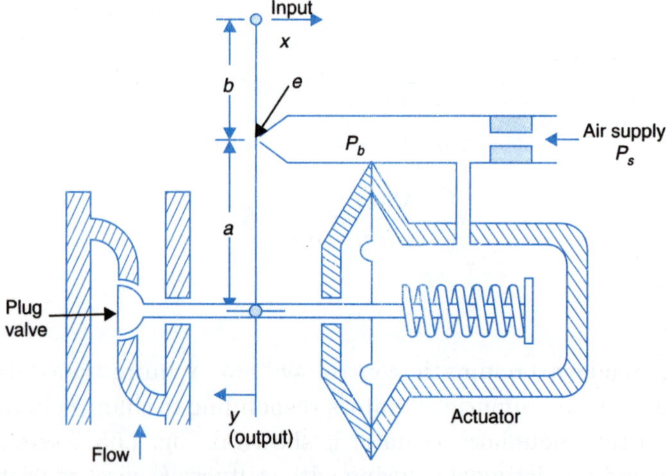

Fig. 4.56. Pneumatic position control system.

The geometry of the flapper for small input and output displacements is shown in Fig. 4.57. From this figure, we have

$$E(s) = \frac{a}{a+b} X(s) - \frac{b}{a+b} Y(s) \qquad ...(4.65)$$

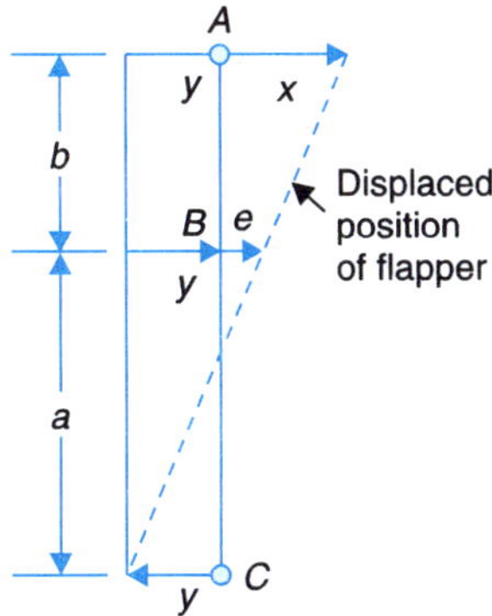

Fig. 4.57. Geometry of flapper movement.

Using eqns. (4.64) and (4.65) the block diagram of the system is drawn in Fig. 4.58. All the variables are incremental values around the operating point.

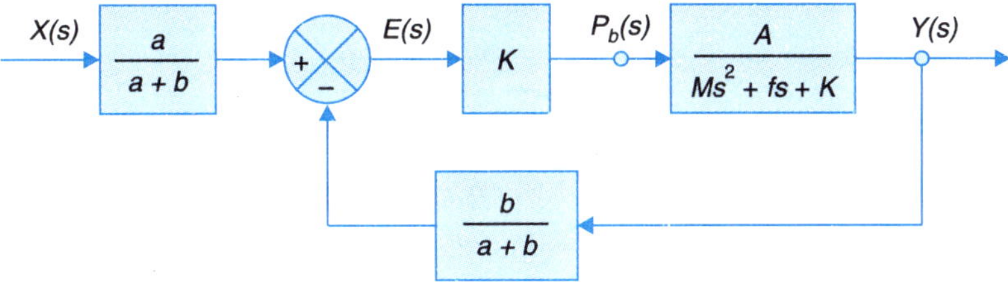

Fig. 4.58. Block diagram of the system shown in Fig. 4.56.

Example 4.3 :

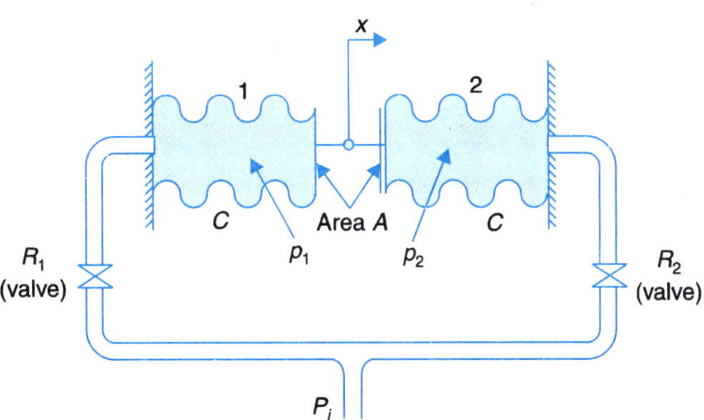

Fig. 4.59. Variables are incremental values around the steady values (representing the operating point).

Consider the pneumatic pressure system of Fig. 4.59 wherein the two bellows are assumed to be identical. The bellows output (mechanical) end are connected by a link from which the mechanical (linear) movement x is picked up; x being proportional to the differential force of the bellows. The zero position of the output motion (x) corresponds to steady input pressure. Relate the changes in input pressure P_i to output motion x in form of the transfer function $X(s)/P_i(s)$. Bellows stiffness is K_s. What is x (steady state) for unit step change in input pressure ?

[**Hint :** Pressure changes in bellows would correspond to differential equation (2.36).]

Solution. For the valve and bellows 1, we can write from eqn. (2.36)

$$R_1C(dp_1/dt) + 1 = p_i$$

or
$$\frac{P_1(s)}{P_i(s)} = \frac{1}{R_1Cs + 1} \quad ...(i)$$

Similarly for the valve and bellows 2

or
$$\frac{P_2(s)}{P_i(s)} = \frac{1}{R_2Cs + 1} \quad ...(ii)$$

Mechanical movement (x) is given by

$$A(p_1 - p_2) = Kx \text{ ; } K = \text{constant of proportionality} \quad ...(iii)$$

or
$$A[P_1(s) - P_2(s)] = KX(s) \quad ...(iv)$$

Substituting from Eqs. (i) and (ii) in Eq. (iv)

$$A\left[\frac{1}{R_1Cs + 1} - \frac{1}{R_2Cs + 1}\right]P_i(s) = KX(s)$$

or
$$\frac{X(s)}{P_i(s)} = (A/K)\left[\frac{(R_2 - R_1)\,Cs}{(R_1Cs + 1)\,(R_2Cs + 1)}\right] \quad ...(v)$$

For unit step input $\qquad P_i(s) = 1/s$

Then
$$X(s) = (A/K)\frac{1}{s}\left[\frac{(R_2 - R_1)s}{(R_1Cs + 1)(R_2Cs + 1)}\right] \quad ...(vi)$$

From final value theorem, we get

$$x(\infty) = \underset{x \to 0}{sX(s)} = 0$$

It is found that steady state value of x is zero for unit step input pressure. Physically it is visualized that in steady state pressures p_1 and p_2 become equal and bellows being identical, the net mechanical force is zero and so $x(\infty) = 0$.

Example 4.4 : The schematic diagram of a pneumatic controller is shown in Fig. 4.60. Capital letters with super-bars are indicate of the operating point (set point) quantities and small case letters stand for small changes around this point.

Draw the block diagram of the controller for small changes deriving the transfer function of each block. Therefrom find the overall transfer function $P_c(s)/E(s)$, where e is the small actuating error signal shown in the figure. Assume that $R_i >> R_d$.

At the operating point the relay nozzle combination is characterized by the relationship

$$p_c = Kx \; ; K > 0$$

What kind of controller results if

(i) $R_i = 0$ (ii) $R_d = 0$

Low resistance values are achieved by making the respective pipes wide enough.

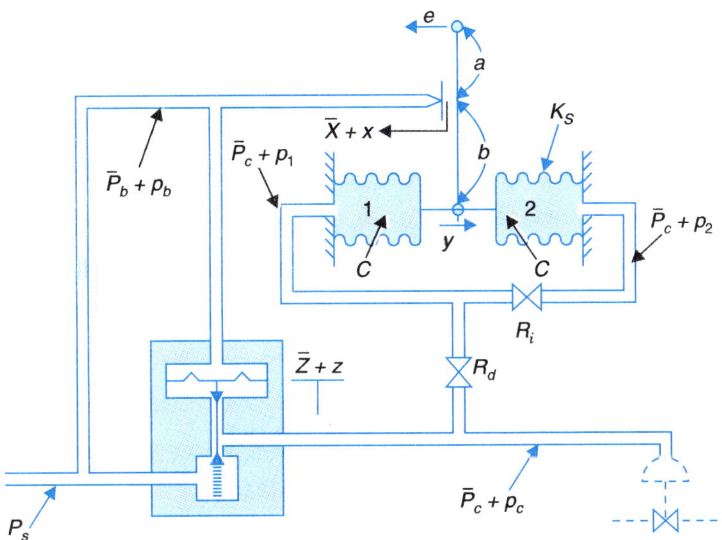

Fig. 4.60. Pneumatic controller—variables indicated are incremental changes about the operating point.

Solution. The electric analog of the bellows and pneumatic flow into these is drawn in Fig. 4.61. Assuming $R_i >> R_d$, it can be written that

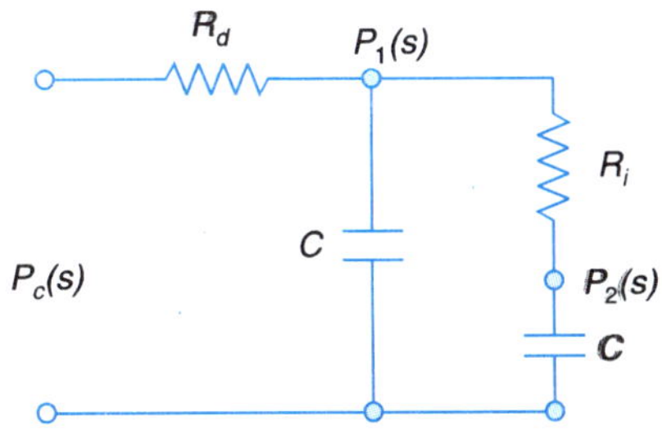

Fig. 4.61

$$\frac{P_1(s)}{P_c(s)} = \frac{1}{R_d Cs + 1} = \frac{1}{\tau_d s + 1} \quad ...(i)$$

and $$\frac{P_2(s)}{P_c(s)} = \frac{1}{R_i Cs + 1} = \frac{1}{\tau_i s + 1} \; ; \text{assuming } R_i >> R_d \quad ...(ii)$$

The force-balance equation for the two bellows gives

$$(P_1 - P_2)A = K_s y \quad ...(iii)$$

For the lever (positive x is in positive direction of e)

$$x = \frac{b}{a+b}e - \frac{a}{a+b}y \qquad \text{...(iv)}$$

For the pneumatic relay

$$p_c = Kx \ (K > 0) \qquad \text{...(v)}$$

The block diagram of Fig. 4.62 can now be drawn.

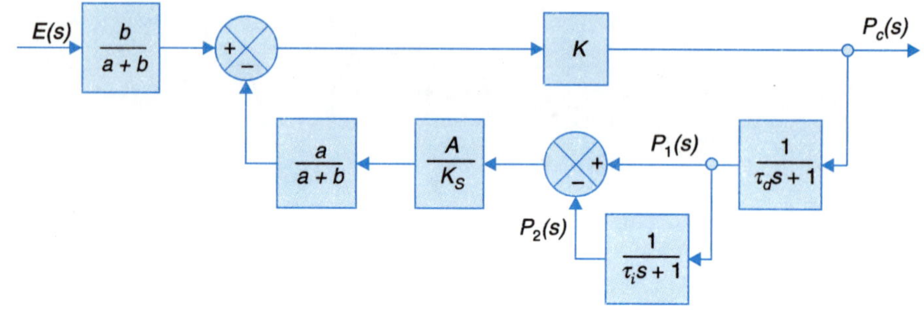

Fig. 4.62

It is easily established from the block diagram that

$$\frac{P_c(s)}{E(s)} = \frac{b/(a+b)E(s)}{1 + K\left(\dfrac{a}{a+b}\right)\left(\dfrac{A}{K_s}\right)\left(\dfrac{\tau_i s}{\tau_i s + 1}\right)\left(\dfrac{1}{\tau_d s + 1}\right)} \qquad \text{...(vi)}$$

Magnitude of second term in denominator >> 1 (for practical controller)

The transfer function of eqn. (*vi*) for frequencies of interest then approximates to

$$\frac{P_c(s)}{E(s)} = \frac{bK_S(\tau_i s + 1)(\tau_d s + 1)}{aA\tau_i s} = \frac{bK_s}{aA}\left(\frac{\tau_i + \tau_d}{\tau_i} + \frac{1}{\tau_i s} + \tau_d s\right) \qquad \text{...(vii)}$$

As $\tau_i >> \tau_d$, we can write

$$\frac{P_c(s)}{E(s)} = K_p\left(1 + \frac{1}{\tau_i s} + \tau_d s\right); \ K_p = \frac{bK_s}{aA} \qquad \text{..(viii)}$$

The controller is thus a proportional-plus-integral-plus-derivative controller (to be discussed in Section 5.7).

(*i*) If $R_i = 0$

$$\frac{P_c(s)}{E(s)} = K_p(1 + \tau_d s) \text{ ; proportional-plus-derivative controller} \qquad \text{...(ix)}$$

(*ii*) If $R_d = 0$

$$\frac{P_c(s)}{E(s)} = K_p\left(1 + \frac{1}{\tau_i s}\right) \text{ ; proportional-plus-integral controller} \qquad \text{...(x)}$$

PROBLEMS

4.1 A two-phase servomotor has rated voltage applied to its reference winding. The torque-speed characteristic of the motor with 115 volts, 50 Hz applied to control winding is shown in Fig P-4.1. The moment of inertia of the motor is 1×10^{-5} kg-m^2 and friction is negligible. Find the transfer function connecting the shaft position θ with the control winding voltage v_c.

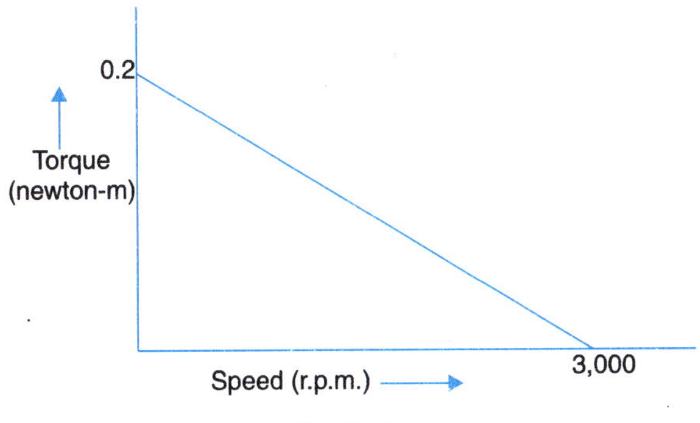

Fig. P-4.1

4.2 The schematic diagram of a servo system is shown in Fig. P-4.2 (given on next page). The two-phase servomotor develops a torque in accordance with the equation

$$T_M = -K_1\dot{\theta}_M + K_2 v_C$$

where $K_1 = 0.5 \times 10^{-3}$ newton-m per rad/sec

$K_2 = 2 \times 10^{-3}$ newton-m/volt

Draw the block diagram of the system indicating the transfer function of each block. Calculate the value of amplifier gain K_A such that the system has a steady-state error $(\theta_R - \theta_C)$ of 0.1 deg for an input velocity of 1 rad/sec. What will be the control field voltage under this condition ?

Given:

Load inertia, $J_L = 2.5.$ kg-m^2

Coefficient of load friction, $f_L = 250$ newton - m per rad/sec

Motor to load gear ratio, $(\dot{\theta}_L / \dot{\theta}_M) = 1/50$

Motor to synchro gear ratio, $(\dot{\theta}_C / \dot{\theta}_M) = 1/1$

Sensitivity of synchros, $K_s = 100$ volts/rad

The motor inertia and friction are negligible.

4.3 For the hydraulic actuator with lever feedback shown in Fig. P-4.3 (given on next page) draw the block diagram giving transfer function of each block. Assume the actuator to be an ideal integrator. Find the transfer function $Y(s)/W(s)$. What is the approximate transfer function for large loop gain ?

4.4 A position control system shown in Fig. P-4.4 is used to control the position of a load having a moment of inertia of J_L kg-m^2 and coefficient of viscous friction as f_L, Nm/rad/s. The position of load is controlled by a dc motor having a torque constant of K_T Nm/A. The resistance of armature is R_a ohms. The position of load is read by an optical encoder which is connected to the load through a gear the box of ratio $N_1 : N_2$ as 1 : 1. The output of encoder is fed to a counter which gives the digital value of position at any time. The digital information of position of load is then fed to a D/A converter, which give a analog voltage proportional to the load position. The D/A converter operates at a very fast sampling rate, so that the system can be approximated as continuous-time system. The input to the system is given through a pot of gain K_p V/rad. The input signal and the feedback from D/A converter is then given to a difference amplifier of gain K_A V/V. The output of difference amplifier is fed to the SCR circuitry for adjusting the value of firing angle. The output of rectifier is a dc voltage proportional to the error. This voltage is in turn applied across the armature of d.c. servo motor.

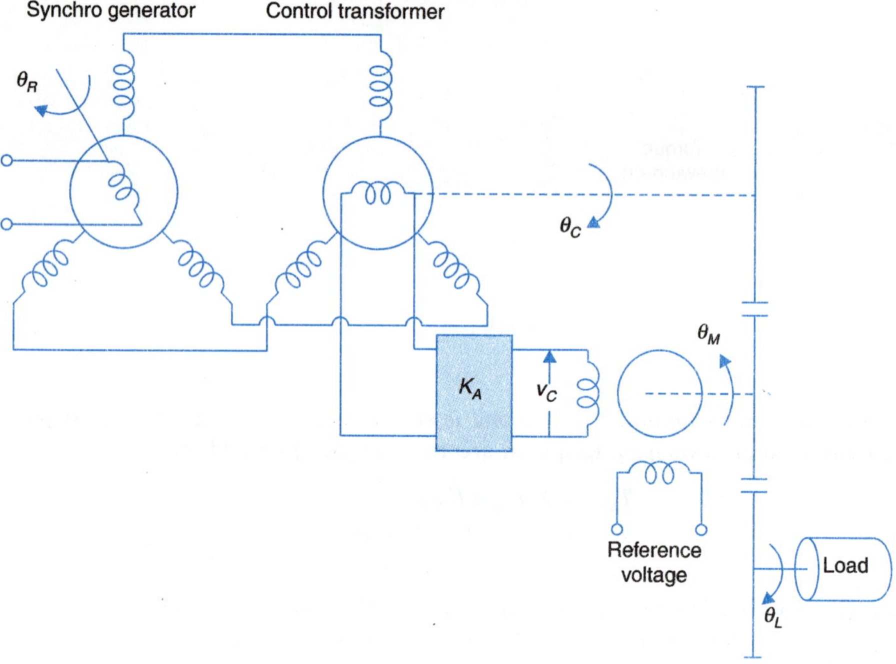

Fig. P-4.2

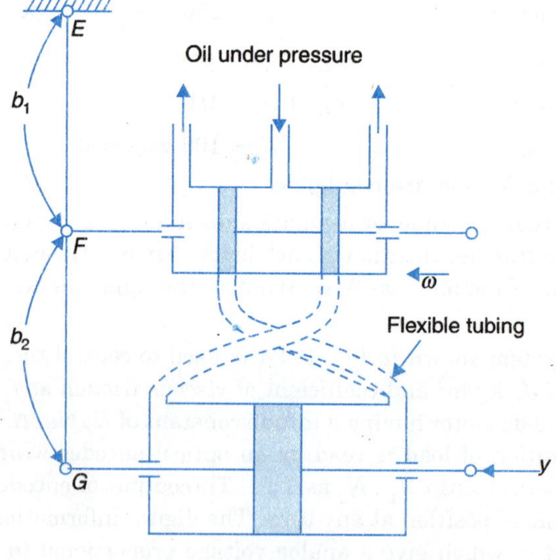

Fig. P-4.3

Given : $K_A = 5$ V/V, $R_a = 1\ \Omega$, $K_T = 1$ N-m/A,
$J_L = 0.5$ kg-m^2, $f_L = 1$ Nm/rad/s, SCR gain = $K_S = 10$ V/V
K_p = potentiometer constant = 10 V/rad.

Combined gain of encoder + counter + D/A converter = $K_X = 10$ V/V

(*a*) Find the transfer function $\theta_L(s)/\theta_R(s)$

(*b*) What is the value of steady-state error of the system for a unit ramp input ?

(*c*) Repeat part (*ii*)if the output of the difference amplifier is modified to

$$K_A v_e + K_A \int v_e\, dt$$

Check the stability of the system.

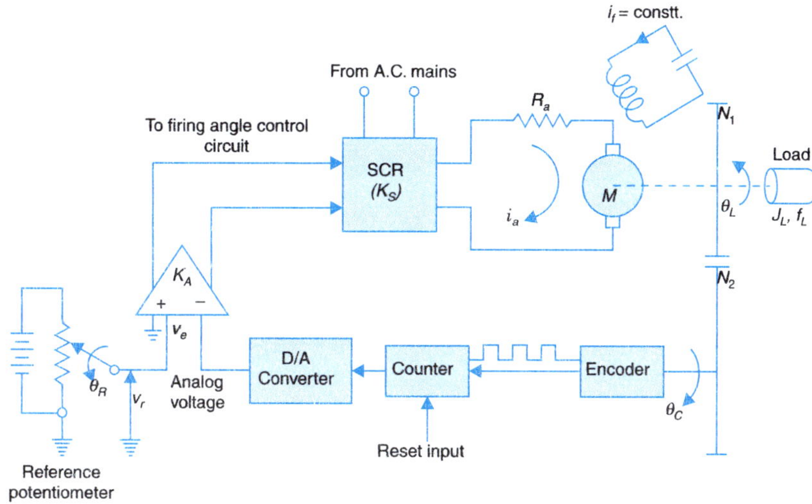

Fig. P-4.4

4.5 A valve piston linkage combination is shown in Fig. P-4.5. The object of this arrangement is to produce an output displacement x which is a function of the speed of the input motion as well as

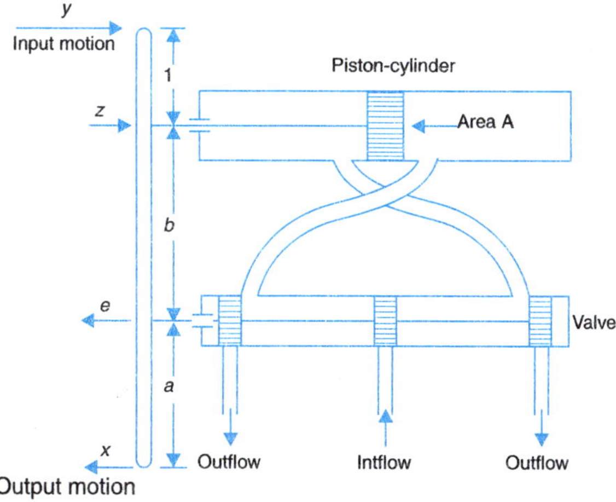

Fig. P-4.5

its amplitude. Find the transfer function $X(s)/Y(s)$ assuming rate of oil-flow to piston to be proportional to valve opening and neglecting oil compressibility, leakage effects and mass of the piston.

4.6. A hydraulic servo system used to control the transverse feed of a machine tool is shown in Fig. P-4.6. Each angular piston of the cam corresponds to a desired reference position x_r such that $y_r = K_r x_r$. The load on the piston is that due to tool reactive force and may be assumed to consist of mass M, friction f and spring with constant K.

Draw the block diagram and obtain the transfer function $Y_c(s)/X_r(s)$, assuming that:

Rate of oil flow to the piston = K_1 × valve opening

Leakage flow across the piston = K_2 × pressure across the piston

The compressibility effect is negligible.

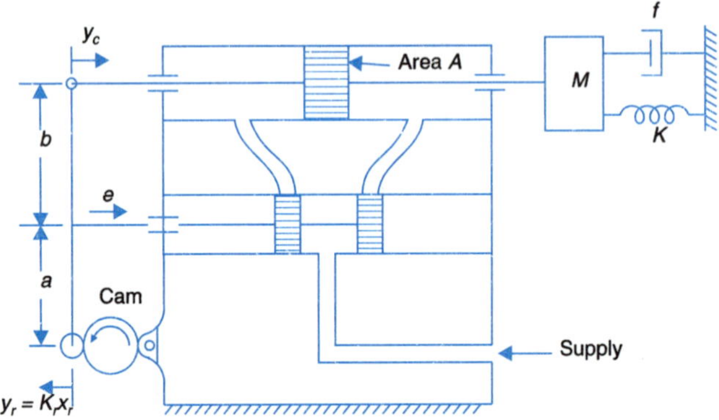

Fig. P-4.6. [From F. Raven, *Automatic Control Engineering*, 2nd ed, 1968, McGraw-Hill, New York. Reproduced with permission.]

4.7 The electrohydraulic position control system shown in Fig. P-4.6 positions a mass M with negligible friction. Assume that the rate of oil flow to the piston is $q = K_x x - K_p p$, where x is the control valve opening and p is the differential pressure. The mass of piston, oil compressibility and leakage are assumed negligible. Draw the signal flow graph of the system and obtain therefrom the transfer function $Y(s)/X_i(s)$. How the transfer function will modify if the mass M is also negligible such that the differential pressure p needed to move the piston is nearly zero ?

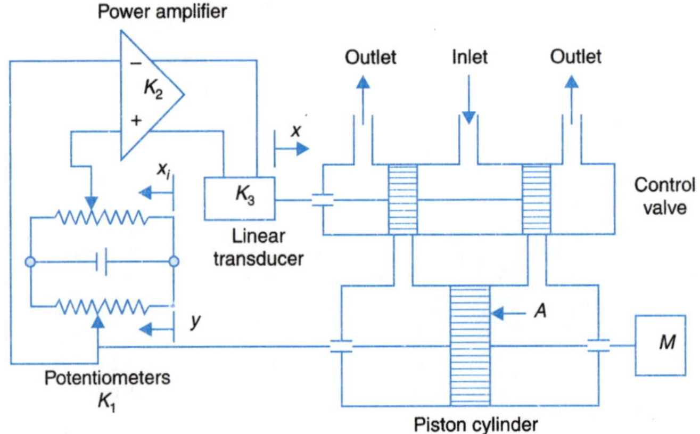

Fig. 4.7

The system constants are given below:

Mass,	$M = 1{,}000$ kg
Control valve constants,	$K_x = 200$ cm^2/sec/cm of valve opening
	$K_p = 5$ cm^3/sec/N/m^3
Potentiometer sensitivity,	$K_1 = 1$ volt/cm
Power amplifier gain,	$K_2 = 500$ mA/volt
Linear transducer constant,	$K_3 = 0.1$ cm/mA
Piston area	$A = 100$ cm^2.

4.8 Fig. P-4.8 shows a bellows with a restrictor. Show that the transfer function of the system is

$$\frac{X(s)}{P_i(s)} = \frac{A/(K+k)}{sT+1}$$

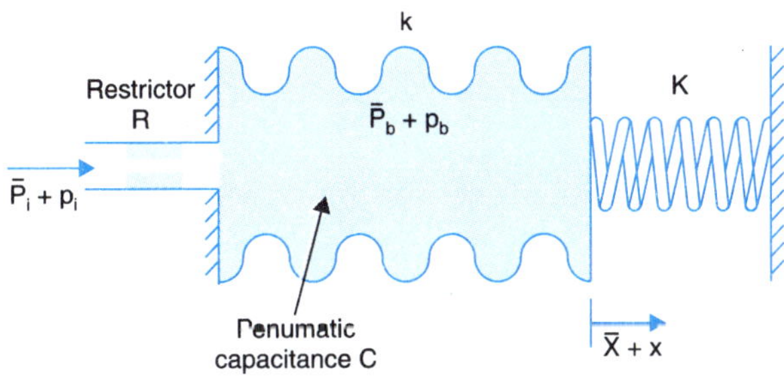

Fig. P-4.8

where $T = RC$; R = pneumatic resistance of restrictor; C = pneumatic capacitance of bellows ; and A = cross-sectional area of base of bellows.

4.9 A position balance pneumatic controller is shown diagrammatically in Fig. 4.9. Determine the transfer function of the system relating the controlled pressure p_c to the movement e, in terms of lever dimensions a and b and the restrictor calibration $T = RC$.

(The supply pressure P_s is assumed to be constant and controlled pressure $p_c = K_p x$, where K_p is a constant).

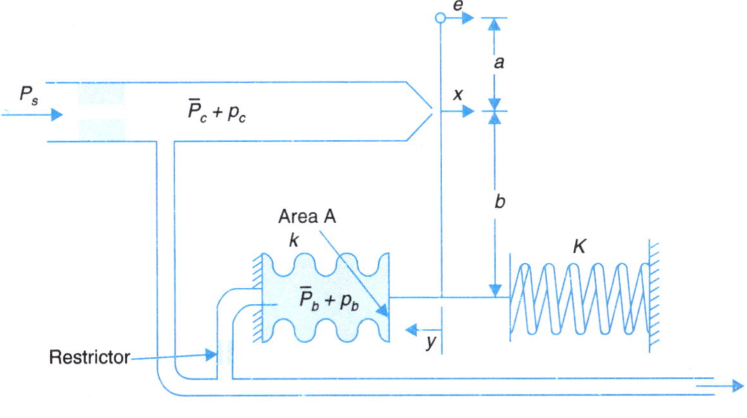

Fig. P-4.9

4.10

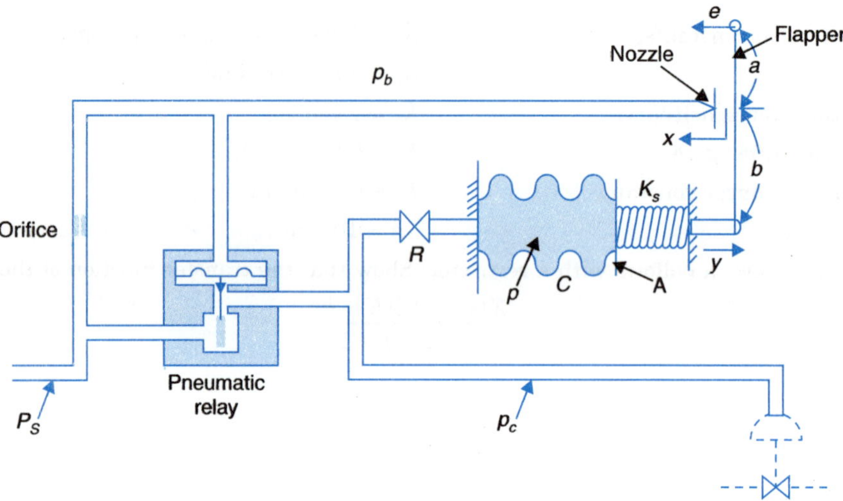

Fig. P-4.10

In the pneumatic controller of Fig. P-4.10 the bellows is spring loaded with spring constant K_s (this includes the effect of the bellows stiffness as well). Draw the block diagram for incremental changes about the steady values (indicated in the figure). Derive therefrom the overall transfer function $P(s)/E(s)$. Comment upon the control law.

4.11 Consider the pneumatic controller of Fig. P-4.11. Draw its block diagram for incremental changes (indicated in the figure) and find the overall transfer function $Y(s)/E(s)$.

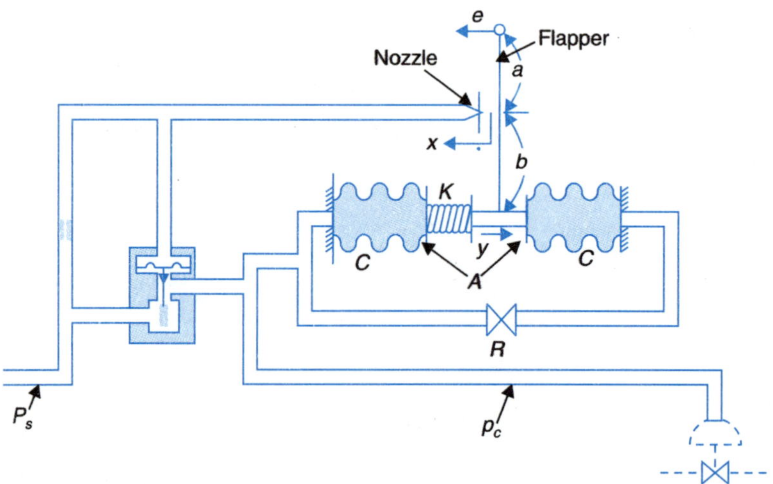

Fig. P-4.11

4.12 Figure P-4.12 shows the pitch circuit of an autopilot system. The control system receives a command signal v from gyroscope. v is some function $f(\theta)$ of θ, the deviation of the aircraft from its set condition in pitch. The signal produces a rotation of the aircraft elevator. The elevator then produces a rate of the pitch (e) of the aircraft which is fed back as shown in the figure.

(*a*) Suggest a suitable hardware for the autopilot system.

(*b*) Draw block diagram of the autopilot system with each block having the transfer function of the corresponding component in the pitch circuit.

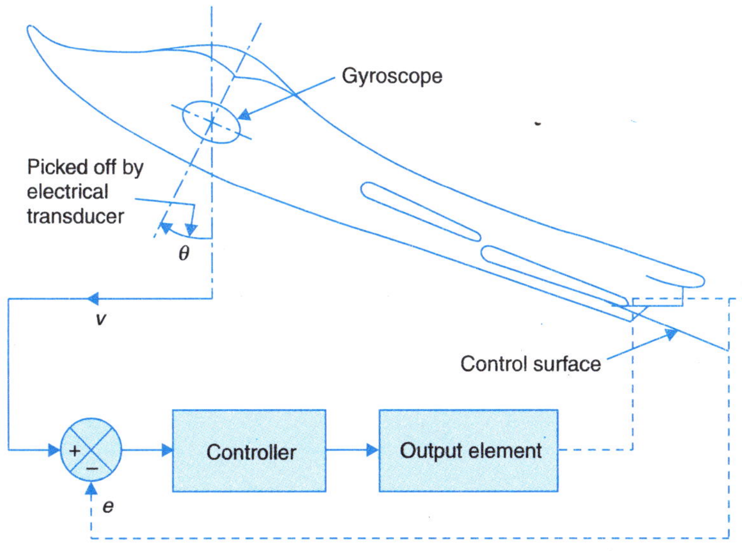

Fig. P-4.12

5

TIME RESPONSE ANALYSIS, DESIGN SPECIFICATIONS AND PERFORMANCE INDICES

5

TIME RESPONSE ANALYSIS, DESIGN SPECIFICATIONS AND PERFORMANCE INDICES

5.1 INTRODUCTION

Control systems are generally called upon to perform both under transient (dynamic) and steady conditions. A feedback control systems has the inherent capability that its parameters can be adjusted to alter both its transient and steady-state behaviour. In order to analyze the transient and steady-state behaviour (these two together are referred to as time response) of control systems, the first step always is to obtain a mathematical model of the system. For any specific input signal a complete time response expression can then be obtained through the Laplace transform inversion (or through convolution integral in case the input is such that we cannot obtain its Laplace transform). This expression yields the steady-state behaviour of the system with the time tending to infinity. In case of simple deterministic signals, steady-state response can also be obtained directly without obtaining the time response expression by use of the final value theorem.

Before proceeding with the time response analysis of a control system, it is necessary to test the stability of the system (the concept of stability shall be explained in Chapter 6). System stability can be tested through indirect tests without actually obtaining the transient response. In case the system happens to be unstable, we need not proceed with its transient response analysis.

Usually the input signals to control systems are not known fully ahead of time. In most cases these signals may be random in nature, *e.g.*, in a radar tracking system, the position and speed of the target to be tracked may vary in a random fashion. It is thus difficult to express the actual input signals mathematically by simple equations. The characteristics of actual signals which severely strain a control system are a sudden shock, a sudden change, a constant velocity and a constant acceleration. System dynamic behaviour for analysis and design is therefore judged and compared under application of standard test signals—an impulse, a step, a constant velocity (a ramp input) and constant acceleration (a parabolic input).

The nature of the transient response is revealed by any one of these test signals as this nature is dependent upon system poles and not upon the type of input.

It is therefore sufficient to analyze the transient response to one of the standard test signals—a step is generally used for this purpose as this signal can be easily generated. Steady state response is then examined to this particular test signal (the step) as well other test signals; the ramp and the parabolic signal. So except for the step test signal the time consuming transient analysis need not be carried out for the ramp and parabolic signals, while their steady state can be quickly determined by the final value theorem as illustrated in Section 5.5 and will also elaborated in other sections of this chapter.

As explained above control systems are inherently time-domain systems subject to time-varying inputs and are to be tested, analyzed and designed by the time-domain test signals like step, ramp and parabolic. The time-domain command signals in a control can also be visualized as (through Fourier Transform) a band of sinusoidal signals of frequencies from dc upwards (control systems are low-pass filters). So another important test signal for control systems is the sinusoidal signal which can be easily generated and its frequency varied. Steady-state sinusoidal response of a control system over a range of frequencies yields a great deal of information about the system; both about its time-domain response and its stability, as there is normally a good correlation between the frequency-domain response and time-domain response of a system. Chapters 8 and 9 will be wholly devoted to steady-state sinusoidal response of control systems.

The time response performance of a control system is measured by computing several time response performance indices as well as steady-state accuracy for the standard input signals described above. These indices give a quantitative method to compare the performance of alternative system configurations or to adjust the parameters of a given system. As a given parameter is varied, various performance indices may change in a conflicting manner. The best parameter choice would thus be the best compromise solution. Certain of the performance indices may be specified as upper or lower bounds in a design.

5.2 STANDARD TEST SIGNALS

Step Signal

The *step* is a signal whose value changes from one level (usually zero) to another level A in zero time. The mathematical representation of the step function is

$$r(t) = Au\,(t)$$

where

$$u(t) = 1\,;\, t > 0 \qquad \text{...(5.1)}$$

$$= 0\,;\, t < 0$$

In the Laplace transform form, $R(s) = A/s$

The graphical representation of a step signal is shown in Fig. 5.1(a).

Ramp Signal

The *ramp* is a signal which starts at a value of zero and increases linearly with time. Mathematically,

$$r(t) = At\,;\, t > 0$$

$$= 0\,;\, t < 0 \qquad \text{...(5.2)}$$

In the Laplace transform form,

$$R(s) = A/s^2$$

The graphical representation of a ramp signal is shown in Fig. 5.1(*b*). From eqns. (5.1) and (5.2), it is seen that a ramp signal is integral of a step signal.

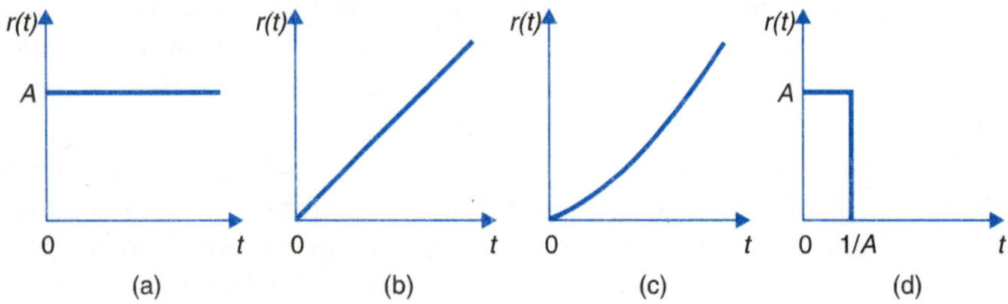

Fig. 5.1. Standard test signals.

Parabolic Signal

The mathematical representation of this signal is

$$r(t) = At^2/2 \; ; \; t > 0$$
$$= 0 \; ; \; t < 0 \qquad \text{...(5.3)}$$

In the Laplace transform form,

$$R(s) = A/s^3$$

The graphical representation of a parabolic signal is shown in Fig. 5.1(*c*). From eqns. (5.2) and (5.3), it is seen that a parabolic signal is integral of a ramp signal.

Impulse Signal

A unit-impulse is defined as a signal which has zero value everywhere except at $t = 0$, where its magnitude is infinite. It is generally called the δ-function and has the following property:

$$\delta(t) = 0 \; ; \; t \neq 0$$

$$\int_{-\varepsilon}^{+\varepsilon} \delta(t)dt = 1$$

where ε tends to zero.

Since a perfect impulse cannot be achieved in practice, it is usually approximated by a pulse of small width but unit area as shown in Fig. 5.1(*d*).

Mathematically, an impulse function is the derivative of a step function *i.e.,*

$$\delta(t) = \dot{u}(t)$$

The Laplace transform of a unit-impulse is

$$\mathcal{L}\delta(t) = 1 = R(s)$$

The impulse response of a system with transfer function $C(s)/R(s) = G(s)$, is given by

$$C(s) = G(s)\,R(s) = G(s)$$

or

$$c(t) = \mathcal{L}^{-1}G(s) = g(t) \qquad \text{...(5.4)}$$

Thus the impulse response of a system, indicated by $g(t)$, is the inverse Laplace transform of its transfer function. This is sometimes referred to as *weighting function* of the system.

Weighting function of a system can be used to find the system's response to any input $r(t)$ (even though its Laplace transform cannot be found) by means of the convolution integral. Thus

$$c(t) = \int_0^t g(t-\tau)r(\tau)d\tau \quad \text{...(5.5)}$$

5.3 TIME RESPONSE OF FIRST-ORDER SYSTEMS

Let us consider the first-order system of Fig. 5.2 with unity fedback. As an example, this can be visualized to be the block diagram of the pneumatic system of Fig. 2.16. Whose dynamics is described by the Laplace transform of eqn. (2.36) wherein one let $T = RC$, $R(s) = \Delta P_r(s)$ and $C(s) = \Delta P_0(s)$.

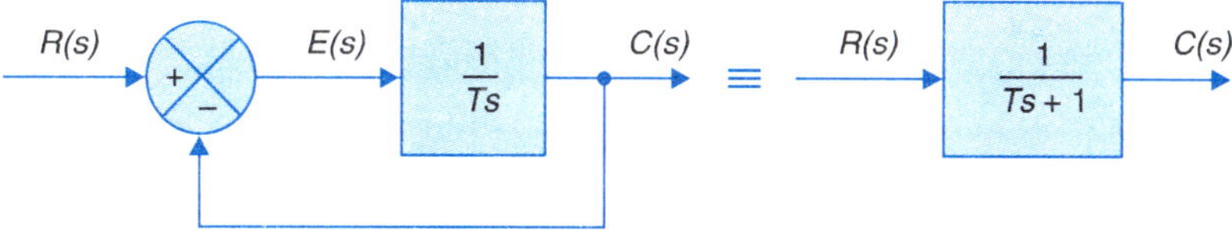

Fig. 5.2. Block diagram of a first-order system.

The transfer function of the system from Fig. 5.2 is

$$\frac{C(s)}{R(s)} = \frac{1}{Ts+1} \quad \text{...(5.6)}$$

In the following sections we shall analyze the system response to unit-step and unit-ramp inputs assuming zero initial conditions.

Response to the Unit-Step Input

For the unit-step input [$R(s) = 1/s$], from eqn. (5.6), the output response is given by

$$C(s) = \frac{1}{s(Ts+1)} = \frac{1}{s} - \frac{T}{Ts+1}$$

Taking the inverse Laplace transform, we get

$$c(t) = 1 - e^{-t/T}$$

which is plotted in Fig. 5.3.

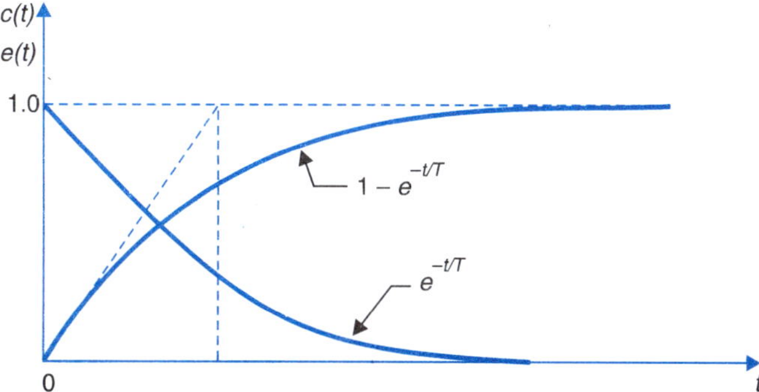

Fig. 5.3. Unit-step response of a first-order system.

It is seen that the output rises exponentially from zero value to the final value of unity. The initial slope of curve at $t = 0$ is given by

$$\left.\frac{dc}{dt}\right|_{t=0} = \left.\frac{1}{T}e^{-t/T}\right|_{t=0} = \frac{1}{T}$$

where T is known as the *time constant* of the system.

The time constant is indicative of how fast the system tends to reach the final value. The speed of response can be quantitatively defined as the time for the output to become a particular percentage of its final value. A large time constant corresponds to a sluggish system and a small time constant corresponds to a fast response as shown in Fig. 5.4.

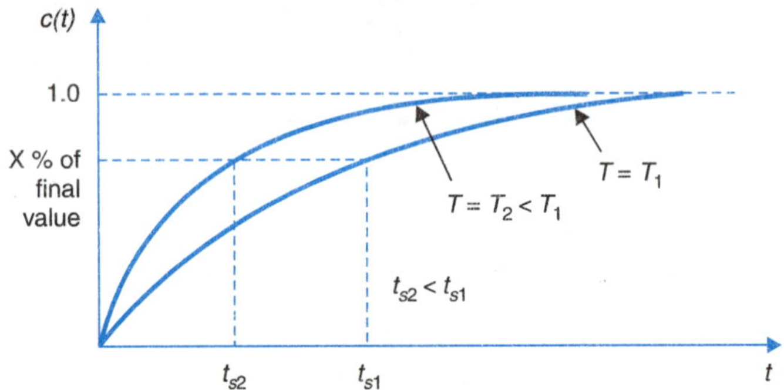

Fig. 5.4. Effect of time constant on system response.

Consider now, the error response of the system which is given by

$$e(t) = r(t) - c(t) = e^{-t/T}$$

which is also plotted in Fig. 5.3. The steady-state error is given by

$$e_{ss} = \lim_{t\to\infty} e(t) = 0$$

Thus this system tracks the unit-step input with zero steady-state error.

Response to the Unit-Ramp Input

From eqn. (5.5), the output response for the unit-ramp input $[R(s) = 1/s^2]$ is given

$$C(s) = \frac{1}{s^2(Ts+1)} = \frac{1}{s^2} - \frac{T}{s} + \frac{T^2}{Ts+1}$$

Taking the inverse Laplace transform of the above equation, we get

$$c(t) = t - T(1 - e^{-t/T})$$

The error signal is

$$e(t) = r(t) - c(t) = T(1 - e^{-t/T})$$

and the steady-state error is given by

$$e_{ss} = \lim_{t\to\infty} e(t) = T$$

Thus the first-order system under consideration will track the unit-ramp input with a steady-state error T, which is equal to the time constant of the system, as shown in Fig. 5.5.

Reducing the system time constant therefore not only improves its speed of response but also reduces its steady-state error to a ramp input.

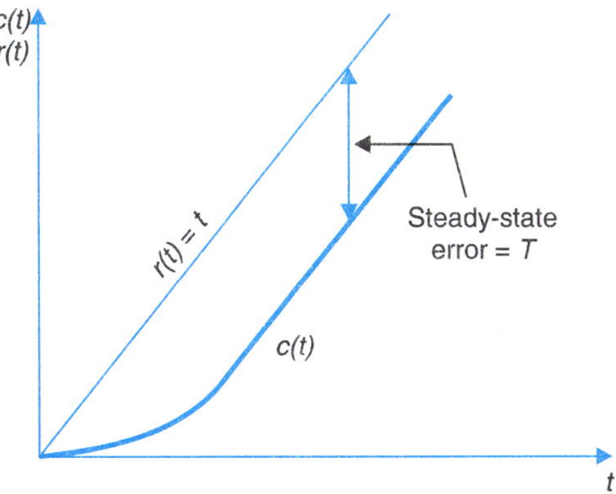

Fig. 5.5. Unit-ramp response of a first-order system.

If we examine the derivative of $c(t)$, *i.e.*,

$$\dot{c}(t) = 1 - e^{-t/T}$$

we find that it is identical to the system response to the unit-step input. The transient response to the ramp input signal thus yields no additional information about the speed of response of the system. We therefore need examine only the steady-state error to the ramp input which can be obtained directly by the final value theorem as given below:

$$e_{ss} = \lim_{t\to\infty} e(t) = \lim_{s\to 0} sE(s) = \lim_{s\to 0} s[R(s) - C(s)]$$

$$= \lim_{s\to 0} s\left[\frac{1}{s^2} - \frac{1}{s^2(Ts+1)}\right] = T$$

This avoids the need to obtain the inverse Laplace transform resulting in a considerable labour saving in higher-order systems.

5.4 TIME RESPONSE OF SECOND-ORDER SYSTEMS

Consider the servomechanism shown in Fig. 5.6, which controls the position of a mechanical load in accordance with the position of the reference shaft. The two potentiometers convert the input and output positions into proportional electrical signals, which are in turn compared and an error signal equal to the difference of the two appears at the leads coming from potentiometer wiper arms.

The error signal (voltage) is

$$v_e = K_p(r - c)$$

where r = reference shaft position in rad; c = output shaft position in rad; and K_p = potentiometer sensitivity in volts/rad.

The error signal is amplified by a factor K_A by the amplifier and is applied to the armature circuit of the d.c. motor whose field winding is excited with a constant voltage or it could also be a PM dc motor. If any error exists, the motor develops a torque which is transmitted to the output shaft through a gear train of ratio $n(n$ = load shaft speed $\dot{c}$/motor shaft speed $\dot{\theta}$). The transmitted torque rotates the output shaft in such a direction so as to reduce the error to zero.

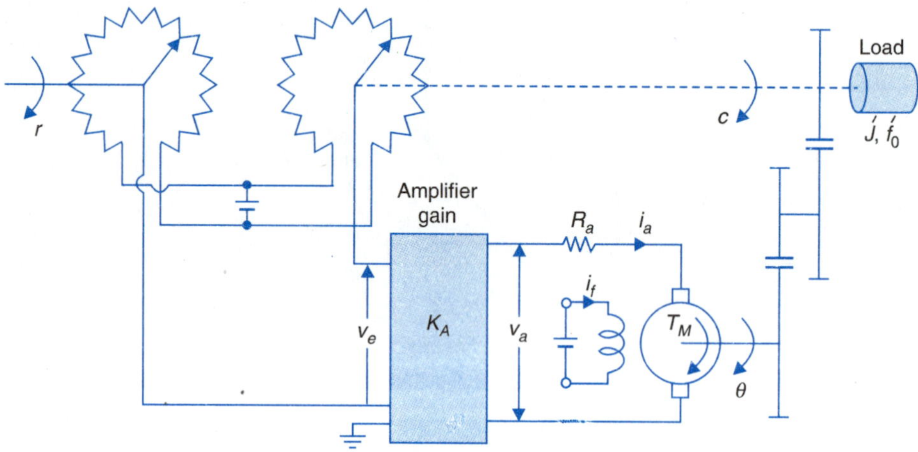

Fig. 5.6. Schematic diagram of position servomechanism.

The block diagram of the system is shown in Fig. 5.7(a). Here J and f_0 are the equivalent inertia and friction at the motor shaft.

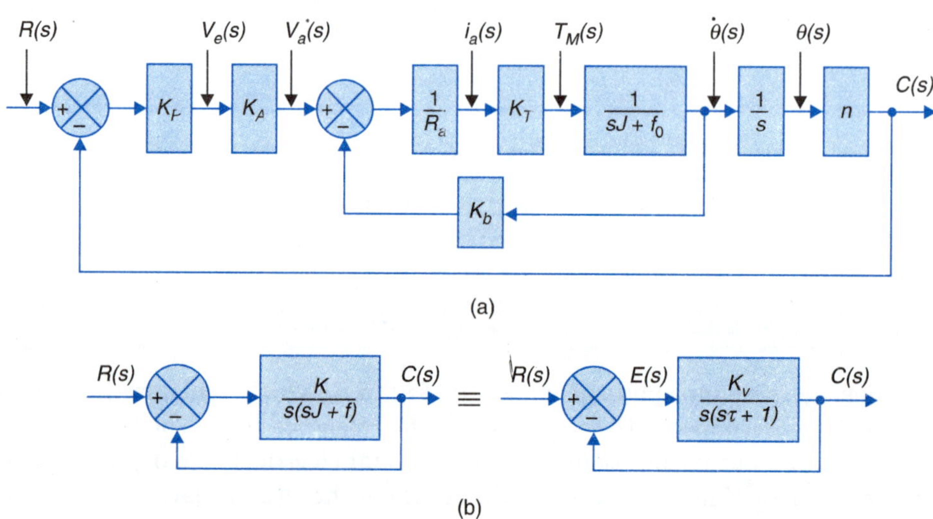

Fig. 5.7. (a) Block diagram of the system shown in Fig. 5.6;
(b) Simplified block diagram of the system.

The inner loop can be reduced to give the motor transfer function as

$$\frac{\theta(s)}{V_a(s)} = \frac{K_T/R_a}{s(sJ + f)}$$

where $f = f_0 + \dfrac{K_T K_b}{R_a}$.

The block diagram can now be simplified to the form of Fig. 5.7(*b*), where

$$K = K_P K_A K_T \frac{n}{R_a}$$

The forward transfer function can be written in the time constant form as

$$G(s) = \frac{K_v}{s(\tau s + 1)}$$

where $K_v = K/f$, $\tau = J/f$.

Fig. 5.7(*b*) is in fact a standard second-order system involving one forward path integration. As shall be explained later in this chapter, this configuration belongs to a general class of systems called type-1.

Response of Second-order System to the Unit-step

The analysis, now onwards, is not restricted to the particular system used as an example above but is valid for a general second-order type-1 system. From Fig. 5.7(*b*), the overall transfer function of the system is

$$\frac{C(s)}{R(s)} = \frac{K_v}{\tau s^2 + s + K_v} = \frac{K_v/\tau}{s^2 + \dfrac{1}{\tau}s + \dfrac{K_v}{\tau}} \qquad ...(5.7)$$

It can also be written in the standard form

$$\frac{C(s)}{R(s)} = \frac{\omega_n^2}{s^2 + 2\zeta\omega_n s + \omega_n^2} = \frac{p(s)}{q(s)} \qquad ...(5.8)$$

where ζ = *damping factor* (or damping ratio) $= \dfrac{1}{2\sqrt{(K_v \tau)}} = \dfrac{f}{2\sqrt{(KJ)}}$;

and ω_n = *undamped natural frequency* $= \sqrt{K_v / \tau} = \sqrt{(K/J)}$.

The time response of any system is characterized by the roots of the denominator polynomial $q(s)$, which in fact are the poles of the transfer function. The denominator polynomial $q(s)$ is therefore called the *characteristic polynomial* and

$$q(s) = 0 \qquad ...(5.9)$$

is called the *characteristic equation*. The characteristic equation of the system under consideration is

$$s^2 + 2\zeta\omega_n s + \omega_n^2 = 0 \qquad ...(5.10)$$

The roots of this characteristic equation are given by

$$(s^2 + 2\zeta\omega_n s + \omega_n^2) = (s - s_1)(s - s_2)$$

For $\zeta < 1$,

$$s_1, s_2 = -\zeta\omega_n \pm j\omega_n\sqrt{(1-\zeta^2)} = -\zeta\omega_n \pm j\omega_d$$

where $\omega_d = \omega_n\sqrt{(1-\zeta^2)}$, is called the *damped natural frequency*.

Most control systems with the exception of robotic control systems are designed with damping factor $\zeta < 1$ to achieve high response speed consistent with other performance measures (explained later in this chapter).

From eqn. (5.7), for the unit-step input, the output response is given by

$$C(s) = \frac{\omega_n^2}{s[s+\zeta\omega_n - j\omega_n\sqrt{(1-\zeta^2)}][s+\zeta\omega_n + j\omega_n\sqrt{(1-\zeta^2)}]} \quad ...(5.11)$$

The Laplace inverse* of eqn. (5.11) is obtained, by the method of residues (refer Appendix I) as

$$c(t) = \left.\frac{\omega_n^2}{s^2 + 2\xi\omega_n s + \omega_n^2}\right|_{s=0}$$

$$+ 2Re\left[\left.\frac{\omega_n^2}{s[s+\zeta\omega_n - j\omega_n\sqrt{(1-\zeta^2)}]}\right|_{s=-\zeta\omega_n - j\omega_n\sqrt{(1-\zeta^2)}} e^{-\zeta\omega_n t}\, e^{-j\omega_n\sqrt{(1-\zeta^2)}t}\right]$$

$$= 1 - \frac{e^{-\zeta\omega_n t}}{\sqrt{(1-\zeta^2)}}\sin\left[\omega_n\sqrt{(1-\zeta^2)}\,t + \tan^{-1}\frac{\sqrt{(1-\zeta^2)}}{\zeta}\right];\ t > 0 \quad ...(5.12)$$

The steady-state value of $c(t)$ is given as

$$c_{ss} = \lim_{t\to\infty} c(t) = 1$$

The time response of an underdamped ($\zeta < 1$) second-order system is plotted from eqn. (5.11) in Fig. 5.8. It is damped sinusoidal. The response reaches a steady-state value of $c_{ss} = 1$, *i.e.*, the steady-state error of this system approaches zero. The time response for various values of ζ plotted against normalized time $\omega_n t$ is shown in Fig. 5.9. The system breaks into continuous oscillations for $\zeta = 0$, as can be seen from eqn. (5.12). The mathematical expression for the time response in this case is given by

$$c(t) = 1 - \cos\omega_n t$$

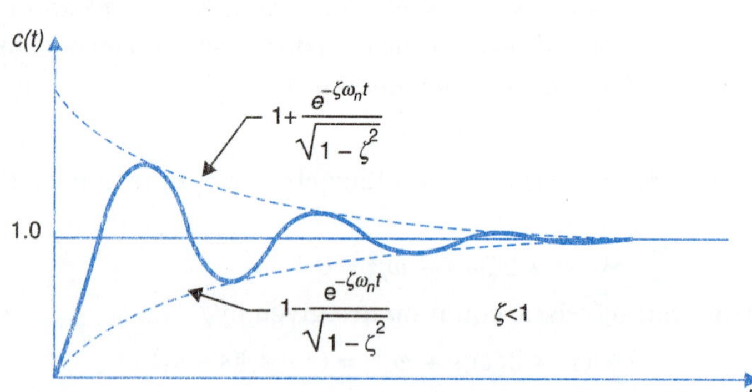

Fig. 5.8. Unit-step response of underdamped second-order system.

*It can also be written down directly from Table I.2, item 17 of Appendix I.

As ζ is increased, the response becomes progressively less oscillatory till it becomes critically damped (just non-oscillatory) for $\zeta = 1$ and becomes overdamped for $\zeta > 1$. Robotic control systems cannot be allowed to have oscillatory response otherwise the end effector would strike against the object that the robot is meant to manipulate. Highest possible speed of response and yet non-oscillatoring response dictates that a robotic control system shall be designed to have a damping factor of $\zeta = 1$ (or close to it but less then unity). For $\zeta = 1$, it easily follows by inverse Laplace transferring eqn (5.12) that

$$c(t) = 1 - e^{-\omega_n t} - \omega_n t\, e^{-\omega_n t} \qquad ...(5.13)$$

where from the characteristic equation (5.10), the roots are

$$s_1 = s_2 = -\omega_n$$

i.e., repeated real negative roots, also see Fig. 5.10.

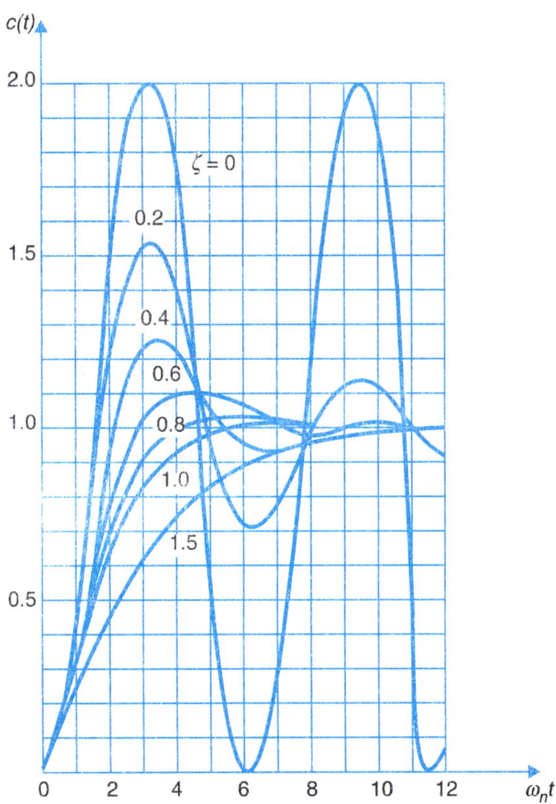

Fig. 5.9. Unit-step response curves of second-order system.

Fig. 5.10 shows the locus of the poles of the second-order system discussed above with ω_n held constant and ζ varying from 0 to ∞. As ζ increases the poles move away from the imaginary axis along a circular path of radius ω_n meeting at the point $\sigma = -\omega_n$ and then separating and travelling along the real axis, one towards zero and the other towards infinity. For $0 < \zeta < 1$, the poles are complex conjugate pair making an angle of $\theta = \cos^{-1} \zeta$ with the negative real axis.

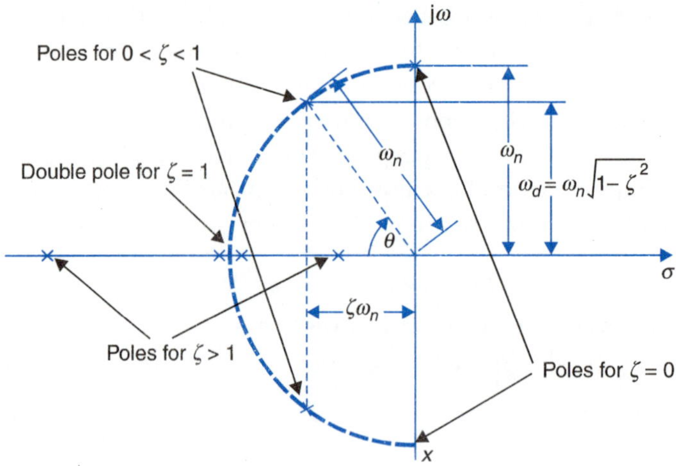

Fig. 5.10. Pole locations for a second-order system.

Time Response Specifications

As pointed out earlier, control systems are generally designed with damping less than one, *i.e.*, oscillatory step response. High-order control systems usually have a pair of complex conjugate poles with damping less than one which dominate over other poles. Therefore the time response of second-and higher-order control systems to a step input is generally of damped oscillatory nature as shown in Fig. 5.11. It is observed from this figure that the step response has a number of overshoots and undershoots with respect to the final steady value. Since the overshoots and undershoots decay exponentially, the peak overshoot is the first overshoot and is the same as the peak of the complete time response. This type of step response is characterized by the following performance indices. The indices are qualitatively related to:

(*i*) How fast the system moves to follow the input ?

(*ii*) How oscillatory it is (indicative of damping) ?

(*iii*) How long does it take to practically reach the final value ?

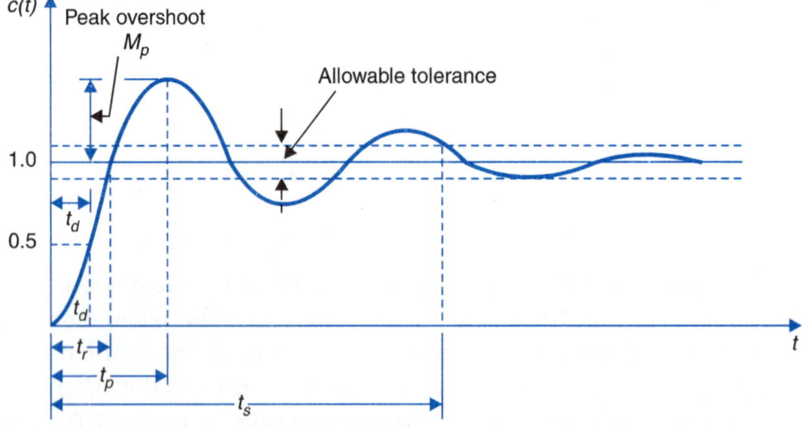

Fig. 5.11. Time response specifications.

It may be noted that various indices are not independent of each other.

1. *Delay time* t_d : It is the time required for the response to reach 50% of the final value in first attempt.

2. *Rise time* t_r : It is the time required for the response to rise from 10% to 90% of the final value for overdamped systems and 0 to 100% of the final value for underdamped systems. Fig. 5.11 shows an underdamped case.

3. *Peak time* t_p : It is the time required for the response to reach the peak of time response or the peak overshoot.

4. *Peak overshoot* M_p : It indicates the normalized difference between the time response peak and the steady output and is defined as

$$\text{Peak percent overshoot} = \frac{c(t_p) - c(\infty)}{c(\infty)} \times 100\% \qquad \text{...(5.14)}$$

In most control systems (except type-0, defined later in this chapter) the steady output for step input is the same as the input. For example, for the second order system of Fig. 5.7(*b*), $c(\infty) = c_{ss} = 1$, *i.e.*, the same as the input.

5. *Settling time* t_s : It is the time required for the response to reach and stay within a specified tolerance band (usually 2% or 5%) of its final value.

6. *Steady-state error* e_{ss} : It indicates the error between the actual output and desired output as t tends to infinity, *i.e.*,

$$e_{ss} = \lim_{t \to \infty} [r(t) - c(t)]$$

From Fig. 5.11 it is seen that by specifying t_d, t_r, t_p, M_p, t_s and e_{ss}, the shape of the unit-step time response curve is virtually fixed. It should however, be noted that these specifications are mutually dependent and must be specified in a consistent manner.

Time Response Specifications of Second-Order Systems

Let us obtain the expressions for the rise time, peak time, peak overshoot and settling time for a second-order system whose block diagram is shown in Fig. 5.7(*b*) and whose dynamics for the underdamped case is described by eqn (5.12).

1. *Rise time* t_r : From eqn. (5.12) the rise time t_r is obtained when $c(t)$ reaches unity for the first time, *i.e.*,

$$c(t_r) = 1 - \frac{e^{-\zeta\omega_n t_r}}{\sqrt{(1-\zeta^2)}} \sin\left[\omega_n \sqrt{(1-\zeta^2)}\, t_r + \tan^{-1}\frac{\sqrt{(1-\zeta^2)}}{\zeta}\right] = 1$$

From the above equation, we get

$$t_r = \frac{\pi - \tan^{-1}\dfrac{\sqrt{(1-\zeta^2)}}{\zeta}}{\omega_n \sqrt{(1-\zeta^2)}} \qquad \text{...(5.15)}$$

For $0 < \zeta < 1$,

$$0 < \tan^{-1}\frac{\sqrt{(1-\zeta^2)}}{\zeta}\,\zeta < \frac{\pi}{2} \qquad \text{...(5.16)}$$

2. *Peak time* t_p : Differentiating eqn. (5.11) with respect to t and equating to zero, we get

$$\frac{dc(t)}{dt} = \frac{\zeta\omega_n}{\sqrt{(1-\zeta^2)}} e^{-\zeta\omega_n t} \sin\left[\omega_n\sqrt{(1-\zeta^2)}\,t + \tan^{-1}\frac{\sqrt{(1-\zeta^2)}}{\zeta}\right]$$

$$- \frac{e^{-\zeta\omega_n t}}{\sqrt{(1-\zeta^2)}}\,\omega_n\sqrt{(1-\zeta^2)}\cos\left[\omega_n\sqrt{(1-\zeta^2)}\,t + \tan^{-1}\frac{\sqrt{(1-\zeta^2)}}{\zeta}\right] = 0 \qquad ...(5.17)$$

With reference to Fig. 5.12, eqn. (5.17) may be written as

$$\sin[\omega_n\sqrt{(1-\zeta^2)}\,t + \phi]\cos\phi - \cos[\omega_n\sqrt{(1-\zeta^2)}\,t + \phi]\sin\phi = 0$$

or

$$\sin[\omega_n\sqrt{(1-\zeta^2)}\,] = 0 \qquad ...(5.18)$$

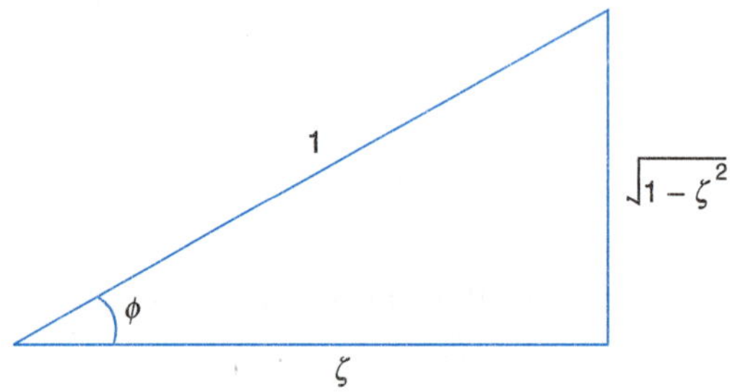

Fig. 5.12

Therefore, the time to various peaks is given by

$$\omega_n\sqrt{(1-\zeta^2)}\,t = 0,\ \pi,\ 2\pi,\ 3\pi, ...$$

Since the peak time corresponds to the first overshoot,

$$t_p = \frac{\pi}{\omega_n\sqrt{(1-\zeta^2)}} \qquad ...(5.19)$$

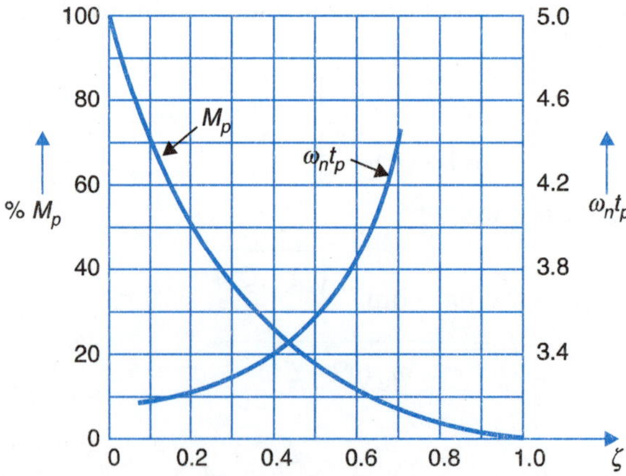

Fig. 5.13. M_p and $\omega_n t_p$ versus ζ for a second-order system.

The first undershoot will occur at $t = 2\pi/\omega_n\sqrt{(1-\zeta^2)}$, the second over shoot at $t = 3\pi/\omega_n\sqrt{(1-\zeta^2)}$ and so on. A plot of normalized peak time $\omega_n t_p$ versus ζ is given in Fig. 5.13.

3. Peak overshoot M_p : From eqn. (5.11) and Fig. 5.11,

$$M_p = c(t_p) - 1$$

$$= -\frac{e^{-\zeta\omega_n t_p}}{\sqrt{(1-\zeta^2)}} \sin\left[\omega_n\sqrt{(1-\zeta^2)}t_p + \tan^{-1}\frac{\sqrt{(1-\zeta^2)}}{\zeta}\right]$$

$$= e^{-\pi\zeta/\sqrt{(1-\zeta^2)}}$$

Therefore the peak percent overshoot

$$= 100\, e^{-\pi\zeta/\sqrt{(1-\zeta^2)}}\ \% \qquad \text{...(5.20)}$$

As seen from Fig. 5.13, the peak overshoot is a monotonically decreasing function of damping ζ and is independent of ω_n. It is, therefore an excellent measure of system damping.

4. *Settling time* t_s : From Fig. 5.8, it is observed that the time response $c(t)$ given by eqn. (5.12) for $\zeta < 1$, oscillates between a pair of envelopes before reaching steady-state. The transient is comprised of a product of an exponentially decaying term $[\exp(-\zeta\omega_n t)]/\sqrt{(1-\zeta^2)}$ and a sinusoidally oscillating term $\sin[\omega_n\sqrt{(1-\zeta^2)}t + \phi]$. The time constant of the exponential envelopes is $T = 1/\zeta\omega_n$. It may be noted that this time constant is equal to 2τ where τ is the motor time constant in Fig. 5.7(*b*).

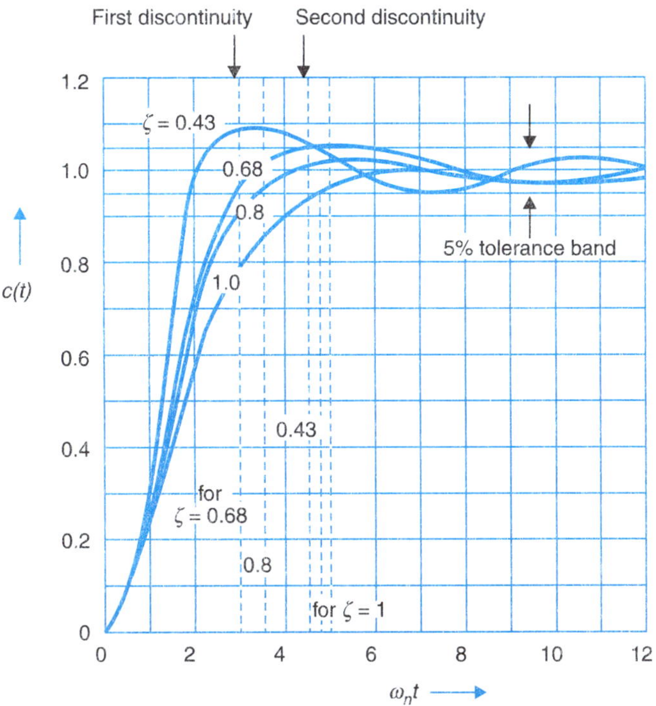

Fig. 5.14. Settling time for various values of ζ.

Fig. 5.14 shows the unit-step response versus normalized time $\omega_n t$ for various values of ζ as well as the tolerance band. It is observed from this figure that as the damping ζ is reduced from the value of unity (corresponding to the critical damping case), the normalized settling time decreases monotonically till the first overshoot just touches the upper limit of the tolerance band. As the damping is decreased further, $\omega_n t_s$ suddenly jumps up (*i.e.,* it has a discontinuity and then increases slightly with further decrease in damping. The plot of $\omega_n t_s$ versus ζ is shown in Fig. 5.15. The plot has another discontinuity corresponding to the first undershoot touching the lower limit of the tolerance band and similar discontinuities occur corresponding to each overshoot and undershoot, respectively touching the upper and lower limits of the tolerance band.

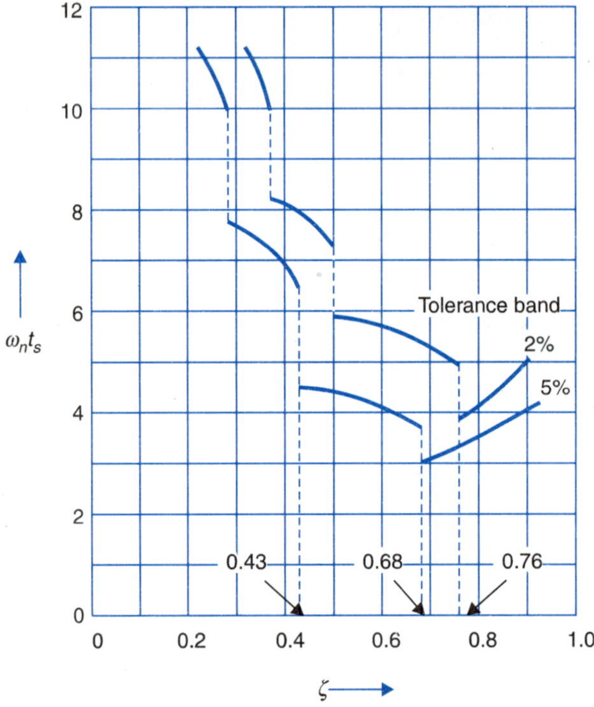

Fig. 5.15. Normalized settling time versus ζ.

The normalized settling time $\omega_n t_s$ is the least at the first discontinuity ($\zeta = 0.76$ for 2% tolerance band and $\zeta = 0.68$ for 5% tolerance band). This fact justifies why the control systems are normally designed to be underdamped. This argument is further strengthened by the observation that rise time reduces monotonically as the damping is decreased.

In fact many control systems are designed to have even lower damping. Justification of this lies in the fact that almost all practical systems possess some coulomb friction and other nonlinearities (backlash, binding in gears and linkages, etc.). The presence of these nonlinearities tends to introduce steady-state error. As these nonlinearities do not appear in the linear mathematical model, therefore in order to compensate for error introduced by them, the linear system is designed with somewhat higher gain and hence lower damping since $\zeta = (1/2)\sqrt{(K_v \tau)}$.

As already pointed out robotic control systems are designed with damping factor close to but more than unity. Further there are constraints on their natural frequency to avoid resonance conditions in their structural arrangement. (see Eq. (5.80)).

For relatively lower values of damping, the time response is oscillatory and an approximate expression for settling time can be easily obtained by finding the time when the envelope of the damped sinusoidal shown in Fig. 5.8 enters the tolerance band. It is sufficiently accurate to assume that the overshoots do not go outside the tolerance band after this time.

Thus, considering only the exponential decaying envelope for a tolerance band of 2%, the settling time is given by

$$\frac{e^{-\zeta\omega_n t_s}}{\sqrt{(1-\zeta^2)}} = 0.02$$

$$e^{-\zeta\omega_n t_s} \approx 0.02 \text{ (for low values of } \zeta)$$

or

$$t_s \approx \frac{4}{\zeta\omega_n} = 4T \qquad \text{...(5.21)}$$

where T is the time constant of the exponential term [in fact $T = 2\tau$, where τ is the time constant of the open-loop transfer function $G(s)$ in Fig. 5.7(*b*)]. Approximate value of t_s is $3/\zeta\omega_n$ or $3T$ for 5% tolerance band.

It is seen that the settling time is inversely proportional to ω_n.

5. *Steady-step error* e_{ss} : From eqn. (5.12), for unit-step input

$$e_{ss} = \lim_{t\to\infty} [1 - c(t)] = 0$$

Thus the type of second-order system under consideration has zero steady state error to unit-step input.

From eqn. (5.8), the response to a unit-ramp input [$r(t) = t$; $R(s) = 1/s^2$], is given by

$$c(t) = \mathcal{L}^{-1}\left[\frac{\omega_n^2}{s^2(s^2 + 2\zeta\omega_n s + \omega_n^2)}\right]$$

$$= t - \frac{2\zeta}{\omega_n} + \frac{e^{-\zeta\omega_n t}}{\omega_n\sqrt{(1-\zeta^2)}} \sin[\omega_n\sqrt{(1-\zeta^2)}t + \phi] \text{ for } \zeta < 1$$

$c(t)$ for various values of ζ is plotted in Fig. 5.16. The steady-state error for unit-ramp input is given by

$$e_{ss} = \lim_{t\to\infty} [t - c(t)] = \frac{2\zeta}{\omega_n} = \frac{1}{K_v}$$

From Fig. 5.16, it is observed that the nature of the transient response to the ramp input is similar to that of the step input, *i.e.*, it is damped oscillatory and yields no new information about the transient response of the system. As pointed out earlier, it is therefore sufficient to test the transient response of the system for a step input only. Ramp input response, of course, gives new information about its steady-state behaviour which may be evaluated directly by the final value theorem.

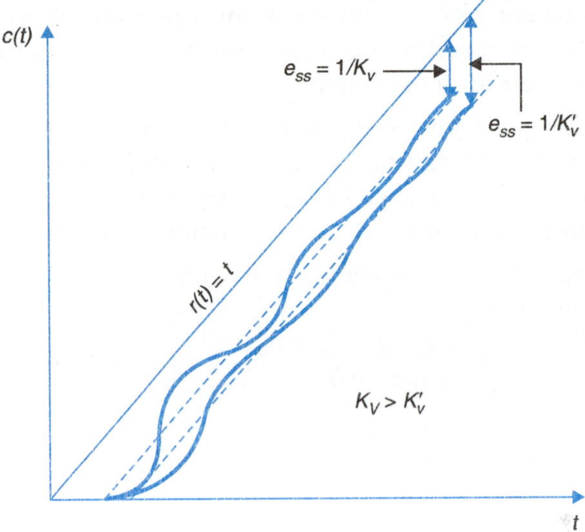

Fig. 5.16. Unit-ramp response of second-order system.

For the system under consideration,

$$e_{ss} = \lim_{s \to 0} s\left[\frac{1}{s^2} - C(s)\right]$$

Substituting for $C(s)$ from eqn. (5.7), we get

$$e_{ss} = 2\zeta/\omega_n = 1/K_v \quad \text{...(5.22)}$$

The result will be further illustrated in the following section on steady-state errors.

5.5 STEADY-STATE ERRORS AND ERROR CONSTANTS

Steady-state errors constitute an extremely important aspect of system performance, for it would be meaningless to design for dynamic accuracy if the steady output differed substantially from the desired value for one reason or another. The steady-state error is a measure of system accuracy. These errors arise from the nature of inputs, type of system and from nonlinearities of system components such as static friction, backlash, etc. These are generally aggravated by amplifier drifts, aging or deterioration.

As discussed in the introduction to this chapter, the steady-state performance of a stable control system is generally judged by its steady-state error to step, ramp and parabolic inputs.

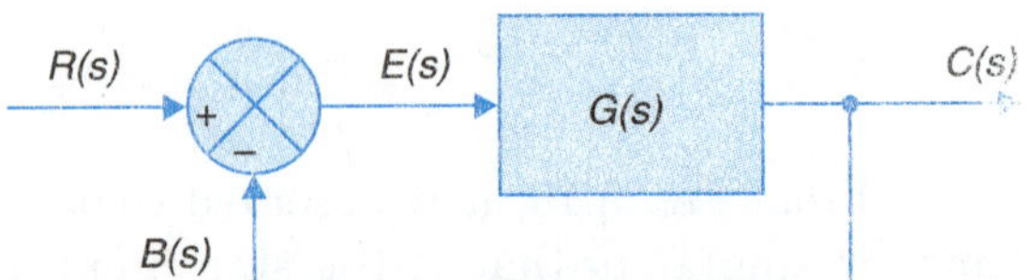

Fig. 5.17. Unit feedback system.

Consider a unit feedback system shown in Fig. 5.17. The input is $R(s)$, the output $C(s)$, the feedback signal $B(s)$ and the difference between input and output is the error signal $E(s)$.

From Fig. 5.17, we see that

$$\frac{C(s)}{R(s)} = \frac{G(s)}{1+G(s)}$$

$$C(s) = E(s)G(s)$$

Therefore
$$E(s) = \frac{1}{1+G(s)} R(s) \qquad ...(5.23)$$

The steady-state error e_{ss} may now be found by use of the final value theorem as follows.

$$e_{ss} = \lim_{t\to\infty} e(t) = \lim_{s\to 0} sE(s) = \lim_{s\to 0} \frac{sR(s)}{1+G(s)} \qquad ...(5.24)$$

Equation (5.24) shows that the steady-state error depends upon the input $R(s)$ and the forward transfer function $G(s)$. The expression for steady-state errors for various types of standard test signals are derived below:

1. *Unit-step Input*

$$\text{Input} \quad r(t) = u(t)$$

$$R(s) = 1/s$$

From eqn. (5.24)

$$e_{ss} = \lim_{s\to 0} \frac{1}{1+G(s)} = \frac{1}{1+G(0)} = \frac{1}{1+K_p} \qquad ...(5.25)$$

where $K_p = G(0)$ is defined as the *position error constant.*

2. *Unit-ramp (Velocity) Input*

$$\text{Input} \quad r(t) = t \text{ or } \dot{r}(t) = 1$$

$$R(s) = 1/s^2$$

From eqn. (5.24)

$$e_{ss} = \lim_{s\to 0} \frac{1}{s + sG(s)} = \lim_{s\to 0} \frac{1}{sG(s)} = \frac{1}{K_v} \qquad ...(5.26)$$

where $K_v = \lim_{s\to 0} sG(s)$ is defined as the *velocity error constant.*

3. *Unit-parabolic (Acceleration) Input*

$$\text{Input} \quad r(t) = t^2/2 \text{ or } \ddot{r}(t) = 1$$

$$R(s) = 1/s^3$$

From eqn. (5.24)

$$e_{ss} = \lim_{s\to 0} \frac{1}{s^2 + s^2G(s)} = \lim_{s\to 0} \frac{1}{s^2G(s)} = \frac{1}{K_a} \qquad ...(5.27)$$

where $K_a = \lim_{s\to 0} s^2G(s)$ *is defined as the acceleration error constant.*

Types of Feedback Control System

The open-loop transfer function of a unity feedback system can be written in two standard forms—the *time-constant form* and the *pole-zero form*. In these two forms, $G(s)$ is as given below.

$$G(s) = \frac{K(T_{z1}s + 1)(T_{z2}s + 1)\ldots}{s^n (T_{p1}s + 1)(T_{p2}s + 1)\ldots} \quad \text{(time-constant form)} \quad \ldots(5.28)$$

$$= \frac{K'(s + z_1)(s + z_2)\ldots}{s^n (s + p_1)(s + p_2)\ldots} \quad \text{(pole-zero form)} \quad \ldots(5.29)$$

The gains in the two forms are related by

$$K = K' \frac{\prod_i z_i}{\prod_j p_j} \quad \ldots(5.30)$$

With the gain relation of eqn. (5.30) for the two forms of $G(s)$, it is sufficient to obtain steady-state errors in terms of the gains of any one of the forms. We shall use the time constant form in the discussions below.

Equation (5.28) involves the term s^n in the denominator which corresponds to number of integrations in the system. As s tends to zero, this term dominates in determining the steady-state error. Control systems are therefore classified in accordance with the number of integrations in the open-loop transfer function $G(s)$ as described below:

1. *Type-0 System*

If $n = 0$, the steady-state errors to various standard inputs, obtained from eqns. (5.25), (5.26), (5.27), and (5.28) are

$$e_{ss}(\text{position}) = \frac{1}{1+G(0)} = \frac{1}{1+K} = \frac{1}{1+K_p} \quad \ldots(5.31)$$

$$e_{ss}(\text{velocity}) = \lim_{s\to 0} \frac{1}{sG(s)} = \infty$$

$$e_{ss}(\text{acceleration}) = \lim_{s\to 0} \frac{1}{s^2G(s)} = \infty$$

Thus a system with $n = 0$ or no integration in $G(s)$ has a constant position error, infinite velocity and acceleration errors. The position error constant is given by the open-loop gain of the transfer function in the time-constant form.

2. *Type-1 System*

If $n = 1$, the steady-state errors to various standard inputs are

$$e_{ss}(\text{position}) = \frac{1}{1+G(0)} = \frac{1}{1+\infty} = 0$$

$$e_{ss}(\text{velocity}) = \lim_{s\to 0} \frac{1}{sG(s)} = \frac{1}{K} = \frac{1}{K_v} \quad \ldots(5.32)$$

$$e_{ss}(\text{acceleration}) = \lim_{s\to 0} \frac{1}{s^2G(s)} = \infty$$

Thus a system $n = 1$ or with one integration in $G(s)$ has zero position error, a constant velocity error and an infinite acceleration error at steady-state.

3. *Type-2 System*

If $n = 2$, the steady-state errors to various standard inputs are

$$e_{ss}(\text{position}) = \frac{1}{1+G(0)} = 0$$

$$e_{ss}(\text{velocity}) = \lim_{s \to 0} \frac{1}{sG(s)} = 0$$

$$e_{ss}(\text{acceleration}) = \lim_{s \to 0} \frac{1}{s^2 G(s)} = \frac{1}{K} = \frac{1}{K_a} \qquad ...(5.33)$$

Thus a system with $n = 2$ or two integrations in $G(s)$ has a zero position error, zero velocity error and a constant acceleration error at steady-state.

Steady-state errors for various inputs and systems are summarized in Table 5.1.

Table 5.1. Steady-State Errors for Various Inputs and Systems Types

Type of input	*Steady-state error*		
	Type-0 system	*Type-1 system*	*Type-2 system*
Unit-step	$1/(1 + K_p)$	0	0
Unit-ramp	∞	$1/K_v$	0
Unit-parabolic	∞	∞	$1/K_a$
	$K_p = \lim_{s \to 0} G(s)$	$K_v = \lim_{s \to 0} sG(s)$	$K_a = \lim_{s \to 0} s^2 G(s)$

For nonunity feedback systems (Fig. 5.18) the difference between the input signal $R(s)$ and feedback signal $B(s)$ is the actuating error signal $E_a(s)$ which is given by [refer eqn. (5.20)]

$$E_a(s) = \frac{1}{1+G(s)H(s)} R(s) \qquad ...(5.34(a))$$

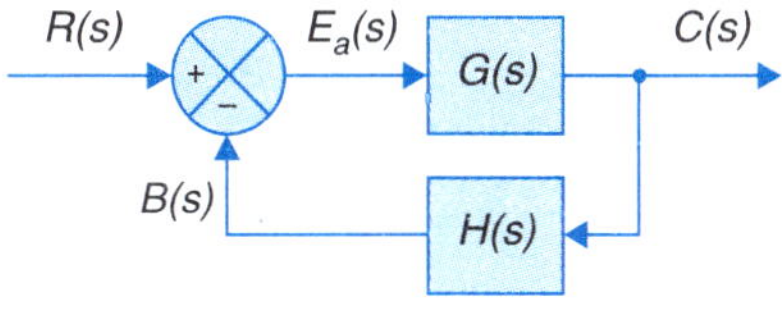

Fig. 5.18

Therefore, the steady-state actuating error is

$$e_{ass} = \lim_{s \to 0} \frac{sR(s)}{1+G(s)H(s)} \qquad ...(5.34(b))$$

The error constants for nonunity feedback systems may be obtained by replacing $G(s)$ by $G(s)H(s)$ in Table 5.1.

The error constants K_p, K_v and K_a describe the ability of a system to reduce or eliminate steady-state errors. As the type of system becomes higher (*i.e.*, increasing number of integrations), progressively more steady-state errors are eliminated. Although it appears that there is no limit to the number of integrations, types-0, -1 and -2 are the most commonly employed systems in practice. Systems of type higher than 2, *i.e.*, with more than two integrations are not employed in practice because of two reasons.

(*i*) These are more difficult to stabilize (this point will be discussed in detail in Chapter 9).

(*ii*) The dynamic errors for such systems tend to be larger than those for types-0, -1 and -2, although their steady-state performance is desirable.

One of the disadvantages of error constants is that they do not give information on the steady-state error when inputs are other than the three basic types-step, ramp and parabolic. Another difficulty is that the error constants fail to indicate the exact manner in which error function change with time. The dynamic error may be evaluated using the dynamic error coefficients—the concept generalized to include inputs of almost any arbitrary function of time (refer Problem 5.13).

5.6 EFFECT OF ADDING A ZERO TO A SYSTEM

The expressions for time response specifications established in Section 5.4 are valid only for the second-order closed-loop transfer function of the form given in eqn. (5.8). Let a zero at $s = -z$ be added to this transfer function. Then we have

$$\frac{C(s)}{R(s)} = \frac{(s+z)(\omega_n^2 / z)}{s^2 + 2\zeta\omega_n s + \omega_n^2} \quad \text{...(5.35(a))}$$

Note that multiplication term in the numerator of above expression has been adjusted so that *steady-state gain* $\frac{C}{R}(0)$ of the system is unity. This gives the steady-state value of output $c_{ss} = 1$ when input is unit step. Thus the system will track the step input with zero steady-state error.

From eqn. (5.35(*a*)) we have

$$\frac{C(s)}{R(s)} = \frac{\omega_n^2}{s^2 + 2\zeta\omega_n s + \omega_n^2} + \frac{s}{z}\left(\frac{\omega_n^2}{s^2 + 2\zeta\omega_n s + \omega_n^2}\right) \quad \text{...(5.35(b))}$$

Let $c_z(t)$ be the response of the system with a zero at $s = -z$. Then, from eqn. (5.35(*b*)), we have

$$c_z(t) = c(t) + \frac{1}{z}\frac{d}{dt}c(t) \quad \text{...(5.35(c))}$$

where $c(t)$ is the response given by eqn. (5.12).

The effect of added derivative term may be seen by examining Fig. 5.19 (*a*) where a case for a typical value of ζ (less than one) is considered. We see from this figure that the effect of the zero is to contribute a pronounced early peak to the system's response whereby the peak overshoot may increase appreciably. From eqn. (5.35(*c*)) and Fig. 5.19 (*b*) it is seen that the smaller the value of z, *i.e.,* the closer the zero to origin, the more pronounced is the peaking phenomenon. On account of this fact, *the zeros on the real axis near the origin are generally avoided in design.* However, in a sluggish system the artful introduction of a zero at the proper position can improve the transient response.

We further observe from eqn. (5.35(*c*)) that as z increases, *i.e.,* the zero moves farther into the left half of s-plane, its effect becomes less pronounced. For sufficiently large values of z, the effect of zero on transient response may become negligible.

For the closed-loop transfer function of eqn. (5.35(*a*)). peak percentage over-shoot for a unit step input can be read from the log-log graphs of Fig. 5.19(*b*) as function of $(z/\zeta\omega_n)$ for various values of $\zeta \le 1$.

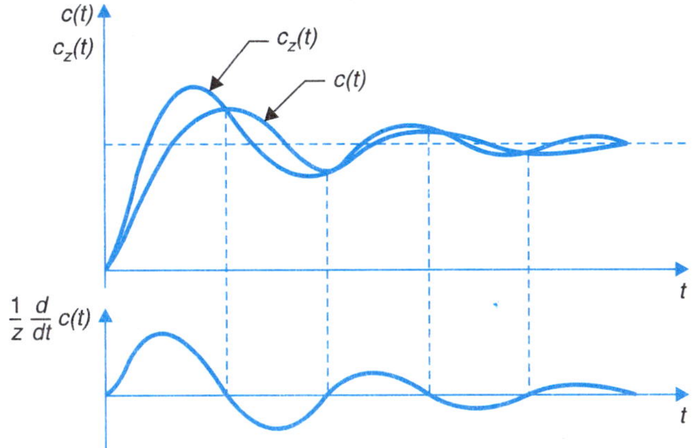

Fig. 5.19(a). Effect of closed-loop zero on unit-step response of a second-order system.

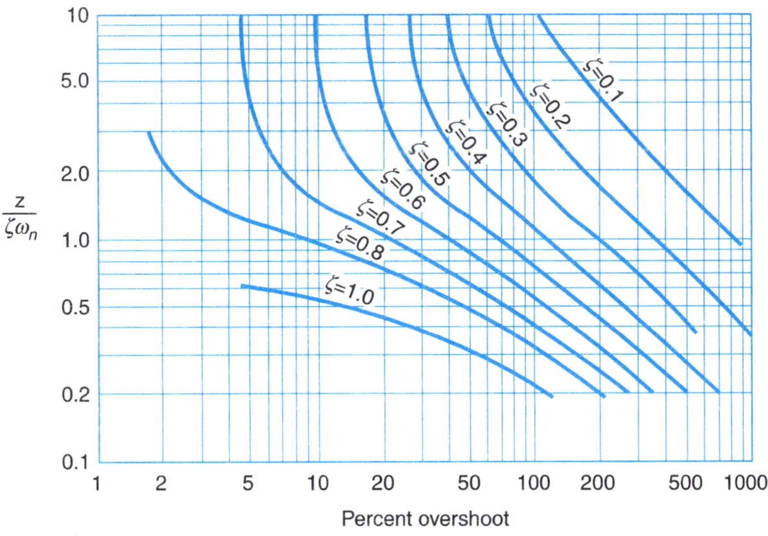

Fig. 5.19(b). Peak overshoot to unit step input of a second-order system with closed-loop zero.

5.7 DESIGN SPECIFICATIONS OF SECOND-ORDER SYSTEMS

A control system is generally required to meet three time response specifications: steady-state accuracy (specified in terms of permissible error e_{ss}), damping factor ζ (or peak overshoot to step input, M_p) and settling time t_s. If the rise time t_r is also specified, it should be consistent with the specification of t_s as both these depend upon ζ and ω_n. Steady-state accuracy requirement is met by a suitable choice of K_p, K_v *or* K_a depending upon the type of the system. As explained earlier, the damping factor ζ sufficiently less than one is preferred in most control systems except for robotic control systems (Section 5.11). For most control systems ζ in the range of 0.7-0.28 (or peak overshoot of 5-40%) is considered acceptable rather desirable from a practical point of view. For this range of ζ, the closed-loop pole locations are restricted to the shaded region of the s-plane as shown in Fig. 5.20.

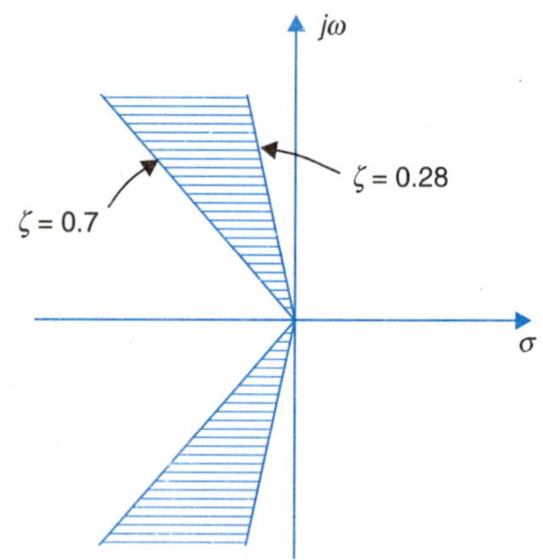

Fig. 5.20. Desirable region of pole locations for a second-order system

Let us now examine the expressions for e_{ss}, ω_n, ζ and t_s for a type-1 second-order system, already derived in this chapter and reproduced below.

$$\zeta = \frac{1}{2\sqrt{(K_v\tau)}};\ \omega_n = \sqrt{(K_v/\tau)} \qquad \text{...(5.36)}$$

$$t_s = 4/(\zeta\omega_n) \text{ (for 2\% tolerance band)} \qquad \text{...(5.37)}$$

$$e_{ss} = \frac{2\zeta}{\omega_n} = \frac{1}{K_v} \text{ (for unit-ramp input)} \qquad \text{...(5.38)}$$

It is obvious from the above equations that only two of the three specifications, *i.e.*, e_{ss}, ζ, t_s can be met exactly by a suitable choice of K_v and τ provided both of these are variable. The third specification, if consistent, can then be met as an upper or lower bound. However, for the practical example of the second-order system of Fig. 5.6, the time constant τ (motor time constant) is generally fixed. Thus K_v or the open-loop gain is the only adjustable parameter of the system (this can be adjusted by varying the amplifier gain). This second-order system can then meet only one of the specifications exactly. Usually this specification is on the allowable steady-state error. In most systems when the open-loop gain is adjusted to meet the steady-state specified accuracy, the ζ reduces below the specified value impairing the system's dynamic response.

We conclude from the above discussion that even to meet two independent specifications a second-order system requires to be modified. This modification termed as *compensation* should allow for high open-loop gains to meet the specified steady-state accuracy and yet preserve a satisfactory dynamic performance. Some of the practical modification schemes are discussed below.

Derivative Error, Compensation

As system is said to possess derivative error compensation when the generation of its output depends in some way on the rate of change of actuating signal. For the system shown in Fig. 5.6, this type of compensation is easily introduced by using an amplifier which provides an output signal containing two terms, one proportional to the derivative of the actuating signal and the other proportional to the actuating signal itself. A controller producing such a signal is called a *proportional plus derivative controller or PD controller*.

When such a controller is introduced in the position control system of Fig. 5.6, the block diagram of Fig. 5.7 gets modified as shown in Fig. 5.21(a). This block diagram can be further reduced to that of Fig. 5.21(b), where

$$K_v = K_p K_A K_T n/(R_a f)$$

$$\tau = J/f$$

$$K_D' = K_D/K_A$$

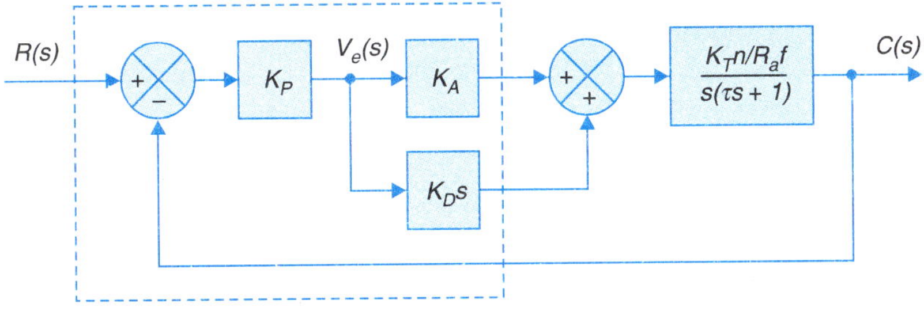

Fig. 5.21. (a) Proportional plus derivative controller for the system shown in Fig. 5.6.

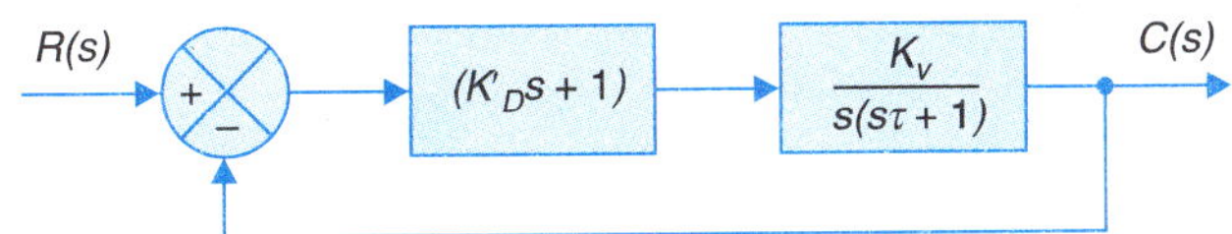

Fig. 5.21. (b) Second-order system with error derivative control.

From Fig. 5.21(*b*) the closed-loop transfer function of the system is given by

$$T(s) - \frac{\frac{K_v}{\tau}(K'_D s + 1)}{s^2 + \left(\frac{1 + K_v K_D}{\tau}\right)s + \frac{K_v}{\tau}}$$

Accordingly, the characteristic equation is modified from that of eqn. (5.7) to the following:

$$s^2 + \left(\frac{1 + K_v K_D}{\tau}\right)s + \frac{K_v}{\tau} = 0$$

The damping and natural frequency of the compensated system are given by

$$\omega_n' = \sqrt{(K_v/\tau)} = \omega_n \qquad \text{...(5.39)}$$

$$\zeta' = \frac{1 + K_v K'_D}{2\sqrt{(K_v \tau)}} = \zeta \frac{K'_D}{2}\sqrt{(K_v / \tau)} \qquad \text{...(5.40)}$$

We observe that compared to the uncompensated system, for the same K_v, the natural frequency of the compensated system remains unchanged, while its damping is increased by $(K'_D/2)\sqrt{(K_v/\tau)}$. If the steady-state error e_{ss} to velocity input is specified (this is generally the case), the system K_v gets fixed (as $e_{ss} = 1/K_v$) and hence the amplifier gain K_A. The closed-loop damping ζ' can now be raised to the desired level by a suitable choice of K'_D in eqn (5.40). $K_D = K'_D K_A$ gives the constant of the derivative term for the amplifier.

An additional advantage of the scheme is that as the damping increases due to compensation with ω_n remaining fixed, the system settling time reduces (as $t_s = 4/\zeta\omega_n$).

Derivative Output Compensation

A system is said to possess a derivative output compensation when the generation of its output depends in some way upon the rate at which the controlled variable is changing. For a servo system, a common way of obtaining this compensation is by means of tachometer feedback.

To obtain this type of compensation for the position control system of Fig. 5.6, a dc. tachometer is coupled to the motor and its output voltage v_t which is proportional to speed (derivative of position), *i.e.*, $v_t = K_t\dot{\theta}$, is fed back negatively to the amplifier input. The system with this modification is resketched in Fig. 5.22(*a*). The new block diagram in which the inner loop due to motor back emf has already been simplified is presented in Fig. 5.22(*b*).

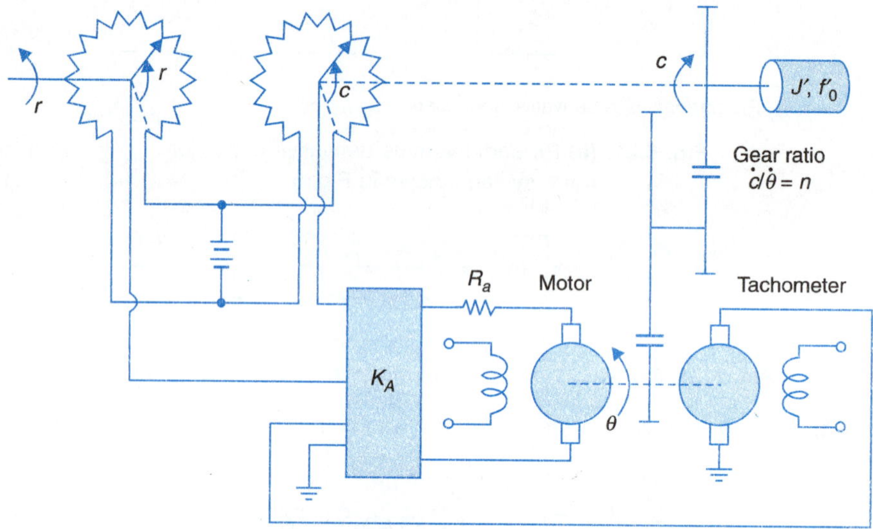

Fig. 5.22. (a) Second-order position control system with tachometer (derivative) feedback.

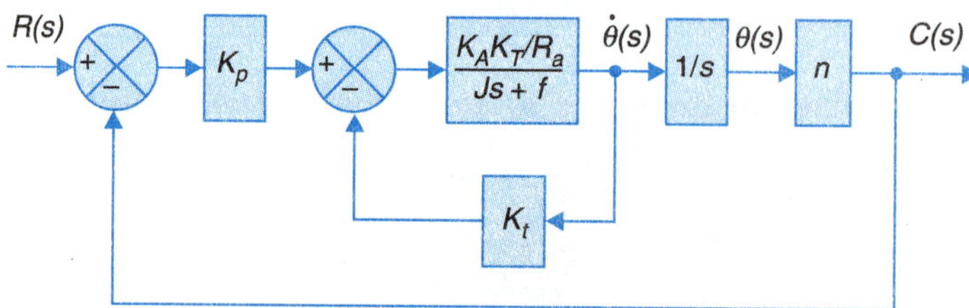

Fig. 5.22. (b) Block diagram of Fig. 5.22(a).

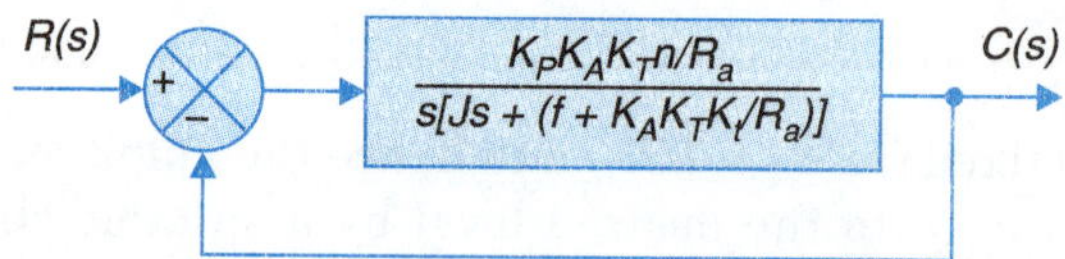

Fig. 5.22. (c) Reduced block diagram of Fig. 5.22(b).

Due to the tachometer feedback, the gain of the inner loop and the effective time constant get reduced. Simplifying the inner loop, we get the block diagram of Fig. 5.22(*c*), from which we can write

$$K_v' = \frac{K_P K_A K_T n}{R_a f + K_A K_T K_t} \qquad \text{...(5.41)}$$

$$\tau' = \frac{JR_a}{R_a f + K_A K_T K_t} \qquad ...(5.42)$$

The natural frequency and damping of the closed-loop can then be expressed as

$$\omega_n' = \sqrt{(K_v' / \tau')} = \sqrt{(K_P K_A K_T n / JR_a)} \qquad ...(5.43)$$

$$\zeta' = (1/2)/\sqrt{(K_v' \tau')} = (R_a f + K_A K_T K_t)/2\sqrt{(K_P K_A K_T n\, JR_a)} \qquad ...(5.44)$$

For a specified K_v', and ζ', we can write from eqns. (5.41) and (5.44)

$$\zeta' K_v' = \frac{1}{2} \sqrt{(K_P K_A K_T n / JR_a)}$$

from which K_A is determined as

$$K_A = 4(\zeta' K_v')^2 (JR_a / K_P K_T n)$$

Using this value of K_A, K_t the tachometer constant is obtained from eqn. (5.41) as

$$K_t = \left(\frac{K_P n}{K_v'} - \frac{R_a f}{K_A K_T} \right)$$

Thus, we notice that by a suitable choice of K_A, K_t, we can simultaneously meet the specification of K_v' and ζ'. On account of the negative derivative feedback, K_A required may sometimes be excessively large. Under such circumstances additional gain outside the derivative loop is very helpful.

Further, it is to be noticed that for the same value of velocity error constant, the system with compensation requires a higher value of K_A and hence it will have a higher natural frequency (eqn. 5.43). Compensation thus increases both the system damping and natural frequency resulting in reduced settling time.

Integral Error Compensation

In an integral error compensation scheme, the output response depends in some manner upon the integral of the actuating signal. This type of compensation is introduced by using a controller which produces an output signal consisting of two terms, one proportional to the actuating signal and the other proportional to its integral. Such a controller is called *proportional plus integral controller or PI controller.*

The block diagram of the system of Fig. 5.6 with proportional plus integral compensation is shown in Fig. 5.23(a), while its simplified block diagram is given in Fig. 5.23(b) where

$$K_v' = K_P K_i K_T n / R_a f$$

$$K_A' = K_A / K_i$$

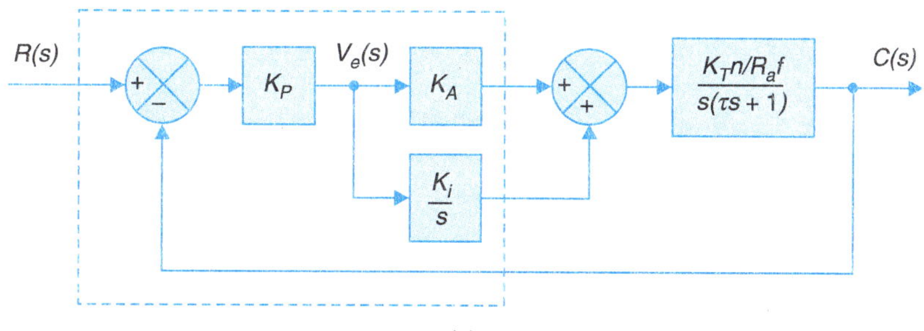

(a)

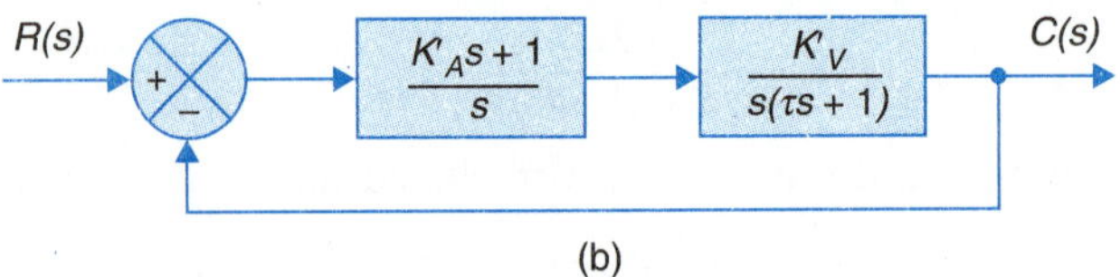

(b)

Fig. 5.23. Proportional plus integral controller for the system shown in Fig. 5.6.

The closed-loop transfer function is easily obtained as

$$\frac{C}{R}(s) = \frac{K_v'(K_A' s + 1)}{(\tau s^3 + s^2 + K_v' K_A s + K_v')} \qquad ...(5.45)$$

We find that the integral error compensation changes a second-order system into a third-order one. The effect of compensation on system dynamics therefore cannot be visualized as easily as in the previous two types of compensation schemes. The techniques to study the dynamics of higher order systems will be taken up later in Chapters 7 and 9. A significant contribution of integral error compensation to the system steady-state performance is, however, obvious as the additional integration in the forward path changes the system from type-1 to type-2 and that the error of velocity input is eliminated or considerably reduced as the practical integration may not be perfect. Thus, we can use integral error compensation to meet high accuracy requirements. However as we shall see in Chapter 6 on stability that a third-order characteristic equation introduces a distinct possibility of system instability.

Proportional Plus Integral Plus Derivative (PID) Controller

It is seen above that error integration in the forward path eliminates steady-state velocity error but it increases system's order making it more susceptible to instability (to be examined in Chapter 6). It also introduces a zero into the forward path so that peak overshoot to step input cannot be easily determined. To increase the damping factor of the dominant poles of a PI controlled system, we take advantage of combining it with derivative error scheme. Such a controller is known as a PID controller and is illustrated in the block diagram of Fig. 5.24.

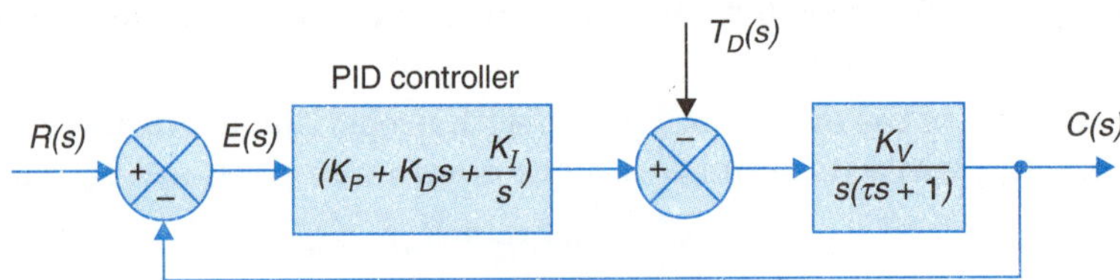

Fig. 5.24

The following expressions can be written down from the block diagram.

$$\frac{C(s)}{R(s)} = \frac{K_v(K_D s^2 + K_P s + K_I)}{\tau s^3 + (1 + K_v K_D)s^2 + (K_v K_P)s + K_v K_I}; \; T_D(s) = 0 \qquad ...(5.46)$$

$$\frac{E(s)}{R(s)} = \frac{s^2(\tau s + 1)}{\tau s^3 + (1 + K_v K_D)s^2 + (K_v K_P)s + K_v K_I}; \; T_D(s) = 0 \qquad ...(5.47)$$

$$\frac{E(s)}{T_D(s)} = -\frac{K_v s}{\tau s^3 + (1 + K_v K_D)s^2 + (K_v K_P)s + K_v K_I}; \; R(s) = 0 \qquad ...(5.48)$$

It is immediately seen from eqns. (5.47) and (5.48) that the steady-state error for velocity input is zero and also the steady-state error caused by disturbance torque (assumed to be step) is zero. These facts are easily established directly from the inclusion of the integral term in the controller.

Effect on dynamic response of the system caused by a PID controller is not obvious as the characteristic equation becomes of third order. Inclusions of two zeros in the numerator $(K_D s^2 + K_P s + K_I)$ helps in the system acquiring desirably dynamic qualities but the third-order characteristic equation causes the system to become unstable for certain combination of gains (K_v, K_P, K_D, K_I). These issues will be examined in details in Chapter 6 and later on in the design chapter 12.

5.8 DESIGN CONSIDERATIONS FOR HIGHER-ORDER SYSTEMS

The results presented in the previous section hold for second-order systems. These results could be extended to higher-order systems with a *dominant pair of complex poles, i.e.,* as long as the real part of these poles is much less than those of the other poles, the time response of higher-order systems can be approximated by that obtained by neglecting all the system poles other than the dominant poles.

For example, let us consider a third-order system with transfer function

$$\frac{C(s)}{R(s)} = T(s) = \frac{\omega_n^2}{(1+sT)(s^2+2\zeta\omega_n s+\omega_n^2)}$$

$$= \frac{\omega_n^2 p}{(s+p)(s^2+2\zeta\omega_n s+\omega_n^2)}\,;\; p = \frac{1}{T} \qquad ...(5.49)$$

The unit-step response is (Laplace inverse of $T(s)/s$; *see* Appendix I)

$$c_p(t) = 1 - K_1 e^{-pt} + K_2 e^{-\zeta\omega_n t}\sin[\omega_n\sqrt{(1-\zeta^2)}\,t - \phi] \qquad ...(5.50)$$

where

$$K_1 = \frac{\omega_n^2}{\omega_n^2 - 2\zeta\omega_n p + p^2}$$

$$K_2 = \frac{1}{\sqrt{1-\zeta^2(1-2\zeta\omega_n/p+\omega_n^2/p^2)}}$$

$$\phi = \tan^{-1}\frac{\sqrt{(1-\zeta^2)}}{-\zeta} + \tan^{-1}\frac{\omega_n\sqrt{(1-\zeta^2)}}{p-\zeta\omega_n}$$

The effect of pole at $s = -p$ will depend on the magnitude of p. If $p << \zeta\omega_n$, the effect of the term e^{-pt} in eqn. (5.49) will last much longer than that of the term whose magnitude is governed by exp $(-\zeta\omega_n t)$. Further, from eqn. (5.50) we observe that

$$\lim_{p\to 0} K_1 = 1; \quad \lim_{p\to 0} K_2 = 0$$

Therefore as the pole at $s = -p$ moves closer to the origin, the response $c_p(t)$ in eqn. (5.50) is dominated by this pole and the effect of the two complex poles at $-\zeta\omega_n \pm j\omega_n\sqrt{(1-\zeta^2)}$ diminishes.

On the other hand, if $\zeta\omega_n << p$, the effect of the term e^{-pt} in eqn. (5.50) will be over much more quickly than the term whose magnitude is governed by exp $(-\ \zeta\omega_n t)$. Again, from eqn. (5.50) we observe that

$$\lim_{p \to \infty} K_1 = 0; \ \lim_{p \to \infty} K_2 = \frac{1}{\sqrt{(1-\zeta^2)}}$$

Therefore, as the pole as $s = -p$ moves far away from the origin, the response $c_p(t)$ in eqn. (5.50) is dominated by the complex poles at

$$-\zeta\omega_n \pm j\omega_n\sqrt{(1-\zeta^2)}.$$

Figure 5.25 shows unit-step response of a third-order system of eqn. (5.49) with $\zeta = 0.3$. It is observed that for $p/\zeta\omega_n = 6$, response of third-order system is, for all practical purposes, the response $c(t)$ of the system with the transfer function

$$T(s) = \frac{\omega_n^2}{s^2 + 2\zeta\omega_n s + \omega_n^2}$$

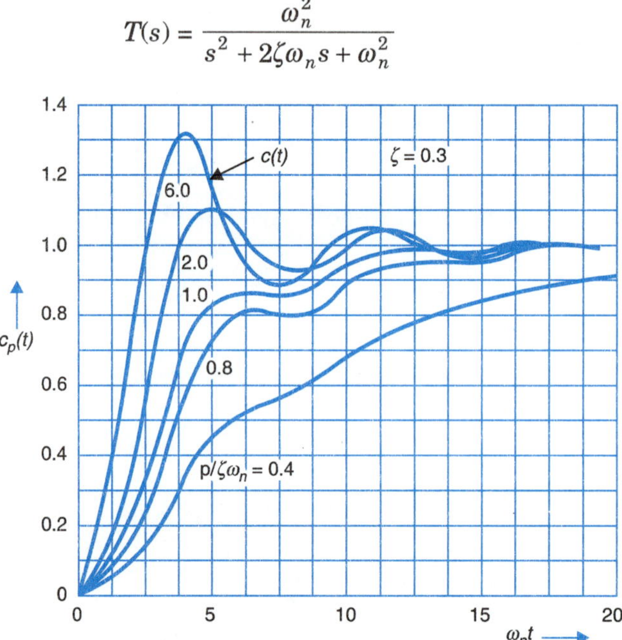

Fig. 5.25. Unit-step response of a third-order system.

The relative location of dominant poles of the third-order system is shown in Fig. 5.26.

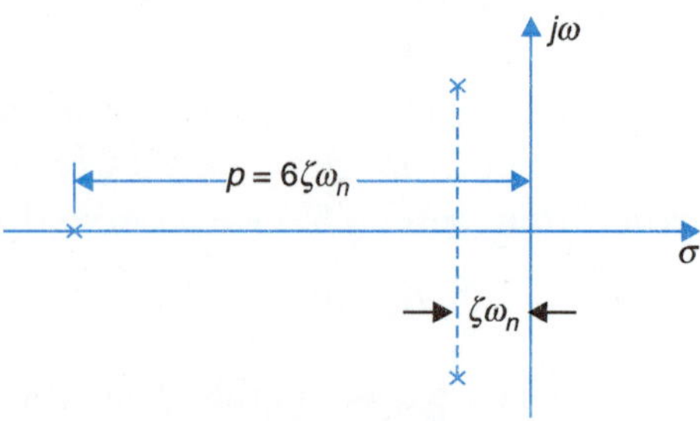

Fig. 5.26. Location of dominant poles of the third-order system.

5.9 PERFORMANCE INDICES

As discussed already, the design of a control system is an attempt to meet a set of specifications which define the overall performance of the system in terms of certain measurable quantities. A number of performance measures have been introduced so far in respect of dynamic response to step input (ζ, M_p, t_r, t_p, t_s etc.) and the steady-state error, e_{ss}, to both step and higher-order inputs. These measures have to be satisfied simultaneously in design and hence the design necessarily becomes a trial and error procedure. If, however, a single performance index could be established on the basis of which one may describe the goodness of the system response, then the design procedure will become logical and straightforward.

Furthermore, in many of the modern control schemes, the system parameters are automatically adjusted to keep the system at an optimum level of performance under varying inputs and varying conditions of operation. Such class of systems is called *adaptive control systems*. These systems require a performance index which is a function of the variable system parameters. Extremum (minimum or maximum) value of this index then corresponds to the optimum set of parameter values. Other desirable features of a performance index are its selectivity, *i.e.*, its power to clearly distinguish between an optimum and non-optimum system, its sensitivity to parameter variations and the ease of its analytical computation or its online analogic or digital determination.

A number of such performance indices are used in practice, the most common being the *integral square error* (ISE), given by

$$\text{ISE} = \int_0^\infty e^2(t)dt \qquad \text{...(5.51)}$$

Apart from the ease of implementation, this index has the advantage that it is mathematically convenient both for analysis and computation. Fig. 5.27(a) and (b) show the system response $c(t)$ and error $e(t)$ respectively to unit-step input. The square error is shown in Fig. 5.27(c) and its integral in Fig. 5.27(d). It is obvious that ISE coverges to a limit as $t \to \infty$. Minimization of ISE by adjusting system parameters is a good compromise between reduction of rise time to limit the effect of large initial error, reduction of peak overshoot and reduction of settling time to limit the effect of small error lasting for a long time.

Consider now the second-order system discussed previously in this chapter from eqn. (5.12), the error response to unit-step is given by

$$e(t) = \frac{e^{-\zeta\omega_n t}}{\sqrt{(1-\zeta^2)}} \sin\left[\omega_n\sqrt{(1-\zeta^2)}t + \tan^{-1}\frac{\sqrt{(1-\zeta^2)}}{\zeta}\right]$$

Therefore

$$\text{ISE} = \int_0^\infty e^2(t)dt = \frac{1}{\omega_n}\int_0^\infty e^2(t_n)dt_n$$

where $t_n = \omega_n t$, *i.e.*, the normalized time.

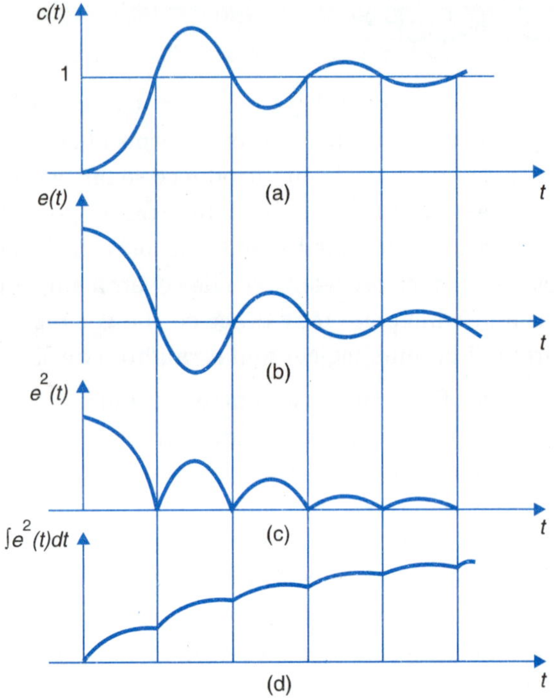

Fig. 5.27. Calculation of integral square error.

Simplification of the above expression leads to the following result:

$$\text{ISE}^* = \frac{1}{2\omega_n}\left(\frac{1}{2\zeta} + 2\zeta\right) \qquad \text{...(5.52)}$$

---

*Details of the derivation are given below:

$$e(t_n) = \frac{e^{-\zeta t_n}}{\sqrt{(1-\zeta^2)}} \sin\left[\sqrt{(1-\zeta^2)}t_n + \phi\right]$$

where $\phi = \tan^{-1}\dfrac{\sqrt{(1-\zeta^2)}}{\zeta}$

$$e^2(t_n) = \frac{e^{-2\zeta t_n}}{(1-\zeta^2)} \sin^2\left[\sqrt{(1-\zeta^2)}t_n + \phi\right]$$

$$= \frac{e^{-2\zeta t_n}}{2(1-\zeta^2)} \left[1 - \cos^2\{\sqrt{(1-\zeta^2)}t_n + \phi\}\right]$$

$$= \frac{e^{-2\zeta t_n}}{2(1-\zeta^2)} \left[1 - (2\zeta^2 - 1)\cos^2\sqrt{(1-\zeta^2)}t_n + 2\zeta\sqrt{(1-\zeta^2)}\sin^2\sqrt{(1-\zeta^2)}t_n\right]$$

$$\frac{1}{\omega_n}\int_0^\infty e^2(t_n)dt = \frac{1}{2\omega_n(1-\zeta^2)}\int_0^\infty e^{-2\zeta t_n}\left[1 - (2\zeta^2 - 1)\cos^2\sqrt{(1-\zeta^2)}t_n\right.$$

$$\left. + 2\zeta\sqrt{(1-\zeta^2)}\sin^2\sqrt{(1-\zeta^2)}t_n\right]dt_n$$

$$= \frac{1}{2\omega_n}\left(\frac{1}{2\zeta} + 2\zeta\right)$$

For a fixed choice of ω_n, ISE given by eqn. (5.52) is minimized for $\zeta = 0.5$. This result corroborates the previously elaborated intuitive design criterion that the damping factor of a control system should be sufficiently less than 1.

For higher-order system, ISE must be computed numerically. Since it is not practicable to integrate up to infinity, the limit ∞ is replaced by T which is chosen sufficiently large so that $e(t)$ for $t > T$ is negligible. Usually T is estimated to be settling time t_s or multiple of settling time, *i.e.*,

$$\text{ISE} = \int_0^{T = nt_s} e^2(t)dt$$

Another easily instrumented performance index is the *integral of the absolute magnitude of error* (IAE), which is written as

$$\text{IAE} = \int_0^T |e(t)|\,dt$$

In order to reduce the weighting of the large initial error and to penalize the small errors occurring later in the response more heavily, the following indices are proposed.

Integral time-absolute error, $\text{ITAE} = \int_0^T t\,|e(t)|\,dt$

Integral time-square error, $\text{ITSE}^* = \int_0^T te^2(t)dt$

Figure 5.28 shows the comparison of the different performance indices of the second-order system having transfer function $C(s)/R(s) = 1/(s^2 + 2s + 1)$.

Inspection of these curves reveals that ISE is not very sensitive to parameter variations since the curve is rather flat near the point where the performance index is minimum (*i.e.*, $\zeta = 0.5$ as already stated). Therefore, the selectivity of this performance index is poor. It, however, has the advantage of being easy to deal with mathematically. The IAE performance index (IAE is minimized for $\zeta = 0.7$ for the second-order system under consideration) gives slightly better selectivity than ISE. The ITAE performance index (ITAE is minimized for $\zeta = 0.707$ for the second-order system under consideration) produces smaller overshoots and oscillations than the IAE and ISE indices. In addition, it is the most sensitive of the three, *i.e.*, it has the

---

*ITSE for the second-order system discussed in this chapter is obtained as below:

$$\text{ITSE} = \int_0^\infty te^2(t)dt = \frac{1}{\omega_n^2}\int_0^\infty t_n e^2(t_n)dt_n$$

$$= \frac{1}{2\omega_n^2(1-\zeta^2)}\int_0^\infty t_n e^{-2\zeta t_n}[1 - (2\zeta^2 - 1)\cos 2\sqrt{(1-\zeta^2)}t_n + 2\zeta\sqrt{(1-\zeta^2)}\sin 2\sqrt{(1-\zeta^2)}t_n]dt_n$$

Integrating each part we get

$$\text{ITSE} = \frac{1}{2\omega_n^2(1-\zeta^2)}\left[\frac{1}{4\zeta^2}(2\zeta^2 - 1)\frac{(2\zeta^2-1)}{4} + 2\zeta\sqrt{(1-\zeta^2)}\frac{1}{2}\zeta\sqrt{(1-\zeta^2)}\right]$$

$$= \frac{1}{2\omega_n^2}\left(\frac{1}{4\zeta^2} + 2\zeta^2\right)$$

For fixed ω_n, ITSE is minimised for $\zeta = 1/(8)^{1/4} = 0.60$

best selectivity. The ITSE index is somewhat less sensitive but is not comfortable computationally.

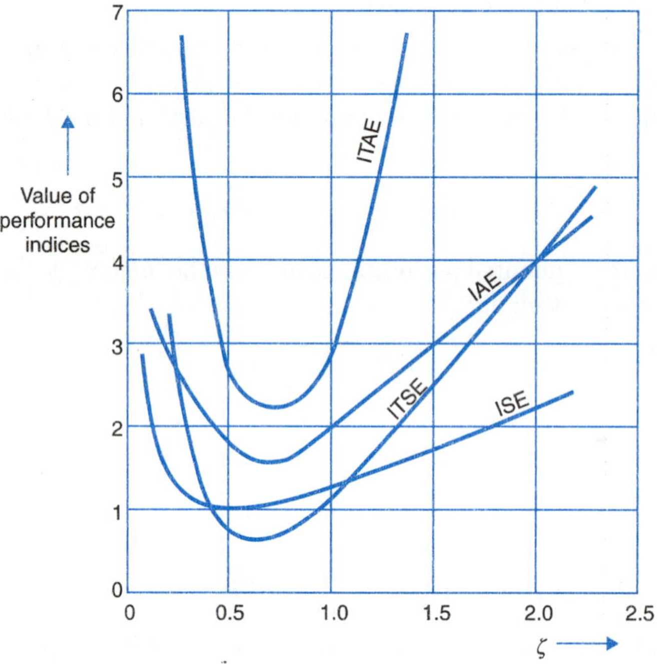

Fig. 5.28. Performance indices for a second-order system.

In practice, relatively insensitive criterion may be more useful and based on this logic. ISE may be the most desirable performance index in some practical applications. Where selectivity is more important, the ITAE performance index is a better choice.

Optimal Control Systems

A control system is optimum when the selected performance index is minimized. The optimum value of the system parameters depends directly on the definition of optimality.

A paper by Graham and Lathrop created a great deal of interest in system optimization using ITAE criterion. They suggested the form that the system transfer function should take, for various order systems, in order to achieve zero steady-state step and ramp error systems and minimize ITAE.

In the case of the zero steady-state step error systems, the general closed-loop transfer function is

$$T(s) = \frac{C(s)}{R(s)} = \frac{a_n}{s^n + a_1 s^{n-1} + \ldots + a_{n-1} s + a_n} \qquad \ldots(5.53)$$

Table 5.2 shows the optimum forms of the closed-loop transfer functions based on the ITAE criterion. The procedure used to produce this table was to vary each coefficient in eqn. (5.53) until the ITAE value became a minimum. Then the successive coefficients were varied in sequence to minimize the ITAE value. These standard forms of transfer functions provide a quick and simple method for synthesizing an optimum dynamic response (other standard forms based on different performance indices are also available).

Table 5.2. Optimum Forms of the Closed-Loop Transfer Functions Based on the ITAE Criterion (Zero Steady-State Step Error Systems)

$\dfrac{C(s)}{R(s)} = \dfrac{a_n}{a^n + a_1 s^{n-1} + \ldots + a_{n-1}s + a_n}$
$s + \omega_n$
$s^2 + 1.4\omega_n s + \omega_n^2$
$s^3 + 1.75\omega_n s^2 + 2.15\omega_n{}^2 s + \omega_n^3$
$s^4 + 2.1\ \omega_n s^3 + 3.4\ \omega_n{}^2 s^2 + 2.7\omega_n{}^3 s + \omega_n^4$
$s^5 + 2.8\ \omega_n s^4 + 5.0\omega_n{}^2 s^3 + 5.5\ \omega_n{}^3 s^2 + 3.4\omega_n{}^4 s + \omega_n^5$
$s^6 + 3.25\omega_n s^5 + 6.60\omega_n{}^2 s^4 + 8.60\omega_n{}^3 s^3 + 7.45\omega_n{}^4 s^2 + 3.95\omega_n{}^5 s + \omega_n^6$

In the case of zero steady-state ramp error systems, the general closed-loop transfer function is

$$T(s) = \frac{C(s)}{R(s)} = \frac{a_{n-1}s + a_n}{s^n + a_1 s^{n-1} + \ldots + a_{n-1}s + a_n} \qquad \ldots(5.54)$$

Table 5.3 gives the optimum forms of the closed loop transfer functions based on the ITAE criterion (refer Problem 5.18)

Table 5.3. Optimum Forms of the Closed-Loop Transfer Functions Based on the ITAE Criterion (Zero Steady-State Ramp Error Systems)

$\dfrac{C(s)}{R(s)} = \dfrac{a_{n-1}s + a_n}{s^n + a_1 s^{n-1} + \ldots + a_{n-1}s + a_n}$
$s^2 + 3.2\omega_n s + \omega_n^2$
$s^3 + 1.75\omega_n s^2 + 3.25\omega_n{}^2 s + \omega_n^3$
$s^4 + 2.41\omega_n s^3 + 4.93\omega_n{}^2 s^2 + 5.14\omega_n{}^3 s + \omega_n^4$
$s^5 + 2.19\omega_n s^4 + 6.50\omega_n{}^2 s^3 + 6.30\omega_n{}^3 s^2 + 5.24\ \omega_n{}^4 s + \omega_n^5$
$s^6 + 6.12\omega_n s^5 + 13.42\omega_n{}^2 s^4 + 17.16\omega_n{}^3 s^3 + 14.14\omega_n{}^4 s^2 + 6.76\omega_n{}^5 s + \omega_n^6$

5.10 ILLUSTRATIVE EXAMPLES

The design concepts introduced in this chapter are further illustrated with the help of following examples.

Example 5.1 : Reconsider the servomechanism shown in Fig. 5.6. The parameters of the system are given as follows:

Sensitivity of error detector,	$K_p = 1\text{V/rad}$
Gain of dc amplifier,	K_A (variable)
Resistance of armature of motor,	$R_a = 5\Omega$
Equivalent inertia at the motor shaft,	$J = 4 \times 10^{-3}$ kg-m^2
Equivalent friction at the motor shaft,	$f_0 = 2 \times 10^{-3}$Nm/rad/s
Torque constant of motor,	$K_T = 1$ Nm/A
Gear ratio,	$n = 1/10$

Solution. The value of the back emf constant K_b is not given in the above list of parameters. However, a definite relationship exists between K_b and K_T (refer Section 2.3). In MKS units $K_b = K_T$.

The open-loop transfer function of the system in the pole-zero form is (refer Fig. 5.7(*a*))

$$G(s) = \frac{K}{s(sJ + f)}$$

where

$$f = f_0 + \frac{K_T K_b}{R_a} = 202 \times 10^{-3}$$

$$J = 4 \times 10^{-3}$$

$$K = K_p K_A K_T \frac{n}{R_a} = \frac{K_A}{50}$$

The open-loop transfer function of the system in the time-constant form is (refer Fig. 5.7(*b*))

$$G(s) = \frac{K_v}{s(s\tau + 1)}$$

where

$$K_v = K/f = \frac{10^3 K_A}{50 \times 202}$$

$$\tau = J/f = 4/202$$

The closed-loop transfer function of the system is

$$\frac{C(s)}{R(s)} = \frac{5K_A}{s^2 + 50.5s + 5K_A}$$

The natural undamped frequency of the system is

$$\omega_n = \sqrt{(5K_A)} \quad \text{...(5.55)}$$

The damping ratio is $\zeta = 25.25/\omega_n$ ...(5.56)

The system under consideration is a type-1 system. Therefore, from Table 5.1 we find that the steady-state error to unit step input is zero and steady-state error to unit-ramp input is

$$e_{ss} = \frac{1}{K_v} = \frac{50 \times 202}{10^3 K_A} = \frac{1.01}{K_A} \quad \text{...(5.57)}$$

Proportional Control

Let us study the effects on the time response when the value of the gain K_A is varied. From eqns. (5.55), (5.57), we note that an increase in K_A increases the natural undamped frequency ω_n but decreases the damping ratio ζ. Therefore, if we choose to improve the steady-state accuracy of the system by increasing the forward gain (eqn. (5.53)), the transient response becomes more oscillatory.

Consider for example, that we desire

$$e_{ss} \text{ (to unit ramp input)} = 0.001$$

Therefore,

$$K_A = 1{,}010$$

$$\omega_n = 71$$

$$\zeta = \frac{25.25}{71} = 0.355$$

This tells us that with $K_A = 1{,}010$, the system becomes quite oscillatory and therefore is not a desirable one.

Proportional plus Derivative Control

Consider that the positional control system of Fig. 5.6 is modified by replacing the dc amplifier with proportional plus derivative control so that the open-loop transfer function now is (Fig. 5.21(*b*))

$$G(s) = \frac{K_v\left(\dfrac{K_D}{K_A}s + 1\right)}{s(\tau s + 1)}$$

The characteristic equation for the system is

$$1 + G(s) = 0 \qquad \text{or} \qquad s^2 + (50.5 + 5\,K_D)s + 5K_A = 0$$

Notice that the derivative control has the effect of increasing the coefficient of s term. It means that the damping of the system is increased. If K_A is fixed by considerations of steady-state accuracy, we can control parameter K_D to achieve desired transient response.

Suppose that keeping $K_A = 1{,}010$, we desire a damping ratio of 0.6. We can now achieve it by adjusting the derivative constant (K_D) as below:

$$50.5 + 5\,K_D = 2 \times 0.6 \times 71 \qquad \text{or} \qquad K_D = 6.94$$

Proportional plus Integral Control

Let us consider the position control system of Fig. 5.6 again. With proportional plus integral control, the open-loop transfer function of the system becomes (Fig. 5.23)

$$G(s) = \frac{5(K_A s + K_i)}{s^2(s + 50.5)}$$

The characteristic equation is

$$s^3 + 50.5s^2 + 5K_A s + 5K_i = 0 \qquad \text{...(5.58)}$$

The effect of K_i on the transient behaviour of the system may be investigated by plotting the roots of eqn. (5.58) as a function of K_i for various values of K_A (refer Section 7.4). The steady-state error to ramp input is of course zero for the system with integral control.

Example 5.2 : A turntable with a moment of inertia of 10 kg-m^2 is used with a proportional error controller in a unity negative feedback system. The controller develops a torque of 60 newton-m per radian of misalignment. The viscous friction is such that the damping factor is 0.3.

(*a*) Draw a signal flow graph of the system and determine therefrom $\frac{\theta_0}{\theta_i}(s)$ and $\frac{\theta_e}{\theta_i}(s)$, where θ_i, θ_e and θ_0 are respectively the input, error and output signals,

(*b*) Determine the steady-state tracking error for a constant velocity input of 0.04 rad/sec.

(*c*) Explain what modification could be made to the system to eliminate error to a ramp input as in part (*b*).

Solution.

(*a*) Signal flow graph is drawn in Fig. 5.29.

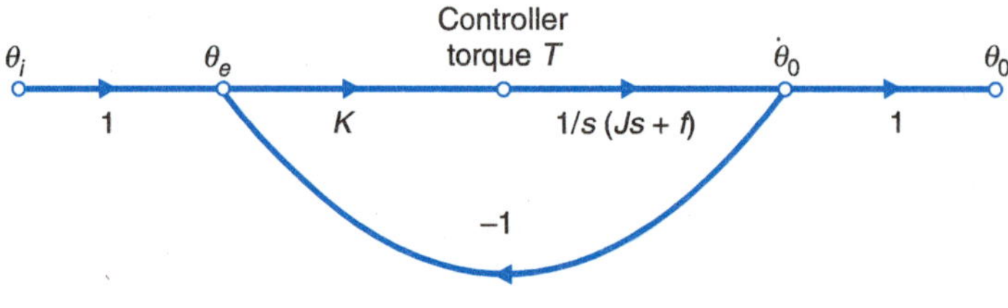

Fig. 5.29

From the signal flow graph

$$\frac{\theta_0}{\theta_i}(s) = \frac{K}{Js^2 + fs + K} = \frac{K/J}{s^2 + (f/J)s + K/J}$$

$$K/J = \omega_n^{\ 2} = 60/10 = 6$$

$$f/J = 2\zeta\omega_n = 2 \times 0.3 \times \sqrt{6} = 0.6\sqrt{6}$$

$$\frac{\theta_0}{\theta_i}(s) = \frac{6}{s^2 + 0.6\sqrt{(6)}s + 6}$$

Again from the signal flow graph

$$\frac{\theta_e}{\theta_i}(s) = \frac{s(Js + f)}{Js^2 + fs + K} = \frac{s(s + 0.6\sqrt{6})}{s^2 + 0.6\sqrt{(6)}s + 6}$$

(*b*)
$$\theta_i(s) = \frac{0.04}{s^2}$$

$$\lim_{t \to 0} \theta_e(t) = \lim_{s \to 0}\left[s \times \frac{0.04}{s^2} \times \frac{s(s + 0.6\sqrt{6})}{s^2 + 0.6\sqrt{(6)}s + 6}\right] = 0.0098 \text{ rad}$$

(*c*) Introduce an integral error term in the controller such that the controller transfer function becomes.

$$\left(K + \frac{K_i}{s}\right) = \frac{Ks + K_i}{s}$$

Since this introduces one more integration in the forward path, the system becomes type-2 and hence the steady-state error to ramp input is reduced to zero.

Example 5.3 : In a d.c. position control servomechanism the load is driven by a motor supplied with constant armature current. The motor field current is supplied from a d.c. amplifier, the input to which is the difference between the voltages obtained from input and output potentiometers.

The load and motor together have a moment of inertia of J = 0.4 kg-m^2 and the viscous friction is f = 2 newton-m/rad/sec. Each potentiometer constant is K_p = 0.6 V/rad. The motor develops a torque of K_T = 2 newton-m per amp of field current. The field time constant is negligible.

(*a*) Make a sketch of the system showing how the hardware is connected.

(*b*) Derive the equation of motion of the system and find the value of the amplifier gain K_A (in amperes output per volt input) to give a natural frequency of 10 rad/sec.

(*c*) A tachogenerator of negligible inertia and friction is connected in the system to improve the damping. Determine the tachogenerator constant (in V/rad/sec) to give critical damping for K_A = 5.

Solution.

(*a*) Figure 5.30 gives the sketch of the position control servomechanism. The switch '*S*' can be closed to provide the tachogenerator feedback.

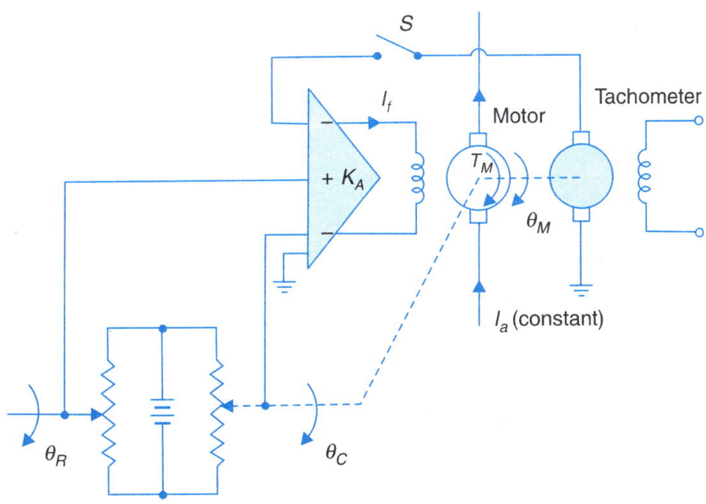

Fig. 5.30

(*b*) Figure 5.31 gives the signal flow graph of the system with switch '*S*' open, *i.e.*, tachogenerator not in the loop.

$$\frac{\theta_C(s)}{\theta_R(s)} = \frac{\theta_M(s)}{\theta_R(s)} = \frac{K_P K_A K_T}{s(Js + f) + K_P K_A K_T}$$

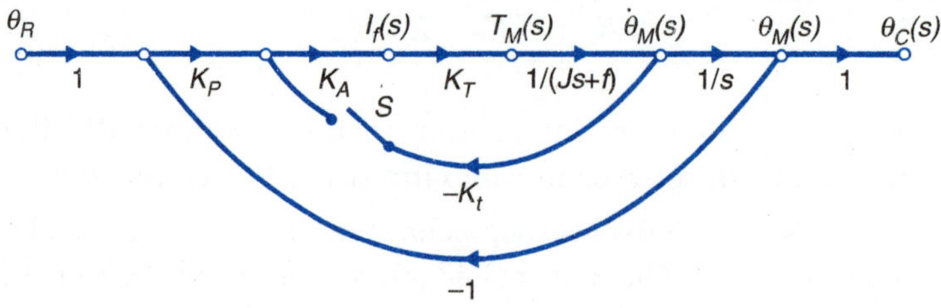

Fig. 5.31

The governing differential equation can be written from the transfer function as

$$J\ddot{\theta}_M(t) + f\dot{\theta}_M(t) + K_PK_AK_T\theta_M(t) = K_PK_AK_T\theta_R(t)$$

The system characteristic equation is

$$s^2 + (f/J)s + K_PK_AK_T/J = 0$$

$$\therefore \qquad \omega_n^2 = \frac{K_PK_AK_T}{J}$$

or $$10^2 = \frac{0.6 \times K_A \times 2}{0.4} \qquad \text{or} \qquad K_A = 33.3 \text{ amp/volt}$$

(*c*) With tachogenerator (switch '*S*' closed)

$$\frac{\theta_M}{\theta_R}(s) = \frac{K_PK_AK_T}{Js^2 + (f + K_AK_tK_T)s + K_PK_AK_T}$$

$$2\zeta\omega_n = (f + K_AK_tK_T)/J \; ; \; \omega_n = \sqrt{(K_PK_AK_T/J)}$$

$$\therefore \qquad \zeta = \frac{1}{2}\frac{f + K_AK_tK_T}{\sqrt{(JK_PK_AK_T)}}$$

or $$1 = \frac{1}{2}\frac{2 + 5 \times 2 \times K_t}{\sqrt{(0.4 \times 0.6 \times 5 \times 2)}} \qquad \text{or} \qquad K_t = 0.11 \text{ V/rad/sec.}$$

Example 5.4 : The system illustrated in Fig. 5.32 is a unity feedback control system with a minor feedback loop (output derivative feedback).

(*a*) In the absence of derivative feedback (a = 0), determine the damping factor and natural fequency. Also determine the steady-state error resulting from a unit-ramp input.

(*b*) Determine the derivative feedback constant of which will increase the damping factor the system to 0.7. What is the steady-state error to unit-ramp input with this setting of the derivative feedback constant ?

(*c*) Illustrate how the steady-state error of the system with derivative feedback to unit-ramp input can be reduced to the same value as in part (*a*), while the damping factor is maintained at 0.7.

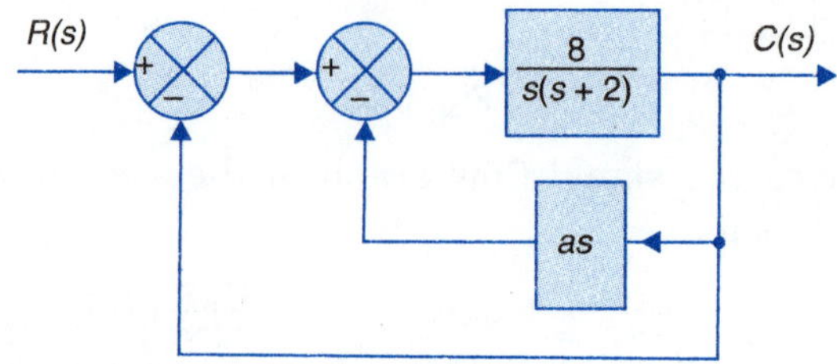

Fig. 5.32

Solution.

(*a*) With $a = 0$, the characteristic equation is

$$s(s + 2) + 8 = 0 \qquad \text{or} \qquad s^2 + 2s + 8 = 0$$

$$\omega_n = \sqrt{8} = 2\sqrt{2} \text{ rad/sec}$$

$$2\zeta\omega_n = 2$$

$$\therefore \qquad \zeta = \frac{1}{2\sqrt{2}} = 0.353$$

System $K_v = 8/2 = 4$

$$\therefore \qquad e_{ss} \text{ (to unit-ramp)} = 1/4 = 0.25$$

(*b*) With derivative feedback, the characteristic equation is

$$1 + G(s) = 1 + \frac{\dfrac{8}{s(s+2)}}{1+\dfrac{8as}{s(s+2)}} = 0$$

or

$$s^2 + (2 + 8a)s + 8 = 0$$

$$\therefore \qquad 2\zeta\omega_n = 2 + 8a$$

$$2 \times 0.7 \times 2\sqrt{2} - 2 + 8a$$

$$a = 0.245$$

System $K_v = 8/(2 + 8a)$

$$\therefore \qquad e_{ss} = (2 + 8a)/8 = 0.495$$

(*c*) Let the gain of 8 in the forward loop be adjusted to a higher value K_A. The new characteristic equation is

$$s^2 + (2 + aK_A)s + K_A = 0$$

$$\therefore \qquad 2\zeta\omega_n = 2 + aK_A$$

$$2 \times 0.7\sqrt{K_A} = 2 + aK_A \qquad \text{...(5.59)}$$

System $K_v = K_A/(2 + aK_A)$

$$\therefore \qquad e_{ss} = (2 + aK_A)/K_A = 0.25 \qquad \text{...(5.60)}$$

Solving eqns. (5.55) and (5.56) we obtain

$$K_A = 31.36; \; a = 0.186$$

Alternative solution is obtained by adding an amplifier of gain K_A between the two summing blocks. The characteristic equation now becomes

$$1 + G(s) = 1 + K_A \frac{8/s(s+2)}{1+8as/s(s+2)} = 0$$

or

$$s^2 + (2 + 8a)s + 8K_A = 0$$

$$\therefore \qquad 2\zeta\omega_n = 2 + 8a$$

$$2 \times 0.7 \times 2\sqrt{2}\sqrt{K_A} = 2 + 8a \qquad \text{...(5.61)}$$

System $K_v = 8K_A/(2 + 8a)$

$\therefore \quad e_{ss} = (2 + 8a)/8K_A = 0.25$ ...(5.62)

Solving eqns. (5.57) and (5.58) we obtain

$$K_A = 3.92 \; ; a = 0.73$$

Note. The second solution requires a smaller gain but a separate amplification stage.

Example 5.5 : Consider the liquid-level system shown in Fig. 5.33. The pump controls the liquid head h by supplying liquid at a rate Q m^3/sec to the tank of cross-sectional area 1 m^2. We shall assume that the flow rate Q is proportional to the error in liquid level (desired level-actual level). Under these assumptions, the system equations are:

(*i*) $Q = \dot{h}$ or $\dfrac{H(s)}{Q(s)} = \dfrac{1}{s}$

(*ii*) $Q = Ke$; e = error in liquid level, K = gain constant

The block diagram representation of the system is given in Fig. 5.34. Let us pose the problem of computing the value of K that minimizes the ISE for unit-step input.

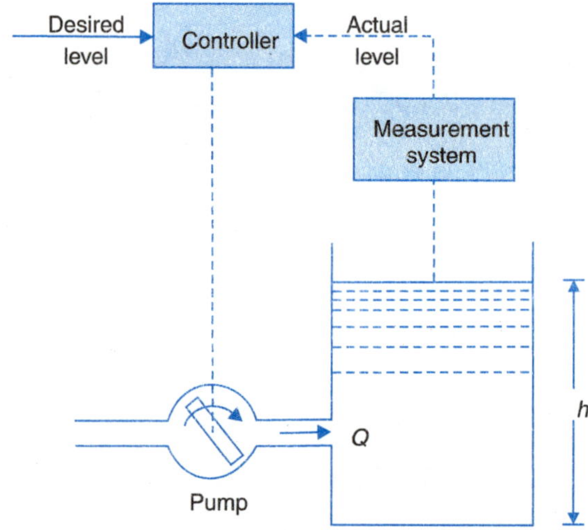

Fig. 5.33. A liquid-level system.

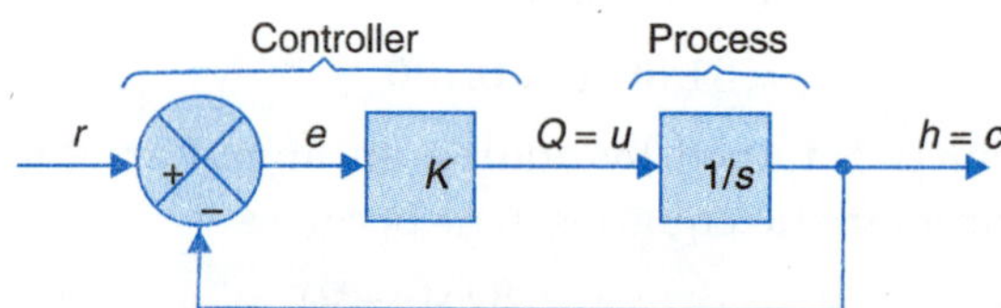

Fig. 5.34. Block diagram representation of Fig. 5.33.

From Fig. 5.34 we have

$$\frac{E(s)}{R(s)} = \frac{s}{s+K}$$

For unit-step input

$$E(s) = \frac{1}{s+K}$$

Therefore

$$e(t) = e^{-Kt} \qquad \text{...(5.63)}$$

$$\text{ISE} = J_e = \int_0^{\infty} e^2(t)dt = 1/(2K) \qquad \text{...(5.64)}$$

Obviously, the minimum value of J_e is obtained as $K \to \infty$. This is an impractical solution resulting in excessive strain on physical components of the system. Increasing the gain means in effect increasing the pump size.

Sound engineering judgement tells us that we must include in our performance index the 'cost' of the control effort. We can do this in many ways. One of the ways is to modify the peformance index J_e so as to include the cost of control effort as a performance measure. For the system under consideration, this cost is proportional to power rating of pump which depends upon kinetic energy required to accelerate the fluid.

Mass of fluid/sec = $Q\rho$; ρ = fluid density

Velocity of fluid = $\frac{Q}{A} = Q$; A = area of cross-section of tank = 1 m^2

Kinetic energy/sec required to accelerate the fluid

$$= \frac{Q^3\rho}{2} = \text{pump power}$$

Therefore, total control effort required = $\int_0^{\infty} \frac{Q^3\rho}{2} dt$

We may limit this control effort, *i.e.*,

$$\int_0^{\infty} \frac{Q^3\rho}{2} dt \le M$$

where M is a constant.

The above inequality may be expressed in the form

$$J_u = \int_0^{\infty} u^3 dt \le N \qquad \text{...(5.65)}$$

where $u = Q$ (Fig. 5.34) and N is the constant determined by M and ρ. J_u is the index based on the cost of control effort.

As per eqn. (5.63)

$$u = Q = K\,e^{-Kt}$$

$$J_u = \int_0^{\infty} K^3 e^{-3Kt} dt = \frac{K^2}{3} \le N$$

Now J_e given by (5.64) is minimized when K has the largest permissible value, *i.e.*,

$$\frac{K^2}{3} = N \quad \text{or} \quad K = \sqrt{3N} \qquad \text{...(5.66)}$$

Hence

$$J_e(\min) = \frac{1}{2K} = \frac{1}{2\sqrt{3N}} \qquad ...(5.67)$$

Example 5.6 : A unity feedback position control system has a forward path transfer function

$$G(s) = K/s$$

For unit-step input, compute the value of K that minimizes ISE.

Solution.

For the system under consideration

$$\frac{E(s)}{R(s)} = \frac{1}{1+G(s)} = \frac{s}{s+K}$$

For unit-step input

$$E(s) = \frac{1}{s+K}$$

Therefore,
$$e(t) = e^{-Kt} \qquad ...(5.68)$$

$$\text{ISE} = \int_0^{\infty} e^2(t)dt = \frac{1}{2K} \qquad ...(5.69)$$

Obviously, the minimum value of ISE is obtained as $K \to \infty$. This is an impractical solution resulting in excessive strain on physical components of the system. A more practical solution results by limiting the second derivative of the output which amounts to limiting the maximum torque of the system.

Let the constraint be

$$|\ddot{c}| \leq M \qquad ...(5.70)$$

where c is the output and M is a constant.

For the system considered here, the output is

$$C(s) = \frac{K}{s(s+K)}$$

$$c(t) = 1 - e^{-Kt}$$

Therefore,
$$\ddot{c}(t) = K^2 e^{-Kt}$$

$$|\ddot{c}(t)|_{\max} = K^2 \leq M \qquad ...(5.71)$$

The maximum value of K satisfying the constraint (5.71) will give the minimum value of ISE. Thus the optimum value of K obtained from eqn. (5.67) is

$$K = \sqrt{M}$$

Therefore, minimum ISE $= 1/2\sqrt{M}$

Example 5.7 : For the system of Example 5.6, compute the value of K that minimizes the following performance index.

$$J = \int_0^{\infty} (e^2 + \lambda\ddot{e})\, dt\ ; \ \lambda \text{ is a positive constant}$$

What is the minimum value of J ?

Solution.

Before solving this problem we note the following:

$$e(t) = 1 - c(t) \qquad \text{or} \qquad \ddot{e}(t) = -\ddot{c}(t)$$

The output acceleration could thus be limited by a single performance index of the type given in this problem.

From eqn. (5.64)

$$\ddot{e} = K^2 e^{-Kt}$$

Therefore,
$$J = \int_0^{\infty} (e^{-2Kt} + \lambda K^4 e^{-2Kt})dt = (1 + \lambda K^4)/2K$$

The minimum value of J is obtained when

$$\frac{\partial J}{\partial K} = -\frac{1}{2K^2} + \frac{3}{2}\lambda K^2 = 0 \qquad \text{or} \qquad K = (3\lambda)^{-1/4}$$

Note. $\dfrac{\partial^2 J}{\partial K^2} = -\dfrac{1}{K^3} + 3\lambda K > 0$

The minimum value of J is $\frac{2}{3}(3\lambda)^{1/4}$.

The constant λ is called the *weighting factor*. When $\lambda = 0$, there is no constraint imposed on the output acceleration. When λ is large, it means more importance is given to the constraint on output acceleration compared to the performance of the system. A suitable value of λ is chosen so that the relative importance of the system performance is contrasted with the importance of the limit on output acceleration.

Figure 5.35 (on next page) gives a plot of the performance index versus K for various values of λ.

5.11 ROBOTIC CONTROL SYSTEMS

A robotic manipulator arm consists of a number of links connected with each other by means of movable joints. The drive at these joints could be provided by any of the following actuators:

1. Hydraulic cylinder
2. Pneumatic cylinder
3. DC motor
4. AC motor
5. Stepper motor

The hydraulic actuator can produce sufficient force to drive the robotic links without need of reduction gearing and this type of actuators are easily applied for robotic position control. But on the other hand hydraulic systems are very cumbersome and messy as these require a great deal of equipment such as pumps, actuators, hose and servo valves etc. In applications where forces must be accurately controlled hydraulic systems prove disadvantageous due to friction of seats and viscosity of oil which has the further complexity of temperature dependence.

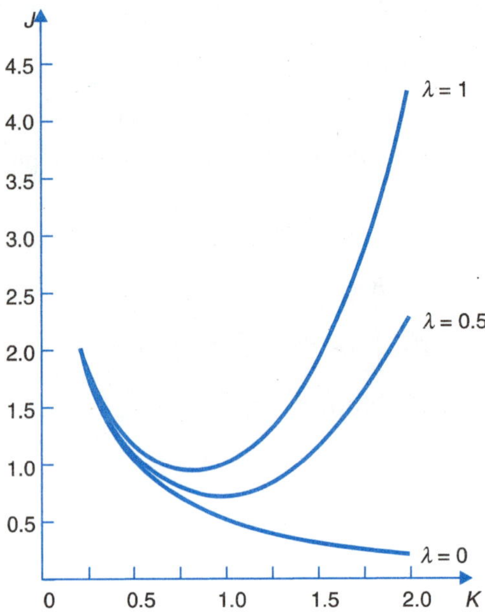

Fig. 5.35. Performance index vs. gain K for Example 5.7.

Pneumatic actuators also possess all the disadvantages of hydraulic systems except that these are cleaner. However, pneumatic actuators are difficult to control accurately due to high friction of seals and compressibility of air (or other pneumatic medium).

The most popular choice of actuator is the electric motor. Although these do not have the power-to-weight ratio of hydraulic or pneumatic actuators, these are easily controlled and interfaced. This makes them attractive for small to medium size manipulators.

Amongst electric motors, ac motors and stepper motors are not very popular. Difficulty of control of former and low torque ability of the latter make their use limited.

DC brush motors (refer Section 4.4) are most straight forward to interface and control, and hence are the most frequently used actuators. The limiting factor on torque output of these motors is the overheating of windings. Brush wear and friction can also be a problem.

Brushless dc motors solve these problems to a great extent. As the windings are on the outside of the motor case, better cooling is possible. This allows for higher sustained torque ratings; brush wear is also eliminated.

Control of a Single Link Manipulator

Let us consider the case of a single link driven by an electric actuator (dc motor) along with inertia link reduction gearing of ratio *i.e.*, link speed/motor speed = $1/n$ where $n > 1$. The actuator inertia reflected through the reduction gearing gets multiplied by factor n^2. The effective link inertia can be expressed as

$$J = D + n^2 J_a \qquad \text{...(5.72)}$$

D = effective link inertia (excluding actuator inertia)

J_a = actuator inertia

J = total (link plus actuator) inertia (on the link side of gearing)

It can be easily seen from eqn. (5.72) that the actuator inertia which is a constant tends to reduce any variation in D, the structural component of inertia.

A dc servomotor is characterized by an actuator gain and effective viscous friction (which also accounts for motor back emf) *expressed in terms of link variables.* These are:

Actuator gain K_m = motor torque in Nm per unit desired speed in rad/s (which is scaled value of motor armature current *i.e.,* the unit are Nm/rad/s)

F = Effective viscous friction in Nm/rad/s

Coulomb friction is ignored in our modelling of the link-actuator mechanism whose block diagram is drawn as in Fig. 5.36 where $\dot{\theta}_d$ = desired speed and $\dot{\theta}_L$ = link speed in rad/s.

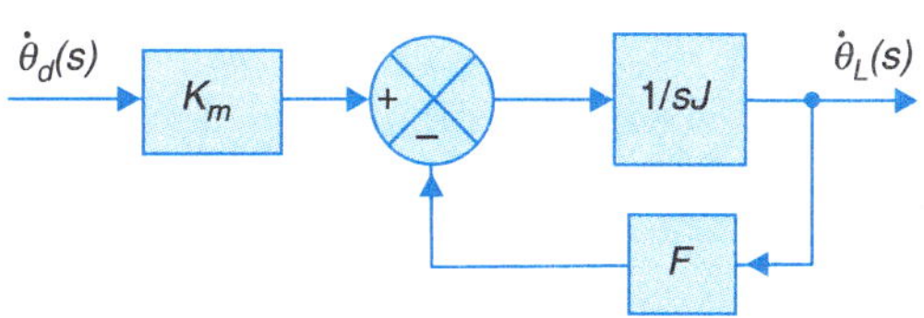

Fig. 5.36

The transfer function corresponding to the block diagram of Fig. 5.36 is

$$\frac{\dot{\theta}_L(s)}{\dot{\theta}_d(s)} = \frac{K_m}{sJ + F} \qquad \text{...(5.73)}$$

The damping factor of the closed-loop actuators can be increased by providing rate feedback from a tachogenerator (refer Section 5.6.) resulting in the block diagram of Fig. 5.37. The transfer function of eqn (5.73) now modifies to

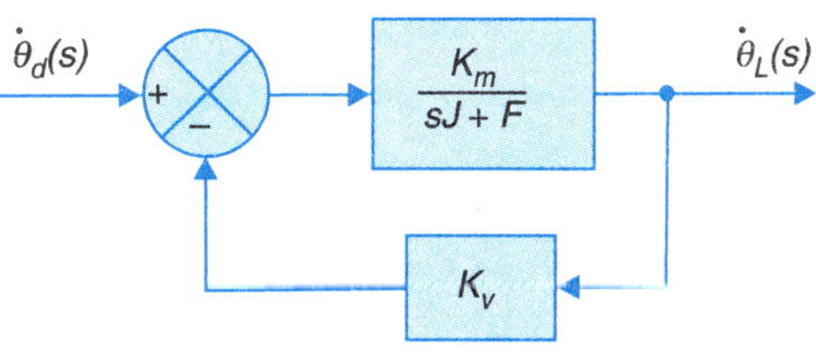

Fig. 5.37

$$\frac{\dot{\theta}_L(s)}{\dot{\theta}_d(s)} = \frac{K_m}{sJ + (F + K_v K_m)} \qquad \text{...(5.74)}$$

For controlling the position (link angle) feedback is provided from the output angle θ_L as in the block diagram of Fig. 5.38 where in an adjustable gain K_e is included in the forward path. The units of K_e are s^{-1}. The closed-loop transfer function of this position control system is

$$\frac{\theta_L(s)}{\theta_d(s)} = \frac{K_e K_m}{s^2 J + s(F + K_v K_m) + K_e K_m} \qquad \text{...(5.75)}$$

$$= \frac{(K_e K_m / J)}{s^2 + s(F + K_v K_m) / J + K_e K_m / J} \qquad \text{...(5.76)}$$

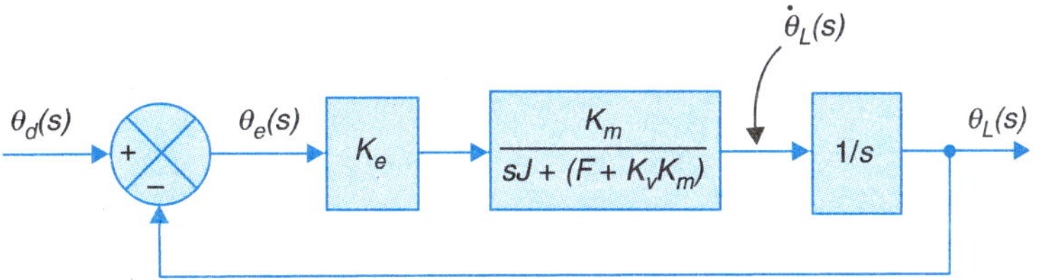

Fig. 5.38

By comparing with eqn. (5.8) it is easily seen that

$$\omega_n = \sqrt{\frac{K_e K_m}{J}} \quad ...(5.77)$$

$$\zeta = \frac{F + K_v K_m}{2\sqrt{JK_e K_m}} \quad ...(5.78)$$

For reasons explained already in Section 5.4 the damping factor of a robotic arm must be unity which is achieved by the condition

$$F + K_v K_m = 2\sqrt{JK_e K_m} \quad ...(5.79)$$

The link has a moment of inertia and structural stiffness, resulting in a structural resonance frequency ω_{struct} whose expression in simple single-link case is K/J where K is the link's structural stiffness. This resonance frequency varies due to variation in link inertia when the end effector carries load. If structural frequency is ω_0 at some value of inertia J_0, then at any other value of J

$$\omega_{struct} = \omega_0 \sqrt{\frac{J_0}{J}} \quad ...(5.80)$$

Typical values of resonance frequency of robot manipulators are 5Hz to 25Hz.

In order to prevent exciting structural oscillations and to ensure structural stability, we must limit ω_n, the natural frequency of link control system, to 0.5 ω_{struct}. From eqn (5.77) and (5.80), we get

$$\sqrt{\frac{K_e K_m}{J}} < 0.5\ \omega_0 \sqrt{\frac{J_0}{J}} \quad \text{or} \quad K_e K_m < \pi^2 f_0{}^2 J_0 \quad ...(5.81)$$

The maximum value of $K_e K_m$ is then given by

$$(K_e K_m)_{max} = \pi^2 f_0{}^2 J_0 \quad ...(5.82)$$

Thus it can be seen that a fixed position gain and variable velocity gain are appropriate for a manipulator of variable inertia.

The velocity feedback gain is selected to provide critical damping given by eqn. (5.79). If gain K_{v0} is selected to provide critical damping with inertia J_0 then at any other value of inertia J, K_v for critical damping is given by

$$K_v = \left[(K_{v0} K_m + F)\sqrt{\frac{J}{J_0}} - F\right]\frac{1}{K_m} \quad ...(5.83)$$

or

$$K_v = (G\sqrt{J} - F)\frac{1}{K_m} \quad ...(5.84)$$

where

$$G = \frac{K_{v0} K_m + F}{\sqrt{J_0}}$$

Steady-state Servo Errors

Due to input command

Examination of the block diagram of Fig. 5.38 reveals that it is a unity feedback system with open-loop transfer function of

$$G(s) = \frac{K_e K_m}{s[sJ + (F + K_v K_m)]} \quad \text{...(5.85)}$$

As this transfer function has one integration, the system is type-1 and so per Table 5.1 its steady-state error ($\theta_e = \theta_d - \theta_L$) to various test signals is given as.

(*i*) Unit step input, $\theta_d(s) = 1/s$; see eqn (5.1)

$$\theta_e(ss) = 0$$

(*ii*) Unit velocity (ramp) input, $\theta_d(s) = 1/s^2$

$$\theta_e(ss) = \frac{1}{\underset{s\to 0}{sG(s)}} = \frac{F + K_v K_m}{K_e K_m} \quad \text{...(5.86)}$$

(*iii*) Unit acceleration (parabolic) input, $\theta_d(s) = 1/s^3$

$$\theta_e(ss) = \infty$$

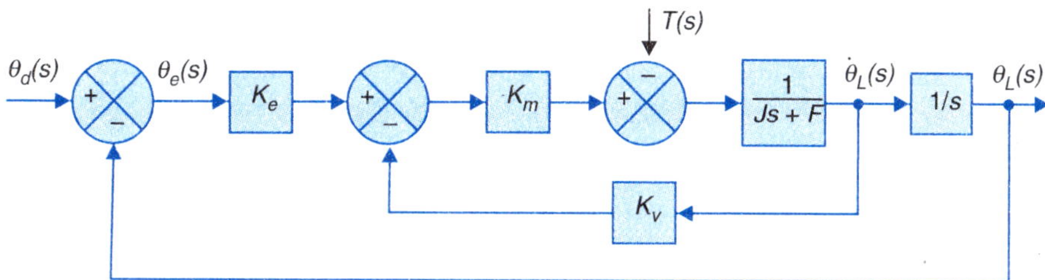

Fig. 5.39

Errors in robotic control system also arise due to disturbance torque, T which appears at the output shaft of reduction gearing. The torque enters the system block diagram at the point indicated in Fig. 5.39. This torque arises due to a combination of the following factors:

1. Error in determining load
2. External forces
3. Coulomb friction
4. Gravity loading

It is easily shown from the block diagram of Fig. 5.39 that the error caused by disturbance torque $T(s)$ is given by

$$\theta_e(s) = -\frac{T(s)}{s^2 J + (F + K_v K_m)s + K_e K_m} \quad \text{...(5.87)}$$

For a unit step disturbance torque the steady state error is given as

$$\theta_e(ss) = \underset{s\to 0}{sG(s)} = -\frac{T}{K_e K_m} \quad \text{...(5.88)}$$

Feedforward to Eliminate Steady-state Errors

A robot is generally required to follow commands of angle, velocity and acceleration for its end effector to move from one Cartesian point to another. Errors due to velocity and acceleration input must therefore be eliminated and also the errors due to known components of the disturbance torque. Error elimination is achieved by means of feedforward technique.

From the block diagram of Fig. 5.39 the expression for error with command input θ_d and disturbance torque T is given as

$$\theta_e(s) = \frac{s[sJ + (F + K_vK_m)]}{s^2J + (F + K_vK_m)s + K_eK_m}\theta_d(s) - \frac{1}{s^2J + (F + K_vK_m)s + K_eK_m}T(s) \qquad ...(5.89)$$

When $\dot{\theta}_d$ is constant velocity input (V/s^2) the steady-state error arises because of the numerator term $(F + K_vK_m)$ in the error expression of eqn. (5.89). This is cancelled out by providing feedforward $\dot{\theta}_d(s)\,(F + K_vK_m)$ as shown in the block diagram of Fig. 5.40. When the input is constant acceleration $\ddot{\theta}_d(s) = A/s^3$, additional error term is contributed by J in the numerator. This is cancelled by providing additional feedforward $\ddot{\theta}_d(s)\,J$. The error due to disturbance is compensated by additional input $T/(K_eK_m)$. The complete block diagram with velocity, acceleration and disturbance torque error compensation is drawn in Fig. 5.40. Effectively this kind of compensation provides additional torque input to the link such that in steady-state $\theta_L = \theta_d$ or $\theta_e = 0$ for velocity, acceleration and disturbance inputs.

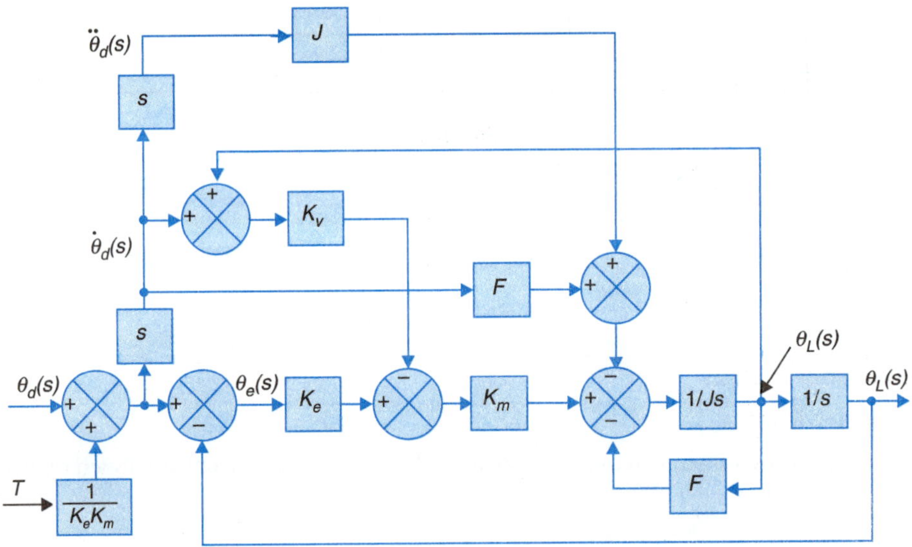

Fig. 5.40. Robotic control system with feedforward compensations.

The compensated control scheme presented above presumes knowledge of J(moment of inertia), F (viscous friction) and T (disturbance torque); but those parameters cannot be estimated accurately in a robot manipulator particularly knowledge of F and T may be far from accurate. More sophisticated control scheme have been devised for which books in robotics may be consulted.

Example 5.8 : A low inertia plotter can be represented by the block diagram of Fig. 5.41. Calculate the value of velocity gain K such that the system is nonoscillatory yet has the lowest possible settling time to step input.

Find the steady-state error for ramp input. Suggest ways of eliminating this error and draw the block diagram of the modified system.

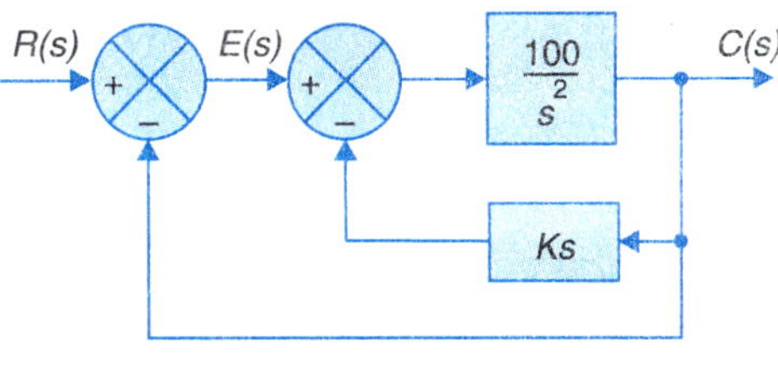

Fig. 5.41

Solution.

$$G(s) = \frac{100/s^2}{1+100K/s} = \frac{100}{s(s+100K)}$$

Characteristic equation is

or $$1 + \frac{100}{s(s+100K)} = 0 \qquad \text{or} \qquad s^2 + 100Ks + 100 = 0$$

which gives

$$\omega_n = 10,\ 2\zeta\omega_n = 100K$$

For lowest possible settling time with nonoscillatary response

$$\zeta = 1 \Rightarrow K = 20/100 = 0.2$$

For ramp input

$$R(s) = V/s^2$$

$$E(s) = \frac{1}{1+G(s)} R(s) = \frac{V/s^2}{1+100/s(s+100K)}$$

$$= \frac{V(s+100K)}{s(s^2+100Ks+100)}$$

$$e(ss) = \underset{s\to 0}{s}\, E(s) = KV = 0.2V$$

For reducing this steady-state error to zero, system is provided with feedforward as in Fig. 5.42.

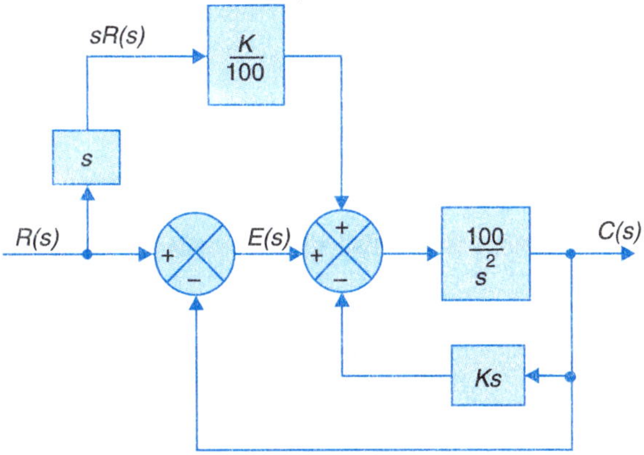

Fig. 5.42

Check

With $R(s) = V/s^2$ consider the steady-state contributed by additional input $sR(s) = V/s$.

$$\left[\frac{KV}{100s} + E(s)\right] G(s) = C(s)$$

But $$C(s) = -E(s)$$

$$\frac{KV}{100s} G(s) = -[1 + G(s)]E(s)$$

or $$E(s) = \frac{(KV/100s)G(s)}{1+G(s)}$$

Substituting value of $G(s)$

$$E(s) = -\frac{KV}{s(s^2+100Ks+100)}$$

$$e(ss) = -KV = -0.2\text{V}, \text{ due to additional input}$$

This cancels out the steady-state error due to $R(s) = V/s^2$.

Hence $e(ss)$ (net) = 0.

Model-Based Robot Control

Several new control strategies other than the conventional control (with feedforward) have been devised for robotic control systems. A simple model-based scheme is presented in Fig. 5.43 in time-domain rather than s-domain, the reason for which will become clear later. The scheme incorporates robot model in form of (F/K_m) and (J/K_m) and also position and velocity errors through gain blocks K_p and K_v respectively. The robot dynamics would follow the trajectory with inputs of θ_d, $\dot{\theta}_d$ and $\ddot{\theta}_d$. Assuming second-order linear dynamics for the robot, we write.

$$\tau = J\ddot{\theta}_0 + F\dot{\theta}_0 \quad ...(5.90)$$

$$\tau = K_m[(J/K_m)\tau' + (F/K_m)\dot{\theta}_0] \quad ...(5.91)$$

$$\tau' = \ddot{\theta}_d + K_v\dot{e} + K_p e \quad ...(5.92)$$

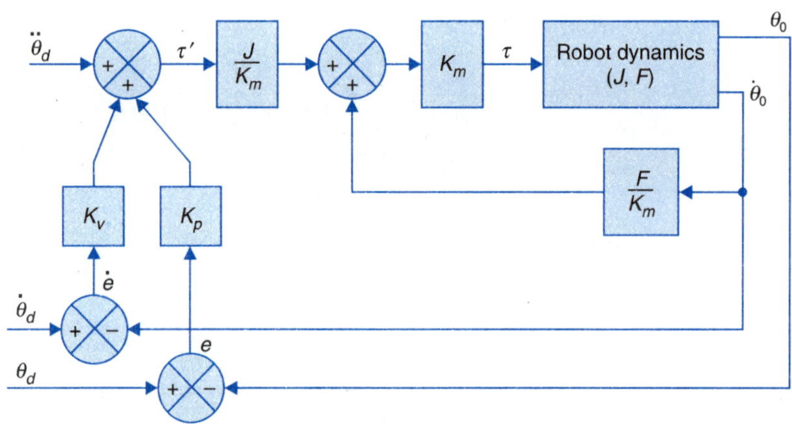

Fig. 5.43. Model-based trajectory following robot control.

Where error and its derivatives are defined as

$$e = \theta_d - \theta_0$$

$$\dot{e} = \dot{\theta}_d - \dot{\theta}_0 \quad ...(5.93)$$

Back substituting eqns. (5.89) and (5.87) we get

$$J(\ddot{\theta}_d + K_v\dot{e} + K_p e) + F\theta_0 = J\ddot{\theta}_0 + F\dot{\theta}_0 \quad ...(5.94)$$

or
$$\ddot{e} + K_v\dot{e} + K_p e = 0 \quad ...(5.95)$$

where
$$\ddot{e} = \ddot{\theta}_d - \ddot{\theta}_0 = \text{second derivative of error}$$

Equation (5.92) gives the system's dynamic performance in error domain. This has resulted in this form because of cancellation of terms of J and K, signals proportional to which have been provided by the model-based control strategy; requiring perfect knowledge of parameters J , f and K_m of the robot and its actuator.

As per error equation (5.95) when excited by initial conditions the error and its derivative, the system dynamics seek zero steady-state error *i.e.,* $e(ss) = 0$. This means that the system follows the trajectory demanded (through command inputs θ_d, $\dot{\theta}_d$ and $\ddot{\theta}_d$).

The gains K_p and K_v are adjusted for critically damped response for which these are to be related as

$$K_v = 2\sqrt{K_p}$$

The error dynamics would be linear as in eqn. (5.95) even though robotic dynamics are nonlinear in which blocks F and J are made correspondingly nonlinear so that these are cancelled out in the error dynamics. This is practically possible with digital control.

However, in actual practice robot parameters J, F are known imperfectly so that have to be estimated on-line and continually adjusted—adaptive control scheme.

5.12 STATE VARIABLE ANALYSIS—LAPLACE TRANSFORM TECHNIQUE

State variable formulations of systems has been presented in Section 2.7 with motivation drawn from physical examples. For a single-input-single-output system we can write the state variable description in the standard form

$$\mathbf{x} = \mathbf{A}\mathbf{x} + \mathbf{B}u \quad \text{...(5.96)}$$

$$y = \mathbf{C}\mathbf{x} \quad \text{...(5.97)}$$

Taking the Laplace transforms of eqn. (5.96) and (5.97), we can write these in the s-domain as

$$s\mathbf{X}(s) - \mathbf{x}(0) = \mathbf{A}\mathbf{X}(s) + \mathbf{B}U(s)\ ;\ \mathbf{x}(0) = \text{initial condition vector} \quad \text{...(5.98)}$$

$$Y(s) = CX(s) \quad \text{...(5.99)}$$

Reorganizing these equations, we get

$$(s\mathbf{I} - \mathbf{A})\,\mathbf{X}(s) = \mathbf{x}(0) + \mathbf{B}U(s)\ ;\ \mathbf{I} = \text{unit diagonal matrix}$$

or
$$\mathbf{X}(s) = (s\mathbf{I} - \mathbf{A})^{-1}\,\mathbf{x}(0) + (s\mathbf{I} - \mathbf{A})^{-1}\,\mathbf{B}U(s) \quad \text{...(5.100)}$$

Taking the inverse Laplace transform of eqn. (5.100) we get the state variable vector in time-domain as

$$\mathbf{x}(t) = \mathcal{L}^{-1}[(s\mathbf{I} - \mathbf{A})^{-1}]\mathbf{x}(0) + \mathcal{L}^{-1}\,[(s\mathbf{I} - \mathbf{A})^{-1}\mathbf{B}U(s)] \quad \text{...(5.101)}$$

We define
$$\mathbf{\Phi}(t) = \mathcal{L}^{-1}\,(s\mathbf{I} - \mathbf{A})^{-1} = \mathcal{L}^{-1}\mathbf{\Phi}(s) \quad \text{...(5.102)}$$

as *state-transition matrix* of the system. Its Laplace transform $\mathbf{\Phi}(s) = (s\mathbf{I} - \mathbf{A})^{-1}$ is called the *resolvant matrix* of the system. From eqn. (5.101) we can obtain the state variable response ($\mathbf{x}(t)$) of the system to initial condition ($\mathbf{x}(0)$) and input ($\mathbf{u}(t)$). This response when used in eqn. (5.97) yields the system output $y(t)$.

We can now obtain the system transfer function from the s-domain state variable model by substituting $\mathbf{X}(s)$ from eqn. (5.100) in eqn. (5.97), while assuming zero initial condition *i.e.,* $\mathbf{X}(0) = 0$. Thus.

$$Y(s) = \mathbf{C}(s\mathbf{I} - \mathbf{A})^{-1}\mathbf{B}U(s)$$

or
$$T(s) = \frac{Y(s)}{U(s)} = \mathbf{C}(s\mathbf{I} - \mathbf{A})^{-1}\mathbf{B} \quad ...(5.103)$$

It can written in the form

$$T(s) = \frac{\mathbf{C}\,\text{adj}\,(s\mathbf{I} - \mathbf{A})\mathbf{B}}{\det\,(s\mathbf{I} - \mathbf{A})\mathbf{B}};\ \text{unique} \quad ...(5.104)$$

It may be noted here that while a system's state-variable model is non-unique, its transfer function is always unique.

The denominator in eqn. (5.104) when set to zero is the system's characteristic equation

$$\det\,(s\mathbf{I} - \mathbf{A}) = 0 \quad ...(5.105)$$

whose roots ($\lambda_1, \lambda_2, ..., \lambda_n$) known as the *eigenvalues* are indeed the poles of the system.

We shall illustrate the concepts and the method of obtaining state-variable response of a system by means of an example.

Example 5.9 : Reconsider the system of Example 5.8 whose modified block diagram identifying the state variables is drawn in Fig. 5.44. Formulate the state variable model of the system and obtain its state transition matrix. Also obtain the output variable for a unit step intput.

Convert the state variable model to transfer function model. Check by means of writing down the transfer function of the system by the signal flow graph technique.

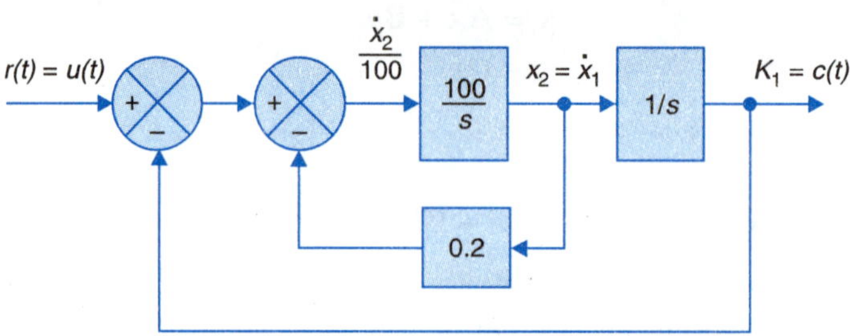

Fig. 5.44

Solution. From Fig. 5.44 we can write down the following state variable equation

$$\dot{x}_1 = x_2$$

$$\dot{x}_2 = -\,100\,x_1 - 20\,x_2 + 100\,r$$

Output
$$y = x_1$$

or
$$\begin{bmatrix} \dot{x}_1 \\ \dot{x}_2 \end{bmatrix} = \begin{bmatrix} 0 & 1 \\ -100 & -20 \end{bmatrix}\begin{bmatrix} x_1 \\ x_2 \end{bmatrix} + \begin{bmatrix} 0 \\ 100 \end{bmatrix} r \quad ...(i)$$

$$y = [1 \quad 0]\begin{bmatrix} x_1 \\ x_2 \end{bmatrix} \quad ...(ii)$$

In Laplace transform from

$$(s\mathbf{I} - \mathbf{A}) = \begin{bmatrix} s & -1 \\ 100 & s+20 \end{bmatrix} \quad ...(iii)$$

Let us take the inverse transform of the resolvant matrix

$$\text{adj}\,(s\mathbf{I}-\mathbf{A}) = \begin{bmatrix} s+20 & 1 \\ -100 & s \end{bmatrix}$$

$$\det\,(s\mathbf{I}-\mathbf{A}) = \begin{bmatrix} s & -1 \\ 100 & s+20 \end{bmatrix} = D(s)$$

$$D(s) = s(s+20) + 100 = 0$$

or $$s^2 + 20s + 100 = 0$$

Its eigenvalues are given as

$s_1, s_2 = -10$; the two eigen values are repeated. This is a critically damped case.

$\mathbf{\Phi}(s) = (s\mathbf{I} - \mathbf{A})^{-1}$ = resolvant matrix

$$= \frac{\text{adj}\,(s\mathbf{I}-\mathbf{A})}{\det\,(s\mathbf{I}-\mathbf{A})} = \begin{bmatrix} s+20 & 1 \\ -100 & s \end{bmatrix}\frac{1}{D(s)} \quad \text{...}(iv)$$

Taking the inverse Laplace transform term by term

$$\mathcal{L}^{-1}\left[\frac{s+20}{D(s)}\right] = \mathcal{L}^{-1}\left[\frac{s+20}{(s+10)^2}\right] = 10\,t\,e^{-10t} + e^{-10t}\ ;\ t > 0$$

$$\mathcal{L}^{-1}\frac{1}{D(s)} = \mathcal{L}^{-1}\frac{1}{(s+10)^2} = e^{-10t}\ ;\ t > 0$$

$$\mathcal{L}^{-1}\frac{-100}{D(s)} = -100t\,e^{-10t}\ ;\ t > 0$$

$$\mathcal{L}^{-1}\frac{s}{D(s)} = \mathcal{L}^{-1}\frac{s}{(s+10)^2} = -10t\,e^{-10t} + e^{-10t}\ ;\ t > 0$$

Thus the state transition matrix is given as

$$\phi(t) = \begin{bmatrix} 10te^{-10t} + e^{-10t} & t\,e^{-10t} \\ -100te^{-10t} & -10te^{-10t} + e^{-10t} \end{bmatrix};\ t > 0 \quad \text{...}(v)$$

For $$\mathbf{X}(s) = [(s\mathbf{I}-\mathbf{A})^{-1}\,\mathbf{B}\,(1/s)]\ ;\ U(s) = 1/s$$

$$= \frac{1}{sD(s)}\begin{bmatrix} s+20 & 1 \\ -100 & s \end{bmatrix}\begin{bmatrix} 0 \\ 100 \end{bmatrix} = 100\begin{bmatrix} 1/s \\ 1 \end{bmatrix}\frac{1}{D(s)} \quad \text{...}(vi)$$

Taking the inverse Laplace transform term by term

$$\mathcal{L}^{-1}\frac{1}{D(s)} = \mathcal{L}^{-1}\frac{1}{(s+10)^2} = te^{-10t}\ ;\ t > 0$$

$$\mathcal{L}^{-1}\frac{1}{sD(s)} = \mathcal{L}^{-1}\frac{1}{s(s+10)^2}$$

$$= \frac{1}{100}\,(1 - 10te^{-10t} - e^{-10t})$$

Thus $$\mathbf{x}(t) = 100\begin{bmatrix} 1 - 10\,te^{-10t} - e^{-10t} \\ 100\,te^{-10t} \end{bmatrix};\ t > 0 \quad \text{...}(vii)$$

The output is given as

$$y(t) = x_1(t) = 1 - 10\,e^{-10t} - e^{-10t}\ ;\ t > 0 \quad \text{...}(viii)$$

The transfer function is

$$T(s) = \frac{\mathbf{C}\,\text{adj}\,(s\mathbf{I} - \mathbf{A})\mathbf{B}}{\det\,(s\mathbf{I} - \mathbf{A})}$$

$$\mathbf{C}\ \text{adj}\ (s\mathbf{I} - \mathbf{A})\ \mathbf{B} = [1 \quad 0]\begin{bmatrix} s+20 & 1 \\ -100 & s \end{bmatrix}\begin{bmatrix} 0 \\ 100 \end{bmatrix} = 100$$

Hence $$T(s) = \frac{100}{s^2 + 20s + 100}$$

Check :

From Fig. 5.44 we can write by means of signal flow graph

$$T(s) = \frac{100/s^2}{1 + \dfrac{100 \times 0.2}{s} + \dfrac{100}{s^2}} = \frac{100}{s^2 + 20s + 100}.$$

5.13 THE APPROXIMATION OF HIGHER-ORDER SYSTEMS BY LOWER-ORDER

Complex systems with higher-order transfer function are studied by approximating to lower-order model. Thus for example a third-order system could be approximated by a second-order system and we can use the performance indices dealt with in the previous section. Out of the several methods available for reducing the order of a system transfer function, one of the simple way to delete a certain insignificant pole of a transfer function, is to note this pole that has a negative real part that is much larger than the other poles. Thus the pole that is expected to affect the transient response insignificantly is deleted.

For example, consider a plant transfer function

$$H(s) = \frac{K}{(s+10)(s^2 + 2s + 2)} \qquad ...(5.106)$$

We can safely neglect the effect of the pole at $s = -10$. However, we must retain the steady-state response, and we reduce the system to

$$L(s) = \frac{(K/10)}{(s^2 + 2s + 2)} \qquad ...(5.107)$$

A second, more complicated system reduction scheme uses an algebraic method. We will let the high-order system be described by the transfer function

$$H(s) = K\frac{b_m s^m + b_{m-1}s^{m-1} + \ldots + b_1 s + 1}{a_n s^n + a_{n-1}s^{n-1} + \ldots + a_1 s + 1}\ ;\ m \le n \qquad ...(5.108)$$

in which the poles are in the left-hand s-plane. The approximate lower-order approximate transfer function is

$$L(s) = K\frac{d_p s^p + d_{p-1}s^{p-1} + \ldots + d_1 s + 1}{c_q s^q + c_{q-1}s^{q-1} \ldots + c_1 s + 1} \qquad ...(5.109)$$

where $p \leq q < n$. To maintain the same steady-state response, the gain constant K is the same for the original and the approximate system. In this method c_i and d_i are selected such that the frequency response of $L(s)$ should be very close to that of $H(s)$. They are mathematically obtained from

$$M^{(k)}(s) = \frac{d^k}{ds^k} M(s) \quad ...(5.110)$$

and

$$\Delta^{(k)}(s) = \frac{d^k}{ds^k} \Delta(s) \quad ...(5.111)$$

where $M(s)$ and Δ are the numerator and denominator polynomials of $H(s)$ and $L(s)$ respectively. We also define

$$M_{2i} = \sum_{k=0}^{2i} \frac{(-1)^{k+i} M^{(k)}(0) M^{(2i-k)}(0)}{k!(2i-k)!} \quad ...(5.112)$$

and a completely indentical of Δ_{2i}. The solutions for the c and d coefficients are obtained by equating

$$M_{2i} = \Delta_{2i} \text{ (same as } \Delta^{2i}) \quad ...(5.113)$$

for i = 1, 2, ... up to the number required to solve for unknown coefficients.

Example 5.10 : Consider a third-order system

$$H(s) = \frac{20}{(s+10)(s^2+2s+2)} = \frac{1}{(1/20)s^3 + (3/5)s^2 + (11/10)s + 1} \quad ...(5.114)$$

Using the second-order model

$$L(s) = \frac{1}{c_2 s^2 + c_1 s + 1} \quad ...(5.115)$$

$M(s) = c_2 s^2 + c_1 s + 1$ and $\Delta(s) = (1/20)s^3 + (3/5)s^2 + (11/10)s + 1$. Then

$$M^0(s) = c_2 s^2 + c_1 s + 1 \quad ...(5.116)$$

and $M^0(0) = 1$. Similarly

$$M^1(s) = 2c_2 s + c_1 \quad ...(5.117)$$

and $M^1(0) = c_1$. Continuing the process we find that,

$$\begin{aligned} M^0(0) &= 1 & \Delta^0(0) &= 1 \\ M^1(0) &= c_1 & \Delta^1(0) &= 11/10, \\ M^2(0) &= 2c_2 & \Delta^2(0) &= 6/5, \\ M^3(0) &= 0 & \Delta^3(0) &= 3/10, \end{aligned}$$

We now equate $M_{2i} = \Delta_{2i}$ for i = 1 and 2. We find for i = 1,

$$M_2 = (-1)\frac{M^0(0)M^2(0)}{2} + \frac{M^1(0)M^1(0)}{1} + (-1)\frac{M^2(0)M^0(0)}{2} \quad ...(5.118)$$

$$= -c_2 + c_1^2 - c_2 = c_1^2 - 2c_2 \quad ...(5.119)$$

Since the equation for Δ_2 is identical, we have

$$\Delta_2 = (-1)\frac{\Delta^0(0)\Delta^2(0)}{2} + \frac{\Delta^1(0)\Delta^1(0)}{1} + (-1)\frac{\Delta^2(0)\Delta^0(0)}{2} \quad ...(5.120)$$

$$= -\frac{3}{5} + \frac{121}{100} - \frac{3}{5} = \frac{1}{100} \quad ...(5.121)$$

Because $M_2 = \Delta_2$ we have

$$c_1^2 - 2c_2 = \frac{1}{100} \quad ...(5.122)$$

Completing the process for $M_4 = \Delta_4$, we obtain

$$c_2^2 = \frac{1}{4} \quad ...(5.123)$$

Then the solution for $L(s)$ is $c_1 = 1.005$ and $c_2 = 0.5$. (The other set of solutions are rejected because they lead to unstable poles). The poles of $H(s)$ are $s = -1 \pm j, -10$, whereas the poles of $L(s)$ are $s = -1.005 \pm 0.995j$. The lower order transfer function is

$$L(s) = \frac{1}{0.5\,s^2 + 1.005\,s + 1} = \frac{2}{s^2 + 2.01\,s + 2} \quad ...(5.124)$$

Since the system is reduced to second-order, it can be analyzed as given in the previous section.

Example 5.11 : A unity feedback central system is to be designed to meet the following specifications

$$30\% > M_{\text{peak}} > 15\%$$
$$t_s < 0.75 \text{ sec.}$$

(*a*) Mark the area in the *s*-plane in which dominant roots of the system's characteristic equation must lie.

(*b*) Determine the smallest value of the third root such that dominance of the complex conjugate roots corresponding to part (*a*) is preserved.

(*c*) The system has unity feedback and its settling time is 0.75 sec. and peak overshoot is 30%. Determine the open loop transfer function of the system.

Solution. (*a*) $M_p = e^{-\pi\xi/\sqrt{(1-\xi^2)}}$

$M_p = 0.15 \Rightarrow \xi = 0.55, \quad \theta = \cos^{-1} 0.55 = 56.6° \approx 57°$

$M_p = 0.30 \Rightarrow \xi = 0.35, \quad \theta = \cos^{-1} 0.35 = 69.5° \approx 70°$

Settling time

$$t_s = \frac{4}{\xi\omega_n} = 0.75$$

$$\omega_n = \frac{4}{0.75\,\xi}, \; \xi\omega_n = \frac{4}{0.75} \quad ...(i)$$

$$\xi = 0.55 \Rightarrow \omega_n = 9.7 \approx 10$$
$$\xi = 0.35 \Rightarrow \omega_n = 15$$

The area on s plane where the complex dominant roots should be in Fig. 5.45

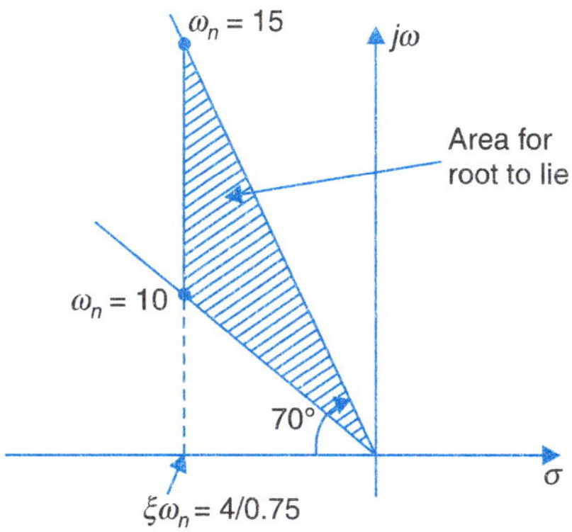

Fig. 5.45

(b) From Eq. (i) $\quad \xi\omega_n = \dfrac{4}{0.75}$

We choose the third root to be at least 6-time further on negative real axis than the real part of the complex conjugate roots. Thus

$$s = -6 \times (\xi\omega_n) = \frac{6 \times 4}{0.75}$$

or $$s = -32$$

(c) $t_s = 0.75$, $M_p = 30\%$ $\Rightarrow$ $\xi = 0.35$, $\xi\omega_n = \dfrac{4}{0.75} = 5.33$, $\omega_n = 15$

Complex conjugate roots term is

$$(s^2 + 2\,\xi\omega_n\, s + \omega_n^{\,2}) = (s^2 + 10.66s + 225)$$

Third root $\quad s = -32$

The closed-loop transfer function can then be written as

$$T(s) = \frac{C(s)}{R(s)} = \frac{32 \times 225}{(s + 32)\,(s^2 + s\,0.66s + 225)}$$

The numerator (32 × 225) assures that $T(s = 0) = 1$, which mean $C(0) = R(0)$ or steady state error is zero, system type-1.

Now $$T(s) = \frac{G(s)}{1 + G(s)} = \frac{32 \times 225}{(s + 32)\,(s^2 + 10.66s + 225)}$$

From which we can write

$$G(s) = \frac{T(s)}{1 - T(s)}$$

Solving yield $$G(s) = \frac{32 \times 225}{s(s^2 + 43s + 566)}.$$

Example 5.11. The open-loop transfer function of a unity feedback system is

$$G(s) = \frac{K}{s(s+2)}$$

The system is required to meet the following specifications.

Peak time, $t_p = 1.0$ sec.

Peak overshoot, $M_p = 10\%$

(*a*) Can we find a value of K to meet both these specifications ? If not, what should be the value of K so that t_p and M_p are violated by the same percentage.

(*b*) Assume that the open-loop pole $s = -2$ can be adjusted. Let it be $s = -\alpha$. Determine the values of K and α in order to meet both the specifications.

(*c*) Suggest a suitable method of adjusting α and draw the block diagram of the modified system.

Solution. (*a*) Characteristic equation of the system

$$1 + G(s) = 0 \quad \text{or} \quad s^2 + 2s + K = 0 \qquad ...(i)$$

Required $$M_p = 0.1 = e^{-\pi\xi/\sqrt{1-\xi^2}} \Rightarrow \xi = 0.6$$

From Eq. (*i*)

$$2\xi\omega_n = 2, \quad \omega_n = \sqrt{K}$$

or $$0.6\sqrt{K} = 1 \Rightarrow K = 2.78$$

Now $$t_p = \frac{\pi}{\omega_n\sqrt{1-\xi^2}} = \frac{\pi}{\sqrt{2.78}\sqrt{1-(0.6)^2}}$$

or $$t_p = 2.36 \text{ sec.} > 1 \text{ ; 2 sec. (specified)}$$

We conclude that with one adjustable variable (K) two specifications can not be met.

(*b*) We now modify the specifications

$$t_p = 1(1 + x) \quad ; \quad x < 1 \qquad ...(ii)$$

$$M_p = 0.1(1 + x) \; ; \quad x < 1 \qquad ...(iii)$$

From Eq. (*i*)

$$\omega_n = \sqrt{K} \; ; \qquad \xi = \frac{1}{\sqrt{K}}$$

Substituting in Eqn. (*ii*) and (*iii*)

$$t_p = \frac{\pi}{\omega_n\sqrt{1-\xi^2}} = \frac{\pi}{\sqrt{K}\sqrt{1-\frac{1}{K}}}$$

or $$t_p = \frac{\pi}{\sqrt{K-1}} = (1 + x) \qquad ...(iv)$$

$$M_p = e^{-\pi\xi/\sqrt{1-\xi^2}}, \; \xi = \frac{1}{\sqrt{K}}$$

or $$M_p = e^{-\pi}/\sqrt{K-1} = 0.1(1 + x) \qquad ...(v)$$

Solving Eqn. (*iv*) and (*v*) we will get K and x. Since these equations are transcendental trial and error numerical solution is needed. We get the solution as

$$K = 4.25 \quad x = 0.75$$

From these values we find

$$t_p = 1.74, \quad M_p = 0.175$$

% change in $t_p = \dfrac{1.74 - 1}{1} \approx 0.75$ or 75% increase

% change in $M_p = \dfrac{0.175 - 0.1}{0.1} = 0.75$ or 75% increase

(c) $$G(s) = \frac{K}{s(s+a)} \quad \text{...(vi)}$$

Characteristic equation

$$s^2 + \alpha s + K = 0$$

$$\omega_n = \sqrt{K}, \quad \alpha = 2\xi\omega_n \quad \text{...(vii)}$$

Specifications : $\xi = 0.6$, $t_p = 1$ sec

$$t_p = \frac{\pi}{\omega_n\sqrt{1-\xi^2}} = \frac{\pi}{\omega_n\sqrt{1-(0.6)^2}} = 1 \text{ sec.} \quad \text{...(viii)}$$

From Eq. (ix)

$$\omega_n = \frac{1 \times 0.8}{\pi} = 3.93 \text{ rad/sec}$$

∴ $$K = \omega_n^2 = 15.4$$

$$\alpha = 2 \times 0.6 \times 3.93$$

or $$\alpha = 4.7$$

(c) To make α adjustable we can use output rate feedback as in the block diagram of Fig. 5.46

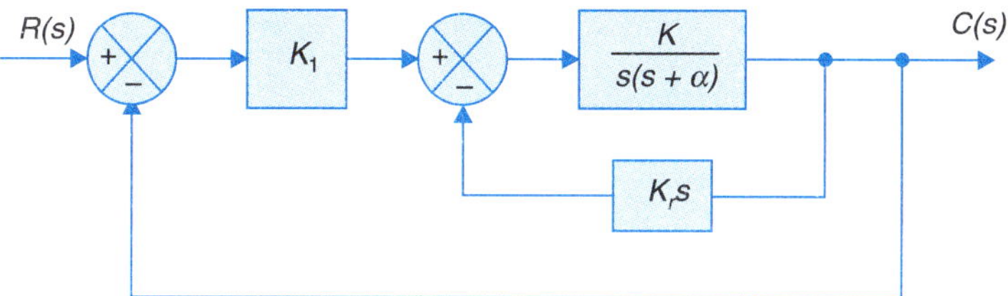

Fig. 5.46

From Fig. 5.46 we can find the forward path transfer function as

$$G(s) = K_1\left[\frac{\dfrac{K}{s(s+\alpha)}}{1+\dfrac{KK_r s}{s(s+\alpha)}}\right]$$

or $$G(s) = \frac{K_1 K}{s[s+(\alpha+KK_r)]} \quad \text{...(ix)}$$

or $$G(s) = \frac{KK}{s(s+\alpha')} \quad \text{...(x)}$$

where $\alpha' = \alpha + KK_r$

We see that α can be adjustable to $\alpha' > \alpha$.

Example 5.12: A unity feedback (negative) system has open-loop transfer function

$$G(s) = \frac{K}{s(s+2)}$$

(*a*) Calculate the value of gain K so that the closed-loop system has a steady-state unit ramp error of 0.1. What are the corresponding damping factor and percentage peak overshoot to unit step input.

(*b*) The system is now modified to include a forward path zero at $s = -6$. What is the new value of K for the same steady-state error as in part (*a*). Also calculate the damping factor and find the percentage peak overshoot corresponding to this damping factor and that found from the curves of Fig. 5.19 (*b*).

Solution. (*a*) Steady-state error to unit ramp input

$$e_{ss} = \lim_{s \to 0} \frac{1}{sG(s)} = \lim_{s \to 0} \frac{(s+2)}{K}$$

or $$e_{ss} = \frac{2}{K} = 0.1 \quad \text{(given)}$$

or $$K = 20$$

Characteristic equation

$$s(s + 2) + K = 0 \quad \text{or} \quad s^2 + 2s + K = 0 \quad \text{or} \quad s^2 + 2s + 20 = 0$$

$$\xi = \frac{2}{2\sqrt{20}} = 0.224$$

$$M_p = e^{-\pi/\xi\sqrt{1-\xi^2}} = 0.486 \quad \text{or} \quad 486\%$$

(*b*) Zero added at $s = -6$

Modified $$G(s) = \frac{K(s+6)}{s(s+2)}$$

Unit ramp input $$e_{ss} = \lim_{s \to 0} \frac{1}{sG(s)} = \frac{2}{6K} = 0.1$$

which gives $$K = 10/3$$

Notice that gain of 6 is contained in $(s + 6)$; so the system's K_v remains unchanged.

Characteristic equation

$$s(s + 2) + K(s + 6) = 0$$

$K = 10/3$ gives

$$s^2 + (16/3)s + 20 = 0$$

$$\omega_n = \sqrt{20}\,,\ \xi = \frac{16}{2 \times 3\sqrt{20}} = 0.6$$

Ignoring the effect of zero

$$M_p = e^{-\pi\xi/\sqrt{1-\xi^2}} \quad \text{at} \quad \xi = 0.6$$

gives

$$M_p = 9.4\%, \text{ considerable reduction}$$

Let us now take into account the effect of zero. For this we will consult the curves of Fig. 5.19 (*b*)

$$\text{Zero, } z = 6, \quad \xi\omega_n = 0.6\sqrt{20}$$

$$\frac{z}{\xi\omega_n} = \frac{6}{0.6\sqrt{20}} = 2.24, \quad \xi = 0.6$$

We obtain from this figure

$$M_p \approx 16\%$$

We find that the effect of zero is to increase M_p from 9.4% to 16%.

Example 5.13. A certain system is described by the differential equation

$$\ddot{y} + b\dot{y} + 4 = r$$

Determine the value of b to satisfy the following specifications.

1. M_p to be as small as possible but no greater than 15%.

2. Rise time t_r to be as small as possible but no greater than 1.2 sec.

If both specifications can not be met simultaneously at least (1) should be met and (2) to be met as closely as possible.

Solution. Since specification (1) must be met, we choose M_p = 15%

$$M_p = e^{-\pi\xi/\sqrt{(1-\xi^2)}} = 0.15$$

Take log natural on both sides

$$-\pi\xi/\sqrt{(1-\xi^2)} = -1.9$$

$$\pi\xi = 1.9\sqrt{(1-\xi^2)}$$

Squaring we get $\quad \pi^2\xi^2 = (1.9)^2(1-\xi^2)$

$$9.87\xi^2 = 3.61(1-\xi^2) \qquad \text{or} \qquad \xi = 0.52$$

From the given differential equation

$$\omega_n = \sqrt{4} = 2 \text{ rad/s}$$

Rise time $\quad t_r = \dfrac{\pi - \phi}{\omega_n\sqrt{1-\xi^2}}, \quad \phi = \tan^{-1}\dfrac{\sqrt{1-\xi^2}}{\xi}, \; \xi = 0.51$

or $\quad t_r = 1.22 \text{ sec.} \quad \phi = 1.036 \text{ rad}$

We find that t_r condition is met.

In this example the two specifications are such that both are met with one parameter ξ (corresponding to b). This is usually not the case.

Example 3.14. The block diagram of a robot joint control is drawn in Fig. 5.47.

Various system parameters are :

J = 10 kg/m², f = 20 Nm/rad/sec.

K_m (motot torque constant) = 2 Nm/V

K_v (velocity feedback constant) = 1 V/rad/sec

K_e(error amplification) in V/degree, to be determined.

(*a*) Calculate the value of K_e for the closed-loop system to have $\xi = 1$.

(*b*) Determine the steady-state error $\theta_e(ss)$ when $\theta_d(s) = 0$ and load torque of 5 Nm is suddenly applied.

(*c*) Determine $\theta_e(ss)$ when $T(s) = 0$ and input θ_d is unit step.

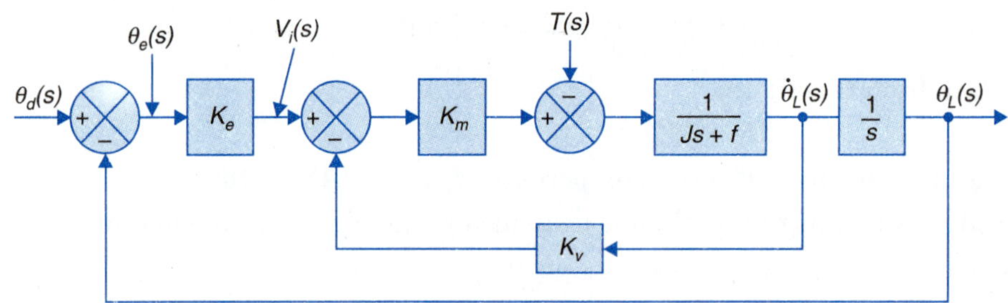

Fig. 5.47

Solution.

(*a*) Reducing the inner loop

$$\frac{\theta_L(s)}{V_i(s)} = \frac{\dfrac{K_m}{Js+f}}{1+\dfrac{K_v K_m}{Js+f}} = \frac{K_m}{Js+(f+K_v\,K_m)} \qquad ...(i)$$

Forward path transfer function

$$G(s) = \frac{\theta_L(s)}{\theta_e(s)} = \frac{K_e K_m}{s[Js+(f+K_v\,K_m)]} \qquad ...(ii)$$

Closed-loop transfer function

$$T(s) = \frac{\theta_L(s)}{\theta_d(s)} = \frac{K_e K_m}{Js^2+(f+K_v\,K_m)s+k_e k_m} \qquad ...(iii)$$

Characteristic equation $s^2 + \dfrac{f+K_v K_m}{J} + \dfrac{K_e K_m}{J} = 0$

Damping factor $\xi = \dfrac{f+K_v K_m}{2J} \times \sqrt{\dfrac{J}{K_e K_m}}$ or $\xi = \dfrac{f+K_v K_m}{2\sqrt{JK_e K_m}} = 1$ ...(*iv*)

which gives the relationship

$$f + K_v K_m = 2\sqrt{JK_e K_m}$$

$$20 + 1 \times 2 = 2\sqrt{10 \times 2\, k_e}$$

We get $K_e = 6.05$

(*b*) $\theta_d(s) = 0,\quad T(s) = \dfrac{10}{s}$

The block diagram can now be drawn in the form of Fig. 3.48.

Now $\theta_e(s) = \theta_d(s) - \theta_L(s) = -\,\theta_L(s)$ as $\theta_d(s) = 0$

So we get from Fig. 3.48

$$\frac{-\theta_e(s)}{-T(s)} = \frac{1}{Js^2+(f+K_v K_m)+K_e K_m} \qquad ...(v)$$

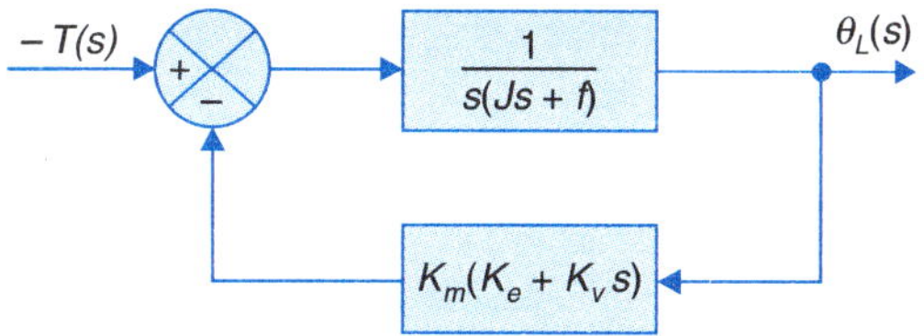

Fig. 5.48

As $$T(s) = \frac{5}{s}$$

$$\theta_e(ss) = \lim_{s \to 0} s\theta_e(s) = \frac{10}{K_e K_m}$$

or $$\theta_e(ss) = \frac{10}{6.05 \times 2} = 0.826 \text{ degree}$$

(*e*) $$T(s) = 0, \; \theta_d(s) = \frac{1}{s}$$

The steady-state error will be zero as there is one integration in the forward path

Thus $$\theta_e(ss) = 0$$

Example 3.15: The block diagram of a unity feedback control system has

$$G(s) = \frac{K}{s(s+8)}$$

Find the value of K such that ITAE performance index is minimized. For this value of K determine the peak overshoot to unit step input.

Solution. From Table 5.2 optimum closed-loop characteristic equation is

$$s^2 + 1.4\,\omega_n s + \omega_n^2 = 0 \qquad ...(i)$$

For the given system

$$\frac{C(s)}{R(s)} = \frac{G(s)}{1+G(s)} = \frac{K}{s^2 + 8s + K} \qquad ...(ii)$$

Comparing the denominator of Eq. (*ii*) with Eq. (*i*), we have

$$\omega_n = \sqrt{K}$$

$$1.4\omega_n = 8$$

or $$1.4\sqrt{K} = 8 \quad \Rightarrow \quad K = 32.7$$

$$\xi = \frac{8}{2\sqrt{K}} = \frac{8}{2\sqrt{32.7}}$$

or $$\xi = 0.7$$

$$M_p = e^{-\pi\xi/\sqrt{(1-\xi^2)}} = 0.046 \quad \text{or} \quad 4.6\%$$

It is noticed that IATE-based system has rather high damping.

Example 3.16. A unity feedback control system has an open-loop transfer function of

$$G(s) = \frac{2(s+8)}{s(s+2)}$$

(*a*) Write the closed-loop transfer function $C(s)/R(s)$ and find there from systems damping factor and natural frequency and unit peak overshoot.

(*b*) Calculate $c(t)$ for unit step input.

(*c*) Calculate e_{ss} for unit ramp input.

(*d*) Remove the zero $s = -10$ and increase the gain to 16. Find now ξ and ω_n.

(*e*) Calculate $c(t)$ for unit step input.

Solution. (*a*)
$$T(s) = \frac{C(s)}{R(s)} = \frac{G(s)}{1+G(s)}$$

or
$$\frac{C(s)}{R(s)} = \frac{2(s+8)}{s^2+4s+16} \qquad \text{...(i)}$$

$$\omega_n = \sqrt{16} = 4, \quad \xi = \frac{4}{2\times 4} = 0.5, \quad M_p = 0.163 \quad \text{or} \quad 16.3\%$$

(*b*)
$$R(s) = \frac{1}{s}, \text{ then}$$

$$C(s) = \frac{2(s+8)}{s(s^2+4s+16)} \qquad \text{...(ii)}$$

$$= \frac{2}{s^2+4s+16} + \frac{16}{s(s^2+4+16)} \qquad \text{...(iii)}$$

Looking up Appendix I Table 1.2 items 15, 17 are reproduced below.

$$\frac{\omega_n^2}{s^2+2\xi\omega_n s+\omega_n^2} \to \frac{\omega_n}{\sqrt{(1-\xi^2)}} e^{-\xi\omega_n t} \sin[\omega_n\sqrt{(1-\xi^2)}]t \;;\; \xi < 1$$

$$\frac{\omega_n^2}{s(s^2+2\xi\omega_n s+\omega\omega_n^2)} \to 1 - \frac{1}{\sqrt{(1-\xi^2)}} e^{-\xi\omega_n t} \sin[\omega_n\sqrt{(1-\xi^2)}\,t+\phi] \;;$$

$$\phi = \tan^{-1}\frac{\sqrt{(1-\xi^2)}}{\xi} \;;\; \xi < 1$$

We can express $C(s)$ as

$$C(s) = \frac{16}{s(s^2+4s+16)} + \frac{1}{8}\,\frac{16}{s^2+4s+16}$$

$$\omega_n = 4, \quad \xi = 0.5, \quad \xi\omega_n = 2$$

$$\sqrt{(1-\xi^2)} = 0.866, \quad \omega_n\sqrt{(1-\xi^2)} = 3.464, \quad \phi = 60°, \quad \frac{\omega_n}{\sqrt{(1-\xi^2)}} = 4.62$$

Taking inverse Laplace transform

$$c(t) = 1 - 1.155e^{-2t}\sin(3.464t + 60°) + 0.155e^{-2t}\sin 3.464t \qquad \text{...(iv)}$$

It can be shown by plotting $e(t)$ that the third term in Eq. (*iv*) caused M_p to be higher than that given by $\xi = 0.5$ because of the presence of zero ($s = -8$)

(*c*) Unit ramp input, $R(s) = \dfrac{1}{s^2}$

$$e_{ss} = \lim_{s\to 0}\frac{1}{sG(s)} = \frac{1}{8} = 0.125$$

(*d*) Zero $s = -8$ removed, then

$$G(s) = \frac{16}{s(s+2)}$$

characteristic equation

$$s^2 + 2s + 16 = 0$$

$$\omega_n = 4, \quad \xi = \frac{2}{2 \times 4} = 0.25, \quad M_p = 0.444 \quad \text{or} \quad 44.4\%$$

It is observed that ξ reduces from 0.5 to 0.25 by removing the zero while M_p increases from 16.3% to 44.4%

(*e*)
$$R(s) = \frac{1}{s}$$

$$\frac{C(s)}{R(s)} = \frac{16}{s^2 + 2s + 16}$$

$$C(s) = \frac{16}{s(s^2 + 2s + 16)}$$

Its Laplace inverse gives

$$C(t) = 1 - 1.033e^{-t} \sin (3.87t + 75.5°) \qquad ...(v)$$

As there is no open-loop zero M_p – 44.4% will be confirmed by $\xi = 0.25$.

Example 5.17: Reconsider the position control system of Fig. 2.54. Determine the transfer function $\theta_o(s)/\theta_d(s)$. Also find the time response of the system for a unit step position input *i.e.* $\theta_d(s) = 1/s$. Given :

$$J = 4 \times 10^{-3} \text{ kg.m}^2 \quad K_A = 60$$

$$f_0 = 2 \times 10^{-3} \text{ Nm/rad/sec.} \ K_t = 0.15 \text{ V/rad}$$

$$K_T = K_b = 0.5 \text{ Nm/A} \qquad n = 1/10$$

$$R_a = 5\ \Omega.$$

Solution. We will proceed by drawing the signal flow graph of the block diagram of Fig. 2.54 and then use Masons gain formula

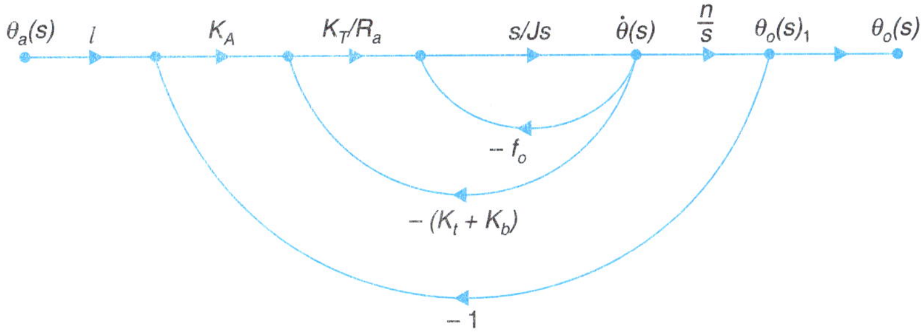

Fig. 5.49

Forward path gain $\quad P_1 = \dfrac{K_A K_T n}{R_a J s^2}$

It is in touch with all the loops, so

$$\Delta_1 = 1$$

Individual loop gains

$$P_{11} = -\frac{f_0}{Js}$$

$$P_{21} = -\frac{K_T(K_t + K_b)}{R_a Js}$$

$$P_{31} = -\frac{K_A K_T n}{R_a Js^2}$$

There are no two or more nontouching loop

$$T(s) = \frac{\theta_o(s)}{\theta_d(s)} = \frac{P_1\Delta}{1-\Delta}$$

or
$$\frac{\theta_o(s)}{\theta_d(s)} = \frac{\dfrac{K_A K_T n}{R_a Js^2}}{1 + \dfrac{f_0}{Js} + \dfrac{K_T(K_t + K_b)}{R_a Js} + \dfrac{K_A K_T n}{R_a Js^2}}$$

$$= \frac{K_A K_T n}{R_a Js^2 + R_a f_0 s + K_T(K_t + K_b)s + K_A K_T n}$$

or
$$\frac{\theta_o(s)}{\theta_d(s)} = \frac{(K_A K_T n / R_a J)}{s^2 + [f_0/J + K_T(K_t + K_b)/R_a J]s + K_A K_T n / R_a J} \quad ...(i)$$

Substituting values

$$\frac{K_A K_T n}{R_a J} = \frac{75 \times 0.5 \times 0.1}{5 \times 4 \times 10^{-3}} = 187.5$$

$$\left[f_0 + \frac{K_T(K_t + K_b)}{R_a}\right]\frac{1}{J} = \left[2 \times 10^{-3} + \frac{0.5(0.15 + 0.5)}{5}\right]\frac{1}{4 \times 10^{-3}} = 16.75$$

In numerical form

$$\frac{\theta_o(s)}{\theta_d(s)} = \frac{187.5}{s^2 + 16.75s + 187.5} \quad ...(ii)$$

Unit step response $\theta_d(s) = \dfrac{1}{s}$

From Eq. (*ii*)
$$\theta_o(s) = \frac{187.5}{s(s^2 + 16.75s + 187.5)} \quad ...(iii)$$

$$\omega_n = \sqrt{187.5} = 13.69$$

$$\xi = \frac{16.75}{2 \times 13.69} = 0.61$$

$$\xi\omega_n = 8.375$$

Laplace transform pair Appendix I, Table I.2, Item 17.

$$\frac{\omega_n{}^2}{s^2 + 2\xi\omega_n s + \omega_n{}^2} \to 1 - \frac{1}{\sqrt{(1-\xi^2)}} e^{-\xi\omega_n t} \sin[\omega_n\sqrt{(1-\xi^2)}\, t + \phi];$$

From eqn. (*iv*), we find ;

$$\phi = \tan^{-1}\frac{\sqrt{(1-\xi^2)}}{\xi}, \xi < 1$$

$$\omega_n\sqrt{(1-\xi^2)} = 10.85 \; ; \qquad \frac{1}{\sqrt{(1-\xi^2)}} = 1.262$$

$$\phi = \tan^{-1}\frac{0.792}{0.61} = 524°$$

By use of Laplace transform pair of eqn. (*iv*), we have

$$\theta_o(t) = 1 - 1.262e^{-8.375t} \sin(10.85t + 52.4°) \qquad ...(iv)$$

This is the well-known response to unit step input of a second-order undamped system. The reader may plot of this response and check its nature. The result of Eq. (*iv*) may be checked that $\theta_o(t = 0) = 0$, $\theta_o(t = \infty) = 1$.

PROBLEMS

5.1 A servomechanism is used to control the angular position θ_0 of a mass through a command signal θ_i. The moment of inertia of moving parts referred to the load shaft is 200 kg-m^2 and the motor torque at the load is 6.88×10^4 newton-m per rad of error. The damping torque coefficient referred to the load shaft is 5×10^3 newton-m per rad/sec.

(*a*) Find the time response of the servomechanism to a step input of 1 rad and determine the frequency of transient oscillation, the time to rise to the peak overshoot and the value of the peak overshoot.

(*b*) Determine the steady-state error when the command signal is a constant angular velocity of 1 revolution/min.

(*c*) Determine the steady-state error which exists when a steady torque of 1,200 newton-m is applied at load shaft.

5.2 In the position control system shown in Fig. P-5.2, the transfer function of the motor is found to be

$$\theta_M(s)/V_C(s) = K_M/s(\tau_m s + 1)$$

where K_m = 15 rad/s/V ; τ_m = 0.15 s. The gear ratios are given as

$$\dot{\theta}_C / \dot{\theta}_L = 1; \; \dot{\theta}_L / \dot{\theta}_M = 1/50$$

The input is given through a pot of sensitivity K_p = 0.1 V/deg. The input and output are given to a d.c. difference amplifier of gain K_A V/V. The output of this amplifier is then modulated by a carrier of 50 Hz. The gain of this modulator is $K_D = s/5$ $V_{rms}/V_{d.c.}$. This signal is then amplified by an a.c. amplifier of gain K_C = 25 V/V and given to the control phase winding of the a.c. servomotor.

(*a*) If the input shaft is driven at a constant speed of π rad/sec, determine the value of the amplifier gain K_A, such that the steady-state error in the position is less than 5 degree. For this value of K_A, determine the damping ratio and the 2% settling time of the system.

(*b*) To improve the system dynamics, the amplifier is modified by introducing an additional derivative term such that its output is given by

$$e_A = K_A e(t) + K_X \dot{e}(t)$$

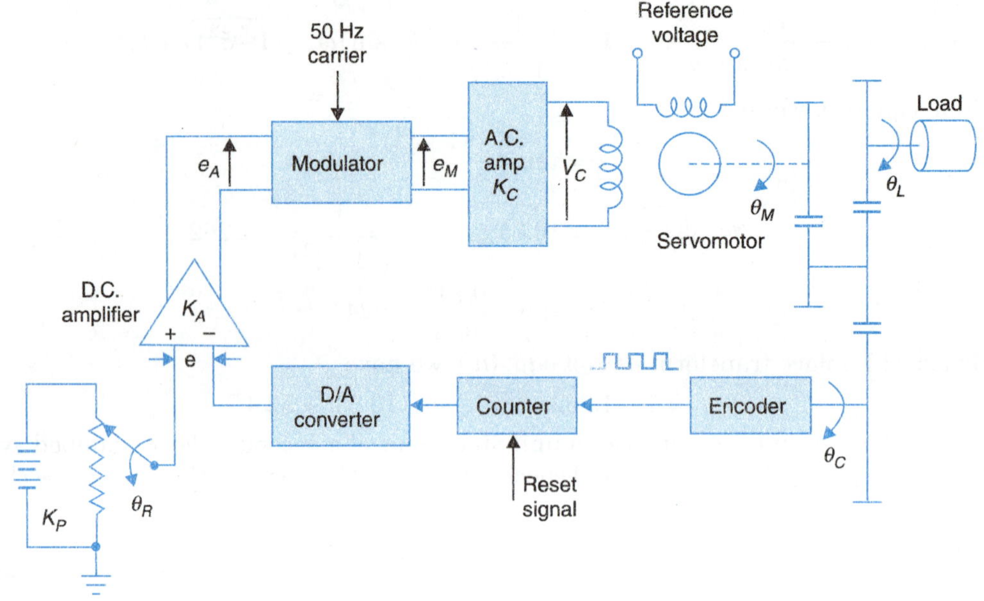

Fig. P-5.2

Determine the value of K_X such that the damping ratio is improved to 0.5. Does this modification affect the steady-state error as in part (*a*) ? Also calculate the new settling time and compare it with that of part (*a*).

5.3 A d.c. position control servomechanism comprises of a permanent magnet d.c. servomotor with constant armature current, potentiometer error detector, a d.c. amplifier and a tachogenerator coupled to motor shaft. A fraction *K* of tachogenerator output is fed back to produce stabilizing effect. The following particulars refer to the system:

Moment of inertia of motor, J_M	$= 2 \times 10^{-3}$ kg-m^2
Moment of inertia of load, J_L	= 5 kg-m^2
Motor to load gear ratio $\dot{\theta}_L / \dot{\theta}_M$	= 1/50
Load to potentiometer gear ratio, $\dot{\theta}_L / \dot{\theta}_C$	= 1
Motor torque constant, K_T	= 2 newton-m/amp
Tachogenerator constant, K_t	= 0.2 volt/rad/sec
Sensitivity of error detector, Kp	= 0.6 volt/rad
Amplifier gain	= K_A amps/volt

Motor and load frictions and motor field time constant are assumed to be negligible.

(*a*) Make a sketch of the system, showing how the hardware is connected.

(*b*) Determine the transfer function of the system.

(*c*) Determine the amplifier gain required and the fraction of the tachogenerator voltage fedback to give an undamped natural frequency of 4 rad/sec and damping of 0.8.

5.4 A position control system shown in Fig. P-5.4 is used to control the position of a load having a moment of inertia of J_L kg-m^2 and coefficient of viscous friction as f_L Nm/rad/s. The position of load is controlled by a motor having a torque constant of K_T Nm/A. The resistance of armature is R_a ohms. The position of load is read by an optical encoder which is connected to the load through a gear box of ratio $N_1 : N_2$ as 1 : 1. The output of encoder is fed to a counter which gives the digital value of position at any time. The digital information of position of load is then fed to a D/A converter, which gives a analog voltage proportional to the load position. The D/A converter operates at a very fast sampling rate, so that the system can be approximated as continuous-time system. The input to the system is given through a pot of gain K_p V/rad. The input signal and the feedback from D/A converter is then given to a difference amplifier of gain K_A V/V. The output of difference amplifier is fed to the SCR rectifier circuitry for fixing the value of fring angle. The output of rectifier is a d.c. voltage proportional to the error. This voltage is in turn applied across the armature of d.c. servo meter.

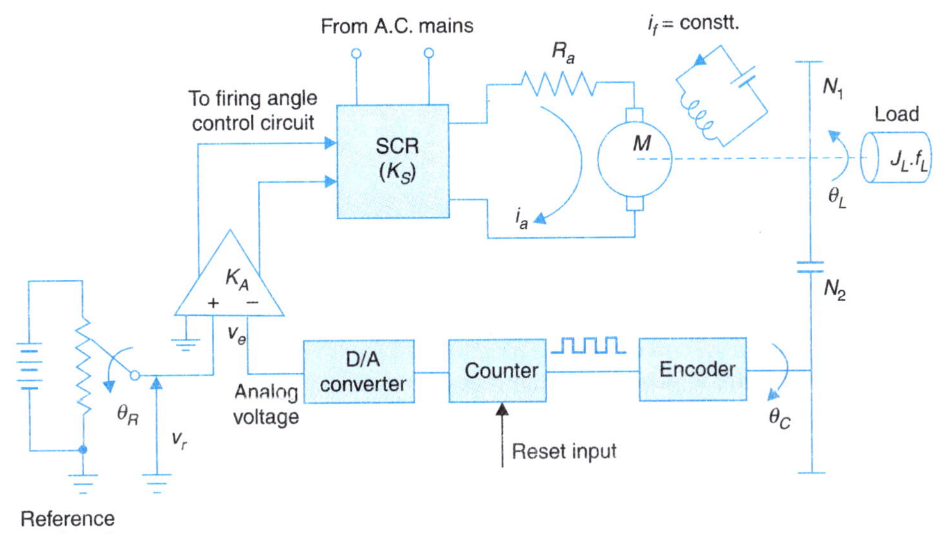

Fig. P-5.4

Given : $K_A = 5$ V/V, $R_a = 1\ \Omega$, $K_T = 1$ N-m/A, J_M and f_M are negligible

$J_L = 0.5$ kg-m^2, $f_L = 2.5$ Nm/rad/s, SCR gain $= K_S = 10$ V/V

$K_P = 10$ V/rad.

Combined gain of encoder + counter + D/A converter $= K_X = 10$ V/rad.

(*a*) Find the transfer function $\theta_L(s)/\theta_R(s)$.

(*b*) What is the value of steady-state error of the system for a unit ramp input ?

(*c*) Repeat part (*ii*) if the output of the difference amplifier is modified to

$$K_A V_e + K_A \int V_e\, dt$$

Compare the steady-state behaviour of the system with that of part (*a*).

5.5 The open-loop transfer function of a unity feedback system is given by

$$G(s) = K/s(Ts + 1)$$

where K and T are positive constants.

By what factor should the amplifier gain be reduced so that the peak overshoot of unit-step response of the system is reduced from 75% to 25% ?

5.6 The following tests are conducted on the position control system of Fig. 5.6 (with gear ratio $n = 1/10$).

(1) Unit-step response is recorded and M_p measured therefrom has a value of 25%.

(2) A constant velocity input of 1 rad/sec produces a steady-state error of 0.04 rad.

(3) With the reference potentiometer shaft held fixed, a torque of 10 newton-m applied to the output shaft produces a steady-state error of 0.01 rad.

From the above data, determine:

(*a*) The natural frequency ω_n.

(*b*) The moment of inertia J referred to the motor shaft.

(*c*) The coefficient of viscous friction f referred to the motor shaft.

5.7 Measurements conducted on a servomechanism show the system response to be

$$c(t) = 1 + 0.2\, e^{-60t} - 1.2\, e^{-10t}$$

when subjected to a unit-step input.

(*a*) Obtain the expression for the closed-loop transfer function.

(*b*) Determine the undamped natural frequency and damping ratio of the system.

5.8 A unity feedback system is characterized by an open-loop transfer function

$$G(s) = K/s(s + 10)$$

Determine the gain K so that the system will have a damping ratio of 0.5. For this value of K determine settling time, peak overshoot and time to peak overshoot for a unit-step input.

5.9 Figure P-5.9 shows a system employing proportional plus error-rate control. Determine the value of the error-rate factor K_e so that the damping ratio is 0.5. Determine the values of settling time, maximum overshoot and steady-state error (for unit-ramp input) with and without error-rate control. Comment upon the effect of error-rate control on system dynamics.

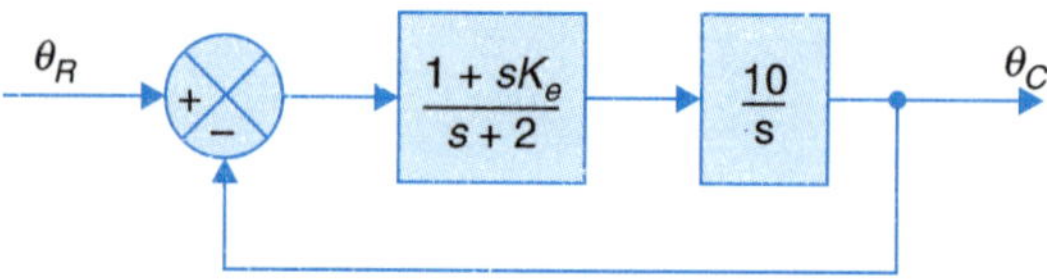

Fig. P-5.9

5.10 A feedback system employing output-rate damping is shown in Fig. P-5.10.

(*a*) In the absence of derivative feedback ($K_0 = 0$), determine the damping factor and natural frequency of the system. What is the steady-state error resulting from unit-ramp input ?

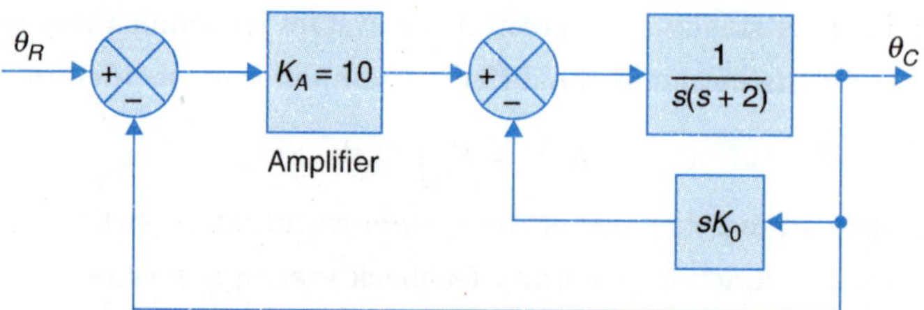

Fig. P-5.10

(*b*) Determine the derivative feedback constant K_0, which will increase the damping factor of the system to 0.6. What is the steady-state error resulting from unit-ramp input with this setting of the derivative feedback constant ?

(*c*) Illustrate how the steady-state error of the system with derivative feedback to unit-ramp input can be reduced to same value as in part (*a*), while the damping factor is maintained at 0.6.

5.11 A unity feedback system is characterized by the open-loop transfer function

$$G(s) = 1/s(0.5s + 1)(0.2s + 1)$$

Determine the steady-state errors for unit-step, unit-ramp and unit-acceleration inputs. Also determine the damping ratio and natural frequency of the dominant roots.

5.12 The open-loop transfer function of a servo system with unity feedback is

$$G(s) = 10/s(0.1s + 1)$$

Evaluate the static error constants (K_p, K_v, K_a) for the system. Obtain the steady-state error of the system when subjected to an input given by the polynomial

$$r(t) = a_0 + a_1 t + \frac{a_2}{2} t^2$$

5.13 For Problem 5.12, evaluate the dynamic error using the dynamic error coefficients.

Hint : For a unity feedback system

$$\frac{E(s)}{R(s)} = \frac{1}{1+G(s)} = \frac{1}{K_1} + \frac{1}{K_2} s + \frac{1}{K_3} s^2 + \ldots$$

Coefficients K_1, K_2, K_3, ... are defined to be dynamic-error coefficients.

$$E(s) = \frac{1}{K_1} R(s) + \frac{1}{K_2} sR(s) + \frac{1}{K_3} s^2 R(s) + \ldots$$

$$e(t) = \frac{1}{K_1} r(t) + \frac{1}{K_2} \dot{r}(t) + \frac{1}{K_3} \ddot{r}(t) + \ldots$$

5.14 A machine tool is required to cut a 30° circular arc of 1 cm radius. The tool moves at a constant feed velocity of 0.1 cm/sec parallel to the *x*-axis, as shown in Fig. P-5.14. A unity feedback servomechanism with open-loop transfer function $G(s) = 10/s(s + 1)$ drives the tool in *y*-direction. Estimate the error when $x = 0.3$.

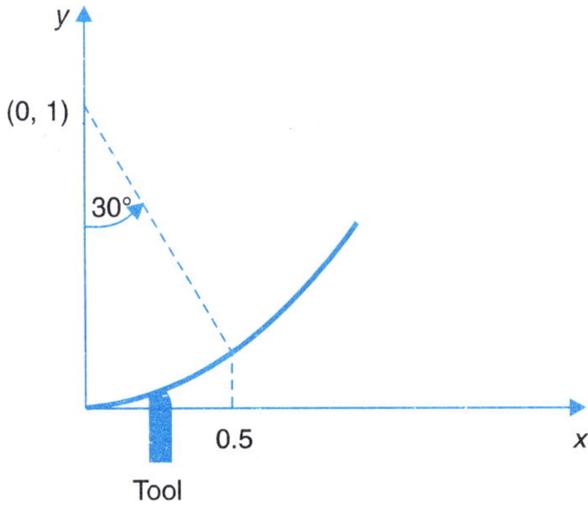

Fig. P-5.14

Hint : The equation of the circular arc is

$$x^2 + (y - 1)^2 = 1$$

which gives

$$y = 1 - \sqrt{(1 - x^3)}, \text{ for } 0 < x < 0.5$$

Since $x = 0.1\,t$

$$y = 1 - \sqrt{[1 - (0.1t)^2]}$$

This is the input to the unity feedback system.

5.15 Consider a unity feedback system with the closed-loop transfer function

$$\frac{C(s)}{R(s)} = \frac{\omega_n^{\,2}}{s^2 + 2\omega_n s + \omega_n^{\,2}}$$

For unit-step input, compute the following performance indices

1. $\text{IAE} = \int_0^\infty |\, e(t) \,|\, dt$ 2. $\text{ITAE} = \int_0^\infty t\,|\, e(t)\,|\, dt$

Calculate the values of ω_n which minimize IAE and ITAE respectively. Are these values of ω_n practical ? It not, then choose suitable values of ω_n and determine the corresponding performance indices.

Hint : (*i*) Since damping factor $\zeta = 1$, the system does not overshoot. Therefore

$$|\, e(t) \,| = e(t)$$

(*ii*) From Appendix I, $\int_0^\infty e(t)\, dt = \lim\limits_{s \to 0} E(s)$;

$$\int_0^\infty te(t)\, dt = \lim_{s \to 0} \left[-\frac{d}{ds} E(s) \right]$$

5.16 A unity feedback system has the forward path transfer function

$$G(s) = K/s(s + \alpha)$$

(*i*) Taking K as constant, determine the value of α which minimize ISE.

(*ii*) Taking α as constant, determine the value of K which minimizes ISE.

5.17 A unity feedback system has an open-loop transfer function

$$G(s) = 1/s(s + 2\zeta)$$

For unit-step input, compute the following:

(*a*) $\text{ISE} = \int_0^\infty e^2(t)\, dt$; (*b*) $\text{ITSE} = \int_0^\infty t\, e^2(t)\, dt$

Calculate the optimal values of $\zeta > 0$ which minimize ISE and ITSE respectively. What are the minimum values of the performance indices ?

5.18 Consider a robot joint control system as shown in Fig. 5.39. Various parameters are:

$$J = 5 \text{ kg-m}^2,\ F = 500 \text{ N-m/rad/s}$$

$$K_m \text{ (mot torque constant)} = 2.5 \text{ N-m/amp}$$

(*a*) Calculate the steady-state error in position of the load, if a sudden disturbance torque of 2N-m is applied with $\theta_d(s) = 0$

(*b*) If a reference step input is now applied, calculate the total error in position.

(*c*) Choose K_e and K_v such that the error in positioning is less than 5 degrees.

5.19 The block diagram of a control system is shown in Fig. P-5.19

(*a*) Identify the state variable of this system.

(*b*) Formulate the state variable model and obtain its state transition matrix. Also obtain the output variable for a unit step input.

(*c*) Convert the state model to transfer function model. Check by writing down the transfer function of the system directly using the signal flow graph technique.

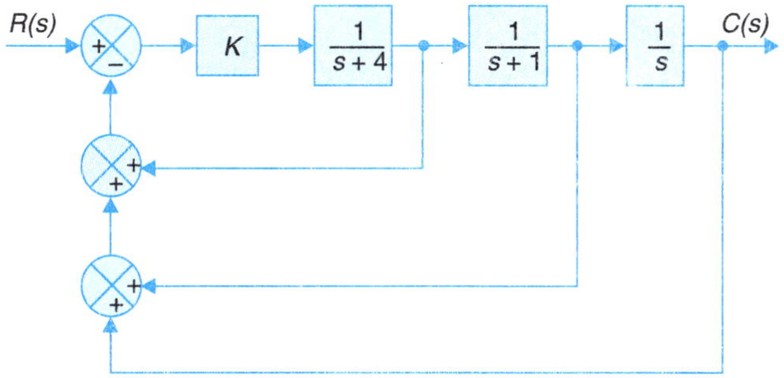

Fig. P-5.19.

6

CONCEPTS OF STABILITY AND ALGEBRAIC CRITERIA

6

CONCEPTS OF STABILITY AND ALGEBRAIC CRITERIA

6.1 THE CONCEPT OF STABILITY

Roughly speaking, stability in a system implies that small changes in the system input, in initial conditions or in system parameters, do not result in large changes in system output. Stability is a very important characteristic of the transient performance of a system. Almost every working system is designed to be stable. Within the boundaries of parameter variations permitted by stability considerations, we can then seek to improve the system performance.

A linear time-invariant system is stable if the following two notions of system stability at satisfied:

(*i*) When the system is excited by a bounded input, the output is bounded.

(*ii*) In the absence of the input, the output tends towards zero (the equilibrium state of the system) irrespective of initial conditions. This stability concept is known as *asymptotic stability*.

The second notion of stability generally concerns a free system relative to its transient behaviour. The first notion concerns a system under the influence of an input. Clearly, if a system is subjected to an unbounded input and produces an unbounded response, nothing can be said about its stability. But if it is subjected to a bounded input and produces an unbounded response, it is by definition, unstable. Actually the output of an unstable system may increase to a certain extent and then the system may break down or become nonlinear after the output exceeds a certain magnitude, so that the linear mathematical model no longer applies.

As we shall see below, the two notions of stability defined above are essentially equivalent in linear time-invariant systems. Simple and powerful tools are available to determine the stability of such systems. For nonlinear systems, because of the possible existence of multiple equilibrium states and other anomalies, the concept of stability is difficult even to define, so that there is no clearcut correspondence between the two notions of stability defined above. For a free stable nonlinear system, there is no guarantee that output will be bounded whenever input is bounded. Also if the output is bounded for a particular bounded input, it may not be bounded for other bounded inputs. Many of the inportant results obtained thus far concern the stability of the nonlinear systems in the sense of the second notion above, *i.e.*, when the system has no input. Here in this chapter, we are concerned with the stability determination of linear time-invariant systems.

Let us observe the physical implication of the two notions of stability defined earlier, by considering a single-input, single-output system with transfer function

$$\frac{C(s)}{R(s)} = G(s) = \frac{b_0 s^m + b_1 s^{m-1} + \ldots + b_m}{a_0 s^n + a_1 s^{n-1} + \ldots + a_n}; m < n \qquad \ldots(6.1)$$

With initial conditions assumed zero, the output of the system is given by

$$c(t) = \mathcal{L}^{-1}[G(s)R(s)]$$

Therefore (see Appendix 1)

$$c(t) = \int_0^{\infty} g(\tau) r(t-\tau) d\tau$$

where $g(t) = \mathcal{L}^{-1}(G(s))$ is the *impulse response* of the system (eqn. (5.4)).

Taking the absolute value on both sides we get

$$| c(t) | = | \int_0^{\infty} g(\tau) r(t-\tau) d\tau |$$

Since the absolute value of integral is not greater than the integral of the absolute value of the integrand,

$$| c(t) | \leq \int_0^{\infty} | g(\tau) r(t-\tau) | d\tau$$

$$\leq \int_0^{\infty} | g(\tau) | | r(t-\tau) | d\tau \qquad \ldots(6.2)$$

The first notion of stability is satisfied if for every bounded input ($| r(t) | \leq M_1 < \infty$), the output is bounded ($| c(t) | \leq M_2 < \infty$). From (6.2), we have for bounded input, the bounded output condition as

$$| c(t) | \leq M_1 \int_0^{\infty} | g(\tau) | d\tau \leq M_2$$

Thus the first notion of stability is satisfied if the impulse response $g(t)$ is absolutely integrable, *i.e.*, $\int_0^{\infty} | g(\tau) | d\tau$ is finite (area under the absolute-value curve of the impulse response $g(t)$ evaluated from $t = 0$ to $t = \infty$ must be finite).

The nature of $g(t)$ is dependent on the poles of the transfer function $G(s)$ which are the roots of the characteristic equation. These roots may be both real and complex conjugate and may have multiplicity of various orders. The nature of response terms contributed by all possible types of roots are given in Table 6.1 and have been illustrated in Fig. 6.1.

Certain observations are easily made from the study of Table 6.1. All the roots which have nonzero real parts [cases (*i*), (*ii*), (*iii*) and (*iv*)], contribute response terms with a multiplying factor of $e\sigma^i$. If $s < 0$ (*i.e.*, the roots have negative real parts), the response terms vanish as $t \to \infty$ and if $\sigma > 0$ (*i.e.*, the roots have positive real parts), the response terms increase without bound. Roots on the $j\omega$-axis with multiplicity two or higher [cases (*vi*) and (*viii*)] also contribute terms which increase without bound as $t \to \infty$.

Table 6.1. Response Terms Contributed By Various Types of Roots

	Type of roots	*Nature of response terms contributed*	
(*i*)	Single root at $s = \sigma$	$Ae^{\sigma t}$	(Fig. 6.1(*a*))
(*ii*)	Roots of multiplicity k at $s = \sigma$	$(A_1 + A_2t + ... + A_kt^{k-1})e^{\sigma t}$	
(*iii*)	Complex conjugate root pair at $s = \sigma \pm j\omega$	$Ae^{\sigma t} \sin(\omega t + \beta)$	(Fig. 6.1(*b*))
(*iv*)	Complex conjugate root pairs of multiplicity k at $s = \sigma \pm j\omega$	$[A_1 \sin(\omega t + \beta_1) + A_2t \sin(\omega t + \beta_2) + ... + A_kt^{k-1} \sin(\omega t + \beta_k)]\, e^{\sigma t}$	
(*v*)	Single complex conjugate root-pair on the $j\omega$ axis (*i.e.*, at $s = \pm j\omega$)	$A \sin(\omega t + \beta)$	(Fig. 6.1(*c*))
(*vi*)	Complex conjugate root pair of multiplicity k on the $j\omega$-axis	$A_1 \sin(\omega t + \beta_1) + A_2t \sin(\omega t + \beta_2) + ... + A_kt^{k-1}\sin(\omega t + \beta_k)$	(Fig. 6.1(*c*))
(*vii*)	Single root at origin (*i.e.*, at $s = 0$)	A	(Fig. 6.1(*d*))
(*viii*)	Roots of multiplicity k at origin	$(A_1 + A^2t + ... + A_kt^{k-1})$	(Fig. 6.1(*d*))

Single root at origin [case (*vii*)] or non-multiple root pairs [case (*v*)] on the $j\omega$-axis contribute response terms which are constant amplitude or constant amplitude oscillation. These observations lead us to the following general conclusions regarding system stability (refer Fig. 6.2).

(1) If all the roots of the characteristic equation have negative real parts, then the impulse response is bounded and eventually decreases to zero.

Therefore, $\int_0^\infty |g(\tau)|\, d\tau$ is finite and the system is *bounded-input, bounded-output* stable.

(2) If any root of the characteristic equation has a positive real part, $g(t)$ is unbounded and $\int_0^\infty |g(\tau)|\, d\tau$ is infinite. The system is therefore unstable.

(3) If the characteristic equation has repeated roots on the jw-axis, $g(t)$ is unbounded and $\int_0^\infty |g(\tau)|\, d\tau$ is infinite. The system is therefore unstable.

(4) If one or more nonrepeated roots of the characteristic equation are on the $j\omega$-axis, then $g(t)$ is bounded but $\int_0^\infty |g(\tau)|\, d\tau$ is infinite. The system is therefore unstable.

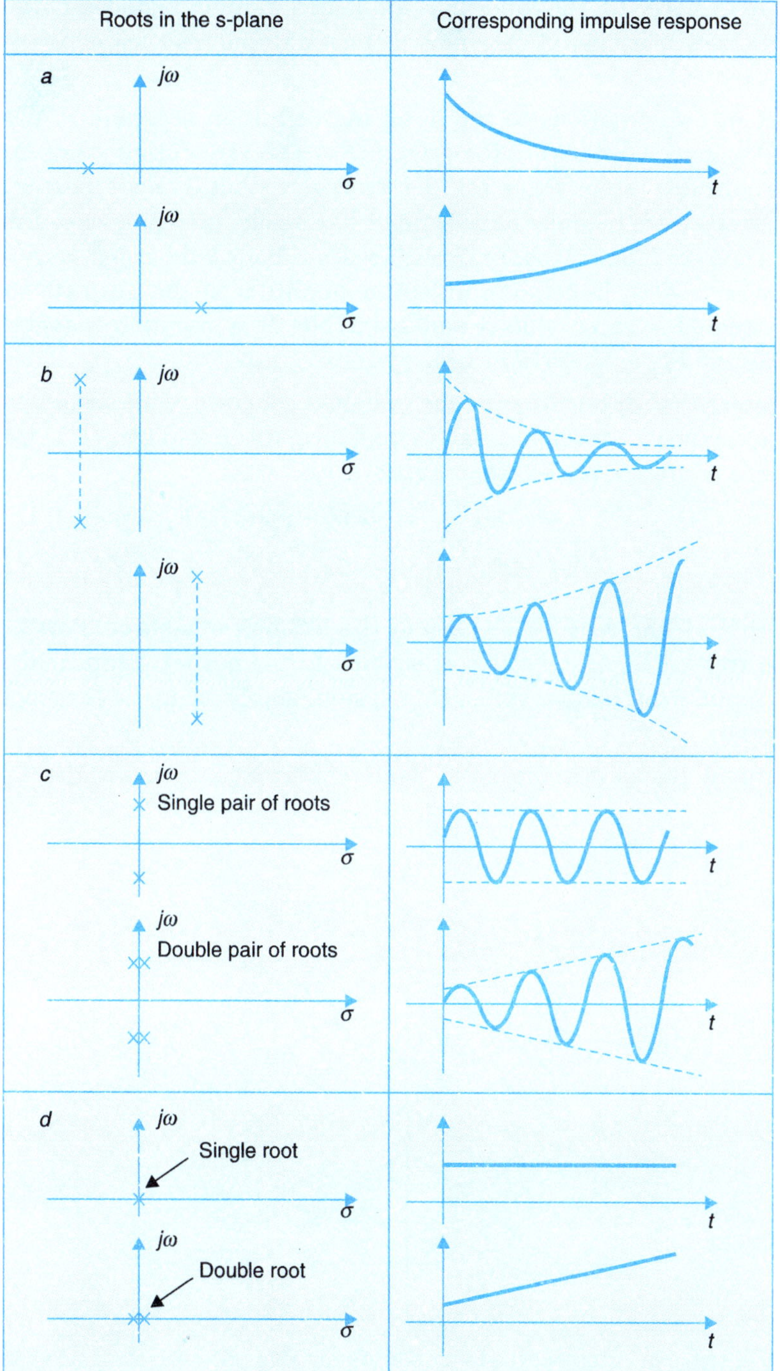

Fig. 6.1. Response terms contributed by various types of roots.

The response to initial conditions is not evident from the model of eqn. (6.1) since the transfer function of a system is derived with the assumption of zero initial conditions. However, as we shall see in Chapter 13, the conditions for bounded-input, bounded-output stability also satisfy the requirement of asymptotic stability of the zero-input response of linear time-invariant systems.

There are a few exceptions to the foregoing definition of stability. When a unit step input ($R(s) = 1/s$) is applied to a perfect integrator ($G(s) = 1/s$), the output is unbounded. However, an integrator is a useful system. From Fig. 6.1 we observe that for nonrepeated poles of $G(s)$ on the $j\omega$-axis, the response is bounded unless input has a pole, matching one of the system poles on the $j\omega$-axis. The zero-input response in such cases is bounded but non-asymptotic ($c(t)$ does not tend to zero as $t \to \infty$). Depending upon the amplitude of the ultimate response, such a system may be treated as acceptable or nonacceptable. This situation is generally referred to as the case of *marginally* or *limitedly* stable system.

This suggests that in our concern for complete generality, we should not be misled by the complexity of occasional special cases. For instance, if $m = n$ in eqn. (6.1) we have (transfer functions with $m > n$ are not physically realizable),

$$G(s) = \frac{b_0 s^n + b_1 s^{n-1} + \ldots + b_n}{a_0 s^n + a_1 s^{n-1} + \ldots + a_n} = \frac{p(s)}{q(s)} = \underbrace{\frac{b_0}{a_0}}_{\text{I}} + \underbrace{\frac{p'(s)}{q(s)}}_{\text{II}}$$

Assume that all the roots of $q(s)$ are in the left half of s-plane. Part I of the transfer function can be treated separately; it corresponds to the output component which depends directly on the input. Whether we call such a system stable or unstable, is again a matter of personal preference.

In a vast majority* of practical systems, the following statements on stability are quite useful (Fig. 6.2):

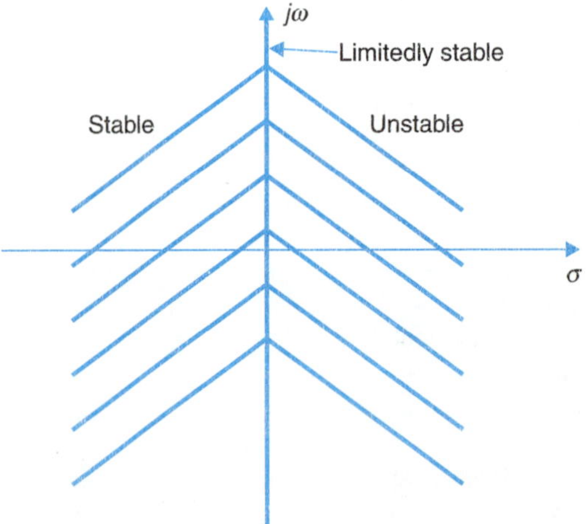

Fig. 6.2. Regions of root-locations for stable, unstable and limitedly stable systems.

*In calculation of $G(s)$ of a system for a particular output, there may be poles of $G(s)$ which are cancelled by the numerator factors in $G(s)$. Such a transfer function does not exhibit all the modes of the systems. However, in vast majority of situations, transfer functions exhibit all natural frequencies; system stability may therefore be determined by examination of poles of the transfer function.

(*i*) If all the roots of the characteristic equation have negative real parts, the system is *stable*.

(*ii*) If any root of the characteristic equation has a positive real part or if there is a repeated root on the *jw*-axis, the system is *unstable*.

(*iii*) If the condition (*i*) is satisfied except for the presence of one or more nonrepeated roots on the *jw*-axis, the system is *limitedly stable*.

In further subdivision of the concept of stability, a linear system is characterized as:

(*i*) *Absolutely stable* with respect to a parameter of the system if it is stable for all values of this parameter.

(*ii*) *Conditionally stable* with respect to a parameter, if the system is stable for only certain bounded ranges of values of this parameter.

It follows from the above discussion that stability can be established by determining the roots of the characteristic equation. Unfortunately, no general formula in algebraic from is available to determine the roots of the characteristic equations of higher than second-order. Though various numerical methods exist for root determination of a characteristic equation, these are quite cumbersome even for third-and fourth-order systems.

However, simple graphical and algebraic criteria have been developed which permit the study of stability of a system without the need of actually determining the roots of its characteristic equation. These criteria answer the question, whether a system be stable or not, in 'yes' or 'no' form.

Relative Stability

In practical systems, it is not sufficient to know that the system is stable but a stable system must meet the specifications on *relative stability* which is a quantitative measure of how fast the transients die out in the system.

Relative stability may be measured by relative settling times of each root or pair of roots. It has been shown in the preceding chapter that the settling time of a pair of complex conjugate poles is inversely proportional to the real part (negative) of the roots. This result is equally valid for real roots. As a root (or a pair of roots) moves farther away from the imaginary axis as shown in Fig. 6.3, the relative stability of the system improves.

6.2 NECESSARY CONDITIONS FOR STABILITY

Certain conclusions regarding the stability of a system can be drawn by merely inspecting the coefficients of its characteristic equation in polynomial form. In the following sections, we shall show that a necessary (but not sufficient) condition for stability of a linear system is that all the coefficients of its characteristic equation $q(s) = 0$, be real and have the same sign. Furthermore, none of the coefficients should be zero.

Consider the characteristic equation

$$q(s) = a_0 s^n + a_1 s^{n-1} + \dots + a_{n-1} s + a_n = 0; \; a_0 > 0 \qquad \text{...(6.3)}$$

It is to be noted that there is no loss of generality in assuming $a_0 > 0$. In case $a_0 < 0$, it can be made positive by multiplying the characteristic equation by -1 throughout.

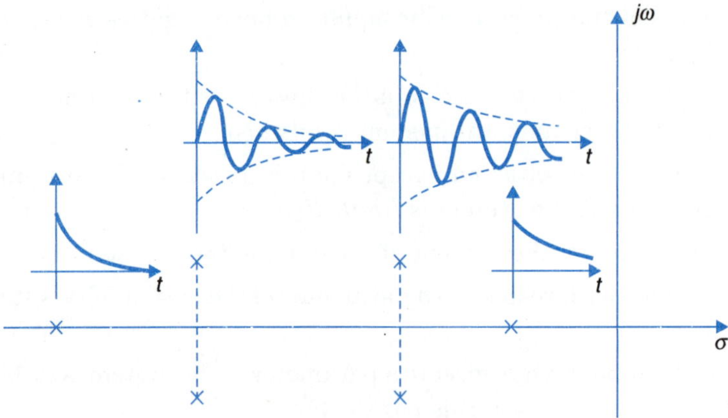

Fig. 6.3. Relative stability for various root-locations in the *s*-plane.

Equation (6.3) may be written in the factored form as

$$q(s) = a_0 \Pi\,(s + s_k)\,\Pi[(s + \sigma_l)^2 + \omega_l^2] \qquad \text{...(6.4)}$$

For the system to be stable, the roots should have negative real parts, which is satisfied if in eqn. (6.4) all s_k and σ_l are positive real. It means that all the factors of eqn. (6.4) have positive terms only. As these factors are multiplied together to get the characteristic equation in polynomial form, all the coefficients of the resulting polynomial must work out to be positive. However, if one or more roots have positive real parts, the coefficients of the characteristic equation may or may not be all positive and hence the above conclusion.

Furthermore, it is to be noted that none of the coefficients can be zero or negative unless one (or more than one) of the following occurs:

(1) one or more roots have positive real parts;

(2) a root (or roots) at origin, *i.e.*, $s_k = 0$ and hence $a_n = 0$;

(3) $\sigma_l = 0$ for some l, which implies the presence of roots on the jw-axis.

We therefore conclude that the absence or negativeness of any of the coefficients of the characteristic equation (with $a_0 > 0$) indicates that the system is either unstable or at most limitedly stable.

From the foregoing discussions, let us prove the following propositions:

1. The positiveness of the coefficients of characteristic equation is necessary as well as sufficient condition for stability of system of first-and second-order.

The characteristic equation of a first order system is

$$a_0 s + a_1 = 0$$

which has a single root

$$s_1 = -\,a_1/a_0 \qquad \text{...(6.5)}$$

It is obvious from eqn. (6.5) that the positiveness of a_0 and a_1 ensures a negative root, *i.e.*, stability.

The characteristic equation of a second-order systems is

$$a_0 s^2 + a_1 s + a_2 = 0$$

which has the roots

$$s_1, s_2 = [-a_1 \pm \sqrt{(a_1^2 - 4a_0a_2)}\,]/2a_0 \qquad ...(6.6)$$

From eqn. (6.6) it is seen that positiveness of a_0, a_1 and a_2 ensures that the roots lie in left half of the s-plane (either both the roots are negative or they have negative real parts) which implies the stability of the system.

2. The positiveness of the coefficients of the characteristic equation ensures the negativeness of real roots but does not ensure the negativeness of the real parts of the complex roots for third-and higher-order systems. Therefore it cannot be a sufficient condition for stability of third-and higher-order systems.

Consider a third-order system with characteristic equation

$$s^3 + s^2 + 2s + 8 = 0 \qquad ...(6.7)$$

Equation (6.7) may be put in the factored form as

$$(s + 2)\left(s - 0.5 + j\frac{\sqrt{15}}{2}\right)\left(s - 0.5 + j\frac{\sqrt{15}}{2}\right) = 0$$

We see that the real part of the complex roots is positive indicating instability of the system even though all the coefficients of the characteristic equation (6.7) are positive.

Therefore, if the characteristic equation of a system is of degree higher than second, the possibility of its instability cannot be excluded even when all the coefficients of its characteristic equation are positive. The first step in analyzing the stability of the system is to examine its characteristic equation. If some of the coefficients are zero or negative it can be concluded that the system is not stable. On the other hand, if all the coefficients of the characteristic equation are positive (or negative), the possibility of stability of system exists and one should proceed further to examine the sufficient conditions of stability.

A. Hurwitz and E.J. Routh independently published the method of investigating the sufficient conditions of stability of a system. The Hurwitz Criterion is in terms of determinants and the Routh Criterion is in terms of array formulation, which is more convenient to handle. We first discuss the Hurwitz Criterion. The Routh Criterion which is derivable from that of Hurwitz, is then presented.

6.3 HURWITZ STABILITY CRITERION

The characteristic equation of an nth order system is

$$q(s) = a_0s^n + a_1s^{n-1} + \ldots + a_{n-1}s + a_n = 0$$

For the stability of this system, it is necessary and sufficient that the n determinants formed from the coefficients $a_0, a_1, \ldots, a_n$ *of the characteristic equation be positive, where these determinants are taken as the principal minors of the following arrangement* (*called the Hurwitz determinant*):

$$\begin{vmatrix} a_1 & a_0 & 0 & 0 & 0 & 0 \cdots 0 & 0 \\ a_3 & a_2 & a_1 & a_0 & 0 & 0 \cdots 0 & 0 \\ a_5 & a_4 & a_3 & a_2 & a_1 & a_0 \cdots 0 & 0 \\ \vdots & \vdots & \vdots & \vdots & \vdots & \vdots & \vdots \\ a_{2n-1} & a_{2n-2} & a_{2n-3} & \cdot & \cdot & \cdot & a_{n-1} + a_n \end{vmatrix} \quad \text{...(6.8)}$$

where the coefficients with indices larger than n or with negative indices are replaced by zeros.

In other words, the necessary and sufficient conditions for stability are

$$\Delta_1 = a_1 > 0 ;$$

$$\Delta_2 = \begin{vmatrix} a_1 & a_0 \\ a_3 & a_2 \end{vmatrix} > 0 ;$$

$$\Delta_3 = \begin{vmatrix} a_1 & a_2 & 0 \\ a_3 & a_2 & a_1 \\ a_5 & a_4 & a_3 \end{vmatrix} > 0 ; \ldots$$

Δ_n = entire arrangement of (6.8) > 0.

When $\Delta_{n-1} = 0$, the system is limitedly stable.

Example 6.1 : Consider a fourth-order system with the characteristic equation

$$s^4 + 8s^3 + 18s^2 + 16s + 5 = 0$$

Solution. The Hurwitz arrangement is given below:

$$\begin{matrix} a_1 \\ a_3 \\ a_5 \\ a_7 \end{matrix} \begin{vmatrix} 8 & 1 & 0 & 0 \\ 16 & 18 & 8 & 1 \\ 0 & 5 & 16 & 18 \\ 0 & 0 & 0 & 5 \end{vmatrix}$$

Therefore, $\Delta_1 = 8 > 0 ;$

$$\Delta_2 = \begin{vmatrix} 8 & 1 \\ 16 & 18 \end{vmatrix} = 128 > 0$$

$$\Delta_3 = \begin{vmatrix} 8 & 1 & 0 \\ 16 & 18 & 8 \\ 0 & 5 & 16 \end{vmatrix} = 1728 > 0$$

$$\Delta_4 = \begin{vmatrix} 8 & 1 & 0 & 0 \\ 16 & 18 & 8 & 1 \\ 0 & 5 & 16 & 18 \\ 0 & 0 & 0 & 5 \end{vmatrix} = 5\Delta_3 > 0$$

Therefore the system under consideration is stable.

6.4 ROUTH STABILITY CRITERION

This criterion is based on ordering the coefficients of the characteristic equation into an array, called the Routh array as given below:

$$q(s) = a_0 s^n + a_1 s^{n-1} + a_2 s^{n-2} + \ldots + a_{n-1} s + a_n = 0$$

Routh Array

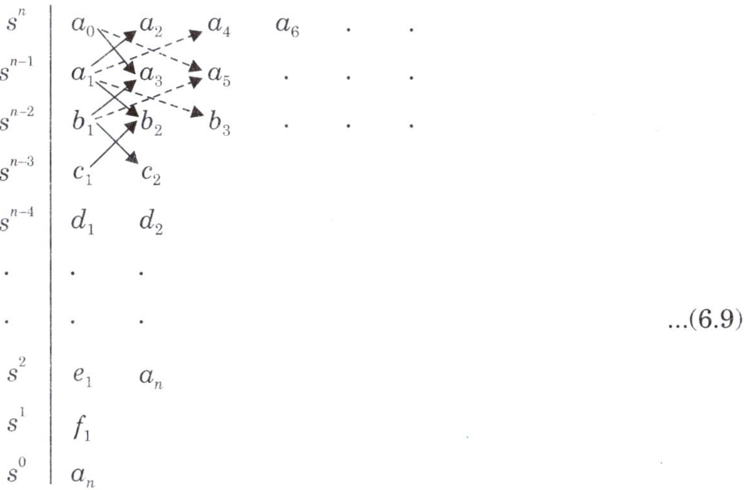

...(6.9)

The coefficients b_1, b_2 ... , are evaluated as follows:

$$b_1 = (a_1a_2 - a_0a_3)/a_1;$$
$$b_2 = (a_1a_4 - a_0a_5)/a_1; \ldots$$

This process is continued till we get a zero as the last coefficient in the third row. In a similar way, the coefficients of 4th, 5th, ..., nth and $(n + 1)$th rows are evaluated, *e.g.*,

$$c_1 = (b_1a_3 - a_1b_2)/b_1;$$
$$c_2 = (b_1a_5 - a_1b_3)/b_1; \ldots$$

and

$$d_1 = (c_1b_2 - b_1c_2)/c_1;$$
$$d_2 = (c_1b_3 - b_1c_3)/c_1; \ldots$$

It is to be noted here that in the process of generating the Routh array, the missing terms are regarded as zero. Also all the elements of any row can be divided by a positive constant during the process to simplify the computational work.

The Routh stability criterion is stated as below.

For a system to be stable, it is necessary and sufficient that each term of first column of Routh array [as given in eqn. (6.9)] of its characteristic equation be positive if $a_0 > 0$. If this condition is not met, the system is unstable and number of sign changes of the terms of the first column of the Routh array corresponds to the number of roots of the characteristic equation in the right half of the s-plane.

The Routh criterion stated above and the Hurwitz criterion are equivalent, as is shown below.

Elements of first column of the Routh array can be interpreted in terms of Hurwitz determinants as follows:

$$b_1 = \frac{a_1a_2 - a_0a_3}{a_1} = \frac{\begin{vmatrix} a_1 & a_0 \\ a_3 & a_2 \end{vmatrix}}{a_1} = \frac{\Delta_2}{\Delta_1}$$

Similarly $$c_1 = D_3/D_2;$$
$$d_1 = D_4/D_3; \ldots$$

Therefore, the condition of positiveness of the Hurwitz determinants corresponds to the condition of positiveness of the elements of the first column of the Routh array.

Example 6.2 : Consider the fourth-order system of Example 6.1 with the characteristic equation

$$s^4 + 8s^3 + 18s^2 + 16s + 5 = 0$$

Solution. The Routh array for this system is given below:

$$\begin{array}{c|ccl}
s^4 & 1 & 18 & 5 \\
s^3 & 8 & 16 & 0 \text{ (for the missing term)} \\
s^2 & \dfrac{8 \times 18 - 1 \times 16}{8} = 16 & \dfrac{8 \times 5 - 1 \times 0}{8} = 5 & \\
s^1 & \dfrac{16 \times 16 - 8 \times 5}{16} = 135 & 0 & \\
s^0 & 5 & &
\end{array}$$

The elements of the first column are all positive and hence the system is stable.

Example 6.3 : Consider the following characteristic equation:

$$3s^4 + 10s^3 + 5s^2 + 5s + 2 = 0$$

Solution. The Routh array is given below:

$$\begin{array}{c|ccc}
s^4 & 3 & 5 & 2 \\
s^3 & 10 & 5 & \\
s^3 & 2 & 1 & \\
s^2 & \dfrac{2 \times 5 - 3 \times 1}{2} = 3.5 & \dfrac{2 \times 2 - 0 \times 3}{2} = 2 & \\
s^1 & \dfrac{3.5 \times 1 - 2 \times 2}{3.5} = -\dfrac{0.5}{3.5} & & \\
s^0 & 2 & &
\end{array}$$

It may be noted that in order to simplify computational work, the s^3-row in the formation of the Routh array has been modified by dividing it by 5 throughout. The modified s^3-row is then used to complete the process of array formation.

Examining the first column of the Routh array, it is found that there are two changes in sign (from 3.5 to – 0.5/3.5 and from – 0.5/3.5 to 2). Therefore the system under consideration is unstable having two poles in the right half of the s-plane.

It is to be noted that the Routh criterion gives only the number of roots in the right half of the s-plane. It gives no information as regards the values of the roots and also does not distinguish between real and complex roots.

Example 6.4 : For the control system of Fig. 5.24 with PID controller find the condition for the system to be stable ($T_D = 0$).

Solution. From eqn. (5.46) the characteristic equation is

$$s^3 + (1 + K_V K_D)s^2 + (K_V K_P)s + K_P K_I = 0$$

Applying Routh criterion

$$\begin{array}{lcc} s^3 & 1 & K_V K_P \\ s^2 & (1+K_V K_D) & K_P K_I \\ s^1 & b_3 & 0 \\ s^0 & K_p K_I & 0 \end{array}$$

Assuming all parameters and variables to be positive the condition of stability is given as

$$b_3 = K_V K_P (1 + K_V K_D) - (K_P K_I) > 0$$

This corroborates the statement made earlier in Section 5.7 that PI and PID controllers, while improving the system's dynamic response and reducing its steady-state error, cause it become conditionally stable. A second-order system with a proportional controller is otherwise absolutely stable.

Example 6.5 : A unity negative feedback control system has an open-loop transfer function consisting of two poles, two zeros and a variable gain K. The zeros are located at –2 and –1; and the poles at –0.1 and +1.

Using Routh stability criterion, determine the range of values of K for which the closed-loop system has 0, 1 or 2 poles in the right-half s-plane.

Solution.

$$G(s) = \frac{K(s + 1)(s + 2)}{(s + 0.1)(s - 1)}$$

The characteristic equation of the system is given as

$$1 + G(s) = 0$$

or

$$(s + 0.1)(s - 1) + K(s + 1)(s + 2) = 0$$

or

$$(1 + K)s^2 + (3K - 0.9)s + (2K - 0.1) = 0 \qquad ...(i)$$

Applying Routh criteria

$$\begin{array}{lcc} s^2 & (1+K) & (2K-0.1) \\ s^2 & (3K-0.9) & 0 \\ s^0 & (2K-0.1) & 0 \end{array}$$

(*i*) No poles in right half s-plane (system stable)

$$K + 1 > 0 \quad \text{or} \quad K > -1$$

$$3K - 0.9 > 0 \quad \text{or} \quad K > 0.3$$

$$2K - 0.1 > 0 \quad \text{or} \quad K > 0.05$$

Hence, $K > 0.3$

(*ii*) 1 pole in right half s-plane (= one change of sign in first column terms)

$$-1 < K < 0.05$$

(*iii*) 2 poles in right half s-plane (= two changes in sign in first column terms)

$$0.05 < K < 0.3$$

Example 6.6 : The characteristic equation of a system in differential equation form is

$$\ddot{x} - (K + 2)\,\dot{x} + (2K + 5)x = 0$$

(*a*) Find the values of K for which the system is

(*i*) stable (*ii*) limitedly stable

(*iii*) unstable

(*b*) For the stable case for what values of K is the system

(*i*) underdamped (*ii*) overdamped.

Solution.

(*a*) $\ddot{x} - (K+2)\,\dot{x} + 2K + 5x = 0$ ...(*i*)

Taking the Laplace transform (assuming zero initial conditions) the characteristic equation in s-domain is

$$s^2 - (K + 2)s + (2K + 5) = 0 \qquad \text{...}(ii)$$

Applying the Routh criterion

s^2	1	$(2K + 5)$
s^1	$-(K + 2)$	0
s^0	$(2K + 5)$	0

(*a*) (*i*) For the system to be stable, there are two conditions

$$-(K + 2) > 0 \text{ and } (2K + 5) > 0$$

or $$K < -2 \text{ and } K > -2.5 \text{ or } -2 > K > -2.5$$

(*ii*) For the system to be limitedly stable

$$K = -2 \text{ and } K = -2.5$$

(*iii*) For the system to be unstable

$$K < -2 \text{ or } K > -2.5$$

(*b*) The roots of the characteristic equation are expressed as

$$s_1, s_2 = \frac{1}{2}\left\{(K+2) \pm \sqrt{[(K+2)^2 - 4(2K+5)]}\right\}$$

For critically damped case

$$(K + 2)^2 - 4(2K + 5) = 0$$

or $$K = 6.47, -2.47$$

As per the stability condition of part (*a*) $K = 6.47$ is unstable. So for critical damping

$$K = -2.47$$

(*i*) Underdamped case

$$-2 > K > -2.47$$

(*ii*) Overdamped case

$$-2.47 > K > -2.5$$

Special Cases

Occasionally, in applying the Routh stability criterion, certain difficulties arise causing the breakdown of the Routh's test. The difficulties encountered are generally of the following types.

Difficulty 1: *When the first term in any row of the Routh array is zero while rest of the row has at least one nonzero term.*

Because of this zero term, the terms in the next row become infinite and Routh's test breaks down. The following methods can be used to overcome this difficulty:

(*a*) Substitute a small positive number ε for the zero and proceed to evaluate the rest of the Routh array. Then examine the signs of the first column of Routh array by letteing $\varepsilon \to 0$.

(*b*) Modify the original characteristic equation by replacing* s by $1/z$. Apply the Routh's test on the modified equation in terms of z. The number of z-roots with positive real parts are the same as the number of s-roots with positive real parts. This method works in most but not all cases.

The following example illustrates these methods.

Example 6.7 : Consider the characteristic equation

$$s^5 + s^4 + 2s^3 + 2s^2 + 3s + 5 = 0$$

Solution. The Routh array is

$$\begin{array}{c|ccc} s^5 & 1 & 2 & 3 \\ s^4 & 1 & 2 & 5 \\ s^3 & \varepsilon & -2 & \\ s^2 & \dfrac{2\varepsilon + 2}{\varepsilon} & 5 & \\ s^1 & \dfrac{-4\varepsilon - 4 - 5\,\varepsilon^2}{2\,\varepsilon + 2} \to -2 & & \\ s^0 & 5 & & \end{array}$$

From the Routh array, it is seen that first element in the third row is 0. This is replaced by ε, a small positive number. The first element in the 4th row is now $(2\varepsilon + 2)/\varepsilon$ which has a positive sign as $\varepsilon \to 0$. (In the Routh stability criterion, we are interested only in the signs of the terms in the first column and not in their magnitudes). The first term of the fifth row is $(-4\varepsilon - 4 - 5\varepsilon^2)/(2\varepsilon + 2)$, which has a limiting value of -2 as $\varepsilon \to 0$. Examining the terms in the first column of the Routh array, it is found that there are two changes in sign and hence the system is unstable having two poles in the right half s-plane.

Consider now the second method of overcoming the difficulty caused by a zero term in the first column of the Routh array.

Replacing s by $1/z$ in the characteristic equation and rearranging, we get

$$5z^5 + 3z^4 + 2z^3 + 2z^2 + z + 1 = 0$$

The Routh array for this equation is

*This transformation maps the left half of the s-plane into the left half of the z plane and the right half of the s-plane into right half of the z-plane.

z^5	5	2	1
z^4	3	2	1
z^3	–4/3	–2/3	
z^2	1/2	1	
z^1	2		
z^0	1		

There are two changes of sign in the first column of the Routh array, which tell us that there are two z-roots in the right half z-plane. Therefore the number of s-roots in the right half s-plane is also two.

Difficulty 2 : *When all the elements in any one row of the Routh array are zero.* This condition indicates that there are symmetrically located roots in the s-plane (pair of real roots with opposite signs and /or pair* of conjugate roots on the imaginary axis and /or complex conjugate roots forming quadrates in the *s-plane*). The polynomial whose coefficients are the elements of the row just above the row of zeros in the Routh array is called an *auxiliary polynomial.* This polynomial gives the number and location of root pairs of the characteristic equation which are symmetrically located in the s-plane. The order of the auxiliary polynomial is always even.

Because of a zero row in the array, the Routh's test breaks down. This situation is overcome by replacing the row of zeros in the Routh array by a row of coefficients of the polynomial generated by taking the first derivative of the auxiliary polynomial. The following example illustrates the procedure.

Example 6.8 : Consider a sixth-order system with the characteristic equation

$$s^6 + 2s^5 + 8s^4 + 12s^3 + 20s^2 + 16s + 16 = 0$$

The Routh array is

s^6	1	8	20	16
s^5	2	12	16	
s^5	1	6	8	
s^4	2	12	16	
s^4	1	6	8	
s^3	0	0		

Since the terms in the s^3-row are all zero, the Routh's test breaks down. Now the auxiliary polynomial is formed from the coefficients of the s^4-row, which is given by

$$A(s) = s^4 + 6s^2 + 8$$

The derivative of the polynomial with respect to s is

$$4s^3 + 12s + 0 = 4s^3 + 12s$$

The zeros in the s^3-row are now replaced by the coefficients 4 and 12. The Routh array then becomes

* There may be more than one such pairs.

s^6	1	8	20	16
s^5	1	6	8	
s^4	1	6	8	
s^3	4	12		
s^3	1	3		
s^2	3	8		
s^1	1/3			
s^0	8			

We see that there is no change of sign in the first column of the new array.

By solving for the roots of auxiliary polynomial

$$s^4 + 6s^2 + 8 = 0$$

we find that the roots are

$$s = \pm j\sqrt{2} \text{ and } s = \pm j2$$

These two pairs of roots are also the roots of the original characteristic equation. Since there is no sign change in the new array formed with the help of the auxiliary polynomial, we conclude that no root of characteristic equation has positive real part. Therefore the system under consideration is limitedly stable.

Application of the Routh Stability Criterion to Linear Feedback System

The Routh stability criterion is frequently used for the determination of the condition of stability of linear feedback control systems. Consider the closed-loop feedback system shown in Fig. 6.4. Let us determine the range of K for which the system is stable.

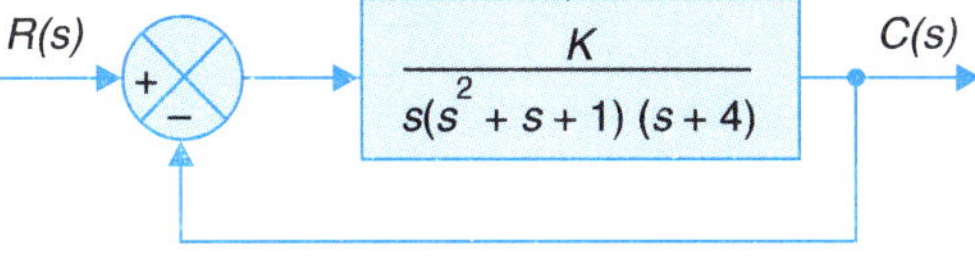

Fig. 6.4

The closed-loop transfer function of the system is

$$\frac{C(s)}{R(s)} = \frac{K}{s(s^2 + s + 1)(s + 4) + K}$$

Therefore, the characteristic equation is

$$s(s^2 + s + 1)(s + 4) + K = 0 \qquad \text{or} \qquad s^4 + 5s^3 + 5s^2 + 4s + K = 0$$

The Routh array for this equation is

$$\begin{array}{c|ccc} s^4 & 1 & 5 & K \\ s^3 & 5 & 4 & 12 \\ s^2 & 21/5 & K & \\ s^1 & \left(\dfrac{84}{5} - 5K\right) \Big/ \dfrac{21}{5} & & \\ s^0 & K & & \end{array}$$

Since for a stable system, the signs of elements of the first column of the Routh array should be all positive, the condition of system stability requires that

$$K > 0$$

and $$(84/5 - 5K) > 0$$

Therefore for stability, K should lie in the range

$$84/25 > K > 0$$

When $K = 84/25$, there will be a zero at the first entry in the fourth row of the Routh array. As explained earlier, this corresponds to the presence of a pair of symmetrical roots which, as shown below, are pure imaginary in this case. Therefore $K = 84/25$, will cause sustained self-oscillations in the closed-loop system.

For $K = 84/25$, the auxiliary polynomial, given by the coefficients of the third row, is

$$(21/5)s^2 + 84/25 = 0$$

which gives the roots as

$$s = \pm j\sqrt{(4/5)} = j\omega_0$$

Hence the frequency of sustained self-oscillations at

$$K = 84/25 \text{ is } \sqrt{(4/5)} \text{ rad/s.}$$

Example 6.9 : Figure 6.5 shows in block diagram form a scheme for controlling (automatically) the direction of travel of a road vehicle (refer also Fig. 1.5). Engine-vehicle dynamics are represented by $G(s)$ and the controller (a compensating network) by $G_c(s)$. Instrumentation of direction sensing and error determination is quite fast and its gain factor is included in K.

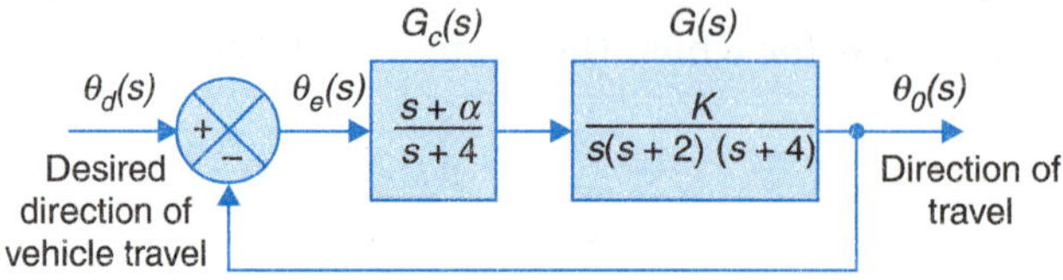

Fig. 6.5

Determine the necessary condition linking a the controller parameter and gain K of the engine-vehicle part for the overall system to be stable.

Suggest suitable values of a and K, while assuring that steady-state error due to unit ramp direction input ($\theta_d(s) = 1/s^2$) is no more than 20%.

Solution. The system's characteristic equation can be written down as below:

$$1 + \left(\frac{s+\alpha}{s+4}\right) \times \frac{K}{s(s+2)(s+4)} = 0$$

$$s^4 + 10s^3 + 32s^2 + (K + 32)s + K\alpha = 0$$

Routh array is formed below

s^4	1	32	$K\alpha$
s^3	10	$(K + 32)$	0
s^2	b_3	$K\alpha$	0
s^1	c_3	0	
s^0	$K\alpha$	0	

where
$$b_3 = \frac{288 - K}{10}$$

$$c_3 = \frac{(288 - K)(K + 32) - 100K\alpha}{b_3}$$

In order for the system to be stable the following three conditions should be satisfied.

$$K < 288$$

$$Ka > 0$$

$$(288 - K)(K + 32) - 100\, K\alpha > 0$$

If we choose $K = 200$

$$\alpha = \frac{88 \times 232}{100 \times 200} \approx 1$$

This choice gives system's velocity error constant as

$$K_v = \frac{K\alpha}{4 \times 2 \times 4} = \frac{K\alpha}{32} = 25/4$$

% velocity error = 4/25 × 100 = 16 (acceptable)

Observe that the controller transfer function becomes

$$G_c(s) = \frac{s + 1}{s + 4}$$

which as we shall see in Chapter 12 is a lead network.

6.5 RELATIVE STABILITY ANALYSIS

Once a system is shown to be stable, we proceed to determine its relative stability quantitatively by finding the settling time of the dominant roots of its characteristic equation. The settling time being inversely proportional to the real part of the dominant roots, the relative stability can be specified by requiring that all the roots of the characteristic equation be more negative than a certain value, *i.e.*, all the roots must lie to the left of the lines $s = -s_1 (s_1 > 0)$. The characteristic equation of the system under study is then modified by shifting the origin of the s plane to $s = -\sigma_1$, *i.e.*, by the substitution

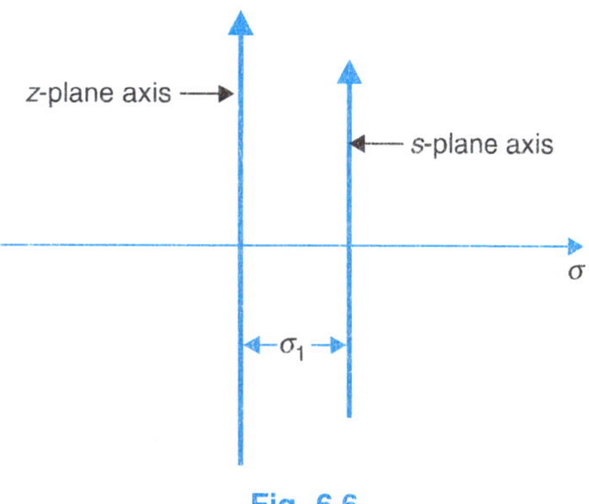

Fig. 6.6

$$s = z - s_1$$

as illustrated in Fig. 6.6. If the new characteristic equation in z satisfies the Routh criterion, it implies that all the roots of the original characteristic equation are more negative than $-\sigma_1$.

Example 6.10 : Consider a third-order system with the characteristic equation

$$s^3 + 7s^2 + 25s + 39 = 0$$

which by the Routh's test can be shown to have all its roots in the left half of the s-plane. Let us check if all the roots of this equation have real parts more negative than –1.

Shift the origin to $s = -1$ by the substitution

$$s = z - 1$$

in the characteristic equation. The characteristic equation in the new variable z is

$$z^3 + 4z^2 + 14z + 20 = 0$$

Forming the Routh array, we have

$$\begin{array}{c|cc} z^3 & 1 & 14 \\ z^2 & 4 & 20 \\ z^1 & 9 & \\ z^0 & 10 & \end{array}$$

As the signs of all the elements of the first column of the Routh array are positive, the roots of the characteristic equation in z lie in the left half z-plane, which implies that all roots of the original characteristic equation is s-domain lie to the left of $s = -1$.

Example 6.11 : The block diagram of Fig. 6.7 is representative of wheel chair velocity control initiated by head nod of a paralyzed patient. Of course for the wheel chair to move from one position in XY-plane to another, an outer position control loop would be needed; this is not being considered here.

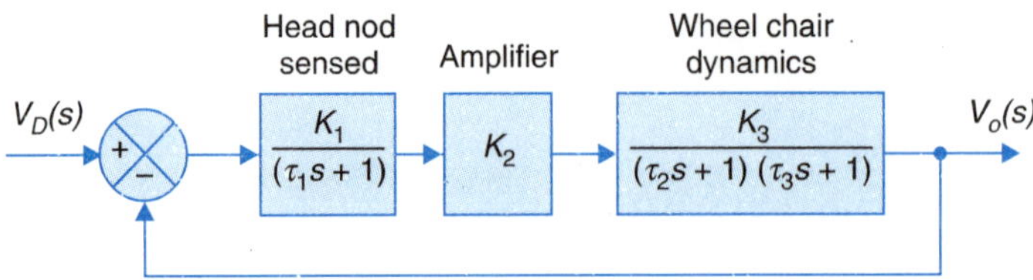

Fig. 6.7

(a) Determine the limiting value of forward gain. K ($= K_1K_2K_3$) for the system to be stable if $\tau_1 = 0.2s$, $\tau_2 = 1s$ and $\tau_3 = 0.4s$.

(b) Calculate the gain K for the system to be stable with a settling time of 4s based on dominant poles.

(c) With K as in part (b) find all the roots of the characteristic equation.

Solution.

(a) The system's characteristic equation can be written as

$$(0.2s + 1)(s + 1)(0.4s + 1) + K = 0$$

or

$$s^3 + 8.5s^2 + 20s + 12.5\,(1 + K) = 0 \qquad \ldots(i)$$

Applying Routh criterion

$$\begin{array}{lll} s^3 & 1 & 20 \\ s^2 & 8.5 & 12.5(1+K) \\ s^1 & \dfrac{8.5 \times 20 - 12.5(1+K)}{8.5} & \end{array}$$

For all terms of the first column to be positive (same sign)

$$8.5 \times 20 - 12.5\,(1 + K) > 0$$

or
$$K < 12.6$$

Limiting value of $K = 12.6$

(*b*) Settling time, $t_s = 4/\zeta\omega_n$

or
$$\zeta\omega_n = 4/4 = 1$$

Thus the real part of dominant root (or complex root pair) should be –1 or more. To ascertain this and to find the corresponding value of K_1, we get

$$s = z - 1$$

in eqn (*i*).

$$(z - 1)^3 + 8.5(z - 1)^2 + 20(z - 1) + 12.5\,(1 + K) = 0$$

or
$$z^3 + 5.5z^3 + 6z + 12.5K = 0 \qquad \text{...}(ii)$$

Applying Routh criterion

$$\begin{array}{lcc} z^3 & 1 & 6 \\ z^2 & 5.5 & 12.5\,K \\ z^1 & \dfrac{5.5 \times 6 - 1 \times 12.5\,K}{5.5} & 0 \\ z^0 & 12.5\,K & \end{array}$$

For the limiting gain of the dominant roots of characteristic equation to be –1 is given as

$$6 \times 5.5 - 12.5K = 0 \qquad \text{or} \qquad K = 2.64$$

For this gain one can find the imaginary part of the dominant roots from the auxiliary equation

$$5.5s^2 + 12.5 \times 2.64 = 0$$

$$s = \pm j\,2.45$$

(*c*) From part (*b*)

$$\zeta\omega_n = 1,\ \omega_n^2 = (1)^2 + (2.45)^2 = 7$$

The dominant roots are given by the factor

$$(s^2 + 2\zeta\omega_n s + \omega_n^2) = (s^2 + 2s + 7)$$

The third root is obtained by dividing out the characteristic equation (*i*) (with $K = 2.64$) by this factor

$$s^3 + 8.5s^2 + 20s + 12.5\,(1 + K) = 0$$

or
$$s^3 + 8.5s^2 + 20s + 45.5 = 0$$

$$\begin{array}{r|l} & s + 6.5 \\ \hline s^2 + 2s + 7 & s^3 + 8.5s^2 + 20s + 45.5 \\ & \underline{s^3 + 2s^2 + 7s} \\ & \quad 6.5s^2 + 13s + 45.5 \\ & \quad \underline{6.5s^2 + 13s + 45.5} \\ & \qquad \times \end{array}$$

Thus the roots of the characteristic equation of (K = 2.64) are

$$s_1, s_2 = 1 \pm j\,2.45;\ \text{dominant roots}$$
$$s_3 = -\,6.5.$$

6.6 MORE ON THE ROUTH STABILITY CRITERION

Some additional useful information regarding the Routh stability criterion which supplements what is already presented in this Chapter, is given below:

1. During the formation of a Routh array, if at any stage a row (not comprising all zeros) having one or more zero elements and/or one or more negative elements appears, the completion of Routh array will always result in sign changes in the first column, *i.e.,* the characteristic polynomial has one or more roots in the right half s-plane. Therefore, for stability determination in such cases, we need not proceed with array formation beyond this stage except when we are interested in determining the number of right half s-plane roots.

2. If the first element in any row of a Routh array happens to be zero, while the rest of the row has at least one non-zero element, the system is unstable as per (1) above. In proceeding with array formation for determining the number of right half s-plane roots, this zero element is replaced by a small number ε. The limit $\varepsilon \to 0$ is then taken to determine the changes in algebraic signs of the first column terms yielding the information regarding the number of right half s-plane roots.

This method, however, fails if the characteristic polynomial has roots on the imaginary axis. In order to establish a modified procedure for such cases, it is necessary to examine the rationale of the method.

If the polynomial under test had slightly different coefficients, the trouble-some first-column zero would instead be some small non-zero number. As long as the polynomial has no roots on the imaginary axis, sufficiently small perturbations of its coefficients will not alter the number of right half s-plane roots. Rather than to actually perturb the polynomial coefficients, the effect of such perturbation is indicated by replacing the zero term with ε. Therefore for a polynomial with no imaginary-axis roots, $\varepsilon \to 0$ from positive or negative side would give identical results for the number of right half s-plane roots. ε is generally taken to be positive for convenience.

For the case where the characteristic equation has imaginary-axis roots, replacement of the first-column zero element by ε would cause the imaginary-axis roots to move into either left or right half of the s-plane, with the result the roots of characteristic polynomial in the right half s-plane cannot be correctly determined.

Therefore to apply the limit method for a characteristic polynomial with the imaginary-axis roots, one must first extract these roots before the test is applied on the remainder polynomial having no imaginary-axis roots.

For example, consider the polynomial

$$q(s) = s^6 + s^5 + 3s^4 + 3s^3 + 3s^2 + 2s + 1$$

The Routh array is

$$\begin{array}{c|cccc} s^6 & 1 & 3 & 3 & 1 \\ s^5 & 1 & 3 & 2 & \\ s^4 & \varepsilon & 1 & 1 & \\ s^3 & \dfrac{3\varepsilon-1}{\varepsilon} & \dfrac{2\varepsilon-1}{\varepsilon} & & \\ s^2 & \dfrac{-2\varepsilon^2+4\varepsilon-1}{3\varepsilon-1} & 1 & & \\ s^1 & \dfrac{4\varepsilon^2-\varepsilon}{2\varepsilon^2-4\varepsilon+1} & & & \\ s^0 & 1 & & & \end{array}$$

As $\varepsilon \to 0$, the element of s^1 row tend to zero. This indicates that there are symmetrically located roots in the s-plane. We therefore need to examine the auxiliary polynomial to find out the possibility of the imaginary-axis roots. If no such roots exist, the usual procedure of replacing the all-zero row by coefficients of the derivative of the auxiliary polynomial is adopted. If the imaginary-axis roots are found to exist, the original polynomial is divided out by the auxiliary polynomial and test is performed on the remainder polynomial.

For the example under consideration, the auxiliary equation is (left $\varepsilon \to 0$ in s^2-row)

$$s^2 + 1 = 0$$

yielding two roots on the imaginary axis. Dividing the original polynomial $q(s)$ by $(s^2 + 1)$, we get

$$q'(s) = s^4 + s^3 + 2s^2 + 2s + 1$$

The Routh array for this polynomial is

$$\begin{array}{c|ccc} s^4 & 1 & 2 & 1 \\ s^3 & 1 & 2 & \\ s^2 & \varepsilon & 1 & \\ s^1 & \dfrac{2\varepsilon-1}{\varepsilon} & & \\ s^0 & 1 & & \end{array}$$

As $\varepsilon \to 0$, there are two sign changes in the first-column elements. This indicates that there are two roots in the right s-plane.

6.7 STABILITY OF SYSTEMS MODELLED IN STATE VARIABLE FORM

We have seen in Section 2.7, that a single-input-single-output system can be expressed in state-variable form as (see eqns. 2.95(a), (b))

$$\dot{\mathbf{x}} = \mathbf{Ax} + \mathbf{b}u \quad \text{...(6.10)}$$

$$y = \mathbf{Cx} \quad \text{...(6.11)}$$

The system's characteristic equation is given by (eqn. 5.105)

$$|\,\lambda\mathbf{I} - \mathbf{A}\,| = 0$$

Once the characteristic equation is found out, the system stability can be determined by applying the Routh criterion. This is illustrated in the example below.

Example 6.12 : The state-variable model of open-loop system is described by

$$\mathbf{A} = \begin{bmatrix} 0 & 1 & 0 \\ 0 & 0 & 1 \\ 0 & -3 & 2 \end{bmatrix} ; \mathbf{B} = \begin{bmatrix} 0 \\ 0 \\ 1 \end{bmatrix}$$

(*a*) Check the stability of the system,

(*b*) The system's loop is now closed by a state feedback

$$u(t) = -\mathbf{Kx}(t)$$

where $\mathbf{K} = [K_1, K_2, K_3]$ is the feedback matrix of constant gains. Determine the constraints on the elements of **K** for the system to be stable.

Solution. The system's characteristic equation is

$$(a)\ |\ \lambda\mathbf{I} - \mathbf{A}\ | = \begin{bmatrix} \lambda & -1 & 0 \\ 0 & \lambda & -1 \\ 0 & 3 & \lambda+2 \end{bmatrix} = \lambda \begin{bmatrix} \lambda & -1 \\ 3 & \lambda+2 \end{bmatrix} = \lambda\,(\lambda^2 + 2\lambda + 3) = 0$$

As there is a root at origin ($\lambda = 0$), the system is limitedly stable.

(*b*) The system block diagram with state feedback is drawn in Fig. 6.8 wherein the input is reorganized as r and u is constructed from the state feedback. The overall model of the closed-loop system is found below.

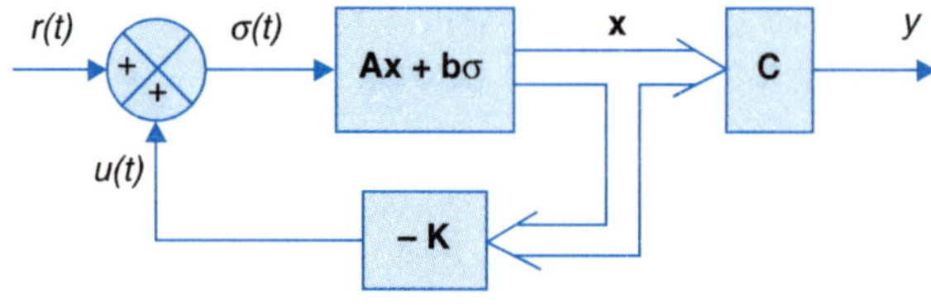

Fig. 6.8

$$\sigma = r - u = r - \mathbf{Kx}$$

$$\dot{\mathbf{x}} = \mathbf{Ax} + \mathbf{b}\sigma = \mathbf{Ax} + \mathbf{b}\,(r - \mathbf{Kx}) = (\mathbf{A} - \mathbf{bK})\,\mathbf{x} + \mathbf{b}r$$

or

$$\dot{\mathbf{x}} = \mathbf{A'x} + \mathbf{b}\,r$$

Stability is now to be determined from the closed-loop system matrix

$$\mathbf{A}' = \mathbf{A} - \mathbf{bK}$$

$$= \begin{bmatrix} 0 & 1 & 0 \\ 0 & 0 & 1 \\ 0 & -3 & -2 \end{bmatrix} - \begin{bmatrix} 0 \\ 0 \\ 1 \end{bmatrix} [K_1 \quad K_2 \quad K_3]$$

$$= \begin{bmatrix} 0 & 1 & 0 \\ 0 & 0 & 1 \\ -K_1 & -(K_3 + 3) & -(K_3 + 2) \end{bmatrix}$$

The characteristic equation now becomes

$$\begin{vmatrix} \lambda & -1 & 0 \\ 0 & \lambda & -1 \\ K_1 & (K_2+3) & \lambda+(K_3+2) \end{vmatrix} = 0$$

or
$$\lambda^3 + (K_3 + 2)\,\lambda^3 + (K_2 + 3) + K_1 = 0$$

Applying Routh criterion

$$\begin{array}{lll} \lambda^3 & 1 & (K_2+2) \\ \lambda^2 & (K_3+2) & K_1 \\ \lambda^1 & \dfrac{(K_3+2)(K_2+3)-K_1}{(K_3+2)} & 0 \\ \lambda^0 & K_1 & 0 \end{array}$$

Conditions for the system to be stable are:

$$K_1 > 1 \quad \text{...}(i)$$

$$K_3 > -2 \quad \text{...}(ii)$$

$$\frac{(K_3+2)(K_2+3)-K_1}{(K_3+2)}$$

or
$$K_2 > (K_1 - 3K_3 - 6)/(K_3 + 2) \quad \text{...}(iii)$$

PROBLEMS

6.1 Find the roots of the characteristic equations for systems whose open-loop transfer functions are given below. Locate the roots in the s-plane and indicate the stability of each system.

(*a*) $G(s)H(s) = \dfrac{1}{(s+2)(s+4)}$

(*b*) $G(s)H(s) = \dfrac{1(s+3)}{s(s+3)(s+8)}$

(*c*) $G(s)H(s) = \dfrac{9}{s^2(s+2)}$.

6.2 The characteristic equation of a servo system is given by

$$a_0s^4 + a_1s^3 + a_2s^2 + a_3s + a_4 = 0$$

Determine the conditions which must be satisfied by the coefficients of the characteristic equation for the system to be stable.

6.3 By means of the Routh criterion, determine the stability of the systems represented by the following characteristic equations. For systems found to be unstable, determine the number of roots of the characteristic equation in the right half s-plane.

(*a*) $s^4 + 2s^3 + 8s^2 + 4s + 3 = 0$

(*b*) $s^4 + 2s^3 + s^2 + 4s + 2 = 0$

(*c*) $s^5 + s^4 + 3s^3 + 9s^2 + 16s + 10 = 0$

(*d*) $s^6 + 3s^5 + 5s^4 + 9s^3 + 8s^2 + 6s + 4 = 0$.

6.4 The characteristic equations for certain feedback control systems are given below. In each case, determine the range of values k for the system to be stable.

(*a*) $s^3 + 2ks^2 + (k + 2)s + 4 = 0$

(*b*) $s^4 + 4s^3 + 13s^2 + 36s + k = 0$

(*c*) $s^4 + 20ks^3 + 5s^2 + 10s + 15 = 0$.

6.5 The open-loop transfer function of a unity feedback control system is given by

$$G(s) = \frac{K}{(s+2)(s+4)(s^2+6s+25)}$$

By applying Routh criterion, discuss the stability of the closed-loop system as a function of K. Dertermine the values of K which will cause sustained oscillations in the closed-loop system. What are the corresponding oscillation frequencies ?

6.6 A system oscillates with frequency w, if it has poles at $s = \pm j\omega$ and no poles in the right half s-plane. Determine values of K and a, so that the system shown in Fig. P-6.6 oscillates at a frequency 2 rad/sec.

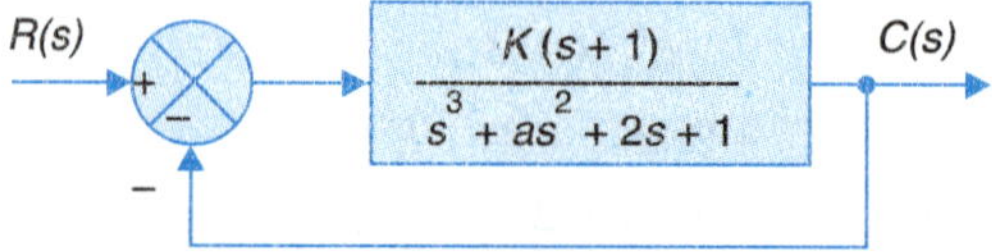

Fig. P-6.6

6.7 Consider the open-loop transfer function of a unity feedback control system;

$$G(s) = \frac{K(s+10)}{s(s+\alpha)(s+2)}$$

Using Routh criterion find the region of the (K, a)-plane wherein the values of (K, a) corresponds to a stable system. Note that K and a are positive real constant.

6.8 For the feedback control system of Fig. P-6.8

(*a*) Find the limiting values of K for system to be stable.

(*b*) For the value of K found in part (*a*) find the magnitude of the imaginary roots.

(*c*) For the value of K equal to 1/2 of that in part (*a*) find the real part of the least damped root. If the root is a complex conjugate pair, find also its imaginary part.

(*d*) For part (*c*) find the remaining root (s). Comment upon the dominance of the least-damped root.

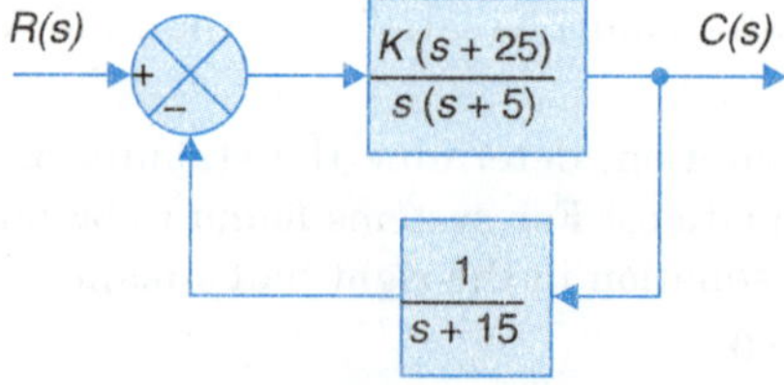

Fig. P-6.8

6.94 A unity feedback control system is characterized by the open-loop transfer function

$$G(s) = \frac{K(s+13)}{s(s+3)(s+7)}$$

(*a*) Using the Routh criterion, calculate the range of values of K for the system to be stable.

(*b*) Check if for $K = 1$, all these roots of the characteristic equation of the above system have damping factor greater than 0.5.

Note. Part (*b*) requires actual determination of the roots.

6.10 Determine whether the largest time constant of the characteristic equation given below is greater than, less than, or equal to 1.0 sec.

$$s^3 + 4s^2 + 6s + 4 = 0.$$

6.11 A feedback system has an open-loop transfer function of

$$G(s)H(s) = \frac{Ke^{-s}}{s(s^2 + 5s + 9)}$$

Determine by use of the Routh criterion, the maximum value of K for the closed-loop system to be stable.

[**Hint :** For low frequencies $e^{-s} \approx (1 - s)$]

6.12 A process is represented by the following state equations

$$\dot{x}_1 = x_1 - 4x_2$$

$$\dot{x}_2 = 8x_1 + u$$

The process dynamics is modified by state feedback

$$u = -k_1x_1 - k_2x_2$$

where k_1 and k_2 are real constants. Sketch the region of (k_1, k_2) for the system to be stable.

6.13 Determine the range of values of K ($K > 0$) such that the characteristic equation

$$s^3 + 3(k + 1)s^2 + (7K + 5)s + (4K + 7) = 0$$

has roots more negative than $s = -1$.

7

THE ROOT LOCUS TECHNIQUE

7

THE ROOT LOCUS TECHNIQUE

7.1 INTRODUCTION

It has been discussed earlier that the possibility of unstable operation is inherent in all feedback control systems because of the very nature of the feedback itself. An unstable system, obviously cannot perform the control task required of it. Therefore, while analyzing a given system, the very first investigation that needs to be made is, whether the system stable. However, the determination of stability of a system, is necessary but not sufficient, for a stable system with low damping is still undersirable. In an analysis problem one must, therefore, proceed to determine not only the absolute stability of a system but also its relative stability (peak overshoot, settling time, etc.) As discussed in the preceding chapter, the relative stability is directly related to the location of the closed-loop poles of a system. The Routh's criterion gives a satisfactory answer to the question of stability but its adoption to determine the relative stability is tedious and requires trial and error procedure even in the analysis problem.

Consider now a design problem in which the designer is required to achieve the desired performance for a system by adjusting the location of its closed-loop poles in the s-plane by varying one or more system parameters. The Routh's criterion, obviously does not help much in such problems. For determining the location of the closed-loop poles, one may resort to the classical techniques of factoring the characteristic polynomial and determining its roots, since the closed-loop poles are the roots of the characteristic equation. This technique is very laborious when the degree of the characteristic polynomial is three or higher. Furthermore, repeated calculations are required as a system parameter is varied for adjustments.

A simple technique, known as the *root locus techniques,* for finding the roots of the characteristic equation, introduced by W.R. Evans, is extensively used in control engineering practice. This technique provides a graphical method of ploting the locus of the roots in the

s-plane as a given system parameter is varied over the complete range of values (may be from zero to infinity). The roots corresponding to a particular value of the system parameter can then be located on the locus or the value of the parameter for a desired root location can be determined from the locus. The root locus is a powerful technique as it brings into focus the complete dynamic response of the system and futher, being a graphical technique, an approximate root locus sketch can be made quickly and the designer can easily visualize the effects of varying various system parameters on root locations. The root locus also provides a measure of sensitivity of roots to the variation in the parameter being considered. It may further be pointed out here that the root locus technique is applicable for single as well as multiple-loop systems.

In this chapter, we shall discuss the concepts underlying the root locus technique, the general procedure for sketching root locus plots, and the analysis of feedback systems by use of this technique. The use of this technique for design problems will be considered in Chapter 10.

7.2 THE ROOT LOCUS CONCEPTS

To understand the concepts underlying the root locus technique, consider the simple second-order system shown in Fig. 7.1. The open-loop transfer function of this system is

$$G(s) = \frac{K}{s(s+a)}$$

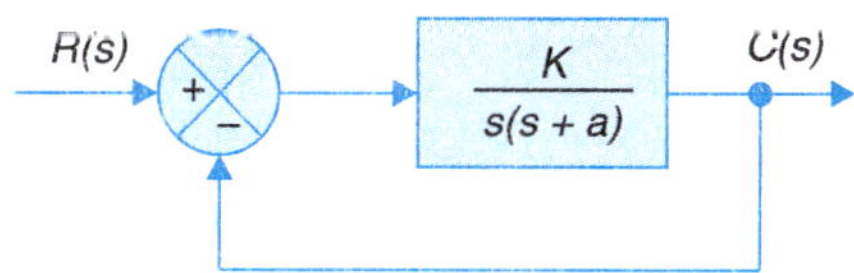

Fig. 7.1. A second-order system.

where K and a are constants.

The open-loop transfer function has two poles–one at origins $s = 0$ and the other at $s = -a$. The closed-loop transfer function of this system is

$$\frac{C(s)}{R(s)} = \frac{K}{s^2 + as + K} \qquad ...(7.1)$$

From eqn. (7.1), the characteristic equation of the system is

$$s^2 + as + K = 0 \qquad ...(7.2)$$

The second-order system under consideration is always stable for positive values of a and K but its dynamic behaviour is controlled by the roots of eqn. (7.2) and hence in turn by the magnitudes of a and K, since the roots are given by

$$s_1, s_2 = -\frac{a}{2} \pm \sqrt{\left[\left(\frac{a}{2}\right)^2 - K\right]} \qquad ...(7.3)$$

From eqn. (7.3), it is seen that as any of the system parameters (a or K) varies, the roots of the characteristic equation change. Let us consider the commonly occurring case of variable gain K, while the parameter a is held fixed. As K is varied from zero to infinity, the two roots (s_1, s_2) describe loci in the s-plane, Root locations for various ranges of values of K are:

(1) $0 \le K < a^2/4$, the roots are real and distinct.

when $K = 0$, the two roots are $s_1 = 0$, $s_2 = -a$ *i.e.*, they coincide with the open-loop pole of the system.

(2) $K = a^2/4$, the roots are real and equal in value, *i.e.*, $s_1 = s_2 = -a/2$.

(3) $a^2/4 < K < \infty$, the roots are complex conjugate with real part $= -a/2$, *i.e.*, unvarying real part.

The root locus with varying K is plotted in Fig. 7.2. These loci given the following information about the system behaviour.

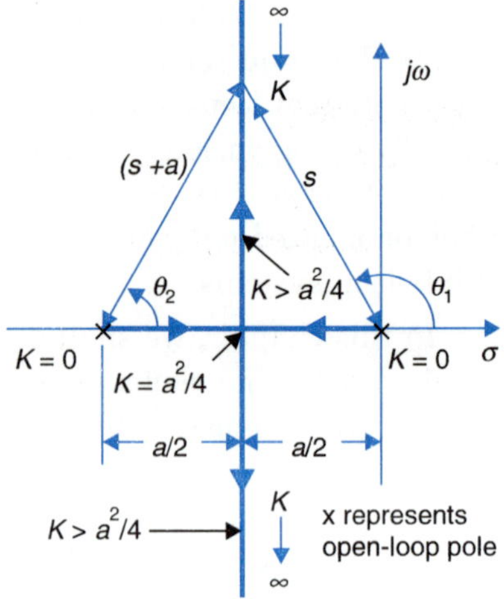

Fig. 7.2. Root loci of $s^2 + as + K = 0$ as a function of K.

1. The root locus plot has two branches starting at the two open-loop poles ($s = 0$, and $s = -a$) for $K = 0$,

2. As K is increased from 0 to $a^2/4$, the roots move towards the point $(-a/2, 0)$ from opposite directions. Both the roots lie on the negative real axis which corresponds to an overdamped system. The two roots meet at $s = -a/2$ for $K = a^2/4$. This point corresponds to a critically-damped system. As K is increased further ($K > a^2/4$), the roots break away from the real axis, become complex conjugate and since the real part of both the roots remains fixed at $-a/2$, the roots move along the line $\sigma = -a/2$ and the system becomes underdamped.

3. For $K > a^2/4$, the real parts of the roots are fixed, therefore the settling time is nearly constant.

The root locus shown in Fig. 7.2 has been drawn by the direct solution of the characteristic equation. This procedure becomes highly tedious for higher-order systems. Evans developed a graphical technique by use of which the root locus for third-and higher-order systems can be constructed as easily as for a second order system. The characteristic equation of any system is given by

$$\Delta(s) = 0 \qquad \text{...(7.4)}$$

where $\Delta(s)$ is the determinant of the signal flow graph of the system and is given by eqn. (2.86) which is reproduced below:

$$\Delta(s) = 1 - \sum_m P_{m1} + \sum_m P_{m2} - \sum_m P_{m3} + \ldots$$

where P_{mr} = gain product of mth possible combination of r non-touching loops of the graph. Thus, the characteristic equation can always be written in the form

$$1 + P(s) = 0 \qquad \text{...(7.5)}$$

For the single-loop system shown in Fig. 7.3

$$P(s) = G(s)H(s)$$

where $G(s)H(s)$ is the open-loop transfer function in block diagram terminology or loop transmittance in signal flow graph terminology.

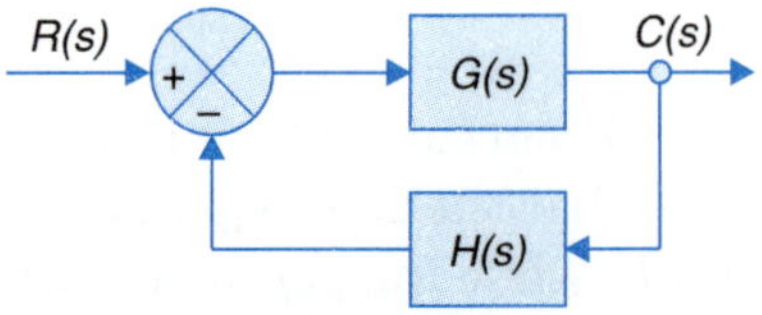

Fig. 7.3. Single-loop feedback system.

From eqn. (7.5) it is seen that the roots of the characteristic equation (*i.e.*, the closed-loop poles of the system) occur only for those values of s, where

$$P(s) = -1 \qquad \text{...(7.6)}$$

Since s is a complex variable, eqn. (7.6) is converted into the two conditions given below:

$$| P(s) | = 1 \quad ...(7.7)$$

$$\angle P(s) = \pm 180° (2q + 1) ; q = 0, 1, 2, ... \quad ...(7.8)$$

Equations (7.7) and (7.8) imply that the roots of $1 + P(s) = 0$ are those values of s at which the magnitude of $P(s)$ equals 1 and the angle of $P(s)$ equals $\pm 180° (2q + 1)$; $q = 0, 1, 2, ...$ These two conditions are not independent and one implies the other. Therefore, a plot of the points in the complex plane satisfying the *angle criterion* or eqn. (7.8) is the root locus. The value of gain corresponding to a root, *i.e.*, a point on the root locus, can be determined from the *magnitude criterion* of eqn. (7.7).

The root locus can be quickly drawn by checking the angle criterion at various points of the s-plane. The root locus technique is thus much easier to apply as compared to the direct solution of the characteristic equation.

Consider for example the feedback system of Fig. 7.3 with

$$G(s) = K\frac{(s+b)}{s(s+a)} \quad ...(7.9)$$

$$H(s) = 1$$

Its characteristic equation is

$$1 + G(s) = 0 \quad ...(7.10)$$

The open-loop transfer function $G(s)$ has two poles at $s = 0, -a$ and a zero at $s = -b$ as shown in Fig. 7.4. The angle criterion for the root locus is

$$\angle\left[\frac{(s+b)}{s(s+a)} = \pm 180° (2q+1) ; q = 0, 1, 2, ...\right] \quad ...(7.11)$$

With $s = (\sigma + j\omega)$

$$\angle\left[\frac{(s+b)}{s(s+a)}\right] = \angle\left[\frac{(\sigma + j\omega + b)}{(\sigma + j\omega)(\sigma + j\omega + a)}\right]$$

Therefore $\tan^{-1}\left(\dfrac{\omega}{\sigma + b}\right) - \tan^{-1}\left(\dfrac{\omega}{\sigma}\right) - \tan^{-1}\left(\dfrac{\omega}{\sigma + a}\right) = -\pi$

or $$\tan^{-1}\left(\frac{\omega}{\sigma}\right) + \tan^{-1}\left(\frac{\omega}{\sigma + a}\right) = \pi + \tan^{-1}\left(\frac{\omega}{\sigma + b}\right)$$

Taking tan on both sides, we have

$$\frac{\dfrac{\omega}{\sigma} + \dfrac{\omega}{\sigma + a}}{1 - \dfrac{\omega^2}{\sigma(\sigma + a)}} = \frac{\dfrac{\omega}{\sigma + b} + \tan \pi}{1 - (\tan \pi)\left(\dfrac{\omega}{\sigma + b}\right)}$$

Simplifying we get

$$(\sigma + b)^2 + \omega^2 = b^2 - ab \quad ...(7.12)$$

Equation (7.12) is the equation for root locus in the s-plane. It is a circle centred at $(-b, 0)$, *i.e.*, at the zero of the open-loop transfer function and of radius $\sqrt{(b^2 - ab)}$. The locus is drawn in Fig. 7.4. The sections of the real axis which lie on the root locus and the direction of arrows for increasing K are easily determined by the rules given in the next section.

As K is varied, the least-damped (minimum damping factor) complex conjugate poles are obtained by drawing OP tangential to the circular locus. By geometry

$$OP = \sqrt{[b^2 - (b^2 - ab)]} = \sqrt{(ab)}$$

$$\xi_{\min} = \cos\theta = \sqrt{(ab)}\,/\,b = \sqrt{(a/b)}$$

The magnitude criterion can be used to determine the value of K for any particular location of roots.

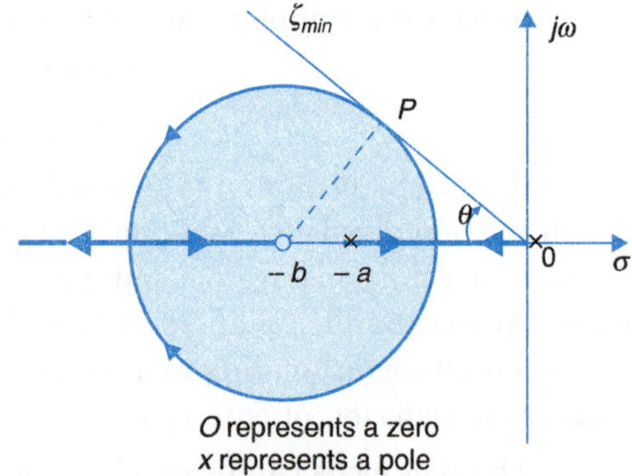

Fig. 7.4. Root locus plot of $1 + K(s + b)/s(s + a) = 0$.

7.3 CONSTRUCTION FOOT LOCI

Consider the basic feedback system of Fig. 7.3. Its characteristic equation is

$$1 + G(s)H(s) = 0$$

The open-loop transfer function $G(s)H(s)$ is generally known in the factored form as it is obtained by modelling the transfer function of individual components comprising the system. Therefore $G(s)H(s)$ [or $P(s)$ in case of multiple-loop systems] can generally be expressed in either of the factored forms given below:

$$1 + G(s)H(s) = 1 + \frac{K'\prod_{i=1}^{m}(\tau_{zi}s + 1)}{\prod_{j=1}^{n}(\tau_{pj}s + 1)} = 0 \text{ (time-constant form)} \qquad ...(7.13)$$

$$= 1 + \frac{K\prod_{i=1}^{m}(s + z_i)}{\prod_{j=1}^{n}(s + p_j)} = 0 \text{ (pole zero form)} \qquad ...(7.14)$$

These forms are interrelated by the following expression

$$K = K'\left[\prod_{j=1}^{n} p_j \Big/ \prod_{i=1}^{m} z_j\right]$$

where K' = open-loop gain in time-constant form ; K = open-loop gain in pole-zero form; $-z_i = -(1/\tau_{zi})$ $(i = 1, 2, ..., m)$ are the zeros of $G(s)H(s)$; $-p_j = -(1/\tau_{pj})$ $(j = 1, 2, ..., n$ are the poles of $G(s)H(s))$.

In all physically realizable systems $n \geq m$, *i.e.*, the number of poles of $G(s)H(s)$ is more than or equal to the number of its zeros.

The pole-zero form of eqn. (7.14) is more convenient for drawing root locus and will be used as such throughout this book. The open-loop gain parameter* K is considered variable for drawing the root locus of the system. The magnitude and angle criteria given by eqns. (7.7) and (7.8) can now be expressed as

$$\frac{K\prod_{i=1}^{m}|s+z_i|}{\prod_{j=1}^{n}|s+p_j|}=1 \qquad ...(7.15)$$

$$\sum_{i=1}^{m}\angle(s+z_i)-\sum_{j=1}^{n}\angle(s+p_j)=\pm(2q+1)\,180^\circ\;;\;q=0,1,2,... \qquad ...(7.16)$$

The points on root locus can be determined by statisfying the angle criterion through trial and error procedure. The poles and zeros of $G(s)H(s)$ are located on the s-plane on a graph sheet and a trial point s_0 is selected. The lines joining each of poles $(-p_j)$ and zeros $(-z_i)$ to s_0 represent the phasors (s_0+p_j) and (s_0+z_i) respectively as shown in Fig. 7.5 [on next page]. The angles of these phasors can be measured at the pole and zero locations and the angle values are substituted in the angle criterion of eqn. (7.16). The trial point s_0 is suitably shifted till the angle criterion is satisfied to a desired degree of accuracy. A more convenient method is to read the angles at the trial point in anticlockwise direction with reference to the dotted line shown in the figure. The dotted line is drawn parallel to the real axis and to the left of the point s_0. A number of points on the root locus are determined in this manner and a smooth curve drawn through these points gives the root locus with variable K. The value of K for a particular root location s_0 can be obtained from the magnitude criterion of eqn. (7.15) by substituting the phasor magnitudes read to scale from phasor lengths. It is to be noted that the same scale must be employed for both the real and imaginary axes of the complex s-plane.

From the magnitude criterion of eqn. (7.15)

$$K=\frac{\prod_{j=1}^{n}|s_0+p_j|}{\prod_{i=1}^{m}|s_0+z_i|}$$

With reference to Fig. 7.5, we can write

$$K=\frac{\text{Product of phasor lengths from } s_0 \text{ to open-loops poles}}{\text{Product of phasor lengths from } s_0 \text{ to open loops zeros}}$$

Note : The phasor lengths are to be read to scale.

Certain rules have been developed for making a quick approximate sketch of the root locus. This approximate sketch provides a guide for selection of a trial point such that more accurate root locus can be obtained by a few trials.

---

* If the variable parameter is other than the open-loop gain, the characteristic equation can be rearranged such that the prameter of interest appears as a multiplying factor in eqn. (7.14). This technique is illustrated in Sec. 7.4.

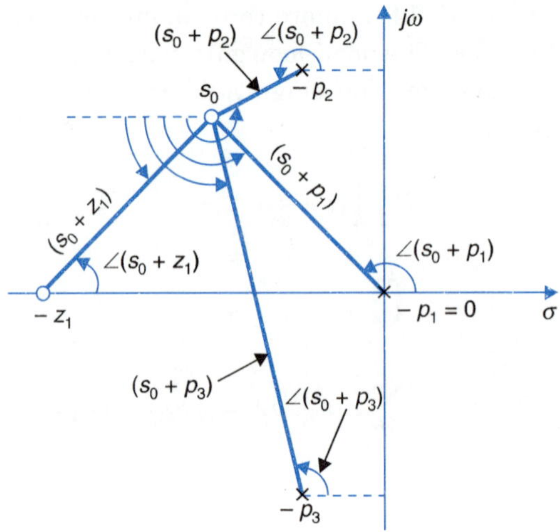

Fig. 7.5. Determining a point on root locus.
× represents a pole; 0 represents a zero.

Further, approximate root locus sketch, as obtained by the rules discussed below, is very useful in visualizing the effects, of variation of system gain K, the effects of shifting pole-zero locations and of bringing in a new set of poles and zeros. Actual root locus plot can then be obtained by MATLAB software; Appendix III.

Construction Rules

Rule 1 : The root locus is symmetrical about the real axis (σ-axis)

We know that the roots of the characteristic equation are either real or complex conjugate or combinations of both. Therefore their locus must be symmetrical about the σ-axis of the s-plane.

Rule 2 : As K increases from zero to infinity, each branch of the root locus originates from on open-loop pole with $K = 0$ and terminates either on an open-loop zero or on infinity with $K = \infty$. The number of branches terminating on infinity equals the number of open-loop poles minus zeros.

The characteristic eqn. (7.14) can be written as

$$\prod_{j=1}^{n}(s+p_j)+K\prod_{i=1}^{m}(s+z_i)=0$$

When $K = 0$, this equation has roots at $-p_j$ ($j = 1, 2, ..., n$) which are the open-loop poles. The root locus branches therefore start at the open-loop poles.

The same characteristic equation can also be written as

$$\frac{1}{K}\prod_{j=1}^{n}(s+p_j)+\prod_{i=1}^{m}(s+z_i)=0$$

As K tends to infinity, the first term of the characteristic equation vanishes and the roots are located at $-z_i$ ($i = 1, 2, ..., m$) which are the open-loop zeros of the system. Therefore m branches of the root locus terminate on the open-loop zeros.

In case $m < n$, the open-loop transfer function has $(n - m)$ zeros at infinity. Examining the magnitude condition (7.15) in the form

$$\frac{\prod_{i=1}^{m} |(s + z_i)|}{\prod_{j=1}^{n} |(s + p_j)|} = \frac{1}{K}$$

we find that this satisfied by $s \to \infty e^{j\phi}$ as $K \to \infty$. Therefore, $(n - m)$ branches of the root locus terminate on infinity.

Rule 3 : A point on the real axis lies on the locus if the number of open-loop poles plus zeros on the real axis to the right of this points is odd.

Consider the open-loop pole-zero configuration shown in Fig. 7.6. Let us examine any point s_0 on the real axis. As this point is joined by phasors to all the open-loop poles and zeros, it is easily seen that (*i*) the poles and zeros on the real axis to the right of this point contribute an angle of 180° each, (*ii*) the poles and zeros to the left of this point contribute an angle of 0° each, and (*iii*) the net angle contribution of a complex conjugate pole or zero pair is always zero. Therefore

$$\angle[G(s)H(s)] = (m_r - n_r)180° = \pm (2q + 1)\ 180°\ ;\ q = 0, 1, 2, ...$$

where m_r = number of zeros on the right of s_0 ; and n_r = number of poles on the right of s_0.

We therefore see that for a point on the real axis, the angle criterion is only met if $(n_r - m_r)$ or $(n_r + m_r)$ is odd and hence the rule. If this rule is satisfied at any point on the real axis, it continues to be satisfied as the point is moved on either side unless the point crosses a real axis pole or zero. By use of this additional fact, the real axis can be divided into segments on-locus and not-on-locus, the dividing points being the real open-loop poles and zeros. The on-locus segments of the real axis must alternate.

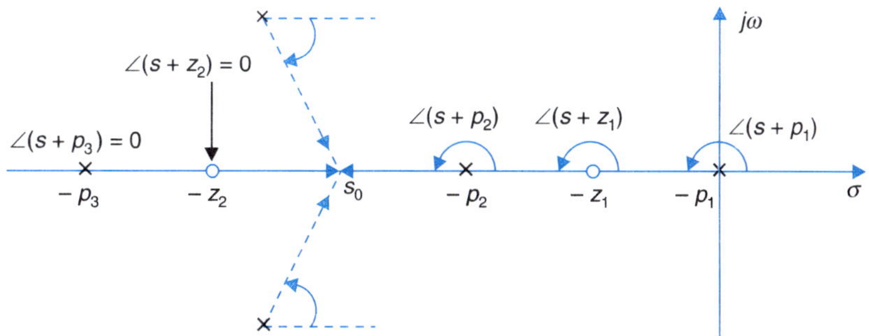

Fig. 7.6. Angle contributions for a point on the real axis.

Example 7.1 : Consider the system shown in Fig. 7.7 with the open-loop transfer function:

The open-loop pole-zero-configuration is shown in Fig. 7.8.

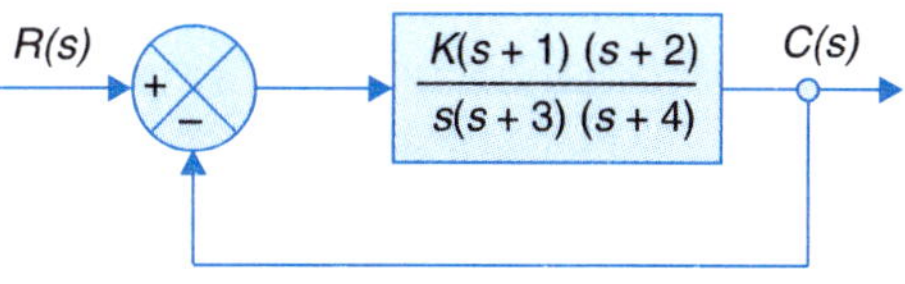

Fig. 7.7

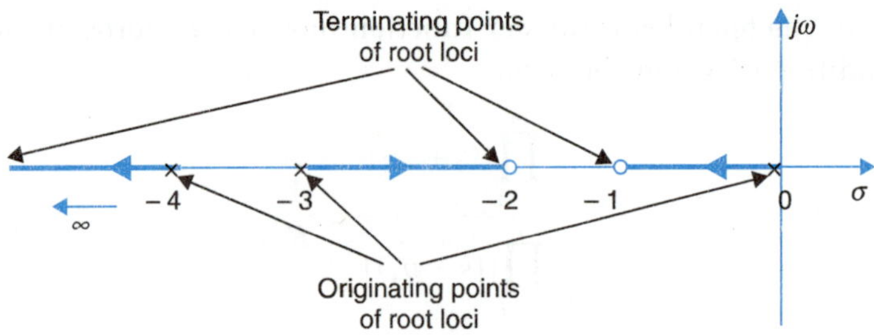

Fig. 7.8. Root locus plot for the system shown in Fig. 7.7.

From the rules described so far, the following information concerning the root locus plot is easily obtained.

1. There are three branches of the root locus as there are three open-loop poles. These branches start with $K = 0$ at each of the poles, $s = 0, -3, -4$.

2. As K increases, the branches leave the open-loop poles and seek open-loop zeros, two branches terminate on the two open-loop zeros ($s = -1, -2$) and one terminates on infinity.

3. The real axis segments between 0 and –1, –2 and –3, –4 and –∞, lie on the root locus (shown by thick lines in Fig. 7.8). This fact is verified by role 3, *e.g.*, for any point on the real axis between –2 and –3, $(n_r + m_r) = 1 + 2 = 3$ is odd and therefore this segment is on the root locus.

If a root locus branch moves along the real axis from an open-loop pole to zero or to infinity, such a branch of the root locus is termed as a *real-root branch*.

Rule 4 : The $(n - m)$ *branches of the root locus which tend to infinity, do so along straight line asymptotes whose angles are given by*

$$\phi_A = \frac{(2q+1)\,180^\circ}{n-m} \; ; q = 0, 1, 2, ..., (n - m - 1).$$

Consider a point on a branch of the root locus (which tends to infinity) at a remote distance from the open-loop poles and zeros. The phasors drawn from such a point to the open-loop poles and zeros essentially make the same angle (say ϕ) to each pole and zero. Therefore, for the point under consideration

$$\angle[G(s)H(s)] = -(n - m)\,\phi$$

For this point to lie on the locus

$$-(n - m)\,\phi = \pm\,(2q + 1)\,180^\circ$$

or

$$\phi = \pm\,\frac{180^\circ\,(2q+1)}{(n-m)} \qquad ...(7.17)$$

The $(n - m)$ branches of the root locus, therefore, tend to infinity along asymptotes having angles given by eqn. (7.17). The number of such asymptotes is equal to number of branches terminating on infinity, *i.e.*, $(n - m)$. Thus the integer q in the above equation can take on values from 0 to $(n - m - 1)$. The angles of asymptotes are therefore given by

$$\phi_A = \frac{(2q+1)\,180^\circ}{n-m} \; ; q = 0, 1, ..., (n - m - 1) \qquad ...(7.18)$$

Rule 5 : The asymptotes cross the real axis at a point known as centroid, determined by the relationship: (sum of real parts of poles – sum of real parts of zeros)/(number of poles – number of zeros).

From eqn. (7.14), the open-loop transfer function is written in the form

$$G(s)H(s) = \frac{K(s+z_1)(s+z_2)\ldots(s+z_m)}{(s+p_1)(s+p_2)\ldots(s+p_n)}\;;\; m \le n$$

$$= \frac{K[s^m + (\sum_{i=1}^{m} z_i)s^{m-1} + \ldots + (\prod_{i=1}^{m} z_i)]}{[s^n + (\sum_{j=1}^{n} p_j)s^{n-1} + \ldots + (\prod_{j=1}^{n} p_j)]}$$

Dividing the denominator by the numerator, we get

$$G(s)H(s) = \frac{K}{[s^{n-m} + (\sum_{j=1}^{n} p_j - \sum_{i=1}^{m} z_i)s^{n-m-1} + \ldots]} \qquad \ldots(7.19)$$

As s tends to infinity, the terms with higher powers of s dominate. Therefore, eqn. (7.19) can be approximate as

$$G(s)H(s)\big|_{s\to\infty} \approx \frac{K}{[s^{n-m} + (\sum_{j=1}^{n} p_j - \sum_{j=1}^{m} z_i)s^{n-m-1}]} \qquad \ldots(7.20)$$

Consider now the following function

$$P(s) = \frac{1}{(s+\sigma_A)^{n-m}}$$

which when expanded, gives

$$P(s) = \frac{1}{[s^{n-m} + (n-m)\sigma_A s^{n-m-1} + \ldots]} \qquad \ldots(7.21)$$

The characteristic equation $1 + P(s) = 0$ has $(n - m)$ root locus branches which are straight lines* passing through the points $s = -\sigma_A$ on the real axis and having angles $[(2q+1)180°]/(n-m)$, $q = 0, 1, 2, \ldots, (n-m-1)$. If σ_A is selected such that

---

*$P(s) = 1/(s + \sigma_A)^{n-m}$ has no zeros and a real pole of multiplicity $(n - m)$ at $s = -\sigma_A$. Consider any point s_0 on a straight line drawn from $-\sigma_A$ at an angle ϕ (see diagram).

Then

$$\angle P(s)\big|_{s=s_0} = -(n-m)\phi$$

For this point to be on the root locus

$$-(n-m)\,\phi = -(2q+1)\,180°$$

or

$$\phi = \frac{(2q+1)\,180°}{(n-m)}\;;\; q = 0, 1, 2, \ldots, (n-m-1)$$

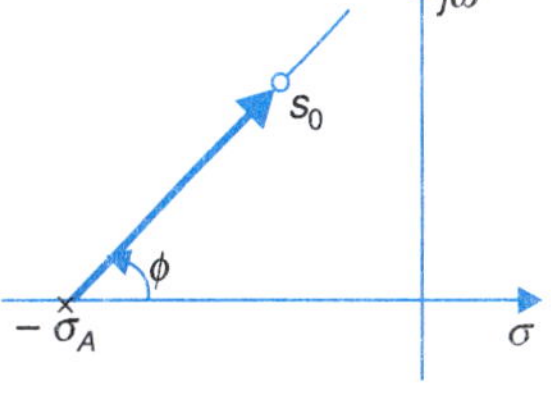

Thus the root locus of $1 + P(s) = 0$ is a set of $(n - m)$ lines drawn from $s = -\sigma_A$ at angles ϕ given by the above expression.

$$(n-m)\sigma_A = \sum_{j=1}^{n} p_j - \sum_{i=1}^{m} z_i \qquad ...(7.22)$$

the function $G(s)H(s)$ behaves in the same manner as $P(s)$ for values of s approaching infinity, because the first two higher-order terms of their denominators are identical. In other words, the general function $G(s)H(s)$ approaches the particular function $P(s)$ as s approaches infinity. Therefore, the branches of the root locus of $1 + G(s)H(s) = 0$ which tend to infinity, approach the straight line root locus branches of $1 + P(s) = 0$. Hence the straight line root loci of $1 + P(s) = 0$ act as asymptotes to the $(n - m)$ root locus branches of $1 + G(s)H(s) = 0$ which tend to infinity.

From eqn. (7.22), the *centroid* of the asymptotes is given by

$$-\sigma_A = \frac{\sum_{j=1}^{n}(-p_j) - \sum_{i=1}^{m}(-z_i)}{n-m} \qquad ...(7.23)$$

Because all the complex poles and zeros occur in conjugate pairs, σ_A is always a real quantity. Therefore eqn. (7.23) may be reduced to the following form:

$$-\sigma_A = \frac{\Sigma \text{ real parts of poles} - \Sigma \text{ real parts of zeros}}{\text{number of poles} - \text{number of zeros}} \qquad ...(7.24)$$

Example 7.2 : Consider a feedback system with the characteristic equation

$$1 + K\frac{1}{s(s+1)(s+2)} = 0$$

The open-loop pole-zero configuration of this system is shown in Fig. 7.9. From the rules described so far, the following information cencerning the root locus is obtained.

1. There are three branches of the root locus.

2. The branches of the root locus originates with $K = 0$ from the open loop poles $s = 0$, -1 and -2.

3. Since there is no open-loop zero in the finite region, all the three branches terminate on infinite, along the asymptotes whose angles with the real axis are given by

$$\phi_A = \frac{(2q+1)\,180°}{3} \;;\; q = 0, 1, 2 = 60°, 180°, 300°$$

4. The centroid of the asymptotes is given by

$$-\sigma_A = \frac{\Sigma \text{ real parts of poles} - \Sigma \text{ real parts of zeros}}{\text{number of poles} - \text{number of zeros}}$$

$$-\sigma_A = \frac{-1-2}{3} = -1.$$

5. The segments of the real axis between 0 and -1, -2 and $-\infty$, lie on the root locus.

From Fig. 7.9, it is seen that out of the three branches of the root locus, one is a real-root branches which originates from $s = -2$ and terminates on $-\infty$. The other two branches originate from $s = 0$ and $s = -1$, and move on the real axis approaching each other as K is increased. These two branches must therefore meet on the real axis. The characteristic equation has a double root at such a point. As the gain K is further increased, the root locus branches break away from the real axis to give a complex conjugate pair of roots. The point which represents a double root is known as *breakaway point* (determination of the breakaway point is discussed in the next rule). The branches which represent complex roots are known as *complex-root branches*.

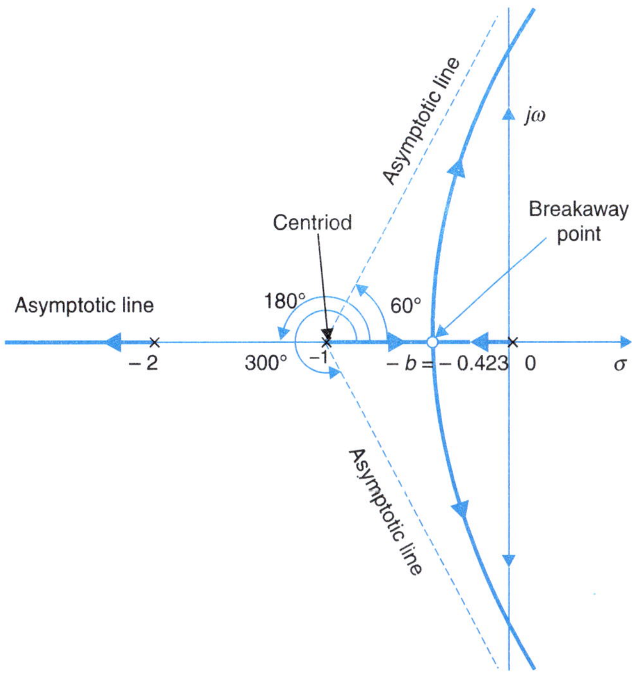

Fig. 7.9. Root locus plot of the equation 1 + K/s(s + 1)(s + 2) = 0.

In Fig. 7.9, the two real-root branches originating from the open-loop poles $s = 0$ and $s = -1$ approaches each other, breakaway at the point $-b$ and then one branch moves to infinity along 60° asymptote and the other movers to infinity along 300° asymptote. The third branch being a real-root branch coincides with the 180° asymptote.

Rule 6 : The breakaway points (points at which multiple roots of the characteristic equation occur) of the root locus are the solutions of $dK/ds = 0$.

Assume that the characteristic equation $1 + G(s)H(s) = 0$ has a multiple root at $s = -b$ of multiplicity r. The

$$1 + G(s)H(s) = (s + b)^r A_1 (s) \qquad ...(7.25)$$

where $A_1(s)$ does not contain the factor $(s + b)$.

Differentiating eqn. (7.25) with respect to s we have

$$\frac{d}{ds}[G(s)H(s)] = (s + b)^{r-1}[rA_1(s) + (s + b)A_1'(s)]$$

where $A_1'(s)$ represents the derivative of $A_1(s)$.

At $s = -b$, the right-hand side of this equation is zero because it has a factor $(s + b)^{r-1}$ and $r \geq 2$. Therefore, at $s = -b$

$$\frac{d}{ds}[G(s)H(s)] = 0 \qquad ...(7.26)$$

It implies that eqn. (7.26) has a root of at least order one at the same location as the multiple root of the original characteristic equation. Thus the breakaway points are the roots of this equation.

In pole-zero form, the characteristic equation may be written as

$$1 + G(s)H(s) = 1 + K\frac{B(s)}{A(s)} = 0 \qquad ...(7.27)$$

Taking the derivative of eqn. (7.27) with respect to s with K as a constant*, we have

$$\frac{d}{ds}[G(s)H(s)] = K\frac{A(s)B'(s) - A'(s)B(s)}{[A(s)]^2} = 0$$

As per eqn. (7.26) the roots of $\frac{d}{ds}[G(s)H(s)] = 0$ are the breakaway points, therefore the breakaway points are also given by the roots of

$$A(s)B'(s) - A'(s)B(s) = 0 \qquad ...(7.28)$$

From eqn. (7.27) we can write

$$K = -A(s)/B(s)$$

Differentiating K with respect to s, we get

$$\frac{dK}{ds} = \frac{A(s)B'(s) - A'(s)B(s)}{[B(s)]^2} \qquad ...(7.29)$$

At the breakaway points, the numerator of eqn. (7.29) is zero as per eqn. (7.28). Therefore, the breakaway points of the original characteristic equation are determined from the roots of

$$\frac{dK}{ds} = 0 \qquad ...(7.30)$$

Equation (7.28) or eqn. (7.30) could be used to determine the breakaway points.

In general, a breakaway point may involve two or more than two branches. If a breakaway point is obtained when r branches of the root locus come together and meet at a point, then that point represents the rth order root of the characteristic equation. Since the characteristic equation can have real as well as complex multiple roots, its root locus can have real as well as complex breakaway points, but because of conjugate symmetry of the root loci, the breakaway points must either be on the real axis or occur in complex conjugate pairs.

From the above discussions, we conclude that the breakaway points are the roots of the equation $dK/ds = 0$. It should, however, be noted that not all the roots of equation $dK/ds = 0$ correspond to the actual breakaway points. The actual breakaway points are those roots of the equation at which the root locus angle criterion is met.

---

*This value of K pertains to the multiple roots of the characteristic equation.

Breakaway Points on the Real Axis

We discuss below different methods of evaluating breakaway points on the real axis.

Method 1 : *An analytical approach.* This method requires the determination of the roots of the equation $dK/ds = 0$ to evaluate the breakaway points. The practical efficacy of the method is limited to third-order systems as in higher-order cases the determination of the roots of $dK/ds = 0$ in itself is a time-consuming job. The method is illustrated through the example below.

Example 7.3 : In the root locus plot of equation $1 + K/[s(s+1)(s+2)] = 0$ shown in Fig. 7.9 we found that there is a breakaway point on the real axis between 0 and –1 as the two real-root branches are oppositely directed on this segment.

From the characteristic equation, the gain K is given by

$$K = -s(s+1)(s+2) \qquad \text{...(7.31)}$$
$$= -(s^3 + 3s^2 + 2s)$$

Differentiating eqn. (7.31), we get

$$\frac{dK}{ds} = -(3s^2 + 6s + 2)$$

The roots of the equation $dK/ds = 0$ are

$$s_{1,2} = \frac{-6 \pm \sqrt{(36-24)}}{6} = -0.423, -1.577$$

Since the breakaway point must lie between 0 and –1, it is clear that $s = -0.423$ corresponds to the actual breakaway point.

Method 2 : *A graphical approach*. It is a more practical method for determining the breakaway points. Along the real axis, the condition $dK/ds = 0$ implies that the gain K is extremized with respect to the real variable $s = \sigma$. A breakaway on the real axis may occur in two ways. First, a breakaway point may result from two real-root branches moving towards each other as K is increased. After the breakaway point these branches become complex-root branches. In this case K is increasing along the real axis from either side to the breakaway point. The condition $dK/ds = 0$, therefore results in maximization of K at the breakaway point. Secondly, a real axis breakaway point may occur with complex-root branches moving towards the real axis and meeting at the breakaway point. These branches then become real-root branches and move in opposite direction along the real axis as K is further increased. In this case K decreases as breakaway point is approached from either side of the real axis. Therefore, the condition $dK/ds = 0$ results in minimization of K at the breakaway point.

As an example of the first case consider the third-order system whose root locus plot is drawn in Fig. 7.9. Fig. 7.10 shows plot of K for this system for various values of s between 0 and –1. The maximum value of K is 0.385 at $s = -0.423$. Thus the breakaway point occurs at $s = -0.423$ with $K = 0.385$.

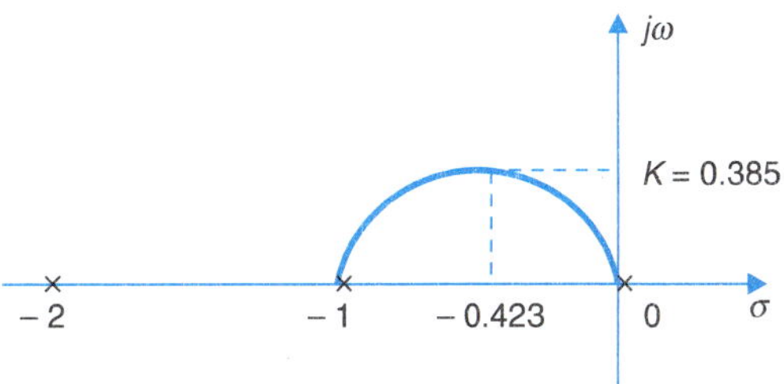

Fig. 7.10. Graphical evaluation of a real axis breakaway point.

An example of the second case is shown in the root locus plot of $1 + K(s+2)/(s^2 + 2s + 2) = 0$ in Fig. 7.11.

Another graphical method of finding a real axis breakaway point is to locate, through use of angle criterion, a few points (two or three such points will do) on the complex-root branches close to the real axis. A smooth curve drawn through these points intersects the real axis at the breakaway point.

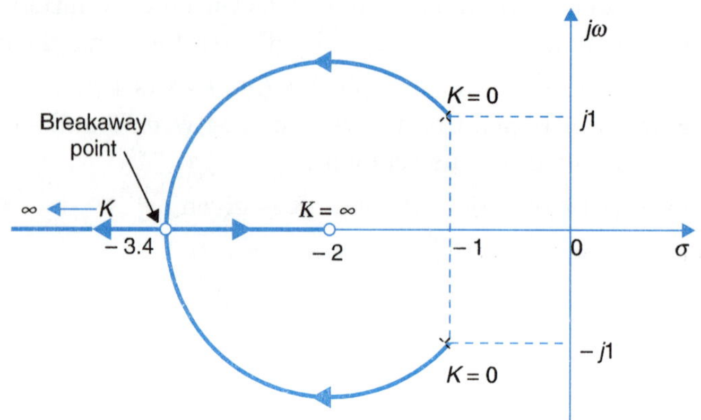

Fig. 7.11. Root locus plot of $1 + K(s + 2)/(s^2 + 2s + 2) = 0$.

Complex Breakaway points

The application of eqn. (7.30) for evaluation of complex breakaway points is illustrated in the following example.

Example 7.4 : The open-loop transfer function of a feedback system is

$$G(s)H(s) = \frac{K}{s(s+4)(s^2+4s+20)}$$

The open-loop pole-zero configuration is shown in Fig. 7.12. From the rules described so far, the following information concerning the root locus plot is obtained.

1. There are 4 branches of root locus originating at $s = 0$, $s = -4$, $s = -2 + j4$ and $s = -2 - j4$ respectively.

2. Since there is no open-loop zero in the finite region, all the four branches terminate on infinity along the asymptotes whose angles with the real axis are

$$\phi_A = \frac{(2q+1)180°}{4} \;; q = 0, 1, 2, 3 = 40°, 135°, 225°, 315°$$

3. The centroid of the asymptotes is given by

$$-\sigma_A = \frac{\Sigma \text{ real parts of poles } - \Sigma \text{ real parts of zeros}}{\text{number of poles } - \text{ number of zeros}}$$

$$= (-4 - 2 - 2)/(4 - 0) = -2$$

4. The points between 0 and –4 on the real axis lie on the root locus.

5. From the characteristic equation we have

$$K = -s(s + 4)\,(s^2 + 4s + 20) = -(s^4 + 8s^3 + 36s^2 + 80s)$$

Therefore $$\frac{dK}{ds} = -(4s^3 + 24s^2 + 72s + 80) = 0$$

the roots of which are found to be at $s = -2$ and $s = -2 \pm j2.45$.

Therefore, there is one breakaway point on the real axis at $s = -2$ and two complex conjugate breakaway points at $s = -2 \pm j2.45$. The rooot locus plot is sketched in Fig. 7.12.

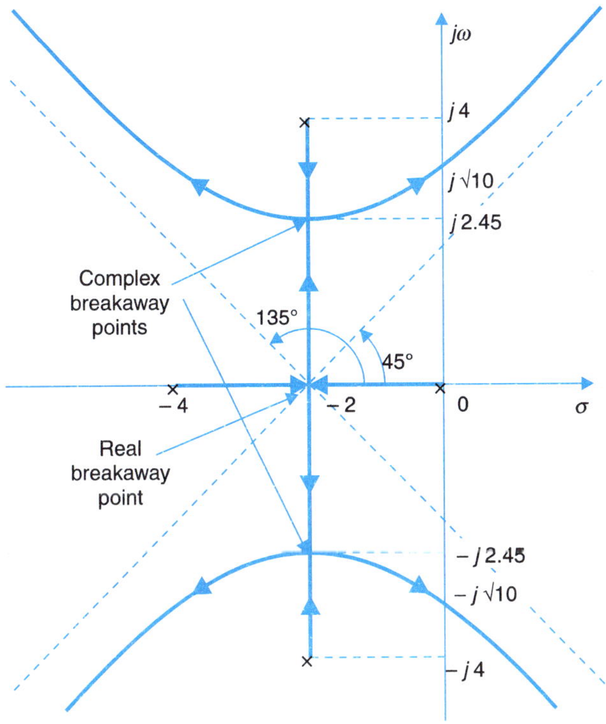

Fig. 7.12. Root locus plot with complex breakaway points.

Breakaway Directions of Root Locus Branches

The root locus branches must approach or leave the breakaway point on the real axis at an angle of ± 180°/r, where r is the number of root locus branches approaching or leaving the point.

The above statement can easily be verified by considering the root locus of Fig. 7.9. It is seen that two root locus branches approach the breakaway point and therefore according to the above statement, the root locus branches must leave the real axis breakaway point at an angle of ± 90°.

Take a point at an angle of 90° to the real axis and very close to the breakaway point. It can be shown by the angle criterion that the point lies on the root locus.

Rule 7 : The angle of departure from an open-loop pole is given by

$$\phi_p = \pm 180° (2q + 1) + \phi \, ; q = 0, 1, 2, \ldots \qquad \ldots(7.32)$$

where ϕ is the net angle contribution, at this pole, of all other open-loop poles and zeros.

Similarly the angle of arrival at an open-loop zeros is given by

$$\phi_z = \pm 180° (2q + 1) - \phi \, ; q = 0, 1, 2, \ldots$$

where ϕ means the net angle contribution at the zero under consideration of all other open-loop poles and zeros.

It easily follows from this rule that the angle of departure from a real open-loop pole or the angle of arrival at a real open-loop zero is always 0° or 180°. The angle of departure or arrival need therefore be calculated only for complex poles and zeros.

Consider the open-loop pole-zero configuration shown in Fig. 7.13. Around a pole p (complex), select a point s on the root locus branch starting from p and located very close to it. The net angle contribution of all other poles and zeros at this point is

$$\phi = \theta_4 - (\theta_1 + \theta_2 + \theta_3 + \theta_5)$$

The angle contribution of the pole p at s is θ_p. In the limit as the point s on the root locus approaches p, θ_p equals the angle of departure of the root locus from the pole p. By the angle criterion, we have

$$\phi - \phi_p = \pm 180° \, (2q + 1) \; ; q = 0, 1, 2, \ldots$$

or

$$\phi_p = \pm \, 180° \, (2q + 1) + \phi \; ; q = 0, 1, 2, \ldots$$

It may be noted that in the limit as the point approaches p, ϕ becomes the net angle contribution of all other open-loop poles and zeros at the pole p.

The rule for the angle of arrival at a zero follows similarly.

Example 7.5 : Let us evaluate the angle of departure of the root locus branch from the pole at $s = -1 + j1$ in Fig. 7.11. The open-loop pole-zero configuration of this figure is reproduced in Fig. 7.14.

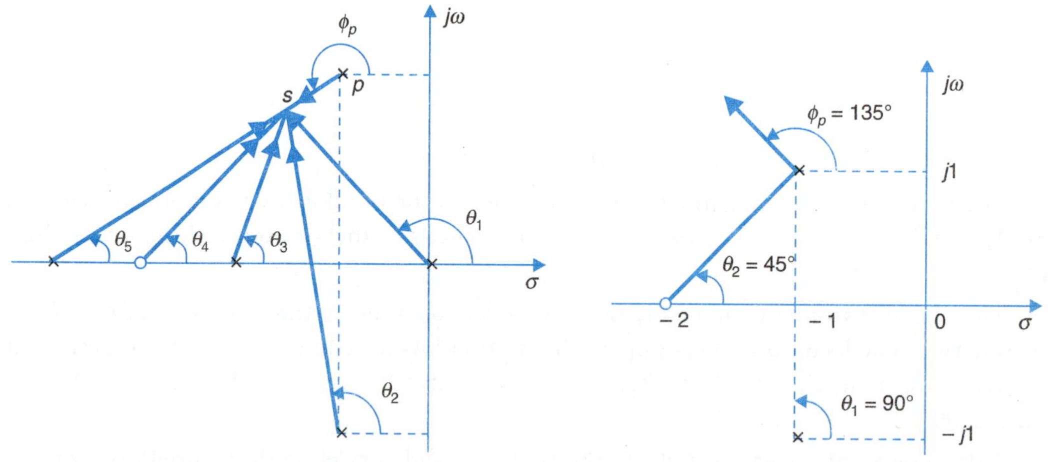

Fig. 7.13. Determination of the angle of departure of root locus.

Fig. 7.14. Determining the angle of departure of root locus branch in Fig. 7.11.

From Fig. 7.14 and eqn. (7.32), the angle of departure from the pole $(-1 + j1)$ is given by

$$\phi_p = \pm(2q + 1)\, 180° + (\theta_2 - \theta_1) = 180° + (45 - 90°) = +\, 135°.$$

Rule 8 : The intersection of root locus branches with the imaginary axis can be determined by use of the Routh criterion.

The rule is simply an application of the Routh criterion discussed in Chapter 6. To illustrate the rule, let us reconsider Example 7.4. The open-loop transfer function of the system considered in this example is

$$G(s)H(s) = \frac{K}{s(s+4)(s^2+4s+20)}$$

Therefore, the characteristic equation of the system is given by

$$s(s+4)(s^2+4s+20) + K = 0 \qquad \text{or} \qquad s^4 + 8s^3 + 36s^2 + 80s + K = 0$$

Application of the Routh criterion to the above equation gives the following Routh array:

$$\begin{array}{c|ccc} s^4 & 1 & 36 & K \\ s^3 & 8 & 80 & \\ s^3 & 1 & 10 & \\ s^2 & 26 & K & \\ s^1 & \dfrac{260-K}{26} & 0 & \\ s^0 & K & & \end{array}$$

For all the roots of the characteristic equation to lie to the left of the imaginary axis, the following conditions should be satisfied:

$$K > 0; \;\; (260 - K)/26 > 0$$

Therefore, the critical value of K (which corresponds to location of roots on the $j\omega$-axis is given by

$$(260 - K)/26 = 0 \qquad \text{or} \qquad K = 260$$

The value of $K = 260$, makes all the coefficients of s^1-row of the Routh array zero. For this value of K the auxiliary equation formed from the coefficient of the s^2-row is given by

$$26s^2 + K = 0$$

For $K = 260$, the roots of the above equation lie on the $j\omega$-axis and are given by

$$s = \pm j\sqrt{10}$$

Thus for the root locus plot shown in Fig. 7.12, the branches intersect the $j\omega$ -axis at $s = \pm j\sqrt{10}$ and the value of K corresponding to these roots is 260.

The rules described above are useful in determining the general configuration of root locus and are summarized in Table 7.1 for easy reference. While each of these rules has already been illustrated through examples, the comprehensive example given below further illustrates the use of all the rules in a single example.

Table 7.1. Rules for Construction of Root Loci of $1 + G(s)H(s) = 0$ where the Open-Loop Transfer Function $G(s)H(s)$ is known in Pole-Zero Form with n = Number of Open-Loop Poles, m = Number of Open-Loop Zeros

S.No.	*Rule*
1.	The root locus is symmetrical about the real axis.
2.	Each branch of the root locus originates from an open-loop pole at $K = 0$ and terminates at $K = \infty$, either on an open-loop zero or on infinity. The number of branches of the root locus terminating on infinity is equal to $(n - m)$, *i.e.*, the number of open-loop poles minus the number of zeros.

(Contd...)

3. Segments of the real axis having an odd number of real axis open-loop poles plus zeros to their right are parts of the root locus.
4. The $(n - m)$ root locus branches that tend to infinity, do so along straight line asymptotes making angles with the real axis given by

$$\phi_A = \frac{180°\,(2q+1)}{(n-m)}\ ;\ q = 0, 1, 2, ..., (n-m-1).$$

5. The point of intersection of the asymptotes with the real axis is at $s = -\sigma_A$ where

$$-\sigma_A = \frac{\Sigma \text{ real parts of poles } - \Sigma \text{ real parts of zeros}}{n-m}.$$

6. The breakaway points of the root locus are determined from the roots of the equation $dK/ds = 0$.

   r braches of root locus which meet at a point, break away at an angle of $\pm 180°/r$.
7. The angle of departure from an open-loop pole is given by

$$\phi_p = \pm\, 180°(2q+1) + \phi\ ;\ q = 0, 1, 2, ...$$

   where ϕ is the net contribution at the pole of all other open-loop poles and zeros.

   Similarly the angle of arrival at an open-loop zero is given by

$$\phi_z = \pm\, 180°\,(2q+1) - \phi\ ;\ q = 0, 1, 2, ...$$

   where ϕ is the net angle contribution at the zero of all other open-loop poles and zeros.
8. The intersection of root locus branches with the imaginary axis can be determined by use of the Routh criterion.
9. The open-loop gain K in pole-zero form at any point s_0 on the root locus is given by

$$K = \frac{\prod_{j=1}^{n} |s_0 + p_j|}{\prod_{i=1}^{m} |s_0 + z_i|} = \frac{\text{Product of phasor lengths* from } s_0 \text{ to open-loop poles}}{\text{Product of phasor lengths* from } s_0 \text{ to open-loop zeros}}$$

Example 7.6 : A feedback control system has an open-loop transfer function

$$G(s)H(s) = \frac{K}{s(s+3)(s^2+2s+2)}$$

Find the root locus as K is varied from 0 to ∞.

Solution. The open-loop poles are located at $s = 0, -3, (-1 + j1)$ and $(-1 - j1)$ while there are no finite open-loop zeros. The pole-zero configuration is shown in Fig.7.15.

Rule 2 tells us that the four branches of the root locus originate at $K = 0$ from the four open-loop poles and terminate at $K = \infty$ on infinity, since there are no finite zeros.

Rule 3 tells us that the root locus exists on the real axis for $-3 \le s \le 0$, shown by thick line on the real axis in Fig. 7.15.

Rule 4 tells us that the four branches tends to infinity along asymptotes whose angles with the real axis are

*Read to scale.

$$\phi_A = \frac{(2q+1)180°}{4-0} \;; q = 0, 1, 2, 3 = 45°, 135°, 225°, 315°$$

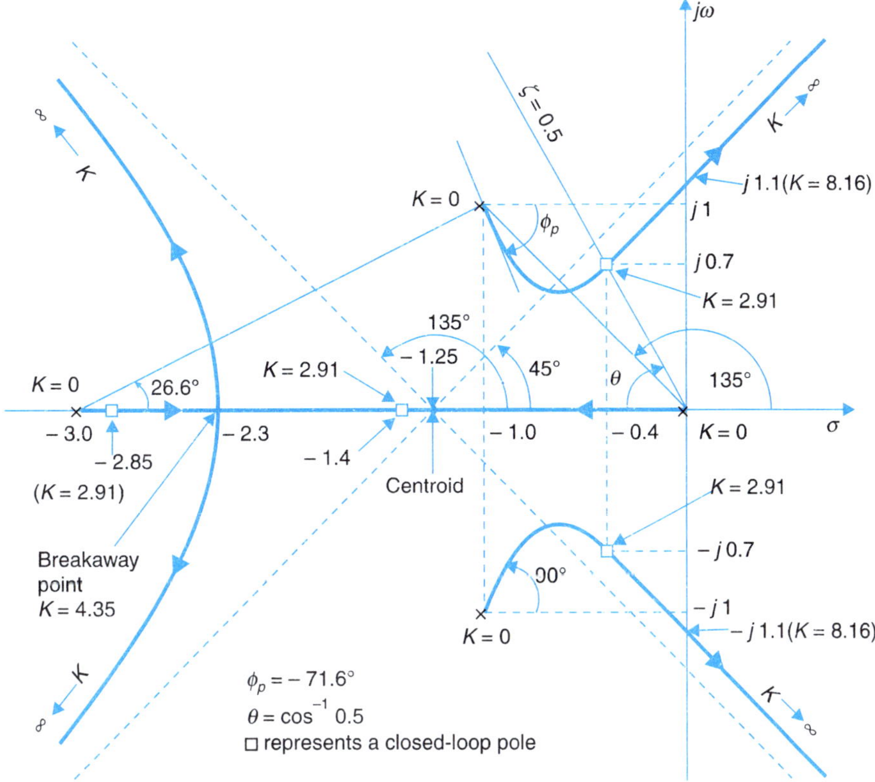

Fig. 7.15. Root locus plot of $1 + K/s(s + 3)(s^2 + 2s + 2) = 0$.

Rule 5 tells us that the centroid is given by

$$-\sigma_A = \frac{-3-1-1}{4} = -1.25$$

The asymptotes are shown by dotted line in Fig. 7.15.

Rule 6 is used below to determine the breakaway points:

From the characteristic equation of the system

$$K = -\,s(s + 3)(s^2 + 2s + 2) = -\,(s^4 + 5s^3 + 8s^2 + 6s)$$

$$\frac{dK}{ds} = -\,4\,(s^3 + 3.75s^2 + 4s + 1.5) = 0$$

The roots of this equation which give the possible breakaway points of the root locus are $s = -\,2.3\,;-\,0.725 \pm j0.365$. A breakaway point must occur at $s = -\,2.3$ as this part of the real axis is on the root locus and the two root locus branches starting from $s = 0$ and $s = -\,3$ are approaching each other. It can be checked that $s = -0.725 \pm j0.365$ are not the breakaway points as the angle criterion is not met at these points. The rule further tells us that two branches break away at an angle of ± 90°.

The value of K at the breakaway points is evaluated as

$$K = \{\,|s\,||s+3\,||s+1+j1\,||s+1-j1|\}_{s=-2.3}$$
$$= A.B.C.D. = 2.3 \times 0.7 \times 1.68 \times 1.68 = 4.54$$

where A, B, C, D are the lengths of the phasors drawn from the open-loop poles to the point $s = -2.3$

Rule 7 tells us that the root locus branch leaves the pole at $s = (-1 + j1)$ at angle ϕ_p given by

$$\phi_p = 180°(2q + 1) + [-135° - 90° - 26.6°]$$
$$\phi_p = 71.6°$$

Rule 8 tells us to check by use of the Routh criterion if the two branches originating from $s = -1 + j1$ and $s = -1 - j1$ will intersect the imaginary axis. The system characteristic equation is

$$s(s + 3)(s^2 + 2s + 2) + K = 0$$

or

$$s^4 + 5s^3 + 8s^2 + 6s + K = 0$$

Routh Array

$$\begin{array}{c|ccc} s^4 & 1 & 8 & K \\ s^3 & 5 & 6 & \\ s^2 & 34/5 & K & \\ s^1 & \dfrac{(204/5 - 5K)}{34/5} & 0 & \\ s^0 & K & 0 & \end{array}$$

Examination of the elements in the first column of the Routh array reveals that the above-mentioned root locus branches will intersect the imaginary axis at a value of K given by

$$(204/5) - 5K = 0$$

from which

$$K = 8.16$$

The auxiliary equation, formed from the coefficient of the s^2-row when $K = 8.16$, is

$$(34/5)s^2 + 8.16 = 0$$

from which

$$s = \pm j1.1$$

Therefore, purely imaginary closed-loop poles of the system are located at $s = \pm j1.1$ as shown in Fig. 7.15.

With the information obtained through the use of these rules, the root locus is sketched in Fig. 7.15 from where it is seen that for $K > 8.16$, the system has two closed-loop poles in the right half of the s-plane and is thus unstable.

Determination of Roots from Root Locus

Two problems of practical significance which can be worked with the root locus technique are explained below.

(*i*) *Determination of roots for a specified open-loop gain.* Along a particular root locus branch, a region is determined by trial and error such that the values of the open-loop gain (calculated by the magnitude criterion) at various points of the region are close to specified value. Further, trial and error will then yield the root location.

The above procedure is repeated for each root locus branch.

(ii) Determination of the open-loop gain for a specified damping of the dominant roots. A damping line (making an angle $\theta = \cos^{-1}\zeta$ with negative real axis) is drawn for the specified damping. In the region where the rough root locus sketch intersects the ζ-line, a trial and error procedure is adopted along the ζ-line to determine the point of intersection accurately. This point is then the desired root, at which the open-loop gain is computed. Roots along other branches can then be obtained for this open-loop gain using the procedure outlined in *(i)* above.

As an example consider the root locus plot in Fig. 7.15. Let the specified damping of the dominant roots be 0.5. A. ζ-line is drawn in the second quadrant at angle of $\theta = \cos^{-1} 0.5$ with the negative real axis. Adopting the trial and error procedure of *(ii)* above, it is found that the point $s = -0.4 + j0.7$ which lies on the ζ-line, satisfies the angle criterion. Therefore the dominant roots of the system are $s_1, s_2 = -0.4 \pm j0.7$. The value of the open-loop gain K at this root is

$$K = \text{product of distances from poles to this point}$$

$$= 0.84 \times 1.86 \times 2.74 \times 0.68 = 2.91$$

The other two roots of the characteristic equation are obtained from the branches originating at the poles at $s = 0$ and $s = -3$. From Fig. 7.15, it is seen that at the breakaway point, the value of K is 4.35. Therefore the points corresponding to $K = 2.91$ must lie on the real-root portions of these branches. Using the trial and error procedure given in *(i)* above, it is found that the points $s = -1.4$ and $s = -2.85$ have open-loop gain $K \approx 2.91$. Thus the closed-loop transfer function of the system under consideration is

$$T(s) = \frac{C(s)}{R(s)} = \frac{2.91}{(s + 0.4 + j0.7)(s + 0.4 - j0.7)(s + 1.4)(s + 2.85)}$$

Some Typical Root Locus Plots

Apart from the root locus plots presented so far, Table 7.2 gives additional root locus plots of some of the commonly occurring pole-zero configurations.

Cancellation of Poles and Zeros

Consider the open-loop transfer function

$$G(s)H(s) = \frac{K(s + z)}{s(s^2 + 2s + 2)} \qquad ...(7.33)$$

Its characteristic equation is

$$s^3 + 2s^2 + (2 + K)s + Kz = 0 \qquad ...(7.34)$$

When $z = 0$, the numerator and denominator of eqn. (7.33) have common factor s such that the open-loop pole at $s = 0$ and the open-loop zero at $s = 0$ cancel each other. After cancelling factor, the resulting open-loop transfer function becomes

$$G(s)H(s) = \frac{K}{s^2 + 2s + 2}$$

The characteristic equation now becomes

$$s^2 + 2s + 2 + K = 0$$

Table 7.2

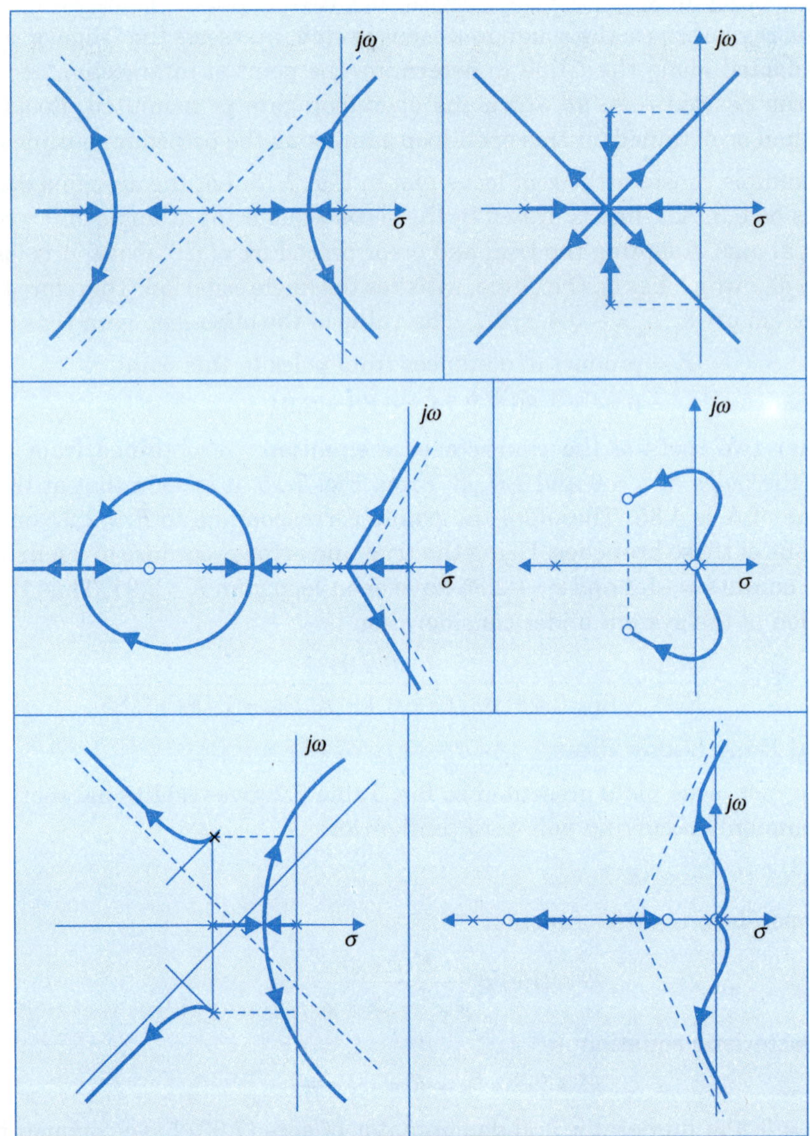

The root locus plot for this characteristic equation given is given in Fig. 7.16. This plot gives two closed-loop poles for the system, while eqn. (7.34) reveals that there must be three closed-loop poles, the missing third pole being the pole at $s = 0$, which is the same as the cancelled open-loop pole. *Therefore in a system where open-loop pole-zero cancellation is adopted, the closed loop poles are the root obtained from root locus plot of the system after pole zero cancellation plus the cancelled open-loop poles.*

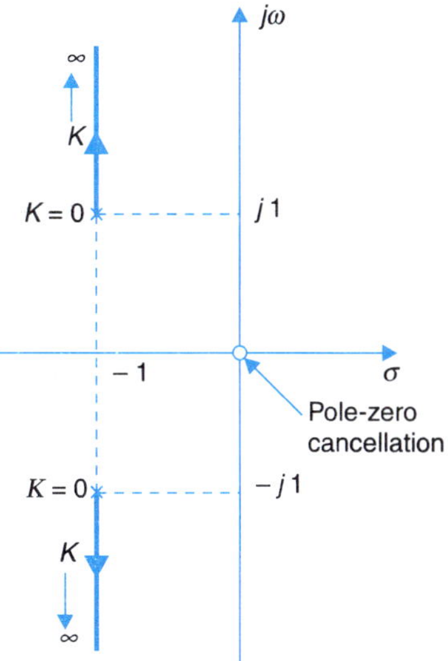

Fig. 7.16. Root locus plot of characteristic equation $(s^2 + 2s + 2 + K) = 0$.

Example 7.7 : An autonomous guided vehicle (AGV) is used to carry payloads to various destinations by an on-board computer. The vehicle having four wheels is powered by two dc servomotors of identical ratings, the two back wheels being free. Each of the motors is independently controlled by a PID scheme.

The basic block diagram of the control scheme is given in Fig. 7.16, wherein the motor is provided with an internal feedback loop.

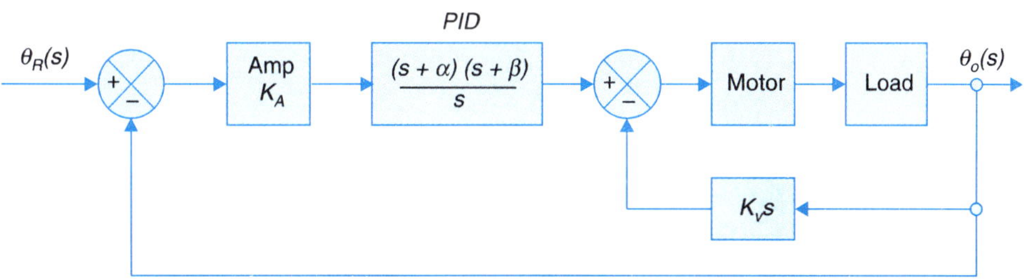

Fig. 7.16

The values of various constants are:

1. PID controller $\alpha = 100, \beta = 275$
2. Motor $K_T = K_b = 0.066\ Nm/A,\ R_a = 2.32\ \Omega$
   $J_m = 0.0002$ kg m^2

Motor inductance and viscous friction can be neglected.

3. Load inertia $J_L = 0.002$ kg m^2 ; directly coupled to motor.
4. Tachometer constant $K_v = 11$ V/rad(s)

Calculate the value of the amplifier gain K_A through root locus method, such that the dominant pole pair has a damping coefficient of $\zeta = 1$.

Also calculate and plot the unit step response of the system. What causes the response to have an overshoot and what is its magnitude. Comment.

Solution. The motor has zero viscous friction and there is no amplifier block in front of motor. So K_b and K_v, constant in V/rad/s add up *i.e.,*

$$K_v' = K_v + K_b = 11 + 0.066 \approx 11.1 \text{ V/rad/s}$$

The overall transfer function of the inner loop is then given as

$$G_m(s) = \frac{(K_T / R_a)}{s(Js + K_T K_v' / R_a)}$$

Let $$f = K_T K_v'/R_a,\ K_T' = K_T/R_a$$

So $$G_m(s) = \frac{K_T'}{s(Js + f)}$$

The overall forward loop transfer function is:

$$G(s) = \frac{K_A K_T' (s+\alpha)(s+\beta)}{s^2(Js+f)} \quad ; K = K_T' K_A / J$$

Substituting values we have

$$\alpha = 10,\ \beta = 50$$

$$f = K_T K_v'/R_a = (0.066 \times 11.1)/2.32 = 0.316$$

$$J = J_m + J_L = 0.0022 \text{ kg m}^2$$

$$\tau = f/J = 0.316/0.0022 \approx 144$$

The forward path transfer function can then be written as

$$G(s) = \frac{K(s+10)(s+50)}{s^2(s+144)}$$

Open loop poles : 0, 0, –144

Open loop zeros : –10, –15

The root locus is drawn to scale in Fig. 7.17 as obtained by MATLAB Control Tool. For illustration we shall find the breakaway point by numerical method given earlier.

Let the breakaway point be at distance d from the origin. At this point

$$K = \frac{d^2 \,|144 - d|}{|10 - d||50 - d|}$$

Searching numerically for d to yield minimum value of K, we get (approximately)

$$d = 17,\ K = 159$$

At this value of ($K = 159$), the dominant poles have

$\zeta = 1$. The corresponding amplifier gain is

$$K_A = KJ/K_T' = \frac{159 \times 0.0022}{(0.066 / 2.32)} \approx 12.3$$

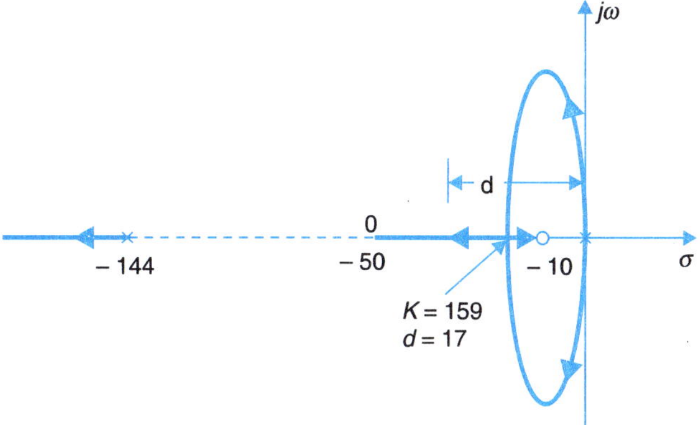

Fig. 7.17

The closed loop poles at $K = 159$ are:

$$-17, -17, -269 \text{ (not shown in figure)}$$

The closed loop transfer function can then be written as

$$\frac{\theta_0(s)}{\theta_R(s)} = \frac{159\,(s+10)\,(s+50)}{(s+17)^2\,(s+269)}$$

Its unit step response ($\theta_R(s) = 1/s$) is plotted in Fig. 7.18. This is obtained by MATLAB Control Tool.

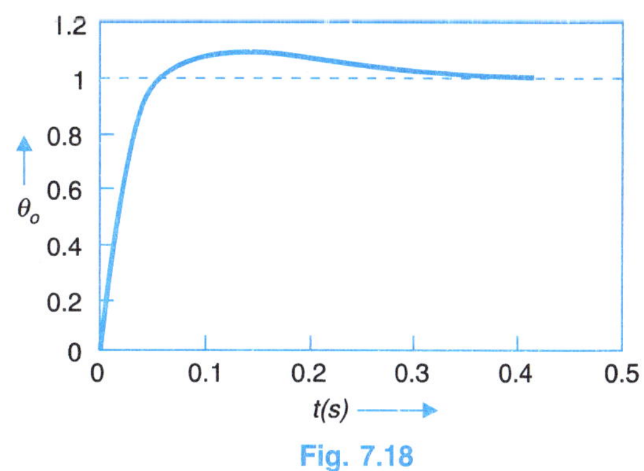

Fig. 7.18

The following observations are made:

* While the dominant poles (repeated real poles) have critical damping ($\zeta = 1$), the step response has an overshoot of about 8%. This results from the presence of the PID controller zeros ($s = -10, -50$). Any adjustment of these zeros cannot eliminate the overshoot totally, but it can be further reduced if the dominant pole pair is made to have $\zeta > 1$ (making $\zeta >> 1$ will not help).
* The step response has a gradually decaying tail which is contributed by the third real pole ($s = -269$).
* It may be noted that but for the PID controller zeros, the system would be absolutely unstable.

Example 7.8 : Consider unity feedback control system with an open-loop transfer function of

$$G(s) = \frac{K(s+1)(s+2)}{(s+0.1)(s-1)}$$

Draw the root loci of the system with gain K as a variable. As an aid to plotting determine: asymptotes, centroid, breakaway point, the gain at which the root locus crosses $j\omega$-axis.

Find from the root locus plot the value of gain K for which a closed-loop system is critically damped.

Solution.

System poles : – 0.1, 1

Zeros : – 1, – 2

These are marked in Fig. 7.19

- The plot has no asymptotes
- To determine the gain at which the root locus crosses the $j\omega$-axis, consider its characteristic equation

$$(s + 0.1)(s - 1) + K(s + 1)(s + 2) = 0$$

or
$$(K + 1)s^2 + (3K - 0.9)s + (2K - 0.1) = 0$$

Applying Routh Hurwitz criterion

$$\begin{array}{ll} s^2(K+1) & (2K-0.1) \\ s^1\,(K-0.3) & 0 \\ s^0(2K-0.1) & 0 \end{array}$$

Hence for stability

$$K > 0.3$$

The root locus crosses the $j\omega$-axis at

$$K = 0.3$$

at a frequency of

$$1.4\,s^2 + 0.5 = 0 \qquad \text{or} \qquad s = \pm j\sqrt{0.5/1.4} = 0.6 \text{ rad/s}$$

Breakaway points

$$B(s) = (s + 1)(s + 2) = s^2 + 3s + 2$$

$$A(s) = (s + 0.1)(s - 1) = s^2 - 0.9s - 0.1$$

From Eqn. (7.28)

$$A(s)\, B'(s) - A'(s)B(s) = 0$$

or
$$(s^2 - 0.9s + 0.1)\,(2s + 3) - (2s - 0.9)(s^2 + 3s + 2) = 0$$

or
$$3.9s^2 + 3.8s - 2.1 = 0$$

or
$$s = -\,1.368,\ 0.394$$

The value of K at the breakaway points is calculated below:

$$K_1 = \left|\frac{(s+0.1)(s-1)}{(s+1)(s+2)}\right|_{s=-1.368} = 12.9$$

$$K_2 = \left|\frac{(s+0.1)(s-1)}{(s+1)(s+2)}\right|_{s=0.394} = 0.09$$

Gain for critical damping

$\zeta = 1$ is achieved when the two roots are equal and negative (real). This happens at the breakaway point in the left half s-plane. The corresponding value of gain is

$$K = 12.9$$

The above information is confirmed by the root locus plot of Fig. 7.19 obtained by means of MATLAB Control Tool.

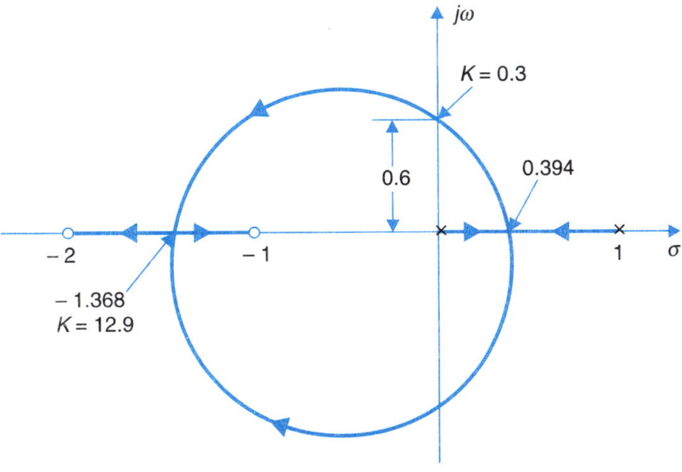

Fig. 7.19

Example 7.9 : A unity feedback control system has an open-loop transfer function of

$$G(s) = \frac{K(s + 4/3)}{s^2(s + 12)}$$

Plot the root locus. Find the value of K for which all the roots are equal. What is the value of these roots ?

Solution.

Poles : $s = 0, 0, -12$

Zeros : $s = -4/3$

$$-\sigma_A = \frac{-12 + 4/3}{3 - 1} = -16/3$$

$$\phi_A = \frac{\pm 180°(2q + 1)}{3 - 1} = \pm 90°$$

$$K = \frac{s^2 + 24s}{(s + 4/3)} + \frac{s^2(s + 12)}{(s + 4/3)} = 0$$

or $$-3s(s + 8)(s + 4/3) + s^2(s + 12) = 0 \quad \text{or} \quad s(s + 4)^2 = 0$$

$$s = 0, s = -4, -4$$

Breakaway point is at $s = -4$

Equal roots (3 nos) are at $s = -4$

$$K(s = -4) = \frac{4 \times 4 \times 8}{(4 - 4/3)} = \frac{128 \times 3}{8} = 48$$

The root locus is plotted in Fig. 7.20.

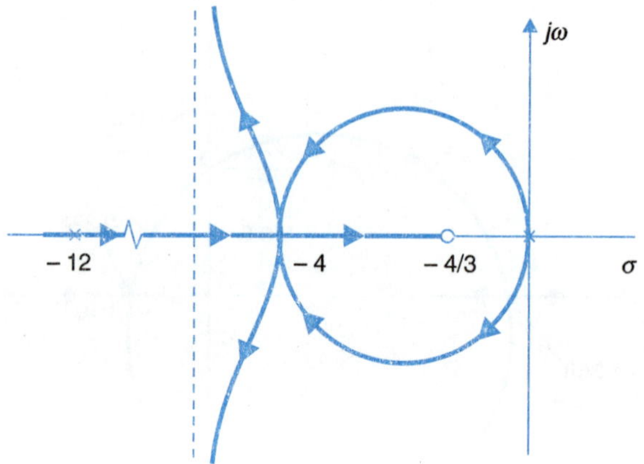

Fig. 7.20

Example 7.10 : In order to overcome the inherent instability of a helicopter (let us say w.r.t. pitch attitude), a stabilizing feedback loop is introduced in its autopilot system as shown in Fig. 7.21. The forward transfer function of the helicopter is estimated to be

$$G(s) = \frac{10(s+0.04)}{(s+0.5)(s^2-0.4s+0.2)}$$

The feedback path transfer function is shown in the figure where the gain of the transfer function (K_h) is adjustable.

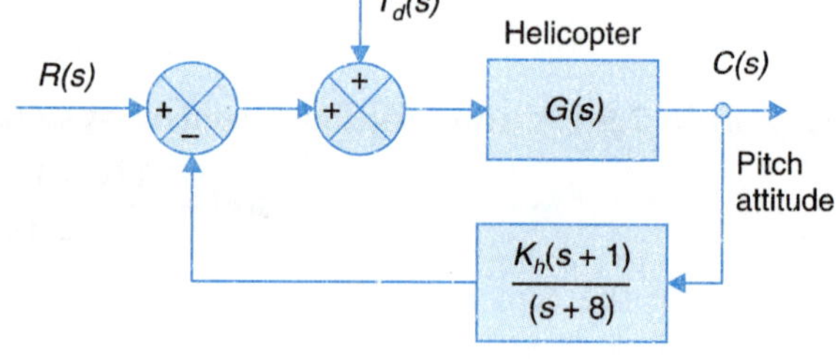

Fig. 7.21

Plot the root locus of the system with variable K_h and find its value for the closed-loop system to have a dominant pole damping of $1/\sqrt{2}$.

With this value of K_h, find the steady state error ($e = r - c$) caused by a wind gust disturbance of $T_d(s) = 1/s$ as shown in the block diagram.

Solution.

$$GH(s) = \frac{10K_h(s+0.04)(s+1)}{(s+0.5)(s^2-0.4s+0.2)(s+8)}$$

Open-loop poles : – 0.5, 0.2 ±j0.2, – 8

Open-loop zeros : – 0.04, – 1

Centroid of asymptotes is given by

$$-\sigma_A = \frac{(-8-0.5+0.2+0.2)-(-1-0.04)}{4-2} = -3.53$$

The root locus plot is drawn in Fig. 7.22 with $K = 10\,K_h$ as variable. We can now find the value of K at which the root locus crosses over to negative half of the s-plans. The system's characteristic equation is

$$(s + 0.5)(s^2 - 0.4s + 0.2)(s + 8) + K(s + 0.04)(s + 1) = 0$$

or
$$s^4 + 8.1s^3 + (K + 0.8)s^2 + (1.04K - 1.43)s + 0.04K + 0.8 = 0$$

Applying Routh's criterion we get the following result.

The root locus crosses the $j\omega$-axis (to left) at $K = 2.385$ at $\pm j\omega = 0.575$ rad/s.

The root locus is plotted to scale in Fig. 7.22. The $\zeta = 1/\sqrt{2}$ line intersects the root locus at two points with K as

$$K_1 = 1.555 \quad \text{or} \quad K_{h1} = 0.156$$
$$K_2 = 0.785 \quad \text{or} \quad K_{h2} = 0.079$$

Steady-state error for unit wind gust disturbance

$$R(s) = 0 \quad \Rightarrow \quad E(s) = -C(s)$$

$$\frac{E(s) = -C(s)}{T_d(s)} = \frac{G(s)}{1 + G(s)\dfrac{K_h(s+1)}{(s+8)}}$$

$$T_d(s) = 1/s$$

$$e_{ss} = \underset{s \to 0}{sE(s)} = \frac{10K_h}{1 + 10K_h} \text{; after substituting for } G(s)$$

This gives
$$e_{ss} = 0.609 \text{ for } K_{h1} = 0.156$$
$$= 0.44 \text{ for } K_{h2} = 0.079.$$

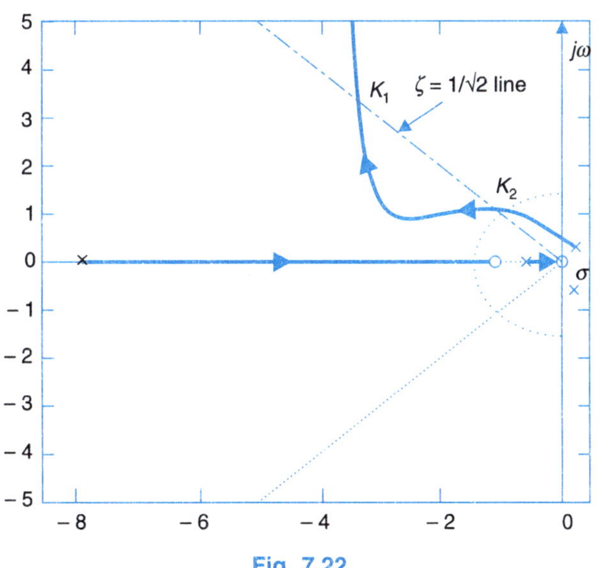

Fig. 7.22

7.4 ROOT CONTOURS

We have discussed so far the root locus technique for the study of feedback system when the open-loop gain K is a variable parameter. This technique can be easily adapted to the cases where a parameter other than K or more than one parameter are to be varied. The corresponding root locus plots are called *root contours* which can be constructed following the rules already enunciated in the previous section.

The basic equation for construction of root locus as already defined in eqn. (7.14) is reproduced below

$$1 + G(s)H(s) = 1 + \frac{K\prod_{i=1}^{m}(s+z_i)}{\prod_{j=1}^{n}(s+p_j)} = 1 + K\left(\frac{s^m + b_1 s^{m-1} + \ldots + b_{m-1}s + b_m}{s^n + a_1 s^{n-1} + \ldots + a_{n-1}s + a_n}\right) \quad \ldots(7.35)$$

where K, the open-loop gain is a variable parameter. If the variable parameter is other than K (say the distance α of an open-loop pole from the $j\omega$-axis), then the first step in construction of root locus is to manipulate the characteristic equation in the form eqn. (7.35) above, with α in place of the gain parameter K. Auxiliary root locus plot may be needed to bring eqn. (7.35) in the factored pole-zero form.

To illustrate the construction of root contours, consider the system shown in Fig. 7.1, whose characteristic equation is

$$s^2 + as + K = 0 \quad \ldots(7.36)$$

The root contours are to be plotted by allowing both a and K to vary. This can be carried out conveniently by rewriting the characteristic equation as

$$1 + a\left(\frac{s}{s^2 + K}\right) = 0 \quad \ldots(7.37)$$

such that a appears as a gain parameter. Equation (7.37) is in the form that its root contours can be drawn for various values of K by allowing a to vary from 0 to ∞. The following information is derived from the root locus construction rules.

The root contours of eqn. (7.37) originate from poles at $s = \pm j\sqrt{K}$ and terminate on zero at $s = 0$ and on $s = \infty$. From eqn. (7.37).

$$a = -(s^2 + K)/s$$

Therefore the roots of $\quad da/ds = -(s^2 - K)/s^2 = 0$

give the breakaway points at $s = \pm\sqrt{K}$. Fig. 7.23 shows that the breakaway point must lie between $s = 0$ and $s = -\infty$. Therefore $s = -\sqrt{K}$ corresponds to the breakaway point. The root contours* for various values of K with a varying from 0 to ∞ are shown in Fig. 7.23.

---

*In this case it can be shown that the complex-root branches are circular.

$$\angle \frac{\sigma + j\omega}{(\sigma + j\omega + j\sqrt{K})(\sigma + j\omega - j\sqrt{K})} = 180^\circ\,(2q + 1)$$

or $$\tan^{-1}(\omega/\sigma) - [\tan^{-1}(\omega + \sqrt{K})/\sigma + \tan^{-1}(\omega - \sqrt{K})/\sigma] = 180^\circ$$

Transposing $\tan^{-1}(\omega/\sigma)$ to the right and taking tan on both sides we get,

$$\frac{(\omega + \sqrt{K})/\sigma + (\omega - \sqrt{K})/\sigma}{1 - (\omega^2 - K)/\sigma^2}$$

which on simplification yields, $\sigma^2 + \omega^2 = K$, *i.e.*, the complex-root branches are parts of a circle centred at (0, 0) and of radius $\sqrt{K}$.

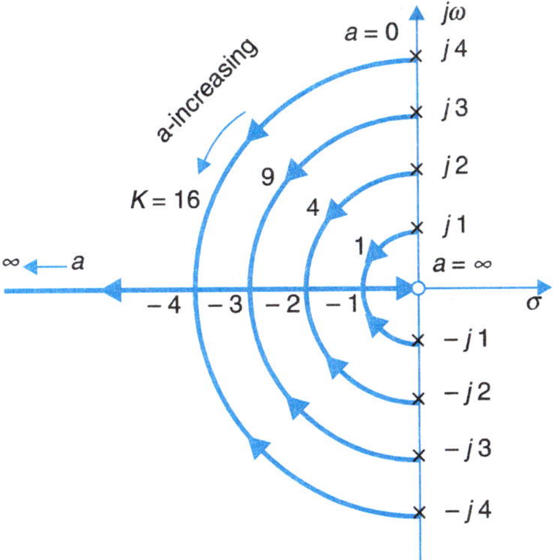

Fig. 7.23. Root contours of the system shown in Fig. 7.1.

Example 7.11 : Consider the feedback control system with an open-loop transfer function

$$G(s)H(s) = \frac{K}{s(s+1)(s+\alpha)}$$

in which the open-loop gain and the pole $s = -\alpha$ are both regarded as variable.

The characteristic equation of the system is

$$1 + \frac{K}{s(s+1)(s+\alpha)} = 0 \quad \text{or} \quad s^2(s+1) + \alpha\,(s+1)s + K = 0 \qquad ...(7.38)$$

It can be manipulated into the form below where α appears as a root locus parameter

$$1 + \frac{\alpha\, s(s+1)}{s^2(s+1) + K} = 0 \qquad ...(7.39)$$

Though eqn. (7.39) is in the form that root locus with respect to the parameter α can be drawn, however, we cannot proceed with it without determining its open-loop poles from the reduced characteristic equation

$$s^2(s+1) + K = 0$$

where K is a variable. The reduced characteristic equation can be obtained from the complete characteristic eqn. (7.38) by putting $\alpha = 0$. The reduced characteristic equation can be rewritten as

$$1 + \frac{K}{s^2(s+1)} = 0 \qquad ...(7.40)$$

The root locus of the reduced characteristic equation with K as a variable parameter is plotted in Fig. 7.24(a). The three roots of eqn. (7.40) for a particular value of K contribute the open-loop poles of eqn. (7.39).

The root contours for various values of K with varying α are drawn in Fig. 7.24(b). The value of α at which the root contours will cross the $j\omega$-axis into the left half s-plane and the system will become stable is obtained by application of the Routh criterion to the characteristic equation (7.38) as given below:

$$s^2(s+1) + \alpha s(s+1) + K = 0 \quad \text{or} \quad s^3 + (\alpha+1)s^2 + \alpha s + K = 0$$

The Routh array is

$$\begin{array}{c|cc} s^3 & 1 & \alpha \\ s^2 & (\alpha+1) & K \\ s^1 & \dfrac{\alpha(\alpha+1)-K}{(\alpha+1)} & 0 \\ s^0 & K & \end{array}$$

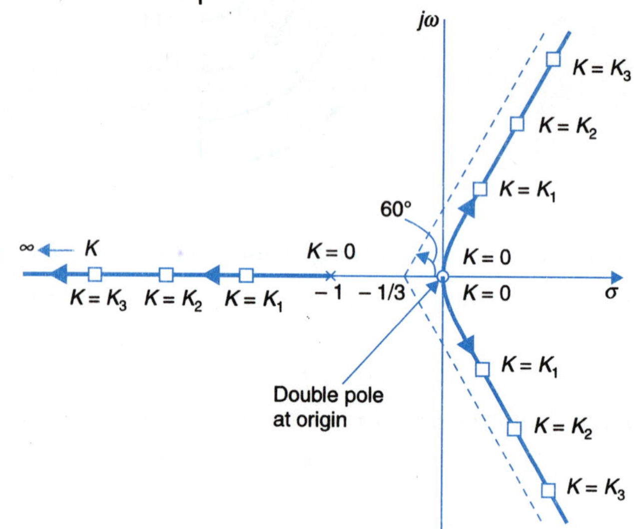

Fig. 7.24. (a) Root locus plot of the auxiliary characteristic equation $s^2(s+1) + K = 0$.

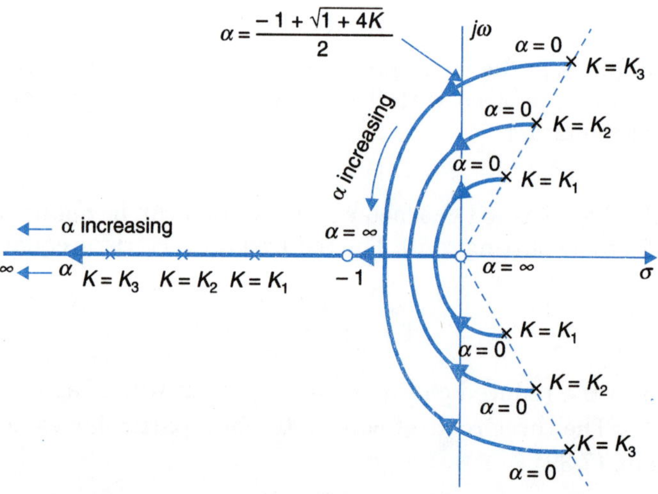

Fig. 7.24. (b) Root contours of $1 + K/s(s+1)(s+\alpha) = 0$.

Therefore, the root contours cross the $j\omega$-axis for

$$\alpha(\alpha+1) - K = 0$$

or $$\alpha = \frac{-1 + \sqrt{(1+4K)}}{2} \text{ (since only positive values of } \alpha \text{ are permitted).}$$

For higher values of α, the root loci cross into the stable region of s-plane as seen from Fig. 7.24(b).

Multiple-Loop System

In the previous examples in this chapter, there has been the implication of one feedback path. Actually the principles that have been discussed so far are equally applicable to a system having any number of feedback paths or closed-loops. As an example, consider the system shown in Fig. 7.25(a) which has two feedback loops.

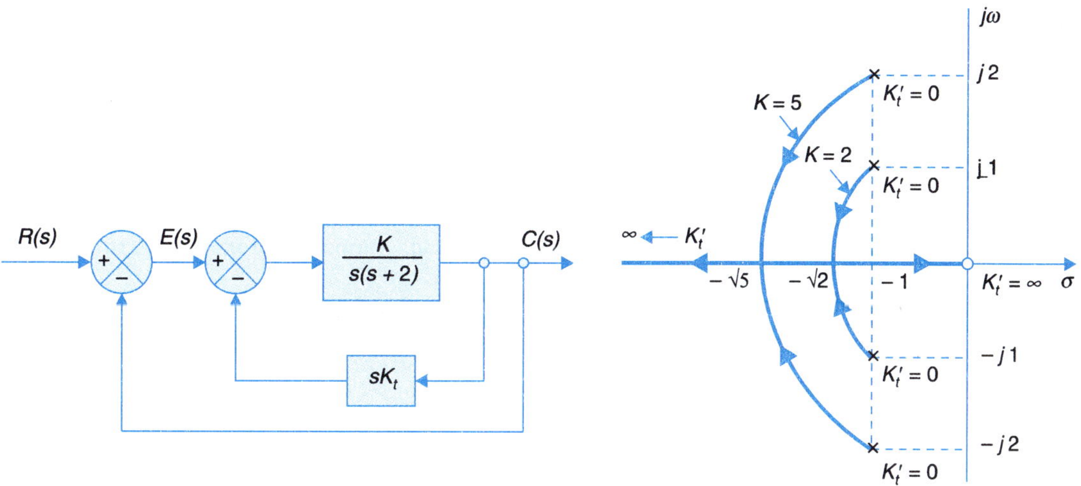

Fig. 7.25. (a) A feedback system with two loops.

Fig. 7.25. (b) Root contours of the system shown in Fig. 7.25(a).

The open-loop transfer function of the system is derived as

$$G(s) = \frac{C(s)}{E(s)} = \frac{K/s(s+2)}{1 + sKK_t/s(s+2)} = \frac{K}{s(s+2) + sKK_t}$$

Therefore the characteristic equation of the system is

$$s(s+2)\,sKK_t + K = 0 \qquad \text{...(7.41)}$$

Equation (7.41) may be rearranged as

$$1 + \frac{K_t Ks}{s(s+2)+K} = 0$$

or $$1 + \frac{K_t Ks}{[(s+1+j\sqrt{(K-1)}][(s+1)-j\sqrt{(K-1)}]} = 0$$

The root contours plotted for various values of K with $K_t' = KK_t$ varying from zero to infinity are shown in Fig. 7.25 (b).

7.5 SYSTEMS WITH TRANSPORTATION LAG

All the feedback systems discussed so far are represented by linear, lumped parameter mathematical models (transfer functions consisting of the ratios of algebraic polynomials). This is valid so long as the time taken for energy transmission is negligible, *i.e.,* the output beings to appear immediately on application of input. This is not quite true of transmission channel—lines, pipes, belt conveyors, etc. In such cases a definite time elapses after application of the input before the output begins to appear. This type of pure time lag is known as *transportation lag* or *dead time*. Linear lumped parameter model is not valid under such situations unless the pure time lag is negligible compared to other lags in the system. For example, the transmission line (pipe) between the hydraulic pump and motor causes a time lag in transportation of oil from pump to motor and vice versa. Similarly, the slow rate of transmission of heat energy by conduction or convection introduces serious transportation lags in process control systems.

The transfer function corresponding to a transportation lag can be easily determined. Consider a component whose output is the same as the input but delayed by T sec. Mathematically this can be expressed as

$$c_0(t) = c_i(t - T)$$

Taking the Laplace transform of the above equation we get

$$C_0(s) = C_i(s)e^{-sT} \qquad \text{or} \qquad C_0(s)/C_i(s) = e^{-sT} \qquad \text{...(7.42)}$$

Thus, the transfer function for a constant transportation lag is given by e^{-sT} (a non-minimum phase function).

At a point on the root locus $(\delta_0 + j\omega_0)$ the contribution of the delay term in the open-loop transfer function is

$$e^{-sT}\Big|_{s = (\delta_0 + j\omega_0)} = \exp(-\delta_0 T) \angle (-j\omega_0 T)$$

This angle contribution will repeat periodically for change in ω_0 by $(2\pi/T)$ and therefore the root locus branches of a system with time delay will repeat periodically. This renders the root locus method as such to be combersome for systems with time delay. However, in this s-domain region of interest it is possible to approximate time delay by one or more poles or zeros and then use the normal techniques for root locus design. Two commonly used approximation techniques are explained below:

$$\text{(1)}\ e^{-sT} \approx \frac{1}{(1 + sT/n)^n} \qquad \text{...(7.43)}$$

This approximation improves in accuracy as n is increased. However, for practical use n can be 1 or 2.

$$\text{(2)}\ e^{-sT} = 1 - Ts + \frac{T^2 s^2}{2!} - \frac{T^3 s^3}{3!} + \ldots \qquad \text{...(7.44)}$$

The series can be suitably truncated for design use and is valid for low values of T. It is most convenient to retain the first two terms only *i.e.,*

$$e^{-sT} \approx (1 - sT) = -T(s - 1/T) \qquad \text{...(7.45)}$$

This method is applied here to analyze the system of Fig. 7.26. This system shows an arrangement of controlling the thickness of a steel plate produced by a rolling mill. A voltage signal corresponding to the desired thickness is the reference input to the system. A thickness gauge provides a feedback voltage signal proportional to the actual thickness of the plate. The error voltage actuates the motor which positions the rolls.

The open-loop transfer function of the system is given by

$$G(s)H(s) = \frac{K_1}{s(\tau_m s + 1)} \qquad ...(7.46)$$

where for the sake of simplicity, we assume a single dominant time constant in the loop, probably that of the motor.

In the transfer function given by eqn. (7.46), we have ignored the fact that a finite time must elapse before a change in thickness of steel plates between rolls reaches the point of measurement at the gauge.

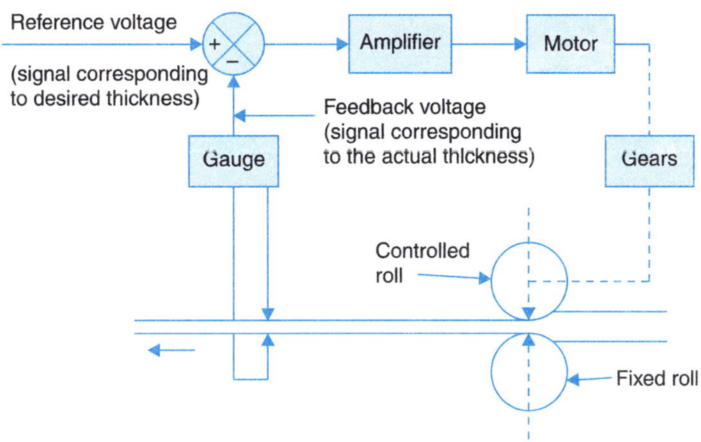

Fig. 7.26. A feedback system with transportation lag.

If this delay in transportation of signal is assumed to be T sec, then the open-loop transfer function of the system becomes.

$$G(s)H(s) = K_1 e^{-sT}/[s(\tau_m s + 1)]$$

The above equation may be rearranged in the following form

$$G(s)H(s) = Ke^{-sT}/[s(s + \alpha)] \qquad ...(7.47)$$

Assuming the transportation lag to be small, we use the approximating $e^{-sT} = (1 - sT)$. Equation (7.47) can now be rearrange in the form

$$G(s)H(s) = \frac{-K(s - 1/T)}{s(s + \alpha)}$$

The characteristic equation becomes

$$1 - K(s - 1/T)/[s(s + \alpha)] = 1 - P(s) = 0 \qquad ...(7.48)$$

Comparing the above equation* with eqn. (7.5), we observe that the angle criterion given by eqn. (7.8) is now modified as

$$\angle P(s) = \pm\, 180°\ (2q)\ ;\ q = 0, 1, 2, ...$$

This leads to following modification in Rules 3, 4 and 7 of Table 7.1.

(*i*) Replace 'odd number' by 'even number'.

(*ii*) Replace '180° (2q + 1)' by '180° (2q)'

The root locus plot for eqn. (7.45) with $\alpha = 2$ and $T = 1$ sec is shown in Fig. 7.27. This plot generally agrees with the exact plot in the region of small values of s where the dominant roots are located.

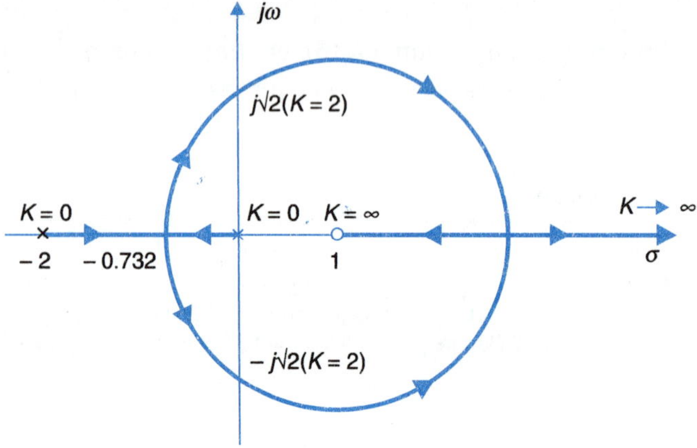

Fig. 7.27. Root locus plot of $1 + Ke^{-s}/[s(s + 2)] = 0$ where e^{-s} is approximated as $(1 - s)$.

7.6 SENSITIVITY OF THE ROOTS OF THE CHARACTERISTIC EQUATION

Earlier in Chapter 3 the sensitivity of the system transfer function $T(s)$ to variation of a parameter has been defined as

$$S_K^T = \frac{d(\ln T)}{d(\ln K)} = \frac{\partial T / T}{\partial K / K}$$

where K is the parameter of interest. Because of the importance of root locus as a design technique, it is useful to define the sensitivity of the roots of the characteristic equation (*i.e.*, the closed-loop poles of the system) to variation of a parameter K. This is defined as

$$S_K^{-r_k} = \frac{\partial(-r_k)}{\partial(\ln K)} = \frac{\partial(-r_k)}{\partial K/K} \qquad ...(7.49)$$

where $-r_k$ is a root of the characteristic equation. In order that the system dynamic response be relatively unaffected by parameter changes, the system must be so designed that the roots of the characteristic equation and in particular the dominant roots are made insensitive to parameter changes, *i.e.*, the root sensitivity $S_K^{-r_k}$ is made less than a specified value.

*Such cases also occur in positive feedback systems.

Example 7.12 : Consider a feedback control system with the characteristic equation

$$1 + \frac{K(s+z)}{s(s+p)} = 0 \qquad \text{...(7.50)}$$

In the case, the parameters of interest are the gain K, the zero $s = -z$ and the pole $s = -p$. An unwanted change in any of these parameters will cause the roots of the characteristic equation to shift. The amount and nature of this shift is determined by the root sensitivity to each of these parameters. Let us consider the root sensitivity to these parameters one by one.

Root Sensitivity To Gain K

Assume that the two open-loop poles are situated at $s = 0$ and $s = -2$ and the zero at $s = -3$. The characteristic equation of the system then becomes.

$$1 + K(s + 3)/s(s + 2) = 0 \qquad \text{...(7.51)}$$

The root locus plot of this equation as K varies from 0 to ∞ is shown in Fig. 7.28. It is seen that part of the locus is a circle with centre at $s = -3$ and radius $= \sqrt{3}$.

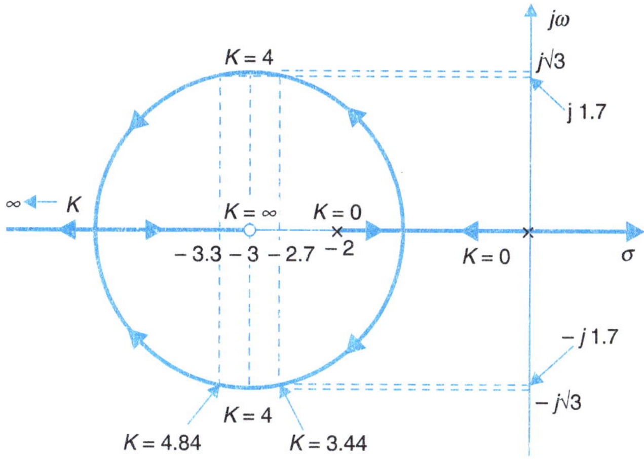

Fig. 7.28. Root locus plot of eqn. (7.48).

Let the nominal system gain be $K = K_0 = 4$. This results in two complex roots, $s_1 = -r_1 = -3 + j\sqrt{3}$ and $s_2 = -r_2 = -3 - j\sqrt{3}$. Since the roots are complex conjugate, the root sensitivity for $-r_1$ is the conjugate of the root sensitivity for $-r_2$. Therefore, we need to evaluate only the root sensitivity of one of the roots, say $-r_1$.

Assume that because of changes in gain K, the root $-r_1$ gets shifted to a new location $s = -3.3 + j1.7$ on the root locus, of course. Then

$$\Delta(-r_1) = (-3.3 + j1.7) - (-3 + j1.73) = -0.3 - j0.03$$

From Fig. 7.28, the value of gain K for this new location of the root (obtained by the magnitude criterion) is 4.84. Therefore

$$\Delta K = K - K_0 = 0.84$$

$$\Delta K/K = 0.84/4 = 0.21$$

From eqn. (7.49), the sensitivity of the root $-r_1$ is

$$S_{K+}^{(-r_1)} = \frac{\Delta(-r_1)}{\Delta K / K} = \frac{-0.3 - j0.03}{0.21} = 1.43 \angle 186°$$

for positive changes of gain.

Considering the new root location at $s = -2.7 + j\,1.7$, the gain corresponding to this location obtained from Fig. 7.28 is 3.44. Therefore

$$\Delta(-r_1) = (-2.7 + j1.7) - (-3 + j1.73) = 0.3 - j0.03$$

$$\Delta K = 0.54$$

$$\Delta K/K = 0.54/4 = 0.135$$

The sensitivity of the root $-r_1$ is now given by

$$S_{K-}^{(-r_1)} = (0.3 - j0.03)/(0.135) = 2.22 \angle -6°$$

for negative change in gain.

It can easily be verified that as the percentage change in gain $\Delta K/K$ decreases, the sensitivity measure $S_{K+}^{(-r_1)}$ and $S_{K-}^{(-r_1)}$ approach equality in magnitude and a difference in angle of 180°. Thus for very small changes in gain $\Delta K/K \to 0$,

$$|\, S_{K+}^{(-r_1)} \,| = |\, S_{K-}^{(-r_1)} \,| \qquad ...(7.52)$$

$$\angle S_{K+}^{(-r_1)} = 180° + \angle S_{K-}^{(-r_1)} \qquad ...(7.53)$$

Usually the desired root sensitivity measure is for small changes in the parameter. Therefore we need to evaluate only one sensitivity measure, say for positive changes in parameter; other sensitivity measure (for negative changes of parameter) can be obtained by use of eqns. (7.52) and (7.53).

Root Sensitivity For Small Parameter Changes

Once the root have been determined for nominal parameter values, the root sensitivity for any particular root for small parameter changes can be computed analytically from the knowledge of the open-loop poles and zeros using the root sensitivity expressions derived below.

It has been shown towards the end of Section 7.4 that the characteristic equation can always be manipulated such that the parameter K of interest appears in the form

$$1 + K\frac{\prod_{i=1}^{m}(s + z_i)}{\prod_{j=1}^{n}(s + p_j)} = 1 + K\frac{B(s)}{A(s)} = 0 \qquad ...(7.54)$$

where $B(s) = \prod_{i=1}^{m}(s + z_i)$ and $A(s) = \prod_{j=1}^{n}(s + p_j)$ are polynomials of degree m and n respectively with the coefficient of their highest degree term is s as unity.

The characteristic eqn. (7.54) can be written as

$$q(s) = A(s) + K\,B(s) = K_1 \prod_{k=1}^{n} (s + r_k) = 0 \qquad ...(7.55)$$

where $-r_k$ $(k = 1, 2, ..., n)$ are the roots of the characteristic equation. In eqn. (7.55) K_1 is given by

$$K_1 = 1 \text{ if } n > m$$
$$= 1 + K \text{ if } n = m$$

Taking logarithm of each side eqn. (7.55), we get

$$\ln q(s) = \ln K_1 + \prod_{k=1}^{n} \ln (s + r_k) \qquad ...(7.56)$$

Differentiating eqn. (7.55) and (7.56) with respect to the parameter K, we have

$$\frac{\partial q(s) / \partial K}{q(s)} = \frac{B(s)}{q(s)} = K_1^{-1} \frac{\partial K_1}{\partial K} - \sum_{k=1}^{n} \frac{\partial(-r_k) / \partial K}{(s + r_k)} \qquad ...(7.57)$$

Each side of eqn. (7.57) must have the same residue at any pole. Taking the residue of both the sides at the pole $s = -r_k$, we have

$$\frac{\partial(-r_k)}{\partial K} = -(s + r_k) \left.\frac{B(s)}{q(s)}\right|_{s=-r_k} = -\frac{B(-r_k)}{q'(-r_k)} \qquad ...(7.58)$$

where $$q'(-r_k) = \left.\frac{dq(s)}{ds}\right|_{s=-r_k}$$

In taking the residue above it has been assumed that the root $-r_k$ is a nonrepeated root of the characteristic equation.

It follows from eqn. (7.54) that

$$B(-r_k) = -\frac{A(-r_k)}{K}$$

Therefore we can write eqn. (7.58) as

$$S_K^{-r_k} = \frac{\partial(-r_k)}{\partial K / K} = \frac{A(-r_k)}{q'(-r_k)} \qquad ...(7.59)$$

The sensitivity of the roots $s = -r_k$ can thus be computed from the expression* given in eqn. (7.59).

The transfer function sensitivity and sensitivities of the roots of the characteristic equation can be correlated as follows.

The transfer function can be expressed as

$$T(s) = \frac{K_2 \prod_{i=1}^{m'} (s + z_i)}{\prod_{k=1}^{n} (s + r_k)} \qquad ...(7.60)$$

*The derivation of this expression is from Horowitz. Reproduced with permission.

where the zeros* of $T(s)$ are contributed by the zeros of $G(s)$ and the poles of $H(s)$ for a single-loop system [*i.e.*, m' = number of zeros of $G(s)$ plus number of poles of $H(s)$]. Taking logarithm of eqn. (7.60) and differentiating it with respect to ln K being the parameter of interest, we get

$$\frac{d(\ln T)}{d(\ln K)} = \frac{\partial(\ln K_2)}{\partial(\ln K)} + \sum_{k=1}^{n} \frac{\partial(-r_k)}{\partial(\ln K)} \frac{1}{(s + r_k)}$$

or
$$S_K^T = \frac{\partial(\ln K_2)}{\partial(\ln K)} + \sum_{k=1}^{n} S_K^{-r_k} \frac{1}{(s + r_k)} \qquad ...(7.61)$$

where it has been assumed that $(\partial z_i/\partial(\ln K)) = 0$, *i.e.*, the zeros of $T(s)$ are independent of the parameter K.

The expression (7.61) can be used to compute the transfer function sensitivity from the root sensitivities for the parameter K. In the particular case where the system gain K_2 is independent of the parameter K, we have

$$S_K^T = \sum_{k=1}^{n} S_K^{-r_k} \frac{1}{(s + r_k)} \qquad ...(7.62)$$

Example 7.13 : In Example 7.8 introduced earlier, the root sensitivity to each of the parameter K, p and z is computed below using the expression (7.59).

Root sensitivity to gain K. From the characteristic eqn. (eqn. (7.51))

$$A(s) = s(s + 2)$$

$$B(s) = (s + 3)$$

$$q(s) = A(s) + KB(s) = s(s + 2) + 4(s + 2)$$

$$\therefore \quad q'(s) = 2s + 6$$

Using eqn. (7.59) the root sensitivity to changes in gain K with nominal gain $K = 4$ is computed below:

$$S_K^{-r_1} = \frac{A(-r_1)}{q'(-r_1)} = \frac{(-3 + j\sqrt{3})(-1 + \sqrt{3})}{2(-3 + j\sqrt{3} + 3)} = -2 = 2\angle 180°$$

For a small change in the nominal gain (say $\Delta K = 0.4$), the new (shifted) root location can be calculated as

$$S_K^{-r_1} \approx \frac{\Delta(-r_1)}{\Delta K / K} \qquad ...(7.63)$$

$$\therefore \quad \Delta(-r_1) = -2 \times 0.4/4 = -0.2$$

---

*$T(s) = G(s)/[1 + G(s)H(s)]$

Let $G(s) = N_G(s)/D_G(s)$ and $H(s) = N_H(s)/D_H(s)$

$$\therefore \; T(s) = \frac{N_G(s)D_H(s)}{D_G(s)D_H(s) + N_G(s)N_H(s)}$$

The zeros of $T(s)$ thus are the roots of $N_G(s)D_H(s)$, *i.e.*, the zeros of $G(s)$ and the poles $H(s)$.

New root location $= (-r_1) + \Delta(-r_1) = (-3.2 + j\sqrt{3})$

If the gain decreases then $\Delta K = -0.4$. Therefore

$$\Delta(-r_1) = -2\,(-0.4/4) = +0.2$$

The new root location $= (-r_1) + \Delta(-r_1) = (-2.8 + j\sqrt{3})$

The expression for root sensitivity (7.59) is not valid for large changes in the parameter K, in which case the roots for changed value of K must be obtained from the root locus plot and the root sensitivity computed from eqn. (7.63).

Root sensitivity to open-loop zero ($s = -z$). The characteristic equation can be rewritten with z as a variable parameter, *i.e.*,

$$1 + 4(s + z)/(s + 2) = 0$$

$$s^2 + 6s + 4z = 0 \qquad \text{or} \qquad 1 + z'/s(s + 6) = 0$$

where $z' = 4z$

The nominal value of $z' = 12$ for which the two roots are $-r_1, -r_2, = -3 \pm j\sqrt{3}$.

Now
$$A(s) = s(s + 6)$$
$$B(s) = 1$$
$$q(s) = s(s + 6) + 12$$
$$q'(s) = 2s + 6$$
$$S_{z'}^{-r_1} = \left.\frac{A(s)}{q'(s)}\right|_{s=-3+j\sqrt{3}} = (2\sqrt{3})\angle 90°$$

Root sensitivity to open-loop poles ($s = -p$). The characteristic equation is written below with p as a variable parameter:

$$1 + 4(s + 3)/s(s + p) = 0 \qquad \text{or} \qquad 1 + ps/(s^2 + 4s + 12) = 0$$

For a nominal value of $p = 2$, the roots of the characteristic equation are $-r_1, -r_2 = -3 \pm j\sqrt{3}$. From the characteristic equation

$$A(s) = s^2 + 4s + 12$$
$$B(s) = s$$
$$q(s) = s(s^2 + 4s + 12) + 2s$$
$$q'(s) = 2s + 6$$
$$S_p^{-r_1} = \left.\frac{A(s)}{q'(s)}\right|_{s=-3+j\sqrt{3}} = 2\angle -120°$$

The root sensitivity measure for parameter variations is useful for comparing several designs. The use of this method as a design technique requires repeated complex calculations but there is obvious direction for adjusting the parameters to reduce or minimize sensitivity. This will be taken up in Chapter 10.

7.7 MATLAB: TOOL FOR DESIGN AND ANALYSIS OF CONTROL SYSTEMS—APPENDIX III.

This software has several Control System Tools; root-locus plotting and design is one such tool. This has made most graphical procedures redundant. These have been presented here as a pedagogical aid for of providing a deep understanding to the reader of the root-locus as a technique.

PROBLEMS

7.1. A unity feedback control system has an open-loop transfer function

$$G(s) = K/s\ (s^2 + 4s + 13)$$

Sketch the root locus plot of the system by determining the following:

(*i*) Centroid, number and angle of asymptotes.

(*ii*) Angle of departure of root loci from the poles.

(*iii*) Breakaway point if any.

(*iv*) The value of K and the frequency at which the root loci cross the $j\omega$-axis.

7.2. Sketch the root locus plot of a unity feedback system with an open-loop transfer function

$$G(s) = K/s(s + 2)(s + 4)$$

Find the range of values of K for which the system has damped oscillatory response. What is the greatest value of K which can be used before continuous oscillations occur ? Also determine the frequency of continuous oscillations.

Determine the value of K so that the dominant pair of complex poles of the system has a damping ration of 0.5. Corresponding to this value of K, determine the closed-loop transfer function in factored form.

7.3. A unity feedback system has an open-loop transfer function

$$G(s) = K(s + 1)/s(s - 1)$$

Sketch the root locus plot with K as a variable parameter and show that the loci of complex roots are part of a circle with (–1, 0) as centre and radius = $\sqrt{2}$.

Is the system stable for all values of K ? If not, determine the range of K for stable system operation. Find also the marginal value of K which causes sustained oscillations and the frequency of these oscillations.

From the root locus plot, determine the value of K such that the resulting system has a settling time of 4 sec. What are the corresponding values of the roots ?

7.4. The signal flow graph of a control system is shown in Fig. P-7.4. Comment upon the stability of this system when the switch S is open.

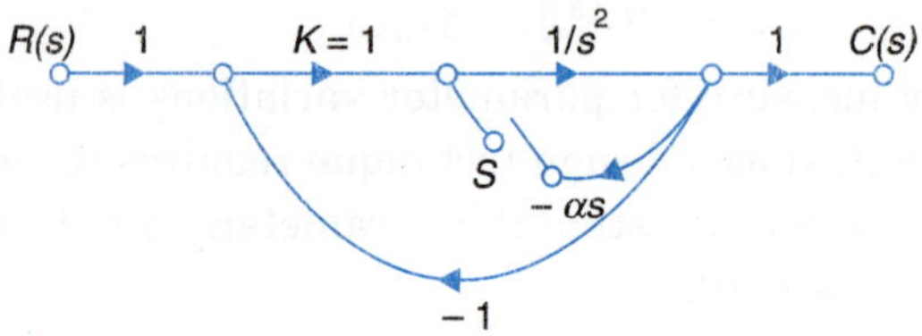

Fig. P-7.4

With the switch S closed, draw the root locus plot of the system with α as a varying parameter. Show that the complex-root branches are part of a circle. From the root locus plot, determine the value of α such that the resulting system has a damping ratio of 0.5. For this value of α, find the overall transfer function in factored form.

7.5. The block diagram of a control system is shown in Fig. P-7.5. Draw the root locus plot of the system with α as varying parameter.

(*a*) Determine the steady-state error to the unit-ramp input, damping ratio and settling time for the system without derivative feedback, *i.e.*,

$$\alpha = 0$$

(*b*) Discuss the effect of derivative feedback on transient as well as steady-state behaviour of the system assuming $\alpha = 0.2$.

(*c*) Determine the value of α for the system to be critically damped.

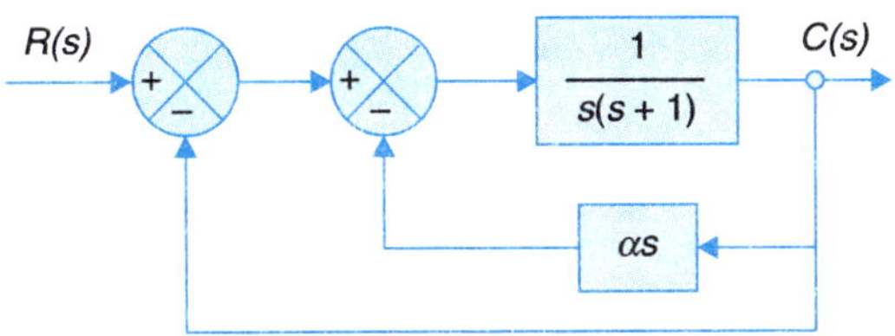

Fig. P-7.5

7.6. A unity feedback system has an open-loop transfer function

$$G(s) = K/s^2(s + 2)$$

(*a*) By sketching a root locus plot, show that the system is unstable for all values of K.

(*b*) Add a zero at $s = -a$ $(0 \le a < 2)$ and show that the addition of zero stabilizes the system.

(*c*) If $a = 1$, sketch the root locus plot and determine approximately the value of K which gives the greatest damping ratio for the oscillatory mode. Find also the value of this damping ratio and the corresponding undamped natural frequency.

7.7. Open-loop transfer function of a unity feedback system is

$$G(s) = K/(s + 2)^3$$

Sketch the root locus plot and determine the following:

(*a*) Static loop sensitivity for which the root locus crosses the $j\omega$-axis and the corresponding frequency of sustained oscillations.

(*b*) The position error constant corresponding to a damping ratio of 0.5. Also determine the peak overshoot, time to peak overshoot and settling time considering the effect of dominant poles only.

(*Note :* The static loop sensitivity is defined to be the gain in pole-zero form.)

7.8. The characteristic equation of a feedback control system is

$$s^4 + 3s^3 + 12s^2 + (K - 16)s + K = 0$$

Sketch the root locus plot for $0 \le K < \infty$ and show that the system is conditionally stable (stable for only a range of gain K). Determine the range of gain for which the system is stable.

7.9. Find the roots of the following polynomial by use of the root locus method.

$$3s^4 + 10s^3 + 21s^2 + 24s + 30 = 0$$

7.10. Figure P-7.10. shows an arrangement of controlling the thickness of steel plates. A signal proportional to the desired steel thickness is the reference input and a signal proportional to the actual thickness is the feedback signal. The error signal actuates the motor which positions the cutter. A definite time T sec elapses before the change in thickness of steel plates reaches the point of measurement. The open-loop transfer function may be assumed to be

$$G(s)H(s) = 100e^{-sT}/s(s + 4)$$

If the steel plate moves at a constant velocity of 1 m/sec, find the maximum permissible distance d before sustained oscillations are produced.

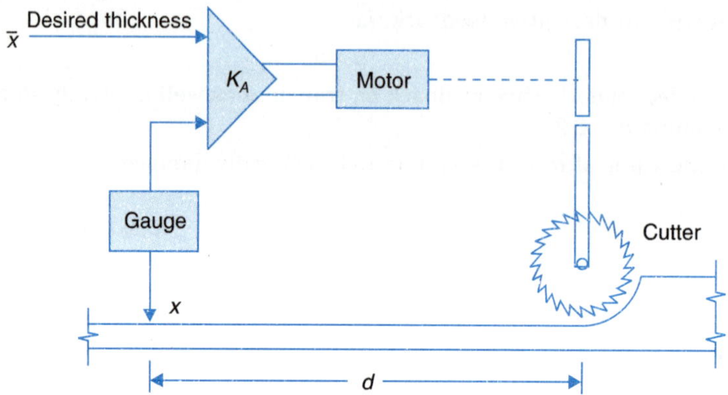

Fig. P-7.10

7.11. The block diagram of Fig. P-7.11 shows a feedback system using a proportional plus integral controller.

(*a*) Draw the root contours of the system for $a = 0, 1, 2$ and 4.

(*b*) Find the range of values of a for which the system is stable.

(*c*) Determine the steady-state error for unit-parabolic input.

(*d*) Determine the damping ratio of the dominant roots for various values of $a(K_A = 5)$ and show that the system is very sensitive to the value of a.

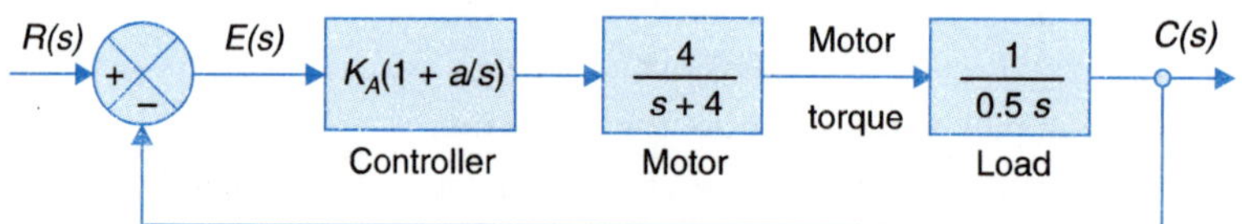

Fig. P-7.11

7.12. A unity feedback system has an open-loop transfer function

$$G(s) = 25/s(s + 4)$$

(*a*) Determine the settling time and damping ratio of the system.

(*b*) The system dynamics is to be improved by cascading a compensating network having the transfer function

$$G_c(s) = (as + 1)/(bs + 1)$$

in the forward path as shown in Fig. P-7.12. By drawing root contours, discuss the effect of varying a and b on system dynamics. Also determine suitable values of a and b, so that the settling time is less than 1 sec and the dominant roots have damping ratio greater than 0.5.

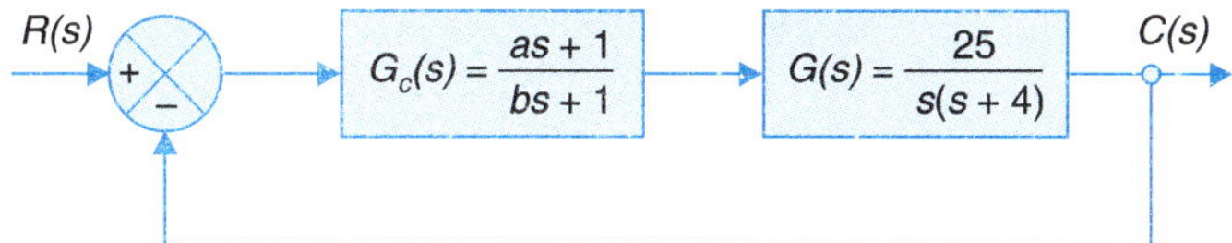

Fig. P-7.12

7.13 Determine the root sensitivity of the complex roots of the system of Problem 7.5 to variations in derivative feedback constant $\alpha = 0.2$.

7.14 Determine and compare the root sensitivity of the dominant roots of the system of Problem 7.6 to variations in (*i*) gain $K = 5$, (*ii*) open-loop pole at $s = -2$, (*iii*) open-loop zero at $s = -1$.

7.15 The root locus of a second order, unity feedback control system consisting of an amplifier of gain K and a load is given in Fig. P-7.15.

For $K = 0$ the roots of the characteristic equation are -1 and -3. At a certain value of K the roots of the characteristic equation are -2 and -2. For $K > 0.5$ the roots of the characteristic equation are complex conjugate. For this system,

(*a*) Determine the open loop transfer function as a function of K.

(*b*) For $K = 6.5$ find the damping ratio and natural frequency of the system. For a unit step input find the maximum peak percentage overshoot, settling time (2% tolerance) the steady-state error.

(*c*) It is now desired to increase the damping coefficient to 0.7, while keeping the steady state-error and natural frequency the same as obtained in part (*b*). Draw the complete modified block diagram insert all numerical values.

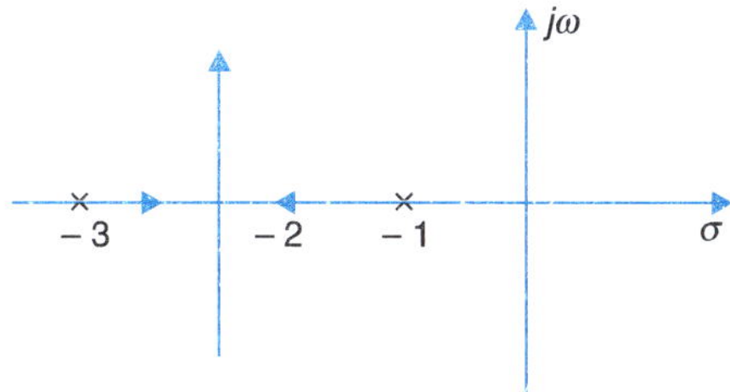

Fig. P-7.15

7.16 Draw the root locus of a unity feedback system with open loop transfer function given by

$$G(s)H(s) = K(s^2 + s + 4)/s\ (s + 4)(s + 10)$$

(*a*) What value of K gives a damping coefficient of 0.5 for the two dominant poles ?

(*b*) What is the approximate value of peak overshoot for this value of K ?

(*c*) Find the value of steady-state error to a unit ramp input for this system. For what value of K is this error minimum ?

(*d*) Find the closed loop transfer function such that the dominant poles have a 2% tolerance band settling time of 4 seconds.

8

FREQUENCY RESPONSE ANALYSIS

8

FREQUENCY RESPONSE ANALYSIS

8.1 INTRODUCTION

Various standard test signals used to study the performance of control systems were discussed in Chapter 5. While the sinusoidal test signal was introduced there, the discussion on the same was postponed till this chapter which is fully devoted to it on account of its importance in control engineering. Consider a linear system with a sinusoidal input

$$r(t) = A \sin \omega t$$

Under steady-state, the system output as well as the signals at all other points in the system are sinusoidal. The steady-state output may be written as

$$c(t) = B \sin (\omega t + \phi)$$

The magnitude and phase relationship between the sinusoidal input and the steady-state ouput of a system is termed the *frequency response.* In linear time-invariant systems, the frequency response is independent of the amplitude and phase of the input signal.

The frequency response test on a system or a component is normally performed by keeping the amplitude A fixed and determining B and ϕ for a suitable range of frequencies. Signal generators and precise measuring instruments are readily available for various ranges of frequencies and amplitudes. The ease and accuracy of measurements are some of the advantages of the frequency response method. Wherever it is not possible to obtain the form of the transfer function of a system through analytical techniques, the necessary information to compute its transfer function can be extracted by performing the frequency response test on the system. The step response test can also be performed easily but the extraction of transfer function from the step response data is quite a laborious procedure. For systems with very large time constants, the frequency response test is cumbersome to perform as the time required for the output to reach steady-state for each frequency of the test signal is excessively long. Therefore, the frequency response test is not recommended for systems with very large time constants.

Further the frequency response test obviously cannot be performed on non-interruptable systems. Under such circumstances a single shot test (step or impulse) is more convenient even though the computation of transfer function from it gets involved.

The design and parameter adjustment of the open-loop transfer function of a system for specified closed-loop performance is carried out somewhat more easily in frequency domain (*i.e.,* through frequency response) than in time domain (*i.e.,* through time response). Further, the effects of noise disturbance and parameter variations are relatively easy to visualize and assess through frequency response. If necessary, the transient response of a system can be obtained from its frequency response through the Fourier integral (discussed in the next section). The correlation is quite tedious to compute except for first and second-order systems. Usually for the sake of the simplicity and ease of analysis, the correlation between frequency and time response of a second-order system is employed as an approximation for higher-order* systems as well. Through the use of this approximate correlation, a satisfactory transient response for higher-order systems can be achieved by the adjustment of their frequency response.

An interesting and revealing comparison of frequency and time domain approaches is based on the relative stability studies of feedback systems. The Routh criterion is a time domain approach which establishes with relative ease the stability of a system, but its adoption to determine the relative stability is involved and requires repeated application of the criterion. The root locus method is a very powerful time domain approach as it reveals not only stability but also the actual time response of the system. On the other hand, the Nyquist criterion (discussed later in Chapter 9) is a powerful frequency domain method of extracting the information regarding stability as well as relative stability of a system without the need to evaluate roots of the characteristic equation.

The frequency response is easily evaluated from the sinusoidal transfer function which can be obtained simply be replacing s with $j\omega$ in the system transfer function $T(s)$. The transfer function $T(j\omega)$ thus obtained, is a complex function of frequency and has both a magnitude and a phase angle. These characteristics are conveniently represented by graphical plots. The various graphical techniques to represent the sinusoidal transfer function $T(j\omega)$ are discussed in detail in this chapter.

8.2 CORRELATION BETWEEN TIME AND FREQUENCY RESPONSE

As discussed earlier, the correlation between time and frequency response has an explicit form only for systems of first- and second-order. Let us first discuss the explicit correlation for second-order systems.

Second-Order Systems

A second-order system of the form shown in Fig. 8.1 has the transfer function

$$\frac{C(s)}{R(s)} = \frac{\omega_n^2}{s^2 + 2\zeta\omega_n s + \omega_n^2}$$

*Such systems should approximately meet the condition for dominance of a pair of complex conjugate closed-loop poles.

where ζ is the damping factor and ω_n is the undamped natural frequency or oscillations.

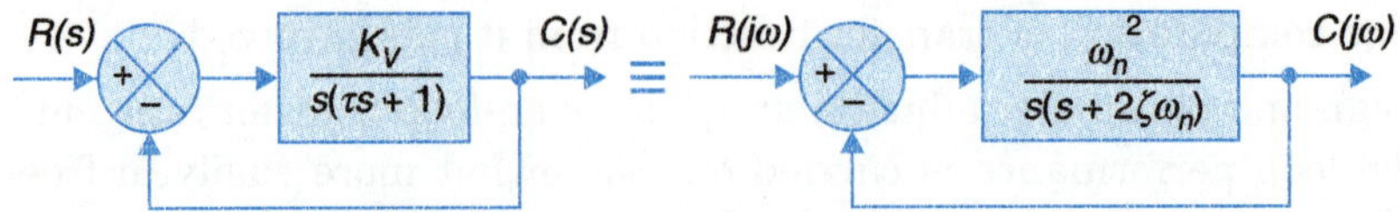

Fig. 8.1. A second-order system.

The sinusoidal transfer function of the system is

$$\frac{C}{R}(j\omega) = T(j\omega) = \frac{\omega_n^2}{(j\omega)^2 + 2\zeta\omega_n(j\omega) + \omega_n^2} = \frac{1}{(1-u^2) + j2\zeta u} \quad \text{...(8.1)}$$

where $u = \omega/\omega_n$ is the normalized driving signal frequency. From eqn. (8.1), we get

$$| T(j\omega) | = M = 1/\sqrt{[(1-u^2)^2 + (2\zeta u)^2} \quad \text{...(8.2)}$$

and

$$\angle T(j\omega) = \phi = -\tan^{-1}[2\zeta u/(1-u^2)] \quad \text{...(8.3)}$$

The steady-state output of the system for a sinusoidal input of unit magnitude and variable frequency ω is given by

$$c(t) = \frac{1}{\sqrt{[(1-u^2)^2 + (2\zeta u)^2]}} \sin\left(\omega t - \tan^{-1}\frac{2\zeta u}{1-u^2}\right)$$

From eqns. (8.2) and (8.3), it is seen that when

$$u = 0, M = 1 \text{ and } \phi = 0$$

$$u = 1, M = \frac{1}{2\zeta} \text{ and } \phi = -\pi/2$$

$$u \to \infty, M \to 0 \text{ and } \phi \to -\pi$$

The magnitude and phase angle characteristic for normalized frequency u for certain values of ζ are shown in Figs. 8.2(*a*) and (*b*) respectively.

The frequency where M has a peak value is known as the *resonant frequency*. At this frequency, the slope of the magnitude curve is zero. Let ω_r be the resonant frequency and $u_r = (\omega_r/\omega_n)$ be the normalized resonant frequency. Then

$$\left.\frac{dM}{du}\right|_{u=u_r} = -\frac{1}{2}\frac{[-4(1-u_r^2)u_r + 8\zeta^2 u_r]}{[(1-u_r^2)^2 + (2\zeta u_r)^2]^{3/2}} = 0$$

which gives

$$4u_r^3 - 4u_r + 8\zeta^2 u_r = 0 \qquad \text{or} \qquad u_r = \sqrt{(1-2\zeta^2)}$$

i.e.,

$$\omega_r = \omega_n\sqrt{(1-2\zeta^2)} \quad \text{...(8.4)}$$

From eqn. (8.2), the maximum value of the magnitude known as the *resonant peak* is given by

$$M_r = \frac{1}{2\zeta\sqrt{(1-\zeta^2)}} \quad \text{...(8.5)}$$

The phase angle ϕ of $T(j\omega)$ at the resonant frequency u_r obtained from eqn. (8.3) is given by

$$\phi_r = -\tan^{-1}[\sqrt{(1-2\zeta^2)}/\zeta]$$

From eqns. (8.4) and (8.5), it is seen that as ζ approaches zero, ω_r approaches ω_n and M_r approaches infinity. For $0 < \zeta \le 1/\sqrt{2}$, the resonant frequency always has a value less than ω_n and the resonant peak has a value greater than 1.

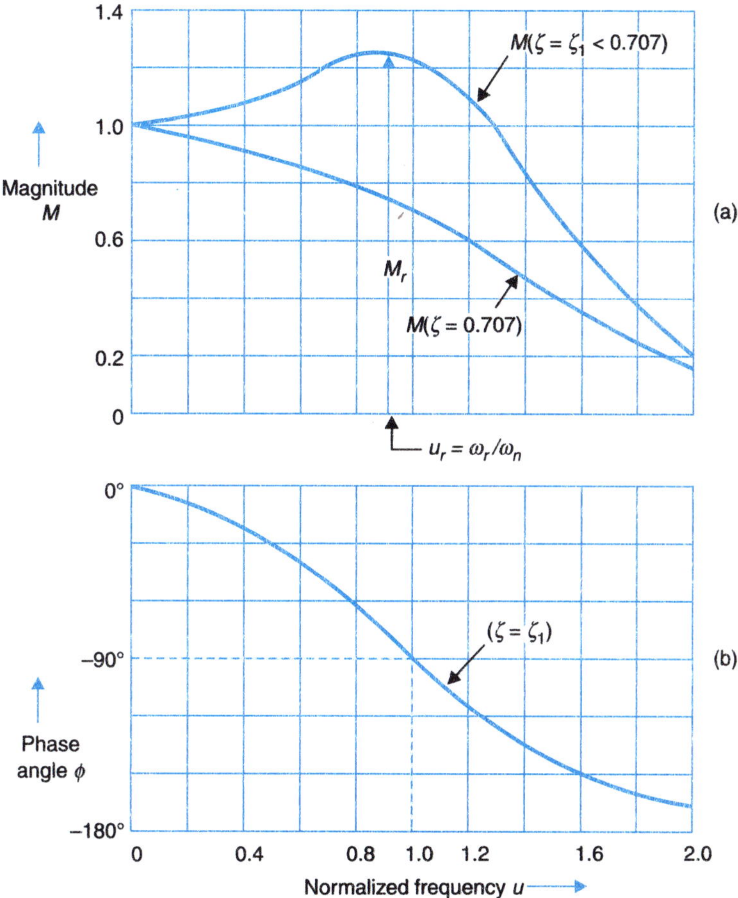

Fig. 8.2. (a) Frequency response magnitude characteristics; (b) Frequency response phase characteristic.

For $\zeta > 1/\sqrt{2}$, it is seen from eqn. (8.4) that dM/du, slope of the magnitude curve does not become zero for any real value of ω. For this range of ζ, the magnitude M decreases monotonically from $M = 1$ at $u = 0$ with increasing u, as shown in Fig. 8.2. It therefore follows that for $\zeta > 1/\sqrt{2}$, there is no resonant peak as such as the greatest value of M equals one.

We then find from eqns. (8.4) and (8.5) and the above discussion that for a second-order system, the resonant peak M_r of its frequency response is indicative of its damping factor ζ for

$0 < \zeta \le 1/\sqrt{2}$, and the resonant frequency ω_r of the frequency response is indicative of its natural frequency for a given ζ and hence indicative of its speed of response (as $t_s = 4/\zeta\omega_n$). M_r and ω_r of the frequency response could thus be used as performance indices for a second-order system.

Re-examining Fig. 8.2, we notice that for $\omega > \omega_r$, M decreases monotonically. The frequency at which M has a value* of $1/\sqrt{2}$ is of special significance and is called the *cut-off frequency* ω_c. The signal frequencies above cut-off are greatly attenuated in passing through a system.

For feedback control systems, the range of frequencies over which M is equal to or greater than $1/\sqrt{2}$ is defined as *bandwidth* ω_b. Control systems being lowpass filters (at zero frequency, $M = 1$), the bandwidth ω_b is equal to cut-off frequency ω_c. The definition of the bandwidth is depicted on typical frequency response of a feedback control system in Fig. 8.3*a*.

In general, the bandwidth of a control system indicates the noise-filtering characteristic of the system. Also, bandwidth gives a measure of the transient response properties as observed below.

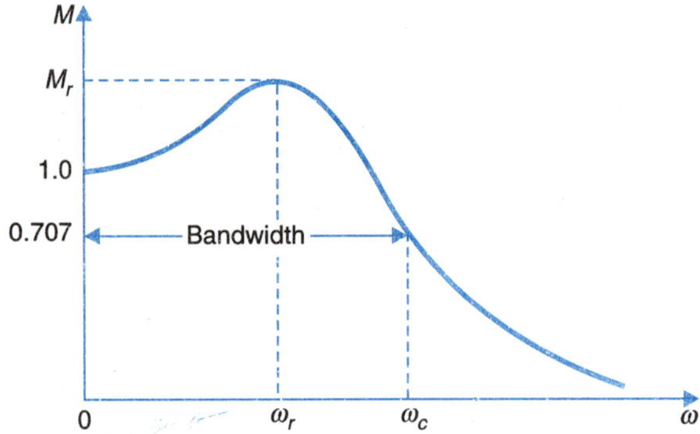

Fig. 8.3. (a) Typical magnification curve of a feedback control system.

The normalized bandwidth $u_b = \omega_b/\omega_n$ of the second-order system under consideration can be readily determined as follows:

$$M = \frac{1}{\sqrt{[(1-u_b^2)^2 + (2\zeta u_b)^2]}} = \frac{1}{\sqrt{2}}$$

or

$$u_b^4 - 2(1 - 2\zeta^2)u_b^4 - 1 = 0$$

Solving for u_b we get

$$u_b = [1 - 2\zeta^2 + \sqrt{(2 - 4\zeta^2 + 4\zeta^4)}\,]^{1/2} \qquad ...(8.6(a))$$

It can be approximated in linear form

$$u_b = -\,1.19\zeta + 1.85 \qquad ...(8.6(b))$$

As the bandwidth must be a positive real quantity, the negative sign in quadratic solution and the negative sign in taking the square-root have been discarded.

We observe from eqn. (8.6) that the normalized bandwidth is a function of damping only; u_b versus ζ is plotted in Fig. 8.3(*b*).

* This value corresponds to – 3db point on the Bode plot of $T(j\omega)$ of the second-order system under consideration. Bode plots are dealt with in later part of this chapter.

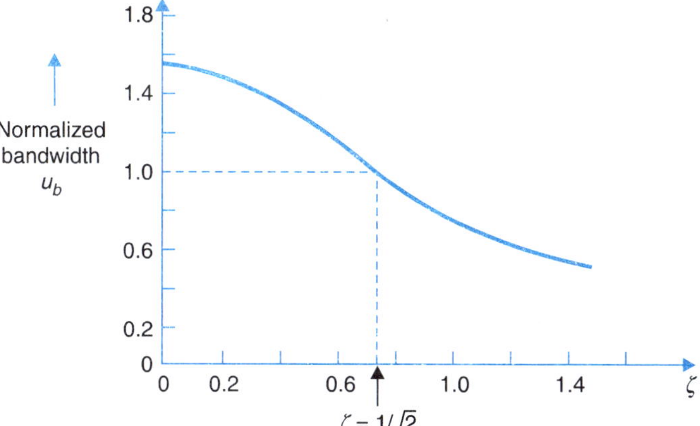

Fig. 8.3. (b) Bandwidth versus damping factor.

The denormalized bandwidth from eqn. (8.6) is given by

$$\omega_b = \omega_n[1 - 2\zeta^2 + \sqrt{(2 - 4\zeta^2 + 4\zeta^4)}]^{1/2} \quad ...(8.7)$$

Let us now consider the step response of a second-order system. The nature of this response has been discussed in detail in Chapter 5. The expressions for the damped frequency of oscillation ω_d and peak overshoot M_p of the step response for $0 \le \zeta \le 1$ are

$$\omega_d = \omega_n \sqrt{(1-\zeta^2)} \quad ...(8.8)$$

$$M_p = \exp\,[-\pi\zeta/\sqrt{(1-\zeta^2)}\,] \quad ...(8.9)$$

M_r, the resonant peak of the frequency response, and M_p, the peak overshoot of the step response for a second-order system, are plotted in Fig. 8.4 for various value of ζ.

The comparison of M_r and M_p plots shows that for $0 < \zeta < 1/\sqrt{2}$, the two performance indices are correlated as both are functions of the system damping factor ζ only. It means that a system with a given value of M_r of its frequency response, must exhibit a corresponding value of M_p if subjected to a step input. For $\zeta > 1/\sqrt{2}$, the resonant peak M_r does not exist and the correlation breaks down. This is not a serious problem as for this range of ζ, the step response oscillations are well-damped and M_p is hardly perceptible.

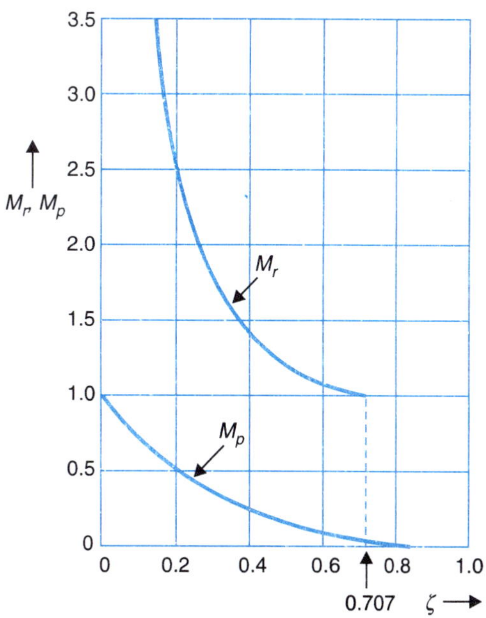

Fig. 8.4. M_r, M_p versus ζ.

Similarly, the comparison of eqn. (8.4) and (8.8) reveals that there exists definite correlation between the resonant frequency ω_r of frequency response and damped frequency of oscillations of the step response. The ratio of these two frequencies is

$$\omega_r/\omega_d = \sqrt{[(1-2\zeta^2)/(1-\zeta^2)]}$$

which is also a function of ζ and is plotted in Fig. 8.5.

It is further observed from eqn. (8.7) that the bandwidth, a frequency domain concept, is indicative of the undamped natural frequency of a system for a given ζ and therefore indicative of the speed of response ($t_s = 4/\zeta\omega_n$), a time domain concept.

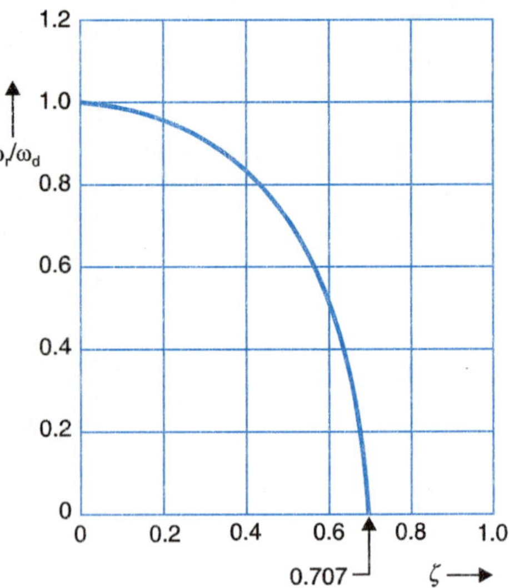

Fig. 8.5. Correlation between ω_r and ω_d.

Higher-order Systems

In general, the transient response and frequency response of linear systems are related through the Fourier integral. This relationship forms an important basis of most design procedures and criteria. Usually the desired time domain behaviour is interpreted in terms of frequency response characteristics. The design is carried out in the frequency domain and the frequency response is then translated back into the time domain. The behaviour in the frequency domain of a given driving function $r(t)$ is given by the Fourier integral as (see Appendix I)

$$R(\omega) = \int_{-\infty}^{\infty} r(t)e^{-j\omega t} dt$$

Similarly, the frequency response $C(j\omega)$ may be translated back into the time domain by the inverse Fourier integral as

$$c(t) = \frac{1}{2\pi}\int_{-\infty}^{\infty} C(j\omega)e^{j\omega t} d\omega$$

However, the correlation between transient and frequency response through the Fourier integral is highly laborious to compute. It is therefore tempting to consider the possibility of using the second-order correlations for higher-order systems as well. Whenever, a higher-order system has a transfer function which is dominated by a pair of complex conjugate poles, it can be approximated by a second-order system whose poles are the dominant poles of the higher-order system. When such an approximation is possible, all the time and frequency domain correlations of a second-order system become valid for use in higher order systems. The design based on these correlations proceeds much faster though it is not exact. Sound engineering skill and judgement are necessary to estimate the accuracy of such a procedure. It is in fact essential to check the exact response after the design is completed.

8.3 POLAR PLOTS

The sinusoidal transfer function $G(j\omega)$ is a complex function and is given by

$$G(j\omega) = Re[G(j\omega)] + jIm[G(j\omega)]$$

or

$$G(j\omega) = |G(j\omega)| \angle G(j\omega) = M \angle \phi \quad ...(8.10)$$

From eqn. (8.10) it is seen that $G(j\omega)$ may be represented as a phasor of magnitude M and phase angle ϕ (measured positively in counter clockwise direction). As the input frequency ω is varied from 0 to ∞, the magnitude M and phase angle ϕ change and hence the tip of the phasor $G(j\omega)$ traces a locus in the complex plane. The locus thus obtained is known as *polar plot.**

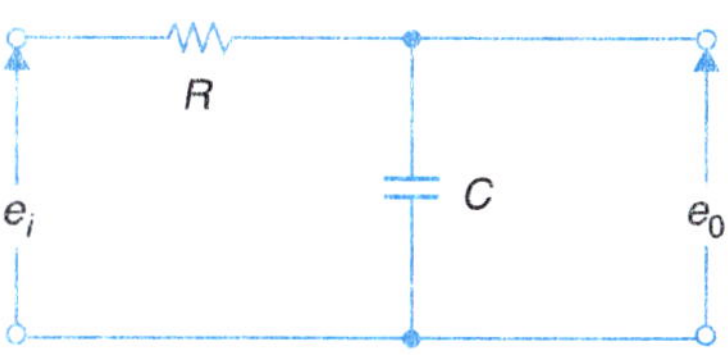

Fig. 8.6. RC filter circuit.

As an example, consider a simple RC filter shown in Fig. 8.6

$$\frac{E_o}{E_i}(s) = G(s) = \frac{1/Cs}{R + 1/Cs} = \frac{1}{1 + Ts}$$

where $T = RC$.

Therefore the sinusoidal transfer function is

$$G(\omega) = \frac{1}{1 + j\omega T} \quad \text{...(8.11)}$$

$$= \frac{1}{\sqrt{(1 + \omega^2 T^2)}} \angle -\tan^{-1}\omega T = M \angle \phi \quad \text{...(8.12)}$$

The polar plot of $G(\omega)$ as in eqn. (8.12) is drawn in Fig. 8.7. When $\omega = 0$, $M = 1$ and $\phi = 0$. Therefore the phasor at $\omega = 0$ has unit length and lies along the positive real axis. As ω increases, M decreases and phase angle increases negatively. When $\omega = 1/T$, $M = 1/\sqrt{2}$ and $\phi = -45°$. As $\omega \to \infty$, M becomes zero and ϕ is $-90°$. This is represented by a phasor of zero length directed along the $-90°$ axis in the complex plane. In fact the locus of $G(\omega)$ (polar plot) can be shown to be a semicircle.

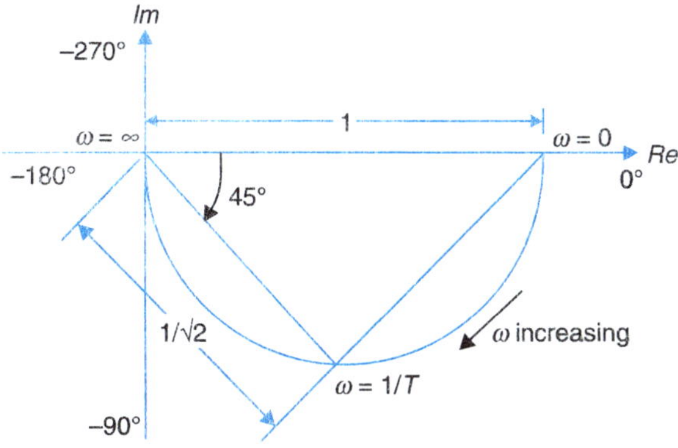

Fig. 8.7. Polar plot of $1/(1 + j\omega T)$.

Consider now the transfer function

$$G(j\omega) = \frac{1}{j\omega(1 + j\omega T)} \quad \text{...(8.13)}$$

---

*Other ways of graphically representing $G(j\omega)$ are discussed in later sections of this chapter.

This transfer function may be rearranged as

$$G(j\omega) = \frac{-T}{1+\omega^2T^2} - j\frac{1}{\omega(1+\omega^2T^2)} \qquad \text{...(8.14)}$$

From eqn. (8.14) we get

$$\lim_{\omega \to 0} G(j\omega) = -T - j\infty = \infty\angle -90°$$

$$\lim_{\omega \to \infty} G(j\omega) = -0 - j0 = 0\angle -180°$$

The general shape of the polar plot of this transfer function is shown in Fig. 8.8. The plot is asymptotic to the vertical line passing through the point (–T, 0).

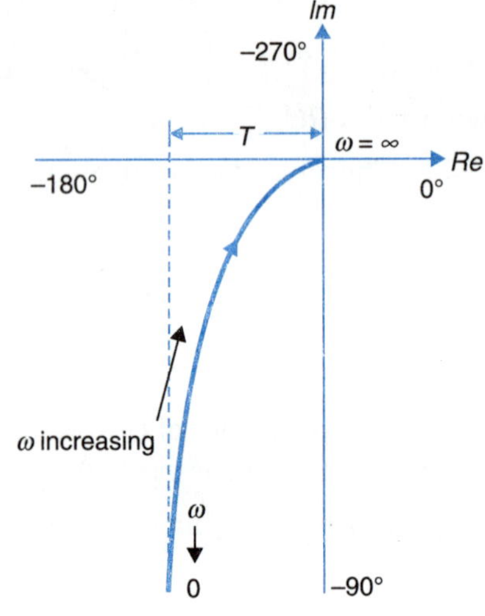

Fig. 8.8. Polar plot of $1/j\omega(1 + j\omega T)$

The major advantage of the polar plot lies in stability study of systems. N. Nyquist (in 1932) related the stability of a system to the form of these plots. Because of his work, the polar plots are commonly referred to as *Nyquist plots*.

The general shapes of the polar plots of some important transfer functions are given in Table 8.1.

From the polar plots of Table 8.1, following observations are made:

(*i*) Addition of a nonzero pole to a transfer function results in further rotation of the polar plot through an angle of –90° as $\omega \to \infty$.

(*ii*) Addition of a pole at the origin to a transfer function rotates the polar plot at zero and infinite frequencies by a further angle of –90°.

The effect of addition of a zero to a transfer function is to rotate the high frequency portion of the polar plot by 90° in counter-clockwise direction.

Inverse Polar Plots

The inverse polar plot of $G(j\omega)$ is a graph of $1/G(j\omega)$ as function of ω. For example, for the RC filter shown in Fig. 8.6.

$$\frac{1}{G(j\omega)} = G^{-1}(j\omega)$$

$$= 1 + j\omega T = \sqrt{(1+\omega^2T^2)}\ \angle \tan^{-1} \omega T$$

The corresponding inverse polar plot is shown in Fig. 8.9.

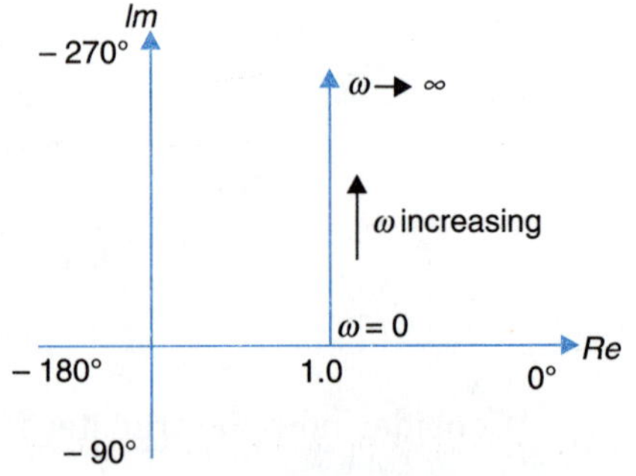

Fig. 8.9. Inverse polar plot of $1/(1 + j\omega T)$.

It will be seen later in Chapter 9 that the inverse polar plots are useful in applying M-criterion and are also valuable in stability study of nonunity feedback systems.

Table 8.1

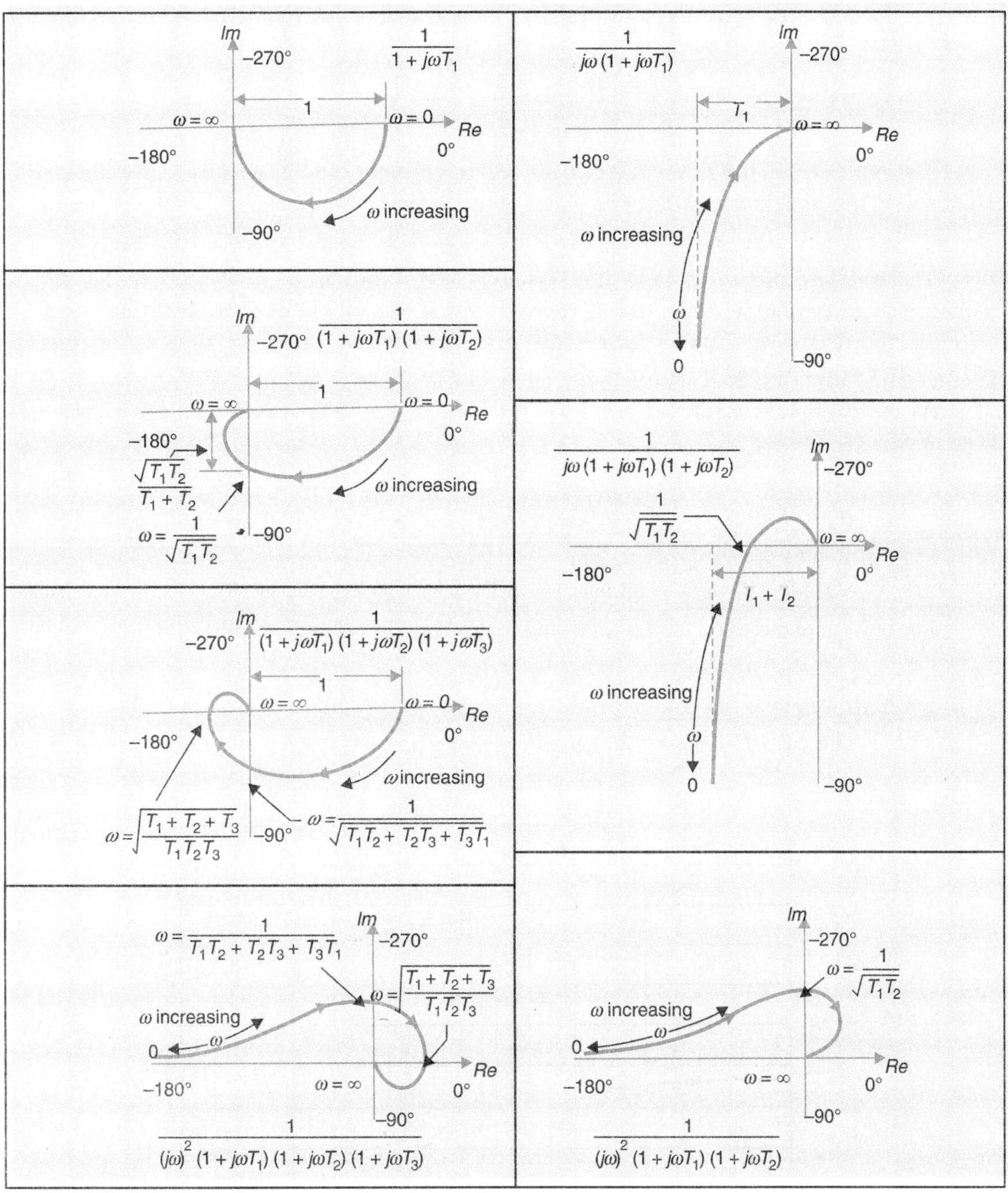

8.4 BODE PLOTS

One of the most useful representations of a transfer function is a logarithmic plot which consists of two graphs, one giving the logarithm of $| G(j\omega) |$ and the other phase angle of $G(j\omega)$ both plotted against frequency in logarithmic scale. These plots are called *Bode plots* in honour of H.W. Bode who did the basic work in this area. The transfer function $G(j\omega)$ is represented by

$$| G(j\omega) | e^{j\phi}$$

Taking natural logarithm of both sides

$$\ln G(\omega) = \ln | G(j\omega) | + j\phi(\omega) \quad ...(8.15)$$

The real part is the natural logarithm of magnitude and is measured in a basic unit called *neper*; the imaginary part is the phase characteristic. Similarly,

$$\log G(j\omega) = \log | G(j\omega) | + \log e^j\phi(\omega) = \log | G(j\omega) | + 0.434 j\phi(\omega)$$

The standard procedure is to plot 20 log $| G(j\omega) |$ and phase angle $\phi(\omega)$ vs. log ω, *i.e.*, frequency on a logarithmic scale. In this representation, the unit of magnitude 20 log $| G(j\omega) |$ is *decibel,* abbreviated as db. The curves are normally drawn on a semilog paper using log scale for frequency and linear scale for magnitude in db and phase angle in degrees.

As an example, consider the *RC* filter shown in Fig. 8.6, whose transfer function as given in eqn. (8.12) is

$$G(j\omega) = \frac{1}{(1+\omega^2 T^2)^{1/2}} \angle -\tan^{-1} \omega T$$

The log-magnitude is

$$20 \log | G(j\omega) | = 20 \log (1 + \omega^2 T^2)^{-1/2} = -10 \log (1 + \omega^2 T^2) \quad ...(8.16)$$

For low frequencies ($\omega << 1/T$), the log-magnitude is approximated as

$$20 \log | G(j\omega) | = -10 \log 1 = 0 \text{ db} \quad ...(8.17)$$

For high frequencies ($\omega >> 1/T$), the log-magnitude is approximated as

$$20 \log | G(j\omega) | = -20 \log \omega T \quad ...(8.18)$$

$$= -20 \log \omega - 20 \log T \quad ...(8.19)$$

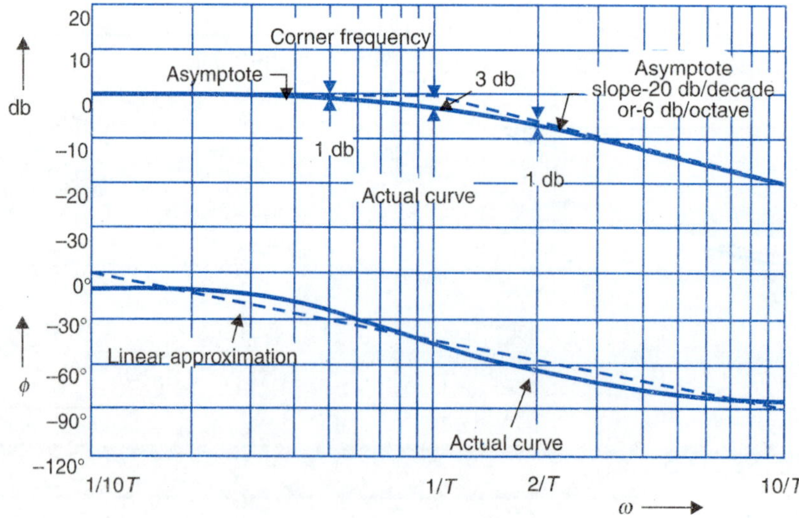

Fig. 8.10. Bode plot of $(1 + j\omega T)^{-1}$.

The logarithmic plot 20 log | $G(j\omega)$ | versus log ω of eqn. (8.17) is a straight line coincident with the horizontal axis. The plot of eqn. (8.19) is also a straight line with a slope –20 db per unit change in log ω. A unit change in log ω means

$$\log (\omega_2/\omega_1) = 1 \qquad \text{or} \qquad \omega_2 = 10\omega_1$$

This range of frequencies is called a *decade.* Thus the slope of eqn. (8.19) is –20 db/decade. The range of frequencies $\omega_2 = 2\omega_1$ is called an *octave*. Since –20 log 2 = –6 db, the slope of eqn. (8.19) could also be expressed as –6 db/octave. Further at $\omega = 1/T$ the plot of eqn. (8.19) has a value of 20 log 1 = 0 db as shown in Fig. 8.10.

Though the straight line approximations of eqn. (8.17) and (8.19) hold good for $\omega << 1/T$ and $\omega >> 1/T$ respectively, with some loss of accuracy these could be extended for frequencies $\omega \le 1/T$ and $\omega \ge 1/T$. Therefore the log-magnitude versus log-frequency curve of $1/(1 + j\omega T)$ can be approximated by two straight line asymptotes, one a straight line at 0 db for the frequency range $0 < \omega \le 1/T$ and the other, a straight line with a slope –20 db/decade (or – 6db/octave) for the frequency range $1/T \le \omega < \infty$. The frequency $\omega = 1/T$ at which the two asymptotes meet is called the *corner frequency* or the *break frequency*. The corner frequency divides the plot in two regions, a low frequency region and a high frequency region.

It is important to note that the log-magnitude plot of $(1 + j\omega T)^{-1}$ shown in Fig. 8.10 is an asymptotic approximation of the actual plot. The actual plot can be obtained from it by applying correction for the errors introduced by asymptotic approximation.

From eqns. (8.16) and (8.17), the error in log-magnitude for $0 < \omega \le 1/T$ is given by

$$-\,10 \log (1 + \omega^2T^2) + 10 \log 1$$

Therefore, the error at the corner frequency $\omega = 1/T$ is

$$-\,10 \log (1 + 1) + 10 \log 1 = -\,3 \text{ db}$$

The error at frequency ($\omega = 1/2\,T$) one octave below the corner frequency is

$$-\,10 \log (1 + 1/4) + 10 \log 1 = -\,1 \text{ db}$$

Similarly, the errors at other frequencies ($\omega < 1/T$) may also be evaluated from this expression.

For $1/T \le \omega < \infty$, the error in log-magnitude given by eqns. (8.16) and (8.18) is

$$-\,10 \log (1 + \omega^2T^2) + 20 \log \omega T$$

From this expression, the error at the corner frequency $\omega = 1/T$ is found to be

$$-\,10 \log\ (1 + 1) + 20 \log 1 = -3 \text{ db}$$

The error at frequency $\omega = 2/T$, *i.e.*, one octave above the corner frequency is

$$-\,10 \log (1 + 4) + 20 \log 2 = 1 \text{ db}$$

Similarly, the errors at other frequencies ($\omega > 1/T$) may be evaluated from this expression. The error caused by the asymptotic approximation of the Bode plot of $(1 + j\omega T)^{-1}$ is plotted in the error-graph of Fig. 8.11 for various frequencies expressed in terms of the corner frequency.

In practice, a sufficiently accurate log-magnitude plot is obtained by correcting the asymptotic plot by –3 db at the corner frequency and by –1 db one octave below and one above the corner frequency, and then drawing a smooth curve through these three points approaching the low and high frequency asymptotes as shown in Fig. 8.10.

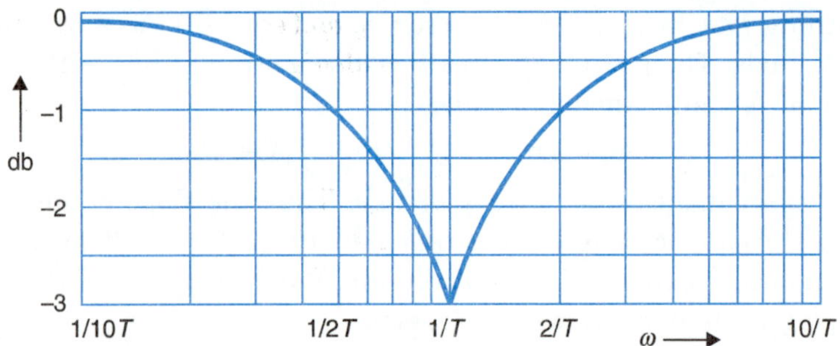

Fig. 8.11. Error in log-magnitude versus frequency of $(1 + j\omega T)^{-1}$.

From eqn. (8.12), the phase angle ϕ of the factor $1/(1 + j\omega T)$ is

$$\phi = \tan^{-1} \omega T$$

At the corner frequency, the phase angle of this factor is

$$\phi = \tan^{-1} (T/T) = -45°$$

At zero frequency, $\phi = 0$ and at infinity, it becomes –90°. Since the phase angle is given by inverse tangent function, the phase characteristic is skew symmetric about the inflection point $\phi = -45°$ as shown in Fig. 8.10.

Phase versus log-frequency plot can also be approximated by a straight line passing through –45° at the corner frequency ($\omega = 1/T$), 0° at $\omega = 1/10T$ and –90° at $\omega = 10/T$ as shown by the dotted line in Fig. 8.10. Such an approximation has a maximum error of about 6°. Using linear approximation, the phase plot of $G(j\omega)$ can be readily obtained. However, in most analysis and design problems a more accurate phase plot is needed and therefore the linear approximation does not find much favour for use.

It is observed that compared to polar plots, Bode plots can be more quickly constructed. In these plots, both low and high frequency regions are brought into focus simultaneously. The main advantage of the Bode plots is conversion of multiplicative factors into additive factors.

Consider a typical transfer function $G(j\omega)$ factored in the time-constant form*

$$G(j\omega) = \frac{K(1+j\omega T_a)(1+j\omega T_b)\ldots}{(j\omega)^r(1+j\omega T_1)(1+j\omega T_2)\ldots\left[1+j2\zeta\left(\frac{\omega}{\omega_n}\right)+\left(j\frac{\omega}{\omega_n}\right)^2\right]\ldots} \quad \ldots(8.20)$$

The transfer function $G(j\omega)$ has real zeros at $-1/T_a$, $-1/T_b$, ..., a pole at the origin of multiplicity r, real poles at $-1/T_1$, $1/T_2$, ... and complex poles at $-\zeta\omega_n \pm j\omega_n \sqrt{(1-\zeta^2)}$, ...

If the transfer function has complex zeros, quadratic terms of the form given in denominator of eqn. (8.20) will appear in the numerator as well.

Note that the constant multiplier K is given by

$$K = \lim_{\omega \to 0} (j\omega)^r G(j\omega)$$

---

*If the transfer function is given in pole-zero form as in eqn. (5.26), it should be rearranged in the time-constant form for construction of its Bode plot.

where r is the number of poles of $G(j\omega)$ at the origin, *i.e.*, r is the system type number. For type-0, type-1 and type-2 systems, $K = K_p$, K_v and K_a respectively.

From eqn. (8.20) the log-magnitude is given by

$$20 \log |G(j\omega)| = 20 \log K + 20 \log |1 + j\omega T_a| + 20 \log |1 + j\omega T_b| + ...$$
$$- 20r \log (\omega) - 20 \log |1 + j\omega T_1| - 20 \log |1 + j\omega T_2| ...$$
$$- 20 \log |1 + j2\zeta(\omega/\omega_n) - (\omega/\omega_n)^2| ... \quad ...(8.21)$$

and the phase angle is given by

$$\angle G(j\omega) = \tan^{-1} \omega T_a + \tan^{-1} \omega T_b + ... -r(90°) - \tan^{-1} \omega T_1$$
$$- \tan^{-1} \omega T_2 - ... - \tan^{-1}\left(\frac{2\zeta\omega\omega_n}{\omega_n^2 - \omega^2}\right) ... \quad ...(8.22)$$

From eqns. (8.21) and (8.22) it is seen that the Bode plots of $G(j\omega)$ may be obtained by adding the Bode plots of the factor of $G(j\omega)$, these factors are found to be of the forms given below:

1. Constant gain K
2. Poles at the origin $1/(j\omega)^r$
3. Pole on real axis $1/(1 + j\omega T)$
4. Zero on real axis $(1 + j\omega T)$
5. Complex conjugate poles $1/[1 + j2\zeta(\omega/\omega_n) - (\omega/\omega_n)^2]$
6. Complex conjugate zeros may also be present.

Let us now consider the plotting of each of these factors and the construction of the Bode plot for a given $G(j\omega)$ as sum of the plots of its individual factors.

Factors of the Form $K/(j\omega)^r$

The log-magnitude of this factors is

$$20 \log \left| \frac{K}{(j\omega)^r} \right| = - 20r \log \omega + 20 \log K \quad ...(8.23)$$

and the phase is

$$\phi(\omega) = - 90° \, r$$

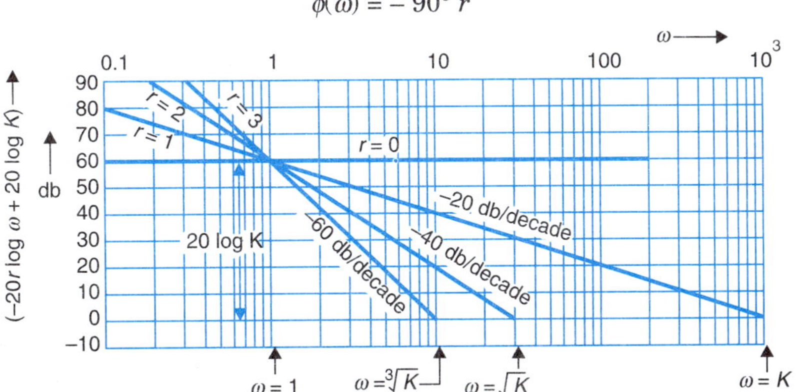

Fig. 8.12. Log-magnitude plot of $K/(j\omega)^r$.

With log ω as abscissa, the plot of eqn. (8.23) is a straight line having a slope of $-20r$ db/decade and passing through 20 log K db when log $\omega = 0$, *i.e.*, $\omega = 1$, as shown in Fig. 8.12. Further, the plot has a value of 0 db at the frequency of

$$20r \log \omega = 20 \log K \qquad \text{or} \qquad \omega = (K)^{1/r}$$

The angle contribution of the factor K is zero and that of $1/(j\omega)^r$ is $-90°r$ at any frequency. The Bode magnitude plot for $r = 0, 1, 2$ and 3 are shown in Fig. 8.12. The phase diagram is simply a horizontal line at an angle of $-90°\ r$.

It also follows from eqn. (8.23) that factor $K/(j\omega)^r$ become 0 db at

$$\omega = \sqrt[r]{K}$$

It means that for (see Fig. 8.12)

$r = 0$; initial 0 db/decade gain = 20 log K, type 0 system

$r = 1$, $\omega = K$, type -1 system

$r = 2$, $\omega = \sqrt{K}$; type-2 system

This result is useful in drawing db Bode plates and in interpretting it.

Pole or Zero on the Real Axis

The pole factor $1/(1 + j\omega T)$ has already been considered in this section. The Bode plot of the zero factor $(1 + j\omega T)$ can be drawn in the same manner as the pole factor but with a slope of +20 db/decade and a phase angle of + $\tan^{-1} \omega T$. The db-corrections can be read from Fig. 8.11 and are to be added to the asymptotic plot. The plot of a zero factor is shown in Fig. 8.13.

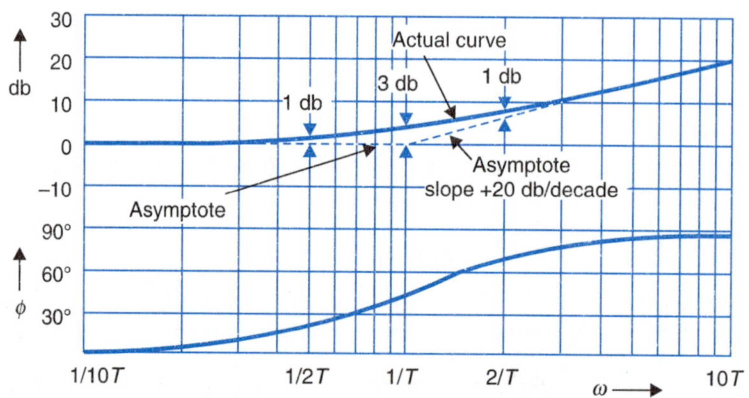

Fig. 8.13. Bode plot of $(1 + j\omega)$.

Complex Conjugate Poles

In normalized form, the quadratic factor for a pair of complex conjugate poles may be written as

$$\frac{1}{(1 + j2\zeta u - u^2)}$$

where $u = \omega/\omega_n$ is the normalized frequency. The log-magnitude of this factor is

$$20\log\left|\frac{1}{1+j2\zeta u-u^2}\right| = -20\log[(1-u^2)^2+(2\zeta u)^2]^{1/2}$$

$$= -10\log[(1-u^2)^2+4\zeta^2u^2]$$

For $u << 1$, the log-magnitude is given by

$$20\log\left|\frac{1}{1+j2\zeta u-u^2}\right| \approx -10\log 1 = 0$$

and for $u >> 1$, the log-magnitude is

$$20\log\left|\frac{1}{1+j2\zeta u-u^2}\right| \approx -20\log u^2 = -40\log u$$

Therefore, the log-magnitude curve of the quadratic factor under consideration, consists of two straight line asymptotes, one horizontal line at 0 db for $u << 1$ and the other, a line with a slope – 40 db/decade for $u >> 1$. The two asymptotes meet on 0-db line at $u = 1$, *i.e.*, $\omega = \omega_n$ which is the corner frequency of the plot. The asymptotic plot and actual plots are shown in Fig. 8.14.

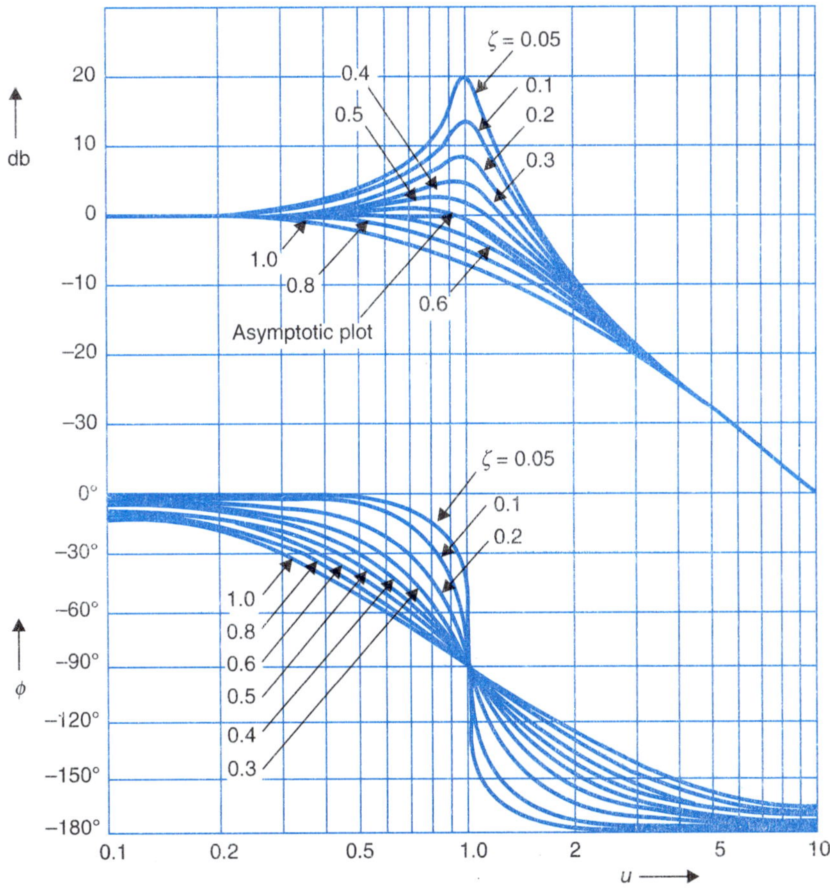

Fig. 8.14. Bode plot of $1/(1 + j2\zeta u - u^2)$.

The error between the actual magnitude and the asymptotic approximation is as given below

For $0 < u << 1$, the error is

$$-10 \log [(1-u^2)^2 + 4\zeta^2u^2] + 10 \log 1$$

and for $1 < u << \infty$, the error is

$$-10 \log [(1-u^2)^2 + 4\zeta^2u^2] + 40 \log u$$

From the above expressions, it is seen that the error is a function of ζ and u. The error versus u curves for different values of ζ are plotted in Fig. 8.15. The corrected log-magnitude curves for various values of ζ are shown in Fig. 8.14.

The phase angle of the quadratic factor $1/(1 + j2\zeta u - u^2)$ is given by

$$\phi = -\tan^{-1}\left(\frac{2\zeta u}{1-u^2}\right) \quad ...(8.24)$$

We find that the phase angle is a function of both u and ζ. The phase angle plots for various values of ζ are given in Fig. 8.14. All these plots have a phase angle of 0° at $u = 0$, – 90° at $u = 1$ and –180° at $u = \infty$. The curves become sharper in going from low frequency range to the high frequency range as ζ decreases, until for $\zeta = 0$ the curve jumps discontinuously from 0° down to –180° at $u = 1$.

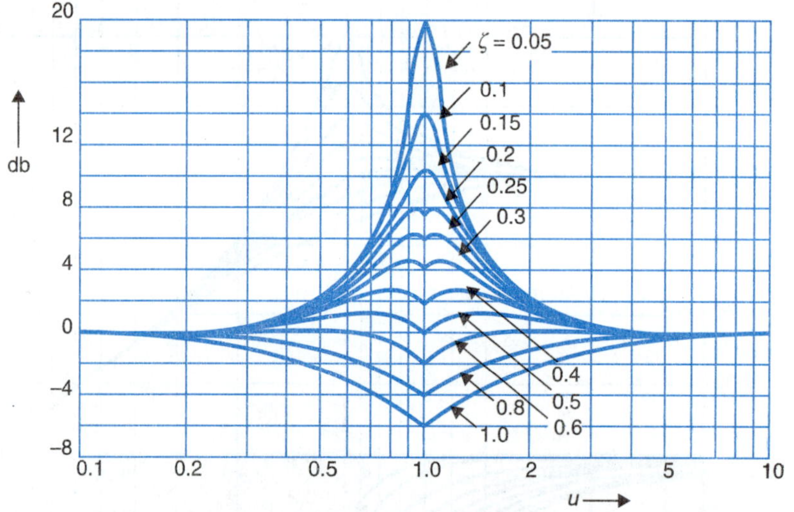

Fig. 8.15. Error in db vs. frequency for asymptotic Bode plot of $1/(1 + j2\zeta u - u^2)$.

In the above discussion, the plots of factors $(j\omega)^r$, *i.e.*, zeros at the origin and $[1 + j2\zeta(\omega/\omega_n) - (\omega/\omega_n)^2]$, *i.e.*, complex zeros, are not considered. The plots of these factors are similar to the plots of poles at the origin $1/(j\omega)^r$ and complex poles $1/[1 + j2\zeta(\omega/\omega_n) - (\omega/\omega_n)^2]$, respectively, but with opposite signs.

General Procedure for Constructing Bode Plots

The following steps are generally involved in constructing the Bode plot for a given $G(j\omega)$.

1. Rewrite the sinusoidal transfer function in the time constant form as given in eqn. (8.20).

2. Identify the corner frequencies associated with each factor of the transfer function.

3. Knowing the corner frequencies, draw the asymptotic magnitude plot. This plot consists of straight line segments with line slope changing at each corner frequency by + 20 db/decade for a zero and – 20 db/decade for a pole (± 20 m db/decade for a zero or pole multiplicity m). For a complex conjugate zero or pole the slope changes by ±40 db/decade (±40 m db/decade for complex conjugate zero or pole of multiplicity m).

4. From the error graphs of Figs. 8.11 and 8.15, determine the corrections to be applied to the asymptotic plot.

5. Draw a smooth curve through the corrected points such that it is asymptotic to the line segments. This gives the actual log-magnitude plot.

6. Draw phase angle curve for each factor and add them algebraically to get the phase plot.

To illustrate the technique, let us draw the Bode plot for the transfer function

$$G(s) = \frac{64(s+2)}{s(s+0.5)(s^2+3.2s+64)}$$

The rearrangement of the transfer function in the time-constant form gives

$$G(s) = \frac{4(1+s/2)}{s(1+2s)(1+0.05s+s^2/64)}$$

Therefore, the sinusoidal transfer function in the time-constant form is given by

$$G(j\omega) = \frac{4(1+j\omega/2)}{j\omega(1+2j\omega)[1+j0.4(\omega/8)-(\omega/8)^2]} \quad ...(8.25)$$

The factors of this transfer function in order of their occurrence as frequency increases, are

1. Constant gain, $K = 4$
2. Pole at origin, $1/j\omega$
3. Pole at $s = -0.5$; corner frequency $\omega_1 = 0.5$
4. Zero at $s = -2$; corner frequency $\omega_2 = 2$
5. Pair of complex conjugate poles with $\zeta = 0.2$, $\omega_n = 8$; corner frequency $\omega_3 = 8$.

The pertinent characteristics of each factor of the transfer function are given in Table 8.2.

Table 8.2 Asymptotic Approximation Table for Construction of Bode Plot of $\dfrac{4(1+j\omega/2)}{j\omega(1+j2\omega)[1+j0.4(\omega/8)-(\omega/8)^2]}$

Factor	*Corner frequency*	*Asymptotic log-magnitude characteristic*	*Phase angle characteristic*
$4/j\omega$	None	Straight line of constant slope –20 db/ decade, passing through 20 log 4 = 12 db point at $\omega = 1$	Constant –90°
$1/(1+j2\omega)$	$\omega_1 = 0.5$	Straight line of the 0 db for $\omega < \omega_1$, straight line of slope –20 db/decade for $\omega > \omega_1$	Phase angle varies, from 0 to – 90°, angle at ω_1 = – 45°

(Contd.)...

$1 + j0.5\omega$	$\omega_2 = 2$	Straight line of 0 db for $\omega < \omega_2$ straight line of slope + 20 db/decade for $\omega > \omega_3$	Phase angle varies from 0 to 90°, angle at $\omega_3 = 45°$
$\left[1 + j0.4\left(\frac{\omega}{8}\right) - \left(\frac{\omega}{8}\right)^2\right]$	$\omega_3 = 8$ ($\zeta = 0.2$)	Straight line of 0 db for $\omega < \omega_3$, straight line of slope – 40 db/decade for $\omega > \omega_3$	Phase angle varies, from 0 to – 180° angle at $\omega_3 = -90°$

Asymptotic log-magnitude plot of the transfer function given in eqn. (8.25) is drawn in Fig. 8.16. The plot is obtained from Table 8.2 following the steps given below:

1. We start with the factor $(4/j\omega)$ corresponding to the pole at origin. Its log-magnitude plot is the asymptote '1', having a slope of –20 db/decade and passing through the point 20 log 4 = 12 db at $\omega = 1$.

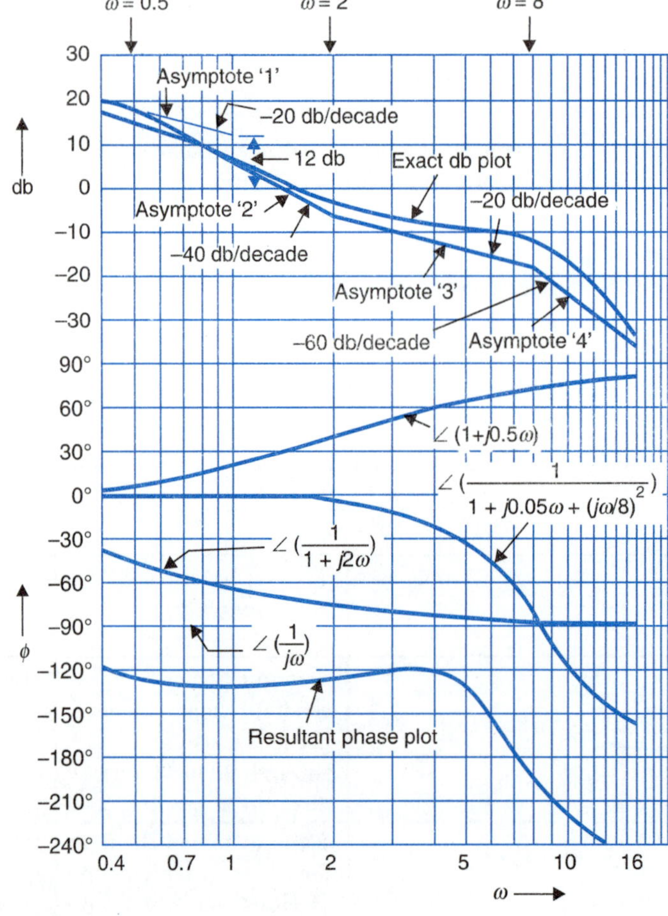

Fig. 8.16. Bode plot of $\frac{4(1+j\omega/2)}{j\omega(1+j2\omega)[1+j0.4(\omega/8)-(\omega/8)^2]}$.

2. Let us now add to the asymptote '1' the plot of the factor $1/(1 + j2\omega)$ corresponding to the lowest corner frequency $\omega = \omega_1 = 0.5$. Since this factor contributes zero db for $\omega \leq \omega_1 = 0.5$, the resultant plot up to $\omega = 0.5$ is the same as that of the asymptote '1'. For $\omega > \omega_1 = 0.5$ this is factor contributes –20 db/decade such that the resultant plot of these two factors is the asymptote '2' of slope $(-20) + (-20) = -40$ db/decade.

3. Above $\omega_2 = 2$, the factor $(1 + j0.5\omega)$ is effective. This gives rise to a straight line of slope +20 db/decade for $\omega > 2$, which when added to asymptote '2' results in asymptote '3' with a slope $(-40 + 20) = -20$ db/decade in the frequency band ω_2 to ω_3.

4. Above ω_3, the plot of $1/[1 + j0.4(\omega/8) - (\omega/8)^2]$ is to be added. This factor is represented by a straight line of –40 db/decade for $\omega > \omega_3 = 8$, which when added to asymptote '3', results in asymptote '4' having a slope of $(-20) + (-40) = -60$ db/decade in the frequency band ω_3 to ∞.

To the asymptotic plot thus obtained, corrections as read from Figs. (8.11) and (8.15) are to be applied to get the actual plot. db corrections are found at each corner frequency and at twice and half its values. Corrections contributed by adjoining corner frequencies are to be added algebraically.

The asymptotes and the exact log-magnitude plot of the transfer function given in eqn. (8.25) are shown in Fig. 8.16.

The determination of the phase angle curve can be simplified by using the following procedure.

1. For the factor $K/(j\omega)^r$, draw a straight line at an angle of $-90°r$.

2. The phase angles of the factor $(1 + j\omega T)^{\pm 1}$ are

$\pm 45°$ at $\omega = 1/T$

$\pm 26.6°$ at $\omega = 1/2T$

$\pm 5.7°$ at $\omega = 1/10T$

$\pm 63.4°$ at $\omega = 2/T$

$\pm 84.3°$ at $\omega = 10/T$

These points are plotted and a smooth curve passing through them gives the phase angle plot of the factor $(1 + j\omega T)^{\pm 1}$.

3. The phase angles of the factor $[1 + j2\zeta(\omega/\omega_n) - (\omega/\omega_n)^2]^{-1}$ are:

(*a*) $-90°$ at $\omega = \omega_n$.

(*b*) Corresponding to the value of ζ given in the factor, a few points of phase angles are read off from Fig. 8.14 or are calculated from the phase relation given in eqn. (8.24). These points are then located and joined to give a phase plot of the quadratic factor.

4. Once the phase curve of each factor of the transfer function has been drawn, the complete phase plot is obtained by adding the individual curves algebrically.

For the example under consideration, the phase angle contributions of the individual factors and the resultant phase angle plot are shown in Fig. 8.16.

8.5 ALL-PASS AND MINIMUM-PHASE SYSTEMS

The cases we have considered so far are termed as *minimum-phase transfer functions, i.e.,* those with all poles and zeros in the left half of the s-plane.

Consider now the special class of transfer function having a pole-zero pattern which is antisymmetric about the imaginary axis, *i.e.,* for every pole in the left half plane, there is a zero in the mirror-image position. A common example of such a transfer function is

$$G(j\omega) = \frac{1 - j\omega T}{1 + j\omega T}$$

whose pole-zero configuration is shown in Fig. 8.17(c). It has a magnitude of unity at all frequencies and a phase angle ($-2\tan^{-1}\omega T$) which varies from 0° to –180° as ω is increased from 0 to ∞. The property of unit magnitude at all frequencies applies to all transfer functions with antisymmetric pole-zero pattern. Physical systems with this property, are called *all-pass systems*.

Consider next the case where the transfer function has poles in the left half s-plane and zeros in both the left-and right-half s-plane. Poles are not permitted to lie in the right-half s-plane because such a system would be unstable. Consider for example the transfer function

$$G_1(j\omega) = \frac{(1 - j\omega T)}{(1 + j\omega T_1)(1 + j\omega T_2)}$$

whose pole-zero pattern is shown in Fig. 8.17(a). This transfer function may be rewritten as

$$G_1(j\omega) = \left[\frac{(1 + j\omega T)}{(1 + j\omega T_1)(1 + j\omega T_2)}\right]\left[\frac{(1 - j\omega T)}{(1 + j\omega T)}\right] = G_2(j\omega)G(j\omega)$$

which now becomes the product of two transfer functions, one $G_2(j\omega)$ having no poles or zeros in the right half of s-plane (Fig. 8.17(b)) and the other, an all-pass transfer function, $G(j\omega)$ of Fig. 8.17(c). It is evident that $G_1(j\omega)$ and $G_2(j\omega)$ have identical curves of magnitude versus frequency but their phase versus frequency curves are different as shown in Fig. 8.18, with $G_2(j\omega)$ having a smaller range of phase angle than $G_1(j\omega)$. It means that for $G_1(j\omega)$, there is no unique relationship between magnitude and phase, as it is always possible to alter the phase curve without affecting the associated magnitude curve by the addition of an all-pass transfer function. A transfer function which has one or more zeros in the right half s-plane is known as *nonminimum phase transfer function*.

In general, if a transfer function has any zeros in the right half s-plane, it is possible to extract them one by one by associating them with all-pass transfer function in a manner shown in Fig. 8.17. Each time this is done, the magnitude curve remains unaltered but the phase lag is reduced, until eventually we are left with a function which contains no zeros in the right half s-plane. Such a transfer function has the least (minimum) phase angle range for a given magnitude curve and is called a *minimum-phase function*. It has a unique relationship between its phase and magnitude curves. Typical phase angle characteristics of minimum and nonminimum-phase transfer functions are shown in Fig. 8.18. It will be seen in Chapter 10 that the larger the phase lags present in a system, the more complex are its stabilization problems. Therefore for control systems, elements with nonminimum-phase transfer functions are avoided as far as possible.

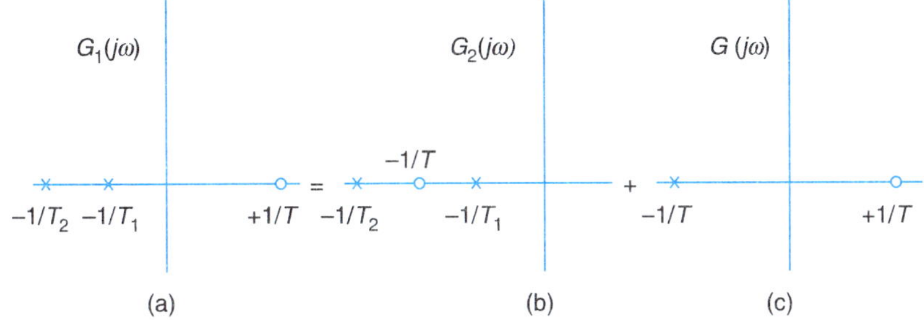

Fig. 8.17. Pole-zero patterns for (a) nonminimum-phase function; (b) minimum-phase function; (c) all-pass function.

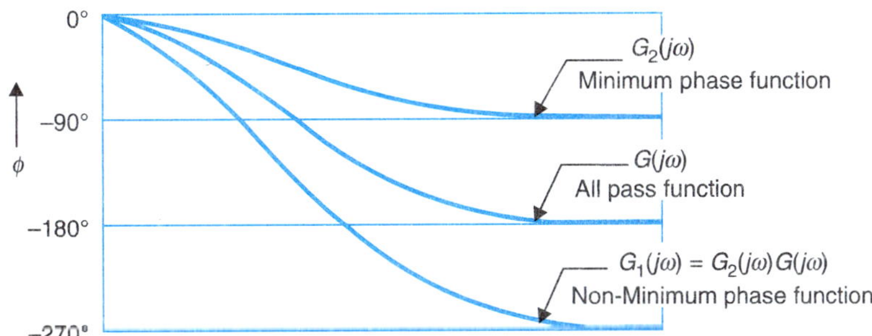

Fig. 8.18. Phase angle characteristics of minimum-phase and nonminimum-phase functions.

A common example of a nonminimum-phase element is transportation lag (see Section 7.5) which has the transfer function

$$G(j\omega) = e^{-j\omega T} = 1 \angle -\omega T \text{ rad} = 1 \angle -57.3\ \omega T \text{ deg}$$

The phase angle characteristics of transportation lag are shown in Fig. 8.19(a) and (b).

Other possible situations where a nonminimum-phase transfer function can arise are, when more than one possible signal paths are available between input and output as in lattice networks and when there is inductive coupling between input and output in addition to a conductive path.

8.6 EXPERIMENTAL DETERMINATION OF TRANSFER FUNCTIONS

The Bode plots are of great value in situations where the transfer function of a given system is unknown. In such cases we proceed to obtain the frequency response data experimentally in the frequency range of interest. The system transfer function within a certain degree of accuracy can then be obtained by fitting an asymptotic log-magnitude plot to the experimental data as per the procedure outlined below.

1. The experimental data is used to plot the exact log-magnitude and phase angle versus frequency curves on a semilog graph sheet.

2. Asymptotes are then drawn on the log-magnitude curve keeping in view the fact that the slopes of these asymptotes must be multiples of ±20 db/decade. The corner frequencies are so adjusted that the db value at the corner frequency on the asymptotic plot differs from the

actual log-magnitude plot by an amount which is in close agreement with the db correction of the kind of factor revealed.

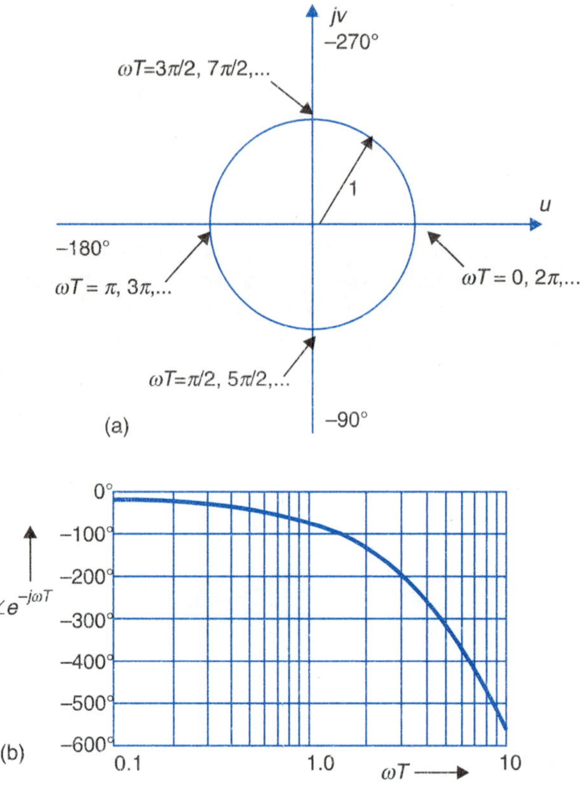

Fig. 8.19. Phase angle characteristics of $e^{-j\omega T}$.

3. If the slope of asymptotic log-magnitude curve obtained in (2) above, changes by –20 m db/decade at $\omega = \omega_1$ it indicates that the factor $1/(1 + j\omega/\omega_2)^m$ exists in the transfer function.

4. If the slope changes by +20 m db/decade at $\omega = \omega_2$, it indicates the presence of the factor $(1 + j\omega/\omega_2)^m$ in the transfer function.

5. The change of slope by –40 db/decade at $\omega = \omega_3$ indicates that either a double pole (*i.e.*, $m = 2$) or pair of complex conjugate poles is present. If the error between the asymptotic and the actual curve is about –6 db, then a factor of the form $1/(1 + j\omega/\omega_3)^2$ is present and if the error is positive then a quadratic factor of the form $1/[1 + j2\zeta(\omega/\omega_3) + (j\omega/\omega_3)^2]$ is present. The corresponding value of ζ is obtained with the help of the error graph of Fig. 8.15.

6. In the low frequency range, the plot is dominated by a factor of the form $K/(j\omega)^r$.

In most practical systems r equals 0, 1 or 2. The value r is determined as follows:

(*a*) If the low frequency asymptote is a horizontal line at x db, it indicates that the transfer function represents a type-0 system with a gain K given by

$$20 \log K = x \qquad \text{or} \qquad K = \frac{1}{20} \log^{-1} x$$

(*b*) If the low frequency asymptote has a slope of –20 db/decade, it indicates the presence of a factor of the form $K/j\omega$ in the transfer function (type-1 system). The frequency at which the asymptote (extended if necessary) intersects the 0-db line, numerically represents the value of K. Also the asymptote (extended if necessary) has a gain of 20 log K at $\omega = 1$.

(*c*) If the low frequency asymptote has a slope of –40 db/decade, then the transfer function has a factor of the form $K/(j\omega)^2$ (type-2 system). The frequency at which this asymptote (extended if necessary) intersects the 0-db line is numerically equal to $\sqrt{K}$. Also the asymptote (extended if necessary) has a gain of 20 log K at $\omega = 1$.

The log-magnitude curves of type-0, type-1 and type-2 systems are shown in Figs. 8.20(*a*), (*b*) and (*c*), respectively.

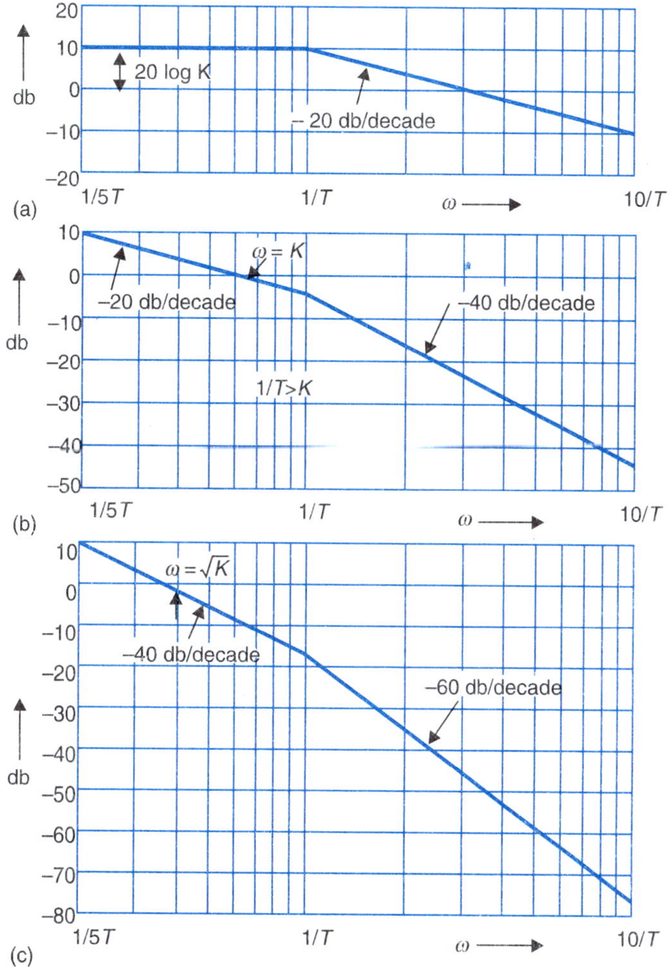

Fig. 8.20. Log-magnitude curves of type-0, type-1, and type-2 systems.

After obtaining the transfer function from the log-magnitude curve, the phase angle curve is constructed from it and is then compared with the one obtained experimentally. If the two curves are in fair agreement and if the phase angles of both the curves at very high frequencies tend to –90° $(q - p)$ where p and q are degrees of the numerator and denominator polynomials respectively of the transfer function, then the transfer function is of minimum-phase type. If the computed phase angle is 180° less negative than the one obtained experimentally, then the transfer function is of nonminimum-phase type and one of the zeros of the transfer function lies in the right half-*s*-plane.

As an illustration, let us derive the transfer function of the system whose experimental log-magnitude and phase-angle curves are shown in Fig. 8.21. First of all, the asymptotes are drawn on the experimentally determined curve as shown.

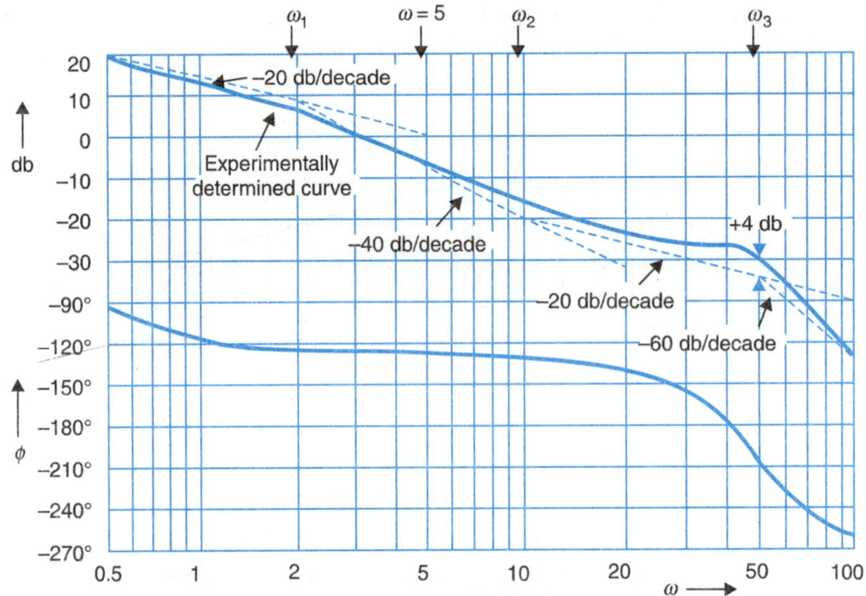

Fig. 8.21. Experimentally obtained log-magnitude and phase characteristics.

The low frequency asymptote has a slope of –20 db/decade and when extended, intersects the 0 db axis at $\omega = 5$. Therefore the asymptote is a plot of the factor $5/(j\omega)$. The corner frequencies are the asymptote is a plot of the factor $5/(j\omega)$. The corner frequencies are found to be located at $\omega_1 = 2$, $\omega_2 = 10$ and $\omega_3 = 50$. At the first corner frequency, the slope of the curve changes by –20 db/decade and at the second corner frequency, it changes by +20 db/decade. Therefore the transfer function has the factors $1/(1 + j\omega/2)$ and $(1 + j\omega/10)$ corresponding to these corner frequencies. At $\omega = \omega_3$, the curve changes by a slope of –40 db/decade. At this frequency the error between actual and approximate plots is +4 db. Therefore, the transfer function has a quadratic factor

$$\frac{1}{1 + j2\zeta(\omega/50) + (j\omega/50)^2}$$

where $\zeta = 0.3$, as obtained from the error graph of Fig. 8.15, corresponding to the error + 4 db.

Thus the transfer function of the system becomes

$$G(j\omega) = \frac{5(1 + j\omega/10)}{j\omega(1 + j\omega/2)[1 + j0.6(\omega/50) + (j\omega/50)^2]}$$

From the experimental phase angle curve shown in Fig. 8.21, it is seen that the phase angle at very high frequencies is –270° which is equal to –90° $(q - p)$ = –90° (4 – 1). Therefore the Bode plot represents a minimum-phase transfer function.

8.7 LOG-MAGNITUDE VERSUS PHASE PLOTS

In the previous sections, we considered polar plots and Bode plots as the graphical representations of the frequency response. An alternative approach to the frequency response representation is to plot the magnitude in db versus phase angle with frequency ω as the running parameter. Normally, the Bode plot is first obtained and then by reading the values of log magnitude and phase angle at different frequencies from these plots the log-magnitude vs. phase angle plot is constructed with the frequencies of various points indicated thereon. The advantages of these plots are that the relative stability of closed-loop control systems can be determined quickly and the compensation can be carried out easily (discussed in Chapter 10).

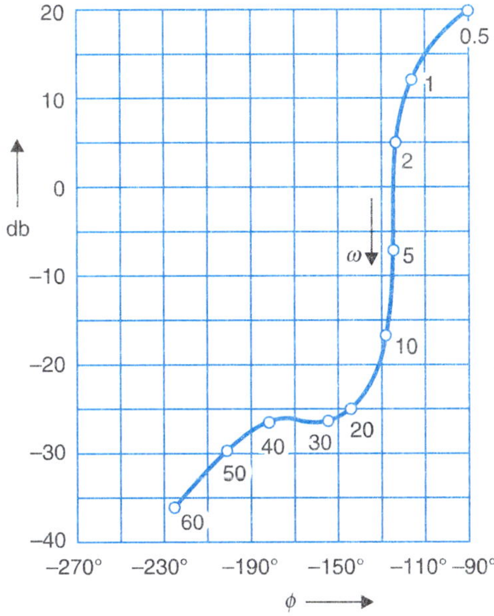

Fig. 8.22. Log-magnitude vs. phase curve obtained from Fig. 8.21.

The log-magnitude versus phase curve for the transfer function

$$G(j\omega) = \frac{5(1 + j\omega/10)}{j\omega(1 + j\omega/2)[1 + j0.6(\omega/50) + (j\omega/50)^2]}$$

is shown in Fig. 8.22 which is obtained by utilizing the Bode plot of Fig. 8.21.

8.8 MATLAB: TOOL FOR DESIGN AND ANALYSIS OF CONTROL SYSTEMS-APPENDIX III

This software has the necessary routines for plotting polar and Bode plots on a PC with a plotter. The above account is presented for the reader to understand the meaning and use of these plots and the procedure of how the software has been written. The student will use the Control System Tools in Control Systems Laboratory equipped with computers, plotters and MATLAB software.

Example 8.1: Bode Magnitude asymptotic plot of a certain system is sketched (not to scale) in Fig. 8.23 with db/decade slopes indicated therein. The asymptote of – 20 db/decade has a value of 0 db at $\omega_2 = 120$. Find the frequency ω_1 at which the two asymptotes intersect.

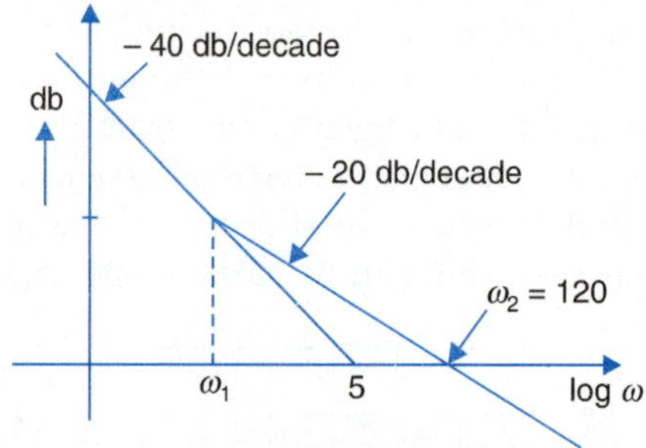

Fig. 8.23

Solution. The gain of – 40 db/decade line is obtained from the frequency $\omega = 5$ at which it intersects the 0 db line. Thus

$$\sqrt{K} = 5, \quad \text{or} \quad K = 25$$

This asymptotes is expressed as

$$\frac{25}{(j\omega)^2}$$

Its magnitude at ω_1 is

$$20 \log \left(\frac{25}{\omega_1^2} \right) \text{ db}$$

Being a corner frequency this magnitude can also be obtained from – 20 db/decade asymptote as

$$20 \log \left(\frac{120}{\omega_1} \right) \text{ db}$$

Equating the two values

$$20 \log \left(\frac{120}{\omega_1^2} \right) = 20 \log \left(\frac{120}{\omega_1} \right)$$

or
$$\frac{25}{\omega_1^2} = \frac{120}{\omega_1}$$

which yields $\omega_1 = 0.21$

Example 8.2: The asymptotic Bode magnitude plot of a system is sketched in Fig. 8.24. Write the expression for its transfer function in frequency domain. Draw it asymptotic phase plot.

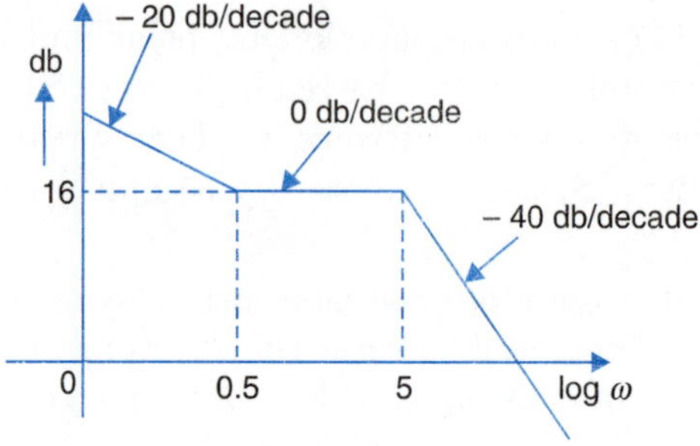

Fig. 8.24

Solution. First asymptote corresponds to $\dfrac{K}{(j\omega)}$

At $\omega = 0.5$

$$20 \log\left(\frac{K}{0.5}\right) = 16 \qquad \text{or} \qquad K = 3.15 \approx 3$$

Break frequencies 0.5, 5

Transfer function $G(j\omega) = \dfrac{3(1 + j2\omega)}{(j\omega)(1 + j0.2\omega)}$

Asymptotic phase angle (ϕ) is plotted in Fig. 8.25.

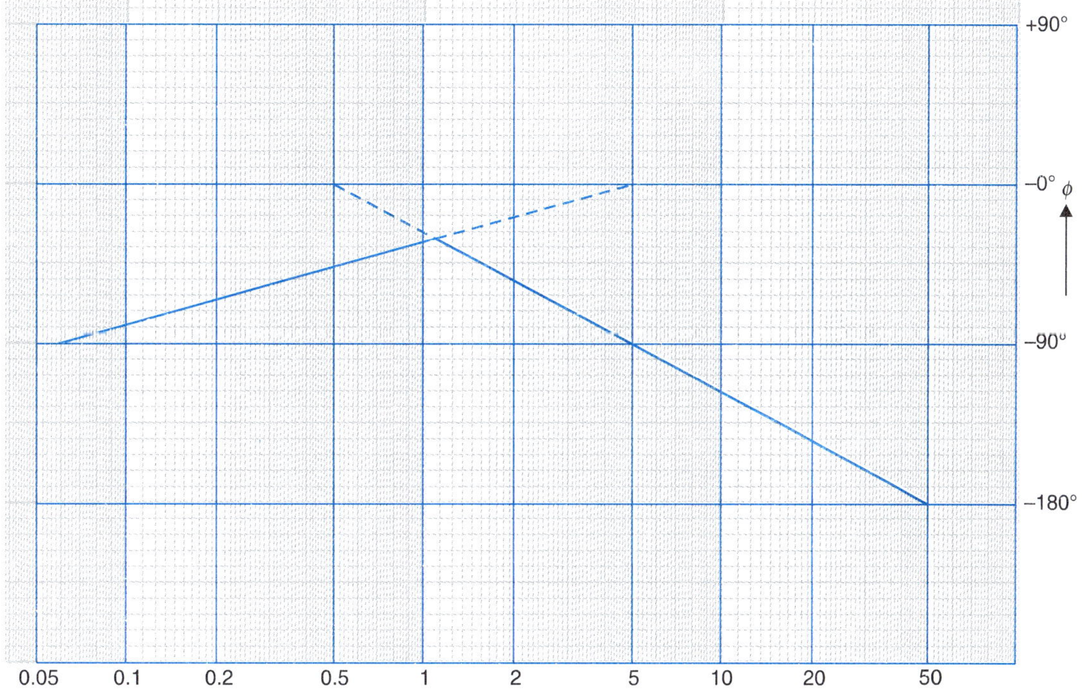

Fig. 8.25. Asymptotic phase plot.

Example 8.3: The frequency response test on a system yielded the following data

ω	db	ϕ
0.1	14	90°
0.5	0	61°
1	5	45°
5	10	0°
10	7.5	– 45°
50	– 19	– 136°
100	– 31	– 180°

Plot the data on a semilog graph sheet. Determine the system's transfer function in frequency domain.

Solution. db-log ω and ϕ-log ω from the given data are plotted in Fig. 8.26. Three asymptotes are fitted to the db-log ω plots.

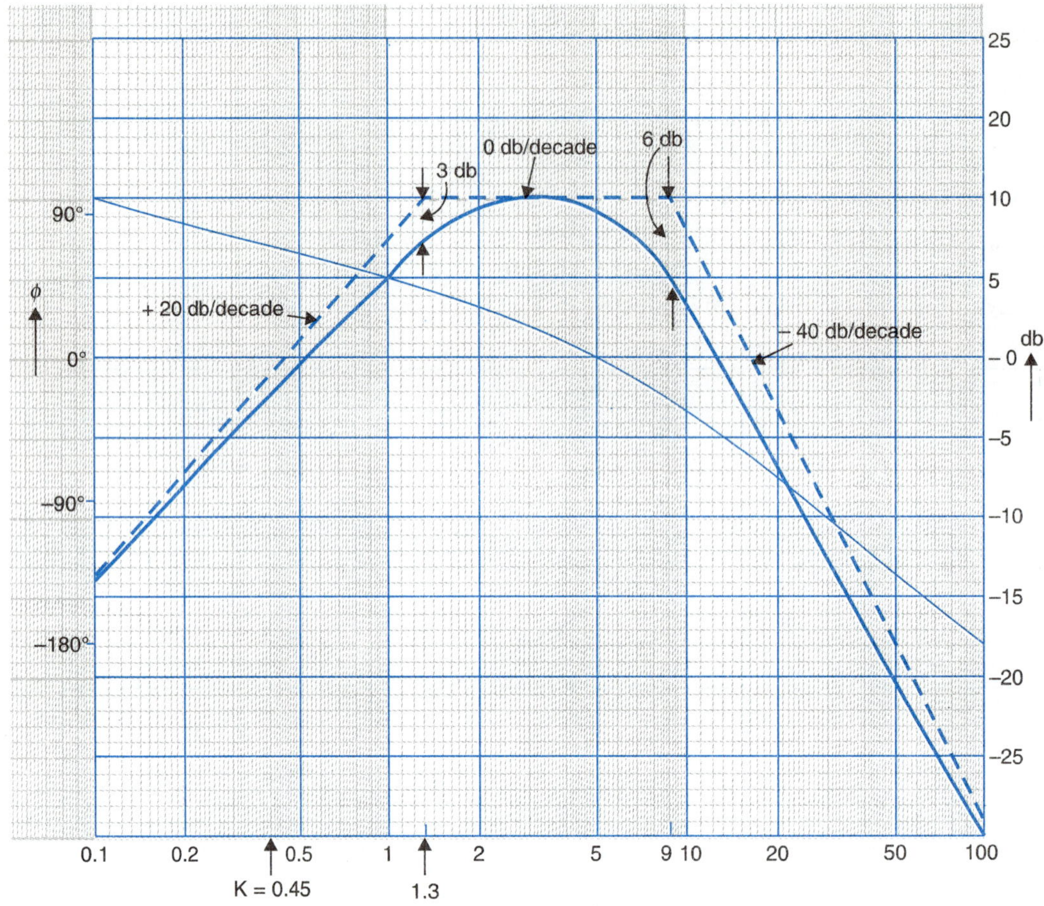

Fig. 8.26

First asymptote + 20 db/decade

Second asymptote 0 db/decade

Third asymptote – 40 db/decade

These asymptotes are confirmed by the fact that :

First break frequency $\omega_1 = 1.3$, db drop from corner to actual curve 3 db

Second break frequency $\omega_2 = 9$, db drop from corner to actual curve 6 db

Initial asymptotic slope is + 20 db and it has a value of 0 db at $\omega = 0.45$. There system gain

$$K = 0.45.$$

From the above result the system transfer function can be written as

$$G(j\omega) = \frac{K(j\omega)}{(1 + j\omega/1.3)(1 + j\omega/9)^2}$$

PROBLEMS

8.1 Sketch the polar plots of the transfer functions given below. Determine whether these plots cross the real axis. If so, determine the frequency at which the plots cross the real axis and the corresponding magnitude $|G(j\omega)|$.

(a) $G(s) = 1/(1 + s)(1 + 2s)$ (b) $G(s) = 1/s(1 + s)(1 + 2s)$

(c) $G(s) = 1/s^2(1 + s)(1 + 2s)$ (d) $G(s) = \dfrac{(1+0.2s)(1+0.025s)}{s^3(1+0.005s)(1+0.001s)}$.

8.2 Sketch the direct and inverse polar plots for a unity feedback system with open-loop transfer function

$$G(s) = 1/s(1 + s)^2$$

Also find the frequency at which $|G(j\omega)| = 1$ and the corresponding phase angle $\angle G(j\omega)$.

8.3 Sketch the Bode plots showing the magnitude the decibels and phase angle in degrees as a function of log frequency for the transfer functions given below. Determine the gain cross-over frequency ω_e (the frequency at which the magnitude curve crosses the 0-db axis) in each case

(a) $G(s) = \dfrac{10}{s(1+0.5s)(1+0.01s)}$ (b) $G(s) = \dfrac{75(1+0.2s)}{s(s^2+16s+100)}$.

8.4 Sketch the Bode plots for the following transfer functions and determine in each case, the system gain K for the gain cross-over frequency ω_c to be 5 rad/sec.

(a) $G(s) = \dfrac{Ks^2}{(1+0.2s)(1+0.02s)}$ (b) $G(s) = \dfrac{Ke^{-0.1s}}{s(1+s)(1+0.1s)}$.

Note : At gain cross-over frequency $|G(j\omega)|_{db} = 0$.

8.5 The frequency response test data of certain elements plotted on Bode diagrams and asymptotically approximated are shown in Fig. P-8.5. Find the transfer function of each element. (Elements are known to have minimum-phase characteristics).

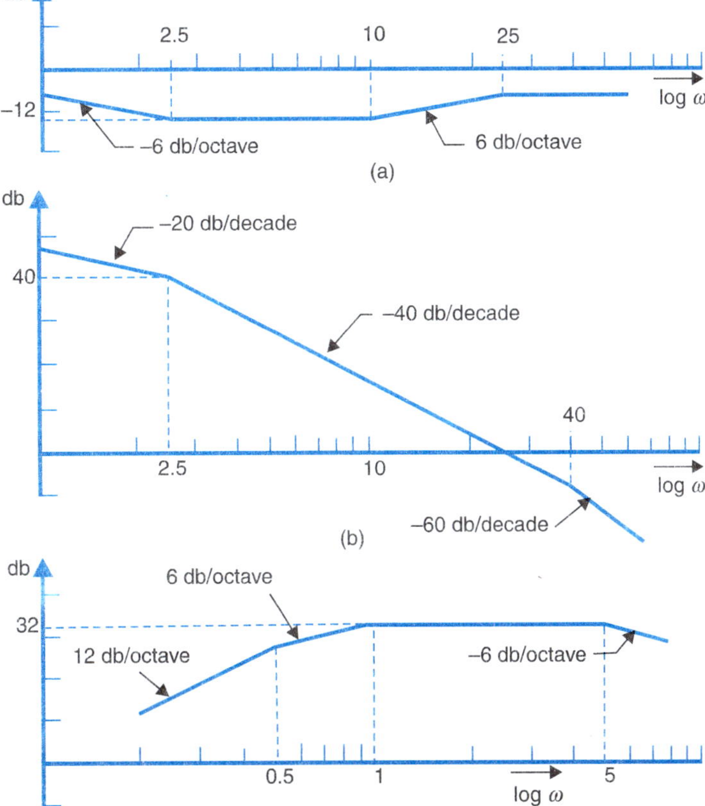

Fig. P-8.5

8.6 The following frequency response test data were obtained for the forward path elements of a unity feedback control system. Plot the data on a semilog graph paper and determine the transfer function of the forward path, using asymptotic approximation. What is the type of the system ?

Gain (db)	34	28	24.6	14.2	8	1.5	– 3.5	– 7.2
Frequency (rad / sec)	0.1	0.2	0.3	0.7	1.0	1.5	2.0	2.5
Gain (db)	–10.0	– 12.5	– 14.7	– 16.0	– 17.5	– 17.5	– 18.0	– 18.5
Frequency (rad / sec)	4.0	5.0	6.0	9.0	20	35	50	100

8.7 Consider the feedback system shown in Fig. P-8.7.

(*a*) Find the value of *K* and *a* to satisfy the following frequency domain specifications:

$$M_r = 1.04$$

$$\omega_r = 11.55 \text{ rad/sec}$$

(*b*) For the values of *K* and *a* determined in part (*a*), calculate the settling time and bandwidth of the system.

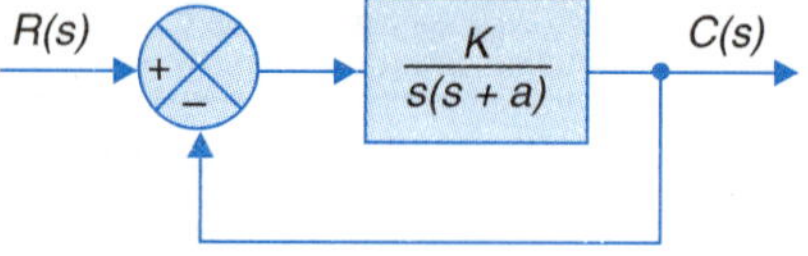

Fig. P-8.7

8.8 The closed-loop transfer function of a feedback system is given

$$T(s) = \frac{1000}{(s + 22.5)(s^2 + 2.45s + 44.4)}$$

(*a*) Determine the resonance peak M_r and resonant frequency ω_r of the system by drawing the frequency response curve.

(*b*) What should be the values of damping ratio ζ and undamped natural frequency ω_n of an equivalent second-order system which will produce the same M_r and ω_r as determined in part (*a*) ?

(*c*) Determine the bandwidth of the equivalent second-order system.

8.9 Unit-step response data of a second-order system is given below. Obtain the corresponding frequency response indices (M_r, ω_r, ω_b) for the system.

t(*sec*)	0	0.05	0.10	0.15	0.20	0.25	0.30	0.35	0.40	0.45	0.50
c(*t*)	0	0.25	0.8	1.08	1.12	1.02	0.98	0.98	1.0	1.0	1.0

8.10 For the systems of parts (*a*) and (*b*) of Problem 8.4 find the phase cross-over frequency. At this frequency $\angle G(j\omega) = -180°$.

9

STABILITY IN FREQUENCY DOMAIN

9

STABILITY IN FREQUENCY DOMAIN

9.1 INTRODUCTION

In earlier chapters, it has been shown that the stability of a system depends upon the location of the roots of its characteristic equation in the s-plane. For a system to be stable, the roots of the characteristic equation should not lie in the right half of the s-plane or on the imaginary axis, for such roots lead to instability or sustained oscillations of the system. The root locus technique discussed in Chapter 7 is an approach to determine the location of the roots in the s-plane, given the open-loop pole-zero configuration.

In this chapter, another equally important and useful technique, the *Nyquist stability criterion,* which relates the location of the roots of the characteristic equation to the open-loop frequency response of a system will be discussed. Unlike the root locus technique, the computation of closed-loop poles is not necessary in this case and the stability study can be carried out graphically from the open-loop frequency response. Therefore, experimentally determined open-loop frequency response can be used directly for the study of stability when the feedback path is closed.

The Nyquist stability criterion is based on a theorem of complex veriables due to Cauchy, commonly known as *principle argument.* In the following section, we shall discuss this theorem and then in the next section, we shall develop the Nyquist stability criterion.

9.2 MATHEMATICAL PRELIMINARIES

Consider a function $q(s)$ that can be expressed as a quotient of two polynomials. Each polynomial may be assumed to be known in the form of product of linear factors as given below:

$$q(s) = \frac{(s-\alpha_1)(s-\alpha_2)\ldots(s-\alpha_m)}{(s-\beta_1)(s-\beta_2)\ldots(s-\beta_n)} \qquad \ldots(9.1)$$

Let s be a complex variable, represented by $s = \sigma + j\omega$ on the complex s-plane. Then the function $q(s)$ is also complex (being a dependent variable) and may be defined as $q(s) = u + jv$ and represented on the complex $q(s)$-plane with coordinates u and v. Equation (9.1) indicates that for every point s in the s-plane at which $q(s)$ is analytic*, we can find a corresponding point $q(s)$ in the $q(s)$-plane. Alternatively, it can be stated that the function $q(s)$ maps the points in the s-plane into the $q(s)$-plane. Since any number of points of in the s-plane, can be mapped into the $q(s)$-plane, it follows that for a contour in the s-plane which does not go through any singular point, there corresponds a contour in the $q(s)$-plane as shown in Fig. 9.1. The region to the right of a closed contour is considered *enclosed* by the contour when the contour is traversed in the clockwise direction**. Thus, the shaded area in Fig. 9.1(*a*) is enclosed by the closed contour.

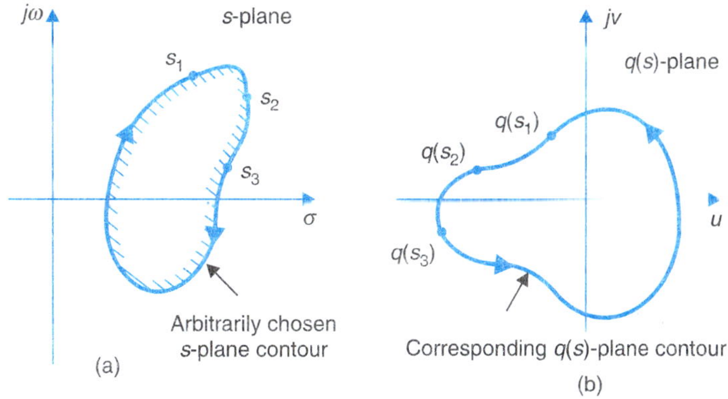

Fig. 9.1. Arbitrarily chosen s-plane contour which does not go through singular points and the corresponding $q(s)$-plane contour.

As will be seen later in the next section that while developing the Nyquist criterion, we are not interested in the exact shape of the $q(s)$-plane contour. An important fact that concerns us is the encirclement of the origin by the $q(s)$-plane contour. To investigate this, consider an s-plane contour which encloses only one of the zeros of $q(s)$, say $s = \alpha_1$, while all the poles and remaining zeros are distributed in the s-plane outside the contour. As discussed above, for any non-singular point s on the s-plane contour, there corresponds a point $q(s)$ on the $q(s)$-plane contour. From eqn. (9.1), the point $q(s)$ is given by

$$|q(s)| = \frac{|(s-\alpha_1)||(s-\alpha_2)|\ldots}{|(s-\beta_1)||(s-\beta_2)|\ldots} \qquad \ldots(9.2)$$

$$\angle q(s) = \angle(s-\alpha_1) + \angle(s-\alpha_2) + \ldots - \angle(s-\beta_1) - \angle(s-\beta_2) - \ldots$$

---

*A function $q(s)$ is analytic in the s-plane provided the function and all its derivatives exist. The points in the s-plane where the function (or its derivatives) does not exist, are called singular points. The poles of a function are singular points.

**This convention is opposite to that usually employed in complex variable theory but is equally applicable and is generally used in control system theory.

From Fig. 9.2(a) it is found that as the point s follows the prescribed path (*i.e.*, clockwise direction) on the s-plane contour, eventually returning to the starting point, the phasor $(s - \alpha_1)$ generates a net angle of -2π, while all other phasors generate zero net angles. Therefore the $q(s)$-phasor undergoes a net phase change of -2π. This implies that the tip of the $q(s)$-phasor must describe a closed contour about the origin of the $q(s)$-plane in the clockwise direction as shown in Fig. 9.2(b). As said earlier, the exact shape of the closed contour in the $q(s)$-plane is not of interest to us, but it is sufficient for us to observe that this contour encircles the origin once. If the contour in the s-plane is so chosen that it does not enclose any zero or pole, the corresponding contour in the $q(s)$-plane then will not encircle the origin. If the s-plane contour encloses two zeros, say at $s = \alpha_1$ and $s = \alpha_2$, the $q(s)$-plane contour encircles the origin twice in the clockwise direction as shown in Fig. 9.3. Generalizing, we can say that for each zero of $q(s)$ enclosed by the s-plane contour, the corresponding $q(s)$-plane contour encircles the origin once in the clockwise direction.

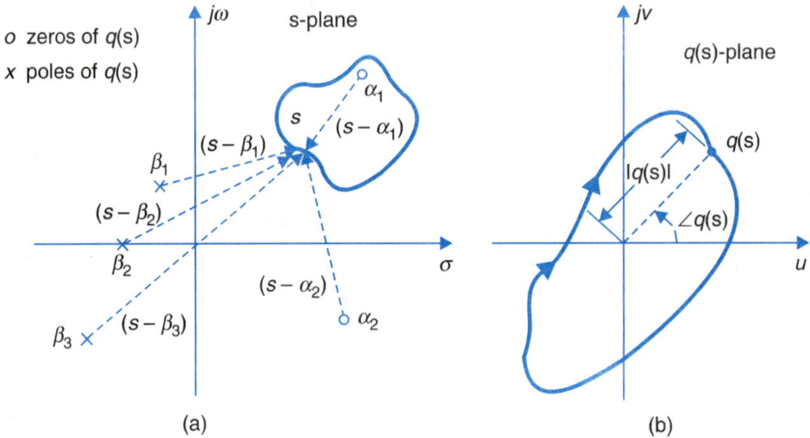

Fig. 9.2. An s-plane contour enclosing a zero of q(s) and the corresponding q(s)-plane contour.

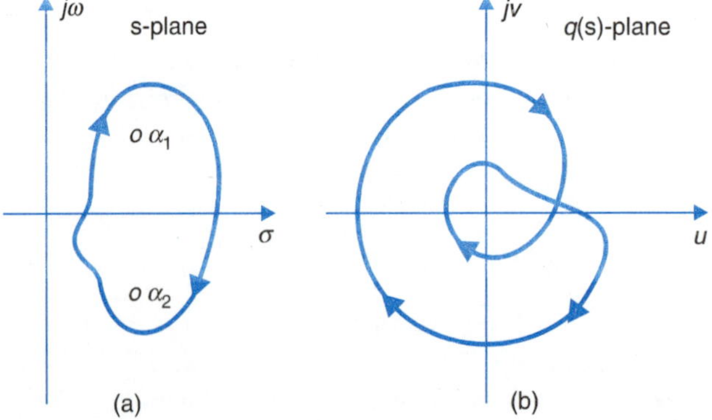

Fig. 9.3. An s-plane contour enclosing two zeros of q(s) and the corresponding q(s)-plane contour.

Consider now the enclosure of a pole of $q(s)$, say at $s = \beta_1$, by the s-plane contour. From an argument similar to the one used in conjunction with Fig. 9.2, it follows that the phasor $(s - \beta_1)$ generates an angle of -2π as s traverses the prescribed path. Since $(s - \beta_1)$ is in the denominator of eqn. (9.1), the $q(s)$-plane contour experiences an angle change of $+2\pi$, which means one counter-clockwise encirclement of the origin. This argument holds for all other poles of $q(s)$.

Thus, if there are P poles and Z zeros of $q(s)$ enclosed by the s-plane contour, then the corresponding $q(s)$-plane contour must encircle the origin Z times in the clockwise direction and P times in the counter-clockwise direction, resulting in a net encirclement of the origin, $(P - Z)$ times in the counter-clockwise direction. For example, in case of 1 zero and 3 poles enclosed by the s-plane contour, the net encirclement of the origin by the $q(s)$-plane contour is $2\pi(3 - 1) = 4\pi$ rad, *i.e.*, two counter-clockwise revolutions, as shown in Figs. 9.4(a) and (b). This relation between the enclosure of poles and zeros of $q(s)$ by the s-plane contour and the encirclements of the origin by the $q(s)$-plane contour is commonly known as *principle of argument*.

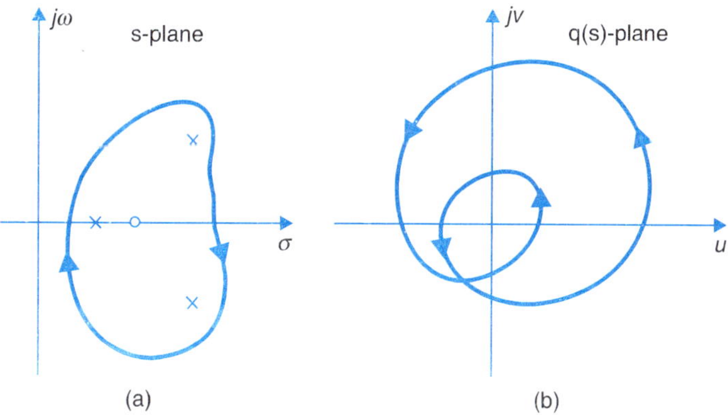

Fig. 9.4. Mapping of the s-plane contour which encloses 1 zero and 3 poles.

9.3 NYQUIST STABILITY CRITERION

Consider the single-loop* feedback system shown in Fig. 9.5. The characteristic equation of the system is

$$q(s) = 1 + G(s)H(s) = 0 \qquad \text{...(9.3)}$$

The standard pole-zero form of the open-loop transfer function $G(s)H(s)$ considered in earlier chapter is

$$G(s)H(s) = K\frac{(s+z_1)(s+z_2)\ldots(s+z_m)}{(s+p_1)(s+p_2)\ldots(s+p_n)}\ ;\ m \le n \qquad \text{...(9.4)}$$

*For the multiple-loop case $q(s) = 1 + P(s)$ as is shown by eqn. (7.5). With $P(s)$ expressed in pole-zero form, all the discussion and results that follow, apply for multiple-loop systems as well.

From eqns. (9.3) and (9.4) we obtain

$$q(s) = 1 + K\frac{(s+z_1)(s+z_2)\ldots(s+z_m)}{(s+p_1)(s+p_2)\ldots(s+p_n)} \quad \ldots(9.5)$$

$$= \frac{(s+p_1)(s+p_2)\ldots(s+p_n) + K(s+z_1)(s+z_2)\ldots(s+z_m)}{(s+p_1)(s+p_2)\ldots(s+p_n)}$$

$$= \frac{(s+z_1')(s+z_2')\ldots(s+z_n')}{(s+p_1)(s+p_2)\ldots(s+p_n)} \quad \ldots(9.6)$$

From eqn. (9.6) it is seen that the zeros of $q(s)$ at $-z_1', -z_2', \ldots, -z_n'$ are the roots of the characteristic equation and the poles of $q(s)$ at $-p_1, -p_2, \ldots, -p_n$ are same as the open-loop poles of the system. For the system to be stable, the roots of the characteristic equation and hence the zeros of $q(s)$ must lie in the left half of the s-plane. It is important to note that even if some of the open-loop poles lie in the right half s-plane, all the zeros of $q(s)$, *i.e.*, the closed-loop poles may lie in the left half s-plane, meaning thereby that an open-loop unstable system may lead to a closed-loop stable operation.

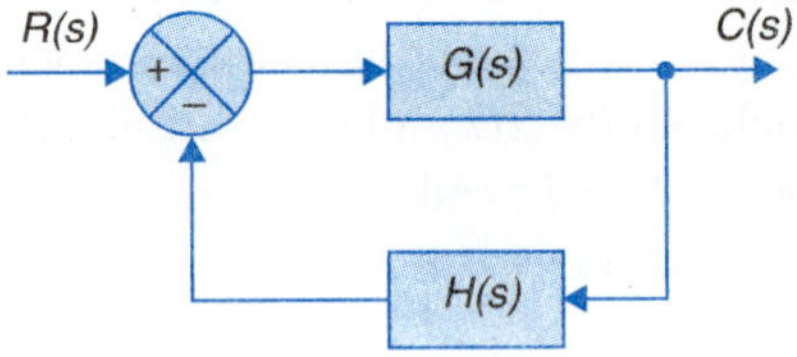

Fig. 9.5. A feedback control system.

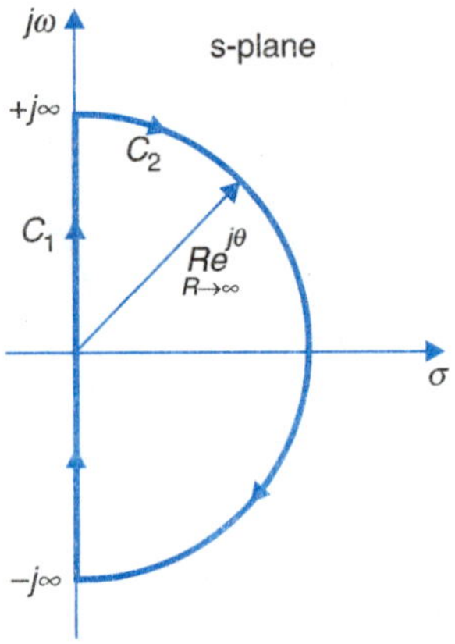

In order to investigate the presence of any zero of $q(s) = 1 + G(s)H(s)$ in the right half s-plane, let us choose a contour which completely encloses this (right) half of the s-plane. Such a contour C, called the *Nyquist contour* is shown in Fig. 9.6. It is directed clockwise and comprises of an infinite line segment C_1 along the $j\omega$-axis and an arc C_2 of infinite radius.

Fig. 9.6. The Nyquist contour.

Along C_1,

$$s = j\omega \text{ with } s \text{ varying from } -j\infty \text{ to } +j\infty$$

and along C_2,

$$s = \underset{R\to\infty}{Re^{j\theta}} \text{ with } \theta \text{ varying from } +\frac{\pi}{2} \text{ to } 0 \text{ to } -\frac{\pi}{2}.$$

The Nyquist contour so defined encloses all the right half s-plane zeros and poles of $q(s) = 1 + G(s)H(s)$. Let there be Z zeros and P poles of $q(s)$ in the right half s-plane. As s moves along the Nyquist contour in the s-plane, a closed contour Γ_q is traversed in the $q(s)$-plane which encloses the origin.

$$N = P - Z \quad \ldots(9.7)$$

times in the counter-clockwise direction.

In order for the system to be stable, there should be no zeros of $q(s) = 1 + G(s)H(s)$ in the right half s-plane, *i.e.*,

$$Z = 0$$

This condition is met if $\quad N = P \quad \ldots(9.8)$

that is, for a closed-loop system to be stable, the number of counter-clockwise encirclements of the origin of the $q(s)$-plane by the contour Γ_q should equal the number of the right half s-plane poles of $q(s)$ which are the poles of open-loop transfer function $G(s)H(s)$. As $G(s)H(s)$ is generally known in the factored form, the number of such poles (P) is easily ascertained by inspection.

In the special case (this is generally the case in most single-loop practical systems) of $P = 0$ (*i.e.*, the open-loop stable system), the closed-loop system is stable if

$$N = P = 0 \qquad ...(9.9)$$

which means that net encirclements, of the origin of the $q(s)$-plane by the Γ_q contour should be zero. It is easily observed that

$$G(s)H(s) = [1 + G(s)H(s)] - 1$$

It therefore follows that the contour Γ_{GH} of $G(s)H(s)$ corresponding to the Nyquist contour in the s-plane is the same as contour Γ_q of $1 + G(s)H(s)$ drawn from the point $(-1 + j0)$. Thus the encirclement of the origin by the contour Γ_q is equivalent to the encirclement of the point $(-1 + j0)$ by the contour Γ_{GH} as shown in Fig. 9.7.

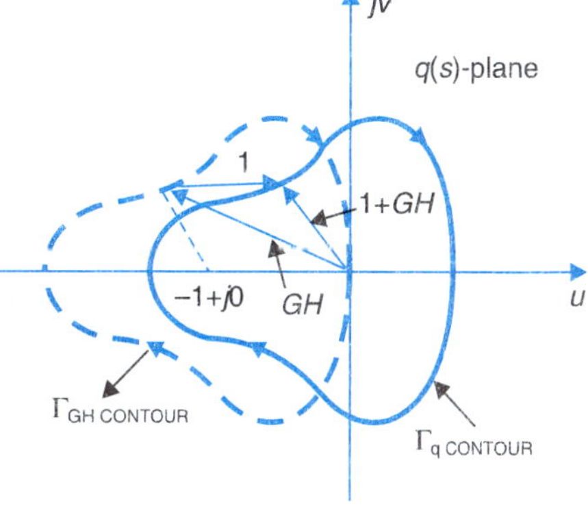

Fig. 9.7

We can now state the Nyquist stability criterion as follows:

If the contour Γ_{GH} of the open-loop transfer function $G(s)H(s)$ corresponding to the Nyquist contour in the s-plane encircles the point $(-1 + j0)$ in the counter-clockwise direction as many times as the number of right half s-plane poles of $G(s)H(s)$, the closed-loop system is stable.

In the commonly occurring case of the open-loop stable system, the closed-loop system is stable if the contour Γ_{GH} of $G(s)H(s)$ does not encircle $(-1 + j0)$ point, i.e., the net encirclement is zero.

The mapping of the Nyquist contour into the contour Γ_{GH} is carried out as follows:

1. The mapping of the imaginary axis is carried out by substitution of $s = j\omega$ in $G(s)H(s)$. This converts the mapping function into a frequency function of $G(j\omega)H(j\omega)$.

2. In physical systems ($m \le n$), $\lim\limits_{\substack{s = Re^{j\theta} \\ R \to \infty}} G(s)H(s)$ = real constant (it is zero if $m < n$). Thus the infinite arc of the Nyquist contour maps into a point on the real axis.

The complete contour Γ_{GH} is thus the polar plot of $G(j\omega)H(j\omega)$ with ω varying from $-\infty$ to ∞. This is usually called the Nyquist plot or locus of $G(s)H(s)$. Further it is important to note that the Nyquist plot is symmetrical about the real axis since $G^*(j\omega)H^*(j\omega) = G(-j\omega)H(-j\omega)$.

Example 9.1 : Consider a feedback system whose open-loop transfer function is given by

$$G(s)H(s) = \frac{K}{(T_1 s + 1)(T_2 s + 1)}$$

The plot* of $G(j\omega)H(j\omega)$ is shown in Fig. 9.8. The infinite semicircular arc of the Nyquist contour maps into origin. It is seen that the plot of $G(j\omega)H(j\omega)$ does not encircle the point $(-1 + j0)$ for any positive values of K, T_1 and T_2. Therefore, the system is stable for all positive values of K, T_1 and T_2.

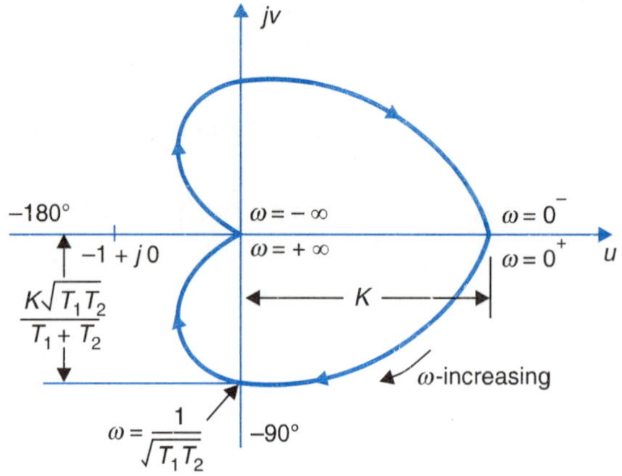

Fig. 9.8. Nyquist plot of $K/(1 + j\omega T_1)(1 + j\omega T_2)$.

Example 9.2 : Consider now, an open-loop unstable system with the transfer function

$$G(s)H(s) = \frac{s+2}{(s+1)(s-1)}$$

Let us determine whether the system is stable when the feedback path is closed.

From the transfer function of the open-loop system, it is observed that there is one open-loop pole in the right half s-plane. Therefore $P = 1$.

The $G(j\omega)H(j\omega)$-locus is sketched in Fig. 9.9 which indicates that the $(-1 + j0)$ point is encircled by this locus once in the counter-clockwise direction. Therefore $N = 1 = P$. Thus $Z = 0$, *i.e.,* there are no zeros of $1 + G(s)H(s)$ in the right half s-plane and hence the closed-loop system is stable.

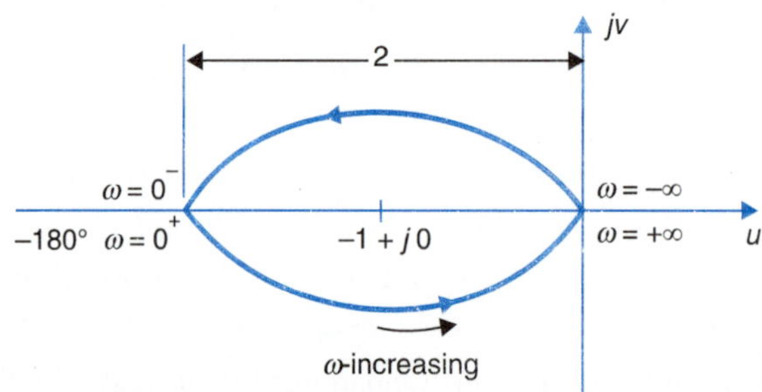

Fig. 9.9. Nyquist plot of $(j\omega + 2)/(j\omega + 1)(j\omega - 1)$.

*Polar plot of this function is given in Table 8.1.

Open-loop Poles on the $j\omega$-Axis

If $G(s)H(s)$ and therefore $1 + G(s)H(s)$ has any poles on the $j\omega$-axis, the Nyquist contour defined earlier (see Fig. 9.6) cannot be used as such since the s-plane contour should not pass through a singularity of $1 + G(s)H(s)$. To study stability in such cases, the Nyquist contour must be modified so as to bypass any $j\omega$-axis poles. This is accomplished by indenting the Nyquist contour around the $j\omega$-axis poles along a semicircle of radius ε, where $\varepsilon \to 0$ as shown in Fig. 9.10.

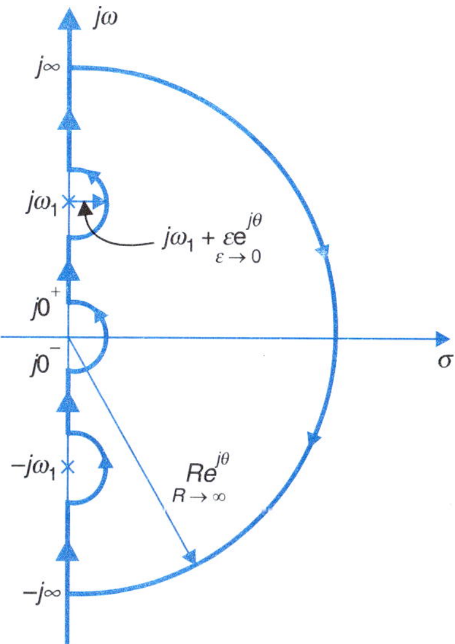

Fig. 9.10. Indented Nyquist contour for $j\omega$-axis open-loop poles.

The following example illustrates the stability study in such cases.

Example 9.3 : Consider a feedback system whose open-loop transfer function is given by

$$G(s)H(s) = \frac{K}{s\,(Ts + 1)}$$

In this example, the open-loop system has a pole at the origin. The Nyquist contour must therefore be indented to bypass the origin as shown in Fig. 9.11(a).

The mapping of the Nyquist contour may be carried out as follows:

1. The semicircular indent around the pole at origin represented by $s = \underset{\varepsilon \to 0}{\varepsilon}\, e^{j\theta}$ (θ varying from $-$ 90° through 0° to + 90°) maps into

$$\lim_{\varepsilon \to 0} [K/(\varepsilon\, e^{-j\theta})(1 + T\varepsilon\, e^{-j\theta})] = \lim_{\varepsilon \to 0} (K/\varepsilon\, e^{-j\theta}) = \lim_{\varepsilon \to 0} (K/\varepsilon)e^{-j\theta}$$

The value K/ε approaches infinity as ε approaches zero and $-\theta$ varies from +90° through 0° to $-$ 90° as s moves along the semicircle. Thus the infinitesimal semicircular indent around the origin maps into a semicircular arc of infinite radius in $G(s)H(s)$-plane as shown in Fig. 9.11(b).

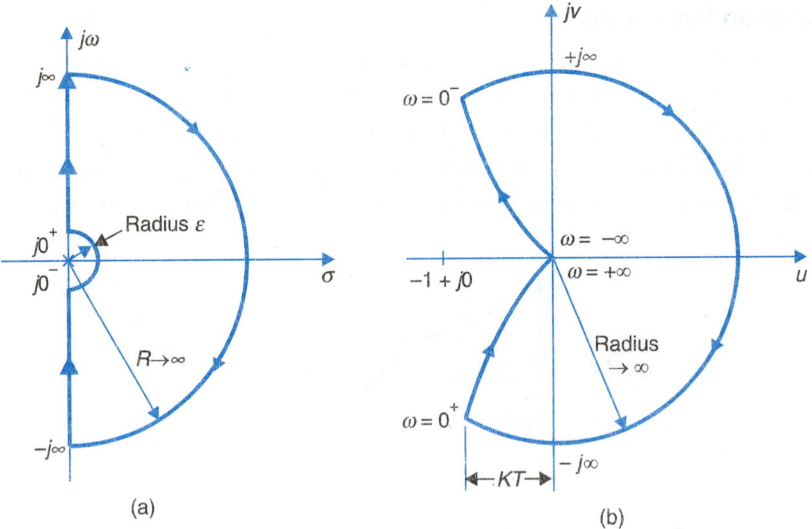

Fig. 9.11. Nyquist contour and $G(s)H(s)$-locus for $G(s)H(s) = K/s(Ts + 1)$.

2. The mapping of the positive imaginary axis ($\omega = 0^+$ to $+\infty$) is obtained by calculating $K/j\omega\,(1 + j\omega T)$ at various values of ω and plotting them in the $G(s)H(s)$-plane. This part of the locus, for the problem under discussion, is sketched in Fig. 9.11(*b*) on the lines indicated in Table 8.1.

3. The infinite semicircular arc of the Nyquist contour represented by $s = \underset{R \to \infty}{Re^{j\phi}}$ (ϕ varying from $+ 90°$ through $0°$ to $- 90°$) is mapped into

$$\lim_{R \to \infty} [K/Re^{j\phi}[T\, Re^{j\phi} + 1)] = \lim_{R \to \infty} [K/TR^2(e^{j2\phi})] = 0e^{-j2\phi}$$

i.e., the origin of the $G(s)H(s)$-plane. The $G(s)H(s)$-locus thus turns at the origin with zero radius from $- 180°$ through $0°$ to $+ 180°$.

4. The mapping of the negative imaginary axis is the mirror image of that for the positive imaginary axis.

The complete Nyquist plot for $G(s)H(s) = K/s(Ts + 1)$ is shown in Fig. 9.11(*b*).

In order to investigate the stability of this system, we first note that number of poles P of $G(s)H(s)$ in the right half s-plane is zero. Examination of Fig. 9.11(*b*) reveals that for all positive values of K and T, the locus of $G(j\omega)H(j\omega)$ does not encircle $(-1 + j0)$ point. Therefore the system under consideration is always stable.

Example 9.4 : Consider a system with an open-loop transfer function

$$G(s)H(s) = \frac{(4s + 1)}{s^2(s + 1)(2s + 1)}$$

In this example, the open-loop system has a double pole at the origin. The Nyquist contour is therefore indented to bypass the origin as shown in Fig. 9.12(*a*). The mapping of the Nyquist contour is obtained as follows:

1. Semicircular indent represented by $s = \lim_{\varepsilon \to 0} \varepsilon e^{j\theta}$ (where θ varies from $-90°$ through $0°$ to $90°$) is mapped into

$$\lim_{\varepsilon \to 0} \left[\frac{4\varepsilon e^{j\theta} + 1}{\varepsilon^2 e^{j2\theta}(\varepsilon e^{j\theta} + 1)(2\varepsilon e^{j\theta} + 1)} \right] = \lim_{\varepsilon \to 0} \left(\frac{1}{\varepsilon^2 e^{j2\theta}} \right) = \infty e^{-j2\theta}$$

$$= \infty (\angle 180° \to \angle 0° \to \angle -180°)$$

This part of the map is an infinite circle shown in Fig. 9.12(*b*).

2. Along the $j\omega$-axis

$$G(j\omega)H(j\omega) = \frac{(1 + j4\omega)}{(j\omega)^2(1 + j\omega)(1 + j2\omega)}$$

For various values of ω, $G(j\omega)H(j\omega)$ is calculated and plotted as shown in Fig. 9.12(*b*).

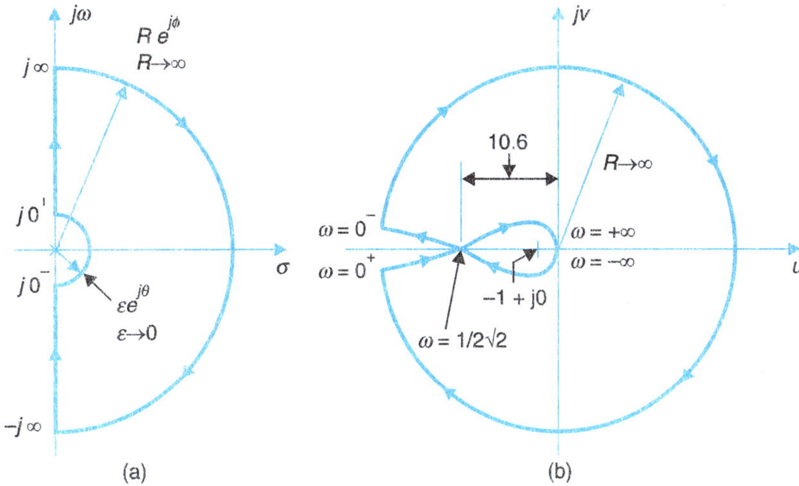

Fig. 9.12. Nyquist contour and the corresponding mapping for $G(s)H(s) = (4s + 1)/s^2(s + 1)(2s + 1)$.

The $G(j\omega)H(j\omega)$-locus intersects the real axis at a point where

$$\angle G(j\omega)H(j\omega) = -180°$$

or $$-180° - \tan^{-1}\omega - \tan^{-1}2\omega + \tan^{-1}4\omega = -180°$$

which gives $$\omega = 1/2\sqrt{2} = 0.354 \text{ rad/sec}$$

Therefore $|G(j\omega)H(j\omega)|_{\omega = 1/2\sqrt{2}} = 10.6$

Further $G(j\omega)H(j\omega)\,|_{\omega \to +j\infty} = 0 \angle -270°$, *i.e.*, the map of $j\omega$-axis ends at $0 \angle -270°$ as $\omega \to +\infty$.

3. The infinite semicircle of the Nyquist contour represented by $s = \lim_{R \to \infty} Re^{j\phi}$ (ϕ varies from $+90°$ through $0°$ to $-90°$) is mapped into

$$\lim_{R\to\infty} \frac{(1+4Re^{j\phi})}{R^2e^{j2\phi}(1+Re^{j\phi})(1+2Re^{j\phi})} = 0e^{-j3}\phi$$

$$= 0(\angle -270^\circ \to \angle 0^\circ \to \angle +270^\circ)$$

The complete mapped plot corresponding to the Nyquist contour of Fig. 9.12(*a*) is shown in Fig. 9.12(*b*). From this plot it is observed that $(-1 + j0)$ point is encircled twice in the clockwise direction. Therefore $N = -2$. From the given transfer function it is seen that no pole of $G(s)H(s)$ lies in the right half s-plane, *i.e.*, $P = 0$. Thus $-2 = 0 - Z$ or $Z = 2$. Hence two zeros of $q(s)$ lie in the right half s-plane from which we conclude that the system is unstable.

Example 9.5 : The system of Example 9.2 is reconsidered here with an integrator and a right hand s-plane pole and a derivative feedback term in the numerator, *i.e.*,

$$G(s)H(s) = \frac{K(s+a)}{s(s-1)}$$

Sketch the Nyquist plot and conclude on closed-loop stability.

Solution.

$$G(s)H(s) = \frac{K(s+a)}{s(s-1)} \qquad ...(i)$$

$$= \frac{K'(s/a+1)}{s(s-1)} \; ; K' = aK \qquad ...(ii)$$

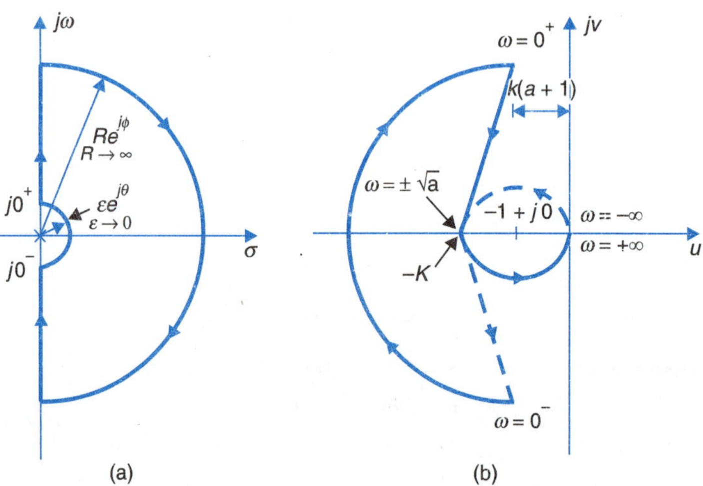

Fig. 9.13

The indented Nyquist contour for the open-loop transfer function is drawn in Fig. 9.13(*a*). The corresponding Nyquist plot for this system is obtained in the following steps:

1. Along the positive imaginary axis

$$G(j\omega)H(j\omega) = \frac{K'(1+j\omega/a)}{j\omega(-1+j\omega)} \; ; \text{for } \omega = 0^+ \text{ to } +\infty \qquad ...(iii)$$

We can write this in form of real and imaginary parts as

$$G(j\omega)H(j\omega) = \frac{K'[(\omega + \omega/a) - j(1-\omega^2/a)]}{-\omega(\omega^2+1)} \quad ...(iv)$$

This part of the plot would cross the axis of reals when the imaginary part of the above expression is zero, that is,

$$1 - \omega^2/a = 0 \qquad \text{or} \qquad \omega = \pm\sqrt{a}\,;\, a > 0$$

At this value of ω

$$|G(j\omega)H(j\omega)|_{\omega^2 = a} = \frac{aK'(1+1/a)}{-(1+a)} = -K' \quad ...(v)$$

At low values of ω, *i.e.*, $\omega = 0^+$ the imaginary part tends to infinite value but real part has a finite value

$$\text{Re}[G(j\omega)H(j\omega)]_{\omega = 0^+} = \frac{aK(1+1/a)}{-1}$$

$$= -K'(a+1)\text{ ; finite negative value}$$

This part of the Nyquist plot is sketched in Fig. 9.13(b)

2. Around the indent at $s = 0$

$$G(s)H(s)\Big|_{\substack{s=\varepsilon e^{j\theta} \\ \varepsilon\to 0,\ \theta\to -90^\circ - 0^\circ - +90^\circ}} = \frac{K'(\varepsilon e^{j\theta}/a+1)}{\varepsilon e^{j\theta}(\varepsilon e^{j\theta}+1)} \quad ...(vi)$$

$$= -\infty e^{j\theta} = \infty e^{-j(180^\circ - \theta)}$$

$$= \infty\angle{-90^\circ} \to \infty\angle{-180^\circ} \to \infty\angle{-270^\circ}$$

It is an infinite semicircular arc in the left-half of $G(s)H(s)$-plane with clockwise direction as in Fig. 9.9(b).

3. Around the infinite circular arc of Nyquist contour

$$G(s)H(s)\Big|_{\substack{s=Re^{j\theta} \\ R\to\infty,\ \theta\to +90^\circ - 0^\circ - -90^\circ}} = \frac{K'(Re^{j\phi}/a+1)}{Re^{j\phi}(Re^{j\phi}+1)} \quad ...(vii)$$

$$= 0e^{-j\theta} = 0\angle{-90^\circ} \to \angle 0^\circ \to \angle{+90^\circ}$$

The map turns around origin from $\angle 90^\circ \to \angle 0^\circ \to \angle 90^\circ$ as sketched in Fig. 9.9(b).

The complete Nyquist plot is sketched in Fig. 9.13(b). Conclusions on stability;

For $\quad K > 1$

the Nyquist plot encircles $(1 + j0)$ once in anticlockwise direction, *i.e.*, $N = 1$. The open-loop transfer function has $P = 1$.

Then $\qquad N = P - Z$

$$1 = 1 - Z \quad \Rightarrow \quad Z = 0 \quad ...(viii)$$

Thus the system is stable.

For $\quad K < 1$

we can easily see that $(-1 + j0)$ will lie beyond $-K$, the crossing point of the plot. So

$$N = -1, P = 1$$

$$N = P - Z$$

$$-1 = 1 - Z \quad \Rightarrow \quad Z = 2 \qquad (ix)$$

The closed-loop is unstable as it has two roots its characteristic equation in the right hand s-plane.

Example 9.6 : Consider now a system with open-loop zero in right-half s-plane. Let

$$G(s)H(s) = \frac{K(s-2)}{(s+1)^2}$$

Investigate in a Nyquist plot stability of the closed-loop system (unity feedback).

Solution : This case corresponds to unindented Nyquist contour of Fig. 9.6. Nyquist plot is obtained in the following steps :

$$j\omega \quad \Rightarrow \quad j0 \to j\infty$$

$$G(s)H(s)\big|_{j\omega} = \frac{K(j\omega - 2)}{(j\omega + 1)^2}$$

It acquires a value of

$$-2K \text{ at } j\omega = j0^+$$

$$0 \angle -90° \text{ at } j\omega = j\infty$$

Crossing the axis of reals

$$G(j\omega)H(j\omega) = \frac{K[(4\omega^2 - 2) + j\omega(5 - \omega^2)]}{(\omega^2 + 1)(\omega^2 + 1)^2}$$

At the crossing, imaginary part is zero *i.e.*,

$$5 - \omega^2 = 0 \quad \text{or} \quad \omega^2 = 5 \quad \text{or} \quad \omega = \pm\sqrt{5}$$

At this value of ω

$$|G(j\omega)H(j\omega)|_{\omega^2 = 5} = K/2$$

The complete Nyquist plot is sketched in Fig. 9.14, examination of which reveals

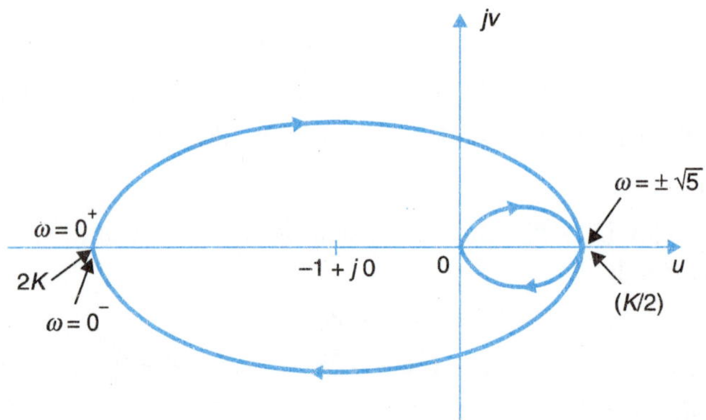

Fig. 9.14

For $\quad K/2 > -1 \quad$ or $\quad K > -2$

$$N = -1, P = 0,$$

$$N = P - Z$$

$$-1 = 0 - Z \quad \text{or} \quad Z = 1$$

The system is therefore unstable for $K > \frac{1}{2}$. It would be stable for $K < \frac{1}{2}$, when

$$N = 0, P = 0 \quad \Rightarrow \quad Z = 0.$$

Example 9.7 : Sketch the Nyquist plot and determine there from the stability of the following open-loop transfer function of unity feedback control systems.

$$(i)\ GH(s) = \frac{K(s+2)}{s^2(s+4)} \qquad (ii)\ GH(s) = \frac{K}{s(s^2+s+4)}$$

If the system is conditionally stable, find the range of K for which the system is stable.

Solution.

$$(i) \qquad GH(j\omega) = \frac{(K/2)(1+j\omega/2)}{(j\omega)^2(1+j\omega/4)}$$

The Nyquist contour is the same as in Fig. 9.13(*a*). Following standard steps the Nyquist plot is sketched in Fig. 9.15.

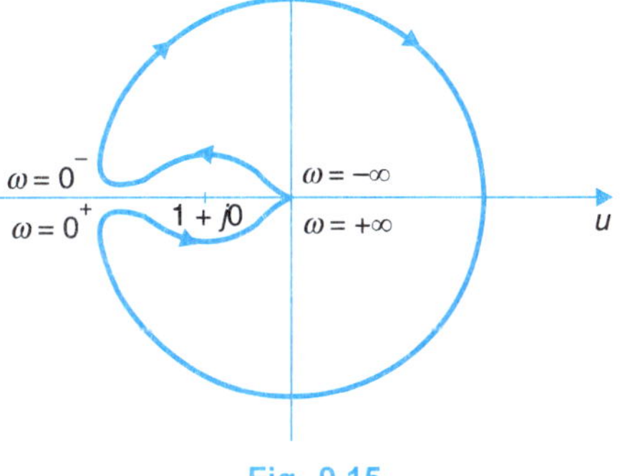

Fig. 9.15

It easily concluded from the Nyquist plot that the system is absolutely stable.

$$(ii) \qquad GH(j\omega) = \frac{K}{j\omega(4-\omega^2+j\omega)}$$

Real axis crossing of Nyquist plot at

$$GH(j\omega) = \frac{-K[\omega + j(4-\omega^2)]}{\omega[\omega^2 + (4-\omega^2)^2]}$$

Equating imaginary part to zero

$$4 - \omega^2 = 0 \quad \text{or} \quad \omega^2 = 4 \quad \text{or} \quad \omega = \pm 2$$

$$|GH(j\omega)|_{\omega^2=4} = -K/4$$

So the Nyquist plot crosses the axis of reals at $\omega = \pm 2$ with an intercept of $-K/4$. The plot is sketched in Fig. 9.16.

From the Nyquist plot

$$K/4 > 1 \quad \text{or} \quad K > 4$$

For this value of K

$$N = P - Z$$

$$-2 = 0 - Z \quad \text{or} \quad Z = 2$$

So the system is unstable. It can be shown to be stable for

$$K < 4$$

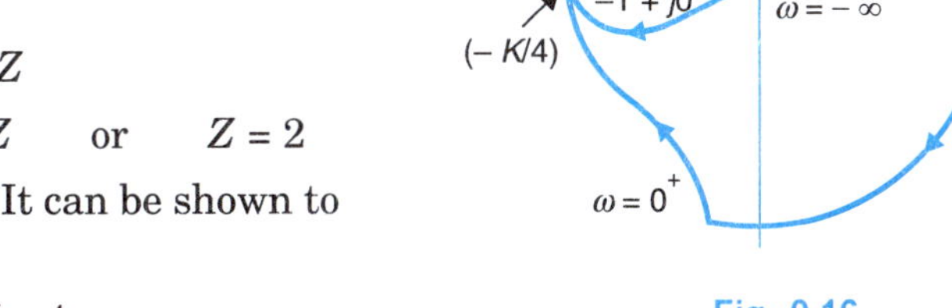

Fig. 9.16

Example 9.8 : The Nyquist plot of a system is shown in Fig. 9.17 for a particular value of gain of the open-loop system. The open-loop system has no right-hand s-plane poles. Is the closed-loop (unity feedback) system stable ? If the answer is no, determine right-half s-plane zeros of the characteristic equation.

The system gain is now adjusted such that the point $(-1 + j0)$ lies at point P. Note that this would correspond to a new scale for the same figure. Determine once again the stability of the system. In the process has the gain been increased or decreased.

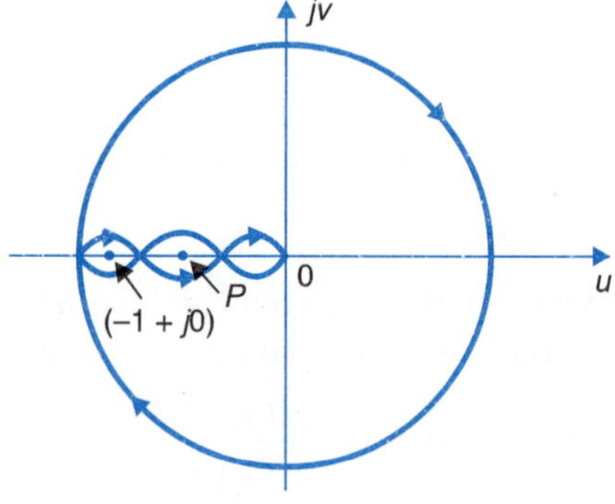

Fig. 9.17

Solution.

$$N = P - Z$$

$$P = 0 \text{ (given)}, N = -2, \text{ from the Nyquist plot}$$

Then $\quad -2 = 0 - Z \quad \text{or} \quad Z = 2$

The closed-loop system is unstable with two roots of the characteristic equation lying in the right half of the s-plane.

When $(-1 + j0)$ lies at point P, indicated, then

$$N = -1 + 1 = 0$$

$$N = P - Z$$

$$0 = 0 - Z \quad \text{or} \quad Z = 0$$

The system is therefore stable. The system is conditionally stable for a range of K. If the gain is further reduced such that $(-1 + j0)$ lies inside the right most loop, it can easily be seen that the system would be stable.

$(-1 + j0)$ shifts to point P by reducing gain.

Nyquist Stability Criterion Applied to Inverse Polar Plots

Occasionally, it is found more convenient* to work with the inverse function $1/G(j\omega)H(j\omega)$ rather than the direct function $G(j\omega)H(j\omega)$. In the following we shall see that the Nyquist stability criterion for direct polar plots can be extended for use to inverse polar plots after minor modification.

Let us consider once again a single-loop feedback system with open-loop transfer function (written in the standard form)

$$G(s)H(s) = K\frac{(s+z_1)(s+z_2)\ldots(s+z_m)}{(s+p_1)(s+p_2)\ldots(s+p_n)} \;;\; m \le n \qquad \ldots(9.10)$$

For the system to be stable none of the roots of the characteristic equation

$$q(s) = 1 + G(s)H(s) = \frac{(s+z_1')(s+z_2')\ldots(s+z_n')}{(s+p_1)(s+p_2)\ldots(s+p_n)} \qquad \ldots(9.11)$$

should lie in the right half s-plane or on the $j\omega$-axis.

*As shall be shown later, inverse polar plots are particularly useful for nonunity feedback systems.

Dividing eqn. (9.11) by eqn. (9.10), we get

$$q'(s) = \frac{1}{G(s)H(s)} + 1 = \frac{(s+z_1{}')(s+z_2{}')\ldots(s+z_n{}')}{(s+z_1)(s+z_2)\ldots(s+z_m)} \quad \ldots(9.12)$$

From eqns. (9.11) and (9.12) it is seen that the zeros of $q'(s)$ are same as zeros of $q(s)$, which are the roots of the characteristic equation. It is further noticed that poles of $q(s)$ are same as the poles of $G(s)H(s)$, while poles of $q'(s)$ are same as the poles of $1/G(s)H(s)$ or the zeros of $G(s)H(s)$.

It can be easily concluded from above that if $1/G(s)H(s)$ has P right half s-plane poles and the characteristic equation has Z right half s-plane zeros; the locus of $1/G(s)H(s)$ encircles the point $(-1 + j0)$ N times in counter-clockwise direction where

$$N = P - Z$$

Since stability implies absence of right half s-plane zeros of the characteristic equation, *i.e.*, $Z = 0$, the Nyquist stability criterion for inverse polar plots immediately follows and can be stated as below:

If the Nyquist plot of $1/G(s)H(s)$, corresponding to the Nyquist contour in the s-plane, encircles counter-clockwise the point $(-1 + j0)$ as many times as are the number of right half s-plane poles of $1/G(s)H(s)$, the closed-loop system is stable. In the special case where $1/G(s)H(s)$ has no poles in the right half s-plane, the closed-loop system is stable provided the net encirclement of $(-1 + j0)$ point by the Nyquist plot of $1/G(s)H(s)$ is zero.

Example 9.9 : Consider a feedback system with an open-loop transfer function

$$G(s)H(s) = K/s(Ts + 1)$$

The inverse polar plot of $G(s)H(s)$ corresponding to the s-plane Nyquist contour of Fig. 9.18(a) is obtained in steps below.

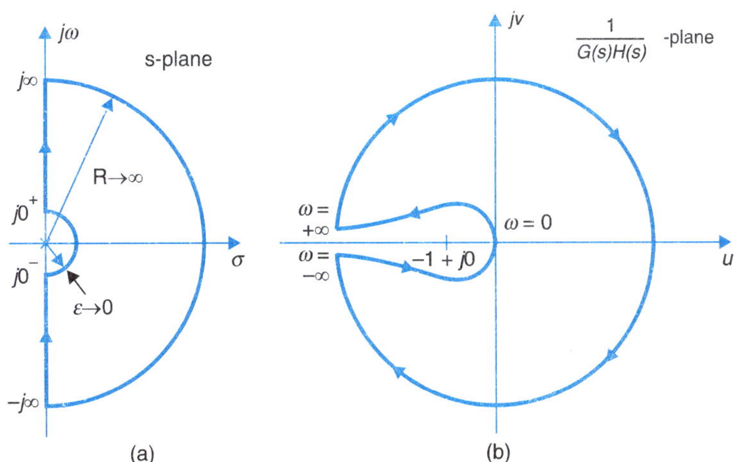

Fig. 9.18. The Nyquist contour and the corresponding plot of 1/G(s)H(s) = s(sT + 1)/K.

1. The semicircular indent around the origin in the s-plane is represented by

$$s = \lim_{\varepsilon \to 0} \varepsilon e^{j\theta}; \text{ where } \theta \text{ varies from } -90° \text{ through } 0° \text{ to } +90°.$$

It is mapped into $1/G(s)H(s)$-plane as

$$\lim_{\varepsilon\to 0}\left[\frac{\varepsilon e^{j\theta}(\varepsilon e^{j\theta}T+1)}{K}\right]=\lim_{\varepsilon\to 0}\frac{\varepsilon}{K}e^{j\theta}=0e^{j\theta}$$

2. Along the $j\omega$-axis $1/G(j\omega)H(j\omega) = j\omega(j\omega T + 1)/K$.

3. The infinite semicircle in the s-plane represented by

$$s=\lim_{R\to\infty} Re^{j\theta};\ \theta \text{ varies from } +90° \text{ through } 0° \text{ to } -90°$$

is mapped into the $1/G(s)H(s)$-plane as

$$\lim_{R\to\infty}\frac{Re^{j\theta}(Re^{j\theta}+1)}{K}=\lim_{R\to\infty}\frac{R^2}{K}e^{j2\theta}$$

which is a circle of infinite radius with angle varying from 180° through 0° to –180°.

The inverse polar plot of $G(s)H(s)$ obtained from the above steps is shown in Fig. 9.18(b). It is found that $(-1 + j0)$ point is not encircled by $1/G(s)H(s)$-locus. Further since $1/G(s)H(s) = s(Ts + 1)/K$ has no poles in the right half s-plane, the system is stable.

9.4 ASSESSMENT OF RELATIVE STABILITY USING NYQUIST CRITERION

In Chapter 7, the need to determine the relative stability of a system, in addition to its absolute stability was clearly established. In this section the use of Nyquist plot shall be extended to extract information regarding relative stability, thus eliminating the need to determine the location of the roots of characteristic equation. The Nyquist plot thus becomes a powerful designer's tool which provides information regarding both stability and relative stability of a feedback system.

Measure of Relative Stability

Measure of relative stability of closed-loop systems which are open-loop stable can be conveniently created through Nyquist plots. The stability information for such systems becomes obvious by inspection of the polar plot of the open-loop function $G(s)H(s)$ since the stability criterion is merely the non-encirclement of $(-1 + j0)$ point. It can be intuitively imagined that as the polar plot gets closer to $(-1 + j0)$ point, the system tends towards instability.

Consider two different systems whose dominant closed-loop poles are shown on the s-plane in Figs. 9.19(a) and (b), respectively. Obviously system A is 'more stable' than system B, since its dominant closed-loop poles are located comparatively away to the left from the $j\omega$-axis. The open-loop frequency response (polar) plots for systems A and B are shown in Figs. 9.19(c) and (d), respectively. The comparison of the closed-loop pole locations of these two systems with their corresponding polar plots reveals that as a polar plot moves closer to $(-1 + j0)$ point, the system closed-loop poles move closer to the $j\omega$-axis and hence the system becomes relatively less stable and vice versa.

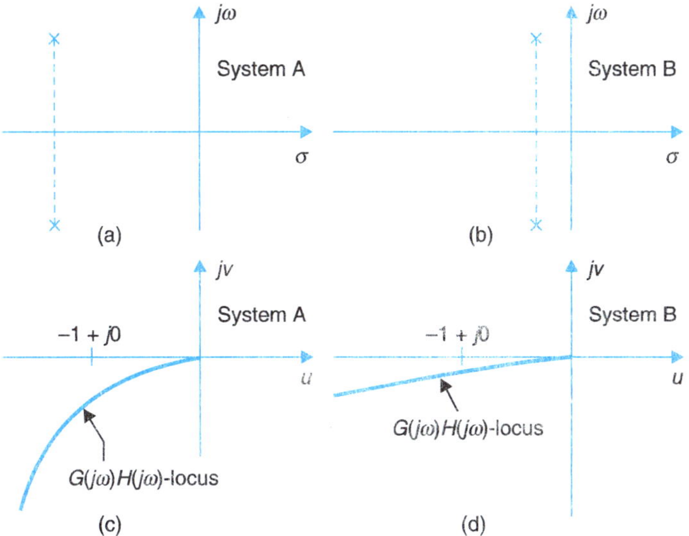

Fig. 9.19. Correlation between the closed-loop s-plane root locations and open-loop frequency response curves.

Figure 9.20 shows a typical $G(j\omega)H(j\omega)$-locus which crosses the negative real axis at a frequency $\omega = \omega_2$ with an intercept of a. Let a unit circle centred at origin [obviously it passes through the point $(-1 + j0)$] intersect the $G(j\omega)H(j\omega)$-locus at a frequency $\omega = \omega_1$ and let the phasor $G(j\omega_1)H(j\omega_1)$ make an angle of ϕ with the negative real axis measured positively in counter-clockwise direction. It is immediately observed that as $G(j\omega)H(j\omega)$-locus approaches $(-1 + j0)$ point, the relative stability reduces. Simultaneously, the value of a approaches unity and that of ϕ tends to zero. The relative stability could thus be measured in terms of the intercept a or the angle ϕ. These concepts are used to define *gain margin* and *phase margin* as practical measures of relative stability. It must be stressed here that gain margin and phase margin concepts are applicable to open-loop stable systems only. Vast majority of practical systems of course lie in this category.

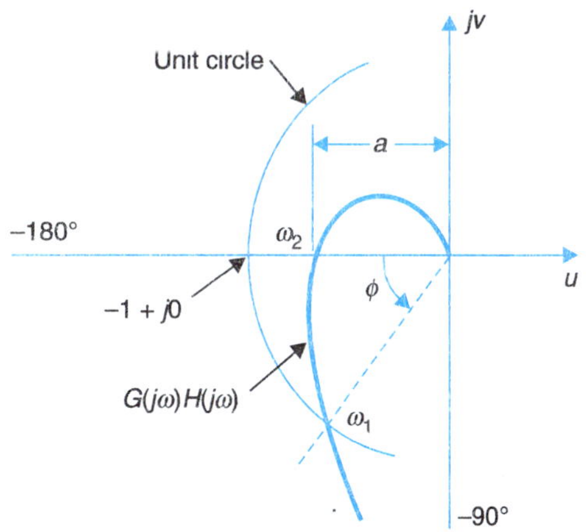

Fig. 9.20. A typical $G(j\omega)H(j\omega)$-locus.

Gain Margin and Phase Margin

Consider a feedback system with the following open-loop transfer function:

$$G(j\omega)H(j\omega) = \frac{K}{j\omega(j\omega T_1 + 1)(j\omega T_2 + 1)} \qquad ...(9.13)$$

Let us investigate the stability of this system for various values of K. The polar plots of $G(j\omega)H(j\omega)$ for various values of K are shown in Fig. 9.21. The point of intersection of the polar plot with the negative real axis can be determined by setting the imaginary part of $G(j\omega)H(j\omega)$ equal to zero. From eqn. (9.13)

$$G(j\omega)H(j\omega) = u + jv = \frac{-K(T_1 + T_2) - jK(1/\omega)(1-\omega^2 T_1 T_2)}{1+\omega^2(T_1^2 + T_2^2) + \omega^4 T_1^2 T_2^2} \quad ...(9.14)$$

Let ω_2 be the frequency at the point of intersection. Then from eqn. (9.14) we have

$$v = \frac{-K(1/\omega_2)(1-\omega_2^2 T_1 T_2)}{1+\omega_2^2(T_1^2 + T_2^2) + \omega_2^4 T_1^2 T_2^2} = 0$$

which gives

$$\omega_2 = 1/\sqrt{(T_1 T_2)} \quad ...(9.15)$$

From eqn. (9.14) the magnitude of the real part of $G(j\omega)H(j\omega)$ at the frequency ω_2 is given by

$$u = \frac{-K(T_1 + T_2)}{1+\omega_2^2(T_1^2 + T_2^2) + \omega_2^4 T_1^2 T_2^2}$$

$$= -\frac{KT_1T_2}{(T_1 + T_2)} \quad ...(9.16)$$

From Fig. 9.21 it is seen that for the system to be stable

$$K \frac{T_1 T_2}{T_1 + T_2} < 1$$

or

$$K < (T_1 + T_2)/T_1 T_2$$

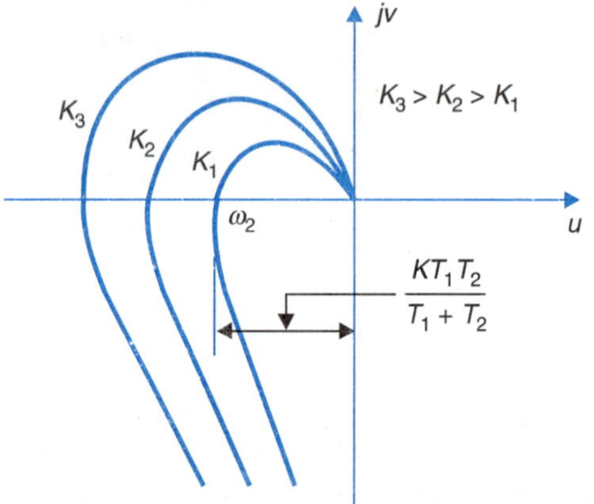

Fig. 9.21. The polar plots of G(jω)H(jω) for various values of gain K.

When the gain K is less than $(T_1 + T_2)/T_1T_2$, the $(-1 + j0)$ point is not encircled by $G(j\omega)H(j\omega)$-plot and the system is stable. When K is equal to $(T_1 + T_2)/T_1T_2$, the plot passes through the point $(-1 + j0)$ which indicates that the system has roots on the $j\omega$-axis. When K is further increased so as to be greater than $(T_1 + T_2)/T_1T_2$, the $G(j\omega)H(j\omega)$-plot encircles the $(-1 + j0)$ point and hence the system is unstable. The margin between the actual gain and critical gain causing sustained oscillations, is a measure of relative stability and is called the gain margin which is defined below.

Gain margin. It is the factor by which the system gain can be increased to drive it to the verge of instability.

In Fig. 9.20, it is seen that at $\omega = \omega_2$ the phase angle $\angle G(j\omega)H(j\omega)$ is 180° and $|G(j\omega)H(j\omega)|$ is a. If the gain of the system is increased by a factor $1/a$, then $|G(j\omega)H(j\omega)|_{\omega = \omega_2}$ becomes $a(1/a) = 1$ and hence the $|G(j\omega)H(j\omega)|$-plot will pass through $(-1 + j0)$ point, driving the system to the verge of instability. Therefore, the gain margin (GM) may be defined as the reciprocal of the gain at the frequency at which the phase angle becomes 180°. The frequency at which the phase angle is 180° is called the *phase cross-over frequency*.

With reference to Fig. 9.20, we have

$$GM = 1/a.$$

where $a = |G(j\omega)H(j\omega)|_{\omega = \omega_2}$.

In decibels the increase in gain for $G(j\omega)H(j\omega)$-plot to pass through $(-1 + j0)$ is given by

$$\text{GM} = -\,20 \log a \text{ db}$$

Since a is less than 1 for stable systems, $\log a$ is negative and hence GM is positive.

In the example considered earlier in this section if $T_1 = 1$, $T_2 = 0.5$, then for $K = 0.75$, the gain margin is given by

$$\text{GM} = \left[\frac{KT_1T_2}{T_1 + T_2}\right]^{-1} = 4$$

In decibels, the gain margin is given by

$$\text{GM} = 20 \log 4 = 12 \text{ db}$$

This value of gain margin indicates that the system gain may be increased by a factor of 4 before the stability limit is reached.

Phase margin. The frequency at which $|\, G(j\omega)H(j\omega) \,| = 1$ is called the *gain cross-over frequency.* It is given by the intersection of the $G(j\omega)H(j\omega)$-plot and a unit circle centred at the origin as shown in Fig. 9.20. At this frequency, the phase angle $\angle G(j\omega_1)H(j\omega_1)$ is equal to $(-\,180° + \phi)$. If an additional phase-lag equal to ϕ is introduced at the gain cross-over frequency, the phase angle $\angle G(j\omega_1)H(j\omega_1)$ will become $-180°$, while the magnitude remains unity. The $G(j\omega)H(j\omega)$-plot will then pass through $(-1 + j0)$ point, driving the system to the verge of instability. This additional phase lag ϕ is known as the phase margin (PM).

The phase margin is thus defined as the amount of additional phase-lag at the gain cross-over frequency required to bring the system to the verge of instability. From Fig. 9.20 it is seen that the phase margin is measured positively in counter-clockwise direction from the negative real axis. The *phase margin is always positive for stable feedback systems.*

The value of phase margin for any system can be computed from

$$\text{Phase margin } \phi = \angle G(j\omega)H(j\omega)\big|_{\omega = \omega_1} + 180° \qquad ...(9.17)$$

where the angle at ω_1, the gain cross-over frequency, is measured negatively.

Gain margin (GM) and phase margin (PM) are frequently used for frequency response specifications by designers. It is important to note once again that these measures of stability are valid for open-loop stable systems only. A large gain margin or a large phase margin indicates a very stable feedback system but usually a very sluggish one. A GM close to unity or a PM close to zero corresponds to a highly oscillatory system. Usually a GM of about 6 db or a PM of 30-35° results in a reasonably good degree of relative stability. In most practical systems a good gain margin automatically guarantees a good phase margin and vice versa. However, the cases where the specification on one does not necessarily satisfy the other, also exist as shown in Figs. 9.22 and 9.23.

In a second-order system with $G(j\omega)H(j\omega) = K/j\omega(j\omega T + 1)$ whose polar plot is shown in Fig. 9.24. GM always remains fixed at infinite value as the plot always reaches the real axis at the origin, while the PM reduces continuously with increasing system gain. In this case, as shall be seen below, PM is the correct measure of relative stability.

Usually it is the phase margin which is specified as a measure of system performance in design.

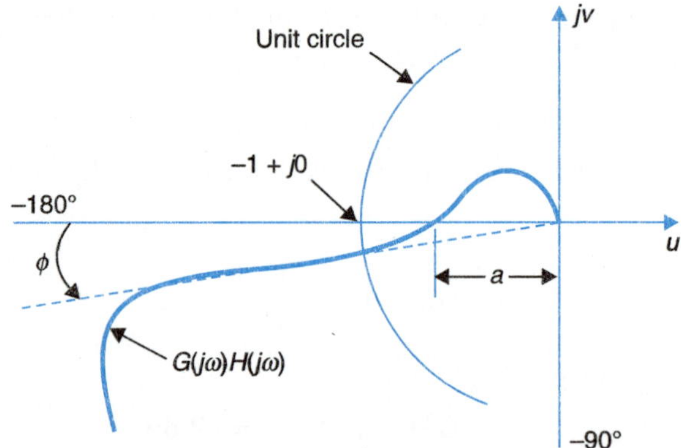

Fig. 9.22. Polar plot of a system with good GM and poor PM.

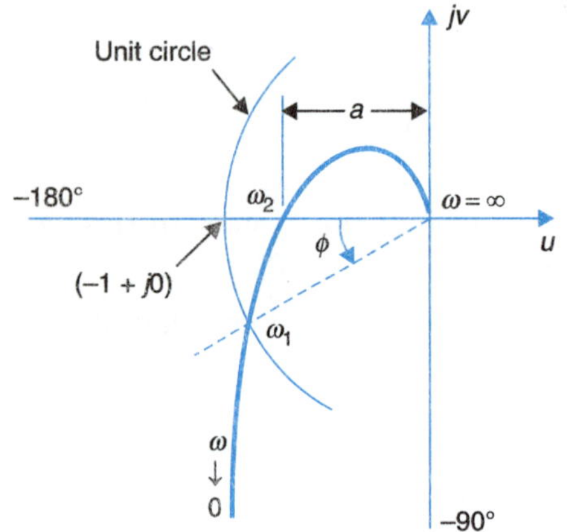

Fig. 9.23. Polar plot of a system with good PM and poor GM.

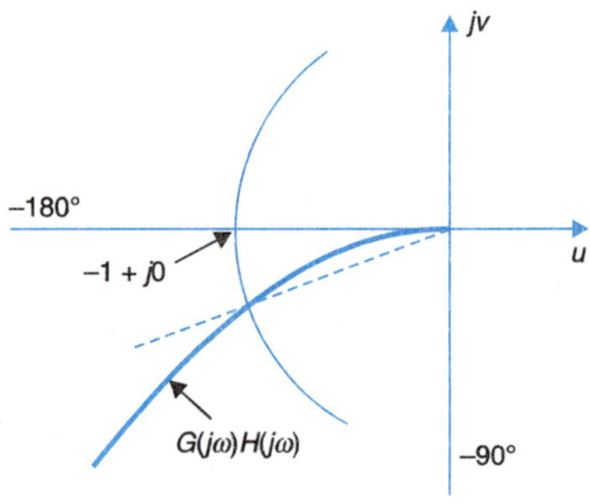

Fig. 9.24. Polar plot of a second-order system with GM always equal to infinity while PM reduces continuously with increasing gain.

Correlation Between Phase Margin and Damping Factor

Consider a unity feedback second-order system with an open-loop transfer function

$$G(s)H(s) = \frac{K}{s(\tau s+1)} = \frac{\omega_n^2}{s(s+2\zeta\omega_n)}$$

where $\omega_n = \sqrt{(K/\tau)}$ and $2\zeta\omega_n = 1/\tau$.

Replacing s by $j\omega$ for obtaining the polar plot, we have

$$G(j\omega)H(j\omega) = \frac{\omega_n^2}{j\omega(j\omega+2\zeta\omega_n)} \qquad ...(9.18)$$

At the gain cross-over frequency $\omega = \omega_1$, the magnitude $|G(j\omega)H(j\omega)| = 1$. Therefore from eqn. (9.18), we have

$$\frac{\omega_n^2}{\omega_1\sqrt{(\omega_1^2+4\zeta^2\omega_n^2)}} = 1 \qquad \text{or} \qquad (\omega_1^2)^2 + 4\zeta^2\omega_n^2(\omega_1)^2 - \omega_n^4 = 0$$

which yields

$$(\omega_1/\omega_n)^2 = \sqrt{(4\zeta^4 + 1)} - 2\zeta^2$$

The phase margin of this system is given by

$$\phi = -90^\circ - \tan^{-1}\left(\frac{\omega_1}{2\zeta\omega_n}\right) + 180^\circ$$

$$= 90^\circ - \tan^{-1}\left[\frac{1}{2\zeta}\{(4\zeta^4 + 1)^{1/2} - 2\zeta^2\}^{1/2}\right]$$

$$= \tan^{-1}\left[2\zeta\left\{\frac{1}{(4\zeta^4 + 1)^{1/2} - 2\zeta^2}\right\}^{1/2}\right] \quad ...(9.19)$$

Equation (9.19) gives a relationship between ζ and ϕ for an underdamped second-order system. The ζ-ϕ relationship is plotted in Fig. 9.25. In the range $\zeta \le 0.7$, a reasonably good linear approximation for ζ-ϕ relationship is given below and is shown by the dotted line in the same figure.

$$\zeta \approx 0.01\ \phi \quad ...(9.20)$$

where ϕ is in degrees.

Equation (9.19) and its approximation of eqn. (9.20) hold good for second-order systems, but can be used for higher-order systems as well, provided the transient response of the system is primarily contributed by a pair of dominant underdamped roots.

Computation of Gain Margin and Phase Margin

GM and PM may be computed by the use of various plots—direct polar plot, inverse polar plot, Bode plot, log-magnitude versus phase angle plot. In relatively simple cases GM and PM may be computed directly.

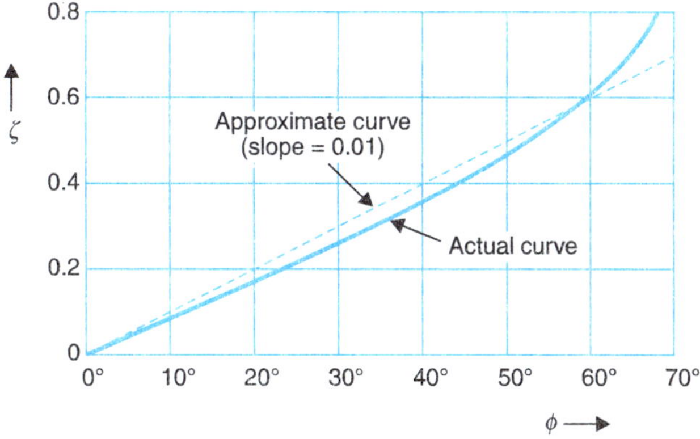

Fig. 9.25. Plot of ζ versus ϕ for a second-order system.

Example 9.10 : Consider a unity feedback system having an open-loop transfer function

$$G(j\omega) = \frac{K}{j\omega(j0.2\omega + 1)(j0.05\omega + 1)}$$

For $K = 1$

$$|G(j\omega)| = \frac{1}{\omega\sqrt{[1+(0.2\omega)^2]}\sqrt{[1+(0.05\omega)^2]}}$$

and $$\angle G(j\omega) = -90° - \tan^{-1} 0.2\omega - \tan^{-1} 0.05\,\omega$$

The Nyquist plot of $G(j\omega)$, for $K = 1$, is shown in Fig. 9.26. From this plot it is found that

$$\text{GM} = 20 \log (1/0.04) = 20 \log 25 = 28 \text{ db}$$

$$\text{PM} = 76°$$

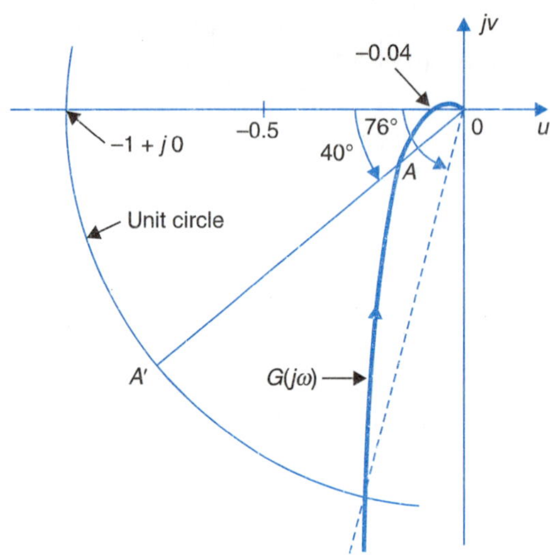

Fig. 9.26. The Nyquist plot of $G(j\omega)$ for $K = 1$.

Let us now use the Nyquist plot for adjustment of system gain for specified GM or PM. Suppose it is desired to find the open-loop gain for

(*i*) a GM to 20 db (*ii*) a PM of 40°

(*i*) For a GM of 20 db, the Nyquist plot should intersect the real axis at $-a$ where

$$20 \log (1/a) = 20 \qquad \text{or} \qquad a = 0.1$$

This is achieved if the system gain is increased by a factor of 0.1/0.04 = 2.5. Thus $K = 2.5$.

(*ii*) A PM of 40° is obtained if the system gain is increased such that point A is shifted to location A' in Fig. 9.26. This is achieved if the system gain is increased by a factor of $0A'/0A$ = 1/0.191 = 5.24. Thus $K = 5.24$.

The above example can also be solved by computation without the need of drawing Nyquist plot (this is easily possible for first-and second-order systems and for those third-order systems which have one open-loop pole at origin and no open-loop zeros).

(*i*) As calculated already, for a GM of 20 db,

$$a = 0.1$$

The Nyquist plot intersects the real axis at a point where

$$G(j\omega) = \frac{K}{j\omega(j0.2\omega + 1)(j0.05\omega + 1)} = \frac{K}{-0.25\omega^2 + j\omega(1 - 0.01\omega^2)} \quad ...(9.21)$$

is real. Setting the imaginary part of eqn. (9.21) equal to zero, we have

$$\omega = \omega_2 = 10 \text{ rad/sec}$$

Now $$|G(j\omega)|_{\omega=\omega_2} = \frac{K}{0.25(10)^2} = a = 0.1$$

which gives $$K = 2.5$$

(*ii*) Let $\omega = \omega_1$ be the gain cross-over frequency. Then for a PM of 40°

$$-90° - \tan^{-1} 0.2\omega_1 - \tan^{-1} 0.05\omega_1 + 180° = 40°$$

or $$\tan^{-1} 0.2\omega_1 + \tan^{-1} 0.05\omega_1 = 50°$$

$$\frac{0.25\omega_1}{1 - 0.01\omega_1^2} = \tan 50° = 1.2$$

or $$0.012\omega_1^2 + 0.25\omega_1 - 1.2 = 0$$

Solving for positive value of ω_1, we get

$$\omega_1 = 4 \text{ rad/sec}$$

Hence $$|G(j\omega)|_{\omega=\omega_1} = \frac{K}{\omega_1\sqrt{[1 + (0.2\omega_1)^2]}\sqrt{[1 + (0.05\omega_1)^2]}} = 1$$

which gives $$K = 5.2.$$

Example 9.11 : Let us determine the gain margin and phase margin of a unity feedback system having an open-loop transfer function

$$G(j\omega) = \frac{10}{j\omega(j0.1\omega + 1)(j0.05\omega + 1)}$$

by use of Bode plot.

The Bode plot of $G(j\omega)$ is shown in Fig. 9.27. From this figure, it is seen that GM = 9.5 db and PM 33°.

Let us now use the Bode plot for adjustment of the system gain for a specified GM or PM. Suppose it is desired to find the open-loop gain for (*i*) a GM of 20 db; and (*ii*) a PM of 24°.

(*i*) A GM of 20 db will be obtained if the log-magnitude plot in Fig. 9.27 is shifted downwards by (20 – 9.5) = 10.5 db. The system gain must therefore be reduced by 10.5 db or by a factor of 3.3. The corresponding open-loop gain is 10/33 ≈ 3.

(*ii*) From Fig. 9.27, it is observed that if the gain cross-over frequency is changed to 9.3 rad/sec, a PM of 24° is obtained. To change the gain cross-over frequency to 9.3 rad/sec, the log-magnitude curve should be raised upwards by 3.5 db or system gain should be raised upwards by 3.5 db or system gain should be increased by a factor of 1.5. Hence the open-loop gain for a PM of 24° is 10 × 1.5 = 15.

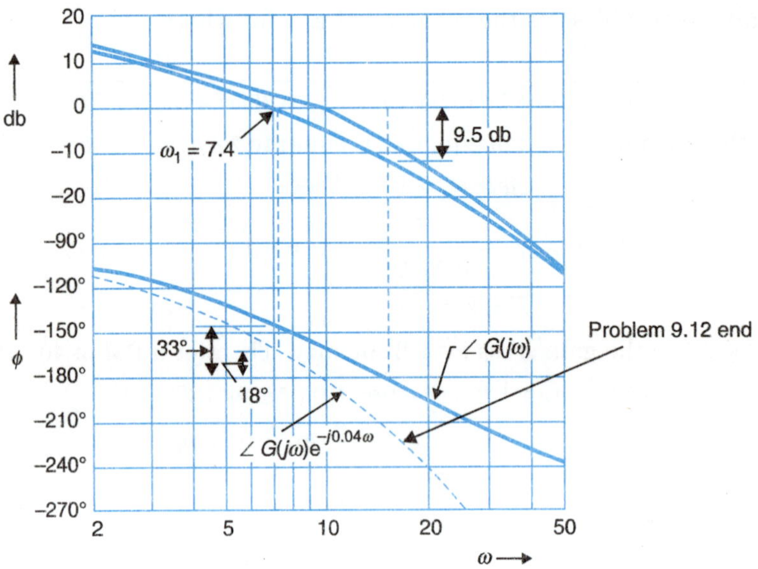

Fig. 9.27. Bode plot of $G(j\omega) = 10/j\omega(j0.1\omega + 1)(j0.05\omega + 1)$.

Systems with Transportation Lag

As was discussed earlier in Chapters 7 and 8, certain elements in control systems are characterized by a dead time or transportation lag. For the block diagram shown in Fig. 9.28, the open-loop transfer function of the system is

$$G(s) = e^{-sT} G_1(s)$$

As $G(s)$ is a transcendental function, analytical approach to a system described by such an equation is difficult. One of the methods of handling a transcendental transfer function is to convert it into a rational function by approximating the exponential term by a polynomial of s so that the usual techniques of analysis can be applied. The approximation suggested in Chapter 7 is

$$e^{-sT} \approx 1 - sT$$

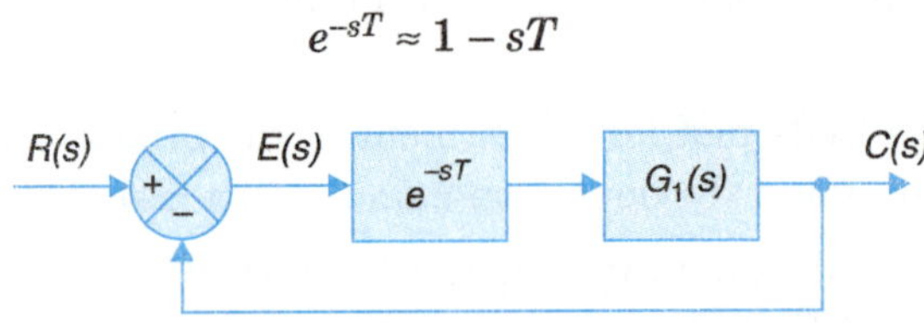

Fig. 9.28. Simple system with transportation lag.

This approximation works fairly well as long as the dead time T is small in comparison to the system time constant and in addition the input time function to dead time element is smooth and continuous.

As was pointed out in Chapter 7, the control systems with dead time could be analyzed with root locus technique without the need to make any approximation but the method is quite complex and time consuming. On the other hand the frequency domain graphical methods provide a simple yet exact approach to handle this problem since the function $e^{-sT}|_{s=j\omega} = e^{-j\omega T}$ is readily interpreted in terms of either the Nyquist or Bode plot.

Assume that the transfer function $G_1(s)$ in Fig. 9.28 is given by

$$G_1(s) = 1/s(s+1) \quad \text{...(9.22)}$$

Then the open-loop transfer function of the system becomes

$$G(s) = e^{-sT}/s(s+1) \quad \text{...(9.23)}$$

The polar plot of e^{-sT} is shown in Fig. 8.19 (*a*), using which the Nyquist plot of $G(j\omega)$ for various values of T is drawn in Fig. 9.29. From this figure, it is observed that the effect of $e^{-j\omega T}$ term in eqn. (9.23) is simply to rotate each point of the $G_1(j\omega)$ plot by an angle of ωT rad in the clockwise direction. Relative stability of the system reduces as T increases. Sufficiently large values of T may drive the system to instability. The effect of e^{-sT} on stability can be determined by the Nyquist stability criterion as follows.

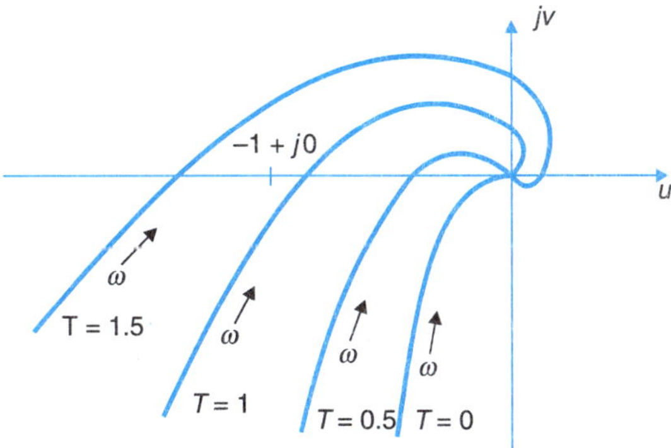

Fig. 9.29. Nyquist plot of $G(s) = e^{-sT}/s(s+1)$.

The characteristic equation of the system shown in Fig. 9.28 is given by

$$1 + G_1(s)e^{-sT} = 0 \qquad \text{or} \qquad G_1(s) = \frac{1}{s(s+1)} = -e^{+sT} \quad \text{...(9.24)}$$

If eqn. (9.24) is satisfied at a particular frequency, the system will exhibit sustained oscillations. The corresponding condition for systems without transportation lag is

$$G_1(j\omega) = -1 \quad \text{...(9.25)}$$

Comparison of eqns. (9.24) and (9.25) reveals that the effect of e^{-sT} is simply to shift the critical stability point from $(-1 + j0)$ to $(-e^{+j\omega T})$, *i.e.*, the critical point now becomes a critical locus.

Example 9.12 : Let us determine with the help of the Nyquist stability criterion, the maximum value of T for the stability of the closed-loop system of Fig. 9.28. From eqn. (9.24), the following condition is satisfied when the system is limitedly stable (*i.e.*, it has sustained oscillations):

$$G_1(j\omega) = \frac{1}{j\omega(j\omega+1)} = -e^{+j\omega T}$$

The polar plots of $G_1(j\omega)$ and $-e^{+j\omega T}$ are shown in Fig. 9.30. These two plots intersect at the point A corresponding to $\omega = 0.75$ rad/sec on $G_1(j\omega)$-plot and $\omega T = 52° \ (\pi/180) = 0.9$ rad/sec on $-e^{+j\omega T}$ plot. For the point A to pertain to the same frequency on both $G_1(j\omega)$ and $-e^{+j\omega T}$ plot, we have

$$0.75\ T = 0.9 \qquad \text{or} \qquad T = 1.2 \text{ sec}$$

From Fig. 9.30 it is observed that the critical point on unit circle ($-e^{j\omega T}$-locus) is enclosed by $G_1(j\omega)$-plot for $T > 1.2$ and is not enclosed for $T < 1.2$ and hence we conclude that the system under discussion is stable if $T < 1.2$ sec.

Transportation lag can be conveniently handled on Bode plot as well without the need to make any approximation. The log-magnitude of transportation lag is

$$20 \log | e^{-j\omega T} | = 0$$

Thus the open-loop log-magnitude plot of a system is unaffected by the presence of upright transportation lag. The lag, of course contributes a phase angle of $-(\omega T \times 180°)/\pi$, thereby causing the modification of the phase plot.

If in Example 9.11, a transportation lag e^{-sT} ($T = 0.04$ sec.) is brought in, the phase plot modifies to that shown dotted in Fig. 9.27. For the given gain, the phase margin reduces to 18°, *i.e.*, as expected, the relative stability of the system reduces due to the presence of transportation lag.

The value of T for the system to be on verge of instability is obtained by setting the phase margin equal to zero, *i.e.*,

$$\angle G_1(j\omega)_{\omega = \omega_1} - \frac{\omega_1 T \times 180°}{\pi} = -180°$$

where $G_1(j\omega)$ is the system without transportation lag and ω_1 is the gain crossover frequency.

We have from Fig. 9.27, $\omega_1 = 7.4$ rad/sec. Thus

$$-153° - \frac{7.4T \times 180°}{\pi} = -180° \qquad \text{or} \qquad T = 0.063 \text{ sec.}$$

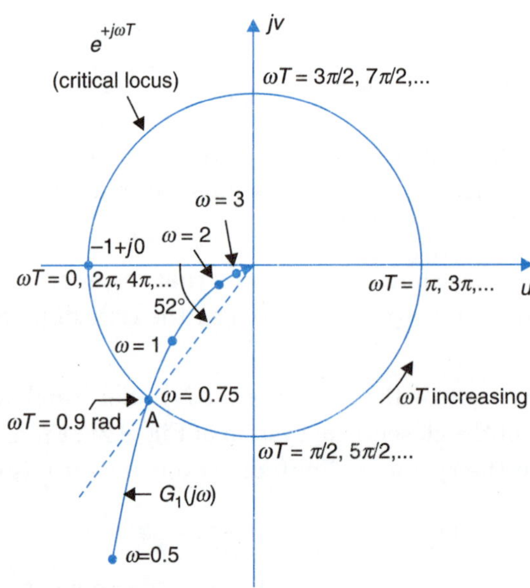

Fig. 9.30. Critical locus and $G_1(j\omega)$-plot for Example 9.8.

Example 9.13 : A unity feedback system has a plant transfer function of

$$G(s) = \frac{K(s+4)}{(s-1)(s-2)}$$

(*a*) For $K = 8$ draw the Bode plots and find therefrom the phase margin and gain margin.

(*b*) What would be the value of K for a phase margin of 30° and what is the corresponding gain margin.

Solution :

$$G(s) = \frac{K(s+4)}{(s-1)(s-2)} = \frac{2K(1+s/4)}{(1-s)(1-s/2)}$$

$$G(j\omega) = \frac{K'(1+j\omega/4)}{(1-j\omega)(1-j\omega/2)} \; ; K' = 2K$$

The corner frequencies are:

Denominator 1, 2

Numerator 4

For $K = 8$ db, db K' = db (16) = 24

The Bode plots of $G(j\omega)$ are drawn in Fig. 9.31 from which are read and following results:

(*a*) $K = 8$

PM = 45°, GM = 8 db

(*b*) PM = 30°

The gain should be reduced by 4 db. Therefore

$$K'(\text{new}) = 24 - 4 = 20 \text{ db or } 10. \; K(\text{new}) = \frac{10}{2} = 5$$

$$\text{GM} = 8 - 4 = 4 \text{ db}$$

Example 9.14 : Fig. 9.32 shows the signal flow graph of a chemical concentration control. The output composition is maintained constant by adjusting the feedflow variable by a controller which operates on the error between the desired and measured (output) concentrations. A conveyer belt brings in the constituent in proper quantities (adjusted by the feedflow value) to the mixer resulting in a time delay. The mixer plant and controller transfer functions are:

$$G_p(s) = \frac{4}{4s+1} \qquad \text{and} \qquad G_c(s) = K_1 + \frac{K_2}{s} \; ;$$

proportional plus integral (PI) controller.

Conveyer delay is indicated in the signal flow diagram wherein $\tau = 1s$.

Investigate the stability of the system via Bode plots and adjust the open-loop gain to get a phase margin of 25° for

(*a*) $K_1 = K_2 = 1$

(*b*) $K_1 = 0.2, K_2 = 0.05$

(*c*) $K_2 = 0$ (proportional controller).

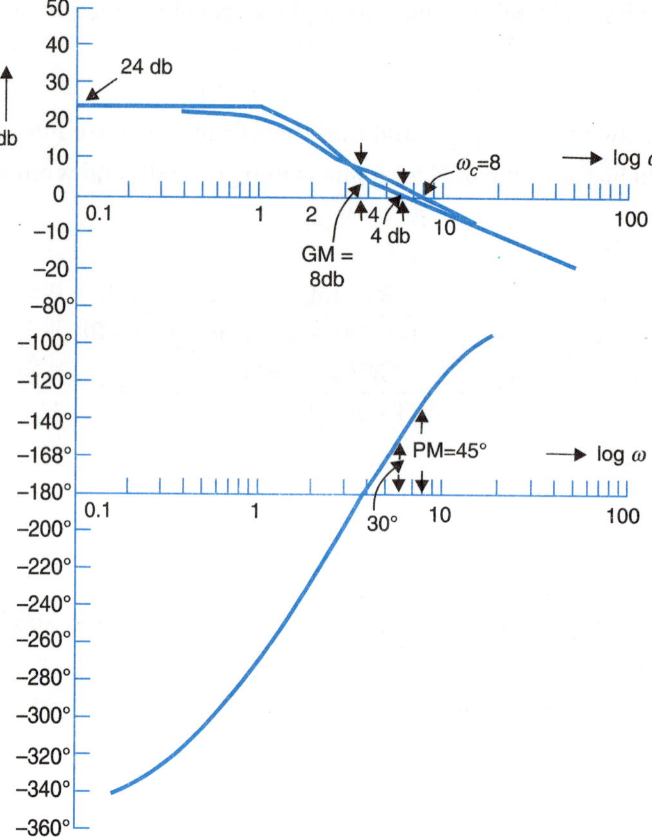

Fig. 9.31

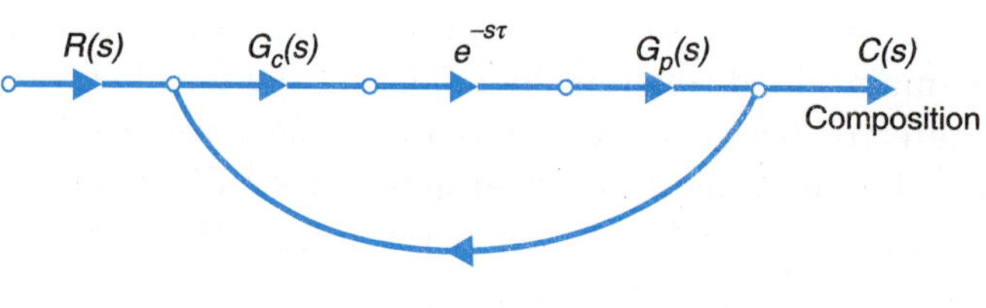

Fig. 9.32

Solution :

The open-loop transfer function is

$$G(s) = G_c(s)\, e^{-s\tau} G_p(s)$$

or
$$G(s) = \frac{4\left(K_1 + \dfrac{K_2}{s}\right)e^{-s\tau}}{(4s+1)} = \frac{4K_2(1 + sK_1/K_2)e^{-s\tau}}{s(4s+1)}$$

(*a*) $\tau = 1s$; $K_1 = K_2 = 1$, then

$$G(s) = \frac{4(1+s)e^{-s}}{s(1+4s)} \qquad ...(i)$$

$$G(j\omega) = \frac{4(1+j\omega)e^{-j\omega}}{j\omega(1+j4\omega)} \qquad ...(ii)$$

Break frequencies:

Denominator $1/4 = 0.25$

Numerator $= 1$

db (4) $= 12$

Bode plots are drawn in Fig. 9.33 (a), from which we find that:

Cross-over frequency, $\omega_c = 1$ rad/s

The system is stable with low phase margin, PM = 2°

Desired PM = 25°

The gain should be reduced by 9 db or by a factor of 2.8.

As per Eqn. (ii), New gain = 4/2.8 = 1.43

(b) $K_1 = 0.2, K_2 = 0.05$

$$K_1 + \frac{K_2}{s} = \frac{K_2}{s}(1 + sK_1/K_2) = \frac{0.05}{s}(1+4s)$$

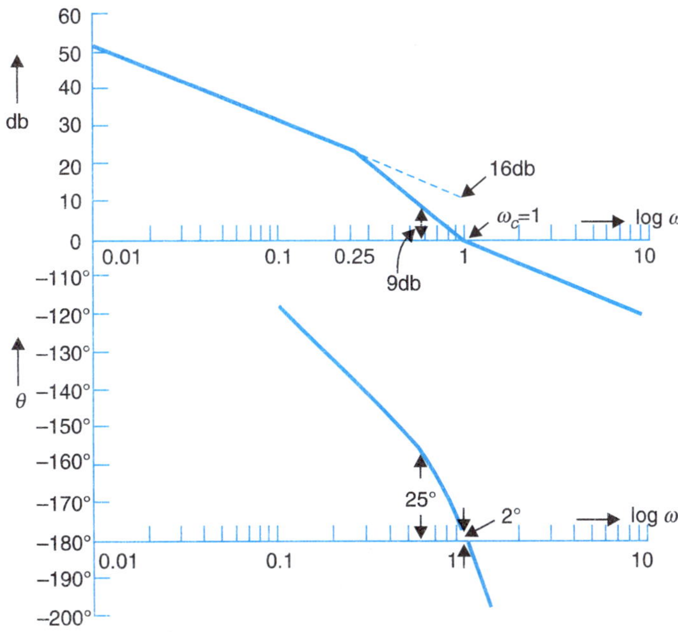

Fig. 9.33(a)

This cancels out the plant pole. Thus

$$G(j\omega) = \frac{0.2e^{-j\omega}}{j\omega}$$

$$\text{db}\,(0.2) = -14$$

The Bode plots are drawn in Fig. 9.33 (*b*). We read from there:

Cross-over frequency, $\omega_c = 3.2$ rad/s

System is stable but PM is too high (it would mean high damping and sluggish system)

For PM = 25°

Gain should be raised by +16 db or by a factor of 6.3.

New gain = 6.3 × 0.2 = 1.26

(*c*) $K_2 = 0$

$$G(j\omega) = \frac{4K_1 e^{-j\omega}}{(1 + j4\omega)}$$

Its Bode plots are drawn in Fig. 9.33 (*c*)

$$\text{db}\,(4) = 12\,;\, K_1 = L$$

Cross-over frequency, $\omega_c = 1$ rad/s

Phase margin = 46°

For PM = 25°

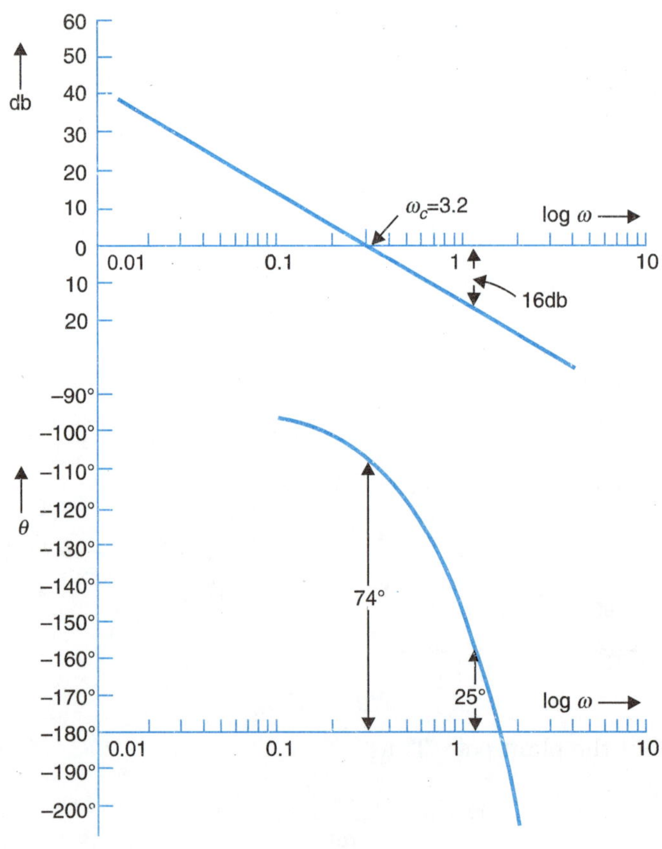

Fig. 9.33(b)

Gain has to be raised by 2db or by a factor of 1.26.

$$\text{New gain } 4K_1 = 4 \times 1.26 = 5 \quad \text{or} \quad K_1 = 1.26$$

The system would be unstable when the gain is raised more than 5 db or by a factor of 1.8

Thus the system would be unstable for

$$4K_1 > 4 \times 1.8 = 7.2 \qquad \text{or} \qquad K_1 > 1.8.$$

Note : The system though is type zero is conditionally stable because of the time delay which contributes to $G(j\omega)$ a lagging angle of $\angle \dfrac{\omega \times 180°}{\pi} = \angle 57.3°\ \omega$, which increases linearly with ω.

9.5 CLOSED-LOOP FREQUENCY RESPONSE

The study of closed-loop frequency response is very useful as it enables us to use the second-order correlations (discussed in Chapter 8) between frequency response and transient response to predict approximately the time response of feedback systems. With the help of these correlations, the time response specifications are first converted into a set of specifications in frequency domain. After design and compensation in frequency domain, the frequency response is translated back to give an approximate time response. Usually the specifications in frequency domain are given in the following terms.

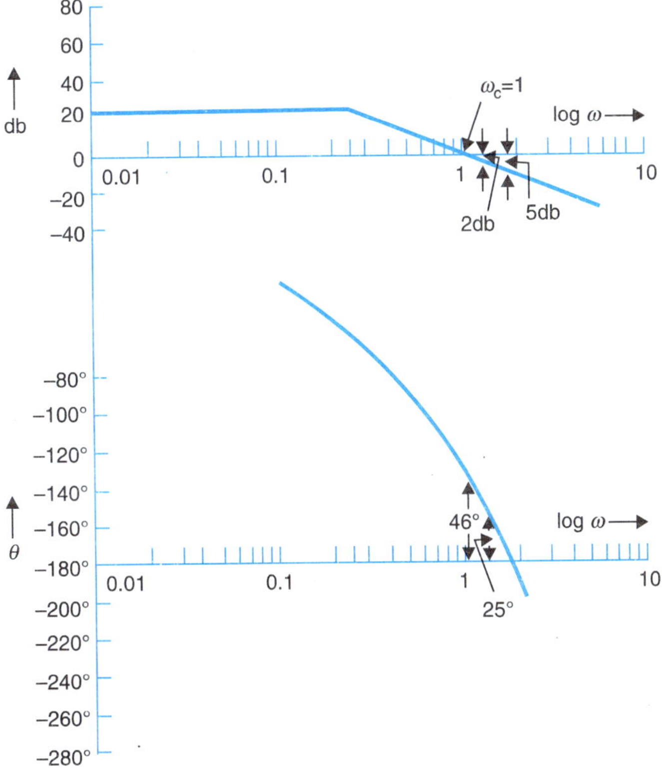

Fig. 9.33(c)

Frequency Domain Specifications

1. *Resonance peak, M_r.* This is the maximum value of M, the magnitude of the closed-loop frequency response. As discussed in Chapter 8, a large resonance peak corresponds to a large overshoot in transient response.

2. *Resonant frequency, ω_r.* This is the frequency at which the resonance peak M_r occurs. This is related to the frequency of oscillation in the step response and thus is indicative of the speed of transient response.

3. *Bandwidth.* As defined already in Chapter 8, it is the range of frequencies for which the system gain is more than –3 db. Such gain is considered adequate to ensure good transmission of signal. The closed-loop system filters out the signal components whose frequencies are greater than the cut-off frequency (frequency of – 3db gain) and transmits those signal components whose frequencies are lower than the cut-off frequency.

The bandwidth information is significant because it measures the ability of a feedback system to reproduce the input signal and also measures its noise rejection characteristics. It is also indicative of rise time in transient response for a given damping factor. A large bandwidth corresponds to small rise time or fast response.

4. *Cutt-off rate.* It is the slope of the log-magnitude curve near the cut-off frequency. The cut-off rate indicates the ability of the system to distinguish the signal from noise.

5. *Gain margin and phase margin.* As defined earlier in this chapter, these are the measures of relative stability and are related to the closeness of the closed-loop poles to the $j\omega$-axis. For a second-order system, the exact correlation between phase margin and ζ is given by eqn. (9.19).

From the frequency domain specifications, we find that the maximum value of M and the frequency at which it occurs are important figures of merit of a system. The frequency domain compensation methods (discussed in Chapter 10) are based upon the knowledge of these two factors. These factors can be evaluated from the closed-loop frequency response. However, this procedure, as is evident, is time consuming.

Graphical techniques are available which help in determining the values of M_r, and ω_r directly from the open-loop frequency response obviating the need of determining complete closed-loop frequency response. These graphical techniques require that constant-M contours be drawn on the complex plane. When these contours are used in conjunction with the opne-loop frequency response, the values of M_r and ω_r are easily obtained.

In the following paragraphs, we shall show that the constant-M and constant-N ($N = \tan \alpha$, α being the phase angle of the frequency response) contours are circles.

Constant-*M* Circles

Consider any point $G(j\omega) = x + jy$, on the polar plot of $G(j\omega)$. The closed-loop frequency response is

$$T(j\omega) = \frac{C(j\omega)}{R(j\omega)} = \frac{G(j\omega)}{1+G(j\omega)} = \frac{x+jy}{1+x+jy} = Me^{j\alpha} \quad \text{...(9.26)}$$

From eqn. (9.26), the magnitude M is given by

$$M = \frac{|x+jy|}{|1+x+jy|} = \left[\frac{x^2+y^2}{(1+x)^2+y^2}\right]^{1/2} \quad \text{or} \quad M^2 = \frac{x^2+y^2}{(1+x)^2+y^2}$$

Rearranging this equation, we get

$$y^2 + \left[x + \frac{M^2}{M^2 - 1}\right]^2 = \frac{M^2}{(M^2 - 1)^2} \quad \text{...(9.27)}$$

Equation (9.27) is the equation of a circle with centre at

$$x_0 = -\frac{M^2}{M^2 - 1}\ ;\ y_0 = 0 \quad \text{...(9.28)}$$

and with radius

$$r_0 = \frac{M}{M^2 - 1} \quad \text{...(9.29)}$$

Using eqns. (9.28) and (9.29) constant-M circles for various values of M can be drawn. Consider Fig. 9.34 in which the $G(j\omega)$-plot and three constant-M circles are plotted. It is observed that M_2-circle is tangent to the $G(j\omega)$-plot. Therefore the maximum value of M is M_2, *i.e.*, $M_r = M_2$.

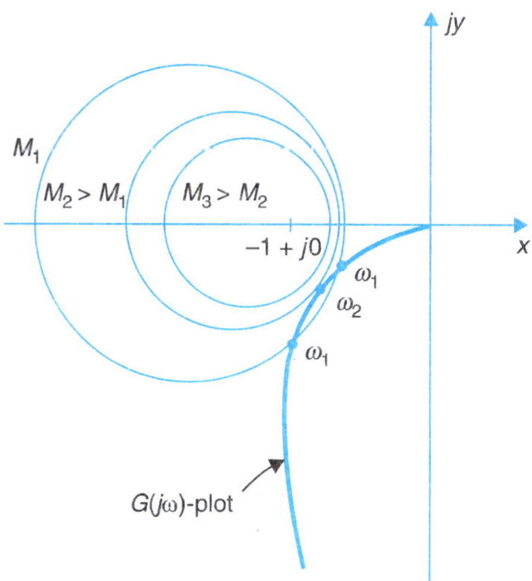

Fig. 9.34. Constant-M circles and $G(j\omega)$-plot.

A family of M-circles for different values of M is constructed in Fig. 9.35.

Constant-N Circles

From eqn. (9.26), the phase angle of $T(j\omega)$ is given by

$$\angle T(j\omega) = \alpha = \angle\left(\frac{x + jy}{1 + x + jy}\right)$$

$$= \tan^{-1}(y/x) - \tan^{-1}[y/(1 + x)] = \tan^{-1}[y/(x^2 + x + y^2)]$$

Therefore $$\tan\alpha = \frac{y}{x^2 + x + y^2} = N \quad \text{...(9.30)}$$

For a constant value of α, $N = \tan\alpha$ is also constant.

Rearranging eqn. (9.30), we get

$$\left(x+\frac{1}{2}\right)^2+\left(y-\frac{1}{2N}\right)^2=\frac{N^2+1}{4N^2} \quad ...(9.31)$$

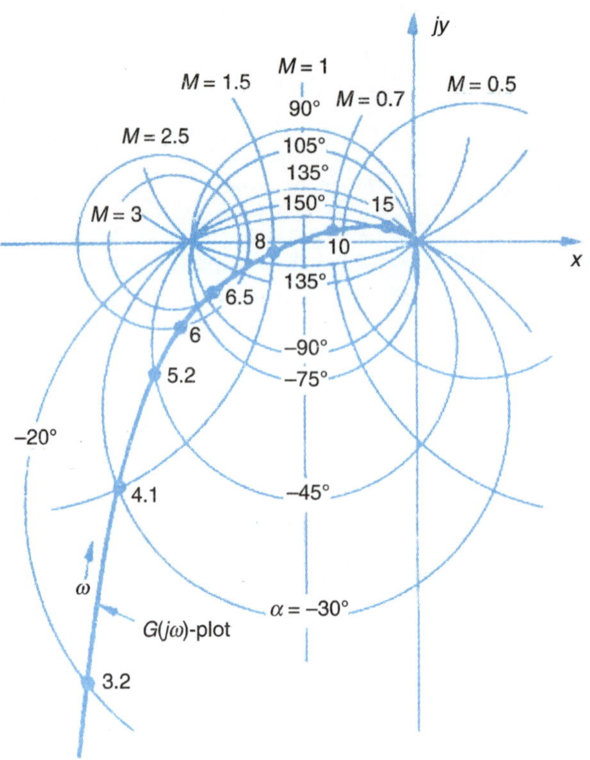

Fig. 9.35. Constant-M and constant-N circles.

This is the equation of a circle with centre at

$$x_0 = -1/2 \; ; \; y_0 = 1/2N \quad ...(9.32)$$

and with radius

$$r_0 = \frac{1}{2N}\,(N^2+1)^{1/2} \quad ...(9.33)$$

For different values of α, a family of N-circles is also constructed in Fig. 9.35. Since eqn. (9.31) is satisfied for $x = 0, y = 0$, and for $x = -1, y = 0$, all the constant-N circles pass through the origin and $(-1 + j0)$ point regardless of the value of N.

Example 9.14 : Let us compute the closed-loop frequency response of a unity feedback system with open-loop transfer function

$$G(j\omega) = \frac{10}{j\omega(1+0.2j\omega)(1+0.05j\omega)}$$

by use of constant-M and constant-N circles.

Figure 9.35 shows $G(j\omega)$-plot superimposed on constant-M and -N circles. At various values of frequencies the magnitude M and phase angle α are read off. With the data thus obtained, the closed-loop frequency response curves are plotted in Fig. 9.36.

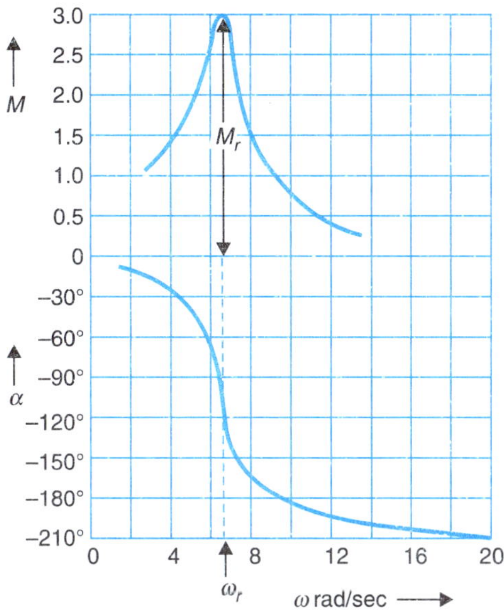

Fig. 9.36. Closed-loop frequency response.

Nonunity Feedback Systems

For a nonunity feedback system, the closed-loop transfer function is given by

$$\frac{C(j\omega)}{R(j\omega)} = T(j\omega) = \frac{G(j\omega)}{1+G(j\omega)H(j\omega)} \qquad \text{...(9.34)}$$

where $G(j\omega)$ is the forward path transfer function and $H(j\omega)$ is the feedback path transfer function.

Equation (9.34) may be written in the form

$$T(j\omega) = \frac{1}{H(j\omega)}\left[\frac{G(j\omega)H(j\omega)}{1+G(j\omega)H(j\omega)}\right]$$

$$= \frac{1}{H(j\omega)}\left[\frac{G_0(j\omega)}{1+G_0(j\omega)}\right] = \frac{1}{H(j\omega)}\,T_0(j\omega)$$

where $G_0(j\omega) = G(j\omega)H(j\omega)$ and $T_0(j\omega) = \dfrac{G_0(j\omega)}{1+G_0(j\omega)}$ ...(9.35)

Equation (9.35) is of the standard form of unity feedback systems. Constant-M and constant-N circles can be used in conjunction with $G_0(j\omega)$ to obtain $T_0(j\omega)$. $T(j\omega)$ can then be obtained by multiplying $T_0(j\omega)$ by $1/H(j\omega)$. This multiplication can be carried out easily by drawing Bode plots for $T_0(j\omega)$ and $H(j\omega)$ and then graphically subtracting the log-magnitude and phase angle of $H(j\omega)$ from that of $T_0(j\omega)$. The resulting Bode plots give the closed-loop frequency response $T(j\omega)$.

The Nichols Chart

Earlier in this section we have seen that the constant-M and constant-N contours on the polar plots are circles. Since it is easier to construct a Bode plot than a polar plot, it is preferable to have constant-M and constant-α contours constructed on logarithmic gain and phase coordinates. N.B. Nichols transformed the constant -M and -N circles to log-magnitude and phase angle coordinates and the resulting chart is known as the *Nichols chart*.

The constant-M contours in log-magnitude phase angle plane may be readily obtained by drawing phasors from the origin to points on constant-M circles in the direct polar plane and then locating these points in log-magnitude phase angle plane. When this is repeated for different values of M, we get a family of constant-M contours in log-magnitude phase angle plane. Similarly a family of constant-α contours may also be plotted in this plane. It is found that constant-M and constant-α contours repeat for every 360° interval and there is a symmetry at every 180° interval. Constant-M and constant-α contours for phase angles from 0° to –210° are shown in Fig. 9.37.

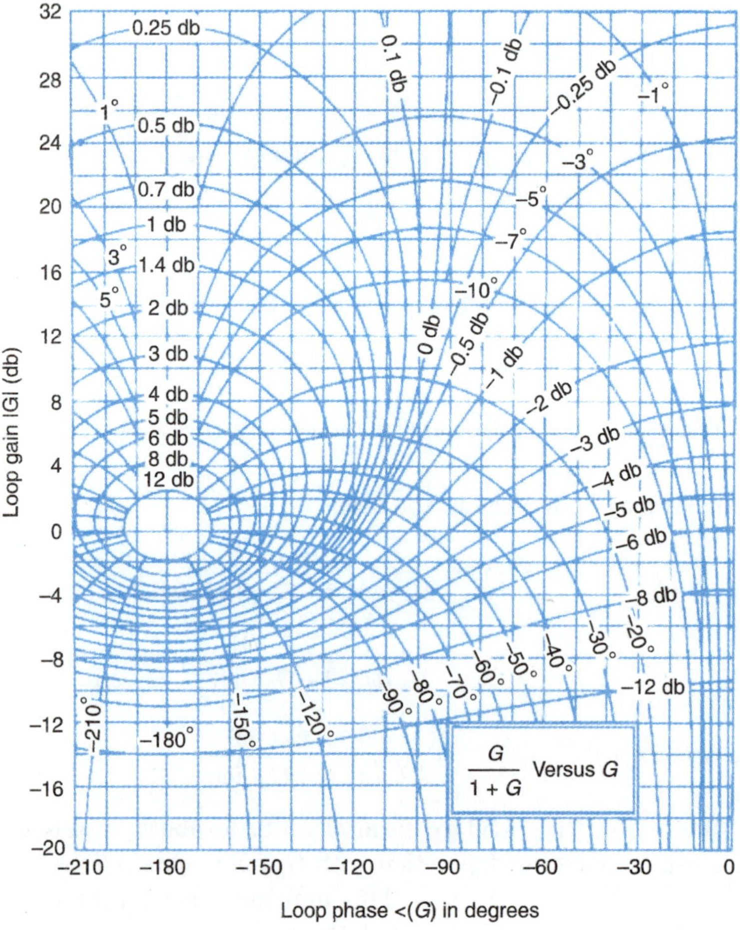

Fig. 9.37. The Nichols chart.

The Nichols chart is very useful for determining the closed-loop frequency response from that of the open-loop. This is accomplished by superimposing the log-magnitude versus phase angle plot of $G(j\omega)$ on Nichols chart. The intersections of the log-magnitude versus phase angle plot and constant-M and -α contours give the magnitude M and phase angle α of the closed-loop frequency response at different frequency points.

Gain adjustments

When a control system is found to be unstable or has poor transient response, the first step is to check if its performance can be modified by adjustment of gain. This adjustment is usually based on a desirable value of M_r. While this adjustment can also be carried out by means of constant-M circles, we shall present here how this is done by means of the Nichols chart which is more convenient and is commonly used.

Gain adjustment by the Nichols chart. The determination of K for specified resonant peak or specified gain and phase margins carried out very conveniently on Nichols chart. For this purpose the log-magnitude versus phase angle [db vs $\angle G(j\omega)$] plot is superimposed on the Nichols chart. Since the gain adjustment has no effect on the phase angle, the db vs $\angle G(j\omega)$ plot merely moves vertically up for increase in gain and down for decrease in gain. The vertical location of the plot is adjusted till it is tangent to the desired M-curve. The db-shift determines the adjustment in gain required to meet the specified M_r. Phase margin or gain margin adjustments are similarly carried out on the Nichols chart. The following example illustrates the procedure for gain adjustment.

Example 9.15 : Let us reconsider the system discussed in Example 9.7. The system has an open-loop transfer function

$$G(j\omega) = \frac{10}{j\omega(j0.1\omega + 1)(j0.05\omega + 1)}$$

The db-$\angle G(j\omega)$ plot for this $G(j\omega)$ is shown in Fig. 9.38. (It is important to note that this plot is conveniently drawn by obtaining the log-magnitude and phase angle data for various values of frequency from the Bode plot of $G(j\omega)$ shown in Fig. 9.27). From Fig. 9.39, it is found that

$$\text{GM} = +\,12 \text{ db}; \text{PM} = +\,33°$$

Gain adjustment for desired GM or PM. Suppose it is desired to find the open-loop gain for (*i*) a GM of 20 db, (*ii*) a PM of 24°.

(*i*) A GM of 20 db is obtained if the plot of Fig. 9.38 is shifted downwards by (20 – 12) = 8 db. The system gain is therefore changed by – 8 db or decreased by a factor of 2.5.

(*ii*) A PM of 24° is obtained if the plot of Fig. 9.38 is raised upwards by 3.5 db or the system gain is increased by a factor of 1.5.

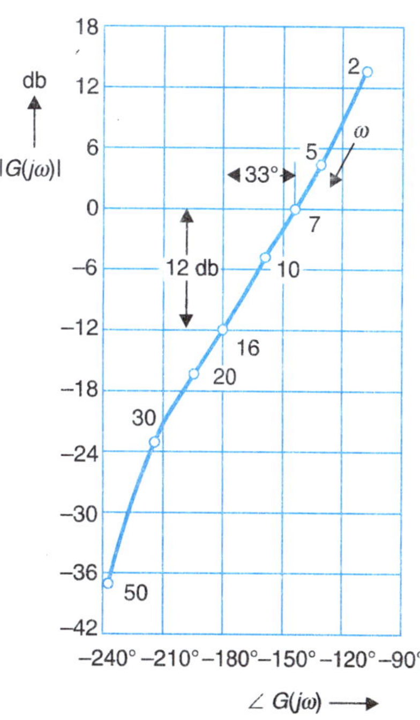

Fig. 9.38. Determination of GM and PM.

Closed-loop frequency response from the Nichols chart. The complete closed-loop frequency response of a system is easily obtained by superimposing the db-$\angle G(j\omega)$ plot on the Nichols chart as shown in Fig. 9.39. The intersections of this plot with M-and α-contours of the Nichols chart determine the closed-loop M and α values for various frequencies from which the closed-loop frequency response of Fig. 9.40 is plotted.

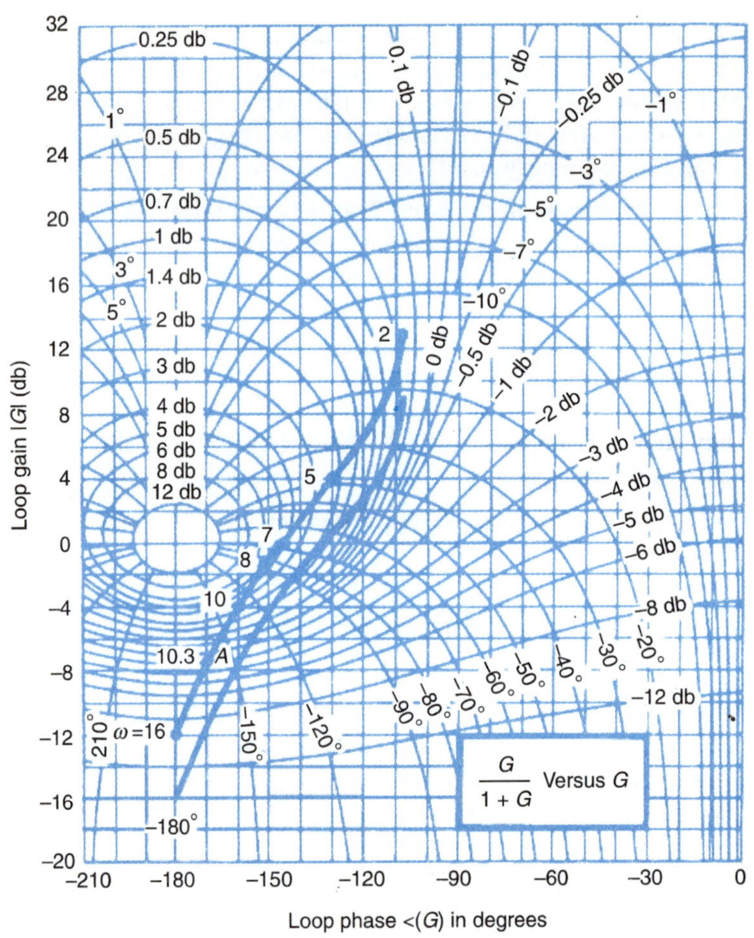

Fig. 9.39. Log-magnitude vs phase angle plot superimposed on the Nichols chart.

From Fig. 9.39 it is seen that db-$\angle G(j\omega)$ plot is tangent to the 5 db M-contour at $\omega = 8$ rad/sec. Therefore for $K = 10$, $M_r = 5$ db and $\omega_r = 8$ rad/sec.

Bandwidth. Bandwidth is defined to be the frequency at which the closed-loop gain is –3 db. From Fig. 9.39 it is observed that db-$\angle G(j\omega)$ plot intersects the –3 db M-contour at the point A. The frequency parameter at this point can be easily determined by transferring the open-loop db data at A to the Bode plot of Fig. 9.27. We find this frequency to be 10.3 rad/sec therefore the bandwidth $\omega_b = 10.3$.

Gain adjustment for desired M_r. Finally, let us investigate the use of db-$\angle G(j\omega)$ plot to obtain the open-loop gain for desired M_r, say 2 db. We find from Fig. 9.39 that the db-$\angle G(j\omega)$

plot must be moved downwards by 3.5 db in order that it be tangent to the 2 db M-contour. Thus for M_r = 2 db, the open-loop gain should be decreased by 3.5 db or by a factor of 1.5 giving the required open-loop gain as 10/1.5 = 6.33.

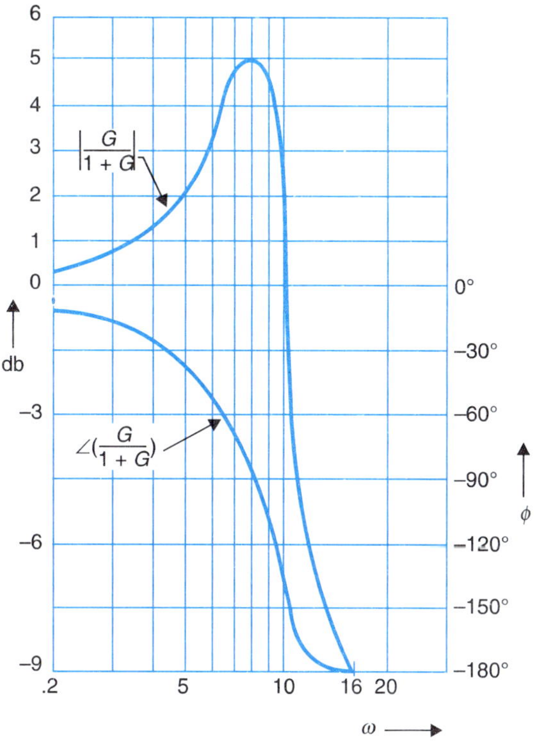

Fig. 9.40. Closed-loop frequency response obtained from Fig. 9.39.

9.6 SENSITIVITY ANALYSIS IN FREQUENCY DOMAIN

Sensitivity considerations often play an important role in the design of control systems. All physical elements have properties that change with environment and age. A good control system should be very insensitive to these parameter variations while being able to follow the command responsively.

The sensitivity of a system's closed-loop transfer function T with respect to the characteristic K of a given element is (refer Section 3.2)

$$S_K^T = \frac{\partial T / T}{\partial K / K}$$

Note that the sensitivity is a function of frequency. For example, the sensitivity of closed-loop transfer function of system of Fig. 3.13 to variations in the parameter K is (eqn. (3.39)

$$S_K^T(s) = \frac{s(s+5)}{s^2 + 5s + 25}$$

Therefore $$S_K^T(j\omega) = \frac{j\omega(j\omega+5)}{(j\omega)^2+5j\omega+25}$$

Figure 9.41 (curve (*i*)) gives a plot of $|S_K^T(j\omega)|$ vs ω. From this figure we observe that sensitivity becomes zero as ω approaches zero. The peak value of $|S_K^T(j\omega)|$ is reached at $\omega = 6$ rad/sec. This means that system of Fig. 3.11 is most sensitive to change in K at this frequency or more generally in this frequency range.

For the two-loop configuration of Fig. 3.14, from eqn. (3.40) we have

$$S_K^T(j\omega) = \frac{j\omega(j\omega+1)}{(j\omega)^2+5j\omega+25}$$

The $|S_K^T(j\omega)|$ vs ω plot is shown in Fig. 9.40 (curve (*ii*)). The two sensitivity curves of this figure give an effective comparison of the two system configurations shown in Figs. 3.13 and 3.14.

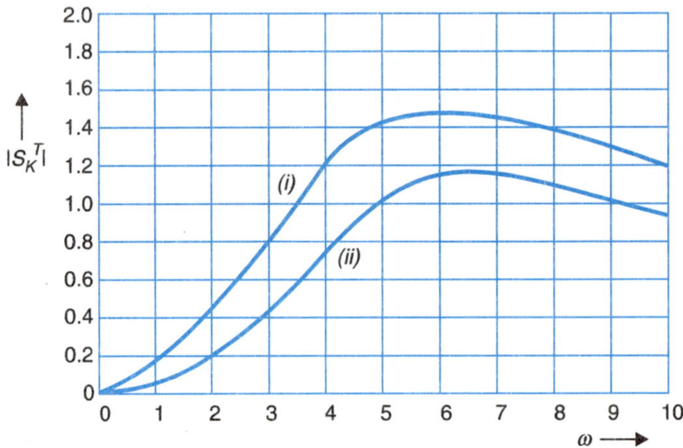

Fig. 9.41. Sensitivity curves for the systems of Figs. 3.11 and 3.12.

Sensitivity of T to Variations in G

The sensitivity of closed-loop transfer function $T(s)$ to variations in open-loop transfer function $G(s)$ is given by the function (eqn. (3.4))

$$S_G^T(j\omega) = \frac{1}{1+G(j\omega)H(j\omega)} \qquad \text{...(9.37)}$$

The function $|S_K^T(j\omega)|$ may conveniently be evaluated from the Nyquist plot of $G(j\omega)H(j\omega)$. The phasor $1 + G(j\omega)H(j\omega)$ is shown in the Nyquist plot of Fig. 9.42. At a frequency ω_1, $|1 + G(j\omega_1)H(j\omega_1)|$ is K_1 and therefore $|S_G^T(j\omega_1)| = 1/K_1$. The maximum sensitivity is given by the circle of radius K_m centred at $(-1 + j0)$ which touches the $G(j\omega)H(j\omega)$ plot. The peak sensitivity is then $|S_G^T(j\omega)| = 1/K_m$ at frequency ω_m.

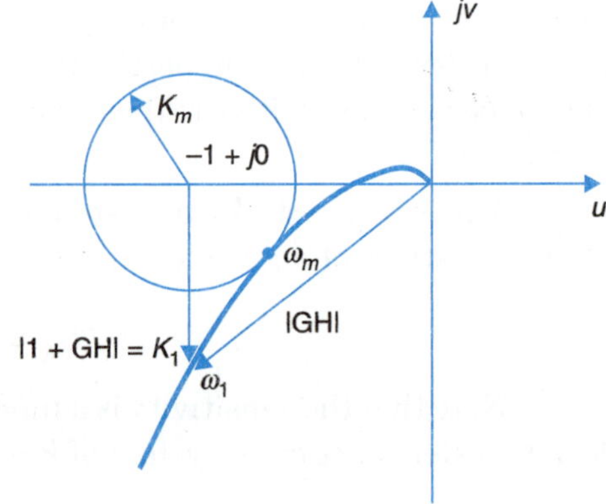

Fig. 9.42. Determination of the sensitivity function from Nyquist plots.

Sensitivity function given by eqn. (9.37) can be evaluated using the Nicholas chart as follows:

$$S_G^T(j\omega) = \frac{1}{1+G(j\omega)H(j\omega)} = \frac{1}{1+G_0(j\omega)} = \frac{G_0^{-1}(j\omega)}{1+G_0^{-1}(j\omega)}$$

This indicates that magnitude and phase of $S_G^T(j\omega)$ can be obtained by plotting $G_0^{-1}(j\omega)$ on the Nichols chart. As an illustration, the function $G^{-1}(j\omega)$ for system of Example 9.15 is plotted on the Nichols chart* in Fig. 9.43. The intersection of $G^{-1}(j\omega)$ curve on the Nichols chart with the constant-M loci gives $|S_G^T(j\omega)|$ at the corresponding frequencies. From Fig. 9.43 it is observed that peak value of $|S_G^T(j\omega)|$ is 7 db at $\omega = 9$ rad/sec.

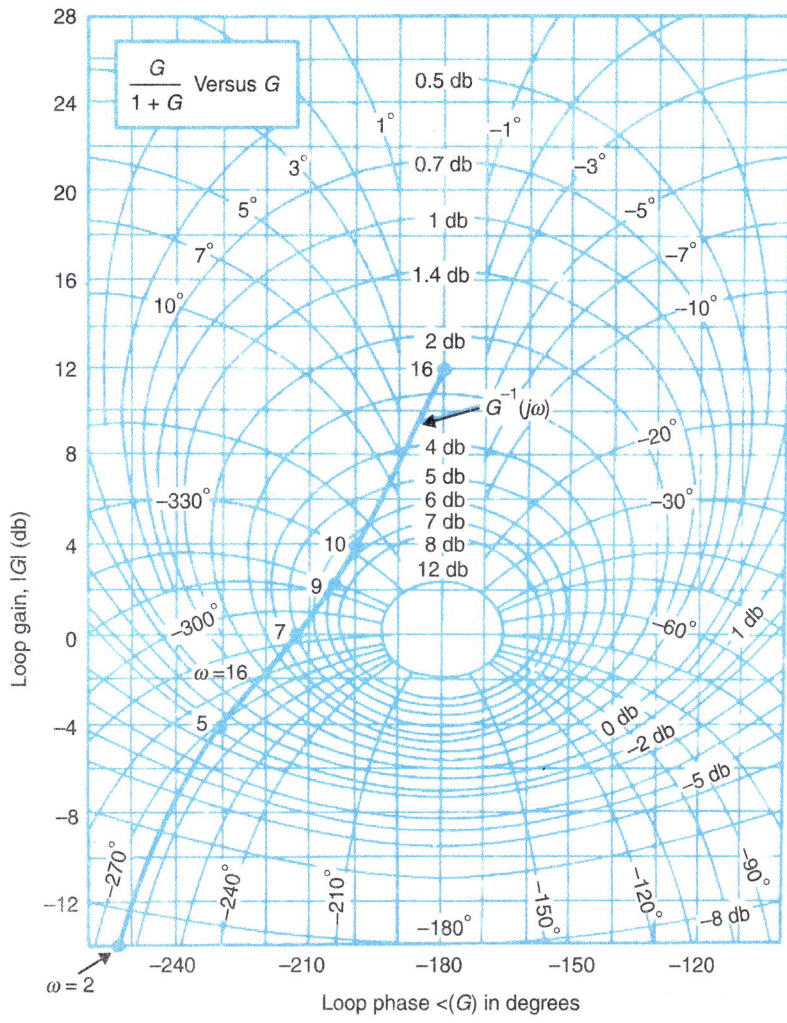

Fig. 9.43. Determination of the sensitivity function from the Nichols chart.

---

*Note that vertical coordinate of the Nichols chart is in decibels; $|G^{-1}(j\omega)|$ and $\angle G^{-1}(j\omega)$ can be obtained from $|G(j\omega)|$ and $\angle G(j\omega)$, since

$$db\,|G^{-1}(j\omega)| = -\,\text{db}\,|G(j\omega)|$$

$$\angle G^{-1}(j\omega) = -\angle G(j\omega)$$

PROBLEMS

9.1. By use of the Nyquist criterion, determine whether the closed-loop systems having the following open-loop transfer functions are stable or not. If not, how many closed-loop poles lie in the right half s-plane ?

(a) $G(s)H(s) = \dfrac{1+4s}{s^2(1+s)(1+2s)}$ (b) $G(s)H(s) = \dfrac{1}{s(1+2s)(1+s)}$

(c) $G(s)H(s) = \dfrac{1}{s^2+100}$.

9.2. Check the stability of systems whose Nyquist plots are shown in Fig. P-9.2.

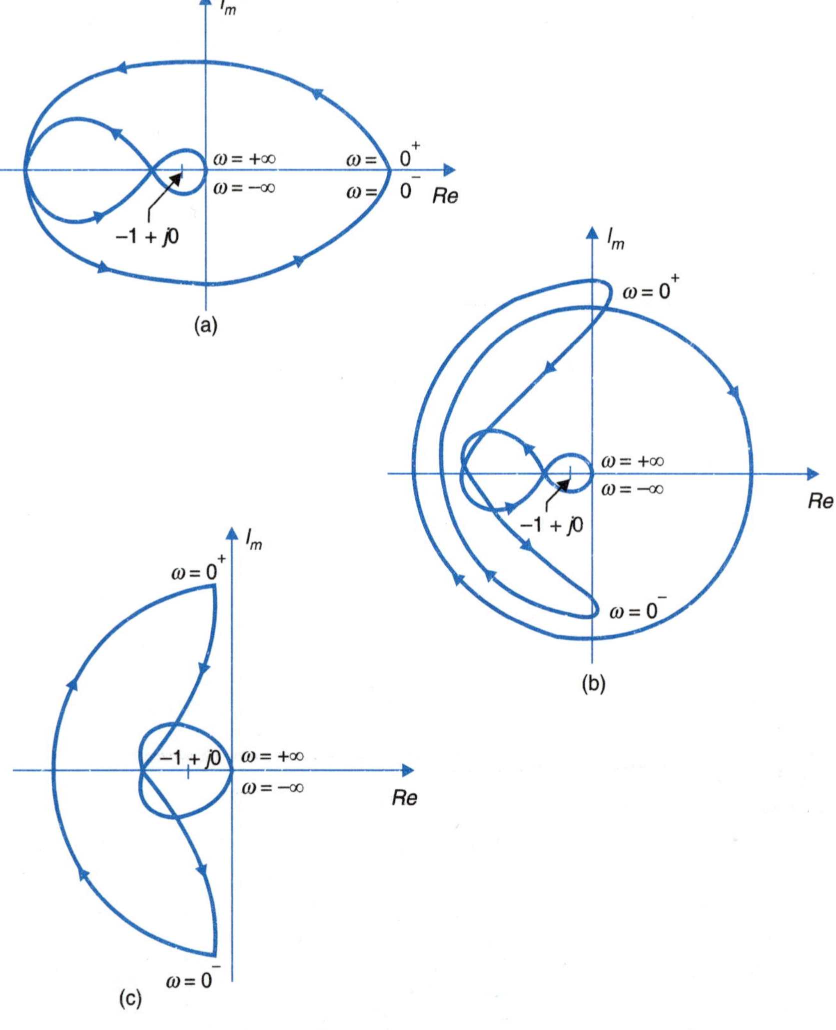

Fig. P-9.2

9.3. Sketch the Nyquist plot for a system with the open-loop transfer function

$$G(s)H(s) = \frac{K(1+0.5s)(s+1)}{(1+10s)(s-1)}$$

Determine the range of values of K for which the system is stable.

9.4. Consider the feedback system shown in Fig. P-9.4. Sketch the Nyquist plot of this system when $G_c(s) = 1$ and determine the maximum value of K for stability.

If $G_c(s)$ is modified to a controller having the transfer function $(1 + 1/s)$, what then is the maximum value of K for stability ? Comment upon the result.

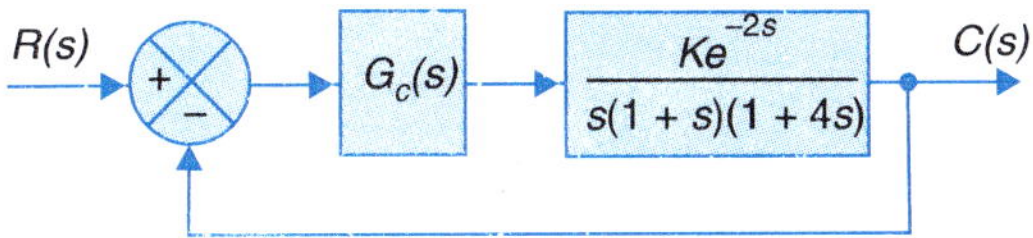

Fig. P-9.4

9.5. Consider a feedback system having the characteristic equation

$$1 + \frac{K}{(s+1)(s+1.5)(s+2)} = 0$$

It is desired that all the roots of the characteristic equation have real parts less than –1. Extend the Nyquist stability criterion to find the largest value of K, satisfying this condition.

[*Hint :* For the above said condition to be satisfied, Nyquist plot of $GH(-1 + j\omega)$ should not encircle $(-1 + j0)$ point.]

9.6. Sketch the Bode plot a closed-loop system which has the open-loop transfer function

$$G(s)H(s) = \frac{2e^{-sT}}{s(1+s)(1+0.5s)}$$

Determine the maximum value of T for the system to be stable.

9.7. Sketch the Bode plot for a unity feedback system, characterized by the open-loop transfer function

$$G(s) = \frac{K(1+0.2s)(1+0.025s)}{s^2(1+0.001s)(1+0.005s)}$$

Show that the system is conditionally stable. Find the range of values of K for which the system is stable.

9.8. The open-loop transfer function of a unity feedback system is given by

$$G(s) = \frac{K}{s(T_1 s + 1)(T_2 s + 1)}$$

Derive an expression for gain K in terms of T_1, T_2 for specified gain margin G_m.

9.9. Phase margin for a second-order system is given by

$$\phi_{pm} = \tan^{-1} 2\zeta \left[\frac{1}{\sqrt{(4\zeta^4 + 1)} - 2\zeta^2} \right]^{1/2}$$

Write the approximate expression for ϕ_{pm} for low values of ζ.

Using the approximate expression, find the value of the gain K such that the system shown in Fig. P-9.9 has a phase margin of ϕ_s, degrees.

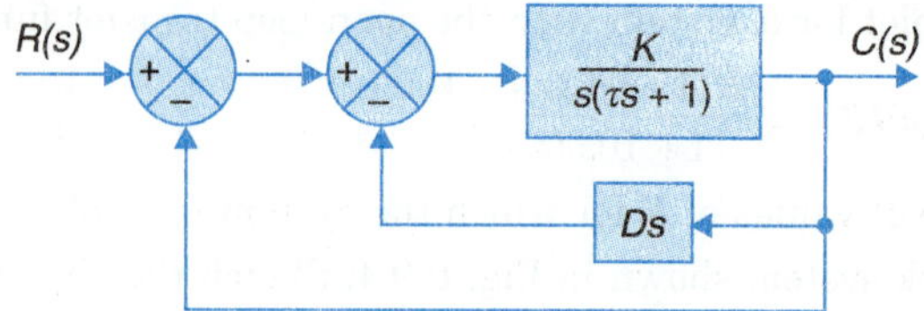

Fig. P-9.9

9.10. The straight line Bode plot of a feedback system is shown in Fig. P-9.10. Derive an expression for the value of ω_c to yield maximum phase margin, in terms of system constants ω_1, ω_2. Determine the maximum phase margin when $m = 1$.

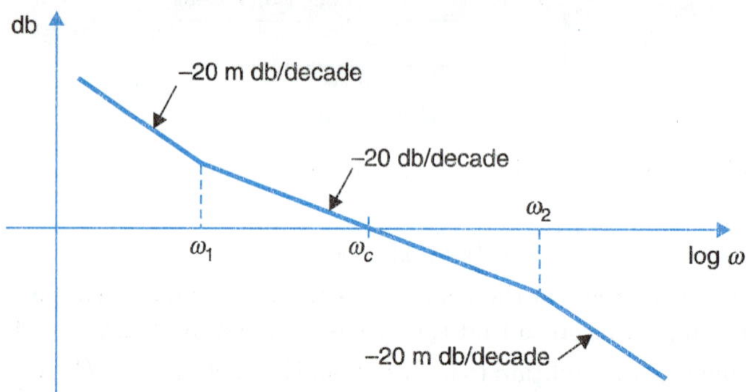

Fig. P-9.10

9.11. The open-loop frequency response of a unity feedback system is given below:

ω	0.8	1.0	1.2	1.4	1.6
$Re[KG(j\omega)]$	– 3.5	– 2.9	– 2.3	– 2.0	– 1.2
$Im[KG(j\omega)]$	– 4.4	– 3.2	– 1.9	– 1.2	– 0.5

Determine the change in gain K required to make the resonant peak $M_r = 1.4$. For this value of gain, determine the phase margin of the system. Evaluate the value of damping ratio using M_r-criterion and phase margin criterion.

9.12. A unity feedback system has the following open-loop frequency response.

ω	2	3	4	5	6	8	10
$\lvert G(j\omega) \rvert$	7.5	4.8	3.15	2.25	1.70	1.00	0.64
$\angle G(j\omega)$	– 118°	– 130°	– 140°	– 150°	– 157°	– 170°	– 180°

(*a*) Evaluate the gain margin and phase margin of the system.

(*b*) Determine the change in gain required so that the phase margin of the system is 60 db.

(*c*) Determine the change in gain required so that the phase margin of the system is 60 deg.

9.13. Sketch the inverse Nyquist plot of a feedback system characterized by the open-loop transfer function

$$G(s) = \frac{K}{s(1 + 0.1s)(1 + s)}$$

Find the value of M_r for $K = 1$. By what factor should the gain K be changed so that M_r is 1.4 ? Determine the value of ω_r for the new setting of gain.

9.14. The open-loop transfer function of a unity feedback system is

$$G(s) = \frac{Ke^{-0.1s}}{s(1+0.1s)(1+s)}$$

By use of Bode plot and/or Nichols chart, determine the following:

(*a*) The value of K so that the gain margin of the system is 20 db.

(*b*) The value of K so that the phase margin of the system is 60 deg.

(*c*) The value of K so that resonant peak M_r of the system is 1 db. What are the corresponding values of ω_r and ω_b ?

(*d*) The value of K so that the bandwidth ω_b of the system is 1.5 rad/sec.

9.15. The open-loop frequency response of a unity feedback system is given below:

ω	1.0	1.5	2.0	2.5	3.0	3.5	4.0
$\dfrac{1}{\lvert G(j\omega)\rvert}$	0.15	0.36	0.55	0.83	1.1	1.43	2.0
$-\angle G(j\omega)$	130°	140°	148°	153°	158°	160°	164°

Sketch the inverse Nyquist plot and determine values of M_r and ω_r.

The configuration of feedback system is now changed to include derivative feedback as shown in Fig. P-9.15. Determine graphically the resultant frequency response and therefrom determine values of M_r and ω_r. Comment upon the effect of derivative feedback on system stability.

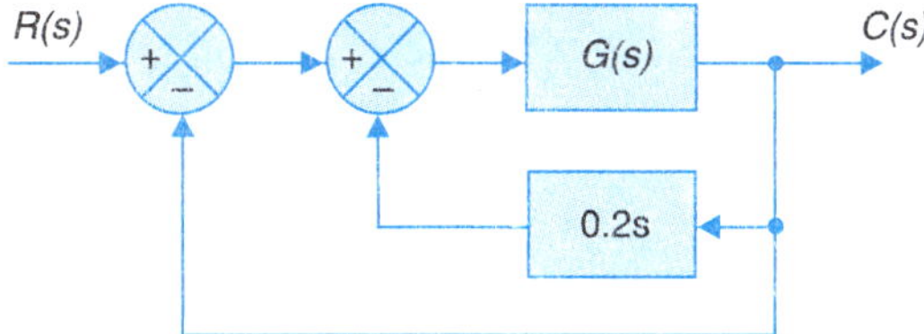

Fig. P-9.15

9.16. The block diagram of a position control system is shown in Fig. P-9.16. Determine the peak sensitivity of the closed-loop transfer function T with respect to G and H, the forward path and feedback path transfer functions respectively. Also determine the frequency at which the peak sensitivity occurs.

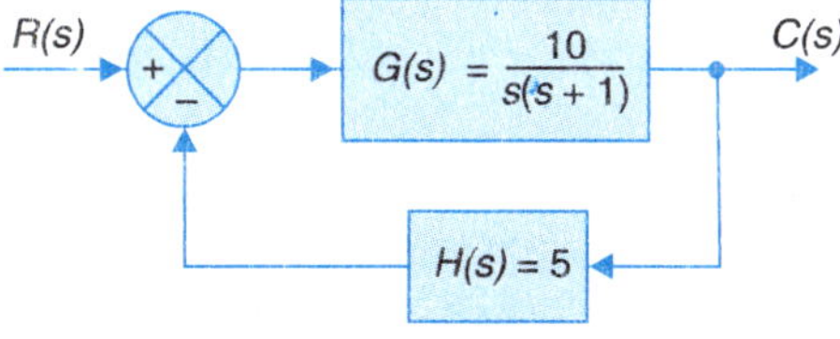

Fig. P-9.16

What will be the bandwidth of the system if it is designed to have

$$|S_G^T(j\omega)| < 1 \,?$$

10

INTRODUCTION TO DESIGN

10

INTRODUCTION TO DESIGN

10.1 THE DESIGN PROBLEM

The design of automatic control systems is perhaps the most important function that the control engineer carries out. Systems analysis presented in the preceding chapters provides the necessary clue to design and construction of practical control systems. While some of the direct design methods can be abstracted from analysis, in most situations the design proceeds on a trial and error basis wherein analysis techniques are repeatedly applied.

Every control system designed for a specific application has to meet certain performance specifications. In Chapter 5 and 8, we discussed two methods of specifying the performance of a control system:

(*i*) By a set of specifications in time domain and/or in frequency domain such as peak overshoot, settling time, gain margin, phase margin, steady-state error, etc.

(*ii*) By optimality of a certain function, *e.g.,* an integral function.

In addition to the performance specifications, some other constraints are always imposed on the control system design. Consider for example the tracking antenna control system discussed in Problem 3.15 (Fig. P-3.15). The objective of the system is to drive large size antenna. The first step in the design is the selection of an actuator to move the antenna. Depending upon the performance specifications, available power supply, space and economical limitations, etc., it could be a d.c. servomotor, a.c. servomotor or hydraulic servomotor. The size of the motor is determined by the inertia, velocity and acceleration ranges of the antenna. Further, since motor's rated speed is usually larger than the required load speed, a gear train of suitable gear ratio is also used. In system of Fig. P-3.15, armature controlled d.c. servomotor with a suitable power amplifier have been taken rather arbitrarily. This collection of devices (components like antenna, gear train, motor, power amplifier) is the *plant* of the control system.

From this discussion, it is evident that the choice of plant components is dictated not only by performance specifications but also by size, weight, available power supply, cost, etc. Therefore the plant generally can not meet the performance specifications. Though the designer is free to choose alternative components, this is generally not done because of cost, availability and other constraints. However, some components of the chosen plant may be easily replaced, *e.g.,* if an electronic amplifier is a component of a plant, its replacement is not a big problem because of the low-cost and wide-range of availability of such amplifiers. Merely by gain adjustment, it may be possible to meet the given specifications on performance of simple control systems. In such cases, gain adjustment seems to be the most direct and simple way of design. However in most practical cases, the gain adjustment does not provide the desired result. As is usually the case, increasing the gain reduces the steady-state error but results in oscillatory transient response or, even in instability. Under such circumstances, it is necessary to introduce some kind of corrective subsystems to force the chosen plant to meet the given specifications. These subsystems are known as *compensators* and their job is to compensate for the deficiency in the performance of the plant. The design problem may therefore be stated as follows:

Given a plant and a set of specifications, design suitable compensators so that the overall system will meet the given specifications.

Approaches to the Design problem

There are basically two approaches to the control system design problem.

1. We select the configuration of the overall system by introducing compensators and then choose the parameters of the compensators to meet the given specifications on performance.

2. For a given plant, we find an overall system that meets the given specifications and then compute the necessary compensators.

The first approach has been used at several places in the preceding chapters. In Chapter 5, we studied control system configurations wherein proportional, derivative and integral controllers were introduced to obtain the desired performance specified in terms of peak overshoot, settling time, steady-state error, etc., or in terms of optimality of a function. In Chapter 7, we discussed the root locus technique for parameter design, while in Chapter 9, we discussed the use of Nyquist and Bode plots for control systems design to achieve the desired performance specified in terms of resonant peak of closed-loop frequency response, bandwidth, gain margin, phase margin, etc.

The primary objective of this chapter is to discuss further the question of design and to present significant design and compensation methods for single-input single-output linear time-invariant systems. We shall assume here fixed system configuration and performance specified in terms of peak overshoot, settling time, gain margin, phase margin, steady-state error, etc., and we shall present the root locus method and frequency domain methods of design, frequently referred to as the *classical method of design*.

Advanced problems in design of control system, shall be taken up in chapters that follows.

10.2 PRELIMINARY CONSIDERATIONS OF CLASSICAL DESIGN

Proper selection of performance specifications is the most important step in control system design. The desired behaviour is specified in terms of transient response measures and steady-state error. The steady-state error is usually specified in terms of error constants for specific inputs, while the transient response measures of relative stability and speed of response may be specified in time or frequency domain or even in both. In time domain, the measure of relative stability is damping factor ζ or peak overshoot M_p, while the speed of response is measured in terms of rise time, settling time or natural frequency. On the other hand, in frequency domain the measure of relative stability is resonant peak M_r or phase margin ϕ_{pm}, while the measure of speed of response is resonant frequency ω_r or bandwidth ω_b.

Once a set of performance specifications has been selected, the designer's next aim is to select a configuration for the overall system. The nature of compensation depends upon the given plant. The compensator may be an electrical, mechanical, hydraulic, pneumatic or other type of device or network. Usually, an electric network serves as compensator in many control systems. (Different types of electric networks used as compensation devices are discussed later in this chapter). The compensator transfer function may be placed in cascade with the plant transfer function (*cascade or series compensation*) or in the feedback path (*feedback or parallel compensation*) as shown in Figs. 10.1 (*a*) and (*b*).

Once the configuration of the overall system has been selected, the next step is to select the structure of the compensator and then to obtain the parameters of the compensator that satisfy the selected specifications in the best possible manner. Parameter search is of course restricted to the feasible domain. Practically most compensators require a trial and error parameter adjustment to achieve at least an acceptable performance if it is not possible to satisfy exactly all the performance specifications.

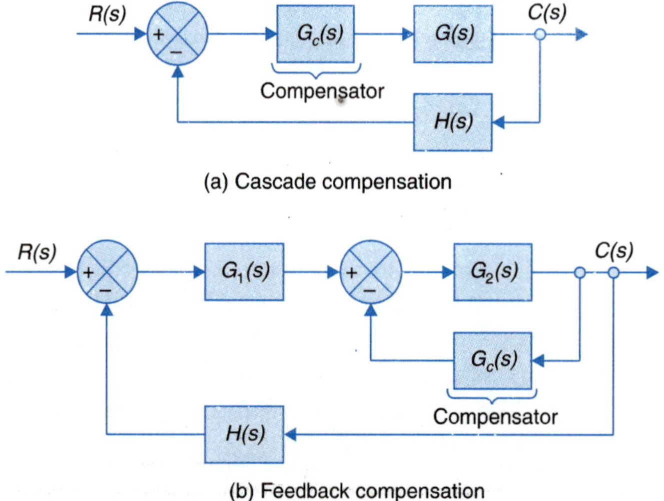

Fig. 10.1. (a) Cascade compensation; and (b) feedback compensation.

A Design Example

Consider the system of Fig. 10.2 which has an open-loop transfer function

$$G(s) = \frac{5K_A}{s(s/2+1)(s/6+1)} \quad ...(10.1)$$

From eqn. (10.1), K_v the velocity error constant of the system is given by

$$K_v = \lim_{s\to 0} sG(s) = 5K_A \quad ...(10.2)$$

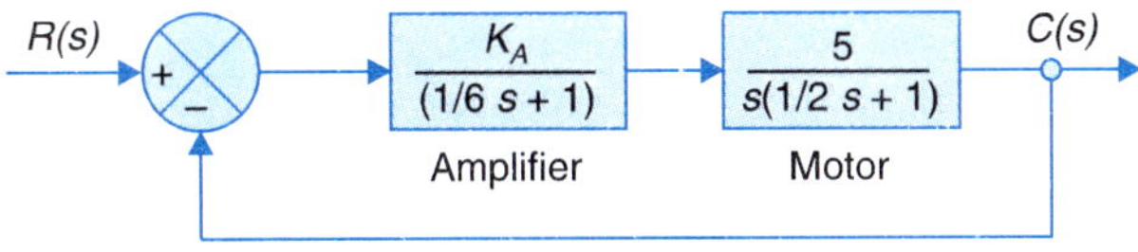

Fig. 10.2. A position control system.

For $K_A = 1$ (*i.e.*, $K_v = 5$) the steady-state error to unit velocity input ($e_{ss} = 1/K_v$) is 0.2, which may be assumed to be an acceptable value.

From eqn. (10.1), we have

$$G(s) = \frac{60K_A}{s(s+2)(s+6)} = \frac{K}{s(s+2)(s+6)} \quad ...(10.3)$$

The root locus plot of the uncompensated system appears in Fig. 10.3. Corresponding to the value $K = 60$ (*i.e.*, $K_v = 5$) on the root locus plot, the dominant pair of root is found to be

$$s_{1,2} = -0.3 \pm j2.8$$

For this root pair, the damping ratio ζ is 0.105 and the undamped natural frequency ω_n is 2.85. The settling time ($= 4/\zeta\omega_n$) for the response is therefore about 13.36 sec.

From the above investigation of the uncompensated system, it is seen that the system has poor relative stability and large settling time.

Assume that the specifications on the transient performance are

$$\zeta = 0.6\ ;$$
$$\text{settling time} < 4 \text{ sec.}$$

A line corresponding to $\zeta = 0.6$ is shown in Fig. 10.3. The intersection of this line with root locus branch determines the value of K which yields the specified ζ. From Fig. 10.3 we find that

$$K = 10.5\ [K_v = 10.5/(2 \times 6) = 0.875,\ i.e.,\ K_A = 0.175];$$
$$\omega_n = 1.26\ ;$$
$$\text{Settling time} = 5.3 \text{ sec;}$$
$$e_{ss} = 1.14.$$

Thus the reduction in gain satisfies the specification on ζ but the steady state error increases well beyond the tolerable of 0.2 and further the settling time specification is not fully satisfied. Therefore we conclude that our purpose is not served by mere given adjustment.

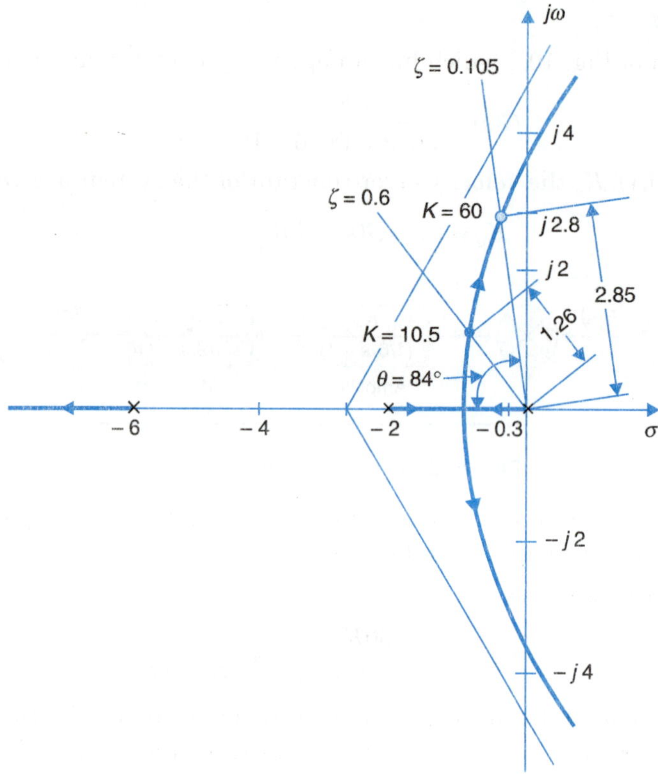

Fig. 10.3. Root locus plot of system shown in Fig. 10.2.

In order to meet the specifications we need to increases ω_n keeping ζ constant (= 0.6) without sacrificing the steady-state performance (*i.e.*, $K_v \geq 5$). To accomplish this goal let us consider the proposition of adding a compensating zero somewhere between the nonzero poles of the open-loop system.

The resulting root locus plot is shown in Fig. 10.4, from which we observe the following effects of the compensating zero.

1. All branches or the root locus now lie completely in the left half s-plane, so that K can be adjusted to any positive value without causing instability.

2. Complex-root branches of the root locus are now bent away from the $j\omega$-axis.

From Fig. 10.4, it is found that for $\zeta = 0.6$;

$$K = 16[K_v = (16 \times 3)/(2 \times 6) = 4];$$

$$\omega_n = 3.4;$$

$$\text{Settling time} = 4/(0.6 \times 3.4) = 1.96 < 4 \text{ sec.}$$

$$e_{ss} = 0.25$$

Thus the addition of a compensating zero helps to satisfy the settling time specifications for $\zeta = 0.6$. The system K_v still falls somewhat short of the specified value of 5. By suitable readjustment of the compensating zero along the negative real axis, one may be able to bring the steady-state error within the specified limit simultaneously satisfying the transient response specifications.

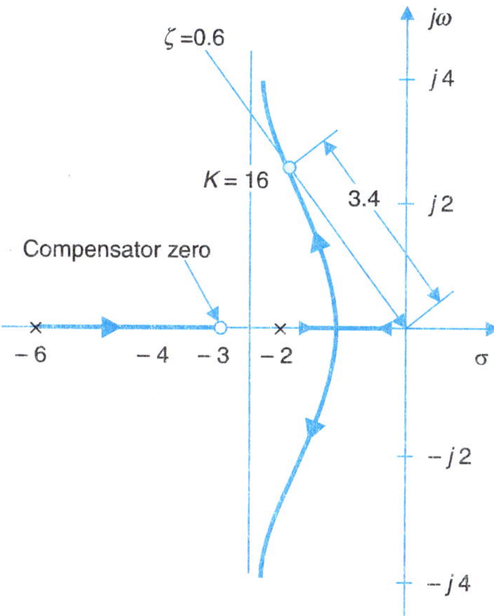

Fig. 10.4. Root locus plot of system shown in Fig. 10.2, compensated through the addition of a zero.

From Figs. 10.3 and 10.4 a further observation is made. Without the compensating zero, the real root in addition to the dominant complex-root pair is always located to the left of the open-loop pole at $s = -6$. On the other hand, in the presence of the compensating zero, this root moves over to the right of $s = -6$ and the dominance condition is therefore weakened. If this root (which is the closed-loop pole) is located close to the open-loop compensating zero, it will contribute a negligible term to the transient response such that the transient response continues to be dominated by the desired closed-loop poles.

The addition of a zero in the open-loop transfer function can be achieved by adding a compensator with the transfer function

$$G_c(s) = (s + z_c) = (s + 1/\tau_c) \qquad ...(10.4)$$

in cascade with the forward transfer function. Since an isolated zero is not physically realizable, we must add a pole along with the compensating zero so as to achieve physical realizability. This pole must of course be added far away from the $j\omega$-axis such that it has relatively negligible effect on the root locus in the region where the two dominant complex closed-loop poles are to occur.

Figure 10.5 shows the locus plot in the presence of both a compensating zero and pole. The root locus near the dominant closed-loop poles gets somewhat modified by the presence of the compensating pole. The effect on the transient response of the system is not very pronounced. For $\zeta = 0.6$, we obtain from root locus

$$K = 67.5 \; [K_v = (67.5 \times 3)/(2 \times 6 \times 10) = 1.7];$$

$$\omega_n = 2;$$

Settling time $= 4/(2 \times 0.6) = 3.33 < 4$ sec.

By adjusting the compensating zero and pole, it may be possible to raise the K_v to the specified value and yet satisfying the transient response specifications.

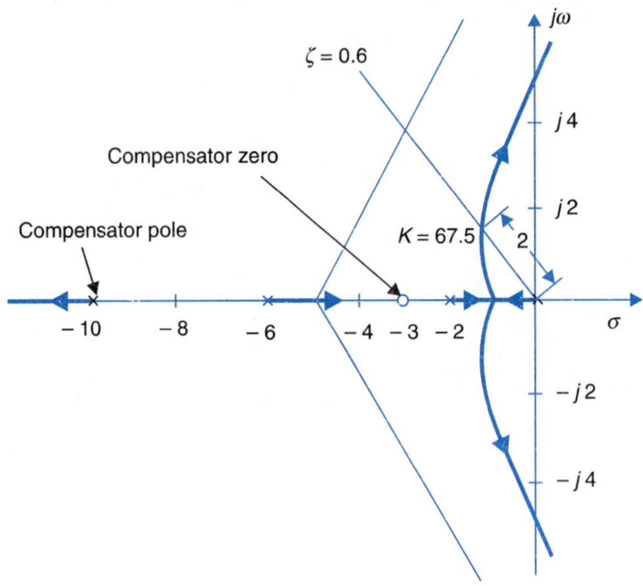

Fig. 10.5. Root locus plot of system shown in Fig. 10.2 with compensating pole and zero.

We see from the above discussion that the compensating pole must be located to the left of the compensating zero. The transfer function of such a compensator then becomes

$$G_c(s) = \frac{(s + z_c)}{(s + p_c)} = \frac{(s + 1/\tau)}{(s + 1/\alpha\tau)}\,;\ \alpha = \frac{z_c}{p_c} < 1,\ \tau > 0 \qquad ...(10.5)$$

Note that $\alpha < 1$ ensures that the pole is located to the left of the zero.

The compensator having a transfer function of the form given in eqn. (10.5) as known as a *lead compensator*. The block diagram of the system with *lead compensator* is shown in Fig. 10.6.

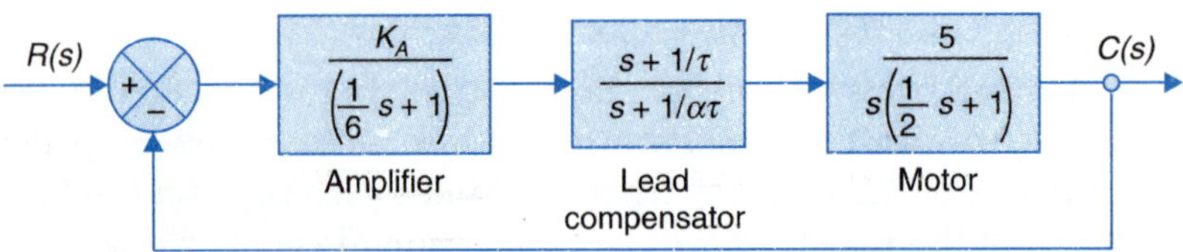

Fig. 10.6. Position control system with lead compensator.

From the foregoing discussion we conclude that a *lead compensator speeds up the transient response and increases the margin of stability of a system. It also helps to increase the system error constant though to a limited extent.*

Let us now consider the case when the system has satisfactory transient response but its steady-state error is quite large. As an example, let us reconsider the system of Fig. 10.2. Assuming $\zeta = 0.6$, we find from the root locus plot of Fig. 10.3 that $K = 10.5$ [*i.e.*, $K_v = 10.5/(2 \times 6) = 0.875$] yields this value of ζ. Such a low K_v means that the steady state error is well beyond the acceptable limit. It is therefore required to raise the system K_v, while ζ is maintained constant at 0.6. To accomplish this, consider the preposition of adding a pole at origin. The velocity error constant of the compensated system is then given by

$$K_v = \lim_{s \to 0} \frac{sG(s)}{s} \to \infty$$

Thus by adding a pole at the origin, the error constant has been increased to infinity and therefore the system now has excellent steady-state response. However, adding a pole at the origin causes the uncompensated type-1 system to become compensated type-2 system, which is inherently closed-loop unstable. This can be seen from the root locus plot of Fig. 10.7, which is obtained by adding a compensating pole at origin to the root locus plot of Fig. 10.3.

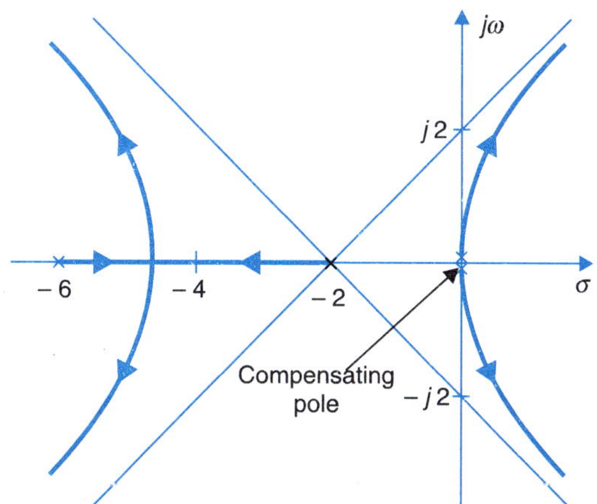

Fig. 10.7. Root locus plot of system shown in Fig. 10.2 with compensating pole at origin.

It is to be noted that addition of a compensating pole at origin to a type-0 system would not change the transient response so drastically as in the case of a type-1 system, still it does degrade the transient response.

This difficulty is remedied by adding a compensating zero very close to the pole at origin and in the left half of the s-plane. The compensated system then continues to be type-2, while the additional zero approximately cancels the effect of the compensating pole on the root locus. The compensated system thus has $K_v = \infty$, while its transient response is not significantly different from that of the uncompensated system.

Figure 10.8 shows the root locus plot of the compensated system with the compensating pole at origin and compensating zero at –0.2. Due to the presence of the compensating zero, an additional real closed-loop pole appears. For $\zeta = 0.6$ which corresponds to $K = 12$, this particular closed-loop pole is located at $s = -0.25$. It contributes an additional term in the transient response which decays slowly but is very small in magnitude on account of its proximity to the open-loop zero. This proximity must be ensured in the compensator design failing which the slowly decaying term will increase the settling time of the system.

The addition of the compensating pole and zero as discussed above can be achieved by a cascade compensator with the transfer function

$$G_c(s) = \frac{(s + z_c)}{s} \qquad ...(10.6)$$

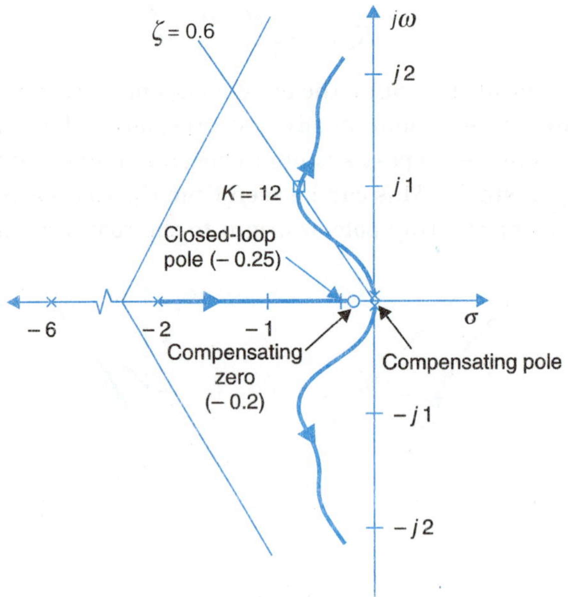

Fig. 10.8. Root locus plot of system shown in Fig. 10.2 with compensating pole and zero.

In actual practice, the considerations of physical realizability require that the pole at origin be shifted on the real axis slightly to the left of origin. The transfer function of the compensator, then becomes

$$G_c(s) = \frac{(s + z_c)}{(s + p_c)}; \frac{z_c}{p_c} = \beta > 1 \qquad \text{...(10.7)}$$

$\beta > 1$ ensures that pole is to the right of zero, *i.e.,* nearer the origin than zero.

The compensator having a transfer function of the form given in eqn. (10.7) is called a *lag compensator*.

From the foregoing discussion we conclude that a *lag compensator improves the steady-state behaviour of a system, while nearly preserving its transient response*.

When both the transient and steady-state response require improvement, a *lag-lead compensator* is required. This is basically a lag and a lead compensator connected in series.

Selection of a Compensator

In general, there are two situations in which compensation is required. In the first case, the system is absolutely unstable and the compensation is required to stabilize it as well as to achieve a specified performance. In the second case, the system is stable but the compensation is required to obtain the desired performance. The systems which are type-2 or higher, are usually absolutely unstable. For these types of systems, clearly lead compensator is required because only the lead compensator increases the margin of stability.

In type-1 and type-0 systems, stable operation is always possible if the gain is sufficiently reduced. In such cases any of the three compensators, *viz.,* lag, lead and lag-lead may be used to obtain the desired performance. The particular choice is based upon factors which are discussed later in this chapter.

10.3 REALIZATION OF BASIC COMPENSATORS

The compensators discussed in earlier sections may be realized by electrical, mechanical, pneumatic, hydraulic or other components. The choice of the type of components to be used depends upon the system structure. Realization by electrical components is quite common in many control systems. In the following, we shall discuss electric network realization of basic compensators and their frequency characteristics.

Lead Compensator

The s-plane representation of the lead compensator is shown in Fig. 10.9. It has a zero at $s = -1/\tau$ and a pole at $s = -1/\alpha\tau$ with the zero closer to the origin than the pole. The general form of the lead compensator is

$$G_c(s) = \frac{(s + z_c)}{(s + p_c)} = \frac{(s + 1/\tau)}{(s + 1/\alpha\tau)} \; ; \alpha = z_c/p_c < 1, \tau > 0 \qquad \text{...(10.8)}$$

$$= \alpha\left(\frac{\tau s + 1}{\alpha\tau s + 1}\right) \qquad \text{...(10.9)}$$

The lead compensator with the transfer function (10.8) can be realized by an electric lead network shown in Fig. 10.10 (a).

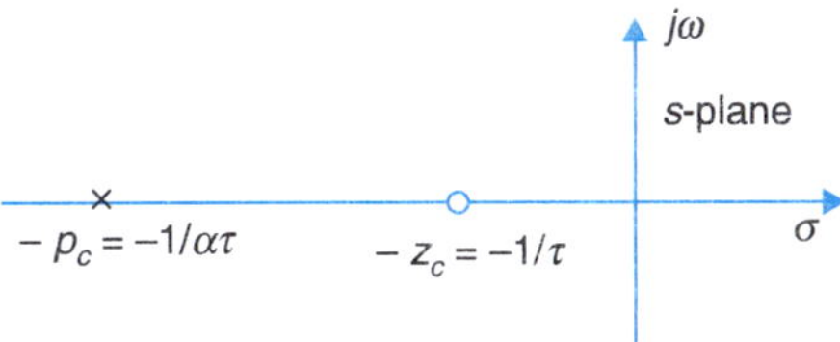

Fig. 10.9. The s-plane representation of lead compensator.

Let us drive the transfer function of this lead network assuming the impedance of the source to be zero and the output load impedance to be infinite. This assumption can be met by suitably designing the impedance levels of the network and the source. From Fig. 10.10 (a)

$$\frac{E_0(s)}{E_i(s)} = \frac{R_2}{R_2 + \dfrac{R_1/sC}{R_1 + 1/sC}} = \frac{(s + 1/R_1C)}{s + \dfrac{1}{[R_2/(R_1 + R_2)]R_1C}} \qquad \text{...(10.10)}$$

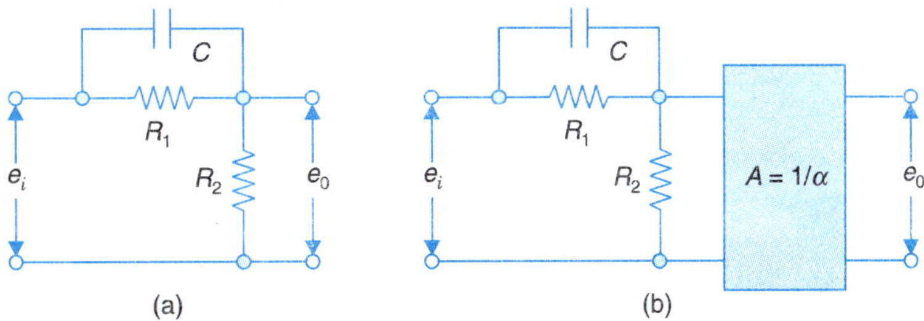

Fig. 10.10. (a) Electric lead network; and (b) phase-lead network with amplifier.

Defining $$\tau = R_1C \text{ and } \alpha = R_2/(R_1 + R_2) < 1 \qquad ...(10.11)$$

we immediately recognize that the network transfer function (10.10) has the same form as that of lead compensator in eqn. (10.8).

It is to be noted that the values of the three network components R_1, R_2 and C are to be determined from the two lead compensator parameters τ and α using eqn. (10.11). Thus there is an additional degree of freedom in the choice of the values of the network components which is used to set the impedance level of the network.

Consider now the sinusoidal transfer function of the lead network. It can be written from eqn. (10.9) as

$$G_c(j\omega) = \alpha\left(\frac{1 + j\omega\tau}{1 + j\omega\alpha\tau}\right) ; \alpha < 1$$

At zero frequency the network has a gain of $\alpha < 1$ or an attenuation of $1/\alpha$. In frequency domain compensation techinique, it is convenient to cancel the d.c. attenuation of the network with an amplification $A = 1/\alpha$. The lead compensator is then visualized as a combination of a network and an amplifier as shown in Fig. 10.10 (*b*). The sinusoidal transfer function of the lead compensator is then given by

$$G_c(j\omega) = \frac{1 + j\omega\tau}{1 + j\alpha\omega\tau} ; \alpha < 1 \qquad ...(10.12)$$

Since $\alpha < 1$, the network output leads the sinusoidal input under steady-state and hence the name lead compensator. The Bode diagram of the lead compensator (with amplifier of gain $A = 1/\alpha$) is given in Fig. 10.11.

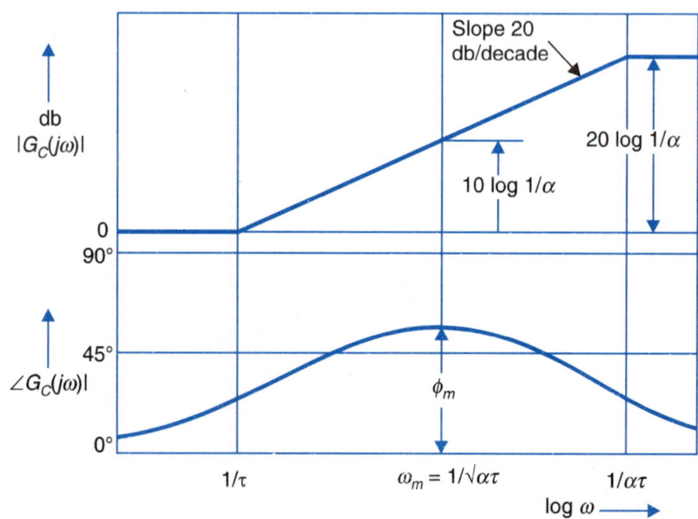

Fig. 10.11. Bode plot of phase-lead network with amplifier of gain A = 1/α.

From eqn. (10.12), the phase-lead of the compensator at any frequency ω is given by

$$\phi = \tan^{-1} \omega\tau - \tan^{-1} \alpha\omega\tau$$

or

$$\tan \phi = \frac{-\omega\tau\,(1 - \alpha)}{1 + \alpha\omega^2\tau^2}$$

Using the condition $d\phi/d\omega = 0$, we find that the maximum phase-lead occurs at frequency

$$\omega_m = 1/(\tau\sqrt{\alpha}) = \sqrt{[(1/\tau)(1/\alpha\tau)]} \quad ...(10.13)$$

As seen from eqn. (10.13) ω_m is the geometric mean of the two corner frequencies of the compensator.

At $\omega = \omega_m$ the maximum phase-lead ϕ_m is given by

$$\tan \phi_m = (1 - \alpha)/(2\sqrt{\alpha}) \qquad \text{or} \qquad \sin \phi_m = (1 - \alpha)/(1 + \alpha)$$

This gives α in terms of ϕ_m as

$$\alpha = \frac{1 - \sin \phi_m}{1 + \sin \phi_m} \quad ...(10.14)$$

The magnitude of $G_c(j\omega)$ at $\omega = \omega_m$, the frequency of maximum phase-lead, is

$$|G_c(j\omega_m)| = \left|\frac{1 + j\omega_m\tau}{1 + j\alpha\omega_m\tau}\right| = 1/\sqrt{\alpha} \quad ...(10.15)$$

Equation (10.14) is useful in computing the α parameter of the network from the required maximum phase-lead. As plot of ϕ_m versus $1/\alpha$ is shown in Fig. 10.12, from which it is observed that to obtain phase-leads more than 60° ($\alpha \approx 0.08$), the network attenuation increases rather sharply, quite out of proportion to the increase in phase-lead. Therefore for phase leads greater than 60°, it is advisable to use two cascaded lead networks with moderate values of α rather than a single lead network with too small a value of α.

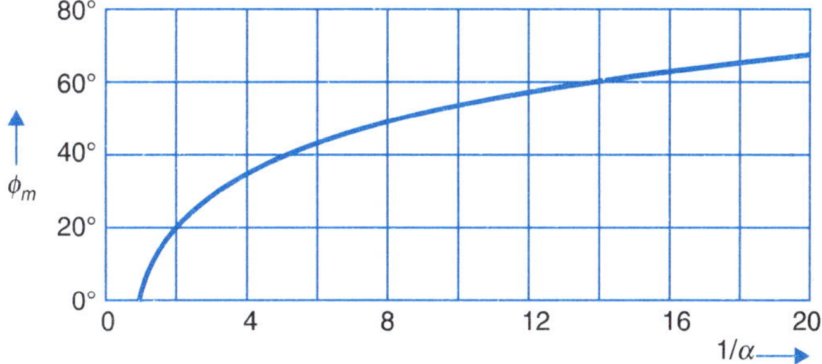

Fig. 10.12. ϕ_m for a lead network.

An important consideration governing the choice of α is the inherent noise in control systems. The nature of this noise is such that the noise signal frequencies are higher than the control signal frequencies. We easily see from Fig. 10.11 that in a lead network the high frequency noise signals are amplified by a factor $1/\alpha > 1$, while the (low frequency) control signals undergo unit amplification (0 db gain). Thus the signal/noise ratio at the output of the lead compensator is poorer than at its input. To prevent the signal/noise ratio at the output from deteriorating excessively, it is recommended that the value of α should not be less than 0.07. A common choice is $\alpha = 0.1$.

Lag Compensator

The s-plane representation of lag compensator is shown in Fig. 10.13 (a). It has a pole at $-1/\beta\tau$ and a zero at $-1/\tau$ with the zero located to the left of the pole on the negative real axis. The general form of the transfer function of the lag compensator is

$$G_c(s) = \frac{(s+z_c)}{(s+p_c)} = \frac{(s+1/\tau)}{(s+1/\beta\tau)};\ \beta = \frac{z_c}{p_c} \qquad ...(10.16)$$

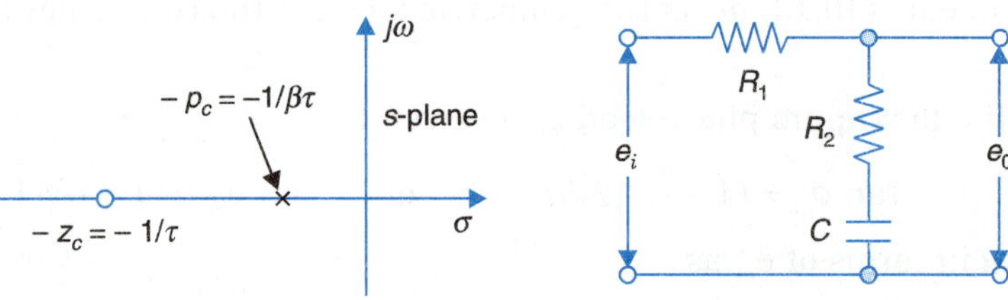

Fig. 10.13. (a) The s-plane representation of lag compensator; and (b) electric lag network.

The realization of the transfer function of eqn. (10.16) is achieved with an electric lag network shown in Fig. 10.13 (*b*) from which we can write

$$\frac{E_0(s)}{E_i(s)} = \frac{R_2 + \dfrac{1}{sC}}{R_1 + R_2 + \dfrac{1}{sC}} = \frac{1}{\dfrac{R_1+R_2}{R_2}}\left[\frac{s+(1/R_2C)}{s+\dfrac{1}{\left(\dfrac{R_1+R_2}{R_2}\right)R_2C}}\right] \qquad ...(10.17)$$

Comparing eqns. (10.16) and (10.17), we get

$$\tau = R_2C,\ \beta = (R_1+R_2)/R_2 > 1$$

Therefore, the transfer function of the network becomes

$$G_c(s) = \frac{1}{\beta}\left(\frac{s+1/\tau}{s+1/\beta\tau}\right) = \frac{1}{\beta}\left(\frac{s+z_c}{s+p_c}\right);\ \beta = z_c/p_c > 1 \qquad ...(10.18)$$

$$= \frac{\tau s+1}{\beta\tau s+1} \qquad ...(10.19)$$

It is to be noted that compared to the form of transfer function (10.16) $G_c(s)$ realized by the network has a multiplicative factor of $1/\beta$. As in the case of lead network realization, we have an additional degree of freedom in the lag network realization also, which is used for impedance matching.

The sinusoidal transfer function of the lag network is given by

$$G_c(j\omega) = \frac{(1+j\omega\tau)}{(1+j\beta\omega\tau)} \qquad ...(10.20)$$

Since $\beta > 1$, the steady-state output has a lagging phase angle with respect to the sinusoidal input and hence the name lag network. The Bode diagram of the lag network is drawn in Fig. 10.14. The maximum phase lag ϕ_m and the corresponding frequency ω_m are obtained from eqns. (10.14) and (10.13) respectively by replacing α with β.

From Fig. 10.14 it is observed that the lag network has a d.c. gain of unity while it offers a high frequency gain of $1/\beta$. Since $\beta > 1$, it means that the high frequency noise is attenuated in passing through the network whereby the signal to noise ratio is improved, in contrast to the lead network. A typical choice of β is 10.

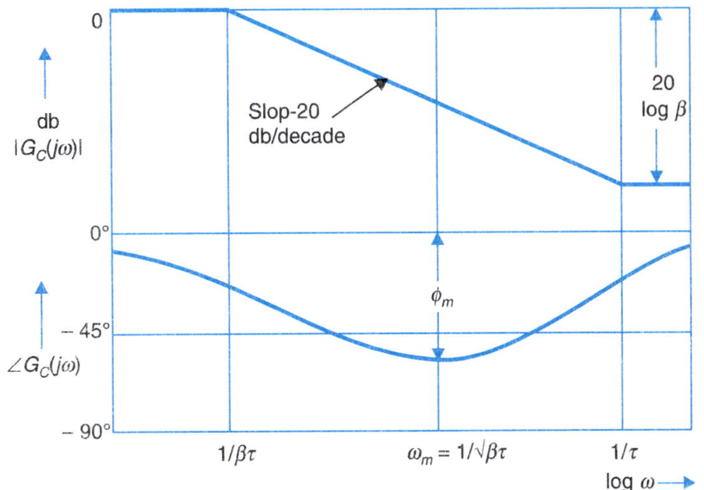

Fig. 10.14. Bode plot of phase-lag network.

Lag-lead Compensator*

As discussed earlier, the lag-lead compensator is a combination of a lag compensator and a lead compensator. The lag-section has one real pole and one real zero with the pole to the right of zero. The lead-section also has one real pole and one real zero but the zero is to the right of the pole. The general form of this compensator is

$$G_c(s) = \underbrace{\left(\frac{s+1/\tau_1}{s+1/\beta\tau_1}\right)}_{\text{Lag section}}\underbrace{\left(\frac{s+1/\tau_2}{s+1/\alpha\tau_2}\right)}_{\text{Lead section}};\ \beta > 1, \alpha < 1 \qquad \text{...(10.21)}$$

The eqn. (10.21) can be realized by a single electric lag-lead network shown in Fig. 10.15. From this figure, the transfer function of the network is given by

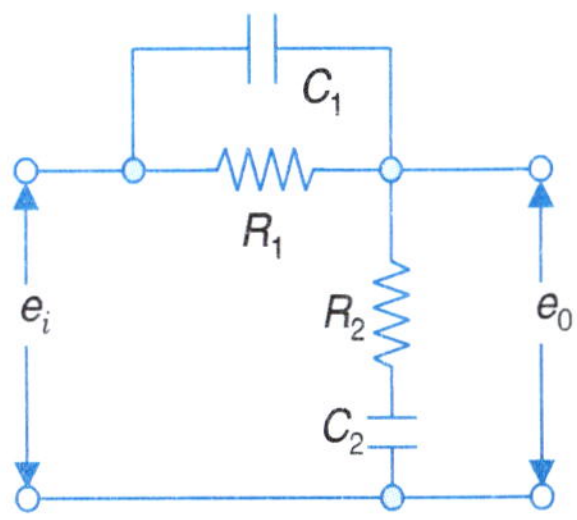

Fig. 10.15. Electric lag-lead network.

$$\frac{E_0(s)}{E_i(s)} = \left[\frac{R_2 + 1/sC_2}{R_2 + \dfrac{1}{sC_2} + \dfrac{R_1/sC_1}{R_1 + 1/sC_1}}\right]$$

$$= \left[\frac{\left(s+\dfrac{1}{R_1C_1}\right)\left(s+\dfrac{1}{R_2C_2}\right)}{s^2 + \left(\dfrac{1}{R_1C_1}+\dfrac{1}{R_2C_1}+\dfrac{1}{R_2C_2}\right)s + \dfrac{1}{R_1R_2C_1C_2}}\right] \qquad \text{...(10.22)}$$

*When the forward path transfer function has complex poles close to the $j\omega$-axis, phase lead, phase-lag or phase lag-lead networks are not effective. In such cases compensation may be achieved by Bridged-T networks.

Comparing eqns. (10.21) and (10.22), we have

$$R_1C_1 = \tau_1 \quad ...(10.23)$$

$$R_2C_2 = \tau_2 \quad ...(10.24)$$

$$R_1R_2C_1C_2 = \alpha\beta\tau_1\tau_2 \quad ...(10.25)$$

$$\frac{1}{R_1C_1} + \frac{1}{R_2C_1} + \frac{1}{R_2C_2} = \frac{1}{\beta\tau_1} + \frac{1}{\alpha\tau_2} \quad ...(10.26)$$

From eqns. (10.23), (10.24) and (10.25), it is found that

$$\alpha\beta = 1$$

It means that a single lag-lead network does not permit us an independent choice of α and β. Keeping this in view, the transfer function of lag-lead compensator may be written as

$$G_c(s) = \frac{(s + 1/\tau_1)(s + 1/\tau_2)}{(s + 1/\beta\tau_1)(s + \beta/\tau_2)}\,;\ \beta > 1 \quad ...(10.27)$$

$$= \left(\frac{s + z_{c_1}}{s + p_{c_1}}\right)\left(\frac{s + z_{c_2}}{s + p_{c_2}}\right);\ \beta = z_{c_1}/p_{c_1} = p_{c_2}/z_{c_2} > 1 \quad ...(10.28)$$

where $\tau_1 = R_1C_1$, $\tau_2 = R_2C_2$ and $\beta > 1$ such that

$$\frac{1}{R_1C_1} + \frac{1}{R_2C_1} + \frac{1}{R_2C_2} = \frac{1}{\beta\tau_1} + \frac{\beta}{\tau_2} \quad ...(10.29)$$

The s-plane representation of a lag-lead compensator is shown in Fig. 10.16. The sinusoidal transfer function of lag-lead compensator is given by

$$G_c(j\omega) = \frac{(1 + j\omega\tau_1)(1 + j\omega\tau_2)}{(1 + j\omega\beta\tau_1)(1 + j\omega\tau_2/\beta)} \quad ...(10.30)$$

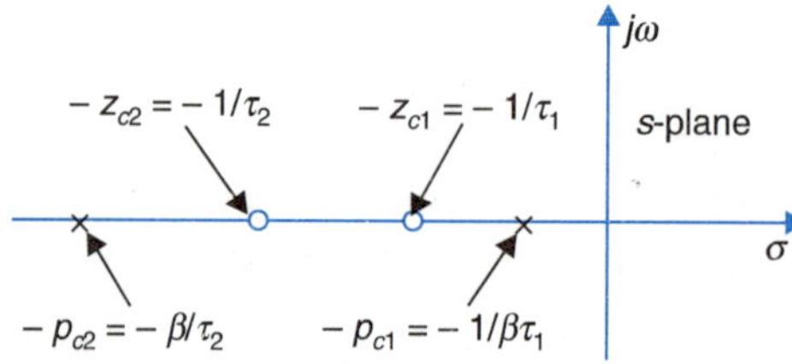

Fig. 10.16. The s-plane representation of lag-lead compensator.

The corresponding Bode plot is given in Fig. 10.17.

10.4 CASCADE COMPENSATION IN TIME DOMAIN

Cascade compensation in time domain is conveniently carried out by the root locus technique. In this method of compensation, the original design specifications on dynamic response are converted into ζ and ω_n of a pair of desired complex conjugate closed-loop poles based on the assumption that the system will be dominated by these two complex poles and therefore its dynamic behaviour can be approximated by that of a second-order system. The desired complex pole pair can be located in the s-plane as shown in Fig. 10.18.

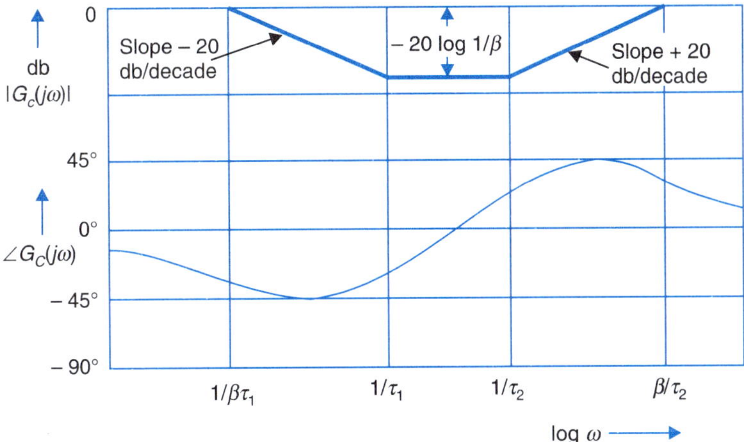

Fig. 10.17. Bode plot of lag-lead network.

A compensator is now designed so that the least-damped complex poles of the resulting transfer function correspond to the desired dominant poles and all other closed-loop poles are either located very close to the open-loop zeros or are relatively far away from the $j\omega$-axis. This ensures that the poles other than the dominant poles make a negligible contribution to the system dynamics.

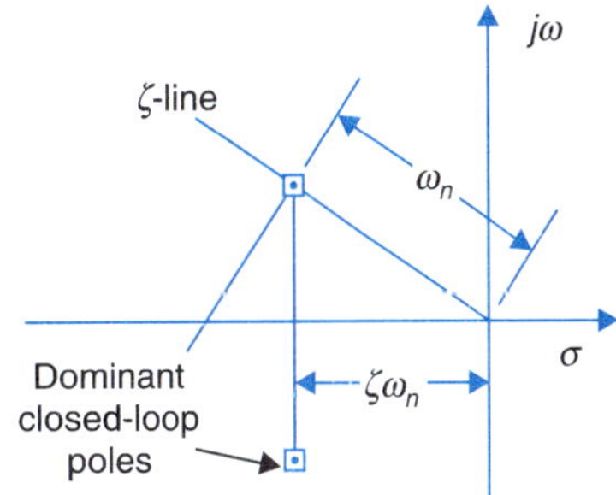

Fig. 10.18. Locating the desired dominant closed-loop poles.

Lead Compensation

Consider a unity feedback system with a forward path unalterable transfer function $G_f(s)$. Let the system dynamics response specifications be translated into the desired location s_d for the dominant complex closed-loop poles as shown in Fig. 10.19.

If the angle criterion at s_d is not met, *i.e.*, $\angle G_f(s_d) \neq \pm 180°$, the uncompensated root locus with variable open-loop gain will not pass through the desired root location, indicating the need for compensation. The lead compensator $G_c(s)$ has to be so designed that the compensated root locus passes through s_d. In terms of the angle criterion this requires that

$$\angle G_c(s_d)\, G_f(s_d) = \angle G_c(s_d) + \angle G_f(s_d) = \pm 180°$$

or

$$\angle G_c(s_d) = \phi = \pm 180° - \angle G_f(s_d) \qquad \text{...(10.31)}$$

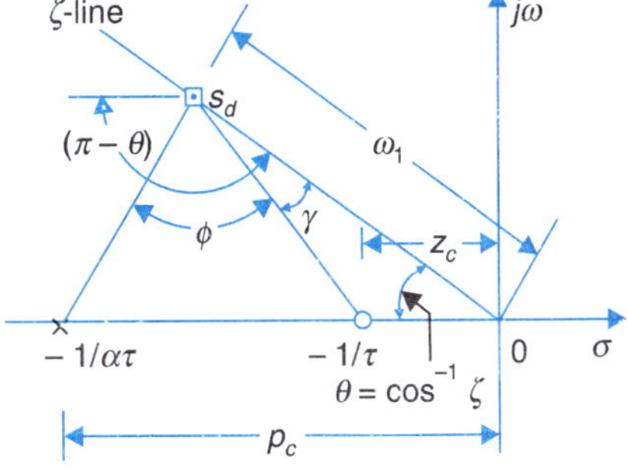

Fig. 10.19. Angle contribution of lead compensator.

Thus for the root locus of the compensated system to pass through the desired root location, the lead compensator pole-zero pair must contribute at s_d an angle ϕ given by eqn. (10.31) and shown in Fig. 10.19.

For a given angle ϕ required for lead compensation, there is no unique location for the pole-zero pair. From the point of view of lead network attenuation (as we shall see shortly, this is a narrow point of view), the best compensator pole-zero location is the one which gives the largest value of $\alpha = z_c/p_c$. The compensator zero is located by drawing a line from s_d making an angle γ with $0s_d$ as shown in Fig. 10.19. The compensator pole is then located by drawing further the requisite angle ϕ to be contributed at s_d by the pole-zero pair. It readily follows from the geometry of this figure that

$$z_c = \omega_n\left[\frac{\sin\gamma}{\sin(\pi-\theta-\gamma)}\right] \quad \text{and} \quad p_c = \omega_n\left[\frac{\sin(\gamma+\theta)}{\sin(\pi-\theta-\gamma-\phi)}\right]$$

$$\alpha = \frac{\sin\gamma\,\sin(\pi-\theta-\gamma-\phi)}{\sin(\pi-\theta-\gamma)\,\sin(\gamma+\phi)}$$

The angle γ for largest α is obtained from the condition

$$\frac{d\alpha}{d\gamma} = 0$$

which gives

$$\gamma = \frac{1}{2}(\pi-\theta-\phi) \qquad \text{...(10.32)}$$

Though the above method of locating the lead compensator pole-zero yields the largest value of α, it does not guarantee the dominance of the desired closed-loop poles in the compensated root locus. The dominance condition must be checked before completing the design.

With compensator pole-zero so located, the system gain at s_d is computed to determine the error constant. If the value of the error constant so obtained is unsatisfactory, the above procedure is repeated after readjusting the compensator pole-zero location while keeping their angle contribution fixed at ϕ. In this readjustment the dominance condition is not allowed to be violated. It may be noted that by adjusting the compensator pole-zero location, it is not always possible to meet any arbitrarily specified value of the error constant. In the above design procedure if $(\gamma + \phi) > (\pi - \theta)$, the compensator pole lies on the positive real axis and therefore a single lead network becomes unrealizable. In such cases, realization can be achieved by using two or more identical networks in tandem.

Example 10.1: Let us first consider the example of a unity feedback type-2 system with

$$G_f(s) = K/s^2$$

This system has zero steady-state error for both the step and ramp inputs. It can be seen from its root locus plot that the closed-loop poles always lie on the $j\omega$-axis. As explained earlier, the lead compensator is the only choice for this system.

It is desired to compensate the system as to meet the following transient response specifications.

Settling time ≤ 4 sec

Peak overshoot for step input $\leq 20\%$

These specifications imply that

$$t_s = \frac{4}{\zeta\omega_n} \leq 4\,\text{sec} \text{ or } \zeta\omega_n \geq 1$$

and

$$\zeta \geq 0.45$$

The desired location ($\zeta = 0.45$ and $\zeta\omega_n = 1$) for the dominant closed-loop poles (s_d) is shown in Fig. 10.20 (a). The angle at (s_d) because of the two poles at origin of the uncompensated system is

$$\angle G_f(s) = -2 \times 117° = -234°$$

Therefore the angle contribution at s_d required of the lead compensator pole-zero pair is

$$\phi = -180° - (-234°) = 54°$$

For the largest value of α, we have from eqn. (10.32)

$$\gamma = \frac{1}{2}(\pi - \theta - \phi) = \frac{1}{2}(180° - \cos^{-1} 0.45 - 54°) = 31.5°$$

Using angles γ and ϕ shown in Fig. 10.20 (a), we obtain

$$z_c = 1/R_1C = 1.15$$

$$p_c = 4.3 = \frac{1}{\left(\dfrac{R_1}{R_1 + R_2}\right)R_1C}$$

$$\alpha = z_c/p_c = 0.268 \text{ (satisfactory)}$$

Choosing $C = 1\mu F$ we obtain

$$R_1 = 870 \text{ k}\Omega \,;\, R_2 = 635 \text{ k}\Omega$$

With the addition of the tandem lead compensator, the open-loop transfer function of the system becomes

$$G(s) = G_f(s)G_c(s) = \frac{K(s + 1.15)}{s^2(s + 4.3)}$$

The root locus for the compensated system is shown in Fig. 10.20 (b) from which the gain at the desired closed-loop pole location is found to be $K = 9.33$. Therefore the open-loop transfer function of the compensated system is

$$G(s) = \frac{9.33(s + 1.15)}{s^2(s + 4.3)}$$

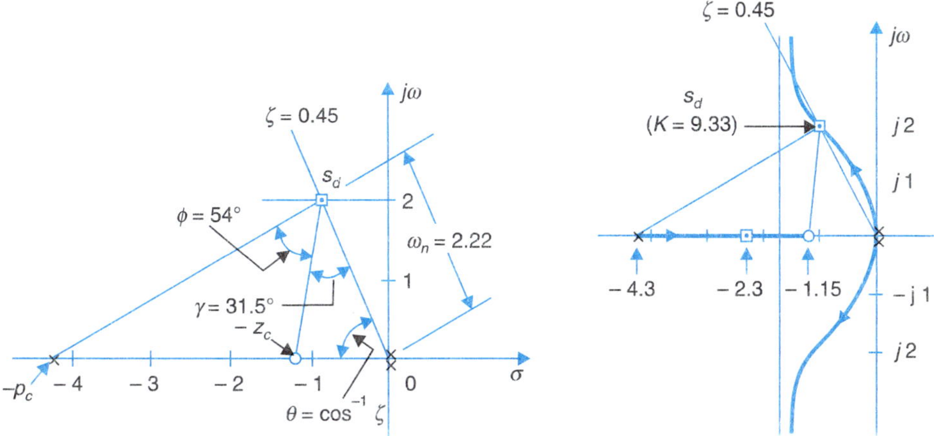

Fig. 10.20. (a) Determination of locations for the compensator pole and zero; (b) Root locus plot of the system with $G(s) = K(s + 1.15)/s^2(s + 4.3)$.

The compensated system block diagram is given in Fig. 10.21.

The acceleration error constant is given by

$$K_a = \lim_{s \to 0} s^2 G(s) = \frac{9.33 \times 1.15}{4.3} = 2.5$$

The steady-state performance of this system is quite satisfactory. In case K_a obtained above is unsatisfactory, the compensator pole-zero pair may be adjusted to achieve a desirable value. The same is of course, not possible for any arbitrarily specified value of K_a.

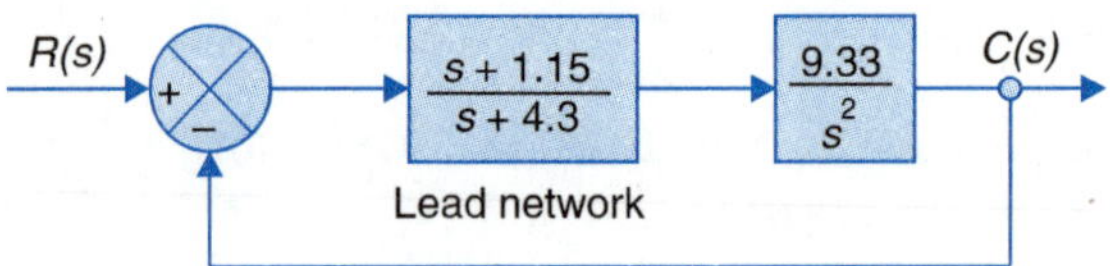

Fig. 10.21. Block diagram of the compensated system.

The compensator designed for the system under discussion results in maximum value of α and hence minimum value of additional required amplifier gain $A = 1/\alpha (= 3.75)$. It is of course essential to check the dominance condition before declaring the compensator design to be complete.

To accomplish this, the third closed-loop pole introduced because of compensator zero is determined. From Fig. 10.20 (*b*), it is found to be at $s = -2.3$. Because of this closed-loop pole, there will be an exponential term in the transient response in addition to the one due to the complex pair of closed-loop poles.

The closed-loop transfer function of the compensated system is obtained as

$$\frac{C(s)}{R(s)} = \frac{9.33(s + 1.15)}{(s + 1 + j2)(s + 1 - j2)(s + 2.3)}$$

For an impulse input,

$$c(t) = \mathcal{L}^{-1}\left[\frac{9.33(s + 1.15)}{(s + 1 + j2)(s + 1 - j2)(s + 2.3)}\right]$$

$$= 3.94e^{-t} \sin (2t - \tan^{-1} 1.83) - 1.84e^{-2.3t}$$

We thus observe that the second term in $c(t)$ contributes appreciably to system dynamics. If this contribution is not tolerable, the compensator zero is shifted further to left and the design procedure repeated. This, of course, means higher network attenuation but as stated earlier, α up to 0.07 is an acceptable value. The above design also illustrates that the best value of α may not result in satisfactory design.

Though the method presented above yields the largest value of α which is a desirable result, it does not guarantee the dominance of desired closed-loop poles in the compensated root locus. No general rule can be laid down which will satisfy this condition for various open-loop pole-zero configurations. However, the following *guidelines* can be of considerable help.

Place the compensating zero on the real axis in the region below the desired closed-loop pole location such that it lies close to the left of any open-loop pole in this region. While giving generally a large value of α, this technique ensures that the closed-loop pole on the real axis caused by the introduction of the compensating zero will be located close to it and would therefore make a negligible contribution to system dynamics. This technique is suggested by *cancellation compensation* wherein the compensating zero is placed exactly in the open-loop pole location completely cancelling the effect of this pole. In case the uncompensated system does not have any open-loop pole in the region below desired closed-loop poles, the dominance condition must be checked by locating the real axis closed-loop pole created by the compensating zero.

Example 10.2: A type-2 system with an open-loop transfer function

$$G_f(s) = \frac{K}{s^2(s+1.5)}$$

is to be compensated to meet the specifications in Example 10.1.

Following the procedure of the previous example, the desired dominant roots are found to lie at $s_d = -1 \pm j2$ (Fig. 10.22). The angle contribution required from a lead compensator is

$$\phi = \pm 180° - \angle G_f(s_d) = \pm 180° - (-2 \times 117° - 75°) = 129°$$

The large value of ϕ here is an indication that a double lead network is appropriate. Each section of a double lead network has then to contribute an angle of 64.5° at s_d.

Let us locate the compensator zero at $s = -1.7$, *i.e.*, in the region below the desired dominant closed-loop pole location and just to the left of the open-loop pole at $s = -1.5$. Join the compensator zero to s_d and locate the compensator pole by making an angle $\phi = 64.5°$ as shown in Fig. 10.22. The location of the pole is found to be at –19.8.

The open-loop transfer function of the compensated system becomes

$$G(s) = \frac{8.30(s+1.7)^2}{s^2(s+1.5)(s+19.8)^2}$$

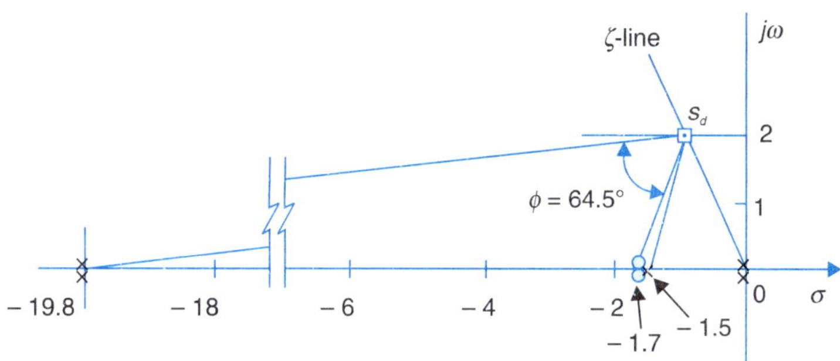

Fig. 10.22. Design of double lead compensator.

By locating the closed-loop poles of the compensated system, it can be easily verified that one closed-loop pole is located very close to the open-loop zero at –1.7 (contributed by the compensating zero) and therefore makes negligible contribution to system dynamics, while the other closed-loop poles are located far to the left of –1 and hence the dominance of the desired closed-loop poles ($-1 \pm j2$) is preserved.

Example 10.3: Consider a type-1 system with an open-loop transfer function of

$$G_f(s) = \frac{K}{s(s+1)(s+4)}$$

The system is to be compensated to meet the following specifications:

Damping ratio $\zeta = 0.5$

Undamped natural frequency $\omega_n = 2$

Using the transient response specifications the desired dominant closed-loop poles are found to lie at

$$s_d = -1 \pm j1.73$$

From Fig. 10.23, the angle contribution required from the lead compensator pole-zero pair is

$$\phi = -180° - \angle G_f(s_d) = -180° - (-120° - 90° - 30°) = 60°$$

Further it is observed that the open-loop pole at $s = -1$ lies directly below the desired closed-loop pole location. Place the compensator zero close to this pole to its left, say at $s = -1.2$. Such a choice of compensator zero generally ensures the dominance condition.

Join the compensator zero to s_d and locate the compensator pole by making an angle of $\phi = 60°$ as shown in Fig. 10.23. The location of the pole is found to be at –4.95.

Open-loop and closed-loop transfer function gains in pole-zero form are equal for unity feedback system. The value of this gain has to such that $C(s)/R(s)|_{s=0} = 1$, as this system is type-1. We then find from the closed-loop poles and zeros that

$$\text{Gain } K = \frac{4 \times 1.35 \times 6.65}{1.2} \approx 30$$

The open-loop transfer function of the compensated system becomes

$$G(s) = \frac{30(s+1.2)}{s(s+1)(s+4)(s+4.95)}$$

The velocity error constant is

$$K_v = \lim_{s \to 0} sG(s) = \frac{30 \times 1.2}{1 \times 4 \times 4.95} = 1.82$$

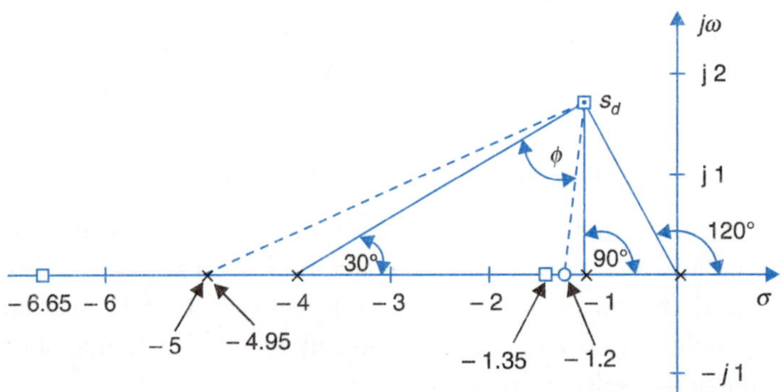

Fig. 10.23. Design of lead compensator for Example 10.3.

It must be understood here that only a marginal increase in K_v above this value can be achieved by a slight readjustment of the compensating zero. Any large shift in the compensating zero would result in violation of the dominance condition.

Let us now check the dominance condition. From the root locus of the compensated system (Fig. 10.23) the closed-loop transfer function in factored form is obtained as

$$\frac{C(s)}{R(s)} = \frac{30(s+1.2)}{(s+1+j1.73)(s+1-j1.73)(s+1.35)(s+6.65)}$$

For an impulse input

$$c(t) = 2.94e^{-t}\sin(1.73t - \tan^{-1} 0.216) - 0.27e^{-1.35t} + 0.89\,e^{-6.65t}$$

In the above equation, the second term is contributed by the closed-loop pole which is created by the introduction of the compensator zero and lies close to this zero. It is observed that the coefficient of this term is small and its contribution to system dynamics is therefore negligible. Equally the contribution of the pole at $s = -6.65$ is negligible as it is located far to the left of the dominant poles. It is thus established that this compensator design guarantees the dominance of the closed-loop poles at s_d.

Lag Compensation

As pointed out earlier in Section 10.2, a lag compensator is used to improve the steady-state behaviour of a system while preserving a satisfactory transient response. This compensation scheme therefore is found useful in systems having satisfactory transient response but unsatisfactory steady-state response.

Consider a unity feedback system with a forward-path transfer function of

$$G_f(s) = \frac{K\prod_{i=1}^{m}(s+z_i)}{s^r \prod_{j=r+1}^{n}(s+p_j)}$$

At a certain value of K, this system has a satisfactory transient response, *i.e.,* its root locus plot passes through (or close to) the desired closed-loop pole location s_d indicated in Fig. 10.24. It is required to improve the system error constant to a specified value K_e^c without impairing its transient response. This requires that after compensation the root locus should continue to pass through s_d, while the error constant at s_d, is raised to K_e^c. To accomplish this, consider adding a lag compensator pole-zero pair with zero the left of the pole. If this pole-zero pair is located close to each other, it will contribute a negligible angle at s_d such that s_d continues to lie on the root locus of the compensated system. Figure 10.24 shows the location of such a lag compensator pole-zero pair. It should be noticed from this figure that apart from being close to each other, the pole-zero pair is also located close to origin, the reason for which will become obvious from the discussion given below.

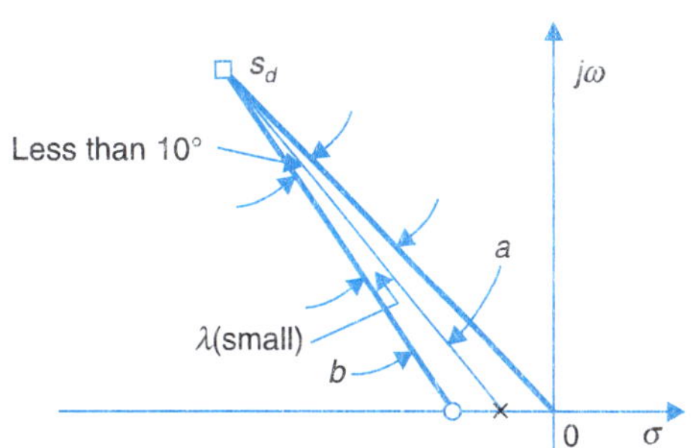

Fig. 10.24. Locating the lag compensator pole-zero.

The gain of the uncompensated system at s_d is given by

$$K^{uc}(s_d) = \frac{|s_d|^r \prod_{j=r+1}^{n}|s_d+p_j|}{\prod_{i=1}^{m}|s_d+z_i|} \qquad \text{...(10.33)}$$

For the compensated system, the system gain at s_d is

$$K^c(s_d) = \frac{|s_d|^r \prod_{j=r+1}^{n} |s_d + p_j|}{\prod_{i=1}^{m} |s_d + z_i|} \frac{a}{b} \qquad ...(10.34)$$

Since the pole and zero are located very close to each other, they are nearly equidistant from s_d, *i.e.*, $a \approx b$. It therefore follows from eqns. (10.34) and (10.35), that

$$K^c(s_d) \approx K^{uc}(s_d)$$

The error constant of the compensated system is given by

$$K_e^{\,c} = K^c(s_d) \frac{\prod_{i=1}^{m} z_i}{\prod_{j=r+1}^{n} p_j} \frac{z_c}{p_c} \approx K^{uc}(s_d) \frac{\prod_{i=1}^{m} z_i}{\prod_{j=r+1}^{n} p_j} \frac{z_c}{p_c}$$

$$\approx K_e^{\,uc} \times \frac{z_c}{p_c}$$

where $K_e^{\,uc}$ = error constant s_d of the uncompensated system. Since the error constant $K_e^{\,c}$ of the compensated system must equal the specified value, we have

$$K_e^{\,c} \approx K_e^{\,uc} \frac{z_c}{p_c} \qquad \text{or} \qquad \beta = \frac{z_c}{p_c} \approx \frac{K_e^c}{K_e^{uc}} \qquad ...(10.35)$$

Thus the β-parameter of the lag compensator is nearly equal to the ratio of the specified error constant to the error constant of the uncompensated system. Any value of $\beta = z_c/p_c > 1$ with $-p_c$ and $-z_c$ close to each other can be realized by keeping the pole-zero pair close to origin. Since the lag compensator does contribute a small negative angle λ at s_d, the actual error constant will somewhat fall short of the specified value if β obtained from eqn. (10.35) is used. Hence for design purpose, we choose a β somewhat larger than that given by this equation.

Further, the effect of the small lag angle λ is to give the closed-loop pole with specified ζ but slightly lower ω_n. This can be anticipated and counteracted by taking the ω_n of s_d to be somewhat larger than the specified value.

In the light of the above discussion, the steps for lag compensator design are summarized below.

1. Draw the root locus plot for the uncompensated system.

2. Translate the transient response specifications into a pair of complex dominant roots and locate these roots on the uncompensated root locus plot. Since the transient performance of the uncompensated system is satisfactory, the desired dominant roots of the closed-loop system will lie on (or close to) the uncompensated root locus.

3. Calculate the gain of the uncompensated system at the dominant root s_d and evaluate the corresponding error constant.

4. Determine the factor by which the error constant of the uncompensated system should be increased to meet the specified value. Choose the β-parameter of lag compensator to be somewhat greater than this factor.

5. Selected zero of the compensator sufficiently close to the origin. As a guide rule, we may construct a line making an angle of 10° (or less) with the desired ζ-line from s_d (see Fig. 10.24). The intersection of this line with the real axis gives location of the compensator zero.

6. The compensator pole can then be located at $-p_c = -z_c/\beta$. It is important to note that the pole-zero pair should contribute an angle λ less than 5° at s_d so that the root locus plot in the region of s_d is not appreciably changed and hence satisfactory transient behaviour of the system is preserved.

Let us illustrate the above steps with the help of an example.

Example 10.4: Let us reconsider the system discussed in Example 10.3 with an open-loop transfer function of

$$G_f(s) = \frac{K}{s(s+1)(s+4)}$$

the system is to be compensated to meet the following specifications:

Damping ratio $\zeta = 0.5$

Settling time $t_s = 10$ sec

Velocity error constant $K_v \geq 5$ sec^{-1}

From the transient response specifications, it follows that

$$\text{undamped natural frequency } \omega_n = \frac{4}{10 \times 0.5} = 0.8 \text{ rad/sec}$$

The desired dominant closed-loop poles are then required to be located at

$$s_d = -\zeta\omega_n \pm j\sqrt{(1-\zeta^2)}\ \omega_n = -0.4 \pm j0.7$$

as shown in Fig. 10.25. These transient response specifications are such that s_d lies on (it may also lie close to) the root locus of the uncompensated system.

At s_d, the gain K^{uc} of the compensated system is given by

$$K^{uc} = 0.8 \times 0.9 \times 3.7 = 2.66$$

The velocity error constant of the uncompensated system is given by

$$K_v^{uc} = \lim_{s \to 0} sG_f(s) = 2.66/4 = 0.666$$

The system has to be compensated to obtain a velocity error constant of $K_v^c \geq 5$. Therefore the β-parameter of the lag network is given by

$$\beta = K_v^c/K_v^{uc} = 5/0.666 = 7.5$$

To counter the effect of a small negative angle contributed by the lag network at s_d, choose a slightly higher value of β, say $\beta = 10$.

From the desired located s_d, draw a line making an angle of say 6° with the desired ζ-line. Its intersection with the real axis determines the compensator zero at $-z_c = -0.1$ as shown in Fig. 10.25. The compensator pole is then located at $-p_c = -z_c/\beta = -0.01$.

Now for the compensated system locate the point s'_d which lies on the $\zeta = 0.5$ line and the compensated root locus using the angle criterion. For the compensated system at s'_d

$$K^c(s'_d) = 2.2$$

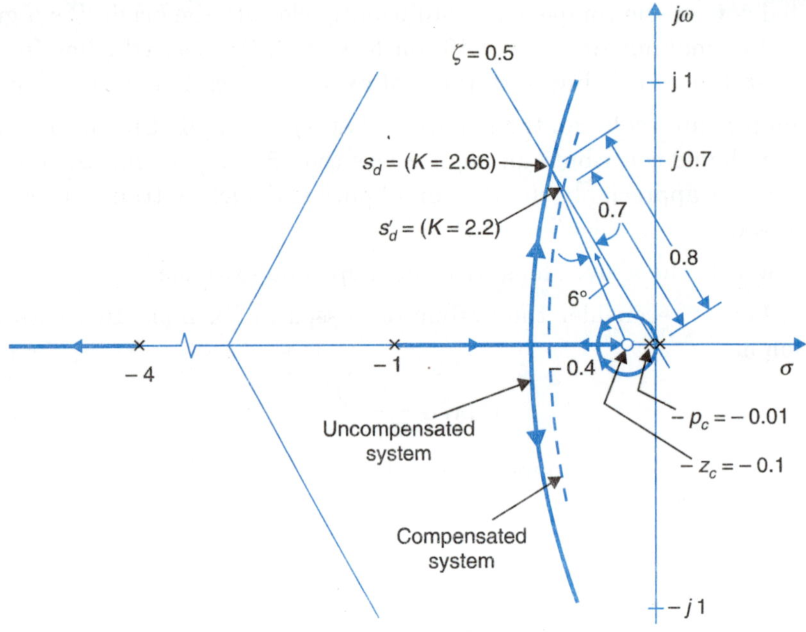

Fig. 10.25. Root locus plots for uncompensated and lag compensated system.

Thus the open-loop transfer function of the compensated system is

$$G(s) = \frac{2.2(s + 0.1)}{s(s + 1)(s + 4)(s + 0.01)}$$

For which we obtain $$K_v = \frac{2.2 \times 0.1}{4 \times 0.01} = 5.5 \text{ (satisfactory)}$$

Comparison of the root locus for the uncompensated and lag compensated systems shows that ω_n has decreased from 0.8 to 0.7 rad/sec. This means a slight increase in the settling time. If this increase is unacceptable, the system must be redesigned by choosing ω_n at s_d to be slightly higher than the specified value 0.8.

PI Controller

In most systems of practical interest the problem is to reduce the steady-state error, while maintaining satisfactory transient response. This is best accomplished by introducing an integration in the forward path for error reduction (or elimination) and a zero to for satisfactory transient response. This lead to a proportional plus integral controller with

$$G_c(s) = K_1 + \frac{K_2}{s} = K_1\left(\frac{s + \alpha}{s}\right) \; ; \; \alpha = K_2/K_1$$

Compensated system has the open-loop transfer function

$$G_c(s)\, G(s) = K_1(s + \alpha)\, \frac{1}{s}\, G(s) = KK_1(s + \alpha).G'(s)$$

where $$G'(s) = \frac{G(s)}{s}$$

This additional integration reduces the steady state error; type -0 steady state step input error is reduced to zero and ramp input error is limited to a finite value ; type-1 ramp input error is also reduced to zero.

PI controller design proceeds conveniently by root locus technique. All the poles (including additional pole at $s = 0$) of $G'(s)$ are located on the s-plane. The transient response specifications lead to the location of s_d, the desired root location. The compensating zero $s = -\alpha$ is located so that the angle criterion at s_d is met. The root locus gain can then be computed at s_d but this does not affect the steady-state error.

The zero of PI controller appears in the closed-loop transfer function which impairs the transient response which was computed on dominant pole pair basis ; see Section 5.6. This problem is more intense in PI controller than in lead network compensation.

Prefilter

To cancel out the closed-loop zero a prefilter

$$G_p(s) = \frac{1}{s+\alpha}$$

is introduced as shown in Fig. 10.26 of the complete compensated system

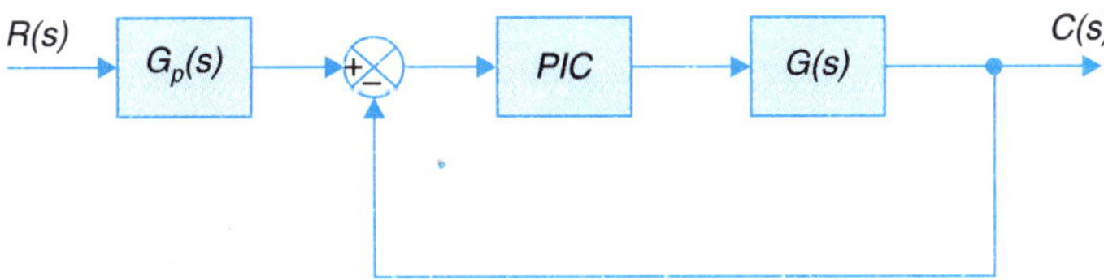

Fig. 10.26. PI compensated system.

The procedure enunciated above is illustrated below through two examples.

Example 10.5: A unity negative feedback system has an open-loop transfer function of

$$G(s) = \frac{K}{(s+4)}$$

Consider a cascade compensator

$$G_c(s) = \frac{s+\alpha}{s}$$

(*a*) Select the values of K and α to achieve

(*i*) Peak overshoot of about 20%.

and (*ii*) Settling time (2% basis) ≈ 1 sec

(*b*) For the values of K and α found in part (*a*) calculate the unit ramp input steady-state error.

Solution. In this simple example we can proceed analytically. Compensated forward path transfer function is

$$G_c(s)\,G(s) = \frac{K(s+\alpha)}{s(s+4)} \qquad ...(i)$$

Its characteristic equation is

$$1 + G_c(s)\,G(s) = 0 \quad \text{or} \quad s(s + 4) + K(s + \alpha) = 0$$

or $$s^2 + (4 + K)s + K\alpha = 0 \qquad ...(ii)$$

Required :

$$M_p = e^{-\pi\xi/\sqrt{(4-\xi^2)}} = 0.2$$

which yields $$\xi = 0.456$$

Settling time $$t_s = \frac{4}{\xi\omega_n} = 1$$

which gives $\omega_n = \dfrac{4}{0.456} = 8.77$ rad/sec.

From the characteristic equation

$$K\alpha = \omega_n{}^2 = (8.77)^2 = 76.9$$

$$K + 4 = 2\xi\omega_n = 8$$

Solving we get $K = 4, \quad \alpha = 19.23$

Transfer function of prefilter

$$G_p(s) = \frac{1}{s + 19.23}$$

(*b*) $$R(s) = \frac{1}{s^2}, \text{ unit ramp}$$

$$e_{ss} = \lim_{s\to 0} \frac{1}{sG_c(s)G(s)} = \lim_{s\to 0} \frac{(s+4)}{K(s+\alpha)} = \frac{4}{K\alpha}$$

or $$e_{ss} = \frac{4}{4 \times 19.23} = 0.05$$

The reader may verify that the compensated system has zero steady-state error to unit step input.

Example 10.6: A unity feedback control system has open-loop transfer function

$$G(s) = \frac{K}{s(s+50)}$$

Design a *PI* controller

$$G_c(s) = \frac{s+\alpha}{s}$$

to meet the following specifications

(*i*) $M_p = 20\%$ and (*ii*) $t_s = 2$ sec.

Solution. Forward path transfer function of the compensated system

$$G_c(s)\,G(s) = \frac{K(s+\alpha)}{s^2(s+50)}$$

Poles 0, 0, – 50 ; Zero – α (to be found). Converting specifications to desired root location (complex conjugate)

$$M_p = 20\% \quad \Rightarrow \quad \xi = 0.456, \text{ angle } \xi-\text{line} = 63°$$

$$t_s = \frac{4}{\xi\omega_n} = 2 \quad \Rightarrow \quad \xi\omega_n = 2,\ \omega_n = 4.4,\ \omega_n\sqrt{1-\xi^2} = 3.9$$

We find $s_d = -2 \pm j3.9$

Objective now is to find the location of zero ($s = -\alpha$) for the compensated root locus to pass through s_d. For this we need to locate poles on the s-plane, which is shown in Fig. 10.27 but not to scale. We will proceed analytically.

We will locate the zero such that the angle criterion is met at s_d.; Therefore

$$2\theta + \theta_3 - \theta_z = -180°$$

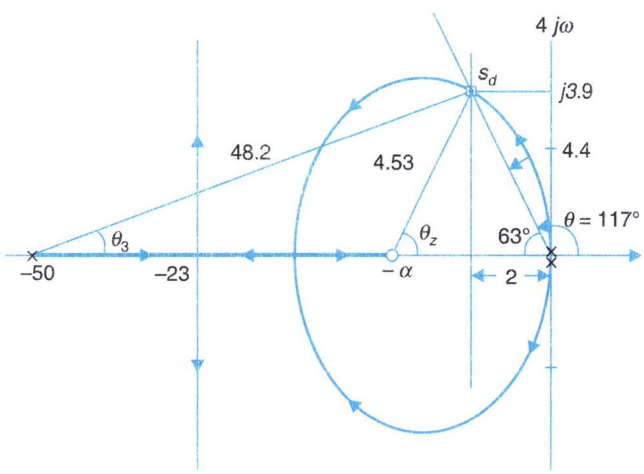

Fig. 10.27

$$\theta = 117°,\ \theta_3 = \tan^{-1}\frac{3.9}{50-2} \approx 5°$$

$$2 \times 117° + 5° - \theta_z = +180°$$

or $\theta_z = 59°$

Now $\tan\theta_z = \dfrac{3.9}{\alpha - 2} = 1.66$

which give $\alpha = 4.35$

The compensator transfer function is the given by

$$G_c(s) = \frac{(s+4.35)}{s}$$

To determine root locus gain (K') at s_d. Distances to s_d from poles and zero are given in Fig. 10.27 From which we get

$$K = \frac{(4.4)^2 \times 48.2}{4.53} = 206$$

Corresponding to this gain these will be a real pole to the left of $s = -4.35$. Dominance condition is met reasonably.

Centroid, $$-\sigma_A = \frac{-0-0-50+4.53}{3-1} = -23$$

$$\text{Angle of asymptotes} = \frac{\pm 180°}{3-1} = \pm 90°$$

Angle of departure from pole at origin

$$\theta_p + 90 + 0 - 0 = \pm 180°$$

$$\theta_p = 90°, -270°$$

Breakaway determination needs the solution of a third order equation.

A rough sketch of root locus is drawn in Fig. 10.27.

Lag-Lead Compensation

In the preceding section, we have seen that the lead compensator is suitable for system having unsatisfactory transient response but it provides only a limited improvement in steady-state response. If the steady-state behaviour is highly unsatisfactory, the lead compensator may not be the answer. On the other hand for systems with satisfactory transient response but unsatisfactory steady-state response, the lag compensator is found to be a good choice.

When both transient and steady-state response are quite unsatisfactory, we must draw upon the combined powers of lag and lead compensators in order to meet the specifications. A more convenient choice is the combined lag-lead compensator introduced earlier in Section 10.3.

The general form of the function of the lag-lead compensator in pole-zero form is

$$G_c(s) = \frac{(s+z_{c_1})(s+z_{c_2})}{\underbrace{(s+p_{c_1})}_{\text{Lag section}}\underbrace{(s+p_{c_2})}_{\text{Lead-section}}};\ \beta = \frac{z_{c_1}}{p_{c_1}} = \frac{p_{c_2}}{z_{c_2}} > 1$$

$$= G_{c_1}(s)G_{c_2}(s)$$

The usual design procedure is to first design the lead-section to meet the specifications on transient response. This design yields a suitable value of β. The error constant is then determined for the lead compensated system. If this is satisfactory the design is complete. If on the other hand, the specified error constant is much higher than that obtained by lead compensation, we then proceed to design the lag-section such that the overall system meets the specifications on steady-state performance. Since β is already determined, the lag-section can increase the error constant by about a factor β. If this increase is sufficient, we proceed to complete the lag-section design as usual. Otherwise, we must redesign the lead-section for a larger value of β. The following example illustrates the procedure.

Example 10.7: Let us again consider the system discussed in Example 10.3. The open-loop transfer function of the uncompensated system is

$$G_f(s) = \frac{K}{s(s+1)(s+4)}$$

This system is now required to be compensated to meet the following specifications:

Damping ratio $\zeta = 0.5$

Undamped natural frequency $\omega_n = 2$

Error constant $K_v \geq 5$

Let us design a lead compensator to meet the transient response specifications (refer Example 10.3). The desired location of the dominant closed-loop poles is

$$s_d = -1 \pm j1.73$$

The angle contribution required of the load compensator pole-zero is $\phi = 60°$. Let the compensator zero be placed at $s = -1$, so as to cancel one of the open-loop poles. The compensating pole is then located by drawing a line at $\phi = 60°$ from the line joining -1 and s_d as shown in Fig. 10.28. The compensating pole is found to be at $s = -4$, giving $\beta = -4/-1 = 4$.

The transfer function of the lead compensated system is

$$G_f(s)G_{c2}(s) = \frac{K^{c2}}{s(s+4)^2} \quad ...(10.36)$$

From Fig. 10.26, the gain K^{c2} at s_d is found to be 23.8. Therefore, from eqn. (10.36)

$$G_f(s)G_{c2}(s) = \frac{23.8}{s(s+4)^2}$$

$$K_v^{c2} = \lim_{s \to 0} sG_f(s)G_{c2}(s) = 1.49$$

This does not meet the specified $K_v \geq 5$. A lag-section using $\beta = 4$ will increase K_v by about four times, which then satisfies the specification on steady-state performance.

The line from s_d making an angle of 10° with the desired β-line intersects real axis at -0.24 which gives the location of the zero of the lag-section. The pole of lag-section is then found to be at $-p_{c_1} = -0.24/4 = -0.06$.

The open-loop transfer function of the lag-lead compensated system then becomes

$$G(s) = \frac{K^c(s+0.24)}{s(s+0.06)(s+4)^2} \quad ...(10.37)$$

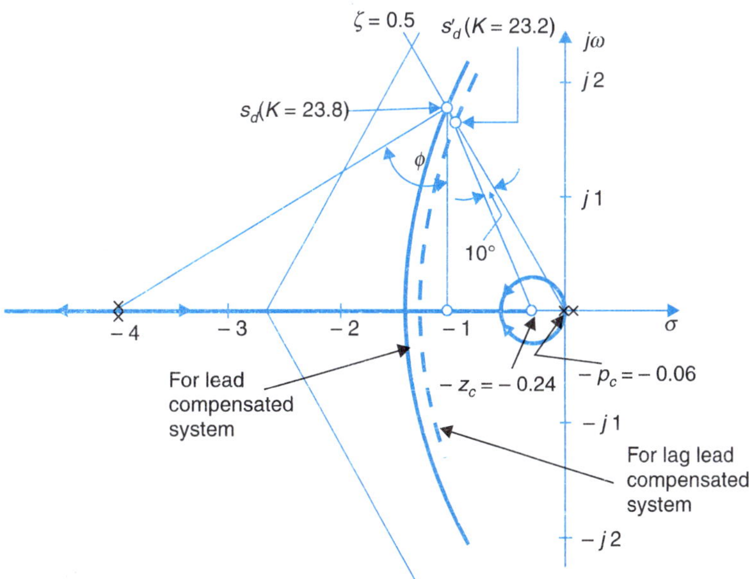

Fig. 10.28. Design of lag-lead compensator.

The root-locus plot for lag-lead compensated system is shown in Fig. 10.26. From this figure, the gain at s'_d (which is slightly shifted from s_d due to introduction of the lag-section) is given by

$$K^c = 23.2$$

Therefore from eqn. (10.37) the open-loop transfer function of the lag-lead compensated system is

$$G(s) = \frac{23.2(s + 0.24)}{s(s + 0.06)(s + 4)^2}$$

$$K_v^c = \frac{23.2 \times 0.24}{0.06 \times 16} = 5.8 \text{ (meet } s \text{ specified value)}$$

If the specifications are very stringent independent lead one lag network compensation is needed as this gives two independently adjustable parameters, α and β. Lag and lead network can be made independent by providing an electronic buffer in between. Such a lead and lag compensator design is illustrated in example below.

Example 10.8: A unity feedback control system is characterized by the open-loop transfer function

$$G(s) = \frac{K}{s(s+3)(s+4)}$$

(*a*) Determine the value of K if 20% peak overshoot to unit step input is desired.

(*b*) For the above value of K determine system K_v (velocity error constant) and settling time.

(*c*) Design a compensator that will give approximately 15% overshoot to unit step input, while settling time is decreased by a factor of 2.5. What is the K_v of the compensated system.

(*d*) For the compensated system as in part (*c*), it is desired to have $K_v \geq 20$. Design a compensator to meet both the specifications.

Solution. (*a*) $M_p = 20\% \Rightarrow \xi \approx 0.45 \Rightarrow \xi$-line angle = 63°

(dominant poles basis)

Open-loop

Poles $s = 0, -3, -9$

No zeroes

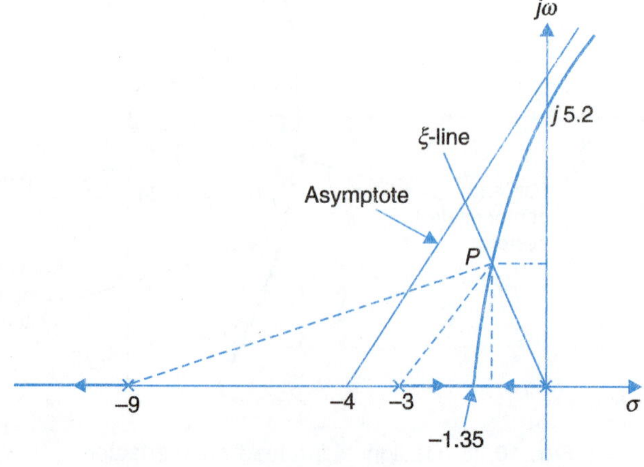

Fig. 10.29 (*a*)

Following all the necessary steps the root locus is drawn Fig. 10.29 (*a*). The root locus intersects the ξ-line at P, corresponding to dominant complex conjugate poles. The third closed-loop pole will lie on real axis beyond $s = -9$ so that the dominance condition is preserved.

At point P, root locus gain

$$K' = \text{product of distances from poles to } P$$
$$= 56$$

(b) For the compensated system

$$K_v = \frac{56}{3 \times 9} = 2.08$$

Real part of point P

$$\xi\omega_n = 1.06$$

Settling time, $$t_s = \frac{4}{\xi\omega_n} = 3.77 \text{ sec.}$$

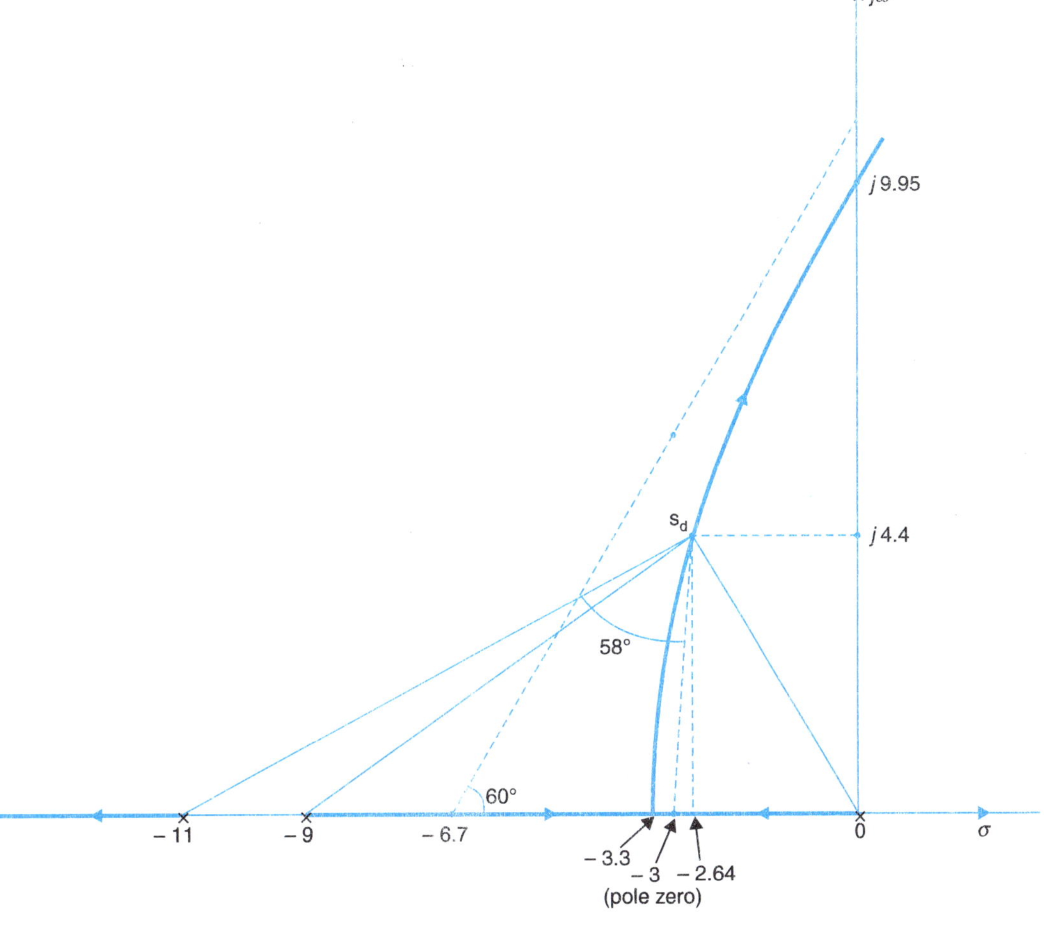

Fig. 10.29 (b)

(c) Peak overshoot specified 15% $\Rightarrow$ $\xi = 0.52$

$$t_s(\text{new}) = \frac{3.77}{2.5} = 1.5 \text{ sec.}$$
$$\omega_n = 2.67$$

$$\xi\omega_n = \frac{4}{1.5} = 2.67,\ \omega_n = \frac{2.67}{0.52} = 5.14$$

$$\omega_n\sqrt{(1-\xi^2)} = 4.39 \approx 4.4$$

We then get the desired root location as

$$s_d = -2.64 \pm j4.4\ ;\ \text{(located in Fig. 10.29 (}b\text{))}$$

If easily found from Fig. 10.29 (*a*) that these roots do not lie on the root locus of the uncompensated system. To make the root locus to pass through s_d, we would need lead compensation.

Total angle at s_d contributed by the poles of $G(s)$ is found to be – 238° from Fig. 10.29 (*b*). The lead compensator pole-zero pair should provide an angle of – 180° – (– 238°) = 58°

Lead Compensator Design

We locate the zero of lead compensator to slightly to left of s_d (– 2.67). We place it at $s = -3$ to cancel out the corresponding pole of $G(s)$. This is called *cecellation compensation*. The compensator pole is then located at an angle of 58° to $s = -3$ to s_d line. The pole is found to lie at $s = -11$. These constructions are indicated in Fig. 10.29 (*b*). The transfer function of the lead compensator is then

$$G_c(s) = \frac{(s+3)}{(s+11)}$$

Thus $$G_c(s)\,G(s) = \frac{K}{s(s+9)(s+11)}$$

Lead compensated root locus is drawn in Fig. 10.29 (*b*) which passes through s_d. Note that breakaway point and $j\omega$-axis crossing are indicated on the figure.

From Fig. 10.29 (*b*) we find at s_d, root locus gain is

$$K = 5 \times 7.8 \times 9.5 = 371$$

$$K_v = \frac{371}{a \times 11} = 3.75$$

Increase in K_v contributed by lead compensator

$$\frac{K_v\ \text{(compensated)}}{K_v\ \text{(uncompensated)}} = \frac{3.75}{2.08} = 1.8 \text{ times}$$

(*d*) It is now specified that $K_v \geq 20$ while the transient performance attained by lead compensation is not impaired. It means that root locus should continue to pass through s_d or pass close by.

For this purpose lag compensation is needed in addition to lead compensation. We shall use an independent lag network so the choice of β parameter is not constrained by the α parameter of the lead compensator. This implies two independent network with a buffer in between.

We need to increase K_v by 20/3.75 ≈ 6 times. Therefore the β parameter of the lag network is

$$\beta = 6$$

The zero of lag is located by drawing a line from s_d at an angle of 10° with ξ-line. This locates the zero at

$$s = -0.9$$

The pole is then located at

$$s = -0.9/6 = -0.15$$

The lag compensator transfer function is then

$$G_c(s)\big|_{\text{lag}} = \frac{(s+0.9)}{(s+0.15)}$$

As the angle contributed by lag compensator at s_d is only $10°$ the root locus will pass very close to s_d.

The open-loop transfer function of the compensated system is

$$G(s)\big|_{\text{Compensated}} = \frac{K'(s+0.9)}{s(s+0.15)(s+9)(s+11)}$$

The reader may verify the specification by root locus plot of the compensated system using MATLAB.

10.5 CASCADE COMPENSATION IN FREQUENCY DOMAIN

Compensation design can also be conveniently carried out using frequency domain methods. Frequency domain specifications are generally given in the following form:

1. Phase margin ϕ_{pm} or resonant peak M_r–indicative of relative stability.
2. Bandwidth ω_b or resonant frequency ω_r–indicative of rise time and settling time.
3. Error constant—indicative of steady-state error.

In case the specifications are given in time domain, we must first translate these into frequency domain to carry out frequency domain compensation. This translation is carried out by using the explicit correlations between the two domains for second-order system. As pointed out earlier in the text, these correlations are valid approximations for higher-order systems dominated by a pair of complex conjugate poles. The correlations given earlier in Chapter 8 and 9 are reproduced below for convenience of use

$$M_r = \frac{1}{2\zeta\sqrt{(1-\zeta^2)}} \qquad ...(10.38)$$

$$\omega_r = \omega_n\sqrt{(1-2\zeta^2)} \qquad ...(10.39)$$

$$\phi_{pm} = \tan^{-1}\left\{2\zeta \Big/ \left[\sqrt{(1+4\zeta^4)} - 2\zeta^2\right]^{1/2}\right\} \qquad ...(10.40)$$

$$\omega_b = \omega_n\,[1 - 2\zeta^2 + \sqrt{(2 - 4\zeta^2 + 4\zeta^2}\,]^{1/2} \qquad ...(10.41)$$

The spcifications in terms of M_r and ω_r are found convenient when compensation is carried out by Nyquist plots, while compensation using Bode plots is usually easier to handle for specified phase margin. Gain cross-over frequency could be used as a rough measure of ω_b. Of course, when Nichols charts are used, any type of specifications can be handled.

After completing the compensation design in frequency domain, one must recheck the time response specifications by computing the exact time response of the compensated system. This is necessary because the time response specifications are converted into frequency response proceeding on the assumption that the compensated system will have a dominant pair of closed-loop poles. Based on the results of this check, it may sometimes be necessary to repeat the complete design process.

From the above discussion it is realized that in frequency domain compensation, direct control on system time performance is lost. This disadvantage is compensated by the advantages of frequency domain methods such as simplicity in analysis and design, ease in experimental determination of frequency response for real systems.

As said earlier, the frequency domain compensation may be carried out using Nyquist polts, Bode plots or Nichols chart. The advantages of the Bode plots are that they are easier to draw and modify. Further the gain adjustments are conveniently carried out and the error constants are always clearly in evidence. We shall discuss below mainly the design procedures using Bode plots, while Nichol charts are used to check the values of M_r, ω_n and ω_b wherever necessary.

Lead Compensation

The lead compensation on Bode plots proceeds by adjusting the system error constant to the desired value. The phase margin of the uncompensated system is then checked. If found satisfactory, the lead compensation is designed to meet the specified phase margin. Figure 10.30 shows the Bode plot of a unity feedback system with an open-loop transfer function $G_f(s) = K_v/s(\tau s + 1)$. With the K_v adjusted to the specified value, let the gain cross-over frequency and the phase margin of the uncompensated system be ω_{c1} and ϕ_1 repectively. Let it be assumed that ϕ_1 falls short of the specified phase margin ϕ_s. Additional phase margin can be provided by a lead network so placed that its two corner frequencies be on either side of ω_{c1}. This location of the network's Bode plot ensures that the phase-lead is provided over the desired region. As is seen from Fig. 10.30, the addition of the phase-lead network (along with necessary amplification to cancel the attenuation of the network) in this region causes the new cross-over frequency to shift to the right to some unknown value ω_{c2}. This fact reduces the contribution to phase margin of the fixed part of the system $G_f(s)$ to some value $\phi_2 < \phi_1$. The phase-lead ϕ_l required at ω_{c2} to bring the phase margin to the specified value is given by

$$\phi_l = \phi_s - \phi_2$$

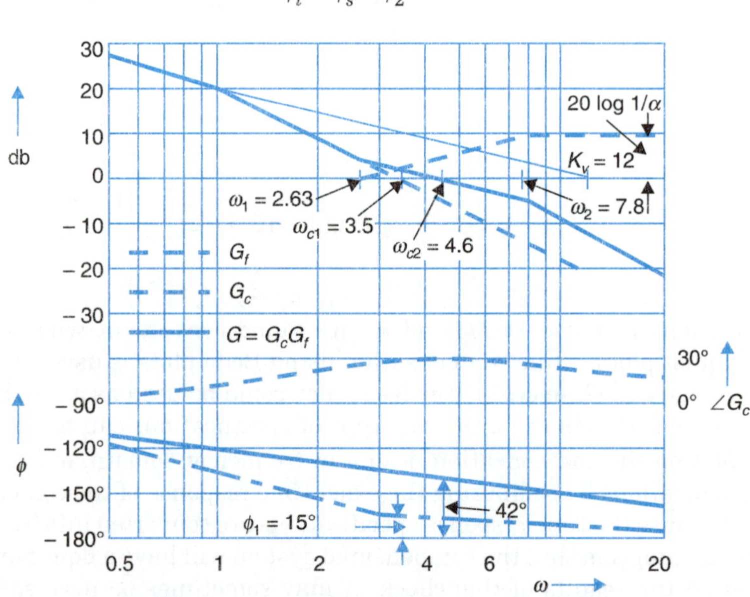

Fig. 10.30. Design of lead compensation for type-1 system.

Since ϕ_2 is unknown, we make a guess on the same as

$$\phi_2 = \phi_1 - \varepsilon$$

Thus the required phase-lead at the new cross-over frequency ω_{c_2} is obtained as

$$\phi_l = \phi_s - \phi_1 + \varepsilon$$

where ε is the unknown reduction in the phase angle $\angle G_f(s)$ on account of the increase in the cross-over frequency. A guess is made on the value of ε depending on the slope in this region of the db-log ω plot of the uncompensated system. For a slope for –40 db/decade, as is in the present example, $\varepsilon = 5°–10°$ is a good guess. The guessed value may have to be as high as 15-20° for a slope of –60 db/decade.

In order to provide a phase-lead ϕ_l at ω_{c2} with the largest value of the network parameter α (this is desirable for signal/noise considerations), the frequency of maximum phase-lead ω_m of the network must be made to coincide with ω_{c_2}. Thus we set

$$\omega_{c2} = \omega_m \qquad \therefore \qquad \phi_m = \phi_l$$

The α-parameter of the network can then be computed from

$$\alpha = \frac{1 - \sin \phi_l}{1 + \sin \phi_l}$$

Since at ω_m the network provides a db-gain of 10 log $(1/\alpha)$ (see Fig. 10.11) the new cross-over frequency $\omega_{c2} = \omega_m$ can be determined as that frequency at which the uncompensated system has a db-gain of – 10 log $(1/\alpha)$.

With $\omega_{c2} = \omega_m$ thus established, the two corner frequencies of the lead network can be computed as

$$\omega_1 = 1/\tau = \omega_m \sqrt{\alpha}\ ;\ \omega_2 = 1/\alpha\tau = \omega_m/\sqrt{\alpha}$$

The log-magnitude and phase plots of the compensated system can now be completely drawn for checking the actual phase margin achieved. How far this falls short of the specified value depends on how good was the guess of ε. If the phase margin achieved is still unsatisfactory, the process is repeated with a higher value of ε. The process usually converges to a staisfactory solution in one or two trials within an error of about 2-3° which is acceptable for all practical purposes. Certain overall observations can be made from the Bode plots of the lead compensated system:

1. The cross-over frequency is increased.

2. This high frequency end of the log-magnitude plot has been raised up by a db-gain of 20 log $(1/\alpha)$.

Since the cross-over frequency is a rough measure of the bandwidth of a closed-loop system, we can say that the lead compensation increases the system bandwidth and hence an improvement in the speed of response of the system results. The actual value of the bandwidth can be determined by transferring the data from Bode plots to Nichols chart. It must of course be noted that too large an increase in system bandwidth so as to include some of the noise frequencies is undersirable and must be guarded against in lead compensation design.

The design procedure for a lead compensator outlined above is quite general and applies to any type and order of system though it has been illustrated through an example of a type-1, second-order system. The complete design procedure is summarized below.

1. Determine the loop gain K to satisfy specified error constant.

2. Using this value of K, determine the phase margin of the uncompensated system.

3. Determine the phase-lead required using the relation

$$\phi_l = \phi_s - \phi_1 + \varepsilon$$

where ϕ_s = specified phase margin; ϕ_l = phase margin of the fixed part of the system (*i.e.*, the uncompensated system); and ε = a margin of safety required by the fact that the cross-over frequency will increase due to compensation.

4. Let $\phi_m = \phi_l$ and determine the α-parameter of the network from

$$\alpha = \frac{1-\sin\phi_m}{1+\sin\phi_m}$$

If the required ϕ_m is more that 60°, it is recommended to use two identical networks each contributing a maximum lead of $\phi_l/2$.

5. Calculate the db-gain 10 log $(1/\alpha)$ provided by the network at ω_m. Locate the frequency at which the uncompensated system has a gain of –10 log $(1/\alpha)$. This is the cross-over frequency $\omega_{c2} = \omega_m$ of the compensated system.

6. Compute the two corner frequencies of the network as

$$\omega_1 = 1/\tau = \omega_m\sqrt{\alpha}\,;\; \omega_2 = 1/\alpha\tau = \omega_m/\sqrt{\alpha}\,.$$

7. Draw the magnitude and phase plots of the compensated system and check the resulting phase margin. If the phase margin is still low, raise the value of ε repeat from step 3 above.

8. Check any additional specifications on system performance, *e.g.*, bandwidth. Redesign for another choice of cross-over frequency ω_{c2}, till this specification is met. It may be noted that the additional specification can only be met if it is consistent.

Example 10.9: Consider a type-1 unity feedback system with an open-loop transfer function

$$G_f(s) = \frac{K_v}{s(s+1)}$$

It is specified that $K_v = 12\ \text{sec}^{-1}$ and $\phi_{pm} = 40°$.

The Bode plots of the system with $K_v = 12$ are drawn in Fig. 10.30. The phase margin of the uncompensated system is found to be $\phi_1 = 15°$. If a lead compensator is used, the phase-lead required at the new cross-over frequency is given by

$$\phi_l = 40° - 15° + 5° = 30° = \phi_m$$

$$\alpha = \frac{1-\sin 30°}{1+\sin 30°} = 0.333$$

The magnitude contribution of the compensating network at ω_m is

$$10 \log (1/0.333) = 4.8 \text{ db}$$

Therefore the frequency at which the uncompensated system has a magnitude of –4.8 db becomes the new cross-over frequency $\omega_{c2} = \omega_m$ when the lead network is added. From the Bode plot of Fig. 10.30, we find

$$\omega_{c2} = 4.6 \text{ rad/sec}$$

Lower corner frequency of the network, $\omega_1 = 1/\tau = \omega_m\sqrt{\alpha} = 2.65$ rad/sec

Upper corner frequency of the network, $\omega_2 = 1/\alpha\tau = \omega_m\sqrt{\alpha} = 8$ rad/sec

The transfer function of the lead network (with amplifier), therefore becomes

$$G_c(s) = \frac{s/2.65 + 1}{s/8 + 1} = \frac{0.377s + 1}{0.125s + 1}$$

The amplification necessary to cancel the lead network attenuation is

$$A = 1/\alpha = 3$$

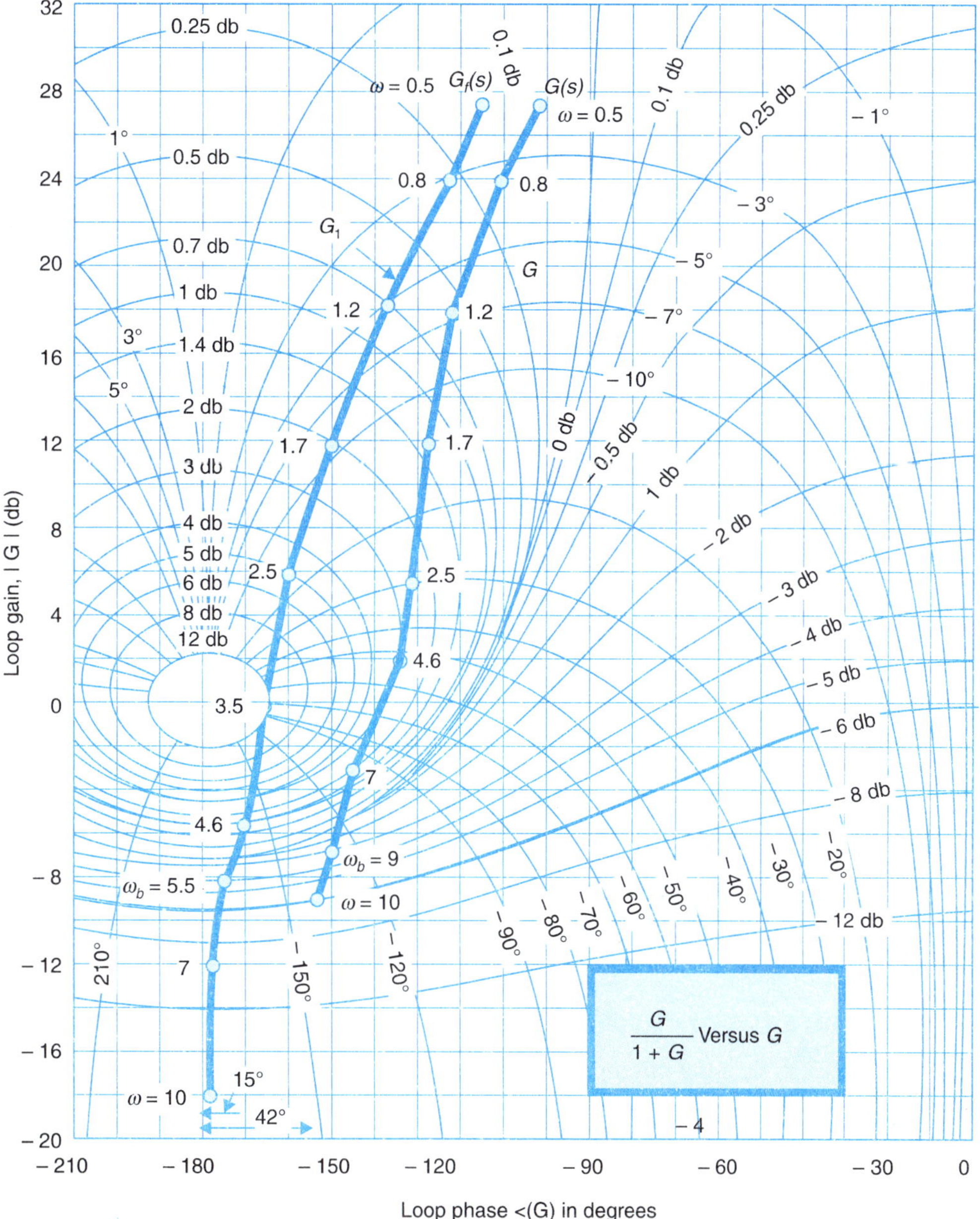

Fig. 10.31. Log-magnitude vs phase angle plots of type-1 system (Example 10.9).

The open-loop transfer function of the compensated system is

$$G(s) = G_f(s)G_c(s) = \frac{12(0.377s + 1)}{s(s + 1)(0.128s + 1)}$$

The log-magnitude vs phase angle for both $G_f(s)$ and $G(s)$ are plotted on Nichols chart in Fig. 10.31, from which we observe the following:

1. Phase margin is increased from 15° to 42°.
2. Bandwidth is increased from 5.5 rad/sec to 9 rad/sec.
3. M_r is reduced from +12 db to +3.0 db.
4. ω_r is increased from 3.5 rad/sec to 4.6 rad/sec.

Thus, in general, the effect of the lead compensator is to increase the margin of stability and speed of response.

Unit step response of the compensated system (as obtained by MATLAB) is drawn in Fig. 10.32 from which we read that

$$\% \text{ peak overshoot} = 30$$

The compensated phase margin of 42° means $\zeta = 0.42$ and $M_p = 25\%$ (on dominant pole basis). So there is good correlation for this method of design.

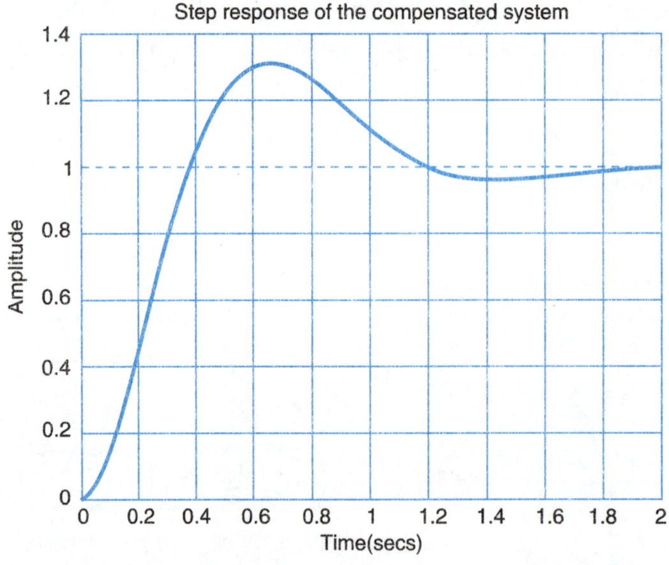

Fig. 10.32. Units step response of the compensated system of Example 10.9.

Example 10.10: Let us design lead compensation for a type-2 system with an open-loop transfer function

$$G_f(j\omega) = \frac{K}{(j\omega)^2(j0.2\omega + 1)}$$

Assume that the system is required to be compensated to meet the following specifications:

1. Acceleration error constant $K_a = 10$
2. Phase margin = 35°

The specification on K_a is met by choosing $K = 10$, such that the open-loop transfer function of the system becomes

$$G_f(j\omega) = \frac{10}{(j\omega)^2\,(j0.2\omega + 1)}$$

The Bode plots of $G_f(j\omega)$ are drawn in Fig. 10.33, from which it is found that the cross-over frequency is $\omega_{c1} = 3.16$ rad/sec and the phase margin is $\phi_1 = -33°$. Further the uncompensated system is absolutely unstable.

Since the required phase margin is $\phi_s = 35°$, the phase-lead needed at the cross-over frequency ω_{c2} of the compensated system is obtained as

$$\phi_1 = 35° - (-33°) + 14° = 82°$$

where $\varepsilon = 14°$ is the estimated reduction in the $\angle G_f(j\omega)$ since $\omega_{c2} > \omega_{c1}$. A large value of ε has been selected in this example as $\angle G_f(j\omega)$ is decreasing at a faster rate since the final slope of the log-magnitude curve is –60 db/decade.

As discussed in Section 10.4, using a single lead network to give such a large phase-lead is not advisable. We shall design a double lead network so that each section has to provide a maximum phase lead to 82/2 = 41°. The attenuation factor of each lead-section is

$$\alpha = \frac{1 - \sin 41°}{1 + \sin 41°} \approx 0.2$$

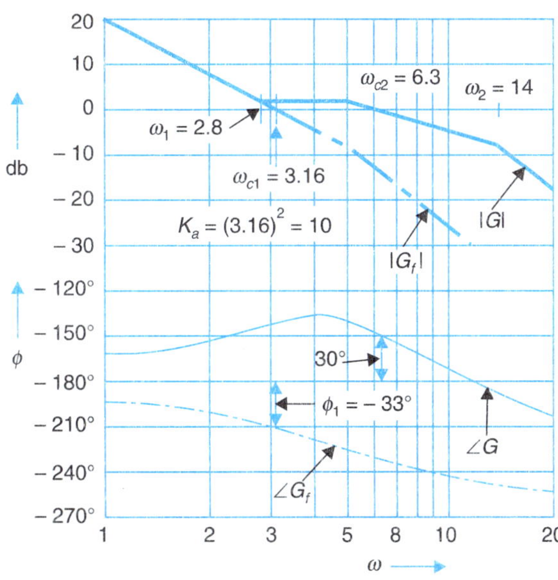

Fig. 10.33. Design of lead compensation for type-2 system.

From the plot of the uncompensated system, it is seen that the gain 2[–10 log (1/0.2)] = –13.2 db occurs at $\omega_{c2} = 6.3$ rad/sec. This should be the cross-over frequency of the compensated system. Choosing $\omega_m = \omega_{c2} = 6.3$ rad/sec, we obtain the network corner frequencies as

$$\omega_1 = (\sqrt{\alpha})\omega_m = 6.3 \times 0.446 = 2.8 \text{ rad/sec};\ \tau_1 = 1/2.8 = 0.358 \text{ sec}$$

$$\omega_2 = (1/\sqrt{\alpha})\omega_m = 2.8/0.2 = 14.0 \text{ rad/sec};\ \tau_2 = 1/14 = 0.077 \text{ sec}$$

Thus the transfer function of each section of the double lead network (with the attenuation cancelled by an amplification $A = 1/\alpha = 5$) is

$$G_c^1(s) = \left(\frac{0.358s + 1}{0.077s + 1}\right)$$

The transfer function of the double lead network becomes

$$G_c(s) = \frac{(0.358s + 1)^2}{(0.077s + 1)^2}$$

The net additional amplifier gain required is $A^2 = 25$ as shown in Fig. 10.34. Cascading this amplifier between the two sections of the lead network provides the isolation needed to prevent the second lead-section from loading the first one.

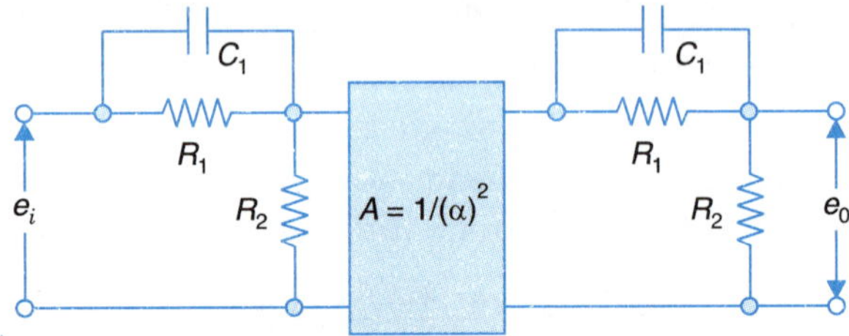

Fig. 10.34. Double lead network.

The open-loop transfer function of the system compensated by the double lead network is given by

$$G(s) = \frac{10(0.358s + 1)^2}{s^2(0.2s + 1)(0.077s + 1)^2}$$

The Bode plot of $G(j\omega)$ is shown in Fig. 10.33 from which it is found that the phase margin at the new cross-over frequency is +30°. This indicates that the system has become stable but the desired phase margin is not yet fully achieved. This is because of the excessive lag of the fixed part of system at the new cross-over frequency. If this phase margin is not acceptable then the compensator should be redesigned with a higher value of ε, say $\varepsilon = 25°$.

Lag Compensation

The frequency response of a phase-lag network has been presented in Fig. 10.14 from which it is seen that the network acts like a low-pass filter attenuating the high frequencies by a db of ($-20 \log \beta$). The phase-lag mainly occurs within and around the two corner frequencies of the lag network. It must be recongnized here that any phase-lag is undesirable at the cross-over frequency of the compensated system. Therefore, it is the attenuation characteristic of the network which is exploited for compensation purposes.

Consider for illustration a unity feedback system with an open-loop transfer function of the form $G_f(s) = K_v/s(\tau_1 s + 1)(\tau_2 s + 1)$ whose Bode plots are given in Fig. 10.34 with K_v adjusted to the specified value. The uncompensated system has a cross-over frequency of ω_{c1} and a phase margin of ϕ_1 which happens to be negative. It is desired to raise the phase margin to a specified value ϕ_s without altering K_v, the error constant. It must be noted here that *lag compensation is only possible if there exists a range of frequencies in which the uncompensated system has a phase angle less negative than* $(-180° + \phi_s)$. Such is not the case with system of type-2 and higher where lead compensation must therefore be used.

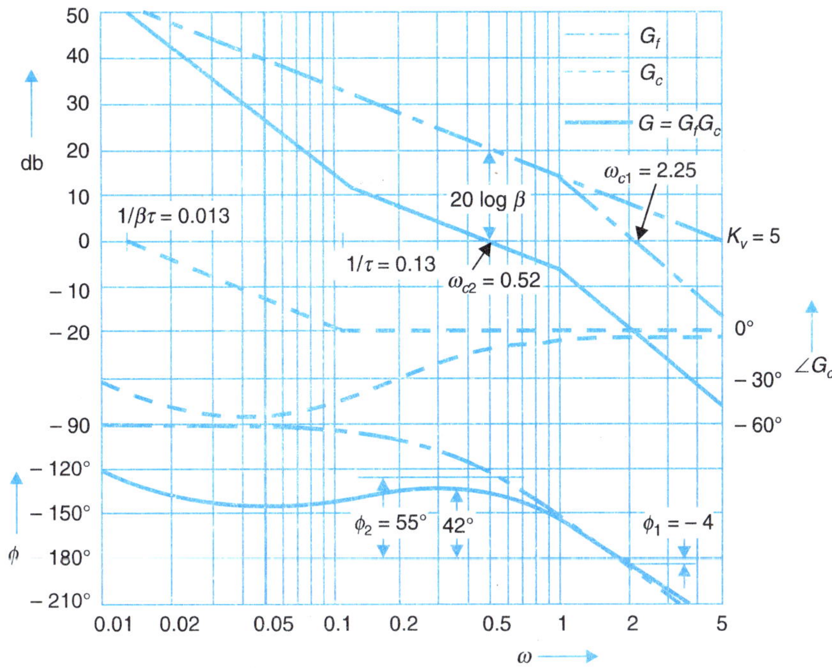

Fig. 10.34. Design of lag compensation for type-1 system.

It is immediately noticed from the Bode plots of the uncompensated system that the desired phase margin can be attained by modifying the db-log ω plot so as to lower the cross-over frequency, while the phase plot is not allowed to alter significantly in the region of the new cross-over frequency. This can be easily achieved by a lag network wherein the high frequency attenuation ($-20 \log \beta$) is utilized to lower the cross-over frequency, while the two corner frequencies of the network are placed sufficiently lower than the desired cross-over frequency so that the phase-lag contribution of the network at this cross-over frequency is made sufficiently small. Usually the upper corner frequency of the network is placed one octave to one decade lower than the cross-over frequency ω_{c2} of the compensated system. To nullify the effect of the small phase-lag contribution of the network which will still be present at ω_{c2}, the compensated system must contribute an angle ϕ_2 at ω_{c2} towards phase margin where,

$$\phi_2 = \phi_s + \varepsilon$$

where ε is allowed a value 5° – 15°.

It shall be noted that the network contributes 0 db in the low frequency region so that K_v remains unaltered by introducing the lag network.

With the above explanation, the lag compensator design procedure may be stated as below.

1. Determine the open-loop gain necessary to satisfy the specified error constant. If the phase margin of the uncompensated system with this gain is unsatisfactory, then design a lag network as per succeeding steps.

2. Find the frequency ω_{c2} where the uncompensated system makes a phase margin contribution of

$$\phi_2 = \phi_s + \varepsilon$$

where ϕ_2 is measured above –180° line. Allow for $\varepsilon = 5°$ to 15° for phase-lag contributed by the network at ω_{c2}.

3. Measure the gain of the uncompensated system at ω_{c2} and equate it to the required high frequency network attenuation (20 log β). Calculate therefrom the β-parameter of the network. This procedure ensures that the compensated cross-over frequency will lie at ω_{c2}.

4. Choose the upper corner frequency ($\omega_2 = 1/\tau$) of the network one octave to one decade below ω_{c_2}, *i.e.*,

$$\omega_2 = 1/\tau = \omega_{c2}/2 \text{ to } \omega_{c2}/10$$

A larger value of τ than that calculated from this rule is undesirable from the point of view of realization as it leads to excessively large size of capacitor for the network.

5. With β and τ determined, the lag compensator design is complete. Draw the frequency response of the compensated system to check the resulting phase margin.

6. If there is any additional specification, check if it is satisfied. Redesign the compensator by choosing another value of ε, if found necessary.

For the system discussed above, the examination of the Bode plots of the lag compensated system reveals that

(*i*) the cross-over frequency is reduced;

(*ii*) the high-frequency end of the log-magnitude plot is lowered by (20 log β) db.

Thus we find the lag compensator reduces the system bandwidth (cross-over frequency being a rough measure of bandwidth) and the additional attenuation of high frequencies improves the signal/noise ratio of the system. If for the specified K_v the reduced bandwidth is unacceptable, lead compensator may be tried.

Example 10.11: Let us reconsider the system discussed in Example 10.4, whose open-loop transfer function is

$$G_f(s) = \frac{K}{s(s+1)(s+4)}$$

The system is to be compensated to meet the following specifications:

Damping ratio $\zeta = 0.4$

Settling time $t_s = 10$ sec

Velocity error constant $K_v \geq 5 \text{ sec}^{-1}$.

Using the relation given in eqns. (10.40 and 41) and $t_s = 3/\zeta\omega_n$ for 5% tolerance band, we obtain the following equivalent specifications in frequency domain:

$$\phi_s = 43° \ ; \ \omega_b = 1.02 \text{ rad/sec} \ ; \ K_v \geq 5 \text{ sec}^{-1}$$

The open-loop transfer function of the uncompensated system may be written as

$$G_f(s) = \frac{K/4}{s(s+1)(0.25s+1)}$$

The specification on K_v is met by choosing $K = 20$. Thus

$$G_f(j\omega) = \frac{5}{j\omega(j\omega + 1)(j0.25\omega + 1)}$$

The Bode plot of $G_f(j\omega)$ is shown in Fig. 10.34 from which it is found that the cross-over frequency $\omega_{c1} = 2.25$ rad/sec and the phase margin $\phi_1 = -4°$. The uncompensated system is, therefore unstable.

It is further seen from the Bode plot that neglecting the phase-lag contribution of the network, the specified phase margin of 43° is obtained, if the cross-over frequency is 0.74 rad/sec. Since this is fairly low, the upper network corner frequency cannot be taken far to its left in order to avoid large time constants. This indicates that phase-lag contribution of the network at the new cross-over frequency will be considerable and may be guessed at 12°. The uncompensated system must, therefore, make a phase margin contribution of

$$\phi_2 = \phi_s + \varepsilon = 43° + 12° = 55°$$

at ω_{c2} which is found to be 0.52 rad/sec.

Placing the upper corner frequency of the compensator two octaves below ω_{c2}, we have

$$\omega_2 = 1/\tau = \omega_{c2}/(2)^2 = 0.13 \text{ rad/sec}$$

To bring the log-magnitude curve down to 0 db at ω_{c2}, the lag network must provide an attenuation of 20 db. Therefore

$$20 \log \beta = 20 \quad \text{or} \quad \beta = 10$$

The lower corner frequency of the network is then fixed at

$$\omega_1 = 1/\beta\tau = 0.013 \text{ rad/sec}$$

The transfer function of the lag network is then

$$G_c(s) = \frac{1}{10}\left(\frac{s + 0.13}{s + 0.013}\right) = \left(\frac{7.7s + 1}{77s + 1}\right)$$

Phase-lag introduced by the lag network at ω_{c2} is

$$(\text{phase-lag})\ \omega_{c2} = \tan^{-1}(7.7\ \omega_{c2}) - \tan^{-1}(77\ \omega_{c2}) = 76° - 88° = -12°$$

Therefore, the safety margin of $\varepsilon = 12°$ is justified.

The open-loop transfer function of the compensated system becomes

$$G(s) = \frac{5(7.7s + 1)}{s(77s + 1)(s + 1)(0.25s + 1)}$$

The Bode plot of $G(j\omega)$ is shown in Fig. 10.34 from where the phase margin of the compensated system is found to be 42°.

The log-magnitude vs. phase angle curves of uncompensated and compensated systems are shown in Fig. 10.35 on Nichols chart. It is observed that addition of a lag-compensator has reduced the bandwidth from 3.4 rad/sec to 1.1 rad/sec.

However, since the reduced value lies in the acceptable range, our design is complete.

Unit step response of the compensated system (as obtained from MATLAB) is drawn in Fig. 10.36.

$$\text{Peak overshoot} \approx 32\%,\ t_s \approx 2s$$

For specified $\zeta = 0.4$, $M_p \approx 27\%$ but that is based on dominant poles. So the figure of 32% is normally acceptable in design work. Observe that t_s is much less than specified.

Lag-lead Compensation

As has been discussed earlier in this section, for a specified error constant, the phase margin can be improved to any desired value by employing lead compensation even though the

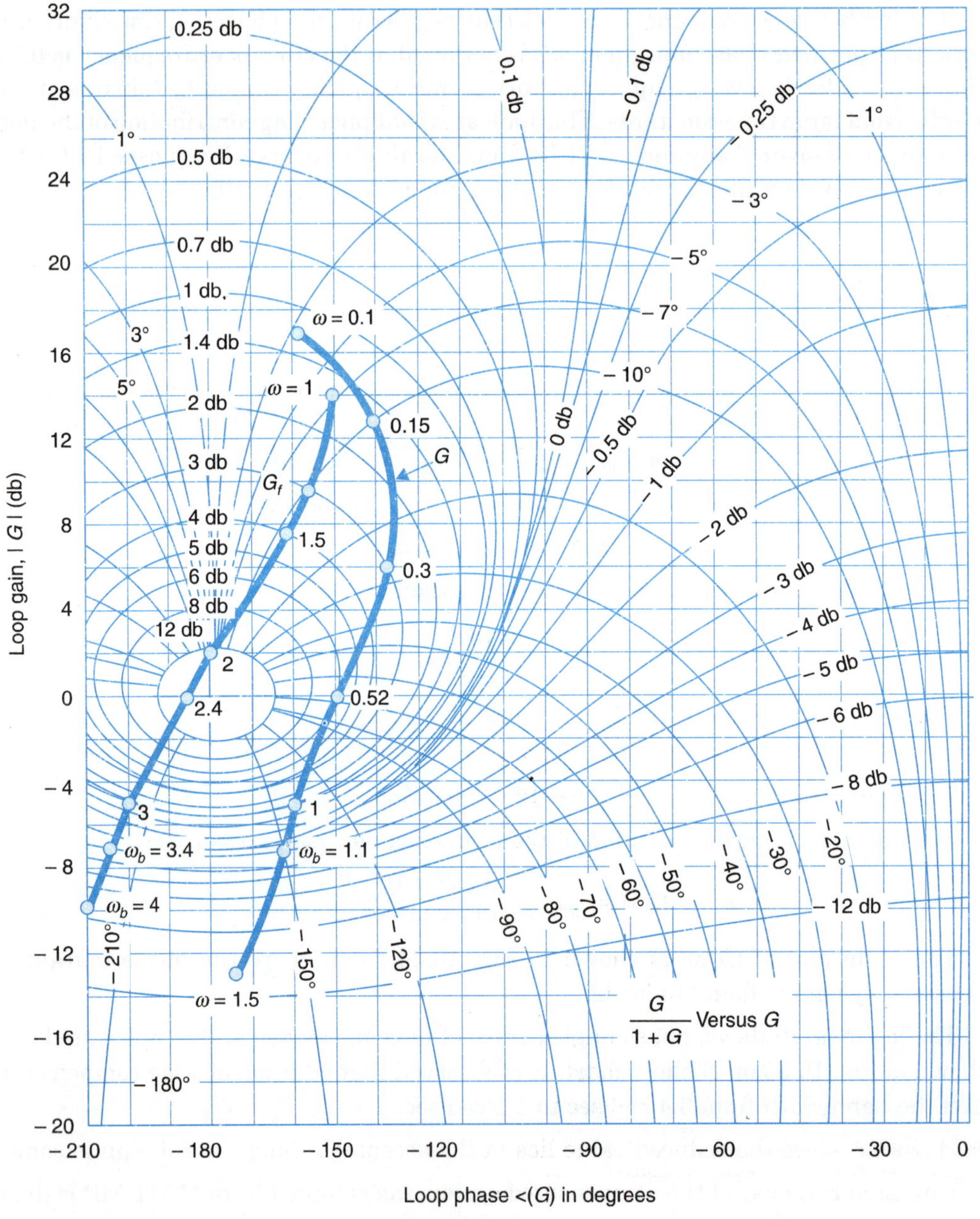

Fig. 10.35. Log-magnitude vs phase angle curves for Example 10.11.

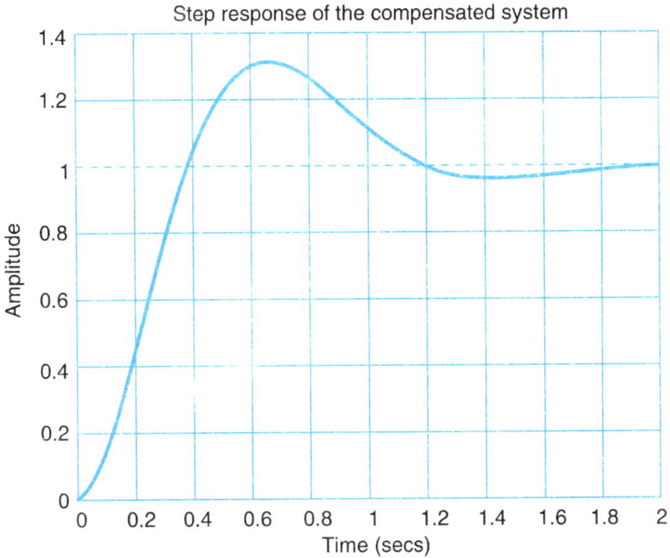

Fig. 10.36. Unit step response of compensated system of Example 10.11.

uncompensated system may be absolutely unstable. The lead compensation results in increased bandwidth and faster speed of response. For high-order systems and for systems with large error constants, large leads are required for compensation, resulting in excessively large bandwidth which is undesirable from noise transmission point of view. For such systems lag compensation is preferred provided the uncompensated system is not absolutely unstable.

The lag compensation, on the other hand, reduces system bandwidth and slows down the speed of response. For large specified error constant and moderately large bandwidth, it may not be possible to meet the specifications through either lead or lag compensation. Under such circumstances we can use a lag-lead compensator wherein for specified error constant, the lag-section is used to provide part of the phase margin and the lead section provides the rest of it as well as gives the desired bandwidth.

To start with this design, we check the phase margin and bandwidth of the uncompensated system for specified error constant. If this bandwidth is smaller than the specified value, lead compensation may be tried. However, if this bandwidth is larger than acceptable, lead compensation would not be desirable, so we try lag compensation provided the uncompensated system is not absolutely unstable. If the lag compensator design results in too low a bandwidth the need for a lag-lead compensation is indicated. A lag-lead compensation design is illustrated through an example below.

From Fig. 10.17, it is seen that a lag-lead network is essentially a band-pass filter having a transfer function

$$G_c(s) = \underbrace{\left[\frac{\tau_1 s + 1}{\beta\tau_1 s + 1}\right]}_{\text{Lag-section}} \underbrace{\left[\frac{\tau_2 s + 1}{(\tau_2/\beta)s + 1}\right]}_{\text{Lead-section}} ; \beta = \frac{1}{\alpha} > 1$$

We therefore notice that once the lag-section has been designed, τ_1 and β get fixed. Therefore τ_2 is the only variable parameter for lead-section design.

Example 10.12: Let us consider a system with an open-loop transfer function

$$G_f(j\omega) = \frac{K}{j\omega(j0.1\omega + 1)(j0.2\omega + 1)}$$

The system is to be compensated to meet the following specifications:

Velocity error constant $K_v = 30 \text{ sec}^{-1}$

Phase margin $\phi_s \geq 50°$

Bandwidth $\omega_b = 12$ rad/sec

It easily follows that $K = 30$ satisfies the specification on K_v. The Bode plot of the open-loop transfer function with this value of K is shown in Fig. 10.37 from which it is found that the uncompensated system has a cross-over frequency of 11 rad/sec and a phase margin of –24°. The uncompensated system is therefore unstable for the specified K_v. It is also observed that the uncopensated system is conditionally stable, *i.e.*, it is stable for a certain range of values of K_v. From the Nichols chart of Fig. 10.38 it is found that the uncompensated system has a bandwidth of 14 rad/sec. If lead compensation is employed, the system bandwidth will increase still further (this increase will be fairly large as $\angle G_f(j\omega)$ is decreasing rapidly near the cross-over frequency) resulting in an undesirable system which will be sensitive to noise.

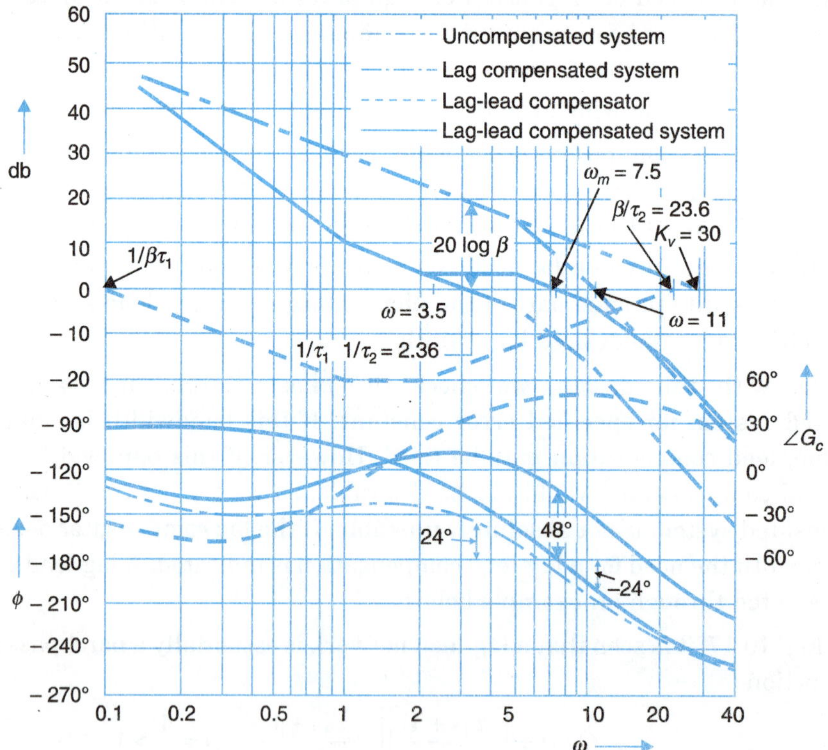

Fig. 10.37. Design of lag-lead compensation for type-1 system.

If lag compensation is attempted, the bandwidth will decrease sufficiently so as to fall short of the specified value of 12 rad/sec, resulting in a sluggish system. This fact can be verified by designing a lag compensator. We thus find that there is need to go in for lag-lead compensation.

Let us now design a lag-lead compensator to overcome the above-mentioned difficulties.

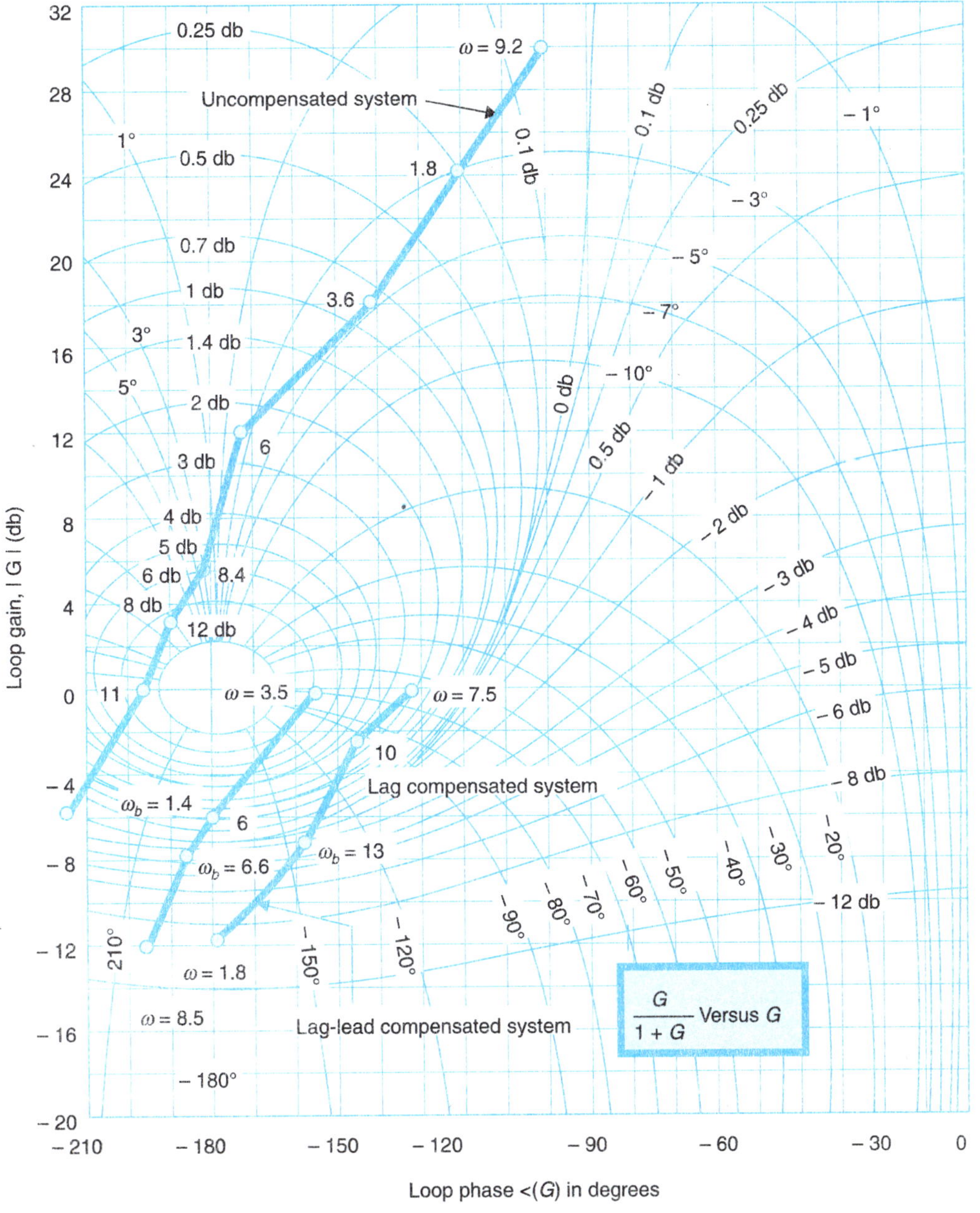

Fig. 10.38. Log-magnitude vs phase angle curves for Example 10.12.

Since a full lag compensator will reduce the system bandwidth excessively, the lag-section of the lag-lead compensator must be designed so as to provide partial compensation only. The lag-section design therefore proceeds by making a choice of the new cross-over frequency due to lag-section compensation only. This choice, of course, must be higher than the cross-over frequency if the system were fully lag compensated. In this example this choice is made as 3.5 rad/sec to start with. It is seen from the Bode plot of the uncompensated system that it must be brought down by 18.5 db for the cross-over frequency to be 3.5 rad/sec. This gives the β-parameter of the lag-section as

$$20 \log \beta = 18.5 \qquad \text{or} \qquad \beta = 8.32 \text{ say } 10$$

Let us now choose τ_1 of the lag-section to be 1 rad/sec. The lag-section design is thus complete and its transfer function is given by $G_{c1}(s) = (s + 1)/(10s + 1)$. It is found that the lag-section compensated system has a phase margin of 24°.

We now proceed to design the lead-section. This design is constrained by the fact that $\alpha = 1/\beta = 0.1$ is already fixed. The maximum lead provided by the lead-section is therefore

$$\phi_m = \sin^{-1}\left(\frac{1 - 1/\beta}{1 + 1/\beta}\right) = 56°$$

To fully utilize the lead effect, we choose the compensated cross-over frequency to coincide with ω_m, which is then obtained as that frequency where the lag-section compensated system has a db of $-10 \log \beta = -10$. This value read off from Bode plot is $\omega_m = 7.5$ rad/sec.

Then $\qquad \omega_m = \sqrt{\beta}/\tau_2 = 7.5 \qquad$ or $\qquad \tau_2 = 0.425$ and $\tau_2/\beta = 0.0425$

The transfer function of the lead-section, therefore, is

$$G_{c2}(s) = \left(\frac{0.425s + 1}{0.0425s + 1}\right)$$

Combining the transfer functions of the lead-and lag-sections, we obtain

$$G_c(s) = \frac{(s + 1)(0.425s + 1)}{(10s + 1)(0.0425s + 1)}$$

The open-loop transfer function of the lag-lead compensated system is

$$G(s) = \frac{30(s + 1)(0.425s + 1)}{s(0.1s + 1)(0.2s + 1)(10s + 1)(0.0425s + 1)}$$

The Bode plot for this transfer function is shown in Fig. 10.35. The phase-margin is found to be 48°.

The log-magnitude vs phase angle curve of the lag-lead compensated system is drawn on Nichols chart in Fig. 10.36 from which the bandwidth of the system is found to be 13 rad/sec.

The design therefore meets the specifications laid down. In case at the first attempt the specifications are not met, must redesign by adjusting the initial choice of β and τ.

Comparision Lead and Lag Compensators

Comparative advantages and application of phase lead and lag compensator carried out at various placed in sections 10.4 and 10.5 are summarized in table below.

Table 10.1. Comparison of Phase Lag and Lead Compensator

Phase Lead Compensator	Phase Lag Compensator
• Bandwidth increases High frequency gain increases	Band width decreases
• Dynamic response becomes faster	Dynamic response slows down
• Susceptable to high frequency noise	High frequency noise suppressed
• No significant decreases in steady-state error	Steady-state error reduced
• Application; When fast dynamics response is required.	Application : When low steady-state error is required.
• Cannot be applied when phase angle of uncompensated system is decreasing rapidly near crossover frequency	Cannot be applied if uncompensated system phase angle in low frequency region is not sufficient to provide requisite phase margin.

Important remark. Compensated system is generally satisfactory when the db slope at gain crossover frequency is 20 db/decade.

OPAMP Compensating Networks

Operational Amplifier circuits for use as compensating networks offer the advantage of having no loading effect, both at input and output. Standard compensating circuits are widely used in industry as control system compensators. These circuits comprise two OPAMPs with input and feedback RC circuits to generate $G_c(s)$, the compensating function, followed by another OPAMP to provide stable gain of desired value with high input and low output resistance.

Some of the commonly used OPAMP circuits are given below :

$$G_c(s) = \frac{E_0(s)}{E_i(s)}$$

PD Compensator

$$G_c(s) = \frac{R_4}{R_3} \cdot \frac{R_2}{R_1} (R_1 C_1 s + 1)$$

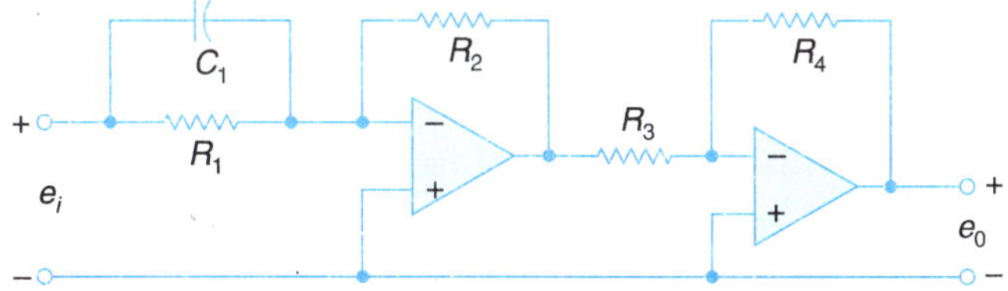

PI Compensator

$$G_c(s) = \frac{R_4}{R_3} \cdot \frac{R_2}{R_1} \cdot \frac{(R_2 C_2 s + 1)}{(R_2 C_2 s)}$$

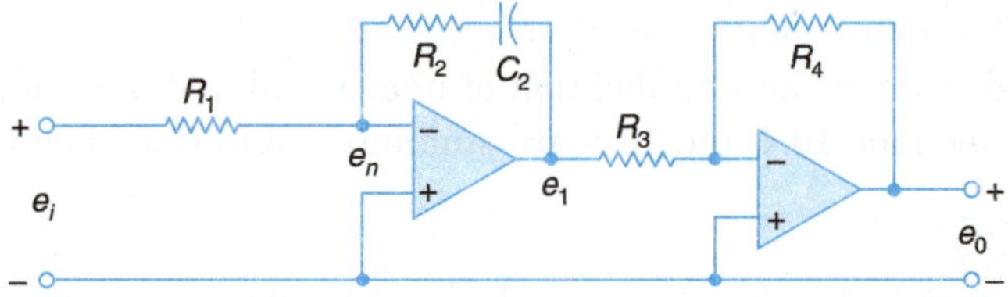

Lead/Lag Compensator

$$G_c(s) = \frac{R_4}{R_3} \cdot \frac{R_2}{R_1} \cdot \frac{(R_1C_1s + 1)}{(R_2C_2s + 1)}$$

Lead if $R_1C_1 > R_2C_2$

Lag if $R_1C_1 < R_2C_2$

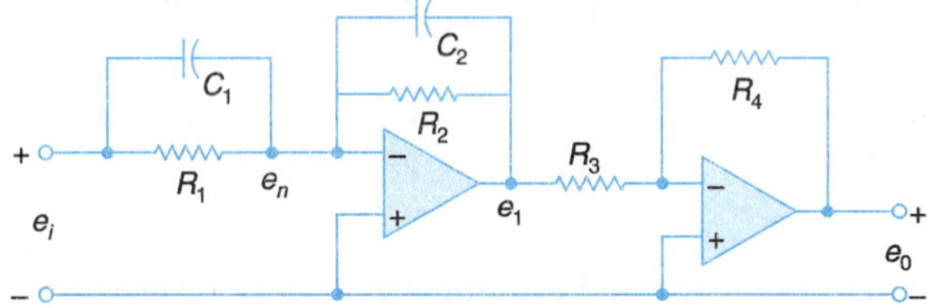

Derivation for Lead/Lag Network

Input circuit admittance

$$Y_i = \frac{1}{R_1} + C_1s = \frac{R_1C_1s + 1}{R_1}$$

Feedback circuit admittance

$$Y_f = \frac{1}{R_2} + C_2s = \frac{R_2C_2s + 1}{R_2}$$

For ideal OPAMP

$$e_n = 0$$

$$i_n \text{ (current into OPAMP)} = 0$$

Writing KCL equation at node n

$$E_i(s)Y_i - E_1(s)Y_f = 0$$

Substituting values and reorganizing yields

$$\frac{E_1(s)}{E_0(s)} = \frac{Y_i}{Y_f} = \frac{R_2}{R_1} \cdot \frac{R_1C_1s + 1}{R_2C_2s + 1}$$

The second OPAMP provides only a gain of value

$$\frac{E_0(s)}{E_1(s)} = \frac{R_4}{R_3}$$

Hence the compensator transfer function is

$$G_c(s) = \frac{E_0(s)}{E_i(s)} = \frac{R_4}{R_3} \cdot \frac{R_2}{R_1} \cdot \frac{R_1C_1s + 1}{R_2C_2s + 1}.$$

10.6 TUNING OF PID CONTROLLERS

Proportional-integral-differential (PID) controllers (compensators) have been introduced briefly in Section 5.7. These are commonly employed in process control industries. Hence we shall present various techniques of tuning PID controllers to achieve certain performance index for system's dynamic response. The technique to be adopted for determining the proportional integral and derivative constants of the controller (called tuning in process control parlance) depends upon the dynamic response of the plant.

In presenting the various tuning techniques we shall assume the basic control configuration of Fig. 10.39, wherein the controller input is the error between the desired output (command, set point, input) and the actual output. This error is manipulated by the controller (PID) to produce a command signal for the plant according to the relationship.

$$U(s) = K_p\,(1 + 1/\tau_i s + \tau_d s) \qquad \text{...(10.42)}$$

or in time domain

$$u(t) = K_p[e(t) + (1/\tau_i)\int_{-\infty}^{t} e dt + \tau_d (de/dt)] \qquad \text{...(10.43)}$$

where K_p = proportional gain

τ_i = integral time constant

τ_d = derivative time constant

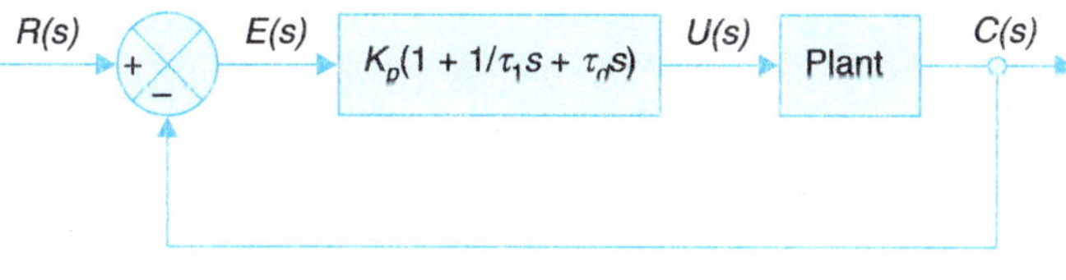

Fig. 10.39

Various tuning sequences and when these are applicable is given in the flow chart of Fig. 10.40.

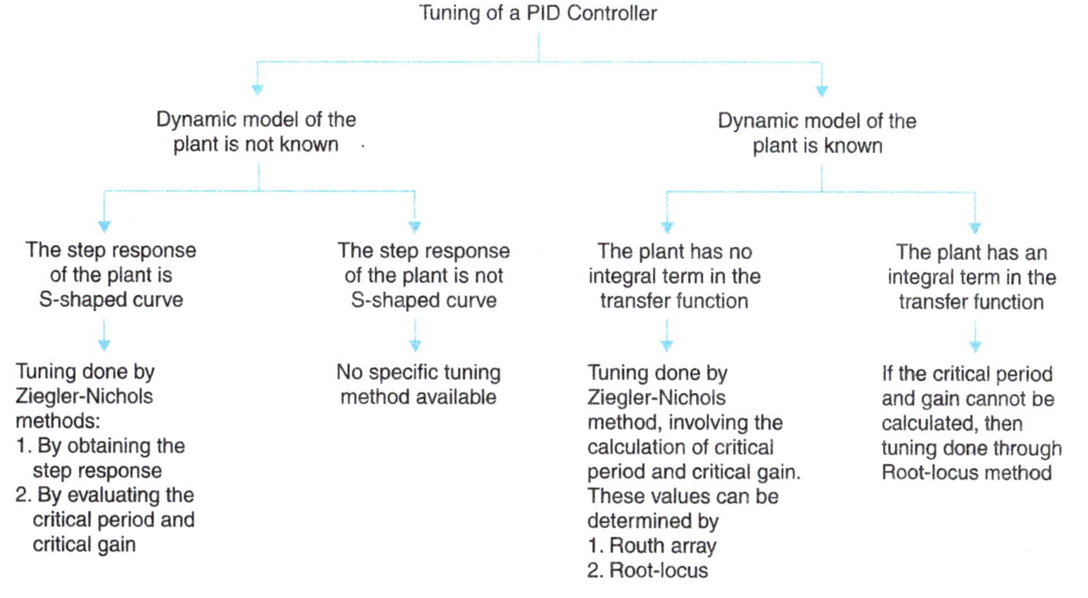

Fig. 10.40

Dynamic Model of the Plant is not Known

When the dynamic model of the process, for which the PID constants are to be found, is not known, its open-loop response for a step input is determined experimentally or by simulation. If this response is S-shaped as in Fig. 10.41 (a), Ziegler-Nichols tuning method is applicable.

The S-shaped response of Fig. 10.41 (a) is characterized by two constants, the dead time L and the time constant T as shown. These constants can be determined by drawing a tangent to the S-shaped curve at the inflection point and finding its intersection with the time axis and the line corresponding to the steady-state value of the output.

From the response of this nature the plant can be mathematically modeled as first order system with a time constant T and delay time L as shown in block diagram form in Fig. 10.41 (b). The gain K corresponds to the steady state value of the output C_{ss}.

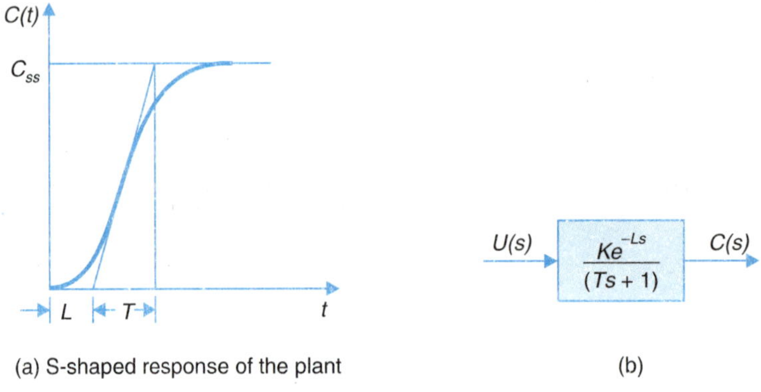

(a) S-shaped response of the plant (b)

Fig. 10.41

The value of K_p, τ_i and τ_d of the controllers can then be calculated as below:

$$\left.\begin{aligned} K_p &= 1.2(T/L) \\ \tau_i &= 2L \\ \tau_d &= 0.5\,L \end{aligned}\right\} \qquad ...(10.44)$$

Example 10.13: Consider the tuning of a PID controller for waste-water treatment where the pH of the waste-water has to be controlled. The block diagram of pH control strategy is shown in Fig. 10.42. The influent is acidic and has a fixed pH. It flows into a continuous stirred tank reactor with a fixed flowrate. The pH of the effluent has to be maintained at a set point in the range of 4 to 7 pH. This is accomplished by controlling the flowrate of the basic reagent with the help of a valve. The opening of the valve is controlled by the PID controller. The gain of the

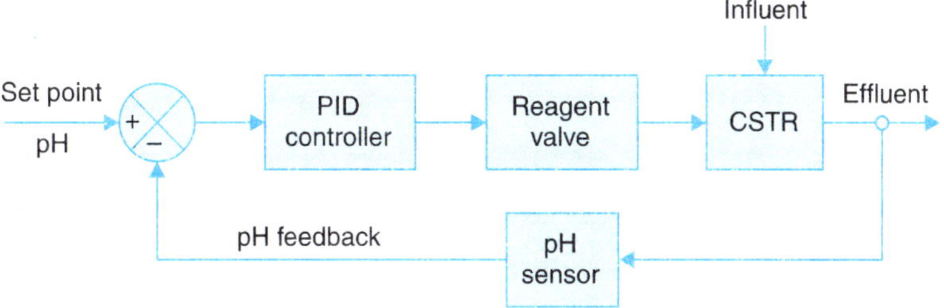

Fig. 10.42. pH Control Strategy.

From the actual setup of pH Control plant (at BITS CfR) the open-loop step response of the plant is depicted in Fig. 10.43 in which the tangent at the inflection point is drawn. From the graph we get

$$L = 19.9s \text{ and } T = 1.8\ s$$

Substituting the values in eqns. 10.44, we get

$$K_p = 0.11,\ \tau_i = 39.8,\ \tau_d = 9.95$$

The PID controller is tuned to these values and the closed-loop step response is obtained which is shown in Fig. 10.44 (on next page). The response has the following performance indices

% overshoot = 26%

Rise time = 0.3 *s*

Settling time = 9.2 *s*

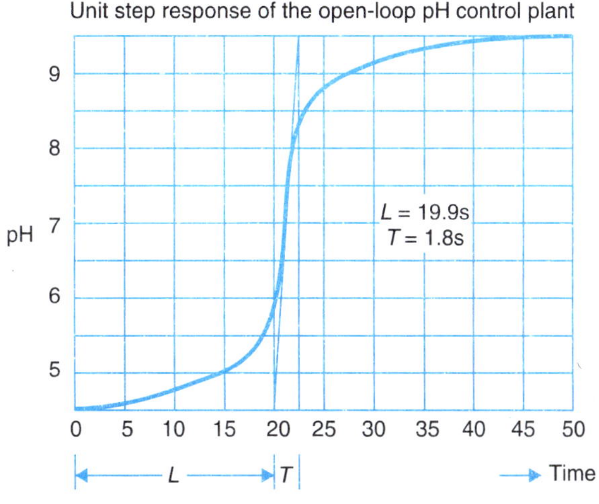

Fig. 10.43

Dynamic Model of the Plant is Known

If the dynamic model of the system is known, the PID controller can be tuned using Ziegler-Nichols method. It is first assumed that the controller has only proportional gain (K_p) term. We then proceed to determine the critical gain, K_{cr} for the closed-loop system (Fig. 10.45) to just get into continuous oscillations. The corresponding period T_{cr} of the oscillations is determined. Knowing these two values the PID controller can be tuned using the following results

$$\left.\begin{aligned} K_p &= 0.6\, K_{cr} \\ \tau_i &= 0.5\, T_{cr} \\ \tau_d &= 0.125\, T_{cr} \end{aligned}\right\} \quad \text{...(10.45)}$$

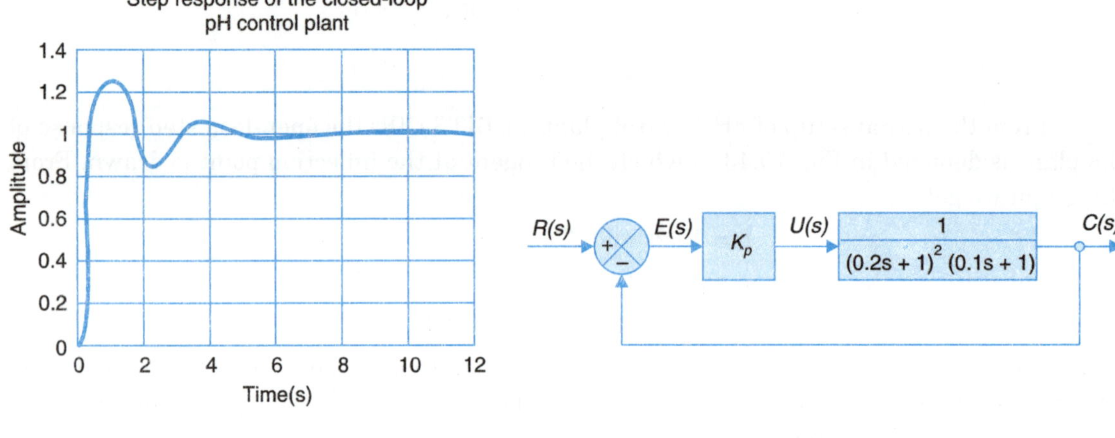

Fig. 10.44 Fig. 10.45

This method is applicable only when K_{cr} exists and the value T_{cr} can be found.

Example 10.14: Consider a third-order process with open-loop transfer function

$$G(s) = C(s)/U(s) = 1/[0.1s + 1)(0.2s + 1)^2]$$

This transfer function has no integral term. With proportional controller the transfer function of the closed loop system becomes

$$C(s)/R(s) = K_p/[(0.1s + 1)(0.2s + 1)^2 + K_p]$$
$$= K_p/(0.004s^3 + 0.08s^2 + 0.5s + 1 + K_p)$$

The critical gain is determined by Routh array for the characteristic equation

$$0.004s^3 + 0.08s^z + 0.5s + 1 + K_p = 0$$

s^3	0.004	0.5
s^2	0.08	$1 + K_p$
s^1	$[0.04 - 0.004(1 + K_p)]/0.08$	0
s^0	$1 + K_p$	0

It is easily seen from the Routh array that the plant would be unstable for

$$0.004(1 + K_p) > 0.04 \text{ or } K_p > 9$$

Hence $K_{cr} = 9$

To find the frequency of oscillations

$$0.08s^2 + (1 + K_p) = 0$$
$$0.08s^2 + 10 = 0$$

or $$s = \pm j\sqrt{(10/0.08)} = \pm j11.2 \text{ rad/s}$$

Therefore $$T_{cr} = 2\pi/11.2 = 0.56 \text{ s.}$$

Substituting these values in eqs. 10.45 gives

$$\begin{aligned} K_p &= 0.6 \times 9 = 5.4 \\ \tau_i &= 0.5 \times 0.56 = 0.28 \\ \tau_d &= 0.125 \times 0.56 = 0.07 \end{aligned} \quad ...(10.46)$$

The block diagram of the closed loop system is drawn in Fig. 10.44.

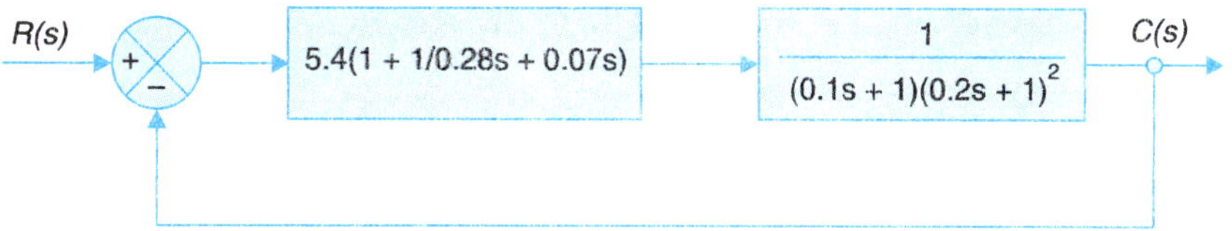

Fig. 10.44

The closed-loop step response of the system is given in Fig. 10.47. Its performance indices are as given below:

% overshoot = 42%

Rise time = 0.2 *s*

Settling time = 1.6 *s*

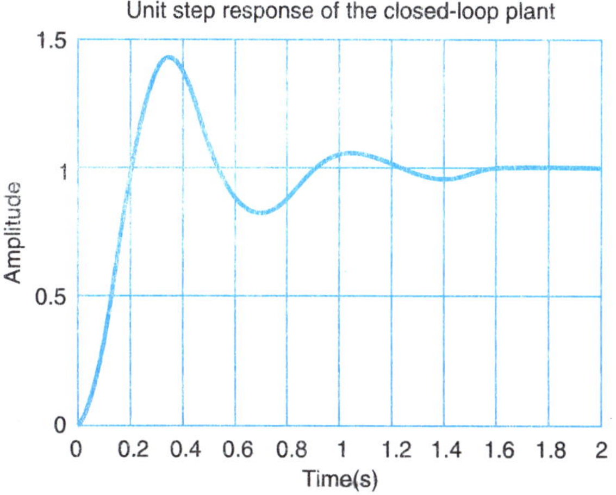

Fig. 10.47

The % overshoot obtained above is rather high. It can be reduced by heuristically fine tuning the controller parameters (PID coefficients). Since the rise time is rather small, the proportional constant K_p can be reduced to get a lower % overshoot, while increasing the rise time to a certain limit. The proportional constant is reduced to 2.16 while keeping τ_i and τ_d same as above to get optimal results. After fine tuning the results obtained are

% overshoot = 26%

Rise time = 0.36 s

Settling time = 1.3 s

K_{cr} and T_{cr} can also be determined by a root locus plot of the transfer function. The root locus of the above system is shown in Fig. 10.48. From the root locus plot the critical gain calculated is same as that determined from the Routh's array.

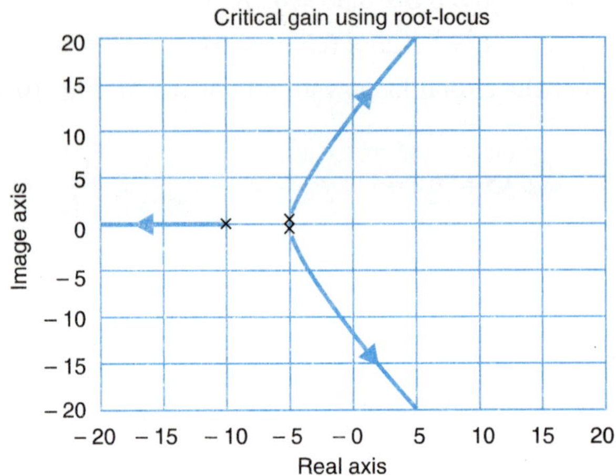

Fig. 10.48

Dynamic model of the plant is known, but

1. *Neither the critical gain exists*
2. *Nor the step response of the system is S-shaped*

In such a situation the plant has to be analysed using the root-locus technique. The block diagram of the plant with a PID controller is as shown in Fig. 10.49.

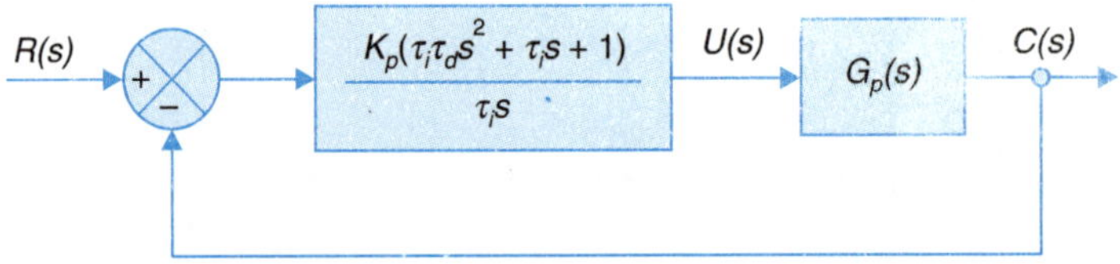

$G_p(s)$: Transfer function of the process

Fig. 10.49

The PID controller induces two zeros and a pole (at origin) in the overall transfer function. The zeros of the controller are,

$$z_1, z_2 = [-\tau_i \pm \sqrt{(\tau_i^2 - 4\tau_i\tau_d)}] / 2\tau_i\tau_d$$

The two zeros have to be placed in such a way that the desired response of the system can be obtained. The steps involved in deciding the values of the controller constants are:

1. Choose any value of τ_i and τ_d and calculate the zeros of the controller.
2. Plot the root locus of the system.

3. From the root locus find the K_p such that the desired closed-loop poles of the system can be obtained.

4. If such a value of K_p does not exist, then start from the step 1 with new values of τ_i and τ_d.

This method will be studied at length in Section 10.8 with examples.

10.7 FEEDBACK COMPENSATION

Other than cascade compensation, compensation can also be achieved by placing a compensating device in an internal feedback path around one or more components of the forward path. Though cascade compensation is quite satisfactory and economical in most cases, feedback compensation may be warranted by the following factors:

1. In a feedback compensator energy transfer is from a high energy level to a lower one thereby obviating the need for feedback amplification.

2. In nonelectrical systems, suitable cascade devices may not be available.

3. Feedback compensation often provides greater stiffness against load disturbances.

Apart from the arguments presented above, it must be mentioned here that availability of components and designer's experience play an important role in selection of a suitable compensation scheme.

It is important to note that the feedback devices may not necessarily be electrical networks. For example, the feedback compensation scheme for the second-order system discussed in Chapter 5 utilizes a tachometer in the feedback path. The net effect of this scheme is to add a zero to the open-loop transfer function, which as discussed in Section 10.2, is the principle of lead compensation.

Since cascade compensation very well illustrates what may be accomplished with simple classical compensation, we shall not discuss feedback compensation in great detail. We discuss below, through an example, a simple and practical scheme of feedback compensation, the output rate compensation, which has already been introduced in Chapter 5.

Example 10.15: Figure 10.50 (*a*) shows the block diagram of a position servo system. The open-loop transfer function of the system is

$$G_f(s) = \frac{K_A K_f K_m}{s^2(s+10)} = \frac{K}{s^2(s+10)}$$

It is easily shown that the system is unstable for all values of K.

Assume that the system is to be compensated to meet the following specifications:-

1. Percent peak overshoot $\leq 10\%$

2. Settling time ≤ 4 sec

Let us utilize tachometer feedback to meet the desired specifications. The block diagram of the system with tachometer feedback is shown in Fig. 10.50 (*b*). Reducing the minor feedback loop, we have

$$\frac{\theta_C(s)}{\theta_E(s)} = G(s) = \frac{K}{s(s^2 + 10s + K\alpha)} \qquad ...(10.47)$$

where $G(s)$ is the forward path transfer function of the major feedback loop.

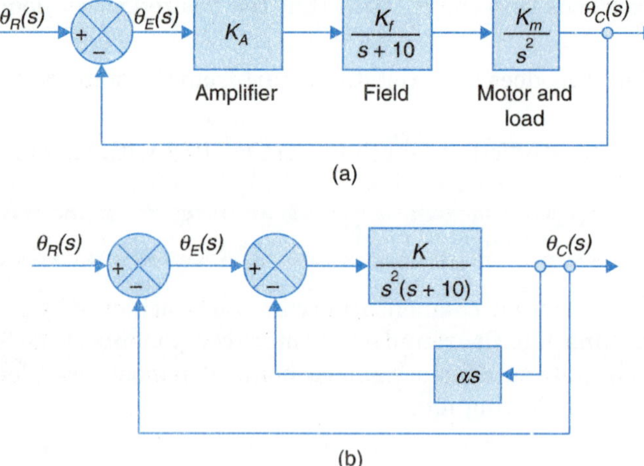

Fig. 10.50. (a) Block diagram of a position servo system; and (b) position servo with tachometer feedback.

The characteristic equation of the system is

$$1+\frac{K}{s^3+10s^2+K\alpha s}=0 \qquad \text{or} \qquad s^3+10s^2+K\alpha s+K=0$$

This characteristic equation can be rewritten in the form

$$1+\frac{K\alpha(s+1/\alpha)}{s^2(s+10)}=0 \qquad \text{...(10.48)}$$

From eqn. (10.50) we have

$$G(s)H(s)=\frac{K'(s+1/\alpha)}{s^2(s+10)} \qquad \text{...(10.49)}$$

where $K' = K\alpha$.

Equation (10.49) shows that the net effect of rate feedback is to add a zero at $s = -1/\alpha$. Now the design procedure calls for a search for suitable location of this additional zero to satisfy the specifications.

From eqn. (5.17) we find that peak overshoot of 10% corresponds to $\zeta = 0.6$. Then

$$\omega_n = 4/\zeta t_s = (4/0.6 \times 4) = 1.67 \text{ rad/sec}$$

The desired dominant roots are then given as

$$s_d = -1 \pm j1.34$$

From Fig. 10.51, the angle contribution of the open-loop poles (s = 0, 0, –10) at s_d is $-2(128°) - 8° = -264°$. Therefore for the point s_d to be on the root locus, the compensating zero should contribute an angle of $\phi = -180° - (-264°) = 84°$. The location of the compensating zero is thus found to be at $s = -1.1$. The open-loop transfer function of the compensated system becomes

$$G(s)H(s)=\frac{K'(s+1.1)}{s^2(s+10)};\ 1/\alpha = 1.1$$

The root locus plot of the compensated system is shown in Fig. 10.51. It is now observed that for all values of K', the system is stable. The value of K' at s_d is found to be 17.4.

The third closed-loop pole is at $s = -8$ which is far away from the imaginary axis, compared to the dominant closed-loop poles. Hence its effect on the transient behaviour is negligible.

Let us now investigate the steady-state behaviour of the compensated system, from Eqn. (10.47), the velocity error constant K_v is given by

$$K_v = \lim_{s \to 0} \left[s \times \frac{K}{s(s^2 + 10s + K\alpha)} \right] = 1/\alpha$$

$$= 1.1; \text{ depends on feedback only}$$

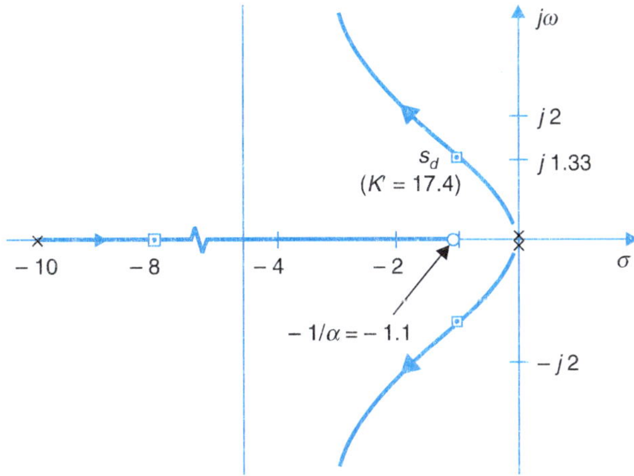

Fig. 10.51. Root locus plot of $K'(s + 1.1)/s^2(s + 10)$.

If this value K_v is acceptable, then the design is complete; otherwise the design procedure above cannot satisfy the steady-state and transient specifications simultaneously. The control over the velocity error constant can be obtained by introducing an amplifier of adjustable gain in the forward path outside the minor feedback loop as shown in Fig. 10.52.

The design procedure to satisfy the steady-state and transient specifications simultaneously, is a trial and error procedure which can be carried out by the root-locus technique but it is conveniently carried out in frequency domain, as discussed below.

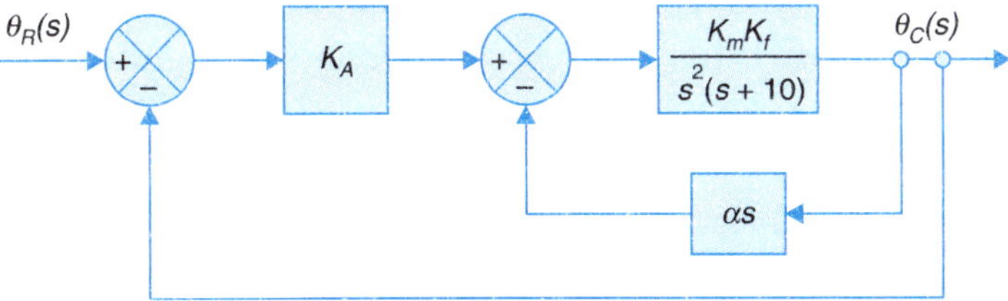

Fig. 10.52. Position servo system with tachometer feedback.

Feedback Compensation in Frequency Domain

Compensation by internal feedback loop can be easily carried out in the frequency domain based on an approximation similar to that made for straight line Bode plots. Consider the basis configuration of Fig. 10.53 with an internal compensating loop. The amplifier K_A is included to provide for the reduction in gain caused by the internal feedback loop. The equivalent transfer function of the inner loop is given by

$$\frac{C(j\omega)}{I(j\omega)} = G(j\omega) = \frac{G_f(j\omega)}{1+G_f(j\omega)H(j\omega)} = \frac{1}{H(j\omega)}\left[\frac{G_f(j\omega)H(j\omega)}{1+G_f(j\omega)H(j\omega)}\right] \quad ...(10.50)$$

Knowing $G_f(j\omega)$ and $H(j\omega)$, we can obtain the bracketed term by plotting $G_f(j\omega)H(j\omega)$ on the Nichols chart. This result must then be added to the Bode plots of $1/H(j\omega)$ to obtain $G(j\omega)$. A trial and error procedure for parameter adjustment then becomes highly tedious and time-consuming. An approximate Bode plot of $G(j\omega)$ can, however, be quickly obtained by the following approximation:

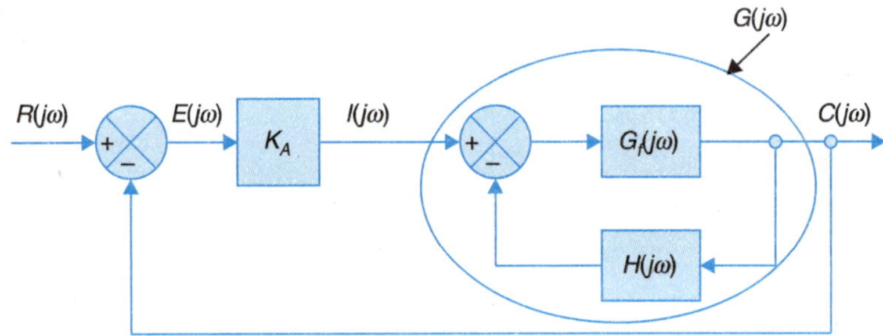

Fig. 10.53. Basic feedback compensation scheme.

$$G(j\omega) \approx G_f(j\omega) \quad \text{for} \quad |G_f(j\omega)\,H(j\omega)| << 1$$

and

$$G(j\omega) \approx \frac{1}{H(j\omega)} \quad \text{for} \quad |G_f(j\omega)\,H(j\omega)| >> 1 \quad ...(10.51)$$

The above approximation is quite valid for the regions of frequencies for which $|G_f(j\omega)\ H(j\omega)| << 1$ (or $>> 1$). As in straight line Bode plots, this approximation can be extended up to frequencies where $|G_f(j\omega)\,H(j\omega)| = 1$. Thus we approximate $G(j\omega)$ as

$$G(j\omega) \approx G_f(j\omega) \quad \text{for} \quad |G_f(j\omega)\,H(j\omega)| \le 1$$

$$\approx \frac{1}{H(j\omega)} \quad \text{for} \quad |G_f(j\omega)\,H(j\omega)| \ge 1 \quad ...(10.52)$$

It must be borne in mind that this approximation is not quite valid in the frequency band around the frequencies at which $|G_f(j\omega)\,H(j\omega)| = 1$. Unlike the straight line Bode plots, there is not direct method of applying corrections at these frequencies.

The condition $|G_f(j\omega)\,H(j\omega)| = 1$ may be interpreted as

$$20 \log |G_f(j\omega)\,H(j\omega)| = 0$$

or

$$20 \log |G_f(j\omega)| = 20 \log \frac{1}{|H(j\omega)|} \quad ...(10.53)$$

This condition will be obtained where the log-magnitude plots of $G_f(j\omega)$ and $1/H(j\omega)$ intersect.

For illustration of the approximation technique, consider the typical Bode plots of $G_f(j\omega)$ and $1/H(j\omega)$ given in Fig. 10.54.

These plots intersect at frequencies ω_1 and ω_2. Thus the approximate log magnitude plot of $G(j\omega)$ coincides with that of $G_f(j\omega)$ for $\omega \le \omega_1$ and $\omega \ge \omega_2$, *i.e.*, in the low and high frequency regions. In the intermediate frequency region, *i.e.*, $\omega_1 \le \omega \le \omega_2$, the approximate plot of $G(j\omega)$ coincides with that of $1/H(j\omega)$. The complete approximate plot of $G(j\omega)$ is shown in thick line in Fig. 10.54. During the design procedure, the approximate plots of $G(j\omega)$ can now be quickly obtained by this technique as the gain and time constants of either $G_f(j\omega)$ or $H(j\omega)$ are adjusted to achieve a desirable performance.

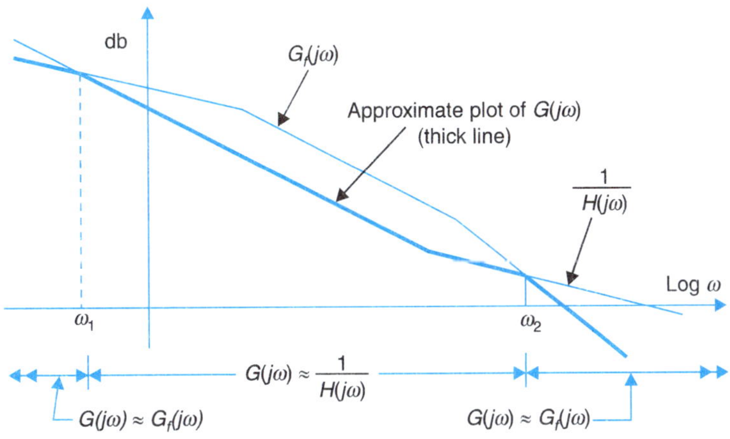

Fig. 10.54. Approximate log-magnitude plot of $G(j\omega)$.

Before we proceed to outline a design procedure, let us dwell upon the choice of a suitable $H(j\omega)$. A direct feedback will reduce the forward gain considerably. A better choice from gain point of view is rate feedback, *i.e.*, $H(s) = \alpha s$. For a typical case let

$$G_f(s) = \frac{K}{s(\tau_1 s + 1)(\tau_2 s + 1)}$$

Then, from Fig. 10.52

$$G(s) = \frac{K}{s(\tau_1 s + 1)(\tau_2 s + 1) + K\alpha s}\ ;\ H(s) = \alpha s$$

$$K_v = \lim_{s \to 0} sK_A G(s) = \frac{KK_A}{1 + K\alpha}$$

It therefore follows that the effect of rate feedback is to reduce the system K_v because of the term $(1 + K\alpha)$ in the denominator. In order to reduce the steady-state errors in the low frequency region, it is desirable to reduce the feedback signal in this range of frequencies. This can be achieved by means of a phase-lead network of the type shown in Fig. 10.55 having a transfer function

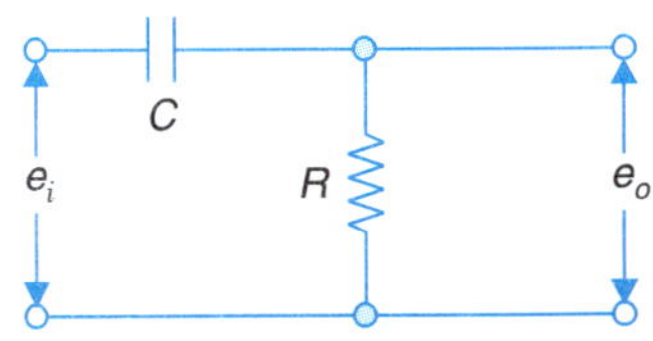

Fig. 10.55. A lead network.

$$\frac{E_o(s)}{E_i(s)} = \frac{Ts}{1+Ts}; T = RC$$

With this network connected in tandem with the rate feedback in the feedback path, we have

$$H(s) = \frac{T\alpha s^2}{1+Ts}$$

The term s^2 corresponds to $(j\omega)^2$ in the numerator and helps to reduce the signal being feedback at low frequencies, thereby improving the system gain and reducing the system errors in this range of frequencies. Now

$$G(s) = \frac{K(1+Ts)}{s[(1+Ts)(1+\tau_1 s)(1+\tau_2 s)+TK\alpha s]}$$

Therefore $K_v = \lim_{s\to 0} sK_A G(s) = KK_A$

We therefore observe that by the introduction of phase-lead network in tandem with rate feedback, the system K_v has now become independent of the feedback.

With the feedback structure so selected we proceed to design the system to meet a set of specifications. This requires the adjustment of two forward path gains K_A and K and two feedback parameters α and T. The design must necessarily be a trial and error procedure, which can be carried out on MATLAB software.

Figure 10.56 shows the Bode plot of

$$\frac{1}{H(j\omega)} = \frac{(1+j\omega T)}{T\alpha(j\omega)^2}$$

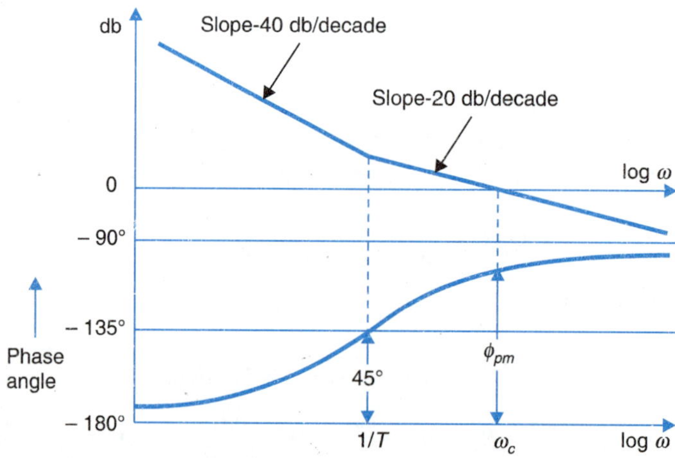

Fig. 10.56. Bode plot of $1/H(j\omega) = (1 + j\omega T)/T\alpha(j\omega)^2$.

It is seen from this figure that if $1/H(j\omega)$ is made effective around the cross-over frequency of $K_A G(j\omega)$, we can obtain any desired phase margin. Depending upon the specified phase margin we can, in fact, decide the system cross-over frequency.

The design procedure is illustrated through a numerical example given below.

Example 10.16: For the system of Fig. 10.51

$$G_f(s) = \frac{K}{s(2s+1)(0.5s+1)}$$

Figure 10.57 shows the Bode plot of $G_f(j\omega)$ for $K = 1$ (it is an initial trial value). If K is increased to 10 to get the desired K_v, it is found that the uncompensated system has a cross-over frequency of 2.2 rad/sec and a phase margin of –36°. This indicates the need for compensation.

Now to provide compensation, consider the proposition of adding an internal feedback loop (output rate element and phase-lead network in tandem) with

$$H(j\omega) = \frac{\alpha T(j\omega)^2}{(1+j\omega T)}$$

Let us design the system such that for the compensated system, $1/H(j\omega)$ is effective over the region of frequencies around the cross-over frequency ω_c. The cross-over frequency of the compensated system can then be approximately determined from the equation

$$-180° + \tan^{-1}\omega_c T + 180° = 35° + \varepsilon$$

where ε allows for the reduction of phase margin caused by the presence of the poles of $G(j\omega)$ other than those contributed by $1/H(j\omega)$.

Choosing $\varepsilon = 25°$, we have $\tan^{-1}\omega_c T = 60°$

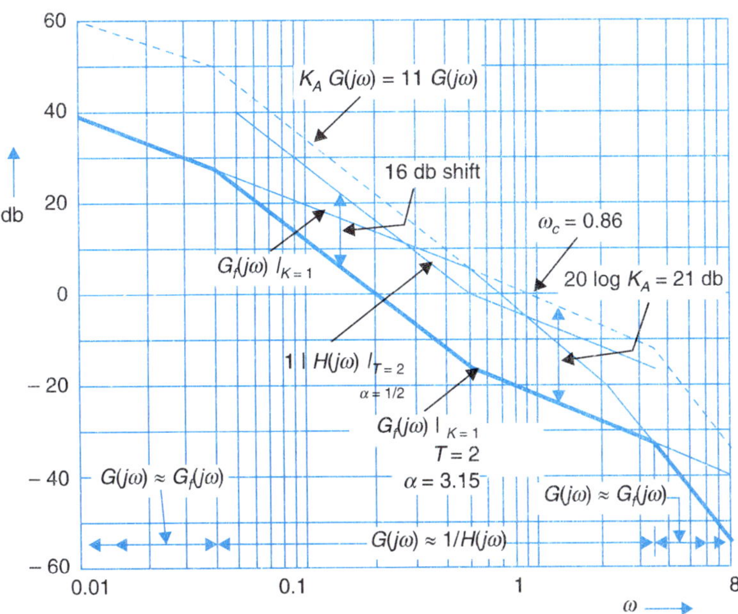

Fig. 10.57. Feedback compensation for Example 10.16.

Why it is necessary to choose such a large value of ε will become clear towards the end of this problem. Now a choice of $T = 2$ sec, gives $\omega_c = 0.86$ rad/sec.

Draw the Bode plot of $1/H(j\omega)$ (for $T = 2$ and $\alpha T = 1$) and shift it relative to the Bode plot of $|G_f(j\omega)|_{K=1}$, such that the frequencies at which the two plots intersect are at least two octaves (*i.e.,* 4 times) away from the expected cross-over frequency $\omega_c = 0.86$. It is found that a downward shift of $1/H(j\omega)$-plot by 16 db satisfies this requirement. Therefore

$$20 \log (1/\alpha T) = -16 \qquad \text{or} \qquad \alpha = 3.15$$

The approximate plot of $G(j\omega)$ with $K = 1$, $T = 2$ and $\alpha = 3.15$ is shown in Fig. 10.57 in thick line. Let the gain K_A be now adjusted so that the cross over frequency of the compensated system lies at $\omega_c = 0.86$. This is achieved by shifting $G(j\omega)$ upwards by 21 db. Therefore

$$20 \log K_A = 21 \qquad \text{or} \qquad K_A = 11$$

The system error constant is then

$$K_v = KK_A = 11$$

The compensated system thus meets the specifications on K_v.

Let us check the specification on phase margin using the approximate Bode plot of $11G(j\omega)$. In factored from

$$11G(j\omega) = \frac{11(1 + j2\omega)}{j\omega(1 + j25\omega)(1 + j0.294\omega)^2}$$

which yields a phase margin of $\phi = 34°$ at the cross-over frequency $\omega_c = 0.86$ rad/sec. The compensation designed therefore meets the specified phase margin as well.

A large choice of ε is a must because of the presence of the factor $(1 + j0.294\omega)^2$ in the denominator. Further because of the strong approximations made, the above design must be finally checked by drawing the exact Bode plots.

10.8 ROBUST CONTROL SYSTEM DESIGN

In the preceding sections of this chapter in designing of a control system we assumed that the model of the plant is completely known and there are no uncertainties or dynamic behaviour affecting the system except for some considerations of inputs like load or random high frequency disturbances. This is usually not the case in most practical system. So we shall now consider the design in the context of uncertainties like

(*i*) Disturbances;

(*ii*) Measurement noise; and,

(*iii*) Unmodelled or imprecisely modelled dynamics

Therefore the design has to seek a control system that functions adequately over a wide range of uncertain parameters. Such a system is said to be Robust *i.e.,* it is durable, hardly and resilient. To be more precise a control system is robust when

(*i*) It has low sensitivities;

(*ii*) It is stable over a wide range of parameter variations; and

(*iii*) The performance stays within prescribed (but practical) limit bounds in presence of changes in the parameters of the controlled system and disturbance inputs.

The two ways of implementing a controller which satisfies the above requirements are outlined below:

(*i*) Adjust the controller parameters by design techniques which take into account the variation in the plant (discussed in the above account) and yet the system (controller plus plant) would give an acceptable performance. Such is Robust Control.

(*ii*) In place of the conventional controller, we design a computer control with on-line measurement of the dynamic response of the system and a suitable measure of its variations from the expected performance. This measure is then used to continuously adjust the parameters of the control algorithm. This is known as Adaptive Control and would be the subject matter of Chapter 15.

Obviously the first choice (Robust Control) is economical, simpler to implement and the controller can be easily returned in the field. So it is still adopted for most plants at present. However, for large complex plants the complexities (and flexibilities) and cost of the second choice (Adaptive Control) can be justified and they have come in the realm of implementation in recent years and are fast becoming a future trend.

As is seen from the earlier discussion robust control design requires the investigation of the control system's sensitivities to plant parameter variations such that the controller settling can be found to meet both the requirements of system performance and low sensitivities.

System Sensitivities to Parameter Perturbations

Differential sensitivities and root sensitivities can be used as a measure of robustness for small perturbations as discussed in Section 3.2 and Section 7.6 respectively. Transfer function sensitivity is defined as

$$S_\alpha^T = \frac{\delta T/T}{\delta\alpha/\alpha} = \frac{\delta T}{\delta\alpha}\left(\frac{\alpha}{T}\right) \qquad ...(10.54)$$

Similarly root sensitivity to a parameter is defined as

$$S_\alpha^{r_i} = \frac{\delta r_i}{\delta\alpha / \alpha} \qquad ...(10.55)$$

where r_i = i^{th} root of the closed-loop system.

When the zeros of $T(s)$ are independent of parameter α, it has been shown in eqn. (7.62) that

$$S_\alpha^T = -\sum_{r=1}^{n} S_\alpha^{r_i}\frac{1}{(s+r_i)} \qquad (10.56)$$

Consider for example a first-order system as shown in Fig. 10.58 where open-loop pole ($-\alpha$) is subject to small perturbations.

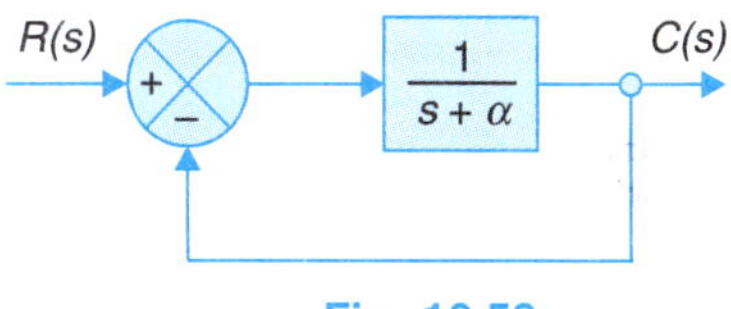

Fig. 10.58

Then $\qquad T(s) = \dfrac{1}{s+\alpha+1}$

Root, $\qquad r_1 = -(\alpha+1)$ and $S_\alpha^{r_1} = -1$

Substituting in eqn. (10.56), we get

$$S_\alpha^T = -S_\alpha^{T_1}\frac{1}{(s+\alpha+1)} = \frac{1}{(s+\alpha+1)} \qquad ...(10.57)$$

Sensitivity in frequency domain can be obtained by substituting $s = j\omega$ and evaluating by its Bode plot for a range of frequencies.

Consider now a type-1 second-order system as depicted in Fig. 10.59. For this system

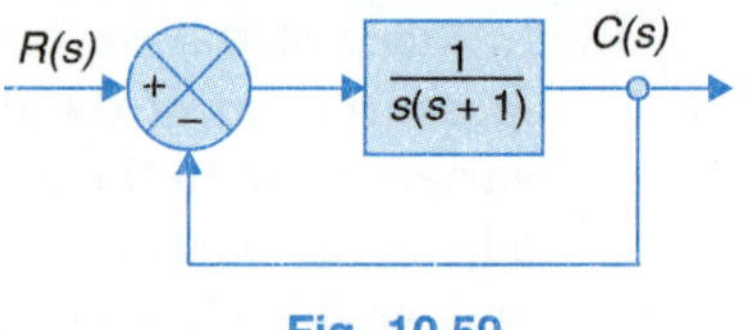

Fig. 10.59

$$T(s) = \frac{K}{s^2 + s + K} \; ; \alpha = 1 \qquad ...(10.58)$$

As in eqn. (3.7*a*)

$$S_K^T = \frac{1}{1 + GH(s)} = \frac{s(s+1)}{s^2 + s + K} \qquad ...(10.59)$$

Comparing this with $T(s)$ eqn. (10.58), we can write

$$T(s) = 1 - S_K^T(s)$$

We now draw the Bode plots of $T(s)$ and $S_K^T(s)$. These are shown in Fig. 10.58 with $K = 1/4$ which corresponds to critical frequency of $\omega_c = 0.5$ rad/s.

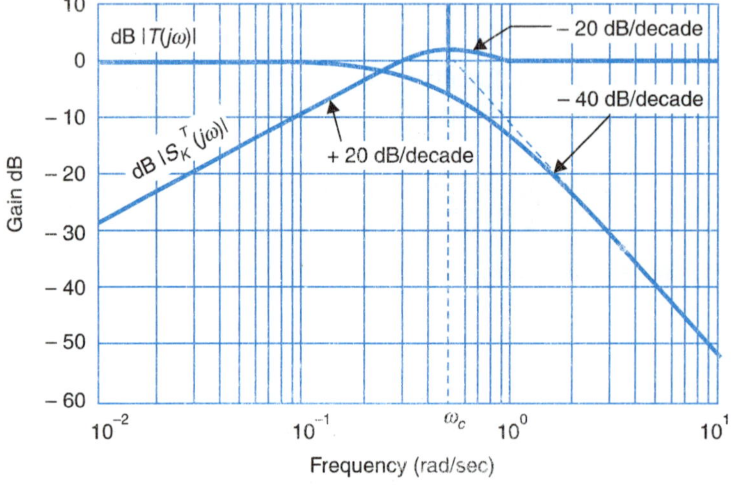

Fig. 10.60

For the closed-loop system

$$\omega_n = \omega_c = \sqrt{K} \; ; K = 1/4$$
$$= 0.5 \text{ rad/s}$$
$$\zeta = 1$$

In the low frequency region, S_K^T is quite low, *e.g.*, at $\omega = 0.1$ db $(|S_K^T|) = -12$. This is quite acceptable, but at this value of K, the system steady-state error to unit velocity input is

$$\theta_{ss} = \frac{1}{(1/4)} = 4$$

This would not be acceptable.

Consider now the option of raising K. This would raise ω_n (increases system bandwidth), but reduces sensitivities at low frequencies. This would be beneficial. Examine now the effect of raising K on system's % overshoot.

From eqn. (10.58),

$$2\zeta\omega_n = 1 \; ; \; \omega_n = \sqrt{K} \qquad \text{or} \qquad \zeta = \frac{0.5}{\sqrt{K}}$$

So the damping decreases with increase in K and % overshoot will increase *e.g.*, at

$$K = 1, \; \zeta = 0.5, \; \% \text{ overshoot} = 16.4\%$$

As the % overshoot increases sharply with K, the system is not robust and hence is unacceptable. A robust system's % overshoot (as well as sensitivity) with parameter variation (gain in this case) should be quite small.

Sensitivity of a Control System

Consider the system shown in Fig. 10.61, where

$$G(s) = \frac{1}{s^2} \; ; \text{ plant} \qquad \ldots(i)$$

$$G_c(s) = b_2 s + b_1 \; ; \text{ PD controller} \qquad \ldots(ii)$$

For this system

$$S_G^T = \frac{1}{1 + G_c G} = \frac{s^2}{s^2 + b_2 s + b_1} \qquad \ldots(iii)$$

$$T(s) = \frac{b_2 s + b_1}{s^2 + b_2 s + b_1} \qquad \ldots(iv)$$

We choose $\zeta = 1$, then $b_2 = 2\omega_n$ where $\omega_n = \sqrt{b_1}$

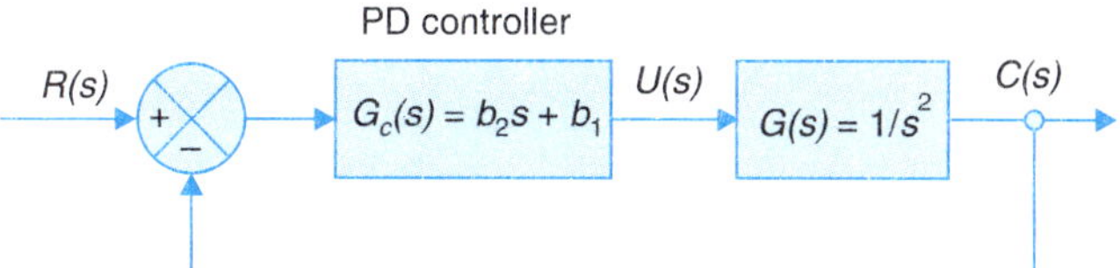

Fig. 10.60

Equations (*iii*) and (*iv*) are then written as

$$S_G^T = \frac{s^2}{(s + \omega_n)^2} \qquad \ldots(v)$$

$$T(s) = \frac{2\omega_n\left(s + \dfrac{\omega_n}{2}\right)}{(s + \omega_n)^2}$$

The Bode plots of $|S_G^T|$ and $|T|$ for $\omega_n = 1$ drawn in Fig. 10.62.

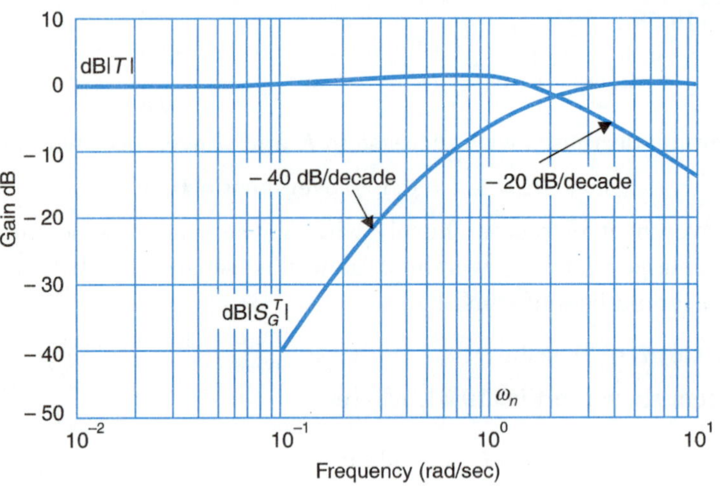

Fig. 10.62

It is seen from these plots that db | T | is flat for $\omega < 0.5\ \omega_n$, rises at 20 db/decade from $\omega = 0.5\omega_n$ to $\omega = \omega_n$ and then falls at 20 db/decade. On the other hand the plot of db $\left|S_G^T\right|$ rises at 40 db/decade for $\omega < \omega_n$ and then remains flat at 0 db. Examination of these plots indicates that sensitivities are important for $\omega < \omega_n$, while the stability margin needs investigation for $\omega > \omega_n$.

Considerations in Design of Robust Control System

For this purpose consider the system of Fig. 10.63 which also shows disturbance input $D(s)$ and the controller $G_c(s)$.

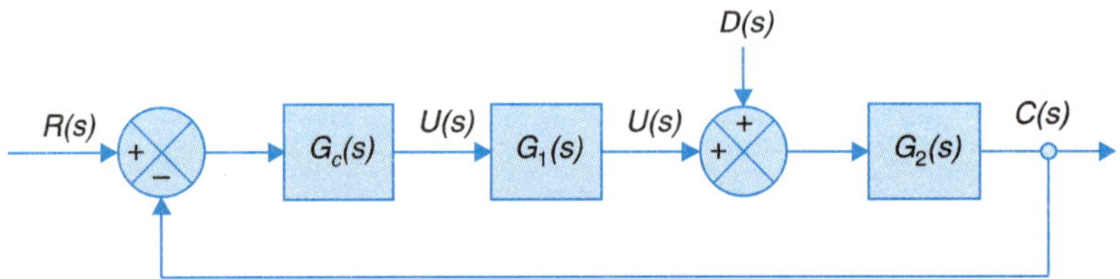

Fig. 10.63

For this system

$$T(s) = \frac{C(s)}{R(s)} = \frac{G_c G_1 G_2}{1 + G_c G_1 G_2} \qquad ...(10.60)$$

and

$$\frac{C(s)}{D(s)} = \frac{G_2}{1 + G_c G_1 G_2} \qquad ...(10.61)$$

These two transfer functions have the same characteristic equation because it is determined by the loop gain and the feedback structure. This is

$$1 + G_c(s)G_1(s)G_2(s) = 1 + L(s) = 0 \qquad ...(10.62)$$

Further it easily follows that

$$S_G^T = \frac{1}{1 + G_c G_1 G_2(s)} \qquad ...(10.63)$$

It can now be visualized from these four expressions that :

(*i*) $T(s)$ should have wide bandwidth for faithful reproduction of expected inputs.

(*ii*) Large loop gain $L(s)$ to minimize sensitivity S.

(*iii*) Large loop gain $L(s)$ should be achieved primarily by $G_c(s)G_1(s)$ (see eqn. (10.63)). This is normally the case as $G_2(s)$ is the plant transfer function which is more or less fixed except for variations of certain parameters.

Large loop gain can mean system instability which should be guarded against by providing suitable stability margin. This consideration is in fact covered by (*i*) above.

The requirements of bandwidth, sensitivity and stability margin may be directly known but may be estimated with a range of variation by the designer from the plant knowledge and its specified performance. No rigid values can be laid down for these requirements as this would lead to internal inconsistency. All can be met only within a range.

In the frequency domain robust design we determine suitable $G_c(j\omega)$ and adjust its parameters on the Bode plot such that:

1. $G_cG(j\omega)$ plot must have adequate phase and gain margin and the cross-over frequency should be high for meeting bandwidth need. A simple rule is that the db-magnitude plot at cross-over should have a slope not more than –20 db/decade.
2. Low frequency gain must be high enough to achieve the desired steady-state accuracy.
3. Maintain accuracy over a bandwidth ω_b by not allowing $|\ G_cG(j\omega)\ |$ to fall below a prescribed level.
4. Achieve disturbance rejection by high gain of $G_c(j\omega)$ over the system bandwidth.

In the root-locus approach the designer has to devise a compensator so as to (*i*) minimize $S_\alpha^{\,r}$, (*ii*) minimize effect of $D(s)$ and simultaneously (*iii*) to maintain the dominant roots within a region so as to give appropriate dynamic response. Let us consider the system shown in Fig. 10.61 with

$$G_2(s) = \frac{1}{s(s+2)},\ G_1(s) = 1,\ G_c(s) = K$$

For this system

$$S_K^r = \frac{dr}{dK}\frac{K}{r} = \frac{ds}{dK}\frac{K(s)}{s}\bigg|_{s=r} \qquad ...(i)$$

The characteristic equation of this system is

$$s(s+2) + K = 0 \qquad ...(ii)$$

or

$$K = -\,s(s+2)$$

$$\frac{dK}{ds} = -\,2(s+1)\ ;\ \frac{ds}{dK} = -\frac{1}{2(s+1)} \qquad ...(iii)$$

Substituting in Eqn. (*i*)

$$S_K^r = -\frac{1}{2(s+1)}\frac{-s(s+2)}{s}\bigg|_{s=r} = \frac{(s+2)}{2(s+1)}\bigg|_{s=r} \qquad ...(iv)$$

For $\zeta < 1$ $$s = -1 \pm j\omega = -1 \pm j\sqrt{(K-1)}\ ; K > 1 \quad ...(v)$$

Then $$S_K^r = \frac{1}{2}\left(1 - \frac{j}{\omega}\right) \Rightarrow \left|S_K^r\right| = \frac{\sqrt{(1+\omega^2)}}{2\omega} \quad ...(iv)$$

From the characteristic equation Eqn. (*ii*)

$$\zeta = \frac{1}{\sqrt{K}} \quad ...(vii)$$

The % overshoot is $$M_p = e^{-\pi\zeta/\sqrt{(1-\zeta^2)}} = e^{-\pi/\sqrt{(K-1)}} \quad ...(viii)$$

With the expression for $\left|S_K^r\right|$ and M_p as function of K, we can find (graphically/numerically) the value of K for which the sensitivity and % overshoot are best possible. The plots of $\left|S_K^r\right|$ and M_p vs K are drawn in Fig. 10.64 from which we find

$$K = 3, \left|S_K^r\right| = 0.62, M_p = 11\% \quad ...(ix)$$

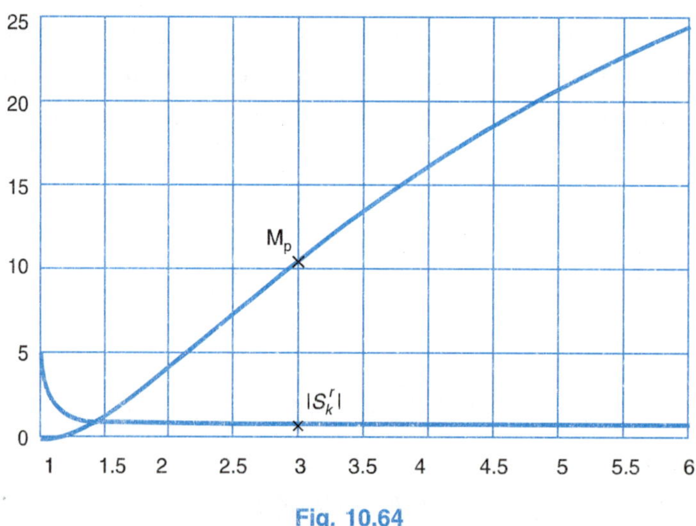

Fig. 10.64

In general, we can use the following guidelines for robust control design with root-locus technique:

1. Draw the root-locus of the system plus compensator (chosen) and adjust the compensator parameter to achieve the desired location for the dominant roots.
2. Maximize the compensator gain to reduce the effect of disturbance inputs.
3. Minimize $\left|S_K^r\right|$, while maintaining the transient response within permitted range.

In the results for $\left|S_K^r\right|$ or M_p as obtained in the above method for finding the gain K of the compensator, the designer may have to allow a smaller value of K for finding an acceptable value of M_P. This is usually the case as the specification on M_P is more stringent than on $|S_K^r|$. The gain reduction has to examined in the context of steady-state errors (and static errors in the low frequency region). If these are violated the designer would then search for suitable $G_c(j\omega)$, which would modify the Bode plots.

Sensitivity and Compensation

Reconsider the design Example 10.6 where a lead compensator was selected with $G_c(s) = \dfrac{(s+z)}{(s+p)}$. The design procedure leads to fixing the values z and p as

$$z = 1/0.377 = 2.65 \text{ rad/s} \; ; \; p = 1/0.128 = 7.8 \text{ rad/s}$$

The compensator yields the desired phase margin of 40°, while preserving the K_v (velocity error constant) of the system at its specified value. This would mean a satisfactory M_P. But one must still check the value of M_P from the unit step response of the system because of the presence of closed-loop zero. This is shown in Fig. 10.65 from which the value of $M_P = 32.01\%$.

We find from this example

$$G(s) = G_f(s)G_c(s) = \frac{12(0.377s+1)}{s(s+1)(0.128s+1)}$$

The closed-loop sensitivity is given as

$$|S_{G_f}^T| = \left|\frac{1}{1+G_cG_f(j\omega)}\right| = \left|\frac{1}{1+G(j\omega)}\right|$$

The value of $|S_{G_f}^T|$ at values of frequency of interest can now be computed directly (or by use of Nichols charts). For the specific case

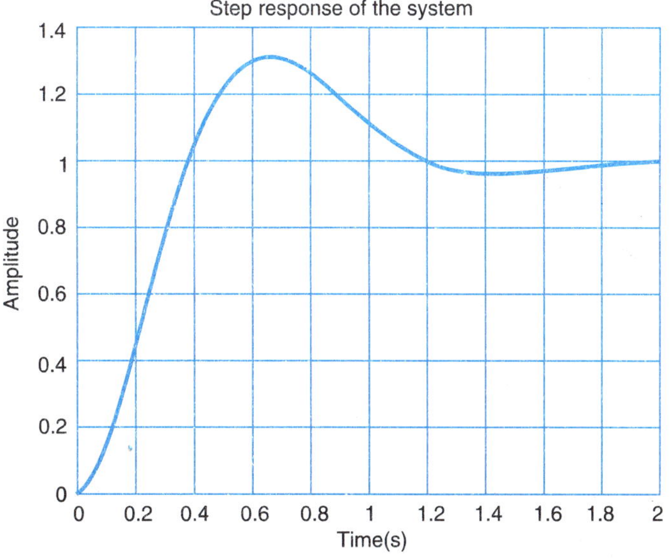

Fig. 10.65. Step response of the system.

$$G(j\omega) = \frac{12(1+j0.377\omega)}{j\omega(1+j\omega)(1+j0.128\omega)}$$

Let us say our frequency of interest is $\omega_1 = \omega_c$ (compensated)/2.5

$= 4.6/2.5 = 1.84$

At this frequency $\left|S_{G_f}^{T}\right| = 0.32$

The effect of the disturbance input can be investigated from

$$\frac{C(s)}{D(s)} = \frac{G_f(s)}{1+G_cG_f(s)} = \frac{G_f(s)}{1+G(s)}$$

or

$$\left|\frac{C(j\omega)}{D(j\omega)}\right| = \left|\frac{G_f(j\omega)}{1+G(j\omega)}\right|$$

With $\left|\frac{1}{1+G(j\omega)}\right| = \left|S_{G_f}^{T}(j\omega)\right|$ already computed we can find

$$\left|\frac{C(j\omega_1)}{D(j\omega_1)}\right| = \left|S_{G_f}^{T}(j\omega_1)\right|\left|G_f(j\omega_1)\right| = 0.9$$

Remark $\left|\frac{C(j\omega)}{D(j\omega)}\right| \approx 1$ is too high. This is not surprising as a lead compensator is prone to respond to noise input. Where a lower value of noise sensitivity is desired, it would be advisable to use lag-lead compensator or a PID controller.

Robust PID Controllers

PID Controllers have already been discussed in Section 10.6 where Ziegler Nichols method for the tuning of the three controller constants were presented and supported by examples. These rules are particularly popular for chemical systems. The controllers set by these rules are found to give robust performance. These rules aim for a percentage overshoot of 25%. These rules may not yield acceptable performance for other systems and particularily some times M_p specification is much more stringent than 25%. For example, for a robotic system the dynamic response is desired to be dead-beat ($\zeta = 1$) or very close to it. It is, therefore necessary to design a robust controller for some systems directly by frequency response or root-locus methods. This is the subject matter for discussion here. Consider a PID Controller

$$G_c(s) = K_p + \frac{K_i}{s} + K_d s = \frac{K_d s^2 + K_p s + K_i}{s} \quad ...(10.64)$$

or

$$G_c(s) = \frac{K_d(s^2 + as + b)}{s} \quad ...(10.65)$$

where $a = K_p/K_d$ and $b = K_i/K_d$. Thus the PID controller has a pole at origin and two zeros which can be placed anywhere in the left-hand half plane.

Consider a controlled system with

$$G_f(s) = \frac{1}{(s+2)(s+3)} \quad ...(i)$$

We will design a PID controller with complex conjugate zeros z, z^*. The root-locus plot for the open-loop system

$$G(s) = G_c(s)G_f(s) = \frac{K_d(s+z)(z+z^*)}{s(s+2)(s+3)} \quad ...(ii)$$

is sketched in Fig. 10.66. Further as the controller introduces a complex conjugate pair of poles at K_d above a certain value, we can write

$$T(s) = \frac{G_c(s)G_f(s)}{1+G_c(s)G_f(s)} = \frac{K_d(s+z)(s+z^*)}{(s+r_1)(s+r_1^*)(s+r_2)} \quad ...(iii)$$

It is seen from the root-locus plot that if K_d is made high enough closed-loop poles r_1 and r_1^* move quite close to open-loop zeros z_1 and z_1^*, thereby cancelling the controller zeros appearing in the closed-loop transfer function which is then reduced to

$$T(s) = \frac{K_d}{s+r_2} \quad ...(iv)$$

Also with large K_d, $r_2 \approx K_d$ and so

$$T(s) = \frac{K_d}{s+K_d} \quad ...(v)$$

It is seen from the approximate closed-loop transfer function of Eqn. (v) that is has a single pole with quite large negative value *i.e.*, a very low time constant $1/K_d$. The step response of the system is therefore dead-beat, fast and also the steady-state error is zero. The value of K_d is however, limited by the allowable magnitude of $U(s)$ (see Fig. 10.63), which can be applied to the fixed part of the controlled system, *i.e.*, $G_f(s)$.

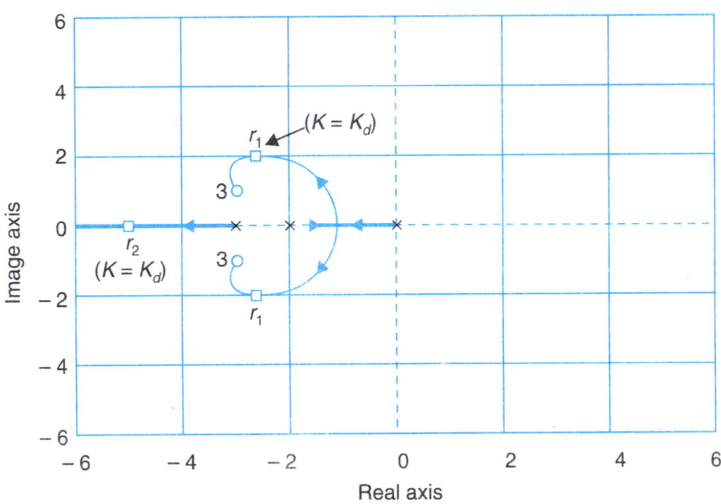

Fig. 10.66

While using the frequency response for PID controller design, we write its transfer function in the form

$$G_c(s) = \frac{K_i\left(\frac{K_d}{K_i}s^2 + \frac{K_p}{K_i}s + 1\right)}{s} = \frac{K_i(\tau s+1)\left(\frac{\tau}{\alpha}s+1\right)}{s} \quad ...(10.66)$$

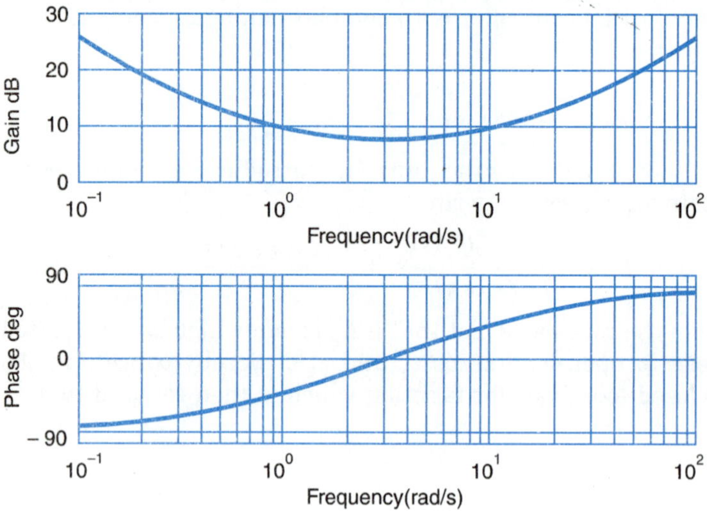

Fig. 10.67. PID controller Bode plots.

wherein the controller constants are adjusted to yield two real zeros. The Bode plot of this $G_c(s)$ is drawn against $\omega\tau$ in Fig. 10.67. It is seen from this figure that Bode plot of PID controller is very similar to that of a lag-lead network as given in Fig. 10.68. Compared to lag-lead network, PID controller has the superiority of having lag behaviour throughout the low frequency end and a lead behaviour throughout the high frequency end (without the attenuation problem of the lead network, high gain in high frequencies which is not a desirable feature when high frequency noise inputs are present).

In case of complex conjugate zeros, the transfer function of the controller has the form (substitute $s = j\omega$ in eqn. (10.65))

$$\frac{K_d\omega_n\left[(1 + j2\zeta(\omega/\omega_n) + (\omega/\omega_n)^2\right]}{j(\omega/\omega_n)} \qquad ...(10.67)$$

Its Bode plot now depends on ζ as well (see Figs. 8.14 and 8.15 for the Bode plots of the numerator). Normally, we require $0.9 < \zeta < 0.7$.

Design Procedure

The design procedure consists of the following steps:

1. Find the closed-loop transfer function $T(s)$ of the controlled system ($G_c(s)$) and the PID controller ($G_c(s)$) for which the constants are to be determined for a robust system.
2. Choose the constants of the characteristic equation of the closed-loop as per the ITAE performance index. Optimum coefficient for minimum ITAE for step input is given in Table 5.2.
3. From step 2 write $T(s)$ in the form of known coefficients except for ω_n. From these get the constants of $G_c(s)$ in terms of ω_n.
4. For specified settling time choose a suitable value of ζ and then find ω_n.
5. $G_c(s)$ is now fully known and so is $T(s)$.

6. Presence of zeros of $G_c(s)$ in $T(s)$ normally does not allow the % overshoot requirements to be met.

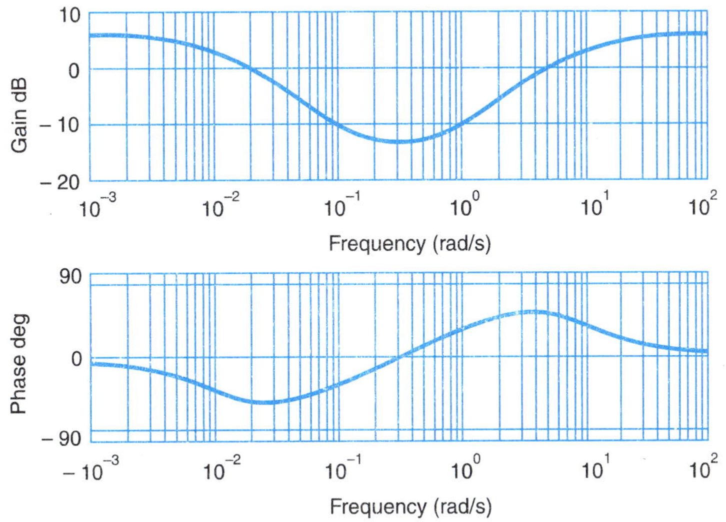

Fig. 16.68. Lag-lead network Bode plots.

7. Determine a prefilter $G_p(s)$ to eliminate the zeros of $T(s)$ as shown in Fig. 10.69.

The above procedure will now be illustrated with an example.

Example 10.17: For the system of Fig. 10.69, select a PID controller to achieve a settling time of 0.8*s* for an ITAE step response. Plot $c(t)$ with $r(t) = u(t)$ with and without a prefilter. Also plot $c(t)$ for unit step disturbance. Discuss the effectiveness of the compensation.

Given $$G_f(s) = \frac{1}{(s+\alpha)^2} = \frac{1}{(s+2)^2}; \ \alpha = 2 \quad \text{or} \quad \tau = 0.5$$

Solution. $$G_c(s) = \frac{K_d s^2 + K_p s + K_i}{s} \qquad ...(i)$$

$$G_f(s) = \frac{1}{(s+2)^2} \qquad ...(ii)$$

With reference to Fig. 10.69,

$$T(s) = \frac{G_c G_f}{1 + G_c G_f} = \frac{K_d s^2 + K_p s + K_i}{s^3 + (K_d + 4)s^2 + (K_p + 4)s + K_i} \qquad ...(iii)$$

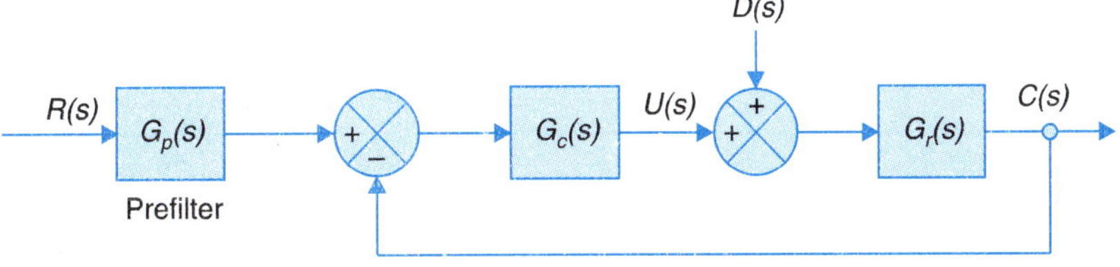

Fig. 10.69

For optimum, ITAE performance, we have from Table 5.2 the characteristic equation as

$$s^3 + 1.75\omega_n s^2 + 2.15\omega_n^2 s + \omega_n^2 \quad ...(iv)$$

We now have to choose ω_n with specified settling time. However, we do not have the value of ζ. We choose $\zeta = 0.8$ for low overshoot based on dominant poles as the zeros would tend to increase it. Therefore $4/\zeta\omega_n = 0.8\ (t_s)$, which yields

$$\omega_n = 4/(0.8 \times 0.8) = 6.25 \text{ rad/s}$$

With this value of ω_n the characteristic equation is given as

$$s^3 + 10.94s^2 + 84.0s + 244 \quad ...(v)$$

Comparing the coefficients of (*iii*) and (*v*), we get

$$K_d = 6.94,\ K_p = 80 \text{ and } K_i = 244$$

Therefore $$G_c(s) = \frac{6.94s^2 + 80s + 244}{s} = \frac{6.94(s^2 + 11.53s + 35.16)}{s} \quad ...(vi)$$

$$T(s) = \frac{6.94(s^2 + 11.53s + 35.16)}{s^3 + 10.94s^2 + 84s + 244} \quad ...(vii)$$

With prefilter $$T_1(s) = T(s)G_p(s) = \frac{244}{s^3 + 10.94s^2 + 84s + 244} \quad ...(viii)$$

Therefore $$G_p(s) = \frac{35.16}{s^2 + 11.53s + 35.16} \quad ...(ix)$$

Let us examine the issues involved. The compensated system is type-1, while the uncompensated system is the type-0. So the compensator by introducing the integral term reduces the steady-state error to step intput to zero and the steady-state error to velocity input to a finite value.

In the compensated system of this example, $K_v = \dfrac{244}{4} = 61$

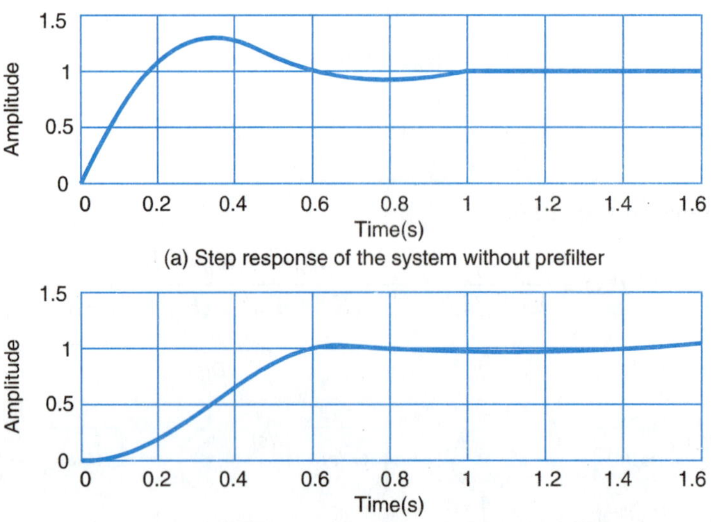

Fig. 10.70

This gain has been built into the system, while maintaining its dynamic response as desired. This would not be possible by merely increasing the gain of the uncompensated systems as this would lead to a very low damping and unacceptable dynamic response. For illustration, if we use $G_c = K_p = 244$, the system (closed-loop) would have $\zeta = 0.25$ (the reader should verify).

The actual dynamic response of the compensated system without and with prefilter to unit step input are plotted in the Figs. 10.70 (*a*) and (*b*) respectively. The results for the same are tabulated in Table 10.1 which bring out the effectiveness of the prefilter in quality of step response in terms of % overshoot and settling time. It is because the prefilter cancels out the troublesome zeros in $T(s)$, which are contributed by the compensator.

Consider now the response of the compensator system to disturbance. From Fig. 10.69 and eqns. (*iii*) and (*v*), removing G_c in numerator,

$$\frac{C(s)}{D(s)} = \frac{G_f}{1+G_cG_f} = \frac{1}{s^3 + 10.94s^2 + 84s + 244} \qquad ...(x)$$

The corresponding unit step disturbance response is plotted in Fig. 10.71 and the results are tabulated in Table 10.1. This value is unaffected by G_p and the zeros of G_c do not appear in $C(s)/D(s)$ as G_c is now in the feedback loop.

Let us now consider the variation of $\alpha = 2$ to 1.5 and see the effect on the compensated system's performance. For eqn. (*iii*)

$$T(s) = \frac{6.94s^2 + 80s + 244}{s^3 + 9.94s^2 + 82.25s + 244} \qquad ...(xi)$$

With prefilter

$$T_1(s) = \frac{244}{s^3 + 9.94s^2 + 82.25s + 244} \qquad ...(xii)$$

Also for disturbance input

$$\frac{C(s)}{D(s)} = \frac{1}{s^3 + 9.94s^2 + 82.25s + 244} \qquad ...(xiii)$$

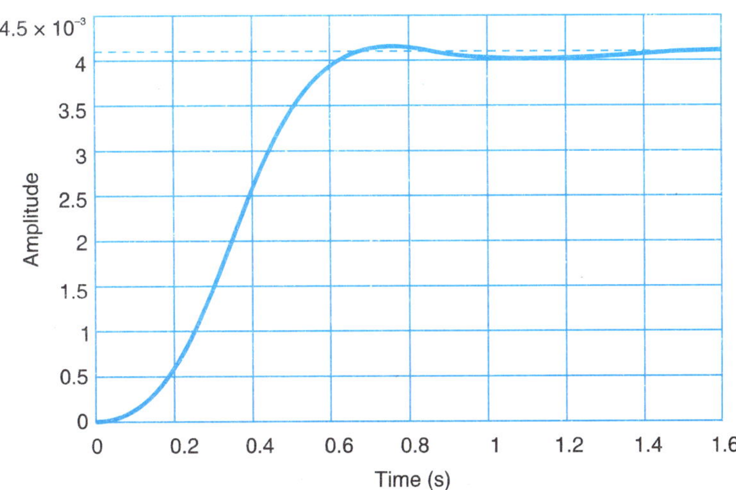

Fig. 10.71. Step response of the system disturbance.

The unit step response for input and disturbances are plotted in Figs. 10.72 (*a*) and (*b*). The results are given in Table 10.1. The system's response satisfies the specified requirements and for a disturbance of 1/*s*, the maximum value of $c(t)$ due to disturbance is 0.4%. This is very favourable design.

Reader should now change α to 2.5 or $\tau = 0.4s$ and obtain step input results. The results would show that this is a very robust system.

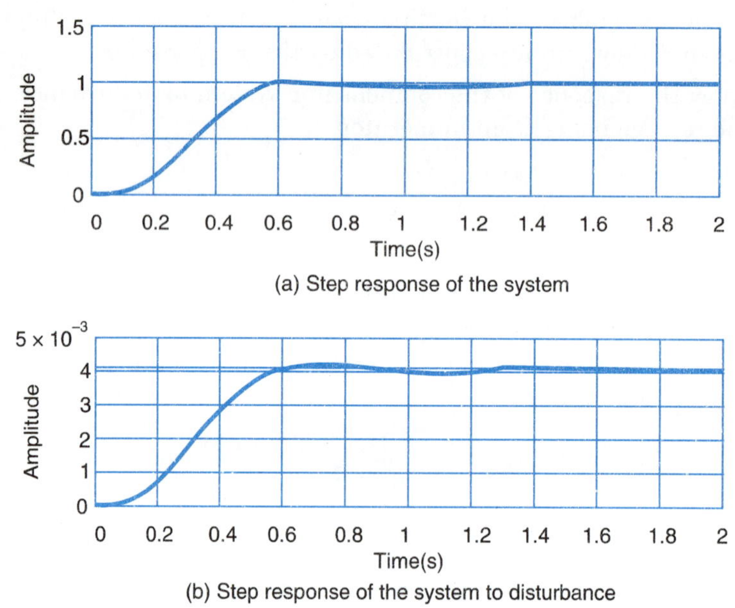

(a) Step response of the system

(b) Step response of the system to disturbance

Fig. 10.72

Table 10.1. Results of the compensated system's performance

Unit step input/ disturbance	$\alpha = 2$ or $\tau = 0.5$		$\alpha = 3$ or $\tau = 0.67$
	$G_p = 1$	$G_p(s)$	$G_p(s)$
% overshoot	29.2%	1.9%	2.4%
Settling time	0.9 *s*	0.57 *s*	0.54 *s*
$\lvert c(t)/d(t) \rvert$ max	0.42%	0.42%	0.42%

To the design procedure given at the start of this subsection we now add a first step

1. Whenever a range of values of plant gain and time constant are given, make a suitable choice of these values and then proceed with the design procedure as given earlier.

In the end the low sensitivity of performance indices to variation in gain and time constant of the plant may then be verified by obtaining requisite step response as illustrated in Example 10.14

Remark: The above presented procedure is an aid, but the robust controller design is basically a trial and error process, wherein the knowledge of the designer about the plant and the techniques of compensation play important role. So we will present another design example.

Example 10.18: Consider the system of Fig. 10.67 with

$$G_f(s) = \frac{1}{(s+2)(s+6)}$$

Design a PID controller ($G_c(s)$) by suitably locating its zeros. Adjust the gain K_d so that the zeros nearly cancel out two of the closed-loop poles. The gain K_d should not exceed the limiting value of 100. Such a controller would not require a prefilter. State why ?

What is the effective time constant of the system and what would be its steady-state error to unit step input ? At what time does the output reach a value of 90% of the input ?

If the smaller pole of $G_f(s)$ changes from –2 to –1.5, find the unit step response of the system.

Solution. For the zero-pole pair cancellation, it would be desirable to choose complex-conjugate zeros for the controller. This would eliminate the smaller pole (at –2) of G_f, thereby speeding up the response.

We choose controller zeros as $z, z^* = -2.5 \pm j3$. The root-locus of the compensated system is drawn in Fig. 10.73. By adjusting the gain K_d such that the zeros nearly cancel out two closed-loop poles, we get $K_d = 87$.

Since the two zeros of the compensated system are cancelled by two of the closed-loop poles, the system thus has only a single pole with quite large negative value, *i.e.*, $(-K_d)$. This means that the system has a very low time constant approximately

$$\frac{1}{K_d} = \frac{1}{87} = 0.01s$$

Hence there is no need for a prefilter for such a controller.

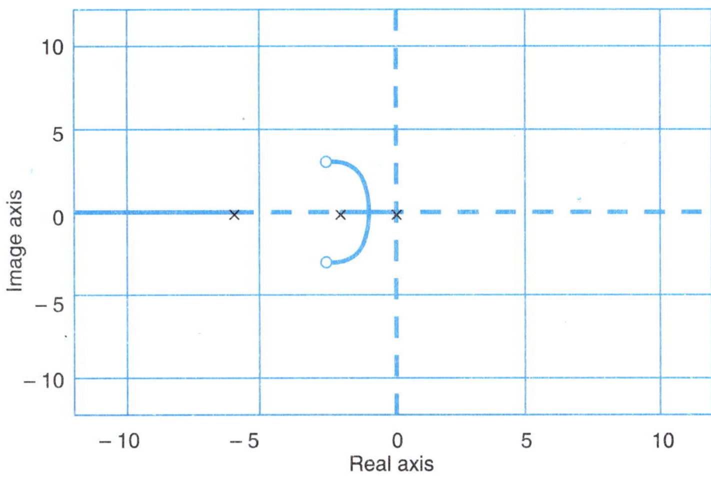

Fig. 10.73

The dynamic response of the system to step input is plotted in Fig. 10.74 (*a*). The results are tabulated in Table 10.2. Let us consider the case where the smaller pole of $G_f(s)$ change from –2 to –1.5. The step response of the system is given in Fig. 10.74 (*b*) and the results are tabulated in Table 10.2.

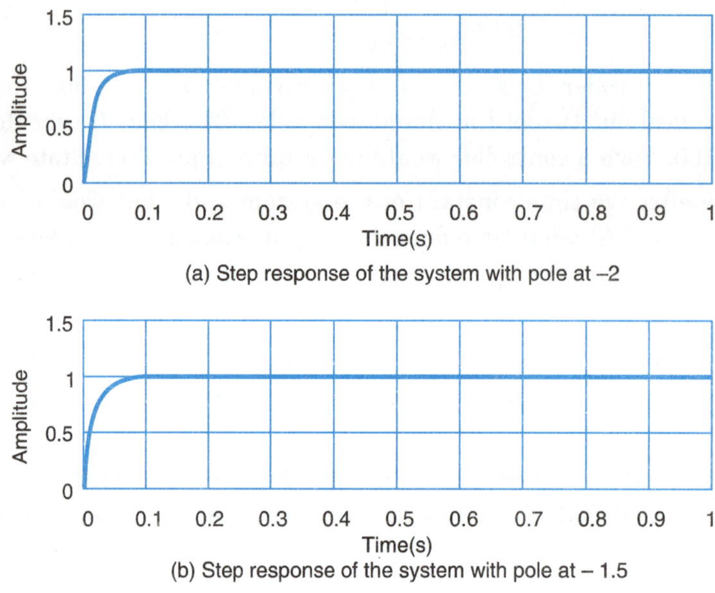

(a) Step response of the system with pole at –2

(b) Step response of the system with pole at – 1.5

Fig. 10.74

Table 10.2. Results of System's Performance

Unit step input	*% overshoot*	*Settling time*	*Steady state error*	*Time for* $\left\|\frac{c(t)}{r(t)}\right\|_{max}$
Pole at –2	1.1%	0.045 *s*	0.0	0.03 *s*
Pole at –1.5	1.3%	0.04 *s*	0.0	0.029 *s*

The results show that the system is robust.

PROBLEMS

10.1. A unity feedback system has an open-loop transfer function of

$$G(s) = \frac{K}{s(s+1)(s+5)}$$

Draw the root locus plot and determine the value of *K* to give a damping ratio of 0.3. A network having a transfer function of 10.(1 + 10*s*)/(1 + 100*s*) is now introduced in tandem. Find the new value of *K* which gives the same damping ratio for the closed-loop response. Compare the velocity error constant and settling time of the original and the compensated system.

10.2. A servomechanism has an open-loop transfer function of

$$G(s) = \frac{10}{s(1+0.5s)(1+0.1s)}$$

Draw the Bode plot and determine the phase and gain margins. A network having the transfer function $(1 + 0.23s)/(1 + 0.23s)$ is now introduced in tandem. Determine the new gain and phase margins. Comment upon the improvement in system response caused by the network.

10.3. A unity feedback control system has an open-loop transfer function of

$$G(s) = 1/s^2$$

Design a suitable compensating network such that a phase margin of 45° is achieved without sacrificing system velocity error constant. Sketch the Bode plot of the uncompensated and compensated systems.

10.4. A unity feedback system has an open-loop transfer function of

$$G(s) = \frac{4}{s(2s+1)}$$

It is desired to obtain a phase margin of 40° without sacrificing the K_v of the system. Design a suitable lag-network and compute the value of network components assuming any suitable impedance level.

10.5. A unity feedback system is characterized by the open-loop transfer function

$$G(s) = \frac{K}{s(s+3)(s+9)}$$

(*a*) Determine the value of K if 20% overshoot to a step input is desired.

(*b*) For the above value of K determine the settling time and K_v (velocity error constant).

(*c*) Design a cascade compensator that will give approximately 15% overshoot to a unit step input, while the settling time is decreased by a factor of 2.5 and $K_v \geq 20$.

10.6. Consider the system shown in Fig. P-10.6. Design a lead compensator for this system to meet the following specifications:

Damping ratio $\zeta = 0.7$

Settling time $t_s = 1.4$ sec

Velocity error constant $K_v = 2$ sec^{-1}.

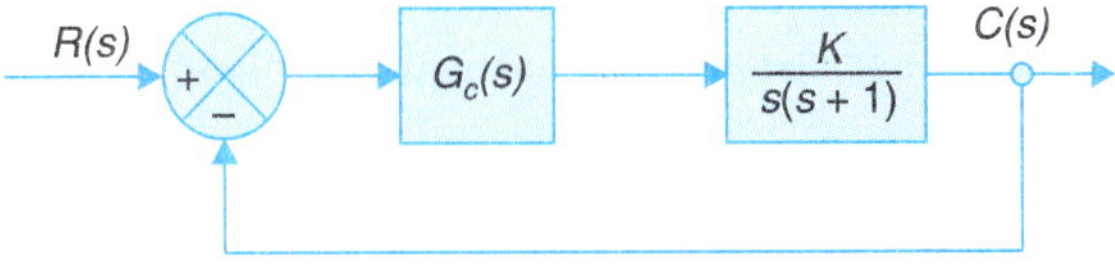

Fig. P-10.6

10.7. Design a phase-lead compensator for the system shown in Fig. P-10.6, to satisfy the following specifications :

(*i*) The phase margin of the system must be greater than 45°.

(*ii*) Steady-state error for a unit step input should be less than 1/15 deg per deg/sec of the final output velocity.

(*iii*) The gain cross-over frequency of the system must be less than 7.5 rad/sec.

10.8. A unity feedback system has an open-loop transfer function

$$G(s) = \frac{K}{s(s+1)(0.2s+1)}$$

Design phase-lag compensation for the system to achieve the following specifications:

Velocity error constant $K_v = 8$

Phase margin $\approx 40°$.

Also compare the cross-over frequency of the uncompensated and compensated systems.

10.9. Figure P-10.9 represents the block diagram of the control system for the pitch rate ($\omega = \dot{\theta}$) control of an aircraft. For improved accuracy, the system is made type-1 by employing an integration gyro which produces an output signal θ_E proportional to the integral of the difference between the command signal $\omega_c = \dot{\theta}_c$ and the pitch rate $\omega = \dot{\theta}$. The command signal is produced by the pilot with the help of a 'control stick'. The error signal adjusts the angle δ_c of the control surfaces of the aircraft by means of a positioning servo. The pitch rate is determined by this angle δ_c and aircraft dynamics.

(*i*) For the transfer functions represented in the block diagram, design a compensator so that the system has a damping factor of 0.6 and a damped natural frequency of 2.4 rad/sec. What is the system error constant for this settling of the controller ?

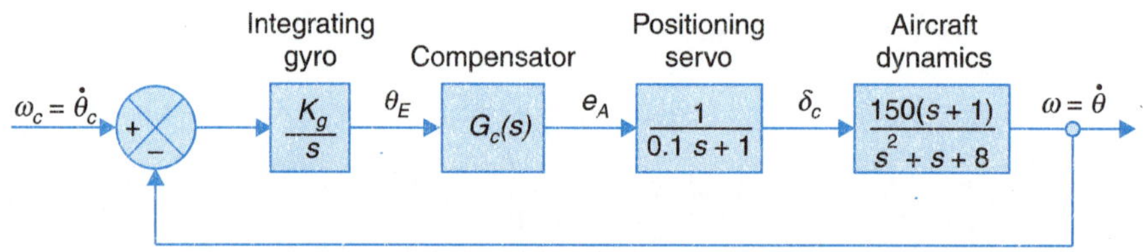

Fig. P-10.9

(*ii*) Redesign the system so that the system velocity error constant of about 0.5 is achieved without impairing its dynamics response.

[**Hint :** Try lead compensation by root locus method. It may be necessary to use more than two lead networks.]

10.10. A unity feedback type-0 system with transportation lag has a forward path transfer function of

$$G(j\omega) = \frac{10e^{-j0.02\omega}}{(1+j0.5\omega)(1+j0.1\omega)(1+j0.05\omega)}$$

Design a suitable compensation scheme so that the system acquires a damping factor of 0.4 without loss of steady-state accuracy. Estimate the bandwidth and settling time of the compensated system.

[**Hint :** Since design is handled more conveniently in the frequency domain in the presence of transportation lag, translate damping factor specification into equivalent phase margin and then proceed.]

10.11. Figure P-10.11 shows a unity feedback system with a forward path transfer function

$$G(s) = \frac{10}{s(s+1)}$$

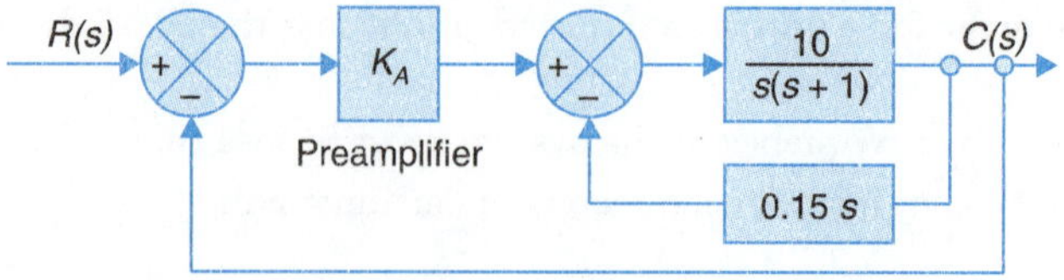

Fig. P-10.11

The system is compensated by means of rate feedback $H(s) = 0.15s$, in the minor feedback loop. Determine the effect of compensation by comparing the phase margins and bandwidths of the compensated and uncompensated schemes. In each case adjust K_A to a value which gives a velocity error constant of $K_v = 10$ for the purpose of comparison.

[**Hint :** Since the system order with internal feedback continues to be second, you can proceed analytically.]

10.12. A position control servomechanism having a forward path transfer function

$$G(s) = \frac{10(1+2s)}{s^2(1+0.2s)}$$

is to be compensated by rate feedback sK_t as indicated in Fig. P-10.12. Extimate a suitable value for K_t so that system has resonant peak $M_r = 1.3$. Estimate also resonant frequency of the compensated system.

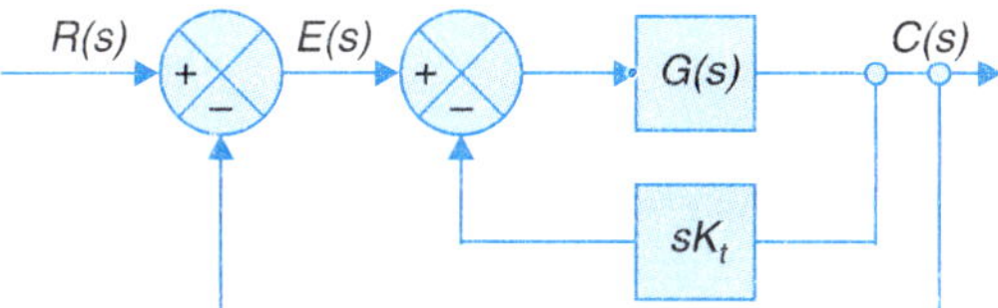

Fig. P-10.12

10.13. A unity feedback system has two transfer functions

$$G_1(s) = \frac{6}{1+0.25s};\ G_2(s) = \frac{4}{s(1+s)(1+0.5s)}$$

in tandem in the forward path. With the given values of gain the system is found to be unstable. It is attempted to stabilize the system by introducing a minor feedback having a transfer function $H(s)$ around $G_2(s)$. Show that the system gets stabilized if

$$H(s) = \frac{8s^2}{(1+2s)}$$

Using approximation techniques in frequency domain determine gain and phase margins of the stabilized system.

10.14. The feedback control system shown in Fig. 10.52 has

$$G_f(s) = \frac{3}{s(s+1)};\ H(s) = \frac{2s^2}{(1+1.25s)}$$

(*a*) Using the approximation technique, find

$$G(s) = \frac{G_f(s)}{1+G_f(s)H(s)}$$

in factored form.

(*b*) With $G(s)$ found in part (*a*), determine the value of K_A, so that the system has a phase margin of 45°.

10.15. Cement is manufactured in large Rotary Kilns where raw lime stone is added and the product clinker emerges after high-temperature processes (calcination, etc.). The clinker is later cooled and estimated for its quality.

The quality of clinker depends on many factors, one of the prime factor being the temperature in the burning zone. This temperature (about 1600°C) is sensed by an optical pyrometer. The temperature is maintained by mainpulating the coal feed rate to the burners. This is accomplished by adjusting the speed of the variable-speed coal feed screws.

The model of the temperature control process is depicted in the Fig. P-10.15. Assume that the transfer function of the clinkerization process is $\frac{1}{(s+1)(5s+1)}$, that of the sensor and coal feed screws is unity. Design a PID control system for the above process by root-locus technique such that the settling time is 3*s* and the steady-state error is zero.

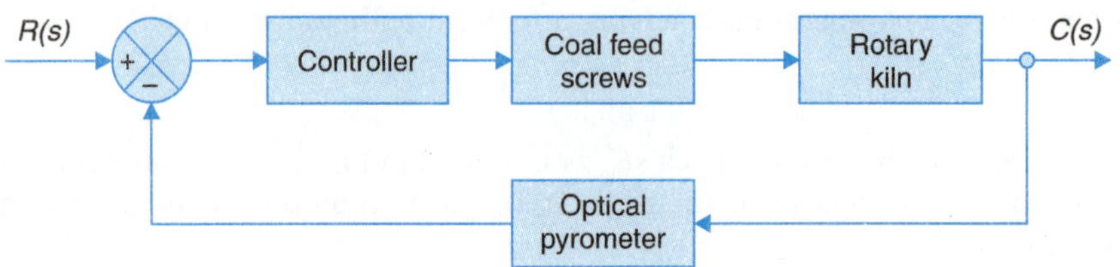

Fig. P-10.15

10.16. Lasers are widely used from drilling holes to vaporizing brain tumors, holography to communications. Mirrors and lenses are used to guide the beam through the laser and to the target. Many such lasers use **Steering Mirrors** that can be quickly moved by a control system to redirect the laser beam. The model of a **Steering Mirror Control** that is to be designed is depicted in Fig. P-10.16.

Assume that the design specifications are

peak overshoot:	20%
settling time:	0.5*s*

(*a*) Find the sensitivity of compensated system and the uncompensated system with respect to changes in $G(s)$. Comment on the stability of the system.

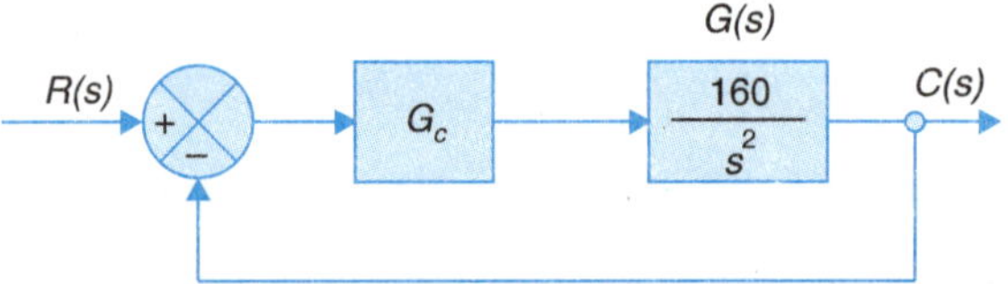

Fig. P-10.16

(*b*) Design a robust control system for the system with the gain K varying between 140 and 180. Do you need a prefilter ? Explain.

(*c*) Plot the unit step response of the compensated system. Comment on the results.

10.17. The position of an industrial robot manipulator used for welding in an automobile industry has to be controlled accurately. The position is maintained by a DC servo motor and the feedback is given by an optical encoder. The block diagram of the process is given in Fig. P-10.17.

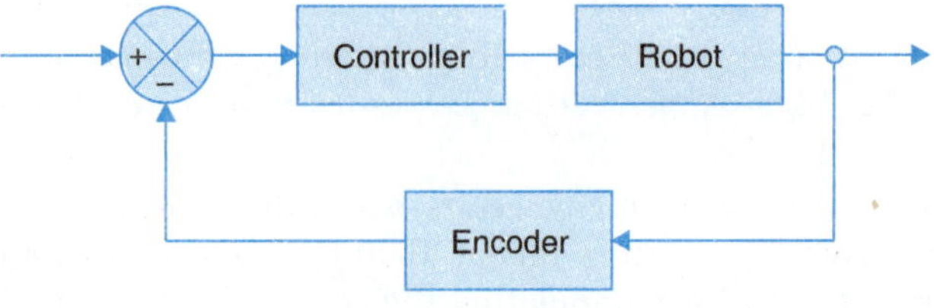

Fig. P-10.17

Given that $G_{robot}(s) = \dfrac{5}{s(2s+5)}$ and the transfer function of optical encoder is unity, design a control system based on ITAE performance. The settling time is desired to be less than 0.5s.

10.18 The prospects of electric vehicles are enormous. The block diagram of one such electrically propelled car having a microcontroller based motor control is shown in Fig. P-10.18. The speed of the vehicle is to be controlled by actuating two AC motor. Consider that the motors are to be controlled individually and that the feedback is given by a tachometer. The AC motor transfer function is $\dfrac{\omega(s)}{V(s)} = \dfrac{20}{(2s+1)^2}$ and that of tachometer is $\dfrac{V_t(s)}{\omega(s)} = 1$.

(a) Design a PID controller based on ITAE performance criterion for the above vehicle such that the peak overshoot is less than 5%.

(b) Plot the step response for a unit step input. Do you need a prefilter ? Comment on your answer.

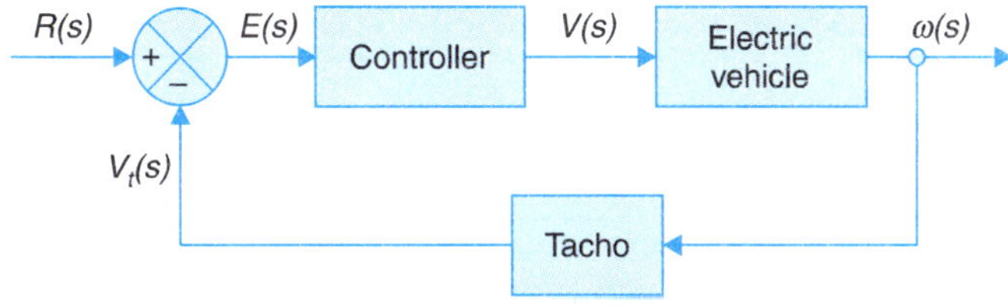

Fig. P-10.18

11

DIGITAL CONTROL SYSTEMS

11

DIGITAL CONTROL SYSTEMS

11.1 INTRODUCTION

In the control systems dealt so far, the signal at every point in the system is a continuous function of time (the independent variable). In particular the controller elements are such that the controller produces continuous time control signals from continuous time input signals. Such a controller is referred to as an *analog* controller*. As the complexity of a control system increases, there arise severe demands of flexibility, adaptability and optimality (Chapter 14) and even demands to account for economic control function complexity. In fact constructing a complex control function may even become technically infeasible, if one is restricted to use only analog elements. The use of a digital computer as a compensator (controller) device has grown during the past two decades as the price of digital computers has reduced and their reliability has improved drastically. Powerful but inexpensive computers, called micro-computers, which use a 16 bit word or 32 bit word, have become readily available. These computers can be equipped with sufficiently large memories to handle a large amount of data in a complex control process. The speed of these computers is also increasing and already systems with a speed as high as 300 MHz are readily available. Besides these general purpose computers, many specific control systems use embedded micro-controller as their heart. These micro-controllers could be either 8 bit or 16 bit depending upon the complexity of the built on the chip itself. This greatly reduces the number of components and the complexity of the controller. A digital controller in which either a special purpose computer or a general purpose computer forms the heart, is therefore an ideal choice for complex control systems. A general purpose computer, if used, lends itself to *time-shared* use for other control functions in the plant or process. A digital controller also has the versatility that its control function can be easily modified by changing a few program instructions or even the entire program and a

*The term stems from the analog computer whose elements produce continuous-time signals.

change in instruction can even be accomplished either manually or automatically under control of a supervisory function. Moreover the computers are able to receive and manipulate several inputs, so a digital computer control system can often be a multivariable system.

Digital controllers used in digital control systems have the inherent characteristic that they accept the data as short duration pulses (*i.e.*, sampled or discrete data) and produce a similar kind of output as control signal. Figure 11.1 shows a simple control scheme employing a digital controller.

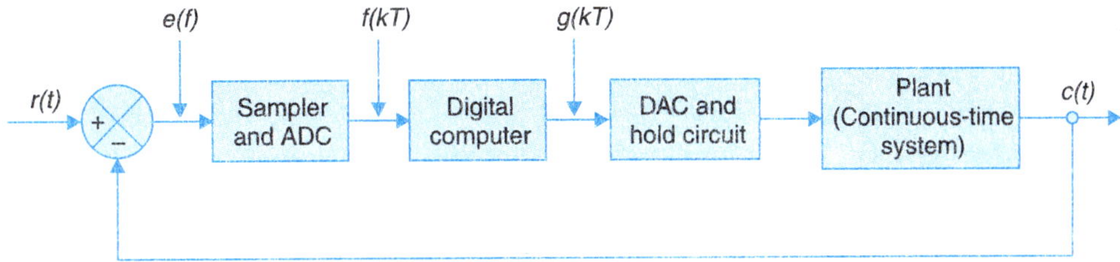

Fig. 11.1. Typical system with digital controller (sampled-data control system).

A sampler and analog-to-digital converter (ADC) is needed at the computer input. The sampler converts the continuous time error signal into a sequence of pulses which are then expressed in numerical code (such as binary code). Numerically coded output data of digital computer are decoded into continuous time signal by digital-to-analog converter (DAC) and hold circuit. This continuous-time signal then controls the plant (continuous-time system). The overall system is *hybrid* in which the signal is in sampled form in the digital controller and in continuous form in the rest of the system. A system of this kind is referred to as a *sampled-data control system*.

Even in relatively simple control schemes, sampling may be warranted from other considerations. In fact sampling is a necessity wherever a high degree of accuracy is a prerequisite. This is the case in most automated machine-tools. For example if it is required to move the table of a drilling machine within an accuracy of 0.01mm over a total distance of 1 m, a resolution of 1 in 100,000 is needed which is impossible to measure using an analog type output transducer say a potentiometer. The problem can only be tackled by employing a digitally coded output sensor like an optical encoder.

Sampled data technique is most appropriate for control systems requiring long distance data transmission. Pulse amplitude modulated* (PAM) data is easily transmitted by means of a carrier over a transmission channel and the data reconstructed at the receiving end. It is well known that pulses may be transmitted with little loss of accuracy; data is analog form will suffer considerable distortion in the transmission channel. Using sampled data technique, more than one channel of information may be sequentially sampled and transmitted through a single transmission system (this technique is known as time-multiplexing), decreasing the cost of transmission installation.

Signal sampling reduces the power demand made on the signal and is therefore helpful for signals of weak power origin.

---

*Pulse width modulation (PWM) is an alternative way of tackling the transmission problem.

The examples cited above were those of systems in which the sampling operation is purposely introduced. however, there is a class of systems where the signals are available in sampled form only. For example in radar tracking system, the signal sent out and received is in form of pulse trains.

The circumstances that lead to the use of sampled-data control systems are summarized below:

1. For using digital computer (or microprocessor) as part of the control loop.
2. For time sharing of control components.
3. Whenever a transmission channel forms part of the control loop.
4. Whenever the output of a control component is essentially in discrete form.

Sampling implies that the signal at the output end of the *sampler* is available in form of short duration pulses each followed by a skip period when no signal is available so that the control system essentially operates open-loop during the skip period. Uniform periodic sampling is illustrated in Fig. 11.2. It is intuitively obvious that if the sampling rate (or frequency) is too low, significant information contained in the input signal may be missed in the output. We shall show in the next article that the minimum sampling rate has a definite relationship with the highest significant signal frequency (*i.e.*, signal bandwidth). With the availability of high speed computers very high sampling rates can be achieved. In fact the sampling period in some cases may be so small that the system can for all practical purposes be approximated as a continuous time system.

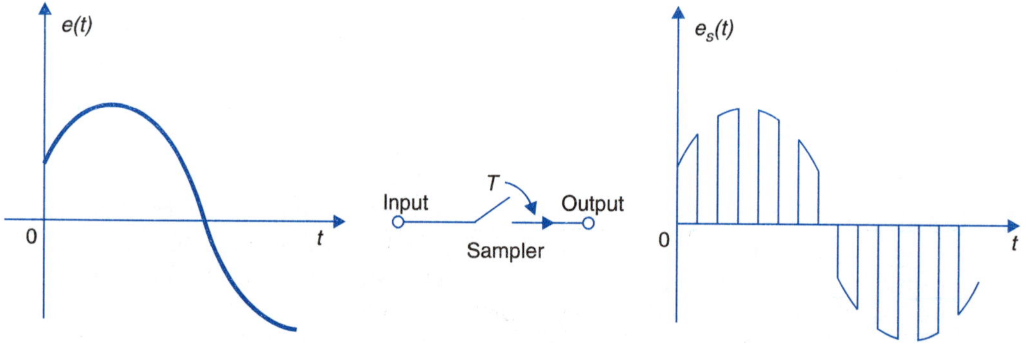

Fig. 11.2. Uniform periodic sampling.

Assuming sample width (time) as fixed, other forms of sampling are:

Multi-order sampling: A particular sampling pattern is repeated periodically.

Multiple-rate sampling: In this case two simultaneous sampling operations with different time periods are carried out on the signal to produce the sampled output.

Random sampling: In this case the sampling instants are random with a particular kind of distribution.

In this book we shall restrict ourselves to uniform periodic sampling.

The mathematical model of a sampled-data control system is essentially in the form of difference equations. The analysis and design of sampled-data systems with linear elements may be effectively carried out by use of the z-transform which was evolved from the Laplace transform as a special form and was later established in its own merit.

11.2 SPECTRUM ANALYSIS OF SAMPLING PROCESS

The sampling process represented in Fig. 11.2 is equivalent to multiplying the signal $e(t)$ with a periodic pulse train $p_{T,}\ \Delta(t)$, shown in Fig. 11.3, to produce the sampled signal

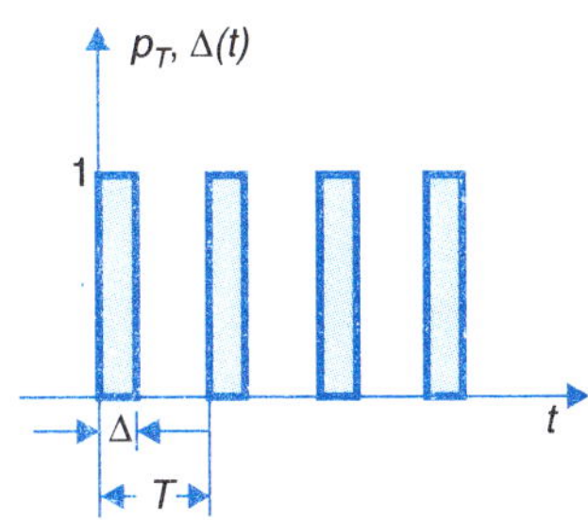

Fig. 11.3 Pulse train $p_T, \Delta(t)$.

The spectral analysis of such a sampled signal is somewhat involved. The pulse train is therefore approximated by an impulse train with each impulse of strength $1 \times \Delta = \Delta$, the pulse area. The sampled signal is consequently approximated as

$$e_s(t) = e(t)\ \Delta\delta_T(t) \qquad ...(11.1)$$

where $\delta_T(t)$ is a unit impulse train. This is illustrated in Fig. 11.4. Notice that the time width (Δ) of the pulse sampler merely modifies the strength of impulses.

Mathematically

$$\delta_T(t) = \qquad ...(11.2)$$

$$\therefore \qquad e_s(t) = \Delta e(t) \sum_{k=-\infty}^{\infty} \delta(t - kT) \qquad ...(11.3)$$

Taking the Fourier transform (see Appendix I) of eqn. (11.3)

$$\mathcal{F}[e_s(t)] = \frac{\Delta}{2\pi} \mathcal{F}[e(t)] * \mathcal{F}\left[\sum_{k=-\infty}^{\infty} \delta(t - kT)\right] \qquad ...(11.4)$$

But
$$\mathcal{F}\left[\sum_{k=-\infty}^{\infty} \delta(t - kT)\right] = \omega_s \sum_{k=-\infty}^{\infty} \delta(\omega - k\omega_s)$$

where $\omega_s = \dfrac{2\pi}{T}$ = sampling frequency.

Equation (11.4) can therefore be written as

$$E_s(\omega) = \frac{\Delta}{2\pi} E(\omega) * \omega_s \sum_{k=-\infty}^{\infty} \delta(\omega - k\omega_s) \qquad ...(11.5)$$

As
$$E(\omega) * \delta(\omega - k\omega_s) = E(\omega - k\omega_s)$$

we have
$$E_s(\omega) = \frac{\Delta}{T} \sum_{k=-\infty}^{\infty} E(\omega - k\omega_s) \qquad ...(11.6)$$

Equation (11.6) gives the frequency spectrum of the impulse sampled signal in terms of the spectrum of the input signal. Let $E(\omega)$, the spectrum of the input signal, be band-limited* signal with a maximum frequency of ω_m as in Fig. 11.5(a). The frequency spectrum of this signal when impulse sampled (see Fig. 11.4(c)) is plotted in Fig. 11.5(b) for $\omega_s > 2\omega_m$ and in Fig. 11.5(c) for $\omega_s < 2\omega_m$.

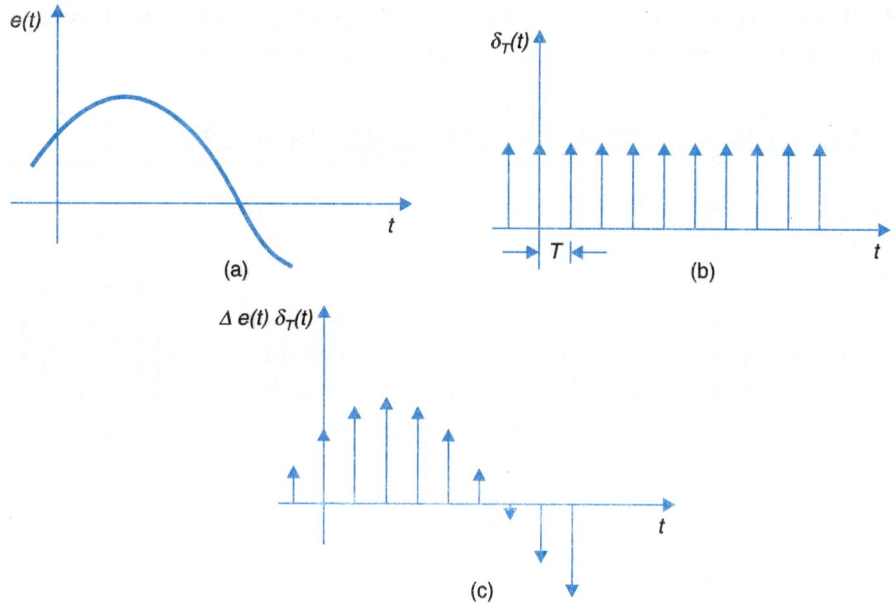

Fig. 11.4. Impulse sampling approximation of pulse sampling.

It is immediately observed from Figs. 11.5(*b*) and (*c*) that so long as $\omega_s \geq 2\omega_m$, the original spectrum is preserved in the sampled signal and can be extracted from it by low-pass filtering (shown in dotted line in Fig. 11.5(*b*)). This is the well-known *Shanon's sampling theorem* according to which the information contained in a signal is fully preserved in the sampled version so long as the sampling frequency is at least twice the maximum frequency contained in the signal.

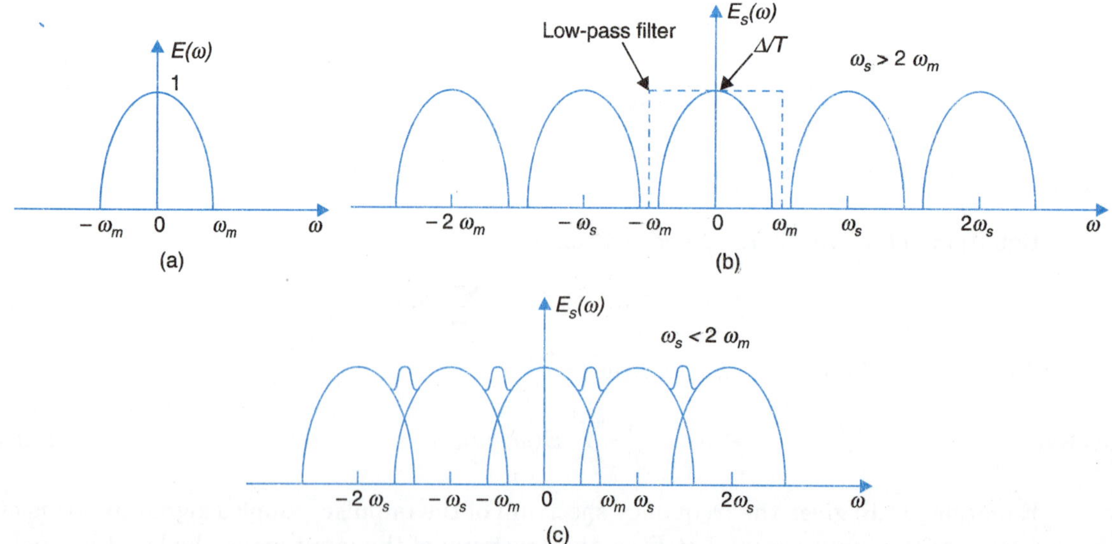

Fig. 11.5. Fourier spectra of input signal and its impulse sampled version.

---

*All signals of practical significance are band-limited in nature as the energy contained in frequencies above a certain figure is of negligible order.

11.3 SIGNAL RECONSTRUCTION

Sampled-data signal which has been modified by a digital controller (or has been transmitted over a channel) must be converted into analog form for use in the continuous part of the system (the controlled plant). This is accomplished by means of various types of *hold circuits* (extrapolators). The simplest hold circuit is the zero-order hold (ZOH) in which the reconstructed signal acquires the same value as the last received sample for the entire sampling period. The schematic diagram of sampler and ZOH is shown in Fig. 11.6, while signal reconstruction is illustrated in Fig. 11.7. The high frequencies present in the reconstructed signal are easily filtered out by the controlled elements of the system which are basically low-pass in frequency behaviour.

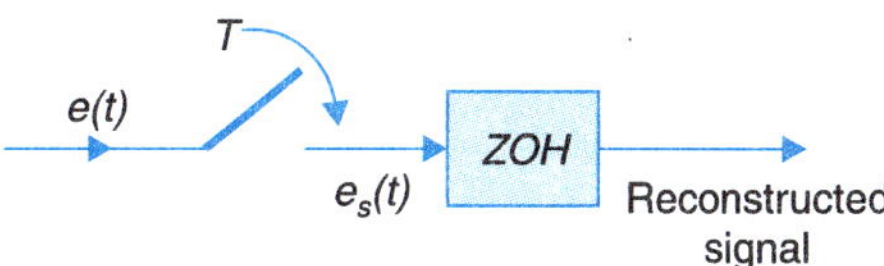

Fig. 11.6. Sampler and zero-order hold.

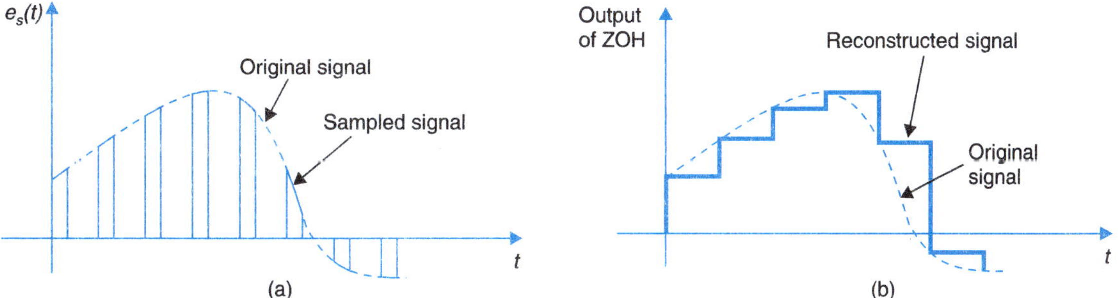

Fig. 11.7. Signal reconstruction by ZOH.

In a first-order hold, the last two signal samples are used to reconstruct the signal for the current sampling period. Similarly higher-order holds can be devised. First or higher-order holds offer no particular advantage over the zero-order hold. The simple zero-order hold when used in conjunction with a high sampling rate provides satisfactory performance.

11.4 DIFFERENCE EQUATIONS

The difference equation model results in sampled-data systems and in numerical analysis of continuous-time systems. Examples of each kind will now be presented here.

Example 11.1: Consider a first-order continuous-time feedback system as in Fig. 11.8. We can immediately write

$$c(t) = A\int_0^t [r(\tau) - c(\tau)]d\tau \quad \text{or} \quad c(t) + A\int_0^t c(\tau)d\tau = A\int_0^t r(\tau)d\tau \qquad ...(11.7)$$

Closed-form solution is not possible if $r(t)$ is an arbitrary function of time. Numerical solution can be obtained by approximating the integral equation (11.7) by a difference equation. Dividing time into discrete intervals of length T, we have

$$c(kT) + A\int_0^{kT} c(\tau)d\tau = A\int_0^{kT} r(\tau)d\tau$$

Approximating integration by discrete summation, we can write

$$c(kT) + A\sum_{m=0}^{k-1} c(mT)T = A\sum_{m=0}^{k-1} r(mT)T \qquad ...(11.8)$$

Also
$$c(kT+T) + A\sum_{m=0}^{k} c(mT)T = A\sum_{m=0}^{k} r(mT)T \qquad ...(11.9)$$

Subtracting (11.8) from (11.9),

$$c[(k+1)T] - c(kT) + ATc(kT) = ATr(kT)$$

Writing $c(kT)$ as $c(k)$ and $r(kT)$ as $r(k)$, we have

$$c(k+1) = (1-AT)\,c(k) + ATr(k) \qquad ...(11.10)$$

which is a first-order difference equation with constant coefficients.

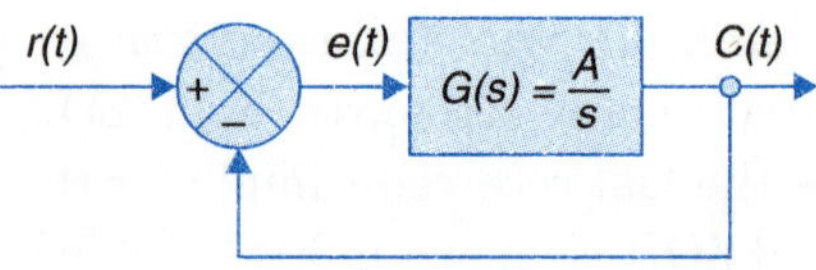

Fig. 11.8

Example 11.2: Introduce now a sampler and ZOH in the forward path of the system of Fig. 11.8 converting it to a sampled-data system of Fig. 11.9.

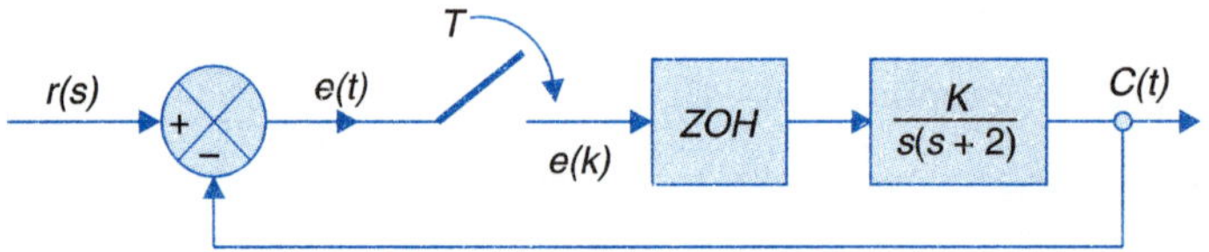

Fig. 11.9. First-order sampled-data system.

Because of the action of sampler and ZOH, the output of ZOH during kth time interval is

$$e_k(t) = e(kT)\ ;\ kT \le t < (k+1)\,T$$

The output $c(t)$ during this period due to action of the integrator (A/s) is given by

$$c(t) = c(kT) + Ae(kT)\,(t - kT)\ ; \quad kT \le t < (k+1)\,T$$

or
$$c[(k+1)T] = c(kT) + ATe(kT)$$

or
$$c(k+1) = c(k) + ATe(k) \qquad ...(11.11)$$

But
$$e(kT) = r(kT) - c(kT)$$

Equation (11.11) can then be written as

$$c(k+1) = (1-AT)\,c(k) + ATr(k) \qquad ...(11.12)$$

which indeed is the same difference equation as (11.10).

Let us obtain the solution of the difference equation (11.12). This can be easily carried out recursively

$$c(1) = (1-AT)c(0) + ATr(0)$$

$$c(2) = (1-AT)c(1) + ATr(1)$$

$$= (1-AT)^2\,c(0) + (1-AT)ATr(0) + ATr(1)$$

$$\cdots \qquad \cdots \qquad \cdots \qquad \cdots$$

$$c(k) = \underbrace{(1 - AT)^k c(0)}_{\text{Zero input response}} + \underbrace{AT \sum_{i=0}^{k-1} (1 - AT)^{k-1-i}\, r(i)}_{\text{Forcing function response}} \qquad ...(11.13)$$

It may be noticed here that the difference equation solution yields $c(k)$, *i.e.*, the values of output $c(t)$ (which is continuous-time function caused by the presence of the continuous part of the system (A/s) at sampling instants only. No information is available about $c(t)$ during the sampling instants. This is satisfactory so long as the sampling rate is sufficiently fast so as not to miss any important output information. In fact cases can exist where parasitic oscillations of frequency multiple of sampling frequency are present in the output but are undetected by the sampled output (see Problem 11.14). Consider now the zero input (*i.e.*, initial condition) response

$$c(k) = (1 - AT)^k c(0) \qquad ...(11.14)$$

The response decays, i.e., the system is stable if

$$|1 - AT| < 1 \qquad \text{or} \qquad 0 < AT < 2$$

or

$$A < \frac{2}{T} \qquad ...(11.15)$$

We therefore conclude that the sampled-data system of Fig. 11.9 is conditionally stable. It may be recalled here that the corresponding continuous time system of Fig. 11.8 is absolutely stable, *i.e.*, stable for all values of A. The instability of the sampled-data system results from the fact that it operates open-loop between sampling instants.

As the difference equation (11.10) approximating the continuous-time system of Fig. 11.8 is identical to eqn. (11.12), it would yield a highly erroneous solution (unstable where the actual system is stable) for $T > 2/A$, which only tells that continuous time must be divided into very much smaller intervals that $T = 2/A$ to be a 'good' approximate solution.

The general from of nth-order linear, constant coefficient difference equation is presented below:

$$\begin{aligned} c(k+n) + a_1 c(k+n-1) + \ldots + a_n c(k) \\ = b_0 r(k+n) + b_1 r(k+n-1) + \ldots + b_n r(k) \end{aligned} \qquad ...(11.16)$$

Often this equation is written in the equivalent from:

$$\begin{aligned} c(k) + a_1 c(k-1) + \ldots + a_n c(k-n) \\ = b_0 r(k) + b_1 r(k-1) + \ldots + b_n r(k-n) \end{aligned} \qquad ...(11.17)$$

The straightforward method of modelling of sampled-data systems gets very messy for higher-order systems. The z-transform technique introduced below is a convenient tool for analysis and design of linear sampled-data systems.

11.5 THE z-TRANSFORM

The one-sided z-transform of a sequence

$$f(k) : \{\ldots 0, 0, f(0), f(1), f(2), \ldots\}$$

which is defined for positive integers k, is defined as the weighted sum

$$Z[f(k)] = F(z) = \sum_{k=0}^{\infty} f(k)z^{-k} \qquad \text{...(11.18)}$$

where z is an arbitrary complex number. $F(z)$, a sum of complex numbers, is also a complex number.

Consider a geometric sequence

$$\begin{aligned} f(k) &= \{... \; 0, 1, a, a^2, ...\} = a^k; \; k \geq 0 \\ &= 0; \; k < 0 \end{aligned} \qquad \text{...(11.19)}$$

where a is any real number.

The z-transform of this sequence is

$$Z[f(k)] = F(z) = \sum_{k=0}^{\infty} a^k z^{-k} = \sum_{k=0}^{\infty} (az^{-1})^k \qquad \text{...(11.20)}$$

The infinite sum (11.20) converges if

$$|az^{-1}| < 1 \qquad \text{or} \qquad |z| > |a|$$

and it diverges (is unbounded) if

$$|z| < |a|$$

In the region of convergence,

$$Z[a^k] = \sum_{k=0}^{\infty} (az^{-1})^k = \frac{1}{1 - az^{-1}} = \frac{z}{z - a}$$

The z-transform of the geometric sequence (11.19) may be compactly expressed as

$$Z[a^k] = F(z) = \begin{cases} \dfrac{z}{z-a} & ; \; |z| > |a| \\ \text{unbounded}; & |z| < |a| \end{cases} \qquad \text{...(11.21)}$$

The convergence and divergence regions of $F(z)$ in (11.21) are shown in Fig. 11.10. On the circle separating the two regions, the z-transform may or may not converge. We should make separate tests for those z which lie on this boundary.

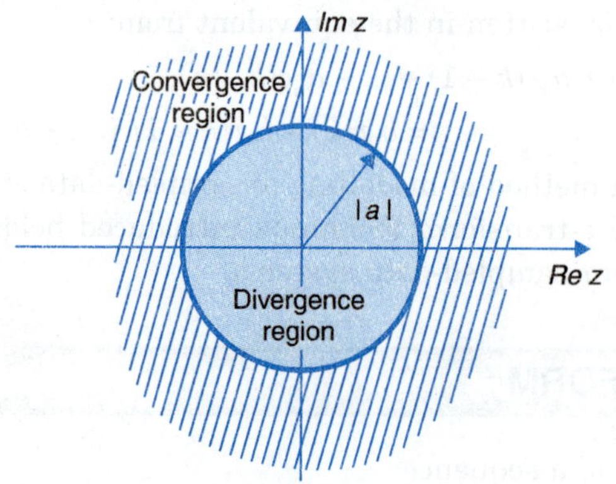

Fig. 11.10. Regions of convergence and divergence for the z-transform.

Consider now the sinusoidal sequence defined by

$$f(k) = \sin k\omega T \quad ; k \geq 0$$
$$= 0 \quad ; k < 0$$

The z-transform is given as

$$Z[\sin k\omega T] = F(z) = \sum_{k=0}^{\infty} (\sin k\omega T) z^{-k}$$

$$= \sum_{k=0}^{\infty} \left(\frac{e^{jk\omega T} - e^{-jk\omega T}}{2j} \right) z^{-k} \quad \text{...(11.22)}$$

Now using the result of eqn. (11.21),

$$Z[(e^{j\omega T})^k] = \frac{z}{z - e^{j\omega T}} \; ; \; |z| > 1 \quad \text{...(11.23)}$$

From eqns. (11.22) and (11.23), we have the following result:

$$Z[\sin k\omega T] = \frac{z \sin \omega T}{\{z - e^{j\omega T}\}\{z - e^{-j\omega T}\}}$$

$$= \frac{z \sin \omega T}{z^2 - 2z \cos \omega T + 1} \; ; \; |z| > 1 \quad \text{...(11.24)}$$

Convergence region of the z-transform lies outside a circle of the unit radius in the complex z-plane.

It can be proved [1] that in general the z-transform of any sequence of numbers $f(k)$ will have a region of convergence specified by

$$|z| > R$$

where the radius of convergence* R depends upon the sequence $f(k)$.

A discrete sequence can be generated from a continuous-time function $f(t)$ by *mathematically sampling* it at time intervals T seconds apart merely by substituting $t = kT$, *i.e.*,

$$f_s(t) = f(kT)$$

This sampling is different from pulse or impulse sampling of a physical signal by suitable hardware wherein the output is a train of pulses or impulses. Mathematical sampling is merely a mathematical contrivance to generate a sequence of values at discrete instants of time. Diagrammatically it is depicted in Fig. 11.11.

Consider a time function

$$f(t) = e^{j\omega t} u(t)$$

Sampling it mathematically

$$f(k) = e^{j\omega kT} u(k) = \{e^{j\omega T}\}^k \; ; \; k \geq 0$$

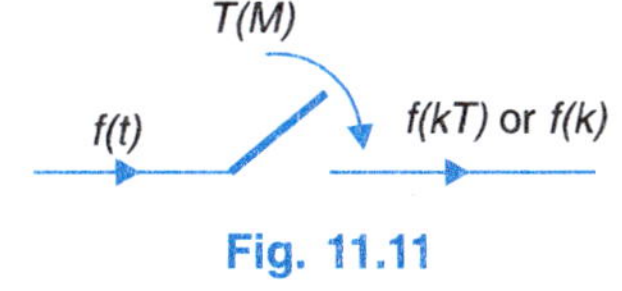

Fig. 11.11

*Note that we will generally not be concerned with the regions of convergence and divergence. These concepts have been introduced for the sake of completeness.

Recognizing $a = e^{j\omega T}$ in (11.21), we can at once write

$$Z[e^{j\omega kT}] = \frac{z}{z - e^{j\omega T}} \text{ for } |z| > 1$$

We can immediately obtain

$$Z[\cos \omega kT] = Z\left[\frac{e^{j\omega kT} + e^{-j\omega kT}}{2}\right]$$

$$= \frac{1}{2}\left[\frac{z}{z - e^{j\omega T}} + \frac{z}{z - e^{-j\omega T}}\right]$$

$$= \frac{z(z - \cos \omega T)}{z^2 - 2z\cos \omega T + 1}; \ |z| > 1$$

The Inverse z-Transform

The inverse z-transformation is a process of determining the sequence which generates a given z-transform. It is denoted by

$$f(k) = Z^{-1}[F(z)]$$

From eqns. (11.21) and (11.24), we have

$$Z^{-1}\left[\frac{z}{z-a}\right] = a^k$$

$$Z^{-1}\left[\frac{z \sin \omega T}{z^2 - 2z \cos \omega T + 1}\right] = \sin k\omega T$$

These equations give the following transform pairs:

$$a^k \leftrightarrow \frac{z}{z-a}$$

$$\sin k\omega T \leftrightarrow \frac{z \sin \omega T}{z^2 - 2z \cos \omega T + 1}$$

For illustration purposes, a few transform pairs are generated below:

(*i*) The process of differentiating a known z-transform pair offers a convenient method for determining a new z-transform pair. For example, consider the known relationship

$$\sum_{k=0}^{\infty} a^k z^{-k} = \frac{z}{z-a}$$

Differentiating both sides with respect to z, we get

$$[- az^{-2} - 2a^2z^{-3} - ...] = -\frac{a}{(z-a)^2}$$

Multiplying both sides by $-z^2$.

$$[a + 2a^2z^{-1} + ...] = \frac{az^2}{(z-a)^2}$$

or
$$\sum_{k=0}^{\infty}(k+1)a^{k+1}z^{-k} = \frac{az^2}{(z-a)^2}$$

$$(k+1)a^{k+1} \leftrightarrow \frac{az^2}{(z-a)^2} \qquad ...(11.25)$$

(*ii*) Consider a discrete unit impulse defined as

$$\delta(k) : \{1, 0, 0, ...\} = 1 \,;\, k = 0$$
$$= 0 \,;\, k \neq 0$$

$$Z[\delta(k)] = \sum_{k=0}^{\infty}\delta(k)z^{-k} = 1$$

Thus the z-transform pair is $\delta(k) \leftrightarrow 1$ ...(11.26)

(*iii*) A discrete unit step function is defined as

$$u(k) : \{1, 1, 1, ...\} = 1;\, k \geq 0$$
$$= 0;\, k < 0$$
$$Z[u(k)] = Z[1^k]$$

Substituting $a = 1$ in eqn. (11.21), we get

$$u(k) \leftrightarrow \frac{z}{z-1} \qquad ...(11.27)$$

Properties of the z-Transform

In this section we will study some of the important properties of the z-transform which can be used to determine $F(z)$ and its inverse.

Linearity

$$Z[af(k) + bg(k)] = \sum_{k=0}^{\infty}[af(k) + bg(k)]z^{-k}$$
$$= a\sum_{k=0}^{\infty}f(k)z^{-k} + b\sum_{k=0}^{\infty}g(k)z^{-k}$$
$$= aF(z) + bG(z) \qquad ...(11.28)$$

Shifting

A sequence $f(k)$ when shifted one interval to the left (advanced), can be written as

$$g(k) = f(k+1) \,;\, k \geq -1$$

This shifted sequence is shown in Fig. 11.12(*b*). Taking the z-transform,

$$Z[f(k+1)] = \sum_{k=0}^{\infty}f(k+1)z^{-k}$$
$$= z\sum_{k=0}^{\infty}f(k+1)z^{-(k+1)}$$

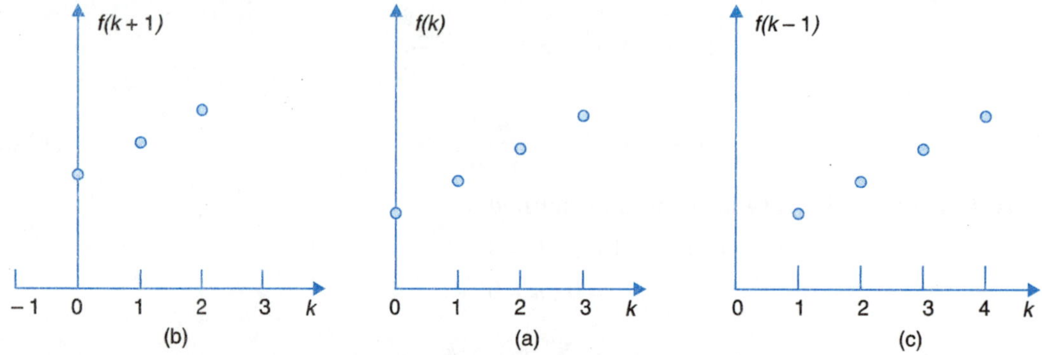

Fig. 11.12. Shifted sequences.

Letting $(k + 1) = m$,

$$Z[f(k+1)] = z \sum_{m=1}^{\infty} f(m) z^{-m}$$

$$= z\left[\sum_{m=0}^{\infty} f(m) z^{-m} - f(0)\right]$$

$$= zF(z) - zf(0) \qquad \text{...(11.29)}$$

In general

$$Z[f(k+n)] = z^n F(z) - \sum_{i=0}^{n-1} f(i)\, z^{n-i};\ k \geq -n \qquad \text{...(11.30)}$$

Consider now the sequence

$$g(k) = f(k-n)\ ;\ k \geq n$$

which is the sequence $f(k)$ shifted n intervals to the right (delayed) as shown in Fig. 11.12(c) for $n = 1$.

Taking the z-transform,

$$Z[f(k-n)] = \sum_{k=0}^{\infty} f(k-n)\, z^{-k} = z^{-n} \sum_{k=0}^{\infty} f(k-n)\, z^{-(k-n)}$$

Letting $(k - n) = m$, we have

$$Z[f(k-n)] = z^{-n} \sum_{m=-n}^{\infty} f(m) z^{-m}$$

As $\qquad f(m) = 0$ for $m < 0$,

$$Z[f(k-n)] = z^{-n} \sum_{m=0}^{\infty} f(m) z^{-m}$$

$$= z^{-n} F(z) \qquad \text{...(11.31)}$$

The above properties are useful in z-transforming difference equations. Consider the difference equation (11.12) reproduced below:

$$c(k+1) - (1 - AT)c(k) = ATr(k)$$

Taking the z-transform of each term (linearity property)

$$Z[c(k+1)] - (1 - AT)\, Z[c(k)] = ATZ\,[r(k)]$$

$$zC(z) - zc(0) - (1 - AT)C(z) = ATR(z)$$

Solving for $C(z)$, we get

$$C(z) = \frac{z}{z-(1-AT)}\,c(0) + \frac{AT}{z-(1-AT)}R(z) \qquad \text{...(11.32)}$$

For the case of zero input

$$C(z) = \frac{z}{z-(1-AT)}c(0)$$

From the z-transform pair of eqn. (11.21), we get

$$c(k) = (1 - AT)^k c(0)$$

which is a result we have earlier obtained by the recursive solution of the difference equation.

Example 11.3: Consider a system described by the difference equation

$$c(k+1) + 2c(k) = r(k)\,;\ c(0) = 0$$

Let us obtain the system's impulse response; that is $c(k)$ satisfying

$$c(k+1) + 2c(k) = \delta(k)$$

Taking the z-transform

$$(z+2)C(z) = 1$$

$$C(z) = \frac{1}{z+2}$$

From (11.21) $\qquad \frac{z}{z+2} \leftrightarrow (-2)^k$

From shifting property (11.30)

$$z^{-1}\left(\frac{z}{z+2}\right) \leftrightarrow (-2)^{k-1}\,;\ k \geq 1$$

Hence $\qquad c(k) = (-2)^{k-1}\,;\ k \geq 1$

Multiplication by k

Let $\qquad Z[f(k)] = F(z)$

Now $\qquad Z[kf(k)] = \sum_{k=0}^{\infty} kf(k)z^{-k} = -z\sum_{k=0}^{\infty} -kf(k)z^{-k-1}$

$$= -z\sum_{k=0}^{\infty} f(k)\frac{d}{dz}z^{-k} = -z\frac{d}{dz}\sum_{k=0}^{\infty} f(k)z^{-k}$$

$$= -z\frac{d}{dz}F(z) \qquad \text{...(11.33)}$$

Example 11.4: Find the z-transform of the discrete ramp function

$$g(k) = k \; ; k \geq 0$$
$$= 0 \; ; k < 0.$$

We can write $$g(k) = ku(k)$$

where $u(k)$ is a discrete step function.

Now $$Z[g(k)] = Z[ku(k)] = -z\frac{d}{dz}U(z)$$

Substituting for $U(z)$ from (11.27),

$$Z[g(k)] = -z\frac{d}{dz}\left(\frac{z}{z-1}\right) = \frac{z}{(z-1)^2}$$

Scale change

Let $$Z[f(k)] = F(z)$$

Consider the inverse z-transform of $F(z/a)$, that is,

$$Z^{-1}\left[F\left(\frac{z}{a}\right)\right] = Z^{-1}\left[\sum_{k=0}^{\infty} f(k)\left(\frac{z}{a}\right)^{-k}\right]$$

$$= Z^{-1}\left[\sum_{k=0}^{\infty} a^k f(k) z^{-k}\right] = a^k f(k)$$

Thus, we have the transform pair

$$a^k f(k) \leftrightarrow F(z/a) \qquad ...(11.34)$$

Initial and final values

(*a*) $$F(z) = \sum_{k=0}^{\infty} f(k) z^{-k} = f(0) + f(1)z^{-1} + f(2)z^{-2} + ...$$

Taking the limit as $z \to \infty$, we obtain

$$f(0) = \lim_{z \to \infty} F(z) \qquad ...(11.35)$$

(*b*) A more useful theorem involves the determination of the final value of the sequence $f(k)$, which is given by

$$\lim_{k \to \infty} f(k) = \lim_{z \to 1} F(z)(z-1) \qquad ...(11.36)$$

provided the limit exists, *i.e.*, the data sequence has a finite value. Consider the z-transform pair

$$F(z) = \frac{z}{z-a} \leftrightarrow f(k) = a^k.$$

The pole of $F(z)$ is located at $z = a$. If $|a| > 1$, that is the pole lies outside a unit circle, the sequence $f(k) = a^k$ tends to increase without bound, *i.e.*, the limit does not exist. Therefore for the final value theorem to be applicable, $F(z)$ must not have any poles in the region outside a unit circle, *i.e.*, it should be analytic for $|z| > 1$.

As an example, consider eqn. (11.32) reproduced below:

$$C(z) = \frac{z}{z-(1-AT)}c(0) + \frac{AT}{z-(1-AT)}R(z)$$

Let $\quad r(k) = u(k)$

$$R(z) = \frac{z}{(z-1)}$$

Now $\quad c(\infty) = \lim_{z\to 1}(z-1)C(z) = \lim_{z\to 1}\left[\frac{z(z-1)}{z-(1-AT)}c(0) + \frac{AT(z-1)}{z-(1-AT)} \times \frac{z}{(z-1)}\right]$

or $\quad c(\infty) = 1$

The properties of the z-transform are summarized in Table 11.1.

Table 11.1. Properties of the (one-sided) z-Transform

Property	*Discrete sequence*	*z-transform*
Linearity	$af(k) + bg(k)$	$aF(z) + bG(z)$
Shifting $n \geq 0$	$f(k+n)$	$z^n F(z) - \sum_{i=0}^{n-1} f(i) z^{n-1}$
	$f(k-n)$	$z^{-n}F(z)$
Multiplication by k^n	$k^n f(k)$	$\left(-z\frac{d}{dz}\right)^n F(z)$
Scaling or multiplication by a^k	$a^k f(k)$	$F(a^{-1}z)$
Convolution*	$\sum_{m=0}^{k} h(k-m)r(m)$	$H(z)R(z)$
Initial value	$f(0) = \lim_{z\to\infty} F(z)$	
Final Value	$f(\infty) = \lim_{z\to 1}(1-z^{-1})F(z)$ $= \lim_{z\to 1}(z-1)F(z)$ if $F(z)$ is analytic for $\lvert z \rvert > 1$	

z-Transform Pairs

Rather than obtaining the z-transform abinitio every time, it is convenient to look up a table of transform pairs given in Table 11.2 (a). Some of the commonly used z-transform pairs along with their Laplace transform are listed in Table 11.2 (b) (see next page). The conversion between Laplace transform to z-Transform and vice versa is dealt with in section 11.8.

* This property is proved in the next article.

Table 11.2. (a) : Laplace and z-Transform (One-sided) Pairs

$f(t)$ $t \geq 0$	$F(s)$	$f(k)/f(kT)$ $k \geq 0$	$F(z)$
$\delta(t)$	1	$\delta(k)$	1
$u(t)$	$\frac{1}{s}$	$u(k)$ or 1	$\frac{z}{z-1}$
		a^k	$\frac{z}{z-a}$
		ka^k	$\frac{az}{(z-a)^2}$
		k^2a^k	$\frac{az(z+a)}{(z-a)^3}$
		$(k+1)a^k$	$\frac{z^2}{(z-a)^2}$
		$\frac{(k+1)(k+2)}{2!a^k}$	$\frac{z^3}{(z-a)^3}$
		$\frac{(k+1)(k+2)(k+3)}{3!a^k}$	$\frac{z^4}{(z-a)^4}$
		$\frac{a^k}{k!}$	$e^{az^{-1}}$

Table 11.2(b) : Laplace and z-Transform (One-sided) Pairs

$f(t)$ $t \geq 0$	$F(s)$	$f(k)/f(kT)$ $k \geq 0$	$F(z)$
t	$\frac{1}{s^2}$	kT	$\frac{Tz}{(z-1)^2}$
t^2	$\frac{1}{s^3}$	$(kT)^2$	$\frac{T^2z(z+1)}{(z-1)^3}$
e^{-at}	$\frac{1}{(s+a)}$	e^{-akT}	$\frac{z}{(z-e^{-aT})}$
te^{-at}	$\frac{1}{(s+a)^2}$	kTe^{-akT}	$\frac{zTe^{-aT}}{(z-e^{-aT})^2}$
$a(1-e^{-at})$	$\frac{a}{s(s+a)}$	$a(1-e^{-akT})$	$\frac{z\{1-e^{-aT}\}}{(z-1)(z-e^{-aT})}$
$\sin \omega t$	$\frac{\omega}{s^2+\omega^2}$	$\sin \omega kT$	$\frac{z \sin \omega T}{z^2 - 2z\cos \omega T + 1}$

...(Contd.)

$\cos \omega t$	$\dfrac{s}{s^2+\omega^2}$	$\cos \omega kT$	$\dfrac{z(z-\cos \omega T)}{z^2-2z\cos \omega T+1}$
$e^{-at}\sin \omega t$	$\dfrac{\omega}{(s+a)^2+\omega^2}$	$e^{-akT}\sin \omega kT$	$\dfrac{ze^{-aT}\sin \omega T}{z^2-2ze^{-aT}\cos \omega T+e^{-2aT}}$
$e^{-at}\cos \omega t$	$\dfrac{(s+a)}{(s+a)^2+\omega^2}$	$e^{-akT}\cos \omega kT$	$\dfrac{z(z-e^{-aT}\cos \omega T)}{z^2-2ze^{-aT}\cos \omega T+e^{-2aT}}$

11.6 THE z-TRANSFER FUNCTION (PULSE TRANSFER FUNCTION)

Linear Discrete Systems (LDS)

Consider a linear time-invariant discrete system represented by the block diagram of Fig. 11.3(a). The system produces an output sequence $c(k)$ for an input sequence $r(k)$. Such a system can be characterized by its response $h(k)$ (also called the *weighting sequence*) to unit discrete impulse

$$\delta(k) = 1; \quad k = 0$$
$$= 0; \quad k \neq 0 \qquad ...(11.37)$$

as shown in Fig. 11.3(b). The system is assumed to be *casual* (*nonanticipative*) so that the output appears only after application of the input. Because of this property, if a delayed unit impulse

$$\delta(k-m) = 1; \quad k = m$$
$$= 0; \quad k \neq m \qquad ...(11.38)$$

is applied to the system, the output sequence gets delayed by m intervals; that is, it now becomes $h\,(k-m)$ as shown in Fig. 11.3(c).

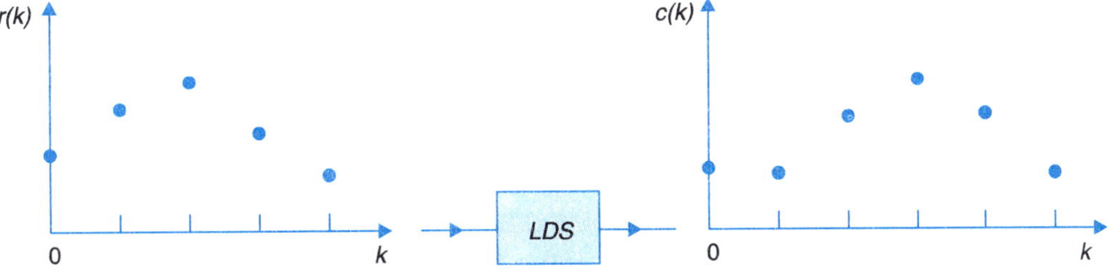

(a) Linear discrete system with arbitrary input

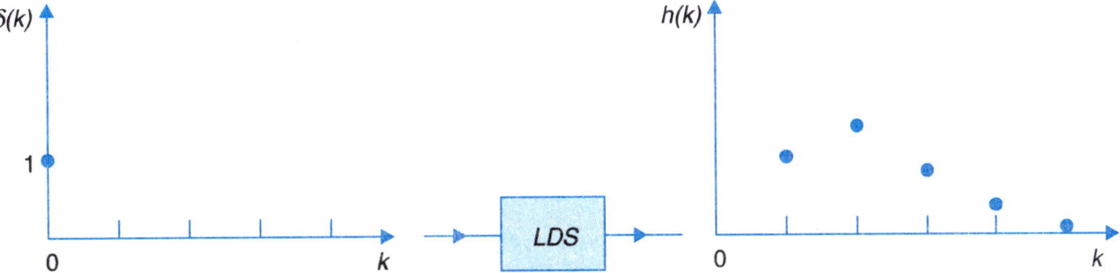

(b) Unit discrete impulse response of LDS

...(Figure 11.3 contd.)

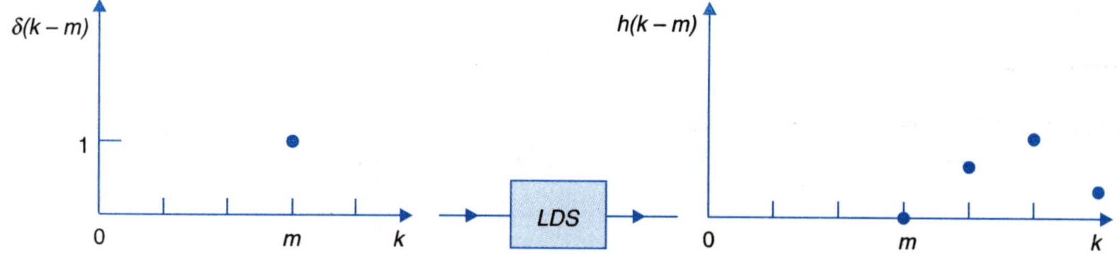

(c) Response of LDS to delayed impulse

Fig. 11.3

The input sequence $r(k)$ can be expressed as sum of impulses as

$$r(k) = r(0)\delta(k) + r(1)\delta(k-1) + r(2)\delta(k-2) + \ldots \qquad \ldots(11.39)$$

The system response to mth impulse $r(m)\delta(k-m)$ is

$$c_m(k) = r(m)h(k-m)$$

Using the principle of superposition, the response to $r(k)$ as expressed in (11.39) can be written as the sum

$$c(k) = \sum_{m=0}^{\infty} c_m(k) = \sum_{m=0}^{\infty} r(m)h(k-m) \qquad \ldots(11.40)$$

The sum (11.40) is known as the *discrete convolution* and is symbolically indicated as

$$c(k) = r(k)\cdot h(k) \qquad \ldots(11.41)$$

The sum (11.40) need not be carried beyond $m = k$ as $h(k-m) = 0$ for $m > k$, the system being casual, that is,

$$c(k) = \sum_{m=0}^{k} r(m)h(k-m)$$

Let $j = k - m$, so that

$$c(k) = \sum_{j=k}^{0} r(k-j)h(j)$$

Reversing the order of summation

$$c(k) = \sum_{j=0}^{k} h(j)r(k-j)$$

Since $r(k) = 0$ for $k < 0$, *i.e.*, the input sequence exists for positive k only, we can write

$$c(k) = \sum_{j=0}^{k} h(j)r(k-j) \qquad \ldots(11.42)$$

It therefore follows that

$$c(k) = r(k)\cdot h(k) = h(k)\cdot r(k) \qquad \ldots(11.43)$$

which means that the convolution sum is *commutative.*

The operation (11.43) to obtain the output sequence is indicated by the block diagram of Fig. 11.14.

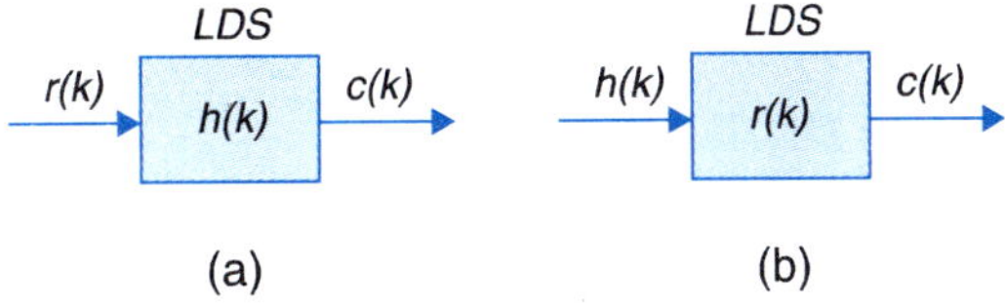

Fig. 11.14

Let us take the z-transform of the convolution sum (11.40)

$$Z[c(k)] = C(z) = \sum_{k=0}^{\infty} c(k) z^{-k}$$

or
$$C(z) = \sum_{k=0}^{\infty} \left[\sum_{m=0}^{\infty} h(k-m)\, r(m) \right] z^{-k}$$

Interchanging the order of summation,

$$C(z) = \sum_{m=0}^{\infty} r(m) \sum_{k=0}^{\infty} h(k-m) z^{-k}$$

Let $j = k - m$, then
$$C(z) = \sum_{m=0}^{\infty} r(m) \sum_{j=-m}^{\infty} h(j) z^{-j-m}$$

$$= \sum_{m=0}^{\infty} r(m)\, z^{-m} \sum_{j=0}^{\infty} h(j) z^{-j}$$

$$= R(z) H(z) \qquad \text{...(11.44)}$$

This is an important result wherein

$$H(z) = \frac{C(z)}{R(z)} \qquad \text{...(11.45)}$$

is interpreted as the z transfer function or pulse transfer function of the LDS as shown by the block diagram of Fig. 11.15. The power of the z-transform lies in the fact that the convolution sum becomes a simple multiplication in the z-domain, parallel to the case of the s-domain in linear continuous-time systems.

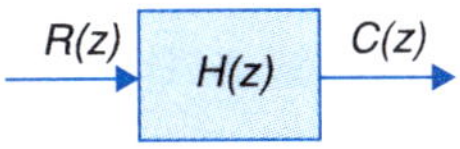

R(z) → H(z) → C(z)

Fig. 11.15. The block diagram of LDS in the z-domain.

As for unit impulse input $\delta(k)$,

$$R(z) = Z[\delta(k)] = 1,$$

the z-transfer function can be interpreted as the z-transform of the unit impulse response, *i.e.*,

$$C(z) = H(z)$$

which is again parallel to the s-domain case.

When two or more LDS are connected in tandem, their individual z-transfer functions are multiplied to obtain the overall transfer function. This is shown by the block diagram of Fig. 11.16.

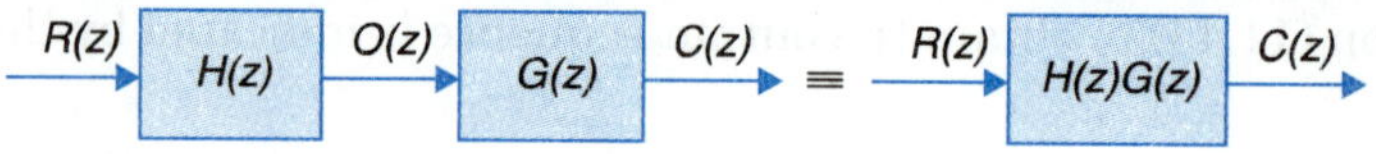

R(z) → H(z) → O(z) → G(z) → C(z) ≡ R(z) → H(z)G(z) → C(z)

Fig. 11.16

Linear Difference Equations (LDE)

Consider a linear, time-invariant discrete system decribed by the difference equation (eqn. (11.17))

$$c(k) + a_n c(k-1) + \dots a_n c(k-n)$$
$$= b_0 r(k) + b_1 r(k-1) + \dots + b_n r(k-n) \qquad \text{...(11.46)}$$

z-transforming each term of eqn. (11.46) by using the shifting theorem of enq. (11.31) (it assumes that $c(k) = 0$ and $r(k) = 0$ for $k < 0$, *i.e.*, zero initial conditions and causal input) and employing the linearity property, we have

$$C(z) + a_1 z^{-1} C(z) + \dots + a_n z^{-n} C(z)$$
$$= b_0 R(z) + b_1 z^{-1} R(z) + \dots + b_n z^{-n} R(z)$$

which can be written in the transfer function form as

$$\frac{C(z)}{R(z)} = \frac{b_0 + b_1 z^{-1} + \dots + b_n z^{-n}}{1 + a_1 z^{-1} + \dots + a_n z^{-n}}$$

Here the z-transfer function of the system described by the difference equation (11.46) is

$$H(z) = \frac{b_0 z^n + b_1 z^{n-1} + \dots + b_n}{z^n + a_1 z^{n-1} + \dots + a_n} \qquad \text{...(11.47)}$$

It is obvious that

$$H(z) = Z[c(k) \,/_{r(k) = \delta(k)}]$$

If instead, the form of difference equation (11.16) is employed, the shifting theorem of eqn. (11.30) will have to be used to obtain the transfer function. The effect of initial conditions is now immediately incorporated. The fact that these alternative forms are dynamically equivalent for zero initial conditions, is illustrated in Example 11.6.

Example 11.5: Determine the impulse response (weighting sequence) of the LDS described by

$$c(k) - \alpha c(k-1) = r(k)$$

Taking the z-transform

$$C(z) - \alpha z^{-1} C(z) = R(z)$$

Note that initial condition have not appeared by the use of the shifting property (11.31) which already assumes zero initial conditions.

Now $$H(z) = \frac{C(z)}{R(z)} = \frac{z}{z-\alpha}$$

The weighting sequence is given by

$$h(k) = Z^{-1}[H(z)] = Z^{-1}\left[\frac{z}{z-\alpha}\right]$$

Looking up Table 11.2, $h(k) = \alpha^k ; k \geq 0$

$$= 0 \; ; k < 0.$$

Example 11.6: Find the z transform of the output for the LDS described by

$$c(k+2) + a_1 c(k+1) + a_2 c(k)$$
$$= b_0 r(k+2) + b_1 r(k+1) + b_2 r(k) \qquad \text{...(11.48)}$$

Taking the z-transform by use of (11.30), we have

$$(z^2 + a_1 z + a_2)C(z) - c(0)z^2 - c(1)z - a_1 c(0)z$$
$$= (b_0 z^2 + b_1 z + b_2)R(z) - b_0 r(0)z^2 - b_0 r(1)z - b_1 r(0)z$$

$$C(z) = \frac{b_0 z^2 + b_1 z + b_2}{z^2 + a_1 z + a_2} R(z)$$

or

$$+ \frac{z^2[c(0) - b_0 r(0)] + z[c(1) + a_1 c(0) - b_1 r(0) - b_0 r(1)]}{z^2 + a_1 z + a_2} \qquad \text{...(11.49)}$$

$c(k)$ could be otained by taking the inverse z-transform of eqn. (11.49). Note that the first term of (11.49) incorporates the response due to input $r(k)$ and the second term gives the initial condition response.

If $c(k) = 0$ and $r(k) = 0$ for $k < 0$, we have from eqn. (11.48),

$$k = -2 \; ; c(0) = b_0 r(0)$$
$$k = -1 \; ; c(1) + a_1 c(0) = b_0 r(1) + b_1 r(0)$$

Hence eqn. (11.49) reduces to

$$C(z) = \frac{b_0 z^2 + b_1 z + b_2}{z^2 + a_1 z + a_2} R(z)$$

or

$$H(z) = \frac{C(z)}{R(z)} = \frac{b_0 z^2 + b_1 z + b_2}{z^2 + a_1 z + a_2}$$

which is of the same form as (11.47).

Note that $H(z)$ could be obtained directly by z-transforming the given difference equation ignoring initial conditions.

11.7 THE INVERSE z-TRANSFORM AND RESPONSE OF LINEAR DISCRETE SYSTEMS

For obtaining the dynamic response of LDS via the z-transform method, the final step is to obtain the inverse transform of the output. We shall study here basically two methods. The third method of inversion integral is beyond the scope of this book [2].

Power Series Method

Recalling the definition of the z-transform,

$$Z[f(k)] = F(z) = \sum_{k=0}^{\infty} f(k) z^{-k}$$

and the fact that $F(z)$ can mostly be expressed in the ratio of polynomial form

$$F(z) = \frac{b_0 z^m + b_1 z^{m-1} + \ldots + b_m}{z^n + a_1 z^{n-1} + \ldots a_n};\ m \le n \text{ (for causal systems)} \qquad \ldots(11.50)$$

we can immediately recognize the sequence $f(k)$ if $F(z)$ can be written in form of series with increasing powers of z^{-1}. This is easily carried out by dividing the numerator by the denominator.

Example 11.7: Determine the first few terms of the sequence $f(k)$ when

$$F(z) = \frac{z^2 + z}{z^2 + 2z + 1}$$

Dividing out the numerator by the denominator, we obtain

$$F(z) = 1 + 3z^{-1} + 5z^{-2} + 7z^{-3} + 9z^{-4} + \ldots$$

Hence

$$f(0) = 1$$
$$f(1) = 3$$
$$f(2) = 5$$
$$f(3) = 7$$
$$f(4) = 9$$
$$\ldots \quad \ldots$$

It may be recognized that while the power series method gives $f(k)$ for any k, if does not help us to get the general form of $f(k)$. It is therefore useful in numerical studies and not analytical studies.

It is also to be noted that $f(0) = 0$ if in the ratio polynomial $m < n$ (it immediately follows from the initial value theorem).

Partial Fraction Expansion Method

The inverse z-transform of $F(z)$; $m \le n$, as in (11.50), can be obtained by partial fractioning it. For $m = n$ case, divide out the numerator polynomial by the denominator polynomial so as to separate the constant term, *i.e.*,

$$F(z) = d_0 + \frac{N(z)}{M(z)}$$

where the polynomial $N(z)$ is of order less than that of $M(z)$. Assuming distinct poles of $F(z)$, we can expand $F(z)$ into partial fractions as (see Appendix I)

$$F(z) = d_0 + \frac{A_1}{z - a_1} + \frac{A_2}{z - a_2} + \ldots + \frac{A_n}{z - a_n} \qquad \ldots(11.51)$$

Now

$$\frac{1}{z - a_i} = z^{-1}\left(\frac{z}{z - a_i}\right) \leftrightarrow (a_i)^{k-1};\ k \ge 1$$

Hence taking the inverse z-transform of (11.51), we have

$$f(k) = d_0\delta(k) + [A_1(a_1)^{k-1} + A_2(a_2)^{k-1} + \ldots + A_n(a_n)^{k-1}];\ k \ge 1 \qquad \ldots(11.52)$$

In case of repeated roots, we get factors of the type

$$\frac{1}{(z-a)^2} = \frac{1}{a} z^{-1}\left[\frac{az}{(z-a)^2}\right] \leftrightarrow (k-1)a^{k-2};\ k \ge 1$$

If $\frac{N(z)}{M(z)}$ is of the form

$$\frac{zN'(z)}{M(z)}$$

It can be expanded as

$$\frac{zN'(z)}{M(z)} = z\left[\frac{A_1'}{z-a_1} + \frac{A_2'}{z-a_2} + \ldots + \frac{A_n'}{z-a_n}\right]$$

$$= A_1'\frac{z}{z-a_1} + A_2'\frac{z}{z-a_2} + \ldots + A_n'\frac{z}{z-a_n}$$

The inverse z-transform now has the form

$$f(k) = A_1'(a_1)^k + A_2'(a_2)^k + \ldots + A_n'(a_n)^k \,;\, k \geq 0$$

Both the methods are illustrated in Example 11.8.

Example 11.8: Find the inverse z-transform of

$$\frac{4z^2 - 2z}{z^3 - 5z^2 + 8z - 4}$$

$$\frac{4z^2 - 2z}{z^3 - 5z^2 + 8z - 4} = \frac{4z^2 - 2z}{(z-1)(z-2)^2} = \left.\frac{4z^2 - 2z}{(z-2)^2}\right|_{z=1} \frac{1}{(z-1)}$$

$$+ \left.\frac{4z^2 - 2z}{(z-1)}\right|_{z=2} \frac{1}{(z-2)^2} + \left.\frac{d}{dz}\left[\frac{4z^2 - 2z}{z-1}\right]\right|_{z=2} \frac{1}{(z-2)}$$

$$= \frac{2}{(z-1)} + \frac{12}{(z-2)^2} + \frac{2}{(z-2)}$$

Taking the inverse z-transform, we have

$$f(k) = 2(1)^{k-1} + 2(2)^{k-1} + 6(k-1)(2)^{k-1};\, k \geq 1$$

Alternatively

$$z\left[\frac{4z-2}{(z-1)(z-2)}\right] = z\left[\left.\frac{4z-2}{(z-2)^2}\right|_{z=1} \frac{1}{z-1} + \left.\frac{4z-2}{z-1}\right|_{z=2} \frac{1}{(z-2)^2} + \left.\frac{d}{dz}\left(\frac{4z-2}{z-1}\right)\right|_{z=2} \frac{1}{z-2}\right]$$

$$= 2\frac{z}{z-1} + 6z^{-1}\frac{z^2}{(z-2)^2} - 2\frac{z}{z-2}$$

Taking the inverse z-transform, we get

$$f(k) = 2(1)^{k-1} + 6(k-1)(2)^{k-1} + 2(2)^{k-1} \,;\, k \geq 1$$

It can be easily checked that the two forms of $f(k)$ are equivalent.

Example 11.9: The z-transform for the output in Example 11.2 was given in eqn. (11.32). Assuming zero initial conditions, we have

$$C(z) = \frac{AT}{z-(1-AT)} R(z)$$

For discrete step input,

$$R(z) = \frac{z}{z-1}$$

$$\therefore \qquad C(z) = \frac{ATz}{[z-(1-AT)](z-1)} = \frac{z}{z-1} - \frac{z}{z-(1-AT)}$$

Taking the inverse z-transform

$$c(k) = 1 - (1 - AT)^k$$

Example 11.10: Solve the difference equation

$$x(k+2) - 3x(k+1) + 2x(k) = 4^k \,;\, x(0) = 0,\, x(1) = 1$$

Taking the z-transform of the difference equation,

$$[z^2X(z) - 3x(1) - z^2x(0)] - 3[zX(z) - zx(0)] + 2X(z) = \frac{z}{z-4}$$

$$(z^2 - 3z + 2)X(z) = z^2x(0) + z[x(1) - 3x(0)] + \frac{z}{z-4}$$

or

$$X(z) = \frac{z^2x(0) + z[x(1) - 3x(0)]}{z^2 - 3z + 2} + \frac{z}{(z-4)(z^2 - 3z + 2)}$$

Substituting the initial conditions

$$X(z) = \frac{z}{(z-1)(z-2)} + \frac{z}{(z-1)(z-2)(z-4)}$$

$$= \left[-\frac{z}{z-1} + \frac{z}{z-2}\right] + \left[\frac{1}{3}\frac{z}{z-1} - \frac{1}{2}\frac{z}{z-2} + \frac{1}{6}\frac{z}{z-4}\right]$$

Taking the inverse z-transform,

$$x(k) = \underbrace{\left[-1 + (2)^k\right]}_{\text{Initial condition response}} + \underbrace{\left[\frac{1}{3} - \frac{1}{2}(2)^k + \frac{1}{6}(4)^k\right]}_{\text{Forcing function response}}$$

11.8 THE z-TRANSFORM ANALYSIS OF SAMPLED-DATA CONTROL SYSTEMS

Sampled-data control systems are in fact hybrid systems as they have both discrete-time and continuous-time signals. Special techniques are therefore required for analysis of these systems.

Analysis of Sampler and Zero-Order Hold

Figure 11.17 shows a sampler with zero-order hold (ZOH). As ZOH holds the input signal value for a period T, it means that for a short duration (Δ) input pulse, it produces an output pulse of duration T, the sampling period. This is illustrated in Fig. 11.18.

The ZOH output pulse appearing at kT instant can be expressed as

$$i(kT)[u(t - kT) - u(t - \overline{k+1}T)]$$

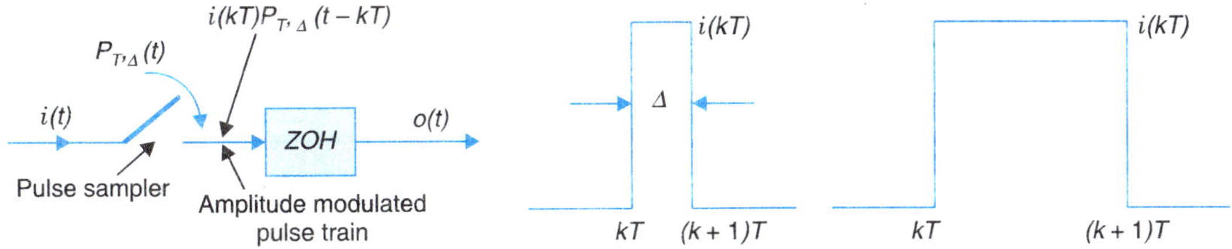

Fig. 11.17. Pulse sampler with ZOH.

Fig. 11.18. ZOH input and output pulses.

We can therefore write the ZOH output as

$$O(t) = \sum_{k=0}^{\infty} i(kT)[u(t - kT) - u(t - \overline{k+1}T)] \qquad ...(11.53)$$

Taking the Laplace transform, we have

$$O(s) = \sum_{k=0}^{\infty} i(kT)\left[\frac{e^{-skT} - e^{-s(k+1)T}}{s}\right]$$

$$= \left(\frac{1 - e^{-sT}}{s}\right)\sum_{k=0}^{\infty} i(kT)e^{-skT} \qquad ...(11.54)$$

We can give a new input-ouput interpretation to eqn. (11.54). Taking the inverse Laplace transform of the infinite sum, we get (see Table 1.1 in Appendix I)

$$\mathcal{L}^{-1}\left[\sum_{k=0}^{\infty} i(kT)e^{-skT}\right] = \sum_{k=0}^{\infty} i(kT)\delta(t - kT) = i(t)\delta_T(t) \qquad ...(11.55)$$

Thus the output $o(t)$ of pulse sampler and ZOH can be produced by *impulse sampled* $i(t)$ when passed through a transfer function

$$G_0(s) = \frac{1 - e^{-sT}}{s} \qquad ...(11.56)$$

This is illustrated in Fig. 11.19. Figure 11.17 and 11.19 represent equivalent operations, but the operation of Fig. 11.19 offers ease of analysis.

It may be noted here that Figs. 11.17 and 11.19 are exactly equivalent and no approximation of any kind has been made here.

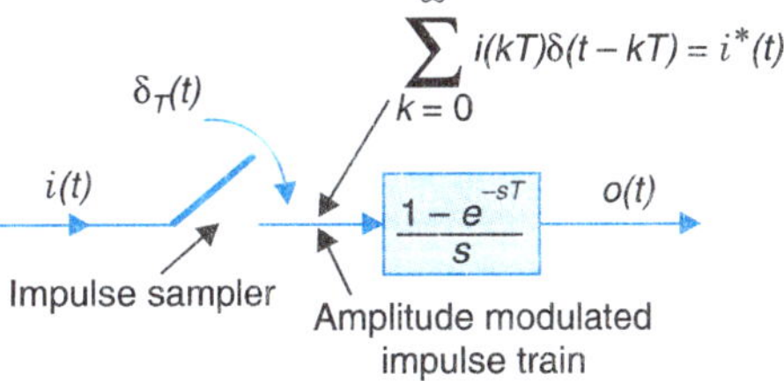

Fig. 11.19. Equivalent representation of pulse sampler and ZOH.

Analysis of Systems with Impulse Sampling

In sampled-data systems, our interest generally is to obtain the values of the output at sampling instants. Also we have seen above how pulse sampling and ZOH can be replaced by impulse sampling and $G_0(s)$. Consider now a linear continuous system fed from an impulse sampler as shown in Fig. 11.20 (*a*). The output signal $c(t)$ is read off at discrete synchronous sampling instants (kT) by means of a mathematical sample $T(M)$ or what we can also call a read-out sampler.

The continuous output of the system in the s-domain is given by (Fig. 11.20 (b))

$$C(s) = H(s)R^*(s)$$

where $R^*(s) = \mathcal{L}[r^*(t)]$ = the Laplace transform of the impulse sampled input signal.

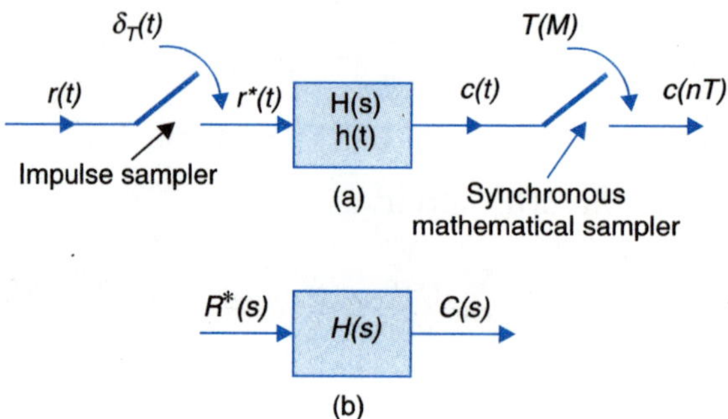

Fig. 11.20. Linear continuous system with impulse sampled input.

$\mathcal{L}^{-1}[H(s)] = h(t)$ = the impulse response of the linear continuous system.

Now $$r^*(t) = \sum_{k=0}^{\infty} r(kT)\delta(t - kT)$$

Using superposition

$$c(t) = \sum_{k=0}^{\infty} r(kT)h(t - kT)$$

$$\therefore \qquad c(nT) = \sum_{k=0}^{\infty} r(kT)h(nT - kT) \qquad ...(11.57)$$

We can at once recognize it as discrete convolution of eqn. (11.40).

Taking the z-transform of both sides (see eqn. (11.44)),

$$C(z) = R(z)H(z) \qquad ...(11.58)$$

where $$H(z) = Z[h(kT)]$$

$$= Z\{\mathcal{L}^{-1}[H(s)] \,|_{t = kT}\} \qquad ...(11.59)$$

The z-transforming operation of (11.59) is commonly indicated as

$$H(z) = Z\,[H(s)] \qquad ...(11.60)$$

It follows from eqn. (11.58) that the time-domain block diagram of Fig. 11.20 (a) becomes the z-domain block diagram of Fig. 11.21.

We could of course directly draw the z-domain block diagram of Fig. 11.21 from the s-domain block diagram of Fig. 11.20 (b) which implies that

$$C(s) = H(s)R^*(s) \longleftrightarrow C(z) = H(z)R(z) \qquad ...(11.61)$$

or $$Z[H(s)R^*(s)] = H(z)R(z) \qquad ...(11.62)$$

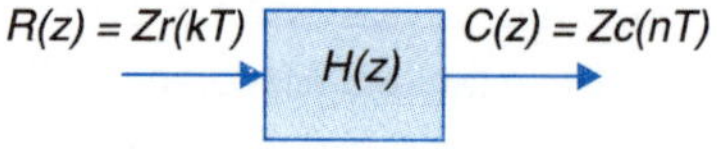

R(z) = Zr(kT) → H(z) → C(z) = Zc(nT)

Fig. 11.21. The z-transform equivalent of Fig. 11.20 (a).

To Obtain $H(z)$

The direct approach to obtain

$$H(z) = Z[H(s)]$$

involves the following steps.

1. Find $h(t) = \mathcal{L}^{-1}[H(s)]$
2. Find $h(kT)$
3. Find $Z[h(kT)]$

Example 11.11:

$$H(s) = \frac{a}{s(s+a)} = \frac{1}{s} - \frac{1}{s+a}$$

$$h(t) = \mathcal{L}^{-1}\left[\frac{1}{s} - \frac{1}{s+a}\right] = \left[1 - e^{-aT}\right]u(t)$$

$$h(kT) = 1 - e^{-akT};\ k \geq 0$$

$$= 0\ ;\ k < 0$$

The z-transform of this equation using Table 11.2 is

$$H(z) = \frac{z}{z-1} - \frac{z}{z-e^{-aT}} = \frac{z\{1-e^{-aT}\}}{(z-1)\{z-e^{-aT}\}}$$

Single factor building blocks of the Laplace and z-transform pairs are given in Table 11.2 (b) Expanding any $H(s)$ into partial fraction, $H(z)$ can be found by the use of this table.

Consider now the case of a continuous system $H(s)$ with continuous input $r(t)$ as in Fig. 11.22 (a). Imagine that we are interested to read the values of the continuous output at sampling instants *i.e.*, imagine a mathematical sampler at the output state. We can equivalently represent it as a block $H(s)R(s)$ with impulse input $\delta(t)$ as in Fig. 11.22 (b). Now the input and therefore the output does not change by imagining a fictitious impulse sampler through which $\delta(t)$ is applied to $H(s)R(s)$ as in Fig. 11.22 (c). On application of (11.58), we can at once write

$$C(z) = Z[H(s)R(s)]Z[\delta(k)] = Z[H(s)R(s)] = HR(z) \quad \text{...(11.63)}$$

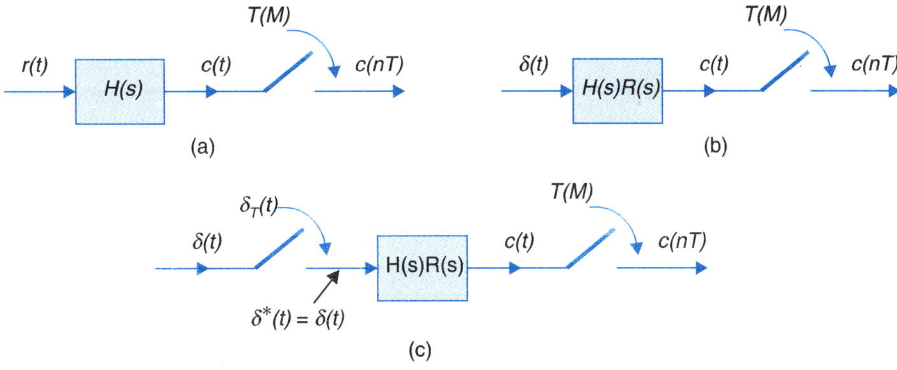

Fig. 11.22

The following s-domain and z-domain relationship follows from above:

$$C(s) = H(s)R(s) \longleftrightarrow C(z) = HR(z) \quad ...(11.64)$$

or

$$Z[H(s)R(s)] = HR(z) \quad ...(11.65)$$

When impulse sampled input is applied to two (or more) s-domain transfer functions in tandem as in Fig. 11.23 (a), we get the z-domain transfer function as in Fig. 11.23 (b) where

$$H(z) = H_1H_2(z) \neq H_1(z)H_2(z) \quad ...(11.66)$$

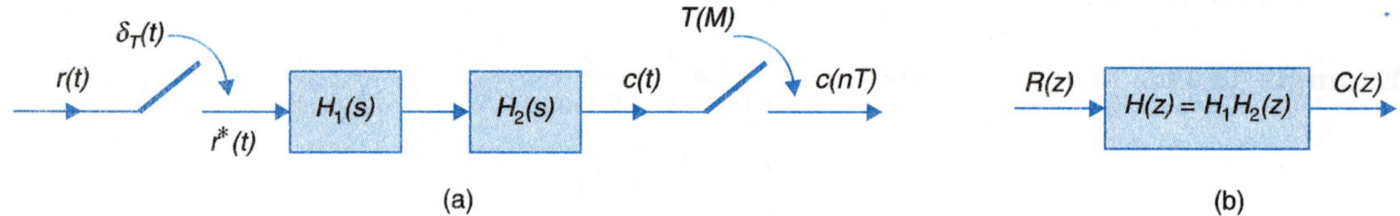

Fig. 11.23

However, if the blocks of Fig. 11.23 (a) are separated by an impulse sampler as in Fig. 11.24, it obviously follows that

$$H(z) = H_1(z)H_2(z) \quad ...(11.67)$$

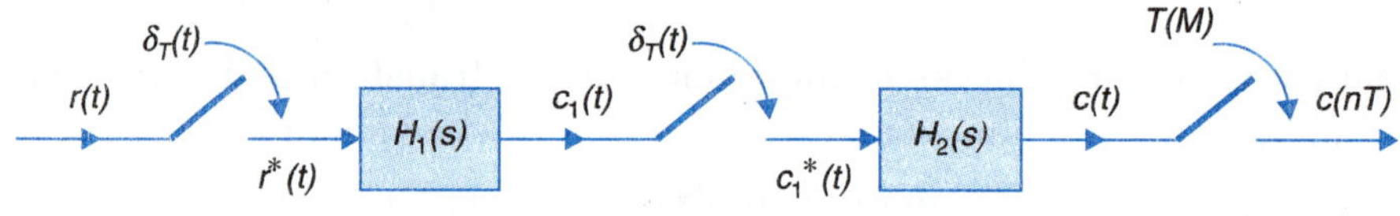

Fig. 11.24

The z-Transform Analysis of Sampled-Data Systems

Consider an open-loop system wherein a linear continuous part $G(s)$ is fed from a sampler and ZOH as shown in Fig. 11.25 (a). Its equivalent impulse sampled system is shown in Fig. 11.25 (b), while the corresponding z-domain block diagram is drawn in Fig. 11.25 (c). The z-transfer function of the system is given by

$$Z[G_0(s)C(s)] = Z\left[\frac{1-e^{-sT}}{s}G(s)\right] = Z\left[\frac{G(s)}{s} - \frac{e^{-sT}G(s)}{s}\right]$$

Let

$$\mathcal{L}^{-1}\left[\frac{G(s)}{s}\right] = g_1(t)$$

$$\mathcal{L}^{-1}\left[\frac{e^{-sT}G(s)}{s}\right] = g_1(t-T)$$

Hence

$$Z\left[\frac{e^{-sT}G(s)}{s}\right] = Z[(g_1(kT-T)] = z^{-1}Z[g_1(kT)]$$

$$= z^{-1}Z\left[\frac{G(s)}{s}\right]$$

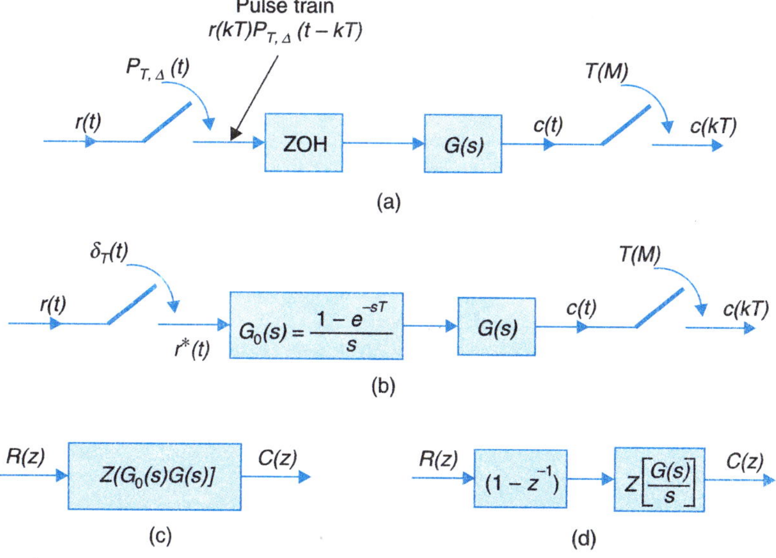

Fig. 11.25

It therefore follows that

$$Z[G_0(s)G(s)] = (1 - z^{-1})Z\left[\frac{G(s)}{s}\right] \qquad ...(11.68)$$

as depicted in Fig. 11.25 (*d*).

For example, if $$G(s) = \frac{a}{s+a}$$

then $$Z\left[\frac{G(s)}{s}\right] = Z\left[\frac{a}{s(s+a)}\right]$$

This result has already been obtained in Example 11.11. We have therefore

$$Z[G_0(s)G(s)] = \frac{z\{1-e^{-aT}\}}{\{z-e^{-aT}\}}$$

In the absence of ZOH in Fig. 11.25 (*a*), the linear continuous part of the system is fed from a train of pulses as shown in Fig. 11.26 (*a*). The *z*-transform though no longer valid can still be used if the pulse train is approximated as an impulse train as in Fig. 11.26 (*b*). It immediately follows that

$$C(z) = Z[G(s)]\, \Delta R(z) = \Delta G(z)R(z) = G'(z)R(z) \qquad ...(11.69)$$

It shall be assumed now onwards that wherever an impulse approximation of a pulse sampler is employed, the gain Δ is already absorbed in the transfer function.

Consider now the basic sampled-data feedback control system whose block diagram is depicted in Fig. 11.27 (*a*). In terms of impulse sampling this diagram can be redrawn as in Fig. 11.27 (*b*). In the *z*-transform form we can write

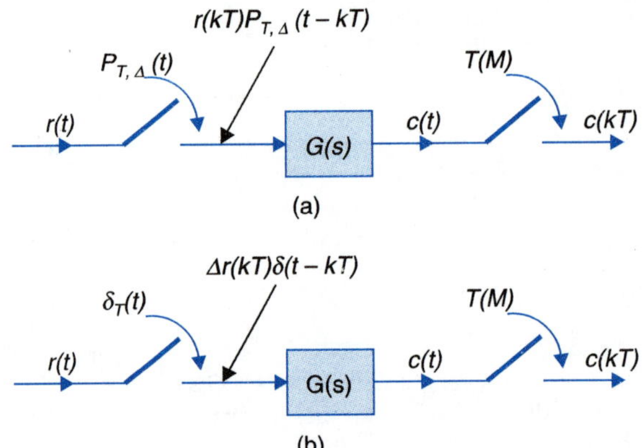

Fig. 11.26

$$C(z) = Z[G_0(s)G(s)]E(z) \qquad \text{...(11.70)}$$

$$B(z) = Z[G_0(s)G(s)H(s)]E(z) \qquad \text{...(11.71)}$$

$$e(t) = r(t) - b(t)$$

or
$$e(kT) = r(kT) - b(kT)$$

or
$$E(z) = R(z) - B(z) \qquad \text{...(11.72)}$$

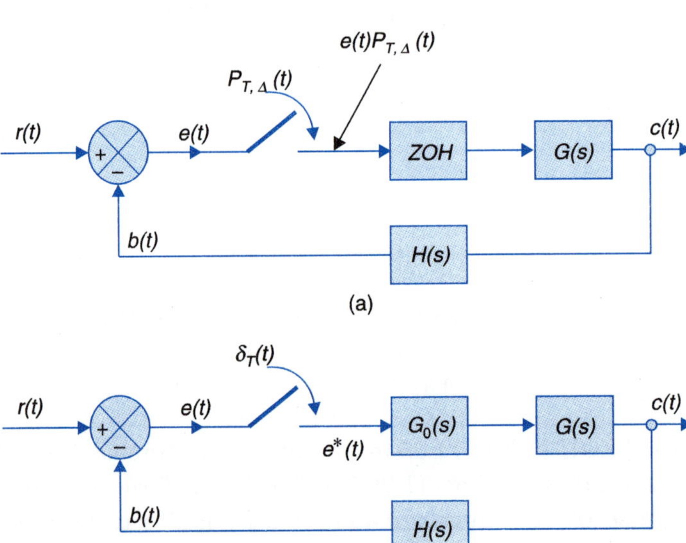

Fig. 11.27

From eqns. (11.70)-(11.72) it immediately follows that

$$\frac{C(z)}{R(z)} = \frac{Z[G_0(s)G(s)]}{1 + Z[G_0(s)G(s)H(s)]} = \frac{G_0G(z)}{1 + G_0GH(z)} \qquad \text{...(11.73)}$$

Having become familiar with the technique, from now onwards we may directly draw the block diagrams of sampled-data systems in terms of impulse sampler and $G_0(s)$.

Consider now a sampled-data system with sampler and ZOH in the feedback path as shown in Fig. 11.28. The signal $e(t)$ is contributed by impulse sampled $c(t)$ and continuous input signal $r(t)$. In s-domain, we can write

$$E(s) = -G_0(s)H(s)C^*(s) + R(s)$$
$$C(s) = -G_0(s)G(s)H(s)C^*(s) + G(s)R(s)$$

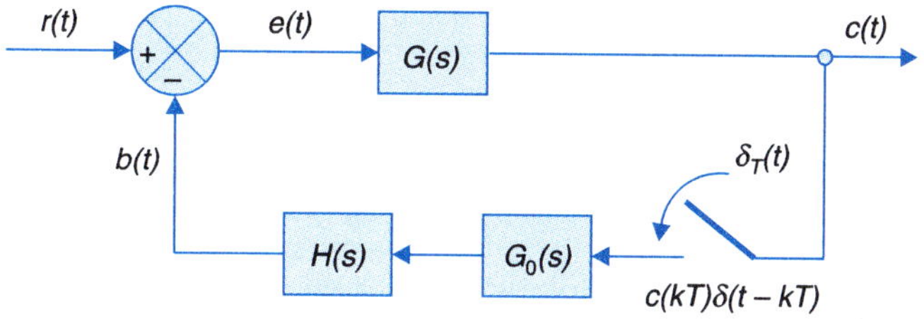

Fig. 11.28

Carrying out z-transformation using eqns. (11.58) and (11.62), we get

$$C(z) = Z[-G_0(s)H(s)G(s)]C(z) + Z[R(s)G(s)]$$

or $$C(z)[1 + G_0HG(z)] = RG(z)$$

$$\therefore \quad C(z) = \frac{RG(z)}{1 + G_0HG(z)} \qquad \text{...(11.74)}$$

Since in the z-transform analysis of sampled-data systems we are dealing merely with the discrete values of the impulse at the impulse sampler output, we can afford to be careless to the extent so as to write $c(nT)$ as the sampler output in Fig. 11.28 rather than the rigidly correct expression of $c(kT)\,\delta(t - kT)$. This is what we shall write from now onwards.

The z-Transfer Function of a Digital Computer

Let Fig. 11.25 (a) be modified by the inclusion of a digital computer or digital network which modifies the strength of the pulses from the sampler in accordance with a difference equation of the type (11.46). This is shown by Fig. 11.29 (a). The action of the digital computer in the z-domain is therefore that of a z-transfer function

$$D(z) = \frac{V(z)}{R(z)}$$

of the type of (11.47). The rest of the system behaves in the manner already illustrated in Fig. 11.25. Thus the system can be represented by the block diagram of Fig. 11.29 (b). It may be noted that $D(z)$ is an independent block representing the digital computer (or network).

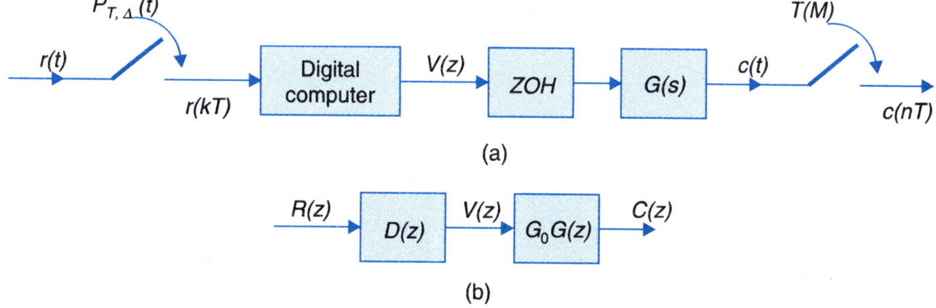

Fig. 11.29. Sampled-data system with digital computer.

If a digital computer is included in the forward path of the basic sampled data control system, we can represent it in the form of a block diagram of Fig. 11.30. It easily follows from this figure that

$$C(z) = D(z)Z[G_0(s)G(s)]E(z)$$

$$B(z) = D(z)Z[G_0(s)G(s)H(s)]E(z)$$

$$E(z) = R(z) - B(z)$$

$$\frac{C(z)}{R(z)} = \frac{D(z)Z[G_0(s)G(s)]}{1 + D(z)Z[G_0(s)G(s)H(s)]} = \frac{D(z)G_0G(z)}{1 + D(z)G_0GH(z)} \quad ...(11.75)$$

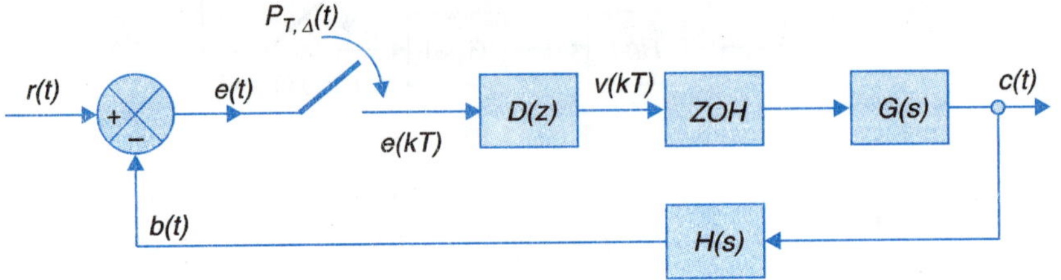

Fig. 11.30

Inclusion of $D(z)$, the digital computer transfer function, provides us with the technique of compensating a sampled-data control system. This will be discussed later in this chapter.

Example 11.12: For the sampled-data control system shown in Fig. 11.31. find the output $c(k)$ for $r(t)$ = unit step.

$$\frac{C(z)}{R(z)} = \frac{Z[G_0G(s)]}{1 + Z[G_0GH(s)]}$$

$$Z[G_0G(s)] = (1 - z^{-1})Z\left[\frac{1}{s(s+1)}\right]$$

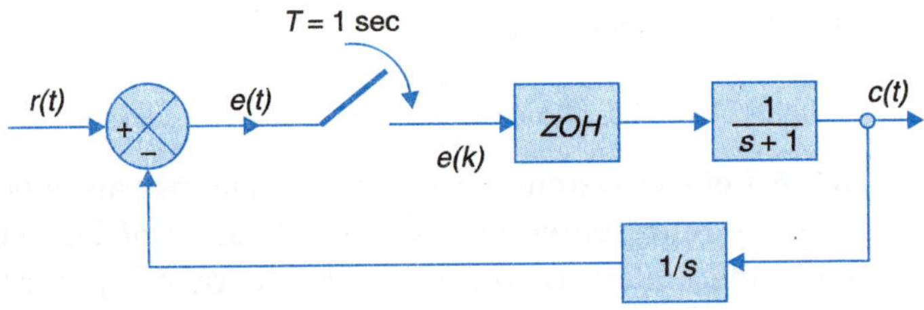

Fig. 11.31

Consulting Table 11.3,

$$Z\left[\frac{1}{s(s+1)}\right] = \frac{z(1-e^{-1})}{(z-1)(z-e^{-1})}$$

Therefore,
$$Z[G_0G(s)] = \frac{1-e^{-1}}{z-e^{-1}}$$

Now
$$Z[G_0GH(s)] = (1 - z^{-1})Z\left[\frac{1}{s^2(s+1)}\right]$$
$$= (1 - z^{-1})Z\left[\frac{1}{s^2} - \frac{1}{s} + \frac{1}{s+1}\right]$$
$$= (1 - z^{-1})\left[\frac{z}{(z-1)^2} - \frac{z}{z-1} + \frac{z}{z-e^{-1}}\right]$$
$$= \frac{e^{-1}z - 2e^{-1} + 1}{(z-1)(z-e^{-1})}$$

It then follows:
$$\frac{C(z)}{R(z)} = \frac{(1-e^{-1})(z-1)}{z^2 - z + (1-e^{-1})}$$

For unit step input,
$$R(z) = \frac{z}{z-1}$$
$$C(z) = \frac{(1-e^{-1})z}{z^2 - z + (1-e^{-1})}$$
$$= \frac{0.632z}{(z - 0.5 - j0.62)(z - 0.5 + j0.62)}$$
$$= \frac{-j0.51z}{(z - 0.5 - j0.62)} + \frac{j0.51z}{(z - 0.5 + j0.62)}$$

Taking the inverse z-transform, we get
$$c(k) = -j0.51(0.5 + j0.62)^k + j0.51(0.5 - j0.62)^k$$

Converting to polar form and combining,
$$c(k) = -\,1.02(0.785)^k \sin(51.5°k)$$

Example 11.13: Consider the system shown in Fig. 11.32 (a). Derive the difference equation describing the system dynamics when the input voltage applied is piecewise constant, *i.e.*,
$$e(t) = e(kT) \text{ for } kT \le t \le (k+1)\,T\,;\ T = 1 \text{ sec}$$

Also obtain the values of output voltage at sampling instants if
$$e(kT) = kT \text{ for } k \ge 0$$

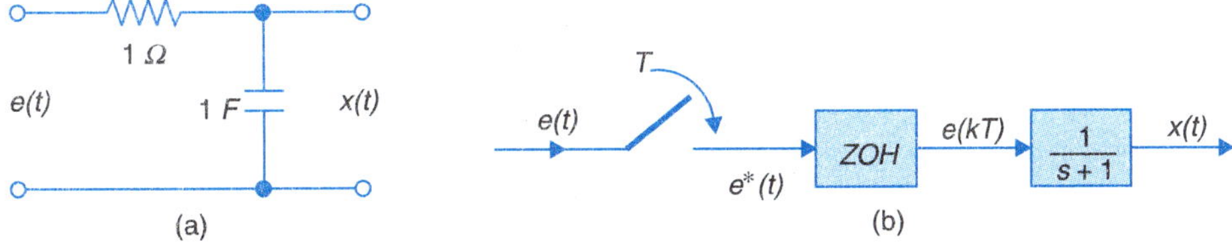

Fig. 11.32

The s-domain transfer function of the system is

$$\frac{1}{s+1}$$

Equivalently we can draw the sampled-data block diagram for the system as in Fig. 11.32 (b). Therefore,

$$X(z) = (1 - z^{-1})Z\left[\frac{1}{s(s+1)}\right]E(z)$$

$$= \frac{(1-z^{-1})z(1-e^{-1})}{(z-1)(z-e^{-1})}E(z) = \frac{1-e^{-1}}{(z-e^{-1})}E(z)$$

or
$$zX(z) - e^{-1}X(z) = (1 - e^{-1})\,E(z)$$

Taking the inverse z-transform

$$x[(k+1)T] = e^{-1}x(kT) + (1 - e^{-1})\,e(kT)$$

This is the desired difference equation

Now for $e(kT) = kT = k$

$$E(z) = \frac{Tz}{(z-1)^2} = \frac{z}{(z-1)^2}$$

$$\therefore \qquad X(z) = \frac{1-e^{-1}}{(z-e^{-1})} \times \frac{z}{(z-1)^2} = (1-e^{-1})\frac{z}{(z-e^{-1})(z-1)^2}$$

$$= \frac{1}{1-e^{-1}}\frac{z}{z-e^{-1}} - \frac{1}{(1-e^{-1})}\frac{z}{z-1} + \frac{z}{(z-1)^2}$$

Taking the inverse z-transform

$$x(k) = \frac{1}{1-e^{-1}}e^{-k} - \frac{1}{1-e^{-1}} + k.$$

11.9 THE z-AND s-DOMAIN RELATIONSHIP

Consider a signal $r(t)$ which has discrete values $r(kT)$ at a sampling rate $1/T$. The z-transform of these discrete values is

$$R(z) = \sum_{k=0}^{\infty} r(kT)z^{-k} \qquad \text{...(11.76)}$$

If the signal $r(t)$ is imagined to be impulse sampled at the same rate, it becomes

$$r^*(t) = \sum_{k=0}^{\infty} r(kT)\delta(t - kT) \qquad \text{...(11.77)}$$

Taking the Laplace transform, we have

$$R^*(s) = \sum_{k=0}^{\infty} r(kT)e^{-ksT} \qquad \text{...(11.78)}$$

If we let $$z = e^{sT} \text{ or } s = \frac{1}{T}\ln z \qquad ...(11.79)$$

we get $$R^*(s)\Big|_{s=\frac{1}{T}\ln z} = \sum_{k=0}^{\infty} r(kT)z^{-k} = R(z) \qquad ...(11.80)$$

Equation (11.80) exposes an interesting result–the z-transform for a set of discrete values can be obtained by imagining these to be impulses, Laplace transforming and then using the transformation (11.79).

The transformation (11.79) maps the s-plane into the z-plane. Consider the mapping of the $j\omega$-axis in the s-plane, *i.e.*,

$$z = e^{j\omega T}$$

$$= e^{j2\pi\omega/\omega_s} = 1\angle 2\pi(\omega/\omega_s) \qquad ...(11.81)$$

where $\omega_s = \frac{2\pi}{T}$, the sampling frequency.

Thus the section $(-j\omega_s/2) - 0 - (+j\omega_s/2)$ of the $j\omega$-axis maps into the unit circle in the anticlockwise direction ($-\pi$, $-\pi/2$, 0, $\pi/2$, π) as shown in Fig. 11.33. In fact every section of the $j\omega$-axis which in integral multiple of ω_s maps into the unit circle. With the direction of mapping indicated, it is then obvious that the left half s-plane can be divided into strips of width ω_s each of which maps into the interior of the unit circle. It suggests that the s-plane stability criterion of the poles of s-transfer function lying in the left half of the s-plane will become the z-plane stability criterion that all the poles of the z-transfer function would lie within the unit circle. This is shown to be so in the next article.

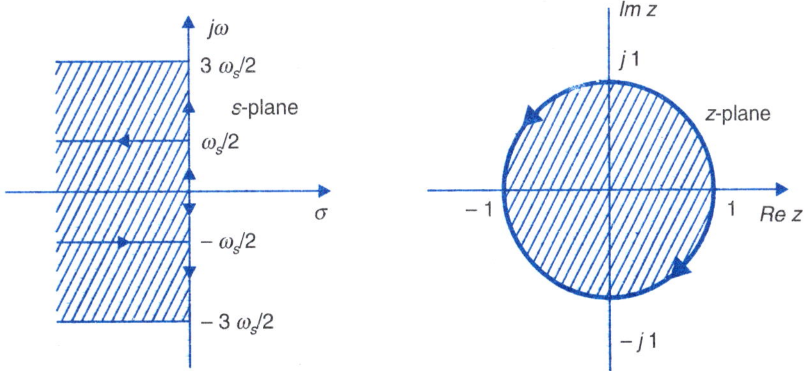

Fig. 11.33. Relationship between the s-and z-domains.

11.10 STABILITY ANALYSIS

Stability Criterion

The input-output relation of a discrete system is given by

$$C(z) = T(z)R(z) \qquad ...(11.82)$$

where $T(z)$ is the z-transfer function of the system, a rational function of two polynomials in z. As in the case of linear continuous systems, the stability is ascertained by examining the system response to impulse input, *i.e.*, $R(z) = 1$. Imagining the poles of $T(z)$ to be simple and distinct, we have

$$C(z) = T(z) = \frac{A_1}{z - a_1} + \frac{A_2}{z - a_2} + \ldots + \frac{A_n}{z - a_n} \qquad ...(11.83)$$

The corresponding discrete time sequence is

$$c(kT) = A_1(a_1)^{k-1} + A_2(a_2)^{k-1} + \ldots + A_n(a_n)^{k-1};\ k \geq 1 \qquad ...(11.84)$$

It is easily seen from eqn. (11.84), that the system response decays to zero if

$$|a_i| < 1\ ;\ i = 1, 2, \ldots, n$$

i.e., all the poles of the system's z-transfer function must lie within the unit circle in the z-plane (as was rightly guessed in Section 11.10).

The presence of complex conjugate pole pair

$$z = a_i \pm jb_i$$

contributes term in the response of the type

$$\alpha_i^k \cos (k\theta_1 + \phi_i)$$

where $\alpha_i = \sqrt{a_i^2 .\ b_i^2}$; and $\phi_i = \tan^{-1} \frac{b_i}{a_i}$ which decays if $\alpha_i < 1$, *i.e.*, the complex pole pair lies within the unit circle.

The presence of real repeated poles contributes additional terms like

$$\frac{1}{(z\ .\ a_i)^2} \longleftrightarrow (k-1)\, a^{k-2};\ k \geq 2$$

which also decay for $|a_i| < 1$.

The criterion of stability that all the poles of the z-transfer function must lie within the unit circle is therefore of general applicability.

As the poles of the closed-loop transfer function are the same as the roots of the system's characteristic equation, the system stability is determined by the roots of (see eqn. (11.73))

$$1 + Z[G_0(s)G(s)H(s)] = 0 \qquad ...(11.85)$$

for systems of Figs. 11.27 and 11.28 and by the roots of (see eqn. (11.75))

$$1 + D(z)Z[G_0(s)G(s)H(s)] = 0 \qquad ...(11.86)$$

for system of Fig. 11.30.

Methods of Stability Analysis

1. *Jury's Stability Test.* It is an algebraic criterion for determining whether or not the roots of the characteristic polynomial lie within a unit circle thereby determining system stability. Consider the characteristic polynomial

$$F_1(z) = a_n z^n + a_{n-1} z^{n-1} + \ldots + a_0 = 0\ ;\ a_n > 0 \qquad ...(11.87)$$

Like the Routh's method, the Jury's test consists of two parts: a simple test for necessary conditions and a second test for sufficiency. The necessary conditions for stability are

$$F_1(1) > 0\ ;\ (-1)^n F_1(-1) > 0 \qquad ...(11.88)$$

The sufficient conditions for stability can be established through one of the important methods.

Method : Prepare a table of coefficients* of the characteristic polynomial as below.

Row	z^0	z^1	z^2	z^3	...	z^{n-k}	...	z^{n-2}	z^{n-1}	z^n
1	a_0	a_1	a_2	a_3	...	a_{n-k}	...	a_{n-2}	a_{n-1}	a_n
2	a_n	a_{n-1}	a_{n-2}	a_{n-3}	...	a_k	...	a_2	a_1	a_0
3	b_0	b_1	b_2	b_3	...	b_{n-k}	...	b_{n-2}	b_{n-1}	
4	b_{n-1}	b_{n-2}	b_{n-3}	b_{n-4}	...	b_{k-1}	...	b_1	b_0	
5	c_0	c_1	c_2	c_3	...	c_{n-k}	...	c_{n-2}		
6	c_{n-2}	c_{n-3}	c_{n-4}	c_{n-5}	...	c_{k-2}	...	c_0		
⋮	⋮	⋮	⋮	⋮	⋮	⋮				
$2n-5$	s_0	s_1	s_2	s_3						
$2n-4$	s_3	s_2	s_1	s_0						
$2n-3$	r_0	r_1	r_2							

where

$$b_k = \begin{vmatrix} a_0 & a_{n-k} \\ a_n & a_k \end{vmatrix}$$

$$c_k = \begin{vmatrix} b_0 & b_{n-1-k} \\ b_{n-1} & b_k \end{vmatrix}$$

$$d_k = \begin{vmatrix} c_0 & c_{n-2-k} \\ c_{n-2} & c_k \end{vmatrix}$$

... ...

... ...

The sufficient conditions for stability are

$$\left.\begin{array}{l} |a_0| < |a_n| \\ |b_0| > |b_{n-1}| \\ |c_0| > |c_{n-2}| \\ \vdots \\ |r_0| > |r_2| \end{array}\right\} (n-1) \text{ constraints} \quad \text{...(11.89)}$$

Example 11.14: Consider the characteristic equation of the second-order system

Solution: $$F_1(z) = a_2 z^2 + a_1 z + a_0 = 0 \; ; \; a_2 > 0$$

The stability constraints are

$$F_1(1) = a_2 + a_1 + a_0 > 0$$

$$(-1)^n F_1(-1) = a_2 - a_1 + a_0 > 0$$

$$|a_0| < a_2.$$

Example 11.15: Consider the characteristic polynomial

Solution: $$F_1(z) = 2z^4 + 7z^3 + 10z^2 + 4z + 1$$

*The singular case where a row of zeros appears is beyond the scope of this book.

$$F_1(1) = 2 + 7 + 10 + 4 + 1 = 24 > 0 \text{ satisfied}$$

$$(-1)^4 F_1(-1) = 2 - 7 + 10 - 4 + 1 = 2 > 0 \text{ satisfied}$$

Row	z^0	z^1	z^2	z^3	z^4
1	1	4	10	7	2
2	2	7	10	4	1
3	– 3	– 10	– 10	– 1	
4	– 1	– 10	– 10	– 3	
5	8	20	20		

Employing stability constraints (11.89)

$|1| < 2$ satisfied

$|-3| > |-1|$ satisfied

$|8| > |20|$ not satisfied

The system is therefore unstable.

Example 11.16: A sampled-data control system of order one with transportation lag is shown in Fig. 11.34. Determine the condition for system stability if $\delta < T$.

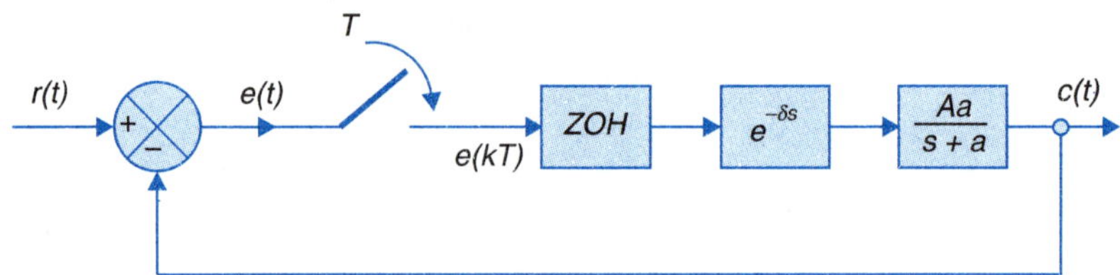

Fig. 11.34

Solution:*

$$G(s) = \left(\frac{1-e^{-sT}}{s}\right)e^{-\delta s}\left(\frac{Aa}{s+a}\right) = [e^{-\delta s} - e^{-(T+\delta)s}]\frac{Aa}{s(s+a)}$$

$$= A[e^{-\delta s} - e^{-(T+\delta)s}]\left(\frac{1}{s} - \frac{1}{s+a}\right)$$

$$g(t) = A[1 - e^{-a(t-\delta)}]u(t-\delta) - A[1 - e^{-a(t-T-\delta)}]\,u(t-T-\delta)$$

$$g(kT) = A[1 - e^{-a(kT-\delta)}]u(kT-\delta) - A[1 - e^{-a(kT-T-\delta)}]\,u(kT-T-\delta)$$

---

*As a time delay which is nonintegral multiple of the sampling period T is encountered, we have to proceed fundamentally to obtain $G(z)$. Obviously

$$G(z) \neq (1 - z^{-1})Z\left[\frac{Aae^{-\delta s}}{s(s+a)}\right]$$

Similar situation exists for time lead (non integral multiple of T (see Problem 11.19).

$$
\begin{aligned}
&= 0 \text{ for} && k = 0\\
&= A[1 - e^{-a(T-\delta)}] && \text{for } k = 1\\
&= A[1 - e^{-a(kT-\delta)}] - A[1 - e^{-a(kT-T-\delta)}] && \text{for } k \geq 2\\
&= Ae^{a\delta}\, e^{-akT}[e^{aT} - 1] && \text{for } k \geq 2
\end{aligned}
$$

$$G(z) = A[1 - e^{-a(T-\delta)}]z^{-1} + Ae^{a\delta}\,(e^{aT} - 1) \sum_{k=2}^{\infty} e^{-akT} z^{-k}$$

$$= A[1 - e^{-a(T-\delta)}]z^{-1} + Ae^{a\delta}\,(e^{aT} - 1)z^{-2} \sum_{k=2}^{\infty} e^{-akT} z^{-(k-2)}$$

Letting $k - 2 = r$, we have

$$G(z) = A[1 - e^{-a(T-\delta)}]z^{-1} + Ae^{a}\delta\,(e^{aT} - 1)z^{-2} \sum_{r=0}^{\infty} e^{-2aT} e^{-arT} z^{-r}$$

$$= A[1 - e^{-a(T-\delta)}]\frac{1}{z} + Ae^{a\delta}\,(e^{aT} - 1)e^{-2aT} \frac{1}{z(z - e^{-aT})}$$

$$G(z) = A\frac{[1 - e^{-a(T-\delta)}](z - e^{-aT}) + e^{a\delta}e^{-aT}(1 - e^{-aT})}{z(z - e^{-aT})}$$

The characteristic equation is

$$
\begin{aligned}
F_1(z) &= 1 + G(z)\\
&= z(z - e^{-aT}) + A[1 - e^{-a(T-\delta)}](z - e^{-aT}) + Ae^{-a(T-\delta)}(1 - e^{-aT})\\
&= z^2 + \{A[1 - e^{-a(T-\delta)}] - e^{-aT}\}z + Ae^{-aT}\,(e^{a\delta} - 1)
\end{aligned}
$$

As per Example 11.14, the conditions of stability for a second-order system are:

(*i*) $1 + A[1 - e^{-a(T-\delta)}] - e^{-aT} + Ae^{-a(T-\delta)} - Ae^{-aT} > 0$

or $$(1 - e^{-aT})(1 + A) > 0$$

This condition is always met for positive values of a, T and A.

(*ii*) $1 - A + Ae^{-a(T-\delta)} + e^{-aT} + Ae^{-a(T-\delta)} - Ae^{-aT} > 0$

or $$(1 + e^{-aT})(1 - A) + 2Ae^{-a(T-\delta)} > 0$$

or $$A < \frac{1 + e^{-aT}}{e^{-aT}(2e^{a\delta} - 1) - 1}$$

(*iii*) $A[e^{-a(T-\delta)} - e^{-aT}] < 1$

For positive values of a, T, A, δ and $\delta < T$, this condition can be written as

$$Ae^{-aT}(e^{a\delta} - 1) < 1$$

or $$A < \left(\frac{e^{aT}}{e^{a\delta} - 1}\right)$$

Conditions (*ii*) and (*iii*) when simultaneously met for specified δ, yield the range of A for the system to be stable. The solution has to be obtained graphically or numerically.

2. *Bilinear Transformation.* We have seen above that in determining stability in the z-domain, we need to find if all the roots of the characteristic equation lie within the unit circle. Standard method of Routh and Nyquist could be applied to this problem if we could find a

complex transformation which maps the interior of the unit circle in the z-plane into the left half of the new plane. The transformation $z = e^{sT}\left[s = \frac{1}{T}\ln z\right]$ can not be used for this purpose because of the periodicity of e^{sT} and the multiple strips of the left half s-plane into which the interior of the unit circle maps. A simple transformation which uniquely maps the interior of the unit circle in the z-plane in to the left half of the r-plane is the *bilinear transformation*

$$r = \frac{z-1}{z+1};\ z = \frac{1+r}{1-r} \qquad \text{...(11.90)}$$

On the unit circle in the z-plane

$$z = e^{j\theta}\ (\theta \text{ varying anticlockwise from } -\pi \text{ through 0 to } +\pi)$$

$$r = \frac{e^{j\theta}-1}{e^{j\theta}+1} = \frac{e^{j\theta/2}-e^{-j\theta/2}}{e^{j\theta/2}+e^{-j\theta/2}}$$

$$= \tanh j\frac{\theta}{2} = j\tan\frac{\theta}{2} = j\omega_r \qquad \text{...(11.91)}$$

where $\omega_r = \tan\frac{\theta}{2}$ varies from $-\infty$ through 0 to $+\infty$.

This mapping is indicated in Fig. 11.35.

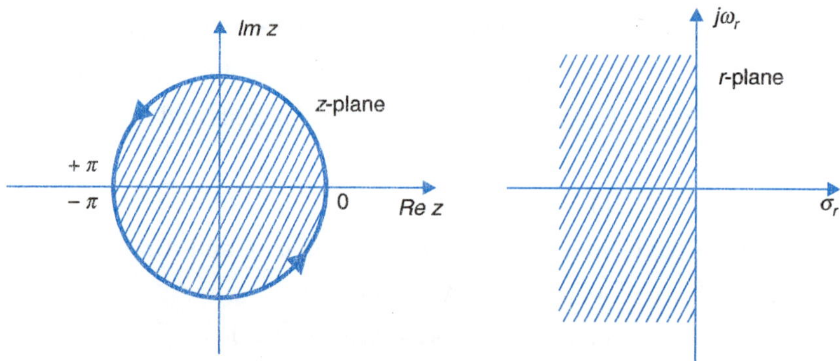

Fig. 11.35

Through use of the bilinear transformation (11.90), the characteristic eqn. (11.87) becomes

$$a_n\left(\frac{1+r}{1-r}\right)^n + a_{n-1}\left(\frac{1+r}{1-r}\right)^{n-1} + \ldots + a_1\left(\frac{1+r}{1-r}\right) + a_0 = 0$$

which can be organized into the form

$$b_n r^n + b_{n-1} r^{n-1} + \ldots + b_1 r + b_0 = 0 \qquad \text{...(11.92)}$$

Now the Routh criterion can be applied to the new characteristic equation (11.92) to determine if all its roots lie in the left half of the r-plane. We could apply the Nyquist criterion as well.

As already shown, the unit circle in the z-plane is the map of sections of width ω_s on the $j\omega$-axis of the s-plane, we can therefore immediately conclude that ω_s-width sections of the $j\omega$-axis in the s-palne map into the whole of the $j\omega_r$-axis in the r-plane.

Heuristically, the Bode plots and gain and phase margin concepts of the frequency domain could be applied once the bilinear transformation has been carried out (see Section 11.12).

Example 11.17: Consider the sampled-data system of Fig. 11.36. Determine its characteristic equation in the z-domain and ascertain its stability via the bilinear transformation.

$$G(z) = Z\left[\frac{5}{s(s-1)(s+2)}\right] = 5Z\left[\frac{1}{2s} - \frac{1}{s+1} + \frac{1}{2(s+2)}\right]$$

$$= 5\left[\frac{z}{2(z-1)} - \frac{z}{z-e^{-1}} + \frac{z}{2(z-e^{-2})}\right] = \frac{5z(0.4z+0.594)}{2(z-1)(z-0.368)(z-0.135)}$$

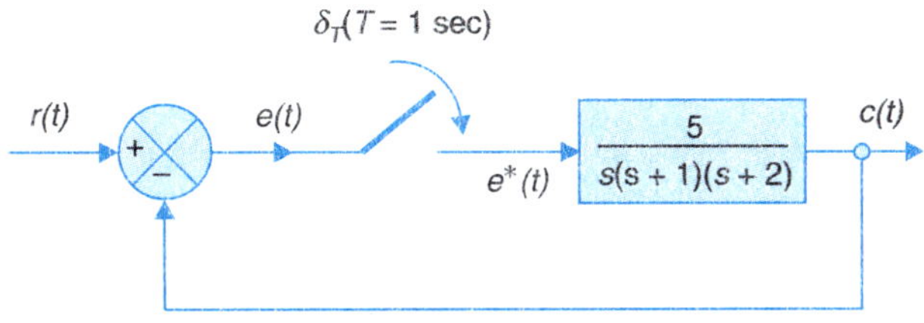

Fig. 11.36

The characteristic equation is

$$1 + G(z) = 2(z - 1)(z - 0.368)(z - 0.135) + 5z(0.4z + 0.594) = 0$$

or

$$z^3 - 0.5z^2 + 2.49z - 0.496 = 0$$

Substituting (11.90), we get

$$\left(\frac{1+r}{1-r}\right)^3 - 0.5\left(\frac{1+r}{1-r}\right)^2 + 2.49\left(\frac{1+r}{1-r}\right) - 0.496 = 0$$

which upon simplification yields

$$3.5r^3 - 2.5r^2 + 0.5r + 2.5 = 0$$

The changes in sign of the characteristic polynomial indicate that the system is unstable. We therefore need not proceed to form the Routh array.

Let us check the stability if the system was linear continuous, *i.e.*,

$$1 + G(s) = 0$$

$$1 + \frac{5}{s(s+1)(s+2)} = 0$$

$$s^3 + 3s^2 + 2s + 5 = 0$$

The Routh array is

s^3	1	2
s^2	3	5
s^1	$\frac{1}{3}$	0
s^0	5	

It indicates a stable system.

By comparison we find that a stable linear continuous system becomes unstable upon introduction of sampling and ZOH in the forward loop. It can therefore be concluded that *sampling has a destabilizing effect on a system.*

3. *The Root Locus Technique.* The root locus technique can be easily adopted to stability analysis and design of sampled-data feedback control systems. The characteristic equation of these systems is of the form

$$1 + F(z) = 0$$

where $F(z)$ is a rational function of z. It can be written in the standard pole-zero form

$$1 + \frac{K\prod(z + z_i)}{\prod(z + p_j)} = 0$$

for drawing the root locus in the z-plane when the gain K is varied.

Examination of the root locus with respect to the unit circle reveals information on system stability and the range of K for the system to be stable can also be determined. This is illustrated by means of Example 11.18.

Example 11.18: Investigate the stability of the system shown in Fig. 11.37 for sampling period $T = 0.4$ sec, 3 sec.

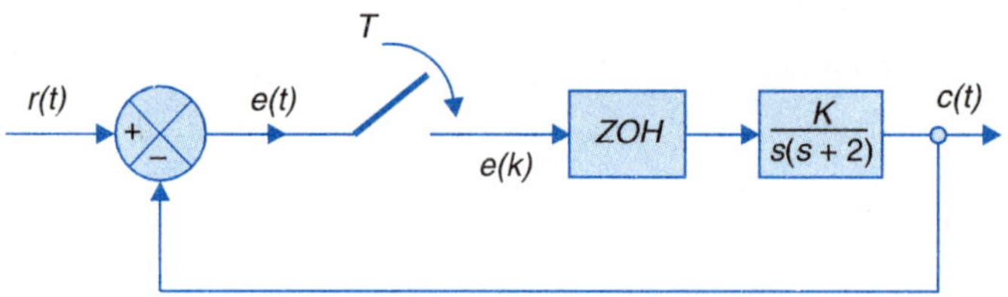

Fig. 11.37

Solution: The characteristic equation of the system is

$$1 + G_0G(z) = 0$$

$$G_0G(z) = Z\left[\frac{1-e^{-sT}}{s}\,\frac{K}{s(s+2)}\right]$$

$$= K(1 - z^{-1})Z\left[\frac{1}{s^2(s+2)}\right]$$

$$= K(1 - z^{-1})Z\left[\frac{1}{2s^2} - \frac{1}{4s} + \frac{1}{4(s+2)}\right] \qquad ...(11.93)$$

$$= \frac{K}{2}(1 - z^{-1})\left[\frac{Tz}{(z-1)^2} - \frac{z}{2(z-1)} + \frac{z}{2(z - e^{-2T})}\right]$$

$$= \frac{K[(2T - 1 + e^{-2T})z - 2Te^{-2T} - e^{-2T} + 1]}{4(z-1)(z - e^{-2T})}$$

Case 1: $T = 0.4$ sec

$$G_0G(z) = \frac{K(z + 0.76)}{16(z-1)(z-0.45)} = \frac{K'(z+0.76)}{(z-1)(z-0.45)}$$

Zero : $z = -0.76$

Poles : $z = 1, z = 0.45$

The root locus is plotted in Fig. 11.38. The two breakaway points are calculated below:

$$1+\frac{K'(z+0.76)}{(z-1)(z-0.45)}=0$$

or
$$K'=\frac{-(z-1)(z-0.45)}{(z+0.76)}$$

$$\frac{dK'}{dz}=\frac{(z^2-1.45z+0.45)+(z+0.76)(-2z+1.45)}{(z+0.76)^2}=0$$

or
$$z^2+1.52z-1.55=0$$

$$z=-2.22,\ 0.7$$

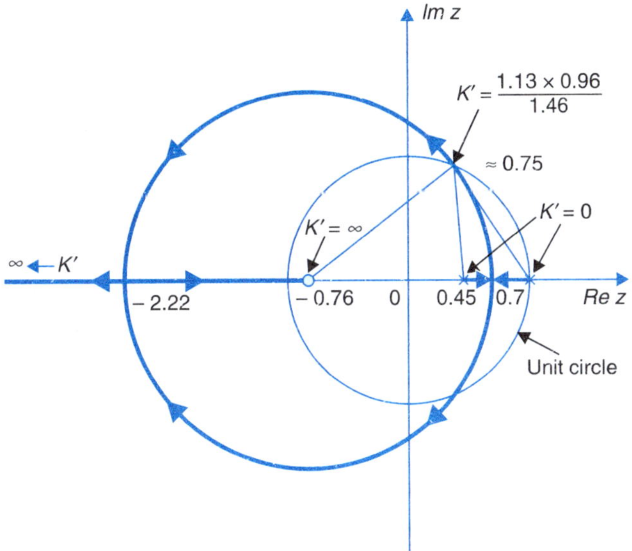

Fig. 11.38. Root locus for $T = 0.4$ sec.

The root locus is a circle centred at $z = -0.76$ (zero) and of radius. 1.46.

The roots of the characteristic equation lie within the unit circle, *i.e.,* the system is stable for

$$K < 16 \times 0.75 = 12$$

Case 2: $T = 3$ sec

$$G_0G(z)=\frac{1.25K(z+0.197)}{(z-1)(z-0.0025)}$$

Zero : $z = -0.197$

Poles : $z = 1, z = 0.0025$

The root locus is plotted in Fig. 11.39. The two breakaway points calculated as before are $z = -0.684$, 0.29.

The root locus is a circle centred at -0.197 (zero) and of radius 0.487. We immediately observe from the root locus that intersection of loci with unit circle occurs at $z = -1$. The marginal value of K is determined as under.

$$1 + \left.\frac{1.25K(z + 0.197)}{(z - 1)(z - 0.0025)}\right|_{z=-1} = 0$$

or

$$K \approx 2$$

We observe in the above example that the system which is stable for $K < 12$ when $T = 0.4$ sec becomes unstable for $K > 2$ when $T = 3$ sec. It means that *increasing the sampling period (or decreasing the sampling rate) reduces the margin of stability*.

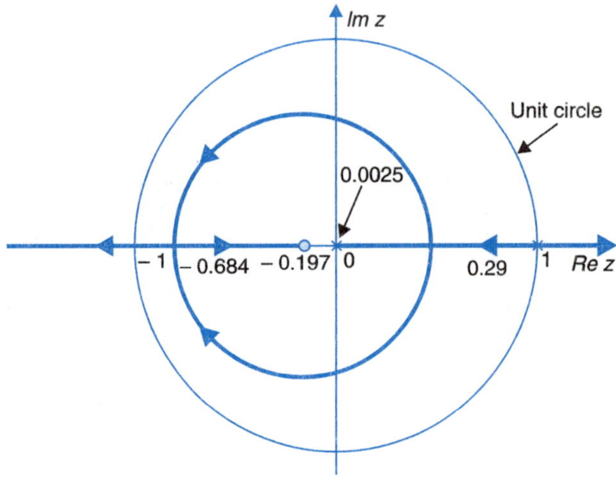

Fig. 11.39. Root locus for $T = 3$ sec.

11.11 COMPENSATION TECHNIQUES

Cascade and feedback compensation techniques valid for linear continuous systems equally apply for linear sampled-data control systems. For want of space, we will not discuss here feedback compensation [2]. There are two ways of accomplishing cascade compensation is sampled-data systems.

1. *Compensation by continuous network:* In this scheme the compensating network is introduced in the continuous part of the system as shown in Fig. 11.40. The overall z-transfer function of the system is

$$\frac{C(z)}{R(z)} = T(z) = \frac{Z[G_0(s)G_c(s)G(s)]}{1 + Z[G_0(s)G_c(s)G(s)]} \qquad \text{...(11.94)}$$

The pole and zero of $G_c(s)$ are adjusted to achieve a desired closed-loop discrete response. As the open-loop poles and zeros (*i.e.*, the poles and zeros of $Z[G_0(s)G_c(s)G(s)]$) are implicity linked to the pole and zero of $G_c(s)$, no direct compensation techniques are possible. Essentially trial and error is the way to proceed. Here this method will not be pursued any further [2].

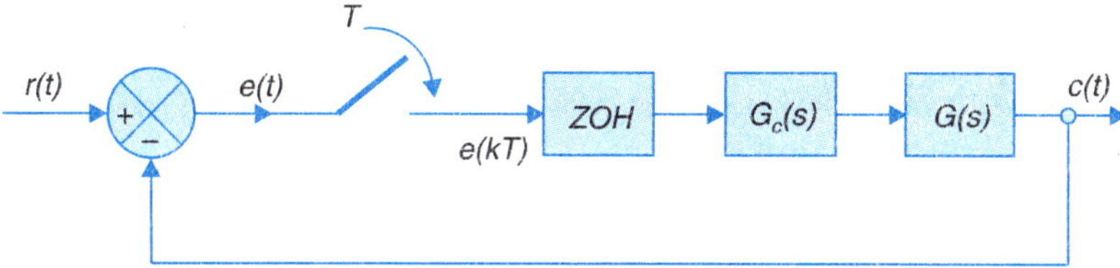

Fig. 11.40. Sampled-data system with linear continuous compensator.

2. *Compensation by digital computer:* The compensation is carried out by a digital controller which for a train of input pulses produces a suitably modified train of output pulses. The scheme is illustrated in Fig. 11.41. The overall z-transfer function of the system is

$$\frac{C}{R}(z) = T(z)\frac{D(z)Z[G_0(s)G(s)]}{1 + D(z)Z[G_0(s)G(s)]} \quad \text{...(11.95)}$$

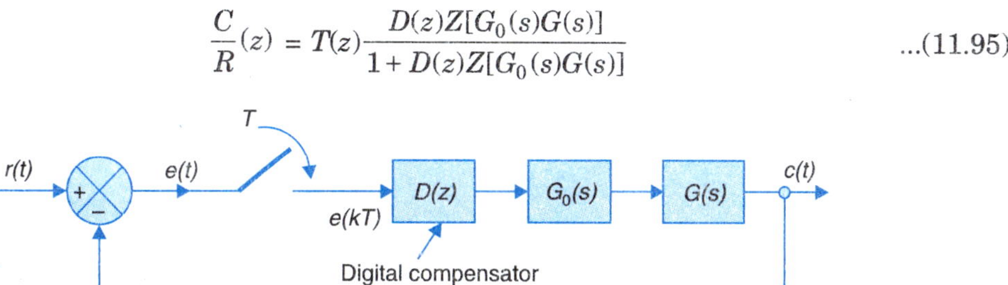

Fig. 11.41. Sampled-data control system with digital compensator.

Since the compensator transfer function $D(z)$ enters the characteristic equation independent of $Z[G_0(s)G(s)]$, it is possible to independently adjust the poles and zeros of $D(z)$ to achieve the desired closed-loop response. The root locus tachnique can be used in design.

Two methods of designing $D(z)$ will be discussed here.

Time-Domain Technique of Designing $D(z)$

Parallel to the case of a continuous system, design specification for a sampled-data system could be laid down as:

1. Zero steady-state error to input $A(kT)^q$ for a given q. For $q = 0, 1, 2$, it corresponds to step, ramp and acceleration inputs in a continuous system.
2. The transient response should settle in finite number of sampling intervals and in as few intervals as possible.

With reference to the compensating scheme of Fig. 11.41, it shall be assumed that the response of all system elements and the overall system is *nonanticipative* (casual). In terms of general transfer function of eqn. (11.50) reproduced below.

$$F(z) = \frac{b_0 z^m + b_1 z^{m-1} + \ldots + b_m}{z^n + a_1 z^{n-1} + \ldots + a_n}$$

it implies that $m \le n$ so that when $F(z)$ is expanded in power series of z^{-1}, no positive power terms of z appear (which would mean appearance of output before input is applied–nonanticipative nature). $F(z)$ can also be written as

$$F(z) = \frac{b_0 z^{-(n-m)} + b_1 z^{-(n-m-1)} + \ldots + b_m z^{-n}}{1 + a_1 z^{-1} + \ldots + a_n z^{-n}}; \; n - m \geq 0 \quad \ldots(11.96)$$

From Fig. 11.41

$$E(z) = R(z) - C(z) = [1 - T(z)]R(z) \quad \ldots(11.97)$$

The steady-state error is given by

$$e(\infty) = \lim_{z \to 1} (1 - z^{-1})[1 - T(z)]R(z) \quad \ldots(11.98)$$

For input of type $A(kT)^q$,

$$R(z) = \frac{B(z)}{(1 - z^{-1})^{q+1}} \quad \ldots(11.99)$$

where $B(z)$ is a finite degree polynomial in z^{-1} (see Table 11.2 for $q = 0, 1, 2$).

$$\therefore \quad e(\infty) = \lim_{z \to 1} (1 - z^{-1})[1 - T(z)] \frac{B(z)}{(1 - z - 1)^{q+1}} \quad \ldots(11.100)$$

It immediately follows from eqn. (11.100) that $e(\infty) = 0$, if

$$1 - T(z) = (1 - z^{-1})^{q+1}$$

or

$$T(z) = 1 - (1 - z^{-1})^{q+1} \quad \ldots(11.101)$$

Now $\quad E(z) = B(z) =$ a finite degree polynomial in z^{-1}.

This further ensures that $e(kT)$ goes to zero in a finite number of sampling intervals which are also the least in number of a specified q.

With $T(z)$ specified as in (11.101), we can obtain $D(z)$ from (11.95) as follows

$$D(z) = \frac{1 - (1 - z^{-1})^{q+1}}{(1 - z^{-1})^{q+1} \, Z[G_0(s)G(s)]} \quad \ldots(11.102)$$

Equation (11.102) also ensures that if $Z[G_0(s)G(s)]$ is non-anticipative, $D(z)$ will also be nonanticipative.

Consider now the system response to three standard inputs wherein the system is compensated with appropriate $D(z)$ (see eqn. (11.97) in each case.

1. Step input (*i.e.*, $q = 0$)

$$E(z) = (1 - z^{-1}) \frac{A}{(1 - z^{-1})} = A$$

Therefore $\quad e(0) = A$

$$e(kT) = 0 \qquad \text{for } k = 1, 2, \ldots$$

2. Ramp input (*i.e.*, $q = 1$)

$$E(z) = (1 - z^{-1})^2 \frac{ATz^{-1}}{(1 - z^{-1})^2} = ATz^{-1}$$

Therefore $\quad e(0) = 0$

$$e(T) = AT$$

$$e(kT) = 0 \qquad \text{for } k = 2, 3, 4, \ldots$$

3. Acceleration input (*i.e.*, $q = 2$)

$$E(z) = (1 - z^{-1})^3 \frac{AT^2 z^{-1}(1 + z^{-1})}{(1 - z^{-1})^3}$$

$$= AT^2 z^{-1} + AT^2 z^{-2}$$

Therefore
$$e(0) = 0$$
$$e(T) = AT^2$$
$$e(2T) = AT^2$$
$$e(kT) = 0 \qquad \text{for } k = 3, 4, \ldots$$

Example 11.19 : In Fig. 11.41, let

$$G(s) = \frac{K}{s(s+2)}$$

Design an appropriate $D(z)$ when ramp input is applied.

From Example 11.17

$$Z[G_0(s)G(s)] = Z\left[\frac{1 - e^{-Ts}}{s} \frac{K}{s(s+2)}\right]$$

$$= \frac{K\,[(2T - 1 + e^{-2T})z^{-1} - (2Te^{-2T} + e^{-2T} - 1)z^{-2}]}{4(1 - z^{-1})(1 - e^{-2T}z^{-1})}$$

From eqn. (11.101)

$$T(z) = 1 - (1 - z^{-1})^2 = 2z^{-1} - z^{-2}$$

It immediately follows from eqn. (11.102) that

$$D(z) = \frac{4(2z^{-1} - z^{-2})(1 - z^{-1})(1 - e^{-2T}z^{-1})}{K(1 - z^{-1})^2[(2T - 1 + e^{-2T})z^{-1} - (2Te^{-2T} + e^{-2T} - 1)z^{-2}]}$$

Frequency-Domain Technique of Designing *D*(*z*)

As already illustrated, use of bilinear transformation permits us to use the Nyquist criterion in the r-plane. Heuristic frequency-domain design criteria of phase and gain margin could be easily extended to the r-plane. This is best illustrated by means of an example.

Example 11.20: Consider the second order sampled-data system of Fig. 11.37. This system is required to meet the following specifications:

$K_v \geq 4$ sec^{-1}; phase margin $\geq 40°$

Bandwidth ≤ 1 Hz(*i.e.*, 6.28 rad/sec)

Design steps are given below:

1. *Selection of sampling frequency:* According to the sampling theorem, the lowest sampling frequency = 2 × bandwidth = 2 Hz. However, to be on the safe side, choose sampling frequency = 2.5 Hz so that the sampling period is $T = 0.4$ sec.

2. *z-transfer function for uncompensated system:* From eqn. (11.93), we have for $T = 0.4$ sec,

$$G_0G(z) = \frac{K(z + 0.76)}{16(z - 1)(z - 0.45)} \qquad \text{...(11.103)}$$

3. *Determination of static gain:* referring to the unity feedback system of Fig. 11.37, the z-transform of the error signal is

$$E(z) = R(z) - C(z) = \frac{R(z)}{1 + G_0G(z)}$$

Using the final value theorem, we have the steady-state error at the sampling instants

$$e_{ss} = \lim_{k\to\infty} e(kT) = \lim_{z\to 1} (1 - z^{-1})E(z)$$

The steady-state error with ramp function input *i.e.,*

$$R(z) = \frac{Tz}{(z-1)^2}$$

is given by

$$e_{ss} = \lim_{z\to 1} \frac{T}{(z-1)[1 + G_0G(z)]} = \frac{1}{K_v}$$

where*

$$K_v = \frac{1}{T} \lim_{z\to 1} [(z-1)G_0G(z)] \qquad \text{...(11.104)}$$

For the system under consideration, we have from eqns. (11.103) and (11.104)

$$K_v = \frac{K}{2} \geq 4$$

Therefore $K = 8$ may be selected as the static gain from the system.

4. *Bilinear transformation:* The z-transfer function of the uncompensated system is

$$G_0G(z) = \frac{0.5(z + 0.76)}{(z-1)(z-0.45)}$$

Using bilinear transformation

$$z = \frac{1+r}{1-r}$$

we get

$$G_0G(r) = \frac{0.8(-r+1)(r/7.35+1)}{r(r/0.38+1)}$$

Letting $r = j\omega_r$, we get

$$G_0G(j\omega_r) = \frac{0.8(-j\omega_r + 1)(j\omega_r/7.35 + 1)}{j\omega_r(j\omega_r/0.38 + 1)}$$

5. *Bode plot for the uncompensated system:* The Bode plot of $G_0G(j\omega_r)$ (*i.e.,* the uncompensated system) is drawn in Fig. 11.42. We immediately find from this figure that

$$\omega_{c1} = 0.5, \ \phi_1 = 14°$$

These results show that compensation is required to increase the phase margin.

$$* \ K_p = \lim_{z\to 1} G_0G(z)$$

$$K_a = \frac{1}{T^2} \lim_{z\to 1} (z-1)^2 G_0G(z)$$

6. *Design of compensation:* We shall design a lag compensator having transfer function

$$D(r) = \frac{1+\tau r}{1+\beta\tau r}\ ;\ \beta > 1$$

in such a way as to realize a phase margin of 40° or more. Following the procedure for designing phase lag compensation for continuous-time systems (Chapter 10) we get

$$D(r) = \frac{1+22.2r}{1+88.8r}$$

7. *Bode plot for the compensated system:* Bode plot for the compensated system is also shown in Fig. 11.42. We find that the required phase margin has been realized.

We can construct the Nicholas plot for the compensated system to verify the satisfaction of bandwidth specification

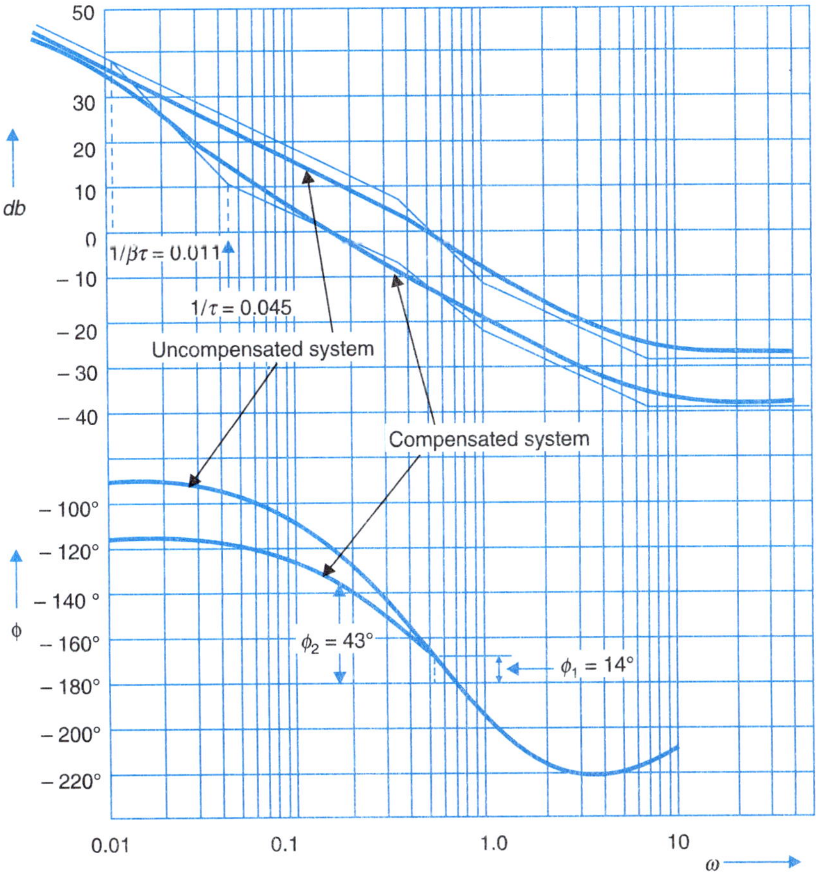

Fig. 11.42. Lag compensator design for a sampled-data control system is the *r*-domain.

8. *Transformation to the z-domain:* Carrying out the reverse transformation

$$r = \frac{z-1}{z+1}$$

we have

$$D(z) = 0.258 \frac{(z - 0.915)}{(z - 0.978)} \quad ...(11.105)$$

9. *Realization of* $D(z)$: The digital controller having the z-transfer function $D(z)$ may be realized in a number of ways:

(*a*) Pulsed-data *RC* network

(*b*) Computer program

(*c*) Digital processor

The general requirement of realizability is that the order of denominator polynomial of $D(z)$ should be greater than/equal to the order of numerator polynomial.

(*a*) Consider first the realization of $D(z)$ of eqn. (11.105) by simple *RC* filter network as shown in Fig. 11.43.

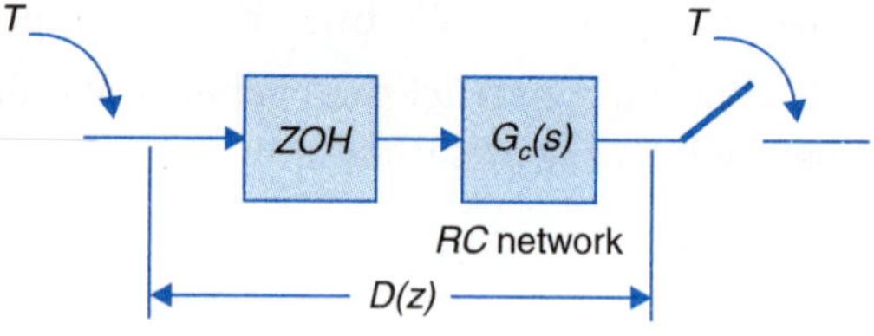

Fig. 11.43

$$D(z) = Z\left[\left(\frac{1 - e^{-Ts}}{s}\right) G_c(s)\right] = 0.258 \frac{(z - 0.915)}{(z - 0.978)}$$

or
$$Z\left(\frac{G_c(s)}{s}\right) = \frac{D(z)}{1 - z^{-1}} = 0.258 \frac{(1 - 0.915z^{-1})}{(1 - z^{-1})(1 - 0.978z^{-1})}$$

$$= \frac{1}{1 - z^{-1}} - \frac{0.775}{1 - 0.978z^{-1}}$$

Note that realization of $D(z)$ by simple *RC* series network is possible when $D(z)$ has simple and real poles lying inside or on the unit circle in the z-plane.

(*b*) The z-transfer function of digital controller is

$$D(z) = \frac{E_2(z)}{E_1(z)} = 0.258 \frac{(z - 0.915)}{(z - 0.978)} = 0.258 \frac{(1 - 0.915z^{-1})}{(1 - 0.978z^{-1})}$$

Therefore

$$(1 - 0.978z^{-1})\, E_2(z) = 0.258\, (1 - 0.915z^{-1})\, E_1(z)$$

or
$$e_2(k) - 0.978e_2(k - 1) = 0.258e_1(k) - 0.246e_1(k - 1)$$

or
$$e_2(k) = 0.258e_1(k) - 0.246e_1(k - 1) + 0.978e_2(k - 1)$$

This difference equation can be solved by a computer program.

(*c*) Realization of $D(z)$ using a special purpose computer by means of a delay network.

PROBLEMS

11.1. For Fig. 11.9, write the difference equation governing the system response for

(*i*) $G(s) = \dfrac{1}{s + 1}$ (*ii*) $G(s) = \dfrac{1}{s^2}$

11.2. Given

$$Z[x(k)] = X(z)$$

find the z-transform of

(i) $y(k) = \sum_{j=0} x(j)$ (ii) $y(k) = e^{-ak}x(k)$

11.3. Find the z-transform of

(i) k^2 (ii) ka^{k-1} ; $k \geq 1$

(iii) $k^2 a^{k-1}$; $k \geq 1$ (iv) $\dfrac{a^k}{k!}$

(v) $\sin h\ \beta k$ (vi) $\cos h\ \beta k$

(vii) $a^k \cos k\pi = (-a)^k$

11.4. Find the z-transforms of the discrete sequences generated by mathematically sampling (at uniform time interval T) the following continuous-time functions

(i) t^2 (ii) e^{-at}

(iii) te^{-at} (iv)* $e^{-at} \sin \omega t$

(v)* $e^{-at} \cos \omega t$

11.5. Find the z-domain transfer function of the following s-domain transfer functions

(i) $\dfrac{a}{(s+a)^2}$ (ii) $\dfrac{s}{s^2 + \omega^2}$

(iii) $\dfrac{a}{(s+b)^2 + a^2}$ (iv) $\dfrac{a}{s^2 - a^2}$

(v) $\dfrac{s+b}{(s+b)^2 + a^2}$

11.6. Find the inverse z-transforms of the following:

(i) $\dfrac{z}{z+a}$ (ii) $\dfrac{1}{z+a}$

(iii) $\dfrac{1}{z-a}$ (iv) $\dfrac{3z^2 + 2z + 1}{z^2 - 3z + 2}$

(v) $\dfrac{3z^2 + 2z + 1}{z^2 + 3z + 2}$ (vi) $\dfrac{2z}{(2z-1)^2}$

(vii) $\dfrac{z - 0.4}{z^2 + z + 2}$ (viii) $\dfrac{z^{-1}}{(1 - az^{-1})^2}$

(ix) $\dfrac{z-4}{(z-1)(z-2)^2}$

11.7. A symmetrical ladder network is shown in Fig. P-11.7. Write the difference equation describing the network behaviour. Also specify the boundary conditions. Using the z-transform techniques solve for current in any loop. Assume each resistor to be 1 ohm.

[*Hint :* The difference equation is

$$3i_{n+1} - i_n - i_{n+2} = 0\ ;\ n = 0, 1, \ldots, 10$$

* You may be use the result of Problem 11.2 (*ii*).

Boundary conditions are

$$i_0 = 4 \; ; \; 3i_{12} = i_{11}$$

The z-transforms of Problem 11.3 (v) and (vi) could be used for inverse transforming.]

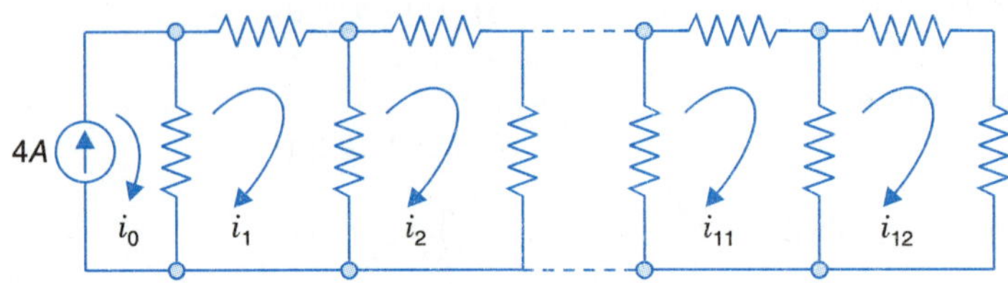

Fig. P-11.7

11.8. The input-output of a sampled-data system is described by the difference equation

$$c(n+2) + 3c(n+1) + 4c(n) = r(n+1) - r(n)$$

Determine the z-transfer function. Also obtain the weighting sequence (discrete impulse response) of the system.

11.9. Solve the difference equation

$$c(k+2) + 3c(k+1) + 2c(k) = u(k) \; ; \; c(0) = 1$$

$$c(k) = 0 \qquad \text{for } k < 0$$

[*Hint* : $c(1)$ needed in the solution can be obtained by letting $k = -1$

$$c(1) + 3c(0) + 2c(-1) = r(-1)$$

or

$$c(1) = -3$$

11.10. For Fig. 11.9, obtain $C(z)/R(z)$ and write therefrom the difference equation governing the system. Check eqn. (11.12).

Repeat for $G(s)$ given in Problem 11.1 and check the difference euqations obtained by the two methods.

11.11. Determine the z-transfer function of two cascaded systems each described by the difference equation

$$c(k) = 0.5c(k-1) + r(k)$$

11.12. The impulse response of a linear continuous system is given by $e^{-at}u(t)$. For impulse sampled unit step input (sampling period = T), find the output at sampled instants.

Repeat for the case where impulse sampler is followed by ZOH. Compare the results.

11.13. Find $C(z)/R(z)$ for the sampled-data closed system of Fig. P-11.13.

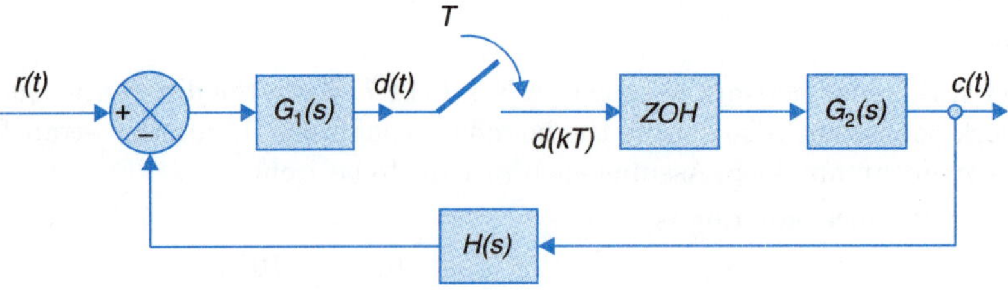

Fig. P-11.13

11.14. A sampled-data control system is shown in Fig. P-11.14. Show that the output of the system at sampling instants is zero.

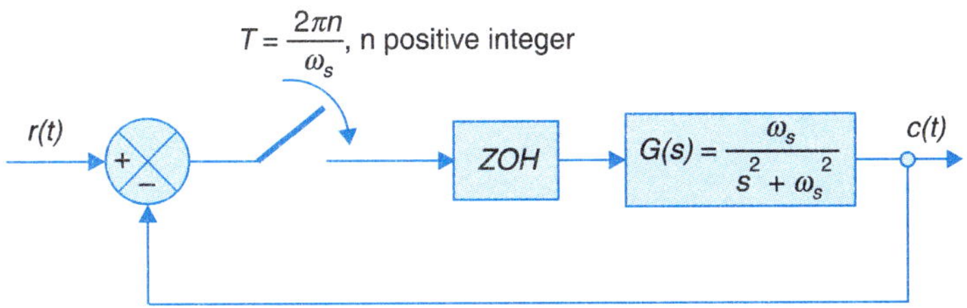

Fig. P-11.14

[*Hint* : $G(z) = 0$ which indicates that the sampled output of the system is zero but the continuous output is not zero.]

11.15. Find $C(z)/R(z)$ for the sampled-data closed-loop system of Fig. P-11.15. Assume both the samples to be of impulse type.

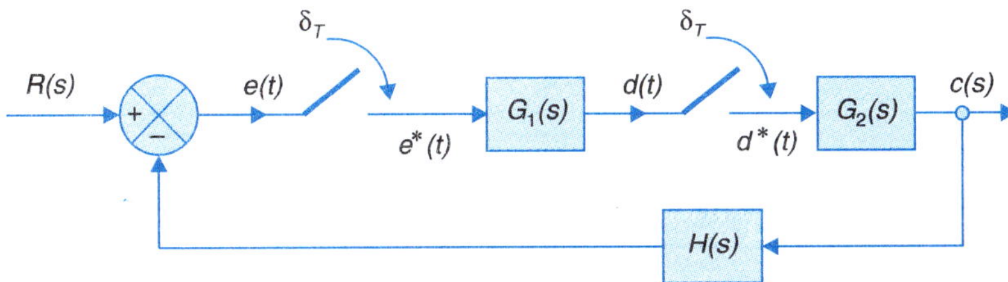

Fig. P-11.15

11.16. For the sampled-data feedback system with a digital network in the feedback path as shown in Fig. P-11.16, find $C(z)/R(z)$.

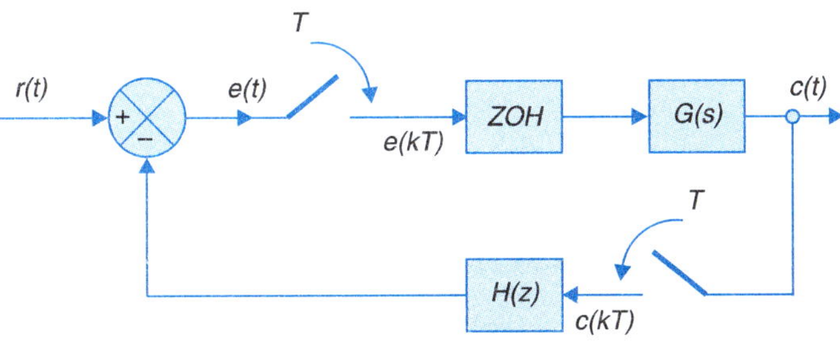

Fig. P-11.16

11.17. For the sampled-data system of Fig. P-11.17, find the response to unit step input

Given: $$G(s) = \frac{1}{s+1}$$

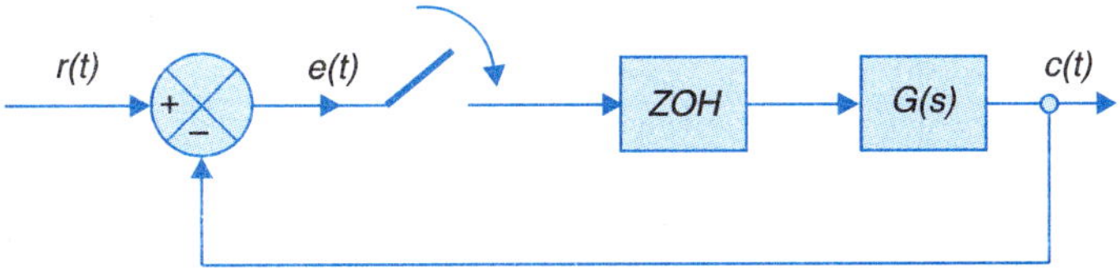

T = 1 sec

r(t) + − e(t) ZOH G(s) c(t)

Fig. P-11.17

11.18. For the system shown in Fig. P-11.18, find the expression for $c(kT)$.

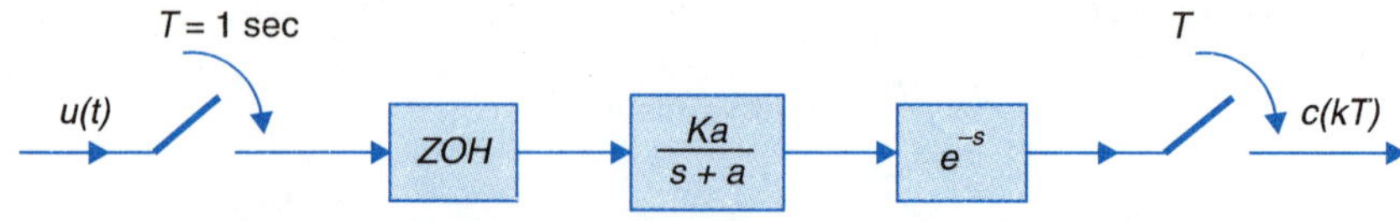

Fig. P-11.18

11.19. Fot the system of Fig. P-11.19, obtain the expression for $c(kT)$ for $r(t)$ = unit step. Given $0 < \Delta < 1$.

[*Hint :*
$$Z\left[\frac{1-e^{-sT}}{s}\,\frac{Kae^{\Delta T}}{s+a}\right] \neq (1-z^{-1})Z\left[\frac{Kae^{\Delta Ts}}{s(s+a)}\right]$$

because ΔT is a submultiple of T. Proceed fundamentally. See Example 11.18.]

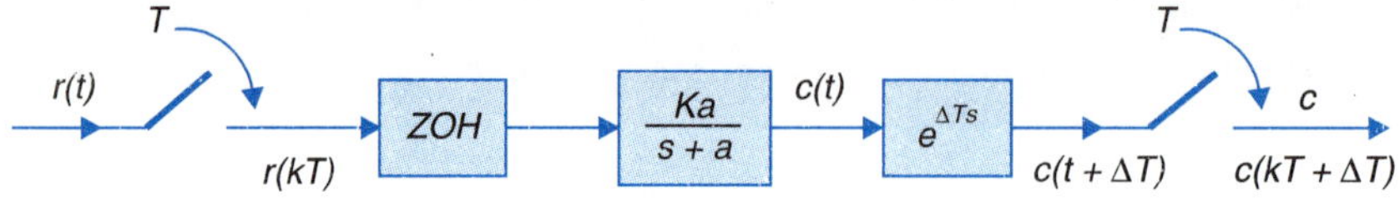

Fig. P-11.19

11.20. In Problem 11.17, obtain the output mid between sampling instants.

11.21. Find the range of K for the system shown in Fig. P-11.21 to be stable.

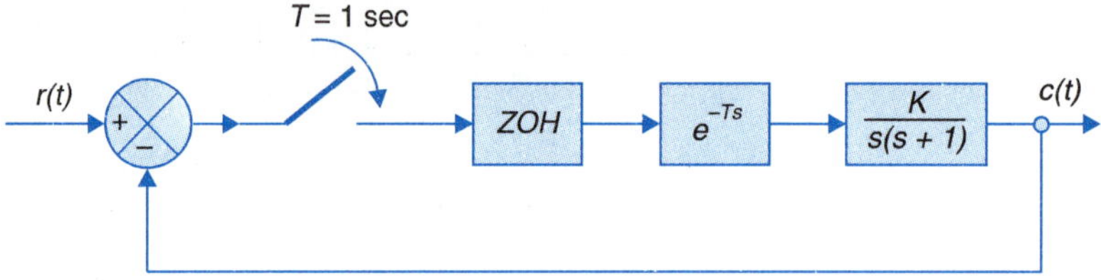

Fig. P-11.21

11.22. Check if all the roots of the following characteristic equation lie within the unit circle.

(*i*) $5z^2 - 2z + 2 = 0$

(*ii*) $z^3 - 0.2z^2 - 0.25z + 0.05 = 0$

(*iii*) $z^4 - 1.7z^3 + 1.04z^2 - 0.268z + 0.024 = 0$

11.23. For the Example 11.20, a compensating digital network with transfer function

$$D(z) = \frac{z+0.5}{z-0.6}$$

is introduced in the forward loop. Obtain the root locus plot of the system for $T = 0.4$ sec. Compare the plot with Fig. 11.40 and find the range of gain K for the system to be stable.

11.24. Design a lead compensator in the r-domain for Example 11.22. What is the corresponding z-domain transfer function ?

11.25. For the $D(z)$ obtained in Problem 11.24, find $G_c(s)$ as per the compensator given in Fig. 11.45 such that

$$(1-z^{-1})Z\left[\frac{G_c(s)}{s}\right] = D(z)$$

Is the s-domain compensator realizable ?

[*Hint :* Partial fractionate $\dfrac{1}{1-z^{-1}}D(z)$ and convert to s-form using Table 11.2(*b*).]

12

STATE VARIABLE ANALYSIS AND DESIGN

12

STATE VARIABLE ANALYSIS AND DESIGN

12.1 INTRODUCTION

In the preceding chapters, we studied several methods of analysis and design of feedback systems such as root locus and frequency response methods. These methods require that the physical system be modelled in the form of a transfer function. Though the transfer function model provides us with simple and powerful analysis and design techniques, it suffers from certain drawbacks, *e.g.*, a transfer function is only defined under zero initial conditions. Further it has certain limitations due to the fact that the transfer function model is only applicable to linear time-invariant systems and there too it is generally restricted to single-input-single-output systems as this approach* becomes highly cumbersome for use in multiple-input-multiple-output systems. Another limitation of the transfer function technique is that it reveals only the system output for a given input and provides no information regarding the internal state of the system. There may be situations where the output of a system is stable and yet some of the system elements may have a tendency to exceed their specified ratings. In addition to this, it may sometimes be necessary and advantageous to provide a feedback proportional to some of the internal variables of a system, rather than the output alone, for the purpose of stabilizing and improving the performance of a system. It is further observed that the classical design methods (root locus and frequency domain methods) based on the transfer function model are essentially trial and error procedures. Such procedures are difficult to visualize and organize even in moderately complex systems and may not lead to a control system which yields an optimum performance in some defined sense.

From the foregoing discussion we feel the need of a more general mathematical representation of a system which, along with the output, yields information about the state of the system variables at some predetermined points along the flow of signals. Such considerations

*Refer Section 2.4.

have led to the development of the state variable approach. It is a direct time domain approach which provides a basis for modern control theory and system optimization. It is a very powerful technique for the analysis and design of linear and nonlinear, time-invariant or time-varying multi-input-multi-output systems. The organization of the state variable approach is such that it is easily amenable to solution through digital computers.

State variable concepts along with some examples have been introduced in Section 2.5.

It will be incorrect to conclude from the above introduction that the state variable approach can completely replace the classical approaches. In fact, the classical approaches provide the control engineer with a deep physical insight into the system and greatly aid the preliminary system design where a complex system is approximated by a more manageable model.

In this chapter our aim is to introduce the state variable approach for linear time-invariant systems in both the continuous-time and discrete-time cases, so as to give a glimpse of its power for the analysis and design of complex systems. It may be pointed out that whereas in transform domain analysis, Laplace transform is needed for continuous-time system and z-transform is needed for discrete-time systems, the state variable approach offers us a way to look at both the continuous-time and discrete-time systems with the same formulation. In the following, we first discuss the state variable techniques for linear continuous-time systems and then show that the techniques are essentially the same for linear discrete-time systems also.

State variable techniques can be fully appreciated if the readers possess a knowledge of the elementary rules of vector and matrix algebra. Only a most elementary knowledge (Appendix II) will prove sufficient for our specific needs.

12.2 CONCEPTS OF STATE, STATE VARIABLES AND STATE MODEL

A mathematical abstraction to represent or model the dynamics of a system utilizes three types of variables called the *input*, the *output* and the *state variables*.

Consider the mechanical system shown in Fig. 12.1 where in mass M is acted upon by the force $F(t)$. The system is characterized by the reactions

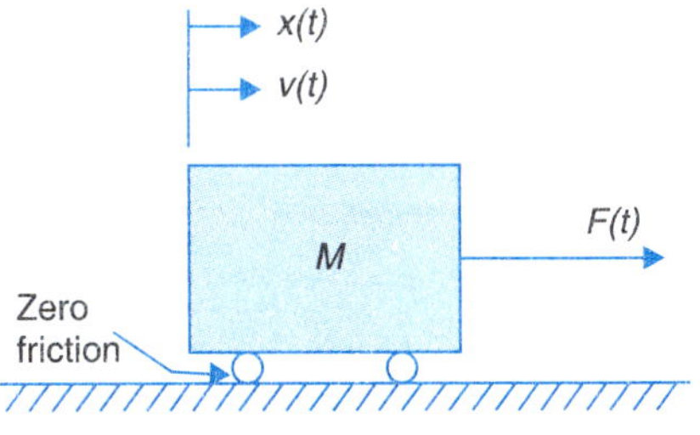

Fig. 12.1. A simple mechanical system.

$$\frac{d}{dt}v(t) = \frac{1}{M}F(t) \quad \text{...(12.1)}$$

$$\frac{d}{dt}x(t) = v(t) \quad \text{...(12.2)}$$

From these relations, we get

$$v(t) = \frac{1}{M}\int_{-\infty}^{t} F(t)\,dt$$

$$= \frac{1}{M}\int_{-\infty}^{t_0} F(t)\,dt + \frac{1}{M}\int_{t_0}^{t} F(t)\,dt$$

$$= v(t_0) + \frac{1}{M}\int_{t_0}^{t} F(t)\, dt \qquad ...(12.3)$$

$$x(t) = \int_{-\infty}^{t} v(t)\, dt = \int_{-\infty}^{t_0} v(t)\, dt + \int_{t_0}^{t} v(t)\, dt$$

$$= x(t_0) + [t - t_0]v(t_0) + \frac{1}{M}\int_{t_0}^{t} d\tau \int_{t_0}^{t} F(t)\, dt \qquad ...(12.4)$$

We observe that the displacement $x(t)$ (output variable) at any time $t \geq t_0$ can be computed if we know the applied force $F(t)$ (input variable) from $t = t_0$ onwards, provided $v(t_0)$ the initial velocity and $x(t_0)$ the initial displacement are known. We may conceive of initial velocity and initial displacement as describing the status or *state* of the system at $t = t_0$. The state of the system of Fig. 12.1 at any time t is given variables $x(t)$ and $v(t)$ which are called the *state variables* of the system.

Let us now look into the precise definitions of state and state variable:

The state of a dynamical system is a minimal set of variables (known as state variables) such that the knowledge of these variables at $t = t_0$ *together with the knowledge of the inputs for* $t \geq t_0$, *completely determines the behaviour of the system for* $t > t_0$.

In state variable formulation of a system, the state variables are usually represented by $x_1(t)$, $x_2(t)$, ..., the inputs by $u_1(t)$, $u_2(t)$ ......; and the outputs by the $y_1(t)$, $y_2(t)$, ... The state space representation may be visualized in block diagram form as shown in Fig. 12.2. For generality, we have depicted a system which has m inputs, p outputs and n state variables. For notational economy, the different variables may be represented by the input vector $\mathbf{u}(t)$, output vector $\mathbf{y}(t)$ and the state $\mathbf{x}(t)$; where

$$\mathbf{u}(t) = \begin{bmatrix} u_1(t) \\ u_2(t) \\ \vdots \\ u_m(t) \end{bmatrix}, \ \mathbf{y}(t) = \begin{bmatrix} y_1(t) \\ y_2(t) \\ \vdots \\ y_p(t) \end{bmatrix}, \ \mathbf{x}(t) = \begin{bmatrix} x_1(t) \\ x_2(t) \\ \vdots \\ x_n(t) \end{bmatrix} \qquad ...(12.5)$$

In Fig. 12.2, broad arrows have been used to represent vector quantities.

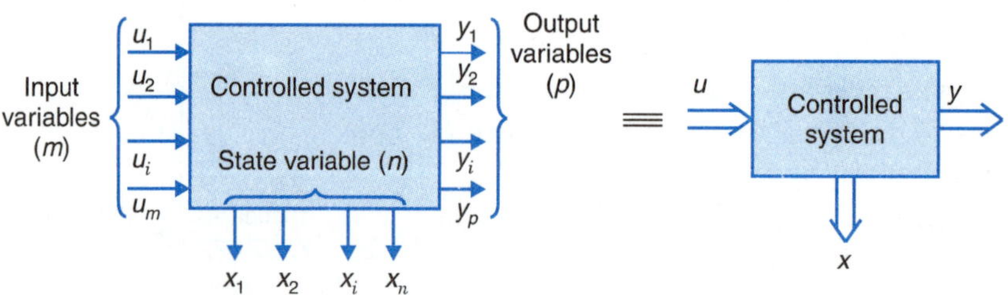

Fig. 12.2. Structure of a general control system.

For the system of Fig. 12.1; the state variable representation is given by two first-order differential equations (12.1) and (12.2), the solution of these equations (eqns. (12.3) and (12.4)) gives the two state variable $x(t)$ and $v(t)$ of the system. For a general system of Fig. 12.2, the state representation can be arranged in the form of n first-order differential equations.

$$\begin{aligned}
\frac{dx_1}{dt} &= \dot{x}_1 = f_1(x_1, x_2, \ldots\ldots, x_n \,;\, u_1, u_2, \ldots\ldots, u_m) \\
&\ \ \vdots \\
\frac{dx_n}{dt} &= \dot{x}_n = f_n(x_1, x_2, \ldots\ldots, x_n \,;\, u_1, u_2, \ldots\ldots, u_m)
\end{aligned} \qquad \ldots(12.6)$$

Integration of equations (12.6) gives

$$x_i(t) = x_i(t_0) + \int_{t_0}^{t} f_i(x_1, x_2, \ldots\ldots, x_n \,;\, u_1, u_2, \ldots\ldots, u_m)\, dt \;;$$

$$i = 1, 2, \ldots, n$$

Thus the n state variables and hence the state of the system can be determined uniquely* at any $t > t_0$ if each state variable is known at $t = t_0$ and all the m control forces are known throughout the interval t_0 to t.

The 'n' differential equations (12.6) may be written in vector notations as

$$\dot{\mathbf{x}}(t) = \mathbf{f}(\mathbf{x}(t), \mathbf{u}(t)) \qquad \ldots(12.7)$$

where $\mathbf{x}$ is $n \times 1$ state vector, $\mathbf{u}$ is $m \times 1$ input vector, both defined earlier in (12.5),

and

$$\mathbf{f}(.) = \begin{bmatrix} f_1(.) \\ f_2(.) \\ \vdots \\ f_n(.) \end{bmatrix} \qquad \ldots(12.8)$$

is $n \times 1$ function vector.

For time-varying systems, the function $\mathbf{f}$ is dependent on time as well and the vector equation may be written as

$$\dot{\mathbf{x}}(t) = \mathbf{f}(\mathbf{x}(t), \mathbf{u}(t), t) \qquad \ldots(12.9)$$

Equation (12.7) and (12.9) are *state equations* for time-invariant and time-varying systems respectively. The state vector $\mathbf{x}$ determines a point (called the state point) in an n-dimensional space, called the *state space*. The curve traced out by the state point from $t = t_0$ to $t = t_1$ in the direction of increasing time is known as the *state trajectory*. For the two-dimensional cases the state space reduces to the state plane or phase plane.

The output $\mathbf{y}(t)$ (Fig. 12.2) can in general be expressed in terms of the state $\mathbf{x}(t)$ and input $\mathbf{u}(t)$ as

$$\mathbf{y}(t) = \mathbf{g}(\mathbf{x}(t), \mathbf{u}(t)) \text{ ; time-invariant systems} \qquad (12.10)$$

$$\mathbf{y}(t) = \mathbf{g}(\mathbf{x}(t), \mathbf{u}(t), t) \text{ ; time-varying systems} \qquad \ldots(12.11)$$

Equations (12.10) and (12.11) are the *output equations* for time-invariant and time-varying systems respectively. It may be noted that the output equation is not a dynamic relation but a static (instantaneous) one. It is called the *read-out* function. We, therefore, need to solve the system state equation and once the system state is known, output can be immediately

---

*It is important to note that a unique solution for eqns. (12.6) exists only if f_i and $\partial f_i/\partial x_j$ are defined and are continuous for $i, j = 1, 2, \ldots, n$. It is being assumed here that these conditions are satisfied.

obtained from the output equation. Solution of the state equations thus provides us information about the system state as well as the system output.

The state and output equations constitute the *state model* of the system.

State Model of Linear Systems

State model of a linear time-invariant system is a special case of the general time-invariant model of eqns. (12.7) and (12.10). Derivative of each state variable now becomes a linear combination of system states and inputs, *i.e.*,

$$\begin{aligned}
\dot{x}_1 &= a_{11}x_1 + a_{12}x_2 + \ldots\ldots + a_{1n}x_n + b_{11}u_1 + b_{12}u_2 + \ldots\ldots + b_{1m}u_m \\
\dot{x}_2 &= a_{21}x_1 + a_{22}x_2 + \ldots\ldots + a_{2n}x_n + b_{21}u_1 + b_{22}u_2 + \ldots\ldots + b_{2m}u_m \\
&\;\;\vdots \qquad\qquad\qquad \vdots \\
\dot{x}_n &= a_{n1}x_1 + a_{n2}x_2 + \ldots\ldots + a_{nn}x_n + b_{n1}u_1 + b_{n2}u_2 + \ldots\ldots + b_{nm}u_m
\end{aligned} \qquad \ldots(12.12)$$

where the coefficients a_{ij} and b_{ij} are constants. In the vector-matrix form, eqns. (12.12) may be written as

$$\dot{\mathbf{x}}(t) = \mathbf{A}\mathbf{x}(t) + \mathbf{B}\mathbf{u}(t)$$

where $\mathbf{x}(t)$ is $n \times 1$ state vector, $\mathbf{u}(t)$ is $m \times 1$ input vector, $\mathbf{A}$ is $n \times n$ *system matrix* defined by

$$\mathbf{A} = \begin{bmatrix} a_{11} & a_{12} & \cdots & a_{1n} \\ a_{21} & a_{22} & \cdots & a_{2n} \\ \vdots & \vdots & & \vdots \\ a_{n1} & a_{n2} & \cdots & a_{nn} \end{bmatrix}$$

and $\mathbf{B}$ is $n \times m$ input matrix defined

$$\mathbf{B} = \begin{bmatrix} b_{11} & b_{12} & \cdots & b_{1m} \\ b_{21} & b_{22} & \cdots & b_{2m} \\ \vdots & \vdots & & \vdots \\ b_{n1} & b_{n2} & \cdots & b_{nm} \end{bmatrix}$$

Similarly, the output variables at time t are linear combinations of the values of the input and state variables at time t, *i.e.*,

$$\begin{aligned}
y_1(t) &= c_{11}x_1(t) + \ldots\ldots + c_{1n}x_n(t) + d_{11}u_1(t) + \ldots\ldots + d_{1m}u_m(t) \\
&\;\;\vdots \\
y_p(t) &= c_{p1}x_1(t) + \ldots\ldots + c_{pn}x_n(t) + d_{p1}u_1(t) + \ldots\ldots + d_{pm}u_m(t)
\end{aligned}$$

where the coefficients c_{ij} and d_{ij} are constants. This set of equations may be put in the vector-matrix form

$$\mathbf{y}(t) = \mathbf{C}\mathbf{x}(t) + \mathbf{D}\mathbf{u}(t)$$

where $\mathbf{y}(t)$ is $p \times 1$ output vector, C is $p \times n$ *output matrix* defined by

$$\mathbf{C} = \begin{bmatrix} c_{11} & c_{12} & \cdots & c_{1n} \\ c_{21} & c_{22} & \cdots & c_{2n} \\ \vdots & \vdots & & \vdots \\ c_{p1} & c_{p2} & \cdots & c_{pn} \end{bmatrix}$$

and $\mathbf{D}$ is $p \times m$ transmission matrix* defined by

$$\mathbf{D} = \begin{bmatrix} d_{11} & d_{12} & \cdots & d_{1m} \\ d_{21} & d_{22} & \cdots & d_{2m} \\ \vdots & \vdots & & \vdots \\ d_{p1} & d_{p2} & \cdots & d_{pm} \end{bmatrix}.$$

The state model of linear time-invariant systems is thus given by the following equations :

$$\dot{\mathbf{x}}(t) = \mathbf{Ax}(t) + \mathbf{Bu}(t) \text{ ; state equation} \tag{12.13a}$$

$$\mathbf{y}(t) = \mathbf{Cx}(t) + \mathbf{Du}(t) \text{ ; output equation} \tag{12.13b}$$

The block diagram representation of the state model is shown in Fig. 12.3.

For system of Fig. 12.1, let us define

$$x_1(t) = x(t)$$

$$x_2(t) = v(t) = \dot{x}(t)$$

$$u(t) = F(t)$$

Equations (12.1) and (12.2) can now be written as

$$\begin{bmatrix} \dot{x}_1 \\ \dot{x}_2 \end{bmatrix} = \begin{bmatrix} 0 & 1 \\ 0 & 0 \end{bmatrix} \begin{bmatrix} x_1 \\ x_2 \end{bmatrix} + \begin{bmatrix} 0 \\ 1/M \end{bmatrix} u \qquad \text{...(12.14a)}$$

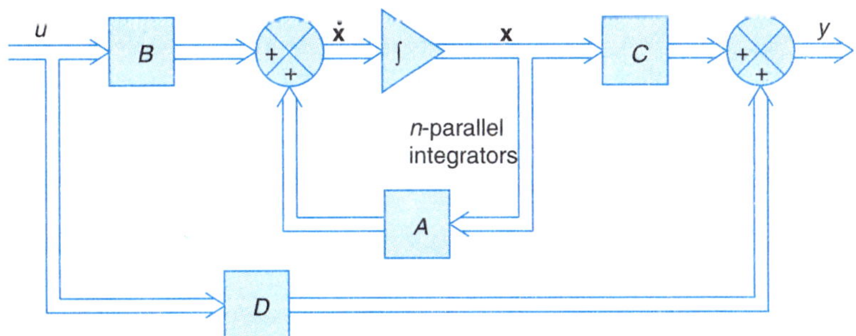

Fig. 12.3. Block diagram representation of the state model of a linear multi-input-multi-output systems.

The output $y = x(t)$ may be expressed as

$$y = [1 \quad 0] \begin{bmatrix} x_1 \\ x_2 \end{bmatrix} \qquad \text{...(12.14b)}$$

Equations (12.14) are of the form of eqns. (12.13).

It may be noted here that the state of a system is not uniquely specified. For example for the system of Fig. 12.1, we have taken displacement $x(t)$ and velocity $v(t)$ as the state variables. We may defined new variables as

$$z_1(t) = 2x(t) + v(t)$$

$$z_2(t) = x(t) + v(t)$$

*Direct coupling of the input to the output is rare in control systems where power amplification is generally desired.

or
$$x(t) = z_1(t) - z_2(t)$$
$$v(t) = -z_1(t) + 2z_2(t)$$

Now
$$\frac{dz_1}{dt} = 2\frac{dx}{dt} + \frac{dv}{dt}$$
$$= 2v(t) + \frac{1}{M}F(t)$$
$$= -2z_1(t) + 4z_2(t) + \frac{1}{M}F(t)$$
$$\frac{dz_2}{dt} = \frac{dx}{dt} + \frac{dv}{dt}$$
$$= v(t) + \frac{1}{M}F(t)$$
$$= -z_1(t) + 2z_2(t) + \frac{1}{M}u(t)$$

We can write from above

$$\begin{bmatrix} \dot{z}_1 \\ \dot{z}_2 \end{bmatrix} = \begin{bmatrix} -2 & 4 \\ -1 & 2 \end{bmatrix} \begin{bmatrix} z_1 \\ z_2 \end{bmatrix} + \begin{bmatrix} \dfrac{1}{M} \\ \dfrac{1}{M} \end{bmatrix} u \quad \text{...(12.14}c\text{)}$$

The output is given by

$$y = x(t) = z_1(t) - z_2(t)$$

or
$$y = [1 \quad -1] \begin{bmatrix} z_1 \\ z_2 \end{bmatrix} \quad \text{...(12.14}d\text{)}$$

We immediately observe that (12.14*c*) and (12.14*d*) given an alternative state variable model of the system previously represented by (12.14*a*) and (12.14*b*). We, therefore, conclude that the state variables of a system are nonunique and the example further brings out the fact that the state variables need not necessarily be the physical variables of the system.

Though the state model of a system is not unique, however, all such models have one characteristic is common for a given system, namely the number of elements in the state vector is equal and minimal. This number n is referred to as the *order* of the system.

The state model for a linear time-varying system is of the same form as defined in eqns. (12.13) except for the fact that the coefficients of the matrices **A, B, C** and **D** are no more constants but are functions of time. In this chapter, we will be mainly concerned with linear time-invariant systems.

State Model for Single-Input-Single-Output Linear Systems

The transfer function analysis deals mainly with single-input-single-output linear time-invariant systems. Here in this chapter we will link the transfer function approach with the state variable approach. If we let $m = 1$ and $p = 1$ in the state model of a multi-input-output linear system, we obtain the following state model for a single-input-single-output linear system:

$$\dot{\mathbf{x}} = \mathbf{A}\mathbf{x} + \mathbf{B}u \quad \text{...(12.15}a\text{)}$$
$$y = \mathbf{C}\mathbf{x} + du \quad \text{...(12.15}b\text{)}$$

where **B** and **C** are now respectively ($n \times 1$) and ($1 \times n$) matrices, d is a constant and u is a scalar control variable. The block diagram representation of this state model is shown in Fig. 12.4.

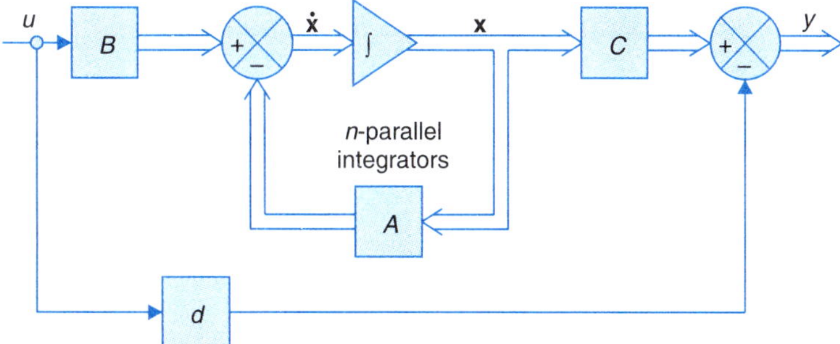

Fig. 12.4. Block diagram representation of the state model for a linear single-input-single-output system.

Linearization of the State Equation

The state equation $\dot{\mathbf{x}} = \mathbf{f}(\mathbf{x}, \mathbf{u})$ of a general time-invariant system can be linearized for small variations about an equilibrium point ($\mathbf{x}_0$, $\mathbf{u}_0$). It is assumed that the system is in equilibrium under the conditions $\mathbf{x}_0$ and $\mathbf{u}_0$, *i.e.*,

$$\dot{\mathbf{x}} = \mathbf{f}(\mathbf{x}, \mathbf{u}) = \mathbf{0}$$

Since the derivatives of all the state variables are zero at the equilibrium point, the system continues to lie at the equilibrium point unless otherwise disturbed.

The state equation can be linearized about the operating point ($\mathbf{x}_0$, $\mathbf{u}_0$) by expanding it into Talyor series and neglecting terms of second-and higher-order. Thus for the ith state equation

$$\dot{x}_i = f_i(\mathbf{x}_0, \mathbf{u}_0) + \sum_{j=1}^{n} \left.\frac{\partial f_i(\mathbf{x}, \mathbf{u})}{\partial x_j}\right|_{\substack{\mathbf{x}=\mathbf{x}_0\\ \mathbf{u}=\mathbf{u}_0}} (x_j - x_{j0}) + \sum_{k=1}^{m} \left.\frac{\partial f_i(\mathbf{x}, \mathbf{u})}{\partial u_k}\right|_{\substack{\mathbf{x}=\mathbf{x}_0\\ \mathbf{u}=\mathbf{u}_0}} (u_k - u_{k0})$$

Recognizing that at the operating point $f_i(\mathbf{x}_0, \mathbf{u}_0) = 0$ and defining the variation about the operating point as

$$\tilde{x}_j = x_j - x_{j0} \qquad \therefore \quad \dot{\tilde{x}}_j = \dot{x}_j$$

$$\tilde{u}_k = u_k - u_{k0}$$

the linearized ith state equation can be written as

$$\dot{\tilde{x}}_i = \sum_{j=1}^{n} \left.\frac{\partial f_i(\mathbf{x}, \mathbf{u})}{\partial x_j}\right|_{\substack{\mathbf{x}=\mathbf{x}_0\\ \mathbf{u}=\mathbf{u}_0}} \tilde{x}_j + \sum_{k=1}^{m} \left.\frac{\partial f_i(\mathbf{x}, \mathbf{u})}{\partial u_k}\right|_{\substack{\mathbf{x}=\mathbf{x}_0\\ \mathbf{u}=\mathbf{u}_0}} \tilde{u}_k$$

The above linearized component equation can be written as the vector matrix equation

$$\dot{\tilde{\mathbf{x}}} = \mathbf{A}\tilde{\mathbf{x}} + \mathbf{B}\tilde{\mathbf{u}} \qquad \text{...(12.16)}$$

where
$$\mathbf{A} = \begin{bmatrix} \frac{\partial f_1}{\partial x_1} & \frac{\partial f_1}{\partial x_2} & \cdots & \frac{\partial f_1}{\partial x_n} \\ \frac{\partial f_2}{\partial x_1} & \frac{\partial f_2}{\partial x_2} & \cdots & \frac{\partial f_2}{\partial x_n} \\ \cdots & \cdots & \cdots & \cdots \\ \frac{\partial f_n}{\partial x_1} & \frac{\partial f_n}{\partial x_2} & \cdots & \frac{\partial f_n}{\partial x_n} \end{bmatrix}$$

$$\mathbf{B} = \begin{bmatrix} \frac{\partial f_1}{\partial u_1} & \frac{\partial f_1}{\partial u_2} & \cdots & \frac{\partial f_1}{\partial u_m} \\ \frac{\partial f_2}{\partial u_1} & \frac{\partial f_2}{\partial u_2} & \cdots & \frac{\partial f_2}{\partial u_m} \\ \cdots & \cdots & \cdots & \cdots \\ \frac{\partial f_n}{\partial u_1} & \frac{\partial f_n}{\partial u_2} & \cdots & \frac{\partial f_n}{\partial u_m} \end{bmatrix}$$

All the partial derivatives in the matrices **A** and **B** defined above (called the *Jacobian matrices*) are evaluated at the equilibrium state ($\mathbf{x}_0$, $\mathbf{u}_0$).

12.3 STATE MODELS FOR LINEAR CONTINUOUS-TIME SYSTEMS

From the discussion of previous section, we know that the state equations of a system are not unique, *i.e.*, there exist more than one set of state variables in terms of which the system behaviour can be completely described. In fact there are infinitely many state models for a given system and any two models are uniquely related. To demonstrate this fact, consider the state model of eqn. (12.13) wherein **x** is an n-dimensional state vector.

Consider now another n-dimensional vector v such that

$$\mathbf{x} = \mathbf{P}v$$

where **P** is any $n \times n$ nonsingular constant matrix.

Since **P** is a constant matrix, it follows that

$$\dot{\mathbf{x}} = \mathbf{P}\dot{v}$$

Substituting **x** and $\dot{\mathbf{x}}$ from above in eqn. (12.13*a*), we get

$$\mathbf{P}\dot{v} = \mathbf{AP}v + \mathbf{Bu}$$

Premultiplying by $\mathbf{P}^{-1}$, we obtain

$$\begin{aligned} \dot{v} &= \mathbf{P}^{-1}\,\mathbf{AP}v + \mathbf{P}^{-1}\mathbf{Bu} \\ &= \tilde{\mathbf{A}}v + \tilde{\mathbf{B}}\mathbf{u} \end{aligned} \qquad \text{...(12.17}a\text{)}$$

where $\tilde{\mathbf{A}} = \mathbf{P}^{-1}\mathbf{AP}$; $\tilde{\mathbf{B}} = \mathbf{P}^{-1}\mathbf{B}$

From eqn. (12.13*b*), we have

$$\begin{aligned} \mathbf{y} &= \mathbf{CP}v + \mathbf{Du} \\ &= \tilde{\mathbf{C}}\mathbf{p} + \mathbf{Du} \end{aligned} \qquad \text{...(12.17}b\text{)}$$

where $\tilde{\mathbf{C}} = \mathbf{CP}$.

Equation (12.17*a*) and (12.17*b*) give another state model for a given system. Since **P** is assumed to be nonunique nonsingular matrix, the state model is also nonunique. It is important to note that the transformation matrix **P** must be nonsingular, *i.e.*, $|\mathbf{P}| \neq 0$. If this were not the case, the inverse transformation would obviously not exist.

This demonstrates that for a given system, infinitely many state models are possible. However, in a particular case the state model resulting from a specific choice of state variables may prove to be simpler and more useful than other choices of state models. Here, judgement has frequently to be depended upon. Some commonly used state variable formulations are discussed below.

State-Space Representation Using Physical Variables

The concept of state-space representation is perhaps best introduced by considering an example. We shall consider state variable formulation for a simple electrical system which is an *RLC* network shown in Fig. 12.5. The network has three energy storage elements: a capacitor C and two inductors L_1 and L_2. History of the network is completely specified by the voltage across the capacitor and currents through the inductors at $t = 0$. If we have a knowledge of initial conditions $v(0)$, $i_1(0)$, $i_2(0)$ and the input signal $e(t)$ for $t \geq 0$, then the behaviour of the network is completely specified for $t \geq 0$. However, if one (or more) of the initial conditions is not known, we are unable to determine the complete response of the network to a given input. Therefore, initial conditions $v(0)$, $i_1(0)$, $i_2(0)$ together with the input signal $e(t)$ for $t \geq 0$ constitute the minimal information needed. It then follows that a *natural* selection of the state variables (but by no means the only selection) would be

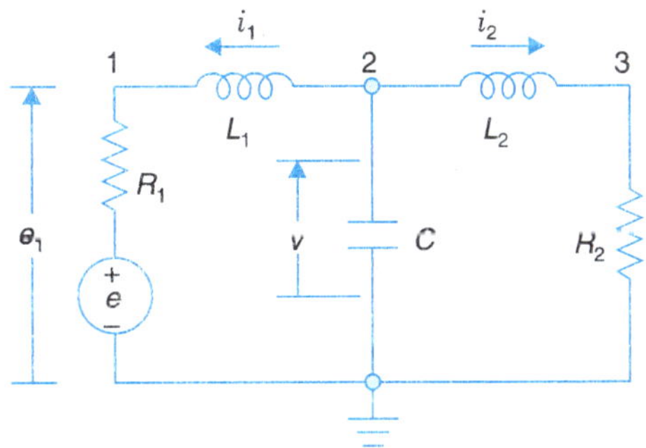

Fig. 12.5. An RLC network.

$$x_1(t) = v(t)$$
$$x_2(t) = i_1(t) \qquad \text{...(12.18)}$$
$$x_3(t) = i_2(t)$$

The differential equations governing the behaviour of the *RLC* network are

$$i_1 + i_2 + C\frac{dv}{dt} = 0$$
$$L_1\frac{di_1}{dt} + R_1 i_1 + e - v = 0 \qquad \text{...(12.19)}$$
$$L_2\frac{di_2}{dt} + R_2 i_2 - v = 0$$

We are interested in expressing the variables $\frac{dv}{dt}, \frac{di_1}{dt}$ and $\frac{di_2}{dt}$ as linear combinations of the variables v, i_1, i_2 and e as required in representation (12.13). For this purpose, eqns. (12.19) may be rewritten as follows :

$$\frac{dv}{dt} = -\frac{1}{C}i_1 - \frac{1}{C}i_2$$

$$\frac{di_1}{dt} = \frac{1}{L_1}v - \frac{R_1}{L_1}i_1 - \frac{1}{L_1}e$$

$$\frac{di_2}{dt} = \frac{1}{L_2}v - \frac{R_2}{L_2}i_2$$

In terms of the state variables defined in (12.18) and the input $u(t) = e(t)$, we have the following equation:

$$\begin{bmatrix} \dot{x}_1 \\ \dot{x}_2 \\ \dot{x}_3 \end{bmatrix} = \begin{bmatrix} 0 & -1/C & -1/C \\ -1/L_1 & -R_1/L_1 & 0 \\ 1/L_2 & 0 & -R_2/L_2 \end{bmatrix} \begin{bmatrix} x_1 \\ x_2 \\ x_3 \end{bmatrix} + \begin{bmatrix} 0 \\ -1/L_1 \\ 0 \end{bmatrix} u \qquad (12.20a)$$

Assume that voltage across R_2 and current through R_2 are the output variables y_1 and y_2 respectively. The output equations are then given by

$$\begin{bmatrix} y_1 \\ y_2 \end{bmatrix} = \begin{bmatrix} 0 & 0 & R_2 \\ 0 & 0 & 1 \end{bmatrix} \begin{bmatrix} x_1 \\ x_2 \\ x_3 \end{bmatrix} \qquad ...(12.20b)$$

This concludes the state-space representation of the *RLC* network. Equations (12.20) provide the state model of system.

Let us consider yet another example in which the electromechanical system of Fig. 12.6*a* is described in state variable form. Figure 12.6(*b*) gives the block diagram of the system which shows that it is a third-order system requiring three state variables to describe its dynamic behaviour. From the block diagram we choose the following set of state variables (in mechanical system, a *natural* choice of state variables is position and speed),

$$x_1 = \theta\,;\, x_2 = \dot{\theta}\,;\text{ and } x_3 = i_a$$

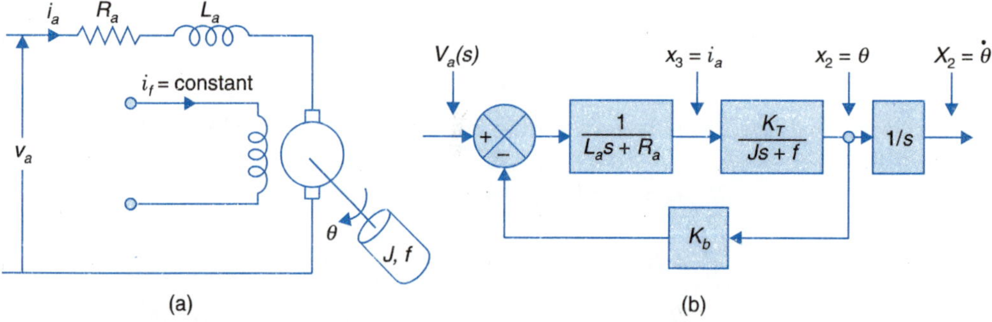

Fig. 12.6. State variables formulation for an electromechanical system.

We can now write down the following set of three first-order differential equations relating the inputs and outputs of the first-order factors $\frac{1}{s}, \frac{K_T}{Js+f}$ and $\frac{1}{L_a s + R_a}$

$$\dot{x}_1 = x_2$$

$$J\dot{x}_2 + fx_2 = K_T x_3$$

$$V_a - K_b x_2 = R_a x_3 + L_a \dot{x}_3$$

These three first-order differential equations can be organized into the vector-matrix form given below which provides the state variable description of the system dynamics.

$$\begin{bmatrix} \dot{x}_1 \\ \dot{x}_2 \\ \dot{x}_3 \end{bmatrix} = \begin{bmatrix} 0 & 1 & 0 \\ 0 & -f/J & K_T/J \\ 0 & -K_b/L_a & -R_a/L_a \end{bmatrix} \begin{bmatrix} x_1 \\ x_2 \\ x_3 \end{bmatrix} + \begin{bmatrix} 0 \\ 0 \\ 1/L_a \end{bmatrix} v_a \quad (12.21a)$$

If the motor angle is regarded as the output, it can be written as

$$y = \theta = x_1 = [1 \quad 0 \quad 0] \begin{bmatrix} x_1 \\ x_2 \\ x_3 \end{bmatrix} \quad ...(12.21b)$$

Example 12.1: For the flow control system of Fig. 12.7 prepare a signal flow graph. Identify suitable state variables and write down the state variable model. Given

$k_a = 25$, $k_p = 1$, $k_d = 0.005$

Motor : $k_m = 5$, $J = 0.05$, $R_a = 1\ \Omega$

Flow $q_i = k_q \theta$, $k_q = 8$, tank area $A = 50\ \text{m}^2$

$q_0 = k_h h$, $k_h = 225$, $k_f = 0.25$.

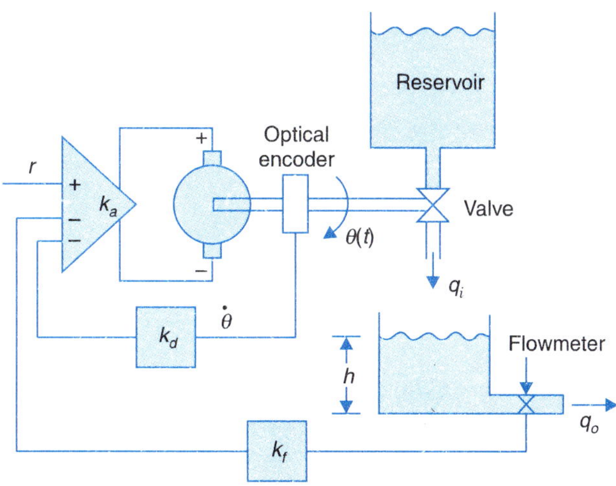

Fig. 12.7

Solution. The signal flow graph of the system is drawn in Fig. 12.8.

We choose the physical variables as the state variables. These are :

$$x_1 = q_0, x_2 = q_i, x_3 = \theta$$

Control variable, $u = r$

We shall now consider the feedback structure imposed on open-loop state model. For the open-loop model, $k_f = 0$ and $k_d = 0$ (The internal motor loop, k_b cannot be opened as it is inherent to the motor)

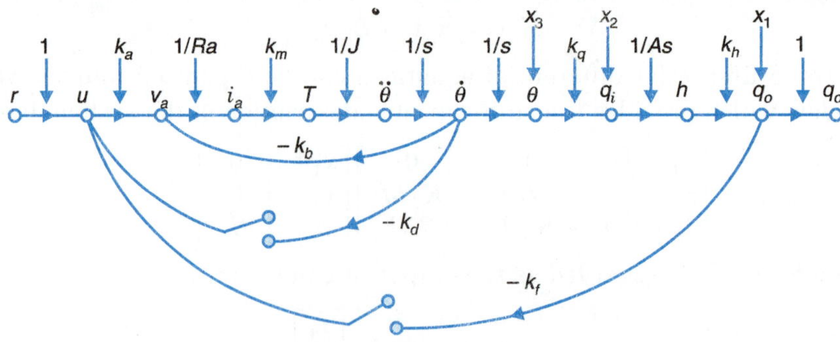

Fig. 12.8

The state variable model equations are written below from the signal flow graph :

$$\dot{x}_i = \frac{k_h}{A} x_2 \qquad \text{...}(i)$$

$$\dot{x}_2 = k_q x_3 \qquad \text{...}(ii)$$

$$\dot{x}_3 = (rk_a - k_b x_3) \frac{k_m}{R_a J} \qquad \text{...}(iii)$$

$$= \left(\frac{k_a k_m}{R_a J}\right) r - \left(\frac{k_b k_m}{R_a J}\right) x_3 \qquad \text{...}(iv)$$

$$y = q_0 = x_1$$

The state equation can be written in matrix form as below :

$$\begin{bmatrix} \dot{x}_1 \\ \dot{x}_2 \\ \dot{x}_3 \end{bmatrix} = \begin{bmatrix} 0 & \frac{k_h}{A} & 0 \\ 0 & 0 & k_q \\ 0 & 0 & -\left(\frac{k_b k_m}{R_a J}\right) \end{bmatrix} \begin{bmatrix} x_1 \\ x_2 \\ x_3 \end{bmatrix} + \begin{vmatrix} 0 \\ 0 \\ \frac{k_a k_m}{R_a J} \end{vmatrix} [u] \; ; u = r \qquad \text{...}(v)$$

$$y = [1\ 0\ 0] \begin{bmatrix} x_1 \\ x_2 \\ x_3 \end{bmatrix} \qquad \text{...}(vi)$$

or $$\dot{\mathbf{x}} = \mathbf{A}\mathbf{x} + \mathbf{b}u \qquad \text{...}(vii)$$

$$y = \mathbf{C}\mathbf{x}$$

Now consider the feedback structure imposed on the closed-loop state model with derivative feedback k_d and the feedback k_f. The feedback matrix is,

$$\mathbf{K}\mathbf{x} = [k_f \;\; 0 \;\; k_d] \begin{bmatrix} x_1 \\ x_2 \\ x_3 \end{bmatrix} \qquad \text{...}(viii)$$

Referring to the Fig. 12.8 with the feedback switches closed

$$u = -\mathbf{K}\mathbf{x} + r$$

Substituting in eqn. (*vii*)

$$\dot{x} = \mathbf{Ax} + \mathbf{b}(-\mathbf{Kx} + r)$$
$$= (\mathbf{A} - \mathbf{bK})\mathbf{x} + \mathbf{b}r$$

Now

$$\mathbf{bK} = \begin{bmatrix} 0 \\ 0 \\ \frac{k_a k_m}{R_a J} \end{bmatrix} [k_f \quad 0 \quad k_d] = \begin{bmatrix} 0 & 0 & 0 \\ 0 & 0 & 0 \\ \frac{k_f k_a k_m}{R_a J} & 0 & \frac{k k_a k_m}{R_a J} \end{bmatrix}$$

Then $(\mathbf{A} - \mathbf{bK})$ is,

$$\begin{bmatrix} 0 & \frac{k_h}{A} & 0 \\ 0 & 0 & k_q \\ -\left(\frac{k_f k_a k_m}{R_a J}\right) & 0 & \frac{-(k_b + k_d k_a)k_m}{R_a J} \end{bmatrix}$$

The state variable matrix can then be written as,

$$\begin{bmatrix} \dot{x}_1 \\ \dot{x}_2 \\ \dot{x}_3 \end{bmatrix} \begin{bmatrix} 0 & \frac{k_h}{A} & 0 \\ 0 & 0 & k_q \\ -\left(\frac{k_f k_a k_m}{R_a J}\right) & 0 & \frac{-(k_b + k_d k_a)k_m}{R_a J} \end{bmatrix} \begin{bmatrix} x_1 \\ x_2 \\ x_3 \end{bmatrix} + \begin{bmatrix} 0 \\ 0 \\ \frac{k_a k_m}{R_a J} \end{bmatrix} [u]$$

$$y = [1 \quad 0 \quad 0] \begin{bmatrix} x_1 \\ x_2 \\ x_3 \end{bmatrix} \qquad \ldots(x$$

Substituting the values we get,

$$\frac{k_h}{A} = \frac{225}{50} = 4.5$$
$$k_q = 8.0$$
$$-\left(\frac{k_f k_a k_m}{R_a J}\right) = -\frac{0.25 \times 25 \times 5}{1 \times 0.01} = -3.125 \times 10^3$$
$$\frac{-(k_b + k_d k_a)k_m}{R_a J} = -\frac{(0.05 + 25 \times 0.05) \times 5}{1 \times 0.01} = -130$$
$$\frac{k_a k_m}{R_a J} = \frac{25 \times 5}{1 \times 0.01} = 12.5 \times 10^3$$

$$\begin{bmatrix} \dot{x}_1 \\ \dot{x}_2 \\ \dot{x}_3 \end{bmatrix} = \begin{bmatrix} 0 & 4.5 & 0 \\ 0 & 0 & 8 \\ -3.125 \times 10^3 & 0 & -130 \end{bmatrix} \begin{bmatrix} x_1 \\ x_2 \\ x_3 \end{bmatrix} + \begin{bmatrix} 0 \\ 0 \\ 12.5 \times 10^3 \end{bmatrix} [u] \qquad \ldots(xi$$

$$y = [1 \quad 0 \quad 0] \begin{bmatrix} x_1 \\ x_2 \\ x_3 \end{bmatrix} \qquad ...(xii)$$

Example 12.2. Consider the system of Fig. 12.9, the classic and fascinating problem of inverted pendulum placed on a cart. The cart of mass M = 1 kg, has to be moved to keep the pendulum, of mass m = 0.05 kg, in the vertical position. The distance between the pivot and the pendulum l = 0.2 m. Express the state variables in terms of the angular rotation $\theta(t)$ and the position of the cart y(t). Assume that $M >> m$ and θ is very small to make the system linear.

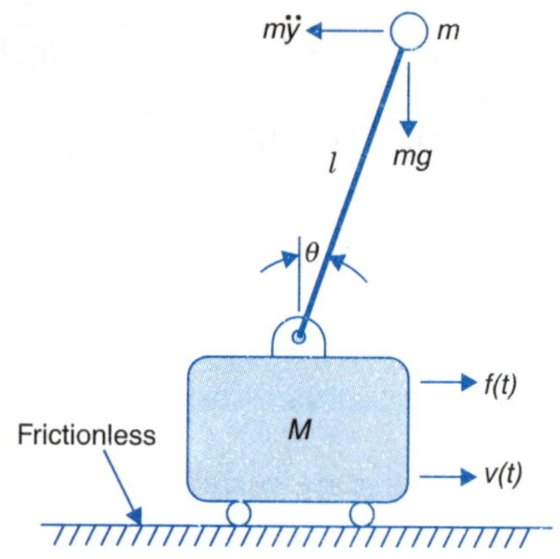

Fig. 12.9

Solution. The differential equations describing the motion of the system can be obtained by writing an expression for the net forces acting in the horizontal direction is,

$$M\ddot{y} = ml\ddot{\theta} - f(t) = 0 \qquad ...(i)$$

where $f(t)$ is the force on the cart. The total moments about the pivot is,

$$ml\ddot{y} + ml^2\ddot{\theta} - mlg\theta = 0 \qquad ...(i)$$

or

$$\ddot{y} = l\ddot{\theta} - g\theta = 0 \qquad ...(ii)$$

These are two second order equations. The state variables (physical) for these are taken as $(x_1, x_2, x_3, x_4) = (y, \dot{y}, \theta, \dot{\theta})$. Then the eqns. (i) and (ii) are written in terms of the state variables as

$$M\dot{x}_2 + ml\dot{x}_4 - f(t) = 0$$

or

$$\dot{x}_2 = -\frac{ml}{M}\dot{x}_4 + \frac{1}{M}f(t) \qquad ...(iii)$$

and

$$\dot{x}_2 + l\dot{x}_4 - gx_3 = 0$$

or

$$\dot{x}_4 = -\frac{1}{l}\dot{x}_2 + \frac{g}{l}x_3 \qquad ...(iv)$$

Substituting eqn. (iv) in eqn. (iii) to eliminate $\dot{x}_4$ we get

$$M\dot{x}_2 + ml\left(-\frac{1}{l}\dot{x}_2 + \frac{g}{l}x_3\right) - f(t) = 0$$

or

$$(M - m)\dot{x}_2 + mgx_3 - f(t) = 0$$

$$M >> m \Rightarrow M\dot{x}_2 + mgx_3 - f(t) = 0 \qquad ...(v)$$

Similarly substituting eqn. (iii) in eqn. (iv) to eliminate $\dot{x}_2$ we get

$$\left(l - \frac{ml}{M}\right)\dot{x}_4 - gx_3 + \left(\frac{1}{M}\right)f(t) = 0$$

or

$$(M - m)l\dot{x}_4 - Mgx_3 + f(t) = 0$$

or

$$Ml\dot{x}_4 - Mgx_3 + f(t) = 0 \qquad ...(vi)$$

Thus the state variable equations can be written as

$$\dot{x}_1 = x_2$$

$$\dot{x}_2 = -\left(\frac{mg}{M}\right)x_3 + \left(\frac{1}{M}\right)f(t)$$

$$\dot{x}_3 = x_4$$

$$\dot{x}_4 = \left(\frac{g}{l}\right)x_3 - \left(\frac{1}{Ml}\right)f(t)$$

Alternatively the equations in the state variable matrix form can be written as

$$\begin{bmatrix}\dot{x}_1\\ \dot{x}_2\\ \dot{x}_3\\ \dot{x}_4\end{bmatrix} = \begin{bmatrix}0 & 1 & 0 & 0\\ 0 & 0 & -\dfrac{mg}{M} & 0\\ 0 & 0 & 0 & 1\\ 0 & 0 & \dfrac{g}{l} & 0\end{bmatrix}\begin{bmatrix}x_1\\ x_2\\ x_3\\ x_4\end{bmatrix} + \begin{bmatrix}0\\ \dfrac{1}{M}\\ 0\\ -\dfrac{1}{Ml}\end{bmatrix}[f(t)]$$

Substituting the value we get

$$\frac{mg}{M} = \frac{0.05 \times 9.8}{1} = 0.49$$

$$\frac{g}{l} = \frac{9.8}{0.2} = 49$$

$$\frac{1}{M} = 1$$

$$\frac{1}{Ml} = \frac{1}{1 \times 0.2} = 5$$

$$\mathbf{A} = \begin{bmatrix}0 & 1 & 0 & 0\\ 0 & 0 & -0.49 & 0\\ 0 & 0 & 0 & 1\\ 0 & 0 & 49 & 0\end{bmatrix};\ \mathbf{B} = \begin{bmatrix}0\\ 1\\ 0\\ -5\end{bmatrix}$$

All the examples discussed above have one common feature; the selected state variables are the physical quantities of the systems which can be measured.

Later in this chapter we will see that in a physical system, in addition to output, other sign variables could be utilized for the purpose of feedback. The implementation of design with state variable feedback becomes straight forward if the state variables are available for feedback. The choice of physical variables of a system as state variables therefore helps in implementation of design.

Other advantage of selecting physical variables for state-space formulation is that the solution of state equation gives time variation of variables which have direct relevance to the physical system. However, with the choice of physical variables, the solution of state equation may become a difficult task. As we shall see later in this section, choice of *canonical* variables which may not even have physical meaning, is helpful in solution of state equation. The system variables of interest may then be obtained through algebraic manipulation of the solution of state equation.

State-space Representation Using Phase Variables

In the following, we discuss an alternate state-space representation of control systems using phase variables as state variables. The phase variables state model is easily determined if the system model is already known in the differential equation/transfer function form.*

The general form of an nth-order linear differential equation relating the output $y(t)$ to the input $u(t)$ of a linear continuous-time system is**

$$y^{(n)} + a_1 y^{(n-1)} + \ldots\ldots + a_{n-1}\dot{y} + a_n y = b_0 u^{(m)} + \ldots\ldots + b_{m-1}\dot{u} + b_m u \qquad \ldots(12.22)$$

where for time-invariant system a_i's and b_j's are constants, m and n are integers with $m \le n$ and $y^{(n)} \triangleq \dfrac{d^n y}{dt^n}$.

The initial conditions are expressed in terms of $y(0)$, $\dot{y}(0)$, ..., $y^{(n-1)}(0)$.

The transfer function obtained from eqn. (12.22) under the assumption of zero initial conditions is

$$T(s) = \frac{Y(s)}{U(s)} = \frac{b_0 s^m + b_1 s^{m-1} + \ldots\ldots + b_{m-1}s + b_m}{s^n + a_1 s^{n-1} + \ldots\ldots + a_{n-1}s + a_n} \qquad \ldots(12.23)$$

When*** $m = n$.

$$T(s) = \frac{b_0 s^n + b_1 s^{n-1} + \ldots\ldots + b_{n-1}s + b_n}{s^n + a_1 s^{n-1} + \ldots\ldots + a_{n-1}s + a_n} \qquad \ldots(12.24)$$

We shall obtain a state model from the transfer function with zero initial conditions and then relax the zero initial conditions to arbitrary inital conditions.

The phase variables are defined as those particular state variables which are obtained from one of the system variables and its derivatives. Often the variable used is the system output and the remaining state variables are then derivatives of the output.

Let us first consider a simple case, where the transfer function does not have zeros. Such a transfer function has the form

$$T(s) = \frac{Y(s)}{U(s)} = \frac{b}{s^n + a_1 s^{n-1} + \ldots\ldots + a_{n-1}s + a_n} \qquad \ldots(12.25)$$

to this transfer function corresponds the differential equation

$$y^{(n)} + a_1 y^{(n-1)} + \ldots\ldots + a_{n-1}\dot{y} + a_n y = bu \qquad \ldots(12.26)$$

---

*Quite often the system dynamics is determined experimentally using standard test signals like a step, impulse or sinusoidal signal. A transfer function is conveniently fitted to the experimental data in some best possible manner. The state model of the system is then obtained from the transfer function.

**Note that there is no loss in generality to assume the coefficient of the highest-order derivative of y to be unity.

***Most of the practical control schemes are realized with $m < n$; we will therefore be mainly concerned with this case. However, for the sake of generality, we will obtain a state model for the transfer function (12.24) ; from the general result, the state model for transfer function (12.23) with $m < n$ may then be obtained by putting appropriate coefficients b_i's equal to zero.

By letting

$$x_1 = y$$
$$x_2 = \dot{y}$$
$$\dots\dots\dots$$
$$x_n = y^{(n-1)} \qquad \dots(12.27)$$

equation (12.26) is reduced to a set of n first-order differential equations given below :

$$\dot{x}_1 = x_2$$
$$\dot{x}_2 = x_3$$
$$\dots\dots\dots$$
$$\dot{x}_{n-1} = x_n$$
$$\dot{x}_n = -a_n x_1 - a_{n-1} x_2 - \dots - a_1 x_n + bu$$

The above equations result in the following state equations :

$$\begin{bmatrix} \dot{x}_1 \\ \dot{x}_2 \\ \vdots \\ \dot{x}_{n-1} \\ \dot{x}_n \end{bmatrix} = \begin{bmatrix} 0 & 1 & 0 & \cdots & 0 \\ 0 & 0 & 1 & \cdots & 0 \\ \vdots & \vdots & \vdots & & \vdots \\ 0 & 0 & 0 & \cdots & 1 \\ -a_n & -a_{n-1} & -a_{n-2} & \cdots & -a_1 \end{bmatrix} \begin{bmatrix} x_1 \\ x_2 \\ \vdots \\ x_{n-1} \\ x_n \end{bmatrix} + \begin{bmatrix} 0 \\ 0 \\ \vdots \\ 0 \\ b \end{bmatrix} u \qquad \dots(12.28)$$

or $$\dot{\mathbf{x}} = \mathbf{A}\mathbf{x} + \mathbf{B}u \qquad \dots(12.29a)$$

It is to be noted that the matrix **A** has very special form. It has all 1's in the upper off-diagonal, its last row is comprised of the negative of the coefficients of the original differential equation and all other elements are zero. This form of matrix **A** is known as the *Bush form or companion form.*

Also note that **B** has the speciality that all its elements except the last are zero. In fact **A** and **B** and therefore the state equation can be written directly by inspection of the linear differential equation.

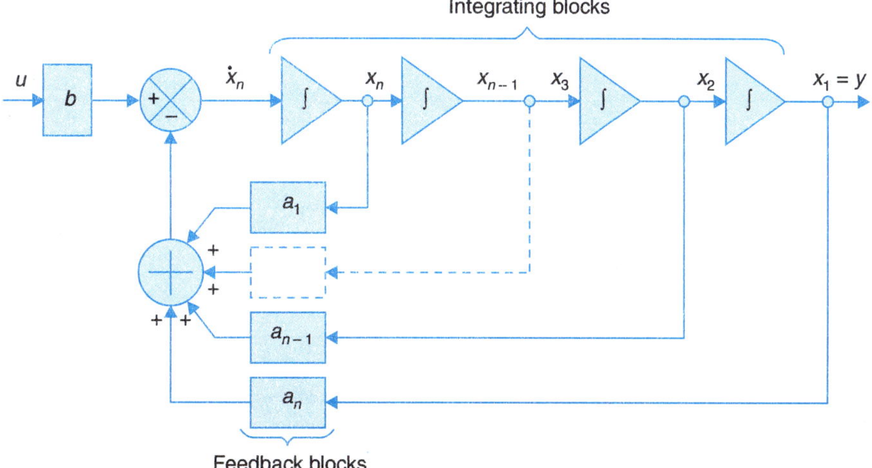

Fig. 12.10. Block diagram representation of the state model given by eqns. (12.29).

The output being $y = x_1$, the output equation is given by

$$y = \mathbf{Cx} \qquad ...(12.29b)$$

where $\mathbf{C} = [1 \quad 0 \quad .... \quad 0]$.

The initial conditions on y give rise to the initial conditions $x_1(0), x_2(0), ..., x_n(0)$ on the state variables as per the definition of the state variables given in eqns. (12.27). Figure 12.10 shows the block diagram representation of the state model derived above.

It is important to note that assuming the initial time equal to zero does not create any loss of generality. Since thè system under the discussion is time-invariant, the charge of state does not depend on initial time but depends only on the length of time during which the control force is applied.

It follows from the above that for the transfer functions with poles only, the derivation of the state model through the differential equation is quite straightforward. However, when a transfer function has zeros as well, the resulting differential equation contains terms which are derivatives of the control force u and the method discussed above can no longer be applied as such. An alternate method using signal flow graphs is presented below.

Phase variable formulations for transfer function with poles and zeros

Let us consider a third order transfer function

$$\frac{Y(s)}{U(s)} = T(s) = \frac{b_0 s^3 + b_1 s^2 + b_2 s + b_3}{s^3 + a_1 s^2 + a_2 s + a_3} \qquad ...(12.30)$$

We identify three state variables x_1, x_2 and x_3. The signal flow graph must have at least three integrators. Equation (12.30) may be rearranged as

$$T(s) = \frac{b_0 + b_1/s + b_2/s^2 + b_3/s^3}{1-(-a_1/s - a_2/s^2 - a_3/s^3)} \qquad ...(12.31)$$

In chapter 2, it was shown that the transfer function and signal flow graphs are related by Mason's gain formula, reproduced below.

$$T(s) = \frac{1}{\Delta}\sum_k P_k \Delta_k \qquad ...(12.32)$$

where P_k = Path gain of the kth forward path; $\Delta = 1 -$ (sum of loop gains of all individual loops) + (sum of gain products of all possible combinations of two non-touching loops) – (sum of gain products of all possible combinations of three non-touching loops) + ... ; Δ_k = the value of Δ for that part of the graph not touching the kth forward path.

Comparing eqn. (12.30) with (12.31) we observe that a signal flow graph of eqn. (12.29) may consist of

(*i*) three feedback loops (touching each other) with gains $-a_i/s$, $-a_2/s^2$ and a_3/s^3 ;

(*ii*) four forward paths which touch the loops and have gains b_0, b_1/s, b_2/s^2 and b_3/s^3

A signal flow graph configuration which satisfies the above requirements is shown in Fig. 12.11. From this figure, we have

$$y = x_1 + b_0 u$$

$$\begin{aligned}\dot{x}_1 &= -a_1(x_1 + b_0u) + x_2 + b_1u \\ &= -a_1x_1 + x_2 + (b_1 - a_1b_0)u \end{aligned} \qquad \text{...(12.33)}$$

$$\dot{x}_2 = -a_2x_1 + x_3 + (b_2 - a_2b_0)u$$

$$\dot{x}_3 = -a_3x_1 + (b_3 - a_3b_0)u$$

or

$$\begin{bmatrix}\dot{x}_1\\ \dot{x}_2\\ \dot{x}_3\end{bmatrix} = \begin{bmatrix}-a_1 & 1 & 0\\ -a_2 & 0 & 1\\ -a_3 & 0 & 0\end{bmatrix}\begin{bmatrix}x_1\\ x_2\\ x_3\end{bmatrix} + \begin{bmatrix}b_1 & -a_1b_0\\ b_2 & -a_2b_0\\ b_3 & -a_3b_0\end{bmatrix}[u] \qquad \text{...(12.34}a\text{)}$$

$$y = [1 \quad 0 \quad 0]\begin{bmatrix}x_1\\ x_2\\ x_3\end{bmatrix} + b_0u \qquad \text{...(12.34}b\text{)}$$

The initial conditions $x_1(0), x_2(0), x_3(0)$ can be obtained in terms of given initial conditions ($y(0)$, $\dot{y}(0)$, $\ddot{y}(0)$, $u(0)$, $\dot{u}(0)$, $\ddot{u}(0)$ from eqn. (12.33)). The result is given below

$$\begin{bmatrix}x_1(0)\\ x_2(0)\\ x_3(0)\end{bmatrix} = \begin{bmatrix}1 & 0 & 0\\ a_1 & 1 & 0\\ a_2 & a_1 & 1\end{bmatrix}\begin{bmatrix}y(0)\\ \dot{y}(0)\\ \ddot{y}(0)\end{bmatrix} + \begin{bmatrix}-b_0 & 0 & 0\\ -b_1 & -b_0 & 0\\ -b_2 & -b_1 & -b_0\end{bmatrix}\begin{bmatrix}u(0)\\ \dot{u}(0)\\ \ddot{u}(0)\end{bmatrix} \qquad \text{...(12.35)}$$

The results given in eqn. (12.33) and (12.34) can easily be generalized for an nth order differential equation

$$y^{(n)} + a_1y^{(n-1)} + ... + a_{n-1}\dot{y} + a_ny = b_0u^{(m)} + ... + b_{m-1}\dot{u} + b_m u \qquad \text{...(12.36)}$$

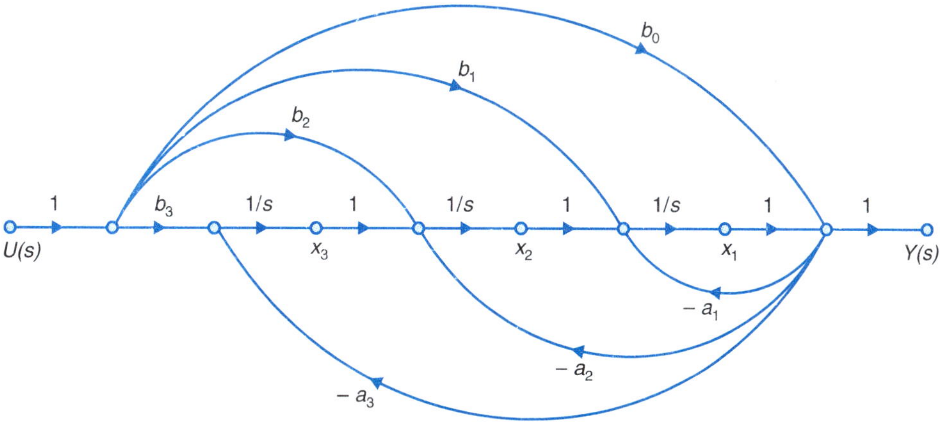

Fig. 12.11

An alternate phase variable formation for the transfer function of eqn. (12.31) may be obtained as follows:

The transfer function $G(s)$ of Fig. 12.12 may be divided into two parts as shown in Fig. 12.13. Thus

$$G(s) = \frac{Y(s)}{U(s)} = \frac{X_1(s)}{U(s)} \cdot \frac{Y(s)}{X_1(s)} \qquad \text{...(12.37)}$$

where $$\frac{X_1(s)}{U(s)} = \frac{1}{s^3 + a_1 s^2 + a_2 s + a_3} \quad \text{...(12.38)}$$

and $$\frac{Y(s)}{X_1(s)} = b_0 s^3 + b_1 s^2 + b_2 s + b_3 \quad \text{...(12.39)}$$

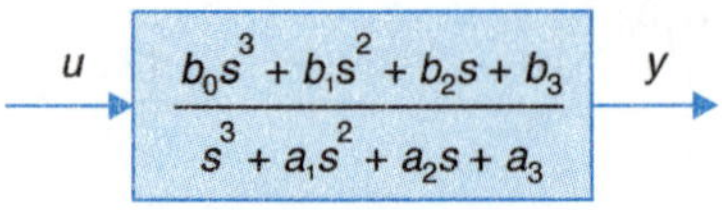

$u \rightarrow \boxed{\dfrac{b_0 s^3 + b_1 s^2 + b_2 s + b_3}{s^3 + a_1 s^2 + a_2 s + a_3}} \rightarrow y$

Fig. 12.12. A third-order system.

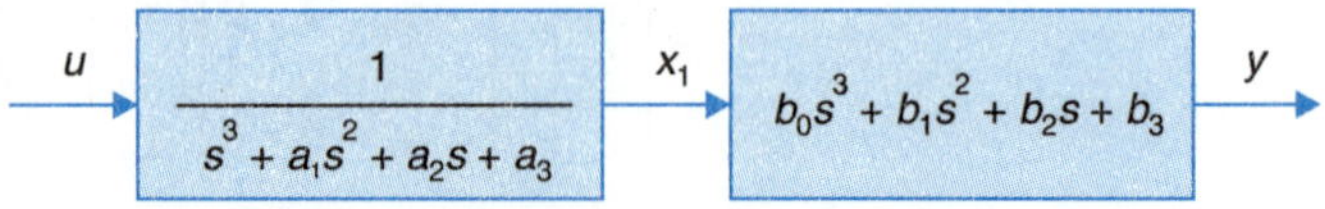

$u \rightarrow \boxed{\dfrac{1}{s^3 + a_1 s^2 + a_2 s + a_3}} \xrightarrow{x_1} \boxed{b_0 s^3 + b_1 s^2 + b_2 s + b_3} \rightarrow y$

Fig. 12.13. An alternate representation of system of Fig. 12.12.

The transfer function of eqn. (12.37) is without zeros and therefore its phase variable representation may be obtained from eqns. (12.28)

$$\begin{bmatrix} \dot{x}_1 \\ \dot{x}_2 \\ \dot{x}_3 \end{bmatrix} = \begin{bmatrix} 0 & 1 & 0 \\ 0 & 0 & 1 \\ -a_3 & -a_2 & -a_1 \end{bmatrix} \begin{bmatrix} x_1 \\ x_2 \\ x_3 \end{bmatrix} + \begin{bmatrix} 0 \\ 0 \\ 1 \end{bmatrix} u \quad \text{...(12.40}a\text{)}$$

From the transfer function of eqn. (12.38), we get

$$\begin{aligned} y &= b_0 \dddot{x}_1 + b_1 \ddot{x}_1 + b_1 \dot{x}_1 + b_3 x_1 \\ &= b_0(-a_3 x_1 - a_2 x_2 - a_1 x_3 + u) + b_1 x_3 + b_2 x_2 + b_3 x_1 \\ &= (b_3 - a_3 b_0) x_1 + (b_2 - a_2 b_0) x_2 + (b_1 - a_1 b_0) x_3 + b_0 u \end{aligned} \quad \text{...(12.40}b\text{)}$$

output equation (12.39*b*) can be written in vector-matrix form as

$$y = [(b_3 - a_3 b_0)\ (b_2 - a_2 b_0)\ (b_1 - a_1 b_0)] \begin{bmatrix} x_1 \\ x_2 \\ x_3 \end{bmatrix} + b_0 u \quad \text{...(12.40}c\text{)}$$

The signal flow graph corresponding to eqns. (12.40*a* and *c*) is drawn in Fig. 12.14.

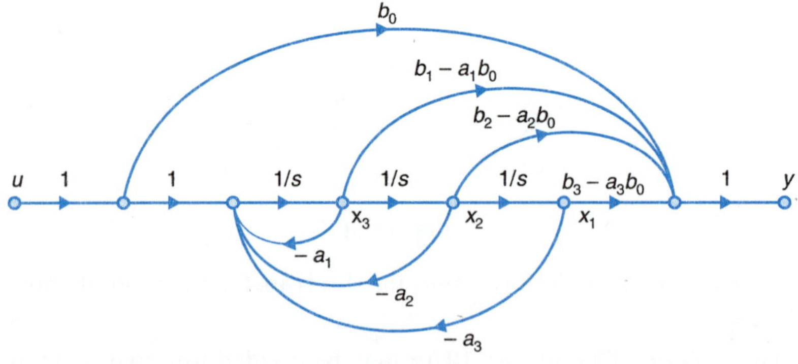

Fig. 12.14

The results of eqns. (12.39*a* and *c*) can be generalized for *n*th order case. It will be shown later in this chapter that this state variable form and its earlier forms given in eqns. 12.28 and 12.34 are completely controllable.

From the results given in (12.33), (12.34) and (12.40) (and their generalized forms) we observe that the phase variable formulation can be obtained by inspection from the transfer function and vice-versa. The reader will also note the similarity of phase variable representation as illustrated by Figs. 12.10 and 12.11 with the *direct programming method* of simulating transfer functions.

A disadvantage of phase variable formation is that the phase variables, in general, are not physical variables of the system and therefore are not available for measurement and control purposes. We have seen that if $G(s)$ has no zeros, the phase variables are given by its output and derivatives of output (eqns. (12.27)). Unfortunately, it becomes difficult to take second or higher derivatives of output. If $G(s)$ have zeros, then the phase variables bear little resemblance to real physical quantities in the system as is seen form eqn. (12.33). Thus, though phase variables are simple to realize mathematically, they are not a practical set a state variables from measurement and control point of view also, phase variables do not offer any advantage. As we shall see shortly, canonical variables are most suitable from analysis point of view.

In spite of these disadvantages, phase variables provide a powerful method of state variable formulation. A link between the transfer function design approach and time-domain design approach is established through phase variables. We shall discuss this link later in this chapter.

State-space Representation Using Canonical Variables

In *canonical-variable* or *normal-form* representation of a system, the matrix, **A** turns out to be a diagonal matrix. This form of state model plays an important role in control theory. Let us derive this model for a general transfer function (12.24), reproduced below :

$$\frac{Y(s)}{U(s)} = T(s) = \frac{b_0 s^n + b_1 s^{n-1} + \dots + b_{n-1}s + b_n}{s^n + a_1 s^{n-1} + \dots + a_{n-1}s + a_n}$$

Assume that the denominator is known in factored form and that the poles of the transfer function, located at $\lambda_1, \lambda_2, \dots, \lambda_n$, are all distinct. The transfer function can then be expanded into partial fractions as (see Appendix I)

$$\frac{Y(s)}{U(s)} = T(s) = b_0 + \sum_{i=1}^{n} \frac{c_i}{s - \lambda_i}$$

where c_i are the residues of the poles at $s = \lambda_i$.

The block diagram model for the transfer function is shown in Fig. 12.15. Defining the output of each integrator to be a state variable, we can write the state equations as

$$\dot{x}_1 = \lambda_i x_i + u; \quad i = 1, 2, \dots, n \tag{12.41a}$$

The output $y(t)$ is given by

$$y = c_1 x_1 + c_2 x_2 + \dots + c_n x_n + b_0 u \qquad \text{...(12.41b)}$$

Equations (12.41) describe the canonical state model of the transfer fuction of eqn. (12.40). This state model can be expressed in the vector-matrix form as

$$\begin{bmatrix} \dot{x}_1 \\ \dot{x}_2 \\ \vdots \\ \dot{x}_n \end{bmatrix} = \begin{bmatrix} \lambda_1 & 0 & 0 & \cdots & 0 \\ 0 & \lambda_2 & 0 & \cdots & 0 \\ \vdots & \vdots & \vdots & & \vdots \\ 0 & 0 & 0 & \cdots & \lambda_n \end{bmatrix} \begin{bmatrix} x_1 \\ x_2 \\ \vdots \\ x_n \end{bmatrix} + \begin{bmatrix} 1 \\ 1 \\ \vdots \\ 1 \end{bmatrix} u \quad \text{(12.42a)}$$

$$y = [c_1 \quad c_2 \quad \ldots \quad c_n] \begin{bmatrix} x_1 \\ x_2 \\ \vdots \\ x_n \end{bmatrix} + b_0 u \quad \text{...(12.42b)}$$

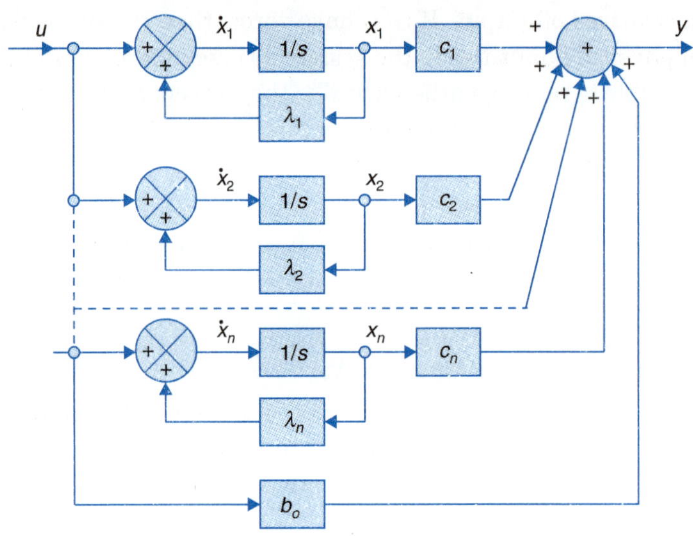

Fig. 12.15. Block diagram of a canonical state model.

It is observed that for the canonical side model described above, *the system matrix* **A** *is a diagonal matrix with the poles of* $T(s)$ *as its diagonal elements*. It is also observed that elements of column vector **B** of the canonical state model are all unity and the elements of the row vector **C** are the residues of the system poles. An alternate canonical state model is given below.

$$\begin{bmatrix} \dot{x}_1 \\ \dot{x}_2 \\ \vdots \\ \dot{x}_n \end{bmatrix} = \begin{bmatrix} \lambda_1 & 0 & 0 & \cdots & 0 \\ 0 & \lambda_2 & 0 & \cdots & 0 \\ \vdots & \vdots & \vdots & & \vdots \\ 0 & 0 & 0 & \cdots & \lambda_n \end{bmatrix} \begin{bmatrix} x_1 \\ x_2 \\ \vdots \\ x_n \end{bmatrix} + \begin{bmatrix} c_1 \\ c_2 \\ \vdots \\ c_n \end{bmatrix} u \quad \text{...(12.43a)}$$

$$y = [1 \quad 1 \quad \ldots \quad 1] \begin{bmatrix} x_1 \\ x_2 \\ \vdots \\ x_3 \end{bmatrix} + b_0 u \quad \text{...(12.43b)}$$

The unique decoupled nature of canonical or normal-form representation is obvious from eqns. (12.43). By decoupled we refer to the fact that in normal form, the n first-order differential equations are completely independent of each other. This decoupling feature, as we shall see later in this chapter, greatly helps in the analysis of the system.

The disadvantage of the canonical form is equally important. The canonical variables, like phase variables, are not real physical variables of the system. This is illustrated through the following example :

Consider the differential equation

$$\dddot{y} + 6\ddot{y} + 11\dot{y} + 6y = \dddot{u} + 8\ddot{u} + 17\dot{u} + 8u$$

Assuming zero initial condition, we get

$$\frac{Y(s)}{U(s)} = T(s) = \frac{s^3 + 8s^2 + 17s + 8}{s^3 + 6s^2 + 11s + 6} \qquad \text{...(12.44)}$$

$$= \frac{s^3 + 8s^2 + 17s + 8}{(s+1)(s+2)(s+3)}$$

$$= 1 - \frac{1}{s+1} + \frac{2}{s+2} + \frac{1}{s+3} \qquad \text{...(12.45)}$$

The block diagram simulation for $T(s)$ of eqn. (12.45) is shown in Fig. 12.16. From eqn. (12.44), we can write the following :

$$\frac{sY(s)}{U(s)} = s + 2 + \frac{1}{s+1} - \frac{4}{s+2} - \frac{3}{s+3} \qquad \text{...(12.46}a\text{)}$$

$$\frac{s^2Y(s)}{U(s)} = s^2 + 2s - 6 - \frac{1}{s+1} + \frac{8}{s+2} + \frac{9}{s+3} \qquad \text{...(12.46}b\text{)}$$

Equations (12.45) and (12.46) may be expressed as

$$\begin{bmatrix} \dfrac{Y(s) - U(s)}{U(s)} \\ \dfrac{sY(s) - sU(s) - 2U(s)}{U(s)} \\ \dfrac{s^2Y(s) - s^2U(s) - 2U(s) + 6U(s)}{U(s)} \end{bmatrix} = \begin{bmatrix} -1 & 2 & 1 \\ 1 & -4 & -3 \\ -1 & 8 & 9 \end{bmatrix} \begin{bmatrix} \dfrac{1}{s+1} \\ \dfrac{1}{s+2} \\ \dfrac{1}{s+3} \end{bmatrix}$$

$$= \begin{bmatrix} -1 & 2 & 1 \\ 1 & -4 & -3 \\ -1 & 8 & 9 \end{bmatrix} \begin{bmatrix} \dfrac{X_1(s)}{U(s)} \\ \dfrac{X_2(s)}{U(s)} \\ \dfrac{X_3(s)}{U(s)} \end{bmatrix}$$

This gives

$$\begin{bmatrix} y - u \\ \dot{y} - \dot{u} - 2u \\ \ddot{y} - \ddot{u} - 2\dot{u} + 6u \end{bmatrix} = \begin{bmatrix} -1 & 2 & 1 \\ 1 & -4 & -3 \\ -1 & 8 & 9 \end{bmatrix} \begin{bmatrix} x_1 \\ x_2 \\ x_3 \end{bmatrix} \qquad \text{...(12.47)}$$

From eqns. (12.47) we find that canonical variables are not physical variables of the system and therefore are not available for measurement and control. (Note that the initial conditions $x_1(0)$, $x_2(0)$, $x_3(0)$, may be obtained in terms of specific initial conditions $y(0)$, $\dot{y}(0)$, $\ddot{y}(0)$, $u(0)$, $\dot{u}(0)$, $\ddot{u}(0)$ from eqns. (12.47).

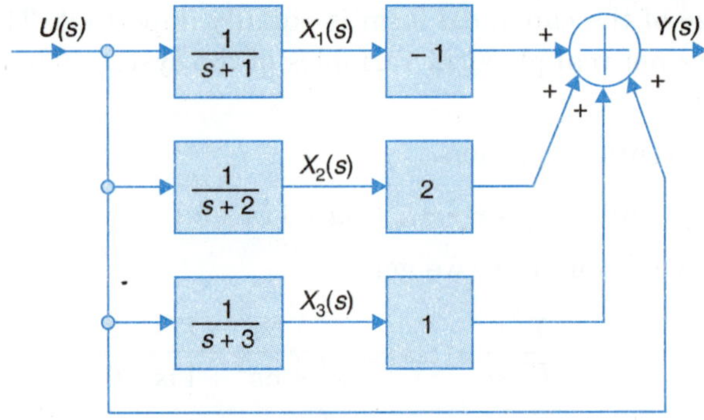

Fig. 12.16. Block diagram for T(s) of eqn. (12.46a).

Example 12.3. There have been significant developments in the application of robotics technology to various fields. One such application is the *Intelligent Wheelchair*. The wheelchair has to move from place to place avoiding obstacles. The simplified block diagram of the controller and the wheelchair is depicted in Fig. 12.17. Write down the canonical state variable form for the complete system, Also draw its block diagram (in state variable form).

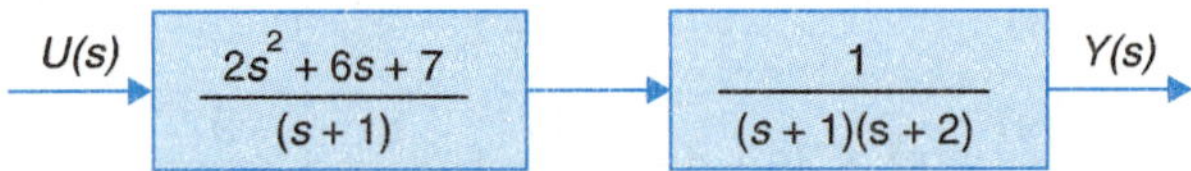

Fig. 12.17

Solution. The transfer function of the complete system is

$$\frac{Y(s)}{U(s)} = \frac{2s^2 + 6s + 7}{(s+1)^2\,(s+2)}$$

Notice that the system has repeated roots. Decomposing the above transfer function by the method of partial fractions yields

$$\frac{Y(s)}{U(s)} = \frac{-1}{(s+1)} + \frac{3}{(s+1)^2} + \frac{3}{(s+2)}$$

We shall define the state variables term by term. Because of repeated root at $s = -1$, the state variable of the second term will feed into the first term.

$$\frac{X_2(s)}{(s+1)} = X_1(s) \;\Rightarrow\; x_2 = \dot{x}_1 + x_1 \;\Rightarrow\; \dot{x}_1 = -x_1 + x_2 \qquad \text{...(i)}$$

$$\frac{U(s)}{(s+1)} = X_2(s) \;\Rightarrow\; u = \dot{x}_2 + x_2 \;\Rightarrow\; \dot{x}_2 = -x_2 + u \qquad \text{...(ii)}$$

$$\frac{U(s)}{(s+2)} = X_3(s) \;\Rightarrow\; u = \dot{x}_3 + 2x_3 \;\Rightarrow\; \dot{x}_3 = -2x_3 + u \qquad \text{...(iii)}$$

Output $$y = -x_1 + 3x_2 + 3x_3 \qquad \text{...(iv)}$$

The state variable model in matrix form is written below from the above eqns.

Jordan block

↓

$$\begin{bmatrix} \dot{x}_1 \\ \dot{x}_2 \\ \dot{x}_3 \end{bmatrix} = \begin{bmatrix} -1 & 1 & 0 \\ 0 & -1 & 0 \\ 0 & 0 & -2 \end{bmatrix} \begin{bmatrix} x_1 \\ x_2 \\ x_3 \end{bmatrix} + \begin{bmatrix} 0 \\ 1 \\ 1 \end{bmatrix} [u] \qquad \ldots(v)$$

$$y = [-1 \quad 3 \quad 3] \begin{bmatrix} x_1 \\ x_2 \\ x_3 \end{bmatrix} \qquad \ldots(vi)$$

The state variable block diagram is drawn in Fig. 12.18.

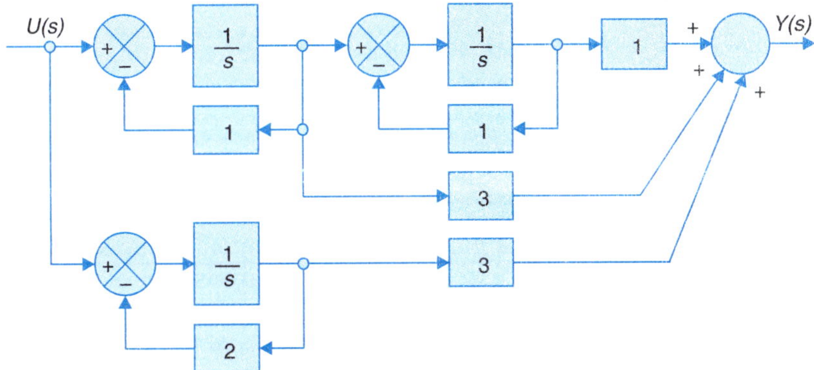

Fig. 12.18

Observation. Notice the dotted block in matrix **A** (eqn. (v)). Therefore, x_1 and x_2, are not in decoupled form (because of repeated root at $s = -1$). This block (dotted) is known as *Jordan block*. This is the simplest decoupled form as far as possible.

Derivation of Transfer Function from State Model

Having answered the question of obtaining the state model for a given transfer function (continuous-time or discrete-time), we next consider the problem of determining the transfer function from a given state model of single-input-single output systems.

The link between transfer function and phase variable formulation has already been demonstrated. For the continuous case the transfer function from a state model using phase variables can be obtained by inspiration from the general results (12.28), (12.34) or (12.40). An alternate method is to draw signal flow graph for the given phase variable model and then obtain the transfer function using Mason's gain formula. Discrete-time case will follow later (page 599).

For a general state model

$$\mathbf{x} = \mathbf{A}\mathbf{x} + \mathbf{B}u \qquad \ldots(12.48a)$$

$$\mathbf{y} = \mathbf{C}\mathbf{x} + du \qquad \ldots(12.48b)$$

the transfer function may be obtained as follows:

Taking the Laplace transform of eqns. (12.50), we have

$$s\mathbf{X}(s) - \mathbf{x}(0) = \mathbf{A}\mathbf{x}(s) + \mathbf{B}U(s)$$

$$Y(s) = \mathbf{C}\mathbf{X}(s) + dU(s)$$

Solving for $Y(s)$, we obtain

$$Y(s) = \mathbf{C}(s\mathbf{I} - \mathbf{A})^{-1}\,\mathbf{x}(0) + \mathbf{C}(s\mathbf{I} - \mathbf{A})^{-1}\,\mathbf{B}U(s) + dU(s)$$

Assuming zero initial conditions, we get the system transfer function as

$$T(s) = \frac{Y(s)}{U(s)} = \mathbf{C}(s\mathbf{I} - \mathbf{A})^{-1}\mathbf{B} + d$$

$$= \frac{\mathbf{C}\,\text{adj}\,(s\mathbf{I} - \mathbf{A})\mathbf{B}}{\det\,(s\mathbf{I} - \mathbf{A})} + d \qquad \text{...(12.49)}$$

An important observation that needs to be made here is that while the state model is nonunique, *the transfer function of the system is unique, i.e.,* the transfer function of eqn. (12.51) must work out to be the same irrespective of which particular state model is used to describe the system.

Setting the denominator of Eq. (12.49), we get the characteristic equation

$$|\, s\mathbf{I} - \mathbf{A} \,| = \mathbf{0}$$

12.4 STATE VARIABLES AND LINEAR DISCRETE-TIME SYSTEMS

State variable methods for the analysis and design of continuous-time systems developed in earlier sections of this chapter can be extended for the analysis and design of discrete-time systems.

In Chapter 11, we discussed analysis and design of discrete-time systems using z-transfer function. There we observed that while dealing with discrete time systems, we often encounter two different situations. The first situation involves systems that are completely discrete with respect to time in the sense that they receive and send out discrete signals only. It is hard to find examples of physical systems of this nature, other than digital computers. Most of the physical systems are of continuous-time type. In socio-economic and some other non-physical systems, a natural discrete unit of time exists and such systems can effectively be modelled using difference equations.

The second situation is that the components of the system are continuous time elements but the signals at certain points of the system are discrete with respect to time because of the sample and hold operations. Both the cases will be studied in this section.

The general form of state model for a multivariable discrete-time system is

$$\mathbf{x}[(k + 1)T] = \mathbf{f}[\mathbf{x}(kT), \mathbf{u}(kT)] \qquad \text{...(12.50}a\text{)}$$

$$\mathbf{y}(kT) = \mathbf{g}[(\mathbf{x}(kT), \mathbf{u}(kT)] \qquad \text{...(12.50}b\text{)}$$

where $\mathbf{x}(kT)$ state vector, $\mathbf{u}(kT)$ = input vector and $\mathbf{y}(kT)$ = output vector.

From these equations we see that given the initial state $\mathbf{x}(0)$ and values of inputs $\mathbf{u}(0)$, $\mathbf{u}(T)$, $\mathbf{u}(2T)$, ..., $\mathbf{u}(kT)$; we can uniquely deduce the evolution of state $\mathbf{x}(T)$, $\mathbf{x}(2T)$, ..., $\mathbf{x}[(k + 1)T]$ and the evolution of the output $\mathbf{y}(0)$, $\mathbf{y}(T)$, ..., $\mathbf{y}(kT)$. Also note that eqn. (12.50a) is a set of first-order difference equations which represents dynamics of discrete-time process and eqn.(12.50b) is an algebraic equation.

For an nth-order linear time-invariant system, eqns. (12.50) reduce to the following form:

$$\mathbf{x}(k + 1) = \mathbf{A}\mathbf{x}(k) + \mathbf{B}\mathbf{u}(k)\,;\; \mathbf{x}(kT) \underline{\underline{\Delta}}\, \mathbf{x}(k) \qquad \text{...(12.51}a\text{)}$$

$$\mathbf{y}(k) = \mathbf{C}\mathbf{x}(k) + \mathbf{D}\mathbf{u}(k) \qquad \text{...(12.51}b\text{)}$$

$\mathbf{x}(k) = n \times 1$ state vector

$\mathbf{u}(k) = m \times 1$ input vector

$\mathbf{y}(k) = p \times 1$ output vector

$\mathbf{A} = n \times n$ system matrix

$\mathbf{B} = n \times m$ input matrix

$\mathbf{C} = p \times n$ output matrix

$\mathbf{D} = p \times m$ transmission matrix

State Models from Linear Difference Equations/z-transfer Functions

Consider a third order system with the following z-transfer function

$$T(z) = \frac{Y(z)}{U(z)} = \frac{b_0 z^3 + b_1 z^2 + b_2 z + b_3}{z^3 + a_1 z^2 + a_2 z + a_3} \qquad \ldots(i)$$

$$= \frac{b_0 + b_1 z^{-1} + b_2 z^{-2} + b_3 z^{-3}}{1 + a_1 z^{-1} + a_2 z^{-2} + a_3 z^{-3}} \qquad \ldots(ii)$$

The implementation of this transfer function by a digital computer can be done in many ways. Two methods are illustrated in terms of state diagrams in the following:

Method 1: The signal flow graph for $T(z)$ of eqn. (i) is shown in Fig. 12.19 (z^{-1} represents a time delay of T sec).

Taking the output of the delay elements as state variables we write

$$\begin{aligned} x_1(k+1) &= x_2(k) \\ x_2(k+1) &= x_3(k) \\ x_3(k+1) &= -a_3 x_1(k) - a_2 x_2(k) - a_1 x_3(k) + u(k) \end{aligned} \qquad \ldots(iii)$$

$$y(k) = (b_3 - a_3 b_0)\, x_1(k) + (b_2 - a_2 b_0)\, x_2(k) + (b_1 - a_1 b_0)\, x_3(k) + b_0 u(k)$$

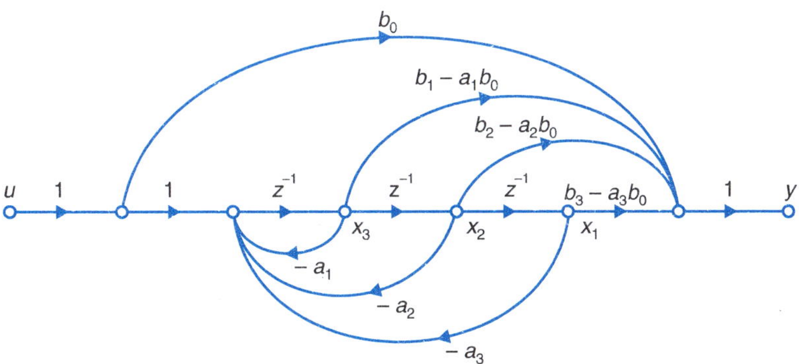

Fig. 12.19

These equations can be written as

$$\mathbf{x}(k+1) = \mathbf{A}\mathbf{x}(k) + \mathbf{B}u(k)$$

$$y(k) = \mathbf{C}\mathbf{x}(k) + b_0 u(k) \qquad \ldots(iv)$$

where

$$\mathbf{A} = \begin{bmatrix} 0 & 1 & 0 \\ 0 & 0 & 1 \\ -a_3 & -a_2 & -a_1 \end{bmatrix}; \mathbf{B} = \begin{bmatrix} 0 \\ 0 \\ 1 \end{bmatrix} \quad ...(v)$$

$$\mathbf{C} = [(b_3 - a_3b_0)\ (b_2 - a_2b_0)\ (b_1 - a_1b_0)]$$

This is phase variable form of state model.

Method 2 : Another state space representation can be obtained by its partial fraction technique. First let us consider $T(z)$ of eqn. (*i*) of method 1 with $b_0 = 1, b_1 = 8, b_2 = 17, b_3 = 8, a_1 = 6, a_2 = 11,$ and $a_3 = 6$.

$$T(z) = \frac{z^3 + 8z^2 + 17z + 8}{z^3 + 6z^2 + 11z + 6} \quad ...(vi)$$

$$= 1 - \frac{1}{z+1} + \frac{2}{z+2} + \frac{1}{z+3} \quad ...(vii)$$

The block diagram for this $T(z)$ is shown is Fig. 12.20. Taking the outputs of delay elements as state variables, we get the following state variable model :

$$\begin{bmatrix} x_1(k+1) \\ x_2(k+1) \\ x_3(k+1) \end{bmatrix} \begin{bmatrix} -1 & 0 & 0 \\ 0 & -2 & 0 \\ 0 & 0 & -3 \end{bmatrix} \begin{bmatrix} x_1(k) \\ x_2(k) \\ x_3(k) \end{bmatrix} + \begin{bmatrix} 1 \\ 1 \\ 1 \end{bmatrix} u(k) \quad ...(viii)$$

$$y(k) = [-1 \quad 2 \quad 1] \begin{bmatrix} x_1(k) \\ x_2(k) \\ x_3(k) \end{bmatrix} + u(k) \quad ...(ix)$$

This is the canonical state model. When some of the poles of $T(z)$ are repeated, we get Jordan canonical form. The method of obtaining this form is identical to that of continuous case.

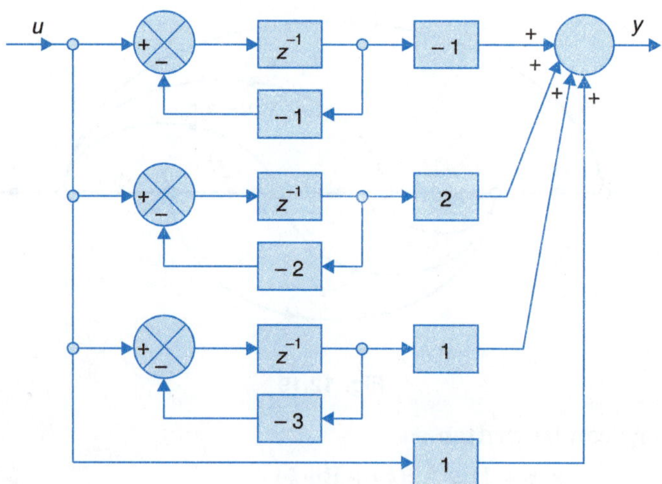

Fig. 12.20

Derivation of z-Transfer Function from Discrete-Time State Model

Taking the z-transform on both sides of eqn. (12.49a) (assuming scalar input),

$$z\mathbf{X}(z) - z\mathbf{x}(0) = \mathbf{A}\mathbf{X}(z) + \mathbf{B}U(z)$$

Solving for $\mathbf{X}(z)$, we get

$$\mathbf{X}(z) = (z\mathbf{I} - \mathbf{A})^{-1} z\mathbf{x}(0) + (z\mathbf{I} - \mathbf{A})^{-1} \mathbf{B}U(z) \quad ...(12.52)$$

Taking the z-transform of eqn. (12.49b) we get (for scalar output, $\mathbf{D} = d$),

$$Y(z) = \mathbf{C}\mathbf{X}(z) + dU(z)$$

Solving for $Y(z)$, we obtain

$$Y(z) = \mathbf{C}(z\mathbf{I} - \mathbf{A})^{-1} z\mathbf{x}(0) + \mathbf{C}(z\mathbf{I} - \mathbf{A})^{-1} \mathbf{B}U(z) + dU(z)$$

Assuming zero initial conditions, we get the system transfer function as

$$T(z) = \frac{Y(z)}{U(z)} = \mathbf{C}(z\mathbf{I} - \mathbf{A})^{-1}\mathbf{B} + d = \frac{C \operatorname{adj}(z\mathbf{I} - \mathbf{A})\mathbf{B}}{|z\mathbf{I} - \mathbf{A}|} + d \quad ...(12.53)$$

Setting the denominator equal to zero, we have the characteristic equation

$$|z\mathbf{I} - \mathbf{A}| = \mathbf{0} \quad ...(12.54)$$

The roots of the characteristic equation are the eigenvalues of matrix **A**.

12.5 DIAGONALIZATION

We observed in earlier sections that the state model of a system is not unique. Some of the state models presented employed physical variables, phase variables and canonical variables. From application point of view, the physical variables for system representation are most useful as the resulting state variables are real physical variables which can be easily measured and used for control purposes. However, the corresponding state model in this case generally not convenient for investigation of system properties and evaluation of time response. The canonical state model wherein **A** is in diagonal form is most suitable for this purpose. It is therefore, useful to study techniques for transforming a general state model into a canonical one. These techniques are often referred to as *diagonalization* techniques.

Consider an nth-order-multi-input-multi-output state model

$$\dot{\mathbf{x}} = \mathbf{A}\mathbf{x} + \mathbf{B}\mathbf{u} \quad ...(12.55a)$$

$$\mathbf{y} = \mathbf{C}\mathbf{x} + \mathbf{D}\mathbf{u} \quad ...(12.55b)$$

Assume that the matrix **A** in this model is nondiagonal. Let us define a new state vector v such that

$$\mathbf{x} = \mathbf{M}\mathbf{v}$$

where **M** is a $(n \times n)$ nonsingular constant matrix. Under this transformation, the original state model modifies to

$$\dot{\mathbf{v}} = \mathbf{M}^{-1} \mathbf{A}\mathbf{M}\mathbf{v} + \mathbf{M}^{-1} \mathbf{B}\mathbf{u} \quad ...(12.56a)$$

$$\mathbf{y} = \mathbf{C}\mathbf{M}\mathbf{v} + \mathbf{D}\mathbf{u} \quad ...(12.56b)$$

If the matrix **M** can be selected such that $\mathbf{M}^{-1}\mathbf{AM}$ is a diagonalized matrix **A**, then the model given by (12.56) is a canonical state model. Under this condition, the matrix **M** is called the *diagonalizing matrix or the modal matrix*. Equations (12.56) may be written as

$$\dot{\mathbf{v}} = \mathbf{\Lambda v} + \tilde{\mathbf{B}}\mathbf{u} \qquad (12.57a)$$

$$\mathbf{y} = \tilde{\mathbf{C}}\mathbf{v} + \mathbf{Du} \qquad ...(12.57b)$$

where $\mathbf{\Lambda} = \mathbf{M}^{-1}\mathbf{AM}$ = a diagonal matrix ; $\tilde{\mathbf{B}} = \mathbf{M}^{-1}\mathbf{B}$; and $\tilde{\mathbf{C}} = \mathbf{CM}$.

The determination of the diagonalizing matrix is facilitated by the use of eigenvectors. With this purpose in mind, the concept of eigenvalues and eigenvectors is briefly reviewed below.

Eigenvalues-and Eigenvectors

Let us now give attention to the equation

$$\mathbf{Ax} = \mathbf{y}$$

and view it as transformation of $n \times 1$ vector **x** to $n \times 1$ vector **y** by $n \times n$ matrix operator **A**. The question which we will like to answer here is : whether there exists a vector **x** such that matrix operator **A** transforms it to a vector $\lambda\mathbf{x}$ (λ is a constant) *i.e.*, to a vector having the same direction in state space as the vector **x**. Such a vector **x** is a solution of the equation

$$\mathbf{Ax} = \lambda\mathbf{x} \qquad ...(12.58)$$

Now our question is about the existence of solution of eqn. (12.58), which may be written as

$$(\lambda\mathbf{I} - \mathbf{A})\mathbf{x} = 0 \qquad ...(12.59)$$

The set of homogeneous eqns. (12.59) have a nontrivial solution if and only if

$$|\,\lambda\mathbf{I} - \mathbf{A}\,| = 0 \qquad ...(12.60a)$$

This equation may be expressed in expanded form as

$$q(\lambda) = \lambda^n + a_1\lambda^{n-1} + a_2\lambda^{n-2} + ... + a_n = 0 \qquad ...(12.60b)$$

The values of λ for which eqn. (12.60*a*) is satisfied are called *eigenvalues* of matrix **A** and eqn. (12.60*b*) is called the *characteristic equation* corresponding to matrix **A**.

From eqn. (12.51) it is observed that the poles of the transfer function $T(s)$ are given by $q(s) = |\,s\mathbf{I} - \mathbf{A}\,| = 0$. This in fact is the same equation as eqn. (12.60*a*) which determines the eigenvalues of **A**. It is therefore concluded that the eigenvalues of the state model and the poles of the system transfer function are the same. All the conclusions established in Chapter 6 on system stability based upon the location of transfer function poles (*i.e.*, roots of the characteristic equation) are therefore valid for eigenvalues of state model. Thus a state model is stable if all its eigenvalues have negative real parts. (In Chapter 13, we shall study stability in details).

Now for $\lambda = \lambda_i$ satisfying eqn. (12.57*a*), we have from eqn. (12.59)

$$(\lambda_i\,\mathbf{I} - \mathbf{A})\mathbf{x} = \mathbf{0} \qquad ...(12.61)$$

Let $\mathbf{x} = \mathbf{m}_i$ be the solution of this equation. The solution $\mathbf{m}_i$ is called the *eigenvector* of **A** associated with eigenvalue λ_i.

As said earlier, solution of (12.61) depends on the rank of matrix $(\lambda \mathbf{I} - \mathbf{A})$. If r is the rank of this matrix, then there are $(n - r)$ independent solutions (eigenvectors).

If the eigenvalues of matrix $\mathbf{A}$ are all distinct, then the rank r of matrix $(\lambda \mathbf{I} - \mathbf{A})$ is $(n - 1)$ and hence we have only one independent eigenvector corresponding to any particular eigenvalue λ_i. This eigenvector, may be obtained by taking cofactors of matrix $(\lambda_i \mathbf{I} - \mathbf{A})$ along any row *i.e.,*

$$\mathbf{m}_i = \begin{bmatrix} C_{k_1} \\ C_{k_2} \\ \vdots \\ C_{k_n} \end{bmatrix}; k = 1, 2, ..., \text{ or } n$$

where C_{k_i} are the co-factors of matrix $(\lambda_i \mathbf{I} - \mathbf{A})$.

Let $\mathbf{m}_1, \mathbf{m}_2, ..., \mathbf{m}_n$ be the eigenvectors corresponding to the eigenvalue $\lambda_1, \lambda_2, ..., \lambda_n$ respectively. Then we have

$$\begin{aligned} \mathbf{AM} &= \mathbf{A}[\mathbf{m}_1 \vdots \mathbf{m}_2 \vdots \ldots \vdots \mathbf{m}_n] \\ &= [\mathbf{Am}_1 \vdots \mathbf{Am}_2 \vdots \ldots \vdots \mathbf{Am}_n] \\ &= [\lambda_1 \mathbf{m}_1 \vdots \lambda_2 \mathbf{m}_2 \vdots \ldots \vdots \lambda_n \mathbf{m}_n] \\ &= \mathbf{M\Lambda} \end{aligned}$$

where
$$\mathbf{\Lambda} = \begin{bmatrix} \lambda_1 & 0 & \cdots & 0 \\ 0 & \lambda_2 & \cdots & 0 \\ \vdots & \vdots & & \vdots \\ 0 & 0 & \cdots & \lambda_n \end{bmatrix}$$
$$= \mathbf{M}^{-1}\mathbf{AM}$$

The matrix constructed by placing the eigenvectors (columns) together is therefore is a diagonalizing or *modal matrix* $\mathbf{M}$ of $\mathbf{A}$, *i.e.*,

$$\mathbf{M} = [\mathbf{m}_1 \vdots \mathbf{m}_2 \vdots \ldots \vdots \mathbf{m}_n]$$

Note that $\mathbf{A}$ and $\mathbf{\Lambda}$ have the same characteristic equation; therefore the eigenvalues are invariant under this transformation.

When $\mathbf{A}$ is expressed in the form given below,

$$\mathbf{A} = \begin{bmatrix} 0 & 1 & 0 & \cdots & 0 \\ 0 & 0 & 1 & \cdots & 0 \\ \vdots & \vdots & \vdots & & \vdots \\ 0 & 0 & 0 & \cdots & 1 \\ -a_n & -a_{n-1} & -a_{n-2} & \cdots & -a_1 \end{bmatrix} \quad ...(12.62)$$

then the modal matrix can be shown to be a special matrix (called the *Vander Monde matrix*)

$$\mathbf{V} = \begin{bmatrix} 1 & 1 & \cdots & 1 \\ \lambda_1 & \lambda_2 & \cdots & \lambda_n \\ \lambda^2 & \lambda_2^2 & \cdots & \lambda_n^2 \\ \vdots & \vdots & & \vdots \\ \lambda_1^{n-1} & \lambda_2^{n-2} & \cdots & \lambda_n^{n-1} \end{bmatrix} \quad ...(12.63)$$

Example 12.4: Consider a matrix **A** given below:

$$\mathbf{A} = \begin{bmatrix} 0 & 1 & 0 \\ 3 & 0 & 2 \\ -12 & -7 & -6 \end{bmatrix}$$

Corresponding to this matrix, the characteristic equation is

$$|\lambda\mathbf{I} - \mathbf{A}| = \begin{bmatrix} \lambda & -1 & 6 \\ -3 & \lambda & -2 \\ 12 & 7 & \lambda+6 \end{bmatrix} = 0$$

or
$$(\lambda + 1)(\lambda + 2)(\lambda + 3) = 0$$

Therefore the eigenvalues of matrix **A** are

$$\lambda_1 = -1,\ \lambda_2 = -2,\ \lambda_3 = -3$$

The eigenvector $\mathbf{m}_i$ associated with $\lambda_1 = -1$ is obtained from co-factors of the matrix

$$(\lambda_1\mathbf{I} - \mathbf{A}) = \begin{bmatrix} -1 & -1 & 0 \\ -3 & -1 & -2 \\ 12 & 7 & 5 \end{bmatrix}$$

The result is (eigenvector has unique direction)

$$\mathbf{m}_1 = \begin{bmatrix} C_{11} \\ C_{12} \\ C_{13} \end{bmatrix} = \begin{bmatrix} 9 \\ -9 \\ -9 \end{bmatrix} \quad \text{or} \quad \mathbf{m}_1 = \begin{bmatrix} 1 \\ -1 \\ -1 \end{bmatrix}$$

The eigenvector $\mathbf{m}_1$ could also be obtained from a solution of the homogeneous equations corresponding to the matrix $(\lambda\mathbf{I} - \mathbf{A})\mathbf{x} = 0$ *i.e.*,

$$-x_1 - x_2 = 0$$
$$-3x_1 - x_2 - 2x_3 = 0$$
$$12x_1 + 7x_2 + 5x_3 = 0$$

Choosing $x_1 = 1$, we get $x_2 = -1$, $x_3 = -1$, which is the same result as obtained by the method of co-factors. Solution of homogeneous equations is computationally more efficient than obtaining co-factors of row when the dimension of **A** is large.

Similarly eigenvectors $\mathbf{m}_2$ and $\mathbf{m}_3$ associated with $\lambda_2 = -2$ and $\lambda_3 = -3$ respectively are

$$\mathbf{m}_2 = \begin{bmatrix} 2 \\ -4 \\ 1 \end{bmatrix}; \quad \mathbf{m}_3 = \begin{bmatrix} 1 \\ -3 \\ 3 \end{bmatrix}$$

The modal matrix **M** obtained by placing the eigenvectors (columns) together is given by

$$\mathbf{M} = [\mathbf{m}_1 \vdots \mathbf{m}_2 \vdots \mathbf{m}_3] = \begin{bmatrix} 1 & 2 & 1 \\ -1 & -4 & -3 \\ -1 & 1 & 3 \end{bmatrix}$$

It can now be verified that

$$\Lambda = [\mathbf{M}^{-1}\mathbf{A}\mathbf{M}] = \begin{bmatrix} -1 & 0 & 0 \\ 0 & -2 & 0 \\ 0 & 0 & -3 \end{bmatrix}$$

which is a diagonal matrix with eigenvalues of **A** as its diagonal elements. In fact Λ could be written directly without the need to compute $\mathbf{M}^{-1}\,\mathbf{AM}$.

Generalized Eigenvectors

In our discussion on eigenvectors so far we have assumed that the eigenvalues of matrix **A** are all distinct. Let us now relax this restriction. Consider the matrix

$$A = \begin{bmatrix} 4 & 1 & -2 \\ 1 & 0 & 2 \\ 1 & -1 & 3 \end{bmatrix}$$

The eigenvalues of this matrix are $\lambda_1 = 1$, $\lambda_2 = 3$, $\lambda_3 = 3$. The eigenvector associated with $\lambda = 1$ may be obtained from the row co-factors of the matrix

$$(\lambda_1\mathbf{I} - \mathbf{A}) = \begin{bmatrix} -3 & -1 & 2 \\ -1 & 1 & -2 \\ -1 & 1 & -2 \end{bmatrix}$$

$$\mathbf{m}_1 = \begin{bmatrix} C_{11} \\ C_{12} \\ C_{13} \end{bmatrix} = \begin{bmatrix} 0 \\ 0 \\ 0 \end{bmatrix}$$

Co-factors along first row give a null solution. Let us take co-factors along the second row.

$$\mathbf{m}_1 = \begin{bmatrix} C_{21} \\ C_{22} \\ C_{23} \end{bmatrix} = \begin{bmatrix} 0 \\ 8 \\ 4 \end{bmatrix}$$

To obtain eigenvectors associated with the repeated eigenvalue at $\lambda = 3$, we construct the matrix

$$(\lambda_2\mathbf{I} - \mathbf{A}) = \begin{bmatrix} \lambda_2 - 4 & -1 & 2 \\ -1 & \lambda_2 & -2 \\ -1 & 1 & \lambda_2 - 3 \end{bmatrix}$$

For $\lambda_2 = 3$, the rank r of 3×3 matrix $(\lambda_2\,\mathbf{I} - \mathbf{A})$ is 2. Therefore one independent eigenvector associated with $\lambda = 3$ can be obtained. It is given by

$$\mathbf{m}_2 = \begin{bmatrix} C_{11} \\ C_{12} \\ C_{13} \end{bmatrix} = \begin{bmatrix} \lambda_2(\lambda_2 - 3) + 2 \\ (\lambda_2 - 3) + 2 \\ -1 + \lambda_2 \end{bmatrix} = \begin{bmatrix} 2 \\ 2 \\ 2 \end{bmatrix}$$

The vector $\mathbf{m}_3$ for the modal matrix

$$\mathbf{M} = [\mathbf{m}_1 \vdots \mathbf{m}_2 \vdots \mathbf{m}_3]$$

may be generated from the independent eigenvector $\mathbf{m}_2$ as follows :

$$\mathbf{m}_3 = \begin{bmatrix} \dfrac{d}{d\lambda_2} C_{11} \\ \dfrac{d}{d\lambda_2} C_{12} \\ \dfrac{d}{d\lambda_2} C_{13} \end{bmatrix}_{\lambda_2 = 3} = \begin{bmatrix} 2\lambda - 3 \\ 1 \\ 1 \end{bmatrix}_{\lambda_2 = 3} = \begin{bmatrix} 3 \\ 1 \\ 1 \end{bmatrix}$$

The vector $\mathbf{m}_3$ is *generalized eigenvector*. The modal matrix **M** is then given by

$$\mathbf{M} = \begin{bmatrix} 0 & 2 & 3 \\ 8 & 2 & 1 \\ 4 & 2 & 1 \end{bmatrix}$$

The modal matrix **M** now transforms **A** to the Jordan matrix, *i.e.*

$$\mathbf{M}^{-1}\,\mathbf{AM} = \begin{bmatrix} 1 & 0 & 0 \\ 0 & 3 & 1 \\ 0 & 0 & 3 \end{bmatrix} = \mathbf{J}$$

↑

Jordan block

It is possible in some rare cases, to have a state model whose **A** matrix has repeated eigenvalue λ_i such that rank of $n \times n$ matrix $(\lambda_i \mathbf{I} - \mathbf{A})$ is less than $(n-1)$, *i.e.*, there are more than one independent eigenvectors corresponding to λ_i. However, we shall assume in our discussion that for the state models we deal with, for each λ_i of multiplicity q, there exists a model matrix **M** such that the Jordan matrix $\mathbf{J} = \mathbf{M}^{-1}\,\mathbf{AM}$ will have a $q \times q$ *Jordan block* corresponding to eigenvalue λ_i.

Linear Transformation of State Vector (Discrete-time Case)

As in the continuous-time case the canonical state modal is equally usefully in the discrete-time system analysis.

The transformation

$$\mathbf{x}(k) = \mathbf{M}v(k)\ ;\ \mathbf{M} = \text{modal matrix}$$

transforms eqns. (12.49) to the following form

$$\mathbf{v}(x+1) = \mathbf{\Lambda} v(k) + \tilde{\mathbf{B}}\mathbf{u}(k)\ ;\ \mathbf{\Lambda} = \mathbf{M}^{-1}\,\mathbf{AM}\ ;\ \tilde{\mathbf{B}} = \mathbf{M}^{-1}\,\mathbf{B} \qquad \text{...(12.64}a\text{)}$$

$$y(k) = \tilde{\mathbf{C}}\mathbf{v}(k) + \mathbf{Du}(k)\ ;\ \tilde{\mathbf{C}} = \mathbf{CM} \qquad \text{...(12.64}b\text{)}$$

The techniques of obtaining modal matrix have already been discussed in this section. If matrix **A** has repeated eigenvalues, it can be transformed to Jordan canonical form.

12.6 SOLUTION OF STATE EQUATIONS

In the preceding two sections, we discussed various state models used to represent a system. In the present section, we shall develop methods for solution of the state equations from which the system transient response can then be obtained.

Let us first review the classical method of solution by considering a first order scalar differential equation.

$$\frac{dx}{dt} = ax\ ;\ x(0) = x_0 \qquad \text{...(12.65)}$$

This equation has the solution

$$x(t) = e^{at}x_0$$

$$= \left(1 + at + \frac{1}{2!}a^2t^2 + ... + \frac{1}{i!}a^it^i + ...\right)x_0 \qquad ...(12.66)$$

Let us now consider the state equation

$$\dot{\mathbf{x}}(t) = \mathbf{A}\mathbf{x}(t)\ ;\ \mathbf{x}(0) = \mathbf{x}_0 \qquad ...(12.67)$$

which represents a homogeneous* linear system (unforced system) with constant coefficients.

By analogy with the scalar case, we assume a solution of the form

$$\mathbf{x}(t) = \mathbf{a}_0 + \mathbf{a}_1t + \mathbf{a}_2t^2 + ... + \mathbf{a}_it^i + ...$$

where $\mathbf{a}_i$ are vector coefficients.

By substituting the assumed solution into eqn. (12.67) we get

$$\mathbf{a}_1 + 2\mathbf{a}_2t + 3\mathbf{a}_3t^2 + ... = \mathbf{A}(\mathbf{a}_0 + \mathbf{a}_1t + \mathbf{a}_2t^2 + ...)$$

The comparison of vector coefficients of equal powers of t, yield

$$\mathbf{a}_1 = \mathbf{A}\mathbf{a}_0$$

$$\mathbf{a}_2 = \frac{1}{2}\mathbf{A}\mathbf{a}_1 = \frac{1}{2!}\mathbf{A}^2\mathbf{a}_0$$

$$\mathbf{a}_i = \frac{1}{i!}\mathbf{A}^i\mathbf{a}_0$$

In the assumed solution, equating $\mathbf{x}(t = 0) = \mathbf{x}_0$, we find that

$$\mathbf{a}_0 = \mathbf{x}_0$$

The solution $\mathbf{x}(t)$ is thus found to be

$$\mathbf{x}(t) = \left(\mathbf{I} + \mathbf{A}t + \frac{1}{2!}\mathbf{A}^2t^2 + ... + \frac{1}{i!}\mathbf{A}^it^i + ...\right)\mathbf{x}_0$$

Each of the terms inside the brackets is an $n \times n$ matrix. Because of the similarity of the entity inside the brackets with a scalar exponential of eqn. (12.66), we call it a *matrix exponential*, which may be written as

$$e^{\mathbf{A}t} = \mathbf{I} + \mathbf{A}t + \frac{1}{2!}\mathbf{A}^2t^2 + ... + \frac{1}{i!}\mathbf{A}^it^i + ... \qquad ...(12.68)$$

The solution $\mathbf{x}(t)$ can now be written as

$$\mathbf{x}(t) = e^{\mathbf{A}t}\mathbf{x}_0 \qquad ...(12.69)$$

---

*The state equation of a linear system is

$$\dot{\mathbf{x}} = \mathbf{A}\mathbf{x} + \mathbf{B}\mathbf{u}$$

If **A** is a constant matrix and **u** is a zero vector (*i.e.*, no control forces are applied to the system) the equation represents a homogeneous linear system. On the other hand, if **A** is a constant matrix and **u** is nonzero vector, the equation represents a nonhomogeneous linear system (*i.e.*, control forces are applied to the system).

From eqn. (12.69) it is observed that the initial state $\mathbf{x}_0$ at $t = 0$, is driven to a state $\mathbf{x}(t)$ at time t. This transition in state is carried out by the matrix exponential $e^{\mathbf{A}t}$. Because of this property, $e^{\mathbf{A}t}$ is known as *state transition matrix* and is denoted by $\Phi(t)$.

Let us now determine the solution of the nonhomogeneous state equation (forced system)

$$\dot{\mathbf{x}}(t) = \mathbf{Ax}(t) + \mathbf{Bu}(t);\ \mathbf{x}(0) = \mathbf{x}_0 \qquad \ldots(12.70)$$

Rewrite this equation in this form

$$\dot{\mathbf{x}}(t) - \mathbf{Ax}(t) = \mathbf{Bu}(t)$$

Multiplying both sides by $e^{-\mathbf{A}t}$, we can write $e^{-\mathbf{A}t}([\dot{\mathbf{x}}(t) - \mathbf{Ax}(t)] = \frac{d}{dt}[e^{-\mathbf{A}t}\,\mathbf{x}(t)] = e^{-\mathbf{A}t}\,\mathbf{Bu}(t)$

Integrating both sides with respect to t between the limits 0 and t, we get

$$e^{-\mathbf{A}t}\mathbf{x}(t)\Big|_0^t = \int_0^t e^{-\mathbf{A}\tau}\mathbf{Bu}(\tau)\,d\tau$$

$$e^{-\mathbf{A}t}\,\mathbf{x}(t) - \mathbf{x}(0) = \int_0^t e^{-\mathbf{A}\tau}\mathbf{Bu}(\tau)\,d\tau$$

Now pre-multiplying both sides by $e^{\mathbf{A}t}$, we have

$$\mathbf{x}(t) = \underbrace{e^{\mathbf{A}t}\mathbf{x}(0)}_{\text{Homogeneous solution}} + \underbrace{\int_{t_0}^t e^{\mathbf{A}(t-\tau)}\,\mathbf{Bu}(\tau)\,d\tau}_{\text{Forced solution}} \qquad \ldots(12.71)$$

If the initial state is known at $t = t_0$ rather than $t = 0$, eqn. (12.71) becomes

$$\mathbf{x}(t) = e^{\mathbf{A}(t-t_0)}\mathbf{x}(t_0) + \int_{t_0}^t e^{\mathbf{A}(t-\tau)}\mathbf{Bu}(\tau)\,d\tau \qquad \ldots(12.72)$$

Properties of the State Transition Matrix

In the above discussion, the state transition matrix has been defined as

$$\Phi(t) = e^{\mathbf{A}t}$$

wherein the initial time has been taken as $t = 0$. In general for a linear time invarient system if the initial time is $t = t_0$, state transition matrix becomes

$$\Phi(t - t_0) = e^{\mathbf{A}(t-t_0)}$$

Since the state transition matrix depends upon $(t - t_0)$, t_0 is conveniently regarded as zero.

Certain useful properties of the state transition matrix $\Phi(t)$ are given below :

1. $\Phi(0) = e^{\mathbf{A}0} = \mathbf{I}$
2. $\Phi(t) = e^{\mathbf{A}t} = (e^{-\mathbf{A}t})^{-1} = [\Phi(-t)]^{-1}$

   or $\Phi^{-1}(t) = \Phi(-t)$
3. $\Phi(t_1 + t_2) = e^{\mathbf{A}(t_1+t_2)} = e^{\mathbf{A}t_1}e^{\mathbf{A}t_2}$

   $= \Phi(t_1)\,\Phi(t_2) = \Phi(t_2)\,\Phi(t_1)$

In terms of the state transition matrix $\Phi(t)$, the solution of the forced system given by eqn. (12.72) can be writen as

$$\mathbf{x}(t) = \boldsymbol{\Phi}(t)\,\mathbf{x}(0) + \int_0^t \Phi(t-\tau)\,\mathbf{Bu}(\tau)\,d\tau \qquad ...(12.73a)$$

If instead the initial state is known at t_0,

$$\mathbf{x}(t) = \boldsymbol{\Phi}(t-t_0)\,\mathbf{x}(t_0) + \int_0^t \Phi(t-\tau)\,\mathbf{Bu}(\tau)\,d\tau \qquad ...(12.73b)$$

where
$$\boldsymbol{\Phi}(t-t_0) = e^{\mathbf{A}(t-t_0)}$$
$$\boldsymbol{\Phi}(t-\tau) = e^{\mathbf{A}(t-\tau)}.$$

Example 12.5. Obtain the time response of the following system

$$\begin{bmatrix}\dot{x}_1\\ \dot{x}_2\end{bmatrix} = \begin{bmatrix}1 & 0\\ 1 & 1\end{bmatrix}\begin{bmatrix}x_1\\ x_2\end{bmatrix} + \begin{bmatrix}1\\ 1\end{bmatrix}u$$

where $u(t)$ is a unit step occurring at $t = 0$ and $x^T(0) = [1 \quad 0]$.

Solution. We have in this case

$$\mathbf{A} = \begin{bmatrix}1 & 0\\ 1 & 1\end{bmatrix};\ \mathbf{B} = \begin{bmatrix}1\\ 1\end{bmatrix}$$

The state transition matrix $\boldsymbol{\Phi}(t)$ is given by

$$e^{\mathbf{A}t} = \mathbf{I} + \mathbf{A}t + \frac{1}{2!}\mathbf{A}^2t^2 + \frac{1}{3!}\mathbf{A}^3t^3 + ...$$

Substituting values of A and collecting terms, we get

$$e^{\mathbf{A}t} = \begin{bmatrix}1+t+0.5t^2+\ldots & 0\\ t+t^2+\ldots & 1+t+0.5t^2+\ldots\end{bmatrix}$$

$$= \begin{bmatrix}\Phi_{11} & \Phi_{12}\\ \Phi_{21} & \Phi_{22}\end{bmatrix}$$

The terms Φ_{11} and Φ_{22} are easily recognized as series expansion of e^t. To recognize Φ_{21}, more terms of the infinite series should be evaluated. In fact Φ_{21} is te^t.

The computation of $e^{\mathbf{A}t}$ by expanding it into a power series in t and then adding the corresponding elements in the matrix terms of the infinite series, is practical only for simple cases. We will shortly discuss alternate methods of computation of state transition matrix

Returning to the example under consideration, we have the state transition matrix as

$$\boldsymbol{\Phi}(t) = \begin{bmatrix}e^t & 0\\ te^t & e^t\end{bmatrix}$$

The time response of the system is given by

$$\mathbf{x}(t) = \boldsymbol{\Phi}(t)\left[\mathbf{x}_0 + \int_0^t \Phi(-\tau)\,\mathbf{Bu}\,d\tau\right]$$

Now with $u = 1$,

$$\boldsymbol{\Phi}(-\tau)\mathbf{B}u = \begin{bmatrix}e^{-\tau} & 0\\ -\tau e^{-\tau} & e^{-\tau}\end{bmatrix}\begin{bmatrix}1\\ 1\end{bmatrix} = \begin{bmatrix}e^{-\tau}\\ e^{-\tau}(1-\tau)\end{bmatrix}$$

Therefore, $$\int_0^t \Phi(-\tau)\mathbf{B}u\,d\tau = \begin{bmatrix} \int_0^t e^{-\tau}\,d\tau \\ \int_0^t e^{-\tau}(1-\tau)\,d\tau \end{bmatrix} = \begin{bmatrix} 1-e^{-t} \\ te^{-t} \end{bmatrix}$$

Then the solution $\mathbf{x}(t)$ is given by

$$\mathbf{x} = \begin{bmatrix} e^t & 0 \\ te^t & e^t \end{bmatrix} \left\{ \begin{bmatrix} 1 \\ 0 \end{bmatrix} + \begin{bmatrix} 1-e^{-t} \\ te^{-1} \end{bmatrix} \right\}$$

$$= \begin{bmatrix} 2e^t - 1 \\ 2e^t \end{bmatrix}.$$

Computation of State Transition Matrix

Computation by Laplace transformation

Let us consider an unforced system whose state equation is

$$\dot{\mathbf{x}} = \mathbf{A}\mathbf{x}, \text{ where } \mathbf{A} \text{ is a constant matrix}$$

Taking the Laplace transform of this equation, we obtain

$$s\mathbf{X}(s) - \mathbf{x}(0) = \mathbf{A}\mathbf{X}(s) \qquad \text{...(12.74)}$$

where $\mathbf{X}(s)$ is the Laplace transform of the unforced response and $\mathbf{x}(0)$ is the initial condition vector. Equation (12.74) may be rearranged as

$$[s\mathbf{I} - \mathbf{A}]\mathbf{X}(s) = \mathbf{x}(0)$$

or $$\mathbf{X}(s) = [s\mathbf{I} - \mathbf{A}]^{-1}\,\mathbf{x}(0)$$

Taking the inverse Laplace transform, we get

$$\mathbf{x}(t) = \mathcal{L}^{-1}[(s\mathbf{I} - \mathbf{A})^{-1}]\,\mathbf{x}(0)$$

where $\mathbf{x}(t)$ is the unforced response of the system. This solution obviously must be identical with the one obtained earlier in eqn. (12.73). The comparison yields a different approach to determine the state transition matrix which is given below:

$$\Phi(t) = e^{\mathbf{A}t} = \mathcal{L}^{-1}[(s\mathbf{I} - \mathbf{A})^{-1}] = \mathcal{L}^{-1}\,\Phi(s) \qquad \text{...(12.75)}$$

where $\Phi(s) = (s\mathbf{I} - \mathbf{A})^{-1}$ is called the *resolvent matrix.*

Let us consider now the response when the control force vector $\mathbf{u}$ is applied. The state equation for this case is

$$\dot{\mathbf{x}} = \mathbf{A}\mathbf{x} + \mathbf{B}\mathbf{u}$$

Performing the Laplace transformation gives

$$s\mathbf{X}(s) - \mathbf{x}_0 = \mathbf{A}\mathbf{X}(s) + \mathbf{B}\mathbf{U}(s);\ \mathbf{x}_0 = \mathbf{x}(0)$$

or $$(s\mathbf{I} - \mathbf{A})\mathbf{X}(s) = \mathbf{x}_0 + \mathbf{B}\mathbf{U}(s)$$

Therefore

$$\mathbf{X}(s) = [(s\mathbf{I} - \mathbf{A})^{-1}]\mathbf{x}_0 + [(s\mathbf{I} - \mathbf{A})^{-1}\,\mathbf{B}\mathbf{U}(s)] \qquad \text{...(12.76)}$$

By inverse Laplace transformation

$$\mathbf{x}(t) = \mathcal{L}^{-1}[(s\mathbf{I} - \mathbf{A})^{-1}]\,\mathbf{x}_0 + \mathcal{L}^{-1}[(s\mathbf{I} - \mathbf{A})^{-1}\,\mathbf{B}\mathbf{U}(s)]$$

$$= \Phi(t)\,\mathbf{x}_0 + \mathcal{L}^{-1}[\Phi(s)\,\mathbf{B}\mathbf{U}(s)] \qquad \text{...(12.77)}$$

Example 12.6. Let us reconsider the system discussed in Example 12.5 in which

$$\mathbf{A} = \begin{bmatrix} 1 & 0 \\ 1 & 1 \end{bmatrix}$$

Then

$$(s\mathbf{I} - \mathbf{A}) = \begin{bmatrix} s & 0 \\ 0 & s \end{bmatrix} - \begin{bmatrix} 1 & 0 \\ 1 & 1 \end{bmatrix} = \begin{bmatrix} s-1 & 0 \\ -1 & s-1 \end{bmatrix}$$

The determinant of $(s\mathbf{I} - \mathbf{A})$ is

$$| s\mathbf{I} - \mathbf{A} | = (s-1)^2$$

Therefore, the resolvant matrix is given by

$$\mathbf{\Phi}(s) = (s\mathbf{I} - \mathbf{A})^{-1} = \frac{1}{(s-1)^2}\begin{bmatrix} (s-1) & 0 \\ 1 & (s-1) \end{bmatrix}$$

$$= \begin{bmatrix} \dfrac{1}{(s-1)} & 0 \\ \dfrac{1}{(s-1)^2} & \dfrac{1}{(s-1)} \end{bmatrix}$$

Hence

$$\mathbf{\Phi}(t) = \mathcal{L}^{-1}[(s\mathbf{I} - \mathbf{A})^{-1}] = \begin{bmatrix} \mathcal{L}^{-1}\left[\dfrac{1}{(s-1)}\right] & 0 \\ \mathcal{L}^{-1}\left[\dfrac{1}{(s-1)^2}\right] & \mathcal{L}^{-1}\left[\dfrac{1}{(s-1)}\right] \end{bmatrix}$$

$$= \begin{bmatrix} e^t & 0 \\ te^t & e^t \end{bmatrix} = e^{\mathbf{A}t}.$$

Example 12.7. Consider a control system with state model

$$\begin{bmatrix} \dot{x}_1 \\ \dot{x}_2 \end{bmatrix} = \begin{bmatrix} 0 & 1 \\ -2 & -3 \end{bmatrix}\begin{bmatrix} x_1 \\ x_2 \end{bmatrix} + \begin{bmatrix} 0 \\ 2 \end{bmatrix}[u]; \quad \begin{bmatrix} x_1(0) \\ x_2(0) \end{bmatrix} = \begin{bmatrix} 0 \\ 1 \end{bmatrix}, \; u = \text{unit step.}$$

Compute the state transition matrix and therefrom find the state response, *i.e.*, $x(t)$ $t > 0$.

Solution. The characteristic equation for this system is

$$| \lambda\mathbf{I} - \mathbf{A} | = \begin{vmatrix} \lambda & -1 \\ 2 & (\lambda+3) \end{vmatrix} = 0$$

or

$$\lambda^2 + 3\lambda + 2 = 0$$

Therefore the eigenvalues of matrix **A** are $\lambda_1 = -1$, $\lambda_2 = -2$. Since the system matrix **A** is in the companion form, we can choose the Vander Monde matrix.

$$\mathbf{M} = \begin{bmatrix} 1 & 1 \\ \lambda_1 & \lambda_2 \end{bmatrix} = \begin{bmatrix} 1 & 1 \\ -1 & -2 \end{bmatrix}$$

as the modal matrix. Then

$$\mathbf{\Lambda} = \mathbf{M}^{-1}\,\mathbf{AM} = \begin{bmatrix} 1 & 1 \\ -1 & -2 \end{bmatrix}\begin{bmatrix} 0 & 1 \\ -2 & -3 \end{bmatrix}\begin{bmatrix} 2 & 1 \\ -1 & -1 \end{bmatrix} = \begin{bmatrix} -1 & 0 \\ 0 & -2 \end{bmatrix}$$

$$\dot{\mathbf{x}} = \mathbf{Ax} + \mathbf{b}u$$

Transforming with modal matrix we get

$$\dot{\mathbf{v}} = \mathbf{M}^{-1}\mathbf{AM}\mathbf{v} + \mathbf{M}^{-1}\mathbf{b}u;\ \mathbf{x} = \mathbf{M}\mathbf{v}$$
$$= \Lambda\mathbf{v} + \mathbf{M}^{-1}\mathbf{b}u$$

The state transition matrix in the transformed domain

$$e^{\wedge t} = \begin{bmatrix} e^{-t} & 0 \\ 0 & e^{-2t} \end{bmatrix}$$

This is easily established by writing out of the decoupled equations

$$v(t) = e^{\wedge t}v(0)$$

Transforming back

$$\mathbf{M}^{-1}\mathbf{x}(t) = e^{\mathbf{\Lambda}t}\mathbf{M}^{-1}\,\mathbf{x}(0)$$

$\Rightarrow$ $$\mathbf{x}(t) = (\mathbf{M}e^{\mathbf{\Lambda}t}\,\mathbf{M}^{-1})\,\mathbf{x}(0)$$

$\therefore$ $$\Phi(t) = \mathbf{M}e^{\mathbf{\Lambda}t}\,\mathbf{M}^{-1}$$

Substituting values we get

$$e^{\mathbf{A}t} = \begin{bmatrix} 1 & 1 \\ -1 & -2 \end{bmatrix}\begin{bmatrix} e^{-t} & 0 \\ 0 & e^{-2t} \end{bmatrix}\begin{bmatrix} 2 & 1 \\ -1 & -1 \end{bmatrix} = \begin{bmatrix} 2e^{-t} - e^{-2t} & e^{-t} - e^{-2t} \\ -2e^{-t} + 2e^{-2t} & -e^{-t} + 2e^{-2t} \end{bmatrix}$$

Writing down the expression for $\mathbf{x}(t)$

$$\mathbf{x}(t) = e^{\mathbf{A}t}\,\mathbf{x}(0) + \int_0^t e^{\mathbf{A}(t-\tau)}\mathbf{b}u(\tau)\,d\tau$$

Computing term by term

$$e^{\mathbf{A}t}\,\mathbf{x}(0) = \begin{bmatrix} e^{-t} - e^{-2t} \\ -e^{-t} + 2e^{-2t} \end{bmatrix}$$

$$e^{-\mathbf{A}\tau}\,\mathbf{b} = \begin{bmatrix} 2e^{\tau} - 2e^{2\tau} \\ -2e^{\tau} + 4e^{2\tau} \end{bmatrix};\ u(\tau) = 1 \text{ for } t > 0$$

$$\int_0^t \begin{bmatrix} 2e^{\tau} - 2e^{2\tau} \\ -2e^{\tau} + 4e^{2\tau} \end{bmatrix} d\tau = \begin{bmatrix} 2e^{\tau} - e^{2\tau} \\ -2e^{\tau} + 2e^{2\tau} \end{bmatrix}_0^t$$

$$= \begin{bmatrix} 2e^{t} - e^{2t} \\ -2e^{t} + 2e^{2t} \end{bmatrix} - \begin{bmatrix} 2-1 \\ -2+2 \end{bmatrix}$$

$$= \begin{bmatrix} 2e^{t} - e^{2t} - 1 \\ -2e^{t} + 2e^{2t} \end{bmatrix}$$

Collecting terms,

$$\mathbf{x}(t) = \begin{bmatrix} e^{-t} - e^{-2t} \\ -e^{-t} + 2e^{-2t} \end{bmatrix} + \begin{bmatrix} 2e^{-t} - e^{-2t} & e^{-t} - e^{-2t} \\ -2e^{-t} + 2e^{-2t} & -e^{-t} + 2e^{-2t} \end{bmatrix}\begin{bmatrix} 2e^{t} - e^{2t} - 1 \\ -2e^{t} + 2e^{2t} \end{bmatrix}$$

Simplifying the above equation we get

$$\mathbf{x}(t) = \begin{bmatrix} 1 - e^{-t} \\ e^{-t} \end{bmatrix}$$

Computation by Techniques Based on the Cayley-Hamilton Theorem

The state transition matrix may be computed using the technique based on the Cayley-Hamilton Theorem. For large systems, this method is far more convenient computationally as compared to the other two methods advanced earlier. To being with, let us state the Cayley-Hamilton Theorem.

Every square matrix **A** satisfies its own characteristic equation. In other words, if

$$q(\lambda) = |\lambda\mathbf{I} - \mathbf{A}| = \lambda^n + a_1 \lambda^{n-1} + \ldots + a_{n-1} \lambda + a_n = 0 \qquad \ldots(12.78)$$

is the characteristic equation of **A**, then

$$q(\mathbf{A}) = \mathbf{A}^n + a_1\mathbf{A}^{n-1} + \ldots. + a_{n-1}\mathbf{A} + a_n\mathbf{I} = 0 \qquad \ldots(12.79)$$

This theorem provides a simple procedure for evaluating the function of a matrix. In the study of linear systems, we are mostly concerned with functions which can be represented as a series of the powers of a matrix. Given an $n \times n$ matrix **A** with the characteristic equation as in eqn. (12.78) above, let $\lambda_1, \lambda_2, \ldots., \lambda_n$ be the eigenvalues of **A**. The matrix polynomial

$$f(\mathbf{A}) = k_0\mathbf{I} + k_1\mathbf{A} + k_2\mathbf{A}^2 + \ldots + k_n\mathbf{A}^n + k_{n+1}\mathbf{A}^{n+1} + \ldots \qquad (12.80)$$

can be computed by consideration of the scalar polynomial

$$f(\lambda) = k_0 + k_1\lambda + k_2\lambda^2 + \ldots.. + k_n\lambda^n + k_{n+1}\lambda^{n+1} + \ldots... \qquad \ldots(12.81)$$

If $f(\lambda)$ is divided by the characteristic polynomial $q(\lambda)$, then we have

$$\frac{f(\lambda)}{q(\lambda)} = Q(\lambda) + \frac{R(\lambda)}{q(\lambda)}$$

or

$$f(\lambda) = Q(\lambda)\, q(\lambda) + R(\lambda) \qquad \ldots(12.82a)$$

where $R(\lambda)$ is the remainder polynomial of the following form :

$$R(\lambda) = \alpha_0 + \alpha_1 \lambda + \alpha_2 \lambda^2 + \ldots\ldots. + \alpha_{n-1} \lambda^{n-1} \qquad \ldots(12.82b)$$

If we evaluate $f(\lambda)$ at the eigenvalues $\lambda_1, \lambda_2, \ldots.., \lambda_n$; then $q(\lambda) = 0$ and we have

$$f(\lambda_i) = R(\lambda_i)\ ;\ i = 1, 2, \ldots.., n \qquad \ldots(12.83)$$

The coefficients $\alpha_0, \alpha_1, \ldots.., \alpha_{n-1}$, can be obtained by successively substituting $\lambda_1, \lambda_2, \ldots..$ λ_n into eqn. (12.83)

Substituting **A** for the variable λ in eqn. (12.82), we get

$$f(\mathbf{A}) = Q(\mathbf{A})\, q(\mathbf{A}) + R(\mathbf{A})$$

Since $q(\mathbf{A})$ is identically zero, it follows that

$$f(\mathbf{A}) = R(\mathbf{A})$$

$$= \alpha_0\mathbf{I} + \alpha_1\mathbf{I} + \ldots. + \alpha_{n-1}\mathbf{A}^{n-1} \qquad \ldots(12.84)$$

which is the desired result.

The formal procedure of evaluation of the matrix polynomial $f(\mathbf{A})$ is given below :

1. Find the eigenvalues of matrix **A**.

2. If all the eigenvalues are distinct, solve n simultaneous equations given by (12.83) for the coefficients $\alpha_0, \alpha_1, \ldots..., \alpha_{n-1}$.

If **A** possesses an eigenvalue λ_k of order m, then only one independent equation can be obtained by substituting λ_k into eqn. (12.83). The remaining $(m - 1)$ linear equations, which must be obtained in order to solve for α_i's, can be found by differentiating both sides of eqn. (12.82). Since

$$\left.\frac{d^j q(\lambda)}{d\lambda^j}\right|_{\lambda=\lambda_k} = 0; \quad j = 0, 1, ..., m-1,$$

it follows that

$$\left.\frac{d^j q(\lambda)}{d\lambda^j}\right|_{\lambda=\lambda_k} = \left.\frac{d^j R(\lambda)}{d\lambda^j}\right|_{\lambda=\lambda_k} \quad ; j = 0, 1, ..., m-1 \qquad ...(12.85)$$

3. The coefficients α_i obtained in step 2 and eqn. (12.85) yield the required result.

Example 12.8. Find $f(A) = A^{10}$ for

$$\mathbf{A} = \begin{bmatrix} 0 & 1 \\ -2 & -3 \end{bmatrix}.$$

Solution. The characteristic equation is

$$q(\lambda) = |\ \lambda\mathbf{I} - \mathbf{A}\ | = \begin{bmatrix} \lambda & -1 \\ 2 & \lambda+3 \end{bmatrix} = (\lambda+1)(\lambda+2) = 0$$

Matrix **A** has distinct eigenvalues $\lambda_1 = -1$, $\lambda_2 = -2$.

Since **A** is of second-order, the polynomial $R(\lambda)$ will be of the following form:

$$R(\lambda) = \alpha_0 + \alpha_1\lambda$$

The coefficients α_0 and α_1 are evaluated from equations

$$f(\lambda_1) = \lambda_1^{10} = \alpha_0 + \alpha_1\lambda_1$$

$$f(\lambda_2) = \lambda_2^{10} = \alpha_0 + \alpha_1\lambda_2$$

The result is $\qquad \alpha_0 = -1022, \alpha_1 = -1023$

From eqn. (12.85), we get

$$f(\mathbf{A}) = \mathbf{A}^{10} = \alpha_0\mathbf{I} + \alpha_1\mathbf{A}$$

$$= \begin{bmatrix} -1022 & -1023 \\ 2046 & 2047 \end{bmatrix}$$

The Cayley-Hamilton technique allows us to attack the problem of computation of $e^{\mathbf{A}t}$ where **A** is a constant $n \times n$ matrix.

The power series for the series $e^{\lambda t}$,

$$e^{\lambda t} = 1 + \lambda t + \frac{\lambda^2 t^2}{2!} + ... + \frac{A^n t^n}{n!} + ...$$

converges for all finite λ and t. It follows from this result that the matrix power series

$$e^{\mathbf{A}t} = \mathbf{I} + \mathbf{A}t + \frac{\mathbf{A}^2 t^2}{2!} + ... + \frac{\mathbf{A}^n t^n}{n!} + ...$$

converges for all **A** and for all finite t. Therefore the matrix polynomial $f(\mathbf{A}) = e^{\mathbf{A}t}$ can be expressed as a polynomial in **A** of degree $(n-1)$ using the techniques presented earlier. This is illustrated below with the help of an example.

Example 12.9. Find the $f(A) = e^{At}$ for

$$\mathbf{A} = \begin{bmatrix} 0 & 1 \\ -1 & -2 \end{bmatrix}.$$

Solution. The characteristic equation is

$$q(\lambda) = |\,\lambda\mathbf{I} - \mathbf{A}\,| = \begin{vmatrix} \lambda & -1 \\ 1 & \lambda+2 \end{vmatrix} = (\lambda + 1)^2 = 0$$

Matrix **A** has eigenvalues $\lambda_1, \lambda_2 = -1$.

Since **A** is of second-order, the polynomial $R(\lambda)$ will be of the following form :

$$R(\lambda) = \alpha_0 + \alpha_1\lambda$$

The coefficients α_0 and α_1 are evaluated from equations (eqns. 12.83) and (12.85)

$$f(-1) = e^{-t} = \alpha_0 - \alpha_1$$

$$\frac{d}{d\lambda} f(\lambda)\bigg|_{\lambda=-1} = te^{-t} = \frac{d}{d\lambda} R(\lambda)\bigg|_{\lambda=-1} = \alpha_1$$

The result is $\qquad \alpha_0 = (1 + t)e^{-t},\ \alpha_1 = te^{-t}$

From eqn. (12.85), we get

$$f(\mathbf{A}) = e^{\mathbf{A}t} = \alpha_0\mathbf{I} + \alpha_1\mathbf{A}$$

$$= \begin{bmatrix} (1+t)e^{-t} & te^{-t} \\ -te^{-t} & (1-t)e^{-t} \end{bmatrix}.$$

Solution of State Equations (Discrete Case)

From the state model (12.49) we can write

$$\mathbf{x}(1) = \mathbf{Ax}(0) + \mathbf{Bu}(0)$$

$$\mathbf{x}(2) = \mathbf{Ax}(1) + \mathbf{Bu}(1)$$

$$= \mathbf{A}^2\mathbf{x}(0) + \mathbf{ABu}(0) + \mathbf{Bu}(1)$$

$$\cdots \quad \cdots \quad \cdots \quad \cdots \quad \cdots$$

$$\mathbf{x}(k) = \mathbf{A}^k\mathbf{x}(0) + \mathbf{A}^{k-1}\mathbf{Bu}(0) + \mathbf{A}^{k-2}\mathbf{Bu}(1) + \ldots + \mathbf{Bu}(k-1)$$

Let us define for $k > 0$

$$\mathbf{\Phi}(k) \triangleq \mathbf{A}^k$$

and $\qquad \mathbf{\Phi}(0) \triangleq \mathbf{I}$ (identity matrix)

Then we have

$$\mathbf{x}(k) = \mathbf{\Phi}(k)\,\mathbf{x}(0) + \sum_{i=0}^{k-1} \mathbf{\Phi}(k - i - 1)\mathbf{Bu}(i) \qquad \ldots(12.86)$$

The matrix $\Phi(k)$ is referred to as the *state transition matrix* for discrete time system (12.49). Properties of the state transition matrix are summarized below :

1. $\mathbf{\Phi}(0) = 1$
2. $\mathbf{\Phi}^{-1}(k) = \mathbf{\Phi}(-k)$
3. $\mathbf{\Phi}(k, k_0) = \mathbf{\Phi}(k - k_0) = A^{(k-k_0)}; k > k_0$.

Example 12.10. Consider the state variable representation of a second order system presented in the Example 12.7.

$$\begin{bmatrix} \dot{x}_1 \\ \dot{x}_2 \end{bmatrix} = \begin{bmatrix} 0 & 1 \\ -2 & -3 \end{bmatrix}\begin{bmatrix} x_1 \\ x_2 \end{bmatrix} + \begin{bmatrix} 0 \\ 2 \end{bmatrix}[u] \qquad ...(i)$$

Find the state response of the system for a unit step input and

$$\begin{bmatrix} x_1(0) \\ x_2(0) \end{bmatrix} = \begin{bmatrix} 0 \\ 1 \end{bmatrix}$$

by using the discrete-time approximation.

Solution. Let the state vector differential equation be written as

$$\dot{\mathbf{x}} = \mathbf{A}\mathbf{x}(t) + \mathbf{B}u(t) \qquad ...(ii)$$

From the basic definition of a derivative, we can determine the value of $\dot{\mathbf{x}}(t)$ when t is divided in small intervals $\Delta t = T$. Thus approximating the derivative as

$$\dot{\mathbf{x}} = \frac{\mathbf{x}(t+T) - \mathbf{x}(t)}{T} \cong \mathbf{A}\mathbf{x}(t) + \mathbf{B}u(t) \qquad ...(iii)$$

Solving for $\mathbf{x}(t + T)$, we have

$$\mathbf{x}(t + T) \cong (T\mathbf{A} + \mathbf{I})\, \mathbf{x}(t) + T\mathbf{B}u(t) \qquad ...(iv)$$

where t is divided into intervals of width T *i.e.* $t = kT$ where $k = 0, 1, 2, 3, ...$. Then eqn. (*iv*) is written as

$$\mathbf{x}[(k + 1)T] \cong (T\mathbf{A} + \mathbf{I})\, \mathbf{x}(kT) + T\mathbf{B}u(kT) \qquad ...(v)$$

Therefore the value of the state vector at $(k + 1)^{st}$ time instant is evaluated in terms of the values of $\mathbf{x}$ and u at the kth time instant. Equation (*v*) can be rewritten as

$$\mathbf{x}(k + 1) \cong \psi(T) + T\mathbf{B}u(k);\ kT \ \Rightarrow\ k \qquad ...(vi)$$

where $\psi(T) = (T\mathbf{A} + \mathbf{I})$ and the symbol T is omitted from the arguments of the variables.

Now we must choose a sufficiently small time interval T so that the approximation of the derivative is reasonably accurate. Usually we choose T to be less than one-half of the smallest time constant of the system. Therefore, sine the smallest time constant of the system this system is 0.5s, we might choose $T = 0.1$s. Alternatively, if a digital computer is used for the calculations and the number of calculations are not important, we would choose $T = 0.05$s in order to obtain greater accuracy. However, we note that as we decrease the increment size the number of calculations increases proportionally. Using $T = 0.1$s, we get

$$\mathbf{x}(k + 1) \cong (0.1\mathbf{A} + \mathbf{I})\, \mathbf{x}(k) + 0.1\mathbf{B}u(t) \qquad ...(vii)$$

Therefore

$$\psi(T) = \begin{bmatrix} 1 & 0.1 \\ -0.2 & 0.7 \end{bmatrix} \qquad ...(viii)$$

and

$$T\mathbf{B} = \begin{bmatrix} 0 \\ 0.2 \end{bmatrix} \qquad ...(ix)$$

Then the discrete-time equation is

$$\mathbf{x}(k + 1) = \begin{bmatrix} 1 & 0.1 \\ -0.2 & 0.7 \end{bmatrix}\mathbf{x}(k) + \begin{bmatrix} 0 \\ 0.2 \end{bmatrix}u(k) \qquad ...(x)$$

Therefore the response at the first instant when $t = T$ or $k = 0$ is obtained as

$$\mathbf{x}(1) \cong \begin{bmatrix} 1 & 0.1 \\ -0.2 & 0.7 \end{bmatrix}\mathbf{x}(0) + \begin{bmatrix} 0 \\ 0.2 \end{bmatrix} = \begin{bmatrix} 0.1 \\ 0.9 \end{bmatrix}$$

where $x(0) = \begin{bmatrix} 0 \\ 1 \end{bmatrix}$ and $u = 1$; $t > 0$. Then the response at the time $t = 2T$ or $k = 1$, is

$$\mathbf{x}(2) \cong \begin{bmatrix} 1 & 0.1 \\ -0.2 & 0.7 \end{bmatrix} \mathbf{x}(1) + \begin{bmatrix} 0 \\ 0.2 \end{bmatrix} = \begin{bmatrix} 0.19 \\ 0.81 \end{bmatrix}$$

The value of the response as $k = 2, 3, 4, \ldots$ is then evaluated in a similar manner. The exact values as calculated from the Example 12.7 are compared with the approximate values of the time response in **Table 12.1**. It is seen that the approximate values match very well with the exact ones.

Table 12.1

Time	*0*	*0.2*	*0.4*	*0.6*	*0.8*
Exact $\mathbf{x}(t)$	$\begin{bmatrix} 0 \\ 1 \end{bmatrix}$	$\begin{bmatrix} 0.18 \\ 0.82 \end{bmatrix}$	$\begin{bmatrix} 0.33 \\ 0.67 \end{bmatrix}$	$\begin{bmatrix} 0.45 \\ 0.55 \end{bmatrix}$	$\begin{bmatrix} 0.55 \\ 0.45 \end{bmatrix}$
Approx $\mathbf{x}(t)$	$\begin{bmatrix} 0 \\ 1 \end{bmatrix}$	$\begin{bmatrix} 0.19 \\ 0.81 \end{bmatrix}$	$\begin{bmatrix} 0.34 \\ 0.66 \end{bmatrix}$	$\begin{bmatrix} 0.47 \\ 0.53 \end{bmatrix}$	$\begin{bmatrix} 0.57 \\ 0.43 \end{bmatrix}$

Computational methods based on discretization of equation (12.110) are available in MATLAB, which have far better computational efficiency and accuracy.

Example 12.11: A discrete-time system has state equation given by

$$\mathbf{x}(k+1) = \begin{bmatrix} 0 & 1 \\ -10 & -7 \end{bmatrix} \mathbf{x}(k)$$

Use Cayley-Hamilton approach to find out its state transition matrix.

Solution. The characteristic equation is

$$|\lambda\mathbf{I} = \mathbf{A}| = \begin{vmatrix} \lambda & -1 \\ 10 & \lambda+7 \end{vmatrix} = (\lambda+2)(\lambda+5) = 0$$

Clearly, the eigenvalues are $\lambda_1 = -2$ and $\lambda_2 = -5$. We have

$$\Phi(k) = \mathbf{A}^k = \alpha_0\mathbf{I} + \alpha_1\mathbf{A}$$

and the equations to be solved are

$$(-2)^k = \alpha_0 - 2\alpha_1$$

$$(-5)^k = \alpha_0 - 5\alpha_1$$

Solving we obtain $\alpha_0 = -\frac{1}{3}(-2)^k - \frac{2}{3}(-5)^k$ and $\alpha_1 = -\frac{1}{3}(-2)^k - \frac{1}{3}(-5)^k$. Then

$$\mathbf{\Phi}(k) = \left[-\frac{1}{3}(-2)^k - \frac{2}{3}(-5)^k\right]\begin{bmatrix} 1 & 0 \\ 0 & 1 \end{bmatrix} + \left[\frac{1}{3}(-2)^k - \frac{1}{3}(-5)^k\right]\begin{bmatrix} 0 & 1 \\ -10 & -7 \end{bmatrix}$$

$$= \begin{bmatrix} -\frac{1}{3}(-2)^k - \frac{2}{3}(-5)^k & \frac{1}{3}(-2)^k - \frac{1}{3}(-5)^k \\ -\frac{10}{3}(-2)^k + \frac{10}{3}(-5)^k & -\frac{8}{3}(-2)^k + \frac{5}{3}(-5)^k \end{bmatrix}.$$

An Efficient Method of Discretization and Solution

Consider the continuous time state equation

$$\dot{\mathbf{x}} = \mathbf{Ax} + \mathbf{Bu} \quad ...(i)$$

Its solution given by eqn. (12.71) is reproduced below

$$\mathbf{x}(t) = e^{\mathbf{A}t}\,\mathbf{x}(0) + \int_0^t e^{\mathbf{A}(t-\tau)}\,\mathbf{Bu}(\tau)d\tau \quad ...(ii)$$

At $t = kT$

$$\mathbf{x}(kT) = e^{\mathbf{A}kT}\,\mathbf{x}(0) + \int_0^{kT} e^{\mathbf{A}(kT-\tau)}\,\mathbf{Bu}(kT)d\tau \quad ...(iii)$$

Assuming $\mathbf{u}(k\tau)$ to remain constant for one time interval T, we can write

$$\mathbf{x}(k+1)T) = e^{\mathbf{A}(k+1)T}\mathbf{x}(0) + \int_0^{(k+1)T} e^{\mathbf{A}((k+1)T-\tau)}\,\mathbf{Bu}(k\tau)d\tau \quad ...(iv)$$

This equation can be written as

$$\mathbf{x}((k+1)T) = e^{\mathbf{A}T}\left[e^{\mathbf{A}kT}\mathbf{x}(0) + \int_0^{kT} e^{\mathbf{A}(kT-\tau)}\mathbf{Bu}(kT)d\tau\right] + \int_{kT}^{(k+1)T} e^{\mathbf{A}(kT+T-\tau)}\mathbf{Bu}(kT)d\tau \quad ...(v)$$

From here onwards we will write $\mathbf{x}(kT)$ as $\mathbf{x}(k)$ and $\mathbf{u}(kT)$ as $\mathbf{u}(k)$.

Recognizing the term within capital brackets as $\mathbf{x}(k)$ and defining $\alpha = kT + T - \tau$, eqn. (v) takes the form

$$\mathbf{x}(k+1) = e^{\mathbf{A}T}\mathbf{x}(k) + \left[\int_0^T e^{\mathbf{A}\alpha}d\alpha\right]\mathbf{Bu}(kT) \quad ...(vi)$$

This result can be expressed as

$$\mathbf{x}(k+1) = \mathbf{A}_d\mathbf{x}(k) + \mathbf{B}_d\mathbf{u}(k) \quad ...(12.87a)$$

Also

$$\mathbf{y}(k) = \mathbf{C}_d\mathbf{x}(k) + \mathbf{u}(k) \quad ...(12.87b)$$

where $\mathbf{A}_d = e^{\mathbf{A}t}$, $\mathbf{B}_d = \left[\int_0^T e^{\mathbf{A}\tau}d\tau\right]\mathbf{B}$

The above equation yield exact result at $t = kT$ provided the input $\mathbf{u}(kT)$ is constant for one time interval T *i.e.* it is piecewise constant.

For computation we can write

$$\mathbf{B}_d = \left[\int_0^T\left(\mathbf{I} + \mathbf{A}\tau + \mathbf{A}^2\frac{\tau^2}{2!}......\right)d\tau\right]\mathbf{B} = \left[T\mathbf{I} + \frac{T^2}{2!}\mathbf{A}^2 + \frac{T^3}{3!}\mathbf{A}^3 + ...\right]\mathbf{B}$$

$$= \mathbf{A}^{-1}\left[\left(T\mathbf{A} + \frac{T^2}{2!}\mathbf{A}^2 + \frac{T^3}{3!}\mathbf{A}^3......\right) + \mathbf{I} - \mathbf{I}\right]\mathbf{B}$$

which indeed is

$$\mathbf{B}_d = \mathbf{A}^{-1}(\mathbf{A}_d - \mathbf{I}) \quad ...(12.87c)$$

$$\mathbf{A}_d = e^{\mathbf{A}T} \quad ...(12.87d)$$

Using equations (12.87) the discrete value of state variables and output can be computed by means of MATLAB programmes.

12.7 CONCEPTS OF CONTROLLABILITY AND OBSERVABILITY

In control problems two basic questions need to be answered in deciding whether or not a control solution exists. These questions may be posed as :

(*i*) Can we transfer the system from any initial state to any other desired state in finite time by application of a suitable control force ?

(*ii*) Knowing the output vector for a finite length of time, can we determine the initial state of the system ?

The answers to these basic questions were conceptualized by Kalman into what is known as controllability and observability advanced below:

A system is said to be completely state controllable if it is possible to transfer the system state from any initial state $\mathbf{x}(t_0)$ *to any desired state* $\mathbf{x}(t)$ *in specified finite time by a control vector* $\mathbf{u}(t)$.

Sometimes it is desired to transfer the system output from an initial value to any other desired value. For a system whose output does not depend upon control vector but depends upon state vector only, the dimensionality of output vector is usually less than that of state vector. Therefore the output control is comparatively easier. In this section, we shall discuss state controllability only. The conditions for output controllability, if desired, may be derived on the similar lines.

A system is said to be completely observable, if every state $\mathbf{x}(t_0)$ *can be completely identified by measurements of the outputs* $y(t)$ *over a finite time interval.*

A system which is not completely observable, implies that some of its state variables are shielded from observation.

As pointed out above, the concepts of controllability and observability play an important role in control engineering. These concepts were originally introduced by *Kalman*. The mathematical tests of controllability and observability developed by Kalman, though elegant do not give a physical feel of the problems involved. The alternative test introduced by *Gilbert* uses canonical state model and provides better physical insight into the problem. In this section, we shall first discuss the Gilbert's method and then Kalman's test without proof. The proof involves concept which have not been introduced in this book.

Controllability

As mentioned earlier, the concept of controllability involves the dependence of state variable of the system on the inputs. Consider a single input linear time invariant system

$$\dot{\mathbf{x}} = \mathbf{A}\mathbf{x} + \mathbf{B}u \qquad ...(12.88)$$

where $\mathbf{x}$ = n-dimensional state vector ; u = control signal (control force) ; $\mathbf{A} = n \times n$ matrix ; and $\mathbf{B} = n \times 1$ matrix.

Let the initial system state be $\mathbf{x}(0)$ and the final desired state be $\mathbf{x}(t_f)$. The system described by eqn. (12.88) is controllable if it is possible to construct a control signal, which in finite time interval $0 < t \le t_f$, will transfer the system state from $\mathbf{x}(0)$ to $\mathbf{x}(t_f)$. First we shall advance the Gilbert's method of testing controllability.

Let us first assume that the eigenvalues of the matrix **A** are all distinct to that it can be transformed into the canonical state variable form

$$\dot{v} = \Lambda v + \tilde{\mathbf{B}} u$$

or
$$\begin{bmatrix} \dot{v}_1 \\ \vdots \\ \dot{v}_n \end{bmatrix} = \begin{bmatrix} \lambda_1 & \cdots & 0 \\ & & \\ 0 & \cdots & \lambda_n \end{bmatrix} \begin{bmatrix} v_1 \\ \vdots \\ v_n \end{bmatrix} + \begin{bmatrix} \tilde{b}_1 \\ \vdots \\ \tilde{b}_n \end{bmatrix} u \qquad \text{...(12.89)}$$

This equation can be written in the component form as

$$\dot{v}_i = \lambda_i v_i + \tilde{b}_i u, \; i = 1, 2, ..., n$$

which has the solution

$$v_i(t) = e^{\lambda_i t} v_i(0) + e^{\lambda_i t} \int_0^t e^{-\lambda_i \tau} \tilde{b}_i u(\tau)\, d\tau$$

The system described by eqn. (12.89) is completely controllable if state variable v_i can be transferred from initial state $v_i(0)$ to a final state $v_i(t_f)$ in a finite time t_f. In other words, the system is controllable if it is possible to construct a control signal $u(t)$ such that the following equation is satisfied :

$$\frac{v_i(t_f) - e^{\lambda_i t_f}\, v_i(0)}{e^{\lambda_i t_f}} = \int_0^t e^{-\lambda_i \tau} \tilde{b}_i u(\tau)\, d\tau$$

There are actually, numerous value of $u(\tau)$ which satisfy this equation provided $\tilde{b}_i \neq 0$ because otherwise the link between input and the corresponding state variable gets broken and hence it is no longer possible to control that particular state variable.

It therefore follows that the necessary condition of complete controllability is simply that the vector $\tilde{\mathbf{B}}$ should not have any zero elements. If any element of this vector is zero, then the corresponding state variable is not controllable. It can be further shown that the condition stated here is in fact both necessary and sufficient.

The result just obtained, can be extended to the case where the control force **u** is an m-dimensional vector. For the system described by

$$\dot{v} = \Lambda v + \tilde{\mathbf{B}} \mathbf{u}$$

where
$$\tilde{\mathbf{B}} = \begin{bmatrix} \tilde{b}_{11} & \tilde{b}_{12} & \cdots & \tilde{b}_{1m} \\ \tilde{b}_{21} & \tilde{b}_{22} & \cdots & \tilde{b}_{2m} \\ \vdots & \vdots & \cdots & \vdots \\ \tilde{b}_{n1} & \tilde{b}_{n2} & \cdots & \tilde{b}_{nm} \end{bmatrix}$$

The necessary and sufficient condition for controllability is that the matrix $\tilde{\mathbf{B}}$ must have no row with all zeros. It is observed from the above equation that if any row of the matrix $\tilde{\mathbf{B}}$ is zero, it is not possible to influence the corresponding state variable by the control forces and hence the particular state variable is uncontrollable. In case **A** has a Jordan block the elements of any row of $\tilde{\mathbf{B}}$ that correspond to the last row of the Jordan block are not all zero.

The Gilbert's method of testing controllability discussed above requires that the system be transformed into canonical state or Jordan canonical form. The test of controllability due to Kalman which can be applied to any state model (canonical or otherwise), is stated below.

A general nth order multi-input linear time-invariant system (with an m-dimensional control vector),

$$\dot{\mathbf{x}} = \mathbf{Ax} + \mathbf{Bu}$$

is completely controllable if and only if the rank of the composite matrix

$$\mathbf{Q}_c = [\mathbf{B} \vdots \mathbf{AB} \vdots \cdots \vdots \mathbf{A}^{n-1}\mathbf{B}] \qquad \text{...(12.90)}$$

is n. Since only matrices **A** and **B** are involved in (12.90), we may say that the pair (**A, B**) is controlled if rank of $\mathbf{Q}_c$ is n.

Example 12.12. Consider the system with state equation

$$\begin{bmatrix} \dot{x}_1 \\ \dot{x}_2 \\ \dot{x}_3 \end{bmatrix} = \begin{bmatrix} 0 & 1 & 0 \\ 0 & 0 & 1 \\ -6 & -11 & -6 \end{bmatrix} \begin{bmatrix} x_1 \\ x_2 \\ x_3 \end{bmatrix} + \begin{bmatrix} 0 \\ 0 \\ 1 \end{bmatrix} u$$

Let us first transform this state equation into canonical form.

The characteristic equation is given by

$$|\mathbf{A} - \lambda \mathbf{I}| = \begin{bmatrix} -\lambda & 1 & 0 \\ 0 & -\lambda & 1 \\ -6 & -11 & -6-\lambda \end{bmatrix} = 0$$

which gives the eigenvalues

$$\lambda_1 = -1, \lambda_2 = -2, \lambda_3 = -3,$$

Choosing Vander Monde matrix as modal matrix, we have

$$\mathbf{M} = \begin{bmatrix} 1 & 1 & 1 \\ \lambda_1 & \lambda_2 & \lambda_3 \\ \lambda_1^2 & \lambda_2^2 & \lambda_3^2 \end{bmatrix} = \begin{bmatrix} 1 & 1 & 1 \\ -1 & -2 & -3 \\ 1 & 4 & 9 \end{bmatrix}$$

$$\tilde{\mathbf{B}} = \mathbf{M}^{-1}\mathbf{B} = \begin{bmatrix} 3 & 5/2 & 1/2 \\ -3 & -4 & -1 \\ 1 & 3/2 & 1/2 \end{bmatrix} \begin{bmatrix} 0 \\ 0 \\ 1 \end{bmatrix} = \begin{bmatrix} 1/2 \\ -1 \\ 1/2 \end{bmatrix}$$

Therefore, the state equation in canonical form is given by

$$\begin{bmatrix} \dot{v}_1 \\ \dot{v}_2 \\ \dot{v}_3 \end{bmatrix} = \begin{bmatrix} -1 & 0 & 0 \\ 0 & -2 & 0 \\ 0 & 0 & -3 \end{bmatrix} \begin{bmatrix} v_1 \\ v_2 \\ v_3 \end{bmatrix} + \begin{bmatrix} 1/2 \\ -1 \\ 1/2 \end{bmatrix} u$$

Since no element of $\tilde{\mathbf{B}}$ is zero, the system is completely controllable.

Let us now test controllability of this system by the Kalman's test. We have

$$\mathbf{B} = \begin{bmatrix} 0 \\ 0 \\ 1 \end{bmatrix}; \quad \mathbf{A} = \begin{bmatrix} 0 & 1 & 0 \\ 0 & 0 & 1 \\ -6 & -11 & -6 \end{bmatrix}$$

Then

$$\mathbf{AB} = \begin{bmatrix} 0 & 1 & 0 \\ 0 & 0 & 1 \\ -6 & -11 & -6 \end{bmatrix} \begin{bmatrix} 0 \\ 0 \\ 1 \end{bmatrix} = \begin{bmatrix} 0 \\ 1 \\ -6 \end{bmatrix}$$

$$\mathbf{A}^2\mathbf{B} = \begin{bmatrix} 0 & 1 & 0 \\ 0 & 0 & 1 \\ -6 & -11 & -6 \end{bmatrix} \begin{bmatrix} 0 \\ 1 \\ -6 \end{bmatrix} = \begin{bmatrix} 1 \\ -6 \\ 25 \end{bmatrix}$$

The composite matrix defined in eqn. (12.90) is given by

$$\mathbf{Q}_c = [\mathbf{B} \vdots \mathbf{AB} \vdots \mathbf{A}^2\mathbf{B}]$$

$$= \begin{bmatrix} 0 & 0 & 1 \\ 0 & 1 & -6 \\ 1 & -6 & 25 \end{bmatrix}$$

It is easily seen that det $\mathbf{Q}_c \neq 0$ *i.e.*, its rank is $r = n = 3$. The system is therefore completely controllable.

Controllable phase variable form

Notice that if the system equation is given by

$$\dot{\mathbf{v}} = \mathbf{A}_c\mathbf{v} + \mathbf{B}_c u \qquad \text{...(12.91)}$$

where v is $n \times 1$ vector and u is a scalar, and

$$\mathbf{A}_c = \begin{bmatrix} 0 & 1 & 0 & \cdots & 0 \\ 0 & 0 & 1 & \cdots & 0 \\ \vdots & \vdots & \vdots & & \vdots \\ 0 & 0 & 0 & \cdots & 1 \\ -a_n & -a_{n-1} & -a_{n-2} & \cdots & -a_0 \end{bmatrix}; \mathbf{B}_c = \begin{bmatrix} 0 \\ 0 \\ \vdots \\ 0 \\ 1 \end{bmatrix} \qquad \text{...(12.92)}$$

then it is easily verified that the rank of the matrix $[\mathbf{B}_c \vdots \mathbf{A}_c\mathbf{B}_c \vdots ... \vdots \mathbf{A}_c^{n-1}\mathbf{B}]$ is n.

Therefore a system described by state equations of the form (12.91) is always controllable.

The converse is also true, *i.e.*, if a linear time-invariant system given by

$$\dot{\mathbf{x}} = \mathbf{Ax} + \mathbf{B}u \qquad \text{...(12.93)}$$

where $\mathbf{x}$ is $n \times 1$ state vector, $\mathbf{A}$ is $n \times n$ system matrix, $\mathbf{B}$ is $n \times 1$ control vector, u is scalar control input, is controllable, then it can be transformed into the form (12.91).

Example 12.13. Consider a linear system described by the differential equation

$$\ddot{y} + 2\dot{y} + y = \dot{u} + u$$

with $\quad x_1 = y, x_2 = \dot{y} - u$

as the state variables, we get the state method

$$\begin{bmatrix} \dot{x}_1 \\ \dot{x}_2 \end{bmatrix} = \begin{bmatrix} 0 & 1 \\ -1 & -2 \end{bmatrix} \begin{bmatrix} x_1 \\ x_2 \end{bmatrix} + \begin{bmatrix} 1 \\ -1 \end{bmatrix} u \qquad \text{...(12.94a)}$$

$$y = x_1 \qquad \text{...(12.94b)}$$

Let us test controllability of this system by Kalman's test.

From eqn. (12.94*a*), we have

$$\mathbf{B} = \begin{bmatrix} 1 \\ -1 \end{bmatrix}; \mathbf{A} = \begin{bmatrix} 0 & 1 \\ -1 & -2 \end{bmatrix}$$

Then

$$\mathbf{AB} = \begin{bmatrix} 0 & 1 \\ -1 & -2 \end{bmatrix}\begin{bmatrix} 1 \\ -1 \end{bmatrix} = \begin{bmatrix} -1 \\ 1 \end{bmatrix}$$

The composite matrix defined in eqn. (12.90) is given by

$$\mathbf{Q}_c = [\mathbf{B} \vdots \mathbf{AB}] = \begin{bmatrix} 1 & -1 \\ -1 & 1 \end{bmatrix}$$

The rank of r of this matrix is 1. The system is therefore not completely controllable. One state of the system is uncontrollable (r out of n states are controllable).

By the methods discussed in Section 12.3, the given differential equation can be transformed to the following controllable phase variable model:

$$\begin{bmatrix} \dot{x}_1 \\ \dot{x}_2 \end{bmatrix} = \begin{bmatrix} 0 & -1 \\ -1 & 2 \end{bmatrix}\begin{bmatrix} x_1 \\ x_2 \end{bmatrix} + \begin{bmatrix} 0 \\ 1 \end{bmatrix} u$$

$$y = [1 \quad 1]\begin{bmatrix} x_1 \\ x_2 \end{bmatrix}$$

Thus state controllability depends on how the state variables are defined for a given system.

Observability

Consider the state model of an nth order single-output linear time-invariant system,

$$\dot{\mathbf{x}} = \mathbf{Ax} + \mathbf{Bu}$$
$$y = \mathbf{Cx}$$

The state equation may be transformed to the canonical form by the linear transformation $\mathbf{x} = \mathbf{M}\mathbf{v}$. The resulting state and output equations are

$$\dot{\mathbf{v}} = \Lambda \mathbf{v} + \tilde{\mathbf{B}}\mathbf{u} \qquad \text{...(12.95}a\text{)}$$

$$y = \tilde{\mathbf{C}}\mathbf{v}$$
$$= \tilde{c}_1 v_1 + \tilde{c}_2 v_2 + \ldots\ldots + \tilde{c}_n v_n \qquad \text{...(12.95}b\text{)}$$

Since diagonalization decouples the states, no state now contains any information regarding any other state, *i.e.*, each state must be independently observable. It therefore follows that for a state to be observed through the output y, its corresponding coefficient in eqn. (12.95b) should be nonzero.

If any particular $\tilde{c}_i$ is zero, the corresponding v_i can have any value without its effect showing up in the output y. Thus the necessary (it is also sufficient) condition for complete state observability is that none of the $\tilde{c}_i$'s (*i.e.*, none of the elements of $\tilde{\mathbf{C}} = \mathbf{CM}$) should be zero.

The result may be extended to the case of multi-input-multi-output systems where the output vector, after canonical transformation is given by

$$\begin{bmatrix} y_1 \\ y_2 \\ \vdots \\ y_p \end{bmatrix} = \begin{bmatrix} \tilde{c}_{11} & \tilde{c}_{12} & \cdots & \tilde{c}_{1n} \\ \tilde{c}_{21} & \tilde{c}_{22} & \cdots & \tilde{c}_{2n} \\ \vdots & \vdots & & \vdots \\ \tilde{c}_{p1} & \tilde{c}_{p2} & \cdots & \tilde{c}_{pn} \end{bmatrix}\begin{bmatrix} v_1 \\ v_2 \\ \vdots \\ v_n \end{bmatrix}$$

or
$$\mathbf{y} = \tilde{\mathbf{C}}\mathbf{v}$$

The necessary condition for complete observability is that none of the columns of the matrix $\tilde{\mathbf{C}}$ be zero. This is the Gilbert's test.

The Kalman's test of observability is as follows:

A general nth order multi-input multi-output linear time-invarient system.

$$\dot{\mathbf{x}} = \mathbf{Ax} + \mathbf{Bu}$$
$$\mathbf{y} = \mathbf{Cx}$$

is completely observable if and only if the rank of composite matrix

$$\mathbf{Q}_0 = [\mathbf{C}^T : \mathbf{A}^T\mathbf{C}^T : \cdots (\mathbf{A}^T)^{n-1}\mathbf{C}^T] \qquad \text{...(12.96)}$$

is n. This condition is also referred as the pair (**A, C**) being observable.

Duality property

Comparing eqn. (12.96) with (12.90), the following observations may be made:

1. The pair (**AB**) is controllable implies that the pair ($\mathbf{A}^T\mathbf{B}^T$) is observable.
2. The pair (**AC**) is observable implies that the pair ($\mathbf{A}^T\mathbf{C}^T$) is controllable.

Thus the concepts of controllability and observability are dual concepts.

Observable phase variable form

If the system equations are given by*

$$\dot{v} = \mathbf{A}_0 v + \mathbf{B}_0 u \qquad \text{...(12.97}a\text{)}$$
$$y = \mathbf{C}_0 v + du \qquad \text{...(12.97}b\text{)}$$

where $\mathbf{v}$ is $n \times 1$ vector, y and u are scalars, and

$$\mathbf{A}_0 = \begin{bmatrix} 0 & 0 & \cdots & 0 & -a_n \\ 1 & 0 & \cdots & 0 & -a_{n-1} \\ 0 & 1 & \cdots & 0 & -a_{n-2} \\ \vdots & \vdots & & \vdots & \vdots \\ 0 & 0 & \cdots & 1 & -a_1 \end{bmatrix}; \quad \mathbf{B}_0 = \begin{bmatrix} \beta_n \\ \beta_{n-1} \\ \beta_{n-2} \\ \vdots \\ \beta_1 \end{bmatrix} \qquad \text{...(12.98)}$$

$$\mathbf{C}_0 = [0 \;\; 0 \;\; \cdots \;\; 0 \;\; 1]$$

It is then easily verified using (12.97) that the system is always observable. The conserve is also true, *i.e.*, linear time-invariant observable system can be transformed into the form (12.98). This result can easily be proved using the duality property. Equations (12.97) are said to be in the *Observable phase variable form.*

Example 12.14. Let us examine the observability of the system given below.

$$\begin{bmatrix} \dot{x}_1 \\ \dot{x}_2 \\ \dot{x}_3 \end{bmatrix} = \begin{bmatrix} 0 & 1 & 0 \\ 0 & 0 & 1 \\ 0 & -2 & -3 \end{bmatrix} \begin{bmatrix} x_1 \\ x_2 \\ x_3 \end{bmatrix} = \begin{bmatrix} 0 \\ 0 \\ 1 \end{bmatrix} u = \mathbf{Ax} + \mathbf{B}u \qquad \text{...(}i\text{)}$$

---

*Note that (12.34) can transformed to (12.97) with the transformation

$$v_i = x_{n+1-i}\,;\; i = 1, 2, \ldots\ldots, n.$$

$$y = [3 \quad 4 \quad 1] \begin{bmatrix} x_1 \\ x_2 \\ x_3 \end{bmatrix} = \mathbf{Cx} \qquad \ldots(ii)$$

The characteristic equation is

$$|\mathbf{A} - \lambda \mathbf{I}| = \begin{bmatrix} -\lambda & 1 & 0 \\ 0 & -\lambda & 1 \\ 0 & -2 & -3-\lambda \end{bmatrix} = 0$$

or $$\lambda(\lambda + 1)(\lambda + 2) = 0$$

Therefore the eigenvalues of matrix **A** are

$$\lambda_1 = 0, \lambda_2 = -1, \lambda_3 = -2$$

The Vander Monde matrix is then

$$\mathbf{M} = \begin{bmatrix} 1 & 1 & 1 \\ 0 & -1 & -2 \\ 0 & 1 & 4 \end{bmatrix}$$

Under the linear transformation $\mathbf{x} = \mathbf{Mv}$, the output is given by

$$y = \mathbf{CMv} = [3 \quad 0 \quad -1] \begin{bmatrix} v_1 \\ v_2 \\ v_3 \end{bmatrix}$$

It is found that the system is not completely observable, since the state variable v_2 is hidden from observation.

Let us apply the Kalman's test to the same system. From eqns. (*i*) and (*ii*)

$$\mathbf{A}^T\mathbf{C}^T = \begin{bmatrix} 0 & 0 & 0 \\ 1 & 0 & -2 \\ 0 & 1 & -3 \end{bmatrix} \begin{bmatrix} 3 \\ 4 \\ 1 \end{bmatrix} = \begin{bmatrix} 0 \\ 1 \\ 1 \end{bmatrix}$$

$$(\mathbf{A}^T)^2\,\mathbf{C}^T = \begin{bmatrix} 0 & 0 & 0 \\ 1 & 0 & -2 \\ 0 & 1 & -3 \end{bmatrix} \begin{bmatrix} 0 \\ 1 \\ 1 \end{bmatrix} = \begin{bmatrix} 0 \\ -2 \\ -2 \end{bmatrix}$$

Therefore the composite matrix defined in eqn. (12.86) is given by

$$\mathbf{Q}_0 = [\mathbf{C}^T \vdots \mathbf{A}^T\,\mathbf{C}^T \vdots (\mathbf{A}^T)^2\,\mathbf{C}^T] = \begin{bmatrix} 3 & 0 & 0 \\ 4 & 1 & -2 \\ 1 & 1 & -2 \end{bmatrix}$$

Since $$\begin{vmatrix} 3 & 0 \\ 4 & 1 \end{vmatrix} \neq 0 \quad \text{and} \quad \begin{vmatrix} 3 & 0 & 0 \\ 4 & 1 & -2 \\ 1 & 1 & -2 \end{vmatrix} = 0$$

the rank of the matrix $\mathbf{Q}_0$ is $r = 2$, while $n = 3$. Hence one of the state variable is unobservable.

Effect of Pole-zero Cancellation in Transfer Function

In the analysis and design of linear time-invariant control systems, we have used transfer functions extensively. Although controllability and observability are concepts of modern control theory, they are closely related to the properties of the transfer function.

For an nth-order system with distinct eigenvalues, let us assume that the input-output transfer function is of the form

$$T(s) = \frac{Y(s)}{U(s)} = \frac{b_0 s^m + b_1 s^{m-1} + \ldots. + b_{m-1}s + b_m}{s^n + a_1 s^{n-1} + \ldots..+ a_{n-1}s + a_n}$$

$$= \frac{K(s-\beta_1)(s-\beta_2)\ldots.(s-\beta_m)}{(s-\lambda_1)(s-\lambda_2)\ldots.(s-\lambda_n)}$$

$$= \sum_{k=1}^{n} \frac{c_k}{(s-\lambda_k)} \; ; c_k \text{ are residue of poles at } s = \lambda_k.$$

Assume that the transfer function has identical pair of pole and zero at $\beta_i = \lambda_i$ thus $c_i = 0$. The effect of this cancellation on controllability and observability properties depends on how the state variable is defined. If state variables are selected so as to get state model of the form (12.43), then $c_i = 0$ will appear in control vector **B** and the state x_i is uncontrollable. On the other hand if state variables are selected so as to get state model of the form (12.42), then $c_i = 0$ will appear in output vector **C** and the state x_i is shielded from observation.

Thus if the input-output transfer function of a linear time-invariant system has pole-zero cancellation, the system will be either not state controllable or unobservable, depending on how the state variables are defined. If the transfer function does not have pole-zero cancellation, the system can always be represented by completely controllable and observable state model.

The reader may note that practically all physical systems are controllable and observable. However, mathematical models of these systems may not be so. Mathematical models may lack these properties, particularly if linearization has been necessary. If this happens, the model is not an accurate representation of the system. Such a model may not given an optimal control solution and if one exists it may not be implementable. One therefore should seek another state model which is controllable and observable.

Controllability and Observability (Discrete Case)

The controllability and observability concepts presented in this Section are carried over directly to the linear discrete-time systems. Kalman's and Gilbert's tests can be used for discrete-time systems as well.

In sampled-data systems, there is an additional requirement. Assume that transfer function of continuous-time plant has a partial fraction expansion which contains the term

$G(s) = \dfrac{\omega}{(s+\sigma)^2 + \omega^2}$. The z-transform of this term is

$$G(z) = Z\left[\frac{\omega}{(s+\sigma)^2+\omega^2}\right] = \frac{z^{-1}e^{-\sigma T}\sin\beta t}{1 - 2z^{-1}e^{-\sigma T}\cos\beta T + z^{-2}e^{-2\sigma T}} \qquad \ldots(12.99)$$

Now if $T = \dfrac{2\pi n}{\omega}$ where n = positive integer, then

$$G(z) = 0$$

The system may even be unstable for $\sigma < 0$, but this fact could not be inferred from the observations of the output. In this situation the system is neither completely controllable nor completely observable. Therefore an additional requirement for complete controllability and abservability of sampled-data system is that if the characteristic roots of the continuous time plant are $-\sigma \pm j\omega$, then $T \neq \dfrac{2\pi n}{\omega}$ for n = positive integer.

12.8 POLE PLACEMENT BY STATE FEEDBACK

Chapter 10 was an exposition of the design of linear control systems using classical control theory techniques, *viz.*, the Bode diagrams, root locus plots, Nicholas chart, etc. We observed that in classical design approach, the form of the compensator is preselected based on the performance of the uncompensated system and the design specifications. The parameters of the compensator are then adjusted to bring the system performance within acceptable bounds. We also observed that compensated system required feedback of only one variable, the output. Of course, in system compensated through inner feedback loop, more than one dynamic variables are fedback.

We shall now consider the case of full state feedback, which would give us greater freedom to satisfy the classical performance indices. In fact the use of full state feedback permits us to place all the poles of the characteristic equation at any desired location.

Consider the single-input-single-output system with nth-order state model

$$\dot{\mathbf{x}} = \mathbf{A}\mathbf{x} + \mathbf{B}u \qquad \text{...(12.100}a\text{)}$$

$$y = \mathbf{C}\mathbf{x} \qquad \text{...(12.100}b\text{)}$$

The state variable feedback for this system is essentially a scalar function which takes the form

$$\sigma = \mathbf{K}\mathbf{x} = [k_1 \quad k_2 \quad ....k_n] \begin{bmatrix} x_1 \\ x_2 \\ \vdots \\ x_n \end{bmatrix} \qquad \text{...(12.101}a\text{)}$$

The block diagram of the system with state variable feedback is shown in Fig. 12.21. From this figure it is observed that

$$u = -\mathbf{K}\mathbf{x} + r \qquad \text{...(12.101}b\text{)}$$

where u is plant input and r is the system input.

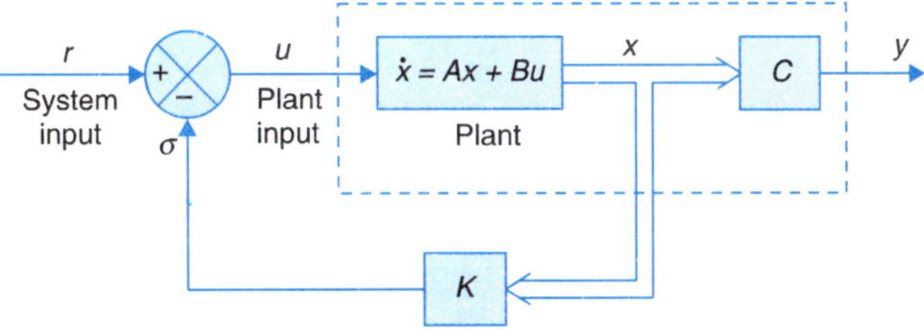

Fig. 12.21. State variable feedback system.

Before presenting a general method of pole placement, we shall consider a simple example of full state feedback.

Example 12.15. Reconsider the system of Fig. 12.8 with full state feedback. Adjust the feedback gains for the system poles to be so placed that if has $\zeta = 0.8$ and $\tau_s = 0.1$ s.

Solution. The signal flow graph of the system with full state feedback is drawn in Fig. 12.22.

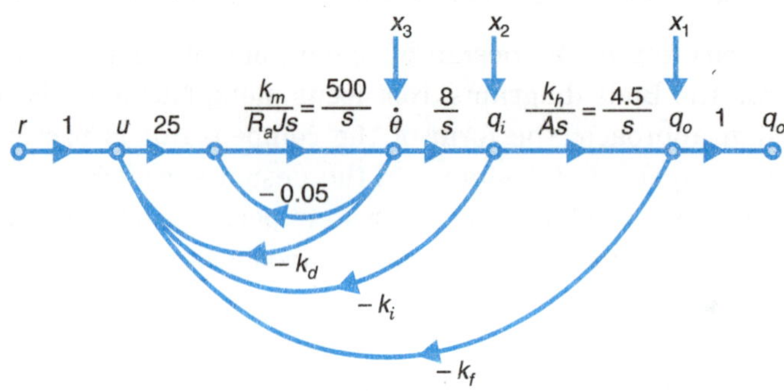

Fig. 12.22

Use of Mason's gain formula gives

$$\frac{q_0(s)}{R(s)} = \frac{25 \times \frac{500}{s} \times \frac{8}{s} \times \frac{4.5}{s}}{1 + \frac{500 \times 0.05 + 25 \times 500 k_d}{s} + \frac{25 \times 500 \times 8 k_i}{s^2} + \frac{25 \times 500 \times 8 \times 4.5 k_f}{s^3}}$$

$$= \frac{4.5 \times 10^5}{s^3 + 3.125 \times 10^5 \, k_d s^2 + 1 \times 10^5 k_i s + 4.5 \times 10^5 k_f} \quad \text{...(i)}$$

We demand the characteristic equation to be

$$(s^2 + 2\zeta\omega_n s + \omega_n^2)\,(s + \zeta\omega_n)$$

with $\zeta = 0.8$

and $$\tau_s = \frac{4}{\zeta\omega_n} = 0.1 \quad \Rightarrow \quad \omega_n = \frac{4}{0.8 \times 0.1} = 50 \quad \text{...(ii)}$$

Substituting the values in eqn. (*ii*)

$$(s^2 + 2 \times 0.8 \times 50s + 2500)\,(s + 0.8 \times 250)$$

or $$s^3 + 120s^3 + 5700s + 10^5 \quad \text{...(iii)}$$

Comparing the eqns. (*i*) and (*iii*) term by term

(1) $$4.5 \times 10^5 \, k_f = 10^5$$

or $$k_f = 0.222$$

(2) $$1.0 \times 10^5 \, k_i = 15.7 \times 10^3$$

or $$k_i = 0.057$$

(3) $$25 + 12.5 \times 10^3 \, k_d = 120$$

or $$k_d = 0.75 \times 10^{-3}$$

Remark: Full state feedback is usually not practical because it is not possible or practical to sense all the states and feed them back. In practical situations, only a few states or combinations of them are possible and the compensator must depend on outputs, inputs and a few state variables for compensation. Also the full state feedback leads to PD-type or PID compensators, with infinite bandwidth, whereas real compensators always have finite bandwidth.

From eqns. (12.100) and (12.101), we get state equation for the feedback system as

$$\dot{\mathbf{x}} = (\mathbf{A} - \mathbf{BK})\mathbf{x} + \mathbf{B}r \qquad ...(12.102)$$

It can be shown that if the pair (**A, B**) is controllable, then by the state feedback given by (12.102), the eigenvalues of (**A** – **BK**) in (12.102) can be arbitrarily assigned.

To prove omit this, let us first transform (12.100*a*) into controllable phase variable form using the transformation, see Appendix IV

$$\mathbf{v} = \mathbf{Px}$$

The resulting phase variables form is

$$\dot{v} = \begin{bmatrix} 0 & 1 & 0 & \cdots & 0 & 0 \\ 0 & 0 & 1 & \cdots & 0 & 0 \\ \vdots & \vdots & \vdots & & \vdots & \vdots \\ 0 & 0 & 0 & \cdots & 0 & 1 \\ -a_n & -a_{n-1} & -a_{n-2} & \cdots & -a_2 & -a_1 \end{bmatrix} \mathbf{v} + \begin{bmatrix} 0 \\ 0 \\ \vdots \\ 0 \\ 1 \end{bmatrix} u$$

$$= \mathbf{A}_c \mathbf{v} + \mathbf{B}_c u$$

Under this transformation, eqn. (12.101*b*) modifies to

$$u = -\mathbf{KP}^{-1}\mathbf{v} + r$$

$$= -\mathbf{K}_c \mathbf{v} + r$$

where

$$\mathbf{K}_c = \mathbf{KP}^{-1} \qquad ...(12.103)$$

Since the eigenvalues are invariant under transformation of state model, the set of eigenvalues of (**A** – **BK**) is the same as that of ($\mathbf{A}_c - \mathbf{B}_c\mathbf{K}_c$).

Let the desired eigenvalues be $\lambda_1, \lambda_2, \ldots\ldots, \lambda_n$ and

$$\lambda_n + \alpha_1\lambda^{n-1} + \alpha_2\lambda^{n-2} + \ldots. + \alpha_n = 0$$

be the corresponding characteristics equation.

If we choose

$$\mathbf{K}_c = [\alpha_n - a_n\, , \alpha_{n-1} - a_{n-1}, \ldots., \alpha_1 - a_1] \qquad ...(12.104)$$

then the controllable phase variable form of the feedback system is

$$\dot{v} = \begin{bmatrix} 0 & 1 & 0 & \cdots & 0 & 0 \\ 0 & 0 & 1 & \cdots & 0 & 0 \\ \vdots & \vdots & \vdots & & \vdots & \vdots \\ 0 & 0 & 0 & \cdots & 0 & 1 \\ -\alpha_n & -\alpha_{n-1} & -\alpha_{n-2} & \cdots & -\alpha_2 & -\alpha_1 \end{bmatrix} \mathbf{v} + \begin{bmatrix} 0 \\ 0 \\ \vdots \\ 0 \\ 1 \end{bmatrix} r$$

The characteristic equation of system matrix in this equation is same as the desired characteristic equation. The required vector **K**, obtained from eqns. (12.103) and (12.104), is

$$\mathbf{K} = \mathbf{K}_c\mathbf{P}$$

$$= [\alpha_n - a_n\, , \alpha_{n-1} - a_{n-1} \cdots, \alpha_1 - a_1]\mathbf{P} \qquad ...(12.105)$$

We have seen that if a system is completely controllable, its eigenvalues can be arbitrarily located by complete state feedback. Thus any unstable system can be stabilized by complete state feedback if all the states are controllable. More restrictively, even if the system is not completely controllable. Is still *stabilizable* as long as the uncontrollable states are stable.

Example 12.16. Consider the position control system of Fig. 12.23 using a d.c. motor in armature control mode. This signal flow graph of the system with usual symbols is drawn in Fig. 12.24(*a*). Being a third-order system three physical variable θ_c, $\dot{\theta}_c$ and i_a are selected as state variables, *i.e.*,

$$\mathbf{x}^T = [\theta_c \quad \dot{\theta}_c \quad i_a]$$

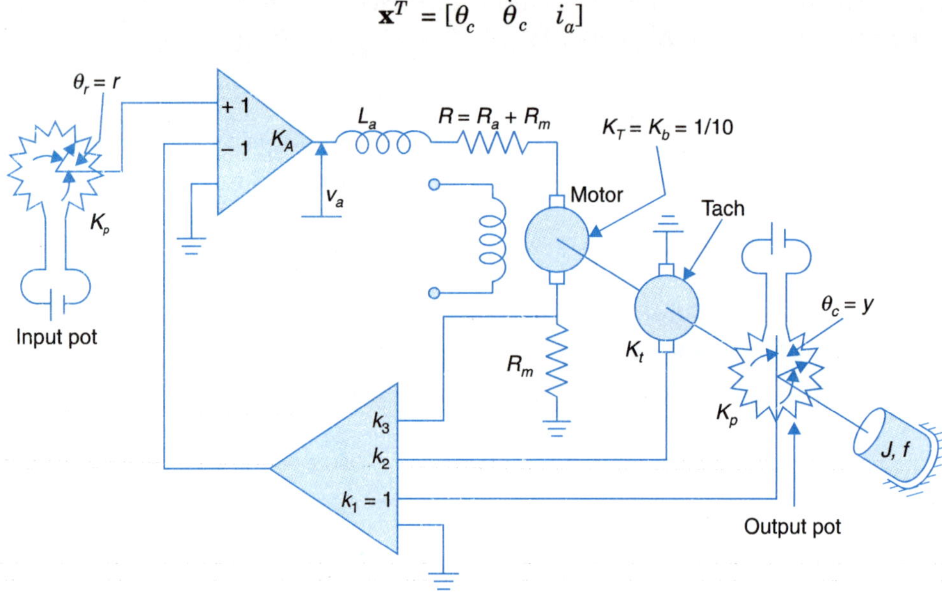

Fig. 12.23. A position control system with state variable feedback.

Full state feedback is employed with position* feedback being obtained from a potentiometer, rate* feedback from a tachometer and current feedback from a current sample across a resistance (small) in series with armature circuit. In the modified signal flow graph of Fig. 12.24(*b*) the outer feedback (position feedback) gain $-k_1 = -K_p$ is reduced to unity by combining potentiometer constant K_p with the gain K_A in the forward path. Three adjustable gains (k'_2, k'_3 and k'_A), one in forward path and two in feedback paths, are now available for achieving design specifications. It must be noted that the inner feedback loop is inherent in the motor operation. Suitable transfer functions for fixed part of the system are shown in Fig 12.24(*b*). The armature circuit time constant has been assumed to be larger than normal for purpose of illustrating the design technique.

*Instead of potentiometer and tachometer, we could use an encoder ; see Fig. P-12.20.

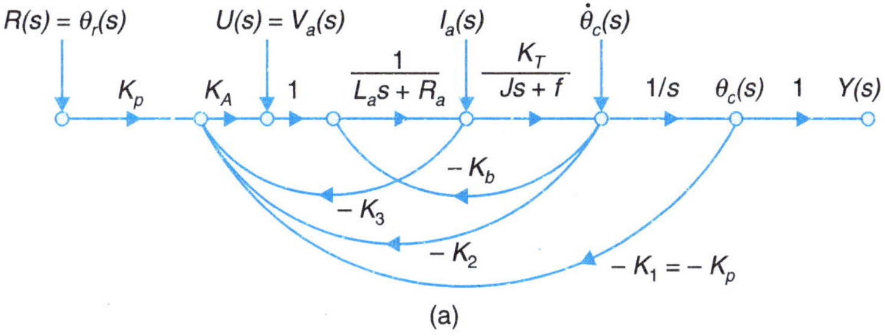

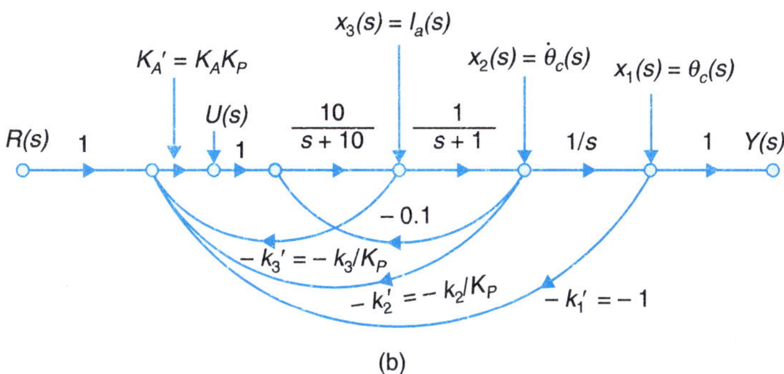

Fig. 12.24. Signal flow graphs of the system shown in Fig. 12.23.

With all the three state feedback loops open, the system is described by

$$\begin{bmatrix} \dot{x}_1 \\ \dot{x}_2 \\ \dot{x}_3 \end{bmatrix} = \begin{bmatrix} 0 & 1 & 0 \\ 0 & -1 & 1 \\ 0 & -1 & -10 \end{bmatrix} \begin{bmatrix} x_1 \\ x_2 \\ x_3 \end{bmatrix} + \begin{bmatrix} 0 \\ 0 \\ 10 \end{bmatrix} u$$

$$= \mathbf{A}\mathbf{x} + \mathbf{B}u \qquad \text{...(12.106)}$$

Let us first test controllability of pair (**AB**) the composite matrix

$$\mathbf{Q}_c = [\mathbf{B} \ \vdots \ \mathbf{AB} \ \vdots \ \mathbf{A}^2\mathbf{B}]$$

$$= \begin{bmatrix} 0 & 0 & 10 \\ 0 & 10 & -110 \\ 10 & -100 & 990 \end{bmatrix}$$

has rank $r = 3$. Therefore, the pair (**A, B**) is controllable. The eigenvalues of **A** in (12.106) as given by the characteristic equation

$$| \lambda \mathbf{I} - \mathbf{A} | = 0$$

or

$$\lambda^3 + 11\lambda^2 + 11\lambda = 0 \qquad \text{...(12.107a)}$$

Note that state model (12.106) is unstable.

Let us assume that for the system under consideration, the following design specifications are to be met for its transient response

$$\zeta = 0.5 \ ; \ \omega_n = 2 \text{ rad/sec}$$

From these specifications, the desired dominant closed-loop poles are found to be at

$$s_d = -1 \pm j\sqrt{3}$$

To ensure the dominance condition the third closed-loop pole should lie at $s \le -10\,\zeta\omega_n$.

Choosing the third pole to be $s_3 = -10\zeta\omega_n = -10$

The denominator of the closed-loop transfer function is obtained as

$$D(s) = (s + 1 + j\sqrt{3})\,(s + 1 - j\sqrt{3})\,(s + 10)$$
$$= s^3 + 12s^2 + 24s + 40$$

In terms of eigenvalues, the desired characteristic equation of the compensated system is

$$\lambda^3 + 12\lambda^2 + 24\lambda + 40 = 0 \qquad \text{...(12.107}b\text{)}$$

Following the procedure given earlier, we have from eqn. (12.107)a and b

$$\mathbf{K}_c = [K_{c1}\; K_{c2}\; K_{c3}] = [(40 - 0)\;(24 - 11)\;(12 - 11)] = [40 \quad 13 \quad 1]$$

From eqn. (12.105), we have

$$\mathbf{K} = \mathbf{K}_c\mathbf{P}$$

Using the result given in Appendix **V**

$$\mathbf{P} = \begin{bmatrix} \mathbf{P}_1 \\ \mathbf{P}_1\mathbf{A} \\ \mathbf{P}_1\mathbf{A}^2 \end{bmatrix}$$

$$\mathbf{P}_1 = [0 \quad 0 \quad 1]\,\mathbf{Q}_c^{-1}$$

$$= [0 \quad 0 \quad 1]\begin{bmatrix} 1.1 & 1 & 0.1 \\ 1.1 & 0.1 & 0 \\ 0.1 & 0 & 0 \end{bmatrix}$$

$$= [0.1 \quad 0 \quad 0]$$

$$\mathbf{P}_1\mathbf{A} = [0 \quad 0.1 \quad 0]$$

$$\mathbf{P}_1\mathbf{A}^2 = [0 \quad -0.1 \quad 0.1]$$

Thus

$$\mathbf{P} = \begin{bmatrix} 0.1 & 0 & 0 \\ 0 & 0.1 & 0 \\ 0 & -0.1 & 0.1 \end{bmatrix}$$

and

$$\mathbf{K} = [40 \quad 13 \quad 1]\,\mathbf{P}$$
$$= [4 \quad 1.2 \quad 0.1]$$
$$= [k_1 \quad k_2 \quad k_3]$$

From Fig. 12.24(b) we find that

$$k_1 = k_1' K_A' = 4$$
$$k_2 = k_2' K_A' = 1.2$$
$$k_3 = k_3' K_A' = 0.1$$

With $k_1' = 1$, we have

$$K_A' = 4\;;\; k_2' = 0.3,\; k_3' = 0.025$$

The above choice of gains meets the design specifications.

Linear Continuous-time System with Sampled Inputs

As has been pointed out earlier, discrete-time systems may arise in various ways. A specific example is that of calculating the response of a continuous-time system by means of a digital computer. Under this situation, a continuous-time system must first be converted into an equivalent discrete-time system.

Consider the continuous-time state equations

$$\dot{\mathbf{x}} = \mathbf{Ax} + \mathbf{Bu} \qquad \text{...(12.108)}$$

we shall impose the following restrictions on the system.

(*i*) The input vector **u** can change only at sampling instants.

(*ii*) The sampling period T is constant at all times.

The continuous-time system with these restrictions is shown in Fig. 12.25. Because of sample and hold operations,

$$u_i(t) = u_i(kT);\ kT \le t < (k+1)\,T$$

$$k = 0, 1, 2, \ldots\ldots$$

$$i = 1, 2, \ldots\ldots, m$$

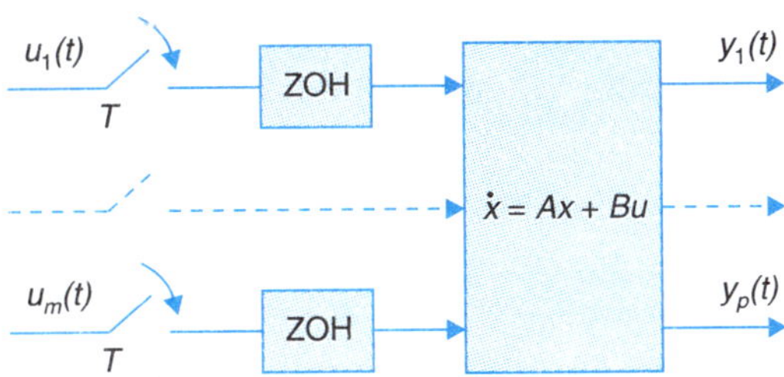

Fig. 12.25. Open-loop sampled-data system.

The solution of eqn. (12.108) in sampled and hold has already been given in eqns. (12.87) *a*, *b*, *c*, *d* which are reproduced below

$$\mathbf{x}(k+1) = \mathbf{A}_d\mathbf{x}(k) + \mathbf{B}_d\mathbf{u}(k) \qquad \text{...(12.109)}$$

$$\mathbf{A}_d = e^{\mathbf{A}T},\ \mathbf{B}_d = \mathbf{A}^{-1}(\mathbf{A}_d - \mathbf{I}) \qquad \text{...(12.110)}$$

Pole Placement by State Feedback (Discrete Case)

The notion of state variable feedback is as powerful in sampled-data systems as it is in continuous-time-systems.

We consider here the system shown in Fig. 12.26. From this figure, we have

$$\dot{\mathbf{x}} = \mathbf{Ax} + \mathbf{B}u \qquad \text{...(12.111}a)$$

$$u = -\mathbf{Kx} + r \qquad \text{...(12.111}b)$$

$$y = \mathbf{Cx} \qquad \text{...(12.111}c)$$

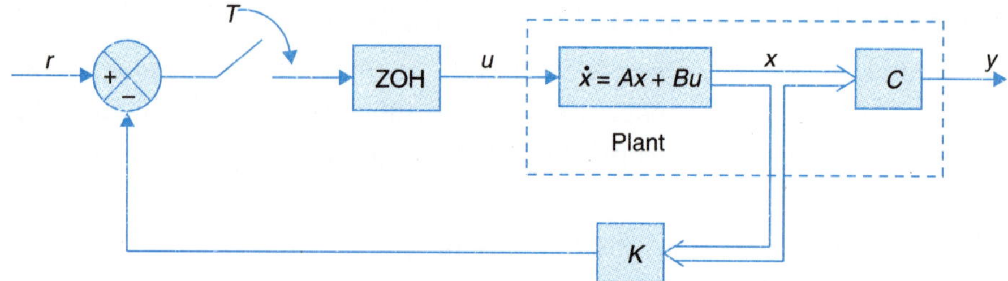

Fig. 12.26. A state variable feedback sampled-data system.

The discrete from the eqn. (12.111*a*) as given by (12.110*a*) is

$$\mathbf{x}(k+1) = \mathbf{A}_d\,\mathbf{x}(k) + \mathbf{B}_d\,u(k)$$

Now $$u(k) = -\mathbf{Kx}(k) + r(k)$$

Therefore $$\mathbf{x}(k+1) = (\mathbf{A}_d - \mathbf{B}_d\,\mathbf{K})\,\mathbf{x}(k) + \mathbf{B}_d\,r(k) \qquad \ldots(12.112)$$

We note that this result is similar to the equivalent result for continuous time systems as seen in (12.102). The method of pole placement discussed earlier in this Section is applicable to the system of Fig. 12.26.

12.9 OBSERVER SYSTEMS

In the problem of pole placement by state feedback we found that the control law is (eqn. (12.101*b*))

$$u = -\mathbf{Kx} + r \qquad \ldots(12.113)$$

As we have seen earlier, to implement this control law, we need to feedback all the state variables. In example (12.16) fortunately the state variables we selected were accessible for feedback. However, in many practical situations, all the state variables are not accessible for measurement and control purposes; only inputs and outputs can be used to drive a device whose output will approximate the state vector. This device is called a *state observer*. The output of observer can then be used to implement the feedback control law. Figure (12.27) shows the overall system structure including the observer. Intuitively, the observer should have the same state equations as the original system and design criterion should be to minimize the difference between the system output $y = \mathbf{Cx}$ and the output $\hat{y} = \mathbf{C}\hat{\mathbf{x}}$ as constructed by the observed vector $\hat{\mathbf{x}}$. Note that this is equivalent to minimization of $\mathbf{x} - \hat{\mathbf{x}}$. Since $\mathbf{x}$ is inaccessible we attempt to minimize $y - \hat{y}$. The difference $(y - \hat{y})$ is multiplied by $n \times 1$ vector $\mathbf{G}$ and fed into the input of the integrator of the observer (Fig. 12.28). The problem is to design $\mathbf{G}$ so that $(y - \hat{y})$ is minimized.

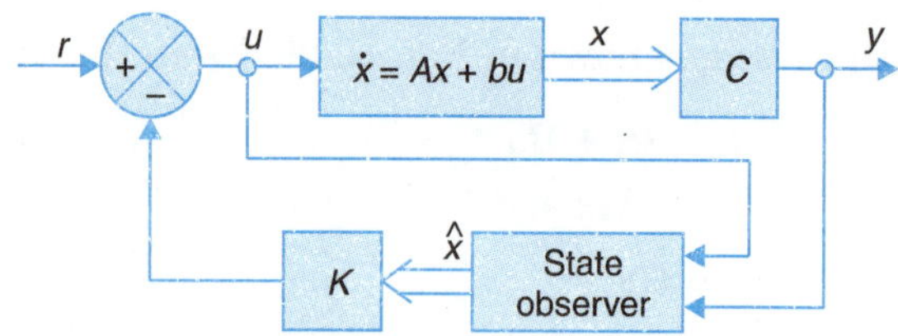

Fig. 12.27. A linear system with state observer.

$$\dot{\hat{\mathbf{x}}} = \mathbf{A}\hat{\mathbf{x}} + \mathbf{G}(\mathbf{C}\mathbf{x} - \mathbf{C}\hat{\mathbf{x}}) + \mathbf{B}u$$

$$= (\mathbf{A} - \mathbf{GC})\hat{\mathbf{x}} + \mathbf{GC}\mathbf{x} + \mathbf{B}u \qquad \ldots(12.114)$$

or

$$\dot{\mathbf{x}} - \dot{\hat{\mathbf{x}}} = (\mathbf{A} - \mathbf{GC})(\mathbf{x} - \hat{\mathbf{x}}) \qquad \ldots(12.115)$$

Define $\tilde{\mathbf{x}} = \mathbf{x} - \hat{\mathbf{x}}$

This gives

$$\dot{\tilde{\mathbf{x}}} = (\mathbf{A} - \mathbf{GC})\tilde{\mathbf{x}} \qquad \ldots(12.116)$$

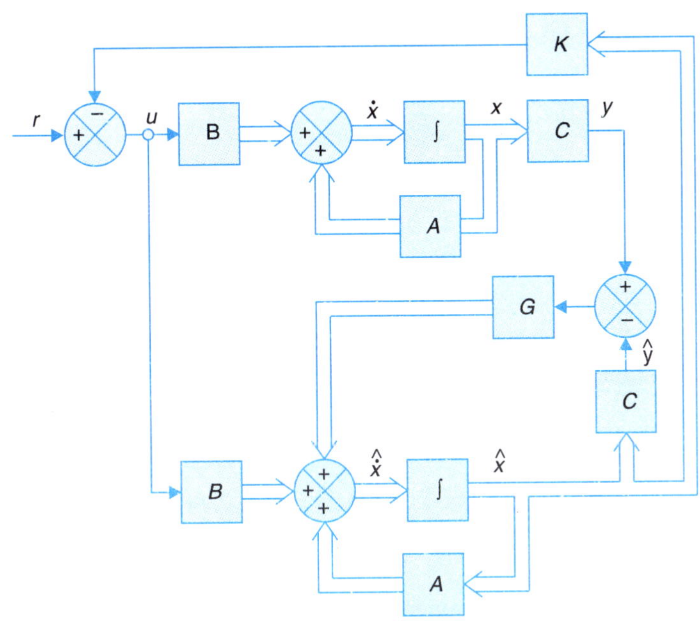

Fig. 12.28. A linear system with state observer.

The vector $\tilde{\mathbf{x}}$ can be made to decay fast if eigenvalues of the matrix $(\mathbf{A} - \mathbf{GC})$ are properly selected. In the following we show that if the pair $(\mathbf{A}, \mathbf{C})$ is observable, then the eigenvalues of the matrix $(\mathbf{A} - \mathbf{GC})$ can be chosen arbitrarily.

We will make use of duality property here. If the pair $(\mathbf{AC})$ is observable, then the pair $(\mathbf{A}^T\mathbf{C}^T)$ is controllable. Using the result proved earlier, we can say that eigenvalues of the matrix $(\mathbf{A}^T - \mathbf{C}^T\mathbf{G}^T)$ can be chosen arbitrarily (replace **A, B** and **K** by $\mathbf{A}^T$, $\mathbf{C}^T$ and $\mathbf{G}^T$ respectively). But the eigenvalues of $((\mathbf{A}^T - \mathbf{C}^T\mathbf{G}^T)$ are same as those of $(\mathbf{A} - \mathbf{GC})^T$ or $(\mathbf{A} - \mathbf{GC})$. This proves the statement.

A simple example given below illustrates the procedure for design of observers.

Example 12.17. Consider a linear system described by the equations

$$\dot{\mathbf{x}} = \begin{bmatrix} 1 & 2 & 0 \\ 3 & -1 & 1 \\ 0 & 2 & 0 \end{bmatrix} \mathbf{x} + \begin{bmatrix} 2 \\ 1 \\ 1 \end{bmatrix} u \qquad \ldots(12.117a)$$

$$= \mathbf{A}\mathbf{x} + \mathbf{B}u$$

$$y = [0 \quad 0 \quad 1]\mathbf{x} \qquad \ldots(12.117b)$$

$$= \mathbf{C}\mathbf{x}$$

It can easily be verified that the state model (12.117) is observable.

It is desired to design a state observer so that eigenvalues of the matrix (**A** – **GC**) are at – 4, – 3 ± *j*1.

Let $$\mathbf{G} = \begin{bmatrix} g_1 \\ g_2 \\ g_3 \end{bmatrix}$$

Then $$(\mathbf{A} - \mathbf{GC}) = \begin{bmatrix} 1 & 2 & -g_1 \\ 3 & -1 & 1-g_2 \\ 0 & 2 & -g_3 \end{bmatrix}$$

The characteristic equation is

$$|\lambda\mathbf{I} - (\mathbf{A} - \mathbf{GC})| = 0$$

or $$\begin{bmatrix} \lambda-1 & -2 & g_1 \\ -3 & \lambda+1 & g_2-1 \\ 0 & -2 & \lambda+g_3 \end{bmatrix} = 0$$

The gives

$$\lambda^3 + g_3\lambda^2 + (2g_2 - 9)\lambda + 2 + 6g_1 - 2g_2 - 7g_3 = 0 \qquad ...(12.118a)$$

The desired characteristic equation is

$$(\lambda + 3 + j1)(\lambda + 3 - j_1)(\lambda + 4) = 0$$

or $$\lambda^3 + 10\lambda^3 + 34\lambda + 40 = 0 \qquad ...(12.118b)$$

Equating the coefficients in eqns. (12.118*a*) and (12.118*b*) we get

$$g_1 = 25.2, g_2 = 21.5, g_3 = 10.$$

PROBLEMS

12.1 For the system shown in Fig. P-12.1, chose $v_1(t)$ and $v_2(t)$ as state variables and write down the state equations satisfied by them. Bring these equations in the vector-matrix form.

12.2 A schematic diagram representing a d.c. motor and load is given in Fig. P-12.2. The field is maintained constant during operation. Assume the motor is operating in the linear region. Determine the state equations in the vector-matrix form for the state variables

(*i*) $\mathbf{x}^T = [\theta \quad \dot{\theta} \quad i_a]$ (*ii*) $[\theta \quad \dot{\theta} \quad \ddot{\theta}]$

Moment of inertia of motor and load = j

Back emf constant = K_b

Coefficient of friction of motor and load = f

Motor torque constant = K_T

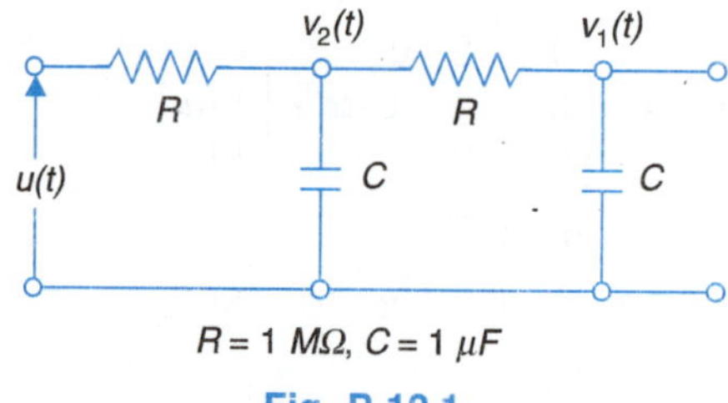

Fig. P-12.1

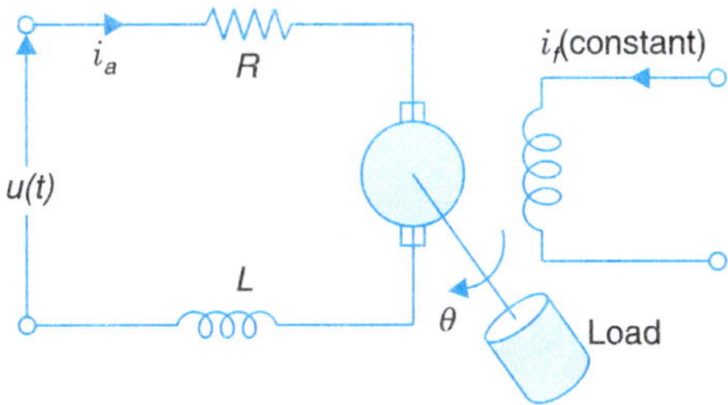

Fig. P-12.2

12.3 Consider the mechanical system depicted in Fig. P-12.3, consisting of two platforms coupled to each other and to ground via springs and dashpot dampers. Choosing suitable state variables, construct a state model of the system.

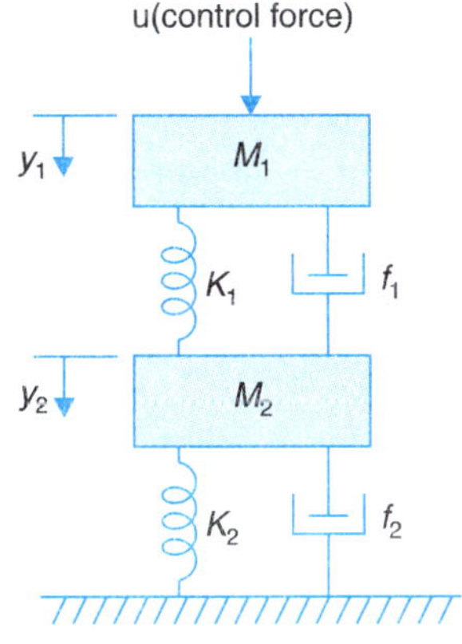

Fig. P-12.3

12.4 Construct the state model for a system characterized by the differential equation

$$\frac{d^3y}{dt^3} + 6\frac{d^2y}{dt^2} + 11\frac{dy}{dt} + 6y = u$$

Given the block diagram representation of the state model.

12.5 A feedback system has a closed-loop transfer function

$$\frac{C(s)}{U(s)} = \frac{10(s+4)}{s(s+1)(s+3)}$$

Construct three different state models for this system and give block diagram representation for each state model.

12.6 A feedback system is characterized by the closed-loop transfer function

$$T(s) = \frac{s^2 + 3s + 3}{s^3 + 2s^2 + 3s + 1}$$

Draw a suitable signal flow graph and therefrom construct a state model of the system.

12.7 Given

$$\mathbf{A}_1 = \begin{bmatrix} \sigma & 0 \\ 0 & \sigma \end{bmatrix}; \mathbf{A}_2 = \begin{bmatrix} 0 & \omega \\ -\omega & \sigma \end{bmatrix}; \mathbf{A} = \begin{bmatrix} \sigma & \omega \\ -\omega & \sigma \end{bmatrix}$$

Compute $e^{\mathbf{A}t}$.

[*Hint* : $e^{\mathbf{A}t} = e^{(\mathbf{A}_1 + \mathbf{A}_2)t} = e^{\mathbf{A}_1 t} . e^{\mathbf{A}_2 t}$; this holds in general if $\mathbf{A}_1\mathbf{A}_2 = \mathbf{A}_2\mathbf{A}_1$.]

12.8 For a system represented by the state equation

$$\dot{\mathbf{x}}(t) = \mathbf{A}\mathbf{x}(t)$$

the response of $\mathbf{x}(t)\begin{bmatrix} e^{-2t} \\ -2e^{-2t} \end{bmatrix}$ when $\mathbf{x}(0) = \begin{bmatrix} 1 \\ -2 \end{bmatrix}$

and $\mathbf{x}(t) = \begin{bmatrix} e^{-t} \\ -e^{-t} \end{bmatrix}$ when $\mathbf{x}(0) = \begin{bmatrix} 1 \\ -1 \end{bmatrix}$

Determine the system matrix **A** and the state transition matrix.

12.9 A linear time-invariant system is characterized by the homogeneous state equation

$$\begin{bmatrix} \dot{x}_1 \\ \dot{x}_2 \end{bmatrix} = \begin{bmatrix} 1 & 0 \\ 1 & 1 \end{bmatrix}\begin{bmatrix} x_1 \\ x_2 \end{bmatrix}$$

(*a*) Compute the solution of the homogeneous equation, assuming the initial state vector

$$\mathbf{x}_0 = \begin{bmatrix} 1 \\ 0 \end{bmatrix}$$

(Employ both the Laplace transform method and the canonical transformation method.)

(*b*) Consider now that the system has a forcing function and is represented by the following nonhomogeneous state equation

$$\begin{bmatrix} \dot{x}_1 \\ \dot{x}_2 \end{bmatrix} = \begin{bmatrix} 1 & 0 \\ 1 & 1 \end{bmatrix}\begin{bmatrix} x_1 \\ x_2 \end{bmatrix} + \begin{bmatrix} 1 \\ 0 \end{bmatrix} u$$

where u is a unit step function.

Compute the solution of this equation assuming initial conditions of part (*a*).

12.10 A linear time-invariant system is described by the following state model.

$$\begin{bmatrix} \dot{x}_1 \\ \dot{x}_2 \\ \dot{x}_3 \end{bmatrix} = \begin{bmatrix} 0 & 1 & 0 \\ 0 & 0 & 1 \\ -6 & -11 & -6 \end{bmatrix}\begin{bmatrix} x_1 \\ x_2 \\ x_3 \end{bmatrix} + \begin{bmatrix} 0 \\ 0 \\ 2 \end{bmatrix} u$$

Transform this state model into a canonical state model and therefrom obtain the explicit solutions for the state vector and output when the control force u is a unit step function and initial state vector is

$$\mathbf{x}_0^T = [0 \quad 0 \quad 2].$$

12.11 A feedback system is represented by a signal flow graph shown in Fig. P-12.11.

(*a*) Construct a state model of the system

(*b*) Diagonalize the coefficient matrix **A** of the state model of part (*a*).

(*c*) Determine the stability of the system.

(*d*) Determine the transfer function $C(s)/U(s)$ from the state model of part (*a*).

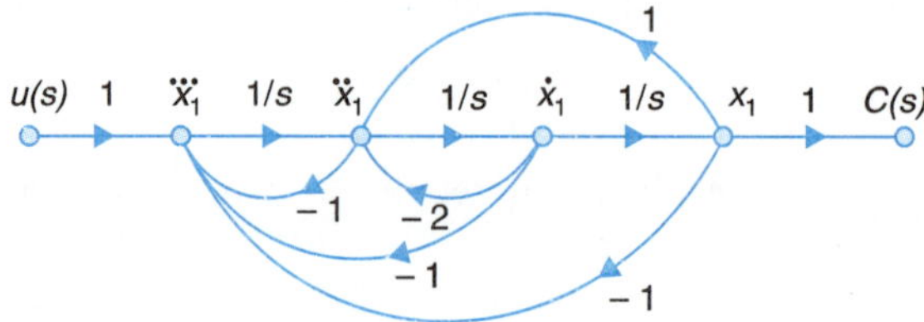

Fig. P-12.11

12.12 For the state equation

$$\dot{\mathbf{x}} = \mathbf{A}\mathbf{x}$$

$$\mathbf{A} = \begin{bmatrix} 0 & 1 & 0 \\ 3 & 0 & 2 \\ -12 & -7 & -6 \end{bmatrix}$$

Find the initial condition vector $\mathbf{x}(0)$ which will excite only the mode corresponding to the eigenvalue with the most negative real part.

[**Hint :** The problem is conveniently handled in the canonical form.]

12.13 Find the transformation matrices **P** which will convert the following equations to diagonal form. Given that the eigenvalues are λ_1, λ_2 and λ_3.

(*i*) $\begin{bmatrix} \dot{x}_1 \\ \dot{x}_2 \\ \dot{x}_3 \end{bmatrix} = \begin{bmatrix} 0 & 1 & 0 \\ 0 & 0 & 1 \\ -a_1 & -a_2 & -a_3 \end{bmatrix} \begin{bmatrix} x_1 \\ x_2 \\ x_3 \end{bmatrix}$ (*ii*) $\begin{bmatrix} \dot{x}_1 \\ \dot{x}_2 \\ \dot{x}_3 \end{bmatrix} = \begin{bmatrix} -a_1 & 1 & 0 \\ -a_2 & 0 & 1 \\ -a_3 & 0 & 0 \end{bmatrix} \begin{bmatrix} x_1 \\ x_2 \\ x_3 \end{bmatrix}$.

12.14 Prove that for

$$\mathbf{A} = \begin{bmatrix} -a_1 & 1 & 0 & \cdots & 0 & 0 \\ -a & 0 & 1 & \cdots & 0 & 0 \\ \vdots & \vdots & \vdots & & \vdots & \vdots \\ -a_{n-1} & 0 & 0 & \cdots & 0 & 1 \\ -a_n & 0 & 0 & \cdots & 0 & 0 \end{bmatrix}$$

with distinct eigenvalues $\lambda_1, \lambda_2, \ldots\ldots, \lambda_3$ the diagonalizing matrix is given by

$$\mathbf{M} = \begin{bmatrix} 1 & \cdots & 1 \\ \lambda_1 + a_1 & \cdots & \lambda_n + a_1 \\ \lambda_1^2 + a_1\lambda_1 + a_2 & \cdots & \lambda_n^2 + a_1\lambda_n + a_2 \\ \vdots & & \vdots \\ \lambda_1^{n-1} + a_1\lambda_1^{n-2} + \ldots\ldots + a_{n-2}\lambda_2 + a_{n-1} & \cdots & \lambda_n^{n-1} + a_1\lambda_n^{n-2} + \ldots\ldots + a_{n-2}\lambda_n + a_{n-1} \end{bmatrix}$$

12.15 Consider a state model

$$\dot{\mathbf{x}} = \mathbf{A}\mathbf{x} + \mathbf{B}u$$

where $$\mathbf{A} = \begin{bmatrix} 0 & 1 & 0 \\ 0 & 0 & 1 \\ -40 & -34 & -10 \end{bmatrix};\ \mathbf{B} = \begin{bmatrix} 0 \\ 0 \\ 1 \end{bmatrix}$$

(*i*) Show that the eigenvalues of **A** are $-3 \pm j1, -4$.

(*ii*) Suggest a suitable transformation matrix **M** so that

$$\mathbf{M}^{-1}\mathbf{A}\mathbf{M} = \Lambda = \begin{bmatrix} -3+j1 & 0 & 0 \\ 0 & -3-j1 & 0 \\ 0 & 0 & -4 \end{bmatrix}$$

(*iii*) Suggest a suitable transformation matrix **Q** so that

$$\mathbf{Q}^{-1}\Lambda\mathbf{Q} = \begin{bmatrix} -3 & 1 & 0 \\ -1 & -3 & 0 \\ 0 & 0 & -4 \end{bmatrix}$$

[**Hint:** $$\mathbf{Q} = \begin{bmatrix} 1/2 & -j/2 & 0 \\ 1/2 & j/2 & 0 \\ 0 & 0 & 1 \end{bmatrix}$$

12.16 Write the state equations of the system shown in Fig. P-12.16, in which x_1, x_2 and x_3 constitute the state vector. Determine whether the system is completely controllable and observable.

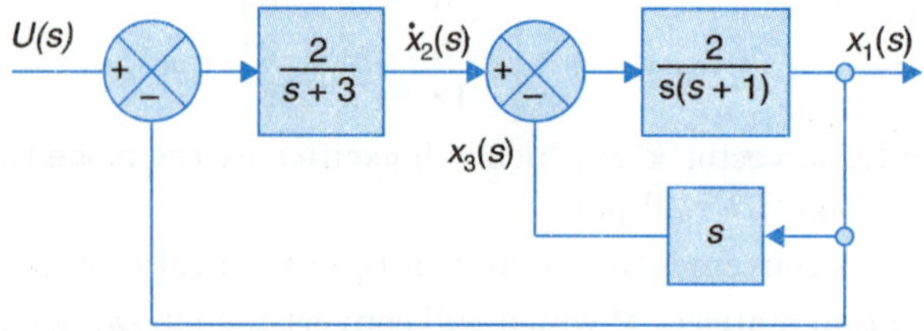

Fig. P-12.16

12.17 Block diagram representation of a linear time-invariant system is given in Fig. P 12.17. Check whether the system is completely observable.

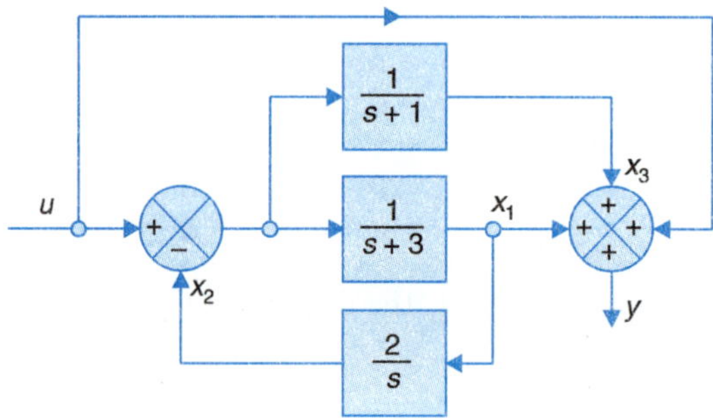

Fig. P-12.17

12.18. Consider the system with state variable feedback shown in Fig. P-12.18. Determine the values of K and $\mathbf{K}^T$ so that the system meets the following specifications:

Peak overshoot $\leq 15\%$

Setting time ≤ 4 sec

Velocity error constant ≥ 1.5

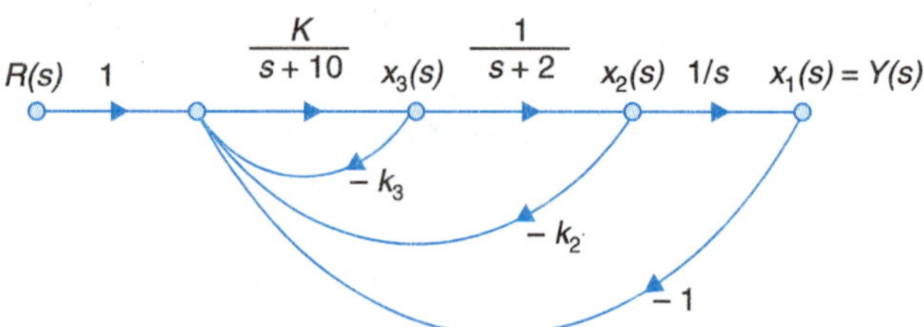

Fig. P-12.18

12.19. Find the forward path gain K_A and the $\mathbf{K}^T$ for the system shown in Fig. P-12.19 such that the closed-loop transfer function is

$$\frac{24}{[(s+1)^2 + (\sqrt{3})^2](s+6)}$$

Draw the root locus plot and comment upon the stability of the compensated system.

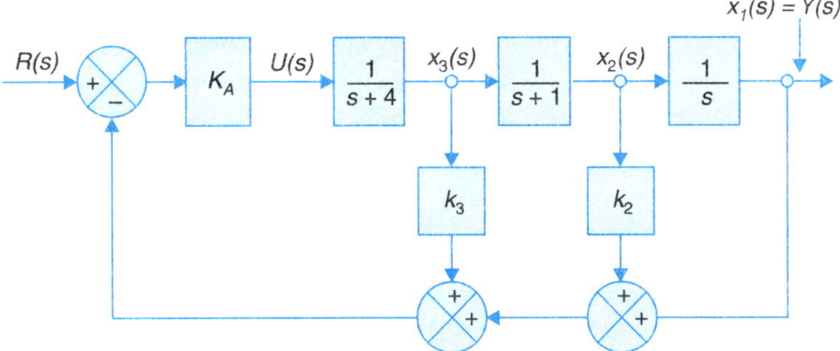

Fig. P-12.19

12.20. Consider the position control system of Fig. P-12.20 employing a d.c. motor in armature control mode with state variable defined on the diagram. Incomplete state feedback ($x_3 = i_a$ is not fedback) is used to achieve the location of the dominant closed loop poles at $-1 + j\sqrt{3}$. Obtain the values of the gains K_A and K_2. Find the location of the third closed loop pole and check if the dominance condition is met. It may be noted that because of incomplete state feedback, the control over one of the closed loop poles is lost.

Given :

Motor field resistance,	$R_a = 100\ \Omega$
Motor field inductance,	$L_a = 10^{-4}$ H
Motor torque constant,	$K_T = 1$ N-m/amp
Moment of inertia referred to motor shaft,	$J = 1$ kg-m^2
Coefficient of friction referred to motor shaft,	$f = 1$ N-m/rad/s

Amplifier gain, K_A V/V

Gain of the encoder + counter + D/A converter = K_2 V/rad/s.

[*Hint* : Find the transfer function of the given system and compare the co-efficients of denominator polynomial of the transfer function with the required characteristic equation.]

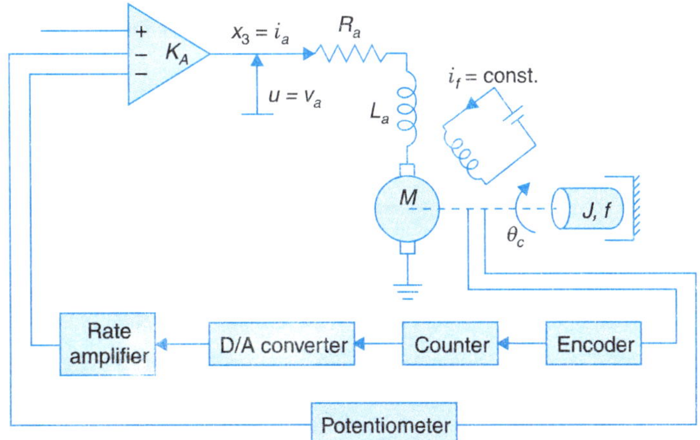

Fig. P-12.20

12.21 Consider a linear system

$$\dot{\mathbf{x}} = \begin{bmatrix} 0 & 1 \\ 0 & -5 \end{bmatrix}\mathbf{x} + \begin{bmatrix} 0 \\ 100 \end{bmatrix}u$$

$$y = [1 \quad 0]\mathbf{x}$$

The feedback controller for the system is given by

$$u = [-k_1 \quad -k_2]\begin{bmatrix} x_1 \\ x_2 \end{bmatrix} + r$$

Assume that the states x_1 and x_2 are not accessible for feedback. An observer system is to be designed to reconstruct **x**. The structure of the overall system is as shown in Fig. 12.28. Design the feedback matrix **G** so that $\mathbf{x} - \hat{\mathbf{x}}$ will delay as fast as e^{-10t}.

12.22 A discrete-time system has the transfer function

$$T(z) = \frac{4z^3 - 12z^2 + 13z - 7}{(z-1)^2(z-2)}$$

Determine the state model of the system in

(*i*) phase variable form (*ii*) Jordan canonical form.

12.23 A discrete-time system is described by the state equation

$y(k + 2) + 5y(k + 1) + 6y(k) = u(k)$

$y(0) = y(1) = 0$; $T = 1$ sec

(*i*) Determine a state model in canonical form.

(*ii*) Find the state transition matrix.

(*iii*) For input $u(k) = 1$ for $k \geq 0$, find the output $y(k)$.

12.24 A continuous-time plant is described by the state equation

$$\dot{\mathbf{x}} = \begin{bmatrix} 0 & 1 \\ -2 & -2 \end{bmatrix}\mathbf{x} + \begin{bmatrix} 0 \\ 1 \end{bmatrix}u \; ; \quad \mathbf{x}(0) + \begin{bmatrix} 1 \\ 0 \end{bmatrix}$$

The state equation is to be solved for $\mathbf{x}(t)$ using digital computer. Obtain suitable recursive relations. Take sampling interval $T = 1$ sec.

12.25 Show that if there is pole-zero cancellation in the input-output transfer function of a linear time-invariant discrete-time system, the system will be either uncontrollable or unobservable.

12.26 The linear continuous-time plant of a sampled-data system is described by the state equation

$$\dot{\mathbf{x}} = \begin{bmatrix} 0 & 1 \\ -4 & 0 \end{bmatrix}\mathbf{x} + \begin{bmatrix} 0 \\ 2 \end{bmatrix}u$$

Determine the value of sampling period T which make the system uncontrollable.

13

LIAPUNOV'S STABILITY ANALYSIS

13

LIAPUNOV'S STABILITY ANALYSIS

13.1 INTRODUCTION

Consider a general time-invariant system described by the state equation

$$\dot{\mathbf{x}} = \mathbf{f}(\mathbf{x}, \mathbf{u}) \qquad \text{...(13.1)}$$

If the input vector **u** is constant, it is possible to write the above equation in the form

$$\dot{\mathbf{x}} = \mathbf{F}(\mathbf{x}) \qquad \text{...(13.2)}$$

A system represented by an equation of this form is called an *autonomous system.* For such a system, consider the points in the phase-space at which the derivatives of all the state variables are zero. Such points are called *singular points*. These are in fact *equilibrium points* already defined in Chapter 12. If the system is placed at such a point, it will continue to lie there if left undisturbed (the derivatives of all the phase variables being zero, the system state remains unchanged).

For studying the system dynamic response at an equilibrium (singular) point to small perturbation, the system is linearized (using linearization techniques presented in Chapter 12) at that point. The linearized model of system of eqn. (13.2) may be written as

$$\dot{\mathbf{x}} = \mathbf{A}\mathbf{x} \qquad \text{...(13.3)}$$

For this linear autonomous system, the equilibrium states are given by those states $\mathbf{x}_e$ satisfying $\mathbf{A}\mathbf{x}_e = \mathbf{0}$. From this we see that $\mathbf{x}_e = \mathbf{0}$ will be the only solution of the above equation provided that the determinant of **A** is nonzero. Equivalently, if all the eigenvalues of the system are different from 0, then the origin is the only singular point.

In general, if $\mathbf{x}_e$ is a singular point, it is convenient to shift the origin of coordinates to $\mathbf{x}_e$. To achieve this, we define new phase variables as

$$\tilde{\mathbf{x}} = \mathbf{x} - \mathbf{x}_e \qquad \text{...(13.4)}$$

The system of eqn. (13.2) in terms of new phase variables is represented as

$$\dot{\tilde{\mathbf{x}}} = \mathbf{F}(\tilde{\mathbf{x}}) \quad \text{...(13.5)}$$

with the equilibrium point at $\tilde{\mathbf{x}} = \mathbf{0}$.

The problem of stability for linear-time-invariant systems was studied in Chapter 6. There we observed that stability may be defined as per the following two notions:

(*i*) For free system : A system is stable with zero input and arbitrary initial conditions if the resulting trajectory tends towards the equilibrium state.

(*ii*) For forced system : A system is stable if with bounded input, the system output is bounded.

These two notions were found to be essentially equivalent for linear time-invariant systems. In nonlinear systems, unfortunately, there is no definite correspondence between the two notions. Many important results have been obtained for free systems. (Our discussion will be limited to this class of nonlinear systems). Even for this class of systems, the concept of stability is non clear-cut. The linear autonomous systems (with nonzero eigenvalues) have only one equilibrium state and their behaviour about the equilibrium state completely determines the qualitative behaviour in the entire state-plane. In nonlinear systems, on the other hand, system behaviour for small deviations about the equilibrium point may be different from that for large deviations. Therefore, *local stability* does not imply stability in the overall state-plane and the two concepts should be considered separately. Secondly, in a nonlinear system with multiple equilibrium states, the system trajectories may move away from one equilibrium point and tend to other as time progresses. Thus it appears that in case of nonlinear systems, it is simpler to speak of system stability relative to the equilibrium state rather than using a general term 'stability of a system'.

There are many types of stability definitions in the literature. We shall concentrate on three of these : stability, asymptotic stability and asymptotic stability in-the-large.

Consider an autonomous system described by the state equation (13.2). Assume that the system has only one equilibrium point. This is the case with all properly designed systems. Futhermore, without loss of generality, let the origin of state space be taken as the equilibrium point.

The system of eqn. (13.2) is *stable* at the origin if, for every initial state $x(t_0)$ which is sufficiently close to origin, $\mathbf{x}(t)$ remains near the origin for all t. It is *asymptotically stable* if $\mathbf{x}(t)$ in fact approaches the origin as $t \longrightarrow \infty$. It is *asymptotically stable in-the-large* if it is asymptotically stable for every initial state regardless of how near or far it is from the origin.

Following are the mathematically precise definitions of the different types of stability.

The system of eqn. (13.2) is *stable* at the origin if, for every real number $\varepsilon > 0$, there exists a real number $\delta(\epsilon) > 0$ such that $\| \mathbf{x}(t_0) \| \leq \delta$ results in $\| \mathbf{x}(t) \| \leq \varepsilon$ for all $t \geq t_0$.

This definition (from Russian mathematician A.M. Liapunov) of stability uses the concept of vector norm. The Euclidean norm for a vector $\mathbf{x}$ with n components $x_1, x_2, ..., x_n$ is

$$\| \mathbf{x} \| = (x_1{}^2 + x_2{}^2 + ... + x_n)^{1/2}$$

$\| \mathbf{x} \| \leq R$ defines a hyper-spherical region $S(R)$ of radius R surrounding the equilibrium point $\mathbf{x} = \mathbf{0}$. In terms of the Euclidean norm, the above definition of stability implies that for any $S(\varepsilon)$

that we may designate, the designer must produce $S(\delta)$ so that system state initially in $S(\delta)$ will never leave $S(\varepsilon)$. This is illustrated in Fig. 13.1 (*a*).

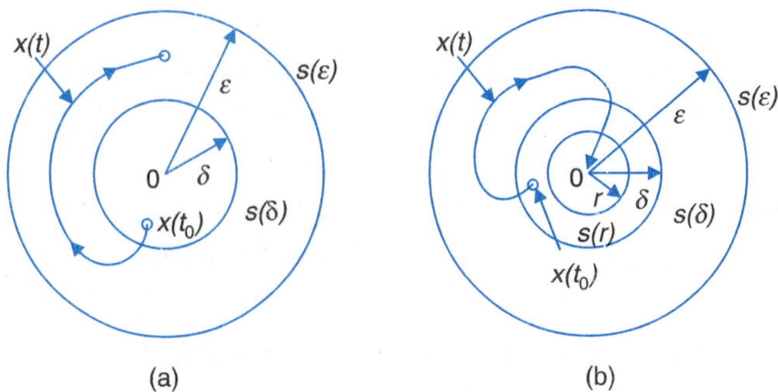

Fig. 13.1. Stability definitions.

Consider for example, a second-order linear autonomous system whose phase-trajectories for various initial conditions are shown in Fig. 13.1 (*b*). For a specified value of ε, we can find a closed phase-trajectory whose maximum distance from the origin is ε. We then select a value of δ which is less than the minimum distance from that curve to the origin. The $\delta(\varepsilon)$, so chosen will satisfy the conditions that guarantee stability.

A system is said to be *locally stable* (or *stable in-the-small*) if the region $S(\varepsilon)$ is small.

Let us now turn to the other two definitions of stability.

The system of eqn. (13.2) is *asymptotically stable* at the origin if

(*a*) it is stable, and

(*b*) there exists a real number $r > 0$ such that $\| \mathbf{x}(t_0) \| \leq r$ results in $\mathbf{x}(t) \to \mathbf{0}$ as $t \to \infty$.

Property (*b*) implies that every motion starting in $S(r)$ converges to origin as $t \to \infty$. This is illustrated in Fig. 13.1 (*b*).

The system of eqn. (13.2) is *asymptotically stable in-the-large* (*globally asymptotically stable*) at the origin if

(*a*) it is stable, and

(*b*) every initial state $\mathbf{x}(t_0)$ results in $\mathbf{x}(t) \to \mathbf{0}$ as $t \to \infty$.

Thus asymptotic stability in-the-large guarantees that every motion will approach the origin.

Limit Cycles

In the definitions of stability of nonlinear systems given above, the system stability has been defined in terms of disturbed steady-state coming back to its equilibrium position or at least staying within tolerable limits from it. This does include the possibility that a disturbed nonlinear system even while staying within tolerance limits, may exhibit a special behaviour of following a closed trajectory or *limit cycle*. The limit cycles describe the oscillations of nonlinear systems. The existence of a limit cycle corresponds to an oscillation of fixed amplitude and period.

As an example, let us consider the well known *Vander Pol's differential equation*

$$\frac{d^2x}{dt^2} - \mu(1-x^2)\frac{dx}{dt} + x = 0 \qquad ...(13.6\ (a))$$

which describes physical situations in many nonlinear systems. By comparing this with the following linear differential equation

$$\frac{d^2x}{dt^2} + 2\zeta\frac{dx}{dt} + x = 0$$

we observe that Vander Pol's equation has damping factor $-(\mu/2)(1-x^2)$ which depends upon x. If it is assumed that initially $|x| >> 1$, then the damping factor has large positive value. The system therefore behaves like an overdamped system with consequent decrease of the amplitude of x. In this process, the damping factor also decreases and so the system state finally enters a limit cycle as shown by the outer trajectory of Fig. 13.2 (a).

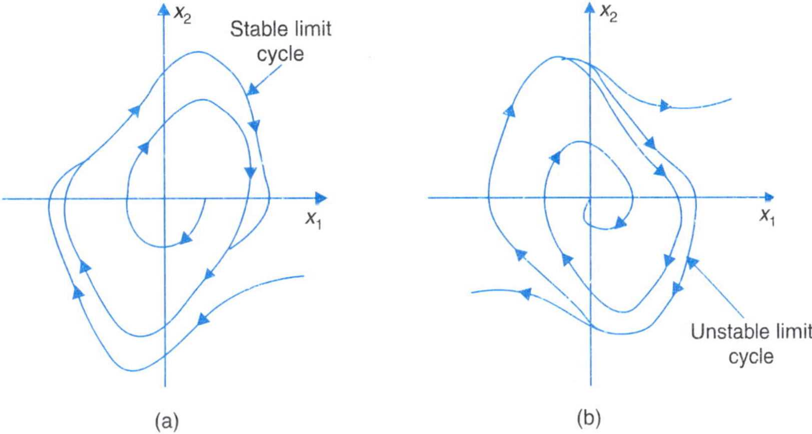

Fig. 13.2. Limit cycle behaviour of nonlinear systems.

On the other hand, if initially $|x| << 1$, the damping is negative hence the amplitude of x increases till the system state again enters the limit cycle as shown by the inner trajectory of Fig. 13.2 (a). The limit cycle shown in Fig. 13.2 (a) is a stable one, since the parts in its neighbourhood converge toward the limit cycle.

If the paths in the neighbourhood of a limit cycle (closed trajectory) diverge away from it, it indicates that the limit cycle is unstable. Consider, for Example, the Vander Pol's equation with the sign of its damping term reversed, *i.e.*,

$$\frac{d^2x}{dt^2} + \mu(1-x^2)\frac{dx}{dt} + x = 0 \qquad ...(13.6\ (b))$$

The phase-trajectories for this equation are shown in Fig. 13.2 (b), from which it is observed that an unstable limit cycle occurs.

It is important to note that a limit cycle, in general is an undersirable characteristic of a control system. It may be tolerated only if its amplitude is within specified limits.

It may be noted that in linear autonomous systems, when oscillations occur, the resulting trajectories will be closed curves. The amplitude of the oscillations is not fixed. It changes with the size of the initial conditions. Slight changes in system parameters (shifting the eigenvalues from the imaginary axis of the complex plane) will destroy the oscillations.

In nonlinear systems, on the other hand, there can be oscillations that are independent of the size of initial conditions (see Fig. 13.2) and these oscillations (limit cycles) are usually much less sensitive to system parameter variations. Limit cycles of fixed amplitude and period can be sustained over a finite range of system parameters.

13.2 LIAPUNOV'S STABILITY CRITERION

The general state equation for a nonlinear system can be expressed as

$$\dot{\mathbf{x}} = \mathbf{f}(\mathbf{x}(t), \mathbf{u}(t), t) \qquad \text{...(13.7)}$$

The analytical solution of this equation is rarely possible. If a numerical solution is attempted, the question of stability behaviour can not be fully answered as solutions to an infinite set of initial conditions are needed. Using engineering judgement and prior knowledge of the system, the analyst may restrict solutions to a finite set of initial conditions. A variety of methods have therefore been devised which yield information about the stability and domain of stability of eqn. (13.7) without resorting to its complete solution.

In this text we shall discuss Liapunov's direct method of stability analysis which finds application in optimal control and in particular in adaptive control (Chap 5).

Basic Stability Theorems

The direct method of Liapunov is based on the concept of energy and the relation of stored energy with system stability. Consider an autonomous physical system described as

$$\dot{\mathbf{x}}(t) = \mathbf{f}(\mathbf{x}(t))$$

and let $\mathbf{x}(\mathbf{x}(t_0), t)$ be a solution. Further, let $V(\mathbf{x})$ be the total energy associated with the system. If the derivative $dV(\mathbf{x})/dt$ is negative for all $\mathbf{x}(\mathbf{x}(t_0), t)$ except the equilibrium point, then it follows that energy of the system decreases as t increases and finally the system will reach the equilibrium point. This holds because energy is non-negative function of system state which reaches a minimum only if the system motion stops. These ideas are well illustrated by the following example.

The spring-mass-damper system shown in Fig. 13.3 is governed by the equation

$$\ddot{x}_1 = f\dot{x}_1 + Kx_1 = 0$$

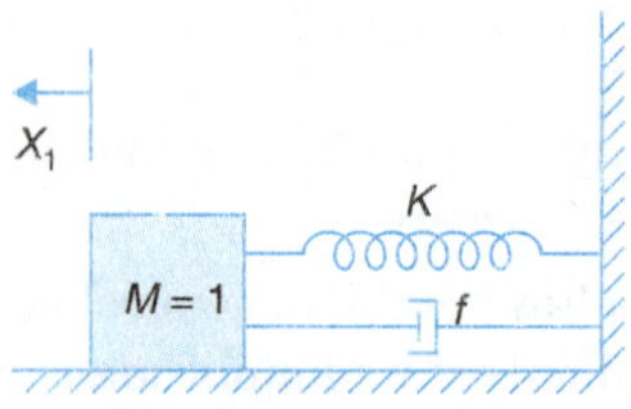

Fig. 13.3. A spring-mass-damper system.

A state model of the system is

$$\dot{x}_1 = x_2 \qquad \text{...(13.8)}$$
$$\dot{x}_2 = -Kx_1 - fx_2$$

At any instant, the total energy V in the system consists of the kinetic energy of the moving mass and the potential energy stored in the spring

$$V(x_1, x_2) = \frac{1}{2}x_2^2 + \frac{1}{2}Kx_1^2$$

Thus $\quad V(\mathbf{x}) > \mathbf{0}$ when $x \neq \mathbf{0}$

$$V(\mathbf{0}) = 0$$

This means that total energy is positive unless the system is at rest at the equilibrium state $\mathbf{x}_e = \mathbf{0}$, where the energy is zero.

The rate of change of energy is given by

$$\frac{d}{dt}V(x_1, x_2) = \frac{\partial V}{\partial x_1}\frac{dx_1}{dt} + \frac{\partial V}{dx_2}\frac{dx_2}{dt}$$
$$= -x_2^2 f$$

Thus dV/dt is negative at all points except where x_2 is zero at which dV/dt is zero. Therefore under positive damping, the system energy can not increase. From eqns. (13.8), we observe that $\dot{x}_2 = -Kx_1$ at the points where $x_2 = 0$; thus the system can not remain in nonequilibrium state for which $x_2 = 0$. Therefore the energy can not remain constant except at the equilibrium point where it will be zero.

A visual analogy may be obtained by considering the surface

$$V = \frac{1}{2}x_2^2 + \frac{1}{2}Kx_1^2$$

This is a cup-shaped surface as shown in Fig. (13.4 (*a*)). The constnat-V loci are ellipses on the surface of the cup. Let (x_{10}, x_{20}) be the initial condition. If one plots trajectory on the surface shown, the representative point $\mathbf{x}(t)$ crosses the constant-V curves and moves towards the lowest point of the cup which is the equilibrium point. Figure 13.4 (*b*) shows the projection of a typical trajectory on the x_1-x_2 plane.

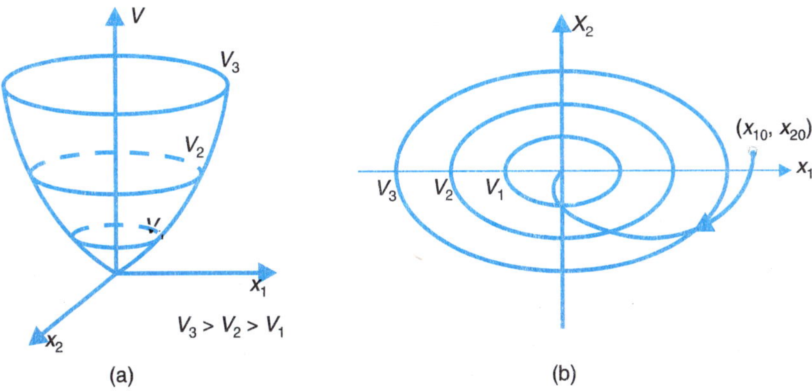

Fig. 13.4. Constant-V curves.

In the example given above, it was easy to associate the energy function V with the given system. However, in general, there is no obvious way of associating an energy function with a given set of equations describing a system. The idea that other non-negative scalar functions of system state can also answer the question of stability was introduced and formalized by the Russian mathematician A.M. Liapunov ; the scalar function is now known as *the Liapunov function* and the method of investigating stability using Liapunov's function as the *Liapunov's direct method.*

The Liapunov's method in its simplest form is given by the following theorem.

Theorem 1: Consider the system

$$\dot{\mathbf{x}} = \mathbf{f}(\mathbf{x}) \; ; \; \mathbf{f}(\mathbf{0}) = \mathbf{0}$$

Suppose there exists a scalar function $V(\mathbf{x})$ which for some real number $\epsilon > 0$, satisfies the following properties for all $\mathbf{x}$ in the region $\| \mathbf{x} \| \le \varepsilon$.

$$\left.\begin{array}{l} (a)\ V(\mathbf{x}) > 0 \,;\, \mathbf{x} \neq \mathbf{0} \\ (b)\ V(\mathbf{0}) = 0 \end{array}\right\} \text{ i.e. } V(\mathbf{x}) \text{ is positive definite scalar function (see Appendix II)}$$

(*c*) $V(\mathbf{x})$ has continuous partial derivatives with respect to all components of $\mathbf{x}$.

(*d*) $\dfrac{dV}{dt} \le 0$ (*i.e., dV/dt* is negative semidefinite scalar function).

Then the system is stable at the origin.

Theorem 2: If the property (*d*) of Theorem 1 is replaced with (d) $dV/dt < 0$, $x \neq 0$ (*i.e., dV/dt* is negative definite scalar function), then the system is asymptotically stable.

It is intuitively obvious since a continuous V function, $V > 0$ except at $\mathbf{x} = 0$, satisfies the condition $dV/dt < 0$, we expect that $\mathbf{x}$ will eventually approach the origin. We shall avoid the rigorous proof of this theorem.

Theorem 3: If all the conditions of Theorem 2 hold and in addition.

$$V(\mathbf{x}) \longrightarrow \infty \text{ as } \| \mathbf{x} \| \longrightarrow \infty$$

then the system is asymptotically stable in-the-large at the origin

Note that the additional requirement for this case is that $V(\mathbf{x})$ must approach ∞ as the distance from point $\mathbf{x}$ to the origin approaches ∞ irrespective of the direction. Essentially, this requirement is to assure that points of constant values of $V(\mathbf{x})$ form closed surfaces in state space. If $V(\mathbf{x})$ does not form closed surfaces, then it is possible for system trajectories to go towards infinity.

Liapunov Functions

The determination of stability via Liapunov's direct method centres around the choice of a positive definite function $V(\mathbf{x})$ called the Liapunov function. unfortunately, there is no universal method for selecting Liapunov function which is unique for a special problem. Some Liapunov functions may provide better answer than others. Several techniques have been devised for systematic construction of Liapunov functions (two such techniques are presented in this chapter) each is applicable to a particular class of systems.

In addition, if a Liapunov function of the required type can not be found, it in no way implies that the system is unstable. (Stability theorems presented in this section merely provide *sufficient conditions* for stability.) It only means that our attempt in trying to estabilish the stability of the system has failed.

For a given function $V(\mathbf{x})$, there is no general method which will easily allow us to ascertain whether it is positive definite. However, if $V(\mathbf{x})$ is in the quadratic form in x_i's, we can use Sylvester's theorem (see Appendix II) to ascertain definiteness of the function.

Inspite of all these limitations, Liapunov's direct method is a most powerful technique available today for stability analysis of nonlinear systems.

Instability

It may be noted that instability in a nonlinear system can be established by direct recourse to the instability theorem of the direct method. The basic instability theorem is presented below:

Theorem 4: Consider the system

$$\dot{\mathbf{x}} = \mathbf{f}(\mathbf{x})\ ;\ \mathbf{f}(\mathbf{0}) = \mathbf{0}$$

Suppose there exists a scalar function $W(\mathbf{x})$ which, for some real number $\in > 0$, satisfies the following properties for all $\mathbf{x}$ in the region $\|\mathbf{x}\| \leq \in$;

(*a*) $W(\mathbf{x}) > 0\ ;\ \mathbf{x} \neq \mathbf{0}$

(*b*) $W(\mathbf{0}) - 0$

(*c*) $W(\mathbf{x})$ has continuous partial derivatives with respect to all components of $\mathbf{x}$.

(*d*) $\dfrac{dW}{dt} \geq 0$

Then the system is unstable at the origin.

Note that it requires as much ingenuity of devise a suitable W function as to devise a Liapunov function V. In stability analysis of nonlinear systems, it is valuable to establish conditions for which the system is unstable. Then the regions of asymptotic stability need not be sought for such conditions and the analyst may not put in fruitless effort.

Example 13.1: Consider a nonlinear system governed by the equations

$$\dot{x}_1 = -x_1 + 2x_1^2 x_2 \qquad ...(i)$$

$$\dot{x}_2 = -x_2$$

We might chose V as

$$V = x_1^2 + x_2^2 \qquad ...(ii)$$

From eqns. (*i*) and (*ii*)

$$\frac{dV}{dt} = \frac{\partial V}{\partial x_1}\frac{dx_1}{dt} + \frac{\partial V}{\partial x_2}\frac{dx_2}{dt}$$

$$= -2x_1^2(1 - 2x_1x_2) - 2x_2^2$$

Thus dV/dt is negative definite if

$$1 - 2x_1x_2 > 0 \qquad ...(iii)$$

Investigation of the inequality (*iii*) reveals that there are curves dividing the stable region from the region where $\dot{V} > 0$. The dividing lines lie in the first and third quadrants and are rectangular hyperbolas as shown in Fig. 13.5. In the second and forth quadrants the inequality is satisfied for all values of x_1 and x_2. Figure 13.5 shows the regions of stability and possible instability. It is easily observed that the orgin of the system is asymptotically stable.

Example 13.2: Consider a nonlinear system described by the equations

$$\dot{x}_1 = x_2$$
$$\dot{x}_2 = -x_2 - x_1^3$$

Clearly origin is the equilibrium point.

A choice of possible Liapunov functions is

$$V = x_1^4 + x_1^2 + 2x_1x_2 + 2x_2^2$$
$$= x_1^4 + (x_1 + x_2)^2$$

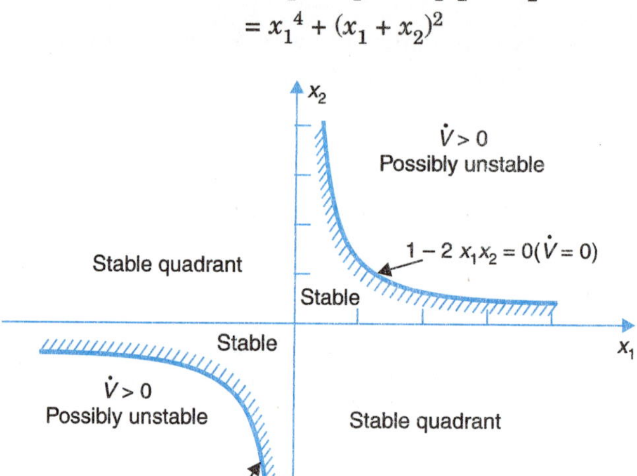

Fig. 13.5. Stability regions for a nonlinear system.

The derivative of the Liapunov function

$$\frac{dV}{dt} = \frac{\partial V}{\partial x_1}\dot{x}_1 + \frac{\partial V}{\partial x_2}\dot{x}_2$$
$$= -2x_1^4 - 2x_2^2$$

is clearly negative definite.

Further $\quad V(\mathbf{x}) \to \infty$ as $\|x\| \to \infty$.

Therefore the origin of the system is asymptotically stable in-the-large.

13.3 THE DIRECT METHOD OF LIAPUNOV AND THE LINEAR SYSTEM

In case of linear systems, the direct method of Liapunov provides a simple approach to stability analysis. It must be emphasized here that compared to the results presented in Chapter 6, no new results are obtained by the use of the direct method for the stability analysis of linear systems. However, the study of linear systems using the direct method is quite useful because it extends our thinking to nonlinear systems.

Consider a linear autonomous system described by the state equation

$$\dot{\mathbf{x}} = \mathbf{A}\mathbf{x} \qquad \text{...(13.9)}$$

The linear system is asymptotically stable in-the-large at the origin if and only if given any symmetric, positive definite matrix $\mathbf{Q}$ (see Appendix II), there exists a symmetric positive definite matrix $\mathbf{P}$ which is the unique solution

$$\mathbf{A}^T\mathbf{P} + \mathbf{P}\mathbf{A} = -\mathbf{Q} \qquad \text{...(13.10)}$$

Proof

To prove the sufficiency of the reult of above theorem, let us assume that a symmetric positive definite matrix $\mathbf{P}$ exists which is the unique solution of eqn. (13.11). Consider the scalar function (Appendix II),

$$V(\mathbf{x}) = \mathbf{x}^T\mathbf{P}\mathbf{x}$$

Note that $V(\mathbf{x}) > \mathbf{0}$ for $\mathbf{x} \neq \mathbf{0}$

and $V(\mathbf{0}) = \mathbf{0}$

The time derivative of $V(\mathbf{x})$ is

$$\dot{V}(\mathbf{x}) = \dot{\mathbf{x}}^T\mathbf{P}\mathbf{x} + \mathbf{x}^T\mathbf{P}\dot{\mathbf{x}}$$

Using eqns. (13.9) and (13.10) we get

$$\begin{aligned}\dot{V}(\mathbf{x}) &= \mathbf{x}^T\mathbf{A}^T\mathbf{P}\mathbf{x} + \mathbf{x}^T\mathbf{P}\mathbf{A}\mathbf{x} \\ &= \mathbf{x}^T(\mathbf{A}^T\mathbf{P} + \mathbf{P}\mathbf{A})\mathbf{x} \\ &= -\mathbf{x}^T\mathbf{Q}\mathbf{x}\end{aligned}$$

Since $\mathbf{Q}$ is positive definite, $\dot{V}(\mathbf{x})$ is negative definite. Norm of $\mathbf{x}$ may be defined as (Appendix II)

$$\|\mathbf{x}\| = (\mathbf{x}^T\mathbf{P}\mathbf{x})^{1/2}$$

Then $V(\mathbf{x}) = \|\mathbf{x}\|^2$

$$V(\mathbf{x}) \to \infty \text{ as } \|\mathbf{x}\| \to \infty$$

The system is therefore asymptotically stable in-the large at the origin (refer Theorem 3).

In order to show that the result is also necessary, suppose that the system is asymptotically stable and $\mathbf{P}$ is negative definite, consider the scalar function

$$V(\mathbf{x}) = \mathbf{x}^T\mathbf{P}\mathbf{x} \qquad \text{...(13.11)}$$

Therefore

$$\begin{aligned}\dot{V}(x) &= -[\dot{\mathbf{x}}^T\mathbf{P}\mathbf{x} + \mathbf{x}^T\mathbf{P}\dot{\mathbf{x}}] \\ &= \mathbf{x}^T\mathbf{Q}\mathbf{x} \\ &> 0\end{aligned}$$

There is contradiction since $V(\mathbf{x})$ given by eqn. (13.11) satisfies instability theorem (refer Theorem 4)

Thus the conditions for the positive definiteness of $\mathbf{P}$ are necessary and sufficient for asymptotic stability of the system of eqn. (13.9).

Hurwitz Criterion and Liapunov's Direct Method

We can prove that the Liapunov's direct method, as applied to linear time-invariant systems, is the same as the Hurwitz stability criterion presented in Chapter 6. Proof will not be advanced here as it is not very instructive.

Example 13.3: Let us determine stability of the system described by the following equation:

$$\dot{\mathbf{x}} = \mathbf{A}\mathbf{x}$$

$$\mathbf{A} = \begin{bmatrix} -1 & -2 \\ 1 & -4 \end{bmatrix}$$

We will first solve eqn. (13.10) for **P** for an arbitrary choice of positive definite real symmetric matrix **Q**. We may choose **Q** = **I**, the identity matrix. Equation (13.10) then becomes

$$\mathbf{A}^T\mathbf{P} + \mathbf{P}\mathbf{A} = -\mathbf{I}$$

or
$$\begin{bmatrix} -1 & 1 \\ -2 & -4 \end{bmatrix}\begin{bmatrix} p_{11} & p_{12} \\ p_{12} & p_{22} \end{bmatrix} + \begin{bmatrix} p_{11} & p_{12} \\ p_{12} & p_{22} \end{bmatrix}\begin{bmatrix} -1 & -2 \\ 1 & -4 \end{bmatrix} = \begin{bmatrix} -1 & 0 \\ 0 & -1 \end{bmatrix} \quad ...(i)$$

Note that we have taken $p_{12} = p_{21}$; this is because solution matrix **P** is known to be a positive definite real symmetric matrix for a stable system.

From eqn. (*i*) we get

$$-2p_{11} + 2p_{12} = -1$$
$$-2p_{11} - 5p_{12} + p_{22} = 0$$
$$-4p_{12} - 8p_{22} = -1$$

Solving for *p*'s, we obtain

$$\mathbf{P} = \begin{bmatrix} p_{11} & p_{12} \\ p_{12} & p_{22} \end{bmatrix} = \begin{bmatrix} \frac{23}{60} & \frac{-7}{60} \\ \frac{-7}{60} & \frac{11}{60} \end{bmatrix} \quad ...(ii)$$

Using Sylvester's theorem (Appendix II) we find that **P** is positive definite. Therefore origin of the system under consideration is asymptotically stable in-the-large.

13.4 METHODS OF CONSTRUCTING LIAPUNOV FUNCTIONS FOR NONLINEAR SYSTEMS

As has been said earlier, the Liapunov theorems give only sufficient conditions on system stability and furthermore there is no unique way of constructing a Liapunov function except in the case of linear systems where a Liapunov function can always be constructed and both necessary and sufficient conditions established. Because of this drawback a host of methods have become available in literature and many refinements have been suggested to enlarge the region in which the system is found to be stable. Since this treatise is meant as a first exposure of the student to the Liapunov's direct method, only two of the relatively simpler techniques of constructing a Liapunov function would be advanced here.

Krasovskii's Method

Consider the system

$$\dot{\mathbf{x}} = \mathbf{f}(\mathbf{x})\ ;\ \mathbf{f}(\mathbf{0}) = \mathbf{0}\ (i.e.,\ \text{singular point at origin})$$

Define a Liapunov function as

$$V = \mathbf{f}^T\mathbf{P}\mathbf{f} \qquad \text{...(13.12)}$$

where $\mathbf{P}$ = a symmetric positive definite matrix.

Now

$$\dot{V} = \dot{\mathbf{f}}^T\mathbf{P}\mathbf{f} + \mathbf{f}^T\mathbf{P}\dot{\mathbf{f}} \qquad \text{...(13.13)}$$

$$\dot{\mathbf{f}} = \frac{\partial \mathbf{f}}{\partial \mathbf{x}}\frac{\partial \mathbf{x}}{\partial t} = \mathbf{J}\mathbf{f}$$

$$\mathbf{J} = \begin{bmatrix} \frac{\partial f_1}{\partial x_1} & \frac{\partial f_1}{\partial x_2} & \cdots & \frac{\partial f_1}{\partial x_n} \\ \frac{\partial f_2}{\partial x_1} & \frac{\partial f_2}{\partial x_2} & \cdots & \frac{\partial f_2}{\partial x_n} \\ \cdots & \cdots & \cdots & \cdots \\ \frac{\partial f_n}{\partial x_1} & \frac{\partial f_n}{\partial x_2} & \cdots & \frac{\partial f_n}{\partial x_n} \end{bmatrix} \text{is Jacobian matrix}$$

Substituting $\dot{\mathbf{f}}$ in (13.13), we have

$$\dot{V} = \mathbf{f}^T\mathbf{J}^T\mathbf{P}\mathbf{f} + \mathbf{f}^T\mathbf{P}\mathbf{J}\mathbf{f}$$

$$= \mathbf{f}^T(\mathbf{J}^T\mathbf{P} + \mathbf{P}\mathbf{J})\,\mathbf{f}$$

Let $\quad \mathbf{Q} = \mathbf{J}^T\mathbf{P} + \mathbf{P}\mathbf{J}$

Since V is positive definite, for the system to be asymptotically stable, $\mathbf{Q}$ should be negative definite. If in addition, $V(\mathbf{x}) \to \infty$ as $\|\mathbf{x}\| \to \infty$, the system is asymptotically stable in-the-large.

Example 13.4: Consider the nonlinear system shown in Fig. 13.6 where the nonlinear element is described as

$$u = g(e)$$

With $\quad r = 0$

$$c = -e$$

$\therefore \quad Ku = -\ddot{e} - \dot{e}$

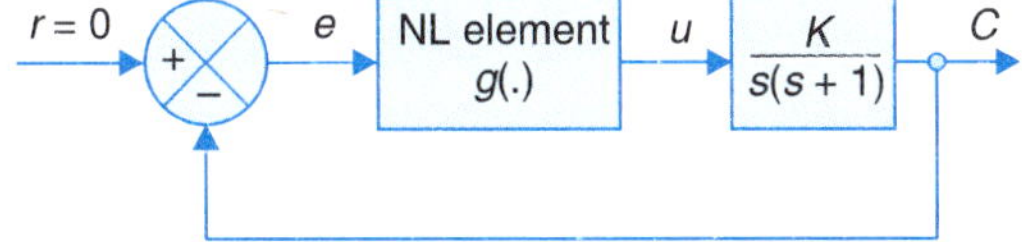

Fig. 13.6

$$x_1 = e$$

$$x_2 = \dot{x}_1$$

$$\dot{x}_1 = x_2$$

$$\dot{x}_2 = -x_2 - Kg(x_1) \qquad ...(i)$$

The state point lies at origin if $g(0) = 0$

$$\mathbf{J} = \begin{bmatrix} 0 & 1 \\ -K\dfrac{dg(x_1)}{dx_1} & -1 \end{bmatrix} \qquad ...(ii)$$

Let
$$\mathbf{P} = \begin{bmatrix} p_{11} & p_{12} \\ p_{12} & p_{22} \end{bmatrix} \qquad ...(iii)$$

For it to be positive definite

$$p_{11} > 0 \qquad ...(iv)$$

$$p_{11}p_{22} - \phi_{12}^2 > 0 \qquad ...(v)$$

Now
$$\mathbf{Q} = \mathbf{J}^T\mathbf{P} + \mathbf{PJ}$$

$$= \begin{bmatrix} 0 & -K\dfrac{dg(x_1)}{dx_1} \\ 1 & -1 \end{bmatrix}\begin{bmatrix} p_{11} & p_{12} \\ p_{12} & p_{22} \end{bmatrix} + \begin{bmatrix} p_{11} & p_{12} \\ p_{12} & p_{22} \end{bmatrix}\begin{bmatrix} 0 & 1 \\ -K\dfrac{dg(x_1)}{dx_1} & -1 \end{bmatrix}$$

or
$$\mathbf{Q} = \begin{bmatrix} -2p_{11}k\dfrac{dg(x_1)}{dx_1} & p_{11} - p_{12} - p_{22}K\dfrac{dg(x_1)}{dx_1} \\ p_{11} - p_{12} - p_{22}K\dfrac{dg(x_1)}{dx_1} & 2(p_{12} - p_{22}) \end{bmatrix} \qquad ...(vi)$$

For the system to be asymptotically stable, **–Q** should be positive definite, *i.e.*,

$$2p_{12}K\frac{dg(x_1)}{dx_1} > 0 \qquad ...(vii)$$

$$-4p_{12}K\frac{dg(x_1)}{dx_1}(p_{12} - p_{22}) - \left(p_{11} - p_{12} - p_{22}K\frac{dg(x_1)}{dx_1}\right)^2 > 0$$

or
$$4p_{12}K\frac{dg(x_1)}{dx_1}(p_{22} - p_{12}) > \left(p_{11} - p_{12} - p_{22}K\frac{dg(x_1)}{dx_1}\right)^2 \qquad ...(viii)$$

Assume that $K > 0$ and choose $p_{12} > 0$. Inequality (*vii*) then yields the condition

$$\frac{dg(x_1)}{dx_1} > 0 \qquad ...(ix)$$

Choose $p_{11} = p_{12}$ and $p_{22} = \beta p_{12}$, where $\beta > 1$. Inequality (*viii*) then gives the condition

$$4(\beta - 1) > K\beta^2\frac{dg(x_1)}{dx_1} \qquad ...(x)$$

The inequalities (*ix*) and (*x*) together constitute the condition under which the system is asymptotically stable.

Take for example

$g(x_1) = x_1^3$ (nonlinearity symmetrically lies in first and third quadrants)

$\therefore \quad \dfrac{dg(x_1)}{dx_1} = 3x_1^2 > 0$ (which is always met)

The condition (x) gives

$$x_1^2 < \frac{4}{3K}\left(\frac{1}{\beta} - \frac{1}{\beta^2}\right) \qquad ...(xi)$$

It can be easily shown that the largest value of x_1 occurs when $\beta = 2$.

Therefore $$x_1^2 < \frac{1}{3K}$$

or $$-\frac{1}{\sqrt{3K}} < x_1 < \frac{1}{\sqrt{3K}} \qquad ...(xii)$$

This region of asymptotic stability is illustrated in Fig. 13.7.

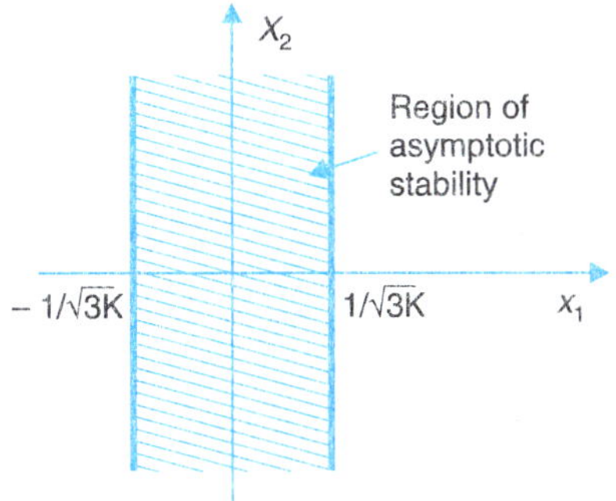

Fig. 13.7. Region of asymptotic stability for Example 13.4 with $g(e) = e^3$.

Variable Gradient Method

The quadratic form approach used so far to Liapunov function formulation is too restrictive. Here we shall advance the variable gradient method which provides considerable flexibility in selecting a Liapunov function.

For the autonomous system

$$\dot{\mathbf{x}} = \mathbf{f}(\mathbf{x})\ ;\ \mathbf{f}(\mathbf{0}) = \mathbf{0}$$

let $V(\mathbf{x})$ be a candidate for a Liapunov function.

The time derivative of V can be expressed as

$$\dot{V}(\mathbf{x}) = \frac{\partial V}{\partial x_1}\dot{x}_1 + \frac{\partial V}{\partial x_2}\dot{x}_2 + \ldots + \frac{\partial V}{\partial x_n}\dot{x}_n \qquad ...(13.14)$$

which can be expressed in terms of the gradient of V as

$$\dot{V} = (\nabla \mathbf{V})^T \dot{\mathbf{x}} \qquad ...(13.15)$$

where $$\nabla \mathbf{V} = \begin{bmatrix} \dfrac{\partial V}{\partial x_1} = \nabla V_1 \\ \dfrac{\partial V}{\partial x_2} = \nabla V_2 \\ \vdots \\ \dfrac{\partial V}{\partial x_n} = \nabla V_n \end{bmatrix}$$

The Liapunov function can be generated by integrating with respect to time both sides of (13.15)

$$V = \int_0^{\mathbf{x}} \frac{dV}{dt} dt = \int_{\mathbf{0}}^{\mathbf{x}} (\nabla \mathbf{V})^T d\mathbf{x} \quad ...(13.16)$$

The above integral is a line integral whose result is independent of the path. The integral can therefore be evaluated sequentially along the component directions (x_1, x_2 ..., x_n) of the state vector ; that is

$$\begin{aligned} V = \int_{\mathbf{0}}^{\mathbf{x}} (\nabla V)^T d\mathbf{x} &= \int_0^{x_1} \nabla V_1(x_1, 0,..., 0)\, dx_1 \\ &+ \int_0^{x_2} \nabla V_2(x_1, x_2, 0,..., 0) dx_2 + ... \quad ...(13.17) \\ &+ \int_0^{x_n} \nabla V_n(x_1, x_2, ..., x_n)\, dx_n \end{aligned}$$

Let us define

$$\mathbf{e}_1 = \begin{bmatrix} 1 \\ 0 \\ 0 \\ \vdots \\ 0 \end{bmatrix} ; \mathbf{e}_2 = \begin{bmatrix} 0 \\ 1 \\ 0 \\ \vdots \\ 0 \end{bmatrix} ; ...; \mathbf{e}_n = \begin{bmatrix} 0 \\ 0 \\ \vdots \\ \vdots \\ 1 \end{bmatrix}$$

The integral given in (13.17) states that the path starts from the origin and moves along the vector $\mathbf{e}_1$ to x_1. From this point, the path moves in the direction of the vector e_2 to x_2. In this way, the path finally reaches the point (x_1, x_2, ... x_n).

For the scalar function V to be unique, the curl of its gradient must be zero, *i.e.*,

$$\nabla \times (\nabla V) = \mathbf{0} \quad ...(13.18)$$

This results in $\frac{n}{2}(n - 1)$ equations to be satisfied by the components of the gradient vector. These are

$$\frac{\partial \nabla V_i}{\partial x_j} = \frac{\partial \nabla V_j}{\partial x_i} \text{ for all } i, j \quad ...(13.19)$$

To begin with, a completely general form given below is assumed for the gradient vector ∇V.

$$\nabla V = \begin{bmatrix} \nabla V_1 \\ \nabla V_2 \\ \vdots \\ \nabla V_n \end{bmatrix} = \begin{bmatrix} a_{11}x_1 & + & a_{12}x_2 & + & ... & + & a_{1n}x_n \\ a_{21}x_1 & + & a_{22}x_2 & + & ... & + & a_{2n}x_n \\ ... & & ... & & ... & & ... \\ ... & & ... & & ... & & ... \\ ... & & ... & & ... & & ... \\ a_{n1}x_1 & + & a_{n2}x_2 & + & ... & + & a_{nn}x_n \end{bmatrix} \quad ...(13.20)$$

The a_{ij}'s are completely undetermined quantities and could be constant of functions of both state variables and t. It is convenient to choose a_{nn} as a constant.

The procedure to formulate a Liapunov function from which the stability of the equilibrium state $\mathbf{x} = \mathbf{0}$ may be determined is given in the following steps.

1. Form $\dot{V}$ as per eqn. (13.15) with ∇V as in eqn. (13.20). Choose a_{ij}'s to constrain it to be negative definite or at least negative semidefinite.

2. Determine the remaining unknown a_{ij}'s to satisfy the curl equations. (13.19).

3. Recheck $\dot{V}$ in case step (2) has altered it.

4. Determine V by integrating as in eqn. (13.17).

5. Determine the region of stability where V is positive definite.

The above procedure is best illustrated by means of an example.

Example 13.5: Consider the nonlinear system shown in Fig. 13.8. The system is described by the state equations

$$\dot{x}_1 = -3x_2 - f(x_1)$$
$$\dot{x}_2 = -x_2 - f(x_1) \qquad ...(i)$$

Let us consider the special case wherein the nonlinearity can be expressed as

$$f(x_1) = g(x_1)x_1$$

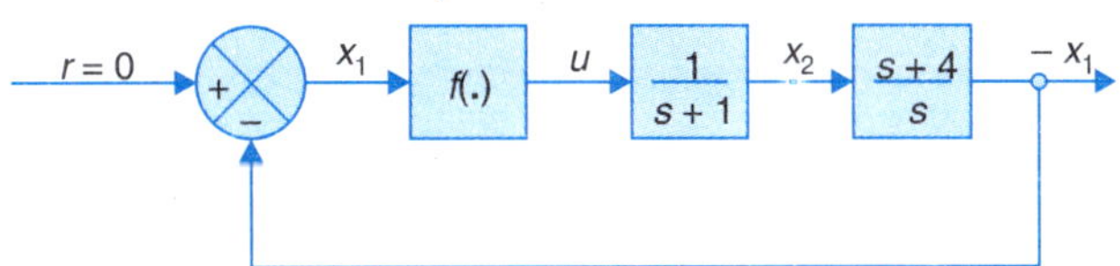

Fig. 13.8

Therefore the state description of the system becomes

$$\dot{x}_1 = -3x_2 - g(x_1)x_1$$
$$\dot{x}_2 = -x_2 + g(x_1)x_1 \qquad ...(ii)$$

It is immediately obvious that the equilibrium point lies at origin.

Assume

$$\nabla V = \begin{bmatrix} a_{11}x_1 + a_{12}x_2 \\ a_{21}x_1 + a_{22}x_2 \end{bmatrix} \qquad ...(iii)$$

Form $\dot{V}$ as per eqn. (13.15)

$$\dot{V} = -x_1^2(a_{11} - a_{21})g(x_1) + x_1x_2[-3a_{11} - a_{12}g(x_1) - a_{21} + a_{22}g(x_1)] - x_2^2(3a_{12} + a_{22}) \qquad ...(iv)$$

One of the ways of keeping $\dot{V}$ negative definite is to choose a_{ij}'s such that

$$(a_{11} - a_{21})g(x_1) > 0$$
$$-3a_{11} - a_{12}g(x_1) - a_{21} + a_{22}g(x_1) = 0 \qquad ...(v)$$
$$3a_{12} + a_{22} > 0$$

These conditions get simplified by making the choice

$$a_{12} = a_{21} = 0 \qquad ...(vi)$$

so that

$$a_{11}g(x_1) > 0$$

$$a_{11} = \frac{a_{22}}{3} g(x_1) \quad \text{...(vii)}$$

$$a_{22} > 0$$

The first of these conditions is satisfied when the other two are met with, because

$$\frac{a_{22}}{3} g^2(x_1) > 0$$

We can therefore choose a_{22} to be any positive constant.

Thus

$$\nabla \mathbf{V} = \begin{bmatrix} \frac{1}{3} a_{22} g(x_1) x_1 \\ a_{22} x_2 \end{bmatrix} \quad \text{...(viii)}$$

$$\dot{\mathbf{V}} = -\frac{1}{3} a_{22} x_1^2 g^2(x_1) - a_{22} x_2^2$$

It may also be noticed that the gradient vector in (*viii*) meets the curl conditions (13.19).

The Liapunov function can now be obtained by taking the line integral of the gradient vector along the path defined in eqn. (13.17). Thus

$$\dot{V} = \frac{1}{3} a_{22} \int_0^{x_1} g(x_1) x_1 dx_1 + a_{22} \int_0^{x_2} x_2 dx$$

$$= \frac{1}{3} a_{22} \int_0^{x_1} g(x_1) x_1 dx_1 + \frac{1}{2} a_{22} x_2^2 \quad \text{...(x)}$$

If $g(x_1) > 0$, *i.e.* $f(x_1) = g(x_1)x_1$ lies in first and third quadrants, V is positive definite. Under this condition the system is asymptotically stable.

Also if

$$\lim_{x_1 \to \infty} \int_0^{x_1} g(x_1) x_1 dx_1 \to \infty$$

the system would be asymptotically stable in-the-large.

Example 13.6: A simple mass, spring and viscous friction system is shown in Fig. 13.9.

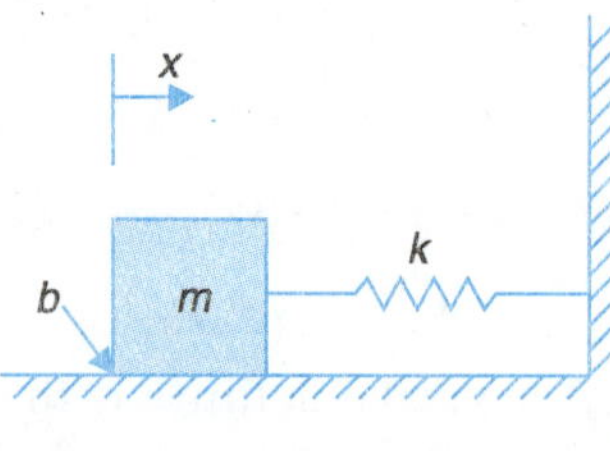

Fig. 13.9

Show that the system is stable.

Solution:

The differential equation governing the system is

$$m\ddot{x} + b\dot{x} + kx = 0$$

The total energy of the system is given by

$$V = \frac{1}{2}m\dot{x}^2 + \frac{1}{2}kx^2$$

where the first term gives the kinetic energy of mass and the second term gives the potential energy stored in spring.

This V can be taken as Lyapunov candidate function as it is nonnegative (*i.e.* positive or zero). Differentiating V with respect to limit, we get

$$\dot{V} = m\dot{x}\ddot{x} + kx\dot{x}$$

$$= -b\dot{x}^2$$

This $\dot{V}$ is always non-positive, since $b > 0$. This implies that energy is always leaving the system unless $\dot{x} = 0$. For checking the possible resting positions, we perform steady-state analysis which yields.

$$kx = 0 \qquad \text{or} \qquad x = 0$$

Thus $x = 0$ is the equilibrium point for this stable system.

Example 13.7: Consider a spring mass system in which the spring and damper are nonlinear. The system is described by the differential equation

$$\ddot{x} + b(\dot{x}) + k(x) = 0.$$

The function $b(.)$ and $k(.)$ are first and third quadrant continuous functions such that

$$\dot{x}b(\dot{x}) > 0 \text{ for } x \neq 0$$

$$xk(x) > 0 \text{ for } x \neq 0$$

show that the system is asymptotically stable.

Solution:

Propose the candidate Lyapunov function as

$$V(x, \dot{x}) = \frac{1}{2}\dot{x}^2 + \int_0^x k(\lambda)d\lambda$$

Differentiating $V(x, \dot{x})$, we get

$$\dot{V}(x, x) = \dot{x}\ddot{x} + k(x)\dot{x} = -\dot{x}b(x)$$

Thus $\dot{V}(.)$ is nonpositive but is only semidefinite, since it is not a function of x but only of $\dot{x}$, For asymptotic stability it is required to show that it is not possible for the system to get stuck with non-zero $\dot{x}$.

To study all trajectories for which $\dot{x} = 0$, we consider

$$\ddot{x} = -k(x)$$

for which $x = 0$ is the only solution. Hence the system will come to rest if $x = \dot{x} = \ddot{x} = 0$.

Example 13.8: A robot manipulator has its dynamics given by

$$\boldsymbol{\tau} = M(\boldsymbol{\theta})\ddot{\theta} + V(\boldsymbol{\theta}, \dot{\theta}) + G(\boldsymbol{\theta}), \text{ where} \qquad ...(i)$$

τ is nxl vector of joint torques.

$M(\boldsymbol{\theta})$ is nxn manipulator mass matrix which is positive definite.

$V(\boldsymbol{\theta}, \dot{\theta})$ is nxl matrix of coriolis and centrifugal forces.

$G(\boldsymbol{\theta})$ is nxl matrix of gravity related terms and $\ddot{\theta}$ is nxl matrix of joint accelerations. It can be shown that the relation

$\frac{1}{2}\dot{\theta}^T \dot{M}(\theta)\dot{\theta} = \dot{\theta}^T V(\theta,\dot{\theta})$ is true for any manipulator with open kinematic chain structure.

This manipulator is controlled by a control law

$$\tau = K_P E - K_d\dot{\theta}\text{, where} \qquad ...(ii)$$

$\boldsymbol{K_p}$ and $\boldsymbol{K_d}$ are nxn diagonal positive definite gain matrices.

$\boldsymbol{E = \theta_d - \theta}$ is the nxl vector of error in position and

$\dot{\boldsymbol{\theta}}$ is nxl vector of joint velocities

Show that the system has global asymptotic stability.

Solution: The resulting closed-loop system dynamics as obtained by equating eqns (*i*) and (*ii*) above is given by

$$M(\boldsymbol{\theta})\ddot{\theta} + V(\boldsymbol{\theta}, \dot{\theta}) + K_d\dot{\theta} + K_P\boldsymbol{\theta} = K_P\boldsymbol{\theta} \qquad ...(iii)$$

Now consider the candidate Lyapunov function

$$v = \frac{1}{2}\dot{\theta}^T M(\theta)\dot{\theta} + \frac{1}{2}E^T K_P E \qquad ...(iv)$$

It can be seen that v is always non-negative

Differentiating v, we get

$$\dot{v} = \frac{1}{2}\dot{\theta}^T \dot{M}(\theta)\dot{\theta} + \dot{\theta}^T M(\theta)\ddot{\theta} - E^T K^P \dot{\theta} \qquad ...(v)$$

$$= \frac{1}{2}\dot{\theta}^T \dot{M}(\theta)\dot{\theta} - \dot{\theta}^T M_d\theta - \theta^T V(\theta,\dot{\theta})$$

$$= \dot{\theta}^T K_d\dot{\theta} \qquad ...(vi)$$

Thus v is nonpositive as $\boldsymbol{K_d}$ is positive definite. Further since v can remain zero only along trajectories that have $\dot{\theta} = 0$ and $\boldsymbol{\theta} = 0$, we see that

$\boldsymbol{K_P E = 0}$ from the dynamical equation of the closed-loop system (eq. (*iii*)). This implies $\boldsymbol{E = 0}$ and proves the global asymptotic stability of the system.

PROBLEMS

13.1 Consider a nonlinear system described by the equations

$$\dot{x}_1 = x_2$$

$$\dot{x}_2 = -(1 - |x_1|)x_2 - x_1$$

Find the region in the state-plane for which the equilibrium state of the system is asymptotically stable.

(*Hint* : A Liapunov function is $V = x_1^2 + x_2^2$)

13.2 Consider the system shown in Fig. P-13.2. Determine the restrictions which must be placed on the nonlinear function $\phi(.)$ in order for the system state with $r(t) = 0$ to be asymptotically stable in-the-large. Use Liapunov's direct method.

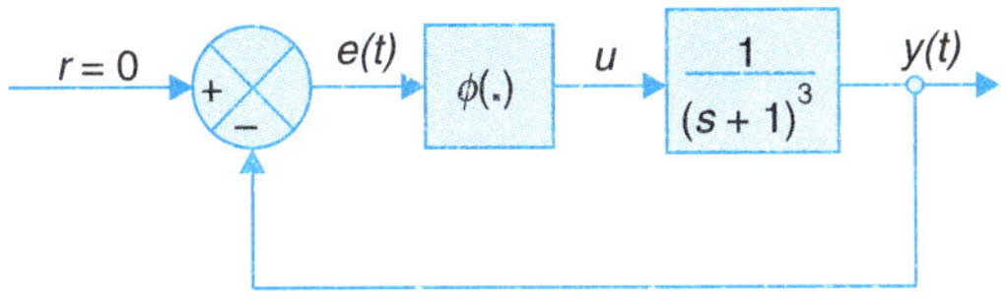

Fig. P-13.2

(*Hint*: Barbashin's result may be used :

The system

$$\dot{x}_1 = x_2$$

$$\dot{x}_2 = x_3$$

$$\dot{x}_3 = -f(x_1) - g(x_2) - ax_3$$

is asymptotically stable in-the-large at the origin provided that

(*i*) $f(0) = 0 = g(0)$ and f, g are differentiable functions

(*ii*) $a > 0$

(*iii*) $\dfrac{f(x_1)}{x_1} \geq \epsilon_1 > 0$ if $x_1 \neq 0$

(*iv*) $\dfrac{ag(x_2)}{x_2} - \dfrac{df(x_1)}{dx_1} \geq \epsilon_2 > 0$ if $x_2 \neq 0$

A Liapunov function is

$$V(x) = aF(x_1) + f(x_1)x_2 + Gx_2 + \frac{(ax_2 + x_3)^2}{2}$$

where $\qquad F(x_1) \underline{\underline{\Delta}} \int_0^{x_1} f(\tau)a\tau \,;\, G(x_2) \underline{\underline{\Delta}} \int_0^{x_2} g(\tau)d\tau$

13.3 Consider a nonlinear system described by the equations

$$\dot{x}_1 = -3x_1 + x_2$$

$$\dot{x}_2 = x_1 - x_2 - x_2^3$$

Investigate the stability of equilibrium state.

(*Hint* : Use Krasovskii's method with **P** as identity matrix.)

13.4 A linear autonomous system is described by the state equation

$$\dot{\mathbf{x}} = \mathbf{A}\mathbf{x}$$

$$\mathbf{A} = \begin{bmatrix} -4K & 4K \\ 2K & -6K \end{bmatrix}$$

Find restrictions on the parameter K to guarantee stability of the system.

13.5 Consider the system of Fig. P-13.2. The nonlinear function is replaced by a linear function $\phi(e) = Ke$. Find the restrictions on the parameter K to guarantee system stability. Use the Liapunov's direct method and the Rough criterion. Compare the results.

13.6 Check the stability of the system described by

$$\dot{x}_1 = x_2$$

$$\dot{x}_2 = -x_1 - b_1 x_2 - b_2 x_2^3 \; ; \; b_1, b_2 > 0$$

(*Hint* : Use the variable gradient method with a_{ij}'s as constants. Also try by choosing $a_{12} = x_1/x_2$; $a_{21} = x_2/x_1$.)

13.7 Check the stability of the system described by

$$\dot{x}_1 = x_2$$

$$\dot{x}_2 = -x_1 - x_1^2 x_2$$

13.8 A linear time-invariant system is described by the equations

$$\dot{\mathbf{x}} = \mathbf{Ax} + \mathbf{Bu}$$

The performance index is

$$J = \int_0^\infty (\mathbf{x}^T\mathbf{Qx} + \mathbf{u}^T\mathbf{Ru})\, dt$$

where **Q** and **R** are constant symmetric, positive definite matrices. The optimal control for the system is

$$\mathbf{u}^* = -\mathrm{R}^{-1}\mathbf{B}^T\mathbf{Px}$$

where **P** is given by the reduced matrix Riccati equation

$$\mathbf{0} = \mathbf{Q} + \mathbf{A}^T\mathbf{P} + \dot{\mathbf{P}}\mathbf{A} - \mathbf{PBR}^{-1}\mathbf{B}^T\mathbf{P}$$

Show that if **P** exists, the closed-loop system is asymptotically stable.

(*Hint* : Choose a Liapunov function $V = \mathbf{x}^T\mathbf{Px}$.)

13.9 Consider the system shown in Fig. P-13.9. The linear plant is described by the equations

$$\dot{\mathbf{x}} = \mathbf{Ax} + \mathbf{B}u$$

$$y = \mathbf{Cx}$$

$$\mathbf{A} = \begin{bmatrix} 0 & 1 \\ -10 & -7 \end{bmatrix}; \mathbf{B} = \begin{bmatrix} 1 \\ -4 \end{bmatrix}; \mathbf{C} = [1 \quad 0]$$

Establish an upper limit on K in

$$0 \le \frac{\phi(e)}{e} \le K$$

which ensures global asymptotic stability of the origin.

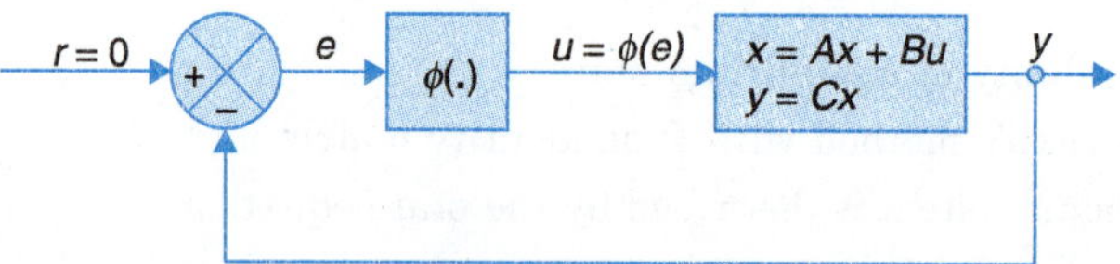

Fig. P-13.9

14

OPTIMAL CONTROL SYSTEMS

14

OPTIMAL CONTROL SYSTEMS

14.1 INTRODUCTION

There are, as discussed in Chapter 10, basically two approaches to the design of control systems. In one approach we select the configuration of the overall system by introducing compensators and then choose the parameters of the compensators to meet the given specifications on performance. In the other approach, for a given plant we find an overall system that meets the given specifications and then compute the necessary compensators.

The classical design method based on first approach mentioned above, has already been discussed at length in earlier chapters. This method relies heavily on the Laplace transform and z-transform. The designer is given a set of specifications in time domain or in frequency domain and system configuration. Peak overshoot, settling time, gain margin, phase margin, steady-state error, etc., are among most commonly used specifications. These design specifications are selected because of convenience in graphical interpretation with respect to root locus or frequency plots. Compensators are selected that give as closely as possible, the desired system performance. In general, it may not be possible to satisfy all the desired specifications. Then, through a trial and error procedure, an acceptable system performance (within certain tolerable error) is achieved. There are generally many designs that can yield this acceptable performance, *i.e.,* the solution is not unique. This trial and error design procedure works satisfactorily for single-input-single-output systems. The gap between the classical design procedure and its application to multi-input-multi-output systems has been bridged recently.

The trial and error uncertainties are eliminated in the parameter optimization method. The point of departure in the parameter optimization procedure is that the performance specifications consist of a single performance index. Integral square error performance index is very common but other performance indices can be used as well (refer Chapter 5). For a fixed system configuration, parameters that minimize the performance index are selected.

The ready availability of the digital computer has led to the development of modern *optimal control theory* which has become the major technique for the design of automatic control systems. This theory is based on the second design approach mentioned above. In this chapter, we attempt to introduce optimal control theory.

14.2 PARAMETER OPTIMIZATION: SERVOMECHANISMS

The parameter optimization method has already been used in Chapter 5 (Examples 5.5-5.7). There we considered a relatively simple case: for fixed configuration of overall system, we determined suitable value of system gain K so as to minimize the given index of performance. The minimization was carried out analytically. The analytical approach of parameter optimization consists of the following steps:

(*i*) Compute the performance index J as a function of the free parameters $K_1, K_2, ..., K_n$ of the system with fixed configuration;

$$J = J(K_1, K_2, ..., K_n) \quad ...(14.1)$$

(*ii*) Determine the solution set K_i of the equations

$$\frac{\partial J}{\partial K_i} = 0; \; i = 1, 2, ..., n \quad ...(14.2)$$

Equations (14.2) give the necessary conditions for J to be minimum. From the solution set of eqns. (14.2), find the subset that satisfies the sufficient conditions which require that the *Hessian matrix* given below is positive definite.

$$\mathbf{H} = \begin{bmatrix} \dfrac{\partial^2 J}{\partial K_1^2} & \dfrac{\partial^2 J}{\partial K_1 \partial K_2} & \cdots & \dfrac{\partial^2 J}{\partial K_1 \partial K_n} \\ \dfrac{\partial^2 J}{\partial K_2 \partial K_1} & \dfrac{\partial^2 J}{\partial K_2^2} & \cdots & \dfrac{\partial^2 J}{\partial K_2 \partial K_n} \\ \cdots & \cdots & \cdots & \cdots \\ \dfrac{\partial^2 J}{\partial K_n \partial K_1} & \dfrac{\partial^2 J}{\partial K_n \partial K_2} & \cdots & \dfrac{\partial^2 J}{\partial K_n^2} \end{bmatrix} \quad ...(14.3)$$

Since $\dfrac{\partial^2 J}{\partial K_i \partial K_j} = \dfrac{\partial^2 J}{\partial K_j \partial K_i}$, the matrix $\mathbf{H}$ is always symmetric.

(*iii*) If there are two or more sets of K_i satisfying the necessary as well as sufficient conditions of minimization given by eqns. (14.2) and (14.3) respectively, then compute the corresponding J for each set. The set that has the smallest J gives the optimal parameters.

The minimization problem will be more easily solved if we can express performance index in terms of transform domain quantities. For quadratic performance index (eqn. (5.51)), this can be done by using the Parseval's theorem which allows us to write

$$\int_0^\infty x^2(t)\, dt = \frac{1}{2\pi j} \int_{-j\infty}^{\infty} X(s) X(-s)\, ds \quad ...(14.4)$$

in which $X(s)$ = Laplace transform of $x(t)$, where $x(t)$ is defined for $t \geq 0$ and $x(t)$ is zero for $t < 0$.

The value of right hand integral in eqn. (14.4) can easily be found from published tables, provided that $X(s)$ can be written in the form $B(s)/A(s)$;

$$B(s) \triangleq b_0 + b_1 s + \ldots + b_{n-1} s^{n-1} \qquad \ldots(14.5)$$

$$A(s) \triangleq a_0 + a_1 s + \ldots + a_n s^n$$

where $A(s)$ has zeros only in the left half of the complex plane.

Results upto fourth-order are given in Table 14.1.

Table 14.1 Tabulation of Definite Integral For Continuous-Time Systems

$$J_n = \frac{1}{2\pi j} \int_{-j\infty}^{j\infty} \frac{B(s)B(-s)}{A(s)A(-s)} ds$$

$$B(s) = \sum_{k=0}^{n-1} b_k s^k$$

$$A(s) = \sum_{k=0}^{n} a_k s^k; \text{ } A(s) \text{ has zeros in left half plane only.}$$

$$J_1 = \frac{b_0^2}{2a_0 a_1}$$

$$J_2 = \frac{b_1^2 a_0 + b_0^2 a_2}{2a_0 a_1 a_2}$$

$$J_3 = \frac{b_2^2 a_0 a_1 + (b_1^2 - 2b_0 b_2) a_0 a_3 + b_0^2 a_2 a_3}{2a_0 a_3(-a_0 a_3 + a_1 a_2)}$$

$$J_4 = \frac{b_3^2(-a_0^2 a_3 + a_0 a_1 a_2) + (b_2^2 - 2b_1 b_3) a_0 a_1 a_4 + (b_1^2 - 2b_0 b_2) a_0 a_3 a_4 + b_0^2(-a_1 a_4^2 + a_2 a_3 a_4)}{2a_0 a_4(-a_0 a_3^2 - a_1^2 a_4 + a_1 a_2 a_3)}$$

For sampled-data systems, the following relation is very useful in analytical design.

$$\sum_{k=0}^{\infty} x^2(kT) = \frac{1}{2\pi j} \oint_{\substack{\text{unit} \\ \text{circle}}} X(z) X(z^{-1}) \frac{dz}{z} \qquad \ldots(14.6)$$

in which $X(z)$ = z-transform of $x(kT)$; where $x(kT)$ is defined for $k \geq 0$ and is 0 for $k < 0$.

The value of right hand side of eqn. (14.6) can be found from published tables, provided that $X(z)$ can be written in the form

$$X(z) = \frac{b_0 z^n + b_1 z^{n-1} + \ldots + b_n}{a_0 z^n + a_1 z^{n-1} + \ldots + a_n} \qquad \ldots(14.7)$$

Results up to third-order are given in Table 14.2.

Table 14.2 Tabulation Of Definite Integral For Sampled-Data Systems

$$J_n = \frac{1}{2\pi j}\oint_{\substack{\text{unit}\\ \text{circle}}} X(z)X(z^{-1})\frac{dz}{z}$$

$$X(z) = \frac{b_0 z^n + b_1 z^{n-1} + \ldots + b_n}{a_0 z^n + a_1 z^{n-1} + \ldots + a_n}$$

$$J_1 = \frac{(b_0^2 + b_1^2)a_0 - 2b_0 b_1 a_1}{a_0(a_0^2 - a_1^2)}$$

$$J_2 = \frac{B_0 a_0 e_1 - B_1 a_0 a_1 + B_2(a_1^2 - a_2 e_1)}{a_0[(a_0^2 - a_2^2)e_1 - (a_0 a_1 - a_1 a_2)a_1]}$$

where

$$B_0 = b_0^2 + b_1^2 + b_2^2$$
$$B_1 = 2(b_0 b_1 + b_1 b_2)$$
$$B_2 = 2b_0 b_2$$
$$e_1 = a_0 + a_2$$

$$J_3 = \frac{a_0 B_0 Q_0 - a_0 B_1 Q_1 + a_0 B_2 Q_2 - B_3 Q_3}{[(a_0^2 - a_3^2)Q_0 - (a_0 a_1 - a_2 a_3)Q_1 + (a_0 a_2 - a_1 a_3)Q_2]a_0}$$

where

$$B_0^2 = b_0^2 + b_1^2 + b_2^2 + b_3^2$$
$$B_1 = 2(b_0 b_1 + b_1 b_2 + b_2 b_3)$$
$$B_2 = 2(b_0 b_2 + b_1 b_3)$$
$$B_3 = 2b_0 b_3$$
$$Q_0 = (a_0 e_1 - a_3 a_2)$$
$$Q_1 = (a_0 a_1 - a_1 a_3)$$
$$Q_2 = (a_1 e_2 - a_2 e_1)$$
$$Q_3 = (a_1 - a_3)(e_2^2 - e_1^2) + a_0(a_0 e_2 - a_3 e_1)$$
$$e_1 = a_0 + a_2$$
$$e_2 = a_1 + a_3$$

Servomechanism or Tracking Problem

In this section we shall be concerned with the parameter optimization in *servomechanisms* or *tracking systems* wherein the objective of design is to maintain the actual output ($c(t)$) of the system as close as possible to the desired output which is usually the reference input ($r(t)$) to the system. We may define an error $e(t)$ by setting $e(t) = c(t) - r(t)$. Loosely speaking, the design objective in a servomechanism or tracking problem is to keep error $e(t)$ 'small'. This objective, as we observed in Chapter 5, can be realized by minimizing the performance index

$$J = \int_0^\infty e^2(t)dt$$

if control $u(t)$ is not constrained in magnitude.

Example 14.1: Referring to the block diagram of Fig. 14.1, consider that $G(s) = 100/s^2$ and $R(s) = 1/s$. Determine the optimal value of parameter K such that

$$J = \int_0^\infty e^2(t)\, dt \text{ is minimum}$$

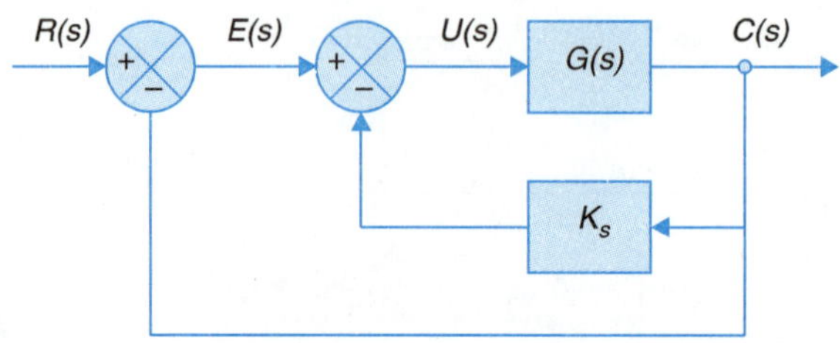

Fig. 14.1. Control system for parameter optimization.

Solution:

From Fig. 14.1,

$$E(s) = \frac{s + 100K}{s^2 + 100Ks + 100}$$

From Table 14.1,

$$J = \int_0^\infty e^2(t)\, dt = \frac{1}{2\pi j}\int_{-j\infty}^{j\infty} E(s)E(-s)\, ds$$

$$= \frac{1 + 100K^2}{200K} \qquad \text{...(14.8)}$$

$$\frac{\partial J}{\partial K} = 0 \text{ gives } K = 0.1$$

It can be checked that $\frac{\partial^2 J}{\partial K^2} > 0.$ Therefore $K = 0.1$ is the optimal value of the free parameter of system of Fig. 14.1. The minimum value of J obtained from eqn. (14.8) is $J_{min} = 0.1$.

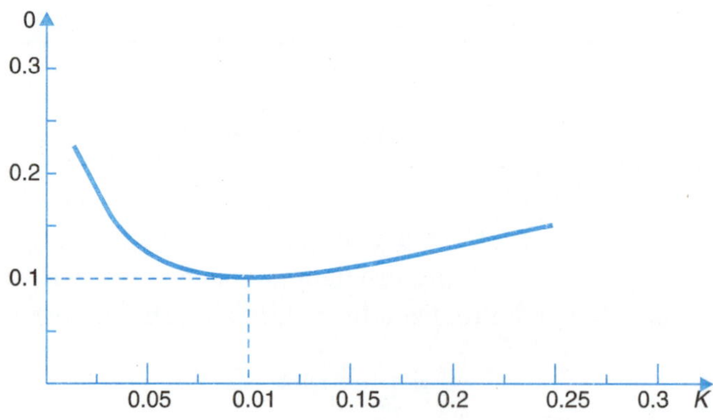

Fig. 14.2. The performance index versus the parameter K.

A curve of performance index as a function of K is shown in Fig. 14.2. It is clear that this system is not very sensitive to changes in K. The sensitivity of an optimum system is given by

$$S_K^{\text{opt}} = \frac{\Delta J / J}{\Delta K / K}$$

where K is the design parameter.

For the system under consideration

$$S_K^{\text{opt}} \approx \frac{0.00045/0.1}{0.01/0.1} = 0.045$$

Example 14.2: Figure 14.3 illustrates a typical sampled-data system. The transfer functions $G_p(s)$ and $G_0(s)$ of the controlled plant and hold circuit respectively are known. The data-processing unit $D(z)$ which operates on the sampled error singal $e_s(t)$ is to be designed.

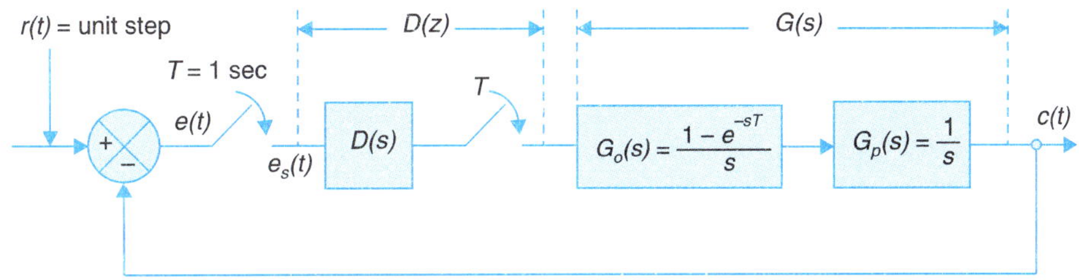

Fig. 14.3. Control system for parameter optimization.

For the design of continuous-time servo systems, we have used the integral square error criterion. This criterion can be effectively utilized for the design of sampled-data servo systems as well. However, for sampled data systems, the *sum square error criterion* which requires the minimization of sum square error

$$J = \sum_{k=0}^{\infty} e^2(kT) \qquad \text{...(14.9)}$$

is more convenient. We need to concern ourselves only with information at sampling instants; therefore we need only the z-transform while the integral square error criterion needs modified z-transform.

Assuming the processing unit $D(z)$ in Fig. 14.3, to be simply an amplifier of gain K, let us find K so that the sum square error given by eqn. (14.9) is minimized (see Problem 14.4).

From Fig. 14.3, we have

$$G(s) = G_0(s)G_p(s) = \frac{1 - e^{-sT}}{s^2}$$

$$r(t) = \text{unit step}$$

Therefore

$$G(z) = (1 - z^{-1})\, Z\left[\frac{1}{s^2}\right] = \frac{1}{z - 1}$$

$$R(z) = \frac{z}{z-1}$$

Now
$$E(z) = R(z) - \frac{G(z)D(z)}{1+G(z)D(z)}R(z)$$

$$= \frac{z}{z-1}\left(1 - \frac{K}{z+K-1}\right)$$

$$= \frac{z}{z+K-1}$$

From Table 14.2, we obtain

$$J = \sum_{i=0}^{\infty} e^2(iT) = \frac{1}{2\pi j} \oint_{\substack{\text{unit}\\\text{circle}}} E(z)E(z^{-1})\frac{dz}{z}$$

$$= \frac{1}{2K - K^2}$$

The necessary condition for J to be minimum is

$$\frac{\partial J}{\partial K} = \frac{-(2-2K)}{(2K-K^2)^2} = 0 \quad \text{or} \quad K = 1$$

The sufficient condition $\frac{\partial^2 J}{\partial K^2} > 0$ is satisfied for $K = 1$. Therefore $K = 1$ is the optimum value of the parameter and corresponding value of performance index is

$$J_{\min} = 1$$

Compensator design subject to constraints

In Chapter 5, we observed that the optimal design of servo systems obtained by minimizing the performance index

$$J = \int_0^{\infty} e^2(t)\,dt \qquad \text{...(14.10)}$$

may be unsatisfactory because it may lead to excessively large magnitudes of some control singals. A more realistic solution to the problem is reached if the performance index is modified to account for the physical constraints like saturation in physical devices. Therefore, a more realistic performance index should be to minimize

$$J = \int_0^{\infty} e^2(t)\,dt \qquad \text{...(14.11)}$$

subject to the constraint

$$\max |u(t)| \le M \qquad \text{...(14.11 a)}$$

for some constant M. The constant M is determined by the linear range of the system plant. Similar constraints could be imposed on other components of the system, *e.g.*, compensators.

However this will make the design extremely complicated. We assume here that these components are of such quality that they are not saturated in the control process. The constraint is therefore imposed only on the plant.

Although the criterion given by (14.11) can be used in the design, it is not convenient to work with. This has already been demonstrated in Example 14.1 wherein evaluation of J became easier because of its quadratic nature. Further, if the criterion given by eqn. (14.11) is used, the resulting optimal system is not necessarily a linear system; *i.e.,* in order to implement the optimal design, nonlinear and/or time-varying devices are required. Therefore, we would very much like to replace the performance criterion given by (14.11) by the following *quadratic performance index*:

$$J = \int_0^{\infty} [e^2(t) + \lambda u^2(t)]\, dt \qquad \text{...(14.12)}$$

where λ, a positive constant, is called the *weighting factor*. If λ is a small positive number, more weight is imposed on the error. As $\lambda \to 0$, the contribution of $u(t)$ becomes less significant and the performance index reduces to (14.10). In this case, the magnitude of $u(t)$ will be very large and the constraint given by (14.11 a) may be violated. If $\lambda \to \infty$, performance criterion given by (14.12) reduces to

$$J = \int_0^{\infty} u^2(t)\, dt \qquad \text{...(14.13)}$$

and the optimal system that minimizes this J is one with $u = 0$. From these two extreme cases, we conclude that if λ is properly chosen, then the constraint of (14.11 a) will be satisfied.

The integral square control index J given by (14.13) may be regarded as a measure of energy required by the control. Since $u^2(t)$ is proportional to power consumed for control, the time integral of $u^2(t)$ is a measure of energy consumed by the system.

For sampled-data servo systems, the performance index J is considered to include the following two summations.

$$J_e = \sum_{k=0}^{\infty} e^2(kT)$$

$$J_u = \sum_{k=0}^{\infty} u^2(kT)$$

The second summation puts a constraint on the control signal. The optimization problem is to find free parameters of the system so that

$$J = J_e + \lambda J_u = \sum_{k=0}^{\infty} [e^2(kT) + \lambda u^2(kT)] \qquad \text{...(14.14)}$$

is minimized.

Example 14.3: Referring to the block diagram of Fig. 14.4, consider that $G(s) = 100/s^2$ and $R(s) = 1/s$. Determine the optimal values of the parameters K_1 and K_2 such that (*i*) $J_e = \int_0^{\infty} e^2(t)\, dt$ is minimized (*ii*) $J_u = \int_0^{\infty} u^2(t)\, dt = 0.1$.

Solution. From Fig. 14.4,

$$E(s) = \frac{s + 100K_1K_2}{s^2 + 100K_1K_2s + 100K_1}$$

$$C(s) = \frac{100K_1}{s(s^2 + 100K_1K_2s + 100K_1)} = \frac{100}{s^2}U(s)$$

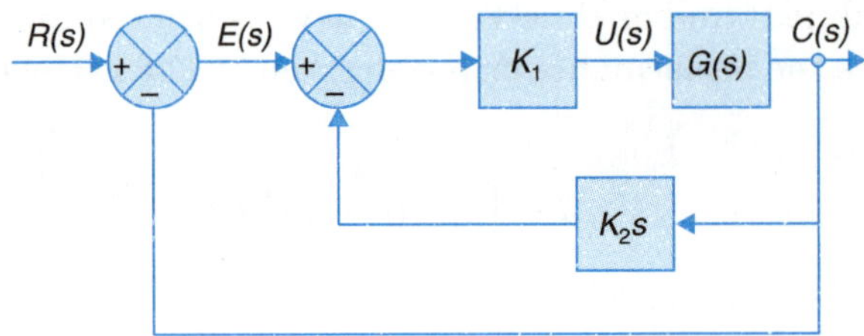

Fig. 14.4. Control system for parameter optimization.

Therefore

$$U(s) = \frac{sK_1}{s^2 + 100K_1K_2s + 100K_1}$$

From Table 14.1,

$$J_e = \int_0^{\infty} e^2(t)\,dt = \frac{1}{2\pi j}\int_{-j\infty}^{j\infty} E(s)E(-s)\,ds$$

$$= \frac{1 + 100K_1K_2^2}{200K_1K_2}$$

$$J_u = \int_0^{\infty} u^2(t)\,dt = \frac{1}{2\pi j}\int_{-j\infty}^{j\infty} U(s)U(-s)\,ds$$

$$= \frac{K_1}{200K_2}$$

The energy constraint on the system is thus expressed by the equation

$$J_u = \frac{K_1}{200K_2} = 0.1 \qquad \text{...(14.15)}$$

The performance index for the system is (eqn. (14.12))

$$J = J_e + \lambda J_u$$

$$= \frac{1 + 100K_1K_2^2}{200K_1K_2} + \lambda\frac{K_1}{200K_2}$$

$$\frac{\partial J}{\partial K_i} = 0 \text{ for } i = 1, 2, \text{ gives}$$

$$\lambda K_1^2 = 1 \qquad \text{...(14.16)}$$

$$100K_1K_2^2 - 1 - \lambda K_1^2 = 0 \qquad \text{...(14.17)}$$

From eqns. (14.15) – (14.17), we get

$$\lambda = 0.25,\ K_1 = 2,\ K_2 = 0.1$$

The Hessian matrix

$$\mathbf{H} = \begin{bmatrix} \dfrac{\partial^2 J}{\partial K_1^2} & \dfrac{\partial^2 J}{\partial K_1 \partial K_2} \\ \dfrac{\partial^2 J}{\partial K_2 \partial K_1} & \dfrac{\partial^2 J}{\partial K_2^2} \end{bmatrix} = \begin{bmatrix} \dfrac{1}{100 K_1^3 K_2} & \dfrac{1-\lambda K_1^2}{200 K_1^2 K_2^2} \\ \dfrac{1-\lambda K_1^2}{200 K_1^2 K_2^2} & \dfrac{1}{K_2} - \dfrac{(100 K_1 K_2^2 - 1)}{100 K_1 K_2^3} \end{bmatrix}$$

For $K_1 = 2, K_2 = 0.1$;

$$\mathbf{H} = \begin{bmatrix} \dfrac{1}{80} & 0 \\ 0 & 5 \end{bmatrix} \text{ is positive definite}$$

Therefore $K_1 = 2, K_2 = 0.1$ satisfy the necessary as well as sufficient conditions for J to be minimum.

14.3 OPTIMAL CONTROL PROBLEMS: TRANSFER FUNCTION APPROACH

Following steps are involved in the solution of an optimal control problem using transfer function apprach:

(*i*) Given a plant with transfer function $G(s)$, find the transfer function of the overall system which is optimal with respect to given performance criterion.

(*ii*) Compute the compensators for the system obtained from step (*i*).

Servomechanism Problem: Continuous-time Systems

The optimal servomechanism or tracking problem for single-input-single-output linear time-invariant systems can be formulated in the following way:

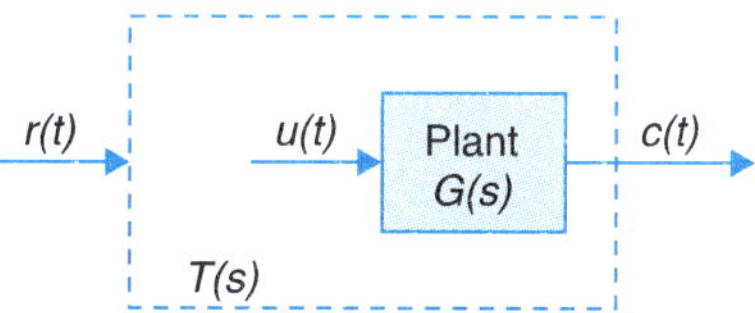

Fig. 14.5. An optimal control problem.

For a given plant (Fig. 14.5) with irreducible proper transfer function $G(s)$, find the transfer function $T^*(s)$ of the overall system that minimizes the quadratic performance index

$$J = \int_0^{\infty} [(r(t) - c(t))^2 + \lambda u^2(t)]\, dt$$

where λ is a positive constant, $u(t)$ is the actuating signal, $c(t)$ is the actual output of the system and $r(t)$ is the desired output (reference input).

In the frequency domain, the performance index is given as

$$J = \frac{1}{2\pi j} \int_{-j\infty}^{j\infty} \{[R(s) - C(s)][R(-s) - C(-s)] + \lambda U(s) U(-s)\} ds \qquad \text{...(14.18)}$$

For the system under consideration

$$\frac{C(s)}{U(s)} = G(s); \ \frac{C(s)}{R(s)} = T(s) \qquad ...(14.19)$$

Therefore

$$U(s) = \frac{T(s)R(s)}{G(s)} \qquad ...(14.20)$$

Substituting eqns. (14.19) and (14.20) in eqn. (14.18), we get

$$J(T(s)) = \frac{1}{2\pi j}\int_{-j\infty}^{j\infty}\left\{[T(s)-1][T(-s)-1]+\lambda\frac{T(s)T(-s)}{G(s)G(-s)}\right\}R(s)R(-s)\,ds \qquad ...(14.21)$$

Now the problem is to find a stable $T^*(s)$ that minimises $J(T(s))$.

Observe that if $T^*(s)$ were the optimum transfer function, the performance index corresponding to $T^*(s)$ must, by definition, have a smaller value than that corresponding to any other transfer function, *i.e.*, for some constant $\in$, the following holds:

$$J(T^*(s)) \le J[T^*(s) + \in T_1(s)] = J\in(T^*(s))$$

where $T_1(s)$ is some arbitrary stable transfer function. The necessary and sufficient conditions for $T^*(s)$ to be the minimizing transfer function are

$$\left.\frac{d}{d\in}J_\varepsilon(T^*(s))\right|_{\varepsilon=0} = 0 \qquad ...(14.22)$$

$$\left.\frac{d^2}{d\in^2}J_\in(T^*(s))\right|_{\in=0} > 0$$

With $T(s) = T^*(s) + \varepsilon T_1(s)$, we have from eqn. (14.21)

$$J\in(T^*(s)) = \frac{1}{2\pi j}\int_{-j\infty}^{j\infty}\left\{[T^*(s)+\varepsilon T_1(s)-1]\,[T^*(-s)+\in T_1(-s)-1]\right.$$

$$\left.+\lambda\frac{[T^*(s)+\in T_1(s)[T^*(-s)+\in T_1(-s)]}{G(s)G(-s)}\right\}R(s)R(-s)\,ds \qquad ...(14.23)$$

Writing the right hand side of this equation in terms of increasing order of $\in$, we get

$$J\in = J_1 + \in(J_2 + J_3) + \in^2 J_4 \qquad ...(14.24)$$

where

$$J_1 = \frac{1}{2\pi j}\int_{-j\infty}^{j\infty}\left\{(T^*(s)-1)(T^*(-s)-1)+\lambda\frac{T^*(s)T^*(-s)}{G(s)G(-s)}\right\}R(s)R(-s)\,ds$$

$$J_2 = \frac{1}{2\pi j}\int_{-j\infty}^{j\infty}\left\{(T^*(-s)-1)+\lambda\frac{T^*(-s)}{G(s)G(-s)}\right\}R(s)R(-s)T_1(s)\,ds$$

$$J_3 = \frac{1}{2\pi j}\int_{-j\infty}^{j\infty}\left\{(T^*(s)-1)+\lambda\frac{T^*(s)}{G(s)G(-s)}\right\}R(s)R(-s)T_1(-s)\,ds$$

$$J_4 = \frac{1}{2\pi j}\int_{-j\infty}^{j\infty}\left\{1+\frac{\lambda}{G(s)G(-s)}\right\}R(s)R(-s)T_1(s)T_1(-s)\,ds$$

Assuming

$$G(s) = \frac{N(s)}{D(s)} \quad ...(14.25)$$

we can write the integrand of J_4 as

$$\left[\frac{N(s)N(-s) + \lambda D(s)D(-s)}{N(s)N(-s)}\right] R(s)R(-s)T_1(s)T_1(-s) \quad ...(14.26)$$

Consider the polynomial

$$P(s) = N(s)N(-s) + \lambda D(s)D(-s)$$

It is clear that $P(s) = P(-s)$. Hence if s_1 is a root of $P(s) = 0$, so is $-s_1$.

We next show that $P(s)$ has no roots on the imaginary axis:

$$\begin{aligned} P(j\omega) &= N(j\omega)N(-j\omega) + \lambda D(j\omega)D(-j\omega) \\ &= |N(j\omega)|^2 + \lambda |D(j\omega)|^2 \end{aligned}$$

Since $\lambda > 0$ and $N(s)$ and $D(s)$ have no common factor, $P(j\omega) \neq 0$ for all ω. Hence $P(s)$ has no roots on the $j\omega$-axis and all the roots are symmetrical with respect to this axis. Consequently, we may write

$$P(s) = D^*(s)D^*(-s) = N(s)N(-s) + \lambda D(s)D(-s) \quad ...(14.27)$$

where $D^*(s)$ consists of all left half s-plane roots of $P(s)$ and $D^*(-s)$ consists of all right half s-plane roots of $P(s)$. The factorization in eqn. (14.27) is called *spectral factorization*.

Let us now look at the integrand (14.26) of J_4. We find that the integrand is symmetrical with respect to the imaginary axis and therefore from eqn. (14.4) we conclude that J_4 is always positive.

One more observation from eqn. (14.24): By substituting $-s$ for s in J_2, it becomes J_3. Thus $J_2 = J_3$.

From (14.22) and (14.24), following necessary and sufficient conditions are obtained for the minimization of performance index:

$$J_2 + J_3 = 0$$

$$J_4 > 0$$

These conditions are satisfied if

$$\frac{1}{2\pi j}\int_{-j\infty}^{j\infty}\left[(T^*(s) - 1) + \lambda \frac{T^*(s)}{G(s)G(-s)}\right] R(s)R(-s)T_1(-s)\, ds = 0 \quad ...(14.28)$$

Due to the requirement of stability, all the poles of $T(s)$, $T^*(s)$ and $T_1(s)$ should be located in the left half of s-plane and corresponding the poles of $T(-s)$, $T^*(-s)$ and $T_1(-s)$ would then be located in the right half of s-plane. One solution of eqn. (14.28) is that all the poles of the integrand are in the right half of s-plane and the line integral is evaluated as a contour integral around the left half of s-plane. Since the contour does not enclose any singularity of integrand, the contour integration is zero.

Thus the condition given by eqn. (14.28) is satisfied if all the poles of the function

$$X = \left[(T^*(s) - 1) + \lambda \frac{T^*(s)}{G(s)G(-s)}\right] R(s)R(-s)$$

lie in the right half of s-plane.

The function X may be written as

$$X = \left[1 + \frac{\lambda}{G(s)G(-s)}\right] T^*(s)R(s)R(-s) - R(s)R(-s)$$

$$= A(s)A(-s)B(s)B(-s)T^*(s) - B(s)B(-s) \quad \text{...(14.29)}$$

where $A(s)$ and $B(s)$ are ratio polynomials with no poles or zeros in the right half s-plane and

$$A(s)A(-s) = \frac{\lambda}{G(s)G(-s)} + 1 \quad \text{...(14.30)}$$

$$B(s)B(-s) = R(s)R(-s) \quad \text{...(14.31)}$$

Dividing eqn. (14.29) by $A(-s)B(-s)$ we get

$$A(s)B(s)T^*(s) - \frac{B(s)}{A(-s)} = \frac{X}{A(-s)B(-s)} \quad \text{...(14.32)}$$

The function $B(s)/A(-s)$ can be expressed in the partial fraction expansion given below:

$$\frac{B(s)}{A(-s)} = C_+(s) + C_-(s) \quad \text{...(14.33)}$$

where $C_+(s)$ is composed of the part involving all the poles in the left half s-plane†.

Equation (14.32) may be written as

$$A(s)B(s)T^*(s) - C_+(s) = \frac{X}{A(-s)B(-s)} + C_-(s) \quad \text{...(14.34)}$$

Since all the poles of the expression on left hand side of eqn. (14.34) are in left half s-plane and all the poles of the expression in the right hand side are in right half s-plane, the two sides of eqn. (14.34) are therefore independent and must each be equal to zero. We obtain as its solution as

$$T^*(s) = \frac{C_+(s)}{A(s)B(s)}$$

Now $$A(s) = \frac{N(s)}{D^*(s)}$$

---

† In $B(s)/A(-s)$, poles in the left half of s-plane are contributed by $B(s)$ only which is given by eqn. (14.31). When $R(s)$ has poles on the $j\omega$-axis, they may be considered to lie on axis parallel to the $j\omega$-axis, shifted by $\pm\ \delta$. The final result is obtained by letting $\delta \to 0$. Exactly same result is obtained if without using δ, we split the $j\omega$-axis poles of $R(s)R(-s)$ equally between $B(s)$ and $B(-s)$ and treat the ones belonging to $B(s)$ as if they were in the left half of s-plane and the ones belonging to $B(-s)$ as if they were in the right half s-plane.

where $N(s)$ is given by eqn. (14.25) and $D^*(s)$ is given by eqn. (14.27).

$$\therefore \qquad T^*(s) = \frac{N(s)C_+(s)}{D^*(s)B(s)} \qquad ...(14.35)$$

Once the transfer function $T^*(s)$ of the overall system is obtained, the designer is free to choose a compensation scheme for the system.

Example 14.4: Consider the linear plant of a system characterized by the transfer function

$$G(s) = 100/s^2$$

The objective of the system is to make the output $c(t)$ follow a unit step input $r(t)$ minimizing

$$J = \int_0^\infty \{(r(t) - c(t))^2 + 0.25u^2(t)\}\, dt$$

where $u(t)$ is the actuating signal of the plant†.

From eqn. (14.25),

$$G(s) = \frac{N(s)}{D(s)} = \frac{100}{s^2}$$

Therefore, $\qquad N(s) = N(-s) = 100;\ D(s) = D(-s) = s^2$

From eqn. (14.27),

$$D^*(s)D^*(-s) = N(s)N(-s) + 0.25D(s)D(-s)$$
$$= 100 + 0.25s^4$$

Let $\qquad D^*(s) = a_0 + a_1 s + a_2 s^2$

Therefore

$$(a_0 + a_1 s + a_2 s^2)(a_0 - a_1 s + a_2 s^2) = 100 + 0.25s^4$$

Comparing the coefficients, we obtain

$$a_0 = 100,\ a_2 = 0.5 \text{ and } a_1 = 10$$

This gives $\qquad D^*(s) = 0.5s^2 + 10s + 100$

$$A(-s) = \frac{D^*(-s)}{N(-s)} = \frac{0.5s^2 - 10s + 100}{100}$$

Given $R(s) = 1/s$

From eqn. (14.31),

$$B(s) = \frac{1}{s}$$

† It may be pointed out here that the choice of the weighting factor λ is quite a problem in design. Its value depends on the linear range of the plant. Here we have arbitrarily taken $\lambda = 0.25$.

We next obtain the partial fraction expansion of $\frac{B(s)}{A(-s)}$:

$$\frac{B(s)}{A(-s)} = \frac{100}{s(0.5s^2 - 10s + 100)} = C_+(s) + C_-(s)$$

This gives $$C_+(s) = \frac{1}{s}$$

The optimal transfer function is (eqn. 14.35)

$$T^*(s) = \frac{100}{0.5s^2 + 10s + 100}$$

$$= \frac{200}{s^2 + 20s + 200} \qquad ...(14.36)$$

Design of compensators

There is a great deal of flexibility in the selection of compensators and hence the final system configuration. Some schemes for the problem under consideration are give below:

(*i*) If the configuration of the compensator is chosen as shown in Fig. 14.6, then the compensator $G_c(s)$ can be obtained from†

$$T^*(s) = \frac{G_c(s)G(s)}{1 + G_c(s)G(s)}$$

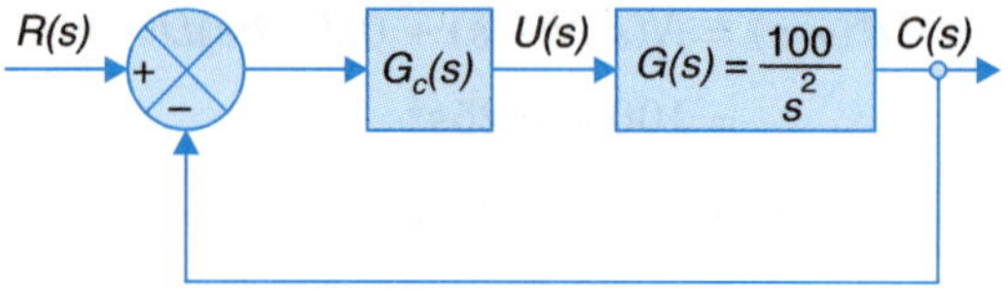

Fig. 14.6. An optimal control system for Example 14.4.

as $$G_c(s) = \frac{T^*(s)}{G(s)[1 - T^*(s)]} \qquad ...(14.37)$$

$$= \frac{20s}{s + 20}$$

(*ii*) Scheme of Fig. 14.7 employing both cascade and feedback compensation realizes the transfer function $T^*(s)$ given by eqn. (14.36).

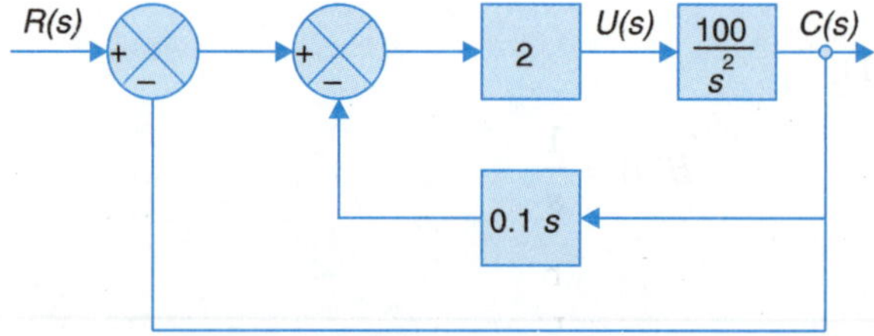

Fig. 14.7. An optimal control system for Example 14.4.

†It may be noted that compensator $G_c(s)$ should not cancel an unstable pole of $G(s)$. If the cancelled or missing pole is a stable one, it will not cause a serious effect on the response of the system.

(*iii*) Given $$G(s) = \frac{100}{s^2} = \frac{C(s)}{U(s)}$$

A state model is given by equations

$$\dot{x}_1 = x_2$$
$$\dot{x}_2 = 100u$$
$$c = x_1$$

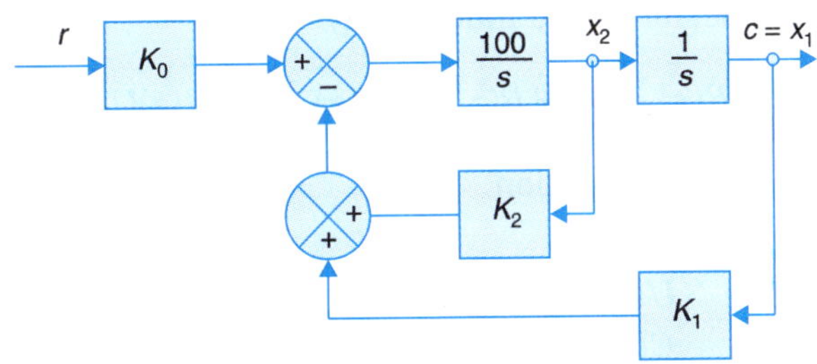

Fig. 14.8. An optimal control system for Example 14.4.

Assuming that the state variables x_1 and x_2 are available for feedback† the overall system may have a configuration shown in Fig. 14.8.

The overall transfer function obtained from Fig.14.8 is

$$T(s) = \frac{100K_0}{s^2 + 100K_2 s + 100K_1}$$

Comparing the coefficients of $T(s)$ with $T^*(s)$, we get

$$K_0 = 2, K_1 = 2 \text{ and } K_2 = 0.2$$

Servomechanism Problem: Discrete-Time Systems

For discrete-time systems, the optimal tracking problem can be formulated in the following way:

For a given plant with irreducible proper transfer function $G_p(s)$ and hold circuit with transfer function $G_0(s)$ (Fig. 14.9), find the transfer function $T^*(z)$ of the overall system that minimizes the quadratic performance index

$$J = \sum_{k=0}^{\infty} \{[r(kT) - c(kT)]^2 + \lambda u^2(kT)\}$$

where λ is a positive constant, $u(t)$ is the actuating signal, $c(t)$ is the actual output of the system and $r(t)$ is the desired output (reference input).

† When some of the states are inaccessible, then we may set the feedback coefficients $K_i = 0$ corresponding to inaccessible state variables x_i and try to adjust remaining coefficients to realize the given transfer function. If this is not possible, we may go for reconstruction of state variables. This has been discussed in Chapter 12.

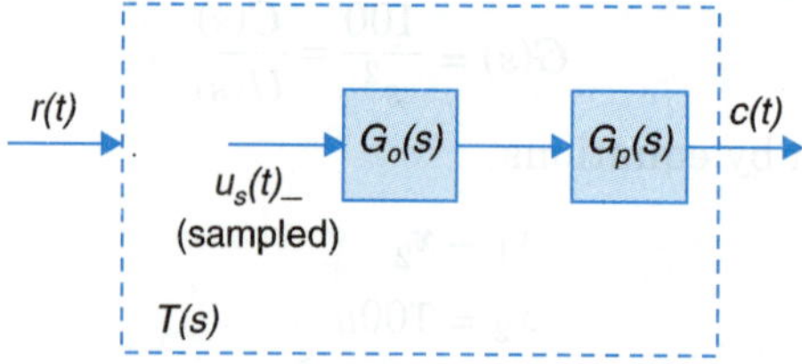

Fig. 14.9. An optimal control problem.

In the z-domain, the performance index is given as

$$J = \frac{1}{2\pi j} \oint_{\substack{\text{unit}\\ \text{circle}}} \left\{[R(z) - C(z)][R(z^{-1}) - C(z)^{-1}] + \lambda U(z)U(z^{-1})\right\} z^{-1} dz \quad ...(14.38)$$

For the system under consideration,

$$\frac{C(z)}{U(z)} = G(z) = \mathcal{Z}[G_0(s)G(s)]$$

$$\frac{C(z)}{R(z)} = T(z)$$

Therefore
$$U(z) = \frac{T(z)R(z)}{G(z)} \quad ...(14.39)$$

Equation (14.38) may be expressed as

$$J(T(z)) = \frac{1}{2\pi j} \oint_{\substack{\text{unit}\\ \text{circle}}} \left[(T(z) - 1)(T(z^{-1}) - 1) + \lambda \frac{T(z)T(z^{-1})}{G(z)G(z^{-1})}\right] R(z)R(z^{-1})z^{-1}\, dz \quad ...(14.40)$$

The procedure for derivation of the optimum form of $T(z)$ is similar to the procedure used for derivation of $T^*(s)$ earlier in this section. Instead of s and $-s$, we have z and z^{-1}. Instead of the imaginary axis, the left half s-plane and the right half s-plane, we have the unit circle, inside and outside of the unit circle respectively.

Following the steps of eqns. (14.21) to (14.28), we have the following result:

A necessary as well as sufficient condition for $T^*(z)$ to minimize J given by eqn. (14.40) is that the function

$$X = \left\{(T^*(z) - 1) + \lambda \frac{T^*(z)}{G(z)G(z^{-1})}\right\} R(z)R(z^{-1})z^{-1} \quad ...(14.41)$$

not have any pole inside the unit circle.

The function X may be written as

$$X = \left[1 + \frac{\lambda}{G(z)G(z^{-1})}\right] T^*(z)R(z)R(z^{-1})z^{-1} - R(z)R(z^{-1})z^{-1}$$

$$= A(z)A(z^{-1})B(z)B(z^{-1})z^{-1}\, T^*(z) - B(z)B(z^{-1})z^{-1} \quad ...(14.42)$$

where $A(z)$ and $B(z)$ are polynomials with no poles or zeros outside the unit circle ($A(z^{-1})$ and $B(z^{-1})$ have no poles and zeros inside the unit circle) and

$$A(z)A(z^{-1}) = 1 + \frac{\lambda}{G(z)G(z^{-1})} \quad ...(14.43)$$

$$B(z)B(z^{-1}) = R(z)R(z^{-1}) \quad ...(14.44)$$

Dividing eqn. (14.42) by $A(z^{-1})B(z^{-1})$, we get

$$A(z)B(z)T^*(z)z^{-1} - \frac{B(z)}{A(z^{-1})} = \frac{X}{A(z^{-1})B(z^{-1})} \qquad \text{...(14.45)}$$

The function $\dfrac{B(z)z^{-1}}{A(z^{-1})}$ can be expressed in the partial fraction form as follows:

$$\frac{B(z)z^{-1}}{A(z^{-1})} = C_+(z) + C_-(z) \qquad \text{...(14.46)}$$

where $C_+(z)$ is composed of the part with all the poles inside the unit circle.

Equation (14.45) may now be written as

$$A(z)B(z)z^{-1}T^*(z) - C_+(z) = \frac{X}{A(z^{-1})B(z^{-1})} + C_-(z)$$

The expression on the left hand side does not have any poles outside the unit circle, while the expression on the right hand side does not have any poles inside the unit circle; the two sides are thus independent and must therefore equal zero. We obtain the solution as

$$T^*(z) = \frac{zC_+(z)}{A(z)B(z)} \qquad \text{...(14.47)}$$

Example 14.5: For the discrete-time system shown in Fig. 14.9, given

$$G_p(s) = \frac{1}{s},\ G_0(s) = \frac{1-e^{-sT}}{s}, r(t) = \text{unit step},\ T = 1 \text{ sec}$$

find optimal transfer function $T^*(z)$ so that output $c(t)$ follows input $r(t)$ minimizing

$$J_e = \sum_{k=0}^{\infty} [r(kT) - c(kT)]^2 \qquad \text{...(14.48)}$$

with

$$J_u = \sum_{k=0}^{\infty} u^2(kT) = 0.5 \qquad \text{...(14.49)}$$

Solution: For the given system

$$G(z) = Z\left[\frac{1-e^{-sT}}{z} \cdot \frac{1}{s}\right] = (1-z^{-1})Z\left[\frac{1}{s^2}\right] = \frac{1}{z-1}$$

$$R(z) = \frac{z}{z-1}$$

The performance index for the system is (eqn. (14.14))

$$J = J_e + \lambda J_u$$

where positive constant λ will be determine from *control effort constraint* given by eqn. (14.49).

From eqn. (14.43),

$$A(z)A(z^{-1}) = 1 + \lambda(z-1)(z^{-1}-1)$$

$$= (1 + 2\lambda) - \lambda z - \lambda z^{-1}$$

Let $\qquad A(z) = a_0 + a_1 z$

Therefore

$$(a_0 + a_1 z)(a_0 + a_1 z^{-1}) = (1 + 2\lambda) - \lambda z - \lambda z^{-1}$$

Comparing the coefficients we obtain

$$a_0^2 + a_1^2 = 1 + 2\lambda$$

$$a_0 a_1 = -\lambda$$

Solving for a_0 and a_1, we get

$$a_0 = \frac{1}{2}(1 - \sqrt{1+4\lambda})$$

$$a_1 = \frac{1}{2}(1 + \sqrt{1+4\lambda})$$

Therefore

$$A(z) = \frac{1}{2}(1 - \sqrt{1+4\lambda}) + \frac{1}{2}(1 + \sqrt{1+4\lambda})z$$

From eqn. (14.44),

$$B(z)B(z^{-1}) = \left(\frac{z}{z-1}\right)\left(\frac{z^{-1}}{z^{-1}-1}\right) \quad \text{...(14.50)}$$

In order to avoid factoring difficulties, we can either introduce the factors $[z - (1 \pm \delta)]$ and let $\delta \to 0$ at the end, or split the poles on unit circle in two parts, half of them belonging to $B(z)$ will be considered to lie inside the unit circle and the remaining half belonging to $B(z^{-1})$ will be considered to lie outside the unit circle.

We can therefore write from eqn. (14.50),

$$B(z) = \frac{z}{z-1},\ B(z^{-1}) = \frac{1}{1-z}$$

We next obtain partial fraction expansion of $\dfrac{B(z)z^{-1}}{A(z^{-1})}$ (eqn. (14.46)).

$$\frac{B(z)z^{-1}}{A(z^{-1})} = \frac{2}{(z-1)[(1-\sqrt{1+4\lambda}) + (1+\sqrt{1+4\lambda})z^{-1}]}$$

$$= C_+(z) + C_-(z)$$

This gives

$$C_+(z) = \frac{1}{z-1}$$

The optimal transfer function is (eqn. (14.47))

$$T^*(z) = \frac{2}{(1-\sqrt{1+4\lambda}) + (1+\sqrt{1+4\lambda})z} \quad \text{...(14.51)}$$

From eqn. (14.39), we obtain the optimum actuating signal

$$U^*(z) = \frac{T^*(z)R(z)}{G(z)} = \frac{2z}{(1-\sqrt{1+4\lambda}) + (1+\sqrt{1+4\lambda})z}$$

The control effort constraint (14.49) may now be expressed in the z-domain.

$$\frac{1}{2\pi j}\oint_{\substack{\text{unit}\\\text{circle}}} *(z)U^*(z^{-1})z^{-1}\,dz = 0.5$$

Using Table 14.2, we get the following equation:

$$\frac{1}{\sqrt{1+4\lambda}} = 0.5$$

This gives $\lambda = 0.75$

From eqn. (14.51), the optimal transfer function is

$$T^*(z) = \frac{2}{3z-1}$$

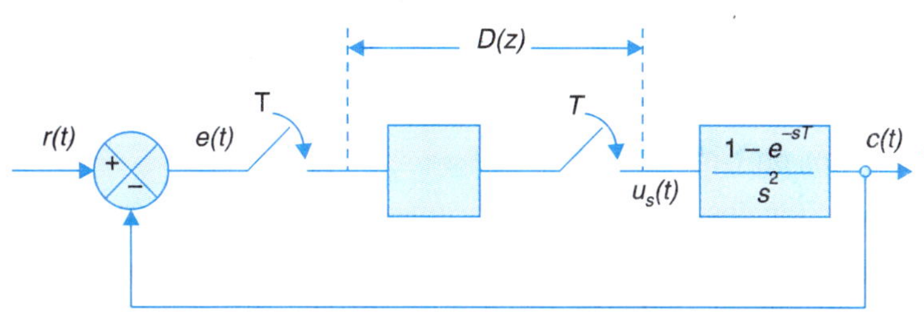

Fig. 14.10. An optimal control system for Example 14.5.

If the configuration of the compensator is chosen as shown in Fig. 14.10, then $D(z)$, the transfer function of data processing unit is obtained from

$$T^*(z) = \frac{D(z)G(z)}{1+D(z)G(z)}$$

as

$$D(z) = \frac{T^*(z)}{G(z)[1-T^*(z)]} = \frac{2}{3} \qquad ...(14.52)$$

Thus the optimum controller in this case is just a gain of 2/3.

Output Regulator Problem

We now introduce another important class of optimal control problems: *the output regulator problem*. It is a special case of the tracking problem in which $r(t) = 0$. For zero input, the output is zero if all the intial conditions are zero. The response $c(t)$ is due to nonzero initial conditions that, in turn, are cuased by disturbances. The primary objective of the design is to damp out the response due to initial conditions quickly without excessive overshoot and oscillations.

Consider for example the system of Problem 3.8 (Fig. P-3.8). The disturbance torque of the sea causes the ship to roll. The response (roll angle $\theta(t)$) to this disturbance is highly oscillatory. The oscillations in the rolling motion are to be damped out quickly without excessive overshoot. If there is no constraint on 'control effort', the controller which minimizes the performance index

$$J = \int_0^\infty (\theta(t) - \theta_d(t))^2\,dt$$

will be the optimum one; θ_d is the desired roll which is clearly zero. Therefore the problem of stabilization of ship against rolling motion is a regulator problem. If, for disturbance torque applied at $t = t_0$, the controller is required to regulate the roll motion within finite time $(t_f - t_0)$, a suitable performance criterion for design of optimum controller is to minimize

$$J = \int_{t_0}^{t_f} \theta^2(t)\, dt \qquad \text{...(14.53)}$$

Limitations Of Transfer Function Approach

Since the responses of regulating systems are due to nonzero initial conditions, the transfer functions do not give useful formulation of such systems.

For the servomechanism or tracking problem in single-input-single-output linear time-invariant systems, the transfer function approach is straight-forward. However, the limitation of this approach for multi-input-multi-output linear time-invariant systems is obvious. The spectral factorization becomes quite complex because of involvement of matrix operations. In addition, the transfer function approach is restricted to systems with quadratic performance index with no constraint on time.

The transfer function design technique becomes ineffective for time-varying and nonlinear systems.

All these limitations of transfer function approach are absent in the *state variable approach*. With the availability of digital computer, the state variable approach has become the most powerful method for solving modern complex control problems. However, the transfer function approach is quite simple for the class of systems for which it is applicable.

14.4 OPTIMAL CONTROL PROBLEMS: STATE VARIABLE APPROACH

Following steps are involved in the solution of an optimal control problem using state variable approach:

(*i*) Given a plant in the form of state equations.

$$\dot{\mathbf{x}}(t) = \mathbf{A}\mathbf{x}(t) + \mathbf{B}\mathbf{u}(t)\text{: Continuous-time systems} \qquad \text{...(14.54 } a\text{)}$$

$$\mathbf{x}(k+1) = \mathbf{A}\mathbf{x}(t) + \mathbf{B}\mathbf{u}(k)\text{: Discrete-time systems} \qquad \text{...(14.54 } b\text{)}$$

where $\mathbf{x} = n \times 1$ state vector; $\mathbf{u} = m \times 1$ control vector; $\mathbf{A} = n \times n$ constant matrix; and $\mathbf{B} = n \times m$ constant matrix, find the control function $\mathbf{u}^*$ which is optimal with respect to given performance criterion.

(*ii*) Realize the control function obtained from step (*i*).

The Stage Regulator Problem

We recall from earlier section that when a system variable $x_1(t)$ (the output) is required to be 'near' zero, the performance measure is (eqn. (14.53)),

$$J = \int_{t_0}^{t_f} {x_1}^2(t)\, dt$$

A peformance index written in terms of two state variables of a system would then be

$$J = \int_{t_0}^{t_f} (x_1^2(t) + x_2^2(t))\, dt$$

Therefore if the state $\mathbf{x}(t)$ of a system described by (14.54 a) is required to be close to $\mathbf{x}_d = \mathbf{0}$, a design criterion would be to determine a control function that minimizes

$$J = \int_{t_0}^{t_f} (\mathbf{x}^T\mathbf{x})\, dt$$

In practical systems, the control of all the states of the system is not equally important. For example, if in addition to roll angle $\theta(t)$ of a ship (Fig. P-3.8), the pitch angle $\phi(t)$ is also required to be zero, the performance index of eqn. (14.53) gets modified to

$$J = \int_{t_0}^{t_f} (\theta^2(t) + \lambda\phi^2(t))\, dt$$

where λ, a positive constant, is the weighting factor. The roll motion contributes much discomfort to the passengers; in the design of passenger ship, the value of λ will be less than one.

A weighted quadratic performance index is

$$J = \int_{t_0}^{t_f} (\mathbf{x}^T\mathbf{Q}\mathbf{x})\, dt$$

where $\mathbf{Q}$ is positive definite, real, symmetric, constant matrix. The simplest form of $\mathbf{Q}$ one can use is a diagonal matrix:

$$\mathbf{Q} = \begin{bmatrix} q_1 & 0 & \cdots & 0 \\ 0 & q_2 & \cdots & 0 \\ \vdots & \vdots & \cdots & \vdots \\ 0 & 0 & \cdots & q_n \end{bmatrix}$$

The ith entry of $\mathbf{Q}$ represents the amount of weight the designer places on the constraint on state variable $x_i(t)$. The larger the value of q_i relative to other values of q, the more control effort is spent to regulate $x_i(t)$.

To minimize the deviation of the final state $\mathbf{x}(t_f)$ of the system from the desired state $\mathbf{x}_d = \mathbf{0}$, a possible performance measure is

$$J = \mathbf{x}^T(t_f)\mathbf{H}\mathbf{x}(t_f)$$

where $\mathbf{H}$ is positive definite, real, symmetric, constant matrix. In the infinite time state regulator problem ($t_f \to \infty$), the final state should approach the equilibrium state $\mathbf{x} = \mathbf{0}$; so the terminal constraint is no longer necessary. The optimal design obtained by minimizing

$$J = \mathbf{x}^T(t_f)\mathbf{H}\mathbf{x}(t_f) + \int_{t_0}^{t_f} \mathbf{x}^T(t)\mathbf{Q}\mathbf{x}(t)\, dt$$

may be unsatisfactory in practice. A more realistic solution to the problem is reached if the performance index is modified by adding a penalty term for physical constraints. One of the ways of accomplishing this is to introduce the following quadratic control term† in the performance index:

$$J = \int_{t_0}^{t_f} \mathbf{u}^T(t)\mathbf{R}\mathbf{u}(t)\, dt$$

† Recall the statement made earlier in this chapter–if the performance index is quadratic, the resulting optimal system is a linear system.

where **R** is a positive definite, real, symmetric, constant matrix. By giving sufficient weight to control terms, the amplitude of controls which minimize the overall performance index may be kept within practical bounds, although at the expense of increased error in $\mathbf{x}(t)$.

Continuous-time systems

We now formulate the *state regulator problem* in continuous-time systems:

Consider the controlled process described by the state equations (14.54 *a*). Find optimal control law $\mathbf{u}^*(t)$, $t \in [t_0, t_f]$, where t_0 and t_f are specified initial and final times respectively, so that the quadratic performance index†

$$J = \frac{1}{2}\mathbf{x}^T(t_f)\mathbf{H}\mathbf{x}(t_f) + \frac{1}{2}\int_{t_0}^{t_f} (\mathbf{x}^T(t)\mathbf{Q}\mathbf{x}(t) + \mathbf{u}^T(t)\mathbf{R}\mathbf{u}(t))\, dt \qquad ...(14.55)$$

is minimized, subject to an initial state

$$\mathbf{x}(t_0) = \mathbf{x}_0$$

If the terminal time t_f is not constrained, the performance index is††

$$J = \frac{1}{2}\int_{t_0}^{\infty} (\mathbf{x}^T(t)\mathbf{Q}\mathbf{x}(t) + \mathbf{u}^T(t)\mathbf{R}\mathbf{u}(t))\, dt \qquad ...(14.56)$$

For this infinite-time state regulator problem, solution exists only if (14.54 *a*) represents a completely controllable system (proved in next section).

In the performance measures given by (14.55) and (14.56), **R** is restricted to be positive definite. This requirement is, as we shall see, a condition for existence of a finite control.

The matrices **Q** and **H** may be positive definite or semidefinite. However, if both of them are simultaneously zero matrices, *i.e.*, $\mathbf{Q} = \mathbf{H} = \mathbf{0}$, then the performance index becomes

$$J = \frac{1}{2}\int_{t_0}^{t_f} \mathbf{u}^T(t)\mathbf{R}\mathbf{u}(t)\, dt$$

which is minimum when $\mathbf{u}(t) = \mathbf{0}$. To exclude this trivial case, we shall assume that **Q** and **H** are not both zero matrices, although we shall allow either **Q** and **H** to be the zero matrix individually.

Discrete-time systems

The state regulator problem in discrete-time systems is formulated as follows:

Consider the controlled process described by state equations (14.56 *b*). Find the optimal control sequence $\mathbf{u}^*(k)$, $k = 0, 1, ..., N-1$, where N is a specified fixed positive integer number, so that the quadratic performance index

$$J = \frac{1}{2}\mathbf{x}^T(N)\mathbf{H}\mathbf{x}(N) + \frac{1}{2}\sum_{k=0}^{N-1} [\mathbf{x}^T(k)\mathbf{Q}\mathbf{x}(k) + \mathbf{u}^T(k)\mathbf{R}\mathbf{u}(k)] \qquad ...(14.57)$$

is minimized, subject to an initial state

$$\mathbf{x}(0) = \mathbf{x}_0$$

† Note that multiplication by a constant 1/2, does not affect the minimization problem. The constant helps in mathematical manipulations as we shall see later.

†† t_0 may be taken as zero in linear time-invariant systems.

Note that eqns. (14.54 *b*) and (14.57) may be the result of a discrete approximation to a continuous-time state regulator problem or the formulation for the state regulator problem in a sampled-data system.

If a process is to be controlled for a large number of stages ($N \to \infty$), then a performance index is

$$J = \frac{1}{2} \sum_{k=0}^{\infty} [\mathbf{x}^T(k)\mathbf{Q}\mathbf{x}(k) + \mathbf{u}^T(k)\mathbf{R}\mathbf{u}(k)] \qquad \text{...(14.58)}$$

For this infinite-time problem, solution exists only if (14.54 *b*) represents a completely controllable system (proved in next section).

In eqns. (14.57) and (14.58), **R** is positive definite, **Q** and **H** are positive definite (or semidefinite). However both **Q** and **H** can not be zero matrices simultaneously.

The Output Regulator Problem

In the state regulator problem, we are concerned with making all the components of the state vector $\mathbf{x}(t)$ small. In the output regulator problem on the other hand, we are concerned with making the components of the output vector small. If the controlled process is observable then, as we shall see, we can reduce the output regulator problem to the state regulator problem.

The output regulator problem is formulated as follows:

Consider an observable controlled process described by the equations

$$\dot{\mathbf{x}}(t) = \mathbf{A}\mathbf{x}(t) + \mathbf{B}\mathbf{u}(t)$$
$$\mathbf{y}(t) = \mathbf{C}\mathbf{x}(t) \qquad \text{...(14.59)}$$

where $\mathbf{x}(t) = n \times 1$ state vector; $\mathbf{u}(t) = m \times 1$ control vector; $\mathbf{y}(t) = p \times 1$ output vector; $\mathbf{A} = n \times n$ constant matrix; $\mathbf{B} = n \times m$ constant matrix; and $\mathbf{C} = p \times n$ constant matrix.

Find optimal control law $\mathbf{u}^*(t)$, $t \in [t_0, t_f]$, where t_0 and t_f are specified initial and final times respectively, so that the quadratic performance index

$$J = \frac{1}{2}\mathbf{y}^T(t_f)\mathbf{H}\mathbf{y}(t_f) + \frac{1}{2}\int_{t_0}^{t_f} (\mathbf{y}^T(t)\mathbf{Q}\mathbf{y}(t) + \mathbf{u}^T(t)\mathbf{R}\mathbf{u}(t))\, dt \qquad \text{...(14.60)}$$

is minimized, subject to an initial state

$$\mathbf{x}(t_0) = \mathbf{x}_0$$

The output regulator problem in discrete-time systems can be formulated on similar lines.

The Tracking Problem

Recall that we discussed in detail the physical aspects of the tracking problem in the earlier section and there we presented its formulation. In the state variable approach, this problem is formulated as follows:

Consider an observable controlled process described by eqn. (14.59). Suppose that the vector $\mathbf{z}(t)$ is the desired output; we assume that the dimension of $\mathbf{z}(t)$ is equal to that of $\mathbf{y}(t)$. Define the error vector $\mathbf{e}(t)$ by setting

$$\mathbf{e}(t) = \mathbf{z}(t) - \mathbf{y}(t)$$

Find optimal control law $\mathbf{u}^*(t)$, $t \in [t_0, t_f]$ where t_0 and t_f are specified initial and final times respectively, so that the quadratic performance index

$$J = \frac{1}{2}\mathbf{e}^T(t_f)\mathbf{H}\mathbf{e}(t_f) + \frac{1}{2}\int_{t_0}^{t_f} (\mathbf{e}^T(t)\mathbf{Q}\mathbf{e}(t) + \mathbf{u}^T(t)\mathbf{R}\mathbf{u}(t))\, dt \qquad ...(14.61)$$

is minimized.

The tracking problem in descrete-time systems can be formulated on the similar lines.

Approaches for minimization of Performance Index

Once the performance index for a system has been chosen, the next task is to determine a control function that minimizes this index. Two approaches for accomplishing this minimization are available:

(*i*) *Minimum Principle of Pontryagin*: This is based on the concepts of calculus of variations.

(*ii*) *Dynamic Programming Developed by Bellman*: This is based on the *principle of invariant imbedding* and leads to the *principle of optimality* that follows the basic laws of nature and does not need complex mathematical development to explain its validity.

For the linear regulator problem, minimization using ordinary differential calculus has been achieved. However, we shall use dynamic programming approach in our discussion. This will help the reader to directly study advanced literature on optimal control theory.

In the following sections, we shall solve various optimal control problems. We shall use the results of Appendix II very often. Many details of the proofs will be omitted. For these details, the reader may refer advanced literature on the subject.

Optimal control problems can be solved with the help of digital computer. However, we shall illustrate the optimal control theory by simple examples which can be solved analytically.

14.5 THE STATE REGULATOR PROBLEM

Discrete-Time Systems

The optimization method based on dynamic programming views the control problem as a multistage decision problem and the control input as a time sequence of decisions. A sampled-data system gives rise to a sequence of transformations of the original state vector; the decision process therefore consists of selecting the transformations at each discrete interval of time.

Consider a process described by state equations (14.54 *b*) with performance index (14.57). This process is called a multistage decision process of N stages where the choice of $\mathbf{u}(k)$ at each sampling instant is considered the decision of interest. If we select from the myriad of possible choices, the choice of decisions $\mathbf{u}(0)$, $\mathbf{u}(1)$, ..., $\mathbf{u}(N-1)$ which minimizes the given performance index J, we have selected the *optimal policy or sequence*.

For a given sampled-data system described by (14.54 *b*), the optimal control sequence is a function of the initial state $\mathbf{x}(0)$ and the number of stages N. We take the decision at the stage corresponding to $k = 0$, to determine optimal path not merely from the given state $\mathbf{x}(0)$ but from all possible states which the system can possess. In general, at the ith stage, we determine optimal path not merely from the state $\mathbf{x}(i)$ which is a function of the initial state

$\mathbf{x}(0)$ and the controls $\mathbf{u}(0), \mathbf{u}(1), \ldots, \mathbf{u}(i-1)$, but from all possible states of the ith stage. Thus we do not regard the control problem as an isolated problem with fixed value of $\mathbf{x}(0)$ and N but rather imbed it within a family of problems. This is the *principle of invariant imbedding* introduced by Bellman.

Another important feature of dynamic programming formulation is that we do not attempt to find all the values of optimal control sequence simultaneously; we find one control value at a time until the entire optimal policy is determined. This sequential decision process is based on the *principle of optimality* which states that an optimal control policy has the property that whatever the initial state and initial decisions are, the remaining decisions must form an optimal policy with respect to the state resulting from the initial decisions. The principle of optimality reduces the N-stage decision process into N single-stage decision processes.

It is permissible to find the last decision as the initial calculation step in multistage decision process, *i.e.,* the calculations of optimal decisions may proceed from the last decision back to the first decision. The system state at the Nth stage is a function of the state at $(N-1)$th stage and the control $\mathbf{u}(N-1)$:

$$\mathbf{x}(N) = \mathbf{f}(\mathbf{x}(N-1), \mathbf{u}(N-1))$$

As per the principle of optimality, regardless of how the state reaches $\mathbf{x}(N-1)$, once $\mathbf{x}(N-1)$ is known then, using $\mathbf{x}(N-1)$ as the initial state for the last stage of state transition, the control $\mathbf{u}(N-1)$ must be so chosen that (eqn. (14.57))

$$J_{N-1,N}(\mathbf{x}(N-1), \mathbf{u}(N-1)) = \frac{1}{2}\mathbf{x}^T(N)\mathbf{H}\mathbf{x}(N) + \frac{1}{2}[\mathbf{x}^T(N-1)\mathbf{Q}\mathbf{x}(N-1) + \mathbf{u}^T(N-1)\mathbf{R}\mathbf{u}(N-1)] \quad \ldots(14.62)$$

is a minimum.

Note that $J_{N-1,N}$ is the performance index of operation during the interval $(N-1)\,\Delta t \le t \le N\,\Delta t$ where Δt is the sampling interval. $J_{N-1,N}$ is also performance index of the one-state process with initial state $\mathbf{x}(N-1)$. Thus optimal policy and minimum performance index of one-stage process are imbedded in the results for the N-stage process.

Equation (14.62) may be written as

$$J_{N-1,N}(\mathbf{x}(N-1), \mathbf{u}(N-1)) = \frac{1}{2}[\mathbf{A}\mathbf{x}(N-1) + \mathbf{B}\mathbf{u}(N-1)]^T\mathbf{P}(0)[\mathbf{A}\mathbf{x}(N-1) + \mathbf{B}\mathbf{u}(N-1)] + \frac{1}{2}[\mathbf{x}^T(N-1)\,\mathbf{Q}\mathbf{x}(N-1) + \mathbf{u}^T(N-1)\mathbf{R}\mathbf{u}(N-1)] \quad \ldots(14.63)$$

where $\mathbf{P}(0) = \mathbf{H}$.

The necessary condition for minimum $J_{N-1,N}$ is

$$\frac{\partial J_{N-1,N}}{\partial \mathbf{u}(N-1)} = 0$$

This gives

$$\mathbf{R}\mathbf{u}^*(N-1) + \mathbf{B}^T\mathbf{P}(0)[\mathbf{A}\mathbf{x}(N-1) + \mathbf{B}\mathbf{u}^*(N-1)] = \mathbf{0}$$

or
$$\mathbf{u}^*(N-1) = -[\mathbf{R} + \mathbf{B}^T\mathbf{P}(0)\mathbf{B}]^{-1}\mathbf{B}^T\mathbf{P}(0)\mathbf{A}\mathbf{x}(N-1)$$
$$= \mathbf{K}(N-1)\mathbf{x}(N-1) \qquad ...(14.64)$$
where
$$\mathbf{K}(N-1) = -[\mathbf{R} + \mathbf{B}^T\mathbf{P}(0)\mathbf{B}]^{-1}\mathbf{B}^T\mathbf{P}(0)\mathbf{A}$$

This control satisfies the sufficient condition for minimum $J_{N-1,N}$ since

$$\frac{\partial^2 J_{N-1,N}}{\partial \mathbf{u}^2(N-1)} = \mathbf{R} + \mathbf{B}^T\mathbf{P}(0)\mathbf{B}$$

is positive definite. This is because **R** is a positive definite matrix and $\mathbf{B}^T\mathbf{P}(0)\,\mathbf{B}$ is positive definite (or semidefinite as $\mathbf{P}(0) = \mathbf{H}$ is positive definite or semidefinite).

The minimum value of performance index obtained from eqns. (14.63) and (14.64) is

$$J^*_{N-1,N}(\mathbf{x}(N-1)) = \frac{1}{2}\mathbf{x}^T(N-1)\left\{[\mathbf{A}+\mathbf{B}\mathbf{K}(N-1)]^T\mathbf{P}(0)\right.$$
$$\left.[\mathbf{A}+\mathbf{B}\mathbf{K}(N-1)] + \mathbf{K}^T(N-1)\mathbf{R}\mathbf{K}(N-1) + \mathbf{Q}\right\}\mathbf{x}(N-1)$$
$$= \frac{1}{2}\mathbf{x}^T(N-1)\mathbf{P}(1)\mathbf{x}(N-1) \qquad ...(14.65)$$
where
$$\mathbf{P}(1) = [\mathbf{A}+\mathbf{B}\mathbf{K}(N-1)]^T\mathbf{P}(0)[\mathbf{A}+\mathbf{B}\mathbf{K}(N-1)] + \mathbf{K}^T(N-1)\mathbf{R}\mathbf{K}(N-1) + \mathbf{Q}$$

Imbedding a two-stage process in the N-stage process and using the principle of optimality, we get the following result:

Regardless of how the state reaches $\mathbf{x}(N-2)$, once $\mathbf{x}(N-2)$ is known then, using $\mathbf{x}(N-2)$, as the initial state for the operation in last two stages, the control sequence $\mathbf{u}(N-2)$, $\mathbf{u}(N-1)$ should be so chosen that

$$J_{N-2,N}(\mathbf{x}(N-2), \mathbf{u}(N-2), \mathbf{u}(N-1)) = \frac{1}{2}\mathbf{x}^T(N)\mathbf{P}(0)\mathbf{x}(N)$$
$$+\frac{1}{2}\sum_{k=N-2}^{N-1}[\mathbf{x}^T(k)\mathbf{Q}\mathbf{x}(k) + \mathbf{u}^T(k)\mathbf{R}\mathbf{u}(k)]$$
$$= \frac{1}{2}\mathbf{x}^T(N-2)\mathbf{Q}\mathbf{x}(N-2) + \frac{1}{2}\mathbf{u}^T(N-2)\mathbf{R}\mathbf{u}(N-2) + J_{N-1,N}$$

is a minimum.

For a two-stage process, whatever the initial state $\mathbf{x}(N-2)$ and initial control $\mathbf{u}(N-2)$, the remaining control $\mathbf{u}(N-1)$ must be optimal with respect to the value of $\mathbf{x}(N-1)$ that results from application of $\mathbf{u}(N-2)$. Therefore the performance index for the two-stage process is (using eqns. (14.54 *b*), (14.65))

$$J_{N-2,N}(\mathbf{x}(N-2), \mathbf{u}(N-2))$$
$$= \frac{1}{2}\mathbf{x}^T(N-2)\mathbf{Q}\mathbf{x}(N-2) + \frac{1}{2}\mathbf{u}^T(N-2)\mathbf{R}\mathbf{u}(N-2) + J^*_{N-1,N}$$

$$= \frac{1}{2}\mathbf{x}^T(N-2)\mathbf{Q}\mathbf{x}(N-2) + \frac{1}{2}\mathbf{u}^T(N-2)\,\mathbf{R}\mathbf{u}(N-2)$$

$$+\frac{1}{2}[\mathbf{A}\mathbf{x}(N-2)+\mathbf{B}\mathbf{u}(N-2)]^T\,\mathbf{P}(1)[\mathbf{A}\mathbf{x}(N-2)+\mathbf{B}\mathbf{u}(N-2)]$$

The minimization procedure (as applied earlier to one-stage process) gives the following result:

$$\mathbf{u}^*(N-2) = -\,[\mathbf{R} + \mathbf{B}^T\mathbf{P}(1)\mathbf{B}]^{-1}\mathbf{B}^T\mathbf{P}(1)\mathbf{A}\mathbf{x}(N-2)$$

$$= \mathbf{K}\,(N-2)\mathbf{x}(N-2)$$

$$J^*(\mathbf{x}(N-2)) = \frac{1}{2}\mathbf{x}^T(N-2)\Big\{[\mathbf{A}+\mathbf{B}\mathbf{K}(N-2)]^T\,\mathbf{P}(1)[\mathbf{A}+\mathbf{B}\mathbf{K}(N-2)]$$

$$+\,\mathbf{K}^T(N-2)\mathbf{R}\mathbf{K}(N-2)+\mathbf{Q}\Big\}\,\mathbf{x}(N-2)$$

$$= \frac{1}{2}\mathbf{x}^T(N-2)\mathbf{P}(2)\mathbf{x}(N-2)$$

By induction, we can write the complete solution for the N-stage process as

$$\mathbf{u}^*(N-j) = \mathbf{K}(N-j)\mathbf{x}(N-j);\; j = 1, 2, ..., N \qquad ...(14.66\ a)$$

$$\mathbf{K}(N-j) = -\,[\mathbf{R} + \mathbf{B}^T\mathbf{P}(j-1)\mathbf{B}]^{-1}\mathbf{B}^T\mathbf{P}(j-1)\mathbf{A} \qquad ...(14.66\ b)$$

$$\mathbf{P}(0) = \mathbf{H} \qquad ...(14.66\ c)$$

$$\mathbf{P}(j) = [\mathbf{A} + \mathbf{B}\mathbf{K}(N-j)]^T\mathbf{P}(j-1)[\mathbf{A} + \mathbf{B}\mathbf{K}(N-j)] \qquad ...(14.66\ d)$$

$$+\,\mathbf{K}^T(N-j)\mathbf{R}\mathbf{K}(N-j) + \mathbf{Q}$$

$$J^*_{N-j,N} = \frac{1}{2}\mathbf{x}^T(N-j)\mathbf{P}(j)\mathbf{x}(N-j) \qquad ...(14.66\ e)$$

Equations (14.66) from the recursive relations for determination of **K** and **P** matrices. Starting with $j = 1$, $\mathbf{K}(N-1)$ is evaluated from eqn. (14.66 *b*) with $\mathbf{P}(0) = \mathbf{H}$, a given matrix. Equation (14.66 *d*) is then solved for $\mathbf{P}(1)$. This constitutes one cycle of procedure, which we then continue by calculating $\mathbf{K}(N-2)$, $\mathbf{P}(2)$ and so on. The matrices $[\mathbf{K}(N-1), \mathbf{K}(N-2), ..., \mathbf{K}(0)]$ are stored. The optimal controller is then realized by determining the appropriate gain settings (eqn. (14.66 (*a*))) as the system transits from stage to stage.

For the N-stage process with specified initial state $\mathbf{x}(0) = \mathbf{x}_0$, the minimum value of performance index is (eqn. (14.66 *e*)

$$J^*_{0,N}(\mathbf{x}_0) = \frac{1}{2}\mathbf{x}_0{}^T\mathbf{P}(N)\mathbf{x}_0 \qquad ...(14.67)$$

The following important points are observed in this result:

(*i*) The optimal control at each stage is a linear combination of the states; thus giving linear state variable feedback control policy.

(*ii*) Feedback is time-varying, although **A, B, R** and **Q** are all constant matrices, *i.e.*, optimal control policy converts a linear time-invariant plant with time-invariant quadratic performance index into a linear time-varying feedback system.

(*iii*) Following the procedure given above, results for optimal control of linear time-varying plant can easily be obtained.

Realization of optimal control policy

Once the optimal control policy has been determined, its realization is the second phase of the optimal control problem.

Since the optimal control is

$$\mathbf{u}^*(k) = \mathbf{K}(k)\mathbf{x}(k)$$

its realization seeks feedback of the state variables. If the plant states are available for measurement, $\mathbf{u}^*(k)$ is implemented by feedback of the state variables through time-varying gain elements; the engineering construction of time-varying functions is done by means of a digital computer.

If the plant states are not available for measurement, then it is possible to construct a physical device—*state observer*, which produces at its output the plant states, when driven by both the plant input and output (discussed in Chapter 12). This is however possible only if the plant equations satisfy the conditions of observability.

It may be noted that it is not necessary that the system (14.54 *b*) be controllable. Assume that some states of the system are uncontrollable. We expect that this will impose problems in the regulator system if the uncontrollable states are unstable because these unstable states will be reflected in the performance index. However the contribution of the unstable uncontrollable states (unstable controllable states are stabilizable) to the performance index is always finite provided that the control interval $[t_0, t_f]$ is finite. We shall see in the next section that we shall require controllability as $t_f \to \infty$ to ensure that the value of performance index is finite.

Example 14.6: The second-order linear system

$$\dot{x}_1(t) = -\,0.2\,x_1(t) + x_2(t) + u(t) \qquad \text{...(14.68)}$$

$$\dot{x}_2(t) = -\,0.5\,x_2(t) + 0.5\,u(t)$$

is to be controlled to minimize the performance index

$$J = \frac{1}{2}\int_0^5 (x_1^2(t) + x_2^2(t) + u^2(t))\,dt \qquad \text{...(14.69)}$$

Let us determine the optimal control policy for this system using the results given by (14.66).

For determination of the optimal control policy using the results (14.66), the system differential equations must be approximated by difference equations and integral in performance index must be approximated by a summation. To do this, divide the time interval $0 \le t \le t_f$ into N equal increments of time duration T. Then from eqn. (14.68),

$$\frac{x_1(t+T) - x_1(t)}{T} \simeq -\,0.2x_1(t) + x_2(t) + u(t)$$

or

$$x_1(k+1) = (1 - 0.2T)x_1(k) + Tx_2(k) + Tu(k)$$

Assuming $T = 1$ sec (this gives $N = 5$), we get

$$x_1(k+1) = 0.8x_1(k) + x_2(k) + u(k) \qquad \text{...(14.70 } a\text{)}$$

Similarly we obtain

$$x_2(k+1) = 0.5x_2(k) + 0.5u(k) \qquad \text{...(14.70 } b\text{)}$$

The performance index given by (14.69) may be approximated to a summation as follows:

$$J = \frac{1}{2}\left\{\int_0^T (x_1^2(t) + x_2^2(t) + u^2(t))\,dt + \int_T^{2T} (x_1^2(t) + x_2^2(t) + u^2(t))\,dt + \ldots + \int_{(N-1)T}^{NT} (x_1^2(t) + x_2^2(t) + u^2(t))\,dt\right\}$$

$$\simeq \frac{1}{2}\left\{[x_1^2(0) + x_2^2(0) + u^2(0)]T + [x_1^2(1) + x_2^2(1) + u^2(1)]\,T + \ldots + [x_1^2(N-1) + x_2^2(N-1) + u^2(N-1)]T\right\}$$

$$= \frac{1}{2}\sum_{k=0}^{4} (x_1^2(k) + x_2^2(k) + u^2(k)) \qquad \ldots(14.71)$$

Comparing eqns. (14.70) and (14.71) with (14.54 (*b*)) and (14.57) respectively, we obtain

$$\mathbf{A} = \begin{bmatrix} 0.8 & 1 \\ 0 & 0.5 \end{bmatrix}, \mathbf{B} = \begin{bmatrix} 1 \\ 0.5 \end{bmatrix}$$

$$\mathbf{H} = \mathbf{O}, \mathbf{Q} = \begin{bmatrix} 1 & 0 \\ 0 & 1 \end{bmatrix}, \mathbf{R} = 1$$

The following results† are obtained from (14.66):

$$\mathbf{K}(4) = [0 \quad 0]$$

$$\mathbf{K}(3) = -[0.355 \quad 0.555]$$

$$\mathbf{K}(2) = -[0.395 \quad 0.677]$$

$$\mathbf{K}(1) = -[0.395 \quad 0.687]$$

$$\mathbf{K}(0) = -[0.395 \quad 0.687]$$

The block diagram of the optimal system is shown in Fig. 14.11.

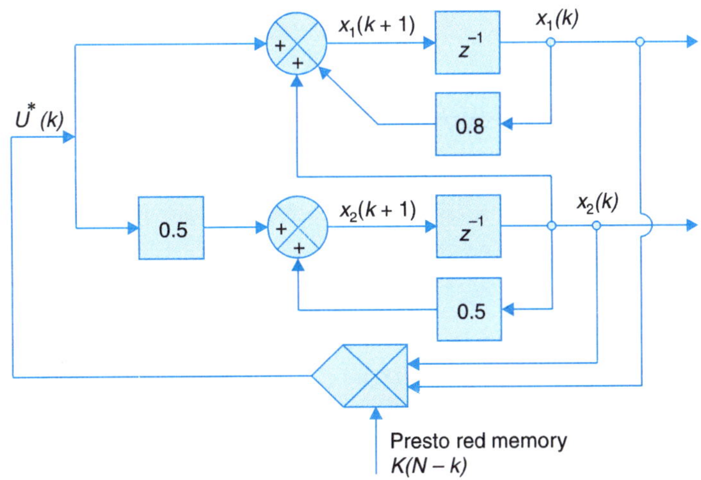

Fig. 14.11 Block diagram of optimal control system for Example 14.6.

† The required calculations may best be carried out using a digital computer.

Continuous-Time Systems

In Example 14.6, we approximated a continuous-time system by a discrete-time system. This approach leads to recursive relations (14.66), ideally suited for digital computer solution. In this section, we consider an alternative approach that leads to nonlinear differential equations.

Consider a process described by the state equations (14.54 a). We are required to find the optimal control law $\mathbf{u}^*(t)$ so that quadratic performance index given by (14.55) is minimized.

Let us use the principle of imbedding to include this problem in a larger class of problems by considering the performance measure

$$J(\mathbf{x}(t)\, t, \underset{t \le \tau \le t_f}{\mathbf{u}}(\tau)) = \frac{1}{2}\mathbf{x}^T(t_j)\mathbf{H}\mathbf{x}(t_f) + \frac{1}{2}\int_t^{t_f}(\mathbf{x}^T(\theta)\mathbf{Q}\mathbf{x}(\theta)$$

$$+ \mathbf{u}^T(\theta)\,\mathbf{R}\mathbf{u}(\theta))\,d\theta;\, t \le t_f$$

The optimal control $\mathbf{u}^*(t)$ is to be found by minimizing J with respect to admissible controls while treating the other variables as constants. Then the optimal value of the performance index is

$$J^*(\mathbf{x}(t)t, \underset{t \le \tau \le t_f}{\mathbf{u}^*}(\tau)) = J^*(\mathbf{x}(t), t)$$

$$= \min_{\substack{\mathbf{u}(\tau) \\ t \le \tau \le t_f}} \left\{\frac{1}{2}\int_t^{t_f}(\mathbf{x}^T(\theta)\mathbf{Q}\mathbf{x}(\theta) + \mathbf{u}^T(\theta)\mathbf{R}\mathbf{u}(\theta))\,d\theta + \frac{1}{2}\mathbf{x}^T(t_f)\,\mathbf{H}\mathbf{x}(t_f)\right\} \quad ...(14.72)$$

*is the minimum value of the performance index for the time interval $\le \tau \le t_f$ with initial state $\mathbf{x}(t)$.

By subdividing the interval, we obtain

$$J^*(\mathbf{x}(t), t) = \min_{\substack{\mathbf{u}(\tau) \\ t \le \tau \le t_f}} \left\{\frac{1}{2}\int_t^{t+\Delta t}(\mathbf{x}^T(\theta)\mathbf{Q}\mathbf{x}(\theta) + \mathbf{u}^T(\theta)\mathbf{R}\mathbf{u}(\theta))d\theta\right.$$

$$\left. + \frac{1}{2}\int_{t+\Delta t}^{t_f}(\mathbf{x}^T(\theta)\mathbf{Q}\mathbf{x}(\theta) + \mathbf{u}^T(\theta)\mathbf{R}\mathbf{u}(\theta))d\theta + \frac{1}{2}\mathbf{x}^T(t_f)\mathbf{H}\mathbf{x}(t_f)\right\}$$

The principle of optimality requires that

$$J^*(\mathbf{x}(t), t) = \min_{\substack{\mathbf{u}(\tau) \\ t \le \tau \le t+\Delta t}} \left\{\frac{1}{2}\int_t^{t+\Delta t}(\mathbf{x}^T(\theta)\mathbf{Q}\mathbf{x}(\theta) + \mathbf{u}^T(\theta)\mathbf{R}\mathbf{u}(\theta))d\theta + J^*(\mathbf{x}(t+\Delta t)\,.\, t+\Delta t)\right\}$$

where $J^*(\mathbf{x}(t + \Delta t), t + \Delta t)$ is the minimum value of the performance index for the process with initial state $\mathbf{x}(t + \Delta t)$.

Expanding $J^*(\mathbf{x}(t + \Delta t), t + \Delta t)$ in Taylor's series about $(\mathbf{x}(t), t)$ and neglecting terms of order higher than first, we get (see Appendix II)

$$J^*(\mathbf{x}(t), t) = \min_{\substack{\mathbf{u}(\tau) \\ t \le \tau \le t+\Delta t}} \left\{\frac{1}{2}\int_t^{t+\Delta t}(\mathbf{x}^T(\theta)\mathbf{Q}\mathbf{x}(\theta) + \mathbf{u}^T(\theta)\mathbf{R}\mathbf{u}(\theta))\,d\theta + J^*(\mathbf{x}(t), t)\right.$$

$$\left. + \left[\frac{\partial J^*(\mathbf{x}(t), t)}{\partial t}\right]\Delta t + \left[\frac{\partial J^*(\mathbf{x}(t), t)}{\partial \mathbf{x}}\right]^T[\mathbf{x}(t+\Delta t) - \mathbf{x}(t)]\right\}$$

$$= J^*(\mathbf{x}(t), t) + \left[\frac{\partial J^*(\mathbf{x}(t), t)}{\partial t}\right]\Delta t$$

$$+ \min_{\mathbf{u}(t)} \left\{\frac{1}{2}(\mathbf{x}^T(t)\mathbf{Q}\mathbf{x}(t) + \mathbf{u}^T(t)\mathbf{R}\mathbf{u}(t))\Delta t + \left[\frac{\partial J^*(\mathbf{x}(t), t)}{\partial \mathbf{x}}\right]^T [\mathbf{A}\mathbf{x}(t) + \mathbf{B}\mathbf{u}(t)]\Delta t\right\}$$

(Note that $\dot{\mathbf{x}}(t) = \dfrac{\mathbf{x}(t+\Delta t) - \mathbf{x}(t)}{\Delta t}$.)

or

$$-\frac{\partial J^*(\mathbf{x}(t), t)}{\partial t}$$

$$= \min_{\mathbf{u}(t)} \left\{\frac{1}{2}(\mathbf{x}^T(t)\mathbf{Q}\mathbf{x}(t) + \mathbf{u}^T(t)\mathbf{R}\mathbf{u}(t)) + \left[\frac{\partial J^*(\mathbf{x}(t), t)}{\partial \mathbf{x}}\right]^T [\mathbf{A}\mathbf{x}(t) + \mathbf{B}\mathbf{u}(t)]\right\} \quad ...(14.73)$$

To carry out the minimization process, we differentiate both sides of above equation with respect to $\mathbf{u}(t)$. This gives

$$\mathbf{0} = \mathbf{R}\mathbf{u}^*(t) + \mathbf{B}^T\frac{\partial J^*}{\partial \mathbf{x}}(\mathbf{x}(t), t)$$

Solving for $\mathbf{u}^*(t)$, we get

$$\mathbf{u}^*(t) = -\mathbf{R}^{-1}\mathbf{B}^T\frac{\partial J^*(\mathbf{x}(t), t)}{\partial \mathbf{x}} \quad ...(14.74)$$

The second derivative of $-\dfrac{\partial J^*(\mathbf{x}(t), t)}{\partial t}$ with respect to $\mathbf{u}(t)$ gives a positive definite matrix $\mathbf{R}$. Therefore $\mathbf{u}^*(t)$ given by eqn. (14.74) satisfies the necessary and sufficient conditions for minimum.

Substituting eqn. (14.74) into eqn. (14.73), we get

$$-\frac{\partial J^*(\mathbf{x}(t), t)}{\partial t} = \frac{1}{2}\mathbf{x}^T(t)\mathbf{Q}\mathbf{x}(t) + \frac{1}{2}\left(\frac{\partial J^*(\mathbf{x}(t), t)}{\partial x}\right)^T \mathbf{B}\mathbf{R}^{-1}\mathbf{B}^T\frac{\partial J^*(\mathbf{x}(t), t)}{\partial \mathbf{x}}$$

$$+ \left(\frac{\partial J^*(\mathbf{x}(t), t)}{\partial \mathbf{x}}\right)^T \mathbf{A}\mathbf{x} - \left(\frac{\partial J^*(\mathbf{x}(t), t)}{\partial \mathbf{x}}\right)^T \mathbf{B}\mathbf{R}^{-1}\mathbf{B}^T\frac{\partial J^*(\mathbf{x}(t), t)}{\partial \mathbf{x}}$$

$$= \frac{1}{2}\mathbf{x}^T(t)\mathbf{Q}\mathbf{x}(t) - \frac{1}{2}\left(\frac{\partial J^*(\mathbf{x}(t), t)}{\partial \mathbf{x}}\right)^T \mathbf{B}\mathbf{R}^{-1}\mathbf{B}^T\frac{\partial J^*(\mathbf{x}(t), t)}{\partial \mathbf{x}} + \left(\frac{\partial J^*(\mathbf{x}(t), t)}{\partial \mathbf{x}}\right)^T \mathbf{A}\mathbf{x} \quad ...(14.75)$$

From eqn. (14.72), we get the boundary condition:

$$J^*(\mathbf{x}(t_f), t_f) = \frac{1}{2}\mathbf{x}^T(t_f)\mathbf{H}\mathbf{x}(t_f) \quad ...(14.76)$$

Earlier in this section we found that in the linear regulator problem for discrete-time systems, the minimum value of the performance index is a time-varying quadratic function of state (eqn. (14.88 (*f*))). Therefore, it is reasonable to assume

$$J^*(\mathbf{x}(t), t) = \frac{1}{2}\mathbf{x}^T(t)\mathbf{P}(t)\mathbf{x}(t) \quad ...(14.77)$$

where $\mathbf{P}(t)$ is a real symmetric positive definite matrix for $t < t_f$ and

$$\mathbf{P}(t_f) = \mathbf{H} \quad ...(14.78)$$

Substituting the assumed solution in eqn. (14.75), we get

$$-\frac{1}{2}\mathbf{x}^T(t)\frac{d\mathbf{P}(t)}{dt}\mathbf{x}(t) = \frac{1}{2}\mathbf{x}^T(t)\mathbf{Q}\mathbf{x}(t) - \frac{1}{2}\mathbf{x}^T(t)\,\mathbf{P}(t)\mathbf{B}\mathbf{R}^{-1}\mathbf{B}^T\mathbf{P}(t)\mathbf{x}(t) + \mathbf{x}^T(t)\mathbf{P}(t)\mathbf{A}\mathbf{x}(t)$$

or $$2\mathbf{x}^T(t)\mathbf{P}(t)\mathbf{A}\mathbf{x}(t) - \mathbf{x}^T(t)\mathbf{P}(t)\mathbf{B}\mathbf{R}^{-1}\mathbf{B}^T\mathbf{P}(t)\mathbf{x}(t) + \mathbf{x}^T(t)\mathbf{Q}\mathbf{x}(t) + \mathbf{x}^T(t)\dot{\mathbf{P}}(t)\mathbf{x}(t) = 0$$

or $$\mathbf{x}^T(t)[2\mathbf{P}(t)\mathbf{A} - \mathbf{P}(t)\mathbf{B}\mathbf{R}^{-1}\mathbf{B}^T(t) + \mathbf{Q} + \dot{\mathbf{P}}(t)]\mathbf{x}(t) = \mathbf{0} \qquad ...(14.79)$$

The only way this equation can be satisfied for an arbitrary $\mathbf{x}(t)$ is that the quantity inside the brackets be equal to zero. However, we know (Appendix II) that in the scalar function $\mathbf{z}^T\mathbf{W}\mathbf{z}$, only the symmetric part of the matrix $\mathbf{W}$,

$$\mathbf{W}_S = \frac{\mathbf{W} + \mathbf{W}^T}{2}$$

is of importance.

Examination of eqn. (14.79) reveals that all the terms within brackets are already symmetric except the first term.

$$\text{Symmetric part of } [2\mathbf{P}(t)\mathbf{A}] = 2\,\frac{\mathbf{P}(t)\mathbf{A} + \mathbf{A}^T\mathbf{P}(t)}{2}$$
$$= \mathbf{P}(t)\mathbf{A} + \mathbf{A}^T\mathbf{P}(t)$$

Therefore in order for eqn. (14.79) to be satisfied, it is necessary that the differential equation

$$\mathbf{P}(t) + \mathbf{Q} - \mathbf{P}(t)\mathbf{B}\mathbf{R}^{-1}\mathbf{B}^T\mathbf{P}(t) + \mathbf{P}(t)\mathbf{A} + \mathbf{A}^T\mathbf{P}(t) = \mathbf{0} \qquad ...(14.80)$$

is satisfied subject to the boundary condition (14.78).

Equation (14.80) is a matrix differential equation of the Riccati type and is thus referred to as *the Riccati equation*. The Riccati equation is nonlinear and for this reason, we usually cannot obtain closed-form solutions; therefore we must compute $\mathbf{P}(t)$ using a digital computer.

Equation (14.80) is a set of n^2 nonlinear equations. Numerical integration is carried out backwards; from $t = t_f$ to $t = t_0$ with the boundary condition $\mathbf{P}(t_f) = \mathbf{H}$. Actually since the $n \times n$ matrix $\mathbf{P}(t)$ is symmetric, we need to integrate only $\frac{n(n+1)}{2}$ differential equations.

Once $\mathbf{P}(t)$ is known for $t_0 \le t \le t_f$, *the optimal control law* is obtained from eqns. (14.74) and (14.77):

$$\mathbf{u}^*(t) = -\,\mathbf{R}^{-1}\mathbf{B}^T\mathbf{P}(t)\mathbf{x}(t)$$
$$= \mathbf{K}(t)\mathbf{x}(t) \qquad ...(14.81)$$

where $$\mathbf{K} = -\,\mathbf{R}^{-1}\mathbf{B}^T\mathbf{P}(t)$$

Example 14.7 : A first-order system is described by the differential equation

$$\dot{x}(t) = 2x(t) + u(t)$$

It is desired to find the control law that minimizes the performance index

$$J = \frac{1}{2}\int_0^{t_f}\left(3x^2 + \frac{1}{4}u^2\right)dt$$

$$t_f = 1 \text{ sec}$$

For this problem, the matrices **A, B, H, Q** and **R** reduce to scalars and are given by

$$\mathbf{A} = 2, \mathbf{B} = 1, \mathbf{H} = 0, \mathbf{Q} = 3, \mathbf{R} = \frac{1}{4}$$

In addition, the matrix $\mathbf{P}(t)$ also reduce to a scalar function of time $p(t)$.

The matrix Riccati equation then becomes the scalar differential equation

$$\dot{p}(t) + 3 - 4p^2(t) + 4p(t) = 0$$

with the boundary condition $p(t_f) = 0$.

The solution $p(t)$ can be obtained by separation of variables.

$$\int_{t_f}^{t} \frac{dp}{4\left(p - \frac{3}{2}\right)\left(p + \frac{1}{2}\right)} = \int_{t_f}^{t} dt$$

or

$$\frac{1}{8}\left\{\ln\left[\frac{p(t) - \frac{3}{2}}{p(t_f) - \frac{3}{2}}\right] - \ln\left[\frac{p(t) + \frac{1}{2}}{p(t_f) + \frac{1}{2}}\right]\right\} = t - t_f$$

This gives

$$p(t) = \frac{\frac{3}{2}(1 - e^8(t - t_f))}{1 + 3e^8(t - t_f)} \qquad \text{...(14.82)}$$

The optimal control law is (eqn. (14.81)),

$$\mathbf{u}^*(t) = -4p(t)x(t) \qquad \text{...(14.83)}$$

The block diagram of the optimal system is shown in Fig. 14.12.

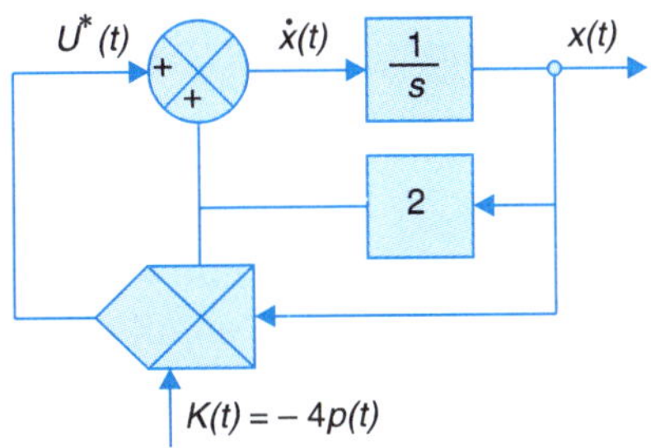

Fig. 14.12. Block diagram of optimal control system of Example 14.7.

14.6 THE INFINITE-TIME REGULATOR PROBLEM

Continuous-Time Systems

The infinite-time state regulator problem may be stated as follows:

Consider the controlled process (eqn. (14.54 (a)))

$$\dot{\mathbf{x}}(t) = \mathbf{A}\mathbf{x}(t) + \mathbf{B}\mathbf{u}(t) \qquad \text{...(14.84)}$$

Find the control law $\mathbf{u}^*(t)$, so that quadratic performance index (eqn. (14.56))

$$J = \frac{1}{2}\int_0^{\infty} (\mathbf{x}^T(t)\mathbf{Q}\mathbf{x}(t) + \mathbf{u}^T(t)\mathbf{R}\mathbf{u}(t))\, dt; \qquad ...(14.85)$$

Q and **R** are positive definite matrices, is minimized, subject to the initial condition

$$\mathbf{x}(0) = \mathbf{x}_0$$

Let us discuss the salient points of the infinite-time state regulator problem.

(*i*) For many applications, the final time t_f has no special significance and there exist no obvious values of t_f which should be specified in the performance index. In such cases, we are satisfied to let $t_f \to \infty$, as is done in eqn. (14.85).

(*ii*) When $t_f \to \infty$, $\mathbf{x}(\infty) \to \mathbf{0}$ for the optimal system to be stable. Therefore the terminal penalty term (in eqn. (14.55)) has no significance; consequently it does not appear in eqn. (14.85). *i.e.*, we have set $\mathbf{H} = \mathbf{0}$ in the general quadratic performance index (14.55).

(*iii*) In the finite-time regulator problem, there is no restriction on the controllability of the plant. This is because J is always finite and instability does not impose any problems in finite-interval control. This may not be so in the infinite-interval case. J can become infinite if

(*a*) one or more states of the plant are uncontrollable,

(*b*) the uncontrollable states are unstable, and

(*c*) the unstable states are reflected in system performance index.

The performance index J would be infinite for all controls; we can not, therefore, distinguish the optimal control from other controls. J will be finite (*i.e.*, the solution to the infinite-time regulator problem will exist) if the states that are not asymptotically stable, are controllable (note that controllable unstable states of the plant can be stabilized by the feedback controller) or broadly, we may say that the controlled process given by eqn. (14.84) should satisfy the conditions of controllability.

(*iv*) Consider the matrix Riccati equation (eqn. (14.80)),

$$\dot{\mathbf{P}}(t) + \mathbf{Q} - \mathbf{P}(t)\,\mathbf{B}\mathbf{R}^{-1}\mathbf{B}^T\mathbf{P}(t) + \mathbf{P}(t)\mathbf{A} + \mathbf{A}^T\mathbf{P}(t) = \mathbf{0}$$

with the boundary condition $\mathbf{P}(t_f) = \mathbf{0}$.

We solve the Riccati equation backward in time with t_f as the starting time and $\mathbf{P}(t_f)$ as the initial condition. At time $t = t_f - \epsilon$, where ϵ is a small positive number, the transient due to initial condition will dominate the solution $\mathbf{P}(t)$; the transient will die out for large values of ϵ resulting in steady-state solution of Riccati equation for some interval of time. This time interval increases as t_f becomes large. As $t_f \to \infty$, the time interval of steady-state solution of Riccati equation grows without bound, *i.e.*, $\mathbf{P}(t) \to$ a constant $\mathbf{P}^0$ for all finite time t.

Consider the system of Example 14.7. For terminal time t_f; the solution of the Riccati equation is (eqn (14.82))

$$p(t) = \frac{\frac{3}{2}(1 - e^8(t - t_f))}{1 + 3e^8(t - t_f)}$$

$p(t)$ is plotted in Fig. 14.13 for t_f = 1 sec and 10 sec.

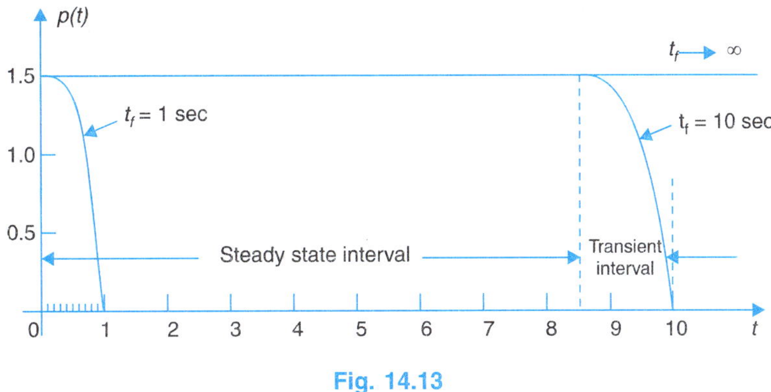

Fig. 14.13

The solution of Riccati equation with infinite terminal time is

$$p^0 = \lim_{t_f \to \infty} p(t) = 1.5$$

The optimal control law is then (eqn. (14.83))

$$u^*(t) = -\,6x(t)$$

The practical implication of this result is obvious.

The optimal law is implemented using time-invariant gain elements in contrast to the finite-time case wherein gain elements must be time-varying to implement the optimal control given by eqn. (14.83).

In general, if the states of the system (14.84) that are not asymptotically stable are controllable, then $\lim_{t_f \to \infty} \mathbf{P}(t)$ tends to a unique positive definite constant matrix $\mathbf{P}^0$ which is the solution of the equation (eqn. (14.80) with $\dot{\mathbf{P}}(t) = \mathbf{0}$)

$$\mathbf{A}^T\mathbf{P}^0 + \mathbf{P}^0\mathbf{A} - \mathbf{P}^0\mathbf{B}\mathbf{R}^{-1}\mathbf{B}^T\mathbf{P}^0 + \mathbf{Q} = \mathbf{0}$$

called the *reduced matrix Riccati equation.*

The solution of the reduced matrix Riccati equation may not be unique. The desired unique answer is obtained by enforcing the requirement that $\mathbf{P}^0$ be positive definite.

Another point to be noted is that computation of $\mathbf{P}^0$ is often convenient by numerical integration of Riccati equation (14.80) for arbitrary large values of t_f till $\mathbf{P}(t)$ converages to a constant matrix $\mathbf{P}^0$.

The optimal control law in the infinite-time regulator problem is (eqn. 14.81))

$$\mathbf{u}^*(t) = -\,\mathbf{R}^{-1}\mathbf{B}^T\mathbf{P}^0\mathbf{x}(t)$$
$$= \mathbf{K}\mathbf{x}(t)$$

where $\mathbf{K} = -\,\mathbf{R}^{-1}\mathbf{B}^T\mathbf{P}^0$

The resulting optimal system is thus a linear time-invariant system.

(v) If a constant positive definite matrix $\mathbf{P}^0$ exists which is the solution of the reduced matrix Riccati equation ($\mathbf{P}^0$ is bound to exist if the plant (14.84) is controllable), the optimal closed-loop system is asymptotically stable if the matrix $\mathbf{Q}$ is positive definite. If $\mathbf{Q}$ is positive semi-definite, the asymptotic stability of the closed-loop regulator is not guaranteed and we should test the closed-loop system for stability.

Practical implications of this stability result are very significant. We guarantee that the optimal system is bound to be stable irrespective of the stability of the open-loop plant. Recall that in the techniques of classical control, the primary role is to achieve system stability; the optimality of performance occupies a secondary role in the design procedure.

(*vi*) For the infinite-time regulator problem, the optimal value of the performance index is obtained as follows:

From eqn. (14.77), we have

$$J^*(\mathbf{x}(t), t) = \frac{1}{2}\mathbf{x}^T(t)\mathbf{P}(t)\mathbf{x}(t)$$

For the infinite-time case, we may write

$$\begin{aligned} J^* &= \int_0^\infty -\frac{1}{2}\frac{d}{dt}(\mathbf{x}^T(t)\,\mathbf{P}^0\mathbf{x}(t))\,dt \\ &= -\frac{1}{2}\mathbf{x}^T(t)\,\mathbf{P}^0\mathbf{x}(t)\Big|_0^\infty \\ &= \frac{1}{2}\mathbf{x}^T(0)\mathbf{P}^0\mathbf{x}(0) \end{aligned}$$

(Note that $\underset{t\to\infty}{\mathbf{x}(t)} = \mathbf{0}$ as per stability result obtained earlier.)

Results of this section are summarised below:

Given the controllable linear time-invariant plant (14.84) and the performance index (14.85), then a unique optimal control exists and is given by

$$\mathbf{u}^*(t) = \mathbf{Kx};\ \mathbf{K} = -\mathbf{R}^{-1}\mathbf{B}^T\mathbf{P}^0 \qquad \text{...(14.86 (a))}$$

where $\mathbf{P}^0$ is a constant positive definite matrix which is the solution of the reduced matrix Riccati equation

$$\mathbf{A}^T\mathbf{P}^0 + \mathbf{P}^0\mathbf{A} - \mathbf{P}^0\mathbf{B}\mathbf{R}^{-1}\mathbf{B}^T\mathbf{P}^0 + \mathbf{Q} = \mathbf{0} \qquad \text{...(14.86 (b))}$$

The minimum value of performance index is

$$J^* = \frac{1}{2}\mathbf{x}_0{}^T\mathbf{P}^0\mathbf{x}_0 \qquad \text{...(14.87)}$$

Example 14.8 : Let us obtain the control law which minimizes the performance index

$$J = \int_0^\infty (x_1^2 + u^2)\,dt$$

for the system

$$\begin{bmatrix}\dot{x}_1\\ \dot{x}_2\end{bmatrix} = \begin{bmatrix}0 & 1\\ 0 & 0\end{bmatrix}\begin{bmatrix}x_1\\ x_2\end{bmatrix} + \begin{bmatrix}0\\ 1\end{bmatrix}u$$

We have for this problem

$$\mathbf{A} = \begin{bmatrix}0 & 1\\ 0 & 0\end{bmatrix},\ \mathbf{B} = \begin{bmatrix}0\\ 1\end{bmatrix},\ \mathbf{Q} = \begin{bmatrix}2 & 0\\ 0 & 0\end{bmatrix},\ \mathbf{R} = [2]$$

The reduced matrix Riccati equation is given as

$$\begin{bmatrix}0 & 0\\ 1 & 0\end{bmatrix}\begin{bmatrix}p_{11} & p_{12}\\ p_{12} & p_{22}\end{bmatrix} + \begin{bmatrix}p_{11} & p_{12}\\ p_{12} & p_{22}\end{bmatrix}\begin{bmatrix}0 & 1\\ 0 & 0\end{bmatrix}$$

$$-\begin{bmatrix} p_{11} & p_{12} \\ p_{12} & p_{22} \end{bmatrix}\begin{bmatrix} 0 \\ 1 \end{bmatrix}\begin{bmatrix} \frac{1}{2} \end{bmatrix}[0 \quad 1]\begin{bmatrix} p_{11} & p_{12} \\ p_{12} & p_{22} \end{bmatrix}+\begin{bmatrix} 2 & 0 \\ 0 & 0 \end{bmatrix}$$

$$= \begin{bmatrix} 0 & 0 \\ 0 & 0 \end{bmatrix}$$

(Note that we have utilized the fact that the $\mathbf{P}^0$ is symmetric.)

Upon simplification we get,

$$\frac{-p_{12}^2}{2} + 2 = 0$$

$$p_{11} - \frac{p_{12}p_{22}}{2} = 0$$

$$\frac{-p_{22}^2}{2} + 2p_{12} = 0$$

The solution of these equations yields the positive definite matrix

$$\mathbf{P}^0 = \begin{bmatrix} 2\sqrt{2} & 2 \\ 2 & 2\sqrt{2} \end{bmatrix}$$

From eqn. (14.86 (*a*)), the optimal control law is given by

$$u^*(t) = -\mathbf{R}^{-1}\mathbf{B}^T\mathbf{P}^0\mathbf{x}(t)$$

$$= -\begin{bmatrix} \frac{1}{2} \end{bmatrix}[0 \quad 1]\begin{bmatrix} 2\sqrt{2} & 2 \\ 2 & 2\sqrt{2} \end{bmatrix}\begin{bmatrix} x_1(t) \\ x_2(t) \end{bmatrix}$$

$$= -x_1(t) - \sqrt{2}\,x_2(t)$$

It can easily be verfied that the closed-loop system is asymptotically stable.

Discrete-Time Systems

Given the Controllable sampled-data system

$$\mathbf{x}(k+1) = \mathbf{A}\mathbf{x}(k) + \mathbf{B}\mathbf{u}(k)$$

Find the control sequence $\mathbf{u}^*(k)$ so that quadratic performance index

$$J = \frac{1}{2}\sum_{k=0}^{\infty}[\mathbf{x}^T(k)\mathbf{Q}\mathbf{x}(k) + \mathbf{u}^T(k)\mathbf{R}\mathbf{u}(k)]$$

is minimized, subject to the initial condition

$$\mathbf{x}(0) = \mathbf{x}_0$$

Results for this infinite-stage discrete-time regulator problem directly follows those of the continuous-time problem. The matrices **P** and **K** in the recursive relations (14.66) are time-invariant when $N \to \infty$. Therefore

$$\mathbf{u}^*(k) = \mathbf{K}\mathbf{x}(k) \quad \text{...(14.88 (a))}$$

$$\mathbf{K} = -[\mathbf{R} + \mathbf{B}^T\mathbf{P}\mathbf{B}]^{-1}\mathbf{B}^T\mathbf{P}\mathbf{A} \quad \text{...(14.88 (b))}$$

$$\mathbf{P} = [\mathbf{A} + \mathbf{B}\mathbf{K}]^T\mathbf{P}[\mathbf{A} + \mathbf{B}\mathbf{K}] + \mathbf{K}^T\mathbf{R}\mathbf{K} + \mathbf{Q} \quad \text{...(14.88 (c))}$$

The minimum value of the performance index is

$$J^* = \frac{1}{2}\mathbf{x}_0^T\mathbf{P}\mathbf{x}_0 \qquad ...(14.88\ (d))$$

The optimal control sequence for the infinite-stage process may be obtained from the algebraic equations given above. An alternative way is to solve the recursive relations (14.66) using digital computer for as many stages as required for $\mathbf{P}(N-k)$ to coverage to a constant matrix.

14.7 THE OUTPUT REGULATOR AND THE TRACKING PROBLEMS

In this section we shall discuss briefly the output regulator problem and the tracking problem (tracking problem has been discussed in Section 14.3 using the transfer function approach). Under certain restrictions, both these problems are reducible to the form of the state regulator problem.

The Output Regulator Problem

The output regulator problem may be stated as follows:

Consider the controlled process

$$\begin{aligned}\dot{\mathbf{x}}(t) &= \mathbf{A}\mathbf{x}(t) + \mathbf{B}\mathbf{u}(t);\ \mathbf{x}(0) = \mathbf{x}_0 \\ \mathbf{y}(t) &= \mathbf{C}\mathbf{x}(t)\end{aligned} \qquad ...(14.89)$$

where $\mathbf{x}(t) = n \times 1$ state vector; $\mathbf{u}(t) = m \times 1$ control vector; $\mathbf{y}(t) = p \times 1$ output vector; $\mathbf{A} = n \times n$ constant matrix; $\mathbf{B} = n \times m$ constant matrix; and $\mathbf{C} = p \times n$ constant matrix.

Find the control law $\mathbf{u}^*(t)$ so that the following quadratic performance index is minimized,

$$J = \frac{1}{2}\mathbf{y}^T(t_f)\mathbf{H}\mathbf{y}(t_f) + \frac{1}{2}\int_0^{t_f}(\mathbf{y}^T(t)\mathbf{Q}\mathbf{y}(t) + \mathbf{u}^T(t)\mathbf{R}\mathbf{u}(t))\,dt \qquad ...(14.90)$$

where $\mathbf{R}$ is a positive definite constant symmetric matrix, $\mathbf{H}$ and $\mathbf{Q}$ are positive definite (or semidefinite) constant, symmetric matrices with the restriction that both of them are not zero matrices at a time.

From the performance index given by (14.90) it is obvious that in the output regulator problem we desire to bring and keep output $\mathbf{y}(t)$ near zero without using an excessive amount of control energy.

Substituting $\mathbf{y}(t) = \mathbf{C}\mathbf{x}(t)$ in eqn. (14.90), we get

$$J = \frac{1}{2}\mathbf{x}^T(t_f)\mathbf{C}^T\mathbf{H}\mathbf{C}\mathbf{x}(t_f) + \frac{1}{2}\int_0^{t_f}(\mathbf{x}^T(t)\mathbf{C}^T\mathbf{Q}\mathbf{C}\mathbf{x}(t) + \mathbf{u}^T(t)\mathbf{R}\mathbf{u}(t))\,dt \qquad ...(14.91)$$

Comparing eqn. (14.91) with eqn. (14.55), we observe that the two indices are identical in form; $\mathbf{H}$ and $\mathbf{Q}$ in eqn. (14.55) are replaced by $\mathbf{C}^T\mathbf{H}\mathbf{C}$ and $\mathbf{C}^T\mathbf{Q}\mathbf{C}$ respectively to get eqn. (14.91). If we assume that the system (14.89) is observable, then $\mathbf{C}$ can not be zero. Therefore $\mathbf{C}^T\mathbf{H}\mathbf{C}$ and $\mathbf{C}^T\mathbf{Q}\mathbf{C}$ will be positive definite (or semidefinite) matrices wherever $\mathbf{H}$ and $\mathbf{Q}$ are positive definite (or semidefinite).

We have the following solution for the output regulator problem which directly follows from the results of Section 14.5.

For the observable system (14.89) with the performance index (14.90), a unique optimal control exists and is given by

$$\mathbf{u}^*(t) = -\mathbf{R}^{-1}\mathbf{B}^T\mathbf{P}(t)\mathbf{x}(t) = \mathbf{K}(t)\mathbf{x}(t) \qquad ...(14.92)$$

where the symmetric positive definite matrix $\mathbf{P}(t)$ is the solution of the matrix Riccati equation

$$\dot{\mathbf{P}}(t) + \mathbf{C}^T\mathbf{Q}\mathbf{C} - \mathbf{P}(t)\mathbf{B}\mathbf{R}^{-1}\mathbf{B}^T\mathbf{P}(t) + \mathbf{P}(t)\mathbf{A} + \mathbf{A}^T\mathbf{P}(t) = \mathbf{0} \qquad ...(14.93)$$

with the boundary condition

$$\mathbf{P}(t_f) = \mathbf{C}^T\mathbf{H}\mathbf{C}$$

The Tracking Problem

Here we shall study a class of tracking problems which are reducible to the form of the output regulator problem.

Consider an observable process described by eqns. (14.89). It is desired to bring and keep output $\mathbf{y}(t)$ close to the desired output $\mathbf{r}(t)$. We define an error vector

$$\mathbf{e}(t) = \mathbf{y}(t) - \mathbf{r}(t)$$

The design objective is to find the control law $\mathbf{u}^*(t)$ so that the performance index

$$J = \frac{1}{2}\mathbf{e}^T(t_f)\mathbf{H}\mathbf{e}(t_f) + \frac{1}{2}\int_0^{t_f}(\mathbf{e}^T(t)\mathbf{Q}\mathbf{e}(t) + \mathbf{u}^T(t)\mathbf{R}\mathbf{u}(t))\,dt \qquad ...(14.94)$$

is minimized.

To reduce this problem to the form of the output regulator problem, we consider only those $\mathbf{r}(t)$ that can be generated by arbitrary initial conditions $\mathbf{z}(0)$ in the system

$$\begin{aligned}\dot{\mathbf{z}}(t) &= \mathbf{A}\mathbf{z}(t)\\ \mathbf{r}(t) &= \mathbf{C}\mathbf{z}(t)\end{aligned} \qquad ...(14.95)$$

The matrices $\mathbf{A}$ and $\mathbf{C}$ are same as those of the plant (14.89).

We now define a new variable

$$\mathbf{w} = \mathbf{x} - \mathbf{z}$$

Then

$$\begin{aligned}\dot{\mathbf{w}} &= \mathbf{A}\mathbf{w} + \mathbf{B}\mathbf{u}\\ \mathbf{e} &= \mathbf{C}\mathbf{w}\end{aligned} \qquad ...(14.96)$$

Applying results of the output regulator problem gives immediately that the optimal control for the tracking problem under consideration is

$$\begin{aligned}\mathbf{u}^*(t) &= -\mathbf{R}^{-1}\mathbf{B}^T\mathbf{P}(t)\mathbf{w}\\ &= \mathbf{K}(t)[\mathbf{x} - \mathbf{z}]\end{aligned} \qquad ...(14.97)$$

where $\mathbf{P}(t)$ is the solution of the Riccati equation (eqn. (14.93))

$$\dot{\mathbf{P}}(t) + \mathbf{C}^T\mathbf{Q}\mathbf{C} - \mathbf{P}(t)\mathbf{B}\mathbf{R}^{-1}\mathbf{B}^T\mathbf{P}(t) + \mathbf{P}(t)\mathbf{A} + \mathbf{A}^T\mathbf{P}(t) = \mathbf{0}$$

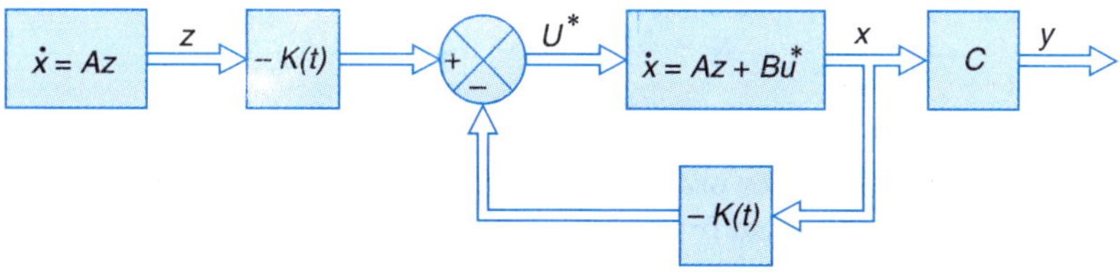

Fig. 14.14. The augmented system for the tracking problem.

with the boundary condition

$$\mathbf{P}(t_f) = \mathbf{C}^T\mathbf{H}\mathbf{C}$$

Figure 14.14 shows the augmented system (14.96) separated into its component systems (14.89) and (14.95) and controlled by the law† (14.97).

14.8 PARAMETER OPTIMIZATION: REGULATORS

We have studied so far the optimal regulator problem wherein (*i*) for a given plant, we find a control function $\mathbf{u}^*$ which is optimal with respect to given performance criterion, and then (*ii*) realize the control function $\mathbf{u}^*(t) = \mathbf{Kx}(t)$.

An optimal solution obtained through steps (*i*) and (*ii*) may not be the best solution in all circumstances. For example, all the elements of matrix **K** may not be free; some gains are fixed by the physical constraints of the system and are therefore relatively inflexible. Similarly if all the states $\mathbf{x}(t)$ are not accessible for feedback, one has to go for a state observer whose complexity is comparable to that of the system itself. It is natural to seek a procedure that relies on the use of feedback from only the accessible state variables, constraining the gain elements of matrix **K** corresponding to the inaccessible state variables to have zero value. Thus, whether one chooses an optimal or suboptimal solution depends on many factors such as cost of equipment, availability of equipment, total space requirement etc., in addition to the performance required out of the system.

In this section, we present a simple method of obtaining the solution of a control problem when some elements of feedback matrix **K** are constrained.

The process considered here may be represented by

$$\dot{\mathbf{x}} = \mathbf{Ax} + \mathbf{Bu} \quad ...(14.98)$$

where $\mathbf{x} = n \times 1$ state vector; $\mathbf{u} = m \times 1$ control vector; $\mathbf{A} = n \times n$ constant matrix; and $\mathbf{B} = n \times m$ constant matrix.

The process is assumed to be completely controllable. The performance index is

$$J = \frac{1}{2}\int_0^\infty (\mathbf{x}^T\mathbf{Qx} + \mathbf{u}^T\mathbf{Ru})\, dt \quad ...(14.99)$$

where **Q** and **R** are, respectively, positive semidefinite and positive definite, real, symmetric, constant matrices.

We know that the optimal control law is linear combination of the state variables, *i.e.*,

$$\mathbf{u} = \mathbf{Kx}(t) \quad ...(14.100)$$

where **K** is $m \times n$ constant matrix (from Section 14.6).

With the linear feedback law of eqn. (14.100), the closed-loop system is described by

$$\begin{aligned}\dot{\mathbf{x}} &= \mathbf{Ax} + \mathbf{BKx} \\ &= (\mathbf{A} + \mathbf{BK})\mathbf{x} \quad ...(14.101)\end{aligned}$$

† Results for the output regulator problem and the tracking problem for the case when final time is not constrained, can easily be obtained from the results of Section 14.6. Identical results immediately follow for the discrete-time systems also.

Substituting for the control vector **u** from eqn. (14.100) in the performance index J of eqn. (14.99), we have

$$J = \frac{1}{2}\int_0^\infty (\mathbf{x}^T\mathbf{Q}\mathbf{x} + \mathbf{x}^T\mathbf{K}^T\mathbf{R}\mathbf{K}\mathbf{x})\,dt$$

$$= \frac{1}{2}\int_0^\infty \mathbf{x}^T(\mathbf{Q} + \mathbf{K}^T\mathbf{R}\mathbf{K})\mathbf{x}\,dt \qquad ...(14.102)$$

This integral is guaranteed to converge because of the fact that a controllable system is stabilizable; the eigenvalues of the matrix **A** + **BK** (eqn. (14.101)) may be assumed to have negative real parts and $\underset{t\to\infty}{\mathbf{x}(t)} = \mathbf{0}$.

Let us postulate the existence of a real symmetric positive definite matrix **P** such that

$$\mathbf{x}^T(\mathbf{Q} + \mathbf{K}^T\mathbf{R}\mathbf{K})\mathbf{x} = -\frac{d}{dt}(\mathbf{x}^T\mathbf{P}\mathbf{x}) \qquad ...(14.103)$$

or

$$\mathbf{x}^T(\mathbf{Q} + \mathbf{K}^T\mathbf{R}\mathbf{K})\mathbf{x} = -\dot{\mathbf{x}}_T\mathbf{P}\mathbf{x} - \mathbf{x}^T\mathbf{P}\dot{\mathbf{x}}$$

Substituting for $\dot{\mathbf{x}}$ and $\dot{\mathbf{x}}^T$ from eqn. (14.101), we have

$$\mathbf{x}^T(\mathbf{Q} + \mathbf{K}^T\mathbf{R}\mathbf{K})\mathbf{x} = -\mathbf{x}^T[(\mathbf{A} + \mathbf{B}\mathbf{K})^T\mathbf{P} + \mathbf{P}(\mathbf{A} + \mathbf{B}\mathbf{K})]\mathbf{x}$$

Since the above equality holds for arbitrary $\mathbf{x}(t)$, we have

$$(\mathbf{A} + \mathbf{B}\mathbf{K})^T\mathbf{P} + \mathbf{P}(\mathbf{A} + \mathbf{B}\mathbf{K}) + \mathbf{K}^T\mathbf{R}\mathbf{K} + \mathbf{Q} = \mathbf{0} \qquad ...(14.104)$$

From this equation we can determine elements of **P** as functions of the elements of the feedback matrix **K**.

By virtue of eqn. (14.103), the performance index is given by

$$J = -\frac{1}{2}\int_0^\infty \frac{d}{dt}(\mathbf{x}^T\mathbf{P}\mathbf{x})\,dt = -\frac{1}{2}\mathbf{x}^T\mathbf{P}\mathbf{x}\Big|_0^\infty$$

$$= \frac{1}{2}\mathbf{x}^T(0)\mathbf{P}\mathbf{x}(0) \qquad ...(14.105)$$

If $K_1, K_2, ..., K_n$ are the free elements of matrix **P**, we have

$$J = J(K_1, K_2, ..., K_n) \qquad ...(14.106)$$

The necessary and sufficient conditions for J to be minimum are given by eqns. (14.2) and (14.3) respectively. Solution set K_i of eqn. (14.106) satisfying these conditions is obtained which gives the suboptimal solution to the control problem. Of course, K_i must satisfy the further constraint that the closed-loop system be asymptotically stable.

If all the parameters of **P** are free, the procedure given above will yield an optimal solution. Procedure of Section 14.6 will also yield the same solution (optimal solution to the regulator problem under consideration is unique). It is more convenient to find **P** from the matrix Riccati equation for the optimal solution.

In the special case where the performance index is independent of the control **u**, we have

$$J = \frac{1}{2}\int_0^\infty \mathbf{x}^T\mathbf{Q}\mathbf{x}\,dt$$

In this case the matrix **P** is obtained from eqn. (14.104) by putting $\mathbf{R} = \mathbf{0}$ resulting in the modified matrix equation

$$(\mathbf{A} + \mathbf{BK})^T\mathbf{P} + \mathbf{P}(\mathbf{A} + \mathbf{BK}) + \mathbf{Q} = \mathbf{0} \quad ...(14.107)$$

Even though **R** is originally assumed to be positive definite, substituting $\mathbf{R} = \mathbf{0}$ is a valid operation here as the positive definiteness of **R** has not been used in this derivation.

Example 14.9: Consider the second-order system of Fig. 14.15 (*a*) wherein it is desired to find optimum ζ which minimizes the integral square error, *i.e.*,

$$J = \int_0^\infty e^2(t)\, dt$$

for the initial conditions $c(0) = 1$, $\dot{c}(0) = 0$.

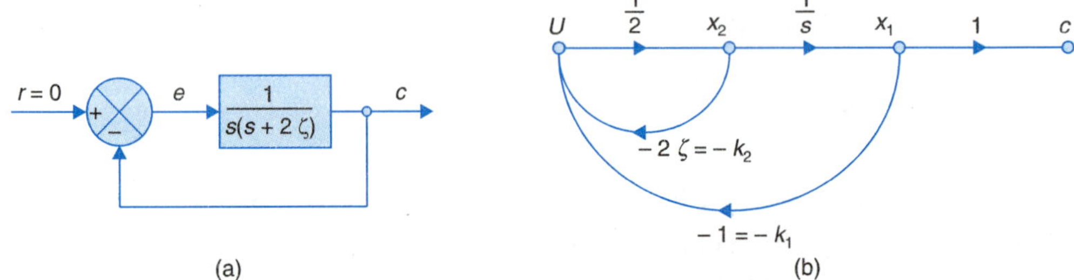

Fig. 14.15 (*a*) A second-order control problem; (*b*) State variable formulation.

The problem is reframed in the state form of Fig. 14.15 (*b*) as one of obtaining feedback control law with the constraint $k_1 = 1$. For this form we have

$$\begin{bmatrix} \dot{x}_1 \\ \dot{x}_2 \end{bmatrix} = \begin{bmatrix} 0 & 1 \\ 0 & 0 \end{bmatrix}\begin{bmatrix} x_1 \\ x_2 \end{bmatrix} + \begin{bmatrix} 0 \\ 1 \end{bmatrix} u; \; x_1(0) = 1, \, x_2(0) = 0$$

$$u = -\begin{bmatrix} k_1 & k_2 \end{bmatrix}\begin{bmatrix} x_1 \\ x_2 \end{bmatrix}$$

Now $\qquad J = \int_0^\infty e^2(t)\, dt = \int_0^\infty x_1^2\, dt$ since $e = -c = -x_1$

Therefore $\qquad \mathbf{Q} = \begin{bmatrix} 2 & 0 \\ 0 & 0 \end{bmatrix}$

Substituting various values in eqn. (14.107) we have

$$\begin{bmatrix} 0 & -1 \\ 0 & -k_2 \end{bmatrix}\begin{bmatrix} p_{11} & p_{12} \\ p_{12} & p_{22} \end{bmatrix} + \begin{bmatrix} p_{11} & p_{12} \\ p_{12} & p_{22} \end{bmatrix}\begin{bmatrix} 0 & 1 \\ -1 & -k_2 \end{bmatrix} + \begin{bmatrix} 2 & 0 \\ 0 & 0 \end{bmatrix} = \begin{bmatrix} 0 & 0 \\ 0 & 0 \end{bmatrix}$$

Solving we get

$$\mathbf{P} = \begin{bmatrix} \dfrac{1 + k_2^2}{k_2} & 1 \\ 1 & \dfrac{1}{k_2} \end{bmatrix}$$

The performance index J is given as (eqn. (14.105))

$$J = \frac{k_2^2 + 1}{2k_2}$$

For J to be minimum, $\frac{\partial J}{\partial k_2} = \left(\frac{1}{2} - \frac{1}{2k_2^2}\right) = 0$; this gives $k_2 = 1$

$$\frac{\partial^2 J}{\partial k_2^2} = \frac{1}{k_2^3} > 0; \text{ this is satisfied for } k_2 = 1$$

Therefore, optimal value of parameter $k_2 = 1$. Since $k_2 = 2\zeta$, we find $\zeta = 0.5$ minimizes integral square error for the given initial conditions.

It can easily be verfied that the suboptimal control derived above results in a closed-loop system which is asymptotically stable.

The control obtained by minimization of J given by (14.105) will vary from one $\mathbf{x}(0)$ to another. However, for practical reasons, it is desirable to have just one control irrespective of what $\mathbf{x}(0)$ is. One way to solve this problem is to assume that $\mathbf{x}(0)$ is a random variable, uniformly distributed on the surface of the n-dimensional unit sphere:

$$E[\mathbf{x}(0)\mathbf{x}^T(0)] = \frac{1}{n}\mathbf{I} \qquad ...(14.108)$$

where E denotes expected value.

We now define a new performance index

$$\tilde{J} = E[J]$$

$$= E\left[\frac{1}{2}\mathbf{x}^T(0)\mathbf{P}\mathbf{x}(0)\right] = \frac{1}{2n}\text{tr}\,[\mathbf{P}] \qquad ...(14.109)$$

where tr[$\mathbf{P}$] = trace of $\mathbf{P}$ = sum of all diagonal elements of matrix $\mathbf{P}$.

For the system of Example 14.9, the parameter k_2 that optimizes $\tilde{J}$ of (14.109) is $\sqrt{2}$. Therefore

$$u = -\begin{bmatrix} 1 & \sqrt{2} \end{bmatrix} \begin{bmatrix} x_1 \\ x_2 \end{bmatrix}$$

is the suboptimal control law that is independent of initial conditions.

PROBLEMS

14.1. Consider the feedback system shown in Fig. P-14.1. The output is required to track unit step. Find the value of α that minimizes integral square error.

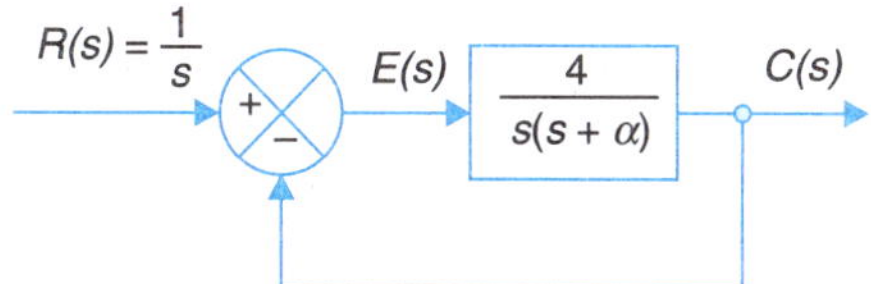

Fig. P-14.1

14.2. Referring to the block diagram of Fig. P-14.2, consider that $G(s) = 100/s^2$ and $R(s) = 1/s$. Determine the optimal values of parameters K_1 and K_2 such that

$$J = \int_0^{\infty} (e^2(t) + 0.25\, u^2(t))\, dt$$

is minimum.

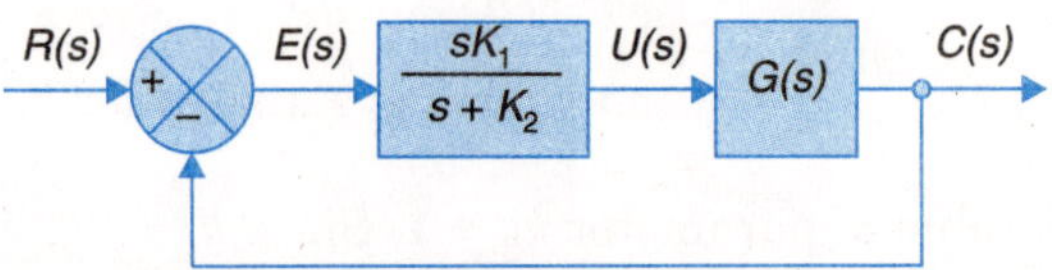

Fig. P-14.2

14.3. Figure P-14.3 illustrates a typical sampled-data system. Find optimal value of K so that

$$J = \sum_{k=0}^{\infty} e^2(kT)$$

is minimized.

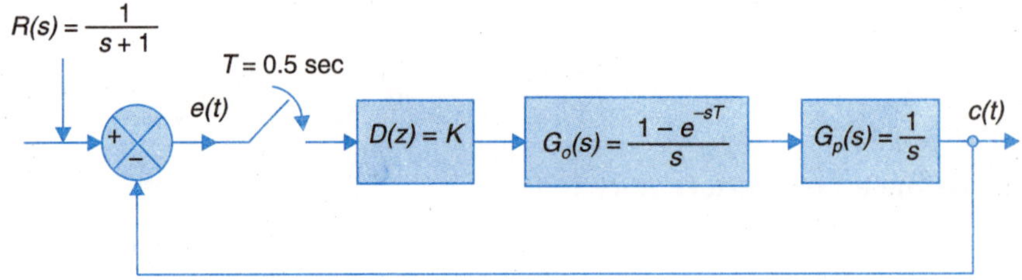

Fig. P-14.3

14.4. Given the system shown in Fig. P-14.4.

(*i*) Find the sum square error

$$J = \sum_{k=0}^{\infty} e^2(kT)$$

as a function of the free parameters K and a.

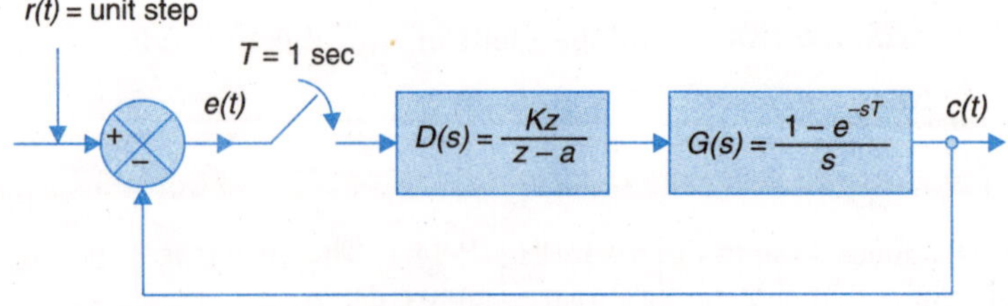

Fig. P-14.4

(*ii*) Determine the range of parameters K and a for which the system is stable.

(*iii*) Determine value of K and a for which J is minimum.

[*Hints*: It may be difficult to solve part (*iii*) analytically. Using digital computer, we can search for a minimum J as K and a are varied within the range determined in part (*ii*).]

14.5. Consider the linear plant of a system characterized by the transfer function

$$G(s) = \frac{10}{s^2}$$

The objective of the system is to make the output $c(t)$ follow a step input $r(t)$ of amplitude 0.5 minimizing

$$J_e = \int_0^\infty (r(t) - c(t))^2 \, dt$$

with

$$J_u = \int_0^\infty u^2(t) \, dt = 2.5$$

where $u(t)$ is the actuating signal of the plant. Find the overall transfer function of optimal system and suggest suitable schemes for implementation of this transfer function.

14.6. For the discrete-data system shown in Fig. P-14.6, given

$$G_p(s) = \frac{1}{s+1}, \; G_0(s) = \frac{1 - e^{-sT}}{s}$$

Find $D(z)$ so that

$$J = \sum_{k=0}^{\infty} [e^2(kT) + u^2(kT)]$$

is minimized.

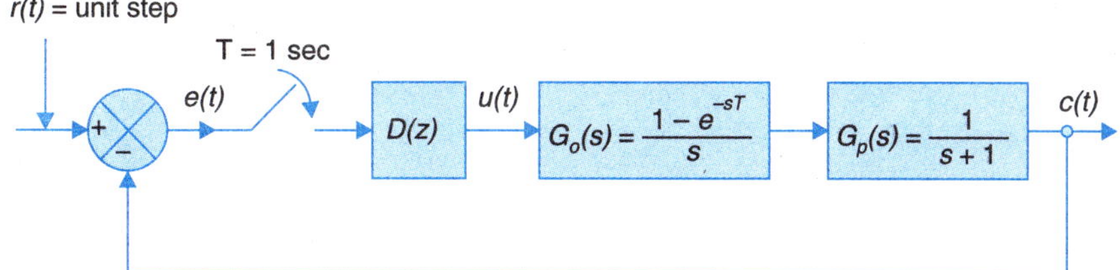

Fig. P-14.6

14.7. Determine the optimal control law for the system described by

$$\mathbf{x} = \begin{bmatrix} 0 & 1 \\ -2 & -3 \end{bmatrix} \mathbf{x} + \begin{bmatrix} 0 \\ 1 \end{bmatrix} u$$

such that the following performance index is minimized

$$J = \int_0^\infty (\mathbf{x}^T \mathbf{x} + u^2) \, dt.$$

14.8. Consider a system described by the equations

$$\begin{bmatrix} \dot{x}_1 \\ \dot{x}_2 \end{bmatrix} = \begin{bmatrix} 0 & 1 \\ 0 & 0 \end{bmatrix} \begin{bmatrix} x_1 \\ x_2 \end{bmatrix} + \begin{bmatrix} 0 \\ 1 \end{bmatrix} u; \; x_1(0) = x_2(0) = 1$$

Choose the feedback law

$$u = -x_1 - Kx_2$$

(*i*) Find the value of K so that

$$J = \frac{1}{2} \int_0^\infty (x_1{}^2 + x_2{}^2) \, dt$$

is minimized.

(*ii*) Find minimum value of J.

(*iii*) Find sensitivity of J with respect to K.

14.9. Consider the plant

$$\begin{bmatrix} \dot{x}_1 \\ \dot{x}_2 \end{bmatrix} = \begin{bmatrix} 1 & 0 \\ -1 & 2 \end{bmatrix} \begin{bmatrix} x_1 \\ x_2 \end{bmatrix} + \begin{bmatrix} 1 \\ 0 \end{bmatrix} u$$

(*i*) Prove that the system is unstable.

(*ii*) Prove that the system is controllable.

(*iii*) Select any values for matrics **Q** and **R** with the constraint that they are positive definite and design a controller for the plant so as to minimize

$$J = \frac{1}{2} \int_0^\infty (\mathbf{x}^T \mathbf{Q} \mathbf{x} + \mathbf{u}^T \mathbf{R} \mathbf{u})\, dt$$

Check that the resulting overall system is stable

14.10. A plant is described by the equations

$$\begin{bmatrix} \dot{x}_1 \\ \dot{x}_2 \end{bmatrix} = \begin{bmatrix} 0 & 1 \\ 0 & 0 \end{bmatrix} \begin{bmatrix} x_1 \\ x_2 \end{bmatrix} + \begin{bmatrix} 0 \\ 1 \end{bmatrix} u;\ x_1(0) = 1,\ x_2(0) = 1$$

Choose the feedback law

$$u = -K[x_1 + x_2]$$

Find the value of K so that

$$J = \frac{1}{2} \int_0^\infty (x_1^2 + x_2^2 + \lambda u^2)\, dt$$

is minimized, when

(*i*) $\lambda = 0$; (*ii*) $\lambda = 1$.

Also determine the values of minimum J in the two cases.

14.11. For the system of Problem P-14.10, find the optimal control law

$$u = -K_1 x_1 - K_2 x_2$$

that minimizes the given performance index for

(*i*) $\lambda = 0$; (*ii*) $\lambda = 1$.

14.12. For the plant described by the equations

$$x_1(k + 1) = 0.8x_1(k) + x_2(k) + u(k)$$

$$x_2(k + 1) = 0.5x_2(k) + 0.5u(k)$$

find the optimal control law that minimizes the performance index

$$J = \frac{1}{2} \sum_{k=0}^{\infty} [x_1^2(k) + x_2^2(k) + u^2(k)].$$

14.13. Consider a system

$$x(k + 1) = 0.368x(k) + 0.632u(k)$$

Find the control sequence so that the following performance index is minimized:

$$J = x^2(N) + \sum_{k=1}^{3} [x^2(k) + u^2(k)]$$

Also find the control sequence when $N \to \infty$.

14.14. The linear discrete system

$$x_1(k + 1) = x_1(k) + x_2(k)$$
$$x_2(k + 1) = x_2(k) + u(k)$$

is to be controlled to minimize the performance measure

$$J = \sum_{k=0}^{2} [4x_1^2(k) + u^2(k)]$$

Obtain the optimal control sequence $[u(0), u(1), u(2)]$; the initial state is

$$x_1(0) = 1, x_2(0) = 0.$$

14.15. It is desired to determine the control law that causes the plant

$$\dot{x}_1 = x_2$$

$$\dot{x}_2 = -x_1 - 2x_2 + u$$

to minimize the performance measure

$$J = 10x_1^2(t_f) + \frac{1}{2}\int_0^{t_f} (x_1^2 + 2x_2^2 + u^2)\, dt$$

The final time t_f is 1 sec.

(*i*) Determine the discrete approximation for the system. Use sampling interval $T = 0.02$ sec.

(*ii*) Determine the optimal control law from the discrete formulation.

(*Hints*: Part (*ii*) may best be solved using digital computer.)

14.16. A first-order system

$$\dot{x} = -x + u$$

is to be controlled to minimize

$$J = \frac{1}{2}\int_0^t (x^2 + u^2)\, dt$$

Find the optimal control law.

14.17. Consider the second-order system

$$\dot{x}_1 = x_2$$
$$\dot{x}_2 = u$$

and the performance index

$$J = \frac{1}{2}\left[x_1^2(t_f) + 2x_2^2(t_f)\right] + \int_0^5 \left[x_1^2(t) + 2x_2^2(t) + x_1(t)x_2(t) + \frac{1}{4}u^2(t)\right] dt$$

The optimal control is

$$u^* = -\mathbf{R}^{-1}\mathbf{B}^T\mathbf{P}(t)\mathbf{x}$$

Obtain **R** and **B** and set up differential equations whose solution will yield the matrix $\mathbf{P}(t)$.

14.18. Consider the system

$$\dot{x}_1 = x_2$$
$$\dot{x}_2 = u$$
$$y = x_1$$

Find the control law which minimizes

$$J = \frac{1}{2}\int_0^\infty (y^2 + u^2)\, dt.$$

14.19. Consider the system shown in Fig. P-14.19. The performance index to be minimized is

$$J = \frac{1}{2}\int_0^\infty ((y - r)^2 + u^2)\, dt$$

Convert this problem to the form of the output regulator problem and find the optimum value of *K*.

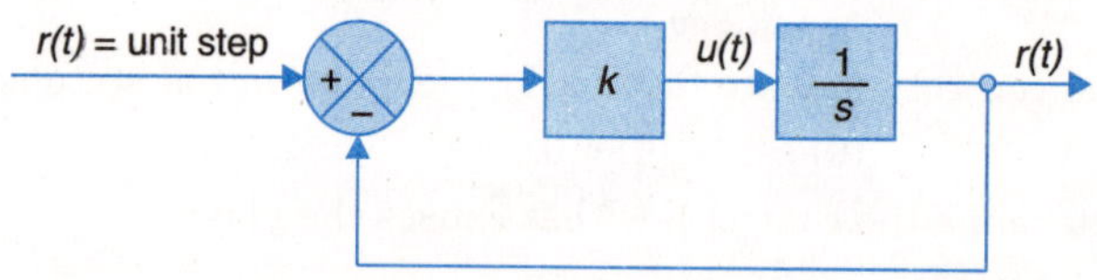

Fig. P-14.19

15

NONLINEAR SYSTEMS

15

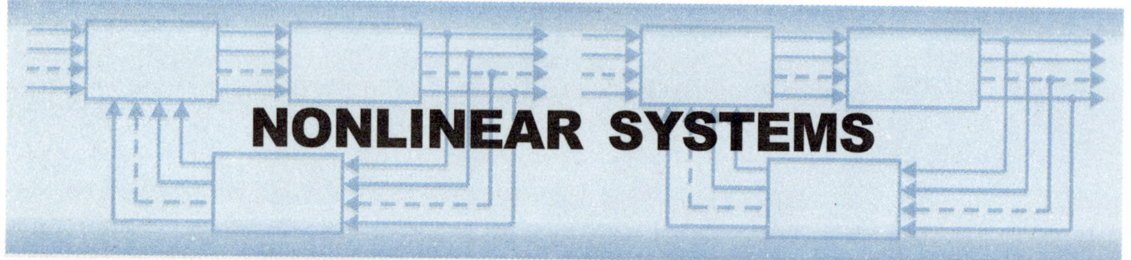

NONLINEAR SYSTEMS

15.1 INTRODUCTION

In the preceding chapters we concerned ourselves mainly with the linear, time-invariant (continuous and discrete) models of systems which, if examined in detail, are always found to be nonlinear to some extent. This does not mean that the powerful techniques of linear theory are not useful as these are restricted in use to the linearized versions of real systems. In fact, for many control applications the linear theory has produced excellent results which are supported experimentally. This again should not mean that we restrict ourselves to linear theory only, which would impose severe burden on system design in two ways: firstly, the constraint of linear operation over wide range demands unnecessarily high quality and therefore high cost components; secondly, the restriction to linear theory may inhibit the designer's curiosity to deliberately introduce nonlinear components or operate the otherwise linear components in nonlinear region with a view to improve system response.

Many practical systems are sufficiently nonlinear so that the important features of their performance may be completely overlooked if they are analyzed and designed through linear techniques as such. For such systems we must necessarily employ special analytical, graphical and numerical techniques which take account of system nonlinearities.

Behaviour of Nonlinear Systems

The most fundamental property of a linear system is the validity of the principle of superposition. It is on account of this property that it can perhaps be guaranteed that a linear system designed to perform satisfactorily when excited by a standard test signal, will exhibit satisfactory behaviour under any circumstances. Furthermore, the amplitude of the test signal is unimportant since any change in input signal amplitude results simply in change of response scale with no change in the basic response characteristics.

In contrast to the linear case, the response of nonlinear systems to a particular test signals is no guide to their behaviour to other inputs, since *the principle of superposition no longer holds*. In fact, the nonlinear system response may be highly sensitive to input amplitude, *e.g.*, a nonlinear system giving its best response for a certain step input may exhibit highly unsatisfactory behaviour when the input amplitude is changed. Further, the Laplace and z-transforms which are simple and powerful tools of linear theory become inapplicable and hence the analysis of nonlinear systems for different time-varying inputs is rendered quite difficult.

The stability of linear systems is determined solely by the location of the system poles (or eigenvalues) and is independent entirely of whether or not the system is driven. Furthermore, the stability of undriven linear systems is independent of the magnitude of the finite initial state.

The situation is not so clear-cut in nonlinear systems. Here the stability is very much dependent on the input and also the initial state. Further, the nonlinear systems may exhibit *limit cycles* which are self-sustained oscillations of fixed frequency and amplitude. Determination of existence of limit cycles is not an easy task as these may depend upon both the type and amplitude of the excitation signal. The stability study of nonlinear systems in fact requires the information about the type and amplitude of anticipated inputs, initial conditions, etc., in addition to the usual requirements of physical and mathematical models of systems.

Even for a stable nonlinear system, the transient and frequency response may exhibit certain peculiar features which are not found in linear systems. To illustrate some of the commonly encountered phenomena, let us consider the spring-mass-damper system of Fig. 15.1(a). If the components are assumed to be linear, the system equation with a sinusoidal forcing function is given by

$$M\ddot{x} + f\dot{x} + Kx = F \cos \omega t \qquad ...(15.1)$$

The well known frequency response curve of this system is shown in Fig. 15.2. This response curve is obtained by measuring the amplitude x of the response, as the frequency of the forcing function is gradually varied keeping its amplitude constant.

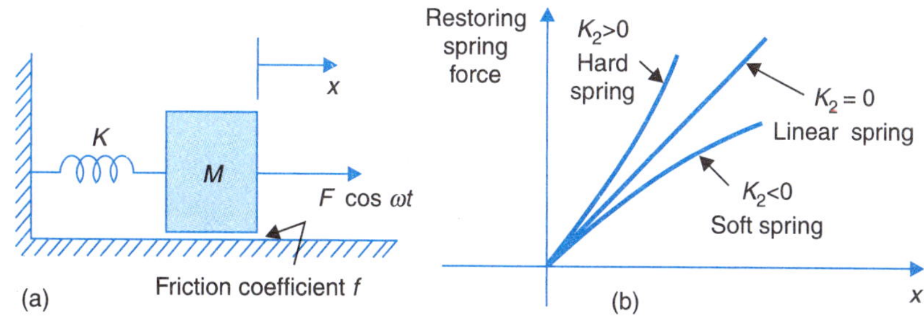

Fig. 15.1. (a) A spring-mass-damper system; (b) Spring characteristics.

Let us now assume that the restoring force of the spring is nonlinear, given by $K_1x + K_2x^3$. The nonlinear spring characteristic is shown in Fig. 15.1 (b). The spring is linear if $K_2 = 0$, while it is called a *hard spring* if $K_2 > 0$ and a *soft spring* if $K_2 < 0$. For the case of the nonlinear spring, eqn. (15.1) modifies to

$$M\ddot{x} + f\dot{x} + K_1x + K_2x^3 = F \cos \omega t \qquad ...(15.2)$$

If an experiment similar to that for the linear case is now performed, the frequency response curve of the form shown in Fig. 15.3 (*a*) may be obtained for the hard spring case, *i.e.*, $K_2 > 0$. The comparison of this figure with Fig. 15.2 reveals that the presence of the nonlinear term K_2x^3 ($K_2 > 0$) in eqn. (15.2) has caused the resonance peak to bend towards higher frequencies. As the input frequency is gradually increased from zero, holding the input amplitude fixed, the measured response follows the curve through the points *A*, *B* and *C*, but at *C* an increment in frequency results in a discontinuous jump down to the point *D*, after which with further increase in frequency, the response curve follows through *DE*. If the frequency is now decreased, the response follows the curve *EDF* with a jump up to *B* occurring at *F* and then the response curve moves towards *A*. It is observed that in certain range of frequencies, the response function is double-valued as seen from Fig. 15.3(*a*). This phenomenon which is peculiar to nonlinear system is known as *jump resonance*.

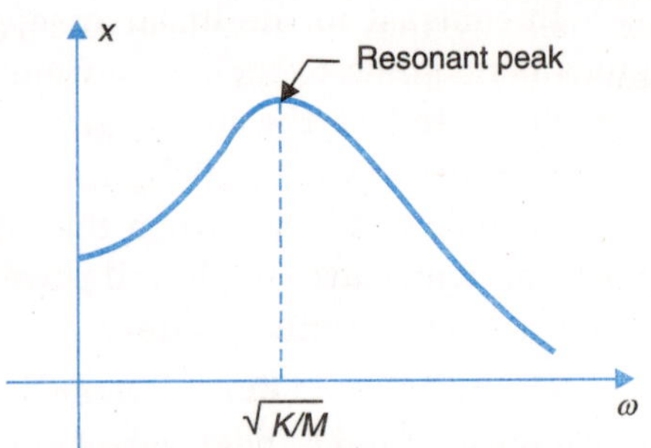

Fig. 15.2. Frequency response curve of spring-mass-damper system.

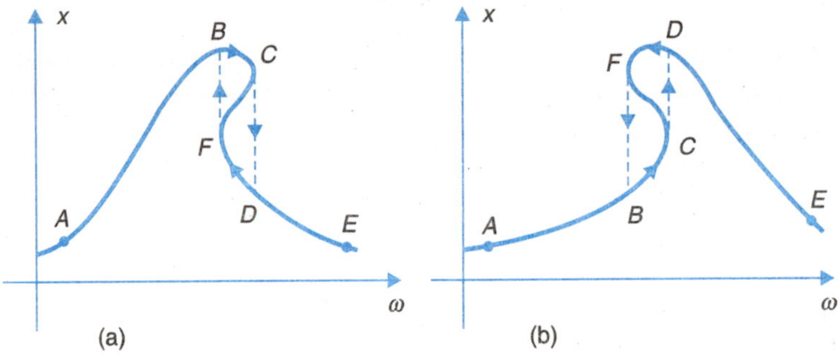

Fig. 15.3. (*a*) Jump resonance in nonlinear systems (hard spring case); (*b*) Jump resonance in nonlinear systems (soft spring case).

For the soft spring case ($K_2 < 0$), the resonant peak bends towards lower frequencies and a similar jump response phenomenon takes place with the difference that there is a jump upward when the frequency is increased and a jump downward when the frequency is decreased as shown in Fig. 15.3 (*b*).

If for the spring-mass-damper system represented by eqn. (15.2), the forcing function is removed and the unforced system is activated by initial conditions only, the response x results in damped oscillations. For the linear case ($K_2 = 0$), the amplitude of successive peaks decreases and the undamped natural frequency of oscillations remains unchanged, while in the nonlinear case, the decrease in amplitude is accompanied by an increase in frequency for $K_2 < 0$ and a decrease in frequency for $K_2 > 0$. As the amplitude of oscillations tends towards zero, the contribution of the non-linear term K_2x^3 becomes negligible and the frequency of oscillations tends towards $\sqrt{(K_1/M)}$, the same as for the linear case. The frequency of oscillations versus amplitude for the three cases is plotted in Fig. 15.4. These plots illustrate the fact that in a nonlinear system the response (the frequency of oscillations in the present case) is sensitive to amplitude.

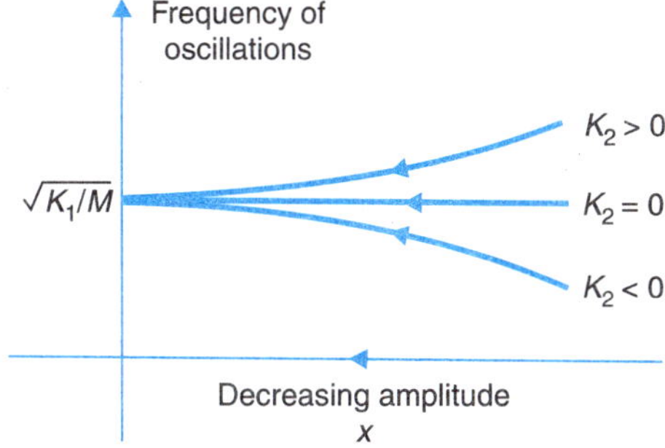

Fig. 15.4. Frequency vs amplitude for the oscillations of spring-mass-damper system with nonlinear spring.

When a linear system is excited by a sinusoidal input of frequency ω, the steady-state output is always sinusoidal of the same frequency. This is not the case in nonlinear systems, where if the input is a sine wave, the output in general is nonsinusoidal containing frequencies (harmonics) which are multiples of the forcing frequency ω. In certain cases, *sub-harmonics*, which are frequencies lower than the forcing frequency, may also exist. It is impossible to extend and justify the concept of frequency response to such situations without drastic modification of its definition and interpretation.

Investigation of nonlinear systems

We discussed above, some of the important phenomena that occur only in nonlinear control systems. These phenomena can not be explained by linear theory. Unfortunately, no single mathematical tool, like the Laplace and z-transforms for linear systems, exists which could analyze the vast variety of nonlinear systems and nonlinear phenomena. Results derived for a class of nonlinear systems cannot be extended to others and no general performance and design criteria exist. Because of this difficulty, an analyst first considers the possibility of approximating a nonlinear system by a linear model. This type of approximation is valid only if the operation is in a restricted range about the operating point. For example, a nonlinear spring having a restoring force of $K_1x + K_2x^3$ may be approximated by a linear one provided the condition $K_2x^3 << K_1x$ is satisfied over the entire operating range of displacement x. This type of approximation gives satisfactory results for systems having small nonlinearities.

Many physical systems are decidedly nonlinear, even within the restricted region about the operating point. For example, a dry friction element produces a damping force of the nature shown in Fig. 15.5 (a). From this figure, it is seen that no single straight line can represent such a curve throughout the range of speeds usually encountered. However, the advantages of linear theory can be extended to this case by piecewise linearization as shown in Fig. 15.5 (b). In such cases, the response obeys a certain linear differential equation in one region of operation and a different one in another region. Depending upon the value of the region parameter (speed in dry friction case), we switch over from one differential equation to the other such that the end conditions of the first are carried over as the initial conditions of the other. Such systems are known as *piecewise linear systems*.

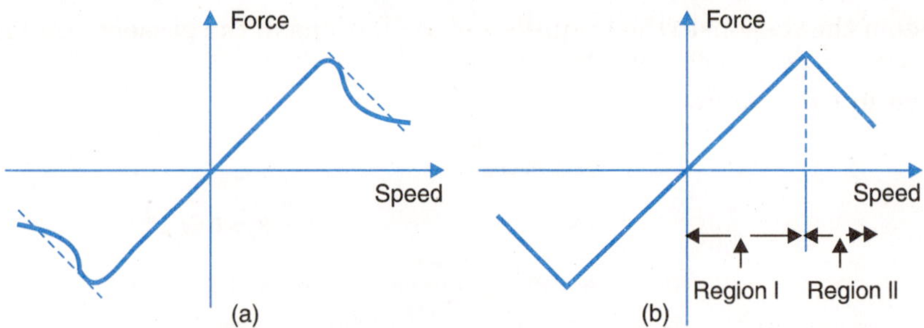

Fig. 15.5. Piecewise linear approximation of dry friction nonlinearity.

At the cost of considerably increased computational work, piecewise linearization could be applied to any nonlinear system by dividing the whole region of operation into small pieces. In fact, this is how we numerically solve a nonlinear differential equation.

In more complicated situations, where the system is distinctly nonlinear, piecewise linearization is cumbersome and time consuming. For such cases, two methods of analysis are available, which have an appreciable degree of generality. One of these is the *phase-plane method*, which is one of the most powerful tools available today. This is basically a graphical method from which information about transient behaviour and stability is easily obtained by constructing phase trajectories. From the point of view of practical utility, the method is restricted to second-order systems, excited by step or ramp inputs. Higher-order systems may first be approximated by their second-order equivalents for investigation by the phase-plane method. Because of this approximation, the results obtained by this analysis should be checked by digital computer simulation.

The other method, known as the *describing function method*, is based on harmonic linearization. We assume here that the input to the nonlinear component is sinusoidal and depending upon the filtering properties of the linear part of the overall system, the output is adequately represented by the fundamental frequency term in the Fourier series. This approximation is not restricted to small signal dynamics. Unfortunately, the approximation is intuitive and has no direct mathematical justification. The degree of accuracy achieved, is heavily dependent upon the filtering property of the linear part of the system. It is no wonder that this linearization may not always work. It usually gives sufficiently accurate information about system stability and limit cycles but does not give a trustworthy information about transient response.

In fact, the phase-plane and describing function methods use complimentary approximations. The phase-plane method retains the nonlinearity as such and uses the second-order approximation of a higher-order linear part, while on the other hand, the describing function method retains the linear part and harmonically linearizes the nonlinearity. The two techniques thus compliment each other. Greater confidence can therefore be placed in results if the system is analyzed through both these techniques.

In order to handle complicated nonlinear systems, digital computers are of course the most powerful tools. However, it is still worthwhile to carry out preliminary design based on suitable approximations by the phase-plane method and/or by the describing function method, as these methods provide the designer with a physical insight into the system behaviour. More detailed analysis and design is then carried out with computer simulation.

We shall devote this appendix to the study of the phase-plane and describing function methods. Powerful methods of stability analysis—the second method of Liapunov have already been presented in Chapter 13.

15.2 COMMON PHYSICAL NONLINEARITIES

In control systems, nonlinearities can be classified as *incidental and intentional.* Incidental nonlinearities are those which are inherently present in the system. The designer strives to design the system so as to limit the adverse effects of these nonlinearities. Common examples of incidental nonlinearities are saturation, dead-zone, coulomb friction, stiction, backlash, etc. The intentional nonlinearities, on the other hand, are those which are deliberately inserted in the system to modify system characteristics. The most common example of this type of nonlinearity is a relay. In the following paragraphs we shall discuss in brief the basic features of the commonly encountered nonlinearities.

Saturation

This is perhaps the most common of all nonlinearities. All practical systems, when driven by sufficiently large signals, exhibit the phenomenon of saturation due to limitations of physical capabilities of their components. Many components such as amplifiers* have output proportional to input in a limited range of input signals. When the input exceeds this range, the output tends to become nearly constant as shown in Fig. 15.6. Though the change over from one range to another is gradual, it is sufficiently accurate in most cases to approximate the saturation phenomenon by straight line segments as shown.

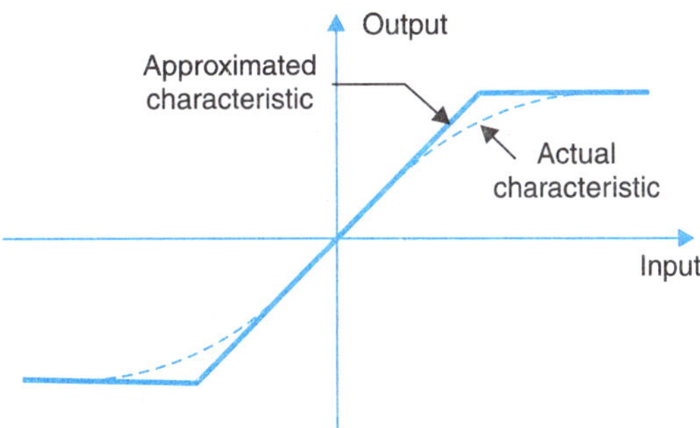

Fig. 15.6. Piecewise linear approximation of saturation nonlinearity.

Friction

Retarding frictional forces exist whenever mechanical surfaces come in sliding contact. The predominant frictional force called the *viscous friction* is proportional to the relative velocity of sliding surfaces, *i.e.*,

$$\text{viscous friction force} = f\dot{x}$$

where f is a constant and x is the relative velocity. Viscous friction is thus linear in nature. Figure 15.7 (a) is the graphical representation of such friction. In addition to the viscous friction,

*Other examples of saturation are torque and speed saturation in electric motors.

there exist two nonlinear frictions. One is the *coulomb friction* which is a constant retarding force (always opposing the relative motion) and the other is the *stiction* which is the force required to initiate motion. The force of stiction is always greater than that of coulomb friction since due to interlocking of surface irregularities, more force is required to move an object from rest than to maintain it in motion. In actual practice, the stiction force gradually decreases with velocity and changes over to coulomb friction at reasonably low velocities as shown in Fig. 15.7 (*b*). It is however sufficiently accurate to regard this change over as sudden as shown in Fig. 15.7 (*c*). The composite characteristics of various frictions are shown in Fig. 15.7 (*d*). It may also be pointed out here that at very high speeds, frictional force will become proportional to square and cube of speed.

Backlash

Another important nonlinearity commonly occurring in physical systems is hysteresis in mechanical transmission such as gear trains and linkages. This nonlinearity is somewhat different from magnetic hysteresis and is commonly referred to as backlash. Backlash in fact is the play between the teeth of the drive gear and those of the driven gear. Consider a gear box as shown in Fig. 15.8 (*a*) having backlash as illustrated in Fig. 15.8 (*b*).

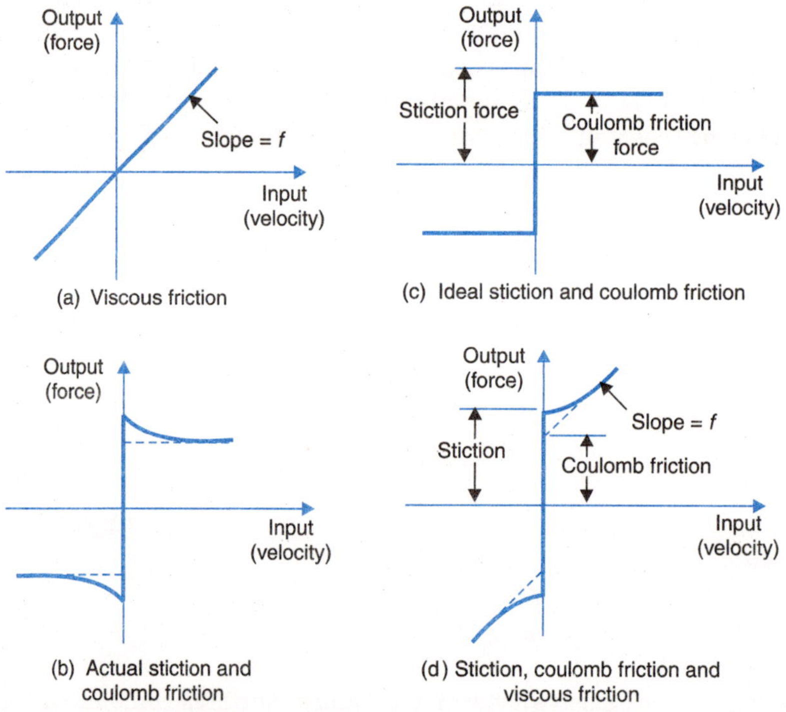

Fig. 15.7. Characteristics of various types of frictions.

Figure 15.8 (*b*) shows the tooth *A* of the drive gear located midway between the teeth B_1, B_2 of the driven gear. Figure 15.8 (*c*) gives the relationship between input and output motions. As the tooth *A* is driven clockwise from this position, no output motion takes place until the tooth *A* makes contact with the tooth B_1 of the driven gear after travelling a distance $b/2$. This output motion corresponds to the segment *pq* of Fig. 15.8 (*c*). After the contact is made the driven gear rotates counter-clockwise through the same angle as the drive gear, if the gear ratio is assumed to be unity. This is illustrated by the line segment qr of Fig. 15.8 (c). As the input motion is reversed, the contact between the teeth A and B_1 is lost and the driven gear immediately becomes stationary based on the assumption that the load is friction-controlled with negligible inertia. The output motion therefore ceases till tooth A has travelled a distance b in the reverse direction as illustrated by the segment rs. After the tooth A establishes contact with the tooth B_2, the driven gear now moves in clockwise direction as shown by the segment st. As the input motion is reversed the driven gear is again at standstill for the segment tu and then follows the drive gear along uq. This completes one cycle of the output motion.

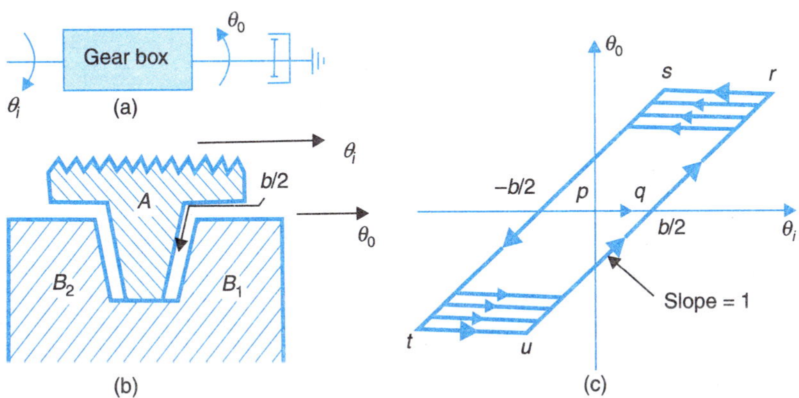

Fig. 15.8 Backlash nonlinearity.

It is easily seen from Fig. 15.8 (c) that for a given input, the output is multivalued. Which particular output will result for a given input depends upon the history of the output. This type of nonlinearity has thus inherent memory and is referred to as *memory type* nonlinearity. The width of the input-output curve equals the total backlash b while its height corresponds to the limits of the input angle θ_i.

In a servo system, the gear backlash may cause sustained oscillations or chattering phenomenon and the system may even turn unstable for large backlash. Backlash can be reduced by using high quality gears. It can be almost eliminated by using spring loaded split gear as the driven gear. Such gears are employed in instrumentation systems.

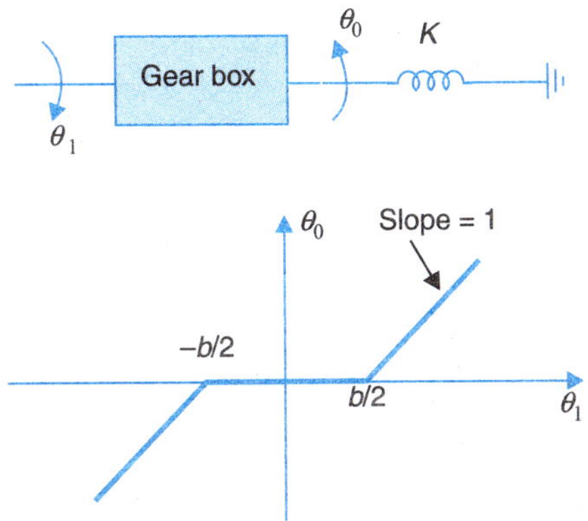

Fig. 15.9. Dead-zone nonlinearity.

Dead-zone

Figure 15.9 shows the case when gears with backlash drive a torsional spring load. The drive and the driven gears must now move together except in the region ± $b/2$ where the spring remains untwisted and there is no contact between the teeth of the drive and driven gears. Figure 15.9 also illustrates the relationship between input and output motions which exhibit the phenomenon of dead-zone.

In addition to the example cited above dead-zone nonlinearity occurs in many other devices which are insensitive to small signals.

Relay

A relay is a nonlinear power amplifier which can provide large power amplification inexpensively and is therefore deliberately introduced in control systems. A relay-controlled system can be switched abruptly between several discrete states which are usually off, full forward and full reverse. Relay-controlled systems find wide appllications in the control field.

A simple relay servo system is shown in Fig. 15.10. The error signal controls the relay current which moves the solenoid so as to make the contact in one direction or the other. The characteristic of such an ideal relay is shown in Fig. 15.11 (*a*). In practice a relay has a definite amount of dead-zone as shown in Fig. 15.11 (*b*). This dead-zone is caused by the fact that the relay coil requires a finite amount of current to actuate the relay. Further, since a larger coil current is needed to close the relay than the current at which the relay drops out, the relay characteristic always exhibits hysteresis. Relay characteristic with hysteresis alone is illustrated in Fig. 15.11 (*c*), while the more practical characteristic with both dead-zone and hysteresis is illustrated in Fig. 15.11 (*d*).

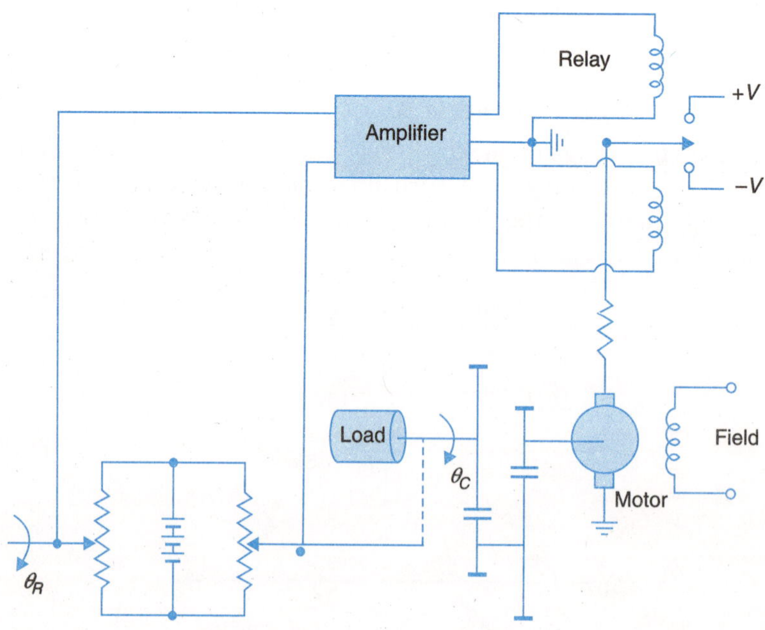

Fig. 15.10. A relay servo system.

In place of an electromechanical relay, SCR switching is now employed. Because of fast switching speeds, such an arrangement exhibits a near ideal characteristic as in Fig. 15.11 (*a*).

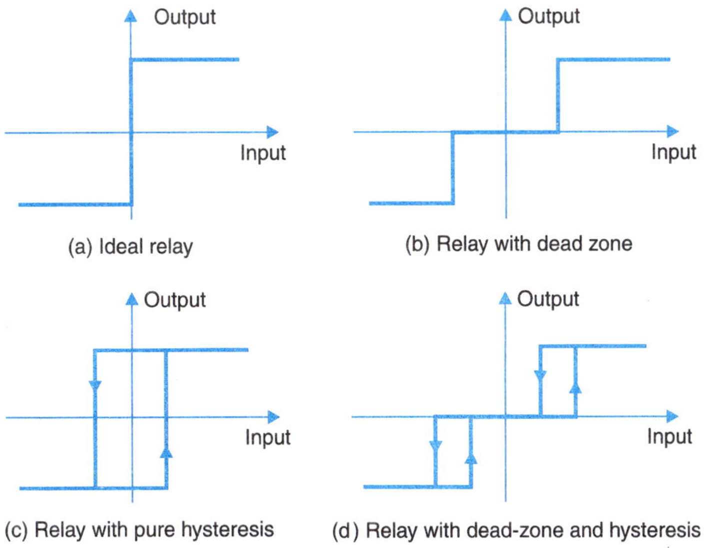

Fig. 15.11. Relay nonlinearity.

Multivariable Nonlinearity

Some nonlinearities such as the torque-speed characteristics of a servomotor, transistor characteristics, etc., are functions of more than one variable. Such nonlinearities are called multivariable nonlinearities. These have been treated through linearization in Chapter 4.

15.3 THE PHASE-PLANE METHOD: BASIC CONCEPTS

Consider an unforced linear spring-mass-damper system whose dynamics is described by

$$M\frac{d^2x}{dt^2} + f\frac{dx}{dt} + Kx = 0 \qquad \text{...(15.3)}$$

Let this system be activated by initial conditions only.

Equation (15.3) may be written in the standard form

$$\frac{d^2x}{dt^2} = 2\zeta\omega_n\frac{dx}{dt} + \omega_n^2 x = 0 \qquad \text{...(15.4)}$$

where ζ and ω_n are familiar quantities; the damping factor and undamped natural frequency of the system.

Let the state of this system be described by two variables, the displacement x and the velocity dx/dt, *i.e.*, in state variable notation

$$x_1 = x;\ x_2 = dx/dt$$

The state variables defined in this fashion (refer Chapter 12) are called *phase variables*. With this definition, the scalar differential equation (15.4) can be written in the state variable form,

$$\frac{dx_1}{dt} = x_2 \quad \text{...(15.5)}$$

$$\frac{dx_2}{dt} = -\omega_n^2 x_1 - 2\zeta\omega_n x_2 \quad \text{...(15.6)}$$

These equation may then be solved for phase variables x_1 and x_2. The time response plots of x_1, x_2 for various values of damping with initial conditions $x_1(0) = x_1^0$ and $x_2(0) = 0$ are shown in Fig. 15.12 (*a*). When the differential equations describing the dynamics of the system are nonlinear, it is in general not possible to obtain a closed form solution of x_1, x_2. For example, if the spring force is nonlinear $(K_1x + K_2x^3)$, eqns. (15.5) and (15.6) take the form

$$\frac{dx_1}{dt} = x_2$$

$$\frac{dx_2}{dt} = -\frac{f}{M}x_2 - \frac{K_1}{M}x_1 - \frac{K_2}{M}x_1^3$$

In such situations, a graphical method known as the *phase-plane method* is found to be very helpful. It provides essentially the same information as obtained from time response curves. Further, this method provides a visual display of system response from any set of initial condition but the method is restricted to second-order systems. Time response and phase plot could be obtained speedily by numerical integration using powerful software tools which are now available.

The coordinate plane with axes that correspond to the dependent variable $x_1 = x$ and its first derivative $x_2 = \dot{x}$ is called the *phase-plane*. The curve described by the state point (x_1, x_2) in the phase-plane with time as running parameter is called a *phase-trajectory*. A phase-trajectory can be easily constructed by graphical or analytical techniques to be discussed later. For the linear mass-spring-damper system under discussion, the phase-trajectories for different values of ζ corresponding to the time response curves of Fig. 15.12 (*a*) are shown in Fig. 15.12 (*b*).

In Fig. 15.13, trajectories of the critically-damped linear system for a given initial condition of displacement (curve *A*), an initial condition of velocity (curve *B*) and initial condition of both displacement and velocity (curve *C*) are shown in thick lines. Trajectories shown in dotted lines correspond to other initial conditions. Such a family of trajectories is called a *phase-portrait*. It is observed that in a phase-portrait, larger or smaller initial conditions produce geometrically similar trajectories.

The power of the phase-plane method lies in the fact that an easily constructable family of trajectories provides a graphical picture of the total system response such that a glance at the phase-portrait is sufficient to answer many important questions regarding system behaviour. More complete information about system behaviour, if required, can then be obtained by performing simple graphical constructions on the phase-portrait.

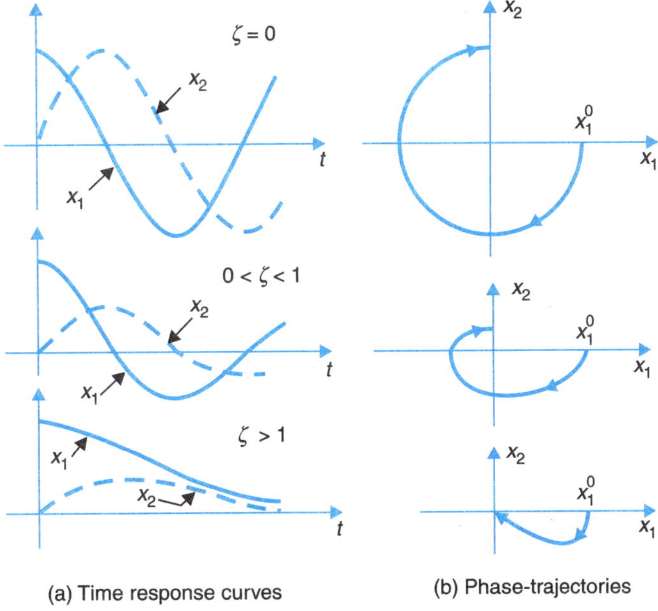

Fig. 15.12. Time response curves and phase-trajectories for linear spring-mass-damper system.

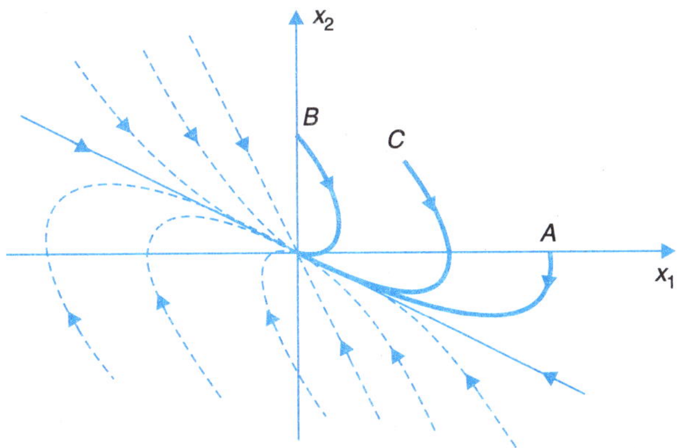

Fig. 15.13. Phase-portrait for a critically-damped system.

Our discussion will be limited to autonomous systems (defined in the next section) only.

15.4 SINGULAR POINTS

Consider a general time-invariant system described by the state equation

$$\dot{\mathbf{x}} = \mathbf{f}(\mathbf{x}, \mathbf{u}) \qquad ...(15.7)$$

If the input vector **u** is constant, it is possible to write the above equation in the form

$$\dot{\mathbf{x}} = \mathbf{F}(\mathbf{x}) \qquad ...(15.8)$$

A system represented by an equation of this form is called an *autonomous system.* For such a system, consider the points in the phase-space at which the derivatives of all the state variables are zero. Such points are called *singular points*. These are in fact *equilibrium points* already defined in Chapter 12. If the system is placed at such a point, it will continue to lie there if left undisturbed (the derivatives of all the phase variables being zero, the system state remains unchanged).

For studying the system dynamic response at an equilibrium (singular) point to small perturbation, the system is linearized (using linearization techniques presented in Chapter 12) at that point. The linearized model of system of eqn. (15.8) may be written as

$$\dot{\mathbf{x}} = \mathbf{A}\mathbf{x} \qquad ...(15.9)$$

For this linear autonomous system, the equilibrium states are given by those states $\mathbf{x}_e$ satisfying $\mathbf{A}\mathbf{x}_e = \mathbf{0}$. From this we see that $x_e = \mathbf{0}$ will be the only solution of the above equation provided that the determinant of **A** is nonzero. Equivalently, if all the eigenvalues of the system are different from 0, then the origin is the only singular point.

In general, if x_e is a singular point, it is convenient to shift the origin of coordinates to $\mathbf{x}_e$. To achieve this, we define new phase variables as

$$\tilde{\mathbf{x}} = \mathbf{x} - \mathbf{x}_e \qquad ...(15.10)$$

The system of eqn. (15.8) in terms of new phase variables is represented as

$$\dot{\tilde{\mathbf{x}}} = \mathbf{F}(\tilde{\mathbf{x}}) \qquad ...(15.11)$$

with the equilibrium point lying at $\tilde{\mathbf{x}} = \mathbf{0}$.

Consider now, a linearized autonomous second-order system described by equation of the form (15.9). The examination of phase-plane trajectory can best be done in the canonical form of representation. Using the linear transformation

$$\mathbf{x} = \mathbf{M}\mathbf{z}; \ \mathbf{M} \text{ is a modal matrix} \qquad ...(15.12)$$

equation (15.9), being of second-order, may be converted into the canonical form

$$\begin{bmatrix} \dot{z}_1 \\ \dot{z}_2 \end{bmatrix} = \begin{bmatrix} \lambda_1 & 0 \\ 0 & \lambda_2 \end{bmatrix} \begin{bmatrix} z_1 \\ z_2 \end{bmatrix} \qquad ...(15.13)$$

where λ_1 and λ_2 are eigenvalues of matrix **A**, which are assumed to be distinct.

It should be noticed that the transformation given by eqn. (15.12) merely changes the coordinate system from (x_1, x_2) to (z_1, z_2) having the same origin, as shown in Fig. 15.14. The new coordinate system (z_1, z_2) in general may not be rectangular.

Equation (15.13) may be written in the component form as

$$\dot{z}_1 = \lambda_1 z_1 \qquad ...(15.14)$$

$$\dot{z}_2 = \lambda_2 z_2 \qquad ...(15.15)$$

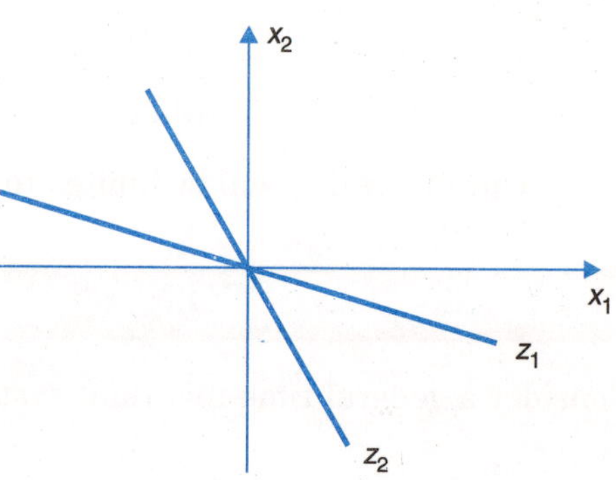

Fig. 15.14. Change of coordinate axes by linear transformation.

The trajectory traced out by the representative point P in the (z_1, z_2)-plane has at any instant a resultant velocity vector associated with the motion as shown in Fig. 15.15. The slope of the velocity vector is given by

$$\tan\theta = \frac{dz_2/dt}{dz_1/dt} = \frac{dz_2}{dz_1} = \frac{\lambda_2 z_2}{\lambda_1 z_1}$$

which upon integration gives

$$z_2 = c(z_1)^{\lambda_2/\lambda_1} \quad ...(15.16)$$

where c is the constant of integration.

Nodal Point

Eigenvalues (λ_1, λ_2) are both real and negative as shown in Fig. 15.15 (*a*). The nature of trajectories is shown in Fig. 15.15 (*b*). They all converge to origin, which is then called a nodal point. This is a stable nodal point. The corresponding trajectories appear as in Fig. 15.16 in x_1x_2 plane. When λ_1, λ_2 are both real and positive, the result is an unstable nodal point with portrait depicted in Fig. 15.17.

Saddle Point

Both eigenvalues are real, equal and negative of each other as in Fig. 15.18 (*a*): The corresponding phase portraits are drawn in Figs. 15.18 (*b*) and (*c*). The origin in this case a saddle point which is always unstable, one eigenvalue being negative.

Focus Point

Let the eigenvalues be

$$\lambda_1, \lambda_2 = \sigma \pm j\omega$$
$$= \text{complex conjugate pair}$$

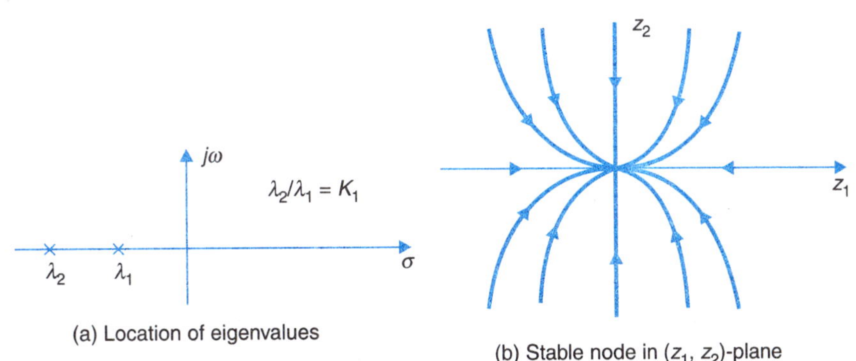

(a) Location of eigenvalues

(b) Stable node in (z_1, z_2)-plane

Fig. 15.15

The phase portraits for negative and positive real parts are drawn in Figs. 15.19 (*a*) and (*b*). The origin is the focus point and is stable/unstable for negative/positive real parts of the eigenvalues. Notice that certain transformation has been carried out for (x_1, x_2) to (y_1, y_2) to present the trajectory in form of a true spiral.

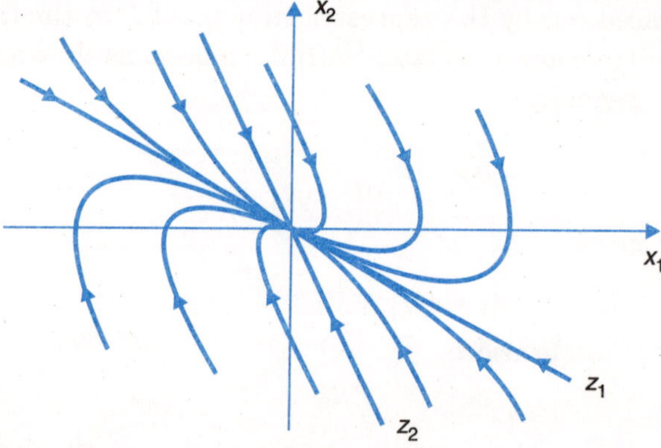

Fig. 15.16. Stable node in (x_1, x_2)-plane.

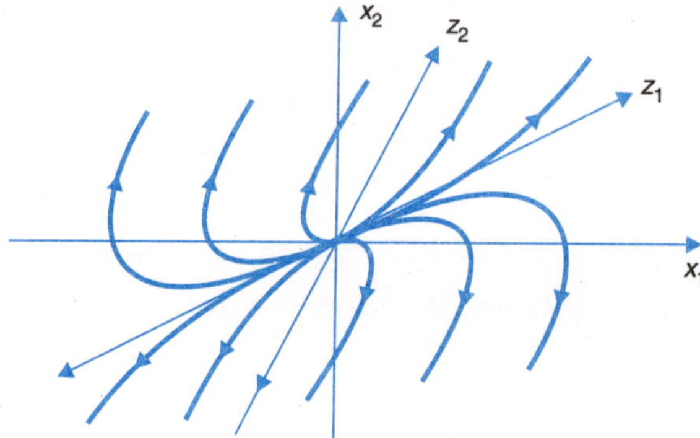

Fig. 15.17. Unstable node in (x_1, x_2)-plane.

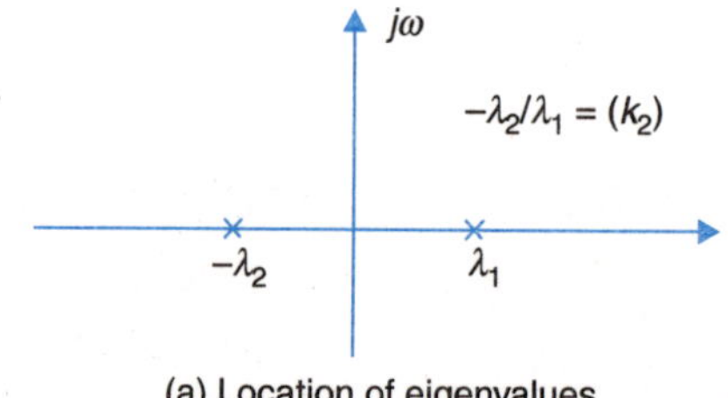

(a) Location of eigenvalues

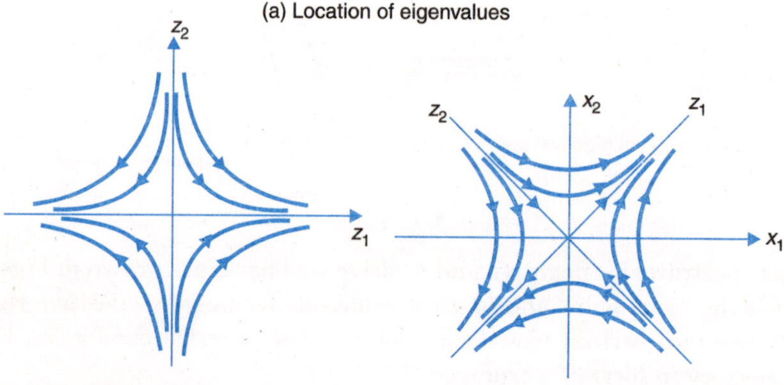

(b) Saddle point in $(z_1 - z_2)$-plane

(c) Saddle point in (x_1, x_2)-plane

Fig. 15.18

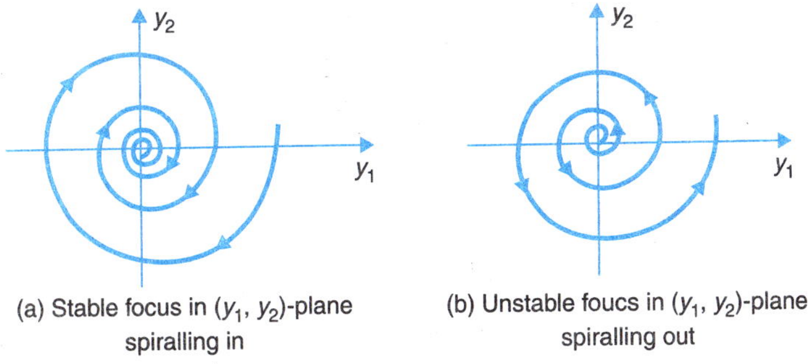

(a) Stable focus in (y_1, y_2)-plane spiralling in

(b) Unstable foucs in (y_1, y_2)-plane spiralling out

Fig. 15.19. (*a*) and (*b*)

Centre or Vortex Point

$$\lambda_1, \lambda_2 = \pm j\omega$$
$$= \text{on the imaginary axis}$$

The phase portrait has closed paths trajectories shown in Figs. 15.20 (*a*) and (*b*) and the origin is called a centre. The system is limitedly stable (see Chap. 6).

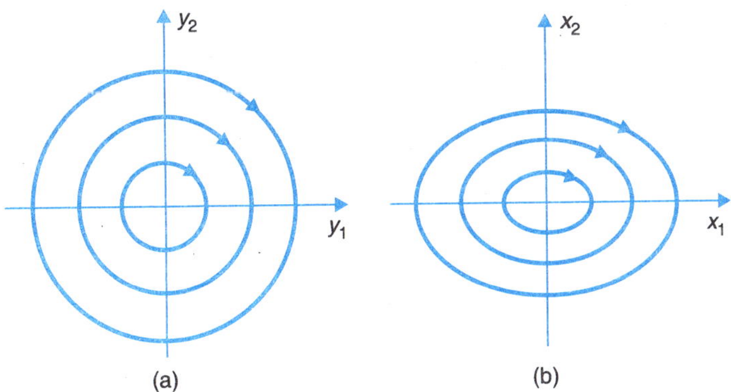

Fig. 15.20. (a) Centre in (y_1, y_2)-plane; and (b) Centre in (x_1, x_2)-plane.

Example 15.1 :

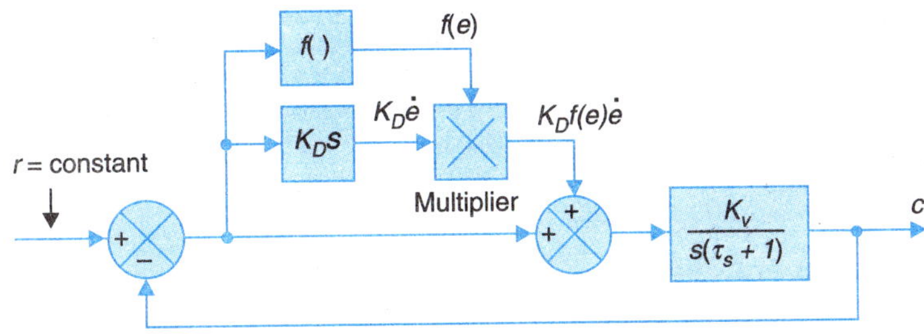

Fig. 15.21. Nonlinear proportional plus derivative controller for second-order system.

In Section 5.7 proportional plus derivative controller was introduced for compensating a second-order system. It was observed there that with the introduction of the derivative term, the system damping increases, settling time reduces while the system K_v is maintained. If in addition a low rise time is required, the design specifications become contradictory. Intentional nonlinearity can be used to advantage in meeting the contradictory requirements of low rise time and low settling time with fixed K_v. This is easily achieved by making the derivative term become dependent upon the magnitude of error as shown in Fig. 15.21. By suitably shaping the nonlinear function $f(e)$, the derivative term $K_D f(e)\,\dot{e}$ is kept low (or even made negative) under large error conditions giving a small rise time. This term is made to become large for small errors so that the system settles is short time. The differential equation describing the nonlinear system is

$$[e + K_D f(e)\,\dot{e}]\,K_v = \tau\ddot{c} + \dot{c}$$

$$e = r - c$$

If r = constant

$$\dot{e} = -\,\dot{c};\ \ddot{e} = -\ddot{c}$$

$$\therefore \qquad \tau\ddot{e} + [1 + Kf(e)]\dot{e} + K_v e = 0 \qquad \text{...[15.17 (a)]}$$

where $$K = K_D K_v$$

Defining the state variables as $x_1 = e$, $x_2 = \dot{e}$, the state equation corresponding to eqn. [15.17 (*a*)] is

$$\dot{x}_1 = x_2$$

$$\dot{x}_2 = -\frac{K_v}{\tau}x_1 - \frac{1 + Kf(x_1)}{\tau}x_2 \qquad \text{...[15.17 (b)]}$$

Letting $\dot{x}_1 = \dot{x}_2 = 0$, the only equilibrium point is found to be at $(x_1 = 0, x_2 = 0)$.

Linearizing eqn. [15.17 (*b*)] about the equilibrium point by using eqn. (12.16) of Section 12.2, we have

$$\dot{\tilde{\mathbf{x}}} = \mathbf{A}\tilde{\mathbf{x}}$$

where $$\mathbf{A} = \begin{bmatrix} 0 & 1 \\ -\dfrac{K_v}{\tau} & -\dfrac{1 + Kf(0)}{\tau} \end{bmatrix} \qquad \text{...(15.18)}$$

The eigenvalues of **A** are given by

$$\lambda_1, \lambda_2 = -\left(\frac{1 + Kf(0)}{2\tau}\right) \pm \sqrt{\left\{\frac{1}{4}\left[\frac{1 + Kf(0)}{\tau}\right]^2 - \frac{K_v}{\tau}\right\}}$$

A desirable design feature could be that the system rises and settles fast without oscillations, for which the equilibrium point must be a stable node. For $f(0) > 0$, stable node, *i.e.*, $\lambda_1, \lambda_2 < 0$; is achieved by the condition

$$f(0) > \left[\frac{2\sqrt{(K_v\tau)} - 1}{K}\right] \qquad \text{...[15.19 (a)]}$$

The stable node requirement fixes the point $f(0)$ on the $f(e)$ function.

In case the above inequality is reversed, *i.e.*,

$$f(0) < \left[\frac{2\sqrt{(K_v\tau)} - 1}{K}\right] \qquad \text{...[15.19 (b)]}$$

λ_1, λ_2 are complex conjugate with negative real part and therefore the equilibrium point will exhibit stable focus behaviour.

15.5 STABILITY OF NONLINEAR SYSTEMS

Note: For sake of continuity it is essential to relocate this part from Chapter 13.

The problem of stability for linear time-invariant systems was studied in Chapter 6. There we observed that stability may be defined as per the following two notions:

(*i*) For free system: A system is stable with zero input and arbitrary initial conditions if the resulting trajectory tends towards the equilibrium state.

(*ii*) For forced system: A system is stable if with bounded input, the system output is bounded.

These two notions were found to be essentially equivalent for linear time-invariant systems. In nonlinear systems, unfortunately, there is no definite correspondence between the two notions. Many important results have been obtained for free systems. (Our discussion will be limited to this class of nonlinear systems). Even for this class of systems, the concept of stability is not clear-cut. The linear autonomous systems (with nonzero eigenvalues) have only one equilibrium state and their behaviour about the equilibrium state completely determines the qualitative behaviour in the entire state-plane. In nonlinear systems, on the other hand, system behaviour for small deviations about the equilibrium point may be different from that for large deviations. Therefore, *local stability* does not imply stability in the overall state-plane and the two concepts should be considered separately. Secondly, in a nonlinear system with multiple equilibrium states, the system trajectories may move away from one equilibrium point and tend to other as time progresses. Thus, it appears that in case of nonlinear systems, it is simpler to speak of system stability relative to the equilibrium state rather than using a general term 'stability of a system'.

There are many types of stability definitions in the literature. We shall concentrate on three of these: stability, asymptotic stability and asymptotic stability in-the large.

Consider an autonomous system described by the state equation (15.8). Assume that the system has only one equilibrium point. This is the case with all properly designed systems. Furthermore, without loss of generality, let the origin of state space be taken as the equilibrium point.

The system of eqn. (15.8) is *stable* at the origin if, for every initial state $x(t_0)$ which is sufficiently close to origin, $\mathbf{x}(t)$ remains near the origin for all t. It is *asymptotically stable* if $\mathbf{x}(t)$ in fact approaches the origin as $t \to \infty$. It is *asymptotically stable in-the-large* if it is asymptotically stable for every initial state regardless of how near or far it is from the origin.

Following are the mathematically precise definitions of the different types of stability.

The system of eqn. (15.8) is *stable* at the origin if, for every real number $\varepsilon > 0$, there exists a real number $\delta(\in) > 0$ such that $\| x(t_0) \| \le \delta$ results in $\| \mathbf{x}(t) \| \le \varepsilon$ for all $t \ge t_0$.

This definition (from Russian mathematician A.M. Liapunov) of stability uses the concept of vector norm. The Euclidean norm for a vector $\mathbf{x}$ with n components $x_1, x_2, \ldots, x_n$ is (eqn. (II. 12), Appendix II),

$$\| \mathbf{x} \| = (x_1^2 + x_2^2 + \ldots + x_n)^{1/2}$$

$\| \mathbf{x} \| \leq R$ defines a hyper-spherical region $S(R)$ of radius R surrounding the equilibrium point $\mathbf{x} = 0$. In terms of the Euclidean norm, the above definition of stability implies that for any $S(\varepsilon)$ that we may designate, the designer must produce $S(\delta)$ so that system state initially in $S(\delta)$ will never leave $S(\varepsilon)$. This is illustrated in Fig. 15.22 (*a*).

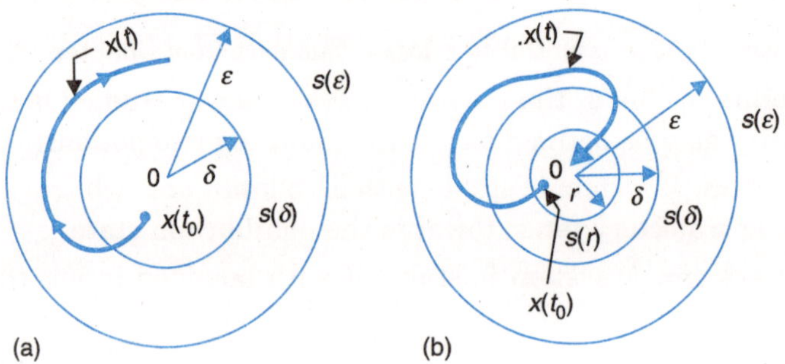

Fig. 15.22. Stability definitions.

Consider for example, a second-order linear autonomous system whose phase-trajectories for various initial conditions are shown in Fig. 15.20 (*b*). For a specified value of ε, we can find a closed phase-trajectory whose maximum distance from the origin is ε. We then select a value of δ which is less than the minimum distance from that curve to the origin. The $\delta(\varepsilon)$, so chosen will satisfy the conditions that guarantee stability. On the other hand, the equilibrium state (saddle point) shown in Fig. 15.18 (*c*) is not stable since for a specified ε, a value of δ can not be produced to satisfy the condition of the definition.

A system is said to be *locally stable* (or *stable in-the-small*) if the region $S(\varepsilon)$ is small.

Let us now turn to the other two definitions of stability.

The system of eqn. (15.8) is *asymptotically stable* at the origin if

(*a*) it is stable, and

(*b*) there exists a real number $r > 0$ such that $\| \mathbf{x}(t_0) \| \leq r$ results in $\mathbf{x}(t) \to 0$ as $t \to \infty$.

Property (*b*) implies that every motion starting in $S(r)$ converges to origin as $t \to \infty$. This is illustrated in Fig. 15.22 (*b*). Note that the singular points in the phase portraits of Figs. 15.16 and 15.19 (*a*) are asymptotically stable.

The system of eqn. (15.8) is *asymptotically stable in-the-large* (*globally asymptotically stable*) at the origin if

(*a*) it is stable, and

(*b*) every initial state $\mathbf{x}(t_0)$ results in $\mathbf{x}(t) \to \mathbf{0}$ as $t \to \infty$.

Thus asymptotic stability in-the-large guarantees that every motion will approach the origin. In fact the signular points in the phase-portraits of Figs. 15.16 and 15.19 (*a*) are asymptotically stable in-the-large.

Limit Cycles

In the definitions of stability of nonlinear systems given above, the system stability has been defined in terms of disturbed steady-state coming back to its equilibrium position or at least staying within tolerable limits from it. This does include the possibility that a disturbed nonlinear system even while staying within tolerance limits, may exhibit a special behaviour of following a closed trajectory or *limit cycle.* The limit cycles describe the oscillations of nonlinear systems. The existence of a limit cycle corresponds to an oscillation of fixed amplitude and period.

As an example, let us consider the well known *Vander Pol's differential equation*

$$\frac{d^2x}{dt^2} - \mu(1 - x^2)\frac{dx}{dt} + x = 0 \quad \text{...[15.20 (a)]}$$

which describes physical situations in many nonlinear systems. By comparing this with the following linear differential equation

$$\frac{d^2x}{dt^2} + 2\zeta\frac{dx}{dt} + x = 0$$

we observe that Vander Pol's equation has damping factor $-(\mu/2)(1 - x^2)$ which depends upon x. If it is assumed that initially $| x | >> 1$, then the damping factor has large positive value. The system therefore behaves like an overdamped system with consequent decrease of the amplitude of x. In this process, the damping factor also decreases and so the system state finally enters a limit cycle as shown by the outer trajectory of Fig. 15.23 (*a*).

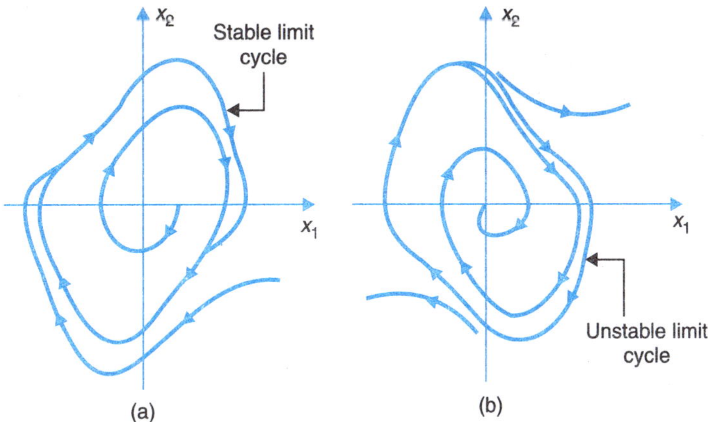

Fig. 15.23. Limit cycle behaviour of nonlinear systems.

On the other hand, if initially $| x | << 1$, the damping is negative hence the amplitude of x increases till the system state again enters the limit cycle as shown by the inner trajectory of Fig. 15.23 (*a*). The limit cycle shown in Fig. 15.23 (*a*) is a stable one, since the parts in its neighbourhood converge toward the limit cycle.

On the other hand, if the paths in the neighbourhood of a limit cycle (closed trajectory) diverge away from it, it indicates that the limit cycle is unstable. Consider, for example, the Vander Pol's equation with the sign of its damping term reversed, *i.e.*,

$$\frac{d^2x}{dt^2} + \mu(1 - x^2)\frac{dx}{dt} + x = 0 \quad \text{...[15.20 (b)]}$$

The phase-trajectories for this equation are shown in Fig. 15.23 (*b*), from which it is observed that an unstable limit cycle occurs.

It is important to note that a limit cycle, in general is an undesirable characteristic of a control system. It may be tolerated only if its amplitude is within specified limits.

It may be noted that the Vander Pol's equation with damping term $+\mu(1 - e^2)$ in fact describes the behaviour of the nonlinear system presented in Example 15.1, if $K_v = \tau = 1$. The identity between the two is established by choosing

$$1 + Kf(e) = \mu(1 - e^2)$$

or

$$f(e) = \frac{\mu(1 - e^2) - 1}{K}$$

For stable node behaviour around origin (equilibrium point)

$$f(0) = \frac{\mu - 1}{K} > \left.\frac{2\sqrt{(K_v\tau) - 1}}{K}\right|_{K_v = \tau = 1} = \frac{1}{K}$$

or

$$\mu > 2$$

and for stable focus behaviour

$$\mu < 2$$

As in the case of Vander Pol's equation, the nonlinear system of Example 15.1 with $f(e)$ defined above will exhibit an unstable limit cycle behaviour as shown in Fig. 15.23 (*b*). Though the nonlinearity makes the transient response more desirable than in a linear case (except for small inputs to which the response is sluggish), the price is paid in terms of unstable behaviour if the system perturbation is large enough to make the initial state point lie outside the limit cycle.

It may be noted that in linear autonomous systems, when oscillations occur, the resulting trajectories will be closed curves (Fig. 15.20 gives trajectories for a second-order system in which oscillations occur). The amplitude of the oscillations is not fixed. It changes with the size of the initial conditions. Slight changes in system parameters (shifting the eigenvalues from the imaginary axis of the complex plane) will destroy the oscillations.

In nonlinear systems, on the other hand, there can be oscillations that are independent of the size of initial conditions (see Fig. 15.23) and these oscillations (limit cycles) are usually much less sensitive to system parameter variations. Limit cycles of fixed amplitude and period can be sustained over a finite range of system parameters.

15.6 CONSTRUCTION OF PHASE-TRAJECTORIES

Consider a second-order autonomous system represented by the following vector differential equation

$$\dot{\mathbf{x}} = \mathbf{F}(\mathbf{x})$$

In the component form, the above equation may be written as two scalar first-order differential equations

$$\dot{x}_1 = f_1(x_1, x_2)$$

$$\dot{x}_2 = f_2(x_1, x_2)$$

Dividing, we get

$$\frac{dx_2}{dx_1} = \frac{f_2(x_1, x_2)}{f_1(x_1, x_2)} \qquad ...(15.21)$$

Equation (15.21) defines the slope of the phase-trajectory. Every point (x_1, x_2) of the phase-plane, has associated with it the slope of the trajectory which passes through that point. The only exception are the singular points at which the trajectory slope is indeterminate. To construct the phase-trajectory, the slope equation must be integrated. Except for simple cases, it is not possible to perform this integration analytically. Therefore the integration must generally be carried out graphically or numerically (may be by use of digital computers).

The direction of phase-trajectory at any point (x_1, x_2) can be ascertained by examining the signs of

$$\Delta x_1 = f_1(x_1, x_2)\ \Delta t$$
$$\Delta x_2 = f_2(x_1, x_2)\ \Delta t$$

in eqn. (15.21) for small positive increment Δt in time. In the commonly met special case, where

$$x_2 = \frac{dx_1}{dt}$$

x_1 increases with time in the upper half of the phase-plane. Since $x_2 = dx_1/dt > 0$ and the state point therefore moves from left to right (→) in this region. In the lower half of the phase-plane $x_2 = dx_1/dt < 0$, x_1 decreases with time and the state point must therefore move from right to left (←).

Further, with this definition of x_2, the trajectories must cross the x_1-axis at 90° since $\left.\frac{dx_2}{dx_1}\right|_{x_2=0} = \left.\frac{f_2(x_1, x_2)}{x_2}\right|_{x_2=0} = \infty$ except at a singular point. In this section we shall present analytical and graphical methods of integrating the slope equation.

Construction by Analytical Method

This is a useful approach for systems described by simple and piecewise linear differential equations. One may obviously raise the question that when time solutions are available by direct integration, where is the necessity of drawing phase-plane plots for such systems? In fact even for such systems, the phase-plane plot provides a powerful qualitative aid for investigating system behaviour and the design of system parameters to achieve a desired response. Furthermore, the existence of limit cycles is sharply brought into focus by the phase-portrait.

We will illustrate the analytical method with the help of the following example:

Example 15.2 : Consider the system shown in Fig. 15.24. The nonlinear elements is an ideal relay whose characteristics are shown in the figure.

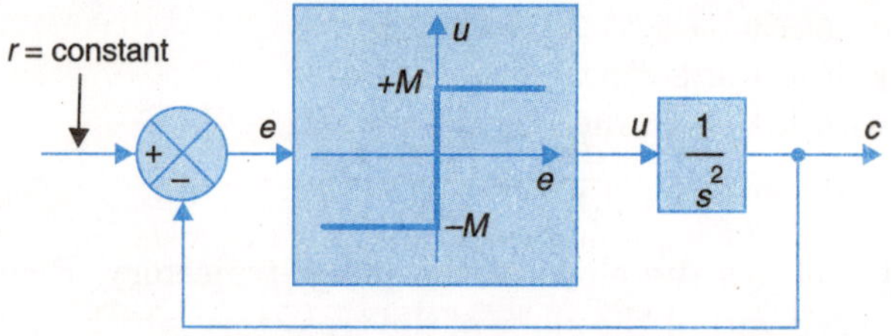

Fig. 15.24. An ON-OFF control system.

From Fig. 15.24 the differential equation describing the dynamics of the system is given by

$$c = u \qquad ...(15.22)$$

But

$$r - c = e$$

$$-\dot{c} = \dot{e}, -\ddot{c} = \ddot{e}\text{ ; as } r = \text{constant}$$

Therefore

$$e = -u \qquad ...(15.23)$$

Choosing the state vector $[x_1\ x_2]^T = [e\ \dot{e}]^T$, we have

$$\dot{x}_1 = x_2$$

$$\dot{x}_2 = \ddot{e} = -u \qquad ...(15.24)$$

The output of the ON/OFF controller is

$$u = M \operatorname{sgn} e$$

$$= M \operatorname{sgn} x_1 \qquad ...(15.25)$$

The state equation are then expressed as

$$\dot{x}_1 = x_2$$

$$\dot{x}_2 = -M \operatorname{sgn} x_1 \qquad ...(15.26)$$

The slope equation is then given by

$$\frac{dx_2}{dx_1} = \frac{M \operatorname{sgn} x_1}{x_2} \qquad ...(15.27)$$

Separating the variables and integrating, we get

$$\int_{x_2(0)}^{x_2} x_2 dx_2 = -M \int_{x_1(0)}^{x_1} \operatorname{sgn} x_1 dx_1$$

Carrying out integration and rearranging, we get

$$x_2^2 = 2Mx_1 - 2M\,x_1(0) + x_2^2(0)\ ;\ x_1 < 0 \qquad ...(15.28)$$

Region I ($x_1 = e$ positive, relay output $+M$)

and

$$x_2^2 = 2M\,x_1 - 2\text{M}x_1\,(0) + x_2^2(0)\ ;\ x_1 < 0 \qquad ...(15.29)$$

Region II ($x_1 = e$ negative, relay output $-M$)

Using these equations the phase portrait is drawn in Fig. 15.25 wherein the centre is at origion. The lower half of Region I trajectory passing through origion is drawn in thick line and the corresponding upper half of Region II trajectory is also drawn in thick line.

If the system is at state point A it follow the trajectory of Region I and switches over to the Region II trajectory that intersects it ; finally closing on to itself. Thus the system oscillates in a limit cycle of magnitude determined by the starting point.

If, however, the switching is made to occur at the crossing of the starting trajectory with the thick line trajectory (as explained above), the system follows this trajectory, it reaches the origion $x_1 = x_2 = 0$ or $e = \dot{e} = 0$ *i.e.*, the output $c = r$ (input). In this traversal the system would take minimum time and this unique trajectory is called **optimum time switching curve.** Suitable logic circuitry has to be added to the controller structure for this kind of switching.

Such systems, are known as **bang-bang** control systems.

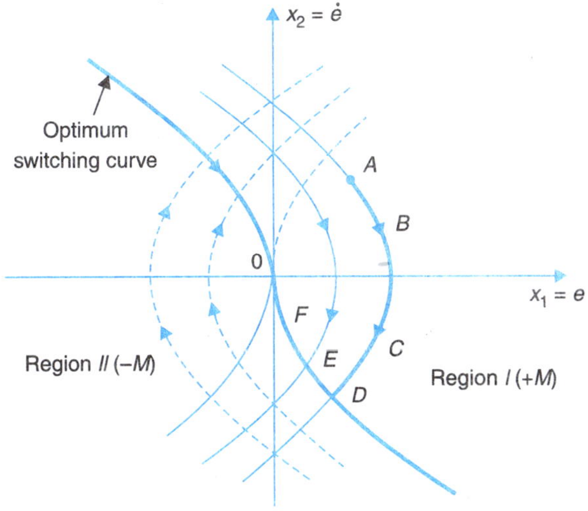

Fig. 15.25. Minimum time trajectory.

Example 15.3 : On lines similar to those of linear systems, consider now the proposition of compensating the ideal relay system of Example 15.2 by means of proportional plus derivative controller ahead of the relay as shown in Fig. 15.26. The modified system is described by

$$\ddot{e} = -u$$
$$u = M \text{ sgn } (e + K_D\dot{e})$$

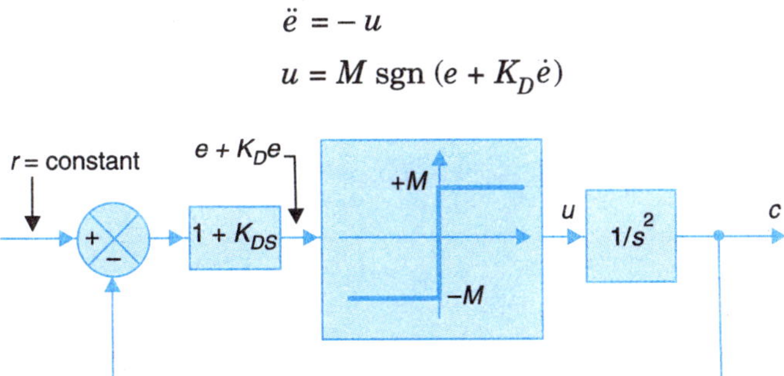

Fig. 15.26. Relay servo system with proportional plus derivative controller.

In the state variable form ($x_1 = e$, $x_2 = \dot{e}$)

$$\dot{x}_1 = x_2$$
$$\dot{x}_2 = -M \text{ sgn } (x_1 + K_D x_2)$$

The slope equation is given by

$$\frac{dx_2}{dx_1} = -\frac{M \operatorname{sgn}(x_1 + K_D x_2)}{x_2}$$

which upon integration gives

$$x_2^2 = -2Mx_1 + 2Mx_1(0) + x_1^2(0)$$

Region I $[(x_1 + K_D x_2) > 0 \,;\, u = +M]$

$$x_2^2 = 2Mx_1 - 2Mx_1(0) + x_2^2(0)$$

Region II $[(x_1 + K_D x_2) < 0 \,;\, u = -M]$

The state-space is divided into two regions by the line

$$x_1 + K_D x_2 = 0$$

This is the equation of the switching line, while the trajectories are parabolas defined by equations given above and are drawn in Fig. 15.27. Any starting state point keeps moving inwards as it switches from one parabole to another till the point moves inside the part A_1A_2 of the switching line. These points are identified by the intersection of the switching line with $\pm M$ trajectories passing through origin.

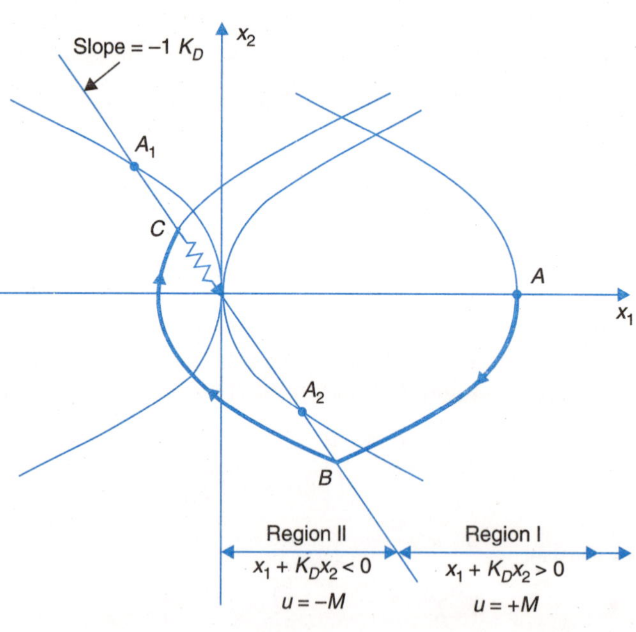

Fig. 15.27

Once the state point reaches inside this line, it will come to origin along this in a chattering fashion (as shown in figure). The relay now immediately switches back and forth; the reader can easily verify this. This is not a desirable feature.

In view of the discussion above, the most desirable switching is along one of the trajectories passing through origin; as shown in Fig. 15.26. This is easy to implement with powerful logic chips being available now.

Construction by Graphical Methods

Graphical methods are no longer of much importance as the trajectories can be constructed by computer methods using powerful integrating software now available. So, only a passing account of graphical methods will be presented.

Isocline method

Trajectory slope at any point is given by

$$\frac{dx_2}{dx_1} = S = \frac{f_2(x_1, x_2)}{f_1(x_1, x_2)} \qquad \text{...(15.30)}$$

Slope at a specific point is expressed as

$$f_2(x_1, x_2) = S_1 f_1(x_1, x_2) \qquad \text{...(15.31)}$$

For a given S_1 this is the equation of an isocline which any trajectory crosses at slope S_1. If isoclines for different values of S are drawn throughout the phase-plane, trjectory starting at any point can be constructed by drawing short lines from one isocline to another at average slope of the two adjoining isoclines as shown in Fig. 15.28.

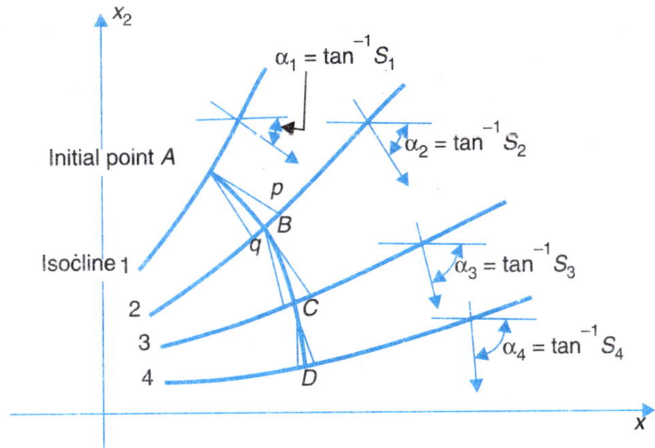

Fig. 15.28. Construction of trajectory by the isocline method.

δ-Method

This method can be used to construct a single trajectory for systems with describing differential equations in the form

$$\ddot{x} + f(x, \dot{x}, t) = 0$$

Convert this equation to the form

$$\ddot{x} + K^2[x + \delta(x, \dot{x}, t)] = 0$$

Let $K = \omega_n$, undamped natural frequency of the system when $\delta = 0$.

Then, the equation is written as

$$\ddot{x} + \omega_n^2 [x + \delta(x, \dot{x}, t)] = 0$$

or

$$\ddot{x} + \omega_n^2 (x + \delta) = 0$$

Choose the state variables as $\quad x_1 = x,\ x_2 = \dot{x}/\omega_n$

giving state equations

$$\dot{x}_1 = \omega_n x_2$$

$$\dot{x}_2 = -\omega_n(x_1 + \delta)$$

Trajectory slope is expressed as

$$\frac{dx_2}{dx_1} = -\frac{x_1 + \delta}{x_2}$$

Consider now Fig. 15.29 wherein for any point $P(x_1, x_2)$, δ is computed and marked on negative side of x_1-axis. The slope of line CP = $x_2/(x_1 + \delta)$. Therefore, the line of slope $-x_2/(x_1 + \delta)$ is at 90° to CP. The trajectory at P is then identified as a short arc at P with centre at C. The process is then repeated for another P shortly away from P (original) on the arc. The completed trajectory can thus be drawn.

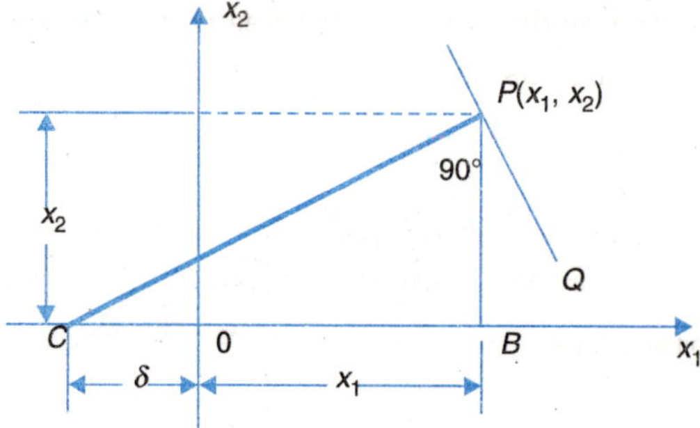

Fig. 15.29. Geometry of the δ-method.

Example 15.4 : Consider the system shown in Fig. 15.30. The nonlinear element is a relay with dead-zone, whose characteristics are shown in the figure.

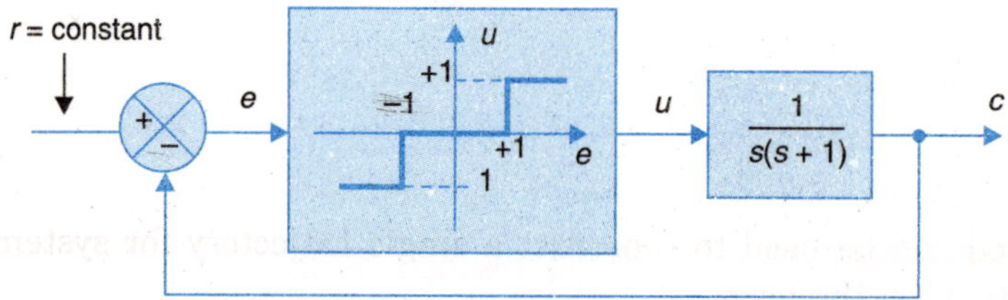

Fig. 15.30. 'ON-OFF' control system with dead-zone.

The differential equation describing the dynamics of the system is given by

$$\ddot{c} + \dot{c} = u$$

$$e = r - c$$

Since r is constant, the first equation can be written as

$$\ddot{e} + \dot{e} = -u$$

The relay output equation is given by

$$u = \phi(e)$$

where $$\phi(e) = \begin{cases} +1 & \text{for } e > 1 \\ 0 & \text{for } -1 < e < 1 \\ -1 & \text{for } e < -1 \end{cases}$$

Choosing the phase variables as $x_1 = e$; $x_2 = \dot{e}$, we obtain the following first-order equations:

$$\dot{x}_1 = x_2$$
$$\dot{x}_2 = -x_2 - \phi(x_1)$$

The slope of trajectory is then given by

$$\frac{dx_2}{dx_1} = -\frac{x_2 + \phi(x_1)}{x_2} = S$$

The trajectory obtained by numerical integration is drawn in Fig. 15.31. Observe that trajectory slope in the dead zone region is $S = -1$ as $u = 0$ and the system is in open-loop mode. Because of dead zone the system comes to rest with a certain error whose magnitude is dependent on the starting point of the trajectory. Maximum error magnitude can be ± 1.

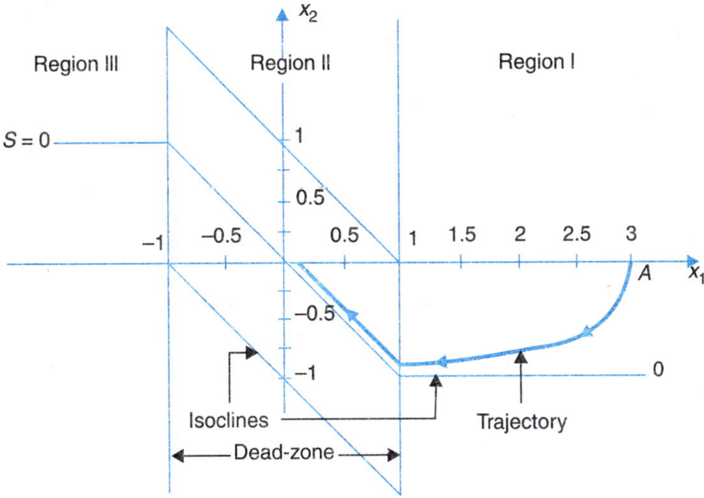

Fig. 15.31. Isoclines and a typical trajectory for the system of Fig. 15.30.

System trajectory with absence of dead zone is plotted in Fig. 15.32, where the system is found to be oscillatory. Dead zone while causing a certain steady error eliminates oscillation.

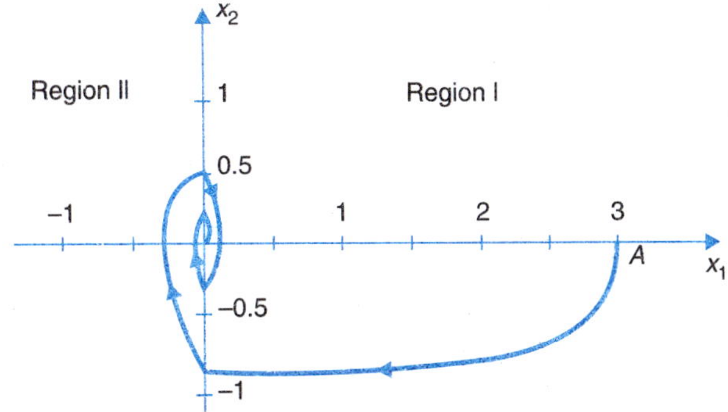

Fig. 15.32. Phase-trajectory for the system of Fig. 15.30 when dead-zone is absent.

15.7 THE DESCRIBING FUNCTION METHOD: BASIC CONCEPTS

The phase-plane method discussed in the preceding sections uses a state model of the system which is obtained from the differential equations governing the system dynamics. This method, though very powerful, is generally restricted to systems described by second-order differential equations. For higher-order systems, therefore the only alternative is to approximate the system by a second-order one, in order to be able to use the phase-plane method for analysis and design.

Certain approximate techniques exist which are capable of determining the behaviour of a wider class of systems than is possible by the phase-plane method. These techniques are known as the *describing function techniques*. To discuss the basic concept underlying these techniques, let us consider the block diagram of a nonlinear system shown in Fig. 15.33 wherein the blocks $G_1(s)$ and $G_2(s)$ represent the linear elements, while the block N represents the nonlinear element.

Let us assume that input x to the nonlinearity is sinusoidal, *i.e.*,

$$x = X \sin \omega t$$

With such an input, the output of the nonlinear element will in general be nonsinusoidal periodic function which may be expressed in terms of Fourier series as follows:

$$y = A_0 + A_1 \sin \omega t + B_1 \cos \omega t + A_2 \sin 2\omega t + B_2 \cos 2\omega t + \ldots$$

It is important to note that in writing the above expression, it has been assumed that the nonlinearity N does not generate subharmonics. Furthermore, if the nonlinearity is assumed to be symmetrical. The average value of y is zero, so that the output y is then given by

$$y = A_1 \sin \omega t + B_1 \cos \omega t + A_2 \sin 2\omega t + B_2 \cos 2\omega t + \ldots$$

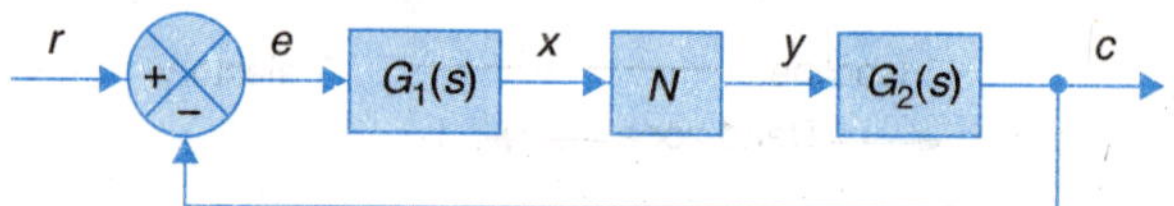

Fig. 15.33. A nonlinear system.

In the absence of an external input (*i.e.*, $r = 0$), the output y of N is fed back to it's input through the linear elements $G_2(s)$ and $G_1(s)$ in tandem. If $G_2(s)G_1(s)$ has low-pass characteristics* (this is usually the case in control systems), it can be assumed to a good degree of accuracy that all the harmonics of y are filtered out in the process such that the input $x(t)$ to the nonlinear element N is mainly contributed by the fundamental component of y, *i.e.*, $x(t)$ remains sinusoidal. Under such conditions the harmonic content of y can be thrown away for the purpose of analysis and the fundamental component of y, *i.e.*,

$$\begin{aligned} y_1 &= A_1 \sin \omega t + B_1 \cos \omega t \\ &= Y_1 \sin (\omega t + \phi_1) \end{aligned}$$

need only be considered.

*Filtering characteristics of the linear part of a nonlinear system improve as the order of the system goes up.

The above procedure heuristically linearizes the nonlinearity, since for a sinusoidal input, only a sinusoidal output of the same frequency is now assumed to be produced. This type of linearization is valid for large signals as well so long as the harmonic filtering condition is satisfied.

Under the above assumption, the nonlinearity can be replaced by a *describing function* $K_N(X, \omega)$ which is defined to be the complex function embodying amplification and phase shift of the fundamental frequency component of y relative to x, *i.e.,*

$$K_N(X, \omega) = (Y_1/X) \angle \phi_1 \quad ...(15.32)$$

when the input to the nonlinearity is

$$x = X \sin \omega t$$

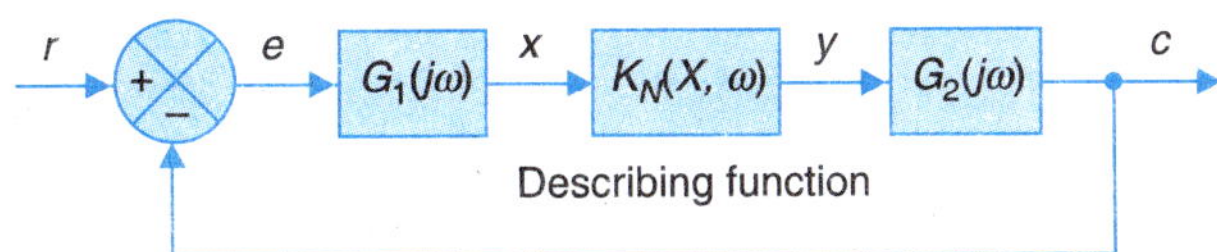

Fig. 15.34. Nonlinear system with nonlinearity replaced by describing function.

As indicated above, the describing function is in general dependent upon the amplitude and frequency of the input.

From Fig. 15.34, it is observed that it is much easier to handle a nonlinear system when the nonlinearity is replaced by its describing function. All the linear theory frequency domain techniques now become applicable. It is important to remind ourselves here that this simplicity has been achieved at the cost of certain limitations, the foremost being the assumption that the linear elements in the feedback path around the nonlinearity have low-pass characteristics. Because of such limitations, the describing function method is mainly used for stability analysis and is not directly applied to the optimization of system design. Furthermore, though the describing function is a frequency domain approach, no general correlation is possible between time and frequency responses.

Before coming to the stability study by the describing function method, it is worthwhile to derive the describing functions of some common nonlinearities.

15.8 DERIVATION OF DESCRIBING FUNCTIONS

As discussed in the previous section, the describing function of a nonlinear element is given by

$$K_N(X, \omega) = (Y_1/X) \angle \phi_1$$

where X = amplitude of the input sinusoid; Y_1 = amplitude of the fundamental harmonic component of the output; and ϕ_1 = phase shift of the fundamental harmonic component of the output with respect to the input.

Therefore, for computing the describing function of a nonlinear element, we are simply required to find the fundamental harmonic component of its output for an input $x = X \sin \omega t$. The fundamental component of the output can be written as

$$y_1 = A_1 \sin \omega t + B_1 \cos \omega t$$

where the coefficients A_1 and B_1 of the Fourier series are

$$B_1 = \frac{1}{\pi}\int_0^{2\pi} y \cos \omega t \, d(\omega t) \qquad \text{...[15.33 (a)]}$$

$$A_1 = \frac{1}{\pi}\int_0^{2\pi} y \sin \omega t \, d(\omega t) \qquad \text{...[15.33 (b)]}$$

The amplitude and phase angle of the fundamental component of the output are given by

$$Y_1 = \sqrt{(A_1^2 + B_1^2)} \qquad \text{...(15.34)}$$

$$\phi_1 = \tan^{-1}(B_1/A_1) \qquad \text{...(15.35)}$$

Illustrative derivations of the describing functions of commonly encountered nonlinearities are given below.

Dead-zone and Saturation

Idealized characteristic of a nonlinearity having dead-zone and saturation and its response to sinusoidal input are shown in Fig. 15.35. The output wave form may be described as follows:

$$y = \begin{cases} 0 & ; 0 \le \omega t \le \alpha \\ K(x - D/2); & \alpha \le \omega t \le \beta \\ K(S - D/2); & \beta \le \omega t \le (\pi - \beta) \\ K(x - D/2); & (\pi - \beta) \le \omega t \le (\pi - \alpha) \\ 0 & ; (\pi - \alpha) \le \omega t \le \pi \end{cases}$$

where $\alpha = \sin^{-1} D/2X$ and $\beta = \sin^{-1} S/X$

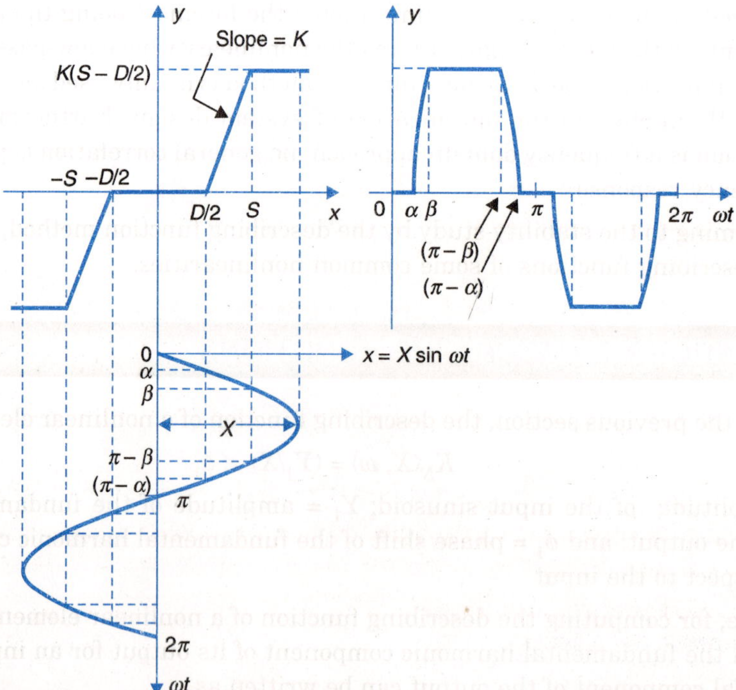

Fig. 15.35. Sinusoidal response of nonlinearity with dead-zone and saturation.

Using eqns. 15.33 (*a*) and 15.33 (*b*) and recognizing that the output has half-wave and quarter-wave symmetries, we have

$$B_1 = 0$$

$$A_1 = \frac{4}{\pi}\int_0^{\pi/2} y \sin \omega t \, d(\omega t)$$

$$= \frac{4}{\pi}\left[\int_\alpha^\beta K(X\sin\omega t - D/2)\sin\omega t\, d(\omega t) + \int_\beta^{\pi/2} K(S - D/2)\sin\omega t\, d(\omega t)\right]$$

$$= \frac{K}{\pi}\left\{2X(\beta-\alpha) - X(\sin 2\beta - \sin 2\alpha) + 4\left[\frac{D}{2}(\cos\beta - \cos\alpha) + \left(S - \frac{D}{2}\right)\cos\beta\right]\right\}$$

$$= \frac{KX}{\pi}\left[2(\beta-\alpha) + (\sin 2\beta - \sin 2\alpha)\right]$$

Therefore, the describing function is given by

$$\frac{K_N(X)}{K} = \begin{cases} 0 & ; X < D/2\,; \alpha = \beta = \pi/2 \\ 1 - \dfrac{2}{\pi}(\alpha + \sin\alpha\cos\alpha) & ; D/2 < X < S\,; \beta = \pi/2 \\ \dfrac{1}{\pi}[2(\beta-\alpha) + (\sin 2\beta - \sin 2\alpha)] & ; X > S \end{cases} \quad \text{...(15.36)}$$

Two special cases immediately follow from eqn. (15.36).

Case I. Saturation nonlinearity ($D/2 = 0,\ \alpha = 0$)

$$\frac{K_N(X)}{K} = \begin{cases} 1 & ; X < S \\ \dfrac{2}{\pi}(\beta + \sin\beta\cos\beta) & \\ = \dfrac{2}{\pi}\{\sin^{-1}(S/X) + (S/X)\sqrt{[1-(S/X)^2]}\} & ; X > S \end{cases} \quad \text{...(15.37)}$$

Case II. Dead-zone nonlinearity ($S \to \infty\,;\ \beta = \pi/2$)

$$\frac{K_N(X)}{K} = \begin{cases} 0 & ; X < D/2 \\ 1 - \dfrac{2}{\pi}(\alpha + \sin\alpha\cos\alpha) & \\ = 1 - \dfrac{2}{\pi}\{\sin^{-1}(D/2X) + (D/2X)\sqrt{[1-(D/2X)^2]}\} & ; X > D/2 \end{cases} \quad \text{...(15.38)}$$

It is found that the describing functions of saturation and dead-zone nonlinearities are frequency-invariant having zero phase shift. In fact all nonlinearities, whose input-output characteristics are represented by a planer graph, would result in describing functions independent of frequency but amplitude dependent. An element described by a nonlinear differential equation on the other hand has both frequency and amplitude dependent describing function. A frequency-invariant describing function having zero phase shift would be produced by a *memoryless* nonlinearity whose output is independent of the history of input as is the case with saturation and dead-zone nonlinearities.

Relay with Dead-zone and Hysteresis

The characteristics of a relay with dead-zone and hysteresis and its response to sinusoidal input are shown in Fig. 15.36.

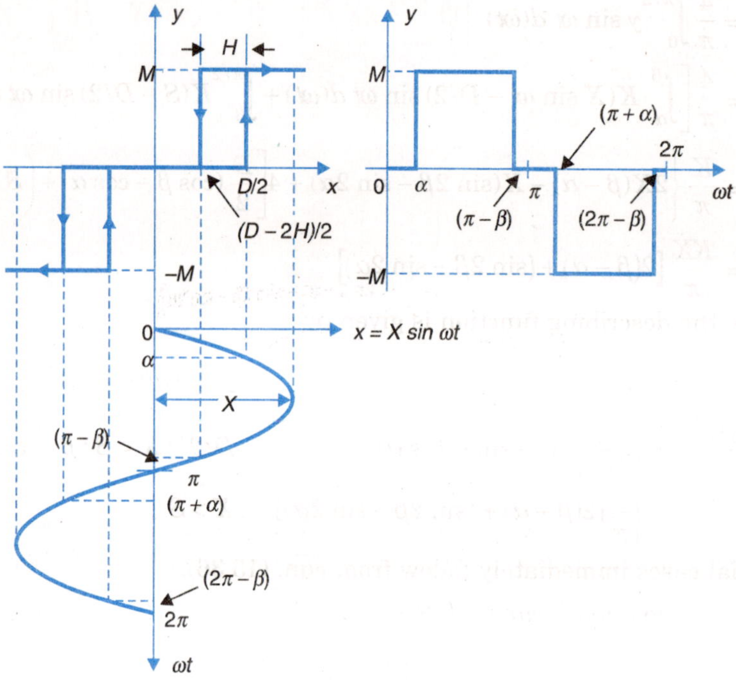

Fig. 15.36. Sinusoidal response of relay with dead-zone and hysteresis.

The output y may be described as follows:

$$y = \begin{cases} 0 & ; 0 \le \omega t \le \alpha \\ +M & ; \alpha \le \omega t \le (\pi - \beta) \\ 0 & ; (\pi - \beta) \le \omega t \le (\pi + \alpha) \\ -M & ; (\pi - \alpha) \le \omega t \le (2\pi - \beta) \\ 0 & ; (2\pi - \beta) \le \omega t \le 2\pi \end{cases}$$

where $\alpha = \sin^{-1} D/2X$; $\beta = \sin^{-1} (D - 2H)/2X$.

Using eqns. [15.33 (a)] and [15.33 (b)] we have

$$B_1 = \frac{2}{\pi} \int_0^{\pi} y \cos \omega t \; d(\omega t)$$

$$= \frac{2}{\pi} \int_{\alpha}^{\pi-\beta} M \cos \omega t \; d(\omega t) = \frac{2M}{\pi} (\sin \beta - \sin \alpha)$$

$$= \frac{2M}{\pi} \left(-\frac{H}{X} \right)$$

$$A_1 = \frac{2}{\pi} \int_0^{\pi} y \sin \omega t \; d(\omega t)$$

$$= \frac{2}{\pi}\int_{\alpha}^{\pi-\beta} M \sin \omega t \, d(\omega t) = \frac{2M}{\pi}(\cos\alpha + \cos\beta)$$

$$= \frac{2M}{\pi}\left\{\sqrt{\left[1-\left(\frac{D}{2X}\right)^2\right]} + \sqrt{\left[1-\left(\frac{D-2H}{2X}\right)\right]^2}\right\}$$

Therefore

$$B_1/X = \frac{2M}{\pi X}\left(-\frac{H}{X}\right) \quad \text{...(15.39)}$$

$$A_1/X = \frac{2M}{\pi X}\left\{\sqrt{\left[1-\left(\frac{D}{2X}\right)^2\right]} + \sqrt{\left[1-\left(\frac{D-2H}{2X}\right)^2\right]}\right\} \quad \text{...(15.40)}$$

$$K_N(X) = \begin{cases} \sqrt{[(A_1/X)^2 + (B_1/X)^2]} \angle \tan^{-1} B_1/A_1 \,; X > D/2 \\ 0 \qquad ; X < D/2 \end{cases} \quad \text{...(15.41)}$$

It is seen that K_N is a function of X, the input amplitude and is independent of frequency. Further, it being a *memory type* nonlinearity (*i.e.,* output is dependent upon the history of input), K_N has both magnitude and angle (a lagging angle which produces the effect of the pole of a linear system). Equation (15.41) is the general expression for the describing function of a relay element, from which the describing functions of various simplified relay characteristics given in Fig. 15.37 are directly obtainable and are given below:

I. *Ideal relay* [Fig. (15.37 (*a*)]. Letting $D = H = 0$ in eqn. (15.41)

$$K_N(X) = 4M/\pi X \quad \text{...(15.42)}$$

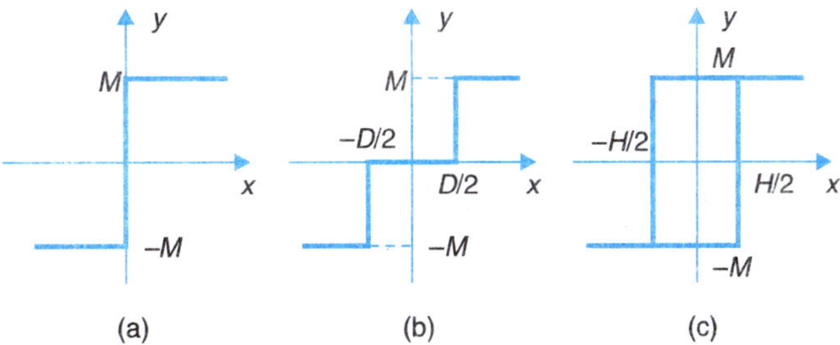

Fig. 15.37. Various simplified relay characteristics.

II. *Relay with dead-zone* [Fig. 15.37 (*b*)]. Letting $H = 0$ in eqn. (15.42)

$$\frac{D}{M} K_N(X) = \begin{cases} 0 & ; X < D/2 \\ \dfrac{4D}{\pi X}\sqrt{[1-(D/2X)^2]} & ; X > D/2 \end{cases} \quad \text{...(15.43)}$$

III. *Relay with hysteresis* [Fig. 15.37 (*c*)]. Letting $H = D$ in eqn. (15.41).

$$\frac{H}{M} K_N(X) = \begin{cases} 0 & ; X < D/2 \\ \dfrac{4H}{\pi X}\angle -\sin^{-1} H/2X & ; X > H/2 \end{cases} \quad \text{...(15.44)}$$

Backlash

The characteristics of a backlash nonlinearity and its response to sinusoidal input are shown in Fig. 15.38. The output may be described as follows :

$$\begin{aligned} y &= x - b/2 \quad ; \quad 0 \le \omega t \le \pi/2 \\ &= X - b/2 \quad ; \quad \pi/2 \le \omega t \le (\pi - \beta) \\ &= x + b/2 \quad ; (\pi - \beta) \le \omega t \le 3\pi/2 \\ &= -X + b/2 \quad ; \quad 3\pi/2 \le \omega t \le (2\pi - \beta) \\ &= x - b/2 \quad ; \quad (2\pi - \beta) \le \omega t \le 2\pi \end{aligned}$$

where $\beta = \sin^{-1}(1 - b/X)$.

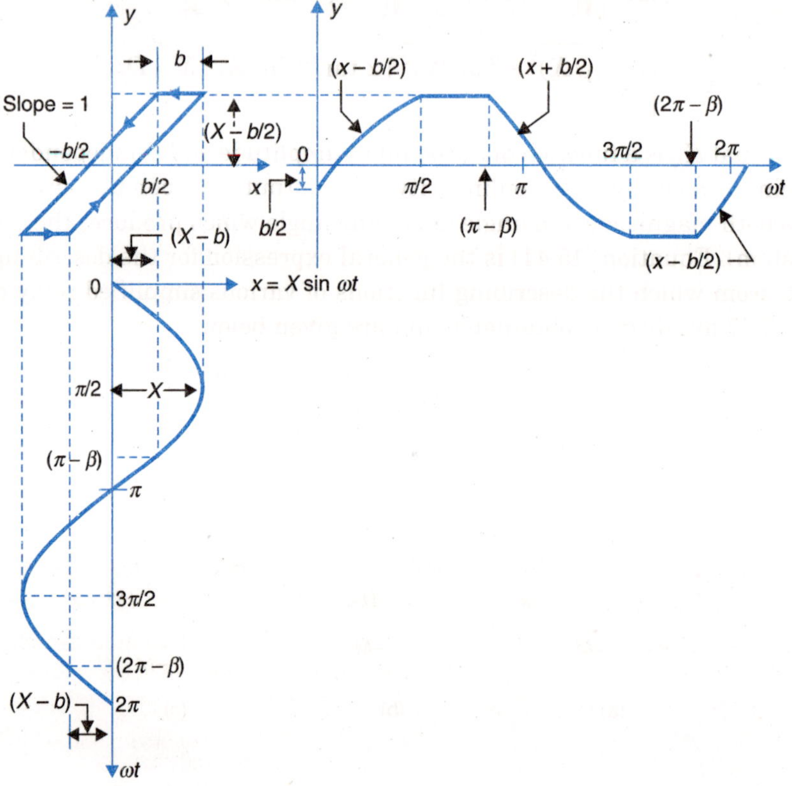

Fig. 15.38. Characteristic and sinusoidal response of backlash nonlinearity.

Using eqns. [15.33 (*a*)] and [15.33 (*b*)], we have

$$B_1 = \frac{2}{\pi}\int_0^{\pi} y \cos \omega t \, d(\omega t)$$

$$= \frac{2}{\pi}\int_0^{\pi/2}\left(X \sin \omega t - \frac{b}{2}\right)\cos \omega t \, d(\omega t) + \frac{2}{\pi}\int_{\alpha}^{\pi-\beta}\left(X - \frac{b}{2}\right)\cos \omega t \, d(\omega t)$$

$$+ \frac{2}{\pi}\int_{\pi-\beta}^{\alpha}\left(X \sin \omega t + \frac{b}{2}\right)\cos \omega t \, d(\omega t)$$

$$= -\frac{X}{\pi}\cos^2\beta$$

$$A_1 = \frac{2}{\pi}\int_0^{\pi} y\sin\omega t\, d(\omega t)$$

$$= \frac{2}{\pi}\int_0^{\pi/2}\left(X\sin\omega t - \frac{b}{2}\right)\sin\omega t\, d(\omega t) + \frac{2}{\pi}\int_{\pi/2}^{\pi/\beta}\left(X - \frac{b}{2}\right)\sin\omega t\, d(\omega t)$$

$$+ \frac{2}{\pi}\int_{\pi-\beta}^{\pi}\left(X\sin\omega t - \frac{b}{2}\right)\sin\omega t\, d(\omega t)$$

$$= \frac{X}{\pi}\left[\left(\frac{\pi}{2}+\beta\right)+\frac{1}{2}\sin 2\beta\right]$$

Therefore,

$$K_N(X) = \begin{cases} 0 & ; X < b/2 \\ \sqrt{[(A_1/X)^2 + (B_1/X)^2]}\angle\tan^{-1} B_1/A_1 & ; X > b/2 \end{cases} \quad ...(15.45)$$

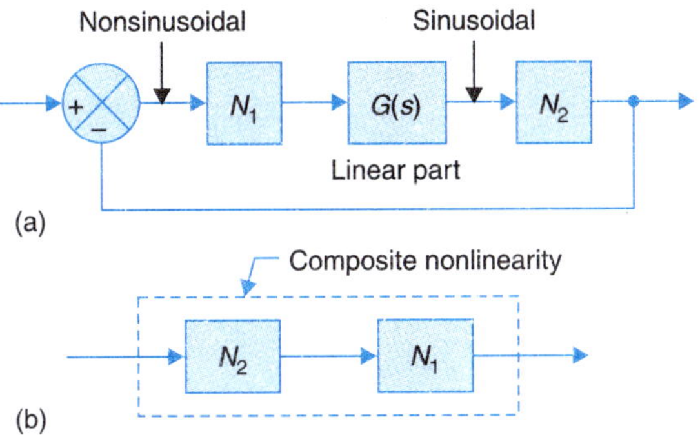

Fig. 15.39. System with two nonlinearities in tandem.

When two nonlinearities are placed in tandem, the resultant describing function cannot be obtained by multiplying the describing functions of the individual nonlinearities.

Consider the closed loop system of Fig. 15.39 (*a*) with two nonlinearities N_1 and N_2 which are in tandem round the loop. The input to N_2 is sinusoidal as it is the output of the linear filter $G(s)$. But the input to N_1 is the nonsinusoidal output of N_2. So while the describing function can be found for N_2, the concept of describing function cannot be applied to N_1 as its input is nonsinusoidal.

In such case N_1, N_2 must be combined into a composite nonlinearity as in Fig. 15.39 (*b*) whose describing function can then be found.

15.9 STABILITY ANALYSIS BY DESCRIBING FUNCTION METHOD

The widest use of describing functions is in stability investigations and prediction of limit cycles in feedback systems.

Consider the system shown in Fig. 15.34 wherein the nonlinearity has been replaced by its describing function. The system having thus been linearized, its characteristic equation can be written as

$$1 + G_1(j\omega)\, G_2(j\omega)\, K_N(X, \omega) = 0$$

Notice that the characteristic equation is independent of the physical location of the nonlinearity in the system. According to the Nyquist stability criterion, the system will exhibit sustained oscillations or limit cycle when

$$G_1(j\omega)\, G_2(j\omega)\, K_N(X, \omega) = -1 \qquad \text{...(15.46)}$$

This condition implies that the plot of $G_1(j\omega)G_2(j\omega)K_N(X, \omega)$ passes through the critical point –1. This extension of the Nyquist criterion is completely heuristic and has no mathematical basis. However, it is a powerful tool because it works.

The condition of eqn. (15.46) can be written in the modified form

$$G_1(j\omega)\, G_2(j\omega) = -1/K_N(X, \omega) \qquad \text{...(15.47)}$$

This means that intersections of the plot $G_1(j\omega)G_2(j\omega)$ with the critical locus determine the amplitude and frequency of the limit cycles.

Consider a simple frequency-invariant describing function wherein the locus of $-1/K_N(X)$ is a single curve in the complex plane as shown in Fig. 15.40. Let the $G(j\omega)$-plot be superimposed on $-1/K_N(X)$-locus. It is assumed here that the system under investigation is open-loop stable. The closed-loop system stability is investigated below.

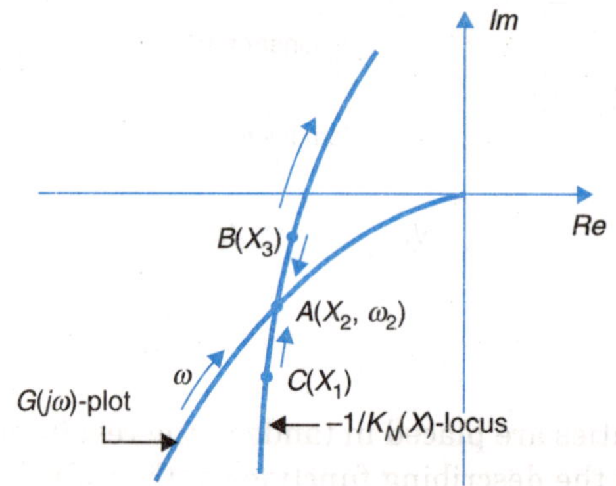

Fig. 15.40. Prediction and stability of limit cycle.

The values of X for which the $-1/K_N(X)$-locus is enclosed by the $G(j\omega)$-plot [*i.e.,* the $-1/K_N(X)$-locus lies in the region to the right of an observer traversing the $G(j\omega)$-plot for positive frequencies in the direction of increasing ω] correspond to unstable conditions. Similarly, the values of X for which the $-1/K_N(X)$-locus is not enclosed by the $G(j\omega)$-plot [*i.e.,* the $-1/K_N(X)$-locus lies in the region to the left of an observer traversing the $G(j\omega)$-plot for positive frequencies in the direction of increasing ω] correspond to stable conditions.

The $-1/K_N(X)$-locus and $G(j\omega)$-plot intersect at the point $(\omega = \omega_2, X = X_2)$ which corresponds to the condition of limit cycle (self-sustained oscillation). The system is unstable for $X < X_2$ and is stable for $X > X_2$.

The stability of the limit cycle can be judged by the *perturbation technique*.

Suppose the system is originally operating at A under the state of a limit cycle. Assume that a slight perturbation is given to the system operating at A, so that the input to the nonlinear element increases to X_3, *i.e.*, the operating point is shifted to B. Since B is in the range of stable operation, the amplitude of input to the nonlinear element progressively decreases and hence the operating point moves back towards A. Similarly, a perturbation which decreases the amplitude of input to the nonlinearity shifts the operating point to C which lies in the range of unstable operation. The input amplitude now progressively increases and the operating point again returns to A. Therefore the system has a stable limit cycle at A. Figure 15.41 (a) shows the case of an unstable limit cycle. For systems having $G(j\omega)$-plots and $-1/K_N(X)$-loci as shown in Figs. 15.41 (b) and (c), there are two limit cycles, one stable and the other unstable.

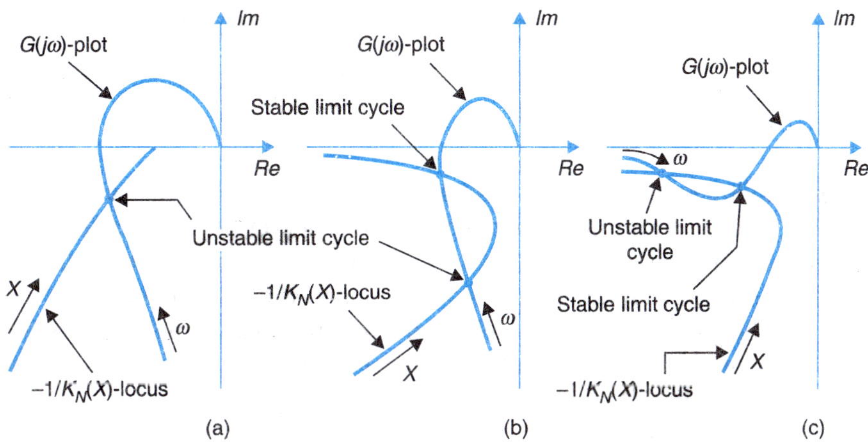

Fig. 15.41. Prediction and stability of limit cycles.

For systems having frequency dependent describing functions, the method of stability analysis is essentially the same except that $-1/K_N(X, \omega)$ has infinitely many loci one for each value of ω. The limit cycles, if any, are determined by the intersections of $-1K_N(X, \omega)$-loci and $G(j\omega)$-plot. Only such intersections qualify for a limit cycle for which the $-1/K_N(X, \omega)$-loci and $G(j\omega)$-plot have a common value of ω.

The concepts of stability analysis are illustrated with the help of two examples given below.

Example 15.6 : Let us investigate the stability of a relay-controlled system shown in Fig. 15.42.

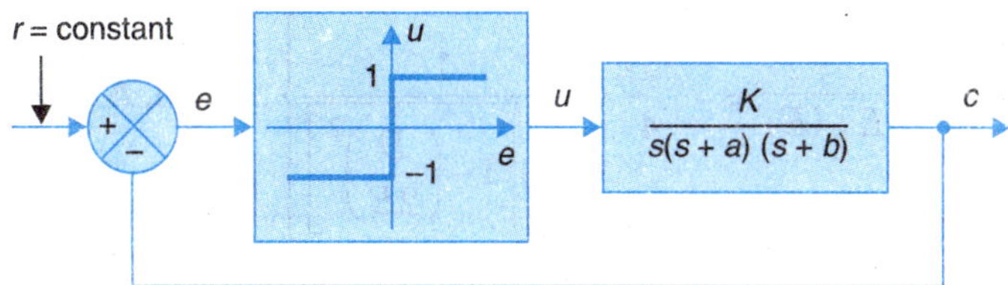

Fig. 15.42. A relay-controlled system.

Using the describing function of an ideal relay given by eqn. (15.42), we have

$$-\frac{1}{K_N(E)} = -\frac{\pi}{4}E\,;\; M = 1 \qquad ...(15.48)$$

where E is the maximum amplitude of the sinusoidal error signal e. Figure 15.43 shows the locus of $-1K_N(E)$ as a function of E and the plots of $G(j\omega)$ for two different values of K.

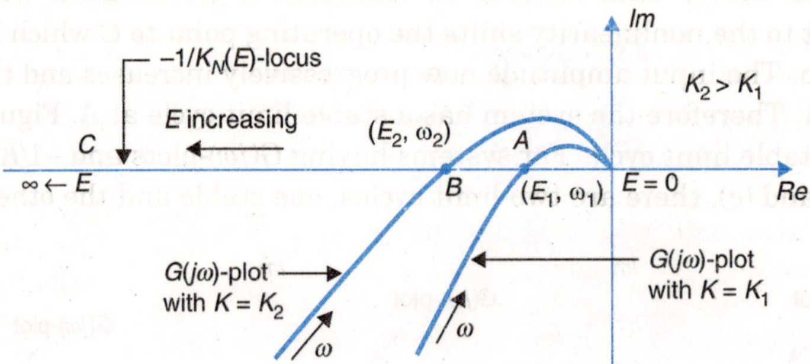

Fig. 15.43. Stability analysis of a system controlled by ideal relay.

Consider the $G(j\omega)$-plot for $K = K_1$ which intersects the $-1/K_N(E)$-locus at A resulting in a limit cycle of amplitude E_1 and frequency ω_1. As an observer traverses the $G(j\omega)$-plot in the direction of increasing ω, the portion OA of $-1/K_N(E)$-locus lies to its right and the portion AC lies to its left. Using the arguments presented previously, we can conclude that the limit cycle is a stable one. As the gain is increased to $K_2 > K_1$, the intersection point shifts to B resulting in a limit cycle of amplitude $E_2 > E_1$ and frequency $\omega = \omega_2$. It should be observed that the system has a limit cycle for all positive values of gain.

Since for the ideal relay, $-1/K_N(E)$-locus is the negative real axis, the frequency of limit cycle can also be determined from the condition

$$\angle G(j\omega_l) = -180°$$

where ω_l is the limit cycle frequency. Knowing ω_l, the amplitude of the limit cycle can be found from

$$\frac{1}{K_N(E)} = \frac{\pi}{4} E = |G(j\omega_l)|$$

Relay with Dead Zone

The describing function of this type of relay is given in eqn. (15.43) which can be written as

$$-\frac{1}{K_N(E)} = -\frac{\pi D}{4}\left\{\frac{1}{\frac{D}{E}\sqrt{\left[1-\left(\frac{D}{2E}\right)^2\right]}}\right\}; M = 1, E > D/2$$

A plot of this function is shown in Fig. 15.44 from which we observe that when the gain K is small enough, the $G(j\omega)$-plot crosses the negative real axis at a point to the right of $-\pi D/4$ such that no intersection takes place between the plots of $G(j\omega)$ and $-1/K_N(E)$ and therefore no limit cycle results. With such a gain the $-1/K_N(E)$-plot lies entirely to the left of the $G(j\omega)$-plot, the system therefore is stable, *i.e.*, it has effectively positive damping.

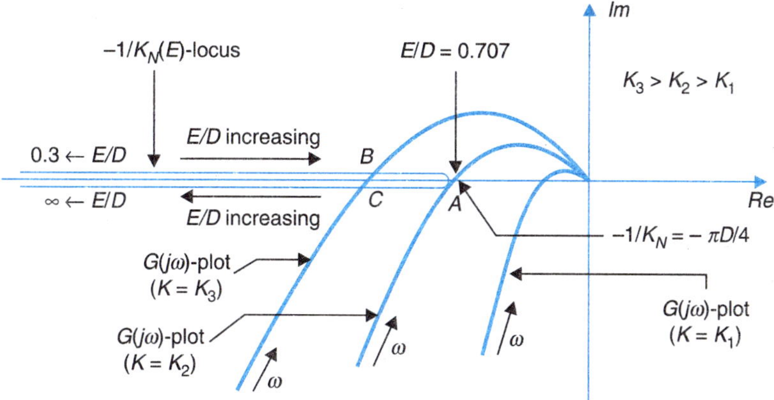

Fig. 15.44. Stability analysis of a system controlled by relay with dead-zone.

If the gain K is now increased to a value K_2 such that the $G(j\omega)$-plot intersects the $-1/K_N(E)$-plot at the point A *i.e.,* on the negative real axis at $-\pi D/4$ then there exists a limit cycle of amplitude $-\pi D/4$ and frequency ω_A. Any perturbation that causes E to increase or decrease corresponding to its value at A takes the operating point to the left of A, so that the system has positive damping causing E to decrease and the operating point returns to A. This is indicative of stable limit cycle but in reality it cannot be so as random disturbance causes E to decrease, the system finally comes to rest *i.e.,* the limit cycles is unstable.

When the gain K is further increased to K_3, the $G(j\omega)$-and $-1/K_N(E)$-plots intersect at two points B and C. By arguments similar to those advanced in the case of ideal relay, it can be shown that the point B represents an unstable limit cycle and C represents a stable limit cycle. It may be noted that though the points B and C lie at the same place on negative real axis, they belong to different values of E/D.

Relay with Hysteresis

The describing function for this type of relay given in eqn. (15.44) can be written as

$$-\frac{1}{K_N(E)} = -\frac{\pi}{4} H\left(\frac{E}{H}\right) \angle \sin^{-1} H/2E;\ E/H > 1/2\text{-}$$

$$= -\frac{\pi H}{4}\left(\frac{E}{H}\right) \angle[-180° + \sin^{-1} H/2E]$$

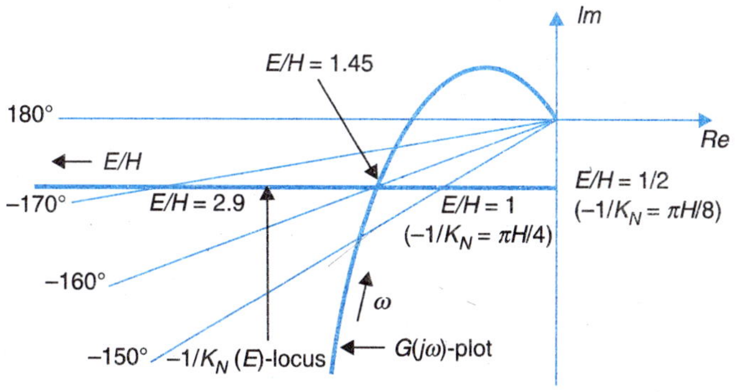

Fig. 15.45. Stability analysis of a system controlled by relay with hysteresis.

This function is plotted in Fig. 15.45, and is found to be a straight line parallel to the real axis. The point of intersection of the $G(j\omega)$-plot with $-1/K_N(E)$ plot determines the amplitude and frequency of the limit cycle which is found to be stable.

Example 15.7 : Consider a third-order system with a saturating amplifier of Fig. 15.46 (*a*) having gain K in its linear region. Determine the largest value of gain K for the system to stay stable. What would be the frequency, amplitude and nature of the limit cycle for a gain of $K = 3$?

Solution. It is convenient to regard the amplifier to have unit gain and the gain K to be attached to the linear part. From eqn. (15.37) for $S = 1$

$$-1/K_N(X) = -1 \qquad ; X < 1$$

$$= \frac{-\pi}{2} \Big/ \left\{\sin^{-1}(1/X) + (1/X)\sqrt{\left[1-(1/X)^2\right]}\right\}; X > 1$$

The plot of $-1/K_N(X)$ thus starts from $(-1 + j0)$ and travels along the negative real axis for increasing X as shown in Fig. 15.46 (*b*). Now for the equation

$$KG(j\omega) = -1/K_N(X)$$

to be satisfied, $G(j\omega)$ must have an angle of

$$\angle G(j\omega) = -90° - \tan^{-1} 2\omega - \tan^{-1} \omega = -180°$$

or $$\frac{2\omega + \omega}{1 - 2\omega^2} = \tan 90° = \infty \qquad \text{or} \qquad \omega = 1/\sqrt{2} \text{ rad/sec}$$

The largest value of K for stability is obtained when $KG(j\omega)$ passes through $(-1 + j0)$, *i.e*,

$$| KG(j\omega) |_{\omega = 1/\sqrt{2}} = 1 \qquad \text{or} \qquad \frac{K}{(1/\sqrt{2})(\sqrt{3})(\sqrt{3}/\sqrt{2})} = 1$$

i.e., $$K = 3/2$$

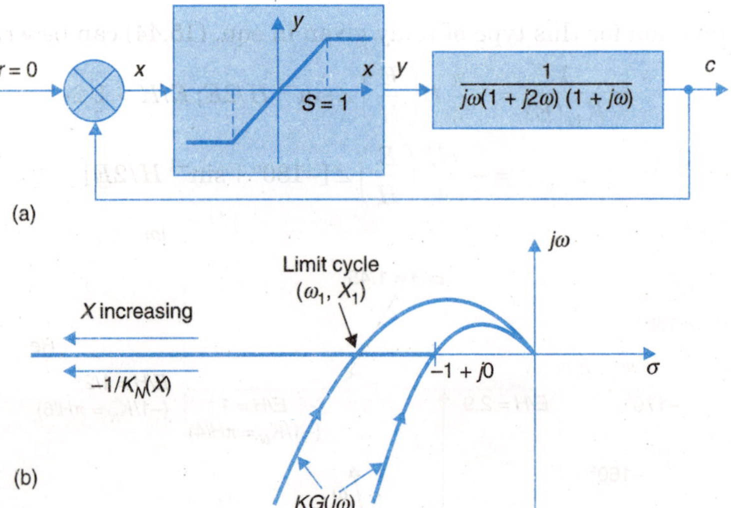

Fig. 15.46. Nonlinear system with saturating amplifier.

Now for $K = 3$, $KG(j\omega)$ plot intersects $-1/K_N(X)$-plot resulting in a limit cycle at (ω_1, X_1) where $\omega_1 = 1/\sqrt{2}$, while X_1 is obtained from

$$\frac{\pi/2}{\sin^{-1} 1/X_1 + 1/X_1\sqrt{\left[1-(1/X_1)^2\right]}} = 3\left(\frac{2}{3}\right) = 2$$

It can be found graphically by plotting eqn. (15.37) that corresponding to $K = 3$ and $K_N(X_1) = 1/2$, we get $X_1 \approx 6.5$.

Stability Analysis by Gain-phase Plots

The stability analysis using describing functions can be carried out by *db* gain-phase plots also as illustrated by the following example.

Example 15.8 : Consider a second-order system with backlash whose block diagram is shown in Fig. 15.47.

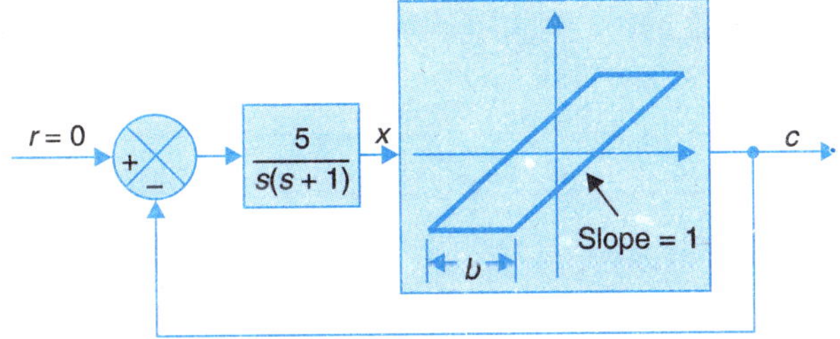

Fig. 15.47. Nonlinear system with backlash.

The describing function of backlash nonlinearity is given by eqn. (15.45), from which the following table is prepared.

Table 15.1

b/X	$\lvert K_N(X) \rvert$	$\angle K_N(X)$	$20 \log [1/\lvert K_N(X)\rvert]$
0	1	0	0
0.2	0.954	− 6.7°	0.4
0.4	0.882	− 13.4°	1.1
0.5	0.838	− 16.5°	1.52
0.6	0.794	− 19.7°	2.0
0.8	0.698	− 26.0°	3.12
1.0	0.592	− 32.5°	4.56
1.2	0.482	− 39.5°	6.36
1.4	0.367	− 46.6°	8.7
1.6	0.248	− 55.2°	12.12
1.7	0.186	− 60.4°	14.62
1.8	0.125	− 66.0°	18.08
1.9	0.064	− 69.8°	23.82
2.0	0	− 90.0°	∞

The plot of the describing function on *db* gain-phase plane is shown in Fig. 15.48.

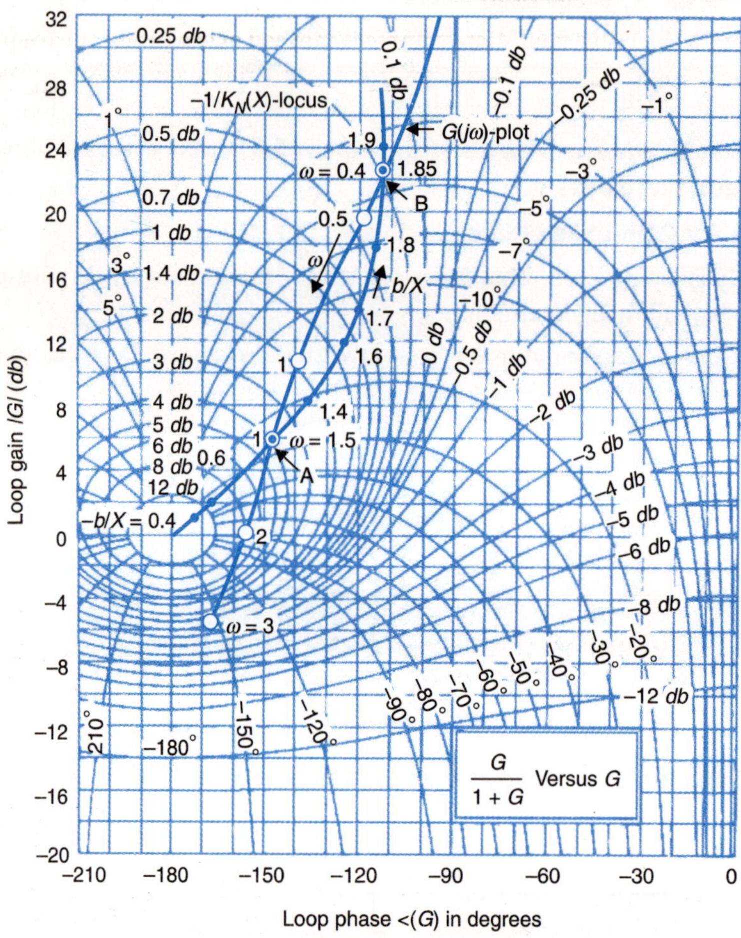

Fig. 15.48. db gain-phase plots of $-1/K_N(X)$ and $G(j\omega)$ of the system shown in Fig. 15.47.

The transfer function of the linear part of the system

$$G(j\omega) = \frac{5}{j\omega(j\omega + 1)}$$

is also plotted in Fig. 15.48. It is observed from this figure that the two plots intersect at points *A* and *B*. Application of the limit cycle stability test* discussed earlier in this section reveals that the point *A* corresponds to a stable limit cycle and the point *B* corresponds to an unstable limit cycle. The stable limit cycle is predicted at $\omega = 1.5$ and $b/X = 1$, while the unstable limit cycle is at $\omega = 0.4$ and $b/X = 1.85$.

---

*Since the plots in *db* gain-phase plane are conformal maps of the corresponding plots in the complex plane, the sense of directions in the complex plane is preserved in the *db* gain-phase plane and hence the same limit cycle stability test applies [*i.e.*, perturbation and enclosure of $-1/K_N(X)$-locus by $G(j\omega)$-plot].

15.10 JUMP RESONANCE

In this section we will illustrate analytico-graphical method of obtaining the frequency response of a nonlinear system and at the same time illustrate that under certain conditions such a frequency response exhibits the phenomenon of jump resonance (refer Section 15.1, Fig. 15.3).

Consider a nonlinear system shown in Fig. 15.49 wherein the nonlinear part has been replaced by its describing function.

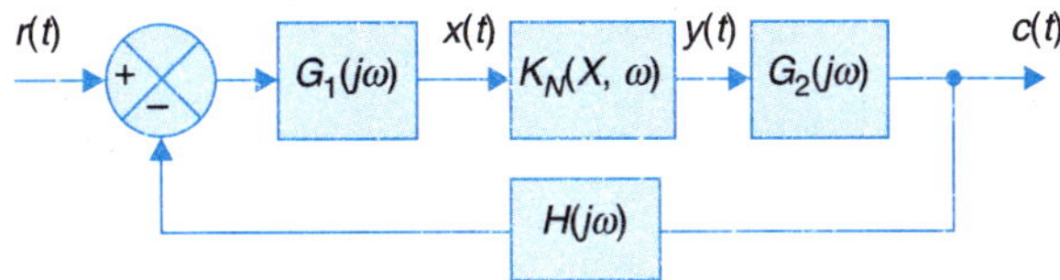

Fig. 15.49. Nonlinear system with nonlinearity replaced by its describing function.

Let $$r(t) = R \sin \omega t$$

$$x(t) = X \sin (\omega t + \phi)$$

It immediately follows that

$$\frac{C(j\omega)}{R(j\omega)} = \frac{G_1(j\omega)K_N(X,\omega)G_2(j\omega)}{1+G_1(j\omega)K_N(X,\omega)G_2(j\omega)H(j\omega)} \quad ...(15.49)$$

We can also write

$$\frac{X(j\omega)}{R(j\omega)} = \frac{G_1(j\omega)}{1+G_1(j\omega)K_N(X,\omega)G_2(j\omega)H(j\omega)} \quad ...(15.50)$$

Further we can write

$$G_1(j\omega) = g_1(\omega) \exp \{j\theta_1(\omega)\}$$

$$G_1(j\omega)G_2(j\omega)H(j\omega) = g_2(\omega) \exp \{j\theta_2(\omega)\}$$

$$K_N(X, \omega) = g_N(X, \omega) \exp \{j\theta_N(\omega)\}$$

In order to simplify the analysis we assume

(*i*) $\theta_N(X, \omega) = 0$

(*ii*) $g_N(X, \omega) = g_N(X)$, *i.e.*, independent of frequency.

So we can write

$$\frac{X(j\omega)}{R(j\omega)} = \frac{g_1(\omega)\exp\{j\theta_1(\omega)\}}{1+g_2(\omega)g_N(X)\exp\{j\theta_2(\omega)\}}$$

In terms of magnitudes, we can write the above expression as

$$\frac{X}{R} = \frac{g_1}{\sqrt{1+2g_N g_2 \cos\theta_2 + g_N^2 g_2^2}}$$

Solving for g_N, we get

$$g_N = \frac{-\cos\theta_2 \pm \sqrt{\cos^2\theta_2 - 1 + (R^2 g_1^2/X^2)}}{g_2}$$

or $$Xg_N(X) = -\frac{X\cos\theta_2(\omega)}{g_2(\omega)} \pm \frac{1}{g_2(\omega)}\sqrt{R^2 g_1^2(\omega) - X^2\sin^2\theta_2(\omega)} \quad ...(15.51)$$

Simultaneously solution of nonlinear eqns. (15.49) and (15.51) for fixed R and a given ω would yield the value X. This can be repeated for different values of ω. The results can then be used to obtain the frequency response of the closed-loop system with fixed R and varying ω, as illustrated in example below.

Example 15.9 : Obtain the frequency response of the system with saturation nonlinearity shown in Fig. 15.50. Given

$$K = 150, S = 2, R = 1.5.$$

Solution. For this particular system we have with reference to Fig. 15.50

$$g_1 = 1, \theta_1 = 0$$

$$g_2(\omega)\frac{150}{\omega\sqrt{1+\omega^2}}, \theta_2(\omega) = -90° - \tan^{-1}\omega.$$

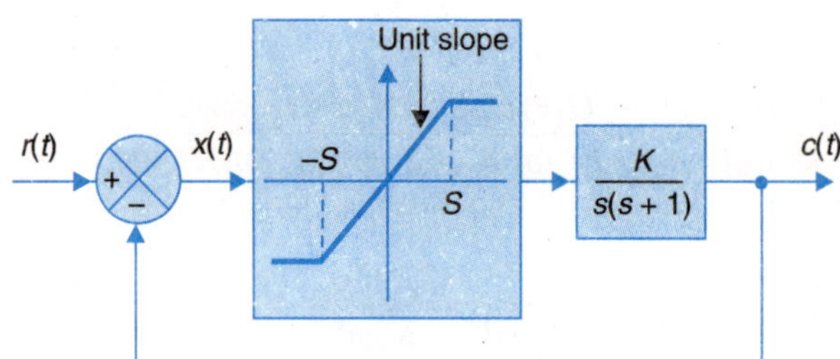

Fig. 15.50. Control system with saturation nonlinearity.

In expression (15.51),

$$\text{RHS} = \frac{X\omega\sqrt{1+\omega^2}\cos(-90° - \tan^{-1}\omega)}{150}$$

$$\pm\frac{\omega\sqrt{1+\omega^2}}{150}\sqrt{2.25 - X^2\sin^2(-90° - \tan^{-1}\omega)}$$

$$\text{LHS} = X\frac{2}{\pi}\left\{\sin^{-1}(2/X) + (2/X)\sqrt{1-(2/X)^2}\right\}; X > S$$

$$= X; \qquad X < S$$

The plots of RHS vs. X with ω as parameter and the plot of LHS vs. X are drawn in Fig. 15.51. The intersection of these plots yields the solution of X for each ω (for sake of clarity the plots are not drawn to scale). We see from these plots that for $\omega \in (\omega_3, \omega_4)$, there are two solutions and outside this range there is only one solution for each ω.

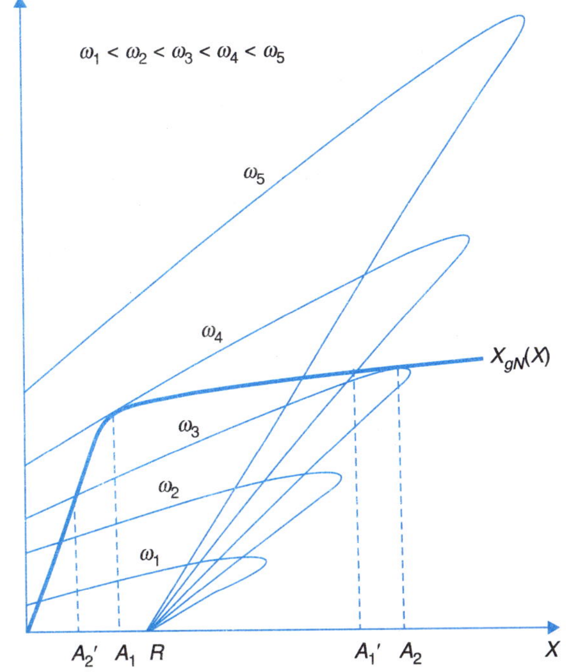

Fig. 15.51

Now
$$\left|\frac{C}{R}(j\omega)\right| = \frac{1}{R}[Xg_N(X)g_2(\omega)]$$

$$= \frac{1}{R}\left[Xg_N(X)\frac{150}{\omega\sqrt{1+\omega^2}}\right]$$

The plot $\left|\frac{C}{R}(j\omega)\right|_{R=1.5}$ vs. ω can be drawn (to scale). It exhibits the jump resonance phenomenon.

PROBLEMS

15.1 A linear second-order servo is described by the equation

$$\ddot{e} + 2\zeta\omega_n\dot{e} + \omega_n^2 e = 0$$

where
$$\zeta = 0.15,\ \omega_n = 1$$
$$e(0) = 1.5,\ \dot{e}(0) = 0$$

Determine the singular points. Construct the phase trajectory, using the method of isoclines.

15.2 A simple servo is described by the following equations:

Reaction torque $= \ddot{\theta}_c + 0.5\dot{\theta}_c$

Drive torque $= 2\ \text{sign}\ (e + 0.5\ \dot{e})$

$$e = \theta_R - \theta_C$$
$$e(0) = 2$$
$$\dot{e}(0) = 0$$

Construct the phase trajectory using the delta method.

15.3 The position control system shown in Fig. P-15.3 has coulomb friction C at the output shaft. Plot the phase-trajectory for a unit step input and zero initial conditions. Calculate and plot the time response of the error and find the value of the steady-state error.

Given

$$\sqrt{(K/J)} = 1.2 \text{ rad/sec}$$
$$C/K = 0.3 \text{ rad}$$

where $K = K_A K_1$

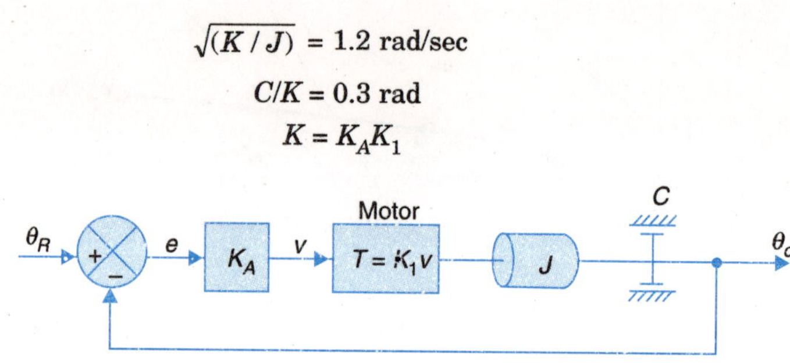

Fig. P-15.3

15.4 A second-order servo containing a relay with dead-zone and hysteresis is shown in the block diagram of Fig. P-15.4. Obtain the phase-trajectory of the system for the initial conditions $e(0) = 0.65$, $\dot{e}(0) = 0$. Does the system have a limit cycle? If so, determine its amplitude and period.

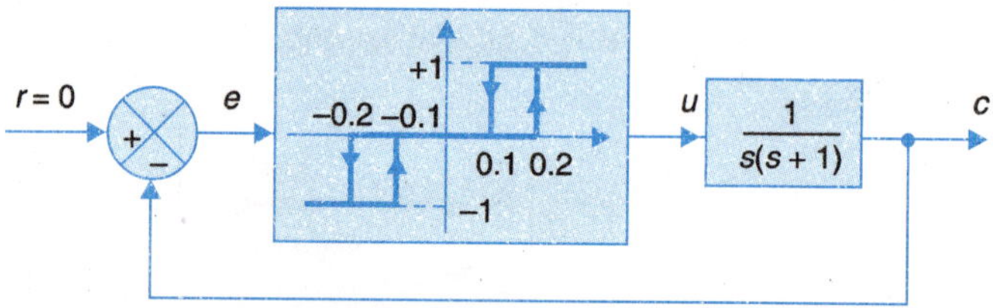

Fig. P-15.4

15.5 For the control system shown in Fig. P-15.5 with tachometer not included in the loop, *i.e.,* with switch S open, plot the phase-trajectory with an initial condition of

$$e(0) = 2, \ \dot{e}(0) = 0$$

Given: $\tau_m = 0.5$ sec.; $K = 8$; $\delta = 0.5$

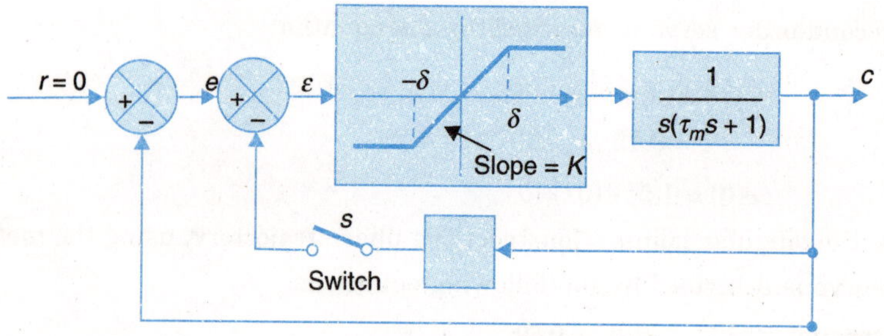

Fig. P-15.5

15.6 Control system shown in Fig. P-15.5, now includes tachometer (K_t = 0.5) in the feedback loop, *i.e.*, switch S closed. Plot the phase-trajectory and comment upon the effect of the tachometer feedback.

15.7 For the nonlinear system shown in Fig. P-15.7, draw the phase-trajectory starting from $(e, \dot{e})$ = (2, 2) in the $(e, \dot{e})$-plane.

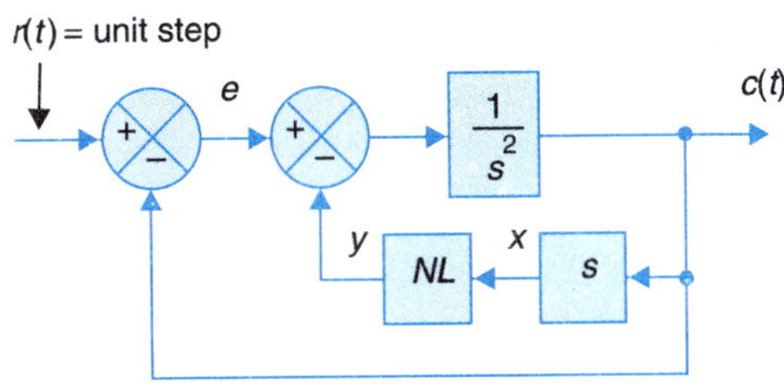

Fig. P-15.7

The characteristics of the nonlinearity (NL) are given below:

$$y = 2x \quad \text{if} \quad |x| \le 1$$
$$y = +2 \quad \text{if} \quad x > 1$$
$$y = -2 \quad \text{if} \quad x < -1$$

15.8 As an alternative to the nonlinear proportional plus derivative compensation of Fig. 15.22, a nonlinear feedback compensation scheme is shown in Fig. P-15.8. Write the differential equation governing the system behaviour. Investigate the nature of the equilibrium point. Would a limit cycle exist? If so, investigate the stability of the limit cycle.

Draw the phase trajectory for $c = 0$, $\dot{c} = 0$ and r = unit step. The δ-method would be more convenient for use here. $\dot{e}$

Given: $K_v = 1$, $\tau = 1$, $b = 1$

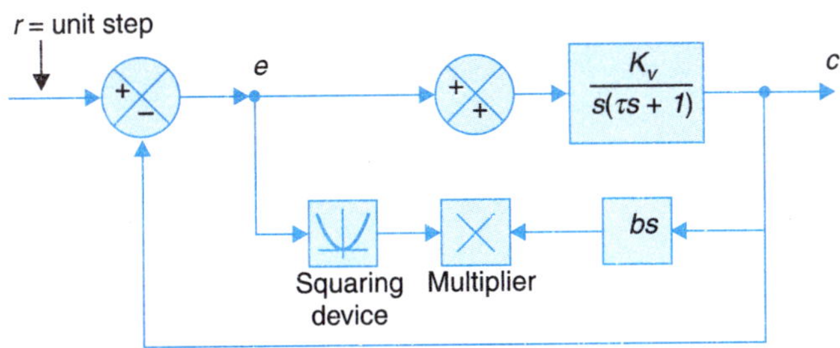

Fig. P-15.8

15.9 The relay control system of Fig. P-15.9 uses a relay with dead-zone and an inner rate feedback loop. Draw a set of isoclines and plot the phase-trajectory for zero input with initial conditions.

$$c(0) = 2,\ \dot{c}(0) = 5.0$$

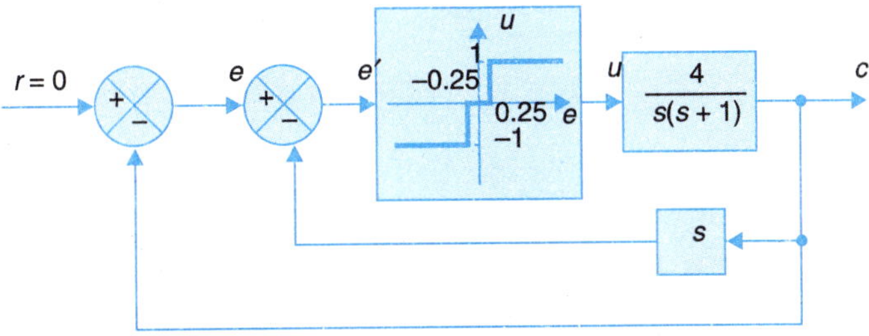

Fig. P-15.9

Also compute and plot time response of the system. Find therefrom the settling time and steady-state error.

15.10 Nonlinear spring characteristic can be approximated by two piecewise linear gains as shown in Fig. P-15.10 ($K_2 > K_1$ for hard spring, $K_2 < K_1$ for soft spring). Derive the expression for its describing function.

15.11 Derive the describing function of the element whose input-output characteristic is shown in Fig. P-15.11. Show that the required describing function equals the sum of the describing functions of relay with dead-zone and amplifier with dead-zone (use the results already derived in this chapter).

Can we generalize the above result for any nonlinearity whose characteristics is the sum of the characteristics of two nonlinearities?

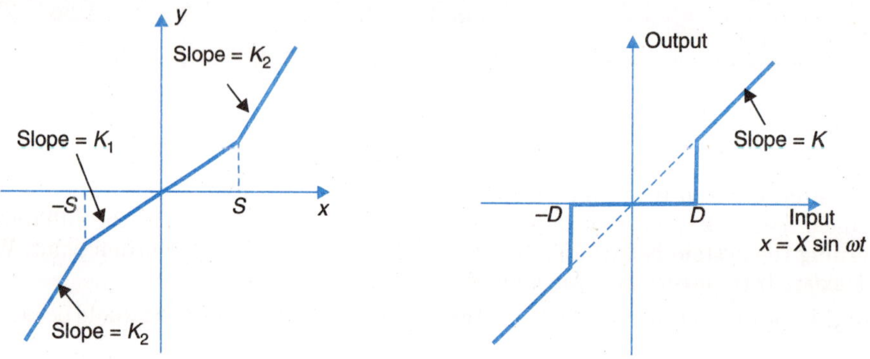

Fig. P-15.10 Fig. P-15.11

15.12 A two-phase servomotor is driven by an amplifier as shown in Fig. P-15.12. The transfer function of the motor is

$$G(s) = \frac{K' e^{-0.1s}}{s(0.1s + 1)}$$

Investigate the stability of the system for $K' = 0.1$. What is the largest value of K' for no limit cycle to exist?

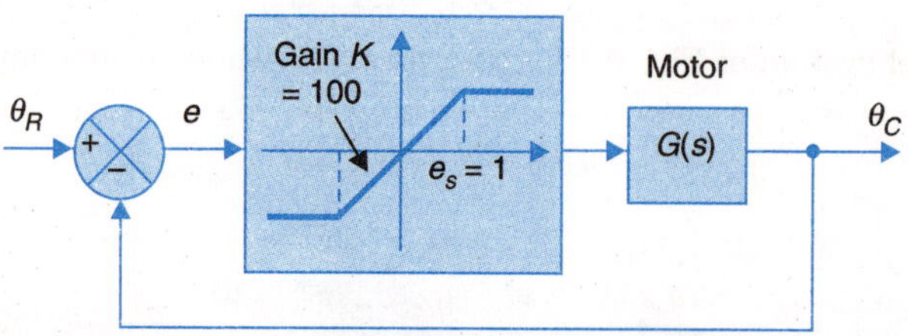

Fig. P-15.12

15.13 An instrument servo system used for positioning a load may be adequately represented by the block diagram shown in Fig. P-15.13 (*a*). The backlash characteristics are shown in Fig. P-15.13 (*b*).

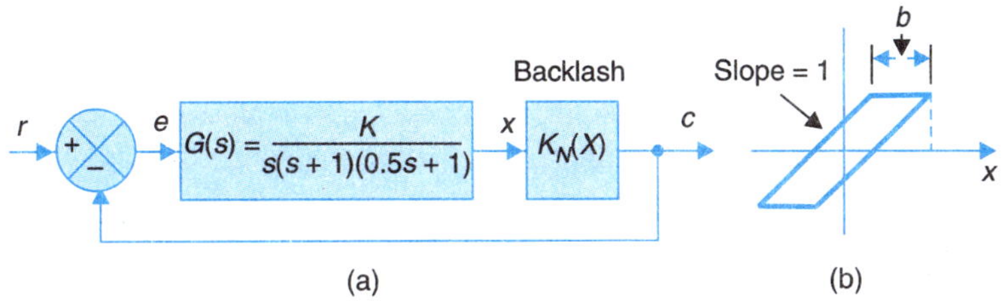

Fig. P-15.13

Using the db gain-phase analysis, show that the system is stable for $K = 1$. If the value of K is now raised to 5, show that limit cycles exist. Investigate the stability of these limit cycles and determine their frequency and b/X.

Given :

b/X	0	0.2	0.4	1	1.4	1.6	1.8	1.9	2.0
$\lvert K_N(X) \rvert$	1	0.954	0.882	0.592	0.367	0.248	0.125	0.064	0
$\angle K_N(X)$	0	−6.7°	−13.4°	−32.5°	−46.6°	−55.2°	−66°	−69.8°	−90°

How is the system behaviour altered if the backlash nonlinearity is present in the feedback path instead of the forward path as in Fig. 15.13 (*a*).

16

ADVANCES IN CONTROL SYSTEMS

16

ADVANCES IN CONTROL SYSTEMS

16.1 INTRODUCTION

Conventional control theory relies on the key assumption of small range operation for the linear model to be valid. When the operation range is large, a linear controller is likely to perform poorly or to be unstable, because the nonlinearities in the system cannot be properly compensated for. For a long time the automatic control of physical processes, in spite of the use of the principle of *feedback,* has been an experimental technique deriving more from art than from scientific basis. The requirement for more complex and higher-performance control systems have been the impulse for the development of a systematic control theory. Even with the development of such a systematic control theory, there is still usually something else missing when a practical control system must be designed; a good knowledge of the dynamic characteristics of the controlled plant.

Another assumption of linear control is that the system is indeed linearizable. However in control systems there are many nonlinearities whose discontinuous nature does not allow linear approximation. These nonlinearities include Coloumb friction, valve hysteresis, reactor deadzones, backlash etc. are often found in control engineering. These effects cannot be derived from linear model and need a nonlinear technique. In designing linear controllers, it is usually necessary to assume that the parameters of the system model are reasonably well known. However, many control problems involve uncertainties in the model parameters. This may be a slow variation of the parameters *e.g.,* fouling of heat exchangers to an abrupt change in parameters *e.g.,* inertial parameters of a robot when a new object is grasped. A linear controller based on inaccurate or obsolete values of the model parameters may exhibit significant performance degradation or even instability. Nonlinearities can be intentionally introduced into the controller part of a control system so that model uncertainties can be tolerated.

To implement high-performance control systems when the plant dynamic characteristics are poorly known or when large and unpredictable variations occur, a new class of control systems called nonlinear control systems have evolved which provide potential solutions. Four classes of nonlinear controllers for this purpose are robust controllers, adaptive controllers fuzzy logic controllers and neural controllers. A brief overview on robust control is given in Chapter 10. In this chapter we will deal with adaptive, fuzzy and neural controllers.

16.2 ADAPTIVE CONTROL

An adaptive controller is a controller that can modify its behaviour in response to changes in the dynamics of the process and the disturbances. Adaptive control can be considered as a special type of nonlinear feedback control in which the stages of the process can be separated into two categories, which can change at different rates. The slowly changing states are viewed as parameters with a fast time scale for the ordinary feedback and a slower one for updating regulator parameters. One of the goals of adaptive control is to compensate for parameter variations, which may occur due to nonlinear actuators, changes in the operating conditions of the process, and non-stationary disturbances acting on the process.

Some of systems to be controlled have parameter uncertainty at the beginning of the control operation for *e.g.,* in robot manipulation the mass and the link lengths are the uncertain parameters. Unless such parameter uncertainty is gradually reduced on-line by an adaptation or estimation technique, it may cause inaccuracy or instability for the control systems. In other systems like a pH control system, the system dynamics may be well known at the beginning, but may experience unpredictable parameter variations (for *e.g.,* due to buffering) as the control operation goes on. Without continuous redesign of the controller, the initial controller may not be able to control the changing plant well. Thus, the basic objective of adaptive controller is to maintain a consistent performance of a system in the presence of uncertainty or unknown variation in the plant parameters.

An adaptive controller is a controller with adjustable parameters and a mechanism for adjusting the parameters. An adaptive control system can be thought of as having two loops. One loop is a normal feedback with the process (plant) and controller. The other loop is a parameter adjustment loop. A block diagram of an adaptive system is shown in Fig. 16.1. The parameter adjustment loop is often slower than the normal feedback loop.

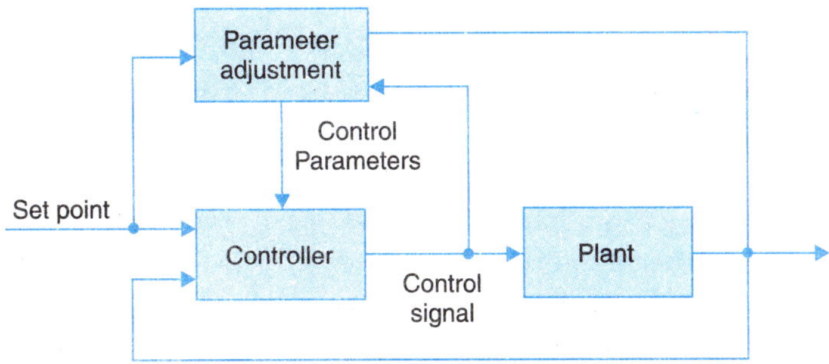

Fig. 16.1. Adaptive controller.

There are two main approaches for designing adaptive controllers. One is known as *Model Reference Adaptive Control* method, while the other is called as *Self-Tuning* method.

Model-Reference Adaptive Control (MRAC)

The Model-Reference Adaptive Control system is an adaptive servo system in which the desired performance is expressed in terms of a reference model, which gives desired response to the reference signal. Such a system can be schematically represented by Fig. 16.2.

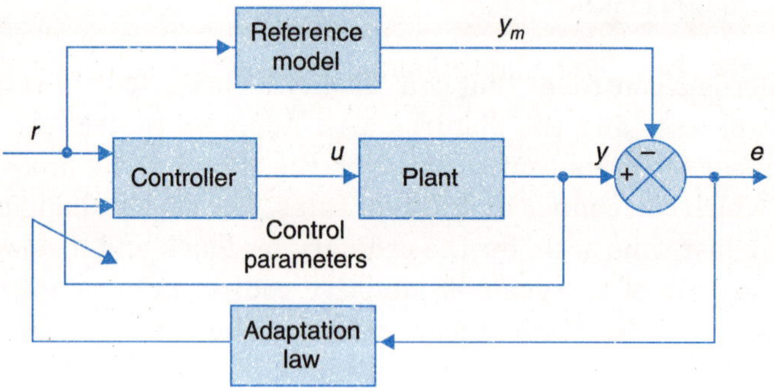

Fig. 16.2. Model Reference Adaptive Controller.

MRAC is composed of four parts : a *plant* containing unknown parameters, *a reference model* for compactly specifying the desired output of the control system, a *feedback control law* containing adjustable parameters. The ordinary feedback loop is known as the inner loop and the parameter adjustment loop is called as the outer loop.

Plant : The plant is assumed to have a known structure, although the parameters are unknown. Consider a linear system, say a servo motor, for which the transfer function is given as

$$G(s) = \frac{k_v}{s(1+\tau s)} \qquad ...(16.1)$$

where k_v is the system gain and τ is the time constant. Here order of the system is known (second order), but system gain k_v and τ time constant are not known. Similarly consider a nonlinear system, say a single link robot manipulator, whose dynamics can be given as

$$ml^2\ddot{\theta} + mgl \cos \theta = \tau \qquad ...(16.2)$$

where m is the link mass, l is the link length, θ is the position, $\ddot{\theta}$ is the acceleration and τ is the joint torque of the manipulator. Here the structure of the dynamics is known but the parameters such as mass and link length are not known (at least not fully known). In such cases, adaptive control provides the solution.

Reference Model : A reference model is used to specify the ideal response of the adaptive control system to the external command. The choice of the reference model has to satisfy two requirements.

1. It should reflect the performance specification in the control tasks, such as rise time, settling time, overshoot or equivalent frequency domain characteristics.
2. The ideal behaviour specfied by the reference model should be achievable for the adaptive control system.

Controller : The controller is usually parameterized by a number of adjustable parameters. This implies that there exists different sets of controller parameter values for which the desired control task is achievable. Usually the control law is linear in terms of the adjustable parameters (linear parameterization). Existing adaptive control designs normally require linear parameterization the controller in order to obtain adaptation mechanisms with guaranteed stability and tracking convergence.

Adaptation Mechanism : The adaptation mechanism is used to adjust the parameters in the control law. In MRAC systems the adaptation law searches for the parameters such that the response of the plant under adaptive control becomes the same as that of the reference model. Thus adaptation mechanism drives the tracking error to zero. This adaptation mechanism is designed to guarantee the stability of the control system as well as convergence of the tracking error to zero. We will use Lyapunov theory to synthesize the adaptation mechanism although other approaches such as hyper-stability and passivity theory exist in the literature.

The design of MRAC usually involves the following three steps:

1. Choose a control law containing variable parameters.
2. Choose an adaptation law for adjusting those parameters.
3. Analyze the convergence properties of the resulting closed loop control system.

We will formally discuss the design methodology of a MRAC using Lyapunov theory after introducing MIT rule, an original approach to model reference adaptive control.

The MIT Rule

We will consider a closed loop system in which the controller has one adjustable parameter θ. The desired closed loop response is specified by a model whose output is y_m. Let e be the error between the output y of the closed loop system and the output y_m of the reference model. The variable control parameter θ is adapted in such a way that the cost function

$$\mathcal{J}(\theta) = \frac{1}{2} e^2 \qquad \text{...(16.3)}$$

is minimized. The given cost function can be minimized if we change the parameters in the direction of the negative gradient of $\mathcal{J}$, generally known as gradient descent approach, in the following manner

$$\frac{d\theta}{dt} = \gamma \frac{\partial \mathcal{J}}{\partial \theta} = -\gamma e \frac{\partial e}{\partial \theta} \qquad \text{...(16.4)}$$

Eqn. (16.4) is known as MIT rule.

Application to a first order control system :

Consider a system described by the model

$$\frac{dy}{dt} = -ay + bu \qquad \text{...(16.5)}$$

where u is the control variable and y is the measured output. Assume that we want to obtain a closed loop system described by

$$\frac{dy_m}{dt} = -a_m y_m + b_m r \qquad \text{...(16.6)}$$

where r is the reference signal and y_m is the reference model response.

Let the controller be given by

$$u(t) = \theta_1 r - \theta_2 y(t) \quad \text{...(16.7)}$$

Let the two controller parameters be chosen as follows:

$$\theta_1 = \theta_1' = \frac{b_m}{b} \quad \text{...(16.8)}$$

$$\theta_2 = \theta_2' = \frac{a_m - a}{b} \quad \text{...(16.9)}$$

If we substitute the control law (16.7) in plant eqn. (16.5), we will get reference model eq. (16.6). This implies that if any adaptation mechanism that drives the controller parameters to the values given by eqns. (16.8, 16.9) will result in perfect model following. In the following we will derive an adaptation mechanism for controller parameters θ_1 and θ_2 using MIT rule assuming that the plant parameters a and b are unknown.

The tracking error is given as $e = y - y_m$. By substituting eqn. (16.7) in eqn. (16.5) we get

$$\frac{dy}{dt} = -(a + b\theta_2)\, y + br \quad \text{...(16.10)}$$

By introducing a differential operator $p = \frac{d}{dt}$, we can write eqn. (16.10) as

$$y = \frac{b\theta_1 r}{p + a + b\theta_2} \quad \text{...(16.11)}$$

Since y_m is independent of controller parameters, the sensitivity derivatives are obtained as

$$\frac{\partial e}{\partial \theta_1} = \frac{\partial y}{\partial \theta_1} = \frac{br}{p + a + b\theta_2} \quad \text{...(16.12)}$$

$$\frac{\partial e}{\partial \theta_2} = \frac{\partial y}{\partial \theta_1} = \frac{b^2\theta_1}{(p + a + b\theta_2)^2} = -\frac{by}{p + a + b\theta_2} \quad \text{...(16.13)}$$

These formulae cannot be used directly because the process parameters a and b are not known. Thus we go for following approximation which is the case for perfect model following.

$$p + a + b\theta_2 \approx p + a_m \quad \text{...(16.14)}$$

where $\theta_2 = \theta_2'$, perfect model following. With this approximation, the update rule for the controller parameters using MIT rule (16.4) can be found as

$$\frac{d\theta_1}{dt} = -\gamma e \frac{\partial e}{\partial \theta_1} = -\gamma e \left(\frac{br}{p + a_m}\right) = -\gamma' e \left(\frac{a_m r}{p + a_m}\right) \quad \text{...(16.15)}$$

where $\gamma' = \frac{\gamma b}{a_m}$. This implies sign of b must be known to update the controller parameter using eqn. (16.15). Similarly the update rule for θ_2 can be derived as

$$\frac{d\theta_2}{dt} = \gamma' e \left(\frac{a_m y}{p + a_m}\right) \quad \text{...(16.16)}$$

The complete MRAC architecture using MIT rule for the first order system is given Fig. 16.3. In this figure s is used in place of p.

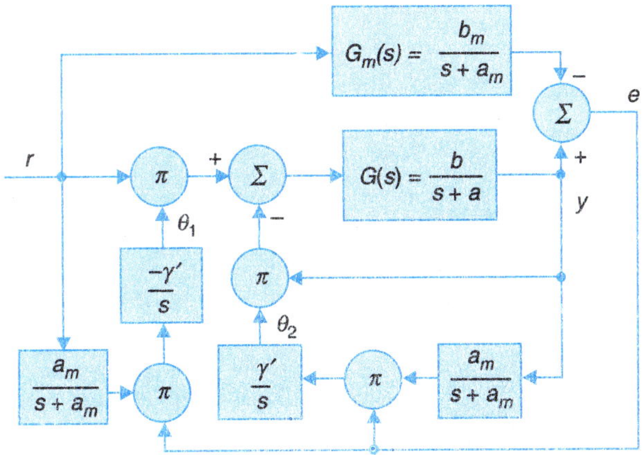

Fig. 16.3. MRAC for first order system. Σ = sum, π = product.

Example 16.1: In this example we will illustrate MRAC for a first order system using MIT rule through simulation. The plant dynamics is given as $\frac{dy}{dt} = -ay + bu$, where $a = 1$ and $b = 2$. The plant response is required to follow the model given by $\frac{dy_m}{dt} = -3y_m + 3r$. The adaptation gain γ is chosen to be equal to 1.5. The initial values of both parameters of the controller are chosen to be zero. The initial conditions of the plant and the model are both zero. Two different reference signals $r(t) = 2$ and $r = 2 \sin(3t)$ are used in the simulation.

Solution.

1. We select the reference signal to be $r(t) = 2$. By implementing control law using MIT rule. Tracking performance and parameter estimation are shown in Fig. 16.4. The results show that although tracking error converges to zero. The controller parameters θ_1 and θ_2 converge to a value of 0.35 and 0.85 respectively. The estimated plant parameter value is given by Eqs. 16.8 and 16.9. The estimated plant parameters comes out to be:

$$\hat{b} = b_m/\theta_1$$
$$= 3/0.85 = 3.53$$
$$\hat{a} = a_m - \hat{b}\theta_2$$
$$= 3 - 3.75*0.35 = 1.76 \quad ...(16.17)$$

It can be clearly seen that the plant parameter estimates $\hat{a}$ and $\hat{b}$ have not converged to the real value of 1 and 2 respectively.

2. We now consider $r(t) = 2 \sin(3t)$. The tracking performance and parameter estimation are shown in Fig. 16.5. In this case also the tracking error converges to zero and the controller parameters θ_1 and θ_2 converge to a value of 1.0 and 1.5 respectively. The estimated plant

parameter value is given by eqns. 16.8 and 16.9. The estimated plant parameters comes out be :

$$\begin{aligned}\hat{b} &= b_m \\ &= 3/1.5 = 2 \\ \hat{a} &= a_m - \hat{b}\theta_2 \\ &= 3 - 2*1 = 1\end{aligned} \quad ...(16.18)$$

It can be clearly seen that the plant parameter estimates $\hat{a}$ and $\hat{b}$ have converged to the real value of 1 and 2 respectively.

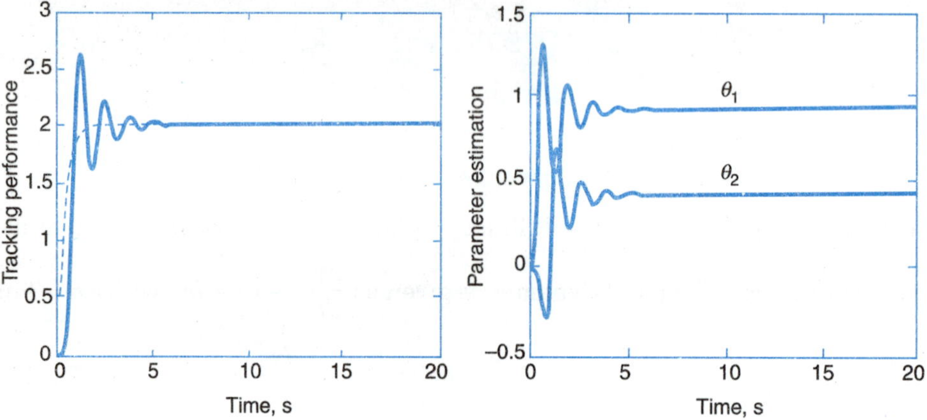

Fig. 16.4. Tracking performance and parameter estimation where $r(t) = 2$.

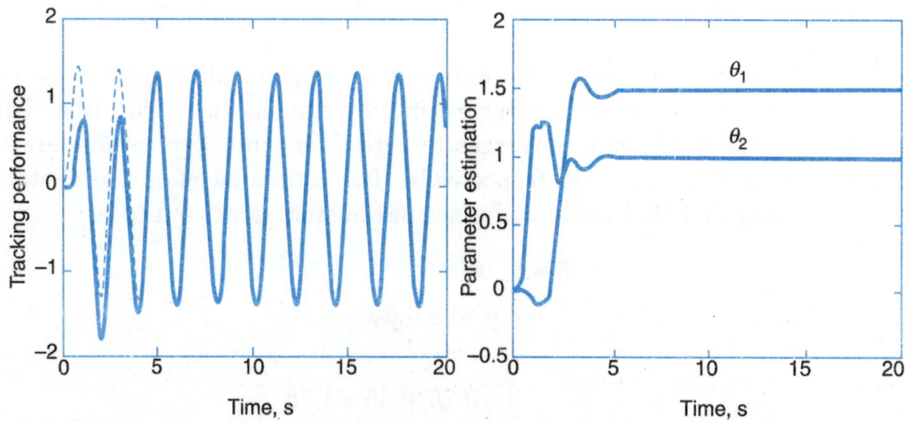

Fig. 16.5. Tracking performance and parameter estimation where $r(t) = 2 \sin(3t)$.

The above example clearly shows that a continuously persisting input is necessary for the convergence of parameter estimation.

MRAC Using Lyapunov Theory

There is no guarantee that an adaptive controller based on the MIT rule will give a stable closed loop system. The Lyapunov theory and Lyapunov function has been discussed in Chapter 13. In this section we are providing Barbalat's lemma for stability analysis.

Barbalat's Lemma

If a scalar function $V(x, t)$ satisfies the following conditions

1. $V(x, t)$ is lower bounded
2. $\dot{V}(x, t)$ is negative semi-definite and
3. $\dot{V}(x, t)$ is uniformly continuous in time

then $\dot{V}(x, t) \to 0$ as $t \to 0$.

First Order System

We again consider the same first order system given section 16.1.2

Reference Model:

$$\frac{dy_m}{dt} = -a_m y_m + b_m r \qquad ...(16.19)$$

where r is the reference signal and $a_m > 0$

Process plant:

$$\frac{dy}{dt} = -ay + bu \qquad ...(16.20)$$

controller: $$u = \theta_1 r - \theta_2 y \qquad ...(16.21)$$

error: $$e = y - y_m \qquad ...(16.22)$$

Before deriving adaptation mechanism, we derive the error dynamics of the closed loop system.

Substitution of control law in the plant dynamics yields

$$\frac{dy}{dt} = -ay + b(\theta_1 r - \theta_2 y) \qquad ...(16.23)$$

$$= -(a + b\theta_2)y + b\theta_1 r \qquad ...(16.24)$$

Subtracting eqn. (16.19) from eqn. (16.24), we get

$$\frac{dy}{dt} - \frac{dy_m}{dt} = \frac{de}{dt} = -a_m e - (b\theta_2 + a - a_m)y + (b\theta_1 - a_m)r \qquad ...(16.25)$$

From the error dynamics given in eqn. (16.25), we can observe that the tracking error will go to zero if $b\theta_2 = a_m - a$ and $b\theta_1 = b_m$. Thus the parameter adjustment rules should achieve this goal.

Usually the Lyapunov function is quadratic in tracking error and controller parameter estimation error since it is expected that the adaptation mechanism will drive both the tracking error and error in estimation of controller parameters to zero. Thus a valid Lyapunov function for the error dynamics eqn. (16.25) can be taken as

$$V(e, \theta_1, \theta_2) = \frac{1}{2}\left(e^2 + \frac{1}{b\gamma}(b\theta_2 + a - a_m)^2 + \frac{1}{b\gamma}(b\theta_1 - b_m)^2\right) \qquad ...(16.26)$$

where $b\gamma > 0$.

This function is zero when e is zero and the controller parameters are equal to the correct values. For a valid Lyapunov function, time derivative of the Lyapunov function must be negative. The derivative is given as

$$\frac{dV}{dt} = e\frac{de}{dt} + \frac{1}{\gamma}(b\theta_2 + a - a_m)\frac{d\theta_2}{dt} + \frac{1}{\gamma}(b\theta_1 - b_m)\frac{d\theta_1}{dt} \qquad ...(16.27)$$

Substituting eqn. (16.25) in the above equation and rearranging we get,

$$\frac{dV}{dt} = -a_m e^2 + \frac{1}{\gamma}(b\theta_2 + a - a_m)\left(\frac{d\theta_2}{dt} - \gamma ye\right) + \frac{1}{\gamma}(b\theta_1 - b_m)\left(\frac{d\theta_1}{dt} + \gamma re\right) \qquad ...(16.28)$$

If the parameters are updated as

$$\frac{d\theta_1}{dt} = -\gamma re \qquad ...(16.29)$$

$$\frac{d\theta_2}{dt} = -\gamma ye \qquad ...(16.30)$$

Then

$$\frac{dV}{dt} = -a_m e^2 \qquad ...(16.31)$$

Thus $\dot{V}$ is negative semidefinite which implies that $V(t) \geq V(0)$. This ensures that e, q_1 and q_2 must be bounded. To ensure that tracking error goes to zero, we compute second time derivative of Lyapunov function.

$$\frac{d^2V}{dt^2} = -2a_m e\frac{de}{dt} \qquad ...(16.32)$$

Substituting eqn. (16.25) in the above equation and rearranging we get,

$$\frac{d^2V}{dt^2} = -2a_m e(-a_m e - (b\theta_2 + a - a_m)y + (b\theta_1 - b_m)r) = f(e, \theta_1, \theta_2, y, r) \qquad ...(16.33)$$

Since all the parameters are bounded and $y = e + y_m$ is bounded, $\ddot{V}$ is also bounded which in turn implies that $\dot{V}$ is uniformly continuous. Now using Barbalat's lemma, we conclude that tracking error converges to zero.

Example 16.2: Let us reconsider the Example 16.1. Here we update the controller parameters using Lyapunov synthesis approach.

Solution.

1. We select the reference signal to be $r(t) = 2$. By implementing control law using MIT rule, tracking performance and parameter estimation are shown in Fig. 16.6. The results show that although tracking error converges to zero. The controller parameters θ_1 and θ_2 converge to a value of 0.43 and 0.93 respectively. The estimated plant parameter value is given by Eqs. 16.8 and 16.9. The estimated plant parameters come out to be:

$$\hat{b} = b_m/\theta_1$$
$$= 3/0.93 = 3.23$$
$$\hat{a} = a_m - \hat{b}\theta_2$$
$$= 3 - 3.23*0.43 = 1.61 \quad ...(16.34)$$

It can be clearly seen that the plant parameter estimates $\hat{a}$ and $\hat{b}$ have not converged to the real value of 1 and 2 respectively.

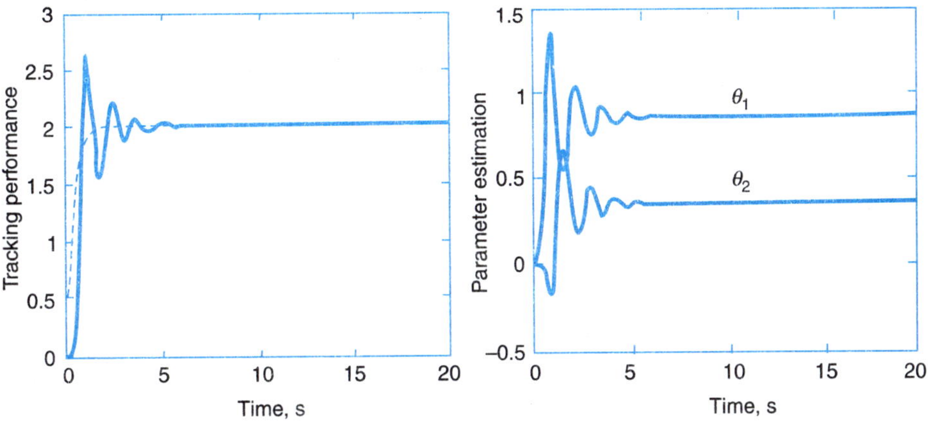

Fig. 16.6. Tracking performance and parameter estimation where $r(t) = 2$.

2. We now consider $r(t) = 2 \sin (3t)$. The tracking performance and parameter estimation are shown in Fig. 16.7. In this case also the tracking error converges to zero and the controller parameters θ_1 and θ_2 converge to a value of 1.00 and 1.5 respectively. The estimated plant parameter value is given by eqns. 16.8 and 16.9. The estimated plant parameters comes out to be:

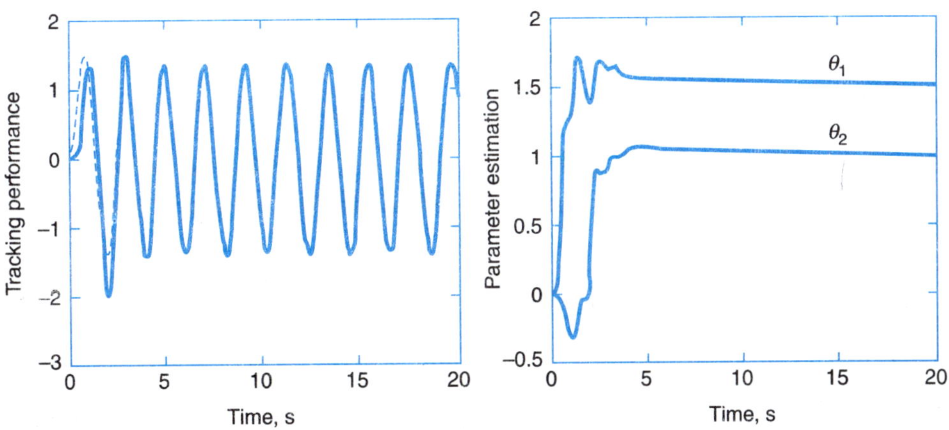

Fig. 16.7. Tracking performance and parameter estimation where $r(t) = 2 \sin (3t)$.

$$\hat{b} = b_m/\theta_1$$
$$= 3/1.5 = 2$$
$$\hat{a} = a_m - \hat{b}\theta_2$$
$$= 3 - 2*1 = 1 \qquad ...(16.35)$$

It can be clearly seen that the plant parameter estimates a and b have converged to the real value of 1 and 2 respectively.

The above example once again substantiates that a continuously persisting input is necessary for the convergence of parameter estimation.

General Higher Order SISO Systems

Plant dynamics of nth order can be described as

$$u = a_n y^{(n)} + a_{n-1} y^{(n-1)} + \ldots + a_0 y \qquad ...(16.36)$$

where the plant parameter vector $\underline{\mathbf{a}} = [a_n a_{n-1} \ldots a_0]^r$ is unknown. It is desired that the closed loop control system has characteristics as

$$\alpha_n y_m^{(n-1)} + \alpha_{n-1} y_m^{(n-1)} + \ldots + \alpha_0 y_m = r \qquad ...(16.37)$$

where the reference model parameters α_n, α_{n-1}, ..., α_0 can be chosen by the designer.

The Control Law:

Let $e = y - y_m$

Define a signal $z(t)$ as follows

$$z(t) = y_m^{\ n} - \beta_{n-1} e^{(n-1)} - \ldots - \beta_0 e \qquad ...(16.38)$$

The positive constants $\beta_1, \beta_2, \ldots, \beta_{n-1}$ in eqn. (16.38) are chosen such that the characteristic polynomial $s^n + \beta_{n-1} s^{n-1} + \ldots + \beta_0$ is a stable (Hurwitz) polynomial *i.e.,* the first column of the Routh's table is positive.

Adding $-a_n z(t)$ to both sides of eqn. (16.36), we get

$$a_n\left[y^{(n)} - z\right] = u - a_n z - a_{n-1} y^{(n-1)} - \ldots -a_0 y \qquad ...(16.39)$$

Let us choose the control law to be

$$u = \hat{a}_n z + \hat{a}_{n-1} y^{(n-1)} + \ldots + \hat{a}_0 y = \mathbf{v}^T \underline{\hat{\mathbf{a}}} \qquad ...(16.40)$$

where $\mathbf{v}(t) = [z(t) y^{(n-1)} \ldots \dot{y}\, y]^T$ and $\underline{\hat{\mathbf{a}}}(t) = [\hat{a}_n\ \hat{a}_{n-1} \ldots \hat{a}_0]$ denotes the estimated parameter vector. The closed loop error dynamics can be derived by substituting eqn. (16.40) in eqn. (16.36) which is as follows :

$$a_n(e^{(n)} + \beta_{n-1} e^{(n-1)} + \ldots + \beta_0 e) = \mathbf{v}^T \underline{\tilde{\mathbf{a}}} \qquad ...(16.41)$$

where $\underline{\tilde{\mathbf{a}}} = \underline{\hat{\mathbf{a}}} - \underline{\mathbf{a}}$.

By observing eqn. (16.41), we can say that the error dynamics is stable when the estimated parameter vector is equal to the actual parameter vector. β_{n-1}, β_{n-2}, ..., β_0 are selected such that error e decays to zero. This control law is a general one for all higher order SISO system and β_{n-1}, β_{n-2}, ..., β_0 are selected by placing the poles properly.

Adaptation Mechanism:

The error dynamics in eqn. (16.41) can be written as

$$e^{(n)} + \beta_{n-1}e^{(n-1)} + \dots + \beta_0 e = \frac{1}{a_n}\mathbf{v}^T\underline{\tilde{\mathbf{a}}} \qquad \dots(16.42)$$

We represent eqn. (16.42) in state space form as discussed in Chapter 12. Selecting the states as

$$x_1 = e$$

$$x_2 = \dot{x}_1 = \dot{e}$$

$$x_3 = \dot{x}_2 = \ddot{e}$$

$$\vdots$$

$$\dot{x}_n = e^{(n)} = -\beta_{n-1}x_{n-1} - \beta_{n-2}x_{n-2} - \dots - \beta_0 x_1 + \frac{1}{a_n}\mathbf{v}^T\underline{\tilde{\mathbf{a}}}$$

We can write the state space form as follows

$$\dot{\mathbf{x}} = \mathbf{Ax} + \mathbf{b}\left[\frac{1}{a_n}\mathbf{v}^T\underline{\tilde{\mathbf{a}}}\right] \qquad \dots(16.43)$$

$$e = \mathbf{Cx} \qquad \dots(16.44)$$

where

$$\mathbf{A} = \begin{bmatrix} 0 & 1 & 0 & \dots & 0 \\ 0 & 0 & 1 & \dots & 0 \\ \vdots & \vdots & \vdots & \vdots & \vdots \\ 0 & 0 & 0 & \dots & 1 \\ -\beta_0 & -\beta_1 & -\beta_2 & \dots & -\beta_{n-1} \end{bmatrix}$$

$$\mathbf{b} = \begin{bmatrix} 0 & 0 & \dots & 1 \end{bmatrix}^T$$

and

$$\mathbf{C} = \begin{bmatrix} 1 & 0 & \dots & 0 \end{bmatrix}$$

Consider the Lyapunov function candidate

$$V(\mathbf{x}, \underline{\tilde{\mathbf{a}}}) = \mathbf{x}^T\mathbf{Px} + \tilde{\mathbf{a}}^T\Gamma^{-1}\underline{\tilde{\mathbf{a}}} \qquad \dots(16.45)$$

where $\boldsymbol{\Gamma}$ and $\mathbf{P}$ are symmetric positive definite constant matrices, and P satisfies

$$\mathbf{PA} + \mathbf{A}^T\mathbf{P} = -\mathbf{Q} \qquad \dots(16.46)$$

for a chosen symmetric positive definite Q.

$$\dot{V} = -\mathbf{x}^T\mathbf{Qx} + 2\underline{\tilde{\mathbf{a}}}^T\mathbf{vb}^T P\mathbf{x} + 2\underline{\dot{\tilde{\mathbf{a}}}}\,\Gamma^{-1}\underline{\tilde{\mathbf{a}}} \qquad \dots(16.47)$$

By selecting the adaptation law to be

$$\underline{\dot{\hat{\mathbf{a}}}} = -\boldsymbol{\Gamma}\mathbf{vb}^T\mathbf{Px} \qquad \dots(16.48)$$

we get $\dot{V} = -\mathbf{x}^T Q\mathbf{x}$. Using Barbalat's lemma we can show convergence of x guaranteeing the convergence of tracking error to zero.

Example 16.3: Mass-spring-damper system

The dynamics of mass-spring-damper system is given as

$$m\ddot{y} + c\dot{y} + ky = u \qquad \dots(16.49)$$

where m = 1 kg, c = 1 N/m.s and k = 1 N/m. Design a MRAC the system response follows the reference model given by

$$\ddot{y}_m + 16.14\ddot{y}_m + y = r \qquad ...(16.50)$$

Select $r(t)$ = 2 and 2 sin (3t).

Solution. $z(t)$ is selected as by placing both the poles at –2.

$$x(t) = \ddot{y}_m - 4\dot{e} - 4e \qquad ...(16.51)$$

The control law is given as

$$u = \hat{a}_2 z(t) + \hat{a}_1 \dot{y} + \hat{a}_0 y \qquad ...(16.52)$$

The system matrix for the closed loop error dynamics is given as

$$\mathbf{A} = \begin{bmatrix} 0 & 1 \\ -4 & -4 \end{bmatrix} b = \begin{bmatrix} 0 \\ 1 \end{bmatrix}$$

Selecting Q to be unity matrix, matrix P can be found to be

$$\mathbf{P} = \begin{bmatrix} \frac{9}{8} & \frac{1}{8} \\ \frac{1}{8} & \frac{5}{32} \end{bmatrix}$$

By selecting

$$\Gamma = \begin{bmatrix} 5 & 0 \\ 0 & 5 \end{bmatrix}$$

the adaptation law can be given as

$$\dot{\hat{a}}_2 = -5z\left[\frac{1}{8}e + \frac{5}{32}\dot{e}\right] \qquad ...(16.53)$$

$$\dot{\hat{a}}_1 = -5\dot{y}\left[\frac{1}{8}e + \frac{5}{32}\dot{e}\right] \qquad ...(16.54)$$

$$\dot{\hat{a}}_0 = -5y\left[\frac{1}{8}e + \frac{5}{32}\dot{e}\right] \qquad ...(16.55)$$

The tracking performance and parameter estimation when $r(t)$ = 2 is shown in Fig. 16.8 while Fig. 16.9 shows the same result for $r(t)$ = 2 sin (3t).

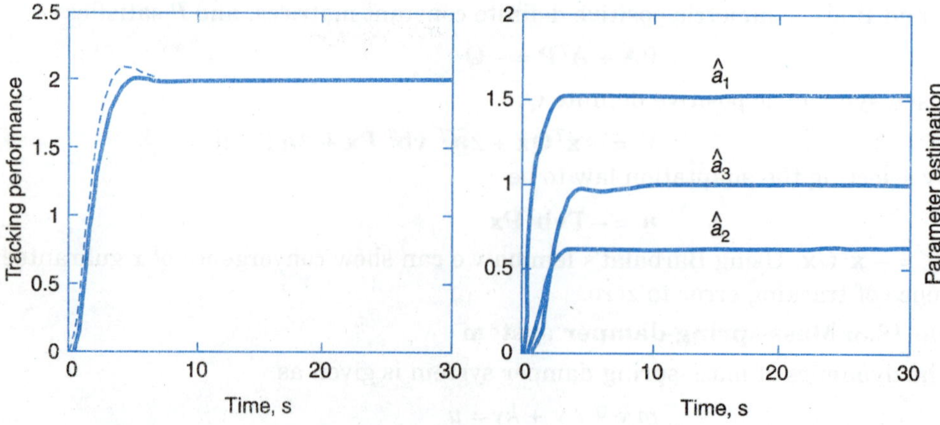

Fig. 16.8. Tracking performance and parameter estimation where $r(t)$ = 2.

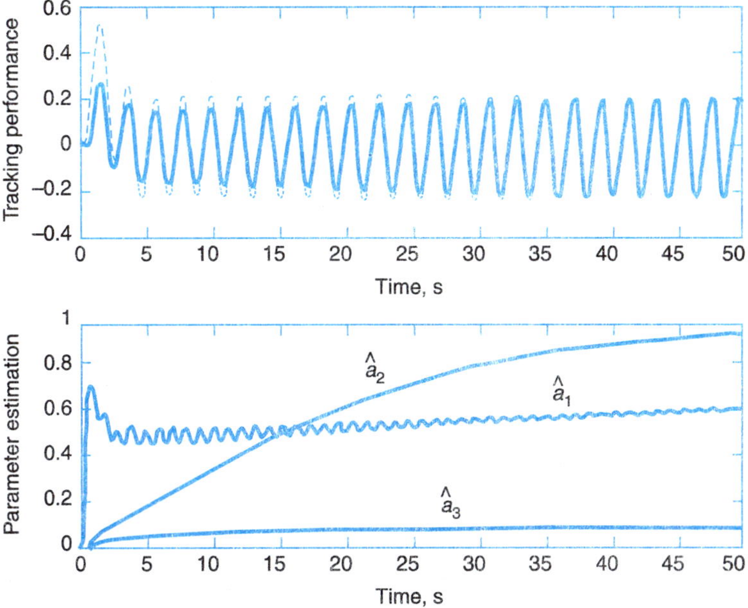

Fig. 16.9. Tracking performance and parameter estimation where $r(t) = 2 \sin (3t)$.

MRAC for A Single Link Manipulator

Dynamics of single link manipulator

$$a\ddot{\theta} + b \cos \theta = u \qquad ...(16.56)$$

The tracking error is defined as

$$e = \theta - \theta^d \qquad ...(16.57)$$

where θ^d is the desired position. Let the control law be given as

$$u = a\left(\ddot{\theta}^d - k_p e - k_d \dot{e}\right) + b \cos\theta \qquad ...(16.58)$$

Substituting eqn. (16.58) in eqn. (16.56), the closed-loop error dynamics is obtained as

$$a\left(\ddot{\theta} - \ddot{\theta}^d\right) + k_d \dot{e} + k_p e = 0$$

$$a\ddot{e} + k_d \dot{e} + k_p e = 0 \qquad ...(16.59)$$

From eqn. (16.59), it is clear that by selecting suitable values of k_d, k_p we can ensure the convergence of tracking error to zero. The above control law is implementable if the manipulator parameters are exactly known. If we assume these parameters to be unknown then the control law becomes

$$u = \hat{a}(\ddot{\theta}^d - k_p e - k_d \dot{e}) + \hat{b} \cos \theta \qquad ...(16.60)$$

$\hat{a}$ and $\hat{b}$ are the estimated parameter. The closed loop error dynamics can be found to be

$$a\ddot{\theta} + b \cos\theta = \hat{a}(\ddot{\theta}^d - k_p e - k_d \dot{e}) + \hat{b} \cos\theta \qquad ...(16.61)$$

$$\hat{a}(\ddot{e} + k_p e + k_d \dot{e}) = \tilde{a}\ddot{\theta} + \tilde{b} \cos\theta \qquad ...(16.62)$$

$$= \begin{bmatrix} \ddot{\theta} & \cos\theta \end{bmatrix} \begin{bmatrix} \tilde{a} \\ \tilde{b} \end{bmatrix} \quad ...(16.63)$$

where $\tilde{a} = \hat{a} - a, \tilde{b} = \hat{b} - b$, $\mathbf{P} = [\hat{a} \quad \hat{b}]^T$ and $\tilde{\mathbf{P}} = \hat{\mathbf{P}} - \mathbf{P}$. Rearranging eqn. (16.63) we get

$$\ddot{e} + k_p e + k_d \dot{e} = \frac{1}{\hat{a}} [\theta \quad \cos\theta] \, \mathbf{P} = \boldsymbol{\Phi}\mathbf{P} \quad ...(16.64)$$

$\hat{a}$ is the estimated value. $\ddot{\theta}$ and $\cos\theta$ are assumed to be available.

Let $\mathbf{X} = [e \quad \dot{e}]^T$. Then eqn. (16.64) is written in state variable form as

$$\dot{\mathbf{X}} = \mathbf{AX} + \mathbf{B}\boldsymbol{\Phi}\tilde{\mathbf{P}} \quad ...(16.65)$$

where $$\mathbf{A} = \begin{bmatrix} 0 & 1 \\ -k_p & -k_d \end{bmatrix}; \mathbf{B} = [0 \quad 1]$$

Taking transpose both the sides, we get

$$\dot{\mathbf{X}}^{\mathbf{T}} = \mathbf{X}^{\mathbf{T}}\mathbf{A}^{\mathbf{T}} + \tilde{\mathbf{P}}^{\mathbf{T}}\boldsymbol{\Phi}^{\mathbf{T}}\mathbf{B}^{\mathbf{T}} \quad ...(16.66)$$

By selecting the Lyapunov function to be

$$V = \mathbf{X}^{\mathbf{T}}\mathbf{P}\mathbf{X} + \tilde{\mathbf{P}}^{\mathbf{T}}\Gamma\tilde{\mathbf{P}} \quad ...(16.67)$$

The time derivative of the Lyapunov function is given as

$$\dot{V} = \dot{\mathbf{X}}^{\mathbf{T}}\mathbf{PX} + \mathbf{X}^{\mathbf{T}}\mathbf{P}\dot{\mathbf{X}} + 2\tilde{\mathbf{P}}^{\mathbf{T}}\Gamma\tilde{\mathbf{P}} \quad ...(16.68)$$

Substituting eqn. (16.66) in the above equation, we get,

$$\dot{V} = X^T(A^TP + PA)X + 2\tilde{P}^T(\Phi^T B^T PX + \Gamma\dot{\tilde{P}}) \quad ...(16.69)$$

We known that $\mathbf{A^TP} + \mathbf{PA} = -\mathbf{Q}$ and if we select the adaptation law to be

$$\dot{\tilde{\mathbf{P}}} = -\boldsymbol{\Gamma}^{-1}\boldsymbol{\Phi}^{\mathbf{T}}\mathbf{B}^{\mathbf{T}}\mathbf{PX} \quad ...(16.70)$$

the eqn. (16.69) reduces to $\dot{V} = -X^TQX$. Using Barbalat's lemma we can show that the system is stable.

Example 16.4: Consider the single link manipulator whose dynamics is given in Eqn. 16.56 where $a = 1$ and $b = 9.81$. Design a MRAC such that manipulator tracks a desired trajectory.

Solution. Let $k_p = 8$ and $k_d = 8$. By selecting $\boldsymbol{\Gamma} = \begin{bmatrix} 0.2 & 0 \\ 0 & 0.25 \end{bmatrix}$, we obtain

$$\mathbf{P} = \begin{bmatrix} 4 & \frac{1}{16} \\ \frac{1}{16} & \frac{7}{16} \end{bmatrix}$$

By applying the control law (16.60) and adaptation law (16.70), the tracking performance and parameter estimation are shown in Fig. 16.10.

Self-Tuning Control

Historically self-tuning control has been implemented mostly in discrete time while MRAC has been implemented in continous time. A general architecture for the self-tuning control is given in Fig. 16.11.

Self-tuning control has two essential components:

1. Parameter estimation, and
2. Control law.

Parameter Estimation: Parameter estimation is a key element in a self-tuner and is performed on-line. The model parameters are estimated based on the measurable process input, process output, and state signals. A number of recursive parameter estimation schemes are employed for self-tuning control. The most popular scheme is the recursive least squares estimation method.

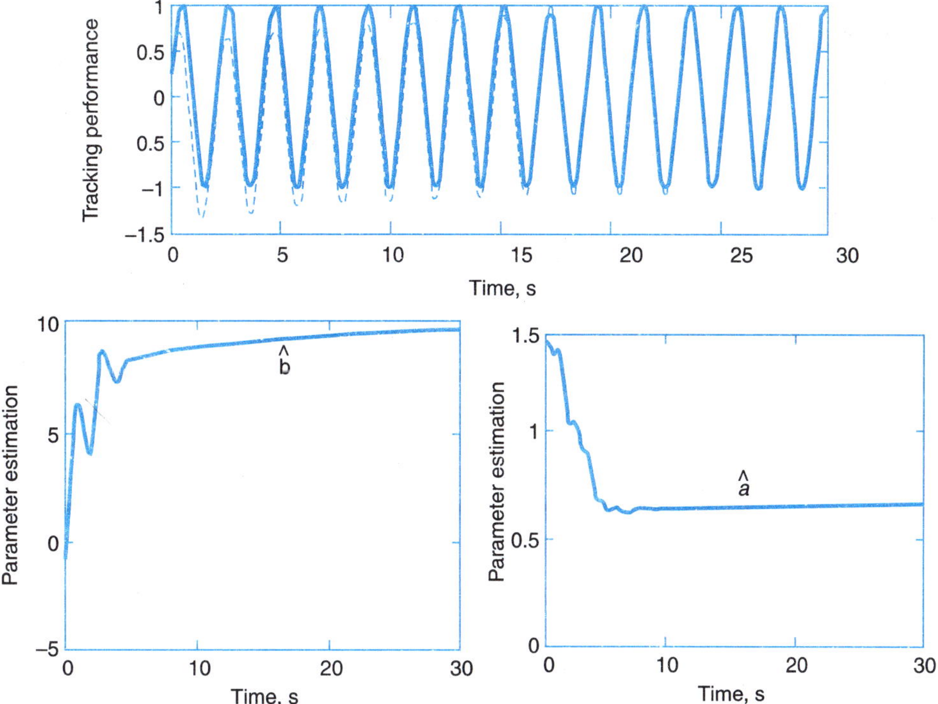

Fig. 16.10. Tracking performance and parameter estimation where $r(t) = \sin(3t)$.

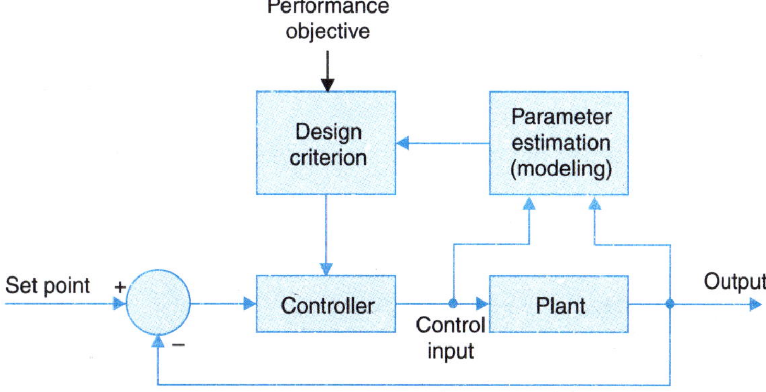

Fig. 16.11. A general configuration of self-tuning controller.

Control Law: The control law is derived based on control performance criterion optimization. Since the parameters are estimated on-line, the calculation of control law is based on a procedure called certainty equivalent in which the current parameter estimates are accepted while ignoring their uncertainties. This approach of designing controller using estimated parameters of the transfer function of the process is known as indirect self-tuning method.

In the following we present the design technique of a simple indirect self-tuner that uses Recursive Least Squares (RLS) estimation technique and pole-placement control law. We will take discrete-time model of the plant although this design can be done in continuous time.

The process model is given as

$$y(t) = -a_1 y(t-1) - \dots - a_n y(t-n) + b_0 u(t-d) + \dots + b_m u(t-d-m) \quad ...(16.71)$$

where y is the output, u is the input of the process and d represents process delay. The model is linear in parameters and can be written as

$$y(t) = \boldsymbol{\varphi}^T \boldsymbol{\theta} \quad ...(16.72)$$

where

$$\boldsymbol{\theta}^T = [a_1 \;\; a_2 \;\; \dots \;\; a_n \;\; b_0 \;\; b_1 \;\; \dots \;\; b_m] \quad ...(16.73)$$

$$\boldsymbol{\varphi}^T(t-1) = [-y(t-1) \dots -y(t-n)\; u(t-d) \dots u(t-d-m)] \quad ...(16.74)$$

The model parameter vector θ is unknown and is estimated using following recursive equations.

$$\hat{\boldsymbol{\theta}}(t) = \hat{\boldsymbol{\theta}}(t-1) + \mathbf{K}(t)[\mathbf{y}(t) - \varphi^T \hat{\boldsymbol{\theta}}(t-1)] \quad ...(16.75)$$

$$\mathbf{K}(t) = \mathbf{P}(t-1)\, \boldsymbol{\varphi}(t)\, [1 + \boldsymbol{\varphi}^T(t)\mathbf{P}(t-1)\, \boldsymbol{\varphi}(t)]^{-1} \quad ...(16.76)$$

$$\mathbf{P}(t) = (\mathbf{I} - \mathbf{K}(t)\, \boldsymbol{\varphi}^T(t))\, \mathbf{P}(t-1) \quad ...(16.77)$$

where $\mathbf{P}(t)$ is known as covariance matrix. In this algorithm the modeling error is calculated by comparing the estimated output with the observed output. The error is then minimized in the least square sense and then $\hat{\boldsymbol{\theta}}(t-1)$ is updated to a new value $\hat{\boldsymbol{\theta}}(t)$.

Using this estimated parameter vector we design control law using pole-placement technique. A general linear controller can be described by

$$Ru(t) = Tr(t) - Sy(t) \quad ...(16.78)$$

where R, S, and T are polynomials in the forward shift operator q. The plant model given in eqn. (16.71) can also be written as

$$A(q)y(t) = B(q)\, u(t) \quad ...(16.79)$$

Elimination of $u(t)$ in plant model using the proposed control law gives the following for the closed loop system:

$$y(t) = \frac{BT}{AR + BS} r(t) \quad ...(16.80)$$

The closed loop characteristic polynomial is thus

$$AR + BS = A_c \quad ...(16.81)$$

The key idea of design is to specify the desired closed-loop characteristic polynomial A_c. The polynomial R and S can then be solved from eqn. (16.81), which is known as *Diophantine equation*. This Diophantine equation determines only the polynomials R and S. Other conditions must be introduced to also determine the polynomial T. One condition may be the model following where plant response y has to follow the model.

$$A_m(q)y_m(t) = B_m(q)r(t) \qquad ...(16.82)$$

This condition can be met if

$$\frac{BT}{AR + BS} = \frac{BT}{A_c} = \frac{B_m}{A_m} \qquad ...(16.83)$$

The selections of R, S and T to satisfy eqn. (16.83) can be done in many ways. We will give two ways in this chapter.

A Pure Feed-forward Solution

If we select $R = BA_m$, $S = 0$ and $T = AB_m$, then eqn. (16.83) is satisfied. Thus the controller thus obtained is a pure feed-forward compensator without feedback. The compensator simply cancels the process dynamics and adds the desired dynamics. But if the process has poles and zeros outside the unit disc, then the system will be unstable.

Error Feedback

If the polynomials S and T are equal, the control law eqn. (16.78) can be written as

$$Ru = S(r - y) = Se \qquad ...(16.84)$$

where e is the control error. This means that the control law is based on feedback from the error only. It is easy to verify that the polynomials $R = B(A_m - B_m)$ and $S = T = AB_m$, satisfy the eqn. (16.83). This choice represents a feedback solution. Here also the closed loop system will remain unstable if there are poles and zeros outside the unit disc.

16.3 FUZZY LOGIC CONTROL

Fuzzy Logic has emerged as one of the active areas of research activity particularly in control applications. Fuzzy logic is a very powerful method of reasoning when mathematical models are not available and input data are imprecise. Its applications, mainly to control, is being studied throughout the world by control engineers. The results of these studies have shown that fuzzy logic is indeed a powerful control tool, when it comes to control systems or processes which are complex. Some studies have also shown that the fuzzy logic performs better when compared to conventional control mechanisms like PID. Wherever a logic in the spirit of human thinking can be introduced, fuzzy logic finds extreme application there. Sacrificing some amount of information we get a more robust summary of the information. What does this realy mean ? Though we are conditioned to think in precise quantities, at a subconscious level, we think and take actions that are fuzzy in nature. And that is the way we perceive the nature and react to it.

Before going into the details, let's look how fuzzy logic has become a house hold jargon. Though fuzzy logic originated in the U.S. some thirty years back, the researchers there were skeptic about its applicability in real world applications and some even scoffed it off as *nothing but probability*. On the other hand, the Japanese watched closely the pioneering work done by

Mamdani and his associates in steam engine control and started applying fuzzy control even to consumer goods like cameras, air conditioners, vacuum cleaners, etc. Thus fuzzy based products became highly competitive due to better performance, high reliability, robustness low power consumption, cheapness etc. One of the landmark success of fuzzy control was the complete automation of the subway train's drive control system in Japan. With fuzzy logic getting a wider acceptance in the recent years, it is predicted that by the end of the decade fuzzy logic will replace most of the conventional logic. Many projects which were nearly impossible earlier are now finding a new way out by *fuzzifying* them.

In the classical control paradigm, much stress is laid on the precision of the input, the intermediate steps that process them, and the modeling of the system in question. In spite of this we observe that many a time such sophisticated classical controllers developed often find it difficult to perform in real world control problems. Because the real world is so complex that regardless of the complexity of our model of the problem and the care taken to design such models, there exist so many parameters that have not been properly accounted for and many more of which we are totally ignorant of. Whereas a fuzzy logic solution is tolerant to the imprecision in the inputs and the model of the system and still produces an output that is desired out of the system. This was put in a more effective way by Lofti A. Zadeh, the father of fuzzy set theory, when he said, *"most applications of fuzzy logic exploit its tolerance for imprecision. Because precision is costly, it makes sense to minimize the precision needed to perform a task."* Thus the applicability of fuzzy logic is indeed promising.

The thinking process involved in the fuzzy realm is not complex. It is simple, elegant, ant easily applicable. The simplicity arises because it eludes mathematics to a great extent and elegance lies in its expressiveness. Even a person who does't know anything about a crane operation can design a fuzzy based controller for it with the help of an expert crane operator.

Fuzzy Sets Vs. Crisp Sets

The main objective behind fuzzy logic is to represent and reason with some particular form of knowledge expressed in a linguistic form. However, when using a language-oriented approach for knowledge representation, one has to build a conceptual framework to tackle its inherent *vagueness*. So let's start with a brief introduction to fuzzy-set concepts.

The traditional (or crisp) sets are based on a two-value logic : objects are either members or not members. Every individual object is assigned a membership value μ of either 1 or 0 that discriminates between members and non-members of the crisp set. For example, the crisp set (Fig. 16.12) *High* in terms of temperature may be defined such that :

$$\mu_{High} = \begin{cases} 0 \ if \ T < 30° \text{ C} \\ 1 \ if \ T \geq 30° \text{ C} \end{cases}$$

where T is the actual temperature. *High* is the *Linguistic Variable* that describes members of the set. The membership function μ_{High} is the measure of "belongingness" to the category of *High*. If T = 50°C, μ_{High} (50°C) = 1, and the temperature is 100% high and definitely not low.

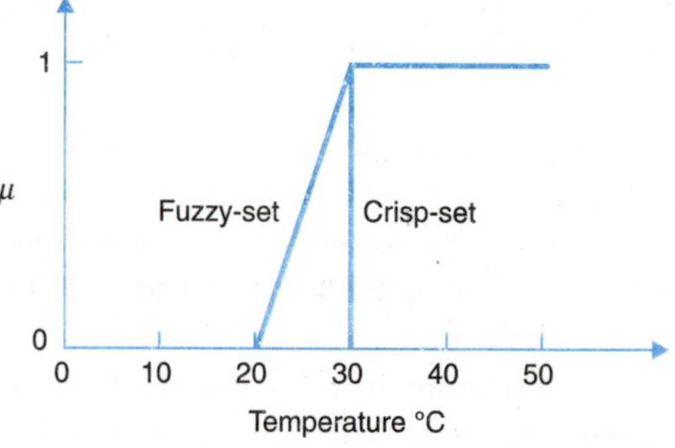

Fig. 16.12. Fuzzy-set and crisp-set definition of linguistic variable High.

Fuzzy-set theory on the other hand, is based on multivalued logic, and deals with the concepts that are not sharply defined. In fuzzy logic membership in the Linguistic variable *High* (Fig. 16.12) may be defined as:

$$\mu_{High} = \begin{cases} 0 & if\ 0°\text{C} \le T < 20°\text{C} \\ \dfrac{(T-20)}{10} & if\ 20°\text{C} \le T \le 30°\text{C} \\ 1 & if\ 30°\text{C} < T \le 40°\text{C} \end{cases}$$

The *Domain* of a fuzzy set is the range of allowable values of the variable. In this example the domain of a fuzzy set *High* is any value of *T* from 0°C to 40°C. The membership value μ_{High} goes from zero (no membership) to unity (complete membership) through intermediate values (partial membership). A temperature of 25°C has a degree of membership of $\dfrac{(25-20)}{10} = 0.50$ in the fuzzy set *High i.e.,* μ_{High} (25°C) = 0.50.

The variable *Temperature* may have many fuzzy sets associated with it (for example *Low, Medium, High* etc.) and the domains of all the *fuzzy* sets constitutes what is known as the *Universe of Discourse.* The fuzzy sets *Low, Medium* and *High* associated with *T* may be defined as shown in Fig. 16.13. Here the universe of discourse is the range of values from 0°C to 40°C. A temperature of 22°C belongs to the set *Low* to a degree of 0 (μ_{Low}(22°C) = 0), to the set *Medium* to a degree of 0.8 (μ_{Medium} (22°C) = 0.8) and to the set *High* to a degree of 0.2 (μ_{High}(22°C) = 0.2). As Fig. 16.13 shows, fuzzy sets always overlap to some degree, indicating that their boundaries are fuzzy or imprecises. A temperature of 22°C is both Medium and High. You might have wanted to call it "The temperature is lightly high". A more precise statement is μ_{Low}(22°C) = 0, μ_{Medium} (22°C) = 0.8, μ_{High}(22°C) = 0.2.

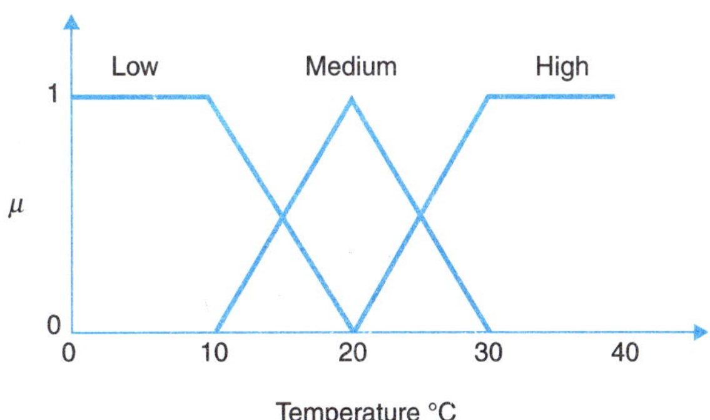

Fig. 16.13. Overlapping sets for temperature.

The traditional set theory and logic always have maintained two inviolate laws : The law of Non-Contradictions and the Law of Excluded Middle. The Law of Non-Contradiction states that an element cannot belong both to sets $\mathcal{A}$ and $\bar{\mathcal{A}}$ (the complement of $\mathcal{A}$). Mathematically this can be written as:

where ϕ is the null set, $$\mathcal{A} \cap \bar{\mathcal{A}} = \phi \qquad ...(16.85)$$

The Law of Excluded Middle declares that an element must belong to either set $\mathcal{A}$ or $\bar{\mathcal{A}}$. There is no middle. Mathematically this can be written as:

$$\mathcal{A}\bar{\mathcal{A}} = \mathcal{U} \qquad ...(16.86)$$

where $\mathcal{U}$ is the universal set.

Fuzzy-set theory and logic flout both these "laws". Hence, fuzzy logic represents a significant paradigm shift. It transcends traditional set theory and logic, and reduces them to special limiting cases.

Fuzzy Set Theory and Operations

In this section, we shall dwell in brief on some of the basic elements of fuzzy set theory that is required in fuzzy control applications. Consider a universe of discourse, $\mathcal{U}$, that consists of elements which shall be denoted by x. A fuzzy set $\mathcal{A}$ can be defined in $\mathcal{U}$ as follows.

$$\mathcal{A} = \{(\mu_A(x), x) \mid x \in \mathcal{U}\} \quad ...(16.87)$$

where $\mu_A(x)$ denotes the degree to which an element x in u belongs to $\mathcal{A}$. For this reason μ_A is known as the fuzzy membership function of $\mathcal{A}$ and $\mu_A(x)$ is known as the membership grade of x in fuzzy set $\mathcal{A}$.

Set Union: The union of two fuzzy sets $\mathcal{A}$ and $\mathcal{B}$ in u is a fuzzy set $\mathcal{A} \cup \mathcal{B}$, is defined as follows,

$$\mathcal{A} \cup \mathcal{B} = \{(\mu_{A \cup B}(x), x) \mid \in \mathcal{U}\} \quad ...(16.88)$$

where,

$$\mu_{A \cup \mathcal{B}}(x) = \max\,(\mu_A(x), \mu_B(x))\ \forall x \in \mathcal{U} \quad ...(16.89)$$

Set Intersection: The intersection of two fuzzy sets $\mathcal{A}$ and $\mathcal{B}$ in u is a fuzzy set $\mathcal{A} \cap \mathcal{B}$, is defined as follows:

$$\mathcal{A} \cap \mathcal{B} = \{(\mu_{A \cap B}(x), x) \mid x \in \mathcal{U}\} \quad ...(16.90)$$

where,

$$\mu_{\mathcal{A} \cap \mathcal{B}}(x) = \min\,(\mu_A(x), \mu_B(x))\ \forall x \in \mathcal{U} \quad ...(16.91)$$

Set Complement: The complement of a fuzzy set $\mathcal{A}$ in u is a fuzzy set $\overline{\mathcal{A}}$, and is defined as follows:

$$\overline{\mathcal{A}} = \{(\mu_{\overline{\mathcal{A}}}(x), x) \mid x \in \mathcal{U}\} \quad ...(16.92)$$

where,

$$\mu_{\overline{\mathcal{A}}}(x) = 1 - \mu_A(x)\ \forall x \in \mathcal{U} \quad ...(16.93)$$

The following example illustrates the above operations.

Consider the fuzzy definition of *High* as given in Section 16.2.2, along with the notion of low Pressure (*Lo-Press*) defined as:

$$\mu_{Lo\text{-}Press} = \begin{cases} 0 & if\ 1.5\ atm < P \le 2.0\ atm \\ (1.5 - P) & if\ 0.5\ atm \le P \le 1.5\ atm \\ 1 & if\ 0.0\ atm \le P \le 0.5\ atm \end{cases}$$

Now consider that the current condition C is 28°C and 1.1 *atm*. Then $\mu_{High}(C) = 0.8$, and $\mu_{High}(C) = 0.4$. It is both *High* temperature **and** *Low* pressure to a degree of $\{\min[\mu_{Low}(C), \mu_{Lo\text{-}Press}(C)]\} = \min\,[0.8, 0.4] = 0.4$. It is either *High* temperature **or** *Low* pressure to a degree of $\{\max[\mu_{High}(C), \mu_{Lo\text{-}Press}(C)]\} = \max\,[0.8, 0.4] = 0.8$. The temperature is not *High* to degree of $(1 - \mu_{High}) = 1 - 0.8 = 0.2$.

Fuzzy Relations : If $\mathcal{X}$ and $\mathcal{Y}$ are universal sets, then a fuzzy relation **R** on $\mathcal{X} \times \mathcal{Y}$ is defined as follows,

$$\mathbf{R} = \{(\mu_R(x, y), (x, y)) \mid (x, y) \in \mathcal{X} \times \mathcal{Y}\} \quad ...(16.94)$$

Composition : The composition of relations is the net effect of applying one relation after another. For the case of two binary relations **P** and **Q**, the composition of their relations as the binary relation **R** is given by,

$$\mathbf{R}(\mathcal{A}, C) = \mathbf{Q}(\mathcal{A}, \mathcal{B}) \text{ o } \mathbf{P}(\mathcal{B}, C) \quad ...(16.95)$$

where

$\mathbf{R}(\mathcal{A}, C)$ is relation between $\mathcal{A}$ and C

$\mathbf{Q}(\mathcal{A}, \mathcal{B})$ is a relation between $\mathcal{A}$ and $\mathcal{B}$

$\mathbf{P}(\mathcal{B}, C)$ is a relation between $\mathcal{B}$ and C

$\mathcal{A}, \mathcal{B}, C$ are fuzzy sets and o is the composition operator. The relation of **R** is the same as applying **P** first, followed by **Q**. In terms of membership functions,

$$\mathbf{R} = \{(\mu_R(a, c), (a, c)) \mid a \in \mathcal{A}, c \in C\} \quad ...(16.96)$$

where,

$$\mu_R(a, c) = \max(\min(\mu_Q(a, b), \mu_P(b, c)))\ b \in \mathcal{B} \quad ...(16.97)$$

Thus the composition is commonly defined by the min-mix matrix product or simply minmax.

Example 16.5 : Let $\mathcal{A} = \{0, 1, 2\}$, $\mathcal{B} = \{-1, 0, 1, 2\}$ and $C = \{0, 1\}$ be three discrete universe of discourses. Let $a \in \mathcal{A}$, $b \in \mathcal{B}$ and $c \in C$. Then "*a is much higher than b*" is a binary fuzzy relation whose membership function may be defined as

$$\mu_P(a, b) = \begin{cases} 1.0 & a - b = 3 \\ 0.7 & a - b = 2 \\ 0.3 & a - b = 1 \\ 0.0 & otherwise \end{cases}$$

The relation may be represented in matrix notation as

		$\mathcal{B}$			
		−1	0	1	2
$\mathcal{A}$	0	0.3	0.0	0.0	0.0
	1	0.7	0.3	0.0	0.0
	2	1.0	0.7	0.3	0.0

and "*b is close to c*" is also a fuzzy relation whose membership function may be defined as

$$\mu_Q(b, c) = \begin{cases} 1.0 & |b - c| = 0 \\ 0.5 & |b - c| = 1 \\ 0.0 & otherwise \end{cases}$$

The relation may be represented in matrix notation as

		C	
		0	1
$\mathcal{B}$	−1	0.5	0.0
	0	1.0	0.5
	1	0.5	1.0
	2	0.0	0.5

The fuzzy set **R** defined as **R** $(\mathcal{A}, C)$ = **Q**$(\mathcal{A}, B)$ o **P**$(\mathcal{B}, C)$, can be calculated from eqn. (16.97). Consider the case $a = 2$ and $c = 0$, then, eqn. (16.97) gives

$$\mu_R(2, 0) = \max [\min (\mu_P(2, -1), \mu_Q(-1, 0)), \min (\mu_{\dot{P}}(2, 0), \mu_Q(0, 0)), \min (\mu_P(2, 1), \mu_Q(1, 0)), \min (\mu_P(2, 2), \mu_Q(2, 0))]$$

$$= \max [\min (1.0, 0.5), \min (0.7, 1.0), \min (0.3, 0.5), \min (0.0, 0.0)]$$

$$= \max [0.5, 0.7, 0.3, 0.0]$$

$$= 0.7$$

The relation **R** may be represented in matrix notation as

		C: 0	1
$\mathcal{A}$	0	0.3	0.0
	1	0.5	0.3
	2	0.7	0.5

Fuzzy Logic Controller Components

A fuzzy Logic based Controller (FLC) is developed using the concept of prevailing conventional controllers. Consider of example an ordinary PID regulator. Its linear behavior can be described by the model :

$$u(t) = k\left[e(t) + \frac{1}{\tau_i}\int^{t} e(t)dt + \tau_d\left(\frac{de(t)}{dt}\right)\right] \qquad ...(16.98)$$

where $u(t)$ is the control signal and $e(t)$ the error signal. PID control can be understood very well from this linear equation. To obtain a good PID regulator, it is also necessary to consider operator interfaces, operational issues like switching smoothly between manual and automatic operations, transients due to parameter changes, the effects of non-linear actuators, wind up of the integral term, maximum and minimum selectors, etc. An operational industrial PID regulator consists of an implementation of the above model and heuristic logic that takes care of these issues.

It can thus be concluded that practical solution to such mundane problems like PID control are not done by theory alone but that heuristics play an import role. The discrete form of equation can be represented as

$$u(k) = u(k - 1) + \delta u(k) \qquad ...(16.99)$$

where

$$\delta u(k) = f(e(k), \delta e(k), ...) \qquad ...(16.100)$$

i.e., incremented controller output is some function of error, change in error and other terms related to present and past status of error.

Once the system behavior is thoroughly studied, comprehensive performance objectives of the controller can be speculated. Then the heuristic rules that describe the equation can be formulated by the help of intelligent trial and error science as well as self institution learned from experience. This forms the basis for the design of a FLC.

The principal design parameters for an FLC are the following:

- fuzzification strategies and the interpretation of a fuzzification operator (fuzzifier)
- data base :
  (*i*) discretization/normalization of universe of discourse,
  (*ii*) fuzzy partition of the input and output spaces,
  (*iii*) completeness,
  (*iv*) choice of the membership function of a primary fuzzy set;
- rule base :
  (*i*) choice of process state (input) variables and control (output) variables of fuzzy control rules,
  (*ii*) source and derivation of fuzzy control rules,
  (*iii*) types of fuzzy control rules,
  (*iv*) consistency interactivity, completeness of fuzzy control rules;
- fuzzy inference mechanism,
- defuzzification strategies and the interpretation of a defuzzification operator (defuzzifier).

The section deals with the design of basic components of a FLC. The above listed components are necessary for the construction of any FLC. Fuzzification implies the process of transforming the crisp values of the inputs to a controller, to the fuzzy domain. The fuzzified values now fire the rule base and generate a so called fuzzified output. As a final step, the fuzzified output is defuzzified to yield the crisp controller output. Fig. 16.14 gives an idea of the construction of a fuzzy logic based controller. Now, let's deal with each of the components in detail.

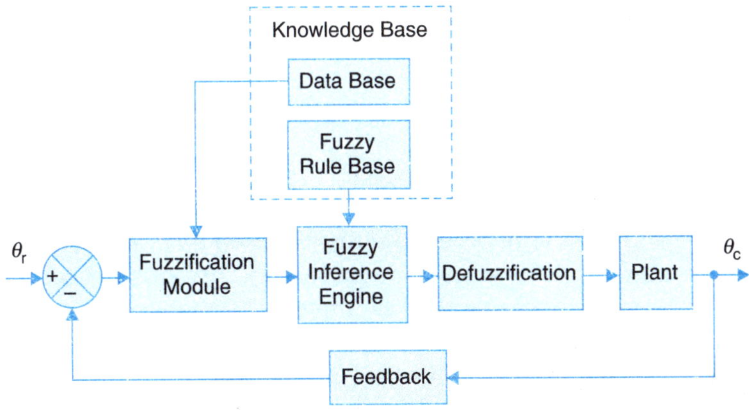

Fig. 16.14. Fuzzy logic based control system.

Fuzzification module

Fuzzification is related to the vagueness and imprecision in a natural language. It is a subjective valuation which transforms a measurement into a valuation of a objective value, and hence it could be defined as a mapping from an observed input space to fuzzy sets in certain input universes of discourse. In fuzzy control applications, the observed data are usually crisp. Since the data manipulation in an FLC is based on fuzzy set theory, fuzzification is necessary in an earlier stage. The fuzzification module (FM) performs the following functions :

- measures the values of input variables.
- Performs a scale transformation (*i.e.*, an input normalization) which maps the physical values of the current process state variables into a normalized universe of discourse. It also maps the normalized value of the control output variable onto its physical domain (*i.e.*, an output denormalization). When a non-normalized domain is used then there is no need for FM. A normalized universe implies the knowledge of input/output space via appropriate scale mappings. Due to normalization, a well-formed term set can be achieved. If this is not the case, or a non normalized universe is used, the terms could be asymmetrical and unevenly distributed in the universe.
- Performs the so called fuzzification which converts a point-wise (crisp), current value of a process state variable into a fuzzy set, in order to make it compatible with the fuzzy set representation of the process state variable in the rule-antecedent.

Consider for example a temperature control system as shown in Fig. 16.15. The temperature of the hot process liquid has to be maintained at temperature T°C by regulating the cold liquid flowrate. For this temperature controller, the error e between the process fluid temperature and its setpoint can be taken as one of the input variables, and can be partitioned into fuzzy sets : Negative-Large (NL), Negative-Small (NS), Zero (ZE), Positive-small (PS) Positive-Larger (PL). Similarly the change in error δe can be taken as the second input and classified as NL, NS, ZE, PS, PL. If the previous process fluid temperature is 40°C, the current temperature is 14.2°C and the set point is 50°C, then the error $e = 50 - 41.2 = 8.8$ and change in error $\delta e = (50 - 41.2) - (50 - 40) = -1.2$. Then the crisp value of error can be converted to a fuzzy set representation based on Fig. 16.16 and is classified as an element of ZE and PS. Similarly the crisp value of change in error based on Fig. 16.17 is classified as an element of NS and NL. In this case the output variable is a value action F_c and can be partitioned into fuzzy sets : NL, NS, ZE, PS, PL. Suppose the control action from an FLC is given as NS, then referring to Fig. 16.22, NS corresponds to –3% change in the controller output to value.

Membership functions

Before going into the detailed discussion on the various components comprising a fuzzy logic control system, it is essential to understand the membership functions. Membership functions form a crucial part in a fuzzy rule base model because they only actually define the fuzziness of a control variable or a process variable. For example, if 50°C represents *hot* for a particular membership function the same value may mean *very hot* for another function. Unfortunately no systematic study and analysis of the effect of membership function on the controller performance exists.

The most popular choices for the shape of the membership functions are triangular, trapezoidal, and bell-shaped functions. These three choices can be explained by the ease with which a parametric, functional description of the membership function can be obtained, stored with minimum use of memory, and manipulated efficiently, in terms of real time requirements by the inference engine.

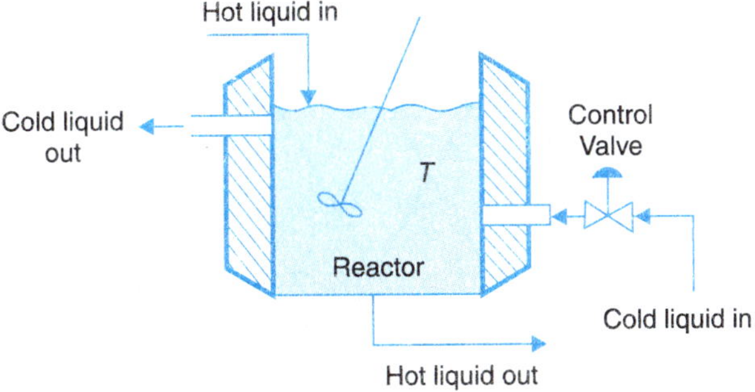

Fig. 16.15. Temperature control system.

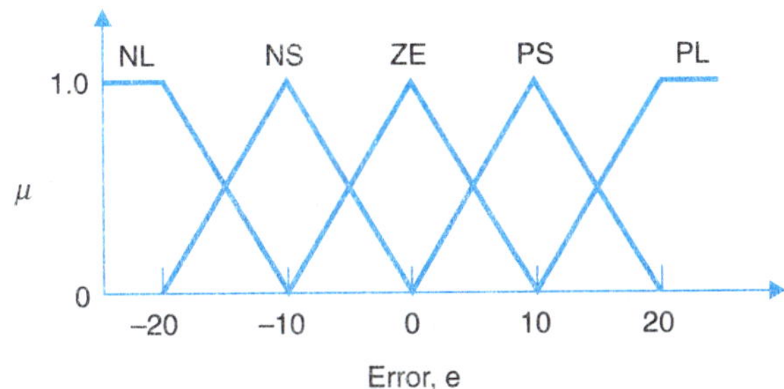

Fig. 16.16. Triangular membership definition of error.

It easy to see that parametric choice of the triangular function is the most economical one. This explains the predominant use of this type of membership functions. Once the shape of the function is selected, one has to map each element of the term set on the domain of the corresponding linguistic variable. This mapping can affect the nature of the fuzzy controller in a number of ways. It has been seen that influence of the crossover points is important. Because if there were no crossovers then there would be discontinuities in the control outputs. Studies has shown that crossover at 0.5 and cross-point ratio at 1 result in a significantly low overshoot faster rise-time and less undershoot. Also a symmetrical shape of the function can lead to a easier inference and defuzzification. The number of quantization levels, influences significantly the overshoot and damping effects. A coarse quantization for large errors and a finer one for smaller errors is the usually followed practice. The above are purely empirical statements and so far no formal analysis tools exist for ascertaining the performance of the fuzzy controller. Below we shall describe the three important class of membership functions.

Triangular membership function

This is one of the most popular among the scientists in this field. Reconsider the case of temperature controller of Fig. 16.15, the membership functions for error e and change in error δe can be defined as shown in Figs. 16.16 and 16.17. The triangular membership function can be generally defined using a left point, center point, and right point. Using these we can calculate the slopes of the sides of the triangle. Once the triangle is thus constructed, membership values can be calculated. The following parameters can be used to adjust the shape of the triangles when there is a need to tune them for better performance.

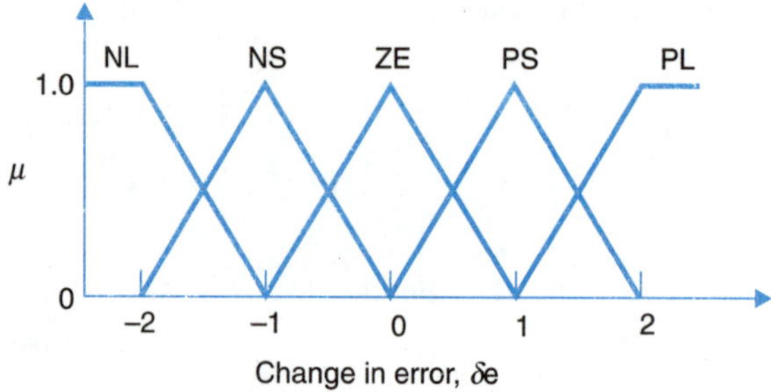

Fig. 16.17. Triangular membership definition of change in error.

Overlap: The overlap is the point of crossover between successive triangles. As the overlap is varied the fuzzification of the input space is changed. Actually, zero overlap is not desirable because there are regions were no strong rule can make a decision. In fact at the point of crossover, there is no rule which is fired. As a result, it is seen that from the point of maximum there is a sudden drop in the output response. There is an improvement in performance when the overlap is increased to 0.5 because in the mid range (at the point of crossover), certain strong rules can fire a valid decision. Finally a further increase in the overlap to say 0.75 results in the degradation of the performance. This is because now the triangles almost merge with each other that there is a clash among them over supremacy in taking decision for a particular situation.

Sensitivity: This is one more area where we can modify the shape of a membership function and observe the effect on the performance of the controller. Sensitivity is actually making the fuzzy engine more sensitive to smaller changes in the input variables. This can be incorporated by making the width of the membership function narrow in the midrange around zero and broader as we move away from zero. So if the system is operating at large values of error and error change, coarse action is taken, but as soon as the values enter within a band the fine control is activated. As a result of this the rule base which previously acted over the entire range now would act only on a narrower range and this small range in turn has all the definitions that were applicable in the large range just multiplied by a proportionality constant.

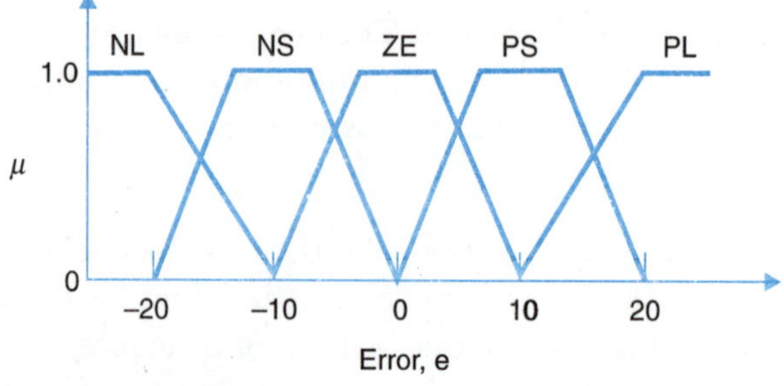

Fig. 16.18. Trapezoidal membership definition of error.

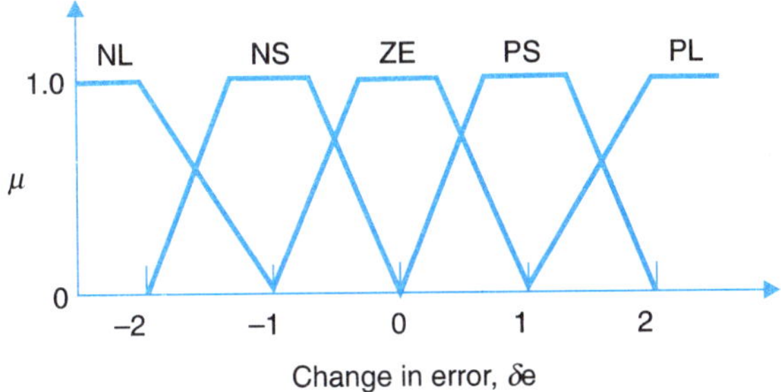

Fig. 16.19. Trapezoidal membership definition of change in error.

Trapezoidal membership function

As the name suggests the shape of this class of membership functions is that of a trapezoid and is shown in Figs. 16.18 and 16.19 for the temperature controller (Fig. 16.15). The maximum membership value of 1.0 occurs over a small range about the central point of the function. Studies show in general that these functions give a better performance than triangular membership functions. Oscillations encountered when triangular functions were used greatly reduces. This can be attributed to the flat top of the membership function. The flat top actually stabilizes the decision over a range and therefore there is not great change in the decision as would happen in a triangular membership function.

Bell shaped function

This choice has the advantage that the desired shape of the fuzzy set can be adapted by three parameters, c alters the point of minimum fuzziness, a alters the spread and b the contrast. Because the decision procedure is too time consuming in the continuous case the fuzzy output sets are calculated at finite quantized intervals of the term set. One example for the membership function that can be used is given below

$$y = \frac{1}{1 + (x - c)^2} \qquad ...(16.101)$$

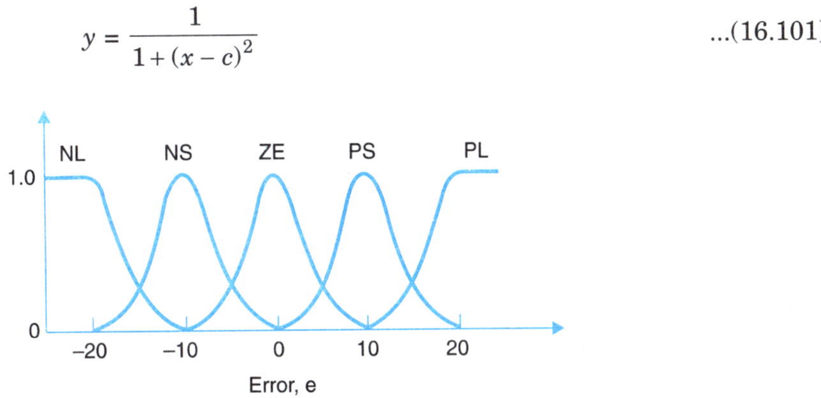

Fig. 16.20. Bell shaped membership definition of error.

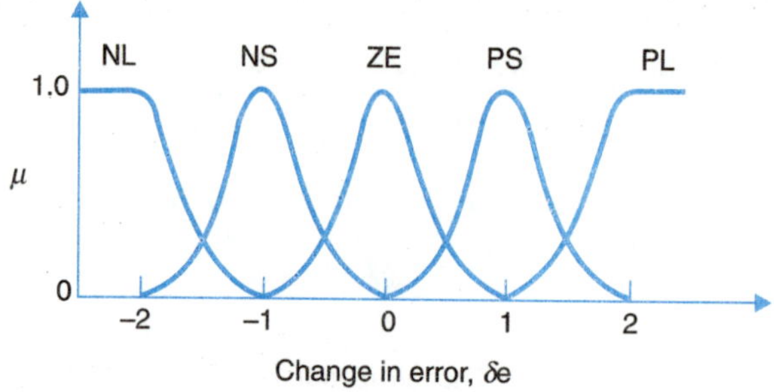

Fig. 16.21. Bell shaped membership definition of change in error.

which gives the membership value of x in that particular set and is shown in Figs. 16.20 and 16.21 for the temperature controller (Fig. 16.15). The set being characterized by its central value c (where the membership value is 1) *i.e.*, say for the linguistic variable *Zero* the value of c is 0, for the variable NS the value of c is –10, and so on. The function for the linguistic variable *Zero* yields a maximum membership value of 1 at $x = 0$ and the membership of x decreases as it deviates from this central value of 0. In the case of bell shaped functions the oscillations are the minimum and the rise time also is greatly reduced.

Knowledge base

The knowledge base of a FLC consists of a database and a rulebase. Some issues relating to the data base are discussed in the next section followed by issue relating to rulebase.

Data Base

The basic function of a database is to provide necessary information for the proper functioning of the FM, the rule base, and the defuzzification module. This information includes :

- fuzzy sets (membership functions) representing the meaning of the linguistic values of the process and the control output variables. These concepts are subjectively defined based on experience and engineering judgement.
- physical domains and their normalized counter parts together with the normalization or denormalization (scaling) factors. If the continuous domains of the process state and control output variables have been discretized then the data base also contains information concerning the quantization look-up tables defining the discretization policy.

Quantization discretizes a universe into a certain number of segments (quantization levels). Each segment labeled as a generic element forms as discrete universe. A fuzzy set is then defined by assigning grade of membership values to each generic element. In the case of an FLC with continuous universes, the number of quantization levels should be large enough to provide adequate approximation and yet be small to save memory space. The choice of quantization levels has an essential influence on how fine a control can be obtained. For example, if a universe is quantized for every ten units of measurement instead of five units, then the controller is half as sensitive to the observed variables. The normalization of a universe requires a discretization of the universe of discourse with each segment mapped into a suitable segment of the normalized universe. The scale mapping can be uniform, non-uniform or both. For example, a universe of discourse [0, 50] can be transformed into a normalized closed interval [0, 1] by using a uniform scaling factor of 0.02.

Rule Base

The basic function of the rule base is to represent in a structured way the control policy of an experienced process operator and/or control engineer in the form of a set of production rules such as

if ⟨Process State⟩ *then* ⟨Control Output⟩

The *if* part of such a rule is called rule antecedent and is a description of a process state in terms of a logical combination of atomic fuzzy propositions. The '*then*' part of the rule is called rule consequent and is again the description of the control output in terms of a logical combination of fuzzy propositions. These propositions state the linguistic values which the control outputs take when the current process state matches the process state description in the rule-antecedent. In our terminology, a fuzzy control rule is a fuzzy conditional statement in which the antecedent is a condition in its application domain and the consequent is a control action for the system under control. Basically, fuzzy control rules provide a convenient way for expressing control policy and domain knowledge. The proper choice of process state variables and control variables is essential to the characterization of the operation of a fuzzy system. The selection of the linguistic variables has a substantial effect on the performance of an FLC. Experience and engineering knowledge play an important role during this section stage. Typically, the linguistic variables in an FLC are the state, state error, state error derivative, state error integral, etc.

The source and derivation of fuzzy control rules are:

- Expert experience and control engineering knowledge: The fomulation of fuzzy control rules can be achieved by means of heuristic approaches. One of the most common approaches involves an introspective verbalization of human expertise. For example verbalization can be the operating manual for a cement kiln.
- Control operator's control actions: A human operator employs-consciously or subconsciously-a set of fuzzy if-then rules to control the process. These if-then rules can be coded as fuzzy rules by employing linguistic variables.
- Fuzzy model of a process: The process dynamics of a controlled process is modeled using the linguistic approach. This model is used to generate fuzzy control rules.
- Based on learning: This refers to self organizing controller (SOC) in which the FLC has the ability to create fuzzy control rules and to modify them based on experience.

Table 16.1. A Typical Rule Base

$e \rightarrow$ δe $\downarrow$	**NL**	**NS**	**NE**	**PS**	**PL**
PL	1.ZE	2.PS	3.PL	4.PL	5.PL
PS	6.NS	7.ZE	8.PS	9.PL	10.PL
ZE	11.NL	12.NS	13.ZE	14.PS	15.PL
NS	16.NL	17.NL	18.NS	19.ZE	20.PS
NL	21.NL	22.NL	23.NL	24.NS	25.ZE

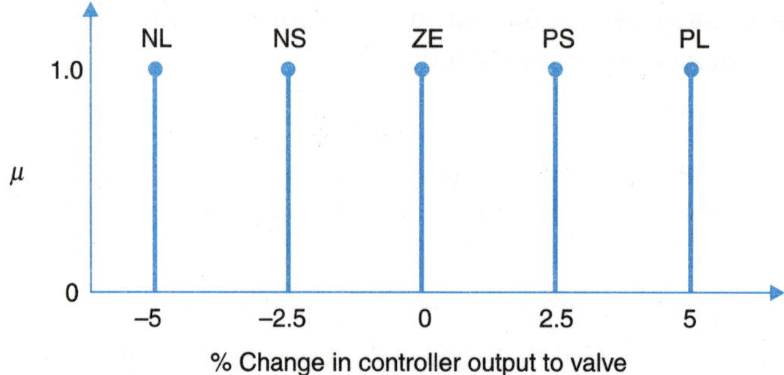

Fig. 16.22. Definitions for the output variable, valve action.

Design of the rule base

In a temperature controller (Fig. 16.15), the error e between the process fluid temperature and its setpoint is one of the input variables, and is partitioned into fuzzy sets : NL, NS, ZE, PS and PL as shown in Fig. 16.16. Similarly the change in error δe is classified as NL, NS, ZE, PS and PL as shown in Fig. 16.17. The output variable of the FLC is the valve action F and is classified in terms of change in controller output to the valve as NL, NS, ZE, PS and PL as shown in Fig. 16.22. The output variable is defined as a singleton, or a set consisting of a single member. The unique member of a singleton is called the support point or the support value of the singleton. The membership function of a singleton vanishes everywhere in the space of the variable except at the support point, where the grade is 1.

The rulebase for the above temperature controller can be imagined to be a two dimensional matrix as summarized in Table 16.1. The rows represent the various linguistic values that change in error δe can take. Like wise the columns indicate the values of the variable error e, can assume. The entries in this matrix would be the control action that has to be taken described in linguistic terms. These rules often are designed to mimic a conventional PID controller in the velocity mode-just as easily, however, they also could incorporate information about process nonlinearities, constraints and so on.

Because the input variables belong to several fuzzy sets, they produce non-zero membership values for a number of different rules. For example, if e is 8°C and δe is 0 °C/s, then, using the memberships as illustrated in Figs. 16.16 and 16.17 $\mu_{PS}(e) = 0.8$ $\mu_{ZE}(e) = 0.2$ $\mu_{ZE}(\delta e) = 1.0$ and all other μ's = 0. Accordingly checking Table 16.1 only the following rules are activated :

- Rule No. 13: if e is ZE and if δe is ZE, then the control action F is ZE
- Rule No. 14: if e is PS and if δe is ZE, then F is PS

Thus the rule base (Table 16.1) for a controller works by monitoring the error and error change of the system. Now we shall see the logic behind building this table using the system response of the process to be controller (for *e.g.*, Fig. 16.23).

Consider the set of rules whose values of e and δe are ZE which correspond to points J, K, L in Fig. 16.23. This indicates that the system is slightly deviated from the setpoint. Therefore the correction provided by the corresponding rules are also of lesser magnitude, just enough to push the state to setpoint. Thus these set of rules represent the steady state behaviour of the process.

If the value of e is PL and δe is ZE (point A in Fig. 16.23), this indicates that the state is below the setpoint and is changing very slowly. Thus a large amount of change is intended to speed up the approach to the set point. If e is PS (point E) which means that process state is below the setpoint and at the same time, δe is ZE, y is changing very slowly. The amount of change which the rules of this type introduce to the previous control output is intended to speed up the approach to the setpoint.

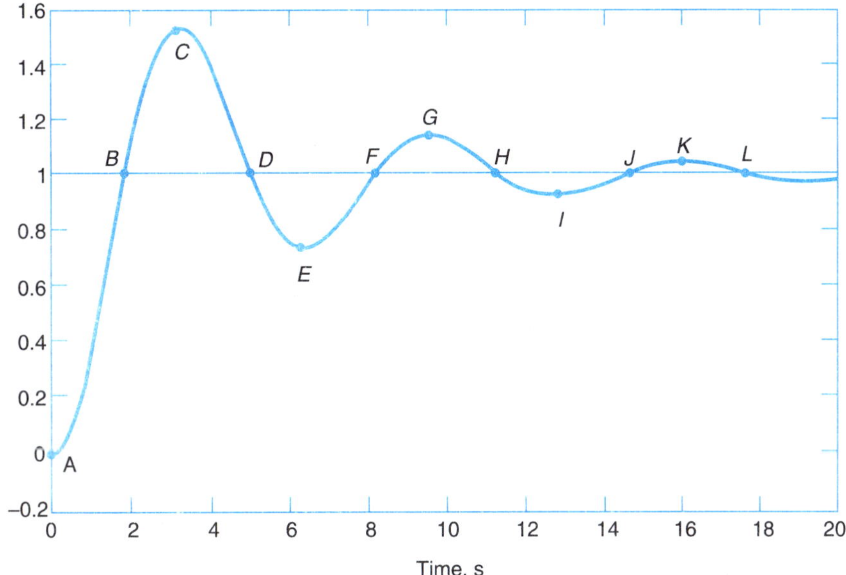

Fig. 16.23. Design of rule base based on system response.

When e is ZE and δe is PL (point D) or PS (point H), process state y, is moving away from the setpoint. Thus a positive change in the previous control output is intended to reverse this trend and make it, instead of moving away from the set point, to start moving towards it. If e is ZE and at the same time, δe is NL (point B) or NS (point F), y is moving away from the set point. Thus a negative change in the previous control output is intended to reverse this trend and make y, instead of moving away from the set point, start moving towards it. The prototype of these fuzzy control rules are summarized in Table 16.2. Better control performance can be obtained by using finer fuzzy partitioned subspaces, for example, with the term set {NL, NM, NS, ZE, PS, PM, PL}.

Fuzzy inference engine

There are two types of approaches employed in the design of the inference engine of a FLC :

- Composition based inference.
- Individual rule-based inference.

In the composition based inference, the fuzzy max-min composition operator is employed. The rulebase matrix represents the fuzzy relation matrix, R. R is then composed with e first and then the resulting fuzzy relation is composed with δe.

$$O = \delta e \; o \; e \; o \; \mathcal{R} \qquad ...(16.102)$$

Table 16.2. Rule Base Design : Prototype of Fuzzy Control Rules

Rule No.	*e*	*δe*	*F*	*Ref. Point*
3.	ZE	PL	PL	D
8.	ZE	PS	PS	H
11.	NL	ZE	NL	C
12.	NS	ZE	NS	G
13.	ZE	ZE	ZE	K
14.	PS	ZE	PS	E
16.	PL	ZE	PL	A
18.	ZE	NS	NS	F
23.	ZE	NL	NL	B

The final result of the above compositions result in O from which the fuzzified output can be extracted.

Reconsider the temperature controller example in section 4.3.3. The rules activated were rule no. 13 and 14. Rule 13 maintains that the valve action should be zero, while rule 14 indicates that it should be positive-small. Both the rules have a claim on how much the valve should change. The truth of the control rule (weight value, an extent to which each active rule is applicable) can be obtained by:

$$\mu(13) = \textit{Truth of Rule No. } 13$$
$$= \mu_{ZE}(e) \cap \mu_{ZE}(\delta e)$$
$$= \min (0.2, 1.0) = 0.2$$
$$\mu(14) = \textit{Truth of Rule No. } 14$$
$$= \mu_{PS}(e) \cap \mu_{ZE}(\delta e)$$
$$= \min (0.8, 1.0) = 0.8$$

The basic function of second type of inference is to compute the overall value of the control output variable based on the individual contributions of each rule in the rule base. Each such individual contribution represents the values of the control output variables as computed by a single rule. The output of the fuzzification module, representing the current crisp values of the process state variables, is matched to each rule-antecedent and a degree of match for each rule is established. Based on this degree of match, the value of the control output variable in the rule-antecedent is modified, *i.e.*, the clipped fuzzy set representing fuzzy value of the control output variable is determined. The set of all clipped fuzzy sets represents the overall fuzzy output.

Defuzzification module

The functions of a defuzzification module (DM) are as follows:

- Performs the so called defuzzification which converts the set of modified control output values into a single point-wise values.
- Performs an output denormalization which maps the point-wise value of the control output onto its physical domain. This step is not needed if non-normalized fuzzy sets are used.

A defuzzification strategy is aimed at producing a non-fuzzy control action that best represents the possibility of an inferred fuzzy control action. There many procedures outlined in the literature for defuzzification which are, center of gravity/area, center of sums, center of largest area, first of maxima, middle of maxima, and height. Of these the center of gravity (COG) is the most efficient in that it gives a defuzzified output which conveys the real meaning of the action that had to be taken at that instant.

The COG strategy generates the center of gravity of the possibility distribution of a control action. In the case of a discrete universe, this method yields

$$F_o = \frac{\sum_{i=1}^{m} \mu(i).\, \omega_i}{\sum_{i=1}^{m} \mu(i)} \qquad \text{...(16.103)}$$

where m is the number of quantization levels of the output, $\mu(i)$ represents the "truth" of the ith rule and ω_i represents the action that i^{th} rule would dictate.

In the temperature controller example considered in the Section 16.2.4, Rule No. 13 would make zero change, while Rule No. 14 would make +2.5% change (Fig. 16.22). The actual output is an average of the active rule outputs (Fig. 16.22), weighted by their corresponding truthes. Thus the output is given by

$$F_o = \frac{0 \times 0.2 + 2.5 \times 0.8}{0.2 + 0.8}$$
$$= 2\%$$

For this particular example, the sum of the truths of activated rules. Rule 13 and 14 is unity : 0.8 + 0.2 = 1.0. This is neither a necessary, nor a desirable, nor common event. It is simply the happenstance of this example.

Case Study

We shall now take up the example of a temperature control in two stirred tanks in series to explain the methodology of designing a fuzzy logic controller. The benchscale process as shown in Fig. 16.24 consists of two stirred tanks connected by a long pipe which introduces time delay. The temperature of the water in the second tank T_2 is controlled by adjusting the heat duty from the electrical heater in the first tank. An additional heater can be introduced to introduce load disturbances manually. The volume of water in each tank is held constant by using overflow lines. The water flowrate through the system is controlled manually and can be adjusted to introduce a second type of disturbance.

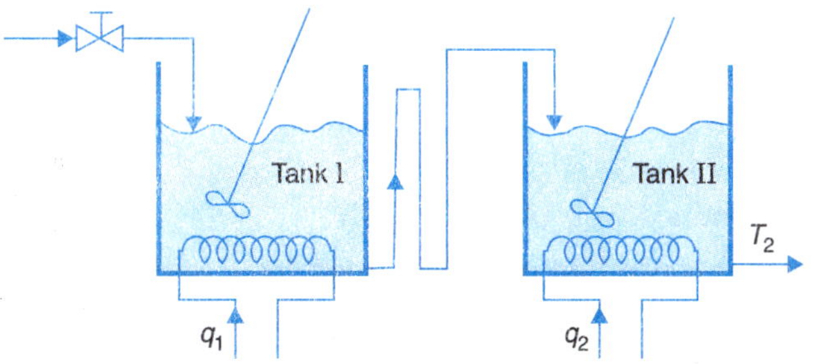

Fig. 16.24. Schematic diagram of the benchj-scale plant.

Assuming constant flowrate of $\omega = 0.05$ kg/s the dynamic model of the heating process is obtained as

$$\frac{T_2(s)}{q_1(s)} = G(s) = \frac{4.15\,\exp(-T_d s)}{(119s+1)(71s+1)} \qquad ...(16.104)$$

where T_2 is the outlet water temperature and q_1 is the control input. The time delay $T_d = 53s$, is determined empirically. Using the techniques of Chapter 11, the digital model is found out with a zero order hold out included in the process. The sampling time is chosen to be 5.3s. From eqn. (11.56) we get

$$G_o(s) = \frac{1-e^{-sT}}{s} \qquad ...(16.105)$$

The approximation time delay term $e^{-T_d s}$ is done by Padé approximation, which is given in the following for a second order approximation:

$$e^{-T_d s} \cong \frac{1 - T_d s/2 + T_d^{\,2} s^2/4}{1 + T_d s/2 + T_d^{\,2} s^2/4} \qquad ...(16.106)$$

The digital model then is given by

$$\frac{T_2(z)}{q_1(z)} = Z[G_o G(s)] \qquad ...(16.107)$$

$$= \frac{0.00658785 z^{-11} + 0.006372325 z^{-12}}{1 - 1.8845096 z^{-1} + 0.8876429 z^{-2}} \qquad ...(16.108)$$

The above eqn. (16.108) is written in the state variable form using the techniques of Chapter 12 as:

$$\mathbf{x}(k+1) = \mathbf{A}\mathbf{x}(k) + \mathbf{B}\mathbf{u}(k) \qquad ...(16.109)$$

where

$$\mathbf{A} = \begin{bmatrix} 1.8845096 & 1 & 0 & 0 & 0 & 0 & 0 & 0 & 0 & 0 & 0 & 0 \\ -0.8876429 & 0 & 1 & 0 & 0 & 0 & 0 & 0 & 0 & 0 & 0 & 0 \\ 0 & 0 & 0 & 1 & 0 & 0 & 0 & 0 & 0 & 0 & 0 & 0 \\ 0 & 0 & 0 & 0 & 1 & 0 & 0 & 0 & 0 & 0 & 0 & 0 \\ 0 & 0 & 0 & 0 & 0 & 1 & 0 & 0 & 0 & 0 & 0 & 0 \\ 0 & 0 & 0 & 0 & 0 & 0 & 1 & 0 & 0 & 0 & 0 & 0 \\ 0 & 0 & 0 & 0 & 0 & 0 & 0 & 1 & 0 & 0 & 0 & 0 \\ 0 & 0 & 0 & 0 & 0 & 0 & 0 & 0 & 1 & 0 & 0 & 0 \\ 0 & 0 & 0 & 0 & 0 & 0 & 0 & 0 & 0 & 1 & 0 & 0 \\ 0 & 0 & 0 & 0 & 0 & 0 & 0 & 0 & 0 & 0 & 1 & 0 \\ 0 & 0 & 0 & 0 & 0 & 0 & 0 & 0 & 0 & 0 & 0 & 1 \\ 0 & 0 & 0 & 0 & 0 & 0 & 0 & 0 & 0 & 0 & 0 & 0 \end{bmatrix}$$

and $$\mathbf{B} = [0 \;\; 0 \;\; 0 \;\; 0 \;\; 0 \;\; 0 \;\; 0 \;\; 0 \;\; 0 \;\; 0 \;\; 0.00658785 \;\; 0.006372325]^T$$

Also $x_1(k) = T_2(k)$ and $u(k) = q_1(k)$

The above system is controlled using the fuzzy logic controller designed as described in the earlier sections. The inputs to the FLC are error ({NL, NM, NS, NZ, PZ, PS, PM, PL}) and change in error ({NL, NM, NS, NZ, PZ, PS, PM, PL}). The heat input, q_1 is the output of the FLC and is defined by singletons. The response of the system using a FLC is compared with that of a PID and is given in Fig. 16.25.

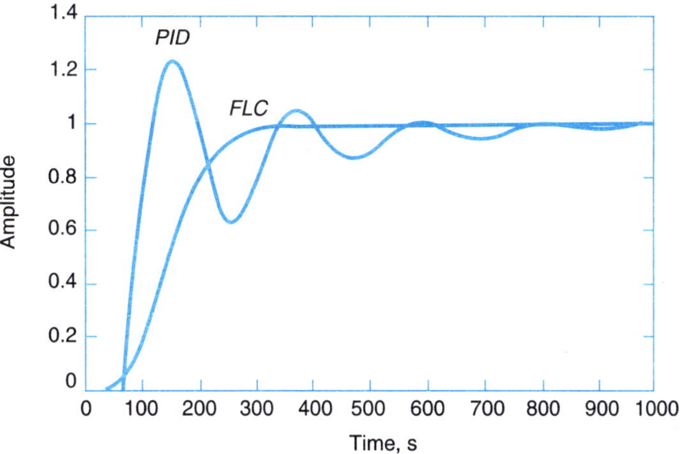

Fig. 16.25. Step input response : *FLC* Vs *PID*.

16.4 NEURAL NETWORKS

Artificial neural networks have emerged from the studies of how brain performs. The human brain is made up of many millions of individual processing elements, called *neurons,* that are highly interconnected. A schematic diagram of a single biological neuron is shown in Fig. 16.26. Information from the outputs of the other neurons, in the form of electrical pulses, are received by the cell at connections called *synapses*. The synapses connect to the cell inputs, *or space dendrites*, and the single output of the neuron appears at the *axon*. An electrical pulse is sent down the axon when the total input stimuli from all of the dendrites exceeds a certain threshold.

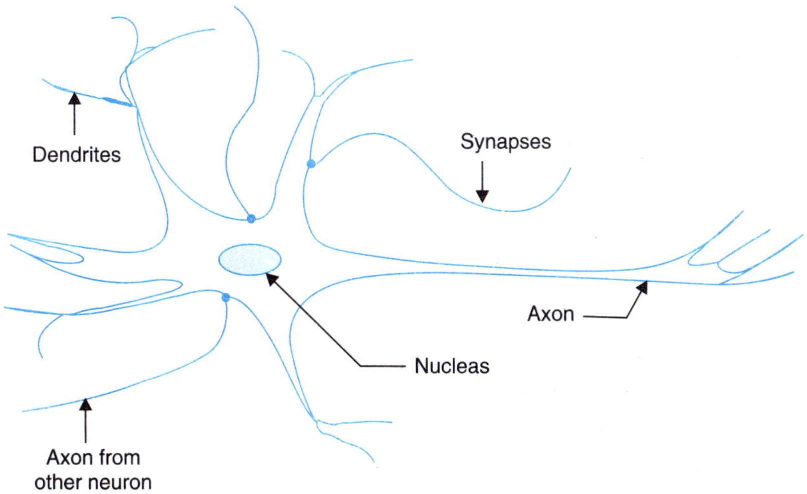

Fig. 16.26. A biological neuron.

Artifical neural networks are made up of simplified individual models of the biological neuron that are connected together to form a network. Information is strored in the network often in the form of different connection strengths, or weights, associated with the synapses in the artificial neuron models. Some of the various properties of neural networks are:

Inherent parallelism in the network architecture due to the repeated use of simple neuron processing elements. This leads to the possibility of very fast hardware implementations of neural networks.

- Capability of *learning* information by example. The learning mechanism is often achieved by appropriate adjustment of the weights in the synapses of the artificial neuron models.
- Ability to generalize to new inputs (*i.e.,* a trained network is capable of providing *sensible* output when presented with input data that has not been used before)
- Robustness to noisy data that occurs in real world applications
- Fault tolerance. In general, network performance does not significantly degenerate if some of the network connections become faulty.

The above properties of neural networks indicate their potential in solving complex, ill-defined problems. Hence, the considerable interest in neural networks that has occurred in recent years is not only due to the significant advances in computer processing power that has enabled their implementation, but also because of the diverse possibility of application areas. Many different types of neural network are available and only the *Multi Layer Network* (MLN) which is most popular in the context of this book, modeling and controls is described. The basic building block of these networks, the artificial neuron model is introduced in the next section.

Artificial Neuron Model

Neural networks represent massively parallel distributed processing capability which makes real-time processing of large volume of data more realizable with the potential for ever-improving performance through dynamical learning. A network of "neuron-like" units operates on data "all at once" rather than "step by step" as in a conventional computer. Neural networks can be considered as a large dimensional non-linear dynamics system, which is defined by a set of I^{st} order non-linear differential equations. It acts as compensation elements in control systems.

A neural network is a system with inputs and outputs and is composed of many simple and similar processing elements. The most commonly used model is depicted in Fig. 16.27 and is based on the model proposed by McCulloch and Pitts in 1943. Each neuron input, x_i, is weighted by the values w_i. A bias, or offset, in the node in characterized by an additional constant input of 1 weighted by the value w_o. The output, y, is obtained by summing the weighted inputs to the neuron and passing the result through a non-linear activation function, $f()$. Various non-linearity are possible and some of these, for *e.g.,* hard limiter, threshold logic, sigmoidal and tanh functions are shown in Fig. 16.28.

The processing elements each have a number of internal parameters called weights. Changing the weights of an element will alter the behaviour of the whole network. The goal here is to choose the weights of the network to achieve a desired input/output relationship. This process is known as training the network. The network can be considered memoryless in the sense that, if the weights are kept constant, the output vector depends only on the current input vector and is independent of past inputs. This may be also called as neural computer.

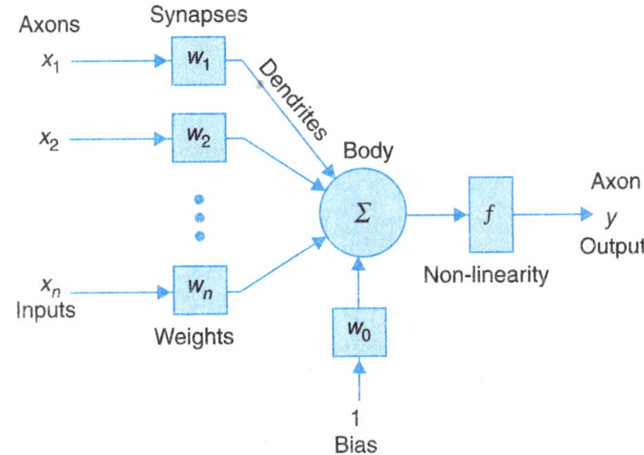

Fig. 16.27. McCulloch-Pitts neuron model.

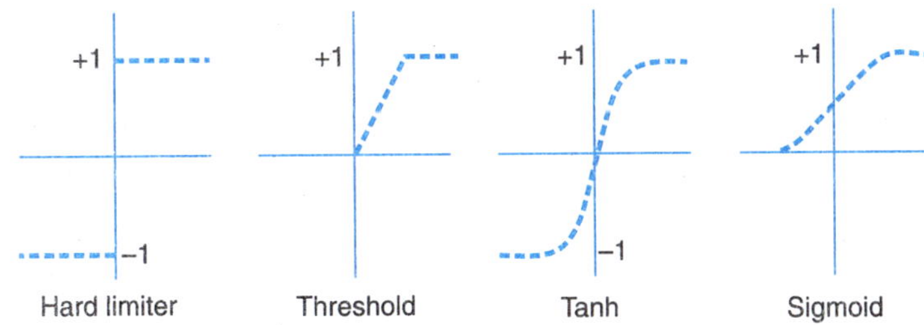

Fig. 16.28. Activation functions, f().

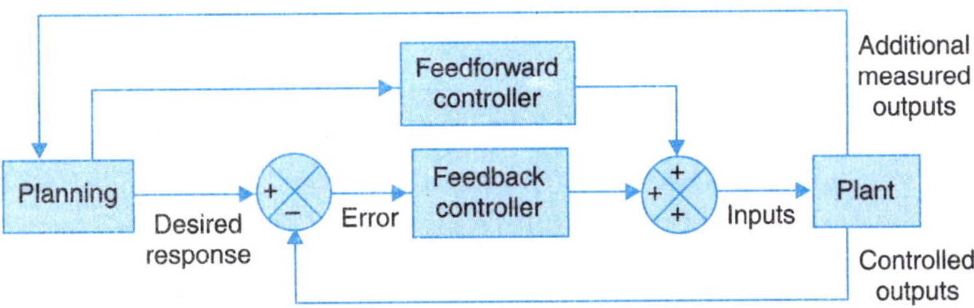

Fig. 16.29. General control systems.

Neural network controller

Plants such as dynamical systems in manufacturing and process industries and other areas including medical applications are typically inaccurately modeled, time varying, and subject to disturbances, but with moderately slow response times in relation to modern computer processing speeds. The niche of interest to artificial neural networks comprises those processes which involve also a degree of non-linearity, which is the main limitation of the otherwise successful conventional control methods, especially where the non-linearity is of an unknown structure, or is very severe.

A neural controller performs a specific form of adaptive control, with the controller taking the form of a nonlinear multi-layer network and the adaptable parameters being the strengths of the interconnections between the neurons. In summary, a controller that is designed as a neural network architecture should exhibit three important characteristics: the utilization of large amounts of sensory information, collective processing capability and adaptation.

A general diagram for a control systems is shown in Fig. 16.29. The feedback and feed-forward controllers and the prefilter can all be implemented as multilayered neural networks. The learning process gradually tunes the weights of the neural network so that the error signal between the desired and actual plant response is minimized. Since the error signal is the input to the feedback controller, the training of the network will lead to a gradual switching from feedback to feed forward action as the error signal becomes small.

In this text, only the implementation of feed-forward controller as a neural network is considered. An immediate consequence of the increased use of feed-forward control action is to speed up the response of the system. In following sections four control learning methods are explained, error-back propagation algorithm described, which is the method used here to adapt the weights in the neural networks that is used as controller. Simulations for a very simple plant to demonstrate the operation of the control learning methods are presented.

Multi-Layer Network

The most popular neural network architecture is the multi-layer perception. The network consists of an input layer, a hidden layer and an output layer as shown in Fig. 16.30. The output and the hidden layer are made up of a number of nodes as described in Section 16.3.2. However the input layer is essentially a direct link to the inputs of the hidden layer and is included by convention. Sigmoidal activation functions for the nodes in the hidden and output layers are the most common choice, although variants on this are also possible. The outputs of each node in a layer are connected to the inputs of all the nodes in the subsequent layer. Data flows through the network in one direction only, from input to output; hence, this type of network is called a feed-forward network.

The network trained in a supervised fashion. This means that during training both the network inputs and target outputs are used. A number of algorithms have been proposed for training the MLN and the most popular is the back-propagation algorithm which will be described in the next section. Briefly, with this algorithm a set of input and corresponding output data is collected that the network is required to learn. An input pattern is applied to the network and an output is generated. This output is compared to the corresponding target output and an error is produced. The error is then propagated back through the network, from output to input, and the network weights are adjusted in such a way as to minimize a cost function, typically the sum of the errors squared. The procedure is repeated through all the data in the training set and numerous passes of the complete training data set are usually necessary before the cost function is reduced to a sufficient value.

An important feature of MLN is that this network can accurately represent any continuous non-linear function relating the inputs and the outputs. Hence MLN exhibits potential for many application including modeling and control of real non-linear processes.

Back Propagation Algorithm

The back-propagation algorithm is central to much current work on learning in neural networks. The algorithm gives a prescription for changing the weights w_{pq} in any feed-forward network to learn a training set of input-output pairs. Since maximum control application use a two layer MLN, we will provide back propagation algorithm. The network is shown in Fig. 16.30. $\underline{x}$ is the $n \times 1$ input vector. $\underline{y}$ is a $m \times 1$ diagonal output vector. The hidden layer consists of h computational units, w_{ij} represents a typical connection weight between output layer and hidden layer while W_{jk} represents a typical connection weight between hidden layer and input layer.

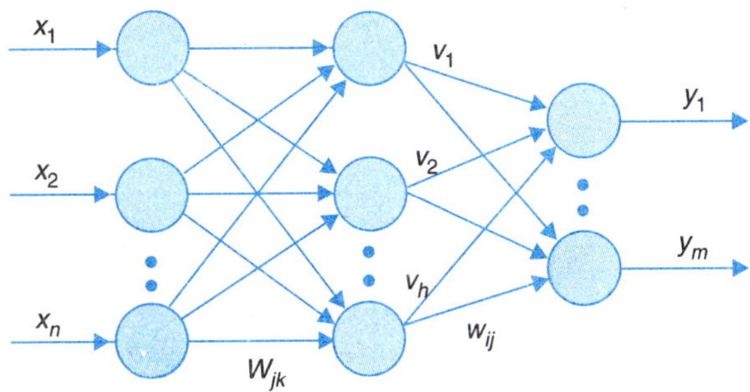

Fig. 16.30. A two layered network.

Forward Propagation

The forward response of such a network is given as follows :

The input of the jth hidden unit is expressed as

$$S_j = \sum_{k=1}^{n} W_{jk} x_k \qquad ...(16.110)$$

Output of the jth hidden unit is given as

$$v_j = f(S_j) \qquad ...(16.111)$$

where f is the squashing function, generally taken as sigmoidal activation

$$f(x) = \frac{1}{1+e^{-x}} \qquad ...(16.112)$$

Finally the input to the ith output unit is

$$q_i = \sum_{i=1}^{h} w_{ij} v_j \qquad ...(16.113)$$

Output y is given as

$$y_i = f(q_i) \qquad ...(16.114)$$

Backward Propagation

The instantaneous error back propagation algorithm is derived following the simple gradient principle where, a typical weight w_{pq} is updated as follows:

$$w_{pq}(t+1) = w_{pq}(t) - \eta \frac{\partial E}{\partial w_{pq}} \qquad ...(16.115)$$

where η is the learning rate and $E = \frac{1}{2}(y^d(t) - y(t))^2$ is the error function to be minimized.

Algorithm

The implementation of error back-propagation algorithm can be given as follows :

1. Initialize the weights to small random values.
2. Choose a pattern x_k ; $k = 1, 2, ..., n$ and apply it to the input layer.
3. Propagate the signal forwards through the network using

$$S_j = \sum_{k=1}^{n} W_{jk} x_k$$

$$v_j = f(S_j)\; j = 1, 2, ..., h$$

$$q_i = \sum_{j=1}^{h} w_{ij} v_j$$

$$y_i = f(q_i)\; ;\; i = 1, 2, ..., m.$$

4. Compute the deltas for the output layer

$$\delta_i = f(q_i)(y_i^d - y_i)$$

by comparing the actual outputs y_i with the desired ones y_i^d for the pattern x_k being considered.

5. Compute the deltas for the hidden layers by propagating the errors backwards

$$\Delta_j = f(S_j) \sum_{i=1}^{n} w_{ij} \delta_i$$

6. Use

$$w_{ij}(t+1) = w_{ij}(t) + \eta \delta_i v_j$$
$$W_{jk}(t+1) = W_{jk}(t) + \eta \Delta_j x_k$$

to update all connections.

7. Go back to step 2 and repeat for the next pattern.

Learning Control Architecture

Figure 16.31 shows the feed-forward controller implemented as a neural network architecture, with its output u driving the plant. The desired response of the plant is denoted by d and its

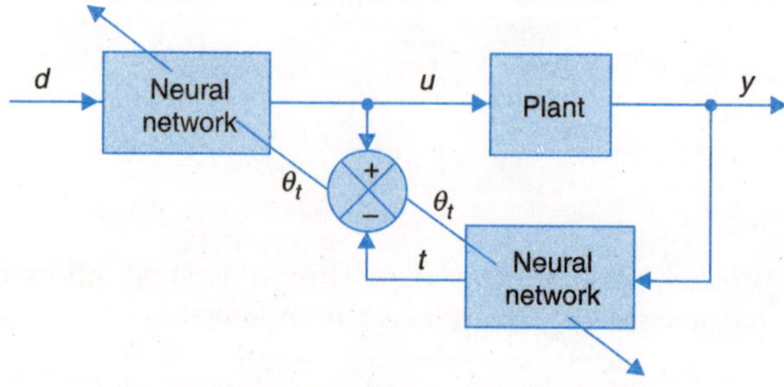

Fig. 16.31. Indirect learning architecture.

actual output by y. The neural controller should act as the inverse of the plant, producing from the desired response d; a signal u that drives the output of the plant to $y \cong d$. The goal of the learning procedure, therefore is to select the values of the weights of the network so that it produces the $d \rightarrow u$ mappings at least over the range of d's in which the plant is to be operated. Four different learning methods are considered.

Indirect Learning Architecture

This architecture is shown in Fig. 16.31. Suppose the feed-forward controller is successfully trained so that the plant output $y \cong d$, then the network used as the feed-forward controller will approximately reproduce the plant input from y (*i.e.*, $t \cong u$). Thus the network can be trained to minimize the error $\varepsilon_1 = u - t$ using this architecture, because if the overall error $\varepsilon_1 = d - y$ goes to 0, so does ε_1.

The positive features of this arrangement would be the fact that the network can be trained only in region of interest since all other signals are generated from d which is the starting point. In addition, it is advantageous to adapt the weights in order to minimize the error directly at the output of the network. Unfortunately, this methods as described is not a valid training procedure bacause minimizing ε_1 does not necessarily minimize ε. For instance simulations with a simple plant shows that the network tends to settle to a solution that maps all d's to a single $u = u_o$, for which ε_1 is zero but obviously ε is not. This training method remains interesting, however because it could be used n conjunction with one of the procedures described below that minimize ε.

Example 16.6 : The above learning architectures were realized using a two-layer architecture with ten hidden neurons plus a fixed-unity input hidden neuron. The hidden neurons have a sigmoid transfer function, $f(x) = \dfrac{1}{(1+e^{-x})}$. Initial weights were selected randomly in the range 0.0 – 0.01, d was taken to be 1.0 and the cases the function of the plant chosen was $y(k+1) = \dfrac{u(k)}{1+y^2(k)}$.

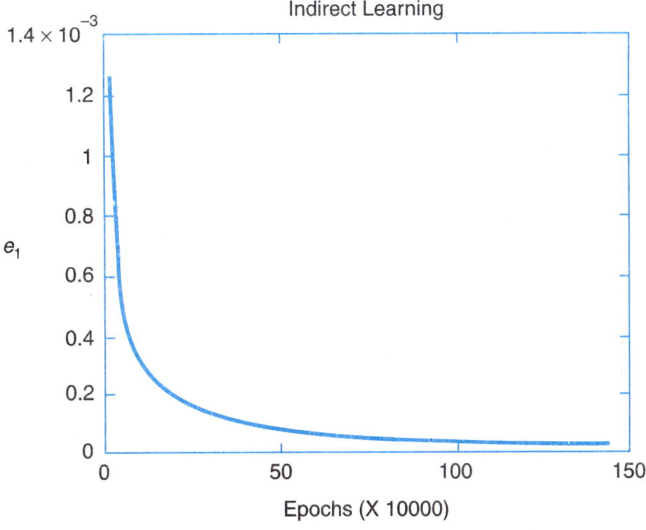

Fig. 16.32. Error.

The plots of the error and the desired and predicted values are given in Figs. 16.32 and 16.33 respectively, From this it is observed that the predicted value reaches the desired value slowly and also it becomes stagnant after thousands of epochs.

General Learning Architecture

The architecture is shown in Fig. 16.34, which provides a method for training the neural controller that does minimize the overall error. The training sequence is as follow : A plant input u is selected and applied to the plant to obtain a corresponding y, and the network is trained to reproduce u at its output from y. The trained network will be able to take a desired response d and produce the appropriate u, making the actual plant output y approach d. This will work if the d happens to be sufficiently close to one of the y's that were used during the training phase. Thus, the success of this method is intimately tied to the ability of the neural network to generalize or learn to respond correctly to inputs it has not specifically been trained for. One of the drawbacks of this method is that the system cannot be selectively trained to respond in regions of interest because which plant inputs u correspond to the desired outputs d is not known. Thus the input space of the plant is uniformly populated with training samples, so that the network can interpolate the intermediate points.

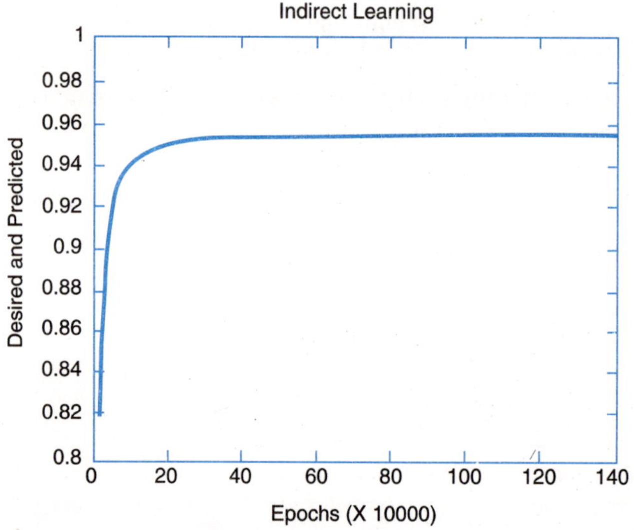

Fig. 16.33. Prediction.

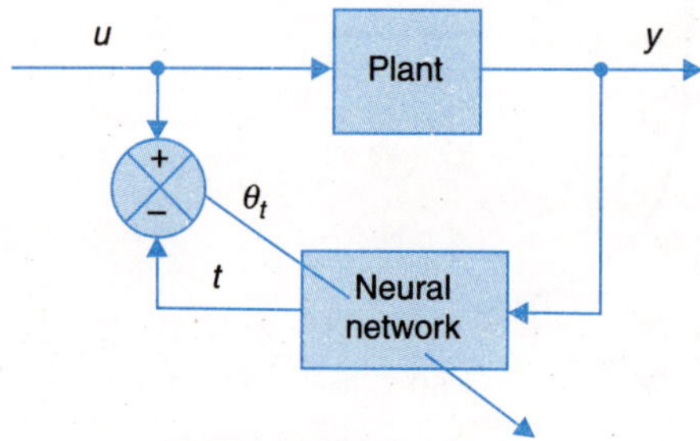

Fig. 16.34. General learning architecture.

Example 16.7 : Reconsider the system of example 16.6. From the Figs. 16.35 and 16.36, it is observed that the error reaches 0, but still the desired output is not obtained. Since with this training method, we cannot place the training samples in the regions of interest, we cannot guarantee what the error will be in these regions.

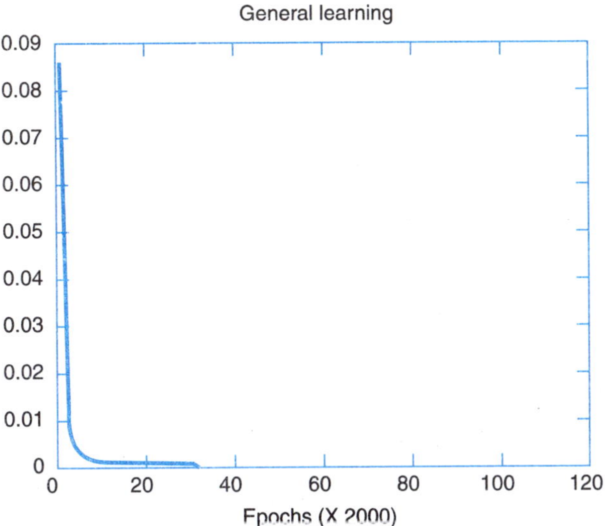

Fig. 16.35. Error.

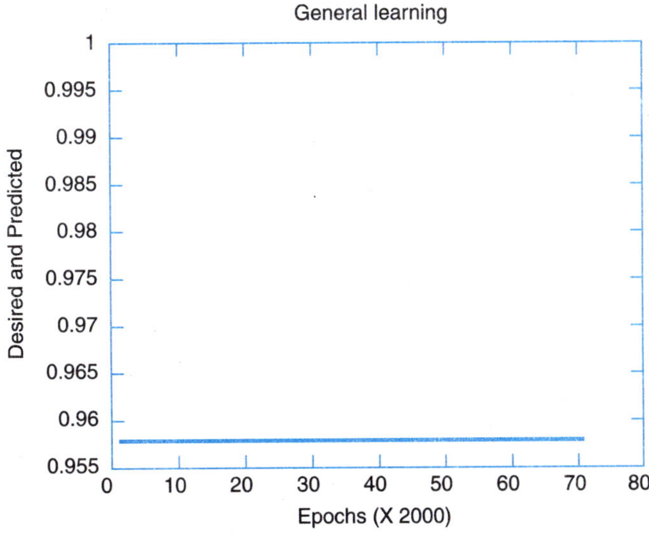

Fig. 16.36. Prediction.

In this case, the general procedure may not be efficient since the network may have to learn the response of the plant over a large operational range than is actually necessary. One solution to this problem is to combine the general method with the specialized procedure described below.

Specialized Learning Architecture

Figure 16.37 shows this architecture which is used for training the neural controller to operate properly in regions of specialization only. The network is trained using desired response d, as input to the network to find the plant input, u that drives the system output, y to the desired d. This is accomplished by using the error between the desired and actual responses of the plant to adjust the weights of the network using steepest descent procedure. During each iteration the weights are adjusted to maximally decrease the error. This architecture can specifically learn in the region of interest, and it may be trained on-line, fine-tuning itself while actually performing useful work. The general learning architecture, on the other hand, must be trained off-line. Feed-forward neural networks are non-dynamical and therefore, input-output stable. Consequently, off-line training of the neural network presents no stability problem for the control systems.

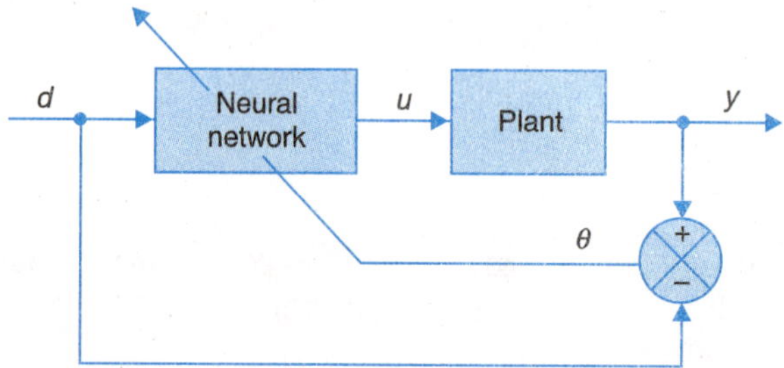

Fig. 16.37. Specialized learning architecture.

Example 16.8 : Let us once again consider the system of example 16.6. The plots of the error and the desired and predicted values are given in Figs. 16.38 and 16.39 respectively. First the network is trained using general learning architecture and then the specialized learning is applied. This is also called hybrid learning and due to this there is a definite advantage. General training will have a tendency to create better initial weights for specialized training. Thus starting with general training can speed the learning process by reducing the number of iterations of the ensuing specialized training. Another advantage of preliminary general learning is that it may result in networks that can adapt more easily if the operating points of the system change or new ones are added.

Forward Inverse Learning Architecture

This architecture is shown in Fig. 16.40 for training the neural controller on-line. Training involves using the desired response d, as input to the network. But here the neural network model $NN1$ is first trained as the plant using the input/output sequence of the plant. Then using another neural network $NN2$, with input as desired response d, output u is obtained that drives the model to $y \cong d$. This architecture is trained using error back propagation. The error found *i.e.*, ε, is back propagated in $NN1$ and also the error $\varepsilon \dfrac{dy}{du}$ in $NN2$. Because of initial training of the neural network mode, this will have a tendency to create better initial weights for this architecture. Thus the learning process is speeded up by this training. Another advantage of this initial training is that this may result in networks that can adopt more easily if the operating points of the system change or new ones are added.

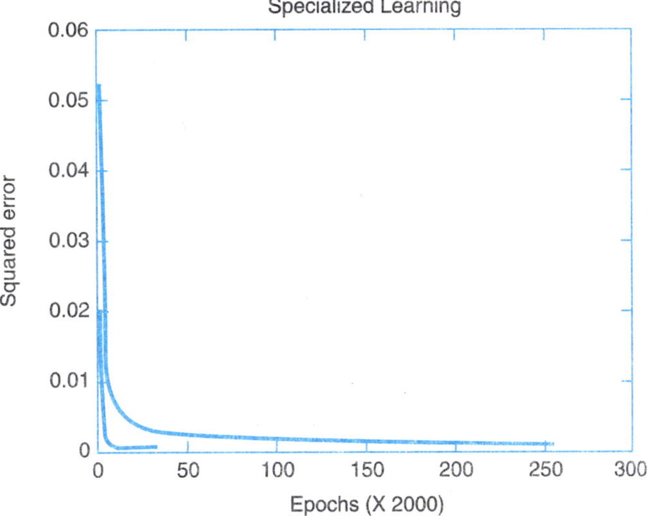

Fig. 16.38. Error.

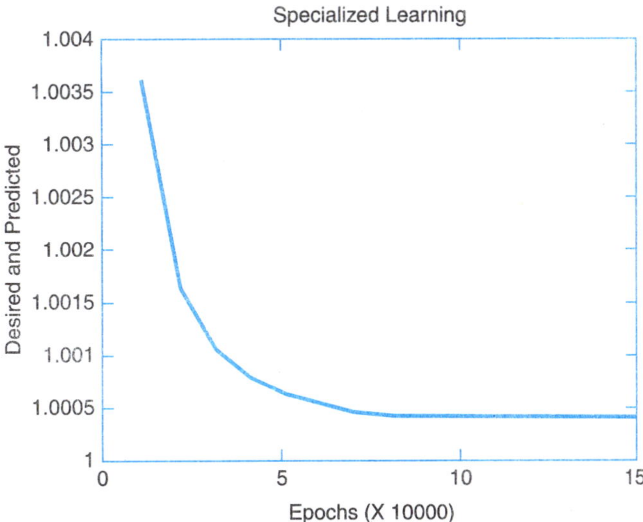

Fig. 16.39. Prediction.

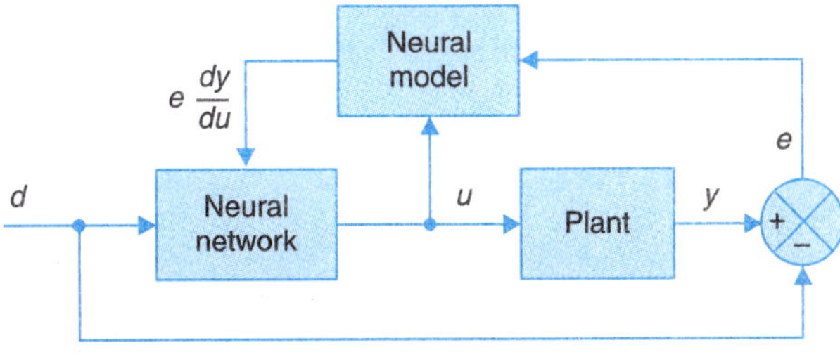

Fig. 16.40. Forward inverse architecture.

Example 16.9 : Reconsider the system described in example 16.6. The Figs. 16.41 and 16.42 show that though the error values first vary, eventually it reaches zeros and so the predicted value also reaches the desired value. This produces more accurate results and it is also less time consuming.

Conclusions

Neural networks were used for controlling physical systems. Specifically, four different methods for using error back propagation to train a feed-forward neural network controller to act as the inverse of the plant. The general learning method attempts to produce the inverse of the plant over the entire state space, but it can be very difficult to use it alone to provide adequate performance in a practical control application. In order to circumvent this problem, error was back-propagated through the plant, which allows to train the network exactly on the operational range of the plant. Also the method of hybrid learning was also explained to gain the advantages of the two different methods and to avoid their potential disadvantages. Forward inverse learning was explored and was found that it was a very accurate learning architecture.

PROBLEMS

16.1. Consider the process $$G(s) = \frac{1}{s(s+a)}$$

where a is an unknown parameter. Determine a controller that can give the closed loop system

$$G(s) = \frac{1}{s(s+a)}.$$

16.2. Consider a position servo described by

$$\frac{dv}{dt} = -av + bu$$

$$\frac{dy}{dt} = v$$

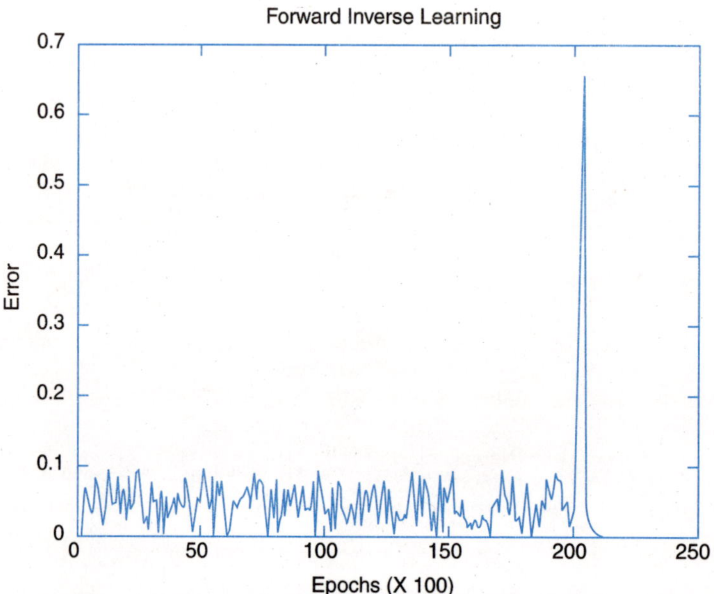

Fig. 16.41. Error.

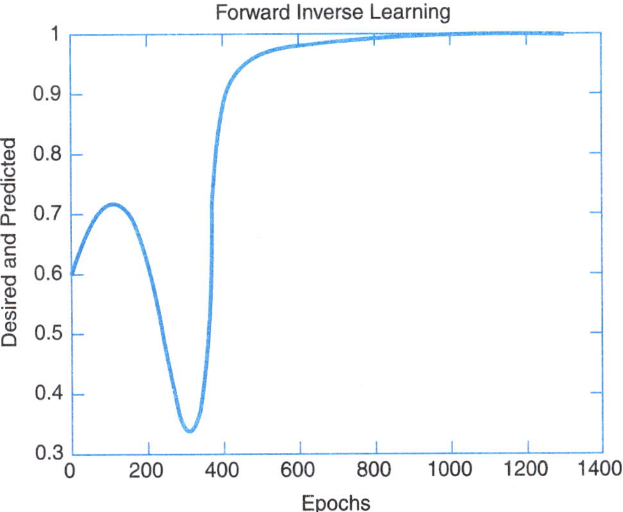

Fig. 16.42. Prediction.

where parameters a and b unknown. Assume that the control law

$$u = \theta_1(u_c - y) - \theta_2 v$$

is used and that it is desired to control the system in such a way that the transfer function from command signal to process output is given by

$$G_m(s) = \frac{\omega^2}{s^2 + 2\zeta\omega s + \omega^2}$$

Determine an adaptive control law that adjusts parameter and so that the desired objective is obtained.

16.3 An integrator

$$G_p(s) = \frac{b}{s}$$

is to be controlled by a zero-order continuous-time controller

$$u(t) = -\, s_0 y(t) + t_0 u_c(t)$$

the desired response model is given by

$$G_m(s) = \frac{b_m}{s + a_m}$$

Derive, using the Lyapunov theory, a parameter update law of an MRAS guaranteeing that the error, $e = y - y_m$ goes to zero. Try the Lyapunov function

$$V(x) = \frac{1}{2}\left(e^2 + \frac{1}{b}(bs_0 - a_m)^2 + \frac{1}{b}(bt_0 - b_m)^2\right).$$

16.4 Consider the problem of adaptation of a feed-forward gain in Example 5.1 when

$$G(s) = \frac{1}{s(s+1)(s+2)}$$

(*i*) Introduce the augmented error and determine an MRAS based on stability theory.

(*ii*) Show that the derived adaptation law in Part (*i*) gives a stable closed loop system.

16.5. A process has the transfer function

$$G(s) = \frac{b}{s(s+1)}$$

where b is a time varying parameter. The system is controlled by a proportional controller

$$u(t) = k(u_c(t) - y(t))$$

It is desirable to choose the feedback gain so that the closed-loop system has the transfer function

$$G_m(s) = \frac{1}{s^2 + s + 1}$$

Design an MRAS that gives the desired result, and investigate the system by simulation. (Compare with problem 11)

16.6. Consider a first-order system with the transfer function

$$G_p(s) = \frac{b}{s + a}$$

where a and b are unknown parameters. Assume that the system is controlled by the control law

$$u = \theta_1 u_c - \theta_2 y$$

Compare by simulation the properties of the systems obtained with the MIT rule and the one derived from Lyapunov theory. Use the same parameter values as in Example [**Hint :** The algorithms are given in Examples]

16.7. Consider a system described by

$$G(s) = \frac{b}{s^2 + a}$$

where a and b are unknown parameters. Find a simple control law that can control the plant well, and derive an adaptive algorithm that gives good performance.

16.8. In sampling a continuous-time process model with $h = 1$ the following pulse transfer function is obtained :

$$H(z) = \frac{2(z+1)}{z^2 - z + 0.25}$$

The design specification states that the discrete-time closed-loop poles should correspond to the continuous-time characteristic polynomial

$$s^2 + 2s + 1$$

(*a*) Design a minimal-order discrete-time indirect self tuning regulator. The controller should have integral action and give a closed-loop system having unit gain in stationary. Determine the Diophantine equation that solves the design problem.

(*b*) Suggest a design that includes direct estimation of the controller parameters. Discuss why a well-working direct self-tuning regulator is more difficult to design for this is an indirect self-tuning regulator.

16.9. Consider the process

$$G_p(s) = \frac{b}{s}$$

where a is an unknown parameter. Assume that the desired closed-loop system is

$$G_p(s) = \frac{b}{s}$$

Consider continuous and discrete-time indirect self-tuning algorithms for the system.

16.10. Consider the system

$$G(s) = G_1(s)G_2(s)$$

where
$$G_1(s) = \frac{b}{s+a}$$

$$G_2(s) = \frac{c}{s+d}$$

where a and b are unknown parameters and c and d are known. Construct discrete-time direct and indirect self-tuning algorithms for partially known system.

16.11. A process has the transfer function

$$G(s) = \frac{b}{s(s+1)}$$

where b is a time-varying parameter. The system is controlled by a proportional controller

$$u(t) = k(u_c(t) - y(t))$$

It is desirable to choose the feedback gain so that the closed-loop system has the transfer function

$$G(s) = \frac{1}{s^2 + s + 1}$$

Construct a continuous-time indirect self-tuning algorithm for the system.

[illegible] equations and discrete-time indirect self-tuning algorithms for the system.

16.10 Consider the system

$$[illegible]$$

where a and b are unknown parameters and d is the known [illegible] [illegible] for the system.

16.11 Propose and [illegible]

$$[illegible]$$

where [illegible]. The [illegible] controlled by a proportional controller [illegible]

[illegible]

$$[illegible]$$

[illegible] and [illegible] for the system.

APPENDICES

APPENDIX I

FOURIER AND LAPLACE TRANSFORMS AND PARTIAL FRACTIONS

I.1 THE FOURIER TRANSFORM

The Fourier transform of a single-valued function $f(t)$ is defined as the integral

$$F(\omega) = \int_{-\infty}^{\infty} f(t)e^{-j\omega t}dt = \mathcal{F}[f(t)] \qquad \text{...(I.1)}$$

where $\mathcal{F}$ is a symbol indicating the Fourier transform operation.

The necessary condition for existence of the Fourier transform is

$$\int_{-\infty}^{\infty} |f(t)|\,dt < \infty \qquad \text{...(I.2)}$$

There are many useful functions, *e.g.*, sine waves, step function etc., which are not absolutely integrable, *i.e.*, they do not satisfy condition (I.2). Many of such functions do, however, possess Fourier transforms if we allow the transforms to include impulse functions.

The inverse Fourier transform is given by

$$f(t) = \frac{1}{2\pi}\int_{-\infty}^{\infty} F(\omega)e^{j\omega t}d\omega = \mathcal{F}^{-1}[F(\omega)] \qquad \text{...(I.3)}$$

where $\mathcal{F}^{-1}$ is the symbol for the inverse Fourier transform operation. Equation (I.3) may be interpreted as decomposition of $f(t)$ in terms of continuum of elementary basic functions exp $(j\omega t)$. $F(\omega)$ is therefore the continuous frequency spectrum of $f(t)$.

Important properties of the Fourier transform are similar counterparts for the Laplace transform presented later.

Fourier Transform of Periodic Functions

A periodic function $f(t)$ of period T can be expanded in terms of *Fourier series* as

$$f(t) = \sum_{n=-\infty}^{\infty} F_n e^{jn\omega_0 t}; \ \omega_0 = \frac{2\pi}{T}$$

where

$$F_n = \frac{1}{T}\int_{-T/2}^{T/2} f(t)e^{-jn\omega_0 t}dt = \frac{1}{T}$$

Therefore

$$F(\omega) = \frac{2\pi}{T}\sum_{n=-\infty}^{\infty}\delta(\omega - n\omega_0) \qquad ...(I.4)$$

I.2 THE LAPLACE TRANSFORM

The single-sided Laplace transform (commonly referred to as the Laplace transform) of a *causal function** ($f(t) = 0$ for $t < 0$) is defined as

$$F(s) = \mathcal{L}[f(t)] = \int_0^{\infty} f(t)e^{-st}dt \qquad (I.5)$$

where $s = \sigma + j\omega$

is complex frequency variable.

In eqn. (I.5), $\mathcal{L}$ is the symbol indicative of the Laplace transforming operation.

The condition for the existance of the Laplace transform of $f(t)$ as

$$\int_0^{\infty} |f(t)|e^{-\sigma t}dt < \infty \qquad ...(I.6)$$

where σ is any real number.

As an example, consider $f(t) = e^{-at}$, $a > 0$. Then

$$\mathcal{L}[f(t)] = \mathcal{L}(e^{-at}) = \int_0^{\infty} e^{-at}e^{-st}dt$$

$$= \int_0^{\infty} e^{-(a+s)t}dt = \frac{1}{s+a}$$

It is easily observed in this case that the condition (I.6) is satisfied if $(a + \sigma) > 0$ or $\sigma > -a$.

The inverse Laplace transform is given by

$$f(t) = \mathcal{L}^{-1}[F(s)] = \frac{1}{2\pi j}\int_{\sigma_1 - j\infty}^{\sigma_1 + j\infty} F(s)e^{st}ds \qquad ...(I.7)$$

where $\sigma_1 > \sigma$, the convergence factor in (I.6).

---

*In majority of engineering applications, the time functions of interest are causal. Noncausal functions can be treated through the double-sided Laplace transform which is similar in this respect to the Fourier transform but has a stronger convergence property.

In eq. (I.7) $\mathcal{L}^{-1}$ is the symbol indicative of the inverse Laplace transforming operation.

Important properties of the Laplace transform are given in Table I.1.

The inverse Laplace transform (I.7) involves complex integration which is quite tedious. However, inverse transformation can be easily carried out by expanding $F(s)$ into partial fractions and then using a table of transform pairs.

This table is based on the fact that the Laplace transform has the uniqueness property *i.e.*, for any given $f(t)$ there is one and only one $F(s)$ and vice-versa. The transform pairs for some of the commonly encountered functions are given in Table I.2.

Table I.1 Properties of the Laplace Transform

Transform pair	$f(t) \leftrightarrow F(s)$
Linearity	$a_1 f_1(t) + a_2 f_2(t) \leftrightarrow a_1 F_1(s) + a_2 F_2(s)$
Scale change	$f\left(\frac{t}{a}\right) \leftrightarrow aF(as); a > 0$
Real translation	$f(t - t_0) \leftrightarrow e^{-st_0} F(s)$
Complex translation	$e^{-at} f(t) \leftrightarrow F(s + a)$
Real differentiation	$\frac{d^n}{dt^n} f(t) \leftrightarrow s^n F(s) - s^{n-1} f(0) - s^{n-2} \frac{df}{dt}(0) - \ldots \frac{d^{n-1} f}{dt^{n-1}}(0)$
Real integration	$\int_{-\infty}^{t} f(t)dt \leftrightarrow \frac{F(s)}{s} + \frac{\left.\int_{-\infty}^{t} f(t)dt\right\|_{t=0}}{s}$
Multiplication by t	$t^n f(t) \leftrightarrow (-1)^n \frac{d^n F(s)}{ds^n}$
Multiplication by $\frac{1}{t}$	$\frac{1}{t} f(t) \leftrightarrow \int_{s}^{\infty} F(s)\, ds$
Real multiplication (complex convolution)	$f_1(t)\, f_2(t) \leftrightarrow F_1(s) * F_2(s)$
Complex multiplication (real convolution)	$f_1(t) * f_2(t) \leftrightarrow F_1(s) F_2(s)$
Initial value theorem	$\lim_{t \to 0} f(t) = \lim_{s \to \infty} sF(s)$
Final value theorem	$\lim_{t \to \infty} f(t) = \lim_{s \to 0} sF(s)$

By use of the real integration property and the final value theorem, we obtain

$$\int_{0}^{\infty} f(t)dt = \lim_{t \to \infty} \int_{0}^{t} f(t)dt = \lim_{s \to 0} s \frac{F(s)}{s} = \lim_{s \to 0} F(s)$$

Parseval's Theorem

Consider a causal signal $f(t)$. Its integral square value (also called energy) is

$$\int_0^{\infty} f^2(t)dt \qquad \text{...(I.8)}$$

this concept is meaningful only if the signal energy is finite. It immediately follows that the signal is then absolutely integrable, *i.e.*,

$$\int_0^{\infty} |f(t)|\,dt < \infty$$

Table I.2 Laplace Transform Pairs

$F(s)$	$f(t)\ ;\ t \geq 0$
1. 1	Unit impulse, $\delta(t)$
2. $1/s$	Unit step, $u(t)$
3. $\dfrac{n!}{s^{n+1}}$	t^n (n = integer)
4. $\dfrac{1}{s+a}$	e^{-at}
5. $\dfrac{1}{(s+a)^n}$	$\dfrac{1}{(n-1)!}t^{n-1}e^{-at}$ (n = integer)
6. $\dfrac{a}{s(s+a)}$	$1-e^{-at}$
7. $\dfrac{1}{(s+a)(s+b)}$	$\dfrac{1}{b-a}(e^{-at}-e^{-bt})$
8. $\dfrac{s}{(s+a)(s+b)}$	$\dfrac{1}{b-a}(be^{-bt}-ae^{-at})$
9. $\dfrac{s}{s(s+a)(s+b)}$	$\dfrac{1}{ab}\left[1+\dfrac{1}{a-b}(be^{-at}-ae^{-bt})\right]$
10. $\dfrac{\omega}{s^2+\omega^2}$	$\sin \omega t$
11. $\dfrac{s}{s^2+\omega^2}$	$\cos \omega t$
12. $\dfrac{\omega}{(s+a)^2+\omega^2}$	$e^{-at}\sin \omega t$
13. $\dfrac{s+a}{(s+a)^2+\omega^2}$	$e^{-at}\cos \omega t$
14. $\dfrac{1}{s^2(s+a)}$	$\dfrac{1}{a^2}(at-1+e^{-at})$

15. $\dfrac{\omega_n^2}{s^2 + 2\zeta\omega_n s + \omega_n^2}$	$\dfrac{\omega_n}{\sqrt{(1-\zeta^2)}} e^{-\zeta\omega_n t} \sin [\omega_n \sqrt{(1-\zeta^2)}]t; \zeta < 1$
16. $\dfrac{s}{s^2 + 2\zeta\omega_n s + \omega_n^2}$	$\dfrac{-1}{\sqrt{(1-\zeta^2)}} e^{-\zeta\omega_n t} \sin [\omega_n \sqrt{(1-\zeta^2)}\, t - \phi]$ $\phi = \tan^{-1} \dfrac{\sqrt{(1-\zeta^2)}}{\zeta}; \zeta < 1$
17. $\dfrac{\omega_n^2}{s(s^2 + 2\zeta\omega_n s + \omega_n^2)}$	$1 - \dfrac{1}{\sqrt{(1-\zeta^2)}} e^{-\zeta\omega_n t} \sin [\omega_n \sqrt{(1-\zeta^2)}t + \phi]$ $\phi = \tan^{-1} \dfrac{\sqrt{(1-\zeta^2)}}{\zeta}; \zeta < 1$

Let us express the signal energy in terms of its Laplace transform. We can write (I.8) as

$$\int_0^\infty f^2(t)dt = \int_0^\infty f(t)\left[\frac{1}{2\pi j}\int_{\sigma-j\infty}^{\sigma+j\infty} F(s)e^{st}ds\right]dt$$

Interchanging the order of integration

$$\int_0^\infty f^2(t)dt = \frac{1}{2\pi j}\int_{\sigma-j\infty}^{\sigma+j\infty} F(s)\left[\int_0^\infty f(t)e^{st}dt\right]ds$$

By definition of the Laplace transform we recognise

$$\int_0^\infty f(t)e^{st}dt = F(-s)$$

Hence

$$\int_0^\infty f^2(t)\,dt = \frac{1}{2\pi j}\int_{\sigma-j\infty}^{\sigma+j\infty} F(s)F(-s)ds$$

Since the signal is absolutely integrable, *i.e.*, $\sigma = 0$, we can write

$$\int_0^\infty f^2(t)\,dt = \frac{1}{2\pi j}\int_{-j\infty}^{j\infty} F(s)F(-s)ds \qquad \text{...(I.9)}$$

This result is known as the Parseval's theorem.

Solution of Differential Equations by the Laplace Transform

The solution of a linear differential equation with constant coefficients (pertaining to a linear system) can be conveniently obtained through the Laplace transform technique in the following three steps :

(1) Using the real differentiation and the linearity properties of Table of I.1 and the transform pairs of Table I.2, transform the differential equation into an algebraic equation in the complex variable s.

(2) Through algebraic manipulation solve for the unknown in terms of the complex variable s.

(3) The time domain solution of the unknown is obtained by inverse Laplace transforming the s-domain solution.

The above procedure is best illustrated by means of an example.

Example : Solve the second-order differential equation

$$\frac{d^2y}{dt^2} + 5\frac{dy}{dt} + 6y = 3t \qquad \text{...(I.10)}$$

for which at $t = 0$, $y = 2$ and $dy/dt = -1$.

Solution. Laplace transforming both sides of eqn. (I.10) using the real differentiation and the linearity properties and the table of transform pairs, we get

$$[s^2Y(s) - 2s + 1] + 5[sY(s) - 2] + 6Y(s) = 3/s^2$$

which gives
$$Y(s) = \frac{2s^3 + 9s^2 + 3}{s^2(s^2 + 5s + 6)} \qquad \text{...(I.11)}$$

Equation (I.11) presents the solution in terms of the s-variable. If the solution in time domain is desired, the inverse Laplace transform of $Y(s)$ is to be obtained.

The function given by eqn. (I.11) does not appear in Table I.2. We may, therefore, expand $Y(s)$ into partial fractions and write it in terms of simple functions of s. Then by linearity property, the inverse Laplace transform may be obtained.

The partial function expansion of $Y(s)$ is

$$Y(s) = \frac{-5/12}{s} + \frac{1/2}{s^2} - \frac{10/3}{s+3} + \frac{23/4}{s+2}$$

Using the table of transform pairs, we get

$$y(t) = -\frac{5}{12} + \frac{1}{2}t - \frac{10}{3}e^{-3t} + \frac{23}{4}e^{-2t} \qquad \text{...(I.12)}$$

which is the required solution in time domain.

We observe from the above example that the final step in solving a differential equation by the Laplace transform is to obtain the inverse Laplace transform of a *rational algebraic function* (a ratio of two polynomials; see eqn. (I.11) by breaking it up into partial fractions. Partial fraction expansion of rational algebraic functions is briefly reviewed in the next section.

I.3 PARTIAL FRACTION EXPANSIONS

As seen above, in obtaining the solution of linear differential equations with constant coefficients (or analysis of linear time-invariant systems) by the Laplace transform method. We encounter rational algebraic fractions that are ratio of two polynomials in s, such as

$$F(s) = \frac{P(s)}{Q(s)}$$

$$= \frac{b_0 s^m + b_1 s^{m-1} + \ldots + b_m}{a_0 s^n + a_1 s^{n-1} + \ldots + a_n} \qquad \text{...(I.13)}$$

In practical systems, the order of polynomial in numerator is equal to or less than that of denominator. In terms of the orders *m* and *n,* rational algebraic fractions are subdivided as follows:

(*i*) Improper fraction if $m \geq n$

(*ii*) Proper fraction if $m < n$

A proper fraction can further be expanded into partial fractions as will be seen shortly. An improper fraction can be separated into a sum of a polynomial in *s* and a proper fraction, *i.e.,*

$$F(s) = \underbrace{\frac{p(s)}{Q(s)}}_{\text{Improper}} = d(s) + \underbrace{\frac{p(s)}{Q(s)}}_{\text{Proper}}$$

This can be achieved by performing a long division.

To obtain the partial fraction expansion of a proper fraction, first of all we factorize the polynomial *Q*(*s*) into *n* first-order factors (see Appendix III). The roots may be real, complex distinct or repeated. Various cases are discussed below.

Partial Fraction Expansion when *Q*(*s*) has Distinct Roots

In this case, eqn. (I.13) may be written as

$$F(s) = \frac{P(s)}{Q(s)} = \frac{P(s)}{(s+p_1)(s+p_2)\ldots(s+p_k)\ldots(s+p_n)} \quad \text{...(I.14)}$$

which when expanded, gives

$$F(s) = \frac{A_1}{s+p_1} + \frac{A_2}{s+p_2} + \ldots + \frac{A_k}{s+p_k} + \ldots + \frac{A_n}{s+p_n} \quad \text{...(I.15)}$$

The coefficient A_k is called the *residue* at the pole $s = -p_k$.

To evaluate A_k multiply *F*(*s*) in eqn. (I.14) by $(s + p_k)$ and let $s = -p_k$.

This gives

$$A_k = (s+p_k)\frac{P(s)}{Q(s)}\bigg|_{s=-p_k} = \frac{P(s)}{\dfrac{d}{ds}Q(s)}\Bigg|_{s=-p_k} \quad \text{...(I.16)}$$

$$= \frac{P(-p_k)}{(p_1-p_k)(p_2-p_k)\ldots(p_{k-1}-p_k)(p_{k+1}-p_k)\ldots(p_n-p_k)} \quad \text{...(I.17)}$$

From eqn. (I.15), we obtain

$$f(t) = \mathcal{L}^{-1}[F(s)] = A_1e^{-p_1t} + A_2e^{-p_2t} + \ldots + A_ke^{-p_kt} + \ldots + A_ne^{-p_nt} \quad \text{...(I.18)}$$

Consider, for example

$$F(s) = \frac{2(s^2+3s+1)}{s(s+1)(s+2)} = \frac{P(s)}{Q(s)} = \frac{A_1}{s} + \frac{A_2}{s+1} + \frac{A_3}{s+2}$$

$$A_1 = sF(s)\big|_{s=0} = \frac{2(s^2+3s+1)}{(s+1)(s+2)}\bigg|_{s=0} = 1$$

$$A_2 = (s+1)F(s)\big|_{s=-1} = \left.\frac{2(s^2+3s+1)}{s(s+2)}\right|_{s=-1} = 2$$

$$A_3 = (s+2)F(s)\big|_{s=-2} = \left.\frac{2(s^2+3s+1)}{s(s+1)}\right|_{s=-2} = -1$$

Partial Fraction Expansion when $Q(s)$ has Complex

Conjugate Roots

Suppose that there is a pair of complex conjugate roots in $Q(s)$, given by

$$s = -a - j\omega \text{ and } s = -a + j\omega$$

Then $F(s)$ may be written as

$$F(s) = \frac{P(s)}{Q(s)} = \frac{P(s)}{(s+a+j\omega)(s+a-j\omega)(s+p_3)(s+p_4)\ldots(s+p_n)} \quad \text{...(I.19)}$$

which when expanded gives

$$F(s) = \frac{A_1}{(s+a+j\omega)} + \frac{A_1{}^*}{(s+a-j\omega)} + \frac{A_3}{s+p_3} + \frac{A_4}{s+p_4} + \ldots + \frac{A_n}{s+p_n} \quad \text{...(I.20)}$$

where A_1 and $A_1{}^*$ are residues at the poles $s = -(a + j\omega)$ and $s = -(a - j\omega)$ respectively. These residues form a complex conjugate pair. From eqn. (I.20), the inverse Laplace transform of $F(s)$ may be obtained as follows.

$$f(t) = \mathcal{L}^{-1}[F(s)] = \mathcal{L}^{-1}\left[\frac{A_1}{s+a+j\omega} + \frac{A_1^*}{s+a-j\omega}\right] + A_3 e^{-p_3 t} + \ldots + A_n e^{-p_n t}$$

$$= 2\text{Re}[A_1 e^{-(a+j\omega)t}] + A_3 e^{-p_3 t} + \ldots + A_n e^{-p_n t} \quad \text{...(I.21)}$$

As per eqn. (I.16), the residue A_1 is given by

$$A_1 = \left.\frac{P(s)}{Q(s)}(s+a+j\omega)\right|_{s=-(a+j\omega)}$$

Consider, for example

$$F(s) = \frac{2s+3}{(s+1)(s^2+4s+5)} = \frac{P(s)}{Q(s)}$$

$$= \frac{2s+3}{(s+2+j1)(s+2-j1)(s+1)}$$

$$= \frac{A_1}{s+2+j1} + \frac{A_1^*}{s+2-j1} + \frac{A_3}{s+1}$$

$$A_1 = (s+2+j1)F(s)\big|_{s=-(2+j1)}$$

$$= \left.\frac{2s+3}{(s+2-j1)(s+1)}\right|_{s=-(2+j1)}$$

$$= -\frac{1}{4} + j\frac{3}{4}$$

$$A_1^* = -\frac{1}{4} - j\frac{3}{4}$$

$$A_3 = (s+1)F(s)\Big|_{s=-1} = \frac{2s+3}{(s^2+4s+5)}\Bigg|_{s=-1} = \frac{1}{2}$$

Partial Fraction Expansion when $Q(s)$ has Repeated Roots

Assume that root p_1 of $Q(s)$ is of multiplicity r and all other roots are distinct. The function $F(s)$ may be written as

$$F(s) = \frac{P(s)}{Q(s)} = \frac{P(s)}{(s+p_1)^r (s+p_{r+1})(s+p_{r+2})\ldots(s+p_n)} \quad \text{...(I.22)}$$

which when expanded, gives

$$F(s) = \frac{A_{1(r)}}{(s+p_1)^r} + \frac{A_{1(r-1)}}{(s+p_1)^{r-1}} + \frac{A_{1(r-2)}}{(s+p_1)^{r-2}} + \ldots + \frac{A_{12}}{(s+p_1)^2}$$
$$+ \frac{A_{11}}{(s+p_1)} + \frac{A_{r+1}}{(s+p_{r+1})} + \frac{A_{r+2}}{(s+p_{r+2})} + \ldots + \frac{A_n}{(s+p_n)} \quad \text{...(I.23)}$$

The coefficients of repeated roots may be obtained using the following relation:

$$A_{1(r-1)} = \frac{1}{i!}\left[\frac{d^i}{ds^i}\left\{(s+p_1)^r \frac{P(s)}{Q(s)}\right\}\right]_{s=-p_1} \quad ; i = 0, 1, 2, \ldots, r-1 \quad \text{...(I.24)}$$

Then from eqn. (I.23) we obtain

$$f(t) = \mathcal{L}^{-1}[F(s)] = \left[\frac{A_{1(r)}}{(r-1)!}t^{r-1} + \frac{A_{1(r-1)}}{(r-2)!}t^{r-2} + \ldots + A_{12}t + A_{11}\right]$$
$$e^{-p_1 t} + A_{r+1}e^{-p_{r+1}t} + \ldots + A_n e^{-p_n t} \quad \text{...(I.25)}$$

Consider, for example

$$F(s) = \frac{1}{s^2(s+1)} = \frac{P(s)}{Q(s)} = \frac{A_{12}}{s^2} + \frac{A_{11}}{s} + \frac{A_3}{s+1}$$

$$A_{12} = s^2 F(s)\Big|_{s=0} = \frac{1}{s+1}\Bigg|_{s=0} = 1$$

$$A_{11} = \frac{d}{ds}(s^2 F(s))\Bigg|_{s=0} = \frac{d}{ds}\left(\frac{1}{s+1}\right)\Bigg|_{s=0}$$

$$= -\frac{1}{(s+1)^2}\Bigg|_{s=0} = -1$$

$$A_3 = 1$$

APPENDIX II

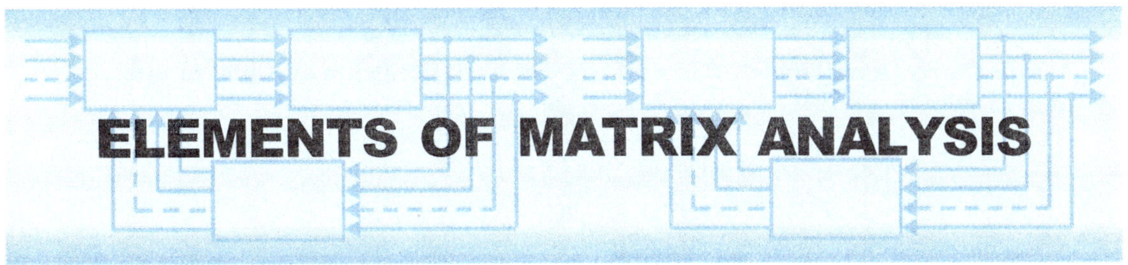

ELEMENTS OF MATRIX ANALYSIS

This appendix is a brief review of some properties of vectors and matrices which has been used in Chapters 12, 13 and 14.

II.1 BASIC DEFINITIONS

Matrix. A matrix is an ordered rectangular array of elements which may be real numbers, complex numbers, functions or operators. The matrix

$$\mathbf{A} = \begin{bmatrix} a_{11} & a_{12} & \cdots & a_{1n} \\ a_{21} & a_{22} & \cdots & a_{2n} \\ \vdots & \vdots & & \vdots \\ a_{m1} & a_{m2} & \cdots & a_{mn} \end{bmatrix} = [a_{ij}] \qquad \text{...(II.1)}$$

is a rectangular array of mn elements.

a_{ij} denotes (i, j)th element, *i.e.,* the element located in ith row and jth column. The matrix defined above has m rows and n columns. This matrix is said to be a rectangular matrix of order $m \times n$.

When $m = n$, *i.e.,* the number of rows is equal to that of columns, the matrix is said to be a *square matrix* of order n.

An $m \times 1$ matrix, *i.e.,* a matrix having only one column is called a *column vector.*

A $1 \times n$ matrix, *i.e.,* a matrix having only row is called a *row vector*.

Diagonal matrix. A diagonal matrix is a square matrix whose elements off the main diagonal are all zeros ($a_{ij} = 0$ for $i \neq j$). The following matrix is a diagonal matrix

$$\mathbf{A} = \begin{bmatrix} a_{11} & 0 & \cdots & 0 \\ 0 & a_{22} & \cdots & 0 \\ \vdots & \vdots & & \vdots \\ 0 & 0 & \cdots & a_{nn} \end{bmatrix}$$

Unit (identity) matrix. A unit matrix **I** is a diagonal matrix whose diagonal elements are all equal to unity ($a_{ij} = 1$ for $i = j$; $a_{ij} = 0$ for $i \neq j$)

$$\mathbf{I} = \begin{bmatrix} 1 & 0 & \dots & 0 \\ 0 & 1 & \dots & 0 \\ \vdots & \vdots & & \vdots \\ 0 & 0 & \dots & 1 \end{bmatrix}$$

Null matrix. A null matrix 0 is a matrix whose elements are all equal to zero.

$$\mathbf{0} = \begin{bmatrix} 0 & 0 & \dots & 0 \\ 0 & 0 & \dots & 0 \\ \vdots & \vdots & & \vdots \\ 0 & 0 & \dots & 0 \end{bmatrix}$$

Determinant of a matrix. For each square matrix, there exists a determinant which is formed by taking the determinant of the elements of the matrix. For example, if

$$\mathbf{A} = \begin{bmatrix} 3 & -2 & 1 \\ -2 & 6 & 4 \\ 1 & 4 & 8 \end{bmatrix} \qquad \text{...(II.2)}$$

then, $\det(\mathbf{A}) = |\,\mathbf{A}\,| = 3\begin{vmatrix} 6 & 4 \\ 4 & 8 \end{vmatrix} - (-2)\begin{vmatrix} -2 & 4 \\ 1 & 8 \end{vmatrix} + 1\begin{vmatrix} -2 & 6 \\ 1 & 4 \end{vmatrix}$

$$= 3(32) + 2(-20) + (-14) = 42 \qquad \text{...(II.3)}$$

Transpose of a matrix. The transpose of matrix **A** denoted by $\mathbf{A}^T$ is the matrix formed by interchanging the rows and columns of **A**. Namely, if **A** is given by (II.1) then $\mathbf{A}^T$ is given by

$$\mathbf{A}^T = \begin{bmatrix} a_{11} & a_{21} & \dots & a_{m1} \\ a_{12} & a_{22} & \dots & a_{m2} \\ \vdots & \vdots & & \vdots \\ a_{1n} & a_{2n} & . & a_{mn} \end{bmatrix} = \left[a_{ji}\right]$$

Note that

$$(\mathbf{A}^T)^T = \mathbf{A}$$

Symmetric matrix. A square matrix is symmetric if it is equal to its transpose, *i.e.,*

$$\mathbf{A}^T = \mathbf{A}$$

For example, matrix **A** given by (II.2) is symmetric.

Skew-symmetric matrix. A square matrix is skew-symmetric if it is equal to its negative transpose, *i.e.,*

$$\mathbf{A}^T = -\mathbf{A}$$

The matrix

$$\mathbf{A} = \begin{bmatrix} 0 & 1 & -3 \\ -1 & 0 & -2 \\ 3 & 2 & 0 \end{bmatrix}$$

is *skew*-symmetric.

Minor. If the ith row and jth column of determinant **A** are deleted, the remaining $(n-1)$ rows and $(n-1)$ columns form a determinant M_{ij}. This determinant is called the minor of the element a_{ij}.

Principal minor. A minor of $|\mathbf{A}|$ whose diagonal elements are also diagonal elements of $|\mathbf{A}|$, is called a principal minor of $|\mathbf{A}|$.

Laplace expansion formula. The value of determinant of a matrix **A** can be obtained by the so-called Laplace expansion formula.

$$|\mathbf{A}| = \sum_{j=1}^{n} (-1)^{i+j} a_{ij} M_{ij} \text{ for any integer } i\,;\ 1 \le i \le n$$

$$= \sum_{i=1}^{n} (-1)^{i+j} a_{ij} M_{ij} \text{ for any integer } j\,;\ 1 \le j \le n$$

where M_{ij} is the minor of element a_{ij}.

Cofactor. The cofactor C_{ij} of element a_{ij} of the matrix **A** is defined as

$$C_{ij} = (-1)^{i+j} M_{ij}$$

Adjoint matrix. The adjoint matrix of a square matrix **A** is found by remaining each element a_{ij} of matrix **A** by its cofactor C_{ij} and then transposing. For example, if **A** is given by (II.2), then

$$adj\ \mathbf{A} = \begin{bmatrix} \begin{vmatrix} 6 & 4 \\ 4 & 8 \end{vmatrix} & -\begin{vmatrix} -2 & 4 \\ 1 & 8 \end{vmatrix} & \begin{vmatrix} -2 & 6 \\ 1 & 4 \end{vmatrix} \\ -\begin{vmatrix} -2 & 1 \\ 4 & 8 \end{vmatrix} & \begin{vmatrix} 3 & 1 \\ 1 & 8 \end{vmatrix} & -\begin{vmatrix} 3 & -2 \\ 1 & 4 \end{vmatrix} \\ \begin{vmatrix} -2 & 1 \\ 6 & 4 \end{vmatrix} & -\begin{vmatrix} 3 & 1 \\ -2 & 4 \end{vmatrix} & \begin{vmatrix} 3 & -2 \\ -2 & 6 \end{vmatrix} \end{bmatrix}^T$$

$$= \begin{bmatrix} 32 & 20 & -14 \\ 20 & 23 & -14 \\ -14 & -14 & 14 \end{bmatrix}^T = \begin{bmatrix} 32 & 20 & -14 \\ 20 & 23 & -14 \\ -14 & -14 & 14 \end{bmatrix} \quad \text{...(II.4)}$$

Singular and nonsingular matrices. A square matrix is called singular if its associated determinant is zero and nonsingular if its associated determinant is non-zero.

Rank of a matrix. A matrix **A** is said to have rank r if there exists an $r \times r$ submatrix of **A** which is nonsingular and all other $q \times q$ submatrices (where $q \ge r + 1$) are singular.

For example, consider a 4×4 matrix

$$\mathbf{A} = \begin{bmatrix} 1 & 2 & 3 & 4 \\ 0 & 1 & -1 & 0 \\ 1 & 0 & 1 & 2 \\ 1 & 1 & 0 & 2 \end{bmatrix}$$

det $\mathbf{A} = 0$

and
$$\begin{vmatrix} 1 & 2 & 3 \\ 0 & 1 & -1 \\ 1 & 0 & 1 \end{vmatrix} \ne 0$$

Hence the rank of matrix **A** is 3.

Conjugate matrix. The conjugate of a matrix **A** (denoted as $\mathbf{A}^*$) is the matrix whose each element is the complex conjugate of the corresponding element of **A**.

Real matrix. If $\mathbf{A} = \mathbf{A}^*$, then the matrix **A** has all real elements and **A** is said to be a real matrix.

II.2 ELEMENTARY MATRIX OPERATIONS

Equality of matrices. Two matrices **A** and **B** are said to be equal if they have the same number of rows and columns and the elements of the corresponding orientations are equal.

Multiplication of a matrix by a scalar. A matrix is multiplied by a scalar K if all the mn elements are multiplied by K, *i.e.*, if **A** is given by eqn. (II. 1), then

$$\mathbf{KA} = \begin{bmatrix} Ka_{11} & Ka_{12} & \dots & Ka_{1n} \\ Ka_{21} & Ka_{22} & \dots & Ka_{2n} \\ \vdots & \vdots & & \vdots \\ Ka_{m1} & Ka_{m2} & \dots & Ka_{mn} \end{bmatrix}$$

Further $\qquad (K\mathbf{A})^T = K\mathbf{A}^T$

Addition and subtraction of matrices. Addition of two matrices **A** and **B** results in a new matrix **C** with its element c_{ij} equal to the sum of the corresponding elements a_{ij} and b_{ij}. We then denote

$$\mathbf{C} = \mathbf{A} + \mathbf{B}$$

The addition of two 3 × 3 matrices is illustrated below:

$$\mathbf{C} = \begin{bmatrix} a_{11} & a_{12} & a_{13} \\ a_{21} & a_{22} & a_{23} \\ a_{31} & a_{32} & a_{33} \end{bmatrix} + \begin{bmatrix} b_{11} & b_{12} & b_{13} \\ b_{21} & b_{22} & b_{23} \\ b_{31} & b_{32} & b_{33} \end{bmatrix}$$

$$= \begin{bmatrix} a_{11}+b_{11} & a_{12}+b_{12} & a_{13}+b_{13} \\ a_{21}+b_{21} & a_{22}+b_{22} & a_{23}+b_{23} \\ a_{31}+b_{31} & a_{32}+b_{32} & a_{33}+b_{33} \end{bmatrix}$$

It is evident that addition of two matrices is possible only if they have the same number of rows and columns. Further

$$(\mathbf{A} + \mathbf{B})^T = \mathbf{A}^T + \mathbf{B}^T$$

Similar arguments apply for subtraction of matrices.

It should be noted that any square matrix **A** may be written as the sum of a symmetric matrix $\mathbf{A}_s$ and a skew-symmetric matrix $\mathbf{A}_{sk}$, as is shown below.

Let $$\mathbf{A} = \mathbf{A}_s + \mathbf{A}_{sk} \qquad \text{...(II.5)}$$

Taking transpose of both sides,

$$\mathbf{A}^T = \mathbf{A}_s^{\;T} + \mathbf{A}_{sk}^{\;T} = \mathbf{A}_s - \mathbf{A}_{sk} \qquad \text{...(II.6)}$$

Solving eqns. (II.5) and (II.6) simultaneously, we obtain

$$\mathbf{A}_s = \frac{\mathbf{A} + \mathbf{A}^T}{2}; \ \mathbf{A}_{sk} = \frac{\mathbf{A} - \mathbf{A}^T}{2}$$

Multiplication of matrices. The multiplication of two matrices is possible only if number of columns of the first matrix is equal to number of rows of the second. Such matrices are called *conformal*. If a matrix **A** is of order $m \times n$ and **B** is of order $n \times q$, then the product **C** is of order $m \times q$. The elements c_{ij} of the product

$$\mathbf{C} = \mathbf{AB}$$

are given by

$$c_{ij} = \sum_{k=1}^{n} a_{ij} b_{kj}$$

In other words the elements c_{ij} are found by multiplying the elements of the ith row of **A** with the element of the jth column of **B** and then summing these element products. For example

$$\begin{bmatrix} a_{11} & a_{12} & a_{13} \\ a_{21} & a_{22} & a_{23} \\ a_{31} & a_{32} & a_{33} \end{bmatrix} \begin{bmatrix} b_{11} & b_{12} & b_{13} \\ b_{21} & b_{22} & b_{23} \\ b_{31} & b_{32} & b_{33} \end{bmatrix} = \begin{bmatrix} c_{11} & c_{12} & c_{13} \\ c_{21} & c_{22} & c_{23} \\ c_{31} & c_{32} & c_{33} \end{bmatrix}$$

where

$$c_{11} = a_{11}b_{11} + a_{12}b_{21} + a_{13}b_{31}$$

$$c_{12} = a_{11}b_{12} + a_{12}b_{22} + a_{13}b_{32}$$

$$\vdots$$

It is important to note that in general matrix multiplication is not *commutative*, *i.e.*,

$$\mathbf{AB} \neq \mathbf{BA}$$

If **AB** = **BA**, we say matrices **A** and **B** commute.

Multiplication is *associative*; *i.e.*,

$$(\mathbf{AB})\mathbf{C} = \mathbf{A}(\mathbf{BC})$$

Multiplication is *distributive* with respect to addition, *i.e.*,

$$\mathbf{A}(\mathbf{B} + \mathbf{C}) = \mathbf{AB} + \mathbf{AC}$$

Multiplication of any matrix by a unit matrix results in the original matrix, *i.e.*,

$$\mathbf{AI} = \mathbf{A}$$

The transpose of the product of two matrices is the product of their transposes in reverse order, *i.e.*,

$$(\mathbf{AB})^T = \mathbf{B}^T\mathbf{A}^T$$

The concept of matrix multiplication assists in the solution of simultaneous linear equations. Consider a set of linear algebraic equations

$$a_{11}x_1 + a_{12}x_2 + \ldots + a_{1n}x_n = c_1$$

$$a_{12}x_1 + a_{22}x_2 + \ldots + a_{2n}x_n = c_1$$

$$\vdots \quad \vdots \quad \vdots \quad \vdots$$

$$a_{m1}x_1 + a_{m2}x_2 + \ldots + a_{mn}x_n = c_m$$

or

$$\sum_{j=1}^{n} a_{ij}x_j = c_i\,;\ i = 1, 2, \ldots, m$$

Using the rules of matrix multiplication defined above, eqns. (II. 7) can be written in the compact notation as

$$\mathbf{Ax} = \mathbf{c} \qquad \text{...(II.8)}$$

where

$$\mathbf{A} = \begin{bmatrix} a_{11} & a_{12} & \cdots & a_{1n} \\ a_{21} & a_{22} & \cdots & a_{2n} \\ \vdots & \vdots & & \vdots \\ a_{m1} & a_{m2} & \cdots & a_{mn} \end{bmatrix}$$

$$\mathbf{x} = \begin{bmatrix} x_1 \\ x_2 \\ \vdots \\ x_n \end{bmatrix}; \ \mathbf{c} = \begin{bmatrix} c_1 \\ c_2 \\ \vdots \\ c_m \end{bmatrix}$$

It is evident that vector-matrix eqn. (II.8) is a useful shorthand representation of the set of eqns. (II.7).

Matrix inversion. The inverse of a square matrix **A** is written as $\mathbf{A}^{-1}$ and is defined by the relation

$$\mathbf{A}^{-1}\mathbf{A} = \mathbf{A}\mathbf{A}^{-1} = \mathbf{I}$$

$\mathbf{A}^{-1}$ is determined by the relation

$$\mathbf{A}^{-1} = \frac{\text{adj } \mathbf{A}}{\det \mathbf{A}} \qquad \text{...(II.9)}$$

For example, if **A** is given by eqn. (II.2), then from eqns. (II.3), (II.4) and (II.9), we get

$$\mathbf{A}^{-1} = \frac{\text{adj } \mathbf{A}}{\det \mathbf{A}} = \frac{1}{42}\begin{bmatrix} 32 & 20 & -14 \\ 20 & 23 & -14 \\ -14 & -14 & 14 \end{bmatrix}$$

The inverse of the product of two nonsingular square matrices of the same order is the product of their inverses in reverse order, *i.e.*,

$$(\mathbf{AB})^{-1} = \mathbf{B}^{-1}\mathbf{A}^{-1}$$

Also

$$(\mathbf{A}^T)^{-1} = (\mathbf{A}^{-1})^T$$

II.3 DEFINITENESS AND QUADRATIC FORMS

In this section, we present the concept of sign definiteness of a scalar function V of a vector **x**.

Norm of a vector. A norm is any function that assigns to every vector **x** of real numbers a real number K satisfying the following four properties :

1. $\| \mathbf{x} \| \geq 0$
2. $\| \mathbf{x} + \mathbf{y} \| \leq \| \mathbf{x} \| + \| \mathbf{y} \|$
3. $\| \alpha\mathbf{x} \| = \alpha \| \mathbf{x} \|$ for every scalar α
4. $\| \mathbf{x} \| = 0$ if and only if $\mathbf{x} = \mathbf{0}$

Examples of norms are

$$\| \mathbf{x} \| = \sum_{i=1}^{n} | x_i |; \ \mathbf{x} \text{ is } n \times 1 \text{ vector} \qquad \text{...(II.10)}$$

$$\| \mathbf{x} \| = \max_i \{| x_i |\} \quad \text{...(II.11)}$$

$$\| \mathbf{x} \| = (\mathbf{x}^T\mathbf{x})^{1/2} \quad \text{...(II.12)}$$

$$\| \mathbf{x} \| = (\mathbf{x}^T\mathbf{A}\mathbf{x})^{1/2} \quad \text{...(II.13)}$$

where **A** is positive definite matrix (definition given later in this section).

The norm in eqn. (II.12) represents magnitude or Euclidean length of **x**. This norm is most commonly used in explicit calculations.

Positive (negative) definite function. A scalar function $V(\mathbf{x})$ is positive (negative) definite if for all **x** such that $\| \mathbf{x} \| \le K$,

(*i*) $V(\mathbf{x}) > 0\ (< 0)\ ;\ \mathbf{x} \ne \mathbf{0}$

(*ii*) $V(\mathbf{0}) = 0$

Positive (negative) semidefinite function. A scalar function $V(\mathbf{x})$ is positive (negative) semidefinite if for all **x** such that $\| \mathbf{x} \| \le K$,

(*i*) $V(\mathbf{x}) \ge 0\ (\le 0)\ ;\ \mathbf{x} \ne \mathbf{0}$

(*ii*) $V(\mathbf{0}) = 0$

In the above definitions if K is made arbitrarily large, the definitions hold in the entire state-space and are said to be global.

Indefinite function. A scalar function $V(\mathbf{x})$ is indefinite if $V(\mathbf{x})$ assumes both positive and negative values within the region $\| \mathbf{x} \| \le K$, no matter how small K is made.

Quadratic form. An expression such as

$$V(x_1, \dots, x_n) = \sum_{i=1}^{n}\sum_{j=1}^{n} k_{ij}x_ix_j$$

involving terms of second degree in x_i and x_j is known as quadratic form of n variables.

Quadratic form can be compactly expressed in the vector-matrix form as

$$V(\mathbf{x}) = \mathbf{x}^T\mathbf{Q}\mathbf{x} \quad \text{...(II.14)}$$

where **Q** is a constant matrix.

In the expanded form

$$V(\mathbf{x}) = [x_1\ x_2 \dots x_n]\begin{bmatrix} q_{11} & q_{12} & \cdots & q_{1n} \\ q_{21} & q_{22} & \cdots & q_{2n} \\ \vdots & \vdots & & \vdots \\ q_{n1} & q_{n2} & \cdots & q_{nn} \end{bmatrix}\begin{bmatrix} x_1 \\ x_2 \\ \vdots \\ x_n \end{bmatrix} \quad \text{...(II.15)}$$

Note that we can always assume **Q** to be symmetric. If **Q** is not symmetric, it can be proved that

$$\mathbf{x}^T\mathbf{Q}\mathbf{x} = \mathbf{x}^T\mathbf{Q}_s\mathbf{x}$$

This easily follows from eqn. (II.5) and property $\mathbf{x}^T\mathbf{Q}\mathbf{x} = \mathbf{x}^T\mathbf{Q}^T\mathbf{x}$, which is evident from eqn. (II.15)

Thus only the symmetric portion of **Q** is of importance. We shall therefore tacitly assume that **Q** is symmetric.

If a scalar function $V(\mathbf{x})$ is of the quadratic form given by eqn. (II.14), the definiteness of $V(\mathbf{x})$ is attributed to matrix $\mathbf{Q}$. Thus we may speak of *definiteness of a matrix*. An important test of definiteness* of a matrix is given by the Sylvester's theorem.

Sylvester's theorem. The necessary and sufficient conditions for a matrix

$$Q = \begin{bmatrix} q_{11} & q_{12} & q_{13} & \cdots & q_{1n} \\ q_{21} & q_{22} & q_{23} & \cdots & q_{2n} \\ \vdots & \vdots & \vdots & & \vdots \\ q_{n1} & q_{n2} & q_{n3} & \cdots & q_{nn} \end{bmatrix} \quad \text{...(II.16)}$$

to be positive definite are that all the successive principal minors of $\mathbf{Q}$ be positive, *i.e.*

$$q_{11} > 0\,; \begin{vmatrix} q_{11} & q_{12} \\ q_{21} & q_{22} \end{vmatrix} > 0; \begin{vmatrix} q_{11} & q_{12} & q_{13} \\ q_{21} & q_{22} & q_{23} \\ q_{31} & q_{32} & q_{33} \end{vmatrix} > 0\,; \ldots ; \det[\mathbf{Q}] > 0 \quad \text{...(II.17)}$$

The matrix $\mathbf{Q}$ is semidefinite if any of the above determinants is zero.

The matrix $\mathbf{Q}$ is negative definite (semidefinite) if the matrix $-\mathbf{Q}$ is positive definite (semidefinite).

If $\mathbf{Q}$ is positive definite, so is $\mathbf{Q}^2$ and $\mathbf{Q}^{-1}$.

It should be noted that the definiteness of a quadratic form scalar function is global.

II.4 MATRIX CALCULUS

1. *Integration*:

(*a*) $\underset{(n \times 1 \text{ vector})}{\mathbf{x} = \mathbf{x}(t)}; \int \mathbf{x}(t)dt = \left[\int x_1(t)\,dt \int x_2(t)\,dt \ldots \int x_n(t)\,dt\right]^T$

(*b*) $\underset{(m \times n \text{ matrix})}{\mathbf{A} = \mathbf{A}(t)}; \int \mathbf{A}(t)\,dt = \begin{bmatrix} \int a_{11}(t)\,dt & \int a_{12}(t)\,dt \ldots & \int a_{1n}(t)\,dt \\ \vdots & \vdots & \vdots \\ \int a_{m1}(t)\,dt & \int a_{m2}(t)\,dt \ldots & \int a_{mn}(t)\,dt \end{bmatrix}$

2. *Differentiation with respect to a scalar (time)*:

(*a*) $\mathbf{x} = \mathbf{x}(t); \dfrac{d\mathbf{x}}{dt} = \left[\dfrac{dx_1}{dt}\,\dfrac{dx_2}{dt} \ldots \dfrac{dx_n}{dt}\right]^T$

(*b*) $\mathbf{A} = \mathbf{A}(t); \dfrac{d\mathbf{A}}{dt} = \begin{bmatrix} \dfrac{da_{11}}{dt} & \dfrac{da_{12}}{dt} & \cdots & \dfrac{da_{1n}}{dt} \\ \vdots & \vdots & & \vdots \\ \dfrac{da_{m1}}{dt} & \dfrac{da_{m2}}{dt} & \cdots & \dfrac{da_{mn}}{dt} \end{bmatrix}$

3. *Differentiation with respect to a vector*:

(*a*) $\underset{(n \times 1 \text{ vector})}{\mathbf{x} = \mathbf{x}(t)}; \underset{(\text{scalar})}{f = f(\mathbf{x})}; \dfrac{\partial f}{\partial \mathbf{x}} = \left[\dfrac{\partial f}{\partial x_1}\,\dfrac{\partial f}{\partial x_2} \ldots \dfrac{\partial f}{\partial x_n}\right]^T \quad \text{...(II.18)}$

*Matrix $\mathbf{Q}$ is positive definite if all the eigenvalues of $\mathbf{Q}$ are positive.

$\dfrac{\partial f}{\partial x}$ is often called the gradient of f with respect to $\mathbf{x}$.

(*b*) $\mathbf{x} = \mathbf{x}(t)$; $\quad \mathbf{f} = \mathbf{f}(\mathbf{x})$

($m \times 1$ vector)

$$\frac{\partial \mathbf{f}}{\partial \mathbf{x}} = \begin{bmatrix} \dfrac{\partial f_1}{\partial x_1} & \dfrac{\partial f_1}{\partial x_2} & \cdots & \dfrac{\partial f_1}{\partial x_n} \\ \vdots & \vdots & & \vdots \\ \dfrac{\partial f_m}{\partial x_1} & \dfrac{\partial f_m}{\partial x_2} & \cdots & \dfrac{\partial f_m}{\partial x_n} \end{bmatrix} \qquad \text{...(II.19)}$$

This is sometimes called the *Jacobian matrix*.

4. *Derivative of product and sum*:

(*a*) $\mathbf{A} = \mathbf{A}(t)$; ($m \times n$ matrix) $\quad \mathbf{B} = \mathbf{B}(t)$ ($n \times q$ matrix)

$\mathbf{F} = \mathbf{AB}$ ($m \times q$ matrix)

$$\frac{d\mathbf{F}}{dt} = \frac{d\mathbf{A}}{dt}\mathbf{B} + \mathbf{A}\frac{d\mathbf{B}}{dt}$$

(*b*) $\mathbf{A} = \mathbf{A}(t)$; ($m \times n$ matrix) $\quad \mathbf{B} = \mathbf{B}(t)$; ($m \times n$ matrix) $\quad \mathbf{F} = \mathbf{A} + \mathbf{B}$

$$\frac{d\mathbf{F}}{dt} = \frac{d\mathbf{A}}{dt} + \frac{d\mathbf{B}}{dt}$$

5. *Derivatives of linear and quadratic forms*:

(*a*) $\dfrac{d\mathbf{x}^T}{d\mathbf{x}} = \mathbf{I}$; $\mathbf{x}$ is $n \times 1$ vector ...(II.20)

(*b*) $\dfrac{d(\mathbf{x}^T\mathbf{b})}{d\mathbf{x}} = \mathbf{b}$; $\mathbf{b}$ is $n \times 1$ vector (II.21)

(*c*) $\dfrac{d(\mathbf{x}^T\mathbf{x})}{d\mathbf{x}} = 2\mathbf{x}$ ...(II.22)

(*d*) $\dfrac{d(\mathbf{x}^T\mathbf{A}\mathbf{x})}{d\mathbf{x}} = (\mathbf{A}^T + \mathbf{A})\mathbf{x}$; $\mathbf{A}$ is $n \times n$ matrix ...(II.23)

(*e*) $\dfrac{d(\mathbf{x}^T\mathbf{A}\mathbf{x})}{d\mathbf{x}} = 2\mathbf{A}\mathbf{x}$; $\mathbf{A}$ is $n \times n$ symmetric matrix ...(II.24)

APPENDIX III

MATLAB: TOOL FOR DESIGN AND ANALYSIS OF CONTROL SYSTEMS

III.1 QUICK REFERENCE

MATLAB provides 20 main categories of functions. The MATLAB **help** displays an on line table of these categories. Typing **help category name** displays the tables within those categories.

Color	Color control models	**matfun**	Matrix functions
Datafun	Data analysis	**ops**	Operators
plotxy	2-D graphics	**plotxyz**	3-D graphics
elmat	matrix manipulation	**polyfun**	Polynomial functions
funfun	Function functions	**sparfun**	Sparse matrix functions
general	General Commands	**specfun**	Specialized math functions
graphics	Graphics functions	**specmat**	Special Matrices
lang	Language constructs	**strfun**	Character string functions

General Purpose Commands

path	control MATLAB's path	**type**	Display M-file contents
which	Locate Functions and Files	**clear**	Clear memory
save	Save Workspace variables	**length**	Length of Vector
size	Size of Matrix	**who**	List Variables
load	Retrieve Variable	**cd**	Change Working Directory
dir	Directory Listing	**quite**	Terminate MATLAB

Language Constructs

function	Add new function	**break**	Terminate Execution Loop
for	Repeat statements	**else**	Used with *if*
if	Conditional Execution	**elseif**	Used with *if*
return	Return to invoking function	**end**	Terminate loops
while	Repeat Conditionally	**menu**	generate Menu
input	Prompt for User input	**pause**	Wait for User response

Matrix and Math Functions

eye	Identity Matrix	**ones**	Ones Matrix
rand	Random numbers	**zeros**	Zeros Matrix
abs	Absolute value	**acos**	Inverse Cosine
exp	Exponential	**angle**	Phase angle
log	Natural Logarithm	**sin**	Sine
cos	Cosine	**conj**	Complex Conjugate
sqrt	Square Root	**rem**	Remainder after division
det	Determinant of Matrix	**rank**	LI rows or cols
norm	Matrix of vector norm	**inv**	Matrix inverse
rref	Row-reduced form	**orth**	Orthogonalization
cross	Vector cross product	**dot**	Vector dot product

Graphics Function

plot	linear plot	**bar**	Bar Graph
hist	Histogram Plot	**title**	Graph Title
xlabel	X-axis label	**ylabel**	Y-axis label
mesh	3-D Mesh Surface	**grid**	Grid lines
axis	Axis scaling	**text**	Text Annotation

Table III.1: Model Conversion Function

Function	*Purpose*
c2d	continuous state-space to discrete state-space
residue	partial fraction expansion
ss2tf	state-space to transfer function
ss2zp	state-space to zero-pole-gain
tf2ss	transfer function to state-space
tf2zp	transfer function to zero-pole-gain
zp2ss	zero-pole-gain to state-space
zp2tf	zero-pole-gain to transfer function

III.2 MODEL CONVERSION

MATLAB allows a number of functions that make it easy to convert from one model form to another and to convert continuous-time systems into discrete-time systems. These conversion functions and their uses are summarized in the Table III.1.

Following are descriptions of each conversion function and example of how to use them.

c2d Function: The *c2d* function converts the continuous-time state-space equation

$$\dot{x} = Ax + Bu \qquad \text{...(III.1)}$$

to the discrete-time space equation

$$x[n + 1] = A_d x[n] + B_d u[n] \qquad \text{...(III.4)}$$

The function has two output matrices as show in the statement:

$$[ad, bd] = c2d(a, b, Ts);$$

The input arguments a and b of the *c2d function* are the matrices A and B of the continuous-time state-space equation that is to be converted, and T_s is the desired sample period. The outputs *ad* and *bd* are the matrices A_d and B_d of the discrete-time equation.

For example, if we take a continuous-time state-space plant equation:

$$\dot{x} = Ax + Bu \qquad \text{...(III.3)}$$

where

$$A = \begin{bmatrix} 0 & 1 \\ -3 & -4 \end{bmatrix}, B = \begin{bmatrix} 0 \\ 1 \end{bmatrix} \qquad \text{...(III.4)}$$

can be converted to a discrete-time state-space equation with a sampling period of 0.1 seconds using the statements

$$a = [0 \quad 1; \quad -3 \quad -4];$$
$$b = [0 \quad 1];$$
$$[ad, bd] = c2d\ (a, b, 0, 1);$$

The values computed by the *c2d function* are the following:

$$ad = \begin{bmatrix} 0.9868 & 0.0820 \\ -0.2460 & 0.6588 \end{bmatrix}, B = \begin{bmatrix} 0.0044 \\ 0.0820 \end{bmatrix} \qquad \text{...(III.5)}$$

Thus the discrete-time state-space equation is

$$x[n + 1] = A_d x[n] + B_d u[n] \qquad \text{...(III.6)}$$

which represents

$$\begin{bmatrix} x_1 \\ x_2 \end{bmatrix}_{n+1} = \begin{bmatrix} 0.9868 & 0.080 \\ -0.2460 & 0.6588 \end{bmatrix} \begin{bmatrix} x_1 \\ x_2 \end{bmatrix}_n + \begin{bmatrix} 0.0044 \\ 0.0820 \end{bmatrix} u_n \qquad \text{...(III.7)}$$

Residue function: The residue function converts the polynomial transfer function:

$$H(s) = \frac{b_0 s^n + b_1 s^{n-1} + \ldots + b_{n-1} s + b_n}{a_0 s^m + a_1 s^{m-1} + \ldots + a_{m-1} s + a_m} \qquad \text{...(III.8)}$$

to the partial fraction:

$$H(s) = \frac{r_1}{s - p_1} + \frac{r_2}{s - p_2} + \ldots + \frac{r_n}{s - p_n} + k(s) \qquad \text{...(III.9)}$$

The function has three output matrices as shown in this statement:

$[r, p, k]$ = residue (b, a);

The input arguments b and a of the residue function are the coefficients of the numerator and denominator polynomials, respectively, of the polynomial transfer function that is to be converted. They must be entered as row vectors in descending powers of s. The outputs include r, a column vector containing the residue values; p, a column vector of the corresponding pole locations; and k, a row vector of the direct terms.

For example, consider a system with transfer function

$$H(s) = \frac{s^2 + 6s + 9}{s^3 + 8s^2 + 20s + 16} \qquad \text{...(III.10)}$$

is computed with the following statements

num = [1 6 9];

den = [1 8 20 16];

$[r, p, k]$ = residue (num, den);

where r, p, k, take following values

$$r = \begin{bmatrix} 0.25 \\ 0.75 \\ 0.5 \end{bmatrix},\ p = \begin{bmatrix} -4.0 \\ -2.0 \\ -2.0 \end{bmatrix},\ k = [\ ] \qquad \text{...(III.11)}$$

Therefore,

$$H(s) = \frac{0.25}{s+4} + \frac{0.75}{s+2} + \frac{0.5}{s+2} \qquad \text{...(III.12)}$$

ss2tf Function: The *ss2tf* function converts the continuous-time, state-space equations:

$$\dot{x} = Ax + Bu \qquad \text{...(III.13)}$$

$$y = Cx + Du$$

to the polynomial transfer function

$$H(s) = \frac{r_1}{s - p_1} + \frac{r_2}{s - p_2} + \ldots + \frac{r_n}{s - p_n} + k(s) \qquad \text{...(III.14)}$$

The function has two output matrices as shown in this statement:

[num, den] = *ss2tf* (a, b, c, d, iu);

The input arguments a, b, c and d of the *ss2tf* function are the matrices A, B, C, and D of the state-space equations corresponding to the iu'th input. In case of single input system iu is 1. The output vectors *num* and *den* contain the coefficients, in descending powers of s, the numerator and denominator of the polynomial transfer function for the iu'th input.

For example, the given state-space equation

$$\begin{bmatrix} \dot{x}_1 \\ \dot{x}_2 \end{bmatrix} = \begin{bmatrix} 0 & 1 \\ -3 & -4 \end{bmatrix} \begin{bmatrix} x_1 \\ x_2 \end{bmatrix} + \begin{bmatrix} 0 \\ 1 \end{bmatrix} u \qquad \text{...(III.15)}$$

$$y = [10 \quad 0]\begin{bmatrix} x_1 \\ x_2 \end{bmatrix} + [0]u \qquad \text{...(III.16)}$$

can be converted to polynomial transfer function

```
a = [0 1 ; – 3 – 4] ;
b = [0 1] ;
c = [10 0] ;
d = 0 ;
iu = 1 ;
[num, den] = ss2tf(a, b, c, d, iu) ;
```

The values computed are the following:

$$num = [0 \quad 0 \quad 10],\ den = [1 \quad 4 \quad 3] \qquad \text{...(III. 17)}$$

Thus, the transfer function is

$$H(s) = \frac{10}{s^2 + 4s + 3} \qquad \text{...(III.18)}$$

ss2zp Function: The *ss2zp* function converts the continuous-time, state-space equations:

$$\dot{x} = Ax + Bu \qquad \text{(III.19)}$$

$$y = Cx + Du$$

to the pole-zero-gain function

$$H(s) = k\,\frac{(s - z_1)(s - z_2)\ldots(s - z_n)}{(s - p_1)(s - p_2)\ldots(s - p_m)} \qquad \text{...(III.20)}$$

The function has three output matrices as shown in this statement.

[z, p, k] = ss2zp(a, b, c, d, iu);

The input arguments *a, b, c* and *d* of the *ss2zp* function are the matrices *A, B, C* and *D* of the state-space equations corresponding to the *iu*'th input, which *iu* is the number of the input for a multi-ouput system, *iu* is 1. The output vectors *z* and *p* are the zeros and the poles respectively, of the zero-pole-gain transfer function for the *iu*'th input, and *k* is the associated gain.

For example the state-space equation

$$\begin{bmatrix} \dot{x}_1 \\ \dot{x}_2 \end{bmatrix} = \begin{bmatrix} 0 & 1 \\ -3 & -4 \end{bmatrix}\begin{bmatrix} x_1 \\ x_2 \end{bmatrix} + \begin{bmatrix} 0 \\ 1 \end{bmatrix} u \qquad \text{...(III.21)}$$

$$y = [10 \quad 0]\begin{bmatrix} x_1 \\ x_2 \end{bmatrix} + [0]u \qquad \text{...(III.22)}$$

can be converted to zero-pole-gain transfer function

```
a = [0  1;  – 3  – 4] ;
b = [0  1] ;
c = [10  0] ;
d = 0 ;
```

```
iu = 1 ;
[z, p, k] = ss2zp(a, b, c, d, iu) ;
```

The values computed are the following:

$$z = [\ ], p = \begin{bmatrix} -1 \\ -3 \end{bmatrix}, k = [10] \qquad \text{...(III.23)}$$

Thus, the transfer function is

$$H(s) = \frac{10}{(s+1)(s+3)} \qquad \text{...(III.24)}$$

III.3 DESIGN AND ANALYSIS FUNCTIONS

MATLAB has several functions for designing and analyzing linear systems both in the time and frequency domains. Some of the functions frequently used are tabulated in Table III.2:

Table III.2 : Design and Analysis Functions

Function	*Purpose*
bode	magnitude and phase-frequency plots
nyquist	Nyquist frequency plots
rlocus	Root-locus plots
step	Unit step time response
lqe	linear-quadratic estimator
lqr	linear-quadratic regulator

Step Response

The format for calling this function is as given below:

```
[y, x, t] = step (num, den) ;
[y, x, t] = step (num, den, t) ;
[y, x, t] = step (a, b, c, d) ;
[y, x, t] = step (a, b, c, d, iu) ;
[y, x, t] = step (a, b, c, d, iu, t) ;
```

The arguments of these functions are dependent on the model you have chosen for the system. For example lets take the step response of a third-transfer function:

$$H(s) = \frac{1}{s(s+1)(s+2)} \qquad \text{...(III.25)}$$

can be generated using the following statements:

```
num = [1] ;
den = poly ([0 – 1 – 2]) ;
k = [0.25 0.4 1.5 6 8] ;        % Range of the gain of
                                % proportional controller
```

```
        t = [0 : 0.2 : 20]';                % Time varies from 0 to
                                            % 20 sec in steps 0.2 secs
    for i = 1 : 5
        [ntc, dtc] = cloop (num*k(i), den) ;
        y(:, i) = step (ntc, dtc, t) ;
        imp (:, i) = impulse (ntc, dtc, t) ;
end
subplot (211) ; plot (t, y(: , 1 : 3)), grid
subplot (212) ; plot (t, y(: , 4 : 5)), grid
pause ;
subplot (211) ; plot(t, y(: , 1 : 3)), grid
subplot (212) ; plot (t, y(: , 4 : 5)), grid
```

The plot is as shown in Fig. III.1

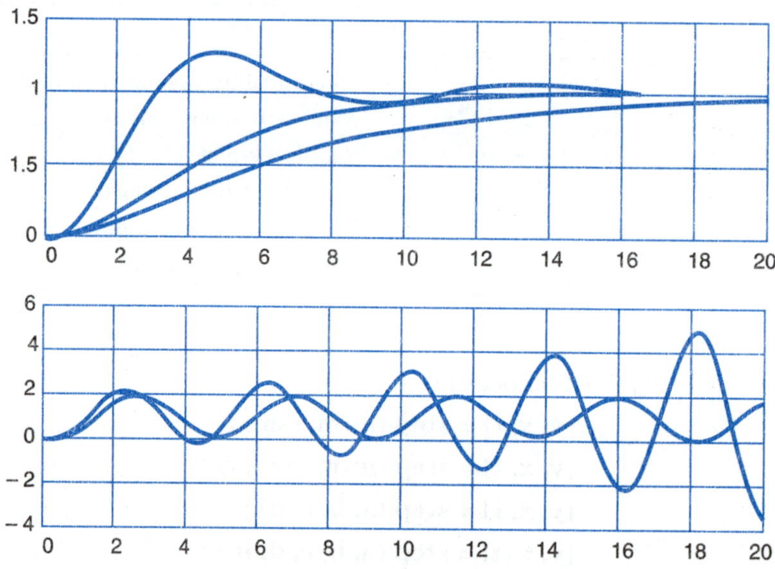

Fig. III.1. Step response plot.

Bode Plots

The bode function calculates the magnitude and phase frequency response of the continuous-time linear-time-invariant systems for use in making *bode* and *Nichols* plots. It can be used in a variety of ways depending on the model representation. The format for Bode Plot is as given below :

```
[mag, phase] = bode (num, gen) ;
[mag, phase] = bode (num, gen, w) ;
[mag, phase] = bode (a, b, c, d) ;
```

[mag, phase] = bode (a, b, c, d, iu) ;

[mag, phase] = bode (a, b, c, d, iu, w) ;

The input arguments are vectors containing the coefficients, in descending power of s, of the numerator and denominator polynomials of the transfer function for which the bode plot is needed. For example the bode-plot for second-order transfer function

$$H(s) = \frac{10}{s^2 + s + 3} \quad \text{...(III.26)}$$

can be generated with statements

num = 10 ;

den = [1 1 3] ;

bode (num, den) ;

title ('Bode Plot for Second-Order System') ;

The plot generated is shown in Fig. III.2.

A bode plot can also be generated using input arguments that represent the continuous state-space system:

$$\begin{bmatrix} \dot{x}_1 \\ \dot{x}_2 \end{bmatrix} = \begin{bmatrix} 0 & 1 \\ -3 & -4 \end{bmatrix} \begin{bmatrix} x_1 \\ x_2 \end{bmatrix} + \begin{bmatrix} 0 \\ 1 \end{bmatrix} u \quad \text{...(III.27)}$$

$$y = \begin{bmatrix} 10 & 0 \end{bmatrix} \begin{bmatrix} x_1 \\ x_2 \end{bmatrix} + [0]u \quad \text{...(III.28)}$$

Thus these statements will produce the same plot as shown in Fig. III.2

a = [0 1 ; - 3 - 1] ;

b = [0 1] ;

c = [10 0] ;

d = 0 ;

bode (a, b, c, d)

Nyquist Plots

The nyquist function is a frequency-domain analysis function that is similar to the bode function in that it uses exactly the same input arguments to produce the frequency response plots. As opposed to bode plot the Nyquist plot gives only one output. It plots the real component of the open-loop transfer function versus the imaginary component for different values of frequency.

The **nyquist** function can be used for continuous-time linear-time-invariant systems. The various forms for its use are shown below:

[re, im, w] = nyquist (num, den) ;

[re, im, w] = nyquist (num, den, w) ;

[re, im, w] = nyquist (a, b, c, d) ;

[re, im, w] = nyquist (a, b, c, d, iu) ;

[re, im, w] = nyquist (a, b, c, d, iu, w) ;

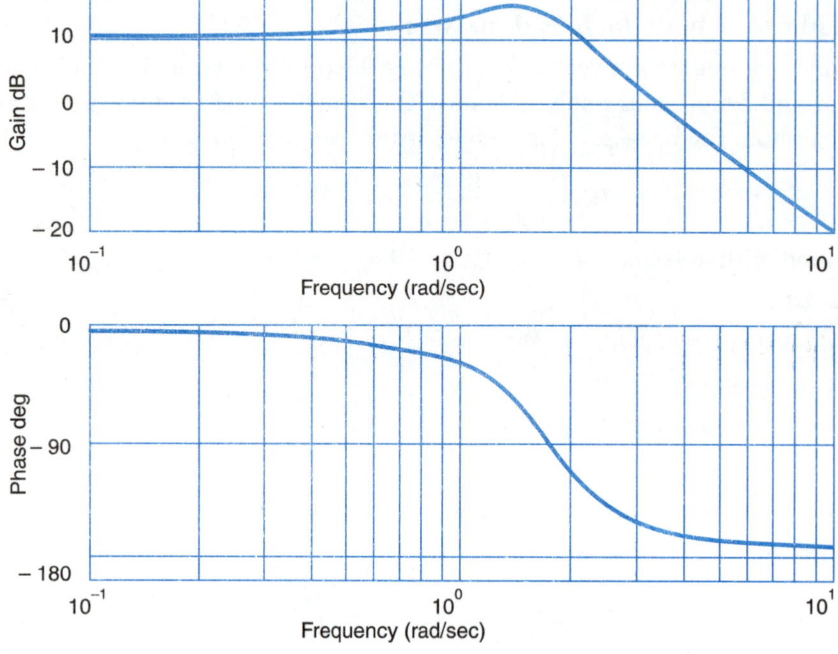

Fig. III.2. Bode plot for second order systems.

The input arguments specify the same thing as that of bode function. The output **re** and **im** are the real and imaginary components of the system response, respectively, and **w** is the vector of frequencies at which the response is computed.

Fro example the **nyquist** plot for the following system function:

$$H(s) = \frac{10}{s^2 + 4s + 3} \qquad \text{...(III.29)}$$

can be produced using the following statements

num = 10 ;

den = [1, 4, 3] ;

nyquist (num, den), title ('Nyquist Plot')

which produces the plot as shown in Fig. III.3.

A nyquist plot can also be generated using the following arguments that represent the continuous state-space system:

$$\dot{x} = Ax + Bu \qquad \text{...(III.30)}$$

$$y = Cx + Du$$

For example the state-space equation taken previously as

$$\begin{bmatrix} \dot{x}_1 \\ \dot{x}_2 \end{bmatrix} = \begin{bmatrix} 0 & 1 \\ -3 & -4 \end{bmatrix} \begin{bmatrix} x_1 \\ x_2 \end{bmatrix} + \begin{bmatrix} 0 \\ 1 \end{bmatrix} u \qquad \text{...(III.31)}$$

$$y = [10 \quad 0] \begin{bmatrix} x_1 \\ x_2 \end{bmatrix} + [0]u \qquad \text{...(III.32)}$$

The statements below will produce the same plot as shown in Fig. III.3:

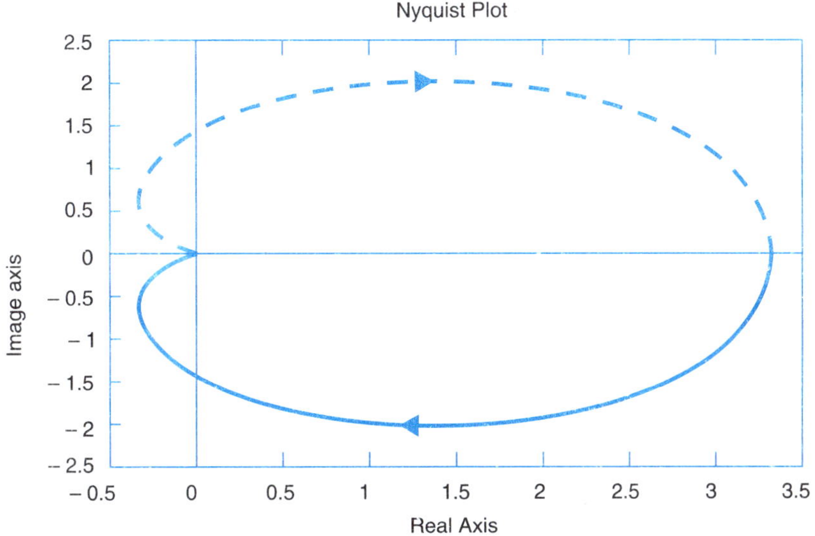

Fig. III.3. Nyquist plot.

```
a = [0 1 ; –3  –1] ;
b = [0 1]' ;
c =[10 0] ;
d = 0 ;
nyquist (a, b, c, d), title ('Nyquist Plot')
```

Root Locus Plots

The root locus is an useful tool for single-input/single-output systems. It helps in assessing the stability and transient response of a system and for determining ways to improve the system performance. The plot shows the location of roots of the characteristic equation of a system. The roots of the characteristic equation determine the stability of the system. By analysing the root-locus the system designer can improve the system performance.

In the equation below $G(s)$ represents the forward path transfer function and $H(s)$ is the feedback path transfer function:

$$\frac{y(s)}{u(s)} = \frac{G(s)}{1 + G(s)H(s)} \qquad ...(III.33)$$

The root locus is obtained by setting the denominator of the closed-loop transfer function equal to zero, which gives the characteristic equation of the system

$$1 + G(s)H(s) = 0 \qquad ...(III.34)$$

$$G(s)H(s) = -1 \qquad ...(III.35)$$

The roots of the characteristic equation are then computed as some parameter, usually the forward path gain, is varied from zero to infinity.

The **rlocus** function can be used to produce root locus plots for both continuous-time and discrete-time systems. The forms for using **rlocus** are

[r, k] = rlocus (num, den) ;

[r, k] = rlocus (num, den, k) ;

[r, k] = rlocus (a, b, c, d) ;

[r, k] = rlocus (a, b, c, d, k) ;

The input arguments **num** and **den** for the first forms of the rlocus function shown are vectors containing the coefficients, in descending powers of s, of the numerator and denominator polynomials of the open-loop transfer function $G(s)H(s)$. The output matrices **r** and **k** are the root locations and the corresponding gains, respectively.

For example, the root locus plot of the unity feedback system with the following forward system function:

$$F(s) = \frac{K(s+6)}{s(s+4)(s^2+4s+8)} \qquad \text{...(III.36)}$$

can be produced by using the following statement:

num = [1 6] ;

p1 = [1 4 0] ;

p2 = [1 4 8] ;

den = conv (p1, p2) ;

rlocus (num, den), title ('Root Locus Plot')

The root locus of the system is as shown in Fig. III.4. In this case, the function automatically selects the value of the gain, K. Similarly the statement rlocus (num, den, k) produces root locus plot using a user-specified input gain vector k. Root Locus plot can also be generated using arguments that represent the continuous-time state-space systems.

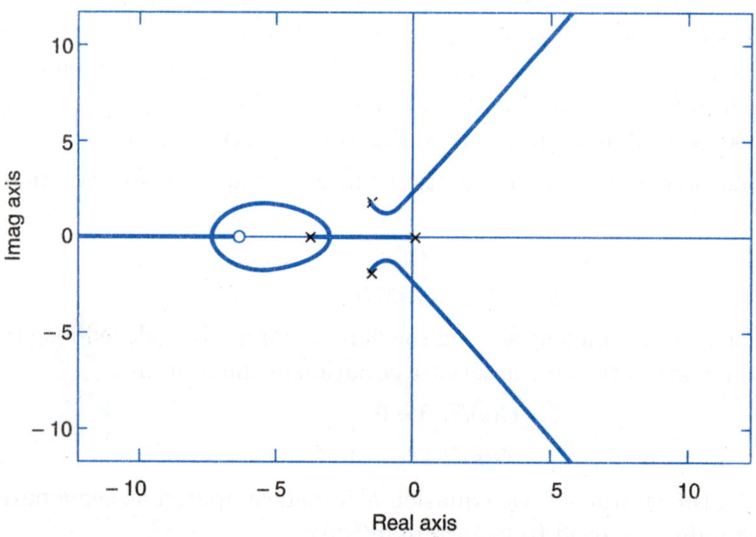

Fig. III.4. Root locus plot.

III.4 CASE STUDY

Engineers design the control systems starting with the performance requirements that the plant or the system needs. Then appropriate controller is chosen and its model along with plant is used to analyze the system. The control system designs are analyzed using such things as root locus, bode plot, step response plot, etc. to evaluate controller configuration and gain values.

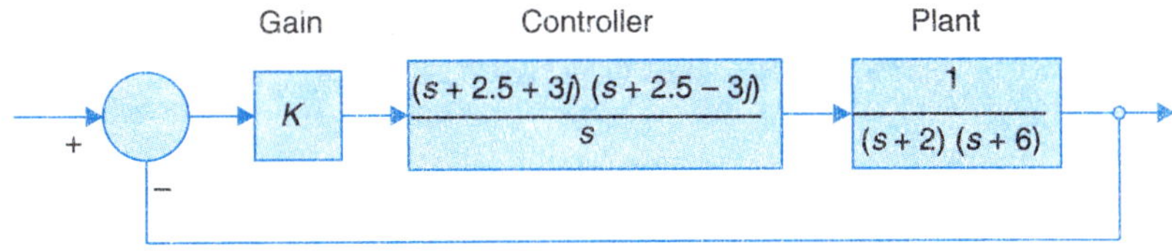

Fig. III.5. Example control system.

Problem Statement

A PID controller is being designed for a system having system function:

$$G(s) = \frac{1}{(s+2)(s+6)} \qquad \text{...(III.37)}$$

The block diagram III-4 shows the gain, controller and the plant with unity feedback path. Select a gain for the control system that will provide a stable well damped response.

MATLAB Solution

We use the root locus technique to see the location of the poles as the gain changes. We can then chose the location of the plant poles that will give a stable response. We use the following MATLAB script determine the gain:

```
ztf = [-2.5-3*j ; -2.5 + 3*j] ;          % zeros of the system
ptf = [0 ; -2 ; -6] ;                    % poles of the system
[numtf, dentf] = zp2tf(ztf, ptf, 1) ;    % zero-pole to transfer fn.
rlocus (numtf, dentf) ;                  % plots the root locus
[K, poles] = rlocusfind (numtf, dentf) ;
K
```

Poles

After plotting the root locus the poles of the system were chosen close to the zeros so that at that gain the poles and zeros cancel. On choosing we got the following values as: $K = 113$ *and Poles* $= (-116, -2.42 + 3j, -2.42 - 3j)$. It can clearly seen that on choosing gain $K = 113$. The two complex poles will be cancelled and the pole $p = 116$ is far of and hence the system will be a stable one with very fast response time.

APPENDIX IV

IV.1 FINAL VALUE THEOREM

The final value of the sequence $f(k)$ is given by

$$\lim_{k\to\infty} f(k) = \lim_{z\to 1} F(z)(z-1) \qquad \text{...(IV.1)}$$

provided the limit exists, *i.e.,* the data sequence has a finite value.

Proof :

$$Z[f(k+1) - f(k)] = \lim_{m\to\infty} \sum_{k=0}^{m} [f(k+1) - f(k)]z^{-k} \qquad \text{...(IV.2)}$$

By the left shifting property, we get

$$zF(z) - f(0) - F(z) = \lim_{m\to\infty} \sum_{k=0}^{m} [f(k+1) - f(k)]z^{-k} \qquad \text{...(IV.3)}$$

Letting $z \to 1$ on both the sides,

$$\lim_{z\to 1} [(z-1)F(z) - f(0)] = \lim_{z\to 1} \lim_{m\to\infty} \sum_{k=0}^{m} [f(k+1) - f(k)]z^{-k} \qquad \text{...(IV.4)}$$

Interchanging the order of the limits on the right hand side, we have

$$\lim_{z\to 1}(z-1)F(z) = f(0) + \lim_{m\to\infty} \sum_{k=0}^{m} [f(k+1) - f(k)]z^{-k} \qquad \text{...(IV.5)}$$

$$= f(\infty) \qquad \text{...(IV.6)}$$

APPENDIX V

To prove that a controllable linear time invariant system state variable model to a controllable phase variable model.

Theorem : Consider a linear time invariant system

$$\dot{\mathbf{x}} = \mathbf{Ax} + \mathbf{Bu}$$
$$\mathbf{y} = \mathbf{Cx} \quad \text{...(V.1)}$$

where $\mathbf{x}$ is $n \times 1$ state vector, $\mathbf{A}$ is as $n \times n$ system matrix, $\mathbf{B}$ is a $n \times 1$ control vector, u is a scalar control input.

If this system is controllable, then it can be transformed into the form

$$\dot{\mathbf{v}} = \mathbf{A_c v} + \mathbf{B_c} u \quad \text{...(V.2)}$$

where v is a $n \times 1$ vector and

$$\mathbf{A}_c = \begin{bmatrix} 0 & 1 & & 0 & . & . & . & 0 \\ 0 & 0 & . & 1 & . & . & . & 0 \\ . & . & & . & & & & . \\ . & . & & \vdots & & & & . \\ . & . & & . & & & & . \\ 0 & 0 & & 0 & . & . & . & 1 \\ -\mathbf{a}_n & -\mathbf{a}_{n-1} & & -\mathbf{a}_{n-2} & . & . & . & -\mathbf{a}_1 \end{bmatrix}; \mathbf{B}_c = \begin{bmatrix} 0 \\ 0 \\ . \\ . \\ . \\ 0 \\ 1 \end{bmatrix}$$

Proof : Let us assume that the transformation exists and is given by

$$\mathbf{v} = \mathbf{Px} \quad \text{...(V.3)}$$

where

$$\mathbf{P} = \begin{bmatrix} p_{11} & p_{12} & \cdots & p_{1n} \\ p_{21} & p_{22} & \cdots & p_{2n} \\ \vdots & \vdots & & \vdots \\ p_{n1} & p_{n2} & & p_{nn} \end{bmatrix} = \begin{bmatrix} \mathbf{P}_1 \\ \mathbf{P}_2 \\ \vdots \\ \mathbf{P}_n \end{bmatrix} \quad \text{...(V.4)}$$

$$\mathbf{P}_i = \begin{bmatrix} p_{i1} & p_{i2} & \cdots & p_{in} \end{bmatrix}; \; i = 1, 2, ..., n$$

In the component form we can write

$$v_1(t) = p_{11}x_1 + p_{12}x_2 + \dots + p_{1n}x_n$$
$$= \mathbf{P}_1\,\mathbf{x}(t)$$

Taking derivative on both sides of this equation, we have

$$\dot{v}_1(t) = \mathbf{P}_1\,\mathbf{x} = \mathbf{P}_1\,\mathbf{Ax} + \mathbf{P}_1\,\mathbf{B}u$$

But $\dot{v}_1 = v_2$ is a function of $\mathbf{x}$ only as per transformation (V.3). Therefore $\mathbf{P}_1\mathbf{B} = \mathbf{0}$ and

$$v_2 = \mathbf{P}_1\,\mathbf{Ax}$$

Taking derivative on both sides once again, we have

$$\dot{v}_2 = v_3 = \mathbf{P}_1\,\mathbf{A}^2\mathbf{x} \text{ and } \mathbf{P}_1\,\mathbf{AB} = \mathbf{0}$$

This process gives

$$\dot{v}_{n-1} = v_n = \mathbf{P}_1\,\mathbf{A}^{n-1}\mathbf{x} \text{ with } \mathbf{P}_1\,\mathbf{A}^{n-2}\mathbf{B} = \mathbf{0}$$

Thus

$$v(t) = \mathbf{Px} = \begin{bmatrix} \mathbf{P}_1 \\ \mathbf{P}_1\mathbf{A} \\ \vdots \\ \mathbf{P}_1\mathbf{A}^{n-1} \end{bmatrix}\mathbf{x} \qquad \text{...(V.5)}$$

and $\mathbf{P}_1$ should satisfy the conditions

$$\mathbf{P}_1\,\mathbf{B} = \mathbf{P}_1\,\mathbf{AB} = \cdots = \mathbf{P}_1\,\mathbf{A}^{n-2}\mathbf{B} = \mathbf{0} \qquad \text{...(V.6)}$$

The transformation (V.3) transforms (V.2) to

$$\dot{\mathbf{v}} = \mathbf{PAP}^{-1}\mathbf{v} + \mathbf{PB}u$$

Comparing this with (V.2), we get

$$\begin{matrix} \mathbf{B} & & \mathbf{PB} \end{matrix}$$

or

$$\begin{bmatrix} 0 \\ 0 \\ \vdots \\ 1 \end{bmatrix} = \begin{bmatrix} \mathbf{P}_1\mathbf{B} \\ \mathbf{P}_1\mathbf{AB} \\ \vdots \\ \mathbf{P}_1\mathbf{A}^{n-1}\mathbf{B} \end{bmatrix}$$

or

$$\mathbf{P}_1[\mathbf{B} \vdots \mathbf{AB} \vdots \cdots \vdots \mathbf{A}^{n-1}\mathbf{B}] = [0 \;\; 0 \cdots 1] \qquad \text{...(V.7)}$$

This gives

$$\mathbf{P}_1 = [0 \quad 0 \quad \cdots \quad 1]\left[\mathbf{B} \vdots \mathbf{AB} \vdots \cdots \vdots \mathbf{A}^{n-1}\mathbf{B}\right]^{-1}$$

The matrix $\mathbf{Q}_c = \left[\mathbf{B} \vdots \mathbf{AB} \vdots \cdots \vdots \mathbf{A}^{n-1}\mathbf{B}\right]$ is nonsingular since the state model is controllable. Therefore, the controllable state model can be transformed to the form (V.2) by the transformation

$$\mathbf{z} = \mathbf{Px}$$

where

$$\mathbf{P} = \begin{bmatrix} \mathbf{P}_1 \\ \mathbf{P}_1\mathbf{A} \\ \vdots \\ \mathbf{P}_1\mathbf{A}^{n-1} \end{bmatrix}; \; \mathbf{P}_1 = [0\; 0 \;\dots\; 1]\mathbf{Q}_c^{-1} \qquad \text{...(V.8)}$$

The state variable form (V.2) is said to be in the *controlled phase variable form*.

Example V.1 : Consider a linear system described by the differential equation

$$\dot{y} = 2\dot{y} + y = \dot{u} + u$$

With $$x_1 = y,\ x_2 = \dot{y} - u$$

as the state variable, we get the state model

$$\begin{bmatrix} \dot{x}_1 \\ \dot{x}_2 \end{bmatrix} = \begin{bmatrix} 0 & 1 \\ -1 & -2 \end{bmatrix} \begin{bmatrix} x_1 \\ x_2 \end{bmatrix} + \begin{bmatrix} 1 \\ -1 \end{bmatrix} u \qquad \text{...[V.9 }(a)]$$

$$y = x_1 \qquad \text{...[V.9 }(b)]$$

Let us test controllability of this system by Kalman's test.

From eqn. [V.9 (*b*)], we have

$$\mathbf{B} = \begin{bmatrix} 1 \\ -1 \end{bmatrix}; \ \mathbf{A} = \begin{bmatrix} 0 & 1 \\ -1 & -2 \end{bmatrix}$$

Then $$\mathbf{AB} = \begin{bmatrix} 0 & 1 \\ -1 & -2 \end{bmatrix} \begin{bmatrix} 1 \\ -1 \end{bmatrix} = \begin{bmatrix} -1 \\ 1 \end{bmatrix}$$

The composite matrix defined in eqn. (12.81) is given by

$$\mathbf{Q}_c = [\mathbf{B} \ \vdots \ \mathbf{AB}] = \begin{bmatrix} 1 & -1 \\ -1 & 1 \end{bmatrix}$$

The rank r of this matrix is 1. The system is therefore not completely controllable. One state of the system is uncontrollable (r out of n states are controllable).

APPENDIX VI

ANSWERS TO PROBLEMS

(*Marked answers are approximate)

2.1. (*a*) $$\frac{M_1 s^2 + (f_1 + f)s + K_1 + K}{\{M_1 M_2 s^4 + [M_1 f_2 + M_2 f_1 + f(M_1 + M_2)]s^3 + [M_2 K_1 + K(M_1 + M_2) + f_1 f_2 + f(f_1 + f_2)]s^2 + K_1(f_1 + f_2) + K(f_1 + f_2)]s + K_1 K\}}$$

(*b*) $$\frac{\dfrac{K}{J_1 J_2}}{s\left[s^3 + \dfrac{f}{J_2}s^2 + K\dfrac{(J_1 + J_2)}{J_1 J_2}s + \dfrac{fK}{J_1 J_2}\right]}$$

2.2. $$F(t) = M_1 \ddot{y}_1 + f\dot{y}_1 + K_1 y_1 + K_{12}(y_1 - y_2)$$
$$0 = M_2 \ddot{y}_2 + K_{12}(y_2 - y_1)$$

2.3. $$F(t) = M_1 \ddot{x}_1 + f_1 \dot{x}_1 + K_1 x_1 + f_{12}(\dot{x}_1 - \dot{x}_2)$$
$$0 = M_2 \ddot{x}_2 + f_2 \dot{x}_2 + K_2 x_2 + f_{12}(\dot{x}_2 - \dot{x}_1)$$

2.4. $$\frac{k_2}{\{RLCMs^4 + L(M + RCf)s^3 + [RM + Lf + RC(2LK + k_1 k_2)]s^2 + (Rf + 2LK + k_1 k_2)s + 2RK\}}$$

2.5. $$\frac{\theta(s)}{\theta_i(s)} = \frac{1}{RCs + 1};\ \theta(t) = \theta_i\{1 - \exp(-t / RC)\}$$

2.6 $\dfrac{C_o(s)}{X(s)} = \dfrac{KC_i / V}{s + Q_o / V}; C_o(t) = \dfrac{KC_i x_o}{Q_o}\{1 - \exp(-Q_o t / V)\}$

where $K = \Delta Q_i / \Delta x$

2.7. $\dfrac{\omega(s)}{E_i(s)} = \dfrac{50}{s + 10.375}; \omega(t) = 481.8\,(1 - \exp(-10.375t))$

2.8. Closed-loop T.F. = $\dfrac{1}{s(0.1s + 1)(0.2s + 1) + 1}$;

Open-loop T.F. = $\dfrac{1}{s(0.1s + 1)(0.2s + 1)}$

2.9. (*a*) $\dfrac{G_1(G_2G_3 + G_4)}{1 + (G_2G_3 + G_4)(G_1 + H_2) + G_1H_1G_2}$

(*b*) $G_4 + \dfrac{G_1G_2G_3}{1 + G_2G_3H_2 + G_2H_1 + (1 - G_1)}$

2.10. (*i*) $\dfrac{G_1G_2G_3}{1 + G_2H_3 + G_3H_2 + G_1G_2G_3H_1}$

(*ii*) $\dfrac{G_3(1 + H_3G_2)}{1 + H_3G_2 + G_3(G_1H_1G_2 + H_2)}$

2.11. $C_1/R_1 = \dfrac{G_1G_2G_3(1 + G_4)}{(1 + G_1G_2)(1 + G_4) - H_1H_2G_1G_4G_5}$

$C_2/R_1 = \dfrac{G_1G_4G_5G_6H_2}{(1 + G_1G_2)(1 + G_4) - H_1H_2G_1G_4G_5}$

2.12. $\dfrac{K_2}{sRC + 1}$

2.13. $\dfrac{(1 + G_4H_2)(G_2 + G_3)G_1}{1 + G_1H_1H_2(G_2 + G_3)}$

2.14. $[G_2G_4G_6(1 + G_5H_2) + G_3G_5G_7(1 + G_4H_1) + G_3G_8G_6 + G_2G_1G_7$

$$\frac{[-G_3G_8H_1G_1G_7 - G_2G_1H_2G_8G_6]}{1 + G_4H_1 + G_5H_2 - H_1G_1H_2G_8 + G_4H_1G_5H_2}$$

2.15. $C_1 = \{[G_1(1 - G_2H_4) + G_3H_4G_4]R_1 + [G_4(1 - G_3H_2) + G_2H_2G_1]R_2\}/\Delta$

$C_2 = \{[G_3(1 - G_4H_1) + G_1H_1G_2]R_1 + [G_2(1 - G_1H_3) + G_4H_3G_3]R_2\}/\Delta$

where $\Delta = 1 - (G_1H_3 + G_2H_4 + G_3H_2 + G_4H_1 + G_1G_2H_1H_2 + G_3G_4H_3H_4) + G_1H_3G_2H_4 + G_3H_2G_4H_1)$

C_1 is independent of R_2 if $H_2 = \dfrac{-G_4}{G_1G_2 - G_3G_4}$

C_2 is independent of R_1 if $H_1 = \dfrac{-G_3}{G_1G_2 - G_3G_4}$

2.16. $\dfrac{X(s)}{U(s)} = \dfrac{\beta_3(s^2 + \alpha_1 s + \alpha_2) + \beta_2 s + \beta_1}{s^2 + \alpha_1 s + \alpha_2}$

3.1. 0.001, – 1, – 10

3.2. (*b*) 0.1, –15 volts, 53 volts (*c*) 25 volts, –30 volts

3.3. $\left| S_G^T \right| = 0.029, \left| S_H^T \right| = 1.02$

3.4. (*a*) 10% (*b*) 0.2%

3.6. (*a*) $\dfrac{5}{s^2(s+6)+10}$ (*b*) 0.5

3.7. 0.04 sec, $K = 0.1$

3.8. (*a*) $\omega_0(t) = 9.62(1 - e^{-130t})$

(*b*) $\omega_0(t) = 9.62(1 - e^{-5t})$

(*c*) For all ω,

$$\left| S_{KA}^{\omega_0(\text{with } fb)} \right| < \left| S_{KA}^{\omega_0(\text{without } fb)} \right|$$

$$\left| S_{\omega_g}^{\omega_0(\text{with } fb)} \right| < \left| S_{\omega_g}^{\omega_0(\text{without } fb)} \right|$$

3.9. $\dfrac{K}{sRC + 1 + K}$; for all ω, $\left| S_R^T \right| \le \left| S_R^G \right|$; time constant reduced by a factor $(1 + K)$ with feedback.

3.10. (*i*) $\dfrac{KK_cC_i / V}{s + (Q_0 + KK_cC_i)/V}$

(*ii*) Steady-state error is reduced by a factor $(Q_0 + KK_cC_i)/Q_0$ with feedback.

(*iii*) Steady-state error is reduced by a factor $(Q_0 + KK_cC_i^0)/Q_0$ with feedback.

3.12. (*a*) $\dfrac{-K_RK_f}{K_2 + K_fK_R}$

3.15. (*a*) 2×10^{-4} (*b*) – 0.01 rad

3.16. (*i*) $\dfrac{K_1 R}{sRC + 1 + RK_2K_x}, \dfrac{K_1}{sRC + 1 + RK_2K_x}$

(*ii*) Steady-state error reduced by a factor $(1 + RK_2K_x)$ with feedback.

3.17. (*a*) $\dfrac{K_G K_B(s\tau_2 + 1)}{s[(s\tau_1 + 1)(s\tau_2 + 1) - K_B K_C(1 - K_T)] + K_G K_B(s\tau_2 + 1)}$

(*b*) 0

(*c*) $\dfrac{K_B(s\tau_2 + 1)}{(s\tau_1 + 1)(s\tau_2 + 1) - K_B K_C(1 - K_T)}$.

3.18. (*a*) $S_K = 1$, $S_K{}^T = \dfrac{1}{1 + KK_t}$

(*b*) For open-loop case, $\Delta\theta(t) = K_2(1 - e^{-t/\tau'})$

For closed-loop case, $\Delta\theta(t)$

$$= \frac{K_2}{1 + K_1K_t}\left[1 - \exp(-(1 + K_1K_t)t/\tau')\right]$$

where $K_1 = 2RK_s{}^2e_0/(Q_0\rho c + Ah)$, $K_2 = Ah/(Q_0\rho c + Ah)$

$\tau' = Mc/(Q_0\tau c + Ah)$

3.19. Open-loop case : $20(1 - e^{-t})$, $50(1 - e^{-0.4t})$

Closed-loop case : $\dfrac{20}{21}(1 - e^{-21t}), \dfrac{20}{20.4}(1 - e^{-20.4t})$

4.1. $\dfrac{2.73}{s(0.0157s + 1)}$

4.2. $K_A = 286.4$, $v_C = 50$ volts

4.4. $\dfrac{10^3}{s^2 + 4s + 10^3}$; 4×10^{-3} rad

4.5. $\dfrac{X(s)}{Y(s)} = \left(\dfrac{sT_1 + 1}{sT_2 + 1}\right)\dfrac{a}{1 + b}$

where $T_1 = \dfrac{A}{K}\left(\dfrac{a + b}{a}\right)$; $T_2 = \dfrac{A}{K(1 + b)}$;

$K = \dfrac{\text{rate of oil–flow}}{\text{valve opening}}$

4.6. $$\frac{Y_c(s)}{X_r(s)} = \frac{\dfrac{bK_rK_1A}{(a+b)K_2}}{Ms^2 + \left(f + \dfrac{A^2}{K_2}\right)s + K + \dfrac{aAK_1}{(a+b)K_2}}$$

4.7. $\dfrac{2}{s^2 + 0.02s + 2}, \dfrac{1}{0.01s + 1}$

4.9. $$\frac{\dfrac{bK_p}{a+b}(sT+1)}{sT + 1 + \dfrac{aAK_p}{(a+b)K}}$$

5.1. (*a*) $\theta_0(t) = 1 - 1.355e^{-12.5t} \sin(13.7t + 0.828)$, 13.7 rad/sec, 0.229 sec, 5.6%

(*b*) 0.44°

(*c*) 1°

5.2. (*a*) $K_A = 0.6$, $\zeta = 0.215$, $t_s = 1.2$ sec

(*b*) $K_D = 0.022$, no effect on steady-state error

$t_s = 0.516$ sec

5.3. (*b*) $\dfrac{6K_A}{s^2 + 100KK_As + 6K_A}$

(*c*) 2.67 amps/volt, 0.024

5.4. (*a*) 0.0916, 2.87°, (*b*) 0.458

(*c*) Integral-error control reduces the steady-state error to zero.

5.5. 20

5.6. (*a*) 20 rad/sec, (*b*) 0.025 kg-m^2, (*c*) 0.4 newton-m/rad/sec

5.7. (*a*) $\dfrac{600}{(s+10)(s+60)}$. (*b*) 24.5 rad/sec, 1.43

5.8. 100. 0.8 sec, 16.2%, 0.364 sec

5.9. $K_e = 0.116$

With feedback : $t_s = 2.53$ sec, $M_p = 16.2\%$, $e_{ss} = 0.2$ rad

Without feedback : $t_s = 4$ sec, $M_p = 35.2\%$, $e_{ss} = 0.2$ rad

5.10. (*a*) 0.316, 3.16 rad/sec, 0.2 rad

(*b*) 1.8, 0.38 rad

(*c*) ($K_A = 36$, $K_0 = 5.2$) yields the desired result

5.11. 0, 1, ∞, 0.567, 1.323 rad/sec

5.12. $K_p = \infty$, $K_v = 10$, $K_a = 0$, $e_{ss} = \infty$

5.13. $e(t) = 0.1\,(a_1 + a_2t)$

5.14. 0.0042

5.15. $IAE = 2/\omega_n$, ITAE $= 3/\omega_n^2$

5.16. (*i*) $\sqrt{K}$ (*ii*) $K \to \infty$

5.17. $ISE = \zeta + \dfrac{1}{4\zeta}$, ITSE $= \zeta^2 + \dfrac{1}{8\zeta^2}$

Optimal ζ to minimize ISE = 0.5, Min. ISE = 1

Optimal ζ to minimize ITSE = 0.595, Min. ITSE = 0.71

5.18. $K_A = 0.77$, $K = 0.15$, $\alpha = 4.4$

6.1. (*a*) $s_1 = s_2 = -3$, (*b*) $s_1 = -0.683$, $s_2 = -7.317$, $s_3 = -3$

(*c*) $s_1 = -3$, $s_2 = 0.5 + j1.66$, $s_3 = 0.5 - j1.66$

6.2. $a_1 > 0$

$a_1a_2 - a_0a_3 > 0$

$(a_1a_2 - a_0a_3)a_3 - a_1^2a_4 > 0$

$a_4 > 0$

6.3. (*a*) Stable (*b*) Two roots in right half *s*-plane

(*c*) Two roots in right half *s*-plane

(*d*) Four roots on the imaginary axis

6.4. (*a*) $\infty > k > 0.528$ (*b*) $36 > k > 0$

(*c*) Always unstable

6.5. $K = 666.25$; $\omega = 4.06$ rad/sec

6.6. $K = 2$; $a = 0.75$

6.7. (*a*) $70 > K > 0$ (*b*) Yes

6.8. Largest time constant is equal to 1 sec

6.9. Maximum value of $K = 7.5$

6.10. $\infty > K > 0.53$

7.1. (*i*) – 4/3 ; 3 ; 60°, 180°, 300°

(*ii*) – 33.7°

(*iii*) No breakaway point

(*iv*) 52, 3.6 rad/sec

7.2* $3.1 < K < 48$, 2.83, rad/sec, 8.34

$$\frac{8.34}{(s + 0.67 + j1.16)(s + 0.67 - j1.16)(s + 4.56)}$$

7.3. No, $1 < K < \infty$, 1, 1 rad/sec, 3, $-1 \pm j\sqrt{2}$

7.4. $\alpha = 1$, $\dfrac{1}{(s + 0.5 + j0.866)(s + 0.5 - j0.866)}$

7.5. (*a*) With $\alpha = 0$, $e_{ss} = 1$, $\zeta = 0.5$, $t_s = 8$ sec

(*b*) With $\alpha = 0.2$, $e_{ss} = 1.2$, $\zeta = 0.6$, $t_s = 6.67$ sec

(*c*) 1.

7.6*. (*c*) 4, 0.2, 1.76 rad/sec

7.7. (*a*) 64, 3.46 rad/sec

(*b*) 1, 16.2%, 1.815 sec, 4 sec

7.8. $35.7 > K > 23.3$

7.9*. $\pm j1.5$, $-1.67 \pm j1.3$

7.10. 4 cm

7.11*. (*b*) $0 \le a < 4$

(*c*)

(*d*) ($a = 0$, $\zeta = 0.316$), ($a = 1$, $\zeta = 0.238$), ($a = 2$, $\zeta = 0.15$), ($a = 4$, $\zeta = 0$)

7.12*. (*a*) 2 sec, 0.4

(*b*) One solutions is $a = 0.124$, $b = 0.05$

7.13. $S_{\alpha+}^{-r} = 0.125 \angle -143.1°$, $-r = -0.6 + j0.8$

7.14. For dominant root $-r = -0.375 + j1.96$,

(*i*) $S_{K+}^{-r} = 1.21 \angle 96.4°$

(*ii*) $S_{p+}^{-r} = 0.96 \angle -133°$

(*iii*) $S_{z+}^{-r} = 0.6 \angle 22.5°$

8.1. (*a*) Does not cross (*b*) 0.71 rad/sec, 0.67 (*c*) Does not cross

(*d*) Crosses the real axis twice at frequencies 16 rad/sec and 375 rad/sec with corresponding amplitude 8.8×10^{-4} and 5.9×10^{-6} respectively

8.2. 0.683 rad/sec, –159°

8.3. (*a*) 4.1 rad/sec (*b*) 0.75 rad/sec

8.4. (*a*) 0.056 (*b*) 28.5

8.5. (*a*) $\dfrac{0.625(1+0.4s)(1+0.1s)}{s(1+0.04s)}$ (*b*) $\dfrac{250}{S(1+0.4s)(1+0.025s)}$

(*c*) $\dfrac{79.8s^2}{(1+2s)(1+s)(1+0.2s)}$

8.6*. $\dfrac{5(1+0.232s)^2}{s(1+2s)}$, Type 1

8.7. (*a*) 475, 26.2 (*b*) 0.305 sec, 25.1 rad/sec

8.8*. (*a*) 2.6, 6.66 rad/sec (*b*) 0.195, 6.93 rad/sec

(*c*) 10.5 rad/sec

8.9*. 1.11, 13.25 rad/sec, 24.6 rad/sec

9.1. (*a*) Two roots in right half *s*-plane

(*b*) Stable

(*c*) Two roots on $j\omega$-axis

9.2. (*a*) Stable (*b*) Unstable (*c*) Stable

9.3. $\infty > K > 6$

9.4. 0.385, unstable for all values of K

9.5. 0.75

9.6. 0.168 sec

9.7. $11.36 \times 10^2 < K < 16.95 \times 10^4$

9.8. $K = \dfrac{1}{G_m}\left(\dfrac{1}{T_1} + \dfrac{1}{T_2}\right)$

9.9. $\zeta = 0.01\ \phi_{pm}$;

$$K = \frac{1}{D^2}\left[2 \times 10^{-2}\phi_s^2\tau - D \pm 2 \times 10^{-2}\sqrt{(10^{-4}\phi^4\tau^2 - \phi_s^2 D\tau}\right]$$

9.10. $\omega_c = \sqrt{(\omega_1\omega_2)}$; 90°

9.11*. Gain to be changed by a factor of 1/3.5, 42.5°, 0.387. 0.394

9.12*. (*a*) 3.86 db, 10°

(*b*) By a factor of 0.156

(*c*) By a factor of 0.147

9.13*. 1, 2 by a factor of 0.77, 1.07 rad/sec

9.14*. (*a*) 0.5 (*b*) 0.446 (*c*) 0.63, 0.5 rad/sec, 1 rad/sec, (*d*) 1.35

9.15*. 2.35, 2.8 rad/sec, 1.15, 2 rad/sec

9.16*. $\left| S_G^T(j\omega) \right|_{\text{peak}} = 7$ at $\omega = 7$ rad/sec,

$\left| S_H^T(j\omega) \right|_{\text{peak}} = 1$ at $\omega = 0$

$\omega_b = 1.6$ rad/sec

10.1*. Uncompensated system : $K = 7.0$, $K_v = 1.4$, $t_s = 11.6$ sec

Compensated system : $K = 6.0$, $K_v = 12.0$, $t_s = 12.7$ sec

10.2*. Uncompensated system : PM = 2.6°, GM = 1.6 db

Compensated system : PM = 37.6°, GM = 18 db

10.3*. One solution is

$G_c(s) = \dfrac{1.61s + 1}{0.214s + 1}$, Lead-network attenuation cancelled by amplification $A = 7.55$

10.4*. One solution is $G_c(s) = \dfrac{10s+1}{80s+1}$,

$C = 10\mu F, R_2 = 1\ M\Omega, R_1 = 7M\Omega$

10.5*. (*a*) 56 (*b*) 3.77 sec, 2.08

(*c*) One solution is $G_c(s) = \dfrac{(s+3)(s+0.9)}{(s+13.6)(s+0.15)}$

10.6*. One solution is $G_c(s) = \dfrac{(s+1.5)^2}{(s+3.5)^2}$

10.7*. One solution is $G_c(s) = \dfrac{0.38s+1}{0.084s+1}$, Lead network attenuation cancelled by amplification

$A = 4.5$

10.8*. One solution is $G_c(s) = \dfrac{6.67s+1}{66.67s+1}$

ω_c (uncompensated) = 2.6 rad/sec

ω_c (compensated) = 0.64 rad/sec

10.9*. One solution is

(*i*) $G_c(s) = \dfrac{(s+1.3)^3}{(s+4.3)^3}$; $K_v = 0.14$

(*ii*) $G_c(s) = \dfrac{(s+2.2)^3}{(s+14)^3}$

10.10*. One solution is $G_c(s) = \dfrac{0.67s+1}{2s+1}$

$\omega_b = 11$ rad/sec, $t_s = 1$ sec

10.11*. Uncompensated system : PM = 18°, $\omega_b = 4.83$ rad/sec

Compensated system : PM = 28°, $\omega_b = 7.42$ rad/sec

10.12*. One solution is

$K_t = 0.038$, $\omega_r = 8$ rad/sec

10.13*. GM = 10 db, PM = 38°

10.14*. (*a*) $\dfrac{3(1+1.25s)}{s(1+5.95s)(1+0.213s)}$

(*b*) 5

10.15. $G_{ac}(j\omega) = \dfrac{(j\omega)^2 R_1 R_2 C^2 + j2\omega(R_1C)+1}{(j\omega)^2 R_1 R_2 C^2 + j\omega C(2R_1 + R_2)+1}$

$$G_{dc}(j\omega) = \alpha\left(\frac{1 + j\omega T}{1 + j\omega\alpha T}\right),$$

$$\alpha = \frac{2R_1}{2R_1 + R_2}, T = \frac{1}{R_1 C\omega_c^2}$$

$$\omega_c^2 = \frac{1}{R_1 R_2 C^2}$$

11.1. (*i*) $c(k + 1) + (1 - 2e^{-T})c(k) = (1 - e^{-T})r(k)$

(*ii*) $2c(k + 2) - (4 - T^2)c(k + 1) + (2 + T^2)c(k) = T^2 r(k + 1) + T^2 r(k)$

11.2. (*i*) $\frac{z}{z-1}X(z)$ (*ii*) $X(e^a z)$

11.3. (*i*) $\frac{z(z+1)}{(z-1)^3}$ (*ii*) $\frac{z}{(z-a)^2}$ (*iii*) $\frac{z(z+a)}{(z-a)^3}$

(*iv*) $e^{az^{-1}}$ (*v*) $\frac{z \sinh \beta}{z^2 - 2z - \cosh \beta + 1}$ (*vi*) $\frac{z(z - \cosh \beta)}{z^2 - 2z \cosh \beta + 1}$

(*vii*) $\frac{z}{z+a}$

11.4. (*i*) $\frac{T^2 z(z+1)}{(z-1)^3}$ (*ii*) $\frac{z}{z - e^{-aT}}$ (*iii*) $\frac{Tze^{-aT}}{(z - e^{-aT})^2}$

(*iv*) $\frac{ze^{aT} \sin \omega T}{e^{2aT}z^2 - 2ze^{aT} \cos \omega T + 1}$ (*v*) $\frac{e^{aT} z[e^{aT} z - \cos \omega T]}{e^{2aT} z - 2ze^{aT} \cos \omega T + 1}$

11.5. (*i*) $\frac{aTe^{-aT}z}{[z - e^{-aT}]^2}$ (*ii*) $\frac{z(z - \cos \omega T)}{z^2 - 2z \cos \omega T + 1}$

(*iii*) $\frac{ze^{bT} \sin aT}{e^{2bT}z^2 - 2ze^{bT} \cos aT + 1}$ (*iv*) $\frac{z \sinh aT}{z^2 - 2z \cosh aT + 1}$

(*v*) $\frac{e^{bT} z[e^{bT} z - \cos aT]}{e^{2bT}z^2 - 2ze^{bT} \cos aT + 1}$

11.6. (*i*) $(-a)^k$ (*ii*) $(-a)^{k-1} u(k - 1)$ (*iii*) $(a)^{k-1}u(k - 1)$

(*iv*) $3\delta(k) - 6u\ (k - 1) + 17(2)^{k-1}u(k - 1)$

(*v*) $3\delta(k) + 2(-1)^{k-1}u(k - 1) - 9(-2)^{k-1}u(k - 1)$ (*vi*) $k\left(\frac{1}{2}\right)^k$

(*vii*) $-\frac{2}{\sqrt{7}}(-\sqrt{2})^k \sin \theta k + \frac{0.8}{\sqrt{7}}(-\sqrt{2})^{k-1} \sin \theta(k - 1)u(k - 1)$

$$\theta = \tan^{-1} \sqrt{7}$$

(*viii*) $\frac{1}{a}ka^k$ (*ix*) $-3u(k - 1) + 3(2)^{k-1}u(k - 1) - (k - 1)(2)^{k-1}u(k - 1)$

11.7*. $i_n = 4 \cosh \beta n - 4.48 \dfrac{\sinh \beta n}{\sinh \beta}; \beta = 0.9624$

11.8. $\dfrac{C(z)}{R(z)} = H(z) = \dfrac{z-1}{z^2+3z+4}$

$$h(k) = -\frac{2}{\sqrt{7}}(-2)^k \sin \theta k + \frac{2}{\sqrt{7}}(-2)^{k-1} \sin \theta(k-1)u(k-1);$$

$$\theta = \tan^{-1} \frac{\sqrt{7}}{3}$$

11.9. $c(k) = \dfrac{1}{6} - \dfrac{3}{2}(-1)^k + \dfrac{7}{3}(-2)^k$

11.11. $\left(\dfrac{z}{z-0.5}\right)^2$

11.12. Without ZOH : $c(k) = \dfrac{1}{(1-e^{-aT})}[1 - \exp\{-aT(k+1)\}]$

With ZOH : $c(k) = \dfrac{1}{a}[1 - \exp(-aTk)]$

11.13. $\dfrac{G_0G_2(z)RG_1(z)}{1 + G_0G_1G_2H(z)}$

11.15. $\dfrac{G_1(z)G_2(z)}{1 + G_1(z)HG_2(z)}$

11.16. $\dfrac{G_0G(z)}{1 + H(z)G_0G(z)}$

11.17. $c(k) = 0.5\,[1 - (-\,0.264)^k]$

11.18. $c(k) = K[1 - \exp\{-\,aT(k-1)\}]u(k-1)$

11.19. $c(k) = K[1 - \exp(-\,aT\Delta)\exp(-\,aTk)]u(k-1)$

11.20. $c(k+0.5) = [0.631 - 0.356\,(-\,0.264)^{k-1}]u(k-1)$

11.21. $0.5 < K < 3.78$

11.22. (*i*) No (*ii*) Yes (*iii*) No

11.23*. $K < 1.15$

11.24*. $G_c(r) = \dfrac{2.8r+1}{0.617r+1}$

$$D(z) = \frac{3.8z - 1.8}{1.617z + 0.383}$$

11.25. $G_c(s) = \dfrac{0.235s - 0.36}{s - 3.6}$

Not realizable

12.1. $\mathbf{A} = \begin{bmatrix} -1 & 1 \\ 1 & -2 \end{bmatrix}; \mathbf{B} = \begin{bmatrix} 0 \\ 1 \end{bmatrix}$

12.2. (*i*) $\begin{bmatrix} \dot{x}_1 \\ \dot{x}_2 \\ \dot{x}_3 \end{bmatrix} = \begin{bmatrix} 0 & 1 & 0 \\ 0 & -f/J & K_T/J \\ 0 & -K_b/L & -R/L \end{bmatrix} \begin{bmatrix} x_1 \\ x_2 \\ x_3 \end{bmatrix} + \begin{bmatrix} 0 \\ 0 \\ 1/L \end{bmatrix} u$

(*ii*) $\begin{bmatrix} \dot{x}_1 \\ \dot{x}_2 \\ \dot{x}_3 \end{bmatrix} = \begin{bmatrix} 0 & 1 & 0 \\ 0 & 0 & 1 \\ 0 & -\dfrac{1}{JL}(K_bK_T + Rf) & -\dfrac{1}{JL}(JR + fL) \end{bmatrix} \begin{bmatrix} x_1 \\ x_2 \\ x_3 \end{bmatrix} + \begin{bmatrix} 0 \\ 0 \\ \dfrac{K_T}{JL} \end{bmatrix} u$

12.3. For state variables

$$\mathbf{x}^T = [y_1 \quad y_2 \quad \dot{y}_1 \quad \dot{y}_2]$$

the state equation is

$$\begin{bmatrix} \dot{x}_1 \\ \dot{x}_2 \\ \dot{x}_3 \\ \dot{x}_4 \end{bmatrix} = \begin{bmatrix} 0 & 0 & 1 & 0 \\ 0 & 0 & 0 & 1 \\ -K_1/M_1 & K_1/M_1 & -f_1/M_1 & f_1/M_1 \\ K_1/M_2 & -\dfrac{K_1 + K_2}{M_2} & f_1/M_2 & -\dfrac{(f_1 + f_2)}{M_2} \end{bmatrix} \begin{bmatrix} x_1 \\ x_2 \\ x_3 \\ x_4 \end{bmatrix} + \begin{bmatrix} 0 \\ 0 \\ 1/M_1 \\ 0 \end{bmatrix} u$$

12.4. For state variables

$\mathbf{x}^T = [y \quad \dot{y} \quad \ddot{y}]$, the stable model is

$$\begin{bmatrix} \dot{x}_1 \\ \dot{x}_2 \\ \dot{x}_3 \end{bmatrix} = \begin{bmatrix} 0 & 1 & 0 \\ 0 & 0 & 1 \\ -6 & -11 & -6 \end{bmatrix} \begin{bmatrix} x_1 \\ x_2 \\ x_3 \end{bmatrix} + \begin{bmatrix} 0 \\ 0 \\ 1 \end{bmatrix} u$$

$$y = x_1$$

12.5. $\dot{\mathbf{x}} = \mathbf{A}\mathbf{x} + \mathbf{B}u$; $c = \mathbf{C}\mathbf{x}$

(*i*) Phase-variable form

$$\mathbf{A} = \begin{bmatrix} 0 & 1 & 0 \\ 0 & 0 & 1 \\ 0 & -3 & -4 \end{bmatrix}; \mathbf{B} = \begin{bmatrix} 0 \\ 0 \\ 1 \end{bmatrix}; \mathbf{C}^T = \begin{bmatrix} 40 \\ 10 \\ 0 \end{bmatrix}$$

(*ii*) Normal form

$$\mathbf{A} = \begin{bmatrix} 0 & 0 & 0 \\ 0 & -1 & 0 \\ 0 & 0 & -3 \end{bmatrix}; \mathbf{B} = \begin{bmatrix} 40/3 \\ -15 \\ 5/3 \end{bmatrix}; \mathbf{C}^T = \begin{bmatrix} 1 \\ 1 \\ 1 \end{bmatrix}$$

(*iii*) Normal form

$$\mathbf{A} = \begin{bmatrix} 0 & 0 & 0 \\ 0 & -1 & 0 \\ 0 & 0 & -3 \end{bmatrix}; \mathbf{B} = \begin{bmatrix} 1 \\ 1 \\ 1 \end{bmatrix}; \mathbf{C}^T = \begin{bmatrix} 40/3 \\ -15 \\ 5/3 \end{bmatrix}$$

12.6. One solution is

$$\begin{bmatrix} \dot{x}_1 \\ \dot{x}_2 \\ \dot{x}_3 \end{bmatrix} = \begin{bmatrix} 0 & 1 & 0 \\ 0 & 0 & 1 \\ -1 & -3 & -2 \end{bmatrix} \begin{bmatrix} x_1 \\ x_2 \\ x_3 \end{bmatrix} + \begin{bmatrix} 0 \\ 0 \\ 1 \end{bmatrix} u$$

$$y = [3 \quad 3 \quad 1] \begin{bmatrix} x_1 \\ x_2 \\ x_3 \end{bmatrix}$$

12.7. $\begin{bmatrix} e^{\sigma t} \cos \omega t & e^{\sigma t} \sin \omega t \\ -e^{\sigma t} \sin \omega t & e^{\sigma t} \cos \omega t \end{bmatrix}$

12.8. $\begin{bmatrix} 0 & 1 \\ -2 & -3 \end{bmatrix}; \begin{bmatrix} 2e^{-t} - e^{-2t} & e^{-t} - e^{-2t} \\ -2e^{-t} + 2e^{-2t} & -e^{-t} + 2e^{-2t} \end{bmatrix}$

12.9. (*a*) $\mathbf{x}(t) = \begin{bmatrix} e^t \\ te^t \end{bmatrix}$ (*b*) $\mathbf{x}(t) = \begin{bmatrix} e^t \\ e^t (t+1) - 1 \end{bmatrix}$

12.10. $\begin{bmatrix} \dot{z}_1 \\ \dot{z}_2 \\ \dot{z}_3 \end{bmatrix} = \begin{bmatrix} -1 & 0 & 0 \\ 0 & -2 & 0 \\ 0 & 0 & -3 \end{bmatrix} \begin{bmatrix} z_1 \\ z_2 \\ z_3 \end{bmatrix} + \begin{bmatrix} 1 \\ -2 \\ 1 \end{bmatrix} u$

$$y = [1 \quad 1 \quad 1] \begin{bmatrix} z_1 \\ z_2 \\ z_3 \end{bmatrix}$$

$$\mathbf{x}(t) = \begin{bmatrix} \frac{1}{3} - e^{-2t} + \frac{2}{3} e^{-3t} \\ 2(e^{-2t} - e^{-3t}) \\ 2(-2e^{-2t} + 3e^{-3t}) \end{bmatrix}, y = x_1(t)$$

12.11. (*a*) $\begin{bmatrix} \dot{x}_1 \\ \dot{x}_2 \\ \dot{x}_3 \end{bmatrix} = \begin{bmatrix} 0 & 1 & 0 \\ 0 & 0 & 1 \\ -1 & 0 & -3 \end{bmatrix} \begin{bmatrix} x_1 \\ x_2 \\ x_3 \end{bmatrix} + \begin{bmatrix} 0 \\ 0 \\ 1 \end{bmatrix} u$

$$c = x_1$$

(*b*) $\lambda_{1,2} = 0.052 \pm j0.565$, $\lambda_3 = -3.104$

$$\Lambda = \begin{bmatrix} \lambda_1 & 0 & 0 \\ 0 & \lambda_2 & 0 \\ 0 & 0 & \lambda_3 \end{bmatrix}$$

(*c*) Unstable (*d*) $\dfrac{1}{s^3 + 3s^2 + 1}$

12.12. $\mathbf{x}^T(0) = [5k \quad -15k \quad 15k]$

where k is a non-zero multiplier

12.13. (*i*) $\mathbf{P} = \begin{bmatrix} 1 & 1 & 1 \\ \lambda_1 & \lambda_2 & \lambda_3 \\ \lambda_1^2 & \lambda_2^2 & \lambda_3^2 \end{bmatrix}$

(*ii*) $\mathbf{P} = \begin{bmatrix} 1 & 1 & 1 \\ \lambda_1 + a_1 & \lambda_2 + a_1 & \lambda_3 + a_1 \\ \lambda_1^2 + a_1\lambda_1 + a_2 & \lambda_2^2 + a_1\lambda_2 + a_2 & \lambda_3^2 + a_1\lambda_3 + a_2 \end{bmatrix}$

12.16. $\begin{bmatrix} \dot{x}_1 \\ \dot{x}_2 \\ \dot{x}_3 \end{bmatrix} = \begin{bmatrix} 0 & 0 & 1 \\ -2 & -3 & 0 \\ 0 & 2 & -3 \end{bmatrix} \begin{bmatrix} x_1 \\ x_2 \\ x_3 \end{bmatrix} + \begin{bmatrix} 0 \\ 2 \\ 0 \end{bmatrix} u$

$y = x_1$

System is completely controllable and observable

12.17. The system is not completely observable

12.18. One solution is

$K = 44,\ k_2 = 0.13,\ k_3 = 0.03$

12.19. $K_A = 24$, $\mathbf{k}^T = [1 \quad 3/8 \quad 1/8]$, system is stable

12.20. $K_A = 4$V/V, $K_2 = 5 \times 10^{-5}$ V/rad/s, third pole at $s = -9999$

12.22. (*i*) $\mathbf{A} = \begin{bmatrix} 0 & 1 & 0 \\ 0 & 0 & 1 \\ 2 & -5 & 4 \end{bmatrix}$; $\mathbf{B} = \begin{bmatrix} 0 \\ 0 \\ 1 \end{bmatrix}$

$\mathbf{C} = [1 \quad -7 \quad 4]$

(*ii*) $\mathbf{A} = \begin{bmatrix} 1 & 1 & 0 \\ 0 & 1 & 0 \\ 0 & 0 & 2 \end{bmatrix}$; $\mathbf{B} = \begin{bmatrix} 0 \\ 1 \\ 1 \end{bmatrix}$

$\mathbf{C} = [2 \quad 1 \quad 3]$

12.23. (*i*) $\mathbf{A} = \begin{bmatrix} -2 & 0 \\ 0 & -3 \end{bmatrix}$; $\mathbf{B} = \begin{bmatrix} 1 \\ 1 \end{bmatrix}$; $\mathbf{C} = [1 \quad -1]$

(*ii*) $\begin{bmatrix} (-2)^k & 0 \\ 0 & (-3)^k \end{bmatrix}$ (*iii*) $\frac{1}{4}(-3)^k - \frac{1}{3}(-2)^k + \frac{1}{12}$

12.24. $\mathbf{x}(k+1) = \begin{bmatrix} 0.6 & 0.232 \\ -0.465 & -0.097 \end{bmatrix} \mathbf{x}(k) + \begin{bmatrix} 0.2 \\ 0.233 \end{bmatrix} u(k)$

12.26. $T = n\pi$ (n = positive integer)

13.1. Stable $K' = 0.114$

13.2*. Unstable limit cycle, $\omega = 0.25$ rad/sec, $b/X = 1.9$

13.3. $x_1^2 + x_2^2 \le 1$

13.4. The origin is asymptotically stable in-the-large.

13.5. $K > 0$

13.6. $-1 < K < 8$

13.7. The origin is asymptotically stable in-the-large.

13.8. The origin is asymptotically stable in-the-large

13.9. Nonlinear function is restricted to first and third quadrants.

14.1. $\alpha = 2$

14.2. $K_1 = 2, K_2 = 20$

14.3. $K = 4.63$

14.4. (*i*) $E(z) = \dfrac{z^3 - (a+1)z^2 + az}{z^3 + z^2(K - a - 2) + z(1 - K) - a}$

Use Table 13.2 to obtain J

(*ii*) $K > 0, -1 + \dfrac{K}{2} < a < 1$

14.5. $T^*(s) = \dfrac{200}{s^2 + 20s + 200}$

14.6. $D^*(z) = \dfrac{1.32(z - 0.368)}{(z + 3.33)}$

14.7. $u^*(z) = -0.236[x_1(t) + x_2(t)]$

14.8. (*i*) $K = 2$

(*ii*) $J_{min} = 1.5$

(*iii*) $S_k^{opt} = 0.1$

14.10. (*i*) $K = \infty, J_{min} = 0.5$

(*ii*) $K = 1, J_{min} = 1$

14.11. (*i*) $K_1 = K_2 = \infty$

(*ii*) $K_1 = 1, K_2 = \sqrt{3}$

14.12. $u^*(k) = -[0.395 \quad 0.687]\begin{bmatrix} x_1(k) \\ x_2(k) \end{bmatrix}$

14.13. $u^*(0) = -0.177x(0)$

$u^*(1) = -0.177x(1)$

$u^*(2) = -0.177x(2)$

$u^*(3) = -0.164x(3)$

$u^*(k) = -0.177x(k)$ when $N \to \infty$

14.14. $u^*(0) = -0.8, u^*(1) = 0, u^*(2) = 0$

14.15. $x_1(k+1) = x_1(k) + 0.02x_2(k)$

$x_2(k+1) = -x_1(k) + 0.96x_2(k) + 0.02u(k)$

$$J = 10x_1^2(50) + 0.01 \sum_{k=0}^{49} [x_1^2(k) + 2x_2^2(k) + u^2(k)]$$

14.16. $u^*(t) = \left[\dfrac{\exp\{-\sqrt{2}(t-1) - \exp\{\sqrt{2}(t-1)\}}{(\sqrt{2}+1)\exp\{-\sqrt{2}(t-1)\} + (\sqrt{2}-1)\exp\{\sqrt{2}(t-1)\}} \right] x(t)$

14.17. $\dot{p}_{11} = 2p_{12}^2 - 2 \quad ; \quad p_{11}(5) = 1$

$\dot{p}_{12} = -p_{11} + 2p_{12}p_{22} - 1 \quad ; \quad p_{12}(5) = 0$

$\dot{p}_{22} = -2p_{12} + 2p_{22}^2 - 4 \quad ; \quad p_{22}(5) = 2$

14.18. $u^*(t) = -[1 \quad \sqrt{2}] \begin{bmatrix} x_1(t) \\ x_2(t) \end{bmatrix}$

14.19. $K = 1$.

15.1. Stable focus

15.3. Steady-state error = – 0.2 rad

15.4. Limit cycle of amplitude 0.2 and time period 3.56 sec. exists

15.4*. $\ddot{e} = (1 - e^2)\dot{e} + e = 0$, stable focus, unstable limit cycle

15.9. $t_s = 2.6$ sec, steady-state error = 0.25

15.10. $K_N(X) = \begin{cases} k_1 & ; \ X < S \\ k_2 + \dfrac{2}{\pi}(k_1 - k_2)\left\{\sin^{-1}\dfrac{S}{X} + \dfrac{S}{X}\sqrt{\left[1 - \left(\dfrac{S}{X}\right)^2\right]}\right\} & ; \ X > S \end{cases}$

15.11. $K_N(X) = \begin{cases} 0 & ; \ X < D \\ \dfrac{2K}{\pi}\left(\dfrac{\pi}{2} - \alpha + \dfrac{\sin 2\alpha}{2}\right) & ; \ X > D \end{cases}$ where $\alpha = \sin^{-1} D/X$.

15.12. Stable $K' = 0.114$

15.13*. Unstable liit cycle, $\omega = 0.25$ rad/sec, $b/X = 1.9$.

BIBLIOGRAPHY

BIBLIOGRAPHY

1 INTRODUCTION

[1] K.J. Astrom and T. Hagglund, *Automatic Tuning of PID Regulartors*. Instrument Society of America, Reasearch Triangle Park, NC, 1988.

[2] L.E. Bayliss, *Living Control System,* English University Press, London 1966.

[3] C.F. Beards, *Vibrations and Control Systems,* Horwood Publishers, Chichester,1988.

[4] J.H. Blacklock, *Automatic Control of Aircraft and Missiles.* John Wiley & Sons, New York, NY, 1965.

[5] P.L. Bogler, *Radar Principles with Applications to Tracking Systems.* John Wiley & Sons, New York, NY, 1990.

[6] B.K. Bose, *Power Electronics and AC Drives*, Prentice-Hall, Englewood Cliffs, NJ, 1986.

[7] E.D. Bukstein, *Basic Servomechanisms,* Holt Rinehert and Winston, New York, NY, 1963.

[8] R.H. Canon, *Dynamics of Physical Systems,* McGraw-Hill, New York, NY, 1967.

[9] E.R. Carson and T. Deutsch, *A Spectrum of Approaches for Controlling Diabetes,* IEEE Control System, pages 25-30, December 1992.

[10] A.B. Corripio, *Tuning of Industrial Control Systems,* Instrument Society of America, Research Triangle Park, NC, 1990.

[11] D.R. Coughanowr, *Process Systems Analysis and Control,* McGraw-Hill, New York, NY, Second Edition, 1991.

[12] J.J. Craig, *Introduction to Robotics and Control*, Addison-Wesley, Reading, M.A, 1986.

[13] P.B. Deshpande and R.H. Ash. *Computer Process Control.* Instrument Society of America, Research Triangle Park, NC, 1989.

[14] R.C. Dorf, *Robotics and Automated Manufacturing,* Reston Publishing Company, Virginia, 1983.

[15] R.C. Dorf, *The Encyclopedia of Robotics,* John Wiley & Sons, New York, Ny, 1988.

[16] R.C. Dorf and A. Kusiak, *Handbook of Automation and Manufacturing,* John Wiley & Sons, New York, NY, 1994.

[17] B. Etkin, *Dynamics of Flight: Stability and Control,* John Wiley & Sons, New York, NY, 1959.

[18] K.S. Fu, R.C. Gonzalez, and C.S.G.Lee, *Robotics: Control Sensing, Vision and Intelligence,* McGraw-Hill, New York, NY, 1967.

[19] L.A. Gould, *Chemical Process Control: Theory and Applications,* Addison Wesley, Reading M.A. 1969.

[20] R. Haberman, *Mathematical: Mechanical Vibrations, Population Dynamics and Traffic Flow.* Prentice-Hall, Englewood Cliffs, NJ, 1977.

[21] S.S. Hacisalihzade, *Control Engineering and Therapeutic Drug Delivery,* IEEE Control Systems, pages 44-46, June 1989.

[22] D.E. Hardt, Modeling and Control of Manufacturing Processes, *Journal of Dynamics Systems, ASME,* pages 291-300, June 1993.

[23] R. Johansson, *System Modeling and Identification,* Prentice-Hall, Englewood Cliffs, NJ, 1993.

[24] C. Klomp et al., Development of An Autonomous Cow Milking Robot Control System, IEEE Control Systems, pages 11-19, October 1990.

[25] W.A. Lynch and J.G. Truxal, *Signals and Systems in Electrical Engineering,* McGraw-Hill, New York, NY, 1962.

[26] O. Mayar, *The Origins of Feedback Control,* MIT Press, Cambridge, MA, 1970.

[27] D. Mclean, *Automatic Flight Control Systems,* Prentice-Hall International, Hemel Hempstead, 1990.

[28] J.H. Milsum, *Biological Control Systems Analysis,* McGraw-Hill, New York, NY, 1966.

[29] M. Mittal et al., Nonlinear Adaptive Control of a Twin Lift Helicopter System, IEEE Control Systems, pages 39-44, April 1991.

[30] G. Newton, L. Gould and J. Kaiser, *Analytical Design of Linear Feedback Controls,* John Wiley & Sons, New York, NY, 1957.

[31] R.P. Paul, *Robot Manipulators: Mathematics, Programming, and Control*, The MIT Press, Cambridge, MA, 1981.

[32] B. Preising and T.C. Hsia, *Robot in Medicine*, IEEE Engineering in medicine and Biology, pages 13-22, June 1991.

[33] M. Rao and H. Qiu, *Process Control Engineering,* Gordon and Breach Science, Amstredam, 1993.

[34] A.P.Sage, *Methodology for Large-scale Systems,* McGraw-Hill, New York, NY, 1977.

[35] D.E. Seborg, T.F. Edgar, and D.A. Mellichamp, *Process Dynamics and Control,* John Wiley & Sons, New York, NY, 1989.

[36] R. Shoureshi, *Intelligent Control Systems, Journal of Dynamic Systems,* ASME, pages 392-400, June 1993.

[37] C.A. Smith and A.B. Corripio, *Principles and Practice of Automatic Process Control,* John Wiley & Sons, New York, NY, 1985.

[38] W. Spong and M. Vidyasagar, *Robotic Dynamics and Control,* John Wiley & Sons, New York, NY, 1989.

[39] G. Stephanopoulos, *Chemical Process Control-An Introdunction to Theory and Practice*, Prentice-Hall, Englewood Cliffs, NJ, 1984.

[40] A.G. Ulsoy, *Control of Machining Process, Journal of Dynamic Systems,* ASME, pages 301-307, June 1993.

[41] P. Varaiya, *Smart Cars on Smart Roads*, IEEE Transactions on Automatic Control, pages 343-350, February 1993.

[42] J.C. West, *Control limitations*, In Proc. IEE, volume 118, pages 225-238, 1971.

[43] R.B. Wilcox, *Analysis and Synthesis of Dynamic Peformance of Industrial Organizations—The Application of Feedback Control Techniques to Organizational Systems*, IRE Transactions on Automatic Control, AC-7:55-67, 1962.

2 MATHEMATICAL MODELING OF PHYSICAL SYSTEMS

[44] R. Bellman, *Introduction to Matrix Analysis,* McGraw-Hill, New York, NY, 1960.

[45] D.P. Campbell, *Process Dynamics,* John Wiley & Sons, New York, NY, 1958.

[46] R. Cannon, *Dynamics of Physical Systems,* McGraw-Hill, New York, NY, 1967.

[47] C.T. Chen, *Linear Systems Theory and Design,* Holt, Rhinehart and Winston, New York, NY, 1984.

[48] C.T. Chen, *Analog & Digital Control Design: Transfer-function, State-space & Algebraic Methods.* Saunders College Publishing Orlando, FL, 1993.

[49] R.V. Churchill, *Operational Mathematics,* McGraw-Hill, New York, NY, 3rd Edition, 1972.

[50] R.V. Churchill, J.W. Brown, and R.F. Verhev, *Complex Variables and Applications,* McGraw-Hill, New York, NY, 3rd Edition, 1976.

[51] C.M. Close and D.K. Frederick, *Modeling and Analysis of Dynamic Systems*, Houghton Mifflin, Boston, 2nd Edition, 1993.

[52] E.A. Coddington and N. Levinson, *Theory of Ordinary Differential Equations,* McGraw-Hill, New York, NY, 1955.

[53] D.R. Coughanowr, *Process Systems Analysis and Control,* McGraw-Hill, New York, NY, 2nd Edition, 1991.

[54] J.J. Craig, *Introduction to Robotics and Control,* Addision-Wesley, Reading, MA, 1986.

[55] B.W. Dickinson, *Mathematics of Controls, Signals and Systems,* Springer-Verlag, New York, NY, 1993.

[56] E.O. Doebelin, *System Modeling and Response,* John Wiley & Sons, New York, NY, 1980.

[57] C.N. Dorny, *Understanding Dynamic Systems,* Prentice-Hall, Englewood Cliffs, NJ, 1993.

[58] R. Haberman, *Mathematical Models: Mechanical Vibrations, Population Dynamics and Traffic Flow,* Prentice-Hall, Englewood Cliffs, NJ, 1977.

[59] F.B. Hilderband, *Methods of Applied Mathematics*, Prentice-Hall, Englewood Cliffs, NJ, 2nd Edition, 1965.

[60] L.B. Jackson, *Signals, Systems and Transforms,* Addison-Wesley, Reading, MA, 1991.

[61] M. Jamshidi, *Computer Aided Analysis and Design of Linear Control Systems*, Prentice-Hall, Englewood, Cliffs, NJ, 1992.

[62] R. Johansson, *System Modeling and Identification*, Prentice-Hall, Englewood Cliffs, NJ, 1993.

[63] T. Kailath, *Linear Systems*, Prentice-Hall, Englewood Cliffs, NJ, 1980.

[64] E. Kamen, *Introduction to Signals and Systems,* Macmillan, New York, NY, 2nd Edition, 1990.

[65] B.C. Kuo, *Automatic Control Systems,* Prentice-Hall, Englewood Cliffs, NJ, 7th Edition 1995.

[66] P. Lancaster and M. Tismenetsky, *The Theory of Matrices,* Academic Press, Orlando, FL, 3rd Edition, 1985.

[67] S. Lefschetz, *Differential Equations: Geometric Theory,* Wiley Interscience, New York, NY, 1957.

[68] W.L. Luyben, *Precess Modeling Simulation and Control for Chemical Engineers,* Chemical engineering Series, McGraw-Hill, New York, NY, 2nd Edition, 1990.

[69] S.J. Mason, *Feedback Theory: Some Properties of Signal Flow Graphs*, Proc. IRE, 41(9): 1144-1156, September 1953.

[70] S.J. Mason, *Feedback Theory: Further Properties of Signal Flow Graphs*, Proc. IRE, 44(7): 920-926, July 1956.

[71] I.J. Nagrath and M. Gopal, *Systems: Modelling and Analysis*, Tata McGraw-Hill, New Delhi, 1982.

[72] B. Noble and J.W. Daniel, *Applied Linear Algebra,* Prentice-Hall Englewood Cliffs, NJ, 3rd Edition, 1988.

[73] K. Ogata, *Modern Control Engineering,* Prentice-Hall, Englewood Cliffs, NJ, 2nd Edition, 1995.

[74] A.V. Oppenheim, A.S. Willsky, and I.T. Young, *Signals and Systems*, Prentice-Hall, Englewood Cliffs, NJ, 1983.

[75] R.P.Paul, *Robot Manipulators: Mathematics, Programming, and Control*, The MIT Press, Cambridge, MA, 1981.

[76] W.J. Rough, *Linear Systems,* Prentice-Hall, Englewood Cliffs, NJ, 1993.

[77] J.L. Shearer, *Dynamics Modeling and Control of Engineering Systems,* Macmillan, New York. NY, 1990.

[78] M. Vidyasagar, *Nonlinear System Analysis,* Prentice-Hall, Englewood Cliffs, NJ, 1993.

[79] C.R. Jr. Wylie, *Advanced Engineering Mathematics,* McGraw-Hall, New York, NY, 2nd Edition, 1960.

[80] D. Younger, *A Simple Derivation of Mason's Gain Formula*, Proc. IEEE, 51(7): 1043-1044, 1963.

3 FEEDBACK CHARACTERISTICS OF CONTROL SYSTEMS

[81] C.T. Chen, *Analog & Digital Control System Design: Transfer-function, State-space & Algebraic Methods,* College Publishing, Orlando, FL, 1993.

[82] C.M. Close and D.K. Feedback, *Modeling and Analysis of Dynamics Systems*, Houghton Mifflin, Boston, 2nd Edition, 1993.

[83] J.J.D'Azzo and C.H. Houpis, *Linear Control System Analysis and Design*, McGraw-Hill, New York, NY, 1988.

[84] R.C. Dorf and R.H. Bishop, *Modern Control Systems*, Addison-Wesley, Reading, MA, 7th Edition, 1986.

[85] C.N. Dorny, *Understanding Dynamic Systems*, Prentice-Hall, Englewood Cliffs, NJ, 1993.

[86] G.F. Franklin, J.D. Powell, and A. Emami-Naeini, *Feedback Control of Dynamic Systems*, Addison-Wesley, Reading, MA, 1986.

[87] K.J. Kurman, *Feedback Control: Theory and Design*, Elsevier Science Publishers, Amsterdam, 1984.

[88] W.R. Perkins, *Sensitivity Function Methods in Control System Education*, In *Proceedings of IFAC Advances in Control Education,* pages 14-22, June 1991.

[89] C.L. Phillips and R.D. Harbor, *Feedback Control Systems*, Prentice-Hall, Englewood Cliffs, NJ, 1988.

[90] C.E. Rohrs, J.L. Melsa, and D. Schultz, *Linear Control Systems*, McGraw-Hill, New York, NY, 1993.

[91] R.T. Stefani, G.H. Hostetter, C.J. Savant Jr., B. Shahian, *Design of Feedback Control Systems.* Saunders College Publishing, Orlando, Fl, 3rd Edition, 1994.

[92] G.J. Thaler, *Automatic Control Systems*, West Publishing Co., St, Paul, MN, 1989.

[93] J.G. Truxal, *Automatic Feedback Systems Synthesis*, McGraw-Hill, New York, NY, 1955.

[94] J. Van De Vegte, *Feedback Control Systems*, Prentice-Hall, Englewood Cliffs, NJ, 3rd Edition, 1993.

4 CONTROL SYSTEMS AND COMPONENTS

[95] W.R. Ahrendt and C.J. Savant Jr, *Servomechanism Practice,* McGraw-Hill, New York, NY, 2nd Edition, 1960.

[96] W.R. Anderson, *Controlling Electrohydraulic Systems*, Marcel Dekker, New York, NY, 1988.

[97] C.W. De Silva, *Control Sensors and Actuators*, Prentice-Hall, Englewood Cliffs, NJ, 1990.

[98] E.O. Doebelin, *Manufacturing Systems*, McGraw-Hill, New York, NY, 3rd Edition, 1983.

[99] R.R. Kadiyala, *A Toolbox for Approximate Linearization of Nonlinear Systems*, IEEE Control Systems, pages 47-56, April 1993.

[100] T.Kenjo, *Stepping Motors and their Microprocessor Controls,* Clarendon Press, Oxford, 1983.

[101] T. Kenjo and S. Nagamori, *Permanent-magnet and Brushless DC Motors,* Clarendon Press, Oxford, 1984.

[102] B.C. Kuo, *Theory and Applications of Step Motors,* West Publishing Co., St. Pual, MN, 1974.

[103] B.C. Kuo, *Incremental Motion Control, Vol. 2: Step Motors and Control Systems,* SRL Publishing Co., Champaign, IL, 1980.

[104] K. Ogata. *Modern Control Systems,* Prentice-Hall Englewood Cliffs, NJ, 2nd Edition, 1995.

[105] S.P. Parker, *Encyclopedia of Engineering,* McGraw-Hill, New York. NY, 2nd Edition, 1993.

[106] E.A. Parr, *Industrial Control Handbook,* volume 1-3. BSP Professional Books, Oxford, 1987.

[107] Y.M. Parr, *Electromagnetic Devices for Motion Control,* Springer-Verlag, New York, NY, 1992.

[108] F.H. Raven, *Automatic Control Engineering*, McGraw-Hill, New York, NY, 5th Edition, 1995.

[109] C.A. Schuler and W.L. McName. *Industrial Electronics and Robotics*, McGraw-Hill, New York, NY, 1986.

[110] M.G. Singh, J.P. Elloy, R. Mezencev, and N. Munro, *Applied Industrial Control,* Pergamon Press, Oxford, 1980.

[111] R.J. Smith, *Electronics: Circuits and Devices.* John Wiley & Sons, New York, NY, 2nd Edition, 1980.

5 TIME RESPONSE ANALYSIS, DESIGN SPECIFICATIONS AND PERFORMANCE INDICES

[112] W.L. Brogan, *Modern Control Systems*, Prentice-Hall Englewood Cliffs, NJ, 2nd Edition, 1985.

[113] C.T. Chen, *Analog & Digital Control System Design: Transfer-function, State-space & Algebraic Methods*, Saunders College Publishing, Orlando, FL, 1993.

[114] R.N. Clark, *Introduction to Automatic Control Systems*, John Wiley & Sons, New York, NY, 1962.

[115] C.M. Close and D.K. Frederick, *Modeling and Analysis of Dynamic Systems*, Houghton Mifflin, Boston, 2nd Edition, 1993.

[116] J.J. Craig, *Introduction to Robotics, Mechanics and Control,* Addison-Wesley, Reading, MA, 1986.

[117] E.J. Davidson, Method for simplifying linear dynamic systems, IEEE Transactions on Automatic Control, Pages 93-101, January 1966.

[118] R.C. Dorf and R.H. Bishop, *Modern Control Systems*, Addison-Wesley. Reading MA, 7th Edition, 1986.

[119] C.N. Dorny, *Understanding Dynamic Systems*, Prentice-Hall Englewood Cliffs, NJ, 1993.

[120] G.F. Franklin, J.D. Powell, and A. Emami-Naeini, *Feedback Control of Dynamic Systems*, Addison-Wesley, Reading, MA, 1986.

[121] D. Graham and R.C. Lathrop, *The Synthesis of Optimal Response: Criteria and Standard Forms*, part 2. Transactions of the AIEE, 72:273-288, November 1953.

[122] T.C. Hsia, *On The Simplification of Linear Systems*, IEEE Transactions on Automatic Control, pages 372-374, June 1972.

[123] B.K. Kuo, *Automatic Control Systems*, Prentice-Hall, Englewood Cliffs, NJ, 7th Edition, 1995.

[124] K. Ogata, *Modern Control Systems*, Prentice-Hall, Englewood Cliffs, NJ, 2nd Edition, 1995.

[125] R.P. Pual, *Robot Manipulators: Mathematics, Programming, and Control*, The MIT Press, Combridge, MA, 1981.

[126] Z.V. Rekasius, *A General Performance Index for Analytical Design of Control Systems*, IRE Transactions on Automatic Control, AC-6: 217-222, 1961.

[127] C.E. Rohrs, J.L. Melsa, and D. Schultz, *Linear Control Systems*, McGraw-Hill, New York, NY, 1993.

[128] W.J. Rugh, *Linear System Theory*, Prentice-Hall Englewood Cliffs, NJ, 1993.

[129] W.C. Schultz and V.C. Rideout, *Control System Performance Measures: Past, Present, and Feature*, IRE Transactions on Automatic Control, AC-6: 22-35, 1961.

[130] S.M. Shinners, *Modern Control System Theory and Design*, John Wiley & Sons, New York, NY, 1992.

[131] G.J. Thaler, *Automatic Control Systems*, West Publishing Co., St, Pual, MN, 1989.

[132] W.A. Wolovich, *Automatic Control System: Basic Analysis and Design*, Saunders College Publishing, Orlando, FL, 1994.

6 CONCEPTS OF STABILITY AND ALGEBRIC CRITERIA

[133] T.S. Change and C.T. Chen, *On The Routh-hurwitz Criterion*, IEEE Transactions on Automatic Control, AC-19: 250-251, March 1974.

[134] R.N. Clark, *The Ourth-Hurwitz Stability Criterion Revisited*, IEEE Control Systems, pages 119-120, June 1992.

[135] M.L. Cohen, *A Set of Stability Constraints on the Denominator of A Sampled Data Filter*, IEEE Transactions on Automatic Control, AC-11: 327-328, April 1966.

[136] C.N. Dorny, *Understanding Dynamics Systems*, Prentice-Hall Englewood Cliffs, NJ, 1993.

[137] G.H. Hotsteller, *Additional Comments On 'On The Routh-Hurwitz Criterion'*, IEEE Transactions on Automatic Control, AC-20: 296-297, February 1975.

[138] A. Hurwitz, *On The Conditions Under Which An Equation has Only Roots With Negative Real Parts.* In *Selected Papers on Mathematical Trends in Control Theory,* pages 70-82. Dover, New York, 1964.

[139] E.I. Jury, *The Number of Roots of A Real Polynomial Inside (or Outside) The Unit Circle Using The Determinant Method*, IEEE Transactions on Automatic Control, AC-10: 1371-372, July 1965.

[140] E.I. Jury and B.D.O. Anderson, *A Simplified Schur-Cohen Test*, IEEE Transactions on Automatic Control, AC-18: 157-163, April 1973.

[141] E.I. Jury and J. Blanchard, *A Stability Test for Linear Discrete Systems in Table Form.* IRE Proceedings, 49(12): 1947-1948, December 1973.

[142] K.J. Khatwani, *On The Routh-Hurwitz Criterion*, IEEE Transactions on Automatic Control, AC-26: 583, 1981.

[143] S.K. Pillai, *The Method of The Routh-Hurwitz Criterion*, IEEE Transactions on Automatic Control, AC-26: 584, April 1981.

[144] R.H. Rible, *A Simplification of Jury's Tabular Form*, IEEE Transactions on Automatic Control, AC-19: 248-250, April 1974.

[145] M.V.C. Rao and A.K. Subramanium, *Elimination of Singular Cases in Jury's Test*, IEEE Transactions on Automatic Control, AC-21: 114-115, February 1976.

[146] C.R. Rohrs, J.L. Melsa, and D. Schultz, *Linear Control Systems,* McGraw Hill, New York, NY, 1993.

[147] E.J. Routh, *Dynamics of a System of Rigid Bodies,* Macmillan, New York NY, 1892.

[148] W.J. Rugh, *Linear System Theory*, Prentice-Hall Englewood Cliffs, NJ, 1993.

[149] S.G. Tzafestas, *Applied Control*, Marcel Dekker, New York, NY, 1993.

7 THE ROOT LOCUS TECHNIQUE

[150] G.A. Bendrikov and K.F. Teodorchick, *The Analytic Theory of Constructing Root Loci*, *Automatic and Remote Control,* pages 340-344, March 1959.

[151] C.S. Chang, *An Analytical Method for Obtaining The Root Locus with Positive and Negative Gain*, IEEE Transactions on Automatic Control, AC-10(1): 92-94, January 1965.

[152] C.F. Chen, *A New Rule for Finding Breaking Points of Root Loci Involving Complex Roots*, IEEE Transactions on Automatic Control, AC-10(7): 373-374, July 1965.

[153] C.T. Chen, *Analog & Digital Control System Design: Transfer-function, State-space & Algebraic Methods,* Saunders College Publishing, Orlando, FL, 1993.

[154] R.C. Dorf and R.H. Bishop, *Modern Control Systems*, Addison-Wesley, Reading, MA, 7th Edition, 1986.

[155] W.R. Evans, *Graphical Analysis of Control Systems*, AIEE Transactions, Part II, 67: 547-551, 1948.

[156] W.R. Evans, Control Systems Synthesis by Root Locus Method, AIEE Transactions, Part II, 69: 66-69, 1950.

[157] W.R. Evans, *Control Systems Dynamics,* McGraw-Hill, New York, NY, 1954.

[158] G.F. Franklin, J.D. Powell, and A. Emami-Naeini, *Feedback Control of Dynamic Systems,* Addison-Wesley, Reading, MA, 1986.

[159] A. Freegosi and Jo Frinstein, *Some Exclusive Properties of The Negative Root Locus*, IEEE Transactions on Automatic Control, AC-14(6): 304-305, June 1969.

[160] J.G. goldberg, *Automatic Controls,* Allyn and Bacon, Boston, 1965.

[161] V. Krishnan, *Semi-analytic Approach to Root Locus*, IEEE Transactions on Automatic Control, AC-11(1): 102-108, January 1966.

[162] B.K. Kuo, *Automatic Control Systems,* Prentice-Hall, Englewood Cliffs, NJ, 7th Edition, 1995.

[163] T.R. Kurfess and M.L. Nagurka, *Understanding The Root Locus Using The Gain Plots*, IRE Control Systems, pages 37-40, August 1991.

[164] C.S. Lorens and R.C. Titsworth, *Properties of Root Locus Asymptotes*, IRE Transactions on Automatic Control, AC-5(1): 71-72, January 1960.

[165] C.C. Macduff, *Theory of Equations,* John Wiley & Sons, New York, NY, 1954.

[166] K. Ogata, *Modern Control Engineering,* Prentice-Hall, Englewood Cliffs, NJ, 2nd Edition, 1995.

[167] S.M. Shinners, *Modern Control Systems Theory and Design,* John Wiley & Sons, New York, NY, 1992.

[168] C.A. Stapelton, *On Root Locus Breakaway Points*, IRE Transactions on Automatic Control, AC-7(4): 88-89, April 1962.

[169] R.T. Stefani, G.H. Hostetter, C.J. Savant Jr., and B. Shahian, *Design of Feedback Control Systems,* Saunders College Publishing, Orlando, FL, 3rd Edition, 1994.

[170] K. Steiglitz, *Analytical Approach to Root Loci*, IRE Transactions on Automatic Control, AC-6(9): 326-332, September 1961.

[171] G.J. Thaler, *Automatic Control Systems,* West Publishing Co., St. Pual, MN, 1989.

[172] J.G. Truxal, *Automatic Feedback Systems Synthesis,* McGraw-Hill, New York, NY, 1955.

[173] H. Ur, *Root Locus Properties and Sensitivity Relations in Control Systems*, IRE Transactions on Automatic Control, AC-5(1): 57-65, January 1960.

[174] C.K. Wojcik, *Analytical Representation of The Root Locus*, ASME Journal on Basic Engineering, 86: 37-43, 1964.

8 FREQUENCY RESPONSE ANALYSIS

[175] C.M. Close and D.K. Frederick, *Modeling and Analysis of Dynamic Systems,* Houghton Mifflin, Boston, 2nd Edition, 1993.

[176] J.J. D'Azzo and C.H. Houpis, *Linear Control System Analysis and Design*, McGraw-Hill, New York, NY, 1988.

[177] W.R. Evans, *The Use of Zeros and Poles for Frequency Response or Transient Response*, ASME Transactions, 76: 1335-1344, 1954.

[178] R.J. Kochenburger, *A Frequency Response Method for Analyzing and Synthesizing Contractor Servomechanisms*, AIEE Transactions, 69: 270-283, 1950.

[179] G.J. Thaler, *Automatic Control Systems,* West Publishing Co., St, Paul, MN, 1989.

9 STABILITY IN FREQUENCY DOMAIN

[180] R.W. Brockett and J.L. Willems, *Frequency Domain Stability Criteria*, part I, IEEE Transactions on Automatic Control, AC-10(7): 255-261, July 1965.

[181] R.W. Brockett and J.L. Willems, *Frequency Domain Stability Criteria*, part II, IEEE Transactions on Automatic Control, AC-10(10): 407-413, October, 1965.

[182] C.T. Chen, *Analog & Digital Control System Design: Transfer-function, State-space & Algebraic Methods,* Saunders College Publishing, Orlando, FL, 1993.

[183] J.J. D'Azzo and C.H. Houpis, *Linear Control System Analysis and Design,* McGraw-Hill, New York, NY, 1988.

[184] R.C. Dorf and R.H. Bishop, *Modern Control Systems,* Addison-Wesley, Reading, MA, 7th Edition, 1986.

[185] G.F. Franklin, J.D. Powell, and A. Emami-Naeini, *Feedback Control of Dynamic Systems,* Addison-Wesley, Reading, MA, 1986.

[186] Aa. Gelb, *Graphical Evaluation of The Sensitivity Function Using The Nichols Chart*, IRE Transactions on Automatic Control, AC-7(7): 57-58, July 1962.

[187] B.K. Kuo, *Automatic Control Systems,* Prentice-Hall Englewood Cliffs, NJ, 7th Edition, 1995.

[188] D.A. Linkens, *CAD for Controls Systems,* Marcel Dekker, New York, NY, 1993.

[189] T.R Natesan, *A Supplement to The Note On The Generalized Nyquist Criterion*, IEEE Transactions on Automatic Control, AC-12(4): 215-216, April, 1967.

[190] K. Ogata, *Modern Control Engineering,* Prentice-Hall Englewood Cliffs, NJ, 2nd Edition 1995.

[191] C.E. Rohrs, J.L. Melsa, and D. Schultz, *Linear Control Systems,* McGraw-Hill, New York, NY, 1993.

[192] W.J. Rough, *Linear Systems,* Prentice-Hall Englewood Cliffs, NJ, 1993.

[193] S.M. Shinners, *Modern Control Systems Theory and Design,* John Wiley & Sons, New York, NY, 1992.

[194] G.J. Thaler, *Automatic Control Systems,* West Publishing Co., St, Pual, MN, 1989.

[195] K.S. Yeung, *A Reformulation of Nyquist's Criterion*, IEEE Transactions on Education, E-28(2): 58-60, February 1985.

10 INTRODUCTION TO DESIGN

[196] K.J. Astrom and B. Wittenmark, *Computer Controlled Systems: Theory and Design,* Prentice-Hall Englewood Cliffs, NJ, 1984.

[197] B. Barnish, *The Robust Root Locus*, Automatica, 26(2): 283-292, 1990.

[198] W.L. Brogan, *Modern Control Systems,* Prentice-Hall, Englewood Cliffs, NJ, 2nd Edition, 1985.

[199] C.T. Chen, *Linear Systems Theory and Design,* Holt, Rhinehart and Winston, New York, NY, 1984.

[200] C.T. Chen, *Analog & Digital Control System Design: Transfer-function, State-space & Algebraic Methods*, Saunders College Publishing, Orlando, FL, 1993.

[201] R.N. Clark, *Introduction to Automatic Control Systems,* John Wiley & Sons, New York, NY, 1962.

[202] D.R. Coughanowr, *Process Systems Analysis and Control,* McGraw-Hill, New York, Second Edition, 1991.

[203] J.J. D'Azzo and C.H. Houpis, *Linear Control System Analysis and Design,* McGraw-Hill, New York, NY, 1988.

[204] E.O. Doebelin, *Control System Principles and Design,* John Wiley & Sons, New York, NY, 1985.

[205] P. Dorato, *Robust Control,* IEEE Press, New York, NY, 1987.

[206] R.C. Dorf and R.H. Bishop, *Modern Control Systems,* Addison-Wesley, Reading, MA, 7th Edition, 1986.

[207] J. C. Doyle, B.A. Francis and A.R. Tannenbaum, *Feedback Control Theory,* Macmillan, 1992.

[208] A.F. D'Souza, *Design of Control Systems,* Prentice-Hall, Englewood Cliffs, NJ, 1988.

[209] W.R. Evans, *Control Systems Synthesis by Root Locus Method*, AIEE Transactions, Part II, 69: 66-69, 1950.

[210] G.F. Franklin, J.D. Powell, and A Emami-Naeini, *Feedback Control of Dynamics Systems,* Addition-Wesley, Reading, MA, 1986.

[211] W.J. Grantham and T.L. Vincent, *Modern Control System Analysis and Design,* John Wiley & Sons, New York, 1993.

[212] M. Green and D.J.N. Limebeer, *Linear Robust Control,* Information and System Sciences Series. Prentice-Hall, Englewood Cliffs, NJ, 1995.

[213] B.C. Kuo, *Automatic Control Systems,* Prentice-Hall. Englewood Cliffs, NJ, 7th Edition, 1995.

[214] K.J. Kurman, *Feedback Control: Theory and Design,* Elsevier Science Publishers, Amsterdam, 1984.

[215] W.L. Luyben, *Process Modeling Sinulation and Control for Chemical Engineers,* Chemical engineering Series, McGraw Hill, New York, second Edition, 1990.

[216] M. Morari, *Robust Process Control,* Prentice-Hall, Englewood Cliffs, NJ, 1991.

[217] K. Ogata, *Modern Control Engineering,* Prentice-Hall, Englewood Cliffs, NJ, 2nd Edition, 1995.

[218] W.J. III Palm, *Control Engineering,* John Wiley & Sons, New York, NY, 1986.

[219] R.M. Phelan, *Automatic Control Systems,* Cornell University Press, London, 7th Edition, 1977.

[220] C.L. Phillips and R.D. Harbor, *Feedback Control Systems,* Prentice-Hall, Englewood Cliffs, NJ, 1988.

[221] F.H. Raven, *Automatic Control Engineering,* McGraw-Hill, New York, NY, 5th Edition, 1995.

[222] C.E. Rohrs. J.L. Melsa, and D. Schultz, *Linear Control Systems,* McGraw-Hill, New York, NY, 1993.

[223] S.M. Shinners, *Modern Control Systems Theory and Design,* John Wiley & Sons, New York, NY, 1992.

[224] M.G. Singh, J.P. Elloy, R. Mezenev, and N. Munro, *Applied Industrial Control,* Pergamon Press, Oxford, 1980.

[225] H.W. Smith and E.J. Davison, *Design of Industrial Regulators*, In Proceedings of IEe, Volume 119, pages 1210-1216, London, August 1972.

[226] R.T. Stefani, G.H. Hostetter, C.J. Savant Jr., and B. Shahian, *Design of Feedback Control Systems,* Saunders College Publishing, Orlando, FL,, 3rd Edition, 1994.

[227] G.J. Thaler, *Automatic Control Systems,* West Publishing Co., St, Paul, MN, 1989.

[228] J.G. Truxal, *Automatic Feedback System Synthesis,* McGraw-Hill, New York, Ny, 1955.

[229] S.G. Tzafestas, *Applied Control,* Marcel Dekker, New York, NY, 1993.

[230] J. Van De Vegte, *Feedback Control Systems,* Prentice-Hall, Englewood Cliffs, NJ, 3rd Edition, 1993.

[231] J.C. Willems and S.K. Miller, *Controllability, Obervability, Pole Allocation and State Reconstruction*, IEEE Transactions on Automatic Control, AC-16(12): 582-595, December 1971.

[232] W.A. Wolovich, *Automatic Control Systems: Basic Analysis and Design,* Saunders College Publishing, Orlando, FL, 1994.

[233] J.G. Ziegler and N.B. Nichols, *Optimum Settings for Automatic Controllers*, ASME Transactions, 64: 759-768, 1942.

[234] J.G. Ziegler and N.B. Nichols, *Process Lags in Automatic Control Circuits*, ASME Transactions, 65: 433-444, 1943.

11 SAMPLED-DATA CONTROL SYSTEMS

[235] J. Ackermann, *Sampled-Data Control Systems,* Springer Verlag, Berlin 2nd Edition, 1985.

[236] K.J. Astrom and B. Wittenmark, *Computer Controlled Systems: Theory and Design,* Prentice-Hall, Englewood Cliffs, NJ, 2nd Edition, 1990.

[237] C.T. Chen, *Analog & Digital Control System Design: Transfer-function, State-space & Algebric Methods,* Saunders College Publishing, Orlando, FL, 1993.

[238] Y. Dote, *Servo Motor and Motion Control Using Digital Signal Processors,* Prentice-Hall, Englewood Cliffs, NJ, 1990.

[239] G.F. Franklin, J.D. Powell, and M.L. Workman, *Digital Control of Dynamic Systems,* Addison-Wesley, Reading MA, 2nd Edition, 1990.

[340] R. Gayakwad and L. Sokoff, *Analog and Digital Control Systems,* Prentice-Hall, Englewood Cliffs, NJ, 1988.

[341] W.J. Grantham and T.L. Vincent, *Modern Control Systems Analysis and Design,* John Wiley & Sons, New York, 1993.

[342] C.H. Houpis and G.B. lamont, *Digital Control System Theory, Hardware, Software,* McGraw-Hill, New York, 1987.

[243] R. Isermann, *Digital Control Systems,* Volume 1-2. Springer Verlag, Berlin, 2nd Edition, 1989.

[244] R.G. Jacquot, *Modern Digital Control Systems,* Marcel Dekker, New York, 1994.

[245] Warwick K. and D. Reeds (eds.), *Industrial Digital Control Systems,* Peter Peregrinus, London, revised Edition, 1988.

[246] P. Katz, *Digital Control Using Microprocessors,* Prentice-Hall, Englewood Cliffs, NJ, 1981.

[247] R. Kuc, *Introduction to Digital Signal Processing,* McGraw-Hill, New York, 1988.

[248] B.C. Kuo, *Digital Control Systems,* Saunders College Publishing, Orlando, FL, 2nd Edition, 1992.

[249] J.R. Leigh, *Applied Digital Control,* Prentice-Hall, Englewood Cliffs, NJ, 1985.

[250] P. Moroney, *Issues in the Implementation of Digital Feedback Compensators,* The MIT Press, Cambridge, 1983.

[251] K. Ogata, *Discrete-Time Control Systems,* Prentice-Hall, Eglewood Cliffs, NJ, 1987.

[252] C.L. Phillips and H.T. Nagle Jr., *Digital Control System Analysis and Design,* Prentice-Hall, Elglewood Cliffs, NJ, 2nd Edition, 1990.

[253] G.V. Rao, *Complex Digital Control Systems,* Van Nostrand Reinhold Co., New York, 1979.

[254] C.E. Rohrs, J.L. Melsa and D. Schultz, *Linear Control Systems,* McGraw-Hill, New York, NY, 1993.

[255] M.S. Satina, A.R. Stubberud, and G.H. Hostetter, *Digital Control System Design,* Saunders College Publishing, Orlando, FL, 2nd Edition, 1994.

[256] N.K. (ed.) Sinha, *Microprocessor-Based Control Systems,* D. Reidel Publishing Co., Dordrecht, 1986.

[257] H.F. Van Landingham, *Introduction to Digital Control Systems,* Macmillan, New York, 1985.

[258] A.M. Zikic, *Practical Digital Control,* Ellis Horwood Publishers, Chichester, 1989.

12 STATE VARIABLE ANALYSIS AND DESIGN

[259] R. Bellman, *Introduction to Matrix Analysis,* McGraw-Hill, New York, NY, 1960.

[260] W.L. Brogan, *Modern Control Systems,* Prentice-Hall, Englewood Cliffs, NJ, 2nd Edition, 1985.

[261] F.M. Callier and C.A. Desoer, *Linear System Theory,* Springer-Verlag, New York, NY, 1991.

[262] C.T. Chen, *Linear Systems Theory and Design,* Holt, Rhinehart and Winston, New York, NY, 1984.

[263] C.T. Chen, *Analog & Digital Control System Design: Transfer-function, State-space & Algebraic Methods,* Saunders College Publishing, Orlando, FL, 1993.

[264] C.M. Close and D.K. Frederick, *Modeling and Analysis of Dynamics Systems,* Houghton Mifflin, Boston, 2nd Edition, 1993.

[265] R.A. Decarlo, *Linear Systems: A State Variable Approach with Numerical Implementation,* Prentice-Hall, Englewood Cliffs, NJ, 1989.

[266] P.M. deRusso, R.J. Roy and C.M. Close, *State Variables for Engineers,* John Wiley & Sons, New York, 1965.

[267] B.W. Dickinson, *Mathematics of Control, Signals and Systems*, Springer-Verlag, New York, NY, 1993.

[268] R.C. Dorf, *Time Domain Analysis and Design of Control Systems,* Addison Wesley, Reading, MA, 7th Edition, 1965.

[269] T.E. Fortman and K.L. Hitz, *An Introduction to Linear Control System,* Marcel Dekker, New York, 1977.

[270] B.Friedland, *Control System Design: An Introduction to State-Space Methods,* McGraw-Hill, New York, 1986.

[271] K. Furuta, A. Sano, and D. Atherton, *State Variable Methods in Automatic Control,* John Wiley & Sons, Chichester, 1988.

[272] E. Kreindler and P.E. Sarachick, *On The Concepts of Controllability and Observability of Linear Systems*, IEEE Transactions on Automatic Control, AC-9: 129-136, 1964.

[273] K. Ogata, *State Space Analysis of Control Systems,* Prentice-Hall, Englewood Cliffs, NJ, 1967.

[274] C.E. Rohrs, J.L. Melsa, and D. Schultz, *Linear Control Systems,* McGraw-Hill, New York, NY, 1993.

[275] W.J. Rugh, *Linear Systems, Theory,* Prentice-Hall, Englewood Cliffs, NJ, 1993.

[276] D.G. Schultz and J.L. Melsa, *State Functions and Linear Control Systems,* McGraw-Hill, New York, NY, 1967.

[277] D.M. Wiberg, *State Space and Linear Systems,* Schaum's Outline Series, McGraw-Hill, New York, NY, 1971.

[278] J.C. Willems and S.K. Mitter, *Controllability, Obervability, Pole Allocation and State Reconstruction*, IEEE Transactions on Automatic Control, AC-16(12): 582-595, December 1971.

[279] L.A. Zadeh and C.A. Desoer, *Linear System Theory: A State Space Approach,* McGraw-Hill, New York, 1963.

13 LIAPUNOV'S STABILITY ANALYSIS

[280] D.P. Atherton, *Nonlinear Control Engineering,* Van Nostrand Reinhold, London, 1975.

[281] D.P. Atherton, *Stability of Nonlinear Systems,* John Wiley & Sons, New York, NY, 1981.

[282] J.J. Craig, *Introduction to Robotics, Mechanics and Control,* Addison-Wesley, Reading, MA, 1986.

[283] W.J. Cunningham, *Introduction to Nonlinear Analysis,* McGraw-Hill, New York, NY, 1958.

[284] C.A. Desoer and M. Vidyasagar, *Feedback Systems: Input Output Properties,* Academic Press, New York, 1975.

[285] J.E. Gibson, *Nonlinear, Automatic Control,* Mcgraw-Hill, New York, NY, 1963.

[286] W. Hahn, *Theory and Application of Lyapunov's Direct Method,* Prentice-Hall, Englewood Cliffs, NJ, 1963.

[287] U. Itkis, *Control Systems of Variable Structure,* John Wiley & Sons, New York, NY, 1976.

[288] C.F. Lin, *Advanced Control Systems Design,* Prentice-Hall, Englewood Cliffs, NJ, 1964.

[289] N. Minorsky, *Theory of Nonlinear Control Systems,* McGraw-Hill, New York, NY, 1969.

[290] M. Mittal et al., *Nonlinear Adaptive Control of a Twin Lift Helicopter System*, IEEE Control Systems, pages 39-44, April 1991.

[291] R.R. Mohler, *Nonlinear Systems: Dynamics and Control,* Volume 1. Prentice-Hall, Englewood Cliffs, NJ, 1994.

[292] J.J.E. Slotine and W.Li, *Applied Nonlinear Control, Prentice-Hall,* Englewood Cliffs, NJ, 1991.

[293] W. Spong and M. Vidyasagar, *Robotic Dynamics and Control,* John Wiley & Sons, New York, NY, 1989.

[294] M. Vidyasagar, *Nonlinear System Analysis,* Prentice-Hall, Englewood Cliffs, NJ, 1993.

14 OPTIMAL CONTROL SYSTEM

[295] Chang, S.S.L., *Synthesis of Optimal Control Systems,* McGraw-Hill, New York, 1961.

[296] Newton, G.C., Jr., L.A. Gould and J.F. Kaiser, *Analytical Design of Linear Feedback Controls,* John Wiley, New York, 1957.

[297] Jury, E.I., and A.G. Dewey, "*A General Formulation of the Total Square Integrals for Continuous Systems*", IEEE Trans. on Aut. Cont., Jan. 1965, AC-10, pp. 119-120.

[298] Jury, E.I., "*A Note on the Evaluation of the Total Square Integral*", IEEE Trans. on Aut. Cont., Jan. 1965, AC-10, pp. 110-111.

[299] Gupta, S.C. and L. Hasdorff, *Fundamentals of Automatic Control,* John Wiley, New York, 1970.

[300] Kirk, D.E., *Optimal Control Theory : An Introduction,* Prentice-Hall, Englewood Cliffs, N.J., 1970.

[301] Wiberg, D.M., *State Space and Linear Systems,* Schaum's Outline Series, McGraw-Hill, New York, 1971.

[302] Cadzow, J.A. and H.R. Martens, *Discrete-time and Computer Control System,* Prentice-Hall, Englewood Cliffs, N.J., 1970.

[303] Schultz, D.G. and J.L. Melsa, *State Functions and Linear Control Systems,* McGraw-Hill, New York, 1967.

[304] Kuo, B.C., *Automatic Control Systems,* 3rd ed., Prentice-Hall, Englewood Cliffs, N.J., 1975.

[305] Athans, M. and P.L. Falb, *Optimal Control: An Introduction to the Thoery and its Applications,* McGraw-Hill, New York, 1966.

[306] Hsu, J.C. and A.U. Meyer, *Modern Control Principles and Applications,* McGraw-Hill, New York, 1968.

[307] Anderson, B.D.O. and J.B. Moore, *Linear Optimal Control,* Prentice-Hall, Englewood Cliffs, N.J., 1971.

[308] Kuo, B.C., *Discrete-Data Control Systems,* Science Tech., Champaign, Illinois, 1974.

[309] Chen, C.T., *Analysis and Synthesis of Linear Control Systems,* Holt, Rinehart and Winston, New York, 1975.

[310] Sage, A.P. and C.C. White, III, *Optimum Systems Control,* Prentice-Hall, Englewood Cliffs, N.J., 1977.

15 NONLINEAR SYSTEMS

[311] Atherton D.P., *Nonlinear Control Engineering*. Van Nostrand Reinhold, London, 1975.

[312] Atherton D.P., *Stability of Nonlinear Systems*. John Wiley & Sons, New York, NY, 1981.

[313] Craig J.J., *Introduction to Robotics, Mechanics and Control.* Addison-Wesley, Reading, MA, 1986.

[314] Cunningham W.J., *Introduction to Nonlinear Analysis*. McGraw-Hill, New York, NY, 1958.

[315] Desoer C.A. and Vidyasagar M., *Feedback Systems: Input Output Properties*. Academic Press, New York, 1975.

[316] Gibson J.E., *Nonlinear, Automatic Control*. McGraw-Hill, New York, NY, 1963.

[317] Hahn W., *Theory and Application of Lyapunov's Direct Method*. Prentice-Hall, Englewood Cliffs, NJ, 1963.

[318] Itkis U., *Control Systems of Variable Structure*. John Wiley & Sons, New York, NY, 1976.

[319] Lin C.F., *Advanced Control Systems Design*. Prentice-Hall, Englewood Cliffs, NJ, 1964.

[320] Minorsky N., *Theory of Nonlinear Control Systems*. McGraw-Hill, New York, NY, 1969.

[321] Mittal et al. Nonlinear adaptive control of a twin lift helicopter system. *IEEE Control Systems*, pages 39-44, April 1991.

[322] Mohler R.R., *Nonlinear Systems: Dynamics and Control*, Volume 1. Prentice-Hall, Englewood Cliffs, NJ, 1994.

[323] Slotine J.J.E., and *Applied Nonlinear Control* W.Li., Prentice-Hall, Englewood Cliffs, NJ, 1991.

[324] Spong W., and Vidyasagar M., *Robotic Dynamics and Control*. John Wiley & Sons, New York, NY, 1989.

[325] Vidyasagar M., *Nonlinear System Analysis*. Prentice-Hall, Englewood Cliffs, NJ, 1993.

16 ADVANCES IN CONTROL SYSTEMS

[326] Astrom, K.J., and B. Wittenmark, *On Self-tuning Regulators*, Automatica, 9:185-199, 1973.

[327] Astrom, K.J., and B. Wittenmark, *Computer Controlled Systems-Theory and Design*, Prentice-Hall, Englewood Cliffs, N.J., second ed., 1990.

[328] Astrom, K.J., and B. Wittenmark. *Adaptive Control*, Addision Wesley, 1995.

[329] Landau, Y.D., *Adaptive Control: The Model Reference Approach*, Marcel Dekker, New York, 1979.

[330] Monopoli, R.V., Model Reference Adaptive Control with an Augmented Error Signal, IEEE Trans. Automat. Contr., AC-19:474-484, 1974.

[331] Narendra, K.S., and L.S. Valavani, *A Comparison of Lyapunov and Hyper-stability Approaches to adaptive control of continuous systems*, IEEE Trans. Automat. Contr., AC-25:243-247, 1980.

[332] Narendra, K.S., and A.M. Annaswamy, *Stable Adaptive Systems*. Prentice-Hall, Englewood Cliffs, N.J., 1989.

[333] Parks, P.C., *Lyapunov Redesign of Model Reference Adaptive Control Systems*, IEEE Trans, Automat. Contr., AC-11:362-365, 1989.

[334] Sastry, S., and M. Bodson, *Adaptive Control: Stability Convergence and Robustness*, Prentice-Hall, Englewood Cliffs, N.J., 1989.

[335] Slotine, J.J.E. and W.Li., *Applied Nonlinear Control*, Prentice-Hall, Englewood Cliffs, N.J., 1991.

[336] Wieslander, J., and B., *Wittenmark An Approach to Adaptive Control Using Real Time Identification*, Automatica, 7:211-217, 1971.

Fuzzy Logic Control

[337] Bernard, J.A., *Use of A Rule Based System for Process Control*, IEEE Control Systems, 8, October 1988.

[338] Dubois, D. and H. Prade, *Fuzzy Sets and Systems: Theory and Applications*, Academic Press, New York, 1980.

[339] Kickett, W. and M.V.N. Lemke, *Application of Fuzzy Controller in a Warm Water Process*, Automatica, 12:301-308, 1976.

[340] King, P.J. and E.H Mamdani, *The Application of Fuzzy Control Systems to Industrial Processes*, Automatica, 13:235-242, 1977.

[341] Larkin, L.I., *A Fuzzy Logic Controller for Aircraft Flight Control,* Industrial Applications of Fuzzy Control, ed. M. Sugeno, North-Holland, pp.87-103, 1985.

[342] Larsen, P.M., *Industrial Applications of Fuzzy Logic Control,* Fuzzy Reasoning and Its Applications, ed. E.H. Mamdani and B.R. Gaines, Academic Press, pp. 335-342, 1981.

[343] Leung, K.S. and N.Lam, *Fuzzy Concepts in Expert Systems*, IEEE Computer, Sept. 1988.

[344] Mamdani, E.H. and Assilian, *A Case Study on the Application of Fuzzy Set Theory to Automatic Control*, Proc. IFAC., 1974.

[345] Sripada, N.R. *et al., AI Application for Process Regulation and Servo Control*, IEE proc., 134(4), July 1987.

[346] Takagi, T., and M. Sugeno, *Fuzzy Identification of Systems and its Applications to Modeling and Control*, IEEE Trans. sys. Man Cybern., SMC-15(1): 116-132, 1985.

[347] Tong, R.M., *A Control Engineering Review of Fuzzy Systems*, Automatica, 13:559-569, 1977.

[348] Tong, R.M, *et al., Fuzzy Control of the Activated Sludge Waste Water Treatment Process*, Automatica, 16:695-501.

[349] Zadeh, L.A., *A Relationale for Fuzzy Control, J. Dynamic Systems, Measurement and Control,* 94(G): 3-4, 1972.

[350] Zadeh, L.A., *Outline of A New Approach to Analysis of Complex Systems and Decision Processes*, IEEE Trans. Sys. Man Cybern, SMC 3:28-44, 1973.

[351] Zadeh, L.A., *The Role of Fuzzy Logic in the Management of Uncertainty in Expert System*, Fuzzy Sets and Systems, 11:199-227, 1983.

[352] Zadeh, L.A., *Making Computers Think Like People*, IEEE spectrum, 21(8): 26-32, 1984.

[353] Zadeh, L.A., *Fuzzy Logic*, IEEE Computer, April, 1988.

[354] Zadeh, L.A., *Knowledge Representation in Fuzzy Logic*, IEEE Trans. Knowledge and Data Eng., 1(1), March 1989.

Neural Control

[355] Albus, J.S., *A New Approach to Manipulator Control: The Cerebellar Model Articulation Controller (CMAC), Trans. of ASME, Journal of Dynamic Systems Measurement and Control,* 97:220-227, September, 1975.

[356] Almedia, L.B., *Back Propagation in Perceptions with Feedback, Neural Computers,* R. Eckmiur and C.V.D. Holsbwa, C.V.D (Eds), Springer Verlag, Hiedelberg, 1988.

[357] Anderson, C.W., *Learning to Control An Inverted Pendulum Using Neural Networks*, IEEE Control Systems Magazine, 31-37, April, 1989.

[358] Astrom, K.J. and B. Wittenmark, *On Self-tuning Regulators*, Automatica, 9:185-199, 1973.

[359] Bavarian, B., *Introduction to Neural Networks for Intelligent Control*, IEEE Control Systems Magazine, 8(2): 3-7, April, 1988.

[360] Beser, B and E. Tulunay, *Preliminary Results on Process Prediction Using Perceptions Which Utilizes Memory*, In *Proc. of Fourth Int. Symp. on Computer and Information Sciences,* Cesme, Izmir, Turkety, volume 1, pages 431-440, 1998.

[361] Bhat, N., and T.J. McAvoy, *Use of Neural Nets for Dynamic Modeling and Control of Chemical Process Systems*, American Control Conference, 1342-1347, 1989. bibitemchu Chu, R. and M. Tenorio, Neural Networks for System Indenfication, American Control Conf., 916-921, 1989.

[362] Gelenbe, E., *Random Neural Networks with Negative and Positive Signals and Product form Solution* In Proc. of Fourth International Symposium on Computer and Information Sciences, Cesme, Izmir, Turkey, volume 1, pages 603-613, 1989. Also in *Neural Computation,* 1(4), 1989.

[363] Guez, A., J.L. Elibert and M. Kam, *Neural Network Architecture for Control*, IEEE Control Systems Magazine, April, 1988.

[364] Hopfield, J.J. and D.W. Tank, *Neural Computation of Decision Optimization Problems*, Biol. Cybern., 52: 1-12, 1985.

[365] Hopfield, J.J., *Neural Networks and Physical Systems with Emergent Collective Computational Abilities*, In Proc. Natl. Acad. Sci., volume 79, pages 2554-2558, 1982.

[366] Hopfield, J.J., *Neurons with Graded Response have Collective Computational Properties Like Those of Two-state Neurons*, In Proc. Natl. Acad. Sci, volume 81, pages 3088-3092, 1984.

[367] Huang, S.C., *Supervised Learning for a Selective Update Strategy for Artificial Neural Networks Masters Thesis* Department of Electrical and Computer Engineering, Univ. of Notre Dame, Notre Dame, IN, August, 1988.

[368] Kraft, D.L. and D.P. Campagna, *A Comparison of CMAC Neural Network and Traditional Adaptive Control Systems*, American Control Conf., 884-889, 1989.

[369] Levin, E., R. Gewirtzman and F.G. Inbar, *Neural Network Architecture for Adaptive System Modeling and Control*, Int Conf. on Neural Networks, Washington D.C., II 311-II 316, June, 1989.

[370] Lippmann, R.P., *An Introduction to Computing with Neural Nets*, IEEE ASSP Magazine 4-22, April, 1987.

[371] Marko, K.A., J. James, J. Dosdall and J. *Murphy, Automotive Control System Diagnostics Using Neural Nets for Rapid Pattern Classification of Large Data Sets*, Int. Conf. on Neural Networks Washington, D.C., II13-II 16, June 1989.

[372] Parks, P.C., *Lyapunov Redesign of Model Reference Adaptive Control Systems*, IEEE Trans. on Automatic Control, AC-11: 362-367, 1966.

[373] Passino, K.M., M.A. Sartori and P.J. Antsaklis, *Neural Computing for Numeric to Symbolic Conversion in Control Systems*, IEEE Control Systems magazine, 44-45, April, 1989.

[374] Psaltis, D., A. Sideris and A. Yamamura, *A Multi-layered Neural Network Controller*, IEEE Control Systems Magazine, 8(2): 17-21, April 1988.

[375] Ramaswamy, G., G.E. Cook, A. Kristinn and G. Karsai, *Neural Networks in GTA Weld Modeling and Control*, American Control Conference, 1989.

[376] Rumelhart, D.E., J.L. McClelland. and the PDP Research Group, *Parallel Distributed Processing, Explorations in the Microstructure of Cognition, vol 1 : Foundations,* A Bradford Book, The MIT Press, Cambridge, MA, 1988.

[377] Shong Lan, M., *Adaptive Control of Unknown Dynamical Systems via Neural Network Approach,* American Control Conference, 910-915, 1989.

[378] Tolat, V.V. and B. Widrow, *And Adaptive Broom Balancer with Visual Inputs*, *Proc.* IEEE Int. Conf, on Neural Networks, San Diego, CA, II 641-II 647, July, 1988.

[379] VerDubin, W.H., *Neural Nets for Diagnoisis and Control Third Annual Experts Systems Conf. and Exposition,* Cobo Center, Detroit, Michigan, 257-267, April,1989

[380] Werbos, P.J., *Back Propagation and Neurocontrol: A Review and Prospectus*, Int. Joint Conf. on Neural Networks, Washington D.C., I 209-I- 216, June, 1989.

[381] Widrow, B. and R. Winter, *Neural Nets for Adaptive Filtering and Adaptive Pattern Recognition* IEEE Computer, 25-29, March, 1988.

[382] Wieland, A. and R. Leighton, *Geometric Analysis of Neural Network Capabilities*, Int. Conf. on Neural Networks, pages III 385-III 391, 1987.

[383] Zafiriov, S.N. and T.J. McAvoy, *Application of Neural Networks on The Detection of Sensor Failure During the Operation of a Control System*, American Control Conf., 1989.

INDEX

INDEX